NATO ASI Series

Advanced Science Institutes Series

A series presenting the results of activities sponsored by the NATO Science Committee, which aims at the dissemination of advanced scientific and technological knowledge, with a view to strengthening links between scientific communities.

The Series is published by an international board of publishers in conjunction with the NATO Scientific Affairs Division

A Life Sciences **B Physics**	Plenum Publishing Corporation London and New York
C Mathematical and Physical Sciences **D Behavioural and Social Sciences** **E Applied Sciences**	Kluwer Academic Publishers Dordrecht, Boston and London
F Computer and Systems Sciences **G Ecological Sciences** **H Cell Biology** **I Global Environmental Change**	Springer-Verlag Berlin Heidelberg New York London Paris Tokyo Hong Kong Barcelona Budapest

NATO-PCO DATABASE

The electronic index to the NATO ASI Series provides full bibliographical references (with keywords and/or abstracts) to more than 30000 contributions from international scientists published in all sections of the NATO ASI Series. Access to the NATO-PCO DATABASE compiled by the NATO Publication Coordination Office is possible in two ways:

- via online FILE 128 (NATO-PCO DATABASE) hosted by ESRIN, Via Galileo Galilei, I-00044 Frascati, Italy.
- via CD-ROM "NATO Science & Technology Disk" with user-friendly retrieval software in English, French and German (© WTV GmbH and DATAWARE Technologies Inc. 1992).

The CD-ROM can be ordered through any member of the Board of Publishers or through NATO-PCO, Overijse, Belgium.

Series I: Global Environmental Change, Vol. 16

The ASI Series Books Published as a Result of Activities of the Special Programme on Global Environmental Change

This book contains the proceedings of a NATO Advanced Research Workshop held within the activities of the NATO Special Programme on Global Environmental Change, which started in 1991 under the auspices of the NATO Science Committee.

The volumes published as a result of the activities of the Special Programme are:

Vol. 1: **Global Environmental Change.** Edited by R. W. Corell and P. A. Anderson. 1991.
Vol. 2: **The Last Deglaciation: Absolute and Radiocarbon Chronologies.** Edited by E. Bard and W. S. Broecker. 1992.
Vol. 3: **Start of a Glacial.** Edited by G. J. Kukla and E. Went. 1992.
Vol. 4: **Interactions of C, N, P and S Biogeochemical Cycles and Global Change.** Edited by R. Wollast, F. T. Mackenzie and L. Chou. 1993.
Vol. 5: **Energy and Water Cycles in the Climate System.** Edited by E. Raschke and D. Jacob. 1993.
Vol. 6: **Prediction of Interannual Climate Variations.** Edited by J. Shukla. 1993.
Vol. 7: **The Tropospheric Chemistry of Ozone in the Polar Regions.** Edited by H. Niki and K. H. Becker. 1993.
Vol. 8: **The Role of the Stratosphere in Global Change.** Edited by M.-L. Chanin. 1993.
Vol. 9: **High Spectral Resolution Infrared Remote Sensing for Earth's Weather and Climate Studies.** Edited by A. Chedin, M.T. Chahine and N.A. Scott. 1993.
Vol. 10: **Towards a Model of Ocean Biogeochemical Processes.** Edited by G. T. Evans and M. J. R. Fasham. 1993.
Vol. 11: **Modelling Oceanic Climate Interactions.** Edited by J. Willebrand and D. L. T. Anderson. 1993.
Vol. 12: **Ice in the Climate System.** Edited by W. Richard Peltier. 1993.
Vol. 13: **Atmospheric Methane: Sources, Sinks, and Role in Global Change.** Edited by M. A. K. Khalil. 1993.
Vol. 14: **The Role of Regional Organizations in the Context of Climate Change.** Edited by M. H. Glantz. 1993.
Vol. 15: **The Global Carbon Cycle.** Edited by M. Heimann. 1993.
Vol. 16: **Interacting Stresses on Plants in a Changing Climate.** Edited by M. B. Jackson and C. R. Black. 1993.

Interacting Stresses on Plants in a Changing Climate

Edited by

Michael B. Jackson

Department of Agricultural Sciences
University of Bristol
AFRC Institute of Arable Crops Research
Long Ashton Research Station
Bristol, BS18 9AF, U.K.

Colin R. Black

Department of Physiology and Environmental Science
University of Nottingham
Sutton Bonington Campus
Loughborough, LE12 5RD, Leicestershire, U.K.

Springer-Verlag
Berlin Heidelberg New York London Paris Tokyo
Hong Kong Barcelona Budapest
Published in cooperation with NATO Scientific Affairs Division

Proceedings of the NATO Advanced Research Workshop on Interacting Stresses on Plants in a Changing Climate, held in Wye College, University of London, U.K., September 13–19, 1992

ISBN-13: 978-3-642-78535-1 e-ISBN-13: 978-3-642-78533-7
DOI: 10.1007/978-3-642-78533-7

Library of Congress Cataloging-in-Publication Data
Interacting stresses on plants in a changing climate / edited by Michael B. Jackson, Colin R. Black. p. cm. – (NATO ASI series. Series I, Global environmental change ; vol. 16)
"Proceedings of the NATO Advanced Research Workshop on Interacting Stresses on Plants in a Changing Climate held in London, U.K., 13-19 September 1992" – T.p. verso. "Published in cooperation with NATO Scientific Affairs Division." Includes index.

1. Vegetation and climate–Congresses. 2. Plants, Effect of stress on–Congresses. 3. Climatic changes–Congresses. I. Jackson, Michael B. II. Black, Colin R., 1948- . III. North Atlantic Treaty Organization. Scientific Affairs Division. IV. NATO Advanced Research Workshop on Interacting Stresses on Plants in a Changing Climate (1992 : London, England) V. Series.
QK754.5.I56 1993 581.2'4–dc20 93-33946

Softcover reprint of the hardcover 1st edition 1993

Typesetting: Camera ready by authors
31/3145 - 5 4 3 2 1 0 - Printed on acid-free paper

FOREWORD

Books dealing with climatic change are commonplace, as are those concerned with effects of environmental stresses on plants. The present volume distinguishes itself from earlier publications by highlighting several interrelated environmental stresses that are changing in intensity as the climate warms in response to the accumulation of 'greenhouse' gases. The stresses examined at the NATO Advanced Research Workshop upon which this book is based include atmospheric pollutants, flooding and submergence, drought and cold. In future, successful farming or landscape management will ultimately depend on strategies that offset the effects of these and other environmental constraints, while exploiting more favourable features. However, the predicted speed of climate change may exceed the rate at which new approaches to farming, forestry, landscape management and genetic conservation can be developed through experience and retroactive response. The alternative is to anticipate future needs and thus identify appropriate management and legislative strategies by research and discussion. The contents of this volume contribute to these vital processes, upon which the productivity of agroecosystems and conservation of natural ecosystems may increasingly depend. Those with any lingering doubts concerning the gravity of the likely future situation are especially encouraged to read the opening chapter.

For convenience, chapters discussing pollution, flooding, drought and cold are grouped in separate sections. However, many authors have taken care to emphasise that interactions between the changing combinations of stresses pose particular problems for plants and plant communities. This is especially clear with respect to atmospheric pollutants and ultra-violet radiation, which may increase susceptibility to other stresses such as frost or drought, while offsetting any beneficial effects of increased temperatures and atmospheric carbon dioxide levels. Such concerns are reiterated in the final section of the book, where the conclusions of four working groups are summarised.

The contribution of molecular biology to understanding plant responses to stress and to the prospects of increasing crop tolerance by genetic engineering offers an attractive and direct approach both to academic studies and to the practical task of improving crop performance in the face of stress. Several chapters deal with these aspects. Progress seems most advanced with respect to cold tolerance, where a combination of physiology, lipid biochemistry and recombinant-DNA technology has generated transformants with increased resilience to cold. The functional link between information at the molecular level and the behaviour of whole plants subjected to stresses other than cold awaits identification of more detailed physiological and biochemical targets for genetic analysis and manipulation.

The Workshop was financed principally by NATO. We gladly acknowledge its generosity, and also thank Mr LV Da Cunha, Programme Director of the NATO Scientific and Environmental Affairs Division for his interest and support during protracted negotiations. Attendance by young scientists was financed by grants from The Annals of Botany Company, Rhône-Poulenc Agriculture and The Federation of European Societies of Plant Physiology (FESPP). These organisations are thanked for their invaluable support. The unstinting efforts of the Local Secretary, Dr Michael Blatt, and his assistant, Dr Alison Wood-Tor, kept the four and a half intensive days of papers and discussion running smoothly, while staff of Withersdane Hall and the Centre for European Agricultural Studies created a country-house atmosphere that proved both relaxing and conducive to informal discussion. On behalf of the Organising Committee, we thank all who were concerned in these highly satisfactory domestic arrangements. We also thank Mr John Whitehead for his excellent work in preparing the manuscripts for publication.

The original intention was to hold the Workshop in Iceland, where both crops and the natural flora are subject to a unique and harsh combination of environmental stresses that would have provided an ideal backdrop to the Workshop. Unfortunately, the highest cost of living in Europe imposed stresses of a financial kind that proved unsurmountable, and the venue was switched to England. Nevertheless, a close link with Iceland was retained. In this regard we thank Dr Bjarni Gudleifsson, a member of the Organising Committee, for a superbly illustrated talk on Iceland's history, flora and landscape, and Mr Jakob Flimann Magnusson of the Icelandic Embassy in London for a memorable evening of Icelandic music and other cultural niceties.

Michael B Jackson
Colin R Black
27 April 1993

SYMPOSIUM ORGANISATION

ORGANISING COMMITTEE

M.B. Jackson, Workshop Director (UK)

C.R. Black, Secretary and Treasurer (UK)

M.R. Blatt, Local Secretary, Wye College (UK)

B.E. Gudleifsson (Iceland)

B.B. Vartapetian (Russia)

J.A.D. Zeevaart (USA)

SPONSORSHIP

PRINCIPAL SPONSOR

The North Atlantic Treaty Organisation, Scientific Affairs Division, Special Programme on the Science of Global Environmental Change

SUPPORTING SPONSORS

The Annals of Botany Company

The Federation of European Societies of Plant Physiology (FESPP)

Rhône-Poulenc Agriculture

PARTICIPANTS AND THEIR AFFILIATIONS

C.J. Andrews,
Plant Research Centre, Agriculture Canada, Ottawa K1A 0C6, Canada

W. Armstrong,
Department of Applied Biology, School of Life Sciences, The University, Hull HU6 7RX, UK

C. Arrese-Igor,
Departamento Producción Agraria, Universidad Pública de Navarra, E-31106 Pamplona, Spain

J.D. Barnes,
Department of Agricultural and Environmental Science, University of Newcastle, Ridley Building, Newcastle upon Tyne NE1 7RU, UK

C. Biel,
IRTA, Centre de Cabrils, Carretera de Cabrils s/n, 08348 Carbrils, Barcelona, Spain

C.R. Black,
Department of Physiology and Environmental Science, Nottingham University, Sutton Bonington Campus, Loughborough LE12 5RD, UK

V.J. Black,
Department of Geography, Loughborough University of Technology, Loughborough LE11 3TU, UK

M.R. Blatt,
Department of Biochemistry and Biological Sciences, Wye College, University of London, Ashford, Kent TN25 5AH, UK

A.M. Boudet,
Centre de Biologie et Physiologie Végétales, Université Paul Sabatier, 118, Route de Narbonne, 31062 Toulouse Cédex, France

R. Brändle,
Universität Bern, Pflanzenphysiologishes Institute, CH-3013 Bern, Altenbergrain 21, Switzerland

N.A. Clarke,
Broom's Barn Experimental Station, Higham, Bury St Edmunds, Suffolk IP28 6NP, UK

R.M.M. Crawford,
Department of Biological and Technical Medicine, Sir Harold Mitchell Building, University of St Andrews, St Andrews, Fife KY16 9AL, UK

W.J. Davies,
Institute of Environmental and Biological Sciences, University of Lancaster, Bailrigg, Lancaster LA1 4YQ, UK

L. van der Eerden,
Research Institute for Plant Protection, PO Box 9060, 6700 GW Wageningen, The Netherlands

H.Z. Enoch,
Soils and Water Institute, Volcani Centre, PO Box 6, Bet Dagan S02SD, Israel

T.W.-M. Fan,
Department of Land, Air and Water Resources, University of California, Hoagland Hall, Davis, California 95616, USA

A.M. Farmer,
English Nature, Northminster House, Peterborough PE1 1UA, UK

B.E. Gudleifsson,
Agricultural Research Institute, Oseyri 2, IS-603 Akureyri, Iceland

G. Güler,
Department of Biology, Faculty of Molecular Biology and Biochemistry, Hacettepe University, Beytepe, Ankara, Turkey

A.D. Hanson,
Institut de Recherche en Biologie Végétale, Université de Montréal, 4101 Sherbrooke Street East, Montréal, Québec H1X 2B2, Canada

W. Hartung,
Lehrstuhl Botanik I, Universität Würzburg, Mittlerer Dallenbergweg 64, D-8700 Würzburg, Germany

T.C. Hsiao,
Department of Land, Air and Water Resources, University of California, Veihmeyer Hall, Davis, California 95616, USA

M.B. Jackson,
Department of Agricultural Sciences, University of Bristol, AFRC Institute of Arable Crops Research, Long Ashton Research Station, Bristol BS18 9AF, UK

C.R. Jensen,
Section of Soil and Water and Plant Nutrition, Royal Veterinary and Agricultural University, Thorvaldsensvej 40, DK-1871 Frederiksberg, Copenhagen, Denmark

T.Ješko,
Department of Plant Physiology, Slovak Academy of Sciences, Dúbravská cesta 14,842 23 Bratislava, Czech and Slovak Federal Republic

B.R. Jordan,
Horticulture Research International, Worthing Road, Littlehampton, West Sussex BN17 6LP, UK

Y.Y. Leshem,
Department of Life Sciences, Bar-Ilan University, Ramat-Gan 52900, Israel

T. Madsen,
Botansk Institute, Åarhus University, Nordlandsvej 68, DK 8420 Risskov, Denmark

T.A. Mansfield,
Institute of Environmental and Biological Sciences, University of Lancaster, Bailrigg, Lancaster LA1 4YQ, UK

S. Mapelli,
Istituto Biosintesi Vegetali, C.N.R., Via Bassini 15, 20133 Milano, Italy

M. Mattana,
Istituto Biosintesi Vegetali, C.N.R., Via Bassini 15, 20133 Milano, Italy

N. Murata,
National Institute for Basic Biology, Myodaiji, Okazaki 444, Japan

J. Murphy,
The Meteorological Office, Hadley Centre, London Road, Bracknell, Berkshire RH12 2SY, UK

P.M. Neumann,
The Lowdermilk Faculty of Agricultural Engineering, Technion City, Haifa 32000, Israel

S. Nogues,
Department of Biology, University of Essex, Wivenhoe Park, Colchester CO4 3SQ, UK

M. Oliveira,
Departamento Fitotecnia, Universidade de Tras-Os-Montes E Alto Douto, 5000 Vila Real, Portugal

J.J. Van Oosten,
Horticulture Research International, Worthing Road, Littlehampton, West Sussex BN17 6LP, UK

M. Pagès
Centro de Investigación y Desarrollo, Jorge Girona Salgado 18-26, 08034 Barcelona, Spain

J.P. Palta,
Department of Horticulture, University of Wisconsin, 1575 Linden Drive, Madison, Wisconsin 53706, USA

M.L. Parry,
Environmental Change Unit, School of Geography, University of Oxford, 1A Mansfield Road, Oxford OX1 3TB, UK

W.H. Patrick Jr,
Wetland Biogeometry Institute, Louisiana State University, Baton Rouge, Louisiana 70803, USA

R.S. Pearce,
Department of Biology and Nutritional Science, The University, Newcastle upon Tyne NE1 7RU, UK

J.W. Radin,
US Department of Agriculture, Agricultural Research Service, Western Cotton Laboratory, Phoenix, Arizona 85040, USA

M.M. Sachs,
USDA-ARS Department of Agronomy UIUC, S108 Turner Hall, 1102 South Goodwin Avenue, Urbana, Illinois 61801, USA

P.H.M. Saglio,
Station de Physiologie Végétale, INRA, BP 81, 33883 Villenave d'Ornon Cédex, France

G. Schenone,
L'energia Elettrica, Direzione Studi e Recherche, Centro Ricerca Termica a Nucleare, 54 via Rubattino, Milano 20134, Italy

R.E. Sharp,
University of Missouri-Columbia, 1-40 Agriculture Building, Columbia, Missouri 65211, USA

G.R. Squire,
Scientific Adviser's Unit, Scottish Office Agriculture and Fisheries Department, 47 Robb's Loan, Edinburgh EH14 1TU, UK

P.L. Steponkus,
Department of Soil, Crops and Atmospheric Sciences, Cornell University, 609 Bradfield Hall, Ithaca, New York 14853, USA

C. Stushnoff,
Department of Horticulture, Colorado State University, Fort Collins, Colorado 80523, USA

J. Summers,
Department of Agricultural Sciences, University of Bristol, AFRC Institute of Arable Crops Research, Long Ashton Research Station, Long Ashton, Bristol BS18 9AF, UK

G. Terry,
Institute of Environmental and Biological Sciences, University of Lancaster, Bailrigg, Lancaster LA1 4YQ, UK

B.B. Vartapetian,
Timiriazev Institute of Plant Physiology, Russian Academy of Sciences, Botanicheskaya 35, Moscow 127276, Russia

L.A.C.J. Voesenek,
Department of Environmental Plant Ecology, University of Nijmegen, Toernooiveld, 6525 ED Nijmegan, The Netherlands

J.A.D. Zeevart,
Michigan State University, MSU/DOE Plant Research Laboratory, East Lansing, Michigan 48824-1312, USA

CONTENTS

SECTION 1

Introduction

THE POTENTIAL EFFECTS OF CLIMATE CHANGE ON WORLD FOOD SUPPLY

Martin Parry and Cynthia Rosenzweig[1]
Environmental Change Unit
1A Mansfield Road
University of Oxford
Oxford OX1 3TB
UK

ABSTRACT

The IPCC Report on potential effects of climate change on agriculture concluded that it is not yet clear whether climate change will increase or decrease the potential for global food production. However, recent work suggests that the net effect on potential production, globally averaged, will tend to be adverse. Projected changes in temperature and precipitation imply increases in potential production in mid latitude regions and, in general, decreases in low latitude areas. Climate change is thus likely to accentuate current inequalities in the world geography of agricultural potential.

BACKGROUND

The impact of climate change on world food production is, as yet, uncertain (Parry *et al.*, 1988; Parry, 1990). This chapter reports the results of a study of the possible effects of climate change on world agriculture[2]. Two main components are considered:

1. *The estimation of potential changes in crop yield.* Potential changes in national grain crop yields were estimated using crop models and a decision support system developed by the US Agency for International Development's International Benchmark Sites Network for Agro-technology Transfer (IBSNAT, 1989). The crops modelled were wheat,

---

[1] Columbia University/Goddard Institute for Space Studies, 2880 Broadway, New York City 10025, USA.

[2] A full version of this study is reported in: Rosenzweig, Parry, Fischer and Frohberg (1993).

NATO ASI Series, Vol. I 16
Interacting Stresses on Plants in a Changing Climate
Edited by M. B. Jackson, C. R. Black

rice, maize, and soybeans. These crops account for more than 85% of the world's traded grains and legumes. The estimated yield changes for 18 countries were interpolated to provide estimates of yield changes for all regions of the world and for all the major crops, by reference to all available published and unpublished information.

2. *Estimation of world food trade responses.* The yield changes were used as inputs into a world food trade model, the Basic Linked System (BLS), developed at the International Institute for Applied Systems Analysis (IIASA) (Fischer *et al.*, 1988). Outputs from simulations by the BLS provided information on food production, food prices and the number of people at risk from hunger.

Climate change scenarios

Scenarios of climate change were developed to estimate effects on crop yields and food trade. A climate change scenario is defined as a physically consistent set of changes in meteorological variables, based on generally accepted projections of CO_2 (and other trace gases) levels. The range of scenarios used is intended to capture the range of possible effects and set limits on the associated uncertainty. The scenarios for this study were created by changing observed data on current climate (1951-80) according to doubled CO_2 simulations of three general circulation models (GCMs). The GCMs used are those from the Goddard Institute for Space Studies (GISS), Geophysical Fluid Dynamics Laboratory (GFDL), and United Kingdom Meteorological Office (UKMO). Mean monthly changes in climate variables from the appropriate gridbox were applied to observed daily climate records to create climate change scenarios for each site.

GCMs currently provide the most advanced means of predicting the potential future climatic consequences of increasing radiatively active trace gases. They have been shown to simulate current temperatures reasonably well, but do not reproduce current precipitation as accurately, and their ability to reproduce current climate varies considerably from region to region (Houghton *et al.*, 1990). They have not yet been validated to project changes in climate variability, such as changes in the frequencies of drought and storms, even though these could affect crop yields significantly.

Rates of future emissions of trace gases, as well as when the full magnitude of their effects will be realised, are not certain. Because other greenhouse gases besides CO_2 (e.g., methane (CH_4), nitrous oxide (N_2O), and the chlorofluorocarbons (CFCs)) are also increasing, an 'effective CO_2 doubling' has been defined as the combined radiative forcing of all greenhouse gases having the same forcing effect as doubled CO_2 (usually defined as 600 ppm). For this study, CO_2 concentrations are estimated to be 555 ppm in 2060 (based on Hansen *et al.*, 1988). The effective CO_2 doubling will occur around the year 2030, if current emission trends continue. The climate change caused by an effective doubling of CO_2 may be delayed by 30 to 40 years or longer, hence the projections for 2060 in this study.

Crop models and yield simulations

Crop models. The IBSNAT crop models were used to estimate how climate change and increasing levels of carbon dioxide may alter the yields of world crops at 112 sites representing both major production areas and vulnerable regions at low, mid and high latitudes. The IBSNAT models employ simplified functions to predict the growth of crops as influenced by the major factors that affect yields, i.e., genetics, climate (daily solar radiation, maximum and minimum temperatures and precipitation), soils and management practices. Models used were for wheat (Ritchie and Otter, 1985; Godwin *et al.*, 1989), maize (Jones and Kiniry, 1986; Ritchie *et al.*, 1989), paddy and upland rice (Godwin *et al.*, 1993) and soybean (Jones *et al.*, 1989).

The IBSNAT models were selected for this study because they have been validated over a wide range of environments (e.g., Otter-Nacke *et al.*, 1986) and are not specific to any particular location or soil type. They are thus suitable for use in large-area studies in which crop growing conditions differ greatly. The validation of the crop models over different environments also improves the ability to estimate effects of changes in climate. Furthermore, because management practices, such as the choice of varieties, planting date, fertiliser application and irrigation, may be varied in the models, they permit experiments that simulate adjustments by farmers and agricultural systems to climate change.

Physiological effects of CO_2. Most plants growing in experimental environments with increased levels of atmospheric CO_2 exhibit increased rates of net photosynthesis (i.e., total photosynthesis minus respiration) and reduced stomatal opening. Experimental effects of CO_2 on crops have been reviewed by Acock and Allen (1985) and Cure (1985). By so doing, CO_2 reduces transpiration per unit leaf area while enhancing photosynthesis. Thus, it often improves water-use efficiency (the ratio of crop biomass accumulation or yield to the amount of water used in evapotranspiration).

The crop models used in this study account for the beneficial physiological effects of increased atmospheric CO_2 concentrations on crop growth and water use (Peart *et al.*, 1989). Ratios were calculated between measured daily photosynthetic and evapotranspiration rates for a canopy exposed to high CO_2 values, based on published experimental results (Cure and Acock, 1986; Allen *et al.*, 1987; Kimball, 1983), and the ratios were applied to the appropriate variables in the crop models on a daily basis. The photosynthesis ratios (555 ppm CO_2/330 ppm CO_2) for soybean, wheat, rice and maize were 1.21, 1.17, and 1.06, respectively. These effects are based on experimental results that may overstate the positive effects of CO_2, because uncertainty exists concerning the extent to which the beneficial effects of increasing CO_2 will be seen in crops growing in variable, windy and pest-infected (weeds, insects and diseases) fields under climate change conditions.

Limitations of crop growth models. The crop models embody a number of simplifications. For example, weeds, diseases and insect pests are assumed to be controlled, there are no problem soil conditions (e.g., high salinity or acidity), and there are no extreme weather events such as heavy storms. The crop models simulate the current range of agricultural technologies available around the world. They do not include induced improvements in such technology, but may be used to test the effects of some potential improvements, such as varieties with higher thermal requirements and the installation of irrigation systems.

Yield simulations. Crop modelling simulation experiments were performed at 112 sites in 18 countries for baseline climate (1951-80) and GCM-doubled CO_2 climate change scenarios, with and without the physiological effects of CO_2. This involved the following tasks:

1. The geographical boundaries were defined for the major production regions in the country studies and the current production of major crops in those regions was estimated.
2. Observed climate data for representative sites within these regions were provided for the baseline period (1951-1980), or for as many years of daily data as were available, and the soil, crop variety, and management inputs necessary to run the crop models at the selected sites were specified.
3. The crop models were validated with experimental data from field trials.
4. The crop models were run with baseline data and climate change scenarios, with and without the direct effects of CO_2 on crop growth.
5. Alterations in agricultural practices that would lessen any adverse consequences of climate change, by simulating irrigated production and other adaptation responses (for example, shifts in planting date and crop varieties) were identified and evaluated.

Deriving estimates of potential yield changes

Aggregation of site results. Crop model results for wheat, rice, maize and soybean from all sites and 18 countries were aggregated by weighting regional yield changes (based on current production) to estimate changes in national yields. The regional yield estimates represent the current mix of rainfed and irrigated production, the current crop varieties, nitrogen management and soils. The study simulated 70-75% of the world production of wheat and maize. Even though only two countries (Brazil and USA) simulated soybean production, their combined output accounts for 76% of the world total. Less of the total world rice production was simulated than for the other crops because India, Indonesia and Vietnam have significant rice production not included in the study. Production data were gathered by scientists participating in the project and from FAO (1988), USDA Crop Production Statistical Division and the USDA International Service.

Yield change estimates for crops and regions not simulated. Changes in national yields of other crops and commodity groups and regions not simulated were estimated on the basis of similarities to modelled crops and growing conditions, previous published and unpublished climate change impact studies and GCM climate change scenarios. Estimates were made of yield changes for the three GCM scenarios, with and without the direct effects of CO_2. The yield changes resulting from the direct effects of CO_2 were based on the mean responses to CO_2 of the different crops used in the crop model simulations.

The primary source of uncertainty in the estimates lies in the sparseness of the crop modelling sites which may not adequately represent the variability of agricultural regions within countries, the variability of agricultural systems within similar agro-ecological zones, or dissimilar agricultural regions. However, since the site results relate to regions that account for about 70% of the world's grain production, the conclusions concerning world production totals contained in this report are believed to be adequately substantiated.

The world food trade model

The estimates of climate-induced changes in yields were used as inputs to a dynamic model of the world food system (the Basic Linked System) in order to assess possible impacts on future levels of food production, food prices and the number of people at risk from hunger. Impacts were assessed for the year 2060, with population growth, technology trends and economic growth projected to that year. Assessments were first made assuming no climate change and subsequently with the climate change scenarios described above. The difference between the two assessments is the climate-induced effect. A further set of assessments examined the efficacy of a number of adaptations at the farm level in mitigating the impacts and the effect on future production of liberalising the world food trade system, and of different rates of growth of economy and population.

The Basic Linked System (BLS) consists of linked national models. The BLS was designed at the International Institute for Applied Systems Analysis for food policy studies, but it also can be used to evaluate the effect of climate-induced changes in yield on world food supply and agricultural prices. It consists of 20 national and/or regional models which cover about 80% of the world food system. The remaining 20% is covered by 14 regional models for the countries which have broadly similar attributes (e.g., African oil exporting countries, Latin American high income exporting countries, Asian low income countries). The grouping is based on country characteristics such as geographical location, income *per capita* and the country's position with regard to net food trade (Fig. 1).

The BLS is a general equilibrium model system, with representation of all economic sectors, empirically estimated parameters and no unaccounted supply sources or demand sinks (see Fischer *et al.*, 1988 for a complete description of the model). In the BLS, countries are linked through trade, world market prices and financial flows. It is a recursively dynamic system: a first round of exports from all countries is calculated for an assumed set of world prices, and international market clearance is checked for each commodity. World prices are then revised, using an optimising algorithm and again transmitted to the national model. Next, these generate new domestic equilibria and adjust net exports. This process is repeated until the world markets are cleared in all commodities. At each stage of the reiteration domestic markets are in equilibrium. This process yields international prices as influenced by governmental and inter-governmental agreements.

The system is solved in annual increments, simultaneously for all countries. Summary indicators of the sensitivity of the world system used in this report include world cereal production, world cereal prices and prevalence of world population at risk from hunger (defined as the population with an income insufficient either to produce or procure their food requirements).

The BLS does not incorporate any climate relationships *per se*. Effects of changes in climate were introduced to the model as changes in the average national or regional yield per commodity as estimated above. Ten commodities are included

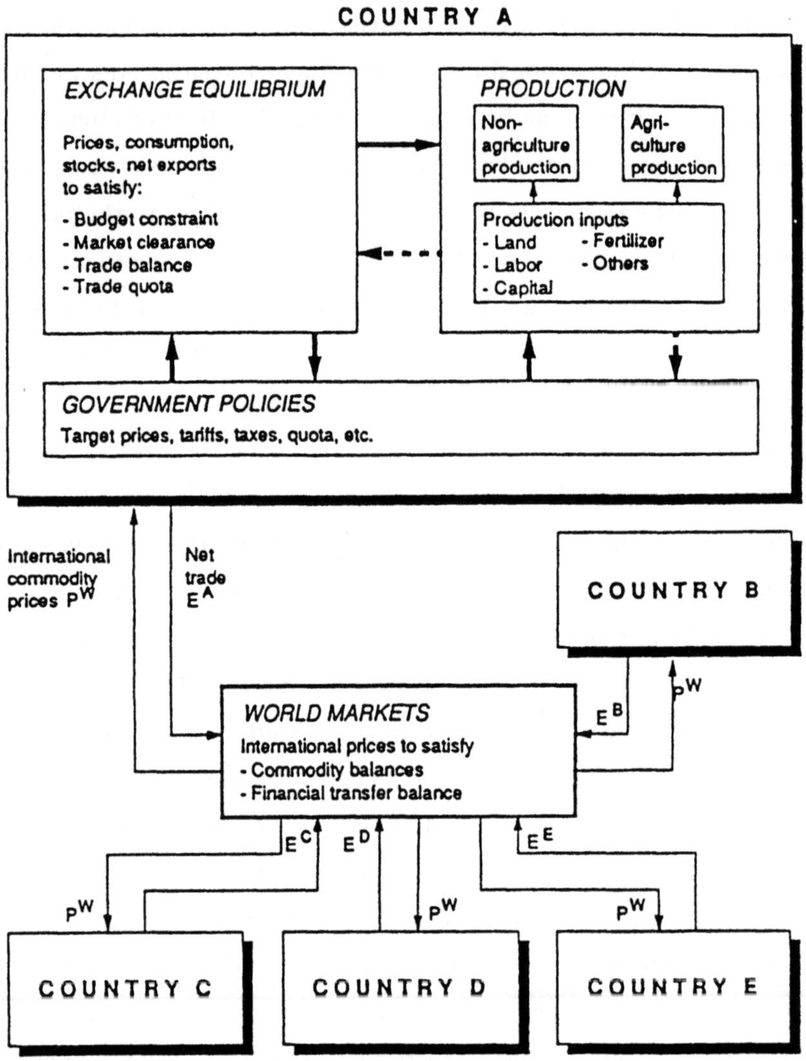

FIGURE 1. The basic linked system: Country details.

in the model: wheat, rice, coarse grains (e.g., maize, millet, sorghum, barley), bovine and ovine meat, dairy products, other animal products, protein feeds, other food, non-food agriculture and non-agriculture. In this context, however, consideration is limited to the major grain food crops.

The set of model experiments

The results described in this chapter consider the following scenarios:

The reference scenario. This involved projection of the agricultural system to the year 2060 with no effects of climate change on yields and with no major changes in the political or economic context of the world food trade. It assumed:

- UN medium population estimates (10.2 billion by 2060; UN, 1989; IBRD/World Bank, 1990).
- 50% trade liberalisation in agriculture introduced gradually by 2020.
- moderate economic growth (ranging from 3.0% year^{-1} in 1980-2000 to 1.1% year^{-1} in 2040-2060).
- technology is projected to increase yields over time (cereal yields for world total, developing countries and developed countries are assumed to increase annually by 0.7%, 0.9% and 0.6%, respectively).
- no changes in agricultural productivity due to climate change.

Three climate change scenarios. These are projections of the world food system including effects on agricultural yields under different climate scenarios (the '2 x CO_2 scenarios' for the GISS, GFDL and UKMO GCMs). The food trade simulations for these three scenarios were started in 1990 and assumed a linear change in yields until the double CO_2 concentration was reached in 2060. Simulations were made both with and without the physiological effects of 555 ppm CO_2 on crop growth and yield for the equilibrium yield estimates. In these scenarios, internal adjustments in the model occur, such as increased agricultural investment, reallocation of agricultural resources according to economic returns and reclamation of additional arable land as an adjustment to higher cereal prices, based on shifts in comparative advantage among countries and regions.

Scenarios including the effects of farm-level adaptations. The food trade model was first run with yield changes assuming no external adaptation to climate change and was then re-run with different climate-induced changes in yield assuming a range of farm-level adaptations. These included such measures as altered planting dates and crop varieties and the use of different amounts of irrigation and fertiliser. Two adaptation levels to cope with potential effects on yield and agriculture were considered. Adaptation Level 1 included those adaptations at the farm level that would not involve any major changes in agricultural practices. It thus took account of changes in planting date, amounts of irrigation, and the choice of crop varieties that are currently available. Adaptation Level 2 encompassed, in addition to the former, major changes in agricultural practices, such as large shifts of planting date, the availability of new cultivars, extensive expansion of irrigation and increased fertiliser application. This level of adaptation would be likely to involve policy changes both at the national and international level and significant costs. However, policy, cost and water resource availability were not studied explicitly.

Scenarios of different future trade, economic and population growth. A final set of scenarios assumed changes to the world tariff structure and different rates of growth of economy and population. As with the previous experiments, these were conducted both with and without climate change impacts. These scenarios included:

- *Full trade liberalisation.* Full trade liberalisation in agriculture introduced gradually by 2020.
- *Lower economic growth.* (ranging from 2.7% year^{-1} in 1980-2000, to 1.0% year^{-1} in 2040-2060). Global GDP in 2060 is 10.3% lower than the reference scenario, 11.2% lower in developing countries and 9.8% lower in developed countries.
- *Low population growth.* UN low population estimates (*c.* 8.6 billion by 2060).

EFFECTS ON YIELDS

Yields without adaptation

Table 1 shows wheat yield changes predicted by the GCM 2 x CO_2 climate change scenario for the countries where crop model simulations were conducted. The results show that climate change scenarios excluding the direct physiological effects of CO_2 predict decreases in simulated crop yields in many cases, while the direct effects of increasing atmospheric CO_2 mitigate the negative effects primarily in mid and high latitudes.

TABLE 1. Current production and changes in simulated wheat yields under GCM 2 x CO_2 climate change scenarios, with and without the direct effects of CO_2.[1]

Country	Current production			Change in simulated yields					
	Yield ($t\ ha^{-1}$)	Area ($ha \times 10^3$)	Prod. ($t \times 10^3$)	GISS[2] (%)	GFDL[2] (%)	UKMO[2] (%)	GISS[3] (%)	GFDL[3] (%)	UKMO[3] (%)
Australia	1.38	11546	15574	-18	-16	-14	8	11	9
Brazil	1.31	2788	3625	-51	-38	-53	-33	-17	-34
Canada	1.88	11365	21412	-12	-10	-38	27	27	-7
China	2.53	29092	73527	-5	-12	-17	16	8	0
Egypt	3.79	572	2166	-36	-28	-54	-31	-26	-51
France	5.93	4636	27485	-12	-28	-23	4	-15	-9
India	1.74	22876	39703	-32	-38	-56	3	-9	-33
Japan	3.25	237	772	-18	-21	-40	-1	-5	-27
Pakistan	1.73	7478	12918	-57	-29	-73	-19	31	-55
Uruguay	2.15	91	195	-41	-48	-50	-23	-31	-35
Former USSR									
winter	2.46	18988	46959	-3	-17	-22	29	9	0
spring	1.14	36647	41959	-12	-25	-48	21	3	-25
USA	2.72	26595	64390	-21	-23	-33	-2	-2	-14
WORLD[4]	2.09	231	482	-16	-22	-33	11	4	-13

[1] Results for each country represent the site results weighted according to regional production. The world estimates represent the country results weighted by national production.
[2] GCM 2 x CO_2 climate change scenario alone.
[3] GCM 2 x CO_2 climate change scenario with direct effects of CO_2.
[4] World area and production x 10^6.

The magnitude of the estimated yield changes varies between crops. Maize production is most negatively affected, probably due to its lower response to the physiological effects of CO_2 on crop growth. Protein feed is least affected because soybean responds significantly to increased CO_2, at least under the scenarios involving smaller increases in temperature (e.g., the GISS and GFDL scenarios).

The differences between countries in yield responses to climate change are related to differences in current growing conditions. At low latitudes, crops are grown nearer the limits of temperature tolerance and global warming may subject them to higher stress. In many mid and high latitude areas, increased temperatures may benefit crops otherwise limited by cold temperatures and short growing seasons in the present climate.

The primary causes of decreases in yield are:

1. *Shortening of the growing period* (especially the grain filling stage) of the crop. This occurred at some sites in all countries.
2. *Decreased water availability.* Depletion of soil water is increased by greater evapotranspiration and, in some cases, a decrease in precipitation in the climate change scenarios. This occurred in Argentina, Brazil, Canada, France, Japan, Mexico and USA.
3. *Poor vernalisation.* Some temperate cereal crops require a period of low temperature in winter to initiate or accelerate the flowering process. Inadequate vernalisation results in low flower bud initiation and ultimately in reduced yields. This caused decreases in winter wheat yields at some sites in Canada and the former USSR.

Fig. 2 shows estimated potential changes in average national crop yields for the GISS, GFDL and UKMO 2 x CO_2 climate change scenarios, allowing for the direct effects of CO_2 on plant growth. Latitudinal differences are apparent in all the scenarios. High latitude changes are less negative or even positive in some cases, while lower latitude regions indicate more detrimental effects of climate change on agricultural yields.

The GISS and GFDL climate change scenarios produced yield changes ranging from +30 to -30%. The GISS scenario is, in general, more detrimental than GFDL for crop yields in parts of Asia and South America, while GFDL results in more negative yields in the USA and Africa and less positive results in the former USSR. The UKMO climate change scenario, which has the greatest warming (5.2 °C global surface air temperature increase), suggests yield declines almost everywhere (up to -50% in Pakistan).

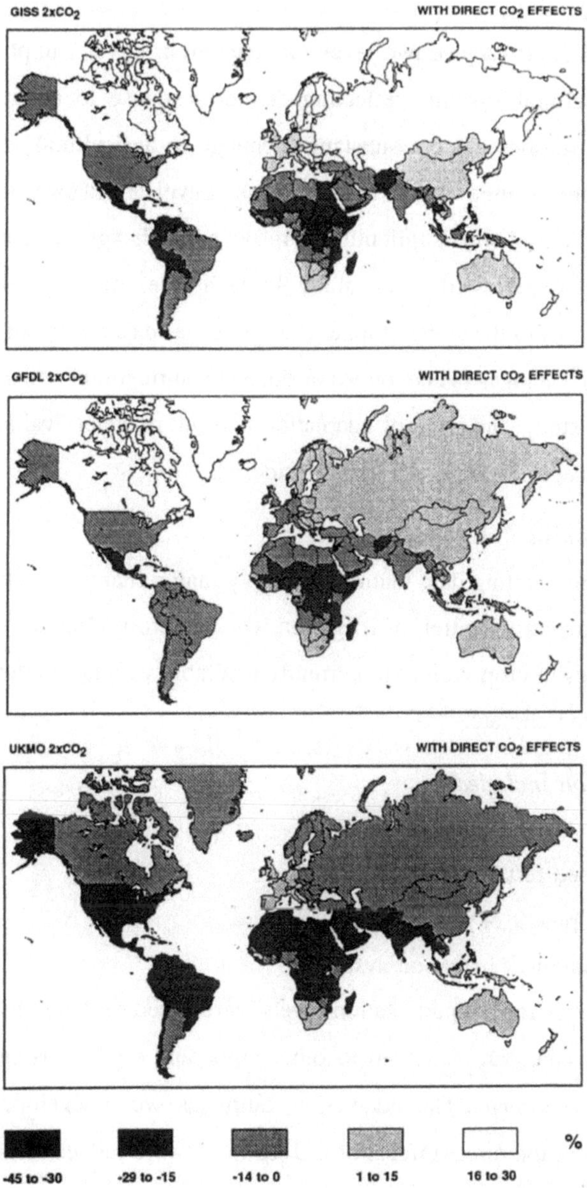

FIGURE 2. *Change in crop yields under the three climate change scenarios, with the direct effects of* CO_2.

Yields with adaptation

The study tested the efficacy of two levels of adaptation: Level 1 implies little change to existing agricultural systems reflecting farmer response to a changing climate, whereas Level 2 implies a more substantial change to agricultural systems possibly requiring resources beyond the farmer's means. Level 2 Adaptation represents an optimistic assessment of world agriculture's response to changed climate conditions as predicted by the GCMs tested in this study. In each case, the adaptations were tested as possible responses to the worst climate change scenario (usually, but not always, the UKMO scenario). Changes in economics or domestic agricultural policies were beyond the scope of this study; the costs of adaptation and future water availability under the climate change scenarios were also not considered.

Level 1 adaptation included:

- Shifts in planting date that do not imply major changes in the crop calendar.
- Additional application of irrigation water to crops already under irrigation.
- Changes in crop variety to currently available varieties better adapted to the projected climate.

Level 2 adaptation included:

- Large shifts in planting date.
- Increased fertiliser application.
- Development of new varieties.
- Installation of irrigation systems.

Yield changes for both adaptation levels were based on crop model simulations, where available, and were extended to other crops and regions using the estimation methods described above. The adaptation estimates were developed only for the scenarios including the direct effects of CO_2 as these were judged to be most realistic. The two levels of adaptation estimates for the UKMO scenario were also examined (Fig. 3). With the high level of global warming projected by the UKMO climate change scenario, neither Level 1 nor Level 2 Adaptation mitigated climate change effects on crop yields in most countries.

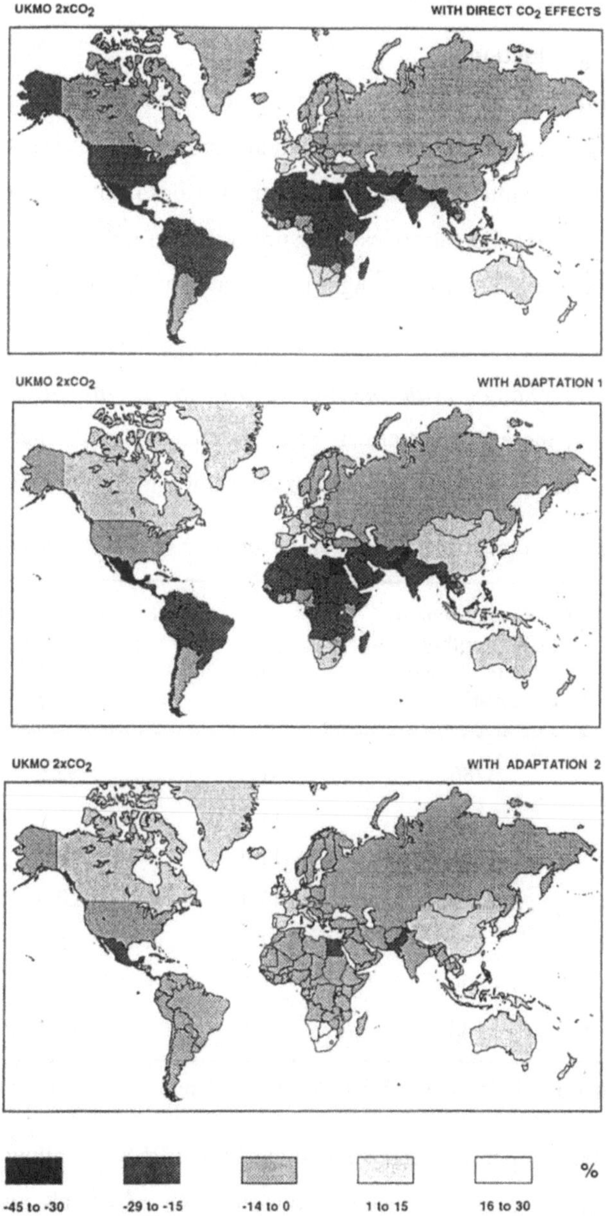

FIGURE 3. *Change in crop yields under two levels of adaptation for UKMO climate change scenarios, all with direct CO_2 effects.*

EFFECTS ON WORLD FOOD TRADE

Effects on food production

The future without climate change. Assuming no effects of climate change on crop yields, but that population growth and economic growth are as stated above, world cereal[1] production is estimated at 3286 million metric tons (mmt) in 2060 as compared with 1795 mmt in 1990. Cereal prices are estimated at an index of 121 (1970 = 100). The number of people at risk from hunger is estimated at about 640 million (530 m in 1990).

Effects of climate change, with internal adjustment in the model, but without adaptation. Under the estimated effects of climate change and atmospheric CO_2 on crop yields, the effects on cereal production are shown in Table 2 and Fig. 4. These

TABLE 2. *Change in cereal production under climate change scenarios in 2060.*

Region	Reference scenario (2060) without climate change (MMT)[1]	Scenario with climate change		
		GISS (%)	GFDL (%)	UKMO (%)
Global	3286	-1.2	-2.8	-7.6
Developed	1449	11.3	5.2	-3.5
Developing	1836	-11.0	-9.2	-10.8
Africa	296	-19.3	-20.9	-11.1
Central & S. America	274	-23.7	-15.7	-5.8
S. & S.E. Asia	690	-9.1	-4.2	-15.5
West Asia	133	-9.8	-14.3	-11.2

[1] Reference scenario is for 2060 assuming no climate change. MMT = million metric tonnes.

estimations are based upon dynamic simulations by the BLS that allow the world food system to respond to climate-induced supply shortfalls of cereals and consequently higher commodity prices through dynamic increases in production factors, cultivated

[1] The estimate for cereals includes wheat, rice, maize, millet, sorghum and minor grains contained in the FAO AGROSTAT database. Rice is included as rice milled equivalent (a factor of 0.67 is used to convert from rice paddy to milled rice).

land, labour and capital and inputs such as fertiliser. World cereal production is estimated to decrease between 1 and 7%, depending on the GCM climate scenario. Under the UKMO scenario, global production is estimated to decrease by more than 7%, while under the GISS scenario (which assumes lower temperature increases) cereal production is estimated to decrease by just over 1%. The largest negative changes occur in developing countries, averaging -9% to -11%. By contrast, in developed countries production is estimated to increase under all but the UKMO scenario (+11% to -3%). Thus existing disparities in crop production between developed and developing countries are estimated to increase.

Effects of climate change under different levels of adaptation

Adaptation level 1. Table 3 and Fig. 5 show the effects of Level 1 adaptation on estimated changes in cereal production. These largely offset the negative climate change yield effects in developed countries, improving their comparative advantage in world markets. In these regions cereal production increases by 4% to 14% over the reference case. However, developing countries are estimated to benefit little from adaptation (-9% to -12%). Averaged global production is altered by between 0% and -5% from the reference case. As a consequence, world cereal prices are estimated to increase by *c.* 10-100%, and the number of people at risk from hunger by *c.* 5-50%. This indicates that Level 1 adaptations would have relatively little influence on reducing the global effects of climate change.

Adaptation level 2. More extensive adaptation virtually eliminates negative cereal yield impacts at the global level under the GISS and GFDL climate scenarios and reduces impacts under the UKMO scenario by one third (Table 3 and Fig. 6). However, the decrease in the comparative advantage of developing countries under these scenarios leads to decreased areas planted to cereals in these areas. Cereal production in developing countries still decreases by about 5%. Globally, however, cereal prices increase by only 5-35%, and the number of people at risk from hunger is altered by between -2% and +20% from the reference case. This suggests that Level 2 adaptations are required to mitigate the negative effects of climate change, but that these still do not eliminate them in developing countries.

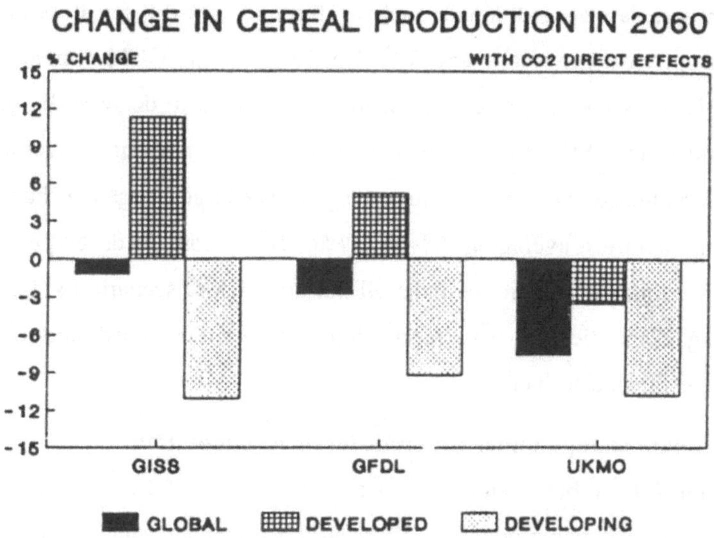

FIGURE 4. *Change in cereal production against reference case for 2060 assuming no climate change (global 3286 MMT, developed 1449 MMT, developing 1836 MMT).*

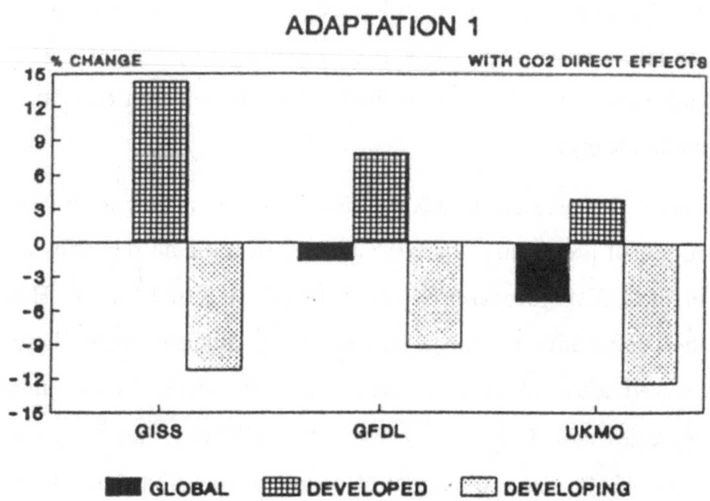

FIGURE 5. *Change in cereal production under climate change scenarios in 2060 and assuming Adaptation Level 1.*

TABLE 3. Effect of climate change on cereal production under two levels of adaptation: Adaptation Level 1 (AD1) and Adaptation Level 2 (AD2).

	Changes in cereal production (% of 2060 reference)		
	GISS	GFDL	UKMO
World			
– without adaptation	-1.2	-2.8	-7.6
– AD1	0.0	-1.6	-5.2
– AD2	1.1	-0.1	-2.4
Developed			
– without adaptation	11.3	5.2	-3.5
– AD1	14.2	7.9	3.9
– AD2	11.0	3.0	0.8
Developing			
– without adaptation	-11.0	-9.2	-10.8
– AD1	-11.2	-9.2	-12.4
– AD2	-6.6	-5.6	-5.7

Net imports of cereals into developing countries will increase under all scenarios. The change in cereal imports is largely determined by the size of the assumed yield changes, the change in relative productivity in developed and developing regions, the change in world market prices and changes in incomes of developing countries. Under the GISS climate scenario, productivity is depressed largely in favour of developed countries, resulting in pronounced increases of net cereal imports into developing countries. Under the UKMO scenario large cereal price increases limit the increase of imports to developing countries. Consequently, despite its beneficial impact for developed countries, the Adaptation Level 1 scenarios show only small improvements in developing countries as compared to the corresponding impacts without such adaptation.

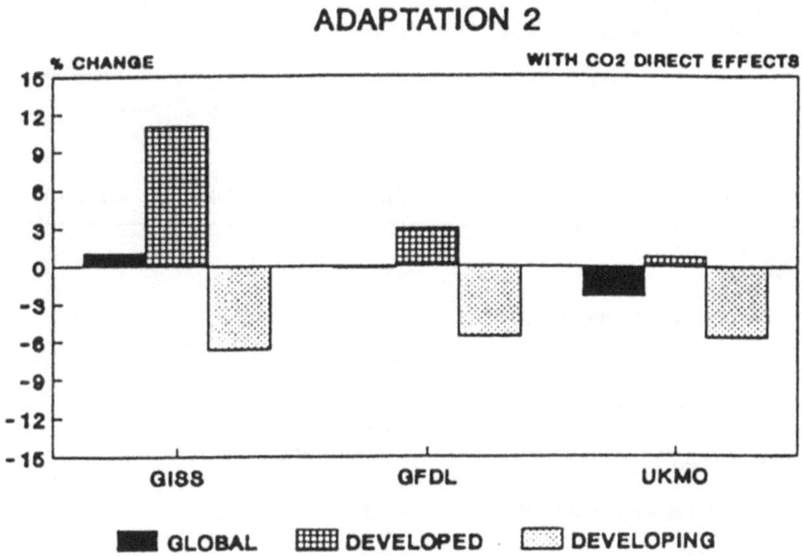

FIGURE 6. Change in cereal production under climate change scenarios in 2060 and assuming Adaptation Level 2.

Effects of climate change assuming full trade liberalisation and lower economic and population growth rates

Full trade liberalisation. Assuming full trade liberalisation in agriculture by 2020 provides for more efficient resource use and leads to a 3.2% higher value added in agriculture globally and a 5.2% higher agriculture GDP in developing countries (excluding China) by 2060 compared to the reference scenario. This policy change results in almost 20% fewer people at risk from hunger. Global cereal production is increased by 70 million mmt, with most of the production increases occurring in developing countries (Table 4). Global impacts due to climate change are slightly reduced under most climatic scenarios, with enhanced gains in production accruing to developed countries, but losses in production being greater in developing countries. Price increases are reduced slightly from what would occur without full trade liberalisation, and the number of people at risk from hunger is reduced by about 100 million.

TABLE 4. Effects of climate change on production assuming full trade liberalisation (REF-FTL).

	Changes in cereal production 2060 (% of REF-FTL)				
	Ref. (2060) (MMT)	REF-FTL (MMT)	GISS (%)	GFDL (%)	UKMO (%)
Global	3286	3356	-0.9	-2.6	-8.2
Developed	1449	1472	12.5	6.5	-3.7
Developing	1836	1884	-11.3	-9.7	-11.6

The reference scenario (2060) assumes no climate change. REF-FTL = Reference 2060 with full trade liberalisation and no climate change. MMT = million metric tonnes. Percent changes are against FTL.

Reduced rate of economic growth. Estimates were also made of impacts under a lower economic growth scenario (10% lower than reference). These are indicated in Table 5. Lower economic growth results in a tighter supply situation, higher prices and more people below the hunger threshold. Prices are 10% higher and the number of people at risk from hunger is 20% greater. The effect of climate change on these trends is generally to reduce production, increase prices and increase the number of people at risk from hunger by about the same ratio as is the case with a higher economic growth rate, but the absolute amounts of change are greater.

Altered rates of population growth. The largest impact of any of the policies considered would result from an accelerated reduction in population growth in developing countries (Table 6). Simulations based on rates of population growth according to UN Low Estimates result in a world population about 17% lower in year 2060 as compared to the UN Mid Estimates used in the reference run. The corresponding reduction in the developing countries (excluding China) would be about 19.5% from 7.3 to 5.9 billion. The combination of higher GDP/capita (about 10%) and lower world population produces an estimated 40% fewer people at risk from hunger in the year 2060 compared to the reference scenario.

TABLE 5. Effects of climate change on production assuming a low rate of economic growth (REF-E).

	Changes in cereal production 2060 (% of REF-E)				
	Ref. (2060) (MMT)	REF-E (MMT)	GISS (%)	GFDL (%)	UKMO (%)
Global	3286	3212	-1.0	-2.7	-7.9
Developed	1449	1428	12.4	6.0	-3.6
Developing	1836	1786	-11.6	-9.7	-11.3

The reference scenario (2060) assumes no climate change. REF-FTL = Reference 2060 with full trade liberalisation and no climate change. MMT = million metric tonnes. Percent changes are against REF-E.

TABLE 6. Effects of climate change on production assuming UN Low Estimate of population growth (REF-P).

	Changes in cereal production 2060 (% of REF-P)				
	Ref. (2060) (MMT)	REF-P (MMT)	GISS (%)	GFDL (%)	UKMO (%)
Global	3286	2929	-0.7	-2.6	-7.1
Developed	1449	1349	10.3	4.8	-3.9
Developing	1836	1582	-10.1	-8.9	-9.9

The reference scenario (2060) assumes no climate change. REF-FTL = Reference 2060 with full trade liberalisation and no climate change. MMT = million metric tonnes. Percent changes are against REF-P.

Even under the most adverse of the three climate scenarios (UKMO), the estimated number of people at risk from hunger is some 10% lower than that estimated for the reference scenario without any climate change. Increases in world prices of agricultural products, in particular of cereals, under the climate change scenarios employing the low population projection are around 75% of those using the UN Mid Estimate.

IMPLICATIONS FOR GRASSLANDS

It is clear that the effects of changes in climate on yields are likely to vary greatly from region to region across the globe. Under the climatic scenarios adopted in this study, the effects on crop yields in mid and high latitude regions appear to be less adverse than those in low latitude regions. Since grass growth is likely to respond in a similar way as C3 food crops to climate change, we may infer the same potential broad-scale effects on grassland productivity.

Fig. 2 suggests that, under the GISS 2 x CO_2 and GFDL 2 x CO_2 scenarios, yield is likely to increase in the northern hemisphere mid-latitude regions of Canada, Northern Europe and Russia and the southern hemisphere regions of southern South America, Southern Africa and Australasia. Mid-latitude regions such as the USA, Southern Europe and China also appear to be favoured. It is possible that grassland yields may increase in these regions. It should be emphasised, however, that while GCMs simulate current temperatures reasonably well, they do not reproduce current precipitation as accurately, and their ability to reproduce current climate varies considerably from region to region. They have not yet been validated to project changes in climate variability, such as changes in the frequency of droughts and storms, even though these could affect the yield significantly.

The second line of inference that may be adopted with reference to grasslands concerns the potential response of grassland production systems to changes in production and prices in the world food system as a result of climate change. The preceding analysis has suggested that a production decrease of 1-5% might occur as a result of climate change from the level that might otherwise be reasonably anticipated in a future without climate change. Estimates of increases in cereal prices resulting from such reduced production vary from 25-150%. In regions where production increases both as a response to possible climate change and is also driven by higher prices, it is plausible to expect a possible increase in comparative advantage of arable agriculture over grassland systems, and thus the extension of the arable area and consequent contraction of grassland. The grassland area might therefore contract while grassland productivity increases. Conversely, the decreases in production in the arable sector indicated for certain regions by the preceding analysis, particularly where these occur in the context of global increases in production (and thus possible reductions in food prices), might be expected to lead to an increasingly competitive position for grassland production over arable production, and the consequent extension of the grassland area, particularly in those regions that may be favoured by changes in climate.

Much, of course, is uncertain. Particularly uncertain is the extent to which the potentially beneficial direct effects of elevated CO_2 on grass growth will be realised. We have also had to speculate about the extent of adaptation at the farm level. Particularly important are likely to be the regional variations in adaptation, since these are likely to enhance the differential effects of climate change on agricultural yields (both for arable crops and grass) between mid-latitude/developed and low-latitude/ developing regions. Adaptation to predicted climate change is predicted to compensate for the detrimental effects of climate change much more successfully in developed areas. When the economic implications of these changes for grassland production are explored, the regional differences in effects remain noticeable. Developed countries are expected to be less affected by climate change than developing economies. The ability of the world food system to absorb impacts decreases with the magnitude of the impact. Adaptation can compensate for lower amounts of climate change, such as under the GISS and GFDL climate scenarios, but not greater ones such as under the UKMO scenario. Prices of agricultural products are broadly exponentially related to the magnitude of the climate change impact.

It should be emphasised that the results reported here are not a forecast of the future. The inferences drawn are most uncertain. We particularly lack information on possible climate change at the regional level, on the effects of technological change on agricultural productivity, on trends in demands (including population growth), and on the wide array of possible adaptations.

REFERENCES

Acock B, Allen LH Jr (1985) Crop responses to elevated carbon dioxide concentrations. In: Strain BR, Cure, JD (eds) Direct effects of increasing carbon dioxide on vegetation. US Department of Energy. DOE/ER-0238. Washington, DC, USA, pp 33-97

Allen LH Jr, Boote KJ, Jones JW, Jones PH, Valle RR, Acock B, Rogers HH, Dahlman RC (1987) Response of vegetation to rising carbon dioxide: Photosynthesis, biomass and seed yield of soybean. Global Biogeochemical Cycles 1: 1-14

Cure JD (1985) Carbon dioxide doubling responses: A crop survey. In: Strain BR, Cure, JD (eds) Direct effects of increasing carbon dioxide on vegetation. US Department of Energy. DOE/ER-0238. Washington, DC, USA, pp 33-97

Cure JD, Acock B (1986) Crop responses to carbon dioxide doubling: A literature survey. Agricultural and Forest Meteorology 38: 127-145

FAO (1988) 1987 Production Yearbook. Food and Agriculture Organization of the United Nations. Statistics Series No. 82 Rome, Italy, pp 351

Fischer G, Frohberg K, Keyzer MA, Parikh KS (1988) Linked national models: A tool for international food policy analysis. Kluwer, Dordrecht, The Netherlands, pp 227

Godwin D, Ritchie JT, Singh U, Hunt L (1989) A user's guide to CERES-Wheat -- V2.10. International Fertiliser Development Center. Muscle Shoals, Alabama, USA

Godwin D, Singh U, Ritchie JT, Alocilja EC (1993) A user's guide to CERES-Rice. International Fertiliser Development Center. Muscle Shoals, Alabama, USA (in press)

Hansen J, Fung I, Lacis A, Rind D, Russell G, Lebedeff S, Ruedy R, Stone P (1988) Global climate changes as forecast by the GISS 3-D model. Journal of Geophysical Research 93: 9341-9364

Houghton JT, Jenkins GJ, Ephraums JJ (1990) Climate change: The IPCC scientific assessment. Cambridge University Press, Cambridge, UK, pp 365

IBRD/World Bank (1990) World population projections. John Hopkins University Press, Baltimore, USA

International Benchmark Sites Network for Agrotechnology Transfer (IBSNAT) Project (1989) Decision support system for agrotechnology transfer version 2.1 (DSSAT V2.1). Department of Agronomy and Soil Science, College of Tropical Agriculture and Human Resources, University of Hawaii, Honolulu, Hawaii, USA

Jones CA, Kiniry JR (1986) CERES-Maize: A simulation model of maize growth and development. Texas A&M Press, College Station, USA, pp 194

Jones JW, Boote KJ, Hoogenboom G, Jagtap SS, Wilkerson GG (1989) SOYGRO V5.42: Soybean crop growth simulation model. Users' guide. Departments of Agricultural Engineering and Agronomy, University of Florida, Gainesville, Florida, USA

Kimball BA (1983) Carbon dioxide and agricultural yield: An assemblage and analysis of 430 prior observations. Agronomy Journal 75: 779-788

Manabe S, Wetherald RT (1987) Large-scale changes in soil wetness induced by an increase in CO_2. Journal of Atmospheric Science 44: 1211-1235

Otter-Nacke S, Godwin DC, Ritchie JT (1986) Testing and validating the CERES-Wheat model in diverse environments. AgGRISTARS YM-15-00407. Johnson Space Center 20244, USA

Parry ML (1990) Climate Change and World Agriculture, Earthscan, London, UK

Parry ML, Carter TR, Konijn NT (eds) (1988) The impact of climatic variations on agriculture. Vol. 1 Assessments in cool temperate and cold regions. Vol. 2 Assessments in semi-arid regions. Kluwer, Dordrecht, The Netherlands, pp 876 and 764

Peart RM, Jones JW, Curry RB, Boote K, Allen LH Jr (1989) Impact of climate change on crop yield in the southeastern USA. In: Smith JB, Tirpak DA (eds) The potential effects of global climate change on the United States. US Environmental Protection Agency, Washington, DC, USA

Ritchie JT, Otter S (1985) Description and performance of CERES-Wheat: A user-oriented wheat yield model. In: Willis WO (ed) ARS Wheat Yield Project. Department of Agriculture, Agricultural Research Service, ARS-38. Washington, DC, USA, pp 159-175

Ritchie JT, Singh U, Godwin D, Hunt L (1989) A user's guide to CERES-Maize -- V2.10. International Fertiliser Development Center, Muscle Shoals, Alabama, USA

Rosenzweig C, Parry ML, Fischer E, Frohberg K (1993) Climate change and world food supply. Research report 3, Environmental Change Unit, Oxford University, UK

United Nations (1989) World Population Prospects 1988. United Nations, New York, USA

MANIFESTATIONS OF MECHANICAL STRESS IN MEMBRANES: A PROSPECTIVE HYPOTHESIS OF ENDOGENOUS EXPRESSION OF EXOGENOUS STRESS SIGNALLING

Y.Y. Leshem[1], R.R. Avtalion[1], R. Agassi[1], H. Gottlieb[2] and D. Bach[3]

Departments of [1]Life Sciences and [2]Chemistry
Bar-Ilan University
Ramat-Gan 52900
Israel

ABSTRACT

Environmental stress is perceived in plants by activation of phospholipase A_2 which releases *sn*-2 polyunsaturated fatty acid from membrane galactolipids which, as a result, become *lyso*-entities. An overall description is presented of how these products of PLA_2 action induce biophysical changes primarily at membrane sites. It is proposed that this stress-signalling mechanism may be either a manifestation of a general adaptation syndrome or serve as a trigger for a detrimental cascade of events culminating in senescence and death.

INTRODUCTION

This chapter presents a hypothesis that environmental stress stimuli are perceived by plants as biophysical and hence mechanical surface property-related changes in membrane bilayers. In his preface to a volume on membranes compiled by one of the authors (Leshem, 1992) A.W. Galston stated:

> *The plasma membrane is at once the window through which the cell senses the environment and the portal through which the environment influences the structure and activities of the cell. Its importance in cellular physiology can thus hardly be overestimated, since constant flow of materials between cell and environment is essential to the well-being of any biological system.*

---

[3] Department of Membrane Research, Weizmann Institute of Science, Rehovoth 72500, Israel

NATO ASI Series, Vol. I 16
Interacting Stresses on Plants in a Changing Climate
Edited by M. B. Jackson, C. R. Black

In keeping with this approach we shall endeavour to show that a pivotal stage in stress perception is the promotion of an endogenous phospholipase A_2 (PLA_2) either by its activation via triggering a Ca^{2+} increment in the cytosol or by *de novo* synthesis. PLA_2 action releases *sn*-2 fatty acyl chains from membrane glycerolipid (inclusive of plasmalemmal *phospho*lipids, chloroplast membrane *galacto*lipids and *sulpho*lipids) bilayers. The predominant (but not sole) fatty acyl species at the *sn*-2 sites are polyunsaturated fatty acids (PUFA) which, in plant membranes, are mainly linoleic or linolenic acids - 18:2 and 18:3 (Mazliak, 1980; Leshem, 1992). Based on previous work by the authors and in other laboratories, and upon further experimental data presented below, a proposed stress sequence resulting from the activation of PLA_2 is envisaged to be as follows. Firstly, de-esterification of the membrane glycerolipid backbone by PLA_2 (Fig. 1) results in the production of two breakdown products:

1. free fatty acids (FFA) - usually PUFA as mentioned above;
2. lyso-phospho-, galacto- or even -sulpholipids.

The consequences of the release of these two groups of substances are discussed below.

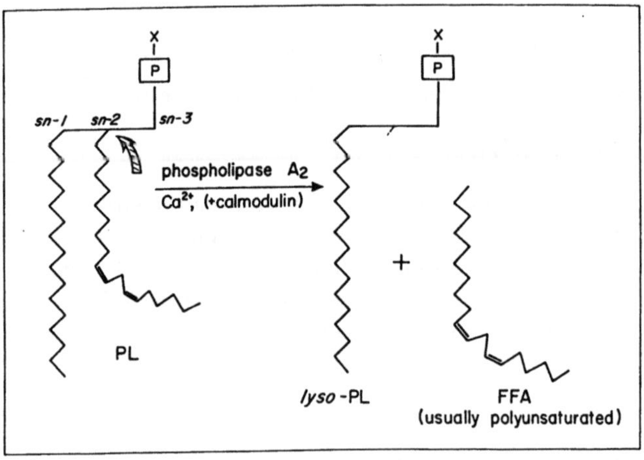

***FIGURE 1**. Mode of action of phospholipase A_2.*

STRESS-ASSOCIATED IMPLICATIONS OF PUFA RELEASE BY PLA_2

a) ***FFAs* per se *act as lytic agents*,** causing further membrane disruption.

b) ***Hydroxylation of PUFA ensues, resulting in the enzymatic (*via *lipoxygenase - LOX) or spontaneous non-enzymatic production of PUFA hydroperoxides*.** The hydroperoxides involved are the 9-linoleic acid hydroperoxide (LAHP) or the 13-LAHP (Leshem *et al.*, 1993), or in certain cases 'double' hydroperoxides (Grechkin *et al.*, 1991). As can be seen from Fig. 2, hydroperoxidation implies a certain degree of straightening of the C=C 'kinks' in the intact PUFAs, with a resulting decrease in molecular radius. This allows more units to be packed into a given surface area and hence membrane rigidity, as expressed by surface tension, is increased. This increase in rigidity (or microviscosity) or, conversely, the decrease in fluidity (η) is regarded as physiologically detrimental.

Is PUFA release a prerequisite for hydroperoxidation? While it is generally accepted that LOX-mediated hydroperoxidation is several orders of magnitude greater on de-esterified PUFAs than from those in the intact membrane glycerolipid (S. Grossman and Y. Sofer, Bar Ilan University, personal communication), there are clear indications that the process may occur in intact phospholipids (PLs) (Leshem *et al.*, 1993). Membrane rigidification can therefore ensue at the level of hydroperoxidation of unesterified galactolipids, as well as of the individual de-esterified acyl chains. Moreover, as pointed out below, FFA can undergo further changes. Direct hydroperoxidation, *not* requiring PLA_2, may in this particular respect contribute more to membrane rigidity than the former type.

c) ***Production of jasmonates from the PUFA*.** Jasmonates as a group (jasmonic acid, methyl jasmonates (MJ) and tuberonic acid are well documented as being associated with environmental stress (Ueda and Kato, 1980; Satler and Thimann, 1981; Sembdner and Gross, 1986; Parthier, 1990; Horton 1991). Under mild stress jasmonate may even be employed as a mechanism for coping with stress; for example, methyl jasmonate curbs transpiration in drought-stressed foliage. However, prolonged exposure induces senescence-associated phenomena. As shown in Fig. 3, jasmonates are formed from an 18C FFA (18:2) precursor, which *via*

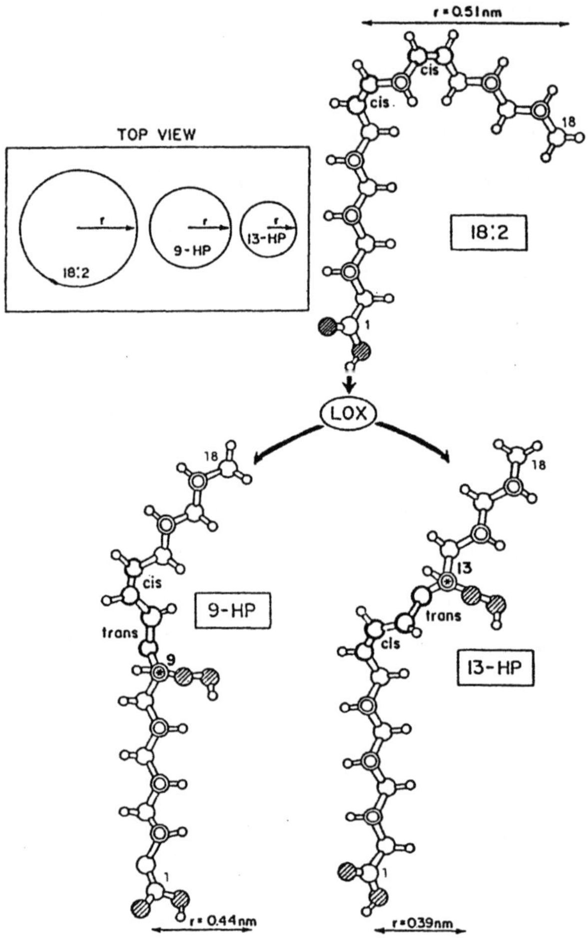

FIGURE 2. Mode of PUFA hydroperoxidation which may be lipoxygenase (LOX)-mediated or spontaneous (after Leshem et al., *1993). The immediate products are linoleic (or linolenic acid)-9, or -13-hydroperoxides in which the* cis-cis *double bond is converted to either* cis-trans *as in the 13-HP, or to* trans-cis *as in the 9-HP. These changes reduce the molecular radius of the 18:2 'backbone'. In the case of the 13-HP, which is believed to be more physiologically active in the membrane, molecular radius is reduced by half and approximates to that of 18:1. This potentially induces greater membrane rigidity by allowing packing of more molecules into a given area.*

hydroperoxidation as shown above, undergoes cyclisation of a 12C fragment and sheds a 6C scission product. In this respect, the plant processes are metabolically parallel to the formation of mammalian prostaglandin which is also a cyclisation product of a PUFA, which in animal systems is 20:4 and not 18:2 or 18:3. Moreover, octadecanoid release and jasmonate formation may induce disease resistance (Farmer and Ryan, 1992).

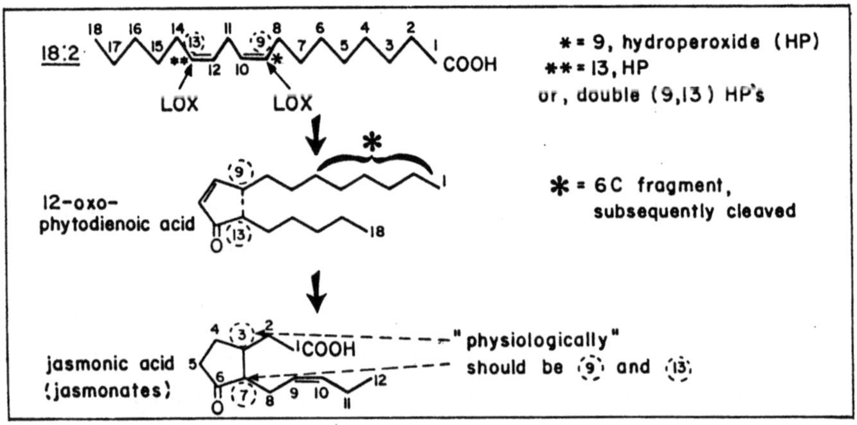

FIGURE 3. Scheme indicating the linoleic acid (18:2) derivation of jasmonate.

d) *Biophysical implication of the presence of 18:2 and 18:3 in membranes.* As will be shown later, incorporation of MJ into phosphatidyl choline bilayers or monolayers has a fluidising effect which may be understood by the chemical structure of the methyl jasmonate molecule, as seen in Plate 1. This plate indicates that MJ is not flat but rather a tri-planar molecule with one of the hairpin bends placed at an angle behind the cyclic 5-membered carbon loop, the continuation of the chain from which carbon 7 to carbon 12 are slanted in front of the loop. It therefore is highly probable that such a molecule serves as a far more effective spacer between individual glycerolipid membrane components than precursor esterified or deesterified PUFA or PUFA hydroperoxides. This spacing effect as of necessity, and as is shown in the present experiments, serves to decrease membrane microviscosity. Moreover, in this respect jasmonates are more effective

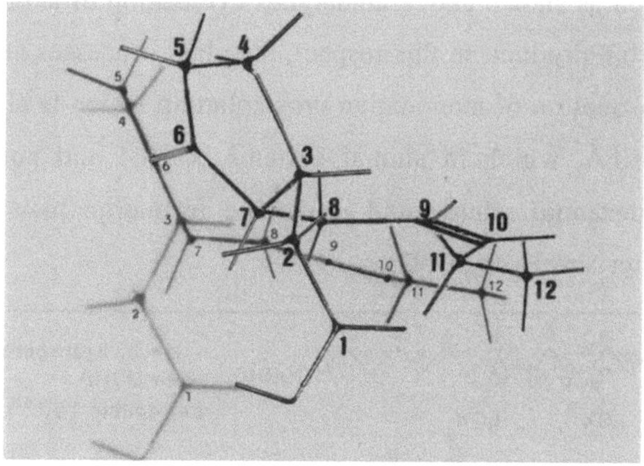

***PLATE 1**. Shadow diagram indicating the tri-planar structure of jasmonate. It is proposed that this structure acts as a 'spacer' which promotes membrane fluidity.*

than abscisic acid (ABA) since, as shown by Satler and Thimann (1981), concentrations one twentieth of those of ABA are equally effective in producing a wide variety of stress-associated manifestations, including reduction of transpiration. Moreover, whereas ABA undergoes rapid deactivation when exposed to sunlight, the UV component of which converts the active *cis* configuration of the sidechain to *trans* (Davies and Jones, 1991), MJ was not noticeably affected by exposure to UV in the present study.

While the above derigidifying action may well account for increased tolerance to drought and possibly cold, and for enhanced guard cell movement and stomatal closure, this is in apparent contradiction to the above property of rigidity. However, in terms of the phospholipase activity of phospholipases as a group, Demel *et al.* (1975) pointed out that optimal activity, especially of PLA_2, occurs between a certain range of bilayer surface tensions, above and below which activity is reduced. The present findings in terms of the overall scheme suggest that, by fluidising membranes, jasmonates make them more amenable to PLA_2 action by bringing the bilayer to a state of surface tension fluidity which ordinarily is non-promotive to phospholytic activity.

e) ***PUFA and ethylene.*** It has been documented that while 1-amino cyclopropane-1- carboxylic acid (ACC) is the ethylene precursor under normal circumstances, linoleic acid can fulfil the same function under conditions of extreme stress, mechanical damage or fungal infection (Leshem *et al.*, 1986). The possibility also exists that ethylene production promoted by PUFA may be indirect by virtue of LAHP serving as a Ca^{2+} ionophore with all the implications of Ca^{2+} increment. Be this as it may, Sridhara and Leshem (1986) have shown in microsomal membranes from pea foliage that, while the increase in ethylene production was minimal over a wide concentration range of saturated stearic acid (18:0), the PUFA, linolenic acid (18:3), significantly enhanced the process.

f) ***Free radical production from de-esterified PUFA.*** This process is an offshoot of the LOX reaction whereby, during the enzyme's action on the 4 pentadiene system, a plethora of oxy free radicals are formed. These induce the PUFA free radical and the superoxide and hydroxyl free radicals. The overall biological importance of free radical involvement in stress and senescence in general and in membrane architecture in particular has been reviewed elsewhere (McKersie *et al.*, 1988; Leshem, 1988; Hendry *et al.*, 1992).

g) ***Effects on membrane stability as a function of sterol esterification.*** Sterols have a modifying effect on membrane fluidity; on the one hand they induce rigidity but on the other hand maintain fluidity under conditions, especially in cold stress, which ordinarily would cause a marked decrease (Shinitzky, 1984). In heat adaptation the converse applies, with sterols maintaining a certain degree of rigidity, the lack of which would result in hyperfluidity and membrane de-stabilisation. Recently it has been shown that, as regards cold acclimation, the sterol-*esters* (i.e., the sterol esterified with fatty acids, especially PUFAs) are probably as important as the sterols (Dyas *et al.*, 1991; Norberg *et al.*, 1991), and serve a clear function in cold adaptation. In keeping with this finding, Leshem (unpublished results) has found that PC monolayers containing sterol esters have lower collapse points and manifest lower surface tensions than PL monolayers containing the corresponding unesterified sterol. The release of PUFA in this respect has a potentially beneficial effect which under certain conditions may counteract some of the above-mentioned harmful ones.

BIOCHEMICAL AND BIOPHYSICAL MANIFESTATIONS OF PLA_2-PRODUCED PHOSPHO- AND GLYCEROLIPIDS

Lyso-phospholipids may be defined as phospholipids from which one of the two fatty acyl components has been removed by phospholipase enzyme action (Fig. 1). Since membranes are dynamic systems and undergo micro- or macro-environmental changes including the degree of saturation or length of fatty acyl side chains, and are subject to phospholytic breakdown, it can be assumed that at any given developmental stage at least some of the major phospholipid species in plant membranes may exist in the form of a *lyso*-phospholipid from which one of the side chains has been removed. Some of the manifestations, in terms of the responses of plant membrane components, of the formation of such *lyso*-products are described below.

a) ***Reduction of molecular radius.*** In a space-filling hexagonal matrix of a given surface area of membrane more *sn*-1 *lyso* glycerolipids (glycerolipids from which the *sn*-2 PUFA tails have been removed by PLA_2) can be packed (Fig. 4). Langmuir isotherm data have concomitantly shown in simulated membranes an increase in the collapse point (the lateral pressure value expressed as $mN\ m^{-1}$ at which the structurally ordered monolayers collapse; Leshem, 1991), reflecting marked rigidification and hence less pliability to withstand the vicissitudes of environmental changes involving heat, cold, osmotic volume, contraction and/or expansion, flooding or drought.

b) ***Ability to complex with abscisic acid by hydrophobic/hydrophilic interactions.*** Fig. 5 shows hydrophobic and hydrophilic sections of membrane PLs and ABA. In the basic non-stressed state, membrane PLs containing the double kinked 18:2 (or triple kinked 18:3) show little if any propensity to conjugate with ABA despite the potential. However, PLA_2-mediated removal of the *sn*-2 chain as seen in Fig. 5 allows the ABA to combine with the *lyso*-product, further affecting the

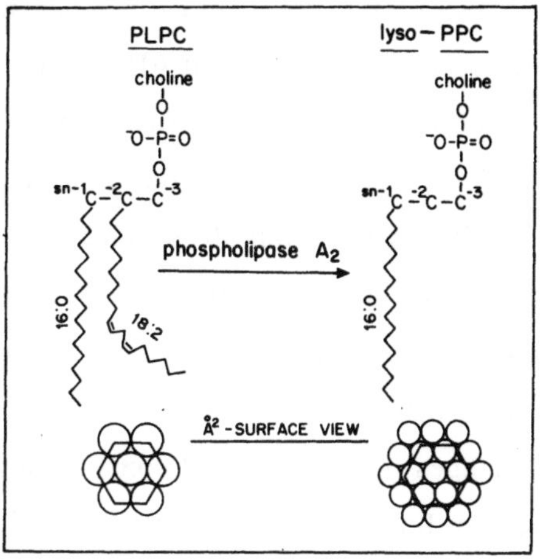

FIGURE 4. Hexagonal packing of an intact PL (left) compared to that of a lyso-PL. Note that more molecules per unit area can be packed in the latter.

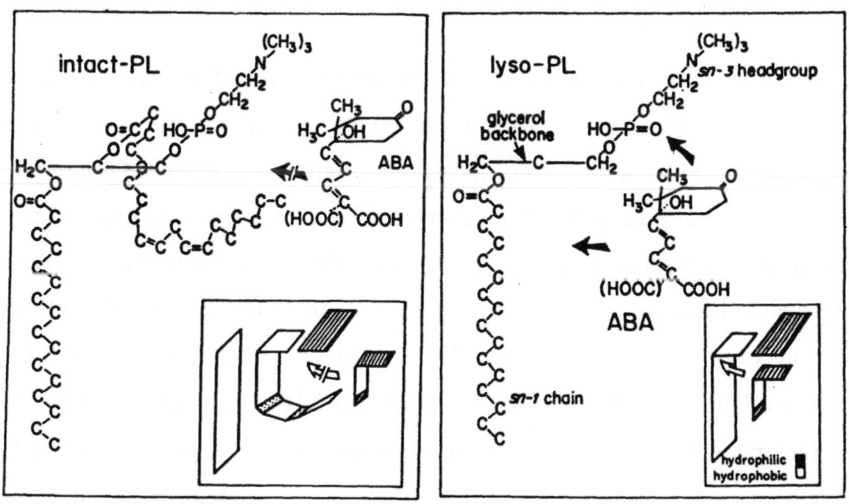

FIGURE 5. Possible mode of binding of abscisic acid to lyso-*phospholipids. The latter are produced after stress-induced phospholipase A_2 action. a: Mode of binding; b: the intact PL precludes such binding.*

motility of *lyso*-GLs. A similar but not identical mechanism has been suggested to account for one aspect of the effectiveness of ABA in inducing stomatal closure where it reportedly combines with a specific species of PL, dipalmitoyl-phosphatidylcholine (DPPC), in which both the *sn*-1 and *sn*-2 acyl tails are saturated; the 'hands off' PUFA lacks this PL species, allowing it to function as a *lyso*-PC in this respect (Leshem *et al.*, 1990).

c) ***Retailoring of membranes to adapt to stress.*** In the adaptation of plants to stress, especially involving temperature, it is believed that one of the regulatory mechanisms involves changes in phospholipid constituency, not by synthesis of new phospholipid molecules but by the 'retailoring' of existing ones. This may be achieved by phospholipase-mediated release of either the *sn*-1 or *sn*-2 fatty acyl chain and reinsertion of a fatty acid residue better equipped to withstand environmental change – increasing saturation at higher temperatures and decreasing saturation at lower temperature; in both cases the trend acts as a homeostatic mechanism to sustain optimal membrane fluidity.

d) ***Effects on membrane stability and interaction with membrane proteins.*** In mitochondrial membranes, Beatrice *et al.* (1980) pointed out that the mechanism involved in catabolic permeability increase, collapse of membrane potential and progressive acceleration of respiration involves *lyso*-phospholipid rather than free PUFA, although the latter may have a synergistic effect. Moreover, in terms of biophysical configurations, de Gier *et al.* (1982) indicated that *lyso*-phospholipids possessing only one fatty acyl tail can result in the physiologically detrimental formation of the non-bilayer hexagonal I (H_1) configuration. On the other hand, these *lyso*-phospholipids, if still located in the bilayer, may be essential for remodelling. *Lyso*-phospholipids may also act as surfactants, disrupt bilayers at concentrations far below those required for membrane dissolution, and aggregate into micelles at much higher Critical Micelle Concentrations than parent phospholipids (Stafford and Dennis, 1988). They are, moreover, involved with membrane fusion and, being cone-shaped, are believed to dominate in membrane surfaces of high curvature, possibly including the highly folded regions of thylakoid membranes and endoplasmic reticulum (Lucy, 1970). *Lyso*-phospholipids, especially *lyso*-PC, also affect membrane-associated proteins such as Na^+/K^+ ATPase (Weltzien, 1979) and plant plasmalemma-contained H^+ATPase (Palmgren *et al.*, 1988). These effects may be caused by a partitioning of substrate out of the bulk phase into a phase of different accessibility to the enzyme, or by inducing membrane surface tension changes which allow enzyme mobility.

The above are the basic tenets of the present hypothesis, many facets of which have clearly been documented. However, several steps have previously been either hypothetical or supported only by indirect evidence. The following section presents direct experimental evidence addressing these questions and endeavours to provide a firmer basis for the above-stated contentions, the two main points being: 1) direct demonstration of the existence of a plant phospholipase A_2 and; (2) membrane-associated biophysical effects of jasmonate as a PUFA scission product.

EXPERIMENTAL DETAILS

1. Immunological identification of a plant PLA_2

Experimentation was based upon the possible existence of an immunological 'cross reaction' between bee venom PLA_2 and the putative plant PLA_2. Positive results would indicate not only the presence of PLA_2 in tested plant protein extracts, but also the existence of some similar immunological determinants shared by both the apian and the plant enzyme.

Preparation of PLA_2 antibodies. Immunisation was performed on two month old female rabbits. Each rabbit was injected with 1.0 mg of bee venom phospholipase A_2 (Merck) emulsified with 1.5 ml of complete Freund's adjuvant. The site of injection was the thigh muscle of the rabbits' hind legs. Three series of injections were made at two week intervals. Three weeks after the final injection, blood was drawn from a cut made in arteries of the rabbits' ears. Serum containing antibodies (AB) was separated from the blood by standard clinical procedures (Johnson, 1986), and samples were taken for protein determination by the method of Lowry *et al.* (1951), and subsequently diluted 1 : 2 with phosphate buffer saline (PBS) pH 7.2 and stored at -20 °C for further use.

Protein extraction from fresh and heat-stressed senescing rose petals. Freshly picked roses (*Rosa hybrida* cv Ariana) of uniform size were aged by placing cuttings in a growth chamber for four days at 30 °C with an 18 hour photoperiod provided by a mixture of incandescent bulbs and fluorescent GRO-LUX tubes providing an intensity of 125 μmol m^{-2} s^{-1}, and relative humidity of approximately 60%. Protein extraction from 75 g each of fresh and senesced flower petals was performed by the ammonium sulphate 'salting out' procedure (Shomer-Ilan *et al.*, 1973). Prior to salting out, care was taken to remove all non-petal flower tissue including anthers, sepals and gynoecia. The protein content of the extract was determined by the method of Lowry *et al.* (1951). The aqueous extract was subsequently lyophilised to a powder which was then hermetically sealed in polyethylene bags and cold-stored. Aliquots of the lyophilised powder served as a possible source of PLA_2 in the Rocket test outlined below.

Rocket immuno-electrophoresis for testing anti PLA_2 sera. An agarose solution containing a 1 : 4 dilution of the rabbit antiserum which, as indicated by the ring test, contained antiPLA_2 AB was poured to form a top layer over a Noble agar-covered microscope slide. The method employed was as described in detail by Avtalion *et al.* (1984), this being a modification of an earlier method (Laurell, 1965). Three round holes, each 3 mm in diameter and separated from each other by 3 mm, were punched through the agarose and agar covering layers to act as origin wells. Using a micropipette, 5 μl bee venom PLA_2 (control) or young and senescing rose petal protein extracts were carefully placed into the wells. The system was subsequently subjected to horizontal electrophoresis employing a 10 V cm^{-1} direct current; the electrophoretic medium was prepared as detailed elsewhere (Avtalion *et al.*, 1984) and comprised tris-veronal buffer, pH 8.5. After a 50 minute separation of electrophoretic bands, the preparation was stained with azocarmine (see Fig. 6 for details).

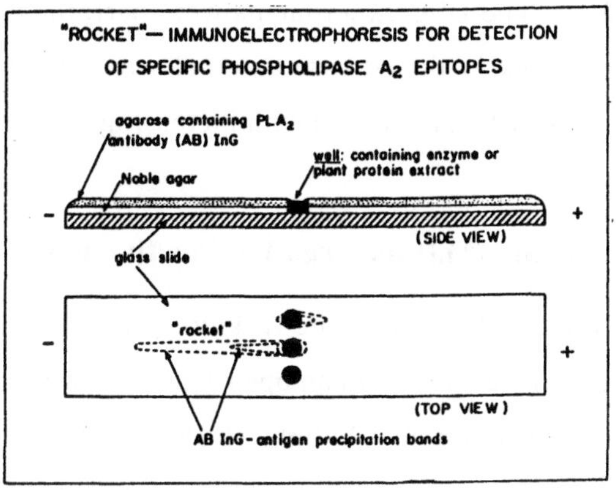

FIGURE 6. Method for rocket immuno-electrophoresis.

2. Effects of jasmonates on membrane fluidity

Differential scanning calorimetry (DSC). DSC is a highly sensitive method which enables membrane phase changes to be monitored. In this method, the timecourse of endothermic or exothermic heat flow is followed as a function of temperature change. The transition between phases, accompanied by heat loss or gain, is manifested by a demarcated-positive or negative-peak which is not accompanied by a temperature change of the tested sample. DPPC was chosen as the simulated membrane system since previous work by our group (Leshem *et al.*, 1990) has indicated its occurrence and putative mode of action in stomatal (and hence transpirational) control. Dipalmitoylphosphatidylcholine (DPPC) was dissolved in chloroform:methanol (2:1 v/v) and methyl jasmonate at the concentrations indicated in Fig. 7 in methanol only. The two solutions were mixed at appropriate concentrations. After mixing, the solvents were removed by a stream of N_2 and the samples placed under high vacuum for three hours. The dry lipid or the lipid-methyl jasmonate mixtures were weighed directly into the aluminium pans of the DSC Instrument and an excess of salt solution (0.15 M NaCl unbuffered or buffered with 10^{-2} M Tris HCl, pH 7.4) was added. The samples were incubated at 60 °C for 30 minutes. The DSC experiments were performed on a Du Pont 990 Thermal Analyser equipped with cell base 2 and at scan rate of 5 °C min^{-1}. Fig. 7 presents thermograms of DPPC alone and with increasing concentrations of methyl jasmonate.

3. Does methyl jasmonate (MJ) undergo UV inactivation?

UV irradiation of MJ. Methyl jasmonate (a kind gift of Takasago International Company, Tokyo, Japan) was placed in an open 1 ml thin-walled transparent glass vial and exposed for four hours to UV light of 150 μmol m^{-2} at 22 °C. This UV-irradiated MJ was compared to non-irradiated MJ. Four hours of UV irradiation sufficed to inactivate ABA completely.

Transpiration studies. Transpiration trials were carried out on two week old wheat (*Triticum aestivum*) plants grown in a growth chamber on a vermiculite : loam (2 : 1) mixture at 22 °C with a 16 hour photoperiod. Irradiance was as detailed in the rose petal experiment. Each treatment comprised four replicates each containing five plants. The treatments applied were controls – aqueous deionised H_2O in 10 ml test tubes – or 22 μM MJ (the optimal concentration for inhibition of transpiration; Satler and Thimann, 1981). Each MJ treatment was either UV-irradiated as described above or non-irradiated. To prevent water loss, the test tubes were sealed with Parafilm. Water loss was measured volumetrically after five days. Upon termination of the experiment, the fresh weight of foliage was assessed and transpiration expressed as a function of foliage fresh weight. This experiment was repeated three times, in each case producing results similar to those presented in Table 1.

Comparison of NMR spectra of UV-irradiated and non-irradiated MJ. NMR spectra were obtained using a Bruker NMR spectrometer (model AM 300 and AC-200) in deuterochloroform Cl_3 solutions; chemical shifts are reported in ppm downfield from internal tetramethylsilane and coupling constants in Hz (Torri *et al.*, 1977; Nishijana *et al.*, 1984). The signals were assigned with the help of

TABLE 1. Effect of methyl jasmonate on transpiration of wheat seedlings. See text for experimental details. Different superscripts indicate statistically different results.

Serial No.	Treatment	Transpiration (ml g^{-1} FW)
1	Control	12.24^a
2	22 μM methyl jasmonate	9.56^b
3	MJ + UV irradiation	9.68^b

two-dimensional correlated spectra, ^1H x ^1H (correlated spectroscopy) and ^{13}C x ^1H; thus ^1H chemical shifts are known even when multiplets overlap in a one-dimensional spectrum. Several of the ^1H signals cannot be identified according to first-order rules and are described as multiplets. The numbering classification was as follows:

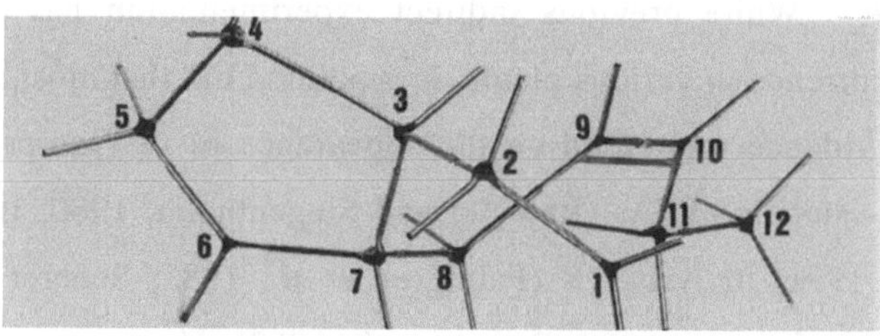

RESULTS AND DISCUSSION

The results of the 'Rocket' electro-immunophoresis clearly indicate that both young and senescing rose petals manifest an immunospecific presence of PLA_2 activity which is far greater in heat-stressed senescing petals than in young petals (Plate 2). It is also of interest that, as opposed to most mammalian, reptile and bacterial PLA_2, the 'rocket tail' observed here is cathodic and that, as compared to young petals, senescent petals produced additional precipitation lines, suggesting the appearance with ageing of additional PLA_2 isozymes.

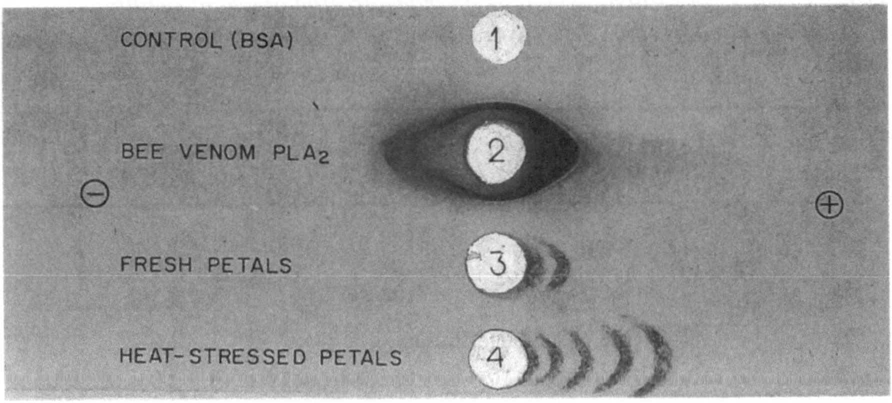

PLATE 2. *Identification of plant phospholipase A_2 in normal and stressed rose petals by 'Rocket' immunoelectrophoresis. Note the marked new epitope bands which appear with senescence and stress. Plant protein extracts were challenged with bee venom antibody immunoglobulin. Heat stress and senesence were obtained by keeping plants at 30 °C for four days.*

We believe these findings are the first direct demonstration of the existence of a plant PLA_2. While previous indirect experimentation has suggested its endogenous occurrence in various plants, it is pointed out that most, if not all, the corroborating evidence was based on the appearance of its *lyso*-products, these being either de-esterified FFAs (Rawyler and Siegenthaler, 1980; Leshem *et al.*, 1984; Leshem, 1991) or *lyso*-PCs (Palmgren *et al.*, 1985; Scherer *et al.*, 1989; Bafor *et al.*, 1991; Scherer, 1992; Banas *et al.*, 1992). However, as discussed in detail by Leshem (1992), these *lyso*-PLs and the pertinent de-esterified FFAs can also be produced by phospholipase B, which hydrolyses both the *sn*-1 and *sn*-2 fatty acyl chain of glycerolipids, or by another phospholytic enzyme, lipolyticacyl-hydrolase, and not necessarily by PLA_2.

Concerning the 'fluidising' action of MJ, Fig. 7 shows that, at low concentrations of MJ, the pretransition characteristic of DPPC disappears and the melting peak broadens. Further addition of the ester causes additional broadening of the melting peak. The total shift of Tm (temperature of the middle of the peak)

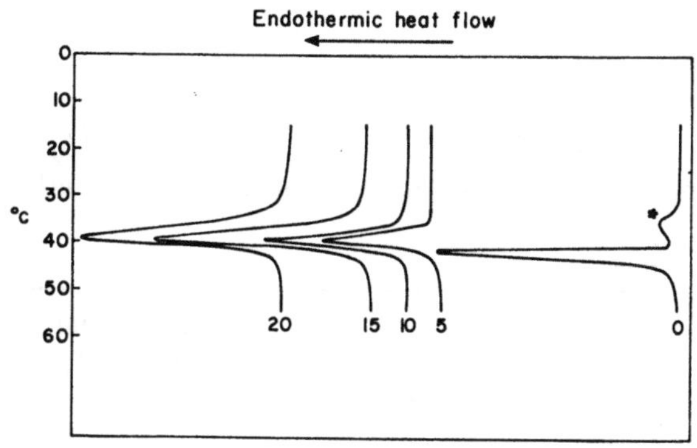

*FIGURE 7. Thermograms of DPPC interacting with methyl jasmonate. The numbers associated with the curves indicate the initial (mole %) concentrations of added methyl jasmonate. * indicates the pre-transition phase.*

is about 2.5 °C and the transition half-width is 5.5 °C, as compared to 1.5 °C for DPPC itself. Within experimental error, no change in the enthalpy of melting was detected due to the interaction of the phospholipid with MJ. The results presented show that MJ accumulates mainly in the head group region of the phospholipid with partial penetration into the hydrocarbon core, as indicated by the broadening of the peak due to the decrease in the size of the cooperative unit. The effect of MJ on DPPC is similar to the one obtained previously reported with abscisic acid by Bach (1986).

Effect of methyl jasmonate on the surface tension of a simulated membrane monolayer

The surface tension of a monolayer comprising a mixture of *sn*-1 palmitoyl, *sn*-2 linoleoyl and phosphatidylcholine (PLPC) was determined with a Wilhelmy Tensiometer and a Du Nöuy platinum rising using a procedure outlined elsewhere (Landau and Leshem, 1988). When applied, 10 µl of MJ was added before readings commenced. The buffer system of the subphase in the hour-glass was 0.01 M HEPES [4-2(hydroxyethyl)-1-piperazine ethane-sulphonic acid] at pH 7.5 containing 10^{-5} M $CaCl_2$. In the case of the non-treated control, an identical quantity of hexane solvent lacking MJ was added to the monolayer. Table 2 presents the results.

TABLE 2. *Surface tension of a PLPC monolayer as affected by methyl jasmonate. Experimental details are given in the text. Numbers in parentheses are standard deviations.*

Treatment	Surface tension ($mN\ m^{-1}$)
Buffer Control	56.6 (0.2)
PLPC	47.1 (1.7)
PLPC + MJ	42.7 (1.5)

Table 2 indicates that the presence of MJ in PLPC monolayers causes a decrease in surface tension of 4.4 mN m^{-1}. This finding complements that of the DSC data (Fig. 7) in another PL species, DPPC, indicating a lowering of transition temperature by 2.8 °C; both experiments indicated a fluidising effect of MJ. In keeping with previous findings (Satler and Thimann, 1981; Horton, 1991), MJ significantly reduced transpiration (Table 1). Moreover, no significant difference existed between the irradiated and non-irradiated MJ treatments. This finding is in keeping with the NMR data (listed below). When viewed together, the two experiments indicate that, in contrast to the phenomenon of ABA inactivation by UV by conversion of the *cis* to the *trans* forms, this does not occur in MJ. The following are the NMR spectra of the non-irradiated MJ and, in the context of the goals of the present research, the ^{1}H-NMR spectrum of UV-irradiated methyl jasmonate was identical to that of the starting material, indicating that no significant chemical transformation had taken place (d, t and m respectively indicate the doublet, triplet and multiplet states):

Methyl jasmonate - NMR ^{13}C (50.3 MHz) (Torri *et al.*, 1977): 218.78 (C-6), 172.51 (C-1), 134.04 (C-10), 125.01 (C-9), 53.97 (C-7), 51.57 (OMe), 38.80 (C-2), 38.03 (C-3), 37.72 (C-5), 27.32 (C-4), 25.50 (C-8), 20.59 (C-11), 14.09 (C-12).

^{1}H (300.1 MHz) (Nishiyama *et al.*, 1984): 5.45 (dtt, 11, 7,1.5, H-99), 5.26 (dtt, 11, 7.5, 1.5, H-10), 3.69 (s, 3H, OMe), 2.71 (m, H-2), 2.36 (m, 3H, H-5, H-8, H-8'), 2.32 (m, 2H, H-2', H-3), 2.24 (m, H-4), 2.10 (ddd, 18.5, 11, 9, H-5'), 2.05 (quintet dt, 7.5, 1.5, 0.5, 2H, H-11, H-11'), 1.90 (dtd, 10, 5, 1.5, H-7), 1.50 (m, H-4'), 0.96 (t, 7.5, 3H, H-12).

CONCLUSIONS

When viewed together with the overall hypothesis presented in the Introduction, the experimental results suggest at least one pathway whereby the manifestations of environmental signals are translated by plants as biochemical, biophysical and mechanical signals which may have far reaching effects. These may, on the one hand, be benign and form part of a plant stress adaptation syndrome; on the other hand, they may essentially be catabolic and cause the plant – initially and primarily at membrane sites – to succumb to environmental constraints and initiate irreversible senescence cascades. To quote Hans Selye who initially coined the biological stress concept over half a century ago, the former response is '*stress without distress*' while the latter is '*stress with distress*'.

ACKNOWLEDGEMENTS

Part of this research was financed by a grant from the USA-Israel Binational Research Foundation. We would like to thank Professor M. Sprecher of the Department of Chemistry, Bar-Ilan University, and Professor Y. Shai, Department of Membrane Research, Weizmann Institute, Israel for their aid and advice.

REFERENCES

Avtalion RR, Shahrabani R, Agassi R, Gringrass L (1984) Sex specific markers in *Tilapias*. Proceedings of the European Mariculture Society 8: 129-138

Bach D (1986) The effect of plant-hormone abscisic acid on model membranes - differential scanning colorimetry and planar bilayer membrane investigations. Biochimica Biophysica Acta 863: 313-317

Bafor M, Smith M, Jonsson L, Stobart K, Stymne S (1991) Ricinoleic acid biosynthesis and triacylglycerol assembly in microsomal preparations from developing castor bean endosperm. Biochemical Journal 280: 507-514

Banas A, Johansson I, Stymne S (1992) Plant microsomal phospholipases with high activities towards oxygenated acyl groups. In: Proceedings of the 10th International Symposium on Metabolism: Structure and utilization of plant lipids. Tunis, pp 1-30

Beatrice MC, Palmer JW, Pfeiffer DR (1980) The relationship between mitochondrial membrane permeability, membrane potential and the retention of Ca^{2+} by mitochondria. Journal of Biological Chemistry 225: 8863-8871

Davies WJ, Jones HG (eds) (1991) Abscisic acid: physiology and biochemistry. Bios Science Publishers. Oxford, UK, pp 280

de Gier J, van Echteld, CJA, van der Steen ATM, Noordam PC, Verkleij AJ, de Kruijff B (1982) Lipid organization and barrier functions of membranes. In: Wintermans JFGN, Kuiper PJC (eds) Biochemistry and metabolism of plant membranes. Elsevier, Amsterdam, The Netherlands, pp 315-325

Demel R, Geurt W, van Kessel A, Zwaal R, Roelofsen B, van Deenen LM (1975) Relation between various phospholipase actions and the interfacial phospholipid pressure in monolayers. Biochimica Biophysica Acta 406: 97-107

Farmer EE, Ryan CA (1992) Octadecanoid derived signals in plants. Trends in Cell Biology 2: 236-241

Grechkin AN, Kuramshin RA, Safonava EY, Yefremov YJ, Latypov SK, Ilyasov AV, Tarchevsky IA (1991) Double peroxidation of gamma-linolenic acid by potato tuber lipoxygenase. Biochimica Biophysica Acta 1081: 79-84

Hendry GAF, Finch-Savage WE, Thorpe PC, Atherton NM, Buckland SM, Nilsson KA, Seel WA (1992) Free radical processes and loss of seed viability during desiccation in the recalcitrant species *Quercus robur* L. The New Phytologist 122: 273-279

Horton RJ (1991) Methyl jasmonate and transpiration in barley. Plant Physiology 96: 1376-1378

Johnson AM (1986) Immunoprecipitation in gels. In: Rose NR, Friedman H, Fahey JL (eds) Manual of clinical laboratory immunology. American Society of Microbiology, Washington, pp 14-24

Landau EM, Leshem YY (1988) Biophysical interactions of membrane anionic phospholipids with pH, calcium and auxins. Journal of Experimental Botany 39: 1689-1697

Laurell CB (1965) Quantitative estimation of proteins by electrophoresis in agarose gel containing antibodies. Analytical Biochemistry 10: 358-361

Leshem YY (1987) Membrane phospholipid catabolism and Ca^{2+} activity in control of senescence. Physiologia Plantarum 69: 551-559

Leshem YY (1988) Plant senescence process and free radicals. Free Radical Biological Medicine 5: 27-38

Leshem YY (1991) Evidence for the presence and mode of action of an endogenous plant phospholipase A_2. In: Quinn PJ, Harwood JL (eds) Plant lipid biochemistry, structure and function. Portland Press, Colchester, UK, pp 53-55

Leshem, YY (1992) Plant membranes: a biophysical approach to structure, development and function. Kluwer Academic Publishers, Dordrecht, The Netherlands, pp 175

Leshem YY, Cojocaru M, Margel S, El-Ani D, Landau EM (1990) A biophysical study of abscisic acid interaction with a phospholipid membrane component. The New Phytologist 116: 487-498

Leshem YY, Halevy AH, Frenkel C (1986) Processes and control of plant senescence. Elsevier, Amsterdam, The Netherlands, pp 214

Leshem YY, Margel S, Aurbach D, Soffer Y (1993) Plant membrane senescence: biophysical parameters of polyunsaturated fatty acid hydroperoxidation as assessed by surface behavior and Fourier transform infra-red spectrometry. Plant Growth Regulation (in press)

Leshem YY, Sridhara S, Thompson JE (1984) Involvement of calcium and calmodulin metabolism in membrane deterioration during senescence of pea foliage. Plant Physiology 75: 329-335

Lowry OH, Rosenbrough NJ, Farr AL, Randall RJ (1951) Protein measurement with the folin phenol reagent. Journal of Biological Chemistry 193: 265-275

Lucy JA (1970) The fusion of biological membranes. Nature (London) 227: 815-817

Mazliak P, (1980) Synthesis and turnover of plant membrane phospholipids. Progress in Phytochemistry 6: 49-102

McKersie BD, Senarata T, Walker MA, Kendall EJ, Hetherington PR (1988) Deterioration of membranes during aging in plants: evidence for free radical mediation. In: Nooden LD, Leopold AC (eds) Senescence and aging in plants. Academic Press, New York, USA, pp 442-465

Nishiyama H, Sakuta K, Itoh K (1984) Conjugate addition of lithiotrimethyl-silylacetate: A simple synthesis of methyl jasmonate via vicinal double alkylation. Tetrahedron Letters 25: 2487-2488

Norberg P, Nilsson R, Liljenberg C (1991) Alterations in lipid composition and phase behavior of oat root plasma membranes after reduction of dehydration tolerance. In: Quinn P, Harwood JL (eds) Plant lipid biochemistry, structure and utilization. Portland Press, London, UK, pp 65-68

Palmgren MJ, Sommarin M, Ulskov P, Jorgensen PJ (1988) Modulation of plasma membrane H^+ ATPase from oat roots by lysophosphatidylcholine free fatty acids and phospholipase A_2. Physiologia Plantarum 74: 11-19

Parthier B (1990) Jasmonates: hormonal regulators or stress factors in leaf senescence? Journal of Plant Growth Regulation 9: 57-63

Rawyler A, Siegenthaler PA (1980) Role of lipids in function of photosynthetic membranes by treatment with lipolytic acyl hydrolase. European Journal of Biochemistry 247: 6970-6977

Satler SO, Thimann KV (1981) Le jasmonate de methyl: nouveau et puissant promoteur de la senescence des feuilles. Comptes Rendus Academie des Sciences, Paris, Series III 293: 735-740

Scherer GFE (1992) Activation of membrane associated protein kinase by lipids and its function in signal transduction. Journal of Experimental Botany 43: 7 (Supplement)

Scherer GFE, André B, Martiny-Baron G (1989) A rapid response to a plant hormone: auxin stimulates phospholipase A_2 *in vivo* and *in vitro*. Biochemical and Biophysical Research Communications 163: 111-117

Sembdner G, Gross D (1986) Plant growth substances of plant and microbial origin. In: Bopp M (ed) Plant growth substances 1985. Springer-Verlag, Berlin, Germany, pp 139-147

Shinitzky M (1984) Membrane fluidity and cellular functions. In: Shinitzky M (ed) Physiology of membrane fluidity. Vol 1. CRC Press, Boca Raton, USA, pp 1-51

Shomer-Ilan A, Avtalion RR, Leshem YY (1973) Further evidence for the presence of an endogenous gonadotropin-like factor 'phytotropin': isolation and mechanism of the active principle. Australian Journal of Biological Sciences 26: 105-112

Sridhara S, Leshem Y (1986) Phospholipid catabolism and senescence of pea foliage membranes: parameters of Ca^{2+}: calmodulin: phospholipase A_2-induced changes. The New Phytologist 102: 5-16

Stafford RE, Dennis EA (1988) Lysophospholipids as surfactants. Colloids and Surfaces 30: 47-64

Torri S, Tanaka H, Kobayashi J (1977) Synthesis of methyl *dl*-jasmonate and its related compounds from methyl (*E*)-and (*Z*)-4,4-dimethoxy-2-butanoates. Journal of Organic Chemistry 42: 3473-3477

Ueda J, Kato J (1980) Isolation and identification of a senescence promoting substance from wormwood (*Artemisia absinthum*). Plant Physiology 66: 246-249

Weltzien HV (1979) Cytolytic and membrane perturbing properties of lysophosphatidylcholine. Biochimica Biophysica Acta 559: 259-287

Wong PYK, Cheung WY (1979) Calmodulin stimulates human platelet phospholipase A_2. Biochemical and Biophysical Research Communications 90: 473-480

SECTION II

Atmospheric Pollution

SIGNIFICANCE OF INCREASING AMBIENT CO_2 FOR PLANT GROWTH AND SURVIVAL, AND INTERACTIONS WITH AIR POLLUTION

H.Z. Enoch and S.J. Honour[1]
Soils and Water Institute
Volcani Center
Agricultural Research Organisation
Bet Dagan
Israel

ABSTRACT

Anthropogenic activities have changed the chemical composition of the global atmosphere. Since the middle of the nineteenth century atmospheric CO_2 has increased by about 25%, whereas global warming has increased mean surface temperature by only 0.6 °C. Throughout the industrialised regions the air pollutants, O_3, SO_2, NO_x and others, are causing increasing damage to vegetation. The temporal and spatial distribution of air pollutants differs by orders of magnitude between the various managed and unmanaged global ecosystems and hence their effect on vegetation varies greatly. The significance of increasing atmospheric CO_2 for plant growth is discussed in terms of its influence on the partitioning of dry matter, leaf growth and development, stem growth and branching, root formation and tillering, growth of the whole plant and flowering. Details are given concerning the influence of CO_2 on the growth and development of plants from different ecosystems. Special attention is given to the interaction of CO_2 concentration, climate and altitude on vegetation in the unmanaged and usually stressed ecosystems of the Arctic, temperate and sub-tropical zones. Some processes which are likely to enhance plant survival at higher CO_2 concentrations are described. Finally, some examples of the ameliorating effects of elevated CO_2 on air pollution damage in plants are outlined.

INTRODUCTION

The object of this chapter is to describe aspects of plant growth in various atmospheric CO_2 concentrations, with and without minor atmospheric constituents ('air pollutants'). Then, in evaluating the significance of these phenomena, we will

---

[1] Institute of Environmental and Biological Sciences, Lancaster University, Lancaster LA1 4YQ, UK.

NATO ASI Series, Vol. I 16
Interacting Stresses on Plants in a Changing Climate
Edited by M. B. Jackson, C. R. Black

attempt to make quantitative statements about the expected changes. These quantitative assessments are based on literature published over the last decade (Acock and Allen Jr, 1985; Idso, 1989; Allen Jr, 1990; Goudriaan *et al.*, 1990). We will restrict our investigation to the largest group of higher plants, the C3 species. Evaluations of the reaction of C4 and CAM plants to elevated CO_2 can be found elsewhere (Nobel, 1991; Ziska *et al.*, 1991).

INFLUENCE OF ATMOSPHERIC CO_2 CONCENTRATION ON PLANTS

Since the beginning of the 19th century it has been known that vegetative growth, development and yield are enhanced when the bulk atmospheric CO_2 concentration increases. This is the principle underlying the common horticultural practice of CO_2 enrichment in greenhouses (Enoch and Kimball, 1986). Where control of CO_2 concentration is employed it is often, for economic reasons, maintained at about 1000 ppm CO_2 (or approximately three times ambient). However, in experiments using 3, 5 and 10-fold increases, it is generally found that yield increases with CO_2 until concentrations of several thousand ppm are reached (see Enoch *et al.*, 1976). Thus, it is likely that well-watered and adequately nourished plants will benefit from future increases in global atmospheric CO_2. The global CO_2 concentration is likely to continue to increase until all fossil fuel is used and/or mankind changes to another energy source. Expected peak values are above 2000 ppm, although these will probably not be reached for several hundred years. Trends in minor

TABLE 1*. Global atmospheric concentration. After Lippold (1992).*

	Carbon dioxide (ppm)	Methane (ppb)	Nitrous oxide (ppb)	Ozone (ppb)
Pre-industrial	280	800	288	10
Present (1991)	355	1740	311	40
Yearly increase (%)	0.5	0.75	0.25	0.5
Atmospheric lifetime (years)	50-200	10	140	0.1

atmospheric constituents are discussed in various publications (Boden *et al.*, 1991; Lippold *et al.*, 1992) and are summarised in Table 1. Our primary consideration will be the reaction of plants to elevated CO_2 levels, and the modifications which air pollutants such as O_3, SO_2 and NO_x may exert on both CO_2-enriched and non-enriched plants.

Temperature interactions. Idso *et al.* (1987) found that CO_2 enrichment increased growth at temperatures above a critical value (about 19 °C for some plants) and had no or negative effects at lower temperatures. Most climate change models predict that the mean surface temperature will increase by about 2-3 °C for each doubling of CO_2 concentration. It is remarkable that the optimal temperature for maximal CO_2 uptake also increases by about 2-3 °C for each doubling of CO_2 concentration (Enoch and Hurd, 1977). This is either a strange coincidence or an adaptation of plants to the various simultaneous variations in CO_2 concentration and temperature that have taken place over millennia. Data for CO_2 concentration and global temperatures deduced from ice cores from Antarctica show that over the last 160,000 years, CO_2 concentration was low during the glacial periods and higher in the warmer interglacial periods. Obviously mankind has only influenced global CO_2 levels significantly during the last few centuries.

Dry matter partitioning. It is convenient to consider the response of various plant organs in turn. Recent information concerning the effect on growth of applying elevated CO_2 to C3 plants reveals that high CO_2 does not always increase root to shoot ratio, although some earlier work seemed to indicate that it does (Enoch and Zieslin, 1988). Some researchers have found an increase in root : shoot ratio (Lakso *et al.*, 1986; Desjardins *et al.*, 1987; Norby and O'Neill, 1988; Ziska *et al.*, 1991) while others have reported no change (Fajer *et al.*, 1991) or a decrease (Oberbauer *et al.*, 1985; Poorter *et al.*, 1988; Nijs *et al.*, 1989; Ziska *et al.*, 1991). Such variation may reflect interspecific differences or varying experimental conditions. For example, at low levels of nutrients there is a tendency towards higher root weight, and *vice versa*. Also light levels, duration of CO_2 enrichment and plant age all affect carbon partitioning.

Leaf growth and development. CO_2 enrichment generally increases the number of leaves per shoot (Poorter *et al.*, 1988), though sometimes no change has been reported (Paez *et al.*, 1983). In most cases, high CO_2 increases branching and tillering (Tissue and Oechel, 1987), possibly by reducing apical dominance (Enoch and Zieslin, 1988). Thus, the overall number of leaves per plant generally increases in response to CO_2 enrichment. If there is no reduction in individual leaf size, this will increase total leaf area per plant, as in soybean (Allen Jr *et al.*, 1982). Many studies have shown a slight increase in the area of individual leaves (Ford and Thorne, 1967, Sionit *et al.*, 1982) in conjunction with a more pronounced increase in leaf weight. Hence the specific leaf area (SLA - leaf area per unit of leaf weight) decreases in response to elevated CO_2 (Cooper and Brun, 1967; Ford and Thorne, 1967; Hurd, 1968; Hofstra and Hesketh, 1975; Acock *et al.*, 1981; Allen Jr *et al.*, 1982; Oberbauer *et al.*, 1985; Desjardins *et al.*, 1987; Mousseau and Enoch, 1989; Fajer *et al.*, 1991; Ziska *et al.*, 1991). The increase in leaf weight is due to the presence of extra layers of palisade cells and more densely packed mesophyll cells (Madsen, 1968; Hofstra and Hesketh, 1975; Rogers *et al.*, 1980).

Stem growth. Elevated CO_2 treatments have been found to increase stem dry weight and stem diameter (Ford and Thorne, 1967; Funsch *et al.*, 1970; Rogers *et al.*, 1980; Mortensen and Sandvik, 1984). Much of the work on stem physiology has been carried out on trees but CO_2 enrichment may also increase the length of flower stems in various herbaceous species (Shaw and Rogers, 1964; Aulenbach, 1965; Jensen, 1983).

Root formation and growth. Elevated atmospheric CO_2 concentrations often enhance root formation and increase root dry weight (Rogers *et al.*, 1980; Allen Jr *et al.*, 1982; Foschner and Reuther, 1984; Mousseau and Enoch, 1989).

Branching and tillering. Increased lateral branching has been found in various herbaceous species including peas, carnations, chrysanthemums, soybean and Campanulas (Goldsberry, 1963; Andersen, 1976; Paez *et al.*, 1980; Luria *et al.*, 1985; Zieslin and Moe, 1985). Many monocotyledonous plants produce more tillers, both in agricultural systems (Ford and Thorne, 1967; Imai and Murata,

1976; Sionit *et al.*, 1982) and natural ecosystems (e.g., *Eriophorum vaginatum*, Tissue and Oechel, 1987).

Whole plant growth. There is wide variation in the results obtained for different species, although total biomass production is often considerably greater at high CO_2 concentrations than at ambient levels (Mousseau and Enoch, 1989). The growth of CO_2-enriched plants is often inhibited at low temperatures (Idso *et al.*, 1987).

Flowering. The effect of CO_2 enrichment on flowering has been extensively reviewed by Enoch and Zieslin (1988). Atmospheric CO_2 may influence flowering, either by acting directly on bud development, floral induction and differentiation, or indirectly *via* branching and stem elongation. Elevated CO_2 (1000-3000 ppm) has also been reported to cause earlier flowering in some species, including flax (Hovland and Dybing, 1973) and cucumber (Enoch *et al.*, 1976). Increasing atmospheric CO_2 concentration generally results in the production of more flowers per plant (Shaw and Rogers, 1964; Lindstroem, 1965; Hovland and Dybing, 1973; Enoch *et al.*, 1976; Penningsfield and Forcthammer, 1980; Sionit *et al.*, 1981; Tsujita *et al.*, 1982; Liu, 1983; van Berkel, 1983). Zieslin *et al.* (1986) also found that CO_2 enrichment of roses enhanced the formation of basal shoots, resulting in the renewal of rose bushes.

Sex expression. The proportions of female to male flowers and of fertile to infertile flowers may both be increased by elevated CO_2 (Imazu *et al.*, 1967). Work by Palloix *et al.* (1985) showed that elevated CO_2 diminishes self-incompatibility in cauliflowers. It has been found (Hicklenton, 1978) that very high CO_2 concentrations (10000-50000 ppm) can induce flowering in short day plants (e.g., Morning Glory, *Pharbitis nil*) even when grown in long day conditions. These and similar experiments suggest either that CO_2 itself acts as a hormone, or that it influences the plant hormone system, possibly by acting as an antidote to endogenously produced ethylene. As ethylene is produced in larger quantities in stressed plants, it is likely that global CO_2 enrichment will play an increasingly significant role in reducing damage to stressed plants.

INFLUENCE OF CO_2 CONCENTRATION ON ECOLOGICAL SYSTEMS

It is impossible to generalise about the effects of elevated CO_2 on plants on a global scale, since the climate and more localised environment which individual species experience are bound also to affect plant growth. We will therefore consider the likely impact of increasing atmospheric CO_2 concentration on vegetation in different climatic zones.

ALTITUDINAL EFFECTS

Although some variation occurs in mountain climates depending on their global location and regional topography, the general effects of altitude on physical quantities such as temperature, radiation and windspeed are remarkably uniform across the world. Thus it is appropriate to consider montane habitats as a distinct mesoclimate (Friend and Woodward, 1990). Predictions of the responses of upland plants to future atmospheric CO_2 concentrations are mostly based on plant responses to present CO_2 levels at various barometric pressures.

Increasing altitude is associated with a decrease in air pressure and density (Barry, 1981; Körner and Cochrane, 1985; Woodward, 1986; Friend and Woodward, 1990). This is particularly important when considering CO_2 uptake by plants, since the CO_2 partial pressure in the atmosphere also decreases with altitude (Woodward, 1986; Körner *et al.*, 1988). Hence the CO_2 partial pressure gradient between the atmosphere and the sites of CO_2 fixation within the cells is decreased. It has been suggested (Körner and Diemer, 1987; Körner *et al.*, 1989; Perry and Borchers, 1990) that plants at high altitudes may therefore be physiologically more sensitive to changes in atmospheric CO_2, and would benefit more from increased global CO_2 levels than lowland plants. There is, however, some debate about whether this is the case and, if it is true, whether such a benefit would be directly due to increased CO_2 levels or an indirect effect acting through changes in temperature. Gale (1972) stated that *'the reduction of the availability of carbon dioxide for photosynthesis with altitude, due to reduced barometric pressure, is only very small. The increased rate of diffusion of CO_2 in air with altitude compensates for the effect of its lower partial pressure.'*

Körner and Diemer (1987) conducted a survey of 12 pairs of altitudinally separated herbaceous plant species, and concluded that an increase in atmospheric CO_2 of 100 ppm would increase carbon assimilation rate by 9% at low elevation and 21% at high elevation. Any reduction in atmospheric pressure also reduces the partial pressure of water vapour, and so evaporation increases. In addition, the general increase in windspeed with altitude (Körner and Cochrane, 1985; Friend and Woodward, 1990) and the smaller leaf size of upland plants reduces boundary layer resistance (Körner and Cochrane, 1985; Körner *et al.*, 1989; Monteith and Unsworth, 1990), thus facilitating transpiration (but also CO_2 uptake).

The characteristic morphology of upland plants is to have small leaves with a dwarf habit. Smith and Geller (1979) showed that small leaves with a low stomatal resistance have similar transpiration rates to larger leaves with a high stomatal resistance. This means that plants at high altitudes offer less resistance to CO_2 uptake by having a low stomatal resistance (Körner and Cochrane, 1985; Woodward, 1986; Körner *et al.*, 1986; Körner and Diemer, 1987). There is also a tendency towards a greater stomatal density at high altitudes (Körner and Cochrane, 1985; Körner *et al.*, 1986; Woodward, 1986; Körner *et al.*, 1989; Friend and Woodward, 1990). Water conservation is rarely a priority in upland plants due to the high levels of precipitation experienced, but the generally shallow soils with their low water holding capacity may negate this effect. The efficiency of CO_2 uptake (assimilation rate) and fixation *via* photosynthesis increases with altitude (Körner and Diemer, 1987; Körner and Renhardt, 1987; Körner and Woodward, 1987; Körner *et al.*, 1988; Körner *et al.*, 1989; Friend and Woodward, 1990).

High altitude plants often exhibit similar morphological and physiological changes as lowland plants grown in elevated CO_2. For example, the leaves are thicker, often because they have more palisade cell layers, and the specific leaf area is lower (Woodward, 1983; Körner and Cochrane, 1985; Körner *et al.*, 1986; Körner and Diemer, 1987; Körner *et al.*, 1989; Friend and Woodward, 1990). Whether such characteristics would be exaggerated still further under conditions of elevated CO_2 is debatable. It is also unlikely that the usual responses of plants to high CO_2 concentrations, such as increased stem height, leaf area and biomass would be advantageous to mountain plants which must resist wind damage.

Although experiments have been carried out on high altitude plants at different CO_2 concentrations, these have tended to involve transplantation of specimens to controlled growth chambers at low altitudes (Woodward, 1986; Honour, unpublished data). The practical difficulties of arranging a system of CO_2 enrichment at high altitudes without affecting other environmental variables has meant that experiments have not adequately reproduced growing conditions in a mountain climate at elevated CO_2 concentration.

CLIMATE ZONE VARIATIONS

Different environmental variables within each of the world's climate zones will modify the effects of elevated atmospheric CO_2 on plant growth.

The polar climate zone

The paucity of literature regarding Arctic vegetation, particularly involving field experiments, means that the probable responses of such plants to elevated global CO_2 concentration is a matter of conjecture. The climatic characteristics of polar regions are low temperatures, limited available water supply when the ground is frozen and frequent strong winds. The low solar angle and seasonal extremes in solar radiation also provide a low photon flux density. During the summer the daylight period lasts for 24 hours, while in winter darkness persists continuously.

If low temperatures reduce the beneficial effect of elevated CO_2 levels (Idso *et al.*, 1987), then it would be expected that little change or possibly a reduction in photosynthetic rates or primary productivity would occur. Tissue and Oechel (1987) found that *Eriophorum vaginatum* grown in the field in open-top chambers in the Arctic at a CO_2 concentration of 680 ppm photosynthesised at the same rate as plants grown in 340 ppm CO_2, when each measurement was made at the appropriate growth concentration. Elevated atmospheric CO_2 reduces stomatal conductance, thereby limiting transpiration and increasing water use efficiency. Paez *et al.* (1983) grew Alaska peas in a controlled environment and found no beneficial effect of CO_2 on well-watered plants. However, drought-stressed plants showed an increase in height, leaf area and dry weight with increasing CO_2 concentration, presumably because CO_2 acted as an antitranspirant.

The tropical and sub-tropical climate zone

The main environmental characteristics of both tropical and sub-tropical climates are high temperature and light intensity. A distinction must be drawn, however, between humid and arid regions.

The humid region. High levels of light, CO_2, temperature and available water would appear to provide ideal conditions for plant growth. Plant nutrients are in adequate supply in the decaying litter on the forest floor. Ziska *et al.* (1991) conducted experiments on nine tropical species, including C3, C4 and CAM plants and both woody and herbaceous species. There is considerable interspecific variation even within, for example, the herbaceous C3 species. In some instances there was up to a threefold increase in leaf area, dry weight and root to shoot ratio. Photosynthesis increased in all C3 species but not in C4 or CAM plants. However, water use efficiency increased in both C3 and C4 plants, but was not measured for CAM species. A similar experiment by Oberbauer *et al.* (1985) indicated similar effects with respect to total weight, leaf area, height and root to shoot ratio, but a significantly larger water use efficiency at elevated CO_2 concentrations for both species studied.

The arid region. Plants growing in tropical or sub-tropical arid zones are usually adapted primarily to water shortage. Idso and Kimball (1992) found that the stomatal conductance of sour orange trees was insensitive to increases in atmospheric CO_2 up to 1000 ppm. This, surprisingly, ensures that stomata remain open, even at high CO_2 concentrations. It was observed that the fine root biomass was increased by raising CO_2 concentration from 400 to 1000 ppm. This phenomenon, in conjunction with a marked decrease in dark respiration, resulted in an almost four-fold enhancement of growth.

Temperate climate zone

The initial part of this chapter, describing the broad effects of CO_2 on plant life, has dealt mainly with the responses of temperate plants. It is likely that temperate vegetation systems together with the humid tropical and subtropical vegetation will respond most to elevated atmospheric CO_2. In the temperate zone, summer temperature is sufficiently high to allow photosynthesis to increase with increasing CO_2. In most temperate regions, water is not the most important limiting factor, although considerable variation in the amount and distribution of precipitation occurs. Maximal response to high CO_2 can be expected where plants are well supplied with nutrients and water, and are growing at relatively high temperatures. Such a response is therefore limited to selected parts of the temperate zone in the summer and the humid parts of the sub-tropical and tropical zone.

PLANT STRESS REDUCTION BY GLOBAL CO_2 ENRICHMENT

When the growth of stressed plants at ambient atmospheric CO_2 concentration is compared with plants in elevated CO_2 (above 600 ppm), it is generally observed that CO_2 enrichment diminishes symptoms of stress. Thus CO_2-enriched plants can survive and even thrive under some level of stress. A few examples are given below.

Salinity stress. Enoch *et al*. (1973) grew tomato plants in nutrient-rich salt solutions containing greater than 1% total salt. At 8000 ppm CO_2 the plants grew better than plants grown at 350 ppm CO_2 in ordinary Hoagland's solution. A more detailed investigation by Schwartz and Gale (1984) confirmed these findings.

Water stress. Stomatal conductance is generally decreased by elevated CO_2. Thus CO_2 acts as an anti-transpirant, reducing water stress and improving water use efficiency. Examples of this phenomenon are given in previous sections dealing with climate zone variations.

Oxygen stress. Atmospheric oxygen inhibits CO_2 uptake by C3 plants (e.g., Long and Drake, 1991) but this inhibition is reduced by elevated CO_2. This phenomenon will be discussed in more detail later.

Ethylene stress. Living plant cells produce ethylene at all times, but stressed plants produce more (Abeles, 1985). Ethylene acts as a hormone and at high concentrations is phytotoxic. Research into the site of ethylene action in plants has revealed that CO_2 can replace the ethylene molecule (Burg and Bars, 1987). Elevated CO_2 therefore competes with ethylene for the reaction site, thus acting as an ethylene inhibitor. Considerable reduction of root growth occurs at ethylene concentrations of 1-10 ppm.

AIR POLLUTANTS

Limited data have been produced concerning the interactive effects of atmospheric CO_2 and air pollution by O_3, NO_x and SO_2. Mulchi *et al.* (1992) observed changes in soybean biomass and seed production which indicated that an additional 150 ppm of CO_2 in the atmosphere played a protective role against the adverse effects of ozone exposure. A similar result was obtained by Heck and Dunning (1967) for tobacco and pinto bean.

In a similar way, elevated CO_2 reduced the sensitivity of various plant species to SO_2 damage (Black, 1982; Carlson and Bazzaz, 1982, 1985). NO_x pollution, a by-product of CO_2 production in many greenhouse systems, has been shown to reduce the growth-enhancing effect of CO_2 enrichment (Caporn, 1988). Little work has been done on CO_2/NO_x interactions with respect to climate change, but it is reasonable to suggest that, since elevated CO_2 reduces stomatal conductance, the entry of all gaseous air pollutants will be reduced. Thus the damage caused by NO_x pollution may be reduced at high CO_2 concentrations.

QUANTITATIVE ASSESSMENT OF RECENT DATA

The results of various investigations are presented in Figs. 1-3 as the log of the ratio of various growth parameters measured under elevated CO_2 concentrations to

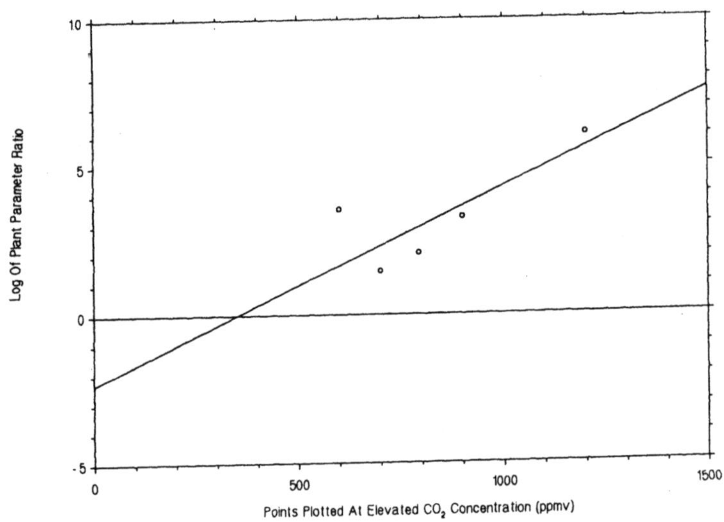

FIGURE 1. *Best linear fit of log plant response ratio at elevated CO_2 relative to the 350 ppmv ambient concentration: net assimilation rate; y = a (x - 350), error^2 0.054, a = 6.573E-004.*

the same parameter measured at the present ambient CO_2 concentration. This form of presentation was chosen in order to compare the dissimilar data in a uniform manner, and is the same approach as used by Kimball in various reviews comparing the yields of CO_2-enriched plants with those growing in ambient CO_2. The variability in the correlations obtained clearly indicates that it is impossible to overcome the variability in the data and make any conclusive statement about the effects of elevated CO_2 on plant growth.

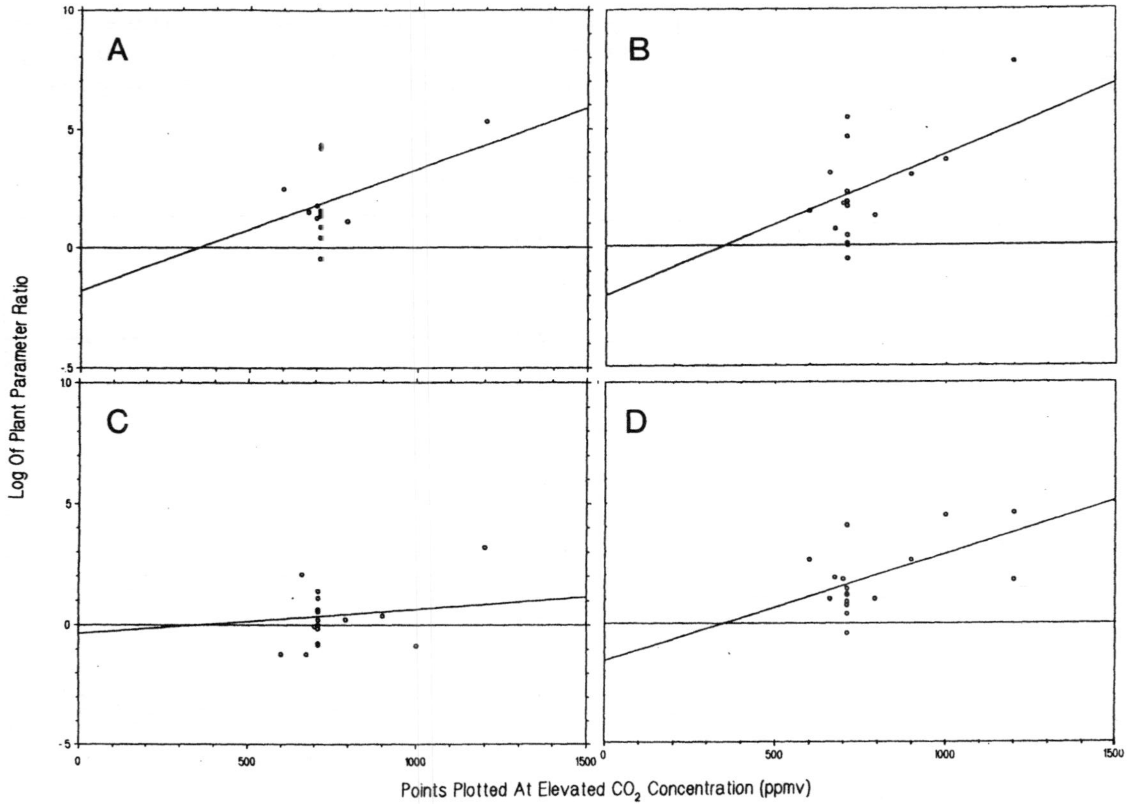

***FIGURE 2.** Best linear fit of log plant response ratio at elevated CO_2 relative to the 350 ppmv ambient concentration. A: Plant weight; $y = a(x - 350)$, $error^2$ 0.248, $a = 5.110E{-}004$. B: Root weight; $y = a(x - 350)$, $error^2$ 0.492, $a = 5.934E{-}004$. C: Shoot to root weight ratio; $y = a(x - 350)$, $error^2$ 0.203, $a = 1.002E{-}004$. D: Shoot weight; $y = a(x - 350)$, $error^2$ 0.298, $a = 4.414E{-}004$.*

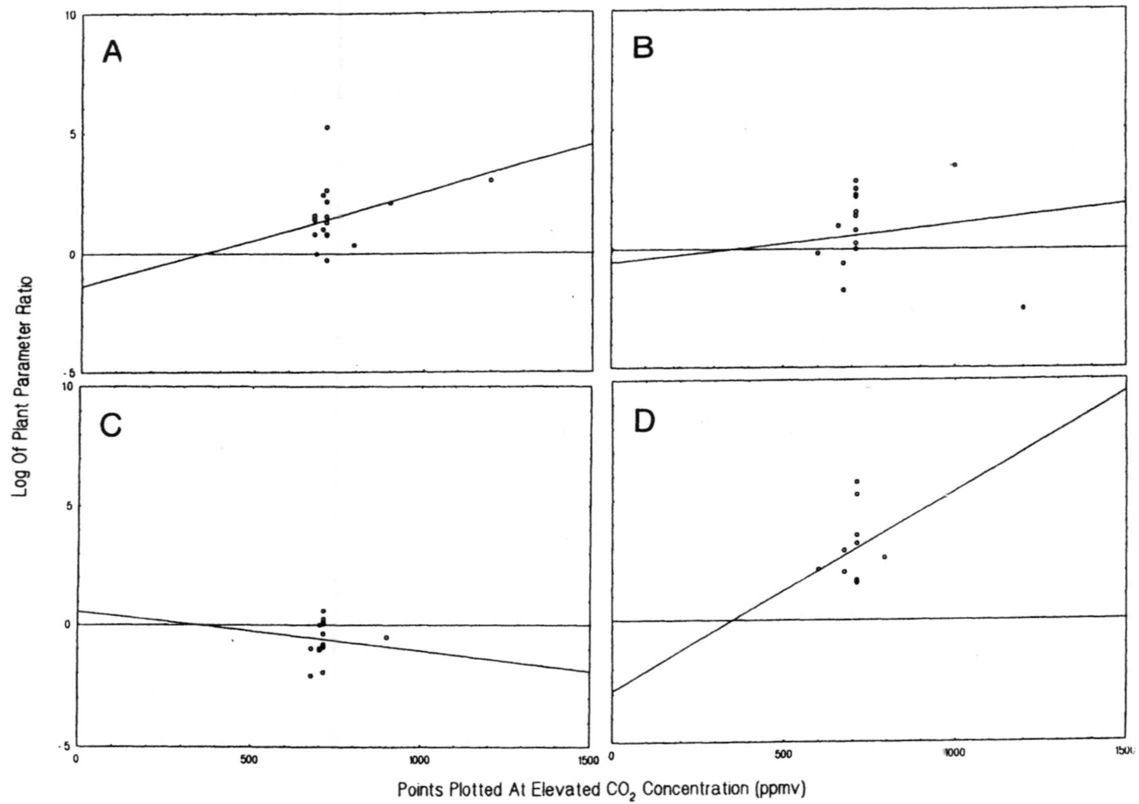

***FIGURE 3**. Best linear fit of log plant response ratio at elevated CO_2 relative to the 350 ppmv ambient concentration. A: Leaf area; y = a (x - 350), error^2 0.253, a = 3.902E-004. B: Net photosynthesis rate; y = a (x - 350), error^2 0.438, a = 1.599E-004. C: Specific leaf area; y = a (x - 350), error^2 0.086, a = -1.643E-004. D: Water use efficiency; y = a (x - 350), error^2 0.203, a = 8.243E-004.*

CONCLUSIONS

The title of this chapter implies an evaluation of changes in the chemical composition of the atmosphere due to anthropogenic activity. The term ‘pollution’ is used to describe substances introduced into the environment by man which cause unwanted changes in the biota. An increase in global CO_2 concentration may be referred to as CO_2 enrichment if it alters plant growth in a manner which is perceived to be beneficial to mankind. Changes in the Earth’s surface temperature are often sensationalised in scientific papers and reports to political bodies, being portrayed as a danger to the wellbeing of the human race (Lippold, 1992). However, it would be difficult to find many laymen in Britain who would object to the 0.6 °C warming that has taken place over the last century or so. The prospect of a slightly warmer environment would be considered a blessing by most people in the temperate zone.

Misleading terms are often used to describe the influence of minor atmospheric constituents on global surface temperature. Such gases are often referred to as ‘greenhouse gases’ and the resulting global warming as ‘the greenhouse effect’. The allusion to greenhouses is misleading for two reasons. Firstly, because the environment inside a greenhouse is not necessarily warmer than outside (a plant enclosure that is warmer is called a ‘hothouse’ by the horticultural community). Secondly, greenhouses without a fossil fuel heat input are only warmer than the outside bulk atmosphere during daylight hours because air is trapped between the ground and the transparent envelope. The trapped air is heated by the solar radiation absorbed by the ground cover and re-emitted by convection to the air. It makes little difference whether the enclosure is covered by glass, which is opaque to long wave radiation, or by polyethylene, which is transparent to long wave radiation (Dayan *et al.*, 1986). The earth is warmer than the moon because some atmospheric gases act as a radiation shield. Thus, the so-called ‘greenhouse gases’ would be more appropriately described as ‘Radiatively Important Gases’ (RIG; Wuebbles and Edmonds, 1991) or ‘Radiation Shield Gases’ (RSG).

Oxygen stress. Plants have changed the global atmosphere by depleting CO_2 and releasing oxygen. Photosynthesis is depressed in C3 plants by the atmospheric abundance of the plant-produced gas, oxygen. In the atmosphere of the early Earth only traces of oxygen occurred and minerals containing iron did not rust. But it is widely believed that 3800 million years ago autotrophic micro-organisms began to photosynthesise, thereby breaking up water molecules and releasing free atmospheric oxygen (Margolis and Lovelock, 1974). Approximately 3000 million years of bacterial photosynthesis have resulted in the modern atmosphere with its high oxygen level. Although the multicellular plants that subsequently developed had to find mechanisms for overcoming the inhibitory effect of oxygen, they still reveal in their photosynthetic characteristics that they have not adapted to the modern atmosphere that has prevailed for about 600 million years. When C3 plants are grown at 350 ppm CO_2 and 2% oxygen, instead of the ambient 21% O_2, their rate of photosynthesis is close to maximal and their photon efficiency is near the theoretical maximum. However, at 21% O_2 the rate of photosynthesis and the photon efficiency are lower at 350 ppm than at elevated CO_2 levels of 500-1500 ppm. It can therefore be said that the atmosphere does not contain a sub-optimal CO_2 concentration, but rather that it contains too large an O_2 concentration. Thus, seen from the plant's point of view, there is no shortage of CO_2 but only a surplus of the plant pollutant, oxygen (which the plants themselves produce). In the bulk atmosphere containing 21% oxygen, elevated CO_2 acts as an antidote to oxygen. Anthropogenic activities such as the combustion of fossil fuels and de-forestation increase atmospheric CO_2 concentration. As we have outlined above, elevated atmospheric CO_2 reduces plant damage induced by salt stress, water stress, heat stress, oxygen stress, ethylene stress and air pollution stress caused by O_3, NO_x and SO_2. Hence mankind is effectively decreasing plant stress by enriching the global environment with CO_2. Stress reduction by elevated CO_2 is a result of either reduced transpiration rates or decreased sensitivity to gaseous air pollutants. Both phenomena are primarily influenced by stomatal behaviour, implying that the role played by stomata in ameliorating or exacerbating stress is a vital area for future study.

ACKNOWLEDGEMENTS

HZE is grateful for the support of the UK Department of the Environment, and SJH acknowledges support from the Natural Environment Research Council.

REFERENCES

Abeles FB (1985) Ethylene and plant development: An introduction. In: Roberts JA, Tucker GA (eds) Ethylene and plant development. Butterworths, London, UK, pp 1-8

Acock B, Allen LH Jr (1985) Crop responses to elevated carbon dioxide concentration. In: Strain BR, Cure JD (eds) Direct effects of increasing carbon dioxide on vegetation. DOE/ER - 0238, US Department of Energy, Washington, D.C., USA, pp 55-97

Acock B, Baker DN, Reddy VR, McKinion JM, Whisler FD, Del Castillo D, Hodges HF (1981) Soybean responses to carbon dioxide: measurement and simulation. Report 004 - Response of vegetation to carbon dioxide. US Department of Energy, Washington, D.C., USA, pp 95

Allen LH Jr (1990) Plant responses to rising carbon dioxide and potential interactions with air pollutants. Journal of Environmental Quality 19: 15-34

Allen LH Jr, Boote KJ, Jones JW, Mishoe JW, Jones PH, Vu CV, Valle R, Campbell WJ (1981) Effects of increased carbon dioxide on photosynthesis and agricultural productivity of soybeans. Report 003 Response of vegetation to carbon dioxide. US Department of Energy, Washington, D.C., USA, pp 131

Allen RG, Gichuki FN, Rosenzweig C (1991) CO_2-induced climatic changes and irrigation-water requirements. Journal of Water Resources Planning and Management 117: 157-178

Alpert P, Warembourg FR, Roy J (1991) Transport of carbon among connected ramets of *Eichornia crassipes (Pontederiaceae)* at normal and high levels of CO_2. American Journal of Botany 78: 1459-1466

Andersen AS (1976) Regulation of apical dominance by ethephon, irradiance and CO_2. Physiologia Plantarum 37: 303-308

Ashenden TW, Baxter R, Rafarel CR (1992) An inexpensive system for exposing plants in the field to elevated concentrations of CO_2. Plant, Cell and Environment 15: 365-372

Aulenbach W (1965) Soils, temperature and CO_2 for cut chrysanthemums. Pennsylvania Flower Growers Bulletin 174

Barry RG (1981) Mountain, weather and climate. Methuen, London, UK, pp 23

Berlekamp J, Overdieck D (1989) Modelling the CO_2 gas exchange of grassland vegetation from experimental data. International Journal of Biometeorology 33: 119-123

Black VJ (1982) Effects of sulphur dioxide on physiological processes in plants. In: Unsworth MH, Ormrod DP (eds) Effects of gaseous pollution in agriculture and horticulture. Butterworths, London, UK, pp 67-91

Boden TA, Kanciruk P, Farrell MP (1990) Trends '90: A compendium of data on global change. ORNL/CDIAC, Oak Ridge Laboratory, Oak Ridge, Tennessee, USA

Bowman WD, Strain BR (1987) Interaction between CO_2 enrichment and salinity stress in the C4 monohalophyte *Andropogon glomeratus*. Plant, Cell and Environment 10: 267-270

Burg SP, Bars EA (1987) Molecular requirements for the biological activity of ethylene. Plant Physiology 42: 144-152

Caporn SMJ (1989) The effects of oxides of nitrogen and carbon dioxide enrichment on photosynthesis and growth of lettuce (*Lactuca sativa* L.). The New Phytologist 111: 473-481

Carlson RW, Bazzaz FA (1982) Photosynthetic and growth responses to fumigation with SO_2 at elevated CO_2 for C3 and C4 plants. Oecologia 54: 50-54

Carlson RW, Bazzaz FA (1985) Plant response to SO_2 and CO_2. In: Winner WE, Mooney HA, Goldstein RA (eds) Sulfur dioxide and vegetation. Stanford University Press, Stanford, USA, pp 313-331

Chang CW (1975) Carbon dioxide and senescence in cotton plants. Plant Physiology 55: 515-519

Cook ER, Cole J (1991) On predicting the response of forests in Eastern North America to future climatic change. Climatic Change 19: 271-282

Cooper RL, Brun WA (1967) Response of soybeans to a carbon dioxide enriched atmosphere. Crop Science 7: 455-467

Cournac L, Dimon B, Carrier P, Lohou A, Chagvardieff (1991) Growth and photosynthetic characteristics of *Solanum tuberosum* plantlets cultivated *in vitro* in different conditions of aeration, sucrose supply, and CO_2 enrichment. Plant Physiology 97: 112-117

Dahlman RC, Reynolds JF, Strain BR (1986) Modelling the response of ecosystems to fossil fuel emissions: carbon dioxide and pollutant interactions. Aïr Pollution Control Association Proceedings 86-95: 1-15

Dayan E, Enoch HZ, Fuchs M, Zipori I (1986) Suitability of greenhouse building types and roof cover materials for growth of export tomatoes in the Besor region of Israel. II. Effect on fresh and dry matter production. Biotronics 15: 71-79

De Kok LJ, Rennenberg H, Kuiper PJC (1991) The internal resistance in spinach leaves to atmospheric H_2S deposition is determined by metabolic processes. Plant Physiology and Biochemistry 29: 463-470

Delmas RA, Servant J, Tathy JP, Cross B, Labat M (1992) Sources and sinks of methane and carbon dioxide exchanges in mountain forest and equatorial Africa. Journal of Geophysical Research 97: 6169-6179

Desjardins Y, Gosselin A, Yelle S (1987) Acclimatization of *ex vitro* strawberry plantlets in CO_2-enriched environments and supplementary lighting. Journal of the American Society of Horticultural Science 112: 846-851

Du Cloux HC, Andre M, Daguenet A, Massimino J (1987) Wheat response to CO_2 enrichment: growth and CO_2 exchanges at two plant densities. Journal of Experimental Botany 38: 1421-1431

Enoch HZ (1987) Global and horticultural carbon dioxide enrichment. In: Newman P (ed) Plant growth and leaf applied chemicals. CRC Press Inc, Boca Raton, Florida, USA, pp 157-167

Enoch HZ, Hurd RG (1977) Effect of light intensity, carbon dioxide concentration and leaf temperature on gas exchange of spray carnation plants. Journal of Experimental Botany 28: 84-95

Enoch HZ, Hurd RG (1979) The effect of elevated CO_2 concentration in the atmosphere on plant transpiration and water use efficiency. A study with potted carnation plants. International Journal of Biometeorology 23: 343-351

Enoch HZ, Rylski I, Spigelman M (1976) CO_2 enrichment of strawberry and cucumber plants grown in unheated greenhouses in Israel. Scientific Horticulture 5: 33-41

Enoch HZ, Sachs JM (1978) An empirical model of CO_2 exchange of a C3 plant in relation to light, CO_2 concentration and temperature. Photosynthetica 12: 150-157

Enoch HZ, Zieslin N (1988) Growth and development of plants in response to carbon dioxide concentrations. Applied Agricultural Research 3: 248-256

Enoch HZ, Zieslin N, Biran Y, Halevy AH, Schwartz M, Kessler B, Shimshi D (1973) Principles of CO_2 nutrition research. Acta Horticulturae 32: 97-118

Fajer ED, Bowers MD, Bazzaz FA (1991) Performance and allocation patterns of the perennial herb, *Plantago lanceolata*, in response to simulated herbivory and elevated CO_2 environments. Oecologia 87: 37-42

Fincher J, Alscher RG (1992) The effect of long-term ozone exposure on injury in seedlings of red spruce (*Picea rubens* Sarg). The New Phytologist 120: 49-59

Ford MA, Thorne GN (1967) Effect of CO_2 concentration on growth of sugarbeet, barley, kale and maize. Annals of Botany 131: 629-644

Foschner W, Reuther G (1984) Photosynthesis and water balance of Pelargonium cuttings during the period of rooting under the influence of various light and CO_2 conditions. Gartenbauwissenschaft 49: 182-190

Friend AD, Woodward FI (1990) Evolutionary and ecophysiological responses of mountain plants to the growing season environment. Advances in Ecological Research 20: 59-124

Funsch RW, Mattson RH, Mowry GR (1970) CO_2-supplemented atmosphere increases growth of *Pinus strobus* seedlings. Forest Science 16: 459-460

Gale J (1972) Availability of carbon dioxide for photosynthesis at high altitudes: theoretical considerations. Ecology 53: 494-497

Gifford RM (1979) Growth and yield of carbon dioxide enriched wheat under water-limited conditions. Australian Journal of Plant Physiology 6: 367-378

Goldsberry KL (1963) Growth of carnation increases with supplementary CO_2. Growers Association Bulletin 164: 1-2

Goudriaan J, van Keulen H, van Laar HH (eds) (1990) The greenhouse effect and primary productivity in European agro-ecosystems. Wageningen, Purdoc, The Netherlands, pp 90

Grout BWW (1988) Photosynthesis of regenerated plantlets *in vitro*, and the stresses of transplanting. Acta Horticulturae 230: 129-135

Grout BWW, Aston MJ (1978) Transplanting of cauliflower plants regenerated from meristem culture. II. Carbon dioxide fixation and the development of photosynthetic ability. Horticultural Research 17: 65-71

Hall DO, Scurlock JMO (1991) Climate change and productivity of natural grasslands. Annals of Botany 67 (Supplement 1): 49-55

Hallgren JE (1984) Photosynthetic gas exchange in leaves affected by air pollutants. In Koziol MJ, Whatley FR (eds) Gaseous air pollutants and plant metabolism. Butterworths, London, UK, pp 147-160

Harley PC, Thomas RB, Reynolds JF, Strain BR (1992) Modelling photosynthesis of cotton grown in elevated CO_2. Plant, Cell and Environment 15: 271-282

Hau L, Hill AC, Soleimani A (1977) Influence of CO_2 on the effects of SO_2 and NO_2 on alfalfa. Environmental Pollution 12: 7-16

Hicklenton PR (1979) Metabolism and energetics in flowering. PhD Thesis, University of British Columbia, Canada

Hocking PJ, Meyer CP (1991) Effects of CO_2 enrichment and nitrogen stress on growth and partitioning of dry matter and nitrogen in wheat and maize. Australian Journal of Plant Physiology 18: 339-356

Hofstra G, Hesketh JD (1975) The effects of temperature and CO_2 enrichment on photosynthesis in soybean. In: Marcelle R (ed) Environmental and biological control of photosynthesis. Dr W. Junk Publishers, The Hague, The Netherlands, pp 71-80

Hovland AS, Dybing CD (1973) Cyclic flowering patterns in flax as influenced by environment and plant growth regulators. Crop Science 13: 380-384

Hunt R, Hand DW, Hannah, MA, Neal AM (1991) Response to CO_2 enrichment in 27 herbaceous species. Functional Ecology 5: 410- 421

Hurd RG (1968) Effect of CO_2 enrichment on the growth of young tomato plants in low light. Annals of Botany 32: 531-542

Idso SB (1989) Carbon dioxide and global change: Earth in transition. IBR Press, Tempe, Arizona, USA, pp 292

Idso SB (1991) The greenhouse effect: questioning the received wisdom. Global Ecology and Biogeography Letters 1: 31-32

Idso SB (1991) A general relationship between CO_2-induced increases in net photosynthesis and concomitant reductions in stomatal conductance. Environmental and Experimental Botany 31: 381-383

Idso SB, Kimball BA (1992) Seasonal fine-root biomass development of sour orange trees grown in atmospheres of ambient and elevated CO_2 concentration. Plant, Cell and Environment 15: 337-341

Idso SB, Kimball BA (1992) Effects of atmospheric CO_2 enrichment on photosynthesis, respiration and growth of sour orange trees. Plant Physiology 99: 341-343

Idso SB, Kimball BA, Andersen HG, Mauney JR (1987) Effects of atmospheric CO_2 enrichment on plant growth: the interactive role of air temperature. Agriculture, Ecosystems and Environment 20: 1-10

Imai K, Murata Y (1976) Effect of carbon dioxide concentration on growth and dry matter production of plants. I. Effects on leaf area, dry matter, tilled dry matter distribution ratio and transpiration. Japanese Journal of Crop Science 45: 598-606

Imazu T, Yabuki K, Oda Y (1967) Studies on the influence of carbon dioxide concentration on the growth, flowering and fruit setting of eggplant (*Solanum melongena* L.). Journal of the Japanese Society for Horticultural Science 36: 222-227

Jensen HEK (1983) Effects of carbon dioxide supplementation and date of cutback on yield and economy of glasshouse roses, *Rosa* L. Tidskrift fur Planteavl 87: 153-166

Kimball BA, Idso SB (1983) Increasing atmospheric CO_2: effects on crop yield, water use and climate. Agricultural Water Management 7: 55-72

Kirkham MB, He H, Bolger TP, Lawlor D, Kanemasu ET (1991) Leaf photosynthesis and water use of big bluestem under elevated carbon dioxide. Crop Science 31: 1589-1594

Körner C (1988) Does global increase of CO_2 alter stomatal density? Flora 181: 253-257

Körner C, Bannister P, Mark AF (1986) Altitudinal variation in stomatal conductance, nitrogen content and leaf anatomy in different plant life forms in New Zealand. Oecologia 69: 577-588

Körner C, Cochrane P (1985) Stomatal responses and water relations of *Eucalyptus pauciflora* in summer along an elevational gradient. Oecologia 66: 443-455

Körner C, Diemer M (1987) *In situ* photosynthetic responses to light, temperature and carbon dioxide in herbaceous plants from low and high altitude. Functional Ecology 1: 179-194

Körner C, Farquhar GD, Roksandic Z (1988) A global survey of carbon isotope discrimination in plants from high altitude. Oecologia 74: 623-532

Körner C, Renhardt U (1987) Dry matter partitioning and root length/leaf area ratios in herbaceous perennial plants with diverse altitudinal distribution. Oecologia 74: 411-418

Körner C, Neumayer M, Pelaez MR, Smeets-Scheel A (1989) Functional morphology of mountain plants. Flora 182: 353-383

Körner C, Woodward FI (1987) The dynamics of leaf extension in plants with diverse altitudinal ranges. II. Field studies in *Poa* species between 600 and 3200 m altitude. Oecologia 72: 279-283

Lakso AN, Reisch BI, Mortensen J, Roberts MH (1986) Carbon dioxide enrichment for stimulation of growth of *in vitro*-propagated grapevines after transfer from culture. Journal of the American Society of Horticultural Science 111: 634-638

Lindstroem RS (1965) Carbon dioxide and its effect on the growth of roses. Proceedings of the American Society of Horticultural Science 87: 521-524

Lippold KW (ed) (1992) Climate change - A threat to global development. Enquete Commission of the German Bundestag, Germany, pp 235

Liu HT (1983) The effects of CO_2 enrichment on growth and cut flower production of *Gerbera jamesonii*, H Bolus Ex. Hook. MSc Thesis, Faculty of Graduate Studies, Guelph, Ontario, Canada

Long SP, Drake BG (1991) Effect of the long-term elevation of CO_2 concentration in the field on the quantum yield of photosynthesis of the C3 sedge, *Scirpus olneyi*. Plant Physiology 96: 221-226

Luria G, Shalma E, Shilo R, Frank M, Frank Y (1985) Influence of cooling transplants and CO_2 enrichment on development and flowering of *Gypsophila*. Hassadeh 65: 956-961 (in hebrew)

Madamanchi NR, Alscher G (1991) Metabolic bases for differences in sensitivity of two pea cultivars to sulfur dioxide. Plant Physiology 97: 88-93

Madsen E (1968) Effect of CO_2 concentration on the accumulation of starch and sugar in tomato leaves. Physiologia Plantarum 21: 168-175

Margolis L, Lovelock JE (1974) Biological modulation of the Earth's atmosphere. Icarus 21: 471-489

Monteith JL, Unsworth MH (1990) Principles of environmental physics (Second edition). Edward Arnold, London, UK, pp 291

Morison JIL (1985) Sensitivity of stomata and water use efficiency to high CO_2. Plant, Cell and Environment 8: 467-474

Morison JIL (1988) Effect of increasing atmospheric CO_2 on plants and their responses to other pollutants, climate and soil factors. Aspects of Applied Biology 17: 113-122

Mortensen LM (1985) Nitrogen oxides produced during CO_2 enrichment. I. Effects on different greenhouse plants. The New Phytologist 101: 103-108

Mortensen LM (1985) Nitrogen oxides produced during CO_2 enrichment. II. Effects on different tomato and lettuce cultivars. The New Phytologist 101: 411-415

Mortensen LM, Sandvik M (1987) Effects of CO_2 enrichment at varying photon flux densities on growth of *Picea abies* (L.) Karst seedlings. Scandinavian Journal of Forestry Research 2: 335-342

Mousseau M, Enoch HZ (1989) Carbon dioxide enrichment reduced shoot growth in sweet chestnut seedlings (*Castanea sativa* Mill.). Plant, Cell and Environment 12: 927-934

Mulchi CL, Slaughter L, Saleem M, Lee EH, Pausch R, Rowland R (1992) Growth and physiological characteristics of soybean in open-top chambers in response to ozone and increased atmospheric CO_2. Agriculture, Ecosystems and Environment 38: 107-118

Nijs I, Impens I, Behaeghe T (1989) Effects of different CO_2 environments on the photosynthesis-yield relationship and the carbon and water balance of a white clover (*Trifolium repens* L. cv. Blanca) sward. Journal of Experimental Botany 40: 353-359

Nobel SP (1991) Achievable productivitites of certain CAM plants: basis for high values compared with C3 and C4 plants. The New Phytologist 119: 183-205

Norby RJ, O'Neill EG (1989) Growth dynamics and water use of seedlings of *Quercus alba* L. in CO_2-enriched atmospheres. The New Phytologist 111: 491-500

Oberbauer SF, Strain BR, Fetcher N (1985) Effect of CO_2-enrichment on seedling physiology and growth of two tropical tree species. Physiologia Plantarum 65: 352-356

Overdieck D (1989) The effects of preindustrial and predicted future atmospheric CO_2 concentration on *Lyonia mariana* L.D. Don. Functional Ecology 3: 569-576

Overdieck D (1990) Direct effects of elevated CO_2 concentration levels on grass and clover in model ecosystems. In Beukema JJ *et al.* (1990) Expected effects of climatic change on marine coastal systems, pp 41-47

Overdieck D, Bossemeyer D, Lieth H (1984) Long-term effects of an increased CO_2 concentration level on terrestrial plants in model-ecosystems. CEC symposium - Interaction between climate and biosphere. Osnabruck, March 21-23, 1983, pp 344-352

Paez A, Hellmers H, Strain BR (1980) CO_2 effects in apical dominance in *Pisum sativum.* Physiologia Plantarum 50: 43-36

Paez A, Hellmers H, Strain BR (1983) CO_2 enrichment, drought stress and growth of Alaska pea plants (*Pisum sativum*). Physiologia Plantarum 58: 161-165

Palloix A, Herve Y, Knox RB, Dumas C (1985) Effect of carbon dioxide and relative humidity on self-incompatibility in cauliflower, *Brassica oleracea.* Theoretical and Applied Genetics 70: 628-633

Pearcy RW, Bjorkmann O (1983) Physiological effects. In: Lemon ER (ed) CO_2 and plants: The response of plants to rising levels of atmospheric carbon dioxide. Westview Press, Boulder, Colorado, USA, pp 65-105

Penningsfeld F, Forcthamer L (1980) Gerbera. Eugen Ulmer Gmbh & Co., Stuttgart, Germany

Penuelas J, Azcon-Bieto J (1992) Changes in leaf $d^{13}C$ of herbarium plant species during the last three centuries of CO_2 increase. Plant, Cell and Environment 15: 85-489

Perry DA, Borchers JG (1990) Climate change and ecosystem responses. The Northwest Environmental Journal 6: 293-313

Poorter H (1992) Interspecific variation in the growth response of plants to an elevated ambient CO_2 concentration. Vegetatio 10: 1-22

Poorter H, Pot S, Lambers H (1988) The effect of an elevated atmospheric CO_2 concentration on growth, photosynthesis and respiration of *Plantago major.* Physiologia Plantarum 73: 553-559

Pospisilova J, Ticha I, Catsky J, Solarova J, Ondrej M (1989) Photosynthesis in regenerants of tobacco transformed by plasmids of *Agrobacterium.* 3. Transplanted plants. Photosynthetica 23: 130-135

Potvin C (1985) Amelioration of chilling effects by CO_2 enrichment. Physiologie Végetale 23: 345-352

Radin JW, Hartung W, Kimball BA, Mauney JR (1988) Correlation of stomatal conductance with photosynthetic capacity of cotton only in a CO_2-enriched atmosphere: mediation by abscisic acid? Plant Physiology 88: 1058-1062

Radoglou KM, Jarvis PG (1990) Effects of CO_2 enrichment on four poplar clones II. Leaf surface properties. Annals of Botany 65: 627-632

Retzlaff WA, DeJong TM, Williams LE (1992) Photosynthesis and growth response of almond to increased atmospheric ozone partial pressures. Journal of Environmental Quality 21: 208-216

Rogers HH, Bingham GE, Cure JD, Heck WW, Heagle AA, Israel DW, Smith JM, Surano KA (1983) Responses of selected plant species to elevated carbon dioxide in the field. Journal of Environmental Quality 12: 569-574

Running SW, Nemani RR (1991) Regional hydrologic and carbon balance responses of forests resulting from potential climate change. Climatic Change 19: 349-368

Sage RF, Sharkey TD, Seemann JR (1989) Acclimation of photosynthesis to elevated CO_2 in five C3 species. Plant Physiology 89: 590-596

Saxe H (1986) Effects of NO, NO_2 and CO_2 on net photosynthesis, dark respiration and transpiration of pot plants. The New Phytologist 103: 185-197

Schwartz M, Gale J (1984) Growth response to salinity at high levels of carbon dioxide. Journal of Experimental Botany 35: 193-196

Shaw RT, Rogers, MN (1964) Interactions between elevated carbon dioxide levels and greenhouse temperatures on the growth of roses, chrysanthemums, carnations, geraniums, snapdragons and African violets. Part 2. Roses, heat plus CO_2. Florists Review 135: 21-22

Sionit N, Hellmers H, Strain BR (1982) Interaction of atmospheric CO_2 enrichment and irradiance on plant growth. Agronomy Journal 74: 721-725

Smith WK, Geller GN (1979) Plant transpiration at high altitudes: theory, field measurements and comparisons with desert plants. Oecologia 41: 109-122

Strain BR, Cure JD (eds) (1986) Direct effects of CO_2 on plants and ecosystems: a bibliography with abstracts. Oak Ridge National Laboratory, Oak Ridge, Tennessee, USA, pp 286

Tissue DT, Oechel WC (1987) Response of *Eriophorum vaginatum* to elevated CO_2 and temperature in the Alaskan tussock tundra. Ecology 68: 401-410

Tsujita MJ, Dutton R, Godzinski B (1982) Gerbera response to carbon dioxide and supplementary lighting. Canadian Florist 77: 39-45

van Berkel N (1983) CO_2 enrichment of gerbera. Vakblad Bloemisterij 38: 47

Warkentin BP (1992) Soil science for environmental quality - how do we know what we know? Journal of Environmental Quality 21: 163-166

Woodward FI (1983) The significance of interspecific differences in specific leaf area to the growth of selected herbaceous species from different altitudes. The New Phytologist 95: 313-323

Woodward FI (1986) Ecophysiological studies on the shrub *Vaccinium myrtillus* L. taken from a wide altitudinal range. Oecologia 70: 580-586

Woodward FI, Rochefort L (1991) Sensitivity analysis of vegetation diversity to environmental change. Global Ecology and Biogeography Letters 1: 7-23

Wray SM, Strain BR (1986) Response of two old field perennials to interaction of CO_2 enhancement and drought stress. American Journal of Botany 73: 1488-1491

Wuebbles DJ, Edmonds J (1991) Primer on greenhouse gases. Lewis Publishers, Chelsea, Michigan, USA, pp 230

Zieslin N, Moe R (1985) *Rosa*. In: Halevy AH (ed) Handbook of Flowering, Vol. 4. CRC Press, Boca Raton, Florida, USA, pp 214-225

Ziska LH, Hogan KP, Smith AP, Drake BG (1991) Growth and photosynthetic response of nine tropical species with long-term exposure to elevated carbon dioxide. Oecologia 86: 383-389

OZONE, SULPHUR DIOXIDE AND NITROGEN OXIDES: SOME EFFECTS ON THE WATER RELATIONS OF HERBACEOUS PLANTS AND TREES

T.A. Mansfield, M. Pearson, C.J. Atkinson[1] and P.A. Wookey[2]
Institute of Environmental and Biological Sciences
Lancaster University
Lancaster LA1 4YQ
UK

ABSTRACT

O_3, SO_2 and NO_2 have all been shown to exert effects on plants which are likely to influence water relations. In this paper a review of evidence (often confusing) from the literature is presented, followed by a description of two phenomena which we have recently studied in detail. It has been known for some time that plants exposed to SO_2, or to mixtures of SO_2 and NO_2, often display differences in stomatal behaviour from those in clean air. We have grown spring barley in SO_2 and NO_2 (24-35 ppb of each gas) and examined the ability of leaves to regulate water loss. The stomata were found to be much less responsive to ABA after exposure. In other respects the stomata of polluted plants appeared to function normally (e.g., they showed similar responses to CO_2). Accumulation of ABA in water-stressed leaves appeared to be unaffected by the pollution treatment. We have concluded that the impairment of stomatal sensitivity to ABA may adversely affect the water relations of polluted plants. In other experiments, young trees of *Fagus sylvatica* (beech) were exposed to realistic doses of ozone in summer. In well-watered trees ozone inhibited stomatal opening, but in trees growing in dry soil, and thus experiencing water stress, stomatal closure was significantly inhibited by ozone. We suggest that ozone may have two deleterious effects on beech: inhibition of CO_2 acquisition when water supplies are adequate, and interference with the protective role of stomatal closure during periods of water stress.

INTRODUCTION

The relationships between plant water relations and the effects of air pollutants are very complex. There is a great deal of experimental evidence showing that soil moisture stress can reduce the sensitivity of plants to gaseous air pollutants. For

[1] Institute of Terrestrial Ecology, Grange-over-Sands, Cumbria LA11 6JU
[2] Horticulture Research International, East Malling, Kent ME19 6BJ

NATO ASI Series, Vol. I 16
Interacting Stresses on Plants in a Changing Climate
Edited by M. B. Jackson, C. R. Black

example, it is well established that the frequency of leaf spots on tobacco caused by ozone increases markedly when crops are well irrigated (e.g., Walker and Vickery, 1961). Experimental manipulations of the water status of tobacco plants have shown that the severity of ozone injury is determined by the extent of stomatal opening (Macdowall, 1965). The increased uptake of ozone through open stomata, and subsequent damage to tissues inside the leaves, is the obvious explanation. The climatic conditions that cause high ambient ozone concentrations are also likely to lead to soil moisture stress. Farmers in such areas thus have a dilemma, for if they irrigate to increase crop production they may well increase the severity of ozone injury. Estimates of the cost effectiveness of irrigation of tobacco must be made in the knowledge that damage caused by ozone may increase.

This kind of protective action of stomata is of less interest to plant physiologists than changes in stomatal behaviour that are *caused* by pollutants. Several of the major air pollutants (O_3, SO_2, NO_2) have been found to influence both the degree of stomatal opening and its diurnal course. Control of stomata by the gaseous composition of the atmosphere is an important part of their normal physiology. In particular, they are sensitive to CO_2 concentration and to water vapour pressure deficit (VPD), and at least some of the pollutant-induced effects reported in the literature are probably not the outcome of direct action on the stomata, but are a result of the changes in intercellular CO_2 concentration that ensue.

EFFECTS OF SO_2 + NO_2 POLLUTION ON RESPONSES OF STOMATA

There is, however, strong evidence that stomata are affected directly by air pollutants. Taylor *et al.* (1981) exposed detached epidermis of *Vicia faba* to different aqueous concentrations of SO_2 (buffered H_2SO_3). They found that low SO_2 concentrations induced stomatal opening and that higher concentrations caused closure. Their data were of particular interest because they helped to explain apparent discrepancies in the literature: some authors had reported that SO_2 caused stomatal opening, while others had observed only closing responses. A review by

Darrall (1989) of a large body of literature led to the suggestion that SO_2 effects on whole plants are indeed concentration- or dose-dependent, with opening responses being more likely when low SO_2 concentrations are applied for short periods of time. Some quite substantial stomatal responses can occur to low SO_2 concentrations; for example, Black and Black (1979) found that 17.5 ppb SO_2 enhanced stomatal opening on whole plants of *Vicia faba.*

Stomatal opening in response to air pollution would have greater physiological impact if it occurred during periods of high evaporative demand, or if it interfered with the stomatal closure that takes place when soil moisture is depleted. As long ago as 1966, Godzik and Piskornik found that leaves of *Aesculus hippocastanum* detached from trees near a coke-producing plant were less able than those from an unpolluted area to control their rate of water loss: the leaves affected by air pollution consequently wilted much more quickly. The simple technique used by Godzik and Piskornik (i.e., detaching a leaf and recording its rate of weight loss to indicate the speed of stomatal closure) has proved very valuable in recent research in our own laboratory. Neighbour *et al.* (1988) found that the technique revealed profound differences in leaves of *Betula pubescens* and *B. pendula* after exposure to low concentrations of $SO_2 + NO_2$ (up to 60 ppb of each gas). The prior exposure to pollution caused the leaves to dry out much more quickly when they were detached from the trees. Studies on the intact trees suggested that the imposition of soil moisture stress had a much more rapid impact on those that had grown in the presence of $SO_2 + NO_2$ pollution. Lucas (1990) found that prior treatment with $SO_2 + NO_2$ at concentrations down to 10 ppb could affect the ability of detached leaves of a grass (*Phleum pratense*) to conserve water.

Abscisic acid (ABA), a well-established inhibitor of stomatal opening, is known to accumulate rapidly in leaves when they are detached from the plant and allowed to wilt. It is this, rather than the loss of turgor *per se* which is believed to determine the speed with which the stomata close. A slower reaction of the stomata on leaves of plants grown in the presence of air pollutants might be the outcome either of reduced biosynthesis of ABA, or a reduced ability of the guard cells to respond to newly produced ABA. In order to study this and other aspects of perturbed stomatal behaviour, we have recently completed a detailed investigation of spring barley (*Hordeum vulgare* L. cv. Klaxon). Seedlings at the one-leaf stage were grown either in relatively unpolluted air (<10 ppb of both SO_2 and NO_2) or in air to which approximately 30 ppb of both SO_2 and NO_2 had been added. The experimental procedures are described by Atkinson *et al.* (1991). It was found that the stomata of well-watered plants were slightly more closed (by about 10-15%) in the presence of the air pollutants, which was in agreement with previous studies on cereals and grasses. However, when leaves were detached and placed either in distilled water or in a solution of 10^{-1} mol m^{-3} ABA, the leaves previously exposed to air pollution showed a considerably reduced response to ABA (Fig. 1). After the application of ABA there was, averaged over three separate experiments, a 68% higher transpiration rate from polluted leaves than from the controls. The leaves supplied with ABA continued to show reduced transpiration during the six hours of observation, as the ABA progressively accumulated in the leaves. Because transpiration of the polluted leaves was greater, their rate of accumulation of ABA must also have been higher, but there is no evidence from the data in Fig. 1 that this produced any increase in the response as time progressed.

The ability of the stomata on leaves of polluted and control plants to respond to CO_2 was also examined by Atkinson *et al.* (1991), and it was found that the closing response was of a similar magnitude when the CO_2 concentration was increased from 200 to 700 μmol mol^{-1}. ABA and CO_2 are two of the most important natural agents that regulate stomatal behaviour, and it appears that prior exposure to SO_2 + NO_2 pollution influences the response to one and not to the other. The stomata of both polluted and control plants also showed a clear diurnal cycle of opening and closing determined by light and darkness. Thus the functioning of the turgor-control mechanism in the guard cells was not impaired, and it appeared that it was the ability to respond to the arrival of ABA in the transpiration stream which was specifically affected.

The current view of the role of ABA in the regulation of leaf water relations is that the stomatal closure which occurs during the initial stages of soil moisture

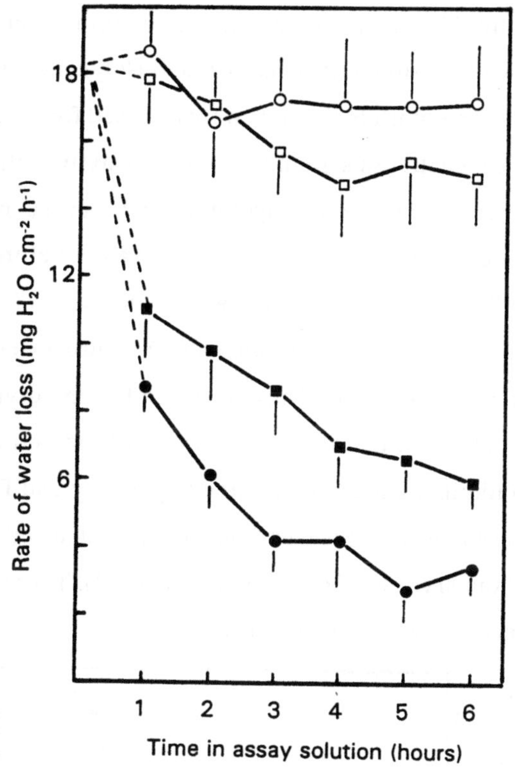

FIGURE 1. *Transpirational water loss from illuminated detached leaves after exposure to ambient air (○ ●) or to SO_2 + NO_2 (□ ■). Leaves were in either distilled water (○ □) or a solution of 10^{-1} mol m^{-3} ABA (● ■). Means ± 1 SE are shown for data obtained on three separate occasions during which each treatment was replicated 10 times. Temperature 25 °C, PAR 200 μmol quanta m^{-2} s^{-1}. Redrawn from Atkinson* et al. *(1991).*

stress is the result of ABA synthesis in roots, and subsequent transport via the xylem to the leaves (Davies and Zhang, 1991). This process appears to be able to achieve very effective regulation of leaf water relations, so that partial stomatal closure takes place without any detectable decline in leaf turgor or water potential. Leaves do, however, possess the ability to biosynthesise ABA when they experience water stress. It has long been known that leaves which are detached and allowed to wilt accumulate large quantities of ABA very rapidly (Wright, 1977). In view of this, we determined whether excised, wilting leaves of barley showed any difference in the ability to accumulate ABA in plants grown in the presence of $SO_2 + NO_2$ pollution. For this experiment, the 'control' plants were exposed to ambient air for which the arithmetic mean SO_2 and NO_2 concentrations were 5 and 18 ppb, respectively. The 'polluted' plants were exposed for 10 days to 33 ppb SO_2 and 46 ppb NO_2, concentrations found in fairly heavily polluted situations in western Europe. At the end of the fumigation period, the second and third leaves were removed from each plant and weighed immediately before being allowed to dry out in a controlled environment chamber. At intervals of 20, 40, 60 and 95 minutes after excision, 12 leaves (6 polluted and 6 control) were weighed, placed in vials and immersed in liquid N_2. Thus, at the end of the drying period, 60 leaf samples had been collected, and these were freeze-dried and analysed for ABA content using the radioimmunoassay described by Zhang and Davies (1987). The ABA content increased rapidly in leaves from both polluted and control plants, and had increased by about six-fold after 95 minutes (Fig. 2). There were, however, no significant differences between treatments.

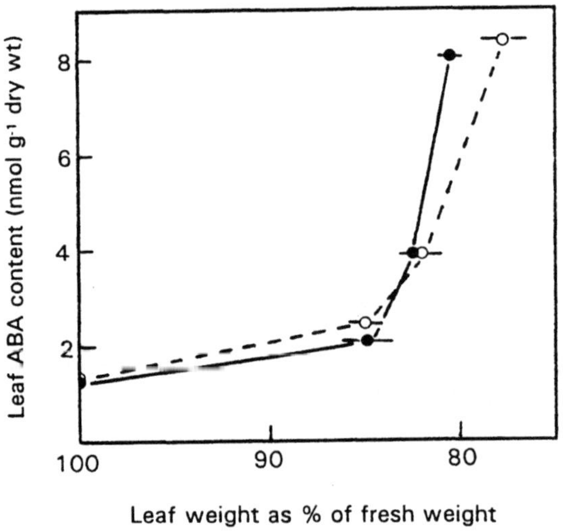

FIGURE 2. ABA contents of control (○) and polluted (●) leaves of barley that were excised and allowed to dry. Means ± 1 SE are shown.

In the light of data shown in Fig. 2, together with the other information that has emerged from our recent studies, we have concluded that:

1. stomata on leaves previously exposed to $SO_2 + NO_2$ pollution show a

reduced rate of response to ABA delivered via the transpiration stream;

2. when leaves from plants polluted with SO_2 + NO_2 are excised and allowed to wilt, their stomata close less rapidly than those on leaves from plants grown in clean air;
3. exposure to SO_2 + NO_2 pollution appears to have no significant effect on the ability of excised, wilting leaves to synthesise ABA.

It was noted earlier that there is a need to explain some apparent discrepancies in the literature concerning the nature of effects of SO_2 (or SO_2 + NO_2) on stomatal behaviour, and the complexity of dose-response relationships was mentioned as one important factor. Our new conclusions suggest that the pretreatment of the plants must also be carefully considered. Well-watered plants with little endogenous ABA may show a limited stomatal closure in response to SO_2 + NO_2, but in the case of plants that are mildly stressed with an elevated concentration of ABA in the xylem fluid (as simulated in the experiment in Fig. 1), the stomata may be more open on the polluted plants.

EFFECTS OF OZONE ON STOMATA: INTERACTIONS WITH WATER STRESS

We have recently concluded a study of effects of ozone on stomatal behaviour which suggests that there may be interactions with plant water relations, and indeed there are close resemblances to the situation found with SO_2 + NO_2. Three year old transplants of *Fagus sylvatica* L. were grown in naturally illuminated fumigation chambers ('Solardomes'; see Lucas *et al.*, 1987) and were subjected to episodes of ozone pollution (concentrations of 30-120 ppb) for 128 days from mid-May. Higher ozone concentrations were applied when incident sunlight was greater, a realistic simulation of the outside environment in the UK and elsewhere in north-west Europe. Determinations of stomatal resistance were made on days after 1 August when the ozone concentration exceeded 100 ppb. Beginning 38 days after 1 August, water was withheld from half of the trees. Older leaves (those of the 'first flush') on the well-watered trees displayed an increase in stomatal resistance due to the ozone treatment, and the magnitude of this effect increased considerably after 20 days (Fig. 3). However, in trees subjected to water stress the pattern of response was quite different. The effect of soil moisture stress on stomatal resistance was considerable, but was less in the polluted trees (Fig. 4). It can be clearly seen that ozone *increased* stomatal resistance in the well-watered trees, but *decreased* it in the trees that were experiencing water stress.

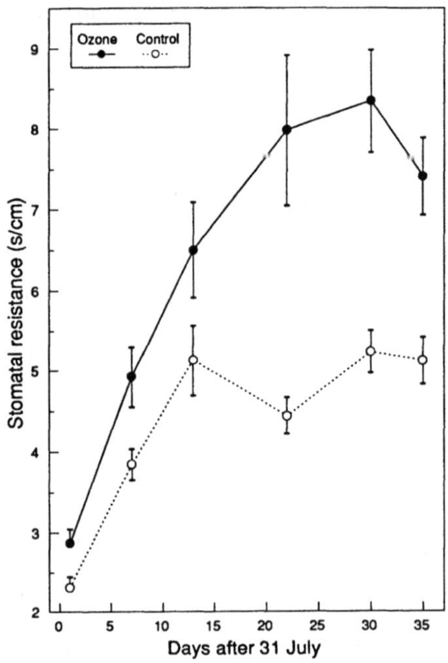

***FIGURE 3**. The effect of episodic ozone pollution on mean stomatal resistance in beech. Data for well-watered trees. Means ± 1 SE are shown.*

There have been many previous reports in the literature that ozone pollution causes an increase in stomatal resistance, and there have also been a few indications that reductions in resistance can occur in some circumstances (e.g., Taylor and Dobson, 1989). However, as far as we know, an experiment of this nature providing a detailed understanding of stomatal behaviour in comparable leaves on plants with and without water stress has not been performed before. We believe that the implications of the data in Fig. 4 could be far-reaching, although we must be cautious in extrapolating our findings on three year old trees in semi-

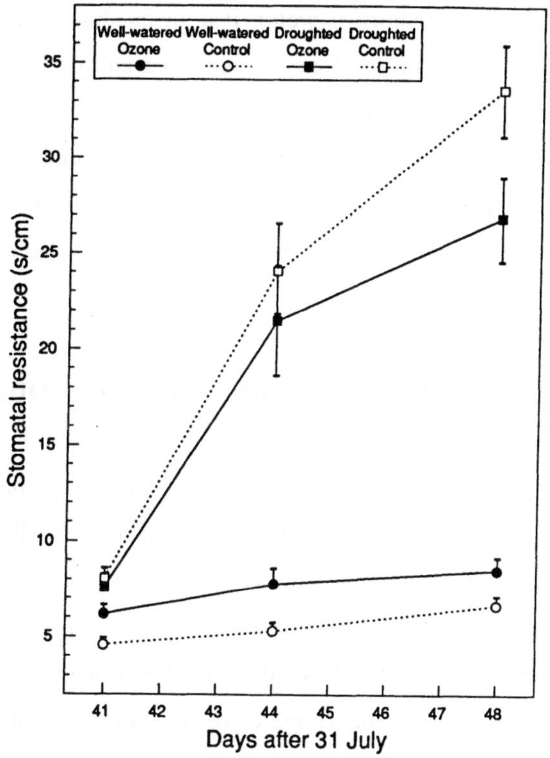

FIGURE 4. The interaction between ozone and water stress on stomatal resistance in beech. Water was withheld from day 39. Means ± 1 SE are shown.

controlled conditions to mature trees in the field. Nevertheless, the data might point to a possible explanation of the damage to beech, and perhaps other deciduous species, which has been reported over a wide area in central Europe (Lonsdale, 1986). It is possible that ozone pollution may interfere with the physiology of beech in two ways:

1. when the trees are well supplied with water, there is an increase in stomatal resistance. This may represent a direct effect of ozone on the stomata, or an indirect effect via intercellular CO_2 concentration if the primary action of ozone is upon carboxylation efficiency (Farage *et al.*, 1991). In either case, the consequence is likely to be interference with carbon acquisition for photosynthesis;

2. when the trees are short of water, it appears that ozone can reduce the ability of the stomata to close. Thus the essential physiological control over water conservation is likely to be disturbed.

Despite the caution that we need to display in interpreting this experiment, we suggest that some detailed observations of the control of water economy and gas exchange of mature beech in the field in areas subject to ozone pollution might prove worthwhile.

AIR POLLUTION AND STIMULUS-RESPONSE COUPLING IN STOMATAL GUARD CELLS

The evidence we have presented here that $SO_2 + NO_2$ pollution may interfere with stomatal responses to ABA is worthy of further discussion. The data in Fig. 4 suggest that the same might apply in the case of ozone pollution, but so far we have no direct evidence that ABA responses *per se* are inhibited by ozone.

Research in the last few years has led to major advances in our understanding of the mode of action of ABA on stomata. A key contribution was by Hartung (1983) who found that ABA does not have to enter the guard cells in order to exert its action upon them. It can apparently act at the outer surface of the plasma membrane. More recently, it has been shown that within a few minutes of the application of ABA to stomata, there is an increase in the concentration of free calcium in the cytosol of the guard cells (McAinsh *et al.*, 1990; Schroeder and Hagiwara, 1990). There is strong evidence that a signal transduction system involving Ca^{2+} is an essential feature of the sequence of events that follows the arrival of ABA at the guard cell surface. The source of the Ca^{2+} is not, however, yet established. The available evidence, though sparse, suggests that it might come both from the extracellular fluid and from intracellular stores (Hetherington and Quatrano, 1991). Drugs that are known to block calcium channels in plasma membranes will, however, only partially inhibit the ABA-induced loss of turgor in guard cells (McAinsh *et al.*, 1991). It is possible that production of inositol triphosphate (IP3) is induced by ABA, and that this then acts as an intracellular signal causing release of Ca^{2+} from inside the cell. IP3 has been shown to be capable of releasing calcium from intracellular stores (such as the vacuole) in plant cells (reviewed by Hetherington and Quatrano, 1991).

These cellular mechanisms involving the perception of ABA molecules at the surface of the plasma membrane, and the subsequent transduction processes, are a vital feature of defence against water stress in plants. Both ozone and mixtures of SO_2 and NO_2 are likely to produce free radicals near the cell surface (Wellburn, 1988) which may damage the plasma membrane and interfere with the action of ABA. It is now believed that functional hormone receptors are located on the external face of the plasma membrane, i.e., in a position likely to be very susceptible to damage. The precise receptors for ABA are, however, still unidentified and hence it is not possible to speculate in more detail about the nature of the damage that pollutants may cause.

ACKNOWLEDGEMENTS

We are grateful to the CEC and to the UK Department of the Environment for financial support.

REFERENCES

Atkinson CJ, Wookey PA, Mansfield TA (1991) Atmospheric pollution and the sensitivity of stomata on barley leaves to abscisic acid and carbon dioxide. The New Phytologist 117: 535-541

Black CR, Black VJ (1979) The effects of low concentrations of sulphur dioxide on stomatal conductance and epidermal cell survival in field beans (*Vicia faba* L.). Journal of Experimental Botany 30: 291-298

Darrall NM (1989) The effect of air pollutants on physiological processes in plants. Plant, Cell and Environment 12: 1-30

Davies WJ, Zhang J (1991) Root signals and the regulation of growth and development of plants in drying soil. Annual Review of Plant Physiology and Plant Molecular Biology 42: 55-76

Farage PK, Long SP, Lechner EG, Baker NR (1991) The sequence of change within the photosynthetic apparatus of wheat following short-term exposure to ozone. Plant Physiology 95: 529-535

Godzik S, Piskornik Z (1966) Transpiration of *Aesculus hippocastanum* leaves from areas of various air pollution. Bulletin of the Polish Academy of Science, Series B 14: 181-184

Hartung W (1983) The site of action of abscisic acid at the guard cell plasmalemma of *Valerianella locusta*. Plant, Cell and Environment 6: 427-428

Hetherington AM, Quatrano RS (1991) Mechanisms of action of abscisic acid at the cellular level. The New Phytologist 119: 9-32

Lonsdale D (1986) Beech health study, 1986. Forestry Commission Research and Development paper 149, UK Forestry Commission, Edinburgh

Lucas PW (1990) The effects of prior exposure to sulphur dioxide and nitrogen dioxide on the water relations of Timothy Grass (*Phleum pratense*) under drought conditions. Environmental Pollution 66: 117-138

Lucas PW, Cottam DA, Mansfield TA (1987) A large-scale fumigation system for investigating interactions between air pollution and cold stress in plants. Environmental Pollution 43: 15-28

Macdowall FDH (1965) Predisposition of tobacco to ozone damage. Canadian Journal of Plant Science 45: 1-12

McAinsh MR, Brownlee C, Hetherington AM (1990) Abscisic acid-induced elevation of guard cell cytosolic Ca^{2+} precedes stomatal closure. Nature (London) 343: 186-188

McAinsh MR, Brownlee C, Hetherington AM (1991) Partial inhibition of ABA induced stomatal closure by calcium channel blockers. Proceedings of the Royal Society, London B243: 195-201

Neighbour EA, Cottam DA, Mansfield TA (1988) Effects of sulphur dioxide and nitrogen dioxide on the control of water loss by birch (*Betula* spp.). The New Phytologist 108: 149-157

Schroeder JI, Hagiwara S (1990) Repetitive increases in cytosolic Ca^{2+} of guard cells by abscisic acid activation of nonselective Ca^{2+} permeable channels. Proceedings of the National Academy of Sciences, USA 87: 9305-9309

Taylor G, Dobson MC (1989) Photosynthetic characteristics, stomatal responses and water relations of *Fagus sylvatica*: impact of air quality at a site in southern Britain. The New Phytologist 113: 265-273

Taylor JS, Reid DM, Pharis RP (1981) Mutual antagonism of sulfur dioxide and abscisic acid in the effect on stomatal aperture in broad bean (*Vicia faba* L.) epidermal strips. Plant Physiology 68: 1504-1507

Walker EK, Vickery LD (1961) Influence of sprinkler irrigation on the incidence of weather fleck on flue-cured tobacco in Ontario. Canadian Journal of Plant Science 41: 281-287

Wellburn AR (1988) Air pollution and acid rain: the biological impact. Longman UK

Wright STC (1977) The relationships between leaf water potential and the levels of abscisic acid and ethylene in excised wheat leaves. Planta 134: 183-189

Zhang J, Davies WJ (1987) ABA in roots and leaves of flooded pea plants. Journal of Experimental Botany 38: 649-659

OZONE/SULPHUR DIOXIDE INTERACTIONS IN TEMPERATE ARABLE CROPS

V.J. Black, J.J. Colls[1] and C.R. Black[1]
Department of Geography
Loughborough University of Technology
Loughborough
Leics LE11 3TU
UK

ABSTRACT

Ozone is a gaseous pollutant of major agricultural significance in many countries, inducing changes in plant growth and altering patterns of resource allocation, water use efficiency and development in a number of crop species. Much of the evidence for these detrimental effects comes from the US NCLAN programme, in which crop plants were exposed to a range of ozone concentrations throughout the growing season. Ambient concentrations of ozone were found to be sufficiently high to cause adverse effects in 13 of the 14 species studied. However, these data cannot be used directly to assess pollutant-induced crop losses in Britain and other European countries because firstly the crops, cultivars and cultivation regimes employed differ between the two continents and secondly climatic factors such as temperature and rainfall, which are known to influence plant responses to pollutants, may also differ markedly. For example, crops in Northern Europe frequently complete their initial growth in winter when cold or waterlogging stress may be prevalent and concentrations of pollutants such as SO_2 are at their highest. In contrast to many areas of the USA, crops in Europe are rarely exposed to regimes of single pollutants, but more commonly to simultaneous or sequential exposure to several gaseous pollutants including ozone, sulphur dioxide and oxides of nitrogen. The presence of these additional pollutants may be an important determinant of plant responses to pollutant regimes in Europe. Since 1985, a European open top chamber network has examined the effects of gaseous pollutants on temperate arable crops. This chapter reviews existing knowledge of the influence of ozone on temperate arable crops, and considers its interactions with other environmental stresses including sulphur dioxide. It also attempts to identify whether sufficient information exists to permit: 1) quantification of the impact of current concentrations of ozone alone and in combination with other pollutants on arable crops in the UK and; 2) prediction of plant responses to changes in the environmental and pollution regimes induced by global climate change.

---

[1] Department of Physiology and Environmental Science, University of Nottingham, Sutton Bonington Campus, Loughborough, Leics LE12 5RD, UK.

NATO ASI Series, Vol. I 16
Interacting Stresses on Plants in a Changing Climate
Edited by M. B. Jackson, C. R. Black

INTRODUCTION

Although predictions of the impact of climate change on global food production remain uncertain, preliminary models suggest that potential agricultural production will increase in mid latitudes but decrease in low latitudes (Parry, this volume). However, these models are based on projected changes in only two of the climatic variables that are known to influence crops, temperature and precipitation. Predicted increases in carbon dioxide, UVB-radiation and gaseous pollutants may also have an impact on agricultural and natural ecosystems. Parry (1990), for example, proposed that a doubling of CO_2 concentration would increase photosynthetic rates in C3 species by 30-100%, while Jordan (this volume) has indicated that increasing fluxes of UVB-radiation may induce deleterious responses ranging from effects on gene expression to whole plant growth. Atmospheric concentrations of nitrogenous air pollutants (NO_x and NH_3), ozone (O_3) and non-methane hydrocarbons (NMHCs) known to have adverse effects on crops are currently rising and are likely to continue to do so in the foreseeable future. It is therefore imperative that any predictions of the effects of global climate change and considerations of air quality guidelines and abatement strategies should be based on realistic models which take account of these key parameters.

However, before these parameters can be included in predictive models, it is necessary to review whether the existing knowledge of the impacts of current and future concentrations of pollutants on crops species is sufficient in quantity and quality to be used to improve our understanding and prediction of the effects of changing climate on temperate agricultural crop production.

OZONE

The main atmospheric pollutant of current concern with respect to arable crops grown in Europe is ozone, a secondary photochemical pollutant produced from the interaction of primary pollutants such as hydrocarbons and oxides of nitrogen under the action of sunlight. Concentrations usually reach a maximum during daylight hours in the summer months, although it has recently been recognised that significant episodes of ozone of several hours' duration can occur in spring and autumn (PORG, 1987). However, other gaseous pollutants may occur concurrently or sequentially with ozone. Depending on the prevailing pollutant emissions and climate, ozone may be supplemented by combinations of SO_2, NO_x, NH_3 and/or NMHCs, giving rise to a spatially and temporally variable pollution climate in Europe.

Experimental fumigations of important agricultural species have shown that ozone has adverse effects on a range of plant growth and yield parameters and underlying biochemical and physiological processes (Heath, 1988; Krupa and Manning, 1988; Wolfenden and Mansfield, 1991). Furthermore, these ozone-induced responses may be moderated in a complex manner by the presence of additional pollutants such as SO_2, NH_3 and NO_x. The direction and degree of this modification is dependent on physical, chemical and biological factors such as the characteristics of the pollutant exposure regime (Ormrod, 1982; Runeckles and Palmer, 1987), the species and cultivar involved and the prevailing and past environmental conditions to which the plants have been exposed (Mansfield and McCune, 1988; van der Eerden *et al.*, this volume).

Many of these effects and dose-response relationships were established for plants exposed for short periods to relatively high concentrations of pollutants under uniform and usually non-stressful growth room conditions. It is now recognised that plant responses to pollutants under these conditions may differ markedly from those induced by long-term exposure under variable and potentially stressful field conditions. Thus, in any attempt to characterise the current responses of arable crops to pollutant regimes in Europe and to predict the future impact of changing pollution and environmental climates, the use of response data from growth room experiments must be treated with caution.

NCLAN PROGRAMME

In order to obtain more valid relationships between seasonal exposure to different concentrations of pollutants, particularly ozone, and the yields of important agricultural crops, a national crop loss assessment network (NCLAN) was operated in the USA between 1980 and 1985 (Preston and Tingey, 1988). In this study 14 major crops were grown in Open Top Chambers (OTCs) at a number of geographical locations. These permitted plants growing in soil in the field and receiving natural radiation to be exposed to an air stream containing either ambient air with its existing complement of pollutants, or to air filtered of the majority of the ambient pollutants. At some sites, enhanced levels of pollutants, usually O_3, were added for seven hours a day to existing ambient pollution concentrations within the chambers. This experimental approach was designed to identify whether current concentrations of pollutants in the USA adversely affected crop yields and whether such impacts would be enhanced by further increases in pollutant concentrations. This study demonstrated that the prevailing ozone concentrations in the USA exerted significant detrimental effects on 13 of the 14 crops studied and might induce estimated yield reductions ranging from zero in barley (*Hordeum vulgare*) to 20% in soybeans (*Glycine max*) (Lesser *et al.*, 1990). Increased injury was observed when the crops were exposed to concentrations of pollutants above those present in the ambient air (Heagle *et al.*, 1988).

The question of whether the NCLAN findings could be extrapolated to predict current and future responses of arable crops to pollution climates in Europe was addressed in 1987 by the United Kingdom Photochemical Oxidants Review Group. Their report (PORG, 1987) concluded that, if the NCLAN data were directly applicable to European conditions, the most sensitive crops would be affected by ozone in the UK only during exceptional summers. However, this extrapolation was largely speculative since the data underpinning the general conclusions of the NCLAN programme exhibited considerable heterogeneity. A number of unidentified factors had apparently modified plant responses to pollutants and contributed to the observed heterogeneity (Heagle *et al.*, 1988). Indeed, according to Rawlings *et al.* (1988), the most serious limitation in the NCLAN study was that environmental conditions had not been adequately monitored.

EUROPEAN OPEN TOP PROGRAMME

The observation that crop responses to pollutants may be modified by a range of factors cast doubt on the feasibility and efficacy of extending the conclusions of the NCLAN programme to assess pollutant-induced crop losses in Europe, with its differing agricultural practices, cultivars grown, prevailing climates and pollution regimes. As Wilson and Sinfield (1988) pointed out, 'The complex mixtures of pollutants that occur in the UK and elsewhere in Europe necessitate different research and policy approaches than those undertaken in the US NCLAN programme, which considers O_3 to be the dominant pollutant'. Therefore, the Commission of the European Communities (CEC) sponsored the establishment of an OTC network in Europe in 1985 which was intended to provide not only a reliable assessment and prediction of crop losses in various pollution climates, but also to take account of the many ecophysiological factors which modify plant responses to air pollutants (Mathy, 1988).

The main hypothesis tested by the CEC programme was that the prevailing concentrations of gaseous air pollutants were sufficient to affect significantly the yield of agricultural crops in European Community countries. Eleven groups in eight countries (including non-EC Switzerland and Sweden) participated, and the experimental period was between 1987 and 1991. Sites in the European network spanned latitudes from 43°N (Pau, France) to 58°N (Ostad, Sweden) and altitudes from 12 m (Roskilde, Denmark) to 900 m (Zimmerwald, Switzerland). Table 1 summarises the mean pollutant concentrations experienced during the growing season at nine of the sites. Pollution climates ranged from the ozone-dominated high-altitude Swiss site to those more characteristic of mixed urban-industrial sources, such as Nottingham in England.

To address the central hypothesis of the CEC programme, comparisons were made between plants grown in charcoal-filtered (CF) and non-filtered (NF) open-top chambers. As in the NCLAN programme, many groups complemented this approach with studies of plant responses to increased concentrations of one pollutant gas, usually ozone. The species grown were either the cereals, oats

TABLE 1. Seasonal mean air pollutant concentrations (ppb) at CEC sites.

Country	O_3	SO_2	NO_2
Belgium (Tervuren)	26	3	10
Switzerland (Zimmerwald)	45	0	0
Germany (Braunschweig)	27	3	10
Denmark (Roskilde)	24	2	7
France (Pau)	23	2	-
England (Nottingham)	14	11	12
England (Ascot)	23	0	8
Ireland (Carlow)	18	4	1
Sweden (Ostad)	20	-	-

(*Avena sativa*), barley (*Hordeum vulgare*) and wheat (*Triticum aestivum*) or the legumes, field (*Vicia faba*) and green beans (*Phaseolus vulgaris*). Varieties were selected according to their local viability, rather than attempting to impose uniformity of experimental material across such a wide range of climatic conditions. Several chamber designs were used; small 1.2 m diameter OTCs enabled many replicate exposures to be made, while the traditional 3 m diameter chambers provided larger samples of plant material for destructive harvests. The CEC protocol did not require a particular type of OTC to be used, but did specify factors such as soil moisture (negligible water stress), pollutant gas measurements (O_3, SO_2 and NO_2 calculated as hourly means) and microclimatic variables (air temperature, vapour pressure deficit and solar radiation). These measures were expected to ensure reasonable comparability between the results obtained from the wide range of cultivars, soils, climates and pollutant regimes employed. In addition to the routine monitoring procedures described above, individual groups made detailed measurements of other factors such as the influence of the OTC on microclimate. Routine analyses of crop performance were made by each group.

Regular non-destructive measurements included plant density, height, leaf and tiller numbers and the timing of flowering and maturity. Destructive harvests taken at least twice during the season (onset of flowering and final harvest) were used to establish leaf areas, numbers of flowers, fruit and seeds and the dry weights of the various plant components.

Although detailed analysis of the CEC network results has still to be completed, two provisional conclusions may be drawn. Firstly, comparison of the results from charcoal filtered (CF) and non-filtered (NF) treatments indicates that the responses of the species examined to current European air pollutant concentrations were small, and variable in both magnitude and sign. Statistically significant yield changes were not detected in most experiments. For example, Fig. 1 shows the ratio of CF to NF yields for all experiments on *Phaseolus vulgaris* and *Triticum aestivum*. A positive value indicates that the CF yield was greater than the NF yield. With *Phaseolus*, there was little consistency between sites, years or cultivars. Three of the four significant results showed a stimulation of yield by unfiltered air, and the non-significant results were equally divided in sign. It does not necessarily follow that the yield changes were economically insignificant, but larger experiments involving greater replication would be needed to improve the sensitivity. This requirement is shown more clearly by the results for *Triticum*, which overwhelmingly indicate that exposure to ambient air pollutants decreased yield, although none of the individual results was statistically significant.

Secondly, the exposure-response experiments conducted over a range of concentrations have shown that ozone may significantly reduce the yield of some crops at concentrations that are likely to occur in Europe, and which were typically 10-15 ppb less than the ambient concentrations prevailing during the NCLAN experiments. For example, Fig. 2 shows the response, expressed as a relative loss of seed weight, of *Triticum aestivum* to treatment period concentrations of ozone ranging from 5 to 100 ppb. In this case, the response was a straightforward function enabling simple estimates to be made of the proportional yield reduction induced by specific ozone concentrations. In contrast, the equivalent results for *Phaseolus vulgaris* cultivars show greater complexity (Fig. 3). Several features of

Phaseolus vulgaris

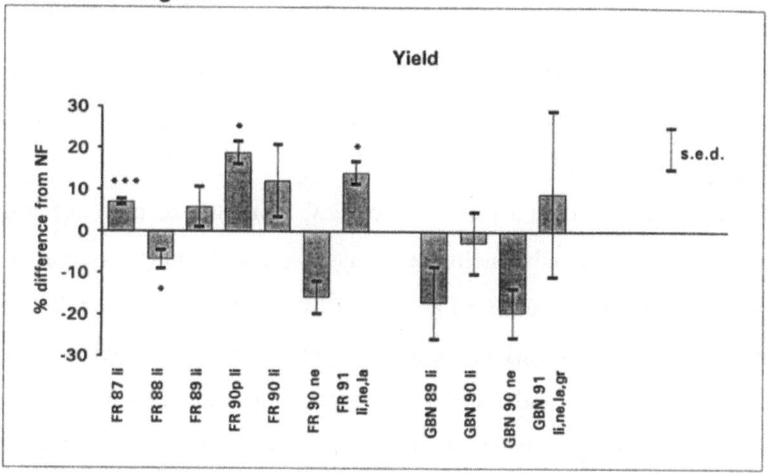

Triticum aestivum

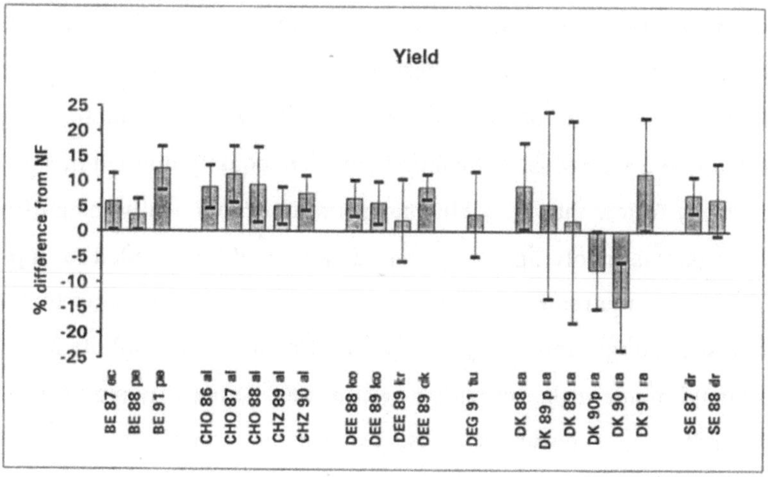

FIGURE 1. *Yield responses of* Phaseolus vulgaris *and* Triticum aestivum *cultivars to a reduction in ambient pollutant concentration in charcoal-filtered (CF) air. A positive response indicates that the yield was higher in CF air than in non-filtered (NF) air.* p *denotes a pot experiment. Key:*

Sites: *France/Pau (FR), Great Britain/Nottingham (GBN), Belgium/Tervuren (BE), Switzerland/Oeschberg (CHO), Switzerland/Zimmerwald (CHZ), Germany/Essen (DEE), Germany/Geissen (DEG), Denmark/Rosskilde (DK), Sweden/Ostad (SE).*

Cultivars: *Lit (li), Nerina (ne), Labrador (la), Groffy (gr), Echo (ec), Pelican (pe), Albis (al), Krako (kr), Okapi (ok), Krako and Okapi (ko), Turbo (tu), Ralle (ra), Drabant (dr).*

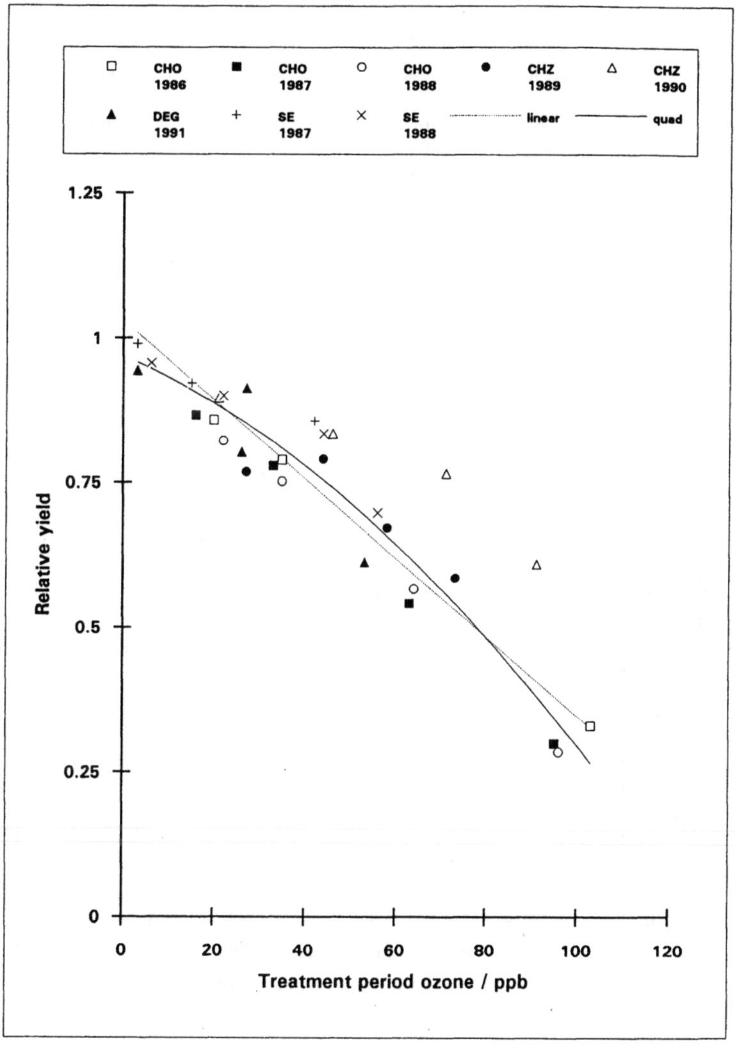

FIGURE 2. Response of grain yield in Triticum aestivum *(various cultivars) to ozone concentration relative to that predicted at 0 ppb. Linear and quadratic models have been fitted. Notation as for Fig. 1.*

this response curve should be noted. Firstly, there is a suggestion of a stimulation of yield as ozone increased from very low concentrations to ambient values. There is currently no accepted physiological explanation for this stimulation, which could reflect an adaptation to ambient ozone concentrations. Secondly, there is a broad concentration band over which yield changes very little with ozone concentration.

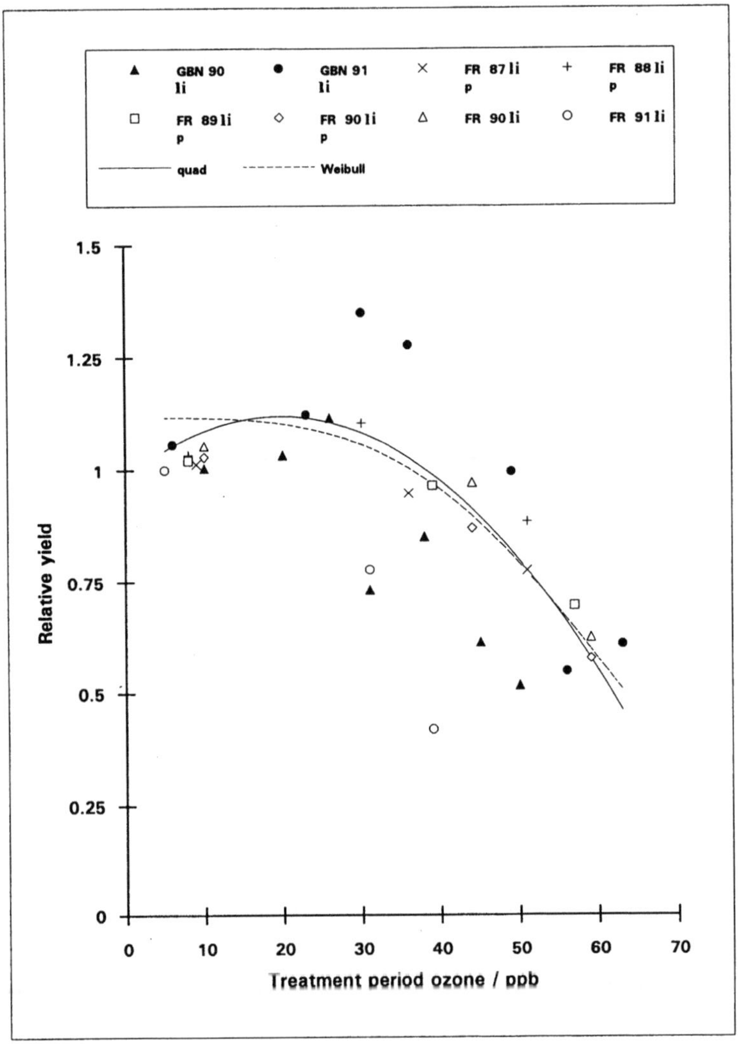

FIGURE 3. Response of grain yield in Phaseolus vulgaris *(various cultivars) to ozone concentration relative to that predicted at 0 ppb. Quadratic and Weibull models have been fitted. Notation as for Fig. 1.*

Thirdly, significant yield losses may occur when the concentration is well above ambient values. One of the practical conclusions of the CEC network was that exposure-response experiments over a range of concentrations can greatly increase our understanding of crop responses at low concentrations, where the CF/NF experiments are weak. These findings will be of value in current attempts to agree standards for critical levels and loads of pollutants, which will eventually influence statutory emission controls for precursor gases.

Considering the inherent variability in the cultivars grown and in the climatic and pollution regimes existing over the area encompassed by the CEC network, the lack of uniformity in plant response to the filtration treatments imposed is hardly surprising. Some groups have already carried out more detailed analyses of the data from the individual sites to gain an understanding of the factors underlying the variable responses observed. Fuhrer *et al.* (1989) and Pleijel *et al.* (1991) have both suggested that the species used in OTC experiments in Switzerland and Sweden exhibited different sensitivities to ozone. For example, although barley was hardly affected by the prevailing pollution climate, grain yield was increased when spring wheat was grown in filtered air as compared with ambient air at both sites. Similar beneficial effects of filtration of ambient air have also been reported by Schenone and Lorenzini (1992) for a range of crops, including wheat, in a rural area of the Po Valley, North Italy.

POLLUTANT X ENVIRONMENT INTERACTIONS

The lack of consistency in the response of individual species to filtration of the ambient air might be explained by variability in the pollution regimes or climate prevailing at different sites. For example, although ozone is thought to have the predominant impact on crops, there is evidence to suggest that responses to O_3 may be greatly modified by the action of one or more of the additional polluting gases found at specific sites within the European network (Table 1). Adaros *et al.* (1991a,b) reported an interactive effect between ozone and sulphur dioxide in both rape (*Brassica napus*) and wheat in the OTCs in Germany. Similarly, Vandermeiren *et al.* (1992) suggested that the presence of both NO_2 and SO_2 at the Belgian sites might alter the response of spring wheat to ozone, sometimes beneficially. Thus, the presence or absence of these additional pollutants may play a significant role in determining the variability of plant responses between different experimental sites.

Vandermeiren *et al.* (1992) demonstrated the potential importance of these additional pollutants when they monitored the composition of filtered and non-filtered air in their OTCs in Belgium. Unfiltered air contained a mixture of four gaseous pollutants, O_3, SO_2, NO_2 and NO, with mean concentrations of 39, 9, 19 and 5 $\mu g\ m^{-3}$ respectively. In the filtered chambers, mean values of 5, 9, 10 and 4 $\mu g\ m^{-3}$ O_3, NO, NO_2 and SO_2 were recorded, whereas the measured maximum hourly concentrations were 67, 98, 46 and 52 $\mu g\ m^{-3}$. Thus, not only would the composition of air in the unfiltered treatments at this site differ from other sites where ozone was the only significant pollutant, but it is also likely that the filtered treatments used as clean air controls exhibited significant variation between sites. Such variability in the composition and pattern of pollutants experienced by plants in filtered and non-filtered chambers at different geographical locations must contribute to the heterogeneity of the responses observed and complicate data analysis and interpretation. Nonetheless, it should be possible to use the very substantial database obtained to further our understanding of the role of differing pollution climates in determining crop responses at present and in the future. The data obtained from the European network clearly show that, in some areas, current ambient concentrations of pollutants have the potential to affect the yield of some arable crops and that this effect would be exacerbated if the frequency, duration and severity of ozone episodes were to increase.

In any discussion of the impact of climatic change, it is important to consider to what extent the response to pollutants of arable crops in Europe may be influenced by prevailing environmental conditions. Some of the data from the European network suggest that crop responses to ozone, and in particular the ozone flux, at some sites are independent of year-to-year fluctuations in climate (Pleijel *et al.*, 1991; Fuhrer *et al.*, 1992) and differences in physical and pollution climates and soil characteristics (Sanders *et al.*, 1992). However, the existence of unexplained variability in the data from both the European network and the NCLAN programme suggests that plant responses to pollutants, including ozone, are likely to be modified significantly by environmental conditions.

Atmospheric carbon dioxide concentration is arguably one of the major environmental variables that has been undergoing consistent long-term change. It has been suggested that rising CO_2 may ameliorate the impact of ozone by inducing changes in stomatal density (Woodward *et al.*, 1991) and conductance (Barnes *et al.*, 1988; Morison, 1989), thereby reducing the flux of damaging gaseous pollutants into plants (Enoch and Honour, this volume). In addition, it has been proposed that increasing atmospheric CO_2 concentration may increase CO_2 assimilation rates and possibly ameliorate reductions in the efficiency of carboxylation induced by ozone exposure (Farage *et al.*, 1991). However, it is now apparent that the responses to these changing variables are much more complex than originally anticipated and that plants may acclimatise to elevated CO_2 concentrations (Van Oosten *et al.*, this volume). Indeed Barnes and Pfirrman (1992) have demonstrated that, although elevated CO_2 concentrations may initially ameliorate the effects of exposure to ozone in radish plants (*Raphanus sativus* L.), this ability is subsequently lost. Furthermore, pollutants are known to alter assimilate partitioning between the roots and shoots, with transport to the above-ground regions being favoured over below-ground organs (Wolfenden and Mansfield, 1991). If the result of increased CO_2 concentration was to amplify this imbalance, then further difficulties for the plant during periods of soil moisture deficit or high evaporative demand could ensue.

DIRECT IMPACT OF GASEOUS POLLUTANTS ON REPRODUCTIVE DEVELOPMENT

Even small increases in pollutant concentrations coupled with small sustained shifts in climatic variables such as temperature could induce complex and far-reaching effects. It is not unreasonable to speculate that in certain areas of Europe daily mean temperatures will rise by 1-2 °C. Such a shift would lead to increased rates of growth and development and the earlier maturation of crops, and therefore alter the timing of critical reproductive stages such as anthesis and fertilisation, bringing these forward into a period coincident with major pollutant episodes. If the buds, flowers, pods and ears are sensitive to pollutant impact, then crops which hold their reproductive organs above the canopy, and thus exposed to maximum pollutant concentrations, would be particularly vulnerable.

Reproductive development is a critical phase in the life cycle of annual plants. Impairment of any step in this developmental sequence has serious implications for both agricultural crops and species in natural and semi-natural communities. The susceptibility of reproductive processes to environmental stresses such as temperature extremes and drought is well established (e.g., Stirling and Black, 1991), but the impact of environmental pollutants has received limited attention (e.g., Feder and Sullivan, 1969; DuBay and Murdy, 1983; Cox, 1983; Kasana, 1991). Furthermore, most studies have involved simultaneous exposure of both vegetative and reproductive components of the plant to pollutants, making it impossible to distinguish direct effects on reproductive development from indirect effects originating from injury to the vegetative organs.

In contrast, recent work with oilseed rape (*Brassica napus* L.) has demonstrated conclusively that reproductive development may be directly affected by exposure to ambient concentrations of gaseous pollutants. For instance, a single 6 h exposure of the reproductive parts to 100 ppb O_3 or 100 ppb O_3 + 30 ppb SO_2 was sufficient to induce both short and long term effects on reproductive development. These effects included reduced pollen germinability and pollen tube growth (Bosac *et al.*, 1993a), both of which are vital to the successful fertilisation of ovules and production of seeds. Significantly increased losses of reproductive sites were also observed 2-5 days after a single exposure to 200 ppb SO_2, 100 ppb O_3 or 100 ppb O_3 + 30 ppb SO_2 (Bosac *et al.*, 1993b). Different racemes exhibited differential sensitivity, with the second and third laterals showing the greatest loss of floral structures following treatment (Fig. 4). Intervarietal differences in sensitivity were also apparent, with cv. Libravo exhibiting a lower susceptibility to reproductive site losses in response to pollutants than cv. Tapidor (Fig. 4). Similar intervarietal differences in sensitivity to pollutants during the vegetative phase are well documented (Taylor *et al.*, 1986).

The extensive losses of reproductive sites observed in rape following pollutant exposure had decreased relative to clean air controls by 26 days after

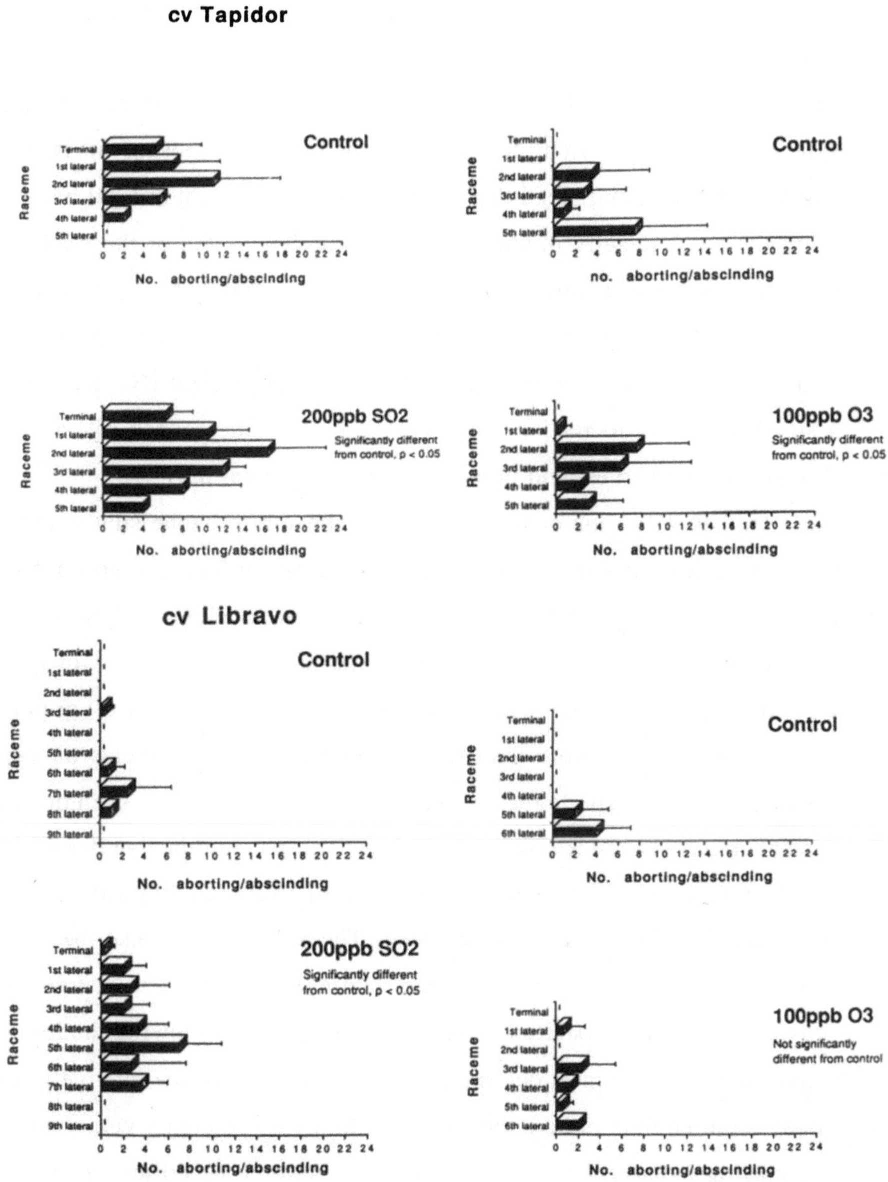

FIGURE 4. Number of reproductive structures aborting or abscinding two days after exposing inflorescences of B. napus *L. to charcoal-filtered air or air containing 200 ppb SO_2 or 100 ppb O_3 for six hours. A, cv. Tapidor; B, cv. Libravo. Single standard errors are shown, n = 4. After Bosac (1992).*

exposure (Bosac, 1992) because the indeterminate growth habit of rape confers considerable compensatory flexibility during reproductive development. Thus, plants subjected to a single pollution episode were able to replace at least some of the lost reproductive sites. Long-term effects of a single exposure on raceme length, mean pod length and mean dry weight of seeds per plant at final harvest, and also on the commercially important harvestable component of the seed, per cent oil content, were also observed (Table 2). More remarkable still was the observation that, although germination of seed from previously exposed plants was unaffected, the vigour of the resulting seedlings was depressed (Bosac, 1992). Thus, a single exposure to realistic pollutant episodes can produce effects that carry forward to final harvest and into the following generation.

These studies with rape demonstrate that exposure to ozone singly and in combination with sulphur dioxide exerts important *direct* effects on reproductive development that are both persistent and of agricultural and ecological significance. It is therefore important to establish whether similar responses occur in other species, not only in crops whose harvestable component is the grain, fruit or pods, but also in native species whose long-term ecological success depends on their reproductive efficacy. If a single realistic exposure to O_3 can produce such diverse and persistent effects in an indeterminate species with high reproductive flexibility, it is important to ascertain the impact of repeated exposures typical of field conditions during reproductive development. The effects may also be more dramatic in species with determinate or semi-determinate reproductive strategies which provide little or no capacity for compensatory growth following the loss of reproductive sites. For instance, determinate species often produce relatively few flowers over a limited time-period, during which they may be highly vulnerable to a single damaging ozone episode. The consequences might be critical in species with small seed banks of limited longevity. An improved understanding of plant responses to pollutants is vital to assist breeding programmes in which cultivars, for example leafless and semi-leafless varieties of *Vicia faba*, are being engineered so that their reproductive structures are held above the crop canopy to improve light penetration to the stipules and expose the bracts and pods to increased solar radiation. Clearly such a strategy, in conjunction with the more open and better ventilated canopy structure, would lead to the increased exposure of reproductive sites to repeated episodes and increased fluxes of atmospheric pollutants.

TABLE 2. *Long-term effects on reproductive development of a single exposure of the inflorescences of* Brassica napus *cv. Libravo to 100 ppb O_3 for six hours. Means±SE. After Bosac (1992).*

Variable	Control	O_3	% Reduction	Significance
Mean raceme length at 71 d (mm)	685 ± 238	452 ± 28	34	NS
Mean pod length at final harvest (mm)	60.7 ± 5.3	55.8 ± 5.2	8	$p < 0.01$
Mean seed dry weight (g plant^{-1})	4.4 ± 0.6	2.4 ± 0.7	45	$p < 0.05$
Oil content of seeds (% dry matter)	25.5 ± 0.5	18.8 ± 0.0	26	$p < 0.05$
Seedling shoot length at 6 d (mm)	29.7 ± 0.8	25.0 ± 0.8	16	$p < 0.05$

MODELLING CROP RESPONSES TO GASEOUS POLLUTANTS

An important question to address is whether sufficient information is available to permit: 1) quantification of the impact on arable crops in Europe of current concentrations of ozone alone and in combination with other pollutants and; 2) prediction of plant responses to potential changes in environmental and pollution regimes induced by global climate change. The data from the NCLAN and CEC programmes indicate that the current ozone climates are adversely affecting certain temperate arable crop species and that such effects are likely to increase if ozone concentrations should rise. However, the impact of the mixtures of pollutants characteristic of many areas of Europe is less well understood. Furthermore, plant responses to allied aspects of climate change are only currently being elucidated, and thus our understanding of the combined impact of changing pollutant regimes and changing climatic factors is also extremely limited.

Is it possible to incorporate our existing knowledge of plant x pollutant x environment interactions into models which would enable tentative predictions of plant responses to a changing climate to be made? Kropff (1987) developed a model describing the short-term physiological responses of *Vicia faba* to SO_2

which could be modified for use with other gases, but unfortunately no mechanistic model is available that successfully simulates the effects of long-term exposure to pollution stress. The model proposed by Kobayashi *et al.* (1990) to simulate the responses of soybeans to ozone is quite limited in that the main parameters used to define the impact of ozone centre on rates of photosynthesis and senescence. This model also does not address the complex action of ozone, either in the presence or absence of other pollutant stresses, on plant processes during both the vegetative and reproductive phases of the life cycle. Dose-response models were developed from the NCLAN data but these were limited because insufficient information was available to permit the development of more realistic models that took account of the interactive influence of important environmental factors (Rawlings *et al.*, 1988).

It may be possible to adapt existing models developed for non-pollution orientated studies to simulate plant x pollutant x environment interactions. For example, at least five major models have been developed to describe the growth and water use of wheat (e.g., AFRCWheat, CERES-Wheat, SWHEAT, DEMETER, WHEAT; GCTE, 1992) and these have been used in attempts to predict the impact of rising atmospheric CO_2 concentration. However, since there were major discrepancies between the predicted values produced by these various models for such key variables as maximum leaf area, dry matter, light interception and water use, it appears premature at present to extend these models to predict the complex impact of environmental pollutants until further development and validation have been carried out. However, it is imperative that modellers addressing plant x global change interactions be aware of the potentially important modifying influence of the pollutant component of the atmospheric environment.

CONCLUSIONS

More information is clearly needed at a number of levels before the impact of changing climate and pollution regimes on temperate arable crops can be predicted with any degree of confidence. The impact of pollutants in the field requires further elucidation using a combination of exposure facilities, including both OTCs (despite the valid criticism that such systems modify the microclimate within the chambers; Davison and Barnes, this volume) and open field exposure systems which permit crops growing in the field to be exposed to experimentally modified climatic and pollution regimes. Allied detailed physiological and biochemical studies under both field and laboratory conditions are also required to improve our understanding of the primary sites and modes of action of pollutants, and establish how these are modified by changing climatic conditions. Only then will our information base be sufficiently strong to permit realistic assessment of the future impact of pollutants. Parallel advances must also be made in the development and validation of mechanistic models that take account of the potential impact of complex pollution regimes in a changing climate.

ACKNOWLEDGEMENTS

We are grateful to the Commission of European Communities, UK Department of the Environment and the NERC/CEGB Special Topic Programme for supporting sections of the work reported here.

REFERENCES

Adaros G, Weigel HJ, Jäger HJ (1991a) Single and interactive effects of low levels of O_3, SO_2 and NO_2 on the growth and yield of spring rape. Environmental Pollution 72: 269-286

Adaros G, Weigel HJ, Jäger HJ (1991b) Concurrent exposure to SO_2 and/or NO_2 alters growth and yield responses of wheat and barley to low concentrations of O_3. The New Phytologist 118: 581-591

Barnes JD, Pfirrman T (1992) The influence of CO_2 and O_3, singly and in combination, on gas exchange, growth and nutrient status of radish (*Raphanus sativus* L.). The New Phytologist 121: 403-412

Barnes JD, Reiling K, Davison AW, Renner CJ (1988) Interaction between ozone and winter stress. Environmental Pollution 53: 235-254

Bosac C (1992) The impact of ozone and sulphur dioxide on reproductive development in oilseed rape. PhD Thesis, Nottingham, UK

Bosac C, Black VJ, Black CR, Roberts JA, Lockwood F (1993a) The impact of O_3 and SO_2 on reproductive development in oilseed rape (*Brassica napus* L.): I Pollen germination and pollen tube growth. The New Phytologist (in press)

Bosac C, Roberts JA, Black VJ, Black CR, (1993b) The impact of O_3 and SO_2 on reproductive development in oilseed rape (*Brassica napus* L.): II Reproductive site losses. The New Phytologist (submitted)

Cox RM (1983) Sensitivity of forest plant reproduction to transported pollutants: *In vitro* sensitivity of pollen to simulated acid rain. The New Phytologist 95: 269-275

Darrall NM (1989) The effects of air pollutants on physiological processes in plants. Plant, Cell and Environment 12: 1-30

DuBay DT, Murdy WH (1983) The impact of sulfur dioxide on plant sexual reproduction: *in vivo* and *in vitro* effects compared. Journal of Environmental Quality 12: 147-149

Farage PK, Long SP, Lechner EG, Baker NR (1991) The sequence of change within the photosynthetic apparatus of wheat following short-term exposure to ozone. Plant Physiology 95: 529-535

Feder WA, Sullivan F (1969) Differential sensitivity of pollen grains to ozone injury. Phytopathology 59: 399

Fuhrer J, Egger A, Lehnherr B, Grandjean A, Tschannen W (1989) Effects of ozone on the yield of spring wheat (*Triticum aestivum* L. cv. Albis) grown in open-top field chambers. Environmental Pollution 60: 273-289

Fuhrer J, Grandjean Grimm A, Tschannen W, Shariat-Madari H (1992) The response of spring wheat (*Triticum aestivum* L.) to ozone at higher elevations. The New Phytologist 121: 211-219

GCTE (1992) Effects of global change on the wheat ecosystem. Report of IGBP Global Change and Terrestrial Ecosystems planning workshop, 22-24 July 1992, Saskatoon, Canada, pp 60

Heagle AS, Kress LW, Temple PJ, Kohut RJ, Miller JE, Heggestad HE (1988) Factors influencing ozone dose-yield response relationships in open-top field chamber studies. In: Heck WW, Taylor OC, Tingey DT (eds) Assessment of crop loss from pollutants. Elsevier Applied Science, London New York, pp 141-179

Heath RL (1988) Biochemical mechanisms of pollutant stress. In: Heck WW, Taylor OC, Tingey DT (eds) Assessment of crop loss from pollutants. Elsevier Applied Science, London New York, pp 259-286

Kasana MS (1991) Sensitivity of three leguminous crops to O_3 as influenced by different stages of growth and development. Environmental Pollution 69: 131-149

Kobayashi K, Miller JE, Flagler RB, Heck WW (1990) Modelling the effects of ozone on soybean growth and yield. Environmental Pollution 65: 33-64

Kropff MJ (1987) Physiological effects of SO_2 on photosynthesis and stomatal regulation of *Vicia faba* L. Plant, Cell and Environment 10: 753-760

Krupa SV, Manning MJ (1988) Atmospheric ozone: Formation and effects on vegetation. Environmental Pollution 50: 101-137

Lesser VM, Rawlings JO, Spruill SE, Somerville MC (1990) Ozone effects on agricultural crops: Statistical methodologies and estimated dose-response relationships. Crop Science 30: 148-155

Mansfield TA, McCune DC (1988) Problems of crop loss assessment when there is exposure to two or more gaseous pollutants. In: Heck WW, Taylor OC, Tingey DT (eds) Assessment of crop loss from pollutants. Elsevier Applied Science, London New York, pp 317-344

Mathy P (1988) The European open-top chambers programme: Objectives and implementation. In: Heck WW, Taylor OC, Tingey DT (eds) Assessment of crop loss from pollutants. Elsevier Applied Science, London New York, pp 505-513

Morison JIL (1989) Plant growth in increased atmospheric CO_2. In: Fantechi R, Ghazi A (eds) Carbon dioxide and other greenhouse gases: Climatic and associated impacts. Kluwer Academic Publishers, London, UK, pp 228-244

Ormrod DP (1982) Air pollutant interactions in mixtures. In: Unsworth MH, Ormrod DP (eds) Effects of gaseous air pollution in agriculture and horticulture. Butterworth Scientific, London, UK, pp 307-331

Parry ML (1990) Climate change and world agriculture. Earthscan Publications, London, UK, pp 157

Pleijel H, Skärby L, Wallin G, Selldén G (1991) Yield and grain quality of spring wheat (*Triticum aestivum* L., cv. Drabant) exposed to different concentrations of ozone in open-top chambers. Environmental Pollution 69: 151-168

PORG (1987) Ozone in the United Kingdom. UK Photochemical Oxidants Review Group Interim Report. Harwell laboratory, UK

Preston EM, Tingey DT (1988) The NCLAN program for crop loss assessment. In: Heck WW, Taylor OC, Tingey DT (eds) Assessment of crop loss from pollutants. Elsevier Applied Science, London New York, pp 45-62

Rawlings JO, Lesser VM, Dassel KA (1988) Statistical approaches to assessing crop losses. In: Heck WW, Taylor OC, Tingey DT (eds) Assessment of crop loss from pollutants. Elsevier Applied Science, London New York, pp 389-415

Runeckles VC, Palmer K (1987) Pretreatment with nitrogen dioxide modifies plant response to ozone. Atmospheric Environment 21: 717-719

Sanders GE, Colls JJ, Clark AG, Galaup S, Bonte J, Cantuel J (1992) *Phaseolus vulgaris* and ozone: results from open-top chamber experiments in France and England. Agriculture, Ecosystems and Environment 38: 31-40

Schenone G, Lorenzini G (1992) Effects of regional air pollution on crops in Italy. Agriculture, Ecosystems and Environment 38: 51-59

Stirling CM, Black CR (1991) Stages of reproductive development in groundnut (*Arachis hypogaea*) most susceptible to environmental stress. Tropical Agriculture 68: 296-300

Vandermeiren K, De Temmerman L, Staquet A, Baeten H (1992) Effects of air filtration on spring wheat grown in open-top field chambers at a rural site. II Effects on mineral partitioning, sulphur and nitrogen metabolism and on grain quality. Environmental Pollution 77: 7-14

Wilson RB, Sinfield AC (1988) Policy implications from crop loss assessment research - a United Kingdom perspective. In: Heck WW, Taylor OC, Tingey DT (eds) Assessment of crop loss from pollutants. Elsevier Applied Science, London New York, pp 515-519

Wolfenden J, Mansfield TA (1991) Physiological disturbances in plants caused by air pollutants. In: Last, F, Whatling R (eds) Acidic deposition - Its nature and impacts. Royal Society of Edinburgh, Edinburgh, UK, pp 117-138

Woodward FI, Thompson BG, McKee IF (1991) The effects of elevated concentrations of carbon dioxide on individual plants, populations communities and ecosystems. Annals of Botany 67: 23-38

PREDISPOSITION TO STRESS FOLLOWING EXPOSURE TO AIR POLLUTION

A.W. Davison and J.D. Barnes
Agricultural and Environmental Science
Ridley Building
The University
Newcastle Upon Tyne
NE1 7RU
UK

ABSTRACT

Most previous research has concentrated on the direct, immediate effects of gaseous air pollution on crops or young trees grown under optimal or unstressed conditions. However, in reality, plants are exposed to cycles and combinations of pollutants and to other stresses such as drought or frost that may occur concurrently or be separated in time. It is known that air pollutants may modify plant responses to environmental stresses and *vice versa*. Examples are presented and the importance of such interactions discussed in relation to predicted changes in climate and atmospheric composition.

INTRODUCTION

In order to survive, plants must withstand a taxing and constantly varying environment. In nature, they contend with diurnal and seasonal variations in temperature and irradiance, periods of water deficit, freezing and wind. They survive the ravages of pests and disease, whilst also competing with other individuals for space, light, nutrients and water. Over the past decade it has become clear that subtle physiological disturbances induced by gaseous air pollutants can modify plant responses to a number of these environmental factors and *vice versa*. Indeed, some authors consider that such interactive effects may be at least as important as the direct effects of gaseous pollutants. In this paper we provide examples of how the common gaseous air pollutants (oxides of nitrogen and sulphur, and ozone) may influence plant responses to freezing and soil water deficit, and *vice versa*. We then discuss problems of interpretation of data from experiments using different techniques of fumigation. Finally, we emphasise the

NATO ASI Series, Vol. I 16
Interacting Stresses on Plants in a Changing Climate
Edited by M. B. Jackson, C. R. Black

need for long-term studies of the impact of rising tropospheric carbon dioxide (CO_2) on perennial species or overwintering crops to be conducted over the full range of environmental conditions that prevail in the field.

CLOSED CHAMBER STUDIES

The vast majority of studies on the response of plants to gaseous air pollutants have been conducted under controlled conditions in closed chambers, of which there are many variants (CEC, 1990). Plants are usually raised under optimal or unstressed conditions, with the result that it is impossible to infer how, or even if, similar levels of pollutants influence plants in the field, where they are subjected to other environmental stresses. However, in a few cases where specific interactions have been investigated the potential for interactive effects has been demonstrated.

Example 1. Pollutant x frost interactions. Exposure to sulphur dioxide (SO_2), SO_2/NO_2 mixtures (Davison and Bailey, 1982; Davison and Barnes, 1986), and to a lesser extent ozone (O_3) (Barnes and Davison, 1988; Lucas *et al.*, 1988; Barnes *et al.*, 1988) may predispose plants to freezing injury. Conversely, in studies where plants have been grown throughout the year in closed chambers outdoors or under controlled conditions indoors and exposed to SO_2, SO_2/NO_2 mixtures or O_3, it has been shown that pollutant effects are exacerbated by low temperatures and/or low photon flux densities (Davis and Wood, 1973; Davies, 1980; Whitmore and Mansfield, 1983). This effect is believed to be a consequence of the low rates of assimilation that occur under sub-optimal conditions. The reduced supply of carbon skeletons, reducing power and ATP decreases the capacity to detoxify the pollutant and/or repair damage (Bell *et al.*, 1979; Jones and Mansfield, 1982). Despite the potential ecological significance of pollutant x frost interactions for overwintering plants (Davison and Barnes, 1986; TERG, 1988), the physiological and biochemical mechanisms underlying such effects remain to be elucidated. One reason is the lack of fundamental information on the mechanisms underlying winter injury, without which an understanding of the interactions with air pollutants is impossible.

Example 2. Pollutant x water stress interactions. There are several ways in which SO_2 and O_3 may alter plant water relations. For instance, both pollutants have a major influence on the partitioning of assimilates between root and shoot, with the shoot generally gaining at the expense of the root (Wolfenden and Mansfield, 1991). This is reflected in a marked decrease in K (allometric root : shoot coefficient) in polluted plants, an effect that would be expected to enhance the sensitivity of plants to soil water deficit. Effects on the relative growth of the root may be exacerbated by direct effects of pollutants on stomatal performance. There are reports that low concentrations of both SO_2 (Neighbour *et al.*, 1988) and O_3 (Barnes *et al.*, 1990a) can increase the basal rate of leaf transpiration through effects on the stomatal apparatus, whilst stomata may not respond as rapidly to induced water stress (Neighbour *et al.*, 1988; Barnes *et al.*, 1990c; Maier-Maercker and Koch, 1991; Barnes and Bahnemann, unpublished). However, the responses of stomata to pollutants are very complex, and higher concentrations of pollutants generally result in partial stomatal closure (Darrall, 1989). In addition, as stomata close in response to increasing soil water deficit, thereby reducing pollutant uptake, water stress may protect the plant against the detrimental effects of air pollutants. Consequently, plant responses to pollutant x water stress interactions may be governed by the sequence of events to which the plant is exposed. However, there are few instances where the net effects of these complex interactions have been fully characterised and none have involved field studies.

Whilst closed chamber studies have the drawback that plants are generally raised under controlled conditions and then exposed to simple combinations of a pollutant and physical stress, this approach has demonstrated an important principle: not only can exposure to pollutants predispose plants to physical stress, but the stress may also modify the impact of the pollutant. In addition, closed-chamber studies allow investigation of the mechanistic basis for the complex and often subtle effects of combinations of pollutants and physical stress.

OPEN-TOP CHAMBERS

On a field scale, most research on the impact of air pollutants (particularly O_3) on crops and young trees has been conducted using open-top chambers (OTCs), of which there are various designs (CEC, 1986). In the USA and Europe there have been important differences in the ways in which these chambers have been employed. In the USA a major, nationally co-ordinated programme (NCLAN) assessed crop losses using standardised techniques and concentrated on annual, non-overwintering crops. It had the aim of generating dose-response models relating yield loss to pollutant exposure (Heck *et al.*, 1982). In contrast, in Europe techniques have varied between countries, although the objectives of the European Open Top Chamber Programme were similar to those of the NCLAN programme (CEC, 1988).

However, the OTC approach to the determination of the impact of ambient air pollutants is fraught with a number of difficulties that limit the interpretation of the data (Last and Mathy, 1986). In particular, OTCs create an environment which differs from that in an unenclosed field. Changes in microclimate and micro-meteorology can influence pollutant uptake, plant development, and the way in which other stresses interactively affect plant physiology and growth (Colls *et al.*, 1988; Unsworth, 1988). The marked 'chamber effects' demonstrated in the majority of OTC studies show the importance of these changes in the physical environment. An example of how subtle such chamber effects can be is provided by a recent study of photoinhibition in watermelon (Fig. 1), in which less photoinhibition (assessed using the ratio of variable to maximal chlorophyll fluorescence) was found inside chambers than outside. This was due to changes in canopy structure associated with increased leaf angles, and to alterations in the quantity and/or quality of the light transmitted to the crop. This is of importance because pollutants such as O_3 may enhance susceptibility to photoinhibition (Eamus and Barnes, 1992).

In most OTC studies, standard agronomic practices of pest and disease control have been used and most have involved irrigation because the chambers reduce the quantity of precipitation received. Both agrochemical use and irrigation may influence plant responses to air pollutants.

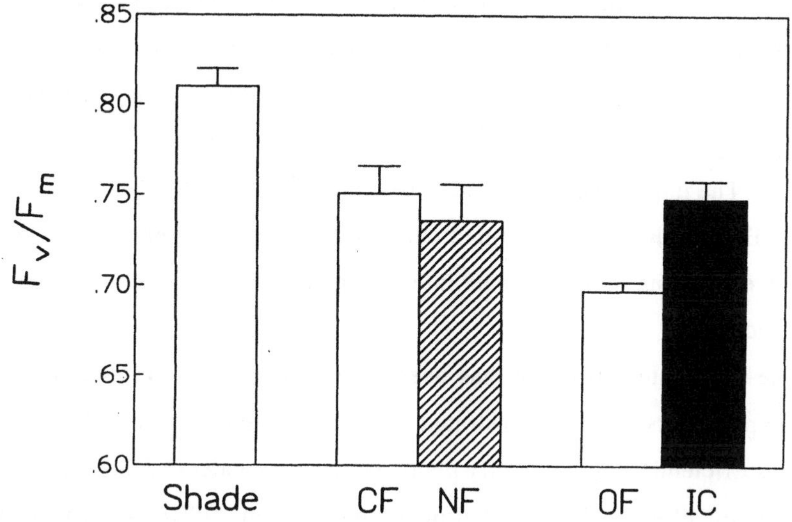

FIGURE 1. Photoinhibition, measured as the ratio of variable to maximal chlorophyll fluorescence (F_v/F_m), in watermelon plants grown in open-top chambers ventilated with charcoal-filtered air (CF) or non-filtered air (NF) and open-field plots (OF). Cumulative PFD (μmol m^{-2}/1000) when leaves were sampled from the open-field, inside chambers (IC) and shade treatments were 11.0, 9.8 and 1.5, respectively. Values represent the means for 10 leaves ± S.E. of four replicate plots/chambers per treatment; the plots/chambers were treated as blocks for ANOVA. Although plants grown in the open-field and in chambers exhibited a significant decline in F_v/F_m ($p<0.001$) compared with plants maintained in shade, there was no significant difference between plants receiving filtered or non-filtered air. However, the decline in F_v/F_m was significantly greater ($p<0.01$) in plants grown in the open-field than in their counterparts raised inside chambers, indicating that photoinhibition was less inside the open-top chambers.

It is well established that certain agrochemicals, notably systemic fungicides such as benomyl (Lorenzini *et al.*, 1987), may protect plants from ozone, but even the wetting agents added to commercial agrochemical formulations may have effects. Recent work at Newcastle (Price, personal communication) has shown that a range of responses may be induced by exposure to O_3 and different agro-chemicals, some of which may even exacerbate pollution damage, for example, cycocel. There appear to have been no field-scale studies to estimate the effects of agrochemical interactions other than attempts to use the antioxidant EDU (ethylenediurea) as an experimental tool (Smith, 1987; Long and Davis, 1991).

Where pest interactions have been investigated, it has been shown that there may be subtle effects of air pollutants on plant biochemistry that may stimulate the growth and survival of phloem-feeding insects (Dohmen *et al.*, 1984). These probably originate from effects on food quality but may also involve defence chemicals. The importance of pollutant/pest interactions in the field is not known, although Bell (TERG, 1988) pointed out that, because of the enormous economic importance of pests, even a small interaction with pollution could be of great significance.

The main interaction that has been investigated using OTCs is with water stress. Experiments have demonstrated that current levels of air pollution in the USA are sufficiently high to modify plant responses to water deficits. For example, the combined effects of O_3 and drought on soybean (*Glycine max*) were investigated as part of the NCLAN programme. Over two successive years a combination of soil water deficit and ambient O_3 reduced yield by 25%, whilst the O_3 effect was less pronounced in irrigated plants (yield reduced by only 5%). Water stress reduced yield by only 4% in filtered air (Heggestad *et al.*, 1985).

Most OTC experiments have involved assessment of the effects of ambient pollution on annual crops grown during the summer months, so there is little information concerning the effects of repeated exposures to pollutants on perennials, or on persistent effects that may influence subsequent responses to environmental stress. However, there are a few notable exceptions. For example, there are reports that ambient air pollution may predispose plants to winter injury. Ashmore *et al.* (1986), working at Ascot, England, reported that winter injury in four cultivars of *Trifolium pratense* was significantly reduced in OTCs receiving filtered air over winter.

Work on forest trees using OTCs in which young conifers were exposed to controlled levels of O_3 over several years has led to the realisation that the effects of O_3 on physiological processes can persist from one growing season to the next. Effects of the pollutant on stomatal performance (Barnes *et al.*, 1990c), photosynthesis (Sasek and Richardson, 1989; Wallin *et al.*, 1990; Eamus *et al.*, 1990), chloroplast membrane lipid composition (Wolfenden and Wellburn, 1991) and carbohydrate content (Alscher *et al.*, 1989; Barnes *et al.*, 1990b) have all been shown to persist for many months after exposure to the pollutant is terminated. This means that pollutants such as O_3, which tends to be episodic and seasonal, may influence responses to environmental stress incurred weeks or months later.

OPEN-FIELD EXPOSURE SYSTEMS

Few experiments have used open-field exposure systems (OFES) to study the direct and indirect effects of gaseous air pollutants, and those that have been undertaken have largely concentrated on the impact of elevated concentrations of SO_2. Although the data obtained are often difficult to interpret, this type of exposure system has some important advantages over OTCs for studies of air pollution effects in the field. Most importantly, the physical environment of the plants is unaffected and it is possible to expose plants to controlled levels of pollutants throughout the year. Consequently, this approach provides a more realistic indication of the scope for interactions. For instance, there are numerous reports of increased susceptibility to frost damage in plants exposed to SO_2 in the field using OFESs (Baker *et al.*, 1982; Holland, personal communication; McLeod, personal communication). Results from our own field exposure system over the past two years have highlighted some of their advantages (see Wilbourn, 1992). At a rural site in the north-east of England a sward of *Lolium perenne* (cv. Melle) and *Trifolium repens* (cv. Grasslands Huia) was exposed to 60-70 ppb O_3 for 7 h d^{-1} for 20-22 days between July and September in each of three consecutive years. The experiment had two treatments in four randomised blocks: on all four blocks two cutting regimes were imposed, a frequent cut to simulate grazing and an infrequent cut to simulate silage production. The difference in cutting regime altered the competitive balance between the two species. Although the ground appeared uniform, two blocks experienced significantly greater soil water deficits than the others, and so there was an inadvertent additional treatment. Destructive harvests clearly demonstrated that:

1. Different management practices influenced the response of the sward to O_3, with the clover declining after exposure in the frequently cut, but not in the infrequently cut plots.
2. There was a significant O_3 x water stress interaction. The clover content of the sward (expressed as a percentage of the total harvested biomass) was reduced by water stress, but there was no significant difference between O_3-fumigated and control plots, in contrast to the marked effects of O_3 on the composition of the less water stressed plots (Fig. 2).
3. Although the clover canopy recovered overwinter in the first two years, it was at the expense of the storage organs, the stolons. The mass of stolons was persistently reduced by summer exposure to ozone.

These results demonstrate the need for long-term studies of the effects of air pollutants on perennial species, and the possible significance of pollutant x stress interactions in the field.

INTERACTIONS OF THE FUTURE?

The continued increase in the atmospheric concentration of CO_2 and the possible associated modifications in climate that this will involve are currently of great concern, but it should not be forgotten that these changes will take place against a shifting background of air pollution. Concentrations of gases such as SO_2 may well continue to fall, but the same combustion sources that produce CO_2 also inject the precursors of O_3 into the atmosphere, with the result that the rise in atmospheric CO_2 concentration is likely to be inextricably linked with increasing frequency and duration of potentially phytotoxic O_3 episodes (Ashmore and Bell, 1991). The potential importance of such interactions is illustrated by work in which O_3 has been shown to reduce progressively with time the effect of elevated CO_2 on growth and photosynthesis (Barnes and Pfirrman, 1992; Barnes and Ollerenshaw,

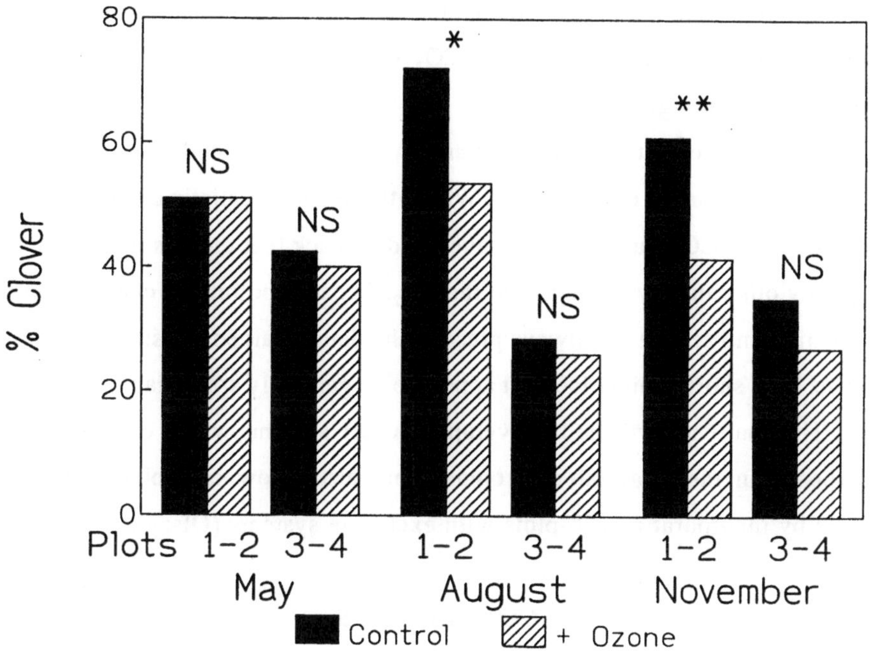

FIGURE 2. Interactive effects of O_3 and water stress on the clover content (expressed as % total living biomass) of a mixed sward. An area sown with a mixture of Lolium perenne *cv. Melle and* Trifolium repens *cv. Grasslands Huia was allowed to establish for one year. Plots were then exposed for three consecutive years to ambient air or ambient air plus 60-70 ppb O_3 for 7 h d^{-1} for 20-22 days each year using a field release system. Two management regimes involving frequent or infrequent cutting were imposed. Plots 3 and 4 experienced a greater soil water deficit than plots 1 and 2. The Figure shows mean data for the different management regimes during the second year of exposure. Mean soil water potentials (MPa) for duplicate blocks measured using soil moisture blocks positioned 15 cm below the soil surface are presented for August (for further details of soil water availability see Wilbourn, 1992). Duplicate plots were treated as blocks for ANOVA. Significant differences between plots receiving ambient air and ambient air plus O_3 are indicated; * $p<0.05$, ** $p<0.01$, NS not significant.*

unpublished). If mechanistically based predictive models are to be realistic, it is essential to learn how CO_2 and O_3interact with other environmental stresses. However, in the numerous experiments presently being conducted on the effects of elevated CO_2, it is often forgotten that there may be large differences in response between controlled and field environments. Current models of the effects of elevated CO_2 on plant growth and yield make little allowance for interactions with pollutants and other environmental factors, with the result that current predictions of the impact of rising CO_2 on plant growth and yield may be extremely misleading.

Field release systems (Hendrey *et al.*, 1988) for elevated CO_2 have already been used or are under development. Whilst these allow realistic estimation of the effects of rising CO_2, such systems have the drawback that there is no control over air quality other than with respect to the CO_2. Because pollutants characteristic of the location and season are always present, the data obtained are site- and season-specific. This shortcoming does not appear to be generally appreciated but must be faced. The challenge will be to develop exposure systems which enable plants to be exposed in their natural environment but also allow control of pollutants, perhaps by incorporating sub-plots with exclusion systems (Olszyk *et al.*, 1986).

CONCLUSIONS

Ambient concentrations of the common gaseous air pollutants (SO_2, SO_2/NO_2 mixtures and O_3) can modify plant responses to a number of environmental stresses and *vice versa.* There is growing evidence to suggest that the effects of these interactions may be as, if not more, important than the direct effects of air pollutants, but a number of technical difficulties are involved in assessing the importance of pollutant x stress interactions in the field. It is vital to understand these complex interactions, not only to define the mechanisms involved in the modification of plant responses to predicted changes in climate and atmosphere, but also to permit the correct interpretation of plant responses to CO_2/pollutant mixtures in controlled environment chambers, OTCs or OFESs. The study of pollutant x stress interactions presents a considerable technological challenge for the future.

ACKNOWLEDGEMENTS

The authors acknowledge funding by the Royal Society, UK Department of the Environment, Natural Environment Research Council, Commission of European

Communities, R.B. Cooke Endowment Trust and University of Newcastle Research Fund. Some of the work reported here was undertaken by Dr. S. Wilbourn, Mr. D. Velissariou, Ms. M. Price and Ms. S. Bahnemann. The authors thank Dr. A.M. Borland for producing the figures.

REFERENCES

Alscher RG, Amundson RG, Cumming JR, Fellows S, Fincher J, Van Leuken P, Weinstein LH (1989) Seasonal changes in the pigments and growth of red spruce as affected by ozone. The New Phytologist 113: 211-223

Ashmore, MR, Bell JNB (1991) The role of ozone in global change. Annals of Botany 67: 39-48

Ashmore MR, Mimmack A, Mepsted R, Bell JNB (1986) Research at Imperial College with open-top chambers 1976-1986. In: Microclimate and plant growth in open-top chambers. Air Pollution Research Report 5, CEC Brussels pp 98-101

Baker CK, Unsworth MH, Greenwood P (1982) Leaf injury on wheat plants exposed in the field in winter to SO_2. Nature (London) 299: 149-151

Barnes JD, Davison AW (1988) The influence of ozone on the winter hardiness of Norway spruce (*Picea abies* L. Karst.). The New Phytologist 108: 159-166

Barnes JD, Eamus D, Brown KA (1990a) The influence of ozone, acid mist and soil nutrient status on Norway spruce I. Plant-water relations. The New Phytologist 114: 713-720

Barnes JD, Eamus D, Brown KA (1990b) The influence of ozone, acid mist and soil nutrient status on Norway spruce II. Photosynthesis, dark respiration and soluble carbohydrate status of trees in late autumn. The New Phytologist 115: 149-156

Barnes JD, Eamus D, Davison AW, Ro-Poulsen H, Mortensen L (1990c) Persistent effects of ozone on needle water loss and wettability in Norway spruce. Environmental Pollution 63: 345-363

Barnes JD, Pfirrmann T (1992) The influence of CO_2 and O_3, singly and in combination, on gas exchange, growth and nutrient status of radish (*Raphanus sativus* L.) The New Phytologist 121: 403-412

Barnes JD, Reiling K, Davison AW, Renner CJ (1988) Interactions between air pollutants and winter stress. Environmental Pollution 53: 235-254

Bell JNB, Rutter AJ, Relton J (1979) Studies on the effects of low levels of SO_2 on the growth of *Lolium perenne* L. The New Phytologist 83: 627-643

CEC (1986) Microclimate and plant growth in open-top chambers. Air Pollution Research Report 5, CEC Brussels, pp 340

CEC (1988) The European Communities research project on open-top chambers: Results on agricultural crops 1987-1988. Air Pollution Research Report 19, CEC Brussels, pp 174

CEC (1990) Environmental research with plants in closed chambers. Payer HD, Pfirrmann T, Mathy P (eds). Air Pollution Research Report 26, CEC Brussels, pp 467

Colls JJ, Sanders GE, Clark AG (1988) Open-top chamber experiments on field grown *Vicia faba* L. at Sutton Bonington (1987-1988) In: The European Communities research project on open-top chambers: Results on agricultural crops 1987-1988. Air Pollution Research Report 19, CEC Brussels, pp 43-73

Darrall NM (1989) The effects of air pollutants on physiological processes in plants. Plant, Cell and Environment 12: 1-30

Davies T (1980) Grasses more sensitive to SO_2 pollution in conditions of low irradiance and short days. Nature (London) 284: 483-485

Davis DD, Wood FA (1973) The influence of environmental factors on the sensitivity of virginia pine to ozone. Phytopathology 63: 371-376

Davison AW, Bailey IF (1982) SO_2 pollution reduces the freezing resistance of ryegrass. Nature (London) 297: 400-402

Davison AW, Barnes JD (1986) Effects of winter stress on pollutant responses. In: How are the effects of air pollution on agricultural crops influenced by the interaction with other limiting factors? In: Proceedings of CEC/COST Workshop Riso, Copenhagen. CEC Brussels, pp 16-32

Dohmen GP, McNeill S, Bell JNB (1984) Air pollution increases *Aphis fabae* pest potential. Nature (London) 307: 52-53

Eamus D, Barnes JD (to be published) Ozone and/or acid mist enhance the susceptibility of beech (*Fagus sylvatica* L.) to photoinhibition. Environmental and Experimental Botany, submitted November 1992

Eamus D, Barnes JD, Davison AW, Ro-Poulsen H, Mortensen L (1990) Persistent stimulation of CO_2 assimilation and stomatal conductance by summer ozone fumigation in Norway spruce. Environmental Pollution 63: 365-379

Heck WW, Taylor OC, Adams R, Bingham G, Miller J, Preston E, Weinstein L (1982) Assessment of crop loss from ozone. Journal of the Air Pollution Control Association 32: 353-361

Heggestad HE, Gish TJ, Lee EH, Bennett JH, Douglas LW (1985) Interaction of soil moisture stress and ambient ozone on growth and yield of soybeans. Phytopathology 75: 472-477

Hendrey GR, Lipfert FW, Kimball BA, Hibman DR, Bhattacharya NC (1988) Free air carbon dioxide enrichment (FACE) facility development II Field tests at Yazoo city MS 1987. Report 046 US Department of Energy, CO_2 Research Division, Office of Energy Research, Washington DC, USA

Jones T, Mansfield TA (1982) The effect of SO_2 on growth and development of seedlings of *Phleum pratense* under different light and temperature environments. Environmental Pollution 27: 57-71

Last F, Mathy P (1986) Stocktaking and implications In: Microclimate and plant growth. Air Pollution Research Report 5, CEC Brussels, pp 318-332

Long RP, Davis DD (1991) Black cherry growth responses to ambient ozone and EDU. Environmental Pollution 70: 241-254

Lorenzini G, Guidi L, Panattoni A (1987) Evaluation of fungicide effects on ozone injury to tobacco plants. Annals of Applied Biology (Suppl.) 110: 102-120

Lucas PW, Cottam DA, Sheppard LJ, Francis BJ (1988) Growth responses and delayed winter hardening in sitka spruce following summer exposure to ozone. The New Phytologist 108: 495-504

Maier-Maercker U, Koch W (1991) Experiments on the control capacity of stomata of *Picea abies* L. Karst. after fumigation with ozone and in environmentally damaged material. Plant, Cell and Environment 14: 175-184

Neighbour EA, Cottam DA, Mansfield TA (1988) Effects of sulphur dioxide and nitrogen dioxide on the control of water loss by birch (*Betula* spp.). The New Phytologist 108: 149-157

Olszyk DM, Kats G, Dawson PJ, Bytnerwicz A, Wolf J, Thompson CR (1986) Characteristics of air exclusion systems vs chambers for field air pollution studies. Journal of Environmental Quality 207: 326-334

Price M (1993) Interaction between ozone and agrochemicals. PhD Thesis, University of Newcastle Upon Tyne, UK

Sasek TW, Richardson CJ (1989) Effects of chronic doses of ozone on loblolly pine: photosynthetic characteristics in the third growing season. Forest Science 35: 745-755

Smith G (1987) Soybean yield in New Jersey relative to ozone pollution and antioxidant application. Plant Disease 71: 121-125

TERG (1988) The Effects of Acid Deposition on the Terrestrial Environment in the United Kingdom. HMSO Publication Centre, London

Unsworth MH (1988) Principles of microclimate and plant growth in open-top chambers. In: Microclimate and plant growth in open-top chambers. Air Pollution Research Report 5, CEC Brussels, pp 16-29

Wallin G, Skärby L, Selldén G (1990) Long-term exposure of Norway spruce, *Picea abies* L. Karst., to ozone in open-top chambers. The New Phytologist 115: 335-344

Whitmore ME, Mansfield TA (1983) Effects of long-term exposures to SO_2 and NO_2 on *Poa pratensis* and other grasses. Environmental Pollution 31: 217-235

Wilbourn S (1992) The effects of ozone on a grass/clover sward. PhD Thesis, University of Newcastle Upon Tyne, UK

Wolfenden J, Mansfield TA (1991) Physiological disturbances in plants caused by air pollutants. In: Last F, Watling R (eds) Acidic deposition - Its nature and impacts. The Royal Society of Edinburgh, Edinburgh, pp 117-138

Wolfenden J, Wellburn AR (1991) Effects of summer ozone on membrane lipid composition during subsequent frost hardening in Norway spruce (*Picea abies* L. Karst.) The New Phytologist 118: 323-329

INFLUENCE OF NITROGENOUS AIR POLLUTANTS ON CARBON DIOXIDE AND OZONE EFFECTS ON VEGETATION

Ludger van der Eerden, Alfred Tonneijck, Wanda Jarosz[1], Sigrid Bestebroer[2] and Thomas Dueck
Research Institute for Plant Protection
PO Box 9060
6700 GW Wageningen
The Netherlands

ABSTRACT

Nitrogenous air pollution can have a major impact on ecosystems. Ambient concentrations of NO_x and NH_3 are increasing in many regions due to the increasing intensity of traffic and livestock farming, respectively. On a more universal scale, concentrations of CO_2 and O_3 are also increasing. This chapter discusses whether nitrogenous air pollutants must be taken into account when evaluating the effects of elevated CO_2 and O_3 on plants. Results reported here indicate that combination effects are generally additive. An exception was the combination of CO_2 and NH_3; plant growth was stimulated by both compounds when applied singly, but not in combination. The growth-stimulating effect of CO_2 was also counteracted by NO_x and O_3. The effect-threshold for O_3 was higher in the presence of elevated CO_2 and NH_3. In more general terms, it appears that the stimulatory effect of elevated CO_2 concentration is reduced by 'conventional' air pollution. This could mean that reductions of NO_x and NH_3 emissions may enhance the effects of CO_2 and O_3, and that in polluted regions the beneficial effect of elevated CO_2 on biomass production may be less than previously assumed. It is suggested that nitrogenous air pollutants must be taken into account when evaluating the true impact of elevated CO_2 and O_3 concentrations.

[1] Environmental Pollution Abatement Centre, 6 Kossutha str. 40-832, Katowice, Poland
[2] Dutch Electricity Supply Company, P.O. Box 9035, 6800 ET, Arnhem, The Netherlands

NATO ASI Series, Vol. I 16
Interacting Stresses on Plants in a Changing Climate
Edited by M. B. Jackson, C. R. Black

INTRODUCTION

Atmospheric CO_2 concentration will probably continue to increase on a worldwide scale in future decades. Current concentrations of ambient O_3 have reached phytotoxic levels (Tonneijck, 1989) and these concentrations might also increase, although on a more localised basis, due to increases in the emission of its precursors (NO_x and hydrocarbons) and the photochemical activity of the atmosphere (increased UV radiation and higher temperatures). Low concentrations of other pollutants are added to this mixture. The excessively high concentrations of SO_2 and HF, which in the past have caused serious damage, have been substantially decreased. However, the concentration of NO_x ($NO + NO_2$) is still increasing in many urban areas, mainly because emission abatement has not completely compensated for the increase in road traffic emissions. The intensification of livestock farming has also tended to increase concentrations of NH_y (NH_3 + NH_4^+). From recent deposition measurements, it appears that the mean deposition velocities for NO_2 and NH_3 are 0.2 and 1.2 cm s^{-1} respectively (Heij and Schneider, 1991), which differs considerably (lower for NO_2 and higher for NH_3) from what was previously expected. Thus, in the rural areas of many countries a substantial part of the atmospheric nitrogen deposition is caused by NH_y. When evaluating the effects of CO_2 and O_3, which tend to be uniformly dispersed over a wide geographic scale, it may be questioned whether pollutants of a more local nature need to be taken into account. This is especially relevant in the case of pollutant interactions.

Generally, exposure to O_3 (Lefohn, 1992) and NO_x (Wellburn, 1990) reduces growth and yield, although low levels of NO_x may act as a fertiliser (Schulze, 1989). Growth stimulation has been shown in response to elevated CO_2 (Kimball, 1983) and NH_y (Van der Eerden *et al.*, 1992), although high concentrations of NH_3 may reduce growth (Van der Eerden, 1982). The concentration threshold at which growth stimulation turns into growth reduction is much higher for NH_3 than for NO_x. This may be due to differences in the mechanisms of detoxification and cellular acidification of the pollutants (Raven, 1988; Pearson *et al.*, 1993). Both

NH_3 and NO_x can also induce secondary effects, such as increased stress sensitivity, changes in flowering and altered competitive ability (Dueck *et al.*, 1990; Wellburn, 1990; Van der Eerden *et al.*, 1991).

Information on combination effects is fragmentary and thus hypotheses on interactions between pollutants must generally be based on results from experiments using pollutants applied individually. Both NO_x (Wellburn, 1990) and NH_3 (Van der Eerden and Pérez-Soba, 1992) can stimulate stomatal opening and thus increase the uptake of other gases. In contrast, exposures to O_3 and CO_2 generally induce stomatal closure, thereby decreasing the uptake of other air pollutants. The content of some polyamines can increase in plants upon exposure to NH_3 (Niederee *et al.*, personal communication). Since polyamines can protect plants from O_3 injury (Ormrod and Beckerson, 1986), antagonism between NH_3 and O_3 can be the result. This may also be true for mixtures of NO_x and O_3, since plants generally respond with less-than-additive effects to this combination (Wellburn, 1990). The growth-stimulating effect of CO_2 will be limited by deficiency of other nutrients. Nitrogen can be one such limiting factor and therefore a nitrogen fertiliser like NH_y (and possibly also NO_x) could have a more-than-additive relationship with CO_2. From this information, it is clearly impossible to assess the impact of the various combinations of these pollutants on plants from existing knowledge. We have therefore started a programme to study the combined effects of CO_2 and O_3, CO_2 and NO_x, CO_2 and NH_y, O_3 and NO_x and O_3 and NH_y. Some of the results are briefly discussed in this chapter.

EXPERIMENTAL PROCEDURES

Fumigation experiments with CO_2, O_3, NH_3, CO_2+O_3 and CO_2+NH_3 were performed in controlled environment chambers (Mooi and Jolink, 1990) and with O_3, NH_3 and O_3+NH_3 in open-top chambers (Van der Eerden *et al.*, 1991). Results for combination effects of O_3 and NO_x and CO_2 and NO_x have been drawn from the literature. In this chapter the notation 'X*Y' indicates the treatments X, Y, X+Y and a control treatment. Fumigation with CO_2 involved raising the CO_2 concentration from 350 to 650 ppm.

In the controlled environment chambers, temperature, relative humidity and irradiance were 17 °C, 75% and 60 W m^{-2} during the day (0800-1900 h), and 11 °C, 90% and 0 W m^{-2} during the night. The O_3 concentrations were 110 $\mu g\ m^{-3}$ from 1000 to 1800 h and 40 $\mu g\ m^{-3}$ during the remaining hours. Concentrations of NH_3 (100 $\mu g\ m^{-3}$) and elevated CO_2 (650 ppm) were kept constant throughout the day and night periods. Saplings of *Fagus sylvatica* and *Pinus sylvestris* were grown in sandy, nutrient-poor soil and were respectively two and three years old when fumigated. The experiment started just after the leaves were fully expanded. Photosynthesis measurements were made during the last two weeks of the exposure period at 60 W m^{-2}. Information on transpiration was obtained by weighing the plants+pots on two dates shortly before final harvest. Plants of *Nicotiana tabacum* cv. Bel W3 (six weeks old) and *Trifolium subterraneum* cv. Geralton (directly after sowing) were also used in these experiments.

In the open-top chambers, plants of *Phaseolus vulgaris* cv. Pros were fumigated for 62 days, from seedling stage until pod maturation. Two identical ranges of six O_3 concentrations were used: during the day (1000-1800 h) the concentrations were 0, 30, 60, 90, 120 and 150 $\mu g\ m^{-3}$. The night-time concentration was one third of the daytime value. One set of O_3 treatments received 40 $\mu g\ m^{-3}$ NH_3 continuously. The plants were grown in pots in nutrient rich soil and were fertilised once. Saplings of *Fagus sylvatica* and *Pinus sylvestris* were also exposed to these pollutant regimes for 15 months. The concentration of CO_2 is expressed in ppm (1000 ppm = 550000 $\mu g\ m^{-3}$); that of the other gases is in $\mu g\ m^{-3}$).

RESULTS

***CO_2*O_3*.** Leaf injury in *Trifolium* caused by O_3 was reduced in the presence of elevated CO_2. This reduction was strongest between 15 and 20 days after the start of the exposure; after 25 days, the difference in injury between the plants in the O_3 and O_3+CO_2 treatment was small and remained constant over the remaining 15 days of exposure. This indicates that O_3-induced leaf injury was both reduced and delayed by elevated CO_2. Exposure to O_3 reduced shoot and root growth and shoot/root ratio, while exposure to elevated CO_2 stimulated shoot and root equally (Fig. 1). An interaction between the impact of CO_2 and O_3 was not found for either shoot or root growth. With respect to leaf injury and shoot growth, the response of *Nicotiana* was very similar to that of *Trifolium*; root growth was not measured. The saplings of *Fagus* did not show any leaf injury due to O_3. Since exposure of *Fagus* began well after the initial development of the foliage, neither treatment influenced leaf biomass production. The younger roots, probably formed during the fumigation, showed an O_3 effect since fewer fine roots (diameter < 2 mm) were formed. This resulted in a decreased specific root length (length per unit of dry matter; SRL). While the CO_2 treatment did not affect SRL, the CO_2+O_3 treatment reduced SRL more than the O_3 treatment alone. The effects of the treatments on SRL and transpiration were comparable, but differed from those on the photosynthesis (Table 1).

***TABLE 1**. Effects of exposure to CO_2, O_3, NH_3 and some mixtures for 90 days on root growth and gas exchange of* Fagus sylvatica. *SE indicates the standard error; n = 10.*

	Roots <2 mm (g dry wt.)		Specific root length (m g^{-1})		Net photo-synthesis (mg CO_2 m^{-2} s^{-1})		Transpiration (g H_2O d^{-1} (g leaf dry wt.$^{-1}$)	
	Mean	SE	Mean	SE	Mean	SE	Mean	SE
Control	1.83	0.18	9.05	0.46	0.12	0.00	20.1	1.94
CO_2	2.40	0.35	9.77	1.31	0.16	0.02	17.0	1.52
O_3	2.55	0.40	7.80	0.73	0.11	0.01	8.07	0.55
NH_3	1.54	0.25	7.22	0.47	0.18	0.02	29.3	2.21
CO_2+O_3	1.10	0.21	5.30	0.22	0.14	0.01	7.14	1.00
CO_2+ NH_3	1.25	0.15	6.88	0.68	0.13	0.01	23.1	3.22

***CO_2*NH_3*.** The CO_2 and control treatments were as in the CO_2*O_3 experiment referred to above. None of the species examined showed any leaf injury in this experiment. Root and shoot growth and root length in *Trifolium* were stimulated by both NH_3 and CO_2, but no significant growth stimulation was found in the

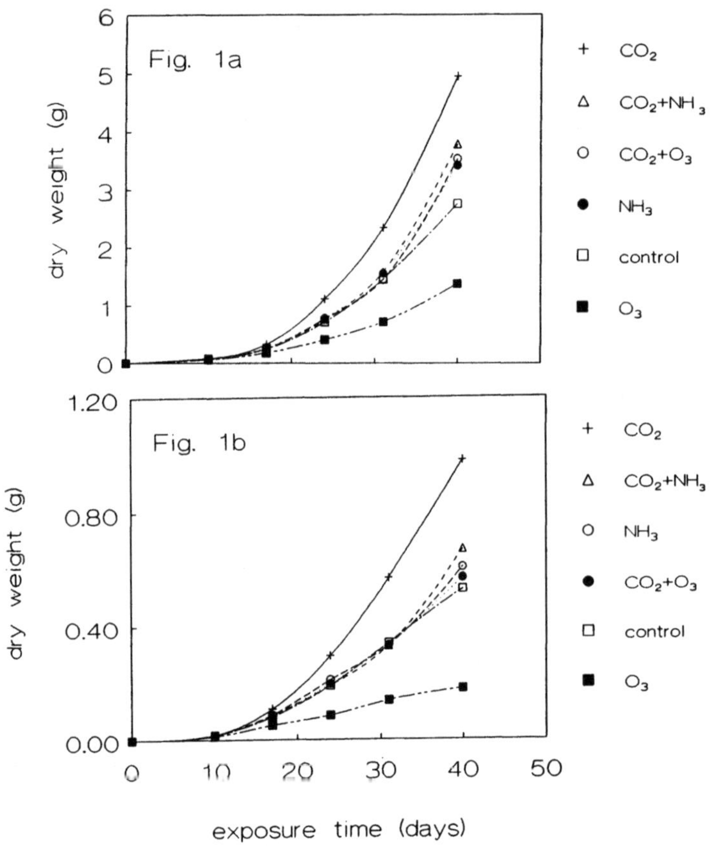

FIGURE 1. *Mean shoot (a) and root (b) dry weight of* Trifolium subterraneum *cv. Geralton following exposure to O_3, NH_3, CO_2 and some mixtures for 40 days.*

the combination treatment (Fig. 1). Thus there was a strong negative interaction between the two compounds in terms of their impact on biomass production. The reaction of *Nicotiana* plants differed somewhat from that of *Trifolium* in that leaf biomass production was stimulated to a similar extent by CO_2 and CO_2+NH_3, but was not influenced by NH_3 alone. Photosynthesis in *Fagus* was stimulated by both compounds individually but not by the mixture of both. However, the variation in both SRL and transpiration was too large to arrive at definite conclusions. *Pinus* was also examined, but no visible injury was detected. Photosynthesis, transpiration and glutamine synthetase activity were measured, and the influence of the various treatments was basically similar to that found in *Fagus* and *Trifolium*, i.e., stimulation by CO_2 and NH_3, but not by CO_2+NH_3.

***O_3*NH_3*.** The effects of this combination have not previously been discussed in the literature, but some research is in progress in our laboratory. Plants of *Fagus, Pinus* and *Phaseolus* were exposed to a range of O_3 concentrations with and without NH_3. The leaves of *Fagus* and needles of *Pinus* were not visibly injured by O_3, even after exposure to 150 $\mu g\ m^{-3}$ for 15 months; only the leaves of *Phaseolus* were visibly injured by O_3. At the highest treatment concentration, injury was visible after five days of exposure and the threshold for leaf injury was between 30 and 60 $\mu g\ m^{-3}$ at the end of the 62 day exposure period. When O_3 concentrations were 60 or 90 $\mu g\ m^{-3}$, NH_3 reduced O_3 injury (Table 2).

Biomass production by the shoots and roots of *Phaseolus* was reduced by O_3 to a similar extent. Shoot and root growth were stimulated by NH_3 without interaction with O_3, although the threshold for reduction of root growth by O_3 may be higher in the presence of NH_3 (Table 2). The fresh weight of the pods which reached a 'marketable' size was strongly influenced by O_3 and NH_3 (Fig. 2). An extensive evaluation of this experiment will be published elsewhere (Tonneijck *et al.*, in preparation), as also applies to the impact of NH_3*O_3 on *Fagus* and *Pinus* (Dueck *et al.*, 1992; Dueck and Bossen, in preparation).

***Additional information from the literature*.** The central theme of this paper is to discuss combination effects of CO_2, O_3, NH_y and NO_x on plants. The results of experiments with CO_2*O_3, CO_2*NH_3 and O_3*NH_3 are presented in other chapters. However, to make this review more complete, it is worth referring to some additional data from the literature. Published data on the effects of O_3*NO_x and

TABLE 2. Effects of exposure to O_3 with and without NH_3 for 62 days on leaf injury and shoot and root growth of Phaseolus vulgaris *cv. Pros. LSD is the least significant difference at $p<0.05$. LSD_{NH3} and LSD_{O3} are meant for comparing treatment means for NH_3 and O_3 respectively. Indices followed by the same letter are not significantly different at $p<0.05$.*

O_3 daytime conc. ($\mu g\ m^{-3}$)	No. injured leaves		Shoot dry wt. (g)			Shoot dry wt. (g)		
	$-NH_3$	$+NH_3$	$-NH_3$	$+NH_3$	mean	$-NH_3$	$+NH_3$	mean
0	0.0 a	0.0 a	11.70	13.16	12.43	2.64	6.10	4.37
30	0.0 a	0.0 a	11.66	13.49	12.57	2.84	5.56	4.20
60	4.0 c	0.9 b	11.26	13.84	12.55	3.97	6.56	5.26
90	13.0 ef	9.9 d	6.83	10.00	8.41	1.69	7.33	4.76
120	9.0 d	10.3 de	5.70	6.47	6.09	2.07	2.31	2.19
150	12.9 ef	13.3 f	4.70	5.89	5.29	1.64	2.20	1.92
Mean			8.64	10.47		2.48	5.09	
LSD_{NH3}			1.40			1.29		
LSD_{O3}						2.43		2.24

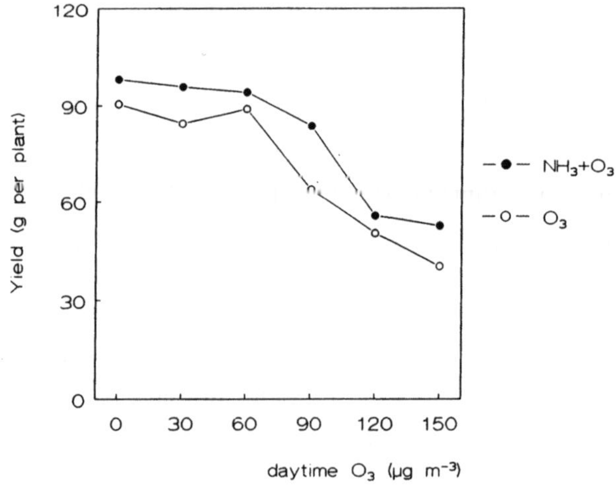

FIGURE 2. Effects of exposure to O_3 with and without NH_3 for 62 days on the marketable yield (g fresh weight[-1] of pods) of Phaseolus vulgaris *cv. Pros.*

CO_2*NO_x will be discussed shortly. To our knowledge, the impact of NO_x*NH_y and combinations of three or all four compounds have not yet been reported. Since the effects of NO differ from those of NO_2 (Wellburn, 1990), and the effects of NH_3 differ from those of NH_4^+ (Van der Eerden *et al.*, 1992), the evaluation of all combinations is extremely complex.

O_3*NO_x. The simultaneous occurrence of elevated concentrations of O_3 and NO_x, characteristic of episodes of photochemical smog, is suspected to be a major cause of crop damage (Van der Eerden *et al.*, 1991). Previous fumigation experiments have been performed with O_3*NO_2, but not with O_3*NO, and most have shown additive effects, although antagonism between the two pollutants has been found in a few cases (Reinert and Gray, 1981; Adaros *et al.*, 1991a; Adaros *et al.*, 1991b; Bender *et al.*, 1991).

CO_2*NO_x. The effects of this combination have not yet been studied within the scope of global climate change, although some relevant information may be gained from the literature concerning CO_2 enrichment in glasshouses. When the flue gases of the heating system are used as a CO_2 source, NO_x (in which NO is dominant) becomes a major contaminant. To evaluate the impact of this contamination, Mortensen (1985) grew tomato (eight cultivars) and lettuce (six cultivars) in CO_2-enriched air (1000 ppm), with and without added NO_x (700-900 ppb). The yield increase induced by elevated CO_2 was completely eliminated by NO_x in most of the tomato cultivars, but not in lettuce. Anderson and Mansfield (1979) and Saxe and Voight Christensen (1984) also found a reduction by NO_x of the growth stimulation induced by CO_2 in tomato and several ornamental species, respectively. Bruggink *et al.* (1988) evaluated this phenomenon in tomato in terms of gas exchange (Fig. 3), using the controlled environment fumigation chambers described earlier. They found that photosynthesis and growth were reduced by NO irrespective of the level of CO_2, while NO did not affect stomatal conductance or cause any visible injury. Contrary to the results of Bruggink *et al.* (1988), who found an additive effect of CO_2 and NO, Saxe (1986) reported that the NO_x effect was slightly reduced in the mixture due to a CO_2-induced increase in stomatal resistance. The available information on the effects of CO_2*NO_x indicates that growth stimulation by CO_2 is reduced by NO_x.

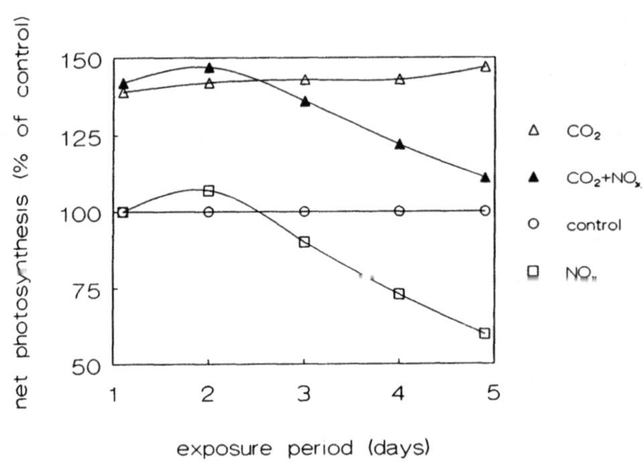

FIGURE 3. Net photosynthesis (% of control) in tomato plants exposed to NO (1150 μg m^{-3}), CO_2 (1000 ppm) and a mixture of both for five days. Control plants were exposed to filtered air (350 ppm CO_2). After Bruggink et al. *(1988).*

DISCUSSION

The results of our experiments are in agreement with data reported in the literature in that O_3 generally reduces growth, whereas CO_2 and NH_3 promote growth. Our results indicate that the responses of plants to pollutant combinations are generally additive. However, this was not true for the combined effects of NH_3 and CO_2; NH_3 and CO_2 both stimulated growth or photosynthesis in all four species examined, but this stimulation disappeared when the gases were combined. It must be emphasised that this antagonistic reaction of NH_3+CO_2 has not been found with NO_x+CO_2. Therefore, the effects of both NO_x and NH_3 should be studied to assess the impact of nitrogenous air pollution on vegetation.

With respect to the impact of pollutants on plants, interactions seem to be an exception rather than a rule. This does not mean that mixtures causing additive effects are of little relevance. For instance, the effect threshold for O_3 is probably higher in the presence of elevated CO_2 and NH_3 (Table 1). Thus, O_3 damage may increase if abatement of CO_2 and NH_3 emission is successful.

The data presented in this paper show that NO_x, O_3 and NH_3 reduce the growth-stimulating effect of CO_2. This may also be true for other pollutants and so the stimulatory effect of CO_2 on biomass production may be less in polluted regions than in cleaner ones. In general, a sound knowledge of pollutants which occur on a local or regional scale is necessary when assessing the effects of pollutants whose concentrations are elevated on a worldwide scale.

Because of the complexity, it is not yet possible to assess the effects of the air pollution 'cocktail', including elevated CO_2, on vegetation in the field. Generalisations and simplifications should be made and a proper philosophy has to be developed to select the relevant air pollution climates, the types of vegetation and effects to be examined and the appropriate exposure conditions. Our results show that the effects of pollutant combinations cannot be neglected in establishing standards. Although standards are aimed at the protection of entire ecosystems, the impact of air pollution on ecosystems remains poorly understood. An alternative is to focus on the most sensitive plant species when information is required urgently. In our experiments, plants of *Trifolium subterraneum*, *Nicotiana tabacum* and *Phaseolus vulgaris* were more sensitive to the pollutant treatments than those of *Pinus sylvestris* and *Fagus sylvatica* with respect to foliar injury. However, more experiments using pollutant combinations are needed to examine other plant species, effects, and environmental conditions before unequivocal conclusions can be drawn. Mechanistic simulation models can become a powerful tool to generalise results from various air pollution experiments (Van de Geijn *et al.*, 1993). However, our knowledge of biochemical and physiological mechanisms is still insufficient to incorporate the impact of air pollution on vegetation in these models. This applies especially to natural vegetation where stress sensitivity and competition are key factors.

From the results presented here, it may be concluded that the effects of NH_y should be studied separately from those of NO_x since these components can cause different effects when present in mixtures. In general, nitrogenous air pollutants must be taken into account when the effects of elevated CO_2 and O_3 need to be evaluated if estimates of CO_2- and O_3-induced effects are not to be inaccurate or even systematically wrong.

ACKNOWLEDGEMENTS

Thanks are due to Dr Gigliola Puppi, who performed the root analysis of *Fagus sylvatica* in the controlled environment experiment.

REFERENCES

Adaros G, Weigel HJ, Jäger HJ (1991a) Concurrent exposure to SO_2 and/or NO_2 alters the growth and yield responses of wheat and barley to low concentrations of O_3. The New Phytologist 118: 581-591

Adaros G, Weigel HJ, Jäger HJ (1991b) Single and interactive effects of low levels of O_3, SO_2 and NO_2 on the growth and yield of spring rape. Environmental Pollution 72: 269-286

Anderson LS, Mansfield TA (1979) The effect of nitric oxide pollution on the growth of tomato. Environmental Pollution 20: 113-121

Bender J, Weigel HJ, Jäger, HJ (1991) Response of nitrogen metabolism in beans (*Phaseolus vulgaris*) after exposure to ozone and nitrogen dioxide, alone and in sequence. The New Phytologist 119: 261-267

Bruggink GT, Wolting HG, Dassen JHA, Bus VGM (1988) The effect of nitric oxide fumigation at two CO_2 concentrations on net photosynthesis and stomatal resistance of tomato (*Lycopersicon lycopersicum* L. cv. Abunda). The New Phytologist 110: 185-191

Dueck ThA, Dorel F, Ter Horst R, Van der Eerden LJ (1991) Effects of ammonia and sulphur dioxide on the frost sensitivity of *Pinus sylvestris*. Water, Air and Soil Pollution 54: 35-49

Dueck ThA, Koopman J, Puppi G, Isopi R (1992) Influence of temperature, ammonia and ozone on *Pinus sylvestris* L. and *Fagus sylvatica* L. In: Tesche M, Feiler S (eds) Air pollution and interactions between organisms in forest ecosystems. Proceedings IUFRO Centennial Congress, 9-11 September, 1992, Tharanolt, Germany, pp 137-142

Heij GT, Schneider T (eds) (1991) Acidification research in The Netherlands. Environmental Science Series no 46. Elsevier, Amsterdam, The Netherlands

Kimball BA (1983) Carbon dioxide and agricultural yield: an assemblage and analysis of 770 prior observations. Agronomy Journal 75: 779-788

Lefohn AL (1992) Surface level ozone exposures and their effects on vegetation. Lewis Publishing Inc, Chelsea MI USA, pp 366

Mooi J, Jolink HHW (1990) Computer controlled plant growth cabinets for fumigation experiments. In: Environmental research with plants in closed chambers. CEC, Air Pollution Research Report 26: 43-51

Mortensen LM (1985) Nitrogen oxides produced during CO_2 enrichment. II. Effects on different tomato and lettuce cultivars. The New Phytologist 101: 411-415

Ormrod DP, Beckerson DW (1986) Polyamines as antioxidants for tomato. Horticultural Science 4: 1070-1071

Pearson J, Fowler D, Stewart GR (1993) Atmospheric ammonia deposition and its effects on plants. Tansley Review. The New Phytologist (in press)

Raven JA (1988) Acquisition of nitrogen by the shoots of land plants: its occurrence and implications for acid-base regulation. The New Phytologist 109: 1-20

Reinert RA, Gray TN (1981) The response of radish to nitrogen dioxide, sulphur dioxide and ozone alone and in combination. Journal of Environmental Quality 10: 240-243

Saxe H (1986) Stomatal-dependent and stomatal-independent uptake of NO_x. The New Phytologist 103: 199-205

Saxe H, Voight Christensen O (1984) Effects of carbon dioxide with and without nitric oxide pollution on growth, morphogenesis and production time of potted plants. Acta Horticulturae 162: 179-186

Schulze ED, Lange OL, Oren R (eds) (1989) Forest decline and air pollution. Ecological Studies 77. Springer Verlag, Berlin, Germany

Tonneijck AEG (1989) Evaluation of ozone effects on vegetation in the Netherlands. In: Schneider T, Lee SD, Wolters GJR, Grant LD (eds) Atmospheric ozone research and its policy implications. Elsevier, Amsterdam, The Netherlands, pp 251-260

Van de Geijn SC, Goudriaan J, Van der Eerden LJ, Rozema J (1993) Problems and approaches to integrating the concurrent impacts of elevated CO_2, temperature, UVB radiation and O_3 on crop production. Proceedings ICSC-Congress. Ames, Iowa USA (in press)

Van der Eerden LJ (1982) Toxicity of ammonia to plants. Agricultural Environment 7: 223-235

Van der Eerden LJ, Blom WF, Van Aalst RM (1990) Schade aan kasgewassen in het Westland (Sept. 1989): een uniek incident? Lucht en Omgeving 1: 26-30

Van der Eerden LJ, Dueck ThA, Berdowski JM, Greven H, Van Dobben HF (1991) Influence of NH_3 and $(NH_4)_2SO_4$ on heathland vegetation. Acta Botanica Neerlandica 4: 281-296

Van der Eerden LJ, Pérez-Soba MGFJ (1992) Physiological responses of *Pinus sylvestris* to atmospheric ammonia. Trees 6: 48-53

Van der Eerden LJ, Lekkerkerk LJ, Smeulders SM, Jansen AE (1992) Effects of atmospheric ammonia and ammonium sulphate on Douglas fir (*Pseudotsuga menziesii*). Environmental Pollution 76: 1-9

Wellburn AR (1990) Why are atmospheric oxides of nitrogen usually phytotoxic and not alternative fertilisers? Tansley Review 24. The New Phytologist 115: 395-429

IMPACT OF AIR POLLUTANTS ON PLANTS IN HOT, DRY CLIMATES

G. Schenone
Laboratory of Terrestrial Ecology
ENELSpA-CRTN
via Rubattino
54 - 20134 Milan
Italy

ABSTRACT

The responses of plants to gaseous air pollutants are known to be influenced by the prevailing exposure conditions, which are largely dependent on climate. In addition, the climate may modify the air pollution pattern of a site by influencing the dispersion mechanisms and photochemical reactions which may lead to highly phytotoxic secondary pollutants, like ozone (O_3). For these reasons the concept of *pollution climate* or *chemical climate* has been introduced. This paper briefly outlines some characteristics of the chemical climate in hot dry places, with especial emphasis on photochemical pollution. The effects of temperature and water stress on the response of plants to air pollutants are then considered, and a short review of the literature on the impact of air pollutants on plants in desert, tropical and mediterranean environments is presented.

CHEMICAL CLIMATE IN HOT DRY PLACES

Native vegetation and crops in hot dry places may be affected by air pollutant emissions from several human activities. The high solar irradiance and temperature in tropical and subtropical climates can promote photochemical reactions which lead to the formation and accumulation of ozone in the boundary layer. There is now well established evidence of relatively high concentrations of O_3 during the dry season over very large tropical areas. For example, aircraft measurements in Brazil gave concentrations of 40-50 nanolitres per litre (nl l^{-1}) in the free troposphere and ~60 nl l^{-1} in the mixing layer over the Cerrado region, characterised by savanna-type ecosystems (Delany *et al.*, 1985). Tropospheric levels of O_3 of 50-90 nl l^{-1} have been measured in Northern tropical Africa during the dry season (December - March) (Marenco *et al.*, 1990). In both cases the production

NATO ASI Series, Vol. I 16
Interacting Stresses on Plants in a Changing Climate
Edited by M. B. Jackson, C. R. Black

of O_3 in the lower and mid-troposphere has been attributed to photochemical reactions which involve precursors originating from biomass burning, in particular nitrogen oxides (NO_x) and hydrocarbons. Although biomass burning is the prevailing source of air pollution in many tropical and subtropical areas, urban and industrial emissions may be very important in developing countries with rapidly increasing populations living in cities. This is the case for the Greater Cairo city, where oxidant concentrations exceeding 100 nl l^{-1} were found on 74% of the measurement days over one year (Nasralla and Shakour, 1981). Another noteworthy case is Mexico City, where the extremely rapid increase of population and motor vehicles has caused great air pollution problems; the high altitude at which the city is located increases the incidence of ultraviolet radiation, facilitating the formation of photochemical oxidants (de Bauer and Krupa, 1990).

The American deserts located within the influence of large urban areas may experience elevated concentrations of O_3 due to transport of polluted air masses; at the Joshua Tree National Monument, located in southeastern California about 100 km east of the Los Angeles Basin, maximum hourly means of 140 nl l^{-1} have been recorded, although O_3 maxima usually occur here at night, because of the pattern of long-range transport of ozone from points south and west (Temple, 1989).

Areas with a Mediterranean climate are characterised by a hot dry summer, while the other seasons are relatively warm and wet. The permanent population density in the Mediterranean region is high, and is increased dramatically during the summer months by the influx of tourists from central and northern Europe. This causes a sharp increase in emissions of pollutants, some of which are important precursors of O_3. For these reasons, the risk to Mediterranean environments from photochemical pollution must be given serious consideration. In the Greater Athens area of Greece, the photochemical smog episodes are called 'nefos' and occur many times in the year, when the O_3 concentration rises for several hours per day above the US Air Quality Standard of 120 nl l^{-1} (Güsten *et al.*, 1988). The available data for Adriatic and Aegean coastal sites show pronounced diurnal variation of the O_3 concentration in urban and industrial areas during summer, with seasonal seven hour averages above 50 nl l^{-1} (Giovanelli *et al.*, 1985; Butkovic *et al.*, 1990); in the same area relatively high values of O_3 have also been measured at rural sites close to large cities (Butkovic *et al.*, 1990). Maximum O_3 concentrations above 120 nl l^{-1} and monthly seven hour mean concentrations up to 70 nl l^{-1} occur at a semirural site in the proximity of the city of Pisa, near the Thyrrenian coast (Lorenzini, 1987). Diurnal summer means above 40-50 nl l^{-1} and hourly peaks well above 100 nl l^{-1} have been measured in Israel (Naveh *et al.*, 1980) and Spain (Martin *et al.*, 1991). Air pollution data which show extremely high concentrations of O_3 at ground level are available for the Mediterranean-type ecosystems in California (Westman, 1985).

ENVIRONMENTAL MODULATION OF THE RESPONSES OF PLANTS TO AIR POLLUTANTS

Temperature. The role of temperature in the response of plants to air pollutants has received little attention to date. The available data show that the modulation may be very different in different species, depending upon their optimal temperature range. One of the ways by which ambient temperature may modify the response of plants to air pollutants is through the regulation of stomatal opening, which in turn influences the uptake of toxic gases from the atmosphere. Generally, stomata tend to open further as temperature increases, following the temperature curve for CO_2 assimilation (Raschke, 1975) and pollutant uptake also increases (Norby and Kozlowsky, 1981; Taylor *et al.*, 1985). In *Phaseolus* the acute sensitivity to ozone and to sulphur dioxide (SO_2) increases steadily above 21-24 °C (Dunning and Heck, 1977). A similar response has been observed in woody plants such as *Fraxinus, Betula* and *Liriodendron* (Norby and Kozlowski, 1981; Taylor *et al.*, 1985). However, for other species the known data are either different or conflicting: Dunning and Heck (1977) observed decreasing O_3 damage in a laboratory experiment as temperature increased in *Nicotiana tabacum* cv Bel-W3, while Biondi *et al.* (1992), found a positive correlation between air temperature and O_3-induced leaf injury in the same species under open-air conditions. The effect of acute SO_2 exposure upon photosynthesis of *Zea mays* is not significantly influenced by increasing temperature from 28 to 35 °C (Taylor *et al.*, 1985). Since the patterns of temperature-dependent plant responses to air pollutants seem quite different in different species, involving to various degrees mechanisms of avoidance and tolerance, generalisations about the influence of exposure temperature on resistance of plants to air pollutants are limited (Norby and Kozlowski, 1981; Taylor *et al.*, 1985). The main research needed on this topic seems to be related to long-term exposures under different temperature regimes, keeping the experimental conditions as close as possible to the field; this should provide the most useful data for modelling and predicting the interactions between temperature change and air pollution-stress in plants.

Water stress. Water stress is one of the more important limiting factors for natural vegetation and crops in hot dry places. Many studies have demonstrated that severe water stress may protect plants from air pollutants (for example, Grzesiak, 1979; Showman, 1991). The reason for this interaction lies mainly in the stomata, which close in response to severe water stress in order to reduce the loss of water by transpiration. As a consequence of the closure of stomata, the uptake of air pollutants decreases, and therefore the damage to the plant is reduced. Thus, the avoidance of one stress also causes avoidance of another. However, these conclusions may only be true for relatively severe water stress. If the water stress is mild, the interaction between the two stresses may not be significant for effects on crops (Heagle, 1989). For desert plants, an experiment performed in an open-air system failed to demonstrate any interaction between effects of ozone and water stress (Bytnerowicz *et al.*, 1988). In at least a few cases, exposure to elevated ozone levels has caused a loss of tolerance to drought: in field-grown soybean the yield reduction due to ambient O_3 was 25% in water-stressed crops and 4% in well-watered ones (Heggestad *et al.*, 1985); in the tropical fibre plant *Hibiscus cannabinus* grown in pots, the O_3 damage was alleviated by mild water stress but enhanced by severe water stress (Kasana, 1992). The variability inherent in the available data on the interactions between water- and air pollutant-stresses results from many factors, including the intensity of water stress (which is not easy to control and quantify) and plant growth conditions. Growth conditions are quite different in the field, where the roots can go deep in order to reach the water table, and in pots, where the roots are restricted. Experiments which involve using large soil columns may help improve our understanding of the mechanisms of interactions between the two stresses (Heggestad and Lee, 1990).

IMPACT OF AIR POLLUTANTS ON PLANTS IN HOT DRY ENVIRONMENTS

Deserts. Most information on the effects of air pollution on desert plants comes from studies performed in North America in areas at risk from industrial sources of SO_2 and NO_x, or from transport of photochemical oxidants from urban areas. The experiments have been performed with plants growing under field conditions or in pots, and in many cases irrigation was applied during exposure to pollutants, to provide optimal sensitivity to air pollutants. Hill *et al.* (1974) fumigated *in situ* 87 species native to the deserts of Utah and New Mexico with short-term episodes of SO_2 and NO_2 using a portable apparatus. They found desert species to be quite resistant to SO_2, with concentrations of 2000 nl l^{-1} being required to induce injury in most of the species tested. Thompson *et al.* (1980) fumigated annual and perennial species of the Mojave desert with different combinations of SO_2 and NO_2 in open-top chambers for several weeks. The annual species were more sensitive than the perennials to SO_2-induced leaf injury. However, in the perennial species, *Ambrosia dumosa*, the production of seeds was strongly impaired even at the lowest concentrations tested (220 nl l^{-1} SO_2 or 110 nl l^{-1} NO_2 and a mixture of both), and a negative flowering response was seen with another perennial, *Encelia farinosa*, at 670 nl l^{-1} SO_2 or 110 nl l^{-1} NO_2. No suggestion of synergism between the two pollutants was noted. Desert plants seem to be more sensitive to SO_2 injury when the prevailing environmental conditions encourage maximum growth rates, possibly due to greater pollutant uptake (Olszyk *et al.*, 1987a). Plants exhibiting Crassulacean Acid Metabolism (CAM) are much more resistant to SO_2 than other desert species, regardless of their water status (Olszyk *et al.*, 1987b). As far as O_3 is concerned, experiments performed with open-top chambers or open-air systems show that some extremely sensitive species, including both annuals (*Camissonia claviformis, Camissonia hirtella, Cryptantha nevadensis*) and perennials (*Rhus trilobata*), may be injured at concentrations as low as 100 nl l^{-1} (Thompson *et al.*, 1984; Bytnerowicz *et al.*, 1988; Temple, 1989). An experiment which included irrigated and non-irrigated plots revealed that water stress did not modify the response of plants to ozone, even though leaf water potential and stomatal conductance were significantly decreased in the non-irrigated plots (Bytnerowicz *et al.*, 1988). Thus, desert plants, like plants of other environments, show great variability in their sensitivity to air pollutants. While CAM plants are in general quite resistant, other species may not be and, even under conditions of low soil moisture typical of desert climates, may be injured at concentrations of ozone which frequently occur in ambient air.

Tropical environments. The available information on the effects of air pollution on tropical plants comes mainly from research in India and Australia and most concerns the effects of SO_2 or fluorine (which will not be considered here), while only a few reports deal with O_3. The sensitivity of tropical trees to SO_2 has been investigated with both field surveys and fumigation experiments. Agrawal and Agrawal (1989) evaluated foliar injury and several biochemical parameters in three woody species around a coal-fired power plant in India: they found that visible injury and the sulphur content of the leaves increased with increasing proximity to the stacks, while chlorophyll and ascorbic acid contents decreased. The changes were greatest in *Mangifera indica*. The average SO_2 concentration measured at the sampling site closest to the source did not exceed 30 nl l^{-1}. Rao and Dubey (1990) analysed four tree species for physiological and biochemical parameters in field conditions at SO_2 levels ranging from 4 to 34 nl l^{-1} and reported that plants with the more efficient antioxidant system (high activities of peroxidase and superoxide dismutase) were more tolerant of SO_2. The most sensitive species were *Azadirachta indica* and *Mangifera indica*. Fumigation experiments performed in laboratory chambers allowed the relative sensitivity to SO_2 of several Indian tree species to be evaluated in terms of visible injury (Farooq and Beg, 1988) and metabolic profile (Beg and Farooq, 1988). It was observed that most of the biochemical constituents examined accumulated in larger quantities following exposure of susceptible species to SO_2, while in the resistant plants the reverse pattern was obtained. These data indicate that the process of defence against acute SO_2 episodes involves avoidance, whereas the field studies discussed earlier indicate that tissue tolerance is the main defence strategy adopted by these tropical trees against chronic exposures to low levels of SO_2.

The relative susceptibility to acute SO_2 injury of 141 Australian tree and shrub species was determined by O'Connor *et al.* (1974); the genera *Eucalyptus* and *Acacia* were the most sensitive, suffering acute leaf injury after a three hour exposure to 1000 nl l^{-1} SO_2; *Casuarina* species appeared to be the most resistant. Plants of *Eucalyptus gomphocephala* were continuously exposed to low levels of SO_2 (40-110 nl l^{-1}) for 126 days in open-top chambers in Australia (Fulford and Murray, 1990). The treatment caused morphogenic changes since the plants grew taller, without an increase in above-ground plant weight, had thinner stems and leaves, and more chlorophyll in healthy leaves than control plants; these results agree with earlier studies of temperate broadleaved trees. A comparison between the tropical pine, *Pinus caribaea*, and the temperate species, *P. virginiana* and *P. sylvestris*, showed that Caribbean pine seedlings are highly sensitive to acute doses of SO_2 (Umbach and Davis, 1986).

Thc available information on the effects of SO_2 on herbaceous tropical plants is slight. The exposure of the herbaceous weed, *Euphorbia hirta*, to coal-smoke effluents from different sources in India caused reductions in growth, stomatal density and pore size (Gupta and Ghouse, 1987a); analogous effects were observed on the Indian crop, *Abelmoschus esculentus* (Gupta and Ghouse, 1987b), but since no information was given on pollutant concentrations at the exposure sites, the above data cannot be compared with other situations. The pasture species, *Medicago truncatula*, a native of the warm temperate areas of Australia, shows growth reductions at concentrations of SO_2 exceeding 255 nl l^{-1} in open-top chamber experiments (Murray and Wilson, 1991).

Little is known about the effects of ambient O_3 on tropical plants. The first report of O_3 damage to a crop in India was published in 1986 and concerns leaf stipple on potato in the Punjab. Although no experimental facilities were available, the appearance of the disease could be prevented by spraying the crops with the antioxidant, ethylenediurea (EDU), or with charcoal dust. Elevated ozone levels were detected at the same site with the bioindicator plant, *Nicotiana tabacum* cv Bel-W3 (Bambawale, 1986). Fumigation experiments performed in closed chambers provided evidence of high relative sensitivity of tropical leguminous crops to ozone pollution, some of which suffered leaf damage and growth reductions at seasonal 7 hour O_3 concentrations of 70 nl l^{-1} (Kasana, 1988). The tropical fibre plant, *Hibiscus cannabinus*, was damaged by six weeks of fumigation at the same O_3 concentration; as already mentioned, it is noteworthy that O_3 damage was alleviated by mild water stress but enhanced by severe water stress (Kasana, 1992). Symptoms of O_3 injury to native vegetation and crops are commonly seen in the Valley of Mexico due to the transport of polluted air masses from Mexico City. In this region *Pinus hartwegii*, *P. leiophylla* and *P. montezumae* show chlorotic banding and mottling of the needles, typical symptoms of O_3 injury in conifers. Severe oxidant injury in the Valley of Mexico has also been noticed on soybean (*Glycine max*) and bean (*Phaseolus vulgaris*) (de Bauer and Krupa, 1990). In conclusion, tropical plants seem to exhibit acute and chronic sensitivities to SO_2 and O_3 comparable to those of temperate plants (Roberts *et al.*, 1983). However, the information available is incomplete. For SO_2, the information comes mainly from India and Australia; for O_3 even fewer data are available, in spite of the occurrence of high concentrations of this pollutant over large tropical areas. In order to begin an assessment of the impact of air pollution on crops in tropical developing countries, the European Communities have sponsored a research programme using open-top chambers under field conditions. The results have not yet been published, but preliminary data from charcoal-filtered versus non-filtered air exposures in Pakistan suggest the occurrence of significant crop losses due to ambient air pollution (JNB Bell, personal communication, 1992).

Mediterranean environments. Mediterranean ecosystems include sclerophyllous and mesophyllous species adapted to summer drought conditions. The strategy of adaptation varies between species, influencing the length of the vegetative period and gas exchange rates. Many species are evergreen and the vegetation period may cover the whole year, with maximum activities in spring and autumn. During the dry summer some species are dormant (e.g., *Euphorbia dendroides*), while others are active and close their stomata during the hot hours of the day to save water (water savers, e.g., *Quercus ilex*); a third group are also active and keep their stomata open all day due to their ability to maintain a high water supply (water expenders, e.g., *Pistacia terebinthus*) (Levitt, 1980). These different mechanisms of adaptation to the Mediterranean climate may cause different patterns of uptake of air pollutants and therefore different sensitivities. The concept of the 'seasonal mean concentration' of pollutants therefore cannot be easily applied to Mediterranean vegetation. The pollutant exposure parameters for these ecosystems should rather be calculated at least on an annual basis, as the growing season extends over the winter.

The threat from photochemical pollutants to Mediterranean ecosystems has long been evident (Naveh *et al.*, 1980), but little research has been conducted to date. Naveh *et al.* (1980) described severe needle damage in *Pinus halepensis* and *P. pinea* in the densely populated coastal region of Israel; the symptom was chlorotic mottle, similar to that occurring in *P. ponderosa* in the USA. The authors attributed the disease to photochemical smog, quoting ozone concentrations of up to 200 nl l^{-1}. Ozone was also indicated as the primary cause of the severe infestation of *P. halepensis* by *Matsucoccus josephi*, scales and Scolytidae beetles. Visible lesions identical to those caused by ozone have also been found on *P. halepensis*, in the Athens basin in Greece. The ozone symptoms here are not easily visible in summer, when the needles are chlorotic, but in autumn the needles regreen and the chlorotic flecks become evident on all but the current year's needles (Velissariou *et al.*, 1992b). Ozone-like injury to the same species has also been noticed in Spain (Velissariou *et al.*, 1992a). Little is known about the sensitivity of other Mediterranean species to O_3. The decline of Greek fir, *Abies cephalonica*, in the mountains facing Athens has been recently related to air pollution (Heliotis *et al.*, 1988) and in Italy symptoms of decline have been observed in several Mediterranean trees (*Pinus pinea*, *P. pinaster*, *P. halepensis*, *Quercus ilex*). The symptoms, similar to the so-called 'new-type forest decline', are quite mild all over the Italian territory, but more severe in the vicinity of strong sources of air pollution (Clauser *et al.*, 1989); the above authors therefore concluded that the main cause of decline appears to be diffused regional atmospheric pollution in central and southern Italy. However, no air quality data are given, and the symptoms described are not specific to air pollution-induced damage.

A survey of seven species in the coastal sage scrub in the Santa Monica mountains in southern California (Westman, 1985) in which field and chamber symptoms were compared suggested that both O_3 and SO_2 were responsible for field injury symptoms, with O_3 being the more frequent cause. The damage increased with proximity to Los Angeles. Westman (1985) concluded that photochemical oxidants in the Santa Monica Mountains cause extensive damage to Mediterranean scrub vegetation, possibly restricting foliar and root growth and 'reducing soil binding function and aggravating soil losses on these steep hillsides'. Some studies have been carried out in Greece on Mediterranean trees and shrubs along gradients of urban pollution. Among the observed effects are ultrastructural modifications of the leaves of sclerophyllous species which lead to a more xeromorphic habit (Christodoulakis and Fasseas, 1990). The sclerophyllous species, *Laurus nobilis* and *Olea europea*, have been classified as 'injury resistant' because, although they show some xeromorphic changes in the heavily polluted urban environment in Greece, the structure of the chloroplast appears intact, indicating that photosynthesis is maintained despite the unfavourable conditions (Christodoulakis and Fasseas, 1990; Christodoulakis and Koutsogeorgopoulou, 1991). As far as agricultural plants are concerned, the evidence for negative effects of ambient ozone is well established for areas with a Mediterranean climate such as California (Olszyk *et al.*, 1988), Italy (Schenone and Lorenzini, 1992) and Spain (Reinert *et al.*, 1992). The available data base mainly concerns effects on herbaceous crops, while for other crops such as fruit trees much less information is available.

CONCLUSIONS

This short overview of the air pollution climate of hot dry places has considered the available evidence concerning the risk to tropical and subtropical environments arising from photochemical air pollution. This risk results from the high solar radiation and air temperature prevailing in such areas in combination with the emission of pollutant precursors from large urban areas or biomass burning. The responses of plants to air pollutants in hot dry climates may be influenced by temperature and water stress, but the role of these factors has not yet been clearly elucidated. A relatively detailed knowledge of the effects of air pollutants on plants is available for the flora of the American deserts, where highly O_3-sensitive species are found. In the tropics and the Mediterranean area, the information is extremely scanty and fragmentary, especially concerning the effects of ozone, despite the potential damage that photochemical air pollution could cause to natural ecosystems, forests and crops in these regions.

REFERENCES

Agrawal M, Agrawal SB (1989) Phytomonitoring of air pollution around a thermal power plant. Atmospheric Environment 23: 763-769

Bambawale OM (1986) Evidence of ozone injury to a crop plant in India. Atmospheric Environment 20: 1501-1503

de Bauer LI, Krupa SV (1990) The Valley of Mexico: Summary of observational studies on its air quality and effects on vegetation. Environmental Pollution 65: 109-118

Beg MU, Farooq M (1988) Sulfur dioxide resistance of Indian trees. I. Experimental evaluation of metabolic profile. Water, Air and Soil Pollution 40: 317-326

Biondi F, Mignanego L, Schenone G (1992) Correlation between environmental parameters and leaf injury in *Nicotiana tabacum* L. cv. 'Bel-W3'. Environmental Monitoring and Assessment 22: 73-87

Butkovic V, Cvitas T, Klasinc L (1990) Photochemical ozone in the Mediterranean. The Science of Total Environment 99: 145- 151

Bytnerowicz A, Olszyk DM, Fox CA, Dawson PJ, Kats G, Morrison CL, Wolf J (1988) Responses of desert annual plants to ozone and water stress in an *in situ* experiment. Journal of Air Pollution Control Association 38: 1145-1151

Christodoulakis NS, Fasseas C (1990) Air pollution effects on the leaf structure of *Laurus nobilis*, an injury resistant species. Bulletin of Environmental Contamination and Toxicology 44: 276-281

Christodoulakis NS, Koutsogeorgopoulou L (1991) Air pollution effects on the leaf structure of two injury resistant species: *Eucalyptus camaldulensis* and *Olea europea* L. Bulletin of Environmental Contamination and Toxicology 47: 433-439

Clauser F, Gellini R, Bussotti F, Cenni E, Bottacci A (1989) New types of damage to forest trees typical of the mediterranean region. European Journal of Forest Pathology 19: 78-83

Delany AC, Haagensen P, Walters S, Wartburg AF, Crutzen PJ (1985) Photochemically produced ozone in the emission from large-scale tropical vegetation fires. Journal of Geophysical Research 90: 2425-2429

Dunning JA, Heck WW (1977) Response of bean and tobacco to ozone: effect of light intensity, temperature and relative humidity. Journal of Air Pollution Control Association 27: 882-886

Farooq M, Beg MU (1988) Sulfur dioxide resistance of Indian trees. II. Experimental evaluation of visible symptoms and SO_2 sorption. Water, Air and Soil Pollution 40: 307-316

Fulford GB, Murray F (1990) Morphogenic changes in *Eucalyptus gomphocephala* exposed to SO_2. Environmental and Experimental Botany 30: 343-347

Giovanelli G, Georgiadis T, Fortezza F, Strocchi V (1985) Transport of photochemical ozone along the western Adriatic shore near a petrochemical plant. Il Nuovo Cimento 8C: 727- 742

Grzesiak S (1979) Influence of sulphur dioxide on the relative rate of photosynthesis in four species of cultivated plants under optimum soil moisture and drought conditions. Bulletin de l'Académie Polonaise des Sciences, Série Science Biologiques 27: 309-321

Gupta MC, Ghouse AKM (1987a) Effects of coal-smoke effluents from different sources on the growth, chlorophyll content, stem anatomy and cuticular traits of *Euphorbia hirta* L. Environmental Pollution 47: 221-229

Gupta MC, Ghouse AKM (1987b) The effects of coal-smoke pollutants on the growth, yield and leaf epidermal features of *Abelmoschus esculentus* Moench. Environmental Pollution 43: 263-270

Güsten H, Heinrich G, Cvitas T, Klasinc L, Ruscic B, Lalas D (1988) Photochemical formation and transport of ozone in Athens, Greece. Atmospheric Environment 22: 1855-1861

Heagle AS (1989) Ozone and crop yield. Annual Review of Phytopathology 27: 397-423

Heggestad HE, Gish TJ, Lee EH, Bennett JH, Douglass LW (1985) Interaction of soil moisture stress and ambient ozone on growth and yield of soybeans. Phytopathology 75: 472-477

Heggestad HE, Lee EH (1990) Soybean root distribution, top growth and yield responses to ambient ozone and soil moisture stress when grown in soil columns in greenhouses. Environmental Pollution 65: 195-207

Heliotis FD, Karandinos MG, Whiton JC (1988) Air pollution and the decline of the fir forest in Parnis National Park, near Athens, Greece. Environmental Pollution 54: 29-40

Hill AC, Hill S, Lamb C, Barrett TW (1974) Sensitivity of native desert vegetation to SO_2 and to SO_2 and NO_2 combined. Journal of Air Pollution Control Association 24: 153

Kasana MS (1988) High relative sensitivities of some leguminous crops to ozone pollution. Environmental Conservation 4: 356-358

Kasana MS (1992) Effects of ozone fumigation on a tropical fibre plant, kenaf (*Hibiscus cannabinus* L.). Agriculture, Ecosystems and Environment 38: 61-70

Levitt J (1980) Responses of plants to environmental stresses, Vol II, Academic Press, New York

Lorenzini G (1987) Towards a 'chemical climate': need for an assessment of economical impact of atmospheric ozone on crop species in Italy. In: Prodi F, Rossi F, Cristoferi G (eds) International Conference on Agrometeorology, Cesena, 1987. Fondazione Cesena Agricultura Publ, Cesena, Italy, pp 409-410

Marenco A, Medale JC, Prieur S (1990) Study of tropospheric ozone in the tropical belt (Africa, America) from STRATOZ and TROPOZ campaigns. Atmospheric Environment 24A: 2823-2834

Martin M, Plaza J, Andrés MD, Bezares JC, Millan MM (1991) Comparative study of seasonal air pollutant behavior in a Mediterranean coastal site: Castellón (Spain). Atmospheric Environment 25A: 1523-1535

Murray F, Wilson S (1991) The effects of SO_2 on the final growth of *Medicago truncatula*. Environmental and Experimental Botany 31: 319-325

Nasralla MM, Shakour AA (1981) Nitrogen oxides and photochemical oxidants in Cairo City atmosphere. Environment International 5: 55-60

Naveh Z, Steinberger EH, Chaim S, Rotmann A (1980) Photochemical oxidants - A threat to Mediterranean forest and upland ecosystems. Environmental Conservation 7: 301-309

Norby RJ, Kozlowski TT (1981) Relative sensitivity of three species of woody plants to SO_2 at high or low exposure temperature. Oecologia 51: 33-36

O'Connor JA, Parbery DG, Strauss W (1974) The effects of phytotoxic gases on native Australian plant species: Part I. Acute effects of sulphur dioxide. Environmental Pollution 7: 7-23

Olszyk DM, Bytnerowicz A, Fox CA, Kats G, Dawson PJ, Wolf J (1987a) Injury and physiological responses of *Larrea tridentata* (DC) Coville exposed *in situ* to sulphur dioxide. Environmental Pollution 48: 197-211

Olszyk DM, Bytnerowicz A, Fox CA (1987b) Sulfur dioxide effects on plants exhibiting Crassulacean Acid Metabolism. Environmental Pollution 43: 47-62

Olszyk DM, Thompson CR, Poe MP (1988) Crop loss assessment for California: Modeling losses with different ozone standard scenarios. Environmental Pollution 53: 303-311

Rao MV, Dubey PS (1990) Explanations for the differential response of certain tropical tree species to SO_2 under field conditions. Water, Air and Soil Pollution 51: 297-305

Raschke K (1975) Stomatal action. Annual Review of Plant Physiology 26: 309-340

Reinert RA, Gimeno BS, Salleras JM, Bermejo V, Ochoa MJ, Tarruel A (1992) Ozone effects on watermelon plants at the Ebro Delta (Spain): Symptomatology. Agriculture, Ecosystems and Environment 38: 41-49

Roberts TM, Darral NM, Lane P (1983) Effects of gaseous air pollutants on agriculture and forestry in the UK. Advances in Applied Biology 9: 1-142

Schenone G, Lorenzini G (1992). Effects of regional air pollution on crops in Italy. Agriculture, Ecosystems and Environment 38: 51-59

Showman RE (1991) A comparison of ozone injury to vegetation during moist and drought years. Journal of Air and Waste Management Association 41: 63-64

Taylor GE, Selvidge WJ, Crumbly IJ (1985) Temperature effects on plant response to sulfur dioxide in *Zea mays*, *Liriodendron tulipifera*, and *Fraxinus pennsylvanica*. Water, Air and Soil Pollution 24: 405-418

Temple P (1989) Oxidant air pollution effects on plants of Joshua Tree National Monument. Environmental Pollution 57: 35-47

Thompson CR, Kats G, Lennox RW (1980) Effects of SO_2 and/or NO_2 on native plants of the Mojave Desert and Eastern Mojave-Colorado Desert. Journal of Air Pollution Control Association 30: 1304-1309

Thompson CR, Olszyk DM, Kats G, Bytnerowicz A, Dawson PJ, Wolf JW (1984) Effects of ozone or sulphur dioxide on annual plants of the Mojave desert. Journal of Air Pollution Control Association 34: 1017-1022

Umbach DM, Davis DD (1986) Severity of SO_2-induced leaf necrosis on Caribbean, Scots, and Virginia pine seedlings. Journal of Air Pollution Control Association 36: 1019-1021

Velissariou D, Davison AW, Barnes JD, Inclán R, Gimeno BS (1992a) The use of Aleppo pine, *Pinus halepensis*, as a bioindicator of ozone stress in Greece and Spain. 7th International Bioindicators Symposium and Workshop on Environmental Health. University of Kuopio, 28 Sept - 3 Oct, 1992, Finland, Abstract Book, 38

Velissariou D, Davison AW, Barnes JD, Pfirrmann T, Maclean DC, Holevas CD (1992b) Effects of air pollution on *Pinus halepensis* (Mill.): Pollution levels in Attica, Greece. Atmospheric Environment 26A: 373-380

Westman WE (1985) Air pollution injury to coastal sage scrub in the Santa Monica Mountains, Southern California. Water, Air and Soil Pollution 26: 19-41

THE MOLECULAR BIOLOGY OF PLANTS EXPOSED TO ULTRAVIOLET-B RADIATION AND THE INTERACTION WITH OTHER STRESSES

Brian R. Jordan
Department of Molecular Biology
Horticulture Research International
Littlehampton
West Sussex BN17 6LP
UK

ABSTRACT

The stratospheric ozone layer acts as a protective shield to remove harmful ultraviolet-B radiation (particularly UV-B:280-320 nm). The progressive global depletion of this ozone layer is, however, leading to increased levels of UV-B radiation reaching the earth's surface. This increase will inevitably continue into the next century with potentially deleterious consequences for agriculture, horticulture and forestry. Studies of UV-B radiation effects on plants have shown a wide range of responses. Some species are particularly sensitive and show signs of damage at moderate increases in UV-B irradiance, while other species show no response. Amongst the more susceptible species are agriculturally important crops from temperate latitudes, such as pea, although tropical crops including rice can also suffer damage from increased UV-B radiation. The specific effects upon plants include inhibition of photosynthesis, loss of fresh and dry weight, changes in assimilate partitioning, alteration in pigment biosynthesis and effects upon flowering and reproduction. These responses would clearly lead to substantial changes in plant communities and to significant losses in agricultural productivity. A major factor that determines the extent of these changes is the interaction between UV-B radiation and other environmental parameters. Thus, any prediction as to the consequences of increased UV-B can only be realistic in the context of interactions with other stresses. One way to predict the plant response is to obtain an understanding of the molecular changes induced by UV-B stress and how they are controlled. In this chapter the molecular biology of UV-B responses is described. Two major UV-B sensitive processes are focussed on, photosynthesis and the biosynthesis of protective pigments. In addition, the influence of other environmental stresses on UV-B induced responses will be discussed.

INTRODUCTION

The stratospheric ozone layer prevents harmful ultraviolet-B radiation (UV-B: 280-320 nm) from reaching the earth's surface. The increase in man-made chloro-

NATO ASI Series, Vol. I 16
Interacting Stresses on Plants in a Changing Climate
Edited by M. B. Jackson, C. R. Black

fluorocarbons (CFCs) and similar compounds is now thought to be destroying this protective ozone layer. Destruction of ozone in the Austral spring is well established (Bowman, 1988) and can account for 50% overall loss or 95% loss at specific altitudes within the ozone column (Kerr, 1988). This depleted level of ozone is then distributed to other latitudes in late spring. In addition to the severe depletion recorded in the Southern hemisphere, the Northern hemisphere ozone has been depleted by 4% over the last decade (Kerr, 1991) and is poised for further destruction. This depletion of the stratospheric ozone layer can be equated to an increase in UV-B radiation of 10% over the same period (Blumthaler and Ambach, 1990). As the concentration of CFCs in the stratosphere will continue to increase for decades, an increase in UV-B radiation will inevitably continue into the next century with potentially deleterious consequences for agriculture, horticulture and forestry.

Studies of UV-B radiation effects on plants have shown a wide range of responses (Tevini and Teramura, 1989). Some species are particularly sensitive and show signs of damage at moderate increases in UV-B irradiance, while other species show no response. These differences are also found between varieties of the same species and this can lead to significant yield loss (Teramura *et al.*, 1990a). Amongst the more susceptible species are agriculturally important crops from temperate latitudes, such as peas and brassicas. Tropical crops, including rice, can also suffer damage from increased UV-B radiation (Teramura *et al.*, 1990b). The specific effects upon plants include inhibition of photosynthesis, loss of fresh and dry weight, changes in assimilate partitioning, alterations in pigment biosynthesis and effects upon flowering and reproduction. These responses resulting from an increase in UV-B would clearly lead to substantial changes in plant communities. Most of the data on plant responses to UV-B radiation have been obtained using controlled environment facilities. The conditions in these facilities are not typical of the natural environment and therefore may not truly reflect the actual response (Caldwell and Flint, 1990). However, it is clear from these studies and the limited field research that a major factor determining the UV-B response is the interaction with other environmental parameters. These results suggest that any prediction of the consequences of increased UV-B can only be realistic in the context of interactions with other stresses. One way to predict the likely response is to obtain an understanding of the molecular changes that are involved and their regulation.

In this chapter the molecular biology involved in UV-B responses is described. The two major processes focussed on are photosynthesis and the biosynthesis of protective pigments. In addition, the review describes how the UV-B responses are influenced by other environmental stresses. For more general accounts of the effects of UV-B radiation on plants a number of excellent reviews are available (Teramura, 1983; Caldwell *et al.*, 1989; Tevini and Teramura, 1989; Tevini *et al.*, 1989; Teramura and Sullivan, 1991).

THE MOLECULAR BIOLOGY OF PLANTS EXPOSED TO UV-B

Photosynthesis. Photosynthesis is one of the most important physiological plant processes and is essential for crop productivity. One of the major sites for UV-B induced change is within the chloroplast with potential impairment of electron transport, photophosphorylation and carbon fixation, with these changes subsequently leading to inhibition of photosynthetic function and decreased efficiency of photosynthesis. UV-B radiation effects on photosynthesis have been extensively studied and clearly demonstrate multiple sites of inhibition (Bornman, 1989; Strid *et al.*, 1990). The diverse processes that are affected reflect both the range of molecules that absorb UV-B radiation and the energetically unstable nature of the photosynthetic apparatus.

Photosystem 2 activity is particularly sensitive to UV-B radiation, while photosystem 1 remains largely uninhibited (Renger *et al.*, 1989; Strid *et al.*, 1990; Prášil *et al.*, 1992). The reaction centre polypeptide D1 is rapidly turned over and the efficiency of this light-dependent process is essential to the correct functioning of photosystem 2 (Barber and Andersson, 1992). Studies by Greenberg *et al.* (1989a) show that the quantum efficiency of degradation of the D1 polypeptide was greatest in the UV-B region and that plastoquinone (in one or more of its redox states) was the photoperception mechanism for the degradation. This explanation relies on a specific excitation of the quinone bound at the Q_B site and alternative suggestions should also be considered (Prášil *et al.*, 1992). Photoinhibition of photosystem 2 is also caused by high levels of visible radiation. However, significant differences exist between UV-B damage and photoinhibition, such as the recovery kinetics (Chow *et al.*, 1992). The degradation rate ($t_{0.5}$ of 5-7 hours) induced by UV-B is also much slower than found during photoinhibition ($t_{0.5}$ of 60 minutes), although it may occur through a common mechanism (Greenberg *et al.*, 1989b) and involve the PEST region on the D1 peptide (see Prášil *et al.*, 1992 for discussion). There is, however, some controversy over the precise protease cleavage sites under normal irradiance and during photoinhibition (Barber and Andersson, 1992). Consequently, the mechanism of UV-B induced damage to the D1 polypeptide and its relationship to other degradation processes is still uncertain. When all detectable UV radiation was filtered out from sunlight the degradation rate of the D1 protein was 30% slower than in sunlight alone. As UV radiation contributes only 4% of the photon flux between 300 and 700 nm, it contributes significantly to the degradation of D1 (Greenberg *et al.*, 1989a). This becomes particularly important with the predicted increase in the proportion of UV-B as a percentage of the total spectral radiation. UV-B and photoinhibition are likely to be synergistic and this may account for photoinhibition in the field being higher than expected under the prevailing light environment (Prášil *et al.*, 1992).

Ribulose 1,5-bisphosphate carboxylase (Rubisco, EC 4.1.1.39) is the primary enzyme of CO_2 fixation in C3 plants. It is also the most abundant protein in leaves and is consequently a major resource of carbon and nitrogen within the leaf. It follows that a change in Rubisco is likely to correlate with a similar change in leaf soluble protein. However, UV-B studies on Rubisco and leaf protein have shown both parallel and opposite responses (see references in Bornman, 1989). Rubisco enzyme activity is also inhibited by UV-B exposure (Vu *et al.*, 1984; Strid *et al.*, 1990; Jordan *et al.*, 1992). This could be caused by protein degradation and or inactivation of the enzyme. Studies by Vu *et al.* (1984) on pea and soybean suggested that a loss of Rubisco protein was the main reason for the inhibition of enzyme activity. In a recent study on pea (Jordan *et al.*, 1992) the Rubisco activity declined more rapidly than did the Rubisco protein at any time point following UV-B exposure. For instance, the Rubisco polypeptides (LSU and SSU) both decreased by 10-15% after one day of UV-B exposure but the enzyme activity declined by almost 40% over the same period. The data on mRNA transcripts (see below) suggest that synthesis of the protein is rapidly switched off. Furthermore, S^{35}-methionine labelling of pea leaf proteins shows that incorporation into Rubisco and other proteins is severely inhibited within hours (Fig. 1; Jordan *et al.*, unpublished observations). As Rubisco polypeptides are relatively stable in mature leaves (Petersen *et al.*, 1973), it seems likely that after UV-B exposure the biosynthesis of Rubisco stops and the polypeptide levels decline relatively slowly as the Rubisco pool gradually depletes (Jordan *et al.*, 1992). The rapid loss of Rubisco enzyme activity is therefore likely to be caused by UV-B damage to the Rubisco holoenzyme. This could be as a direct result of UV-B absorption by aromatic amino acids (phenylalanine, tryptophan and tyrosine) causing ring opening or free radical production (tyrosine). Alternatively, damage could result as a consequence of free radicals created within the cell by UV-B exposure. Free radicals are potent electrophiles and would react with nucleophilic amino acid side chains such as cysteine-SH, lysine and arginine-NH2 and the aromatic amino acids. The modification of proteins by these radicals leads to an increase in hydrophobicity, partial denaturation and loss of catalytic activity (Cadenas, 1989).

Recently, the effect of supplements of UV-B radiation on the gene expression for pea chloroplast proteins has been reported (Jordan *et al.*, 1991; Jordan *et al.*, 1992). This was the first such study and provides a molecular mechanism that can account to some degree for the overall inhibition of photosynthesis by UV-B. A number of important points emerge from this study. Firstly, both nuclear (chlorophyll a/b binding protein, cab, and Rubisco, SSU) and chloroplast encoded (Rubisco LSU and psb A) mRNA transcripts are severely reduced. This reduction can be very rapid (80% reduction after only four hours for cab and Rubisco SSU) or more slowly (psb A mRNA transcripts remain at 35% of the control level after three days of UV-B) and it is likely that the RNA levels reflect their stability

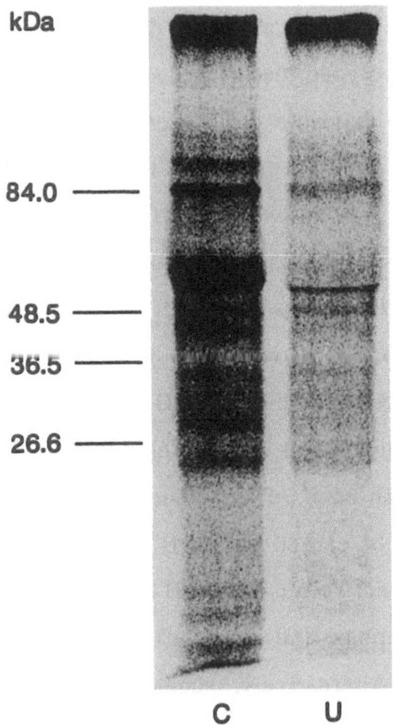

FIGURE 1. The effect of UV-B radiation on S^{35}-methionine incorporation into pea leaf proteins. Pea leaf discs were exposed to UV-B radiation (U) or control conditions (C) for eight hours in the presence of S^{35}-methionine. The labelled proteins were then isolated from the leaf discs, separated by SDS-PAGE and autoradiographed (Jordan et al., *unpublished data).*

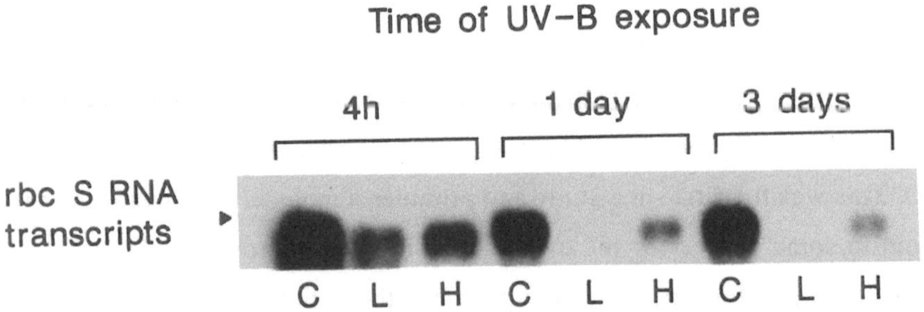

FIGURE 2. The effect of high PAR on mRNA transcripts. Pea seedlings were exposed to UV-B radiation in the presence of low (L: 150 µmol photons m^{-2} s^{-1}) or high (H: 400 µmol photons m^{-2} s^{-1}) irradiance for either four hours, one day or three days. RNA was then isolated from them and untreated control plants (C). The RNA was hybridised with P^{32}-labelled rbc S. From Jordan et al. *(1992).*

(Klaff and Gruissem, 1991). Subsequently, there is a loss of enzyme activity and protein content over a period of days. These losses correlate with a reduction in photosynthetic function. A second feature of the response is that higher photosynthetically active radiation (PAR) ameliorates the reduction in RNA transcripts to some extent (Fig. 2; see below for further discussion). These studies also show that the plants can recover the efficiency of gene expression. Following a four hour exposure to UV-B, the mRNA transcripts continue to decline for approximately 24 hours to a low level and then recover to 60% after three days. The reduction in gene expression and recovery correlate with a similar pattern in quantum yield of photosynthesis (Chow *et al.*, 1992). Repetitive exposure to high UV-B levels is therefore likely to maintain an inhibited and inefficient photosynthetic apparatus in susceptible species. The decline in gene expression is likely to be caused by inhibition of transcription. Three main molecular mechanisms can be envisaged to explain this inhibition:

1. A direct and non-specific effect of UV-B radiation on nuclear or chloroplast DNA, causing pyrimidine dimer formation and inhibition.
2. The changes could be induced through an indirect mechanism. Thus, UV-B absorption by chlorophyll or polyunsaturated acyl-lipids can cause oxidative damage and this is known to be involved in a signal transduction pathway that changes gene expression (Taylor, 1989).
3. The mechanism could be more sophisticated and involve specific photoreceptor molecules (see Ballaré *et al.* (1991) for evidence of a UV-B photoreceptor).

Thus, a UV-B receptor alone or in combination may modulate the control of transcription (see below). Two lines of evidence suggest that the down-regulation of gene expression for chloroplast proteins is not primarily due to non-specific damage to DNA. Firstly, although gene expression for chloroplast proteins is repressed, there is an increase in the expression of the enzymes of the phenylpropanoid pathway (Jordan *et al.*, unpublished observations). Secondly, during de-etiolation of dark-grown seedlings, it is well established that cab gene expression increases substantially from very low levels. Greening pea seedlings continue to express cab mRNA transcripts in a similar manner when exposed to UV-B radiation for a period of seven days (Jordan *et al.*, unpublished observations). These results suggest a specific mechanism regulates the changes in gene expression in response to UV-B radiation. In addition, the developmental stage of the tissue also determines the response. Thus, the effect of UV-B at different developmental stages may reflect the response of individual members of a multi-gene family.

Pigment biosynthesis. A well-characterised response of plants to UV-B radiation is the synthesis of protective pigments that absorb short-wavelength radiation. These pigments are mainly flavonoids (including anthocyanins, flavones and flavanols) and absorb light between 230 and 380 nm (Hahlbrock and Scheel, 1989; Strid and Porra, 1992). Exposure to UV-B induces a rapid and coordinated increase in the enzyme activities of the phenylpropanoid biosynthetic pathway, the flavonoid pigments being deposited in the cell vacuoles. In parsley leaves the biosynthesis and accumulation is specifically localised in the epidermal cells (Schmelzer *et al.*, 1988). This increase is due to dramatic changes in gene expression for enzymes of the phenylpropanoid pathway. Thus, mRNA transcript levels for enzymes such as phenylalanine ammonia lyase (PAL) and chalcone synthase (CHS) increase to maximum levels within hours and then decline to basal amounts. Fig. 3 shows the rapid increase in mRNA transcripts for chalcone synthase after exposure to UV-B radiation. These increases have been shown by run-off transcription assays on isolated nuclei to be caused by changes in transcription (Chappell and Hahlbrock, 1984). The light-dependent regulatory transcription elements within the parsley chalcone synthase gene have now been extensively characterised (Schulze-Lefert *et al.*, 1989; Weisshaar *et al.*, 1991a,b). Four light-dependent regions of the parsley chalcone synthase promoter have been identified by *in vivo* DNA footprinting (Weisshaar *et al.*, 1991b). These sequences contain guanosine residues that show differential reactivity to DMS methylation in control and UV-treated tissue. Furthermore, the altered reactivity to DMS has

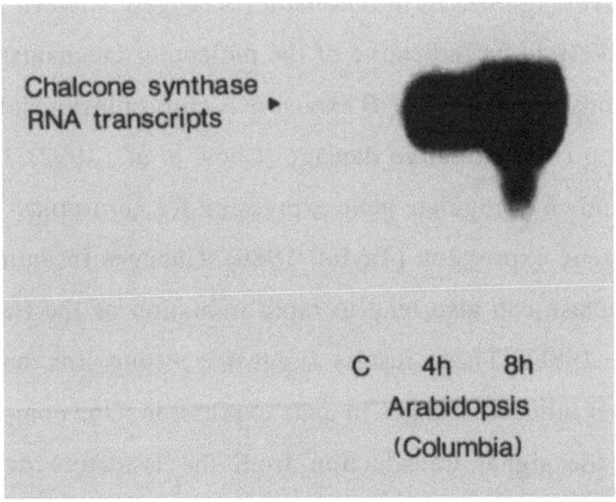

FIGURE 3. The effect of UV-B radiation on chalcone synthase RNA transcripts (Jordan et al., *unpublished data). Six week old* Arabidopsis *plants were exposed to UV-B radiation for four or eight hours. RNA was isolated from them and from untreated control plants (C). The RNA was Northern-blotted and hybridised with a P^{32}-labelled chalcone synthase sequence (generously provided by F. Ausubel).*

a timecourse that correlates to the known transcription kinetics of CHS. Using transient expression assays in isolated parsley protoplasts, it has been shown that boxes 1 and 2 represent the minimal promoter sequence for light-responsive CHS transcription. One of the regions, box 2, contains a heptameric 'core' sequence that is strongly conserved to the G box in a number of light-regulated Rubisco genes, hex in histone promoters and the ABA responsive rab gene promoters. Its presence in developmental (patatin), anaerobic (adh) and pathogenic (rolbc) genes illustrates that these *cis*-elements do not have a specific light/UV responsivity function and that common regulatory *cis*-elements are shared by genes that show regulation by a range of different factors. These studies also indicate that the regulation is dependent on the nature of the *trans*-acting factors binding to the promoters. Recently, several cDNA clones have been isolated that encode proteins that bind to box 2 and specifically to the ACGT motif common to many *cis*-elements (CPRF, common plant regulatory factors: Weisshaar *et al.*, 1991a). One of these, CPRF-1, showed mRNA accumulation that corresponded to CHS expression and supports the suggestion that this *trans*-acting factor is involved in UV light-regulation.

Although the wavelengths used in these studies ranged between 300 and 400 nm, the results are likely to be indicative of the molecular mechanisms involved.

An early consequence of UV-B exposure is chloroplast membrane disruption, probably resulting from oxidative damage (Chow *et al.*, 1992). Oxidation within chloroplasts is known to regulate gene expression for chloroplast proteins without affecting other gene expression (Taylor, 1989). Changes in antioxidants such as glutathione reductase can also lead to rapid induction of the flavonoid pathway (Wingate *et al.*, 1988). These results suggest a strong link between oxidative damage and UV-B induced changes in gene expression. One component that could be involved in the signal transduction from the oxidative damage is methyl jasmonate (Farmer and Ryan, 1992). This compound is produced as a result of lipoxygenase activity and has recently been shown to induce PAL expression (Gundlach *et al.*, 1992).

THE INTERACTION OF OTHER ENVIRONMENTAL STRESSES WITH UV-B INDUCED RESPONSES

The quantity and quality of light has a marked affect on UV-B induced changes in both photosynthesis and pigment biosynthesis. Increased irradiance both before and during exposure to UV-B irradiation has been shown to ameliorate the inhibitory effects on photosynthesis of the UV-B treatment (Warner and Caldwell, 1983; Mirecki and Teramura, 1984). The effects of the increased irradiance could be explained in terms of changes in leaf morphology, increased levels of protective pigments or the alteration of the internal cellular environment (Bornman, 1989; Bornman and Vogelmann, 1991). However, from the work of Jordan *et al.* (1992), the extent of the UV-B induced decrease in mRNA transcripts can be reduced by increased PAR. Thus in some way the higher PAR protects gene expression. These data are consistent with the physiological studies on photosynthesis which show that higher irradiance can afford protection, although the protection of mRNA can be demonstrated after only a few hours (Fig. 2) and so is unlikely to be caused by changes in leaf morphology or pigment content, as was previously suggested (Warner and Caldwell, 1983). The protection by higher PAR may be caused by

increased photorepair and photoprotection mechanisms (Pang and Hays, 1991, and references therein). However, these mechanisms are unlikely to be an explanation at normal field irradiances (Teramura *et al.*, 1990a). Alternatively, the protection may be afforded by the provision of more light-dependent biochemical energy through photosynthesis. The protection by high PAR may be important in moderating the influence of UV-B under natural daylight conditions. However, field studies show that cloud cover is also important in reducing the response to UV-B radiation (Teramura *et al.*, 1990a).

The quality of the light environment can be determined by a variety of photoreceptor mechanisms that include phytochrome, a blue/UV-A receptor, a UV-B receptor, the photosynthetic photosystems and photoperiodic detection. The complexity of the UV-B response within the overall light environment is clearly reflected in a number of studies on flavonoid biosynthesis (Bruns *et al.*, 1986; Batschauer *et al.*, 1991). In parsley tissue culture cells, CHS is strongly induced by UV-light. The increase takes place after a lag of two hours, but if the cells are pretreated with blue light the lag phase can be removed. In addition, red/far-red light treatments which control the phytochrome system will alter both the magnitude and extent of the CHS induction. Thus, three separate light qualities can be involved in this one induction process. The complex role of different photoperception mechanisms is also illustrated during the development of petunia (Koes *et al.*, 1989) and mustard (Batschauer *et al.*, 1991). In petunia seedlings and cell suspension cultures, CHS expression is UV-dependent, while in petals it is controlled by red-light. In mustard, red-light regulates CHS in cotyledons, but blue and UV control the expression in primary leaves. The influence of UV-B therefore changes during development for the expression of genes of the flavonoid pathway as it does for cab gene expression.

Water stress has also been shown strongly to influence UV-B induced responses (Teramura, 1986; Tevini and Teramura, 1989). In cucumber, the adaxial diffusive resistance to water loss increased threefold in the first few days of UV-B treatment and then stomatal function was lost as the leaf resistance rapidly declined. Thus, an increase in UV-B could potentially enhance the susceptibility of cucumber to water stress. In contrast, radish showed no additional response to UV-B (Teramura, 1986). This resistance of radish was accounted for by UV-B and water stress promoting a synergistic increase in flavonoid synthesis and hence UV-B protection. The flavonoid levels decreased with increased UV-B in cucumber and therefore could account for its susceptibility. Sullivan and Teramura (1990) studied the combination of UV-B and water stress on photosynthesis in soybean. Under well watered conditions, increased UV-B radiation reduced photosynthetic capacity, plant dry weight, leaf area and number of pods. The combination of UV-B and water stress did not increase the responses compared to either individual stress. Both UV-B and water stress affected plant growth and photosynthetic gas exchange. Whole leaf gas exchange analysis indicated that stomatal limitations were only significant under the combined stresses of water and supplemental UV-B; UV-B alone had only a minimal influence on stomatal conductance, as has been previously shown (Negash and Björn, 1986). A number of factors may be involved in masking the UV-B response in water stressed plants, in addition to morphology and protective pigmentation. For instance, drought may reduce the plant's content of phosphorus, and phosphorus deficiency reduces UV-B sensitivity in plants. In addition, water stress may delay cell division and therefore reduce growth. As UV-B radiation is particularly damaging during cell division, reduced growth is recognised as a means of UV-B protection. Consequently, reduced growth or phosphorus content resulting from water stress may negate UV-B damage (Sullivan and Teramura, 1990). These results suggest that the influence of increased UV-B radiation on crop productivity may be most inhibitory when there is no shortage of water. The considerable influence of water status has also been demonstrated by a six year field study (Teramura *et al.*, 1990a). Two cultivars of soybean, Essex and Williams, were grown under field conditions with either ambient or supplements of UV-B radiation. Overall, Essex was found to be sensitive to UV-B radiation with yield losses of 20%, while Williams was tolerant. The yield loss was strongly influenced by the prevailing climate throughout the season. Thus the loss of yield in Essex was most severe during seasons when water was readily available. This supports previous findings that water deficit can mask the UV-B response (Murali and Teramura, 1986; Sullivan and Teramura, 1990). However, cv. Williams was only affected by UV-B radiation in a year in which drought and high PAR were most prevalent. This work clearly illustrates the complex interactions between UV-B radiation, water, temperature and light, with water being a major influence.

Another environmental parameter that is likely to increase in the future is the global concentration of CO_2, with most predictions suggesting a rise in concentration from the present levels of 340 to 680 $\mu l\ l^{-1}$ by about 2050. Such an increase in CO_2 would be beneficial to crop productivity in a wide range of C3 plant species. However, this prognosis is based on the consequences of changing only the CO_2 levels and few studies have been made of the interactions between increased CO_2 and increased UV-B, a likely future scenario. From the studies that have been made, it is clear that any beneficial effect of increased CO_2 is likely to be reversed by an increase in UV-B radiation (Teramura *et al.*, 1990b). Similarly, studies of the combination of UV-B and toxic metals, such as cadmium, may reveal interactions that are more damaging than the influence of either stress alone (Bornman and Dubé, 1991).

One further important aspect that has received relatively little attention is the combined influence of UV-B and pathogens (Orth *et al.*, 1990; Panagopoulos *et al.*, 1992, and references therein). In sugar beet there is an accumulative interaction of *Cercospora beticola* and UV-B to give an increased reduction in chlorophyll content, dry weight and lipid peroxidation. Studies on pigment biosynthesis have revealed a complex interaction of UV and fungal elicitors. Both stimuli can activate enzymes of the phenylpropanoid pathway, although UV irradiation results in vacuolar accumulation of flavonoids whereas the fungal elicitors cause the secretion of furanocoumarins. Together the UV irradiation and fungal elicitor result in quantitative changes in each of the individual responses. Thus, elicitor-induced furanocoumarins are reduced and the light-induced accumulation of flavonoids is completely inhibited. These changes are a result of differential responses of enzymes within the general phenylpropanoid pathway to UV and other factors. For instance, an early observation showed that treatment of parsley cells with both fungal elicitor and UV light prevented the UV stimulation of CHS, whereas PAL biosynthesis was increased by UV, elicitor or both (Hahlbrock *et al.*, 1981). Similarly, a number of other enzymes are differentially regulated. Using transient expression analysis of regions of the CHS promoter (Lozoya *et al.*, 1991) it has been demonstrated that elicitor-sensitive regions exist in the promoter and that they are identical to those required for light-inducibility (see further details above). It is also important that other stress factors rapidly down-regulate gene expression for some proteins and activate the expression of other genes. Thus it is well established that the phenylpropanoid pathway is switched on by pathogens and that the expression of genes such as rbcS is switched off (Kombrink and Hahlbrock, 1990). This seems very analogous to the UV-B response at the level of changes in gene regulation. UV radiation may also stimulate pathogen related gene expression (Brederode *et al.*, 1991). Furthermore, during heat shock there is also a rapid reduction in the expression of some genes and the biosynthesis of the heat shock proteins (HSP). The expression of proteins similar to HSP has recently been reported in *Vigna sinensis* seedlings exposed to UV-B radiation (Nedunchezhian *et al.*, 1992).

CONCLUSIONS

It is clear from these few examples that a combination of stress factors can bring about a variety of responses that cannot be predicted on the basis of studying a single environmental parameter. Consequently, unless multiple simultaneous stress factors are studied, it is likely that the environmental impact of UV-B will not be correctly anticipated. It is particularly important to determine the common pathways involved in the signal transduction of these stress responses and elucidate the regulatory mechanisms. This would give the potential to develop long-term strategies to protect susceptible species, either by conventional breeding methods or by genetic engineering.

ACKNOWLEDGEMENTS

I would like to acknowledge the financial support of the Agricultural and Food Research Council and am grateful for the scientific collaboration of colleagues at Horticulture Research International, CSIRO, Canberra and KTH, Stockholm. I would also like to thank all those who provided information to assist in the preparation of this review.

REFERENCES

Ballaré CL, Barnes PW, Kendrick RE (1991) Photomorphogenic effects of UV-B radiation on hypocotyl elongation in wild type and stable-phytochrome-deficient mutant seedlings of cucumber. Physiologia Plantarum 83: 652-658

Barber J, Andersson B (1992) Too much of a good thing: light can be bad for photosynthesis. TIBS 17: 61-66

Batschauer A, Bruno E, Schäfer E (1991) Cloning and characterization of a chalcone synthase gene from mustard and its light-dependent expression. Plant Molecular Biology 16: 175-185

Blumthaler M, Ambach W (1990) Indication of increasing solar ultraviolet-B radiation flux in alpine regions. Science 248: 206-208

Bornman JF (1989) Target sites of UV-B radiation in photosynthesis of higher plants. Journal of Photochemistry and Photobiology 4: 145-158

Bornman JF, Dubé SL (1991) The influence of ultraviolet radiation and cadmium on plants. Current Topics in Plant Biochemistry and Physiology 10: 49-54

Bornman JF, Vogelmann TC (1991) Effect of UV-B radiation on leaf optical properties measured with fibre optics. Journal of Experimental Botany 42: 547-554

Bowman KP (1988) Global trends in total ozone. Science 239: 48-50

Brederode FTh, Linthorst HJM, Bol JF (1991) Differential induction of acquired resistance and PR gene expression in tobacco by virus infection, ethephon treatment, UV light and wounding. Plant Molecular Biology 17: 1117-1125

Bruns B, Hahlbrock K, Schäfer E (1986) Fluence dependence of the ultraviolet-light-induced accumulation of chalcone synthase mRNA and effects of blue and far-red light in cultured parsley cells. Planta 169: 393-398

Cadenas E (1989) Biochemistry of oxygen toxicity. Annual Review of Biochemistry 58: 79-110

Caldwell MM, Flint SD (1990) Plant response to UV-B radiation: comparing growth chamber and field environments. In: Payer HD, Pfirrman T, Mathy P (eds) Environmental research with plants in closed chambers. Air Pollution Research Report 26, Commission of European Communities, pp 264-270

Caldwell MM, Teramura AH, Tevini M (1989) The changing solar ultraviolet climate and the ecological consequences for higher plants. Trends in Ecology and Evolution 4: 363-367

Chappell J, Hahlbrock K (1984) Transcription of plant defence genes in response to UV light or fungal elicitor. Nature (London) 311: 76-78

Chow WS, Strid Å, Anderson JM (1993) Short-term treatment of pea plants with supplementary ultraviolet-B radiation: recovery time-courses of some photosynthetic functions and components. In: Murata N (ed) Proceedings of IX International Congress on Photosynthesis (in press)

Farmer EE, Ryan CA (1992) Octadecanoid precursors of jasmonic acid activate the synthesis of wound-inducible proteinase inhibitors. The Plant Cell 4: 129-134

Greenberg BM, Gaba V, Canaani O, Malkin S, Mattoo AK, Edelman M (1989a) Separate photosensitizers mediate degradation of the 32-kDa photosystem II reaction center protein in the visible and UV spectral regions. Proceedings of the National Academy of Science USA 86: 6617-6620

Greenberg BM, Gaba V, Mattoo AK, Edelman M (1989b) Degradation of the 32 kDa photosystem II reaction centre protein in UV, visible and far red light occurs through a common 23.5 kDa intermediate. Zeitschrift für Naturforschung 44C: 450-452

Gundlach H, Müller MJ, Kutchan TM, Zenk MH (1992) Jasmonic acid is a signal transducer in elicitor-induced plant cell cultures. Proceedings of the National Academy of Science USA 89: 2389-2398

Hahlbrock K, Lamb CJ, Purwin C, Ebel J, Fautz E, Schäfer E (1981) Rapid response of suspension-cultured parsley cells to the elicitor from *Phytophthora megasperma* var. Sojae. Induction of the enzymes of general phenylpropanoid metabolism. Plant Physiology 67: 768-773

Hahlbrock K, Scheel D (1989) Physiology and molecular biology of phenylpropanoid metabolism. Annual Review of Plant Physiology and Plant Molecular Biology 40: 347-369

Jordan BR, Chow WS, Strid A, Anderson JM (1991) Reduction in cab and psbA RNA transcripts in response to supplementary ultraviolet-B radiation. FEBS Letters 284: 5-8

Jordan BR, He J, Chow WS, Anderson JM (1992) Changes in mRNA levels and polypeptide subunits of ribulose 1,5-bisphosphate carboxylase in response to supplementary ultraviolet-B radiation. Plant, Cell and Environment 15: 91-98

Kerr JB (1991) Trends in total ozone at Toronto between 1960 and 1991. Journal of Geophysical Research 96: 20703-20709

Kerr RA (1988) Ozone hole bodes ill for the globe. Science 241: 785-786

Klaff P, Gruissem W (1991) Changes in chloroplast mRNA stability during leaf development. The Plant Cell 3: 517-529

Koes RE, Spelt CE, Mol JNM (1989) The chalcone synthase multigene family of *Petunia hybrida* (V30): differential, light-regulated expression during flower development and UV light induction. Plant Molecular Biology 12: 213-225

Kombrink E, Hahlbrock K (1990) Rapid, systemic repression of the synthesis of ribulose 1,5-bisphosphate carboxylase small-subunit mRNA in fungus-infected or elicitor-treated potato leaves. Planta 181: 216-219

Lozoya E, Block A, Lois R, Hahlbrock K, Scheel D (1991) Transcriptional repression of light-induced flavonoid synthesis by elicitor treatment of cultured parsley cells. The Plant Journal 1: 227-234

Mirecki RM, Teramura AH (1984) Effects of ultraviolet-B irradiance on soybean. Plant Physiology 74: 475-480

Murali NS, Teramura AH (1986) Effectiveness of UV-B radiation on the growth and physiology of field-grown soybean modified by water stress. Photochemistry and Photobiology 44: 215-220

Nedunchezhian N, Annamalainathan K, Kulandaivelu G (1992) Induction of heat shock-like proteins in *Vigna sinensis* seedlings growing under ultraviolet-B (280-320 nm) enhanced radiation. Physiologia Plantarum 85: 503-506

Negash L, Björn LO (1986) Stomatal closure by ultraviolet radiation. Physiologia Plantarum 66: 360-364

Orth AB, Teramura AH, Sisler HD (1990) Effects of ultraviolet-B radiation on fungal disease development in *Cucumis sativus*. American Journal of Botany 77: 1188-1192

Panagopoulos I, Bornman JF, Björn LO (1992) Response of sugar beet plants to ultraviolet-B (280-320 nm) radiation and *Cercospora* leaf spot disease. Physiologia Plantarum 84: 140-145

Pang Q, Hays JB (1991) UV-B-inducible and temperature-sensitive photoreactivation of cyclobutane pyrimidine dimers in *Arabidopsis thaliana*. Plant Physiology 95: 536-543

Peterson LW, Kleinkopf GE, Huffaker RC (1973) Evidence for lack of turnover of ribulose 1,5-diphosphate carboxylase in barley leaves. Plant Physiology 51: 1042-1045

Prášil O, Adir N, Ohad I (1992) Dynamics of photosystem II: mechanisms of photoinhibition and recovery processes. In: Barber J (ed) The photosystems. Structure, function and molecular biology. Elsevier, Amsterdam London New York, pp 295-348

Renger G, Völker M, Eckert HJ, Fromme R, Hohm-Veit S, Gräber P (1989) On the mechanism of photosystem II deterioration by UV-B irradiation. Photochemistry and Photobiology 49: 97-105

Schmelzer E, Jahnen W, Hahlbrock K (1988) *In situ* localization of light-induced chalcone synthase mRNA, chalcone synthase, and flavonoid end products in epidermal cells of parsley leaves. Proceedings of the National Academy of Science USA 85: 2989-2993

Schulze-Lefert P, Dangl JL, Becker-André M, Hahlbrock K, Schulz W (1989) Inducible *in vivo* DNA footprints define sequences necessary for UV light activation of the parsley chalcone synthase gene. The EMBO Journal 8: 651-656

Strid Å, Chow WS, Anderson JM (1990) Effects of supplementary ultraviolet-B radiation on photosynthesis in *Pisum sativum*. Biochimica et Biophysica Acta 1020: 260-268

Strid Å, Porra RJ (1992) Alterations in pigment content in leaves of *Pisum sativum* after exposure to supplementary UV-B. Plant and Cell Physiology 33: 1015-1023

Sullivan JH, Teramura AH (1990) Field study of the interaction between solar ultraviolet-B radiation and drought on photosynthesis and growth in soybean. Plant Physiology 92: 141-146

Taylor WC (1989) Regulatory interactions between nuclear and plastid genomes. Annual Review of Plant Physiology and Plant Molecular Biology 40: 211-233

Teramura AH (1983) Effects of ultraviolet-B radiation on the growth and yield of crop plants. Physiologia Plantarum 58: 415-427

Teramura AH (1986) Interaction between UV-B radiation and other stresses in plants. In: Worrest RC, Caldwell MM (eds) Stratospheric ozone reduction, solar ultraviolet radiation and plant life. Springer-Verlag, Berlin Heidelberg, pp 327-343

Teramura AH, Sullivan JH (1991) Potential impacts of increased solar UV-B on global plant productivity. In: Riklis E (ed) Photobiology. Plenum Press, New York, pp 625-634

Teramura AH, Sullivan JH, Lydon J (1990a) Effects of UV-B radiation on soybean yield and seed quality: a 6-year field study. Physiologia Plantarum 80: 5-11

Teramura AH, Sullivan JH, Ziska LH (1990b) Interaction of elevated ultraviolet-B radiation and CO_2 on productivity and photosynthetic characteristics in wheat, rice, and soybean. Plant Physiology 94: 470-475

Tevini M, Teramura AH (1989) UV-B effects on terrestrial plants. Photochemistry and Photobiology 50: 479-487

Tevini M, Teramura AH, Kulandaivelu G, Caldwell MM, Björn LO (1989) Terrestrial Plants. In: van der Leun JC, Tevini M (eds) United Nations Environment Programme, Environmental Effects Panel Report, pp 25-37

Vu CV, Allen LH Jr, Garrard LA (1984) Effects of enhanced UV-B radiation (280-320 nm) on ribulose-1,5-bisphosphate carboxylase in pea and soybean. Environmental and Experimental Botany 24: 131-143

Warner CW, Caldwell MM (1983) Influence of photon flux density in the 400-700 nm waveband on inhibition of photosynthesis by UV-B (280-320 nm) irradiation in soybean leaves: separation of indirect and immediate effects. Photochemistry and Photobiology 38: 341-346

Weisshaar B, Armstrong GA, Block A, da Costa e Silva O, Hahlbrock K (1991a) Light-inducible and constitutively expressed DNA-binding proteins recognizing a plant promoter element with functional relevance in light responsiveness. The EMBO Journal 10: 1777-1786

Weisshaar B, Block A, Armstrong GA, Herrmann A, Schulze-Lefert P, Hahlbrock K (1991b) Regulatory elements required for light-mediated expression of the *Petroselinum crispum* chalcone synthase gene. In: Jenkins GI, Schuch W (eds) Society for Experimental Biology SEB Symposia Series 45, pp 191-210

Wingate VPM, Lawton MA, Lamb CJ (1988) Glutathione causes a massive and selective induction of plant defense genes. Plant Physiology 87: 206-210

THE ASSESSMENT OF PLANT DAMAGE BY REACTIVE HYDROCARBONS AND THEIR OXIDATION PRODUCTS

G.M. Terry and N.J. Stokes
Institute of Environmental and Biological Sciences
Lancaster University
Lancaster LA1 4YQ
UK

ABSTRACT

Plants produce numerous volatile organic compounds, including aldehydes, ketones, organic acids, alcohols, C_2-C_4 alkanes and alkenes, isoprene and the C_{10}-C_{40} terpenoids. The biogenic alkene isoprene is of particular interest because it is reactive and accounts for almost half of the non-methane hydrocarbons emitted by plants globally. Human activities provide another source of atmospheric hydrocarbons. Propene is an important example, being a reactive hydrocarbon with a large mass emission and high ozone creation potential. The direct effects of non-methane hydrocarbons on plants are currently under investigation. Perhaps of greater concern, however, are the effects of their reaction products. For example, reaction of alkenes with ozone can produce organic hydroperoxides, which are highly reactive and have been implicated in plant damage (enzyme inhibition, mutagenicity), especially in species which are alkene emitters. The study of hydrocarbon pollutants and their oxidation products requires specialised experimental facilities. New facilities at Lancaster are being used to investigate the effects of propene, isoprene, hydrogen peroxide and ozone, all of which are important constituents or precursors of photochemical smog. Fumigation at low pollutant concentrations may produce few visible symptoms of damage, and long-term exposure may be required before pollution effects on growth rate or other gross physiological changes become significant. In contrast, biochemical responses to pollution stress may be detectable within hours or days of fumigation. For example, fumigation of cherry trees over one growing season with 100 ppbv propene with and without 25 ppbv hydrogen peroxide produced no significant changes in photosynthetic rate or stomatal conductance, but quantifiable changes in soluble protein concentration and glutathione reductase activity in leaf extracts were recorded. Bio-assays, particularly those involving antioxidant systems are thus useful indicators of short-term pollution effects involving hydrocarbons and their reaction products, as well as being vital tools in the development of our understanding of the underlying mechanisms of pollution and other stresses in plants.

NATO ASI Series, Vol. I 16
Interacting Stresses on Plants in a Changing Climate
Edited by M. B. Jackson, C. R. Black

INTRODUCTION

The role of atmospheric pollutants in causing damage to plants has been under investigation for some time. Whilst the effects of acid pollutants are beginning to be understood, much less is known about other gaseous primary pollutants such as reactive hydrocarbons and free radical species, and even less is known about the effects of these pollutants in mixtures or acting in combination with their reaction products. Experiments with reactive hydrocarbons require specialist facilities, such as those recently built at Lancaster University, in which at least some of the complex chemical reactions of the troposphere can be reproduced, and in which plants can be grown under controlled conditions. Methods for assessing pollution-induced plant damage must encompass not only the dose-response characteristics but also the effects of both long and short-term exposure periods.

CHEMICAL SPECIES WITH POTENTIAL TO DAMAGE PLANTS

Hydrocarbons found in the troposphere may be biogenic or anthropogenic in origin. Plants produce a wide range of volatile organic compounds, including aldehydes, ketones, organic acids, alcohols, C_2-C_5 alkanes and alkenes, isoprene and the C_{10}-C_{40} terpenoids. Isoprene accounts for almost half of the total of non-methane hydrocarbons emitted by plants globally (Zimmerman *et al.*, 1978), with an emission rate estimated at between 200 and 400 Tg (C-equivalents) per year (Zimmerman *et al.*, 1978; Rasmussen and Khalil, 1988; Turner *et al.*, 1991). Propene is one of the major anthropogenic hydrocarbon pollutants. It is a reactive hydrocarbon with a large mass emission and high ozone creation potential. The direct effects of non-methane hydrocarbons on plants are of interest, and are currently under investigation in our laboratories.

Perhaps of greater interest and relevance, however, are the effects on plants of the reaction products of these hydrocarbons acting singly and in combination with other atmospheric pollutants. For example, recent experimental evidence suggests that visible ozone damage in plants is the result of the reaction of ozone with endogenous unsaturated alkenes such as ethene and isoprene, with the reaction products – a suite of radicals and hydroperoxides – going on to produce the tissue damage (Hewitt *et al.*, 1990; Mehlhorn *et al.*, 1991; Salter and Hewitt, 1992).

Atmospheric reactions involving non-methane hydrocarbons of biogenic or anthropogenic origin can produce a similar suite of products external to the plant, and the more stable of these will be available for plant uptake. Within the troposphere the hydroxyl radical (OH) is the major oxidising species. Table 1 shows a condensed scheme for the oxidation of an alkene initiated by OH. The

TABLE 1*. Principal reactions in the tropospheric oxidation of an alkene by a hydroxyl radical.*

$OH + RCH{=}CH_2 \rightarrow RCH{=}CH_2OH$	(1)
$RCH{=}CH_2OH + O_2 \rightarrow RCH(O_2)CH_2OH$	(2)
$RCH(O_2)CH_2OH + NO \rightarrow RCH(O)CH_2OH + NO_2$	(3)
$RCH(O_2)CH_2OH + HO_2 \rightarrow RCH(OOH)CH_2OH + O_2$	(4)
$RCH(O)CH_2OH + O_2 \rightarrow RC(O)CH_2OH + HO_2$	(5)
$RCH(O)CH_2OH \rightarrow RCHO + CH_2OH$	(6)
$RCHO + hv \rightarrow H + RCO$	(7)
$CH_2OH + O_2 \rightarrow HCHO + HO_2$	(8)
$H + O_2 + m \rightarrow HO_2 + m$	(9)
$NO_2 + hv \rightarrow NO + O$	(10)
$O + O_2 + m \rightarrow O_3 + m$	(11)

scheme shows that the radical formed by reaction with OH in (1) is rapidly oxidised by molecular oxygen to form an alkylperoxyl radical, RO_2 (2). The alkylperoxide rapidly oxidises other atmospheric constituents, commonly nitric oxide when present (3). If the troposphere is unpolluted, however, reaction with the hydroperoxyl radical, HO_2, can occur (4). The alkoxyradical, RO, formed in (3) is oxidised to an aldehyde or ketone plus a free radical (5), or may undergo unimolecular decomposition to produce an aldehyde and a free radical (6). The products from (5) and (6) may then undergo further reaction or decomposition, as in the examples in (7), (8) and (9). An important point to note is that NO_2

produced in (3) is a precursor of ozone formation (10), (11). Gas phase ozonolysis of the alkene will then increase the complexity of the reaction mechanism and the range of reactive products.

LABORATORY EXPERIMENTS

The study of the involvement of reactive alkenes and their oxidation products in plant damage requires specialised experimental facilities. At Lancaster, eight 1 m^3 fumigation chambers have been designed and constructed to allow plants to be exposed to low concentrations of gaseous reactive pollutants (Stokes *et al.*, 1993). The walls and roof of each chamber are lined with 0.05 mm teflon film which is chemically inert and has good light transmission characteristics. Safety systems include automatic shut-down of all gas supply functions in the event of a leak or malfunction of the air extraction system. Ozone, hydrogen peroxide (for *in situ* production of hydroxyl radicals via dissociation) and gaseous hydrocarbons are introduced at known flow rates. Growing conditions (light, temperature, humidity) are monitored and held constant. The concentrations of gases available to plants for uptake are measured as follows: ozone by UV absorption, hydrocarbons by gas chromatography, hydrogen peroxide by chemiluminescence and hydroperoxides by HPLC.

APPROACHES TO THE ASSESSMENT OF PLANT DAMAGE

In investigating the dose-response characteristics of a pollutant or mixture of pollutants, it is important to note that visible symptoms of plant damage, such as leaf mottling or stunting of growth may not occur in the short term, especially with moderate, realistically-low, pollutant concentrations. There may be different dose thresholds, one for acute visible injury and a second for long-term effects on growth parameters such as biomass gain. We therefore feel that it is important to conduct both short-term and long-term fumigation experiments, and to take more than one approach to the assessment of plant damage. In our present work we consider growth rate, gross physiological parameters such as stomatal conductance and photosynthesis measured by gas exchange, and the activities of a number of antioxidant systems in leaf extracts. We are also developing the use of electron spin resonance spectroscopy with the aim of identifying the chemical radicals involved in injury occurring in plant tissues.

ASSESSMENT OF GROWTH RATE AND GROSS PHYSIOLOGICAL CHANGES

Commonly, assessments of growth parameters involve the destructive harvesting of plant parts. Long-term (multi-season) experiments therefore require large numbers of replicate plants, a factor which poses difficulties in compact, laboratory based fumigation systems. The problem can often be overcome by using various non-destructive measurement techniques as indices for monitoring growth (reviewed by Causton, 1985; Lucas *et al.*, 1988). Care must be taken with the statistical analysis of data collected in successive years, since datasets cannot be regarded as fully independent of one another. A further limitation of this approach is that very little is learned about the mechanisms which bring about the observed changes, or of the time-courses involved.

Measurement of gross physiological parameters such as photosynthetic activity and stomatal conductance can be made non-invasively and at frequent intervals using portable infra-red gas analysis equipment with an 'open' type leaf chamber. Repeated measurements can be made on the same leaves, which reduces the variation and allows more sensitive statistical analyses to be applied (Diggle, 1989). The time of onset of pollution-induced changes in these parameters may thus be determined, and in many circumstances this can give an earlier and more sensitive indication of plant injury than could be gained by growth rate measurements alone.

BIOCHEMICAL ASSAY OF DEGRADATION PRODUCTS AND ANTIOXIDANT SYSTEMS

If our aim is to understand the mechanisms underlying the damage caused to plants by atmospheric pollutants, then the identification of those biochemical pathways which are immediately affected by the presence of air pollutants should be helpful. There have been many investigations in which changed cellular conditions in response to single air pollutants have been identified (reviewed by Horsman and Wellburn, 1976). Three categories of assays are commonly carried out:

1. Measurement of the concentrations of degradation products of pollution attack. An example is malondialdehyde – a product of lipid peroxidation – which can be assayed in leaf extracts using the thiobarbituric acid test (Gutteridge and Halliwell, 1990).
2. Measurement of compounds within plant tissues which have known antioxidant activity, the levels of which may be altered when the plant is exposed to pollution or another form of stress. An example is glutathione, a low molecular weight antioxidant which is predominantly present in the reduced form (Halliwell and Foyer, 1978) and whose oxidation restores ascorbate and other compounds to the reduced state.
3. Measurement of enzymes involved in the regeneration of reduced forms of antioxidant molecules. An example is glutathione reductase which catalyses the regeneration of reduced glutathione (Akerboom and Sies, 1981; Smith *et al.*, 1988).

To this category we may also add the measurement of total soluble leaf protein, which may fall in concentration in plant parts which are subject to protein degradation or in which carbon fixation (and hence the concentration of RuBisCo) is reduced. Many such reports of pollution-induced premature senescence may be found in the literature, for example with respect to ozone pollution (Tomlinson and Rich, 1973; Darrall, 1989). Conversely, the synthesis of additional amounts of existing proteins, or of new 'stress proteins' may result from exposure to atmospheric pollutants (Constantinidou and Kozlowski, 1974).

ELECTRON SPIN RESONANCE SPECTROSCOPY

This technique can be used for the detection of any paramagnetic species (i.e., one that has unpaired electrons) and has already begun to be applied to the investigation of pollution damage to plants (Stegmann *et al.*, 1987; Mehlhorn *et al.*, 1990; Runeckles and Vaartnou, 1992). A review of the available literature shows that two types of spectra are obtained with plant tissue, as shown in Fig. 1, both of which are derived from photosynthesising material.

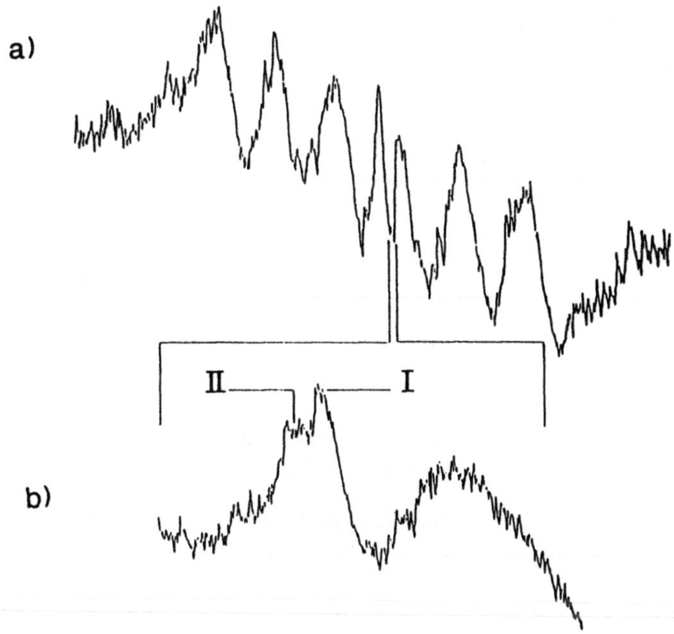

FIGURE 1. ESR spectra from a single piece of excised cherry (*Prunus avium*) leaf showing a) the six line signal obtained from hexa-aquo Mn^{2+} and b) an expanded portion of the spectrum showing two organic radical groups (signals I and II).

Whilst manganese ions are known to be involved in the production of oxygen via the splitting of water in Photosystem II, the divalent ion detected by ESR is not directly involved in this process but is considered to maintain redox balance during photosynthesis and/or to act as a radical scavenger to protect against photo-oxidant damage (Laggner *et al.*, 1988). The organic radicals are thought to be involved in the redox reactions of Photosystems I and II (Runeckles and Vaartnou, 1992).

SOME RECENT RESULTS CONCERNING THE IMPACT OF REACTIVE HYDROCARBONS ON PLANTS

Experiments are currently underway at Lancaster to investigate the effects of reactive hydrocarbons and their products on one year old cherry trees (*Prunus avium*). Fumigation was with propene, with gas-phase hydrogen peroxide, or with a mixture of propene and hydrogen peroxide plus UV light (to generate a supply of OH radicals), as shown in Table 2.

TABLE 2. Cherry tree fumigation conditions.

Chamber	Pollutant	Pollutant concentation (ppb)
1	control	0
2	propene	100
3	propene + hydrogen peroxide + UV	25
4	hydrogen peroxide	25

Fumigation was for an eight week period over the summer growth season. Fortnightly measurements were made of photosynthetic activity, stomatal conductance, transpiration rate and intercellular carbon dioxide concentration using portable infra-red gas analysis equipment. These were carried out on two leaves (at heights of approximately 20 and 45 cm) on each of 15 trees in each fumigation chamber. Measurements were also made of leaf growth rate and stem girth. Considerable variability was encountered in all parameters measured, and it was not possible to deduce definite trends or differences at this early stage.

Measurements were also made of total soluble protein concentrations in leaf extracts from three leaf classes in all treatments at the end of the fumigation period. In plants fumigated with propene and hydrogen peroxide (Chamber 3) and with hydrogen peroxide alone (Chamber 4), significant differences occurred between the protein concentrations seen in young leaves at the top of the plant and mature leaves growing close to stem bases, as compared with the controls (Fig. 2). The highest protein concentration differences were found in the mature lower leaves (significantly different at p=0.05). In addition, mature leaves growing close to the bases of trees fumigated with a mixture of propene and hydrogen peroxide had significantly higher protein concentrations than equivalent leaves from control plants.

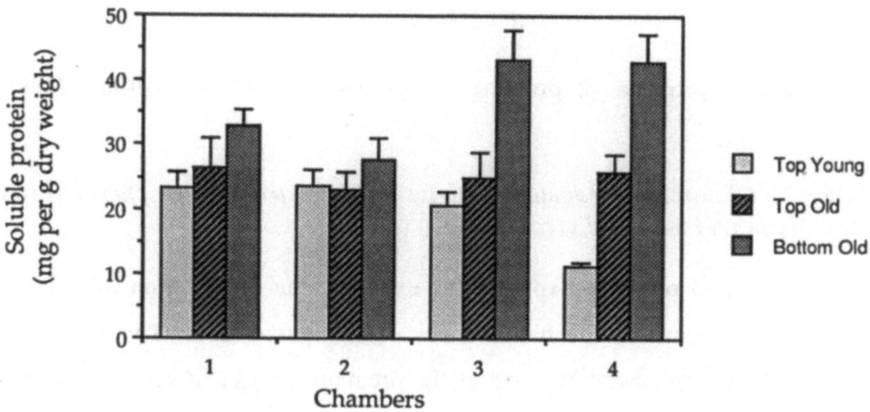

1: control 2: propene 3: propene + hydrogen peroxide 4: hydrogen peroxide

FIGURE 2. *Total soluble protein in leaf extracts from cherry (*Prunus avium*). Mean and standard error, n=7.*

Measurements of glutathione reductase activity made on the same extracts also revealed pollution-induced differences between leaves from different parts of the plant (Fig. 3). Significantly higher activities of this enzyme were found in mature basal leaves fumigated with hydrogen peroxide than in analogous controls. When the data were expressed on a 'per mg protein basis' it was found that the specific activity of glutathione reductase had risen in hydrogen peroxide-fumigated plants.

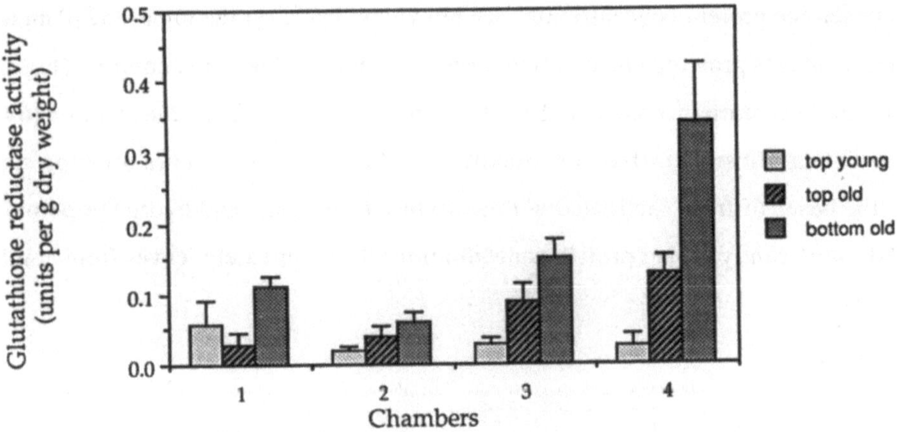

1: control 2: propene 3: propene + hydrogen peroxide 4: hydrogen peroxide

*FIGURE 3. Glutathione reductase activity in leaf extracts from cherry (*Prunus avium*). Mean and standard error, n=7.*

At least two possible explanations exist for the observation of enhanced protein concentrations in the basal leaves of fumigated trees. The first is that protein levels are enhanced because of the premature onset of senescence in the upper parts of the plant, with concomitant changes in source-sink relationships of metabolites prior to abscission of the upper leaves, and an emphasis on the importance of the lower leaves for active metabolism and protein synthesis. The second explanation is that enhanced protein levels are the consequence of increased synthesis of specific proteins in response to pollution-induced stress. Our measurements of changes in glutathione reductase activity following hydrogen peroxide fumigation, and in particular the measured increase in specific activity of the enzyme in some treatments, suggest that *de novo* synthesis or activation of specific proteins may have occurred. However, the extent of the differences in total soluble protein levels between treatments is so large that the involvement of RuBisCo is likely.

Preliminary electron spin resonance spectroscopy measurements have also been made on discs of cherry leaf from plants which had been fumigated as described above. In our experiments, samples from control plants show either a large or a negligible Mn^{2+} signal, whilst samples from fumigated plants show a large range of signals which are never negligible. The spectra obtained from the organic radicals in fumigated trees showed a reduction in the intensity of signal II and an increase in signal I. These results would appear to indicate a pollutant effect upon the photosynthetic apparatus.

CONCLUSIONS

The design and construction of specialist facilities is essential for fumigating plants with reactive hydrocarbons. The coupling of physiological, biochemical and biophysical measurements of pollution-induced effects is already proving successful in our investigation of the impacts of reactive hydrocarbons on plants. In measuring changes in soluble protein content and glutathione reductase activity in leaf extracts we have identified fumigation-induced disturbances of normal plant metabolism which are detectable biochemically before becoming visually apparent. This allows us to identify sub-acute pollution effects sooner than would be possible using gross physiological measurements or growth rate methods alone. Further development of electron spin resonance spectroscopy methods should provide another tool for determining the sequence of events which initiate and propagate damage in plant tissue.

ACKNOWLEDGEMENTS

We acknowledge funding for this project from the Department of the Environment and the Natural Environment Research Council as part of the DoE-NERC Umbrella Programme: Environmental Impact Assessments – Critical Loads and Terrestrial Ecosystems.

REFERENCES

Akerboom TPM, Sies H (1981) Assay of glutathione, glutathione disulphide and glutathione mixed disulphides in biological samples. Methods in Enzymology 77: 373-382

Causton DR (1985) Biometrical, structural and physiological relationships among tree parts. In: Cannell MGR, Jackson JE (eds) Attributes of trees as crop plants. Institute of Terrestrial Ecology, NERC, pp 137-159

Constantinidou HA, Kozlowski TT (1974) Effects of sulphur dioxide and ozone on *Ulnus americana* seedlings II. Carbohydrates, proteins and lipids. Canadian Journal of Botany 57: 176-184

Darrall NM (1989) The effects of air pollutants on physiological processes in plants. Plant, Cell and Environment 12: 1-30

Diggle PJ, Donnelly JB (1989) A selected bibliography on the analysis of repeated measurements and related areas. Australian Journal of Statistics 31: 183-193

Gutteridge JMC, Halliwell B (1990) The measurement and mechanism of lipid peroxidation in biological systems. Trends in Biochemical Sciences 15: 129-135

Halliwell B, Foyer CH (1978) Properties and physiological function of a glutathione reductase purified from spinach leaves by affinity chromatography. Planta 139: 9-17

Hewitt CN, Kok GL, Fall R (1990) Hydroperoxides in plants exposed to ozone mediate air pollution damage to alkene emitters. Nature (London) 344: 56-58

Horsman DC, Wellburn AR (1976) Guide to the metabolic and biochemical effects of air pollutants on higher plants. In: Mansfield TA (ed) Effects of air pollutants on plants. Cambridge University Press, Cambridge, pp 185-199

Laggner P, Mandl R, Schuster A, Zechner M, Grill D (1988) Rapid detection of manganese deficiency in conifer needles by ESR spectroscopy. Angewandte Chemie (International edition, English) 27: 1722-1724

Lucas PW, Cottam DA, Sheppard LJ, Francis BJ (1988) Growth reponses and delayed winter hardening in Sitka spruce following summer exposure to ozone. The New Phytologist 108: 495-504

Mehlhorn H, O'Shea JM, Wellburn AR (1991) Atmospheric ozone interacts with stress ethylene formation by plants to cause visible injury. Journal of Experimental Botany 42: 17-24

Mehlhorn H, Tabner BJ, Wellburn AR (1990) Electron spin resonance evidence for the formation of free radicals in plants exposed to ozone. Physiologia Plantarum 79: 377-383

Rasmussen RA, Khalil MAK (1988) Isoprene over the Amazon Basin. Journal of Geophysical Research 93: 1417-1421

Runeckles V, Vaartnou M (1992) Observations on the *in situ* detection of free radicals in leaves using electron paramagnetic resonance spectrometry. Canadian Journal of Botany 70: 192-199

Salter L, Hewitt CN (1992) Ozone-hydrocarbon interactions in plants. Phytochemistry 31: 4045-4050

Smith IK, Vierheller TL, Thorne CA (1988) Assay of glutathione reductase in crude tissue homogenates using 5,5'-dithiobis(2-nitrobenzoic acid). Analytical Biochemistry 175: 408-413

Stegmann HB, Schuler H, Ruff J (1987) Investigation of damage to forest by EPR spectroscopy *in vivo*. Photochemistry and Photobiology 50: 201-211

Stokes NJ, Lucas PW, Hewitt CN (1993) A controlled environment closed-chamber facility for studies into the effects of reactive hydrocarbons on plant growth and physiology. Atmospheric Environment 27A: 679-683

Tomlinson H, Rich S (1973) Anti-senescent compounds reduce injury and steroid changes in leaves and their chloroplasts. Phytopathology 63: 903-906

Turner DP, Baglio JV, Wones AG, Pross D, Vong R, McVeety BD, Phillips DL (1991) Climate change and isoprene emissions from vegetation. Chemosphere 23: 37-56

Zimmerman P, Chatfield RB, Fishman J, Crutzen PJ, Hanst PL (1978) Estimates of the production of CO and H_2 from the oxidation of hydrocarbon emissions from vegetation. Geophysical Research Letters 5: 679-682

TOO MUCH OF A GOOD THING? LONG-TERM EXPOSURE TO ELEVATED CO_2 DECREASES CARBOXYLATING AND PHOTORESPIRATORY ENZYMES AND INCREASES RESPIRATORY ENZYME ACTIVITY IN SPRUCE

Jean-Jacques Van Oosten, Pierre Dizengremel[1], Eric Laitat[2] and Robert Impens[2]
Horticulture Research International
Worthing Road
Littlehampton
West Sussex BN 17 6LP
UK

ABSTRACT

The long-term effects of an enriched CO_2 atmosphere on the primary carbon metabolism of four year old spruce trees (*Picea abies* L. Karst) were examined. Six key enzymes were studied in one year old needles of trees exposed for two years in open-top chambers to three CO_2 levels (350, 480 and 570 ppmv). The specific activity and quantity of ribulose-1,5-bisphosphate carboxylase/oxygenase (RuBisCO, EC 4.1.1.39) and the specific activities of the photorespiratory enzymes, glycolate oxidase (EC 1.1.3.15) and hydroxypyruvate reductase (HPR, EC 1.1.1.29) showed significant decreases in the CO_2-enriched atmospheres. In contrast, a net increase was found for the specific activities of the mitochondrial enzymes, NAD-malic enzyme (NAD-ME, EC 1.1.1.39) and especially fumarase (EC 4.2.1.2). The carboxylating enzyme, phosphoenolpyruvate carboxylase (PEPC, EC 4.1.1.31), showed a marked decrease in activity. These results clearly demonstrate increases in the activities of enzymes linked to the respiratory process and decreases in CO_2-fixing enzymes as a result of long-term exposure to less than double the present ambient level of CO_2.

---

[1] Laboratoire de Physiologie Vegetale et Forestiere, Universite de Nancy 1, Vandoeuvre, France.

[2] UER de Biologie Vegetale, Faculte des Sciences Agronomiques de Gembloux, Belgique

NATO ASI Series, Vol. I 16
Interacting Stresses on Plants in a Changing Climate
Edited by M. B. Jackson, C. R. Black

INTRODUCTION

The atmospheric CO_2 concentration in Europe has increased from about 270 ppmv before the industrial revolution to 350 ppmv in 1991. It is generally assumed that this increase will continue for the foreseeable future and lead to a doubling of CO_2 during the second half of the 21st century. In this context, the forests which cover one-third of the surface of the earth constitute a major component of carbon cycling by undertaking two-thirds of the global photosynthesis (Kramer, 1981).

The behaviour of trees encountering elevated atmospheric CO_2 is thus of significant interest. As CO_2 is a limiting factor primarily for C3 plants, it is generally assumed that a rise in CO_2 may be beneficial for these plants, stimulating photosynthesis and increasing dry matter production (Kramer, 1981). Many experiments have been conducted on herbaceous plants cultivated in growth cabinets and submitted to short-term exposures (4 weeks - 4 months) to elevated CO_2 concentrations (generally 1000 ppmv). The dry matter production of seedlings is generally enhanced in the first weeks (Clough *et al.*, 1981; Delucia *et al.*, 1985), although the rate of net photosynthesis, after an initial rise, decreases after several weeks of exposure to an enriched CO_2 atmosphere (Delucia *et al.*, 1985; Sasek *et al.*, 1985; Yelle *et al.*, 1989). Experiments on forest trees have generally been performed in growth cabinets or small greenhouses using young seedlings grown in pots, a procedure which is open to criticism (Arp, 1991). When trees were exposed for 4-9 months to 700-1000 ppmv CO_2, slight (Tolley and Strain, 1984; Reekie and Bazzaz, 1989) or marked (O'Neill *et al.*, 1987) increases in biomass have been observed. However, a net decrease in photosynthesis was observed by Reekie and Bazzaz (1989), while Bazzaz and Fajer (1992) concluded that changes induced by a CO_2-rich world would not necessarily benefit plants.

At the biochemical level, accumulation of starch in the leaves of herbaceous plants cultivated under high CO_2 levels is always observed (Cave *et al.*, 1981), and is generally linked to a decrease in RuBisCO activity (Peet *et al.*, 1986; Yelle *et al.*, 1989). Such a decrease in RuBisCO activity is accompanied by a decrease in the amount of enzyme protein in tomato plants exposed to 1000 ppmv CO_2 for nine weeks (Besford *et al.*, 1984, 1990). This particular long-term effect of high CO_2

on the quantity of RuBisCO is analogous to the problem of source-sink relationships within the plant (Stitt *et al.*, 1990). By contrast, no information is available concerning the response of the carbon budget of the leaf tissues of tree species to long-term treatments with elevated CO_2 (for a review of primary carbon metabolism, see Van Oosten, 1992a).

In this paper, for the first time, the effects of two different levels of elevated CO_2 (480 and 570 ppmv) on four year old spruce trees growing in open-top chambers for two years in ambient and enriched CO_2 atmospheres are examined at the level of several key enzymes of primary carbon metabolism.

MATERIALS AND METHODS

Plant material and growth conditions. Open-top chambers (OTCs) were set up in May 1987 to study the physiology of trees grown under controlled stress conditions (Laitat *et al.*, 1991). The station is built under the forest canopy, and was designed to minimise as far as possible any disturbance (soil compaction, destruction of shrubs) so that the forest ecosystem in the chambers and in the control plots can be considered as intact. The chambers are decagonal, 3 m in height and 3 m in diameter, and uniform air distribution is ensured by introducing the inlet pipe into a double-wall towards the base of the chamber. Spruce trees (*Picea abies* (L.) Karst.) are cultivated in the indigenous brown forest soil with moder humus (FAO classification Dystric cambisol with moder humus). After a two year acclimation period, the atmosphere within the chambers was progressively enriched with CO_2 during the following two years. At the time of the measurements, the CO_2 concentrations applied were ambient, ambient + 130 ppmv and ambient + 220 ppmv, or approximately 350, 480 and 570 ppmv CO_2. Measurements were made at weekly intervals after bud burst (from 28/04/91 to 09/06/91). One year old needles were randomly collected from all 30 trees of each chamber, always at 10.30 am to avoid any day to day variation in enzyme activity. The samples were then transferred at 0 °C to the laboratory and two independent enzymatic extractions per sample were performed. RuBisCO measurements were performed on one extract during the last four weeks of sampling.

Extraction and analysis procedures. The extraction medium, protein and enzyme assays and immunological quantitation of RuBisCO protein were all as described by Van Oosten *et al.* (1992). The specific enzyme activities are expressed in nanokatal per milligram of protein (nKat mg^{-1} protein), defined as the amount of enzyme which catalyses one nanomole of substrate per second per milligram of protein. However, RuBisCO activity is expressed in µmol of CO_2 per hour per milligram of protein. The specific activities of the eight enzymes were subjected to a two-factors analysis of variance (ANOVA). The dependent variables were the specific activities of the enzymes, and the protein content. Multiple range comparisons of main effects and interactions were conducted by means of the Student-Newman-Keuls method.

RESULTS

Fig. 1 shows that the protein content of needles from spruce trees grown in OTCs under elevated CO_2 was significantly lower than that of plants grown in ambient air (about 15% less in 570 ppmv CO_2 and over 30% less in 480 ppmv CO_2).

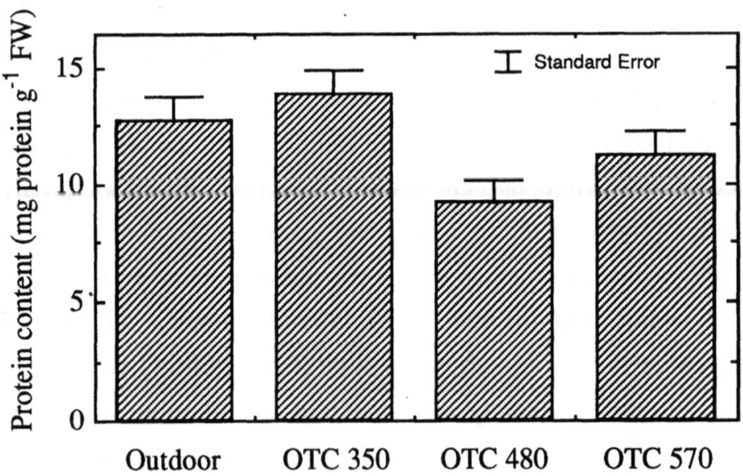

FIGURE 1. Protein content of one year old needles of spruce trees growing outdoors or exposed in open-top chambers to 350, 480 or 570 ppmv CO_2. From Van Oosten et al. *(1992).*

Concerning the enzyme activities, the differences between trees growing outdoors and those in the control OTCs under ambient CO_2 (350 ppmv) were generally small or undetectable (Table 1). The specific activity of RuBisCO was lower in both treatments receiving enriched CO_2 levels than in the control OTCs (40% at 480 ppmv and 27% at 570 ppmv CO_2). The activities of the two photorespiratory enzymes were also decreased by CO_2 enrichment, and it is noteworthy that the decreases were greater at 480 ppmv than at 570 ppmv CO_2. This also applies to the activity of other enzymes such as PFP (see Van Oosten *et al.*, 1992), which only showed a decrease in activity at 480 ppmv CO_2, and to a lesser extent PEPC. The activities of the mitochondrial decarboxylating NAD malic enzyme and of fumarase were slightly increased at elevated CO_2 compared with ambient air.

TABLE 1*. Effect of elevated CO_2 on specific enzyme activities (nkat mg^{-1} protein). OTC, Open top chambers; Control, unchambered ambient plots. 350, 480 and 570, mean CO_2 concentrations in OTCs. M, Mean; SE, Standard error. Multiple range comparisons of the means were conducted using the Student-Newman-Keuls method. Values followed by the same letter are not significantly different at the 0.05 level of probability.*

Enzyme		Specific enzyme activities (nkat mg^{-1} protein)			
		Control	OTC 350	OTC 480	OTC 570
RuBisCO[1]	M	22.56 b	26.01 a	15.65 d	18.93 c
	(SE)	4.98	6.72	4.55	3.43
Glycolate oxidase	M	2.10 a	2.17 a	1.29 b	1.98 a
	(SE)	0.32	0.35	0.23	0.32
Hydroxypyruvate reductase	M	12.39 a	12.17 a	9.23 b	10.28 b
	(SE)	3.43	3.34	3.55	3.50
Phosphoenolpyruvate carboxylase	M	0.71 a	0.75 a	0.41 b	0.47 b
	(SE)	0.14	0.14	0.13	0.11
NAD-malic enzyme	M	2.49 a	2.15 b	2.39 a	2.45 a
	(SE)	0.64	0.65	0.65	0.80
Fumarase	M	2.93 c	2.73 d	3.63 a	3.16 b
	(SE)	1.11	0.91	1.38	1.12

[1] Specific activity expressed in μmol CO_2 h^{-1} mg^{-1} protein. After Van Oosten *et al.* (1992).

The enzyme responsible for RuBisCO activity is known to be quantitatively important in leaf cells and any variation in its activity could be caused by a change in the amount of this protein. A reduced quantity of the RuBisCO protein in needles from trees grown in the enriched CO_2 atmospheres has been observed (Van Oosten *et al.*, 1992). The decreases in the amounts of the protein agree well with the observed decline in RuBisCO activity (Table 1).

DISCUSSION

Needles of four year old spruce trees which had been exposed to enriched CO_2 atmospheres for two years showed changes in the activities of enzymes involved in carbon metabolism. The main results concern the contrasting behaviour of the chloroplast and mitochondrial compartments, with a marked decrease in the specific activities of RuBisCO and enzymes linked to the photorespiratory cycle, while NAD-ME and the Krebs cycle fumarase showed increased activity. The enzyme activity of PEPC also declined.

Short-term exposure of higher plants to elevated CO_2 increases net photosynthesis, whereas long-term treatment often induces a progressive inhibition of photosynthesis (Stitt, 1991), correlated with a decrease in the quantity and activity of RuBisCO (Besford *et al.*, 1984, 1990). In the work reported here, spruce needles from trees exposed to enriched CO_2 atmospheres for extended periods showed a decreased activity of RuBisCO. The question arises as to the possible role of Pi in this decrease since an inhibition of sucrose synthesis (end-product limitation of photosynthesis), which is known to occur under elevated CO_2, could affect chloroplast metabolism by diminishing the availability of Pi (Stitt, 1991). Since Pi is required for both photophosphorylation and regulation of several enzymes of the Calvin cycle (Leegood *et al.*, 1985), the regeneration of RuBP would fall. In addition, a decrease in the ATP/ADP ratio could contribute to a decrease in RubisCO activity by affecting the RubisCO activase (Lilley and Portis, 1990). Moreover, a retention of soluble carbohydrates has been implicated in decreasing gene expression (Sheen, 1990), which could explain both the lower levels of soluble proteins found in the needles of spruce trees exposed to elevated CO_2 (Fig. 1) and the decrease in the quantity of RubisCO (Van Oosten *et al.*, 1992). It is thus suggested that the decline in both the quantity and activity of RuBisCO could depend on both Pi limitation and carbohydrate accumulation (Van Oosten, 1992b). The importance of Pi can easily be demonstrated by the results of Conroy *et al.* (1990), which showed that a lack of Pi led to a decrease in the net photosynthesis of pine species exposed to double the ambient CO_2 level for 49 weeks. However, Pi availability is believed to play an important role in the short-term but not in the longer term, as the activation state of RuBisCO was higher in plants exposed to high CO_2 (Van Oosten *et al.*, 1992b).

The increased fumarase activity observed in spruce needles exposed for two years to enriched CO_2 atmospheres (Table 1) could well indicate an increase in the rate of the respiratory process. In a recent review, Amthor (1991) described other examples where plants showed increased respiration when grown in CO_2-enriched atmospheres. The slight increase in NAD-ME activity in the needles of spruce trees exposed to high CO_2 levels could well be related to a more important contribution by the mitochondrial cyanide-resistant alternative pathway (Lance *et al.*, 1985). This process could serve to drain a carbon overflow (Lambers, 1985), which could be the case when feedback inhibition of sucrose biosynthesis occurs under high CO_2 (Amthor, 1991), as suggested above. The decreased activity of the two enzymes of the photorespiratory cycle (belonging to both the glycolate and glycerate pathways) (Table 1) could diminish the repression exerted by photo-respiration on the Krebs cycle (Douce, 1985), in good agreement with the observed increase in fumarase activity. The PEPC activity was strongly affected by high CO_2 concentrations (Table 1). This result, in addition to the lack of response of PFP, strongly suggests the use of the glycolytic pathway via phosphofructokinase (EC 2.7.1.11) and pyruvate kinase (EC 2.7.1.40). This point requires further clarification.

The effects of CO_2 enrichment on carbon metabolism were more pronounced in needles from spruce trees exposed to 480 ppmv CO_2 than 570 ppmv (Table 1). Differences in light intensity or soil composition cannot explain this difference.

Several authors (Houpis *et al.*, 1988; Reekie and Bazzaz, 1989) have already observed that trees were more strongly affected by CO_2 atmospheres close to 500 ppmv than by higher concentrations of 650-700 ppmv. The particular effect of 500 ppmv CO_2 on tree physiology remains to be established.

CONCLUSIONS

Long-term exposure of spruce trees to high CO_2 increases the activities of respiratory enzymes and decreases the activity of RubisCO, suggesting that acclimation of this coniferous species to enhanced CO_2 results in a reduced cellular carbon balance. It is also important to bear in mind that 'sink' regulation of photosynthetic metabolism usually involves a decrease in the Calvin-cycle enzymes and increases in both glycolytic and respiratory enzymes (Stitt *et al.*, 1990). Finally, the data reported in this paper support the view of Bazzaz and Fajer (1992) that '*These changes* (induced by a CO_2-rich world) *will not necessarily benefit plants*', and it is wise to suspect a problem of sink regulation of photosynthesis caused by long-term exposure of plants to high CO_2.

ACKNOWLEDGMENTS

This work was supported by a CEC doctoral grant (DGXII, STEP/Environment) to the first author. We are also grateful to the SPPS, Region Wallonne and CEC for financial support. We thank Dr R.T. Besford for his comments on this manuscript.

REFERENCES

Amthor JS (1991) Respiration in a future, higher-CO_2 world. Plant, Cell and Environment 14: 13-20

Arp WP (1991) Effects of source-sink relations on photosynthetic acclimation to elevated CO_2. Plant, Cell and Environment 14: 869-875

Bazzaz FA, Fajer ED (1992) Plant life in a CO_2-rich World. Scientific American 266: 68-74

Besford RT, Ludwig LJ, Withers AC (1990) The greenhouse effect: Acclimation of tomato plants growing in high CO_2, photosynthesis and ribulose-1,5-bisphosphate carboxylase protein. Journal of Experimental Botany 41: 925-931

Besford RT, Withers AC, Ludwig LJ (1984) Ribulose bisphosphate carboxylase and net CO_2 fixation in tomato leaves. In: Advances in photosynthetic research III. Martinus Nijhof/Dr W Junk, The Hague, pp 479-482

Cave G, Tolley LC, Strain BR (1981) Effects of carbon dioxide enrichment on chlorophyll content, starch and starch grain structure in *Trifolium subterraneum* leaves. Physiologia Plantarum 51: 171-174

Clough JM, Peet MM, Kramer PJ (1981) Effects of high atmospheric CO_2 and sink size on rates of photosynthesis of a soybean cultivar. Plant Physiology 67: 1007-1010

Conroy JP, Milham PJ, Reed ML, Barlow EW (1990) Increases in phosphorus requirements for CO_2-enriched pine species. Plant Physiology 92: 977-982

Delucia EH, Sasek TW, Strain BR (1985) Photosynthetic inhibition after long-term exposure to elevated levels of atmospheric carbon dioxide. Photosynthesis Research 7: 175-184

Douce R (1985) Mitochondria in higher plants. Academic Press, New York

Houpis JLJ, Surano KA, Cowles S, Shinn JH (1988) Chlorophyll and carotenoid concentrations in two varieties of *Pinus ponderosa* seedlings subjected to long-term elevated carbon dioxide. Tree Physiology 4: 187-193

Kramer PJ (1981) Carbon dioxide concentration, photosynthesis, and dry matter production. Bioscience 31: 29-33

Laitat E, Impens R, Van Oosten JJ (1991) Chambres de culture à ciel ouvert et simulation sur un écosystème forestier des teneurs élevée en CO_2 atmosphérique. Bulletin de Recherche Agronomique Gembloux 26: 21-37

Lambers H (1985) Respiration in intact plants and tissues: its regulation and dependence on environmental factors, metabolism and invaded organisms. In: Douce R, Day DA (eds) Higher plant cell respiration. Springer-Verlag, Berlin, pp 418-465

Lance C, Chauveau M, Dizengremel P (1985) The cyanide-resistant pathway of plant mitochondria. In: Douce R, Day DA (eds) Encyclopedia of plant physiology, Vol. 18. Springer-Verlag, Berlin, pp 202-247

Leegood RC, Walker DA, Foyer C (1985) Regulation of the Benson-Calvin cycle. In: Barber J, Baker NR (eds) Photosynthetic mechanism and the environment. Elsevier, New York, USA, pp 191-258

Lilley RC, Portis AR (1990) Activation of ribulose-1,5-bisphosphate carboxylase/oxygenase (Rubisco) by Rubisco Activase. Effects of some sugar phosphates. Plant Physiology 94: 245-250

O'Neill EG, Luxmoore RJ, Norby RJ (1987) Increases in mycorrhizal colonization and seedling growth in *Pinus echinata* and *Quercus alba* in an enriched CO_2 atmosphere. Canadian Journal of Forest Research 17: 878-883

Peet MM, Huber SC, Patterson DT (1986) Acclimation to CO_2 in cucumbers. Plant Physiology 80: 63-67

Reekie EG, Bazzaz FA (1989) Competition and patterns of resource use among seedlings of five tropical trees at ambient and elevated CO_2. Oecologia 79: 212-222

Sasek TW, Delucia EH, Strain BR (1985) Reversibility of photosynthetic inhibition in cotton after long-term exposure to elevated CO_2 concentrations. Plant Physiology 78: 619-622

Sheen J (1990) Metabolic repression of transcription in higher plants. Plant Cell 2: 1027-1038

Stitt M (1991) Rising CO_2 levels and their potential significance for carbon flow in photosynthetic cells. Plant, Cell and Environment 14: 741-762

Stitt M, von Schaewen A, Willmitzer l (1990) 'Sink' regulation of photosynthetic metabolism in transgenic tobacco plants expressing yeast invertase in their cell wall involves a decrease of the Calvin-cycle enzymes and an increase of glycolytic enzymes. Planta 183: 40-50

Tolley LC, Strain BR (1984) Effects of CO_2 enrichment on growth of *Liquidambar styraciflua* and *Pinus taeda* seedlings under different irradiance levels. Canadian Journal of Forest Research 14: 343-350

Van Oosten J-J (1992a) Apercu de la regulation du metabolisme primaire du carbone chez les vegetaux superieurs de type C3. Annales de Gembloux 98: 63-78

Van Oosten J-J (1992b) Could a higher phosphorus fertilization be required for the plants growing in a CO_2 enriched atmosphere? Proceedings 4th IMPHOS. Louvain-la Neuve, Belgium (in press)

Van Oosten J-J, Afif D, Dizengremel P (1992a) Long-term effects of a CO_2 enriched atmosphere on enzymes of the primary carbon metabolism of spruce trees. Plant Physiology and Biochemistry 30: 541-547

Van Oosten J-J, Laitat E, Dizengremel P, Gerant D (1992b) The effects of CO_2 enrichment on the biochemistry of photosynthesis and photorespiration of spruce trees cultivated in open-top chambers. In: Teller A, Mathy P, Jeffers JNR (eds) Responses of forest ecosystems to environmental changes. Elsevier, London New York, pp 655-656

Yelle S, Beeson RC Jr, Trudel M, Gosselin A (1989) Acclimation of two tomato species to high atmospheric CO_2. II. Ribulose-1,5-bisphosphate carboxylase/oxygenase and phosphoenolpyruvate carboxylase. Plant Physiology 90: 1473-1477

SECTION III

Flooding and Submergence

CLIMATIC CHANGE AND WETLAND VEGETATION IN NORTHERN EUROPE

R.M.M. Crawford
Plant Science Laboratories
Sir Harold Mitchell Building
University of St Andrews
St Andrews
Fife KY16 9AL
Scotland

ABSTRACT

Examination of a number of case histories in Northern Europe from Spitsbergen to Scotland indicates that wetland communities are undergoing marked changes in distribution and species composition as a result of climatic warming. The type of changes that take place will depend on the physiology of the species concerned. This paper examines some case histories from northern habitats and discusses the relative tolerance of plant populations and species in relation to changing water relations and length of the growing season. The potential impact of climatic change is considered in relation to the phenology, physiology and competitive ability of selected wetland species.

INTRODUCTION

Significant climatic warming became noticeable in certain parts of Northern Europe at the beginning of this century. In particular both Finland and Spitsbergen (Kanninen and Anttila, 1992; Jaworowski *et al.*, 1992) began to experience milder winters in the early 1900s. At northern latitudes, even a small amount of extra warmth is likely to have a significant biological impact (French, 1986; Maxwell and Barrie, 1989; Roots, 1989). Sufficient time has therefore elapsed for it to be possible to examine what changes can be observed at these northern latitudes in plant distribution between different sites, as well as in the ecological relationships between species within sites.

A common and almost obvious prediction that is frequently made about plant distribution in relation to climatic warming is that, in the northern hemisphere, southern plants will move north and montane and Arctic species will have their

NATO ASI Series, Vol. I 16
Interacting Stresses on Plants in a Changing Climate
Edited by M. B. Jackson, C. R. Black

range reduced (Maxwell, 1992). Future predictions of the possible extent of species migration are based on their present thermal tolerance and then matched against the temperature change predictions of climatic models. Although these matches between present and future distributions are useful in attempting to translate the abstractions of models into biological terms, they do not examine in sufficient detail either the physiological polymorphism of the species or the role of biological factors such as competition and predation. Even in the most remote regions of the High Arctic many species are represented by a number of different ecotypes occupying sites which, although they may appear to be meteorologically similar, to the plant represent major habitat differences. Projections as to the future distribution of plants that fail to take into account the diversity of response that can exist within plant populations, even in the seemingly harshest environments of the High Arctic, neglect one of the most important features of plant adaptation to changing conditions, namely the capacity of polymorphic species to change their composition. Similarly, environmental tolerance is only one aspect governing the ability of individual species to occupy any particular habitat. In most field studies it is the relative competitive ability of species to maintain their occupancy of present sites against invaders and predators that determines their presence or absence in particular habitats. This distinction between the impact of physical parameters – termed environmental tolerance – and the sum total of physical and biological parameters – termed ecological tolerance – has been extensively discussed for plant communities in many parts of the Earth (Walter, 1960).

Past palynological studies of wetland sites suggest that wetland plants are likely to be much affected by climatic change. The impact of changing climate on swamp and wetland species is likely to be either through increased warmth, reducing water supplies and extending growing seasons – as in areas where plants have hitherto been ice-encased over winter (Kane *et al.*, 1992) – or else the reverse scenario, with increased precipitation (Wigley and Barnett, 1990) giving renewed bog growth and reduced ice cover, allowing greater water movement (Erkamo, 1956). Palynological analyses of mire systems have shown that the bog surfaces of north-western Europe have been very sensitive in the past to frequent changes in climate. In northern Europe in particular, bog-profiles reveal numerous zones where the bog has distinct boundary differences between the type of peat laid down in consecutive wet or dry conditions. These boundaries, normally termed 'recurrence surfaces', have been used to demonstrate the extreme sensitivity of wetland vegetation to changes in climate (Barber, 1981).

This chapter therefore addresses three questions. First, why is northern Europe an important area for studying effects of climate change on wetland vegetation and secondly, what types of change are taking place? Finally, some consideration is given to the types of adaptation that will be needed in wetland plants for survival under changing climatic conditions at northern latitudes.

RELEVANCE OF NORTHERN EUROPE TO STUDIES OF CHANGE IN WETLAND VEGETATION

In many parts of northern Europe, the difference between annual precipitation and potential evaporation is so finely balanced that it takes only minor changes in climate or vegetation cover to cause a significant increase in soil waterlogging (Green, 1964). This feature of northern, cool oceanic climates is therefore a cause of great uncertainty as to the future prospects for wetland vegetation, should there be a significant degree of climatic warming. In more southern regions such as Ireland there is an expectation that increased temperatures will outweigh possible increases in precipitation and that marked reductions in both raised and upland blanket bogs are to be expected (Jeffrey *et al.*, 1992). In Arctic and sub-Arctic regions the position appears much more confused. In some situations there is a prediction that rates of peat decay will accelerate so that 15 years of a warmer arctic climate will be sufficient to dissipate the entire peat accumulation of the last millennium, while other studies present equally convincing arguments that increased warmth will result in greater bog growth (Oechel and Billings, 1992). These contrary predictions are based on different interpretations of the relative effects of climatic warming on the development of the active layer in bogs (the *acrotelm*) and its effect on peat accumulation and decay (Kane *et al.*, 1992). In addition to the uncertainty in relation to peat growth or decline, how the species composition of these wetlands will respond to climatic warming is as yet unexplored. The stability and composition of wetland vegetation is of particular interest, not just for conservation interests, but in a variety of other economically important ways, ranging from resistance to coastal erosion, water purification and the landuse of upland areas. There is therefore a need to go beyond the simplistic predictions based on estimating the degree of climatic shift and the automatic assumption that plants will follow in a commensurate manner. An understanding of the physiology of habitat specialisation in different ecotypes is an essential prerequisite for a more meaningful comprehension of the consequences of global warming.

CASE HISTORIES – SPITSBERGEN

Some of the most marked recent changes in climate are to be found in the very far North in Spitsbergen. Between 1910 and 1925, the five year moving average temperature for January to March at Isfjord Radio (78° 30′N) rose by 9 °C (Jaworowski *et al.*, 1992). This increase in winter temperature has been largely maintained and this century has seen marked retreats in most glaciers, less snow cover and advancement of plant seedlings onto recently uncovered moraines (Fig. 1).

The Natural Environment Research Council (UK) has initiated a special Arctic Research programme in relation to the possible deleterious consequences of climatic warming and one factor under observation is the possible changing role of aeration stress in controlling plant populations in the high Arctic. Any vegetation that has to survive prolonged ice-encasement has to endure very extensive periods of reduced oxygen availability (Andrews and Pomeroy, 1989; Gudleifsson and Larsen, 1991). The duration and severity of this stress varies depending on the depth and continuance of ice cover. This is a factor which is changing rapidly in the high Arctic and is likely to have profound effects on the composition of the vegetation and the animal communities of polar regions.

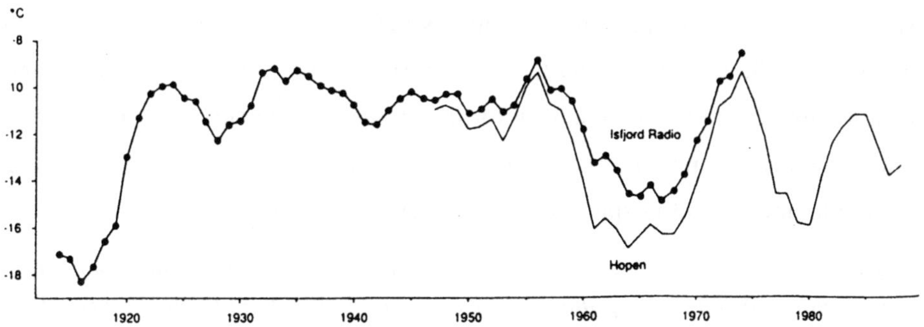

***FIGURE 1**. Five year moving average temperature for January - March for the Spitsbergen weather station, Isfjord Radio (78° 04′ N, 13° 38′ E) and Hopen (76° 04′ N, 25° 04′ E). Reproduced with permission from Jaworowski* et al. *(1992).*

As part of a comparative survey of anoxia tolerance in high arctic species, a preliminary study was carried out in the summer of 1992 at Ny Ålesund (78° 50′ N) in which intact growing plants were subjected to seven days of anoxia in darkness and then observed for three days upon return to air and light for signs of post-anoxic injury. It is during the period of re-exposure to oxygen that tissue injury becomes apparent in anoxia-sensitive species (Crawford, 1992). An unexpected observation from tests carried out *in situ* on intact plants in the middle of the growing season (July, 1992) was the high tolerance of anoxia exhibited by many of the species tested. Intact plants were placed in anaerobe jars from which all oxygen was removed by catalytic reduction with hydrogen liberated from special sachets (Gaspak – Anaerobic Systems, Becton Dickson and Co., Cockeysville, USA). The maintenance of anaerobic conditions was monitored by the inclusion of methylene blue indicator strips and the plants were kept under these conditions in darkness at 20 °C for seven days. Table 1 records those plant species that were able to retain healthy, green and turgid leaves and shoots after 72 hours of exposure to air following anaerobic incubation. The high degree of tolerance of green tissues in such a large number of species was most striking. Anoxia usually kills green leaves in 24 hours and this capacity for prolonged leaf survival is extremely rare. Hitherto, prolonged anoxia tolerance has only been observed in the perennating organs of a number of herbaceous species (Barclay and Crawford, 1982; Crawford, 1992; see also Henzi and Brändle; Summers and Jackson, this volume) and the ability for the whole plant to survive is therefore a remarkable phenomenon. Further research is in progress to investigate the level of anoxia tolerance found in species from contrasting habitats.

TABLE 1. *Tolerance of anoxia in arctic species estimated by subjecting intact green plants to seven days of anoxia in the dark and observing them for three days on return to air and light for signs of irreversible post-anoxic injury.*

ANOXIA TOLERANT	
Saxifraga oppositifolia	(both dryland and wetland forms anoxia-tolerant)
Saxifraga foliosa	
Saxifraga caespitosa	
Salix polaris	(leaves killed by anoxia but stem and buds remain alive)
Draba corymbosa	
Ranunculus sulphureus	
Cardamine nymani	
Luzula confusa	
Juncus biglumis	
Carex misandra	
Huperzia selago	
ANOXIA INTOLERANT	
Saxifraga hieracifolia	
Saxifraga cernua	
Oxyria digyna	
Polygonum viviparum	
Cochlearia officinalis	
Equisetum arvense	

Physiological ecotypes in relation to soil flooding. On morphological criteria alone a number of arctic species show very distinct sub-populations or ecotypes. In the present investigation *Saxifraga oppositifolia* has been studied in detail and it appears that there are distinct populations depending on the lateness and wetness of the habitat. Plants from both dry and wet habitats have a high level of anoxia tolerance but differ in their water use efficiency, respiration rate and photosynthetic activity. The plants from wet habitats, where snow is late to clear, have very little green tissue in relation to plant mass but compensate for this and the late start to the growing season by having very high metabolic rates (respiration and photosynthesis) at the expense of low water use efficiency. The differences between these populations are maintained when plants are kept under cultivation, which suggests that the populations are genetically distinct and that the differences are not due merely to phenotypic plasticity and acclimatisation. The plants from drier sites, which can be only a few metres away, show the converse characteristics, with lower metabolic rates and higher water use efficiency. These plants of drier land have much greater shoot development in terms of green leaves and exhibit a more upright growth form and earlier flowering. Any change in the extent of late snow cover and consequent drying out of the habitats during the brief arctic summer is likely to change the relative proportions of these different ecotypes. Examination of the low-lying coastal site where this study was carried out showed that the largest plants, and presumably the longest established, were the prostrate forms living in the wettest areas. The more tufted plants of the drier sites were much smaller and appeared to be more recent arrivals which may have moved into this habitat in recent years as a consequence of retreating ice cover.

Although morphologically distinct ecotypes between snowfield and fellfield habitats have long been recognised in species such as *Dryas octopetela* (McGraw and Antonovics, 1983), there are few studies of physiological specialisation between adjacent populations in arctic species. It is likely that physiological habitat specialisations between different populations will be found in other species and that *Saxifraga oppositifolia* is not unique in this respect. For this reason, changes in population composition rather than just species migration may be one of the major effects of global warming.

CASE HISTORIES – FINLAND

Similar trends in climatic warming to those recorded in Spitsbergen were observed in Finland in the early years of this century (Hustich, 1947). The vegetation changes that took place in Finland during the climatic warming period of the first half of this century have been documented in great detail both for wetlands and forest trees (Erkamo, 1956). Erkamo paid particular attention to aquatic plants, as is appropriate for Finland given its possession of more water as a percentage of its dry land area than any other country in the World. Erkamo's study is particularly relevant to the present discussion of aeration stress in relation to climatic change as some of the most marked changes were observed in swamp species. The main findings of Erkamo's extensive discussion of recent climatic change in relation to wetland vegetation were:

1. A drying out of wet moorland and increased growth of lichens consistent with a greater depth of the active layer (acrotelm).
2. Increased *Sphagnum* growth in wetter areas.
3. A marked general non-directional spread of some wetland species from refugia, e.g., *Spirodela polyrrhiza* (Lemnaceae), *Carex cyperoides, C. rhynchophysa, C. pseudocyperus, Typha latifolia* and *Zostera marina* (Figs. 2 and 3).

This simultaneous drying out of some moorland areas, with an increased potential for peat growth through renewed *Sphagnum* activity in other wetter localities, matches some of the predictions recently made for Arctic habitats (Kane *et al.*, 1992) where rapid destruction of some organic deposits is predicted for some sites, while renewed peat growth is expected at others.

In Finland the spread of wetland species from localised refugia did not take place in a progressively northern direction in parallel with progressive climatic warming. The species that spread were seen to move out generally into new habitats. Aquatic species reproduce mainly by vegetative means (Crawford, 1992) and increased river activity as a result of climatic warming was suggested by Erkamo as the basis for their spread, either by vegetative fragmentation or by better seed production as a result of climatic warming.

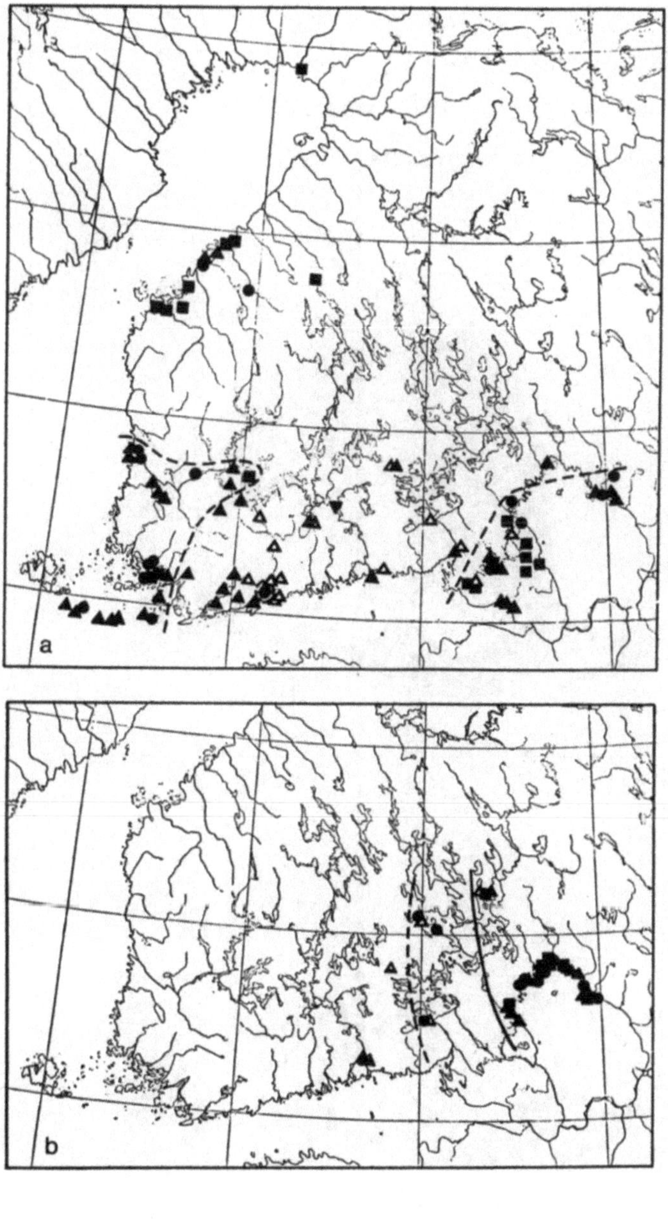

FIGURE 2. *Spread of: (a)* Spirodela polyrrhiza *(Lemnaceae); and (b)* Carex cyperoides *in Finland between 1900 and 1952.* ■ *, at or before 1900;* ● *, new records 1901-1925;* ▲ *, new records 1926-50;* ▼ *, new records 1950-52. Open symbols represent Erkamo's observations; solid symbols, other observations. Reproduced with permission from Erkamo (1956).*

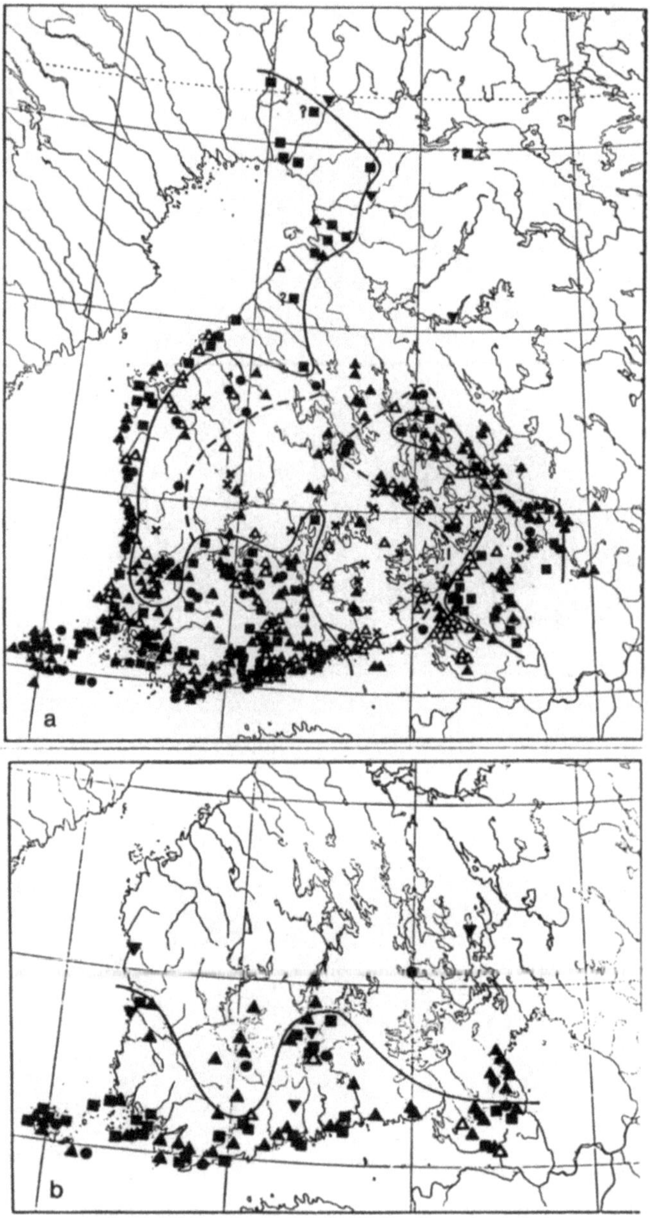

FIGURE 3. Spread of: (a) Typha latifolia; *and (b)* Carex pseudocyperus *in Finland between 1900 and 1952.* ■ *, at or before 1900;* ● *, new records 1901-1925;* ▲*, new records 1926-50;* x, *special survey carried out in 1949;* ▼ *, new records 1950-52. Reproduced with permission from Erkamo (1956).*

PHENOLOGICAL AND PHYSIOLOGICAL ADAPTATIONS TO CHANGING FLOODING CONDITIONS

The examples cited above raise the important question as to why some wetland species spread more readily than others as a consequence of increased flooding? This would appear to be an ecophysiological problem linked to competition and the reproductive biology of these species, and explanations will come only with patience from long-term observations and experiments. Twenty five years ago a marked transect was set up across a flood-prone dune slack near St Andrews, Scotland, and a computer map was made to relate distribution to plant growth and flowering (Crawford and Wishart, 1968). Shortly afterwards the drainage of this area was altered so that the transect was subjected to more frequent flooding. Re-mapping of this dune-slack complex 24 years later provided an opportunity to compare the responses of a number of dune slack communities and their dominant species to a greater flooding frequency in conjunction with generally higher water tables and increased nutrient supplies (Crawford *et al.*, 1993).

The reaction of the species to longer periods of flooding was compared with laboratory studies of the anoxia tolerance of their perennating organs. In areas where flooding was shallow the anoxia-intolerant *Glyceria maxima* expanded rapidly. This species has well developed aerenchyma and shallow rooting rhizomes and does not suffer from anoxia when flooding is shallow. Paradoxically, the anoxia-tolerant species *Filipendula ulmaria* retreated as a result of the shallow flooding that favoured *G. maxima.* Examination of the phenology of the response to flooding in *Filipendula ulmaria* showed that this species combined tolerance of anoxia with a cessation of shoot extension when flooded. Consequently, although not injurious, shallow flooding was sufficient to halt the resumption of growth in the deeper rhizomes of *F. ulmaria.* As a result of this differing phenological response to high water tables, the competitive ability of the two species was radically altered. *Filipendula ulmaria*, although tolerant of anoxia, lost ground to *G. maxima*, with a consequent reduction in species diversity. The spread of *G. maxima* appears to be due to increased access to nutrients and sufficient water to support optimal growth. However, prolonged flooding at depths of 40 cm or more appears to limit further extension of its distribution, most probably by depriving the anoxia-intolerant rhizomes of oxygen.

Two features emerged from this study that were of crucial importance in determining whether wetland species will respond positively to increased flooding, namely (1) phenology and (2) tolerance of anoxia and post-anoxic injury. The ability of plant tissues to withstand interruption of an adequate oxygen supply varies not only with the organ concerned but also with the time of year and ontogeny of the species concerned. Consequently, changes in weather patterns will have diverse effects on the relative tolerance of species to withstand reduced aeration and may alter significantly the competitive ability of certain species to withstand prolonged flooding. Attempts to predict whether or not any particular climate change will adversely affect the ability of plants to withstand low-oxygen stress therefore demands a clearer understanding of the causes of death in plant tissues as a result of oxygen deprivation. Large differences in the speed of death under anoxia suggest fundamentally different sources of cellular injury are involved. Similarly, the distinction between anoxic and post-anoxic injury has important implications for species reactions to changing flooding conditions. A more detailed discussion of these aspects of limits to plant distribution has recently been published (Crawford, 1992).

CONCLUSIONS

1. Wetland plants will spread but are unlikely to move progressively northwards with climatic warming.
2. The pattern of spread will depend on particular adaptations to wetland conditions. Shallow flooding and increased water movement in conjunction with earlier growing seasons will favour species with an early phenological development and lacking anoxia tolerance.
3. Increased flooding in low-lying areas will favour anoxia-tolerant, late developing species.

4. In the Arctic, earlier snow-melt and consequent extension of the growing season and drying out of some Arctic wetland sites will favour plant populations possessing a greater water use efficiency and less reliance on rapid growth and metabolism. Thus, the more flood-tolerant forms, which rely on anoxia tolerance and high rates of photosynthesis to compensate for ultra-short growing seasons, will be restricted to areas where snow persists late into the summer.

REFERENCES

Andrews CJ, Pomeroy MK (1989) Metabolic acclimation to hypoxia in winter cereals. Plant Physiology 91: 1063-1068

Barber KE (1981) Peat stratigraphy and climate change. AA Balkema, Rotterdam

Barclay AM, Crawford RMM (1982) Plant growth and survival under strict anaerobiosis. Journal of Experimental Botany 33: 541-549

Crawford RMM (1992) Oxygen availability as an ecological limit to plant distribution. Advances in Ecological Research 23: 93-185

Crawford RMM, Wishart D (1968) A rapid classification and ordination method and its application to vegetation mapping. Journal of Ecology 56: 385-404

Erkamo V (1956) Untersuchungen über die Pflanzenbiologischen und einige andere Folgeerscheinungen der neuzetlichen Klimaschwankung in Finnland. Annales Botanici Societas-Zoologicae Botanicae Fennicae 'Vanamo' 28: 1-283

French HA (ed) (1986) Impact of climatic change on the Canadian Arctic. Atmospheric Environment Service, Canadian Climate Centre, Downsview, Ontario

Green FHW (1964) The climate of Scotland. In: Burnett JH (ed) The vegetation of Scotland. Oliver and Boyd, Edinburgh, UK, pp 15-35

Gudleifsson G, Larsen A (ed) (1991) Ice encasement injuries in herbage plants. 4th International plant cold hardiness seminar, 1-5 July 1991. Uppsala, Sweden

Hustich I (1947) Climatic fluctuations and vegetation growth in northern Finland during 1890-1939. Nature (London) 160: 478-479

Jaworowski Z, Segalstad TV, Hisdal V (1992) Atmospheric CO_2 and global warming: a critical review. Meddelelser Norsk Polarinstitut 119: 76

Jeffrey DW, Grainger JN, Dowding P, Jones MB, Khan J (1992) The Irish green mantle and its native fauna. In: McWilliams BE (ed) Climate change: studies on the implications for Ireland. Stationary Office, Dublin, pp 69-102

Kane DL, Hinzman LD, Woo M, Everett KR (1992) Arctic hydrology and climate change. In: Chapin III FS, Jefferies RL, Reynolds JF, Shaver GR, Svoboda J, Chu EW (eds) Arctic ecosystems in a changing climate: an ecological perspective. Academic Press, San Diego, USA, pp 35-57

Kanninen M, Anttila P (ed) (1992) The Finnish research programme on climate change. Academy of Finland, Helsinki

Maxwell B (1992) Arctic climate: potential for change under global warming. In: Chapin III FS, Jefferies RL, Reynolds JF, Shaver GR, Svoboda J, Chu EW (eds) Arctic ecosystems in a changing climate: an ecological perspective. Academic Press, San Diego, USA, pp 11-34

Maxwell JB, Barrie LA (1989) Atmospheric and climatic change in the Arctic and Antarctic. Ambio 18: 42-49

McGraw JB, Antonovics J (1983) Experimental ecology of *Dryas octopetela* ecotypes. I. Ecotypic differentiation and life cycle stages of selection. Journal of Ecology 71: 879-897

Oechel WC, Billings WD (1992) Effects of global change on the carbon balance of arctic plants and ecosystems. In: Chapin III FS, Jefferies RL, Reynolds JF, Shaver GR, Svoboda J, Chu EW (eds) Arctic ecosystems in a changing climate: an ecophysiological perspective. Academic Press, San Diego, USA, pp 139-168

Roots EF (1989) Climate change: high-latitude regions. Climate Change 15: 223-253

Studer-Ehrensberger K, Crawford RMM, Studer C (1993) Flood-induced change in a dune slack observed over 24 years. Functional Ecology 7: 156-168

Walter H (1960) Standortslehre (Ökologie unde Geobotanik), Eugen Ulmer Verlag, Stuttgart, Germany

Wigley TML, Barnett TP (1990) Detection of the greenhouse effect in the observations. In: Houghton JJ, Jenkins GJ, Ephraums JJ (eds) Climate change. IPCC Scientific Assessment. Cambridge University Press, Cambridge, UK, pp 245-255

RESPONSE OF COASTAL VEGETATION TO FLOODING AND SALINITY: A CASE STUDY IN THE RAPIDLY SUBSIDING MISSISSIPPI RIVER DELTAIC PLAIN, USA

R.D. DeLaune, S.R. Pezeshki and W.H. Patrick Jr.
Wetland Biogeochemistry Institute
Louisiana State University
Baton Rouge
Louisiana 70803
USA

ABSTRACT

In rapidly subsiding coastal environments such as the Mississippi River deltaic plain, maintenance of the marsh surface in the intertidal zone is important for the survival of marsh vegetation. Coastal wetlands vertically accrete in response to increases in water level through accumulation of organic matter and mineral sediment. In brackish and salt marshes, plant productivity is related to the mineral content of the marsh soil profile. Areas which are not receiving adequate sediment supplies are experiencing marsh deterioration resulting from plant stress caused by flooding and saltwater intrusion. These factors adversely affect normal physiological functions leading to stressed plants characterised by increased anaerobic root respiration, lower carbon fixation and lower productivity. Increases in the intensity of soil reduction restricts root elongation, which reduces the effective rooting zone of wetland vegetation, thereby limiting its access to nutrient sources. Results suggest that coastal marsh vegetation requires periods of drainage or less reducing soil conditions during the growing season for root development in otherwise anaerobic soil zones. Once roots are established, anaerobic root metabolism and oxygen transport mechanisms allow plants to survive adverse soil redox conditions. Increased soil sulphide levels and salinity result in stomatal closure and reduction of photosynthetic rates in brackish and freshwater vegetation. The addition of sediment to a deteriorating marsh reduces plant stress and increases productivity.

INTRODUCTION

Louisiana's Gulf Coast is fringed by a vast expanse of wetlands recognised for their exceptional natural productivity and economic potential. Fresh, brackish and saline marshes and associated water bodies cover some 3.2 million hectares and account for 40% of all the coastal wetlands in the United States. The wetlands are

NATO ASI Series, Vol. I 16
Interacting Stresses on Plants in a Changing Climate
Edited by M. B. Jackson, C. R. Black

located on and influenced by the Mississippi River deltaic plain. The relative elevation of Louisiana coastal marshes has been changing throughout time in response to two fundamentally different groups of factors. Global factors include changes in the volume of the ocean basins resulting from tectonic processes and changes in the total amount of ocean water due to glaciation. Local factors also affect changes in marsh surface-water level relationships. The deltaic plain as a whole is also subject to rapid local and regional subsidence and wetland loss that has been attributed to basement sediment sinking, consolidation of the sediments of the Gulf Coast geosyncline, and local consolidation (Kolb and Van Lopik, 1966). Relict distributary channels and natural levee ridges form an elevated arterial array, which in turn forms hydrologically distinct basins with mature and deteriorating marshes.

Recent analysis of the tide gauges has provided estimates of relative sea-level rise at locations in Louisiana and the Gulf Coast (Gornitz and Lebedeff, 1987; Ramsey and Penland, 1989). These analyses indicate that the most representative rates of relative sea-level rise in Louisiana range between 0.32 and 1.31 cm per year. Mean sea-level rise in Louisiana is reported as being five times the Gulf of Mexico average (Ramsey and Penland *et al.*, 1989). The rapid rate of relative sea-level rise along the Louisiana Gulf Coast is attributed primarily to the compaction and subsidence of Mississippi River deltaic sediment.

MARSH SURFACE-WATER LEVEL RELATIONSHIPS

Marsh surfaces developing in Louisiana are rapidly subsiding. Sediment-deficient coastal environments are maintained in the intertidal zone through plant growth, organic detritus accumulation and limited mineral sediment deposition (DeLaune *et al.*, 1978). The depth of the organic layer is determined by the amount of real and local subsidence, eustatic sea-level change, soil organic matter oxidation and vegetative growth. As plant growth tends to keep pace with the relative rise of sea-level, both organic detritus and mineral sediments are entrapped, resulting in the gradual aggradation of the surface (Fig. 1). The two processes can be viewed as

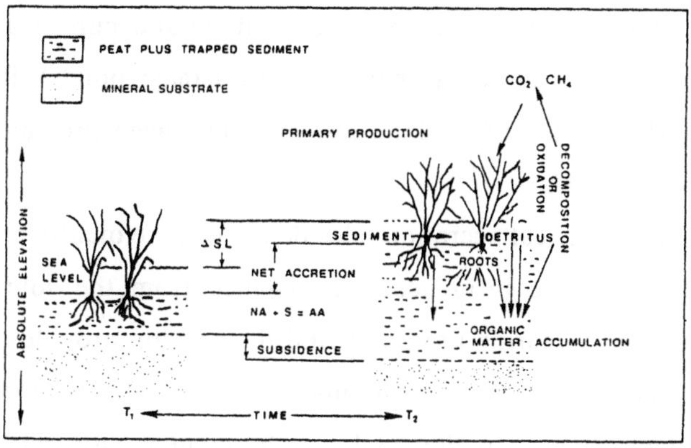

FIGURE 1. Processes governing marsh accretion (ΔSL = change in sea level, NA = net accretion, S = subsidence, AA = absolute accretion). Modified from DeLaune et al. *(1990).*

working in a synergistic manner. Wetland loss can be viewed as the inability of wetlands to maintain themselves. In subsiding environments such as coastal Louisiana, the continued existence of marsh is partially dependent on its ability to maintain its elevation within the tidal range through vertical accretion. This is accomplished through some combination of peat formation and mineral sediment accumulation. The two are interrelated, as the influx of sediments also supplies nutrients for plant growth (DeLaune *et al.*, 1990b). Increased plant growth results in more material available for peat formation, and increases in stem density result in an enhanced ability to further entrap and stabilise sediment.

Many Louisiana Gulf Coast marshes are not accreting rapidly enough or accumulating sufficient organic matter and mineral sediment to maintain marsh surfaces intertidal. From a coast-wide view, it is evident that vertical marsh accretion rates of the order of 0.6-0.8 cm per year (DeLaune *et al.*, 1992) are not sufficient to keep pace with a water level increase of 1.0 cm per year. A continual decrease in the relative elevation of the marsh with respect to the water level results in the eventual conversion of marsh to an open water body. Wetland loss is a direct and rapid response to 'aggradation deficit' in many areas of coastal Louisiana. Wetland loss has been reported to be greater than 100 km^2 per year

(Gagliano *et al.*, 1981). Approximately 75% of this total occurs as a change from emergent marsh to open water environments (bays, canals, ponds). Recent studies have reported a slower rate of wetland loss than had been projected previously (Britsch and Kemp, 1990).

Aggradation or vertical accretion rates depend on the quality and quantity of material available and are influenced by distance from the source and local hydrology. High energy marine processes and tidal action provide an abundance of reworked mineral sediments to salt marshes. The relative contributions of mineral and organic matter to the total marsh segment vary with marsh type (DeLaune *et al.*, 1990a). Mean bulk density integrated over a depth of 45 cm is greater in salt marshes than in fresh and brackish marshes. This is a consequence of the hydrological regime. The mineral sediment content of the salt marsh is directly related to its productivity. When the bulk density of the root zone falls below 0.20 g cm^{-3}, salt marshes will not support appreciable growth of *Spartina alterniflora* (DeLaune *et al.*, 1979). Brackish marsh will not support vegetation grown at bulk densities below 0.20 g cm^{-3} (Fig. 2) (Nyman and DeLaune, unpublished data). As a result of soil bulk density requirements, salt marshes require 1.7 times as much mineral matter as brackish marshes to maintain sufficient vertical accretion in response to coastal submergence (Fig. 3).

Organic matter expressed on a dry weight basis is greater in fresh and brackish marshes and constitutes an increasing fraction of soil solids as the marine influence diminishes inland from the coast. Organic matter is of greatest structural significance in low density, fresh and brackish marsh environments. However, on a unit volume basis, the organic matter is similar in fresh, brackish and salt marshes (Hatton *et al.*, 1983).

Marsh deterioration is a complex problem, seemingly the result of numerous factors with a cumulative effect. Key factors influencing the rate of formation and depth of Louisiana coastal marsh soils include changes in hydrology, local subsidence, eustatic sea-level rise, oxidation, and the compaction of surface peats. The oxidation and decomposition process in surface peats can remove several hundred grams of organic carbon annually through CO_2 and methane emission to the atmosphere (Smith *et al.*, 1983; DeLaune *et al.*, 1983).

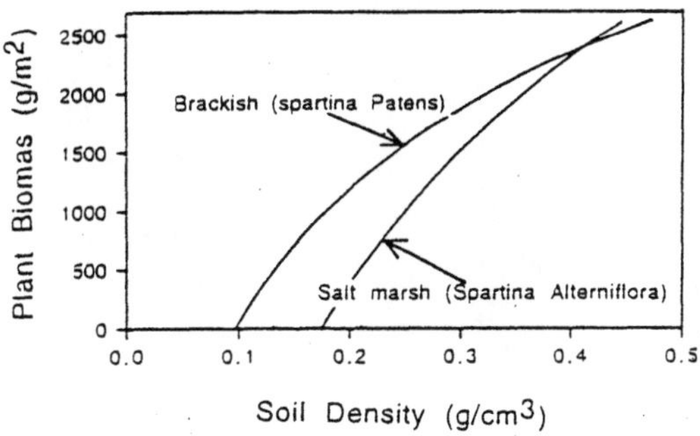

FIGURE 2. *Relationship between soil bulk density and biomass production in brackish (*S. patens*) and salt marshes (*S. alterniflora*) in Louisiana. Modified from DeLaune* et al. *(1979) including unpublished data (Nyman and DeLaune).*

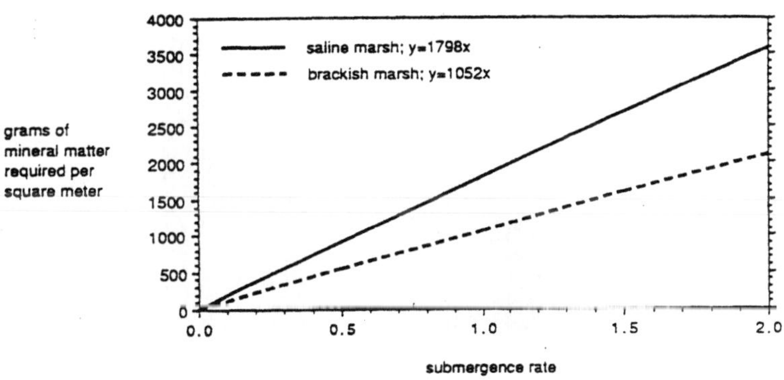

FIGURE 3. *Amount of mineral matter needed to maintain the marsh surface relative to submergence. Modified from Nyman* et al. *(1990).*

Sediment additions to deteriorating Louisiana salt marshes may enhance vegetation production (DeLaune *et al.*, 1990b). For example, raising the marsh surface by approximately 10 cm with the equivalent of 94 kg m^{-2} of sediment more than doubled above-ground plant biomass production (Fig. 4), and plant growth progressively increased in each year following the addition. The enhancement of productivity has persisted for four years after sediment addition. Similarly, the

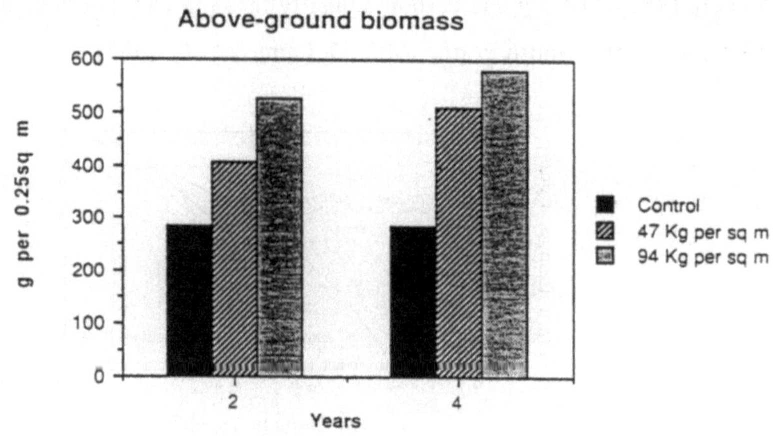

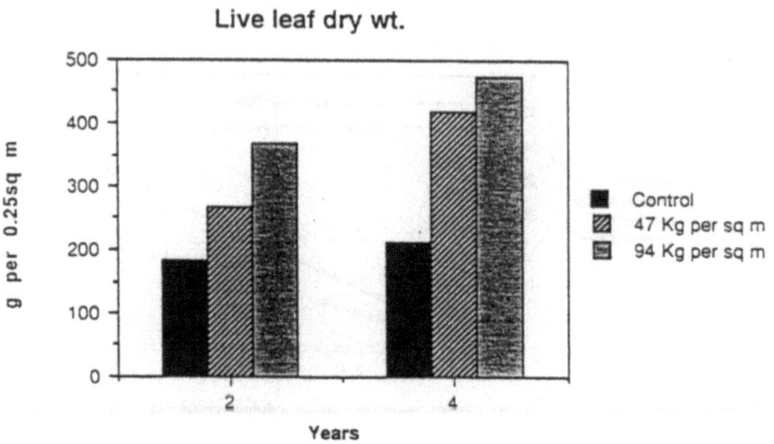

FIGURE 4. *Effect of sediment addition on above-ground biomass and live leaf dry weight four years after adding 47 and 94 kg sediment* m^{-2}. *Two year data modified from DeLaune* et al. *(1990b).*

numbers of regenerating shoots increased greatly compared to control plots. The plant tissue content of Fe, Mn, P and Ca increased substantially, and the uptake of N, Fe, Mn, P, Al, Ca, Mg and K increased in response to the added sediment. Plant growth was enhanced in response to sediment addition, as was evident from the significant increase in leaf area, above-ground biomass production and the regeneration of new culms. Leaf conductance and transpiration rates were significantly greater in sediment-treated plants than in control plants (Pezeshki *et al.*, 1992). The enhanced production of culms per unit area of marsh resulted in an increased leaf area, which allowed a greater capacity for net photosynthesis and contributed to the increased above-ground biomass of sediment-treated plots. These preliminary results suggest that if Mississippi River sediment could be reintroduced and distributed into these rapidly deteriorating salt marshes, plant regeneration and productivity would be enhanced, thus reducing the current high rate of wetland losses in the region.

INFLUENCE OF DEEP PROLONGED FLOODING ON SOIL BIOGEOCHEMICAL PROPERTIES

Deep, prolonged coastal submergence and the resultant increase in the intensity and duration of flooding result in substantial changes in marsh soil redox conditions. In anaerobic soil or sediment the usual terminal electron acceptor is oxygen, with the end product of microbial respiration being carbon dioxide and water. However, in flooded soils, as in coastal wetlands where there is little free oxygen, many soil micro-organisms use alternative compounds as terminal electron acceptors. If an energy source is available to the micro-organisms, nitrate, the higher oxides of manganese, hydrated ferric oxide, sulphate and carbon dioxide are reduced under the strictly anaerobic conditions associated with low potentials. The soil surrounding the roots becomes increasingly reduced biochemically as anaerobic micro-organisms deplete it of these oxidised compounds. Such reducing conditions are influencing wetland macrophyte productivity and distribution along the Louisiana Gulf Coast.

The most striking and easily measured change occurring in Louisiana coastal marsh soil as a result of increased soil wetness is the oxidation-reduction or redox potential (Patrick and DeLaune, 1977). Aerated soils have characteristic redox potentials within the range +400 to +700 mV, but flooded soils along the Louisiana Gulf Coast exhibit potentials as low as -250 to -300 mV. There are many uses or interpretations of redox processes in Louisiana coastal marsh soils; one example is that a knowledge of the redox potential provides an indication of the oxidation-reduction status of various soil components. For example, a redox potential of zero mV indicates that oxygen and nitrate are not likely to be present and that the bioreducible iron and manganese compounds are in a reduced state. At this same potential, however, sulphate is stable in the soil and sulphide is not being formed which is toxic to plants. A redox potential of -150 mV would indicate that sulphide should be present.

The reduction of the inorganic redox systems in Louisiana marsh soils following flooding can be described in both intensity and capacity terms. The intensity factor determines the relative ease of the reduction, whereas the capacity factor denotes the amount of the redox system undergoing reduction (for example, oxygen consumption at the root interface). The capacity factor of a redox system can be best described by the free energy of the reduction, or more commonly by the equivalent electromotive force (EMF) of the reactions. In natural systems where there is biological activity and several redox systems function, such as soils, the oxidation-reduction or redox potential is ordinarily used to denote the intensity of reduction. The capacity factor of the various redox systems can vary depending on soil conditions.

In evaluating the response of flood-tolerant Gulf Coast species to increased flooding, the substrate conditions in which the plants are grown must be properly quantified. Many laboratory and greenhouse conditions used to evaluate plant responses to flooding (for example, growing plants in nutrient solution with oxygen being removed by scrubbing with an inert gas such as N_2) may not be sufficient to quantify the physiological responses of flood-tolerant species (DeLaune *et al.*, 1990c). Many wetland plants can withstand extreme reducing conditions covering a significant portion of the redox scale found in wetland and coastal soils. Investigations of the physiological responses of wetland plants to such conditions should distinguish between plant responses to the absence of oxygen and responses to the intensity and capacity of reduction. Even under strong reducing conditions, the question arises as to whether plants are responding to stress caused by a lack of oxygen in the roots or to toxins (e.g., sulphides) produced by the intense reduction. To illustrate this point, the photosynthetic response of *Spartina alterniflora,* a dominant saltmarsh species, at various soil redox levels representing the anaerobic (no oxygen) range of the redox scale, i.e. < +350 mV is presented (Fig. 5). The results show that net photosynthesis did not respond to the decrease in Eh until soil redox potential reached -100 mV.

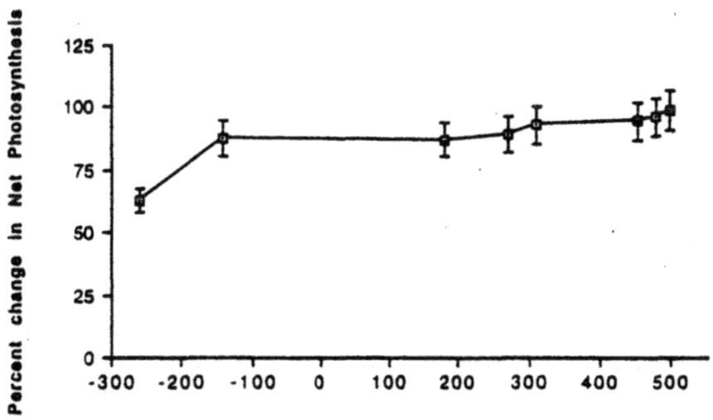

FIGURE 5*. Effect of low soil redox conditions on the photosynthetic activity of* Spartina alterniflora. *Plants were grown in soil suspension in which soil redox level was controlled using techniques described by DeLaune* et al. *(1984).*

Potentially toxic substances are produced in Louisiana marsh soils as a result of flooding and the resultant low redox conditions. Such toxins are of importance to roots that have not been killed by oxygen depletion itself. After prolonged periods of flooding, hydrogen sulphide (H_2S) is produced by the bacterial reduction of sulphates and the dissimilation of sulphur-containing amino acids (Ponnamperuma, 1972; DeLaune *et al.*, 1983). Sulphide has been implicated as a cause of reduced plant growth in Louisiana Gulf Coast salt marshes (DeLaune *et al.*, 1983).

Vegetation found growing along a salinity gradient extending inland from the Louisiana coast was evaluated for sulphide sensitivity. In *Spartina alterniflora,* net photosynthetic rate (P_n) declined significantly when H_2S concentrations exceeded

0.34 mg l^{-1} (Pezeshki *et al.*, 1988). *Panicum hemitomon* Schult and *Spartina patens* (Ait) Muhl. plants from fresh and brackish marshes have also been subjected to hydrogen sulphide under controlled sediment redox conditions (Pezeshki *et al.*, 1991b), and the net carbon assimilation responses of both species to the combined sediment anaerobiosis and hydrogen sulphide concentrations were measured. *Panicum hemitomon* was more sensitive to hydrogen sulphide than *Spartina patens* or *Spartina alterniflora* (Fig. 6). Initiation of reduction in net carbon assimilation

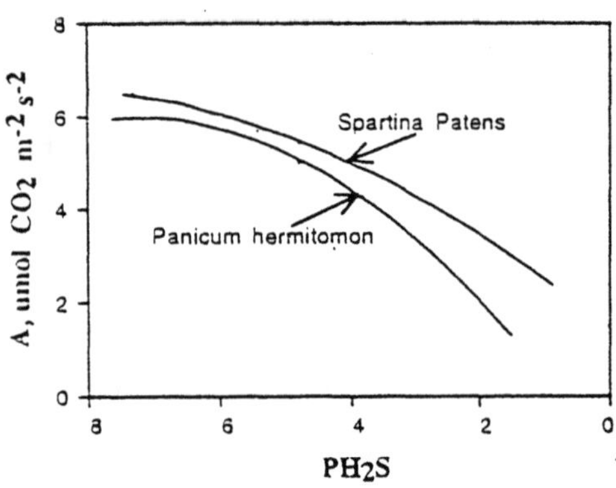

FIGURE 6*. Effect of hydrogen sulphide on carbon assimilation in* Panicum hemitomon *and* Spartina patens. *Modified from Pezeshki* et al. *(1991).*

in *P. hemitomon* began when the H_2S concentrations of the soil solution exceeded 0.22 mg l^{-1}. Reductions in net carbon assimilation in *S. patens* and *S. alterniflora* were noted at H_2S concentrations exceeding 0.34 mg l^{-1}. The reductions in net carbon assimilation observed when these wetland species are exposed to elevated H_2S concentrations suggest that extreme anaerobiosis (low Eh) and elevated sulphide could contribute to the growth reduction of these species under certain conditions. However, based on the H_2S concentrations in fresh marsh soil profile samples collected in the field, the levels were too low to cause adverse effects on *Panicum hemitomon*. In brackish and salt marsh soils containing hydrogen sulphide at concentrations of 3-4 mg l^{-1} in the soil solution, sulphide could be a major factor limiting the growth of *S. patens*. Many soils high in organic matter and with low mineral sediment contents exhibit such high sulphide levels. Mineral sediment serves as a buffer since iron present in the sediment reacts with the sulphide making it less soluble.

PHYSIOLOGICAL RESPONSES OF PLANTS TO INCREASED FLOODING

Plants growing along the Louisiana Gulf Coast use various strategies evolved by wetland plants to cope with soil anaerobiosis. These include anatomical responses (development of aerenchyma systems to allow oxygen diffusion from the aerial parts to the roots) and changes in root metabolism aimed at producing the energy for survival, including accelerated anaerobic fermentation (Mendelssohn *et al.*, 1981; Pezeshki *et al.*, 1991a). Development of aerenchyma and related internal pathways for oxygen diffusion to roots is a major adaptive mechanism in wetland plants, including species growing along the Louisiana Gulf Coast. Aerenchyma development is important because it facilitates the diffusion of oxygen to the roots, allowing some aerobic respiration, reduces the number of respiring cells in the root and helps detoxify the reduced rhizosphere (Pezeshki *et al.*, 1991a). Aerenchyma formation in some species appears to result from the development of hypoxic conditions in the roots, followed by enhanced synthesis and accumulation of ethylene.

The development of morphological and anatomical root adaptations in response to stress conditions is time-dependent. Thus, during the initial period of stress, the energy required for survival is generated through anaerobic metabolism (Jackson and Drew, 1984; Saglio *et al.*, 1988). Wetland plants along the Louisiana Gulf Coast tolerate anaerobic conditions by accelerated ethanol fermentation in the roots. It has been shown that many species, including *S. patens* (Burdick, 1987) and *S. alterniflora* (DeLaune *et al.*, 1984), rely on anaerobic metabolism as a means of surviving anaerobic root conditions. The enzyme involved in catalysing the reaction which produces ethanol, alcohol dehydrogenase (ADH), is found at substantially greater concentrations in the roots of *Spartina alterniflora* under extreme reducing conditions (Mendelssohn *et al.*, 1981; DeLaune *et al.*, 1984; Burdick, 1989).

Normal growth and functioning of roots require more oxygen than is needed for root respiration processes in *Spartina* (Mendelssohn *et al.*, 1981; DeLaune *et al.*, 1983). Under aerated conditions, oxygen diffuses into the roots from soil air spaces *via* the root epidermis. However, when roots are under flooded conditions, the oxygen required must reach the roots through internal paths from the aerial parts (Mendelssohn *et al.*, 1981; Pezeshki *et al.*, 1991a). For example, *Spartina alterniflora* has an extensive aerenchyma system extending from the substomatal cavities to the roots (Morris and Dacey, 1984). A rise and fall pattern of ADH activity in some flood-tolerant plants during extended flooding has been attributed to aerenchyma development in response to hypoxia, which causes a shift from the anaerobic pathway (as indicated by increased ADH activity) to the aerobic pathway as signified by a reduction in ADH activity (Burdick, 1987, 1989). Recent studies have examined various responses of selected populations of *Spartina patens* to changing rhizosphere redox potential (Pezeshki *et al.*, 1991). Rhizotrons were utilised to control redox potential using various gas mixtures. The results indicated that, despite high levels of flood-tolerance, anaerobiosis caused severe inhibition of root elongation in some populations. Fig. 7 illustrates root elongation responses

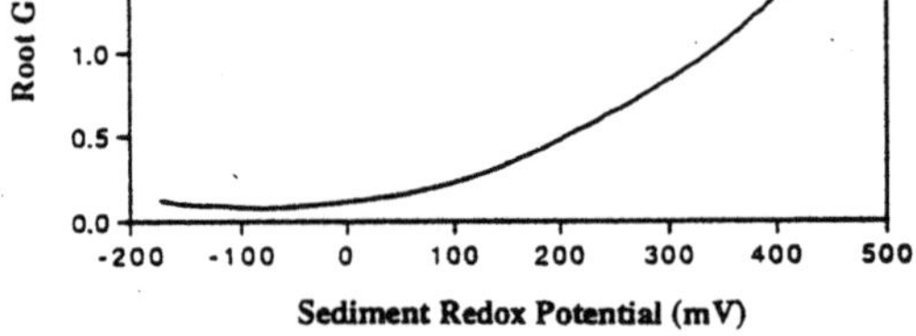

FIGURE 7. Schematic relationship between root growth of Spartina patens *and sediment redox conditions. Modified from Pezeshki and DeLaune (1990).*

to changes in soil redox conditions; low soil redox potentials resulted in dramatic reductions of root elongation, which resumed following an increase in redox potential. Such a response indicates the existence of adaptive mechanisms which allow rapid root growth in the intermittently aerobic soil conditions. Such mechanisms include extensive aerenchyma development under hypoxic conditions. The roots of plants grown in the rhizotrons under low redox conditions showed a three-fold increase in root ADH activity as compared to roots growing in oxidised soil (Pezeshki *et al.*, 1991).

The above research demonstrates that the oxygen supply mechanism in some instances does not produce sufficient oxygen for aerobic respiration in the apical and expanding root tissue under low soil redox conditions. An adequate oxygen supply is essential for root elongation in wetland plants. Due to increased flooding, soils on the Louisiana Gulf Coast have low soil redox potentials. As a result, many species have shallow rooting systems which reflect the decrease in soil oxidation reduction potential. The low soil redox conditions inhibit root growth and development, resulting in stressed plants. A certain amount of soil drainage is necessary during the growing season for roots to penetrate the soil profile once they have become established. Oxygen transport mechanisms and anaerobic root metabolism allow plants to function when the soil redox level decreases once the soil is again flooded.

Increased flooding resulting from subsidence can also affect carbon fixation by vegetation growing along the Gulf Coast. Pezeshki *et al.* (1989a) subjected *Spartina alterniflora* Lois. plants from a Louisiana salt marsh under controlled environmental conditions to fluctuating levels of soil redox potential representing a broad range of reduction normally associated with oxygen deficient environments. Carbon assimilation rates declined significantly (15-20%) when soil redox level decreased rapidly to below -200 mV. The results indicated the adverse effects of extreme anoxia on carbon assimilation by *S. alterniflora*, a possible reflection of the limited ability of this species to maintain root oxygenation during a rapid, intense reduction in soil redox potential.

RESPONSE OF COASTAL VEGETATION TO SALINITY INCREASE

Saltwater intrusion along the Louisiana coast can also impact on the survival and growth of many wetland species. Evidence suggests adverse effects of high salinity on the photosynthesis and growth of brackish and freshwater species. In a laboratory study of the response to salinity of *S. patens*, a dominant brackish marsh species, elevated salinity in the range of 22 ppt caused reductions in net carbon assimilation rates of up to 60% (Pezeshki *et al.*, 1987a).

The impact of saltwater intrusion on Louisiana freshwater marshes is likely to be greater than on brackish marshes because freshwater species are more salt-sensitive. For instance, net carbon assimilation rates were reduced by 74 and 76% when *Panicum hemiitomon*, a dominant freshwater marsh species, was exposed to saltwater containing 5 ppt salt (Pezeshki *et al.*, 1987b). Salinity of 12 ppt resulted in tissue death within four days of salt exposure. Saltwater treatment also caused a substantial decrease in net carbon assimilation in *Sagitaria lancifolia*, another important Gulf Coast freshwater marsh species. Stomatal limitation of photosynthesis, however, was relatively small, ranging between 12 and 18%. This finding indicates that, in addition to diffusional limitations, photosynthetic capacity was adversely affected through metabolic pathways (Pezeshki *et al.*, 1987d). In *Scirpus olneyi*, an important intermediate-brackish marsh species, salinities of 7 and 10 ppt caused immediate reduction in carbon fixation (Table 1). Some photosynthetic recovery was noted in treated plants, including those which survived the 10 ppt saltwater treatment. The resultant recovery indicated that some photosynthetic acclimation occurred in this species, especially at these salinity concentrations.

Tree species found near the coast also responded to salinity by stomatal closure and reduction of net photosynthesis. Coastal and bottomland tree species were more susceptible to salinity than flooding alone, suggesting a synergistic effect of combined flooding and salt on photosynthetic processes. In *Nyssa aquatica* seedlings, carbon assimilation rates declined following salt application in the absence of flooding (Pezeshki *et al.*, 1989b). In green ash (*Fraxinus pennsylvanica*), net carbon assimilation was reduced by as much as 86% in response to soil salinity (Pezeshki and Chambers, 1986). This study indicated a high level of sensitivity of green ash, a flood tolerant species, to soil salinity, but did not examine the effect of combined stresses and their possible interaction on carbon assimilation in this species. Studies on baldcypress (*Taxodium distichum*) seedlings showed 34-58% reductions in net assimilation rate when plants were exposed to saline water containing salt concentrations ranging from 2-7 ppt (Pezeshki *et al.*, 1987c). The impact was evident for all salinity levels and was closely related to the salinity of the floodwater. In tupelo-gum (*Nyssa aquanica*) seedlings subjected to floodwater containing 3 ppt salt, substantial reductions in net assimilation were observed which persisted for the entire 12 weeks of the experiment. Furthermore, the interaction of salt and flooding was significant (Pezeshki *et al.*, 1989b). Salinity caused stomatal closure as well as a reduction of carbon assimilation rates throughout the day, altering normal gas exchange patterns, and subsequently reducing the net diurnal carbon fixation of Gulf Coast tree species (Pezeshki *et al.*, 1987c; Pezeshki, 1990).

TABLE 1. *Long-term effect (29 weeks) of salinity on* Scirpus olneyi.

Salt concentration (ppt)	Net photosynthesis (mol CO_2 m^{-2} s^{-1})	Live culm dry wt. (g)
Control	11.6	17.4
7	9.5 (18)	7.8 (56)
10	5.8 (50)	1.2 (93)

Figures in brackets are the % reduction compared with the relevant control value.

The salinity responses of vegetation growing along the Louisiana Gulf Coast agree with studies reporting the effects of salt stress on coastal plants from other wetland regions (Longstreth and Strain, 1977; Drake and Gallagher, 1984; Pearcy and Austin, 1984). The responses range from a decline in gas exchange rates under low salt concentrations to complete tissue death under high salinity, especially in freshwater marsh species. The overall inhibitory effect of salinity on gas exchange is reduced carbon fixation, leading to inhibition of growth and productivity in both marsh and forest species.

CONCLUSIONS

The Louisiana Gulf Coast is experiencing rapid rates of water level increase resulting primarily from the compacting of Mississippi River deltaic sediments. Vertical marsh accretion is not keeping pace with water level increases. Leeving of the Mississippi River has resulted in an insufficient distribution of sediment supply into the coastal marshes. Correspondingly, marsh vegetation is being subjected to increases in salinity and flooding, resulting in increased plant stress. Intense soil reduction, sulphide and salinity result in decreased photosynthetic activity and plant productivity. Reductions in the productivity of wetland marsh species found growing along the Louisiana Gulf Coast will affect estuarine carbon cycling as well as organic matter pools, important governing factors for vertical marsh accretion (DeLaune *et al.*, 1978). Organic carbon accumulation is essential in maintaining intertidal marshes. Smith *et al.* (1983) reported that approximately 300 g C m^{-2} is accumulated through processes of vertical marsh accretion in brackish marshes in Louisiana. The source of this organic carbon is mainly from carbon fixed by marsh macrophytes. Reductions in primary production by flooding and salinity will adversely affect the ability of these marshes to remain intertidal, causing further marsh degradation. Similar effects on net primary productivity, carbon allocation and nutrient cycling are expected in lowland coastal forests where the effect of these stresses are beginning to appear. Results may depict what other coastal areas may experience if sea-level increases continue as predicted.

ACKNOWLEDGEMENT

This work was supported by the United States Geological Survey, grant number 14-08-001-23411.

REFERENCES

Britsch LD, Kemp III EB (1990) Land loss rates: Mississippi River Deltaic Plain. Technical report GL-90-2. US Army Engineer District, New Orleans, Louisiana, USA, pp 35

Burdick DM (1988) The relationship between root anatomic and metabolic responses to soil water-logging in coastal grass *Spartina patens*. PhD Dissertation, Louisiana State University, Baton Rouge, USA, pp 138

Burdick DM (1989) Rooting aerenchyma development in *Spartina patens* in response to flooding. American Journal of Botany 76: 777-780

Burdick DM, Mendelssohn IA (1989) Waterlogging response in dune swale and marsh populations of *Spartina patens* under field conditions. Oecologia 74: 321-329

DeLaune RD, Buresh RJ, Patrick Jr WH (1979) Relationship of soil properties to standing crop biomass of *Spartina alterniflora* in a Louisiana marsh. Estuarine Coastal Marsh Science 8: 477-487

DeLaune RD, Patrick Jr WH, Buresh RJ (1978) Sedimentation rate determined by ^{137}Cs dating in a rapid accreting salt marsh. Nature (London) 275: 532-537

DeLaune RD, Patrick Jr WH, Smith CJ (1992) Marsh aggradation and sediment distribution along rapidly submerging Louisiana Gulf Coast. Environmental and Geological Water Science 20: 57-64

DeLaune RD, Patrick Jr WH, von Breemen N (1990a) Processes governing marsh formation in a rapidly subsiding coastal environment. Cateria 17: 277-288

DeLaune RD, Pezeshki SR, Pardue JH, Whitcomb HH, Patrick Jr WH (1990b) Some influences of sediment addition to a deteriorating salt marsh in the Mississippi River Deltaic Plain: A pilot study. Journal of Coastal Research 6: 181-188

DeLaune RD, Pezeshki SR, Pardue JH (1990c) An oxidation-reduction buffer for evaluating the physiological response of plants to root oxygen stress. Environmental and Experimental Botany 30: 243-247

DeLaune RD, Smith CJ, Patrick Jr WH (1983) Relationship of marsh elevation, redox potential and sulfide to *Spartina alterniflora* productivity. Soil Science Society of America Journal 47: 930-935

DeLaune RD, Smith CJ, Tolley MD (1984) The effect of sediment redox potential on nitrogen uptake, anaerobic root respiration and growth of *Spartina alterniflora* Loisel. Aquatic Botany 18: 223-230

Drake BG, Gallagher JL (1984) Osmotic potential and turgor maintenance in *Spartina alterniflora*. Oecologia 62: 368-375

Gagliano SW, Meyer-Arendt KJ, Wicker KM (1981) Land loss in the Mississippi River Deltaic Plain. Gulf Coast Association Geological Society 31: 295-300

Gornitz V, Lebedeff S (1987) Global sea-level changes during the past century. In: Nummedal D, Pilkey W, Howard JD (eds) Sea level fluctuation and coastal erosion: SEPM Special Publication 4: 3-16

Hatton RS, DeLaune RD, Patrick Jr WH (1983) Sedimentation, accretion and subsidence in marshes of Barataria Basin, Louisiana. Limnology and Oceanography 28: 494-502

Jackson MB, Drew MC (1984) Effects of flooding on growth and metabolism of herbaceous plants. In: Kozlowski TT (ed) Flooding and plant growth. Academic Press Inc. New York, pp 47-128

Kolb CR, Van Lopik JW (1966) Depositional environments of the Mississippi River Deltaic plain, southeastern Louisiana. In: Deltas and their geologic framework. Houston Geological Society, pp 17-61

Longstreth DJ, Strain BR (1977) Effects of salinity and illumination on photosynthesis of *Spartina alterniflora*. Oecologia 31: 191-199

Mendelssohn IA, McKee KL, Patrick Jr WH (1981) Oxygen deficiency in *Spartina alterniflora* roots: Metabolic adaptation to anoxia. Science 214: 439-441

Morris JT, Dancey JWH (1984) Effects of O_2 on ammonium uptake and root respiration by *Spartina alterniflora*. American Journal of Botany 71: 979-985

Nyman JA, DeLaune RD, Patrick Jr WH (1990) Wetland soil formation in the rapidly subsiding Mississippi River deltaic plain: mineral organic matter relationships. Estuarine Coastal Shelf Science 31: 57-59

Patrick Jr WH, DeLaune RD (1977) Chemical and biological redox systems affecting nutrient availability in the coastal wetlands. Geoscience and Man 18: 131-137

Pearcy RW, Austin SL (1984) Effects of salinity on growth and photosynthesis of three California tidal marsh species. Oecologia 62: 68-73

Pezeshki SR (1990) A comparative study of response of *Taxodium distichum* and *Nyssa aquatica* to soil anaerobiosis and salinity. Forest Ecology and Management 33/34: 531-541

Pezeshki SR, Chambers JL (1986) Effect of soil salinity on stomatal conductance and photosynthesis of green ash (*Fraxinus pennsylvanica*) seedlings. Canadian Journal of Forest Research 16: 569-573

Pezeshki SR, DeLaune RD (1990) Influence of sediment oxidation-reduction potential on root elongation in *Spartina patens*. Acta Oecologia 11: 377-383

Pezeshki SR, DeLaune RD, Pan SZ (1991b) Relationship of soil hydrogen sulfide level to net carbon assimilation of *Panicum hemitomon* and *Spartina patens*. Vegetatio 95: 159-166

Pezeshki SR, DeLaune RD, Pardue JH (1992) Sediment addition enhances transpiration and growth of *Spartina alterniflora* in deteriorating Louisiana Gulf Coast salt marshes. Wetlands Ecology and Management 1: 185-189

Pezeshki SR, DeLaune RD, Patrick Jr WH (1987a) Response of *Spartina patens* to increasing levels of salinity in rapidly subsiding marshes of the Mississippi River Deltaic Plain. Estuarine Coast Shelf Science 24: 389-399

Pezeshki SR, DeLaune RD, Patrick Jr WH (1987b) Response of the freshwater marsh species, *Panicum hemitomon*, to increased salinity. Freshwater Biology 17: 195-200

Pezeshki SR, DeLaune RD, Patrick Jr WH (1987c) Physiological responses of baldcypress (*Taxodium distichum* L.) to increases in flooding salinity in Mississippi River Deltaic Plain. Wetlands 7: 1-10

Pezeshki SR, DeLaune RD, Patrick Jr WH (1987d) Effect of soil anaerobiosis and salinity on photosynthesis of *Sagittaria lancifolia*. Marine Ecology Progress Series 41: 87-91

Pezeshki SR, DeLaune RD, Patrick Jr WH (1989a) Effect of fluctuating rhizosphere redox potential on carbon assimilation of *Spartina alterniflora*. Oecologia 80: 132-135

Pezeshki SR, Matthews SW, DeLaune RD (1991a) Root cortex structure and metabolic response of *Spartina patens* to soil redox conditions. Environmental and Experimental Botany 31: 291-297

Pezeshki SR, Pan SZ, DeLaune RD, Patrick Jr WH (1988) Sulfide induced toxicity: inhibition of carbon assimilation in *Spartina alterniflora*. Photosynthetica 22: 437-442

Pezeshki SR, Patrick Jr WH, DeLaune RD, Moser EB (1989b) Effects of waterlogging and salinity interaction on *Nyssa aquatica* seedlings. Forest Ecology and Management 27: 41-51

Ponnamperuma FN (1972) The chemistry of submerged soil. Advances in Agronomy 24: 29-96

Ramsey KE, Penland S (1989) Sea-level rise and subsidence in Louisiana and the Gulf of Mexico. Gulf Coast Association of Geological Associates 39: 491-500

Saglio PH, Drew MC, Pradet A (1988) Metabolic acclimation to anoxia induced by low (2-4 kPa partial pressure) oxygen pretreatment (hypoxia) in root tips of *Zea mays*. Plant Physiology 86: 61-66

Smith CJ, DeLaune RD, Patrick Jr WH (1983) Carbon dioxide emission and carbon accumulation in coastal wetlands. Estuarine Coastal Shelf Science 17: 21-29

FLOOD TOLERANT AND FLOOD SENSITIVE PLANTS UNDER PRIMARY AND SECONDARY ANOXIA

B.B. Vartapetian
Institute of Plant Physiology
Russian Academy of Sciences
Moscow 127276
Russian Federation

ABSTRACT

The elucidation of the mechanisms of plant adaptation to anaerobic environments and the causes of plant injury and death in oxygen-free conditions are two fundamental problems which have been studied extensively during the last 20 years. The high sensitivity of root cell ultrastructure to oxygen deficiency and facilitated oxygen transport from shoots to roots of flood-tolerant plants indicate that adaptation of plants to oxygen shortage in the rooting medium is achieved by the avoidance of anaerobiosis in root cells. The second main strategy is metabolic adaptation. This is realised at the molecular level in the total absence of oxygen. Both strategies are discussed using results obtained mostly from experiments with rice. In addition, the responses of flood-sensitive plants such as wheat, pea and maize under primary and secondary anoxia are considered.

INTRODUCTION

With respect to their capacity to inhabit flooded, anaerobic soils, higher plants can be divided in two contrasting ecological types: (i) tolerant species that can survive for many days or weeks in the absence of oxygen and even maintain a state of active metabolism. As a consequence, wetland species flourish in anaerobic soils; (ii) sensitive species that are damaged and killed even by short periods without oxygen. The growth of these plants on flooded soils is thus largely precluded. This applies both to germinating seeds (primary anoxia) and plants that experience anoxia after preliminary growth in normal aerobic conditions (secondary anoxia).

Taking into account the importance of the problem from the botanical point of view, and in relation to its impact on natural and agricultural vegetation and the prospect of global climate change, it is not surprising that during the last two decades scientists in several countries have been exploring the mechanisms of plant

NATO ASI Series, Vol. I 16
Interacting Stresses on Plants in a Changing Climate
Edited by M. B. Jackson, C. R. Black

adaptation and injury in anaerobic environments. Our laboratory has been involved in such studies using electron microscopy to investigate plant cell ultrastructure, in particular of mitochondria, and carrying out related functional studies of this organelle. The focus on the mitochondrion is explained by the fact that it is the principal source of energy for aerobic organisms and damage to this organelle during periods of oxygen shortage inevitably results in a dramatic loss of energy provision. In addition, the study of mitochondria under anoxia is interesting because, in oxygen-free media, their ultrastructure undergoes specific re-arrangements which have diagnostic significance. Results obtained in this field are reviewed below.

PLANT ADAPTATION TO OXYGEN-FREE ENVIRONMENTS

Avoidance Strategy. The capacity of flood-tolerant, wetland species to inhabit anaerobic soils gave grounds for some authors to suggest that the roots of wetland species are more resistant to anoxia than those of dryland species (Soldatenkov and Chirkova, 1963; McManmon and Crawford, 1971; Crawford, 1980). In particular, it was demonstrated that detached roots of dryland plants (french bean, tomato, potato) are killed by 24-36 hours of anoxic treatment, while under the same anoxic conditions the detached roots of flood-tolerant wetland plants (*Scirpus silvaticus* and *Glyceria aquatica*) could survive six days of anoxic incubation (Soldatenkov and Chirkova, 1963). It was suggested that the roots of flood-tolerant plants are biochemically adapted to oxygen-free environments due to the unusual nature of their energy metabolism (Crawford, 1980).

In contrast, the results of electron microscopic investigations revealed, surprisingly, that the cellular ultrastructure of roots of rice (*Oryza sativa*) and some natural wetland species (*Alisma plantago-aquatica, Lycopus europaeus, Glyceria fluitans, Carex leporina*) is super-sensitive rather than super-resistant to anoxia (Vartapetian, 1973; Vartapetian 1982; Vartapetian and Andreeva, 1986). For instance, it appeared that, while the mitochondria from roots of flooding-sensitive pumpkin (*Cucurbita pepo*) were destroyed by 12-24 hours of anoxic incubation, the adventitious roots of wetland rice suffered irreversible damage after only 4-6 hours of anoxia (Vartapetian, 1973). This unexpected hypersensitivity of flood-tolerant roots to oxygen deficiency was confirmed using another criterion, the capacity of roots to regrow after anaerobic incubation (Webb and Armstrong, 1983). It was shown that the roots of rice are more vulnerable under anoxia than those of pea and pumpkin. Using a biochemical criterion for root resistance to anoxia, that of the capacity for protein synthesis, it was demonstrated that the roots of flood-tolerant wetland *Glyceria* are more sensitive to oxygen stress than those of dryland, flood-intolerant *Pisum* (ApRees and Wilson, 1984; ApRees *et al.*, 1987).

Thus, the paradoxical results first obtained from electron microscopic investigations and then confirmed in physiological and biochemical studies give solid support for the conclusion that the success of wetland species cannot depend upon the metabolic characteristics of the roots, since they are seen to be more, rather than less sensitive to oxygen deprivation when compared with dryland plants. The more likely explanation is to be found at the whole plant level in the process of oxygen translocation from shoots to roots, which gives rise to the avoidance of root anaerobiosis. Large gas spaces (aerenchyma) facilitate oxygen translocation in the roots of flood-tolerant plants. The formation of aerenchyma in roots, resulting from the degradation of much internal root tissue, also aids root survival by significantly decreasing the number of respiring cells and thereby decreasing the demand for oxygen and ensuring more economical use of oxygen transported from shoots. Aerenchyma may also be important for removing, by upward transport, CO_2 and other potentially harmful volatile compounds accumulated in reduced anaerobic soils and roots. In this context the role of wetland rice fields and rice aerenchyma in the emission into the atmosphere of the greenhouse gas, methane, should also be mentioned. Global annual emission of methane from rice fields is about 25-60 million tonnes, which are transported upwards from the roots mainly through aerenchyma (Neue *et al.*, 1990). In addition to the internal, downward diffusive transport of molecular oxygen through aerenchyma, Raskin and Kende (1985) demonstrated that, in the case of deepwater rice, oxygen is also transported externally due to mass flow along a thin gaseous layer localised between hydrophobic leaf surfaces and the surrounding water.

Metabolic adaptation. The second main strategy of plant adaptation to oxygen-free environments is metabolic adaptation by the shoot tissues of some plant species. Anaerobically germinating rice seed is an excellent example. Rice seeds can germinate under conditions of total absence of oxygen and will germinate even in a vacuum (Costes and Vartapetian, 1978). It appears that the cells of anaerobically growing rice coleoptiles (but no other organ of the seedling) contain all the usual subcellular structures, including structurally intact mitochondria. On establishing contact with oxygen they are able to oxidise Krebs cycle substrates and sustain electron transport that is coupled with the phosphorylation of ADP to ATP (Vartapetian, 1978). Similar observations have been made with anaerobically germinating seeds of *Echinochloa crus-galli* (Kennedy *et al.*, 1983).

The adenylate energy charge of anaerobically growing rice coleoptiles can reach 0.8 (Pradet and Prat, 1976). This rather high value is close to the energy charge of coleoptiles growing in the presence of oxygen (0.9). As a result of such an active anaerobic energy metabolism in rice coleoptiles, the synthesis of certain proteins (Maslova *et al.*, 1975; Mocquot *et al.*, 1981; Atwell and ApRees, 1986), nucleic acids (Mocquot *et al.*, 1977; Aspart *et al.*, 1983) and lipids (Vartapetian *et al.*, 1978) takes place. Thus, we have to consider germinating rice seeds to be neither obligatory anaerobes, as suggested by Aleshin *et al.* (1971), nor obligatory aerobes, as proposed by Neunylov (1979). They are facultative anaerobes, a rare phenomenon among higher plants.

AEROBICALLY GROWN RICE SEEDLINGS UNDER TOTAL ANOXIA

To find whether super-resistance to anoxia by rice is also a feature of aerobically grown seedlings, the latter were transferred to anoxic medium after 5-6 days of aerobic germination. After three or even five days of anoxia, no destructive changes were observed in the ultrastructure of mitochondria in the coleoptiles and leaves. The mitochondria maintained a dense matrix and rows of parallel cristae

(Vartapetian *et al.*, 1974b, c). The roots also possessed intact mitochondria with rows of parallel cristae after 48 hours of anoxia. However, 72 hours without oxygen brought about degradation of mitochondrial ultrastructure. Nevertheless, it should be noted that the mitochondrial fine structure of the seminal roots of seedlings appeared to be much more resistant to anoxia than that of the adventitious roots of established plants (Vartapetian, 1973).

However, the high resistance of the shoots of intact seedlings to anoxia was lost when seedling organs were detached (Vartapetian *et al.*, 1976), since marked degradation of mitochondria was observed after 24-48 hours in excised coleoptiles and leaves. Feeding detached organs with glucose substantially increased their resistance to anoxia to a level similar to that of intact seedling organs. These results were interpreted as evidence that the cause of injury to detached organs under anoxia is carbon deficiency rather than oxygen starvation *per se*. Organ excision presumably interrupts assimilate transport from the grain (Vartapetian *et al.*, 1976, 1977). Thus, it is concluded that intact, germinating rice seedlings are highly resistant to anoxia, with the coleoptile and leaves showing the most resistance, but with the roots also possessing greater resistance than the adventitious roots of mature plants.

FLOOD-SENSITIVE PLANTS UNDER SECONDARY ANOXIA

Intact wheat, pumpkin and maize seedlings and their detached organs revealed much greater sensitivity to anoxia than rice. After only 90 minutes of anoxia, four day old intact wheat (*Triticum aestivum*) seedlings exhibited destructive changes in mitochondrial ultrastructure in root cells. However, these changes were reversible because restoration of mitochondrial fine structure occurred during post-anoxic aeration. Surprisingly, an additional two to three hours of continuous anaerobic incubation also resulted in some restoration of ultrastructure (Vartapetian and Zakhmilova, 1990). After 4-5 hours of anoxic incubation, a further wave of destructive changes was registered and these were irreversible after six hours of anoxia. The ultrastructure of the leaves appeared to be more resistant to anoxia, although early destructive but reversible changes of mitochondria were observed after only 30 minutes of anoxia. Unlike the roots, destructive changes of leaf mitochondria were reversible after 9, 17 and 24 hours of anoxic incubation (Vartapetian and Zakhmilova, 1990). Full, irreversible injury of mitochondria in leaves took place only after 48 hours of anaerobiosis.

As already mentioned, elucidation of the cause of plant cell injury under anoxia is one of the fundamental problems in the study of plant anaerobiosis. It has been suggested that one cause of anoxic injury is the accumulation of toxic end-products of fermentation such as ethanol (McManmon and Crawford, 1971; Chirkova *et al.*, 1974; Crawford, 1980). However, this suggestion has not enjoyed solid experimental confirmation. On the contrary, fermentative ethanol production may play a positive role in the survival of anoxia, as has been demonstrated in the following experiments. The ultrastructure of mitochondria from detached pumpkin roots was irreversibly damaged after 12-24 hours of anoxic incubation. However, mitochondrial damage could be prevented for 48 or even 72 hours by submerging the roots in 1 to 3% glucose, which maintained or even accelerated fermentation (Vartapetian *et al.*, 1977). Similar results were obtained with detached rice seedling organs (Vartapetian *et al.*, 1976) and with pea, pumpkin and rice roots (Webb and Armstrong, 1983) supplied with exogenous glucose. In addition, it was shown that the concentration of exogenous ethanol needed to damage the plants was one or two orders of magnitude higher than that found in plant tissues under natural or experimental anaerobiosis (Jackson *et al.*, 1982; Crawford and Vartapetian, 1984). Thus, the results obtained demonstrate a beneficial rather than a negative role of fermentation for plant survival under anaerobic conditions.

The above conclusion made on the basis of electron microscopical and physiological experiments was directly confirmed at the molecular level in biochemical studies by Saglio *et al.* (1980). They showed that a drastic decrease in energy charge from 0.9 to about 0.2 in anoxic maize root tips could be restored to 0.6 by glucose feeding, with the acceleration of fermentation being associated with the accumulation of ethanol and lactate. The role of energy metabolism in plant survival under anoxia was demonstrated in recent experiments with detached pea roots by Andreev *et al.* (1991). It was shown that irreversible damage to the fine structure of mitochondria and oxidative phosphorylation under anoxia coincided with a marked inhibition of fermentation rate, as measured by ethanol, lactate and alanine accumulation.

Under conditions of soil flooding, the oxygen supply situation can be complicated by gas transport from the shoots. During the past 20 years this problem has been extensively investigated in several laboratories (Armstrong, 1979, 1980; Armstrong *et al.*, 1991). Summarising the results obtained in our laboratory using polarographic (Vartapetian and Nuritdinov, 1976), chemiluminescence (Vartapetian *et al.*, 1974a) and electron microscopy approaches (Vartapetian *et al.*, 1980), one can state that, in contrast to flood-tolerant plants, oxygen transport in flood-sensitive plants such as pumpkin and cotton does not support normal aerobic metabolism in the roots. This is because the quantity of oxygen transported from the shoots at room temperature amounts to no more than a few percent of the root's requirements for normal aerobic respiration. The poor oxygen supply reflects the limited amount of aerenchyma present in these plants. Nevertheless, at lower soil temperatures, the physiological role of oxygen transported from the shoots can be more significant for root metabolism in these plants (Nuritdinov and Vartapetian, 1981), because the amounts required for respiration are much smaller.

The significance of oxygen transported from the shoots of some other mesophytes (e.g., maize, sunflower, wheat, tomato) can increase markedly in response to oxygen shortage. In these plants, the original roots are killed by soil flooding, but replacement adventitious roots typically contain large amounts of aerenchyma. Thus, oxygen transport from shoots is facilitated and survival of the roots is enhanced (Konings, 1982; Drew *et al.*, 1985).

FLOOD-SENSITIVE PLANTS UNDER PRIMARY ANOXIA

In order to complete the comparison of tolerance to anoxia by representatives of the two contrasting ecological types, we now consider the resistance of germinating seeds of flood-sensitive plants, such as wheat, pea and maize, to primary anoxia. The seeds of these plants, in contrast to those of rice, do not germinate without oxygen and germination fails even in water that is only partially depleted of oxygen. Nevertheless, it was surprising to find that the formation of mitochondria can take place during 48 or 72 hours of anoxic imbibition (Vartapetian, 1991) in the embryos of flood-sensitive pea, wheat and maize. Parallel with the anaerobic assembly of mitochondrial membranes, an acceleration of embryo respiration occurred when anaerobically imbibed seeds were transferred to aerobic conditions. Marked incrases in ATP indicated that the oxidative phosphorylation machinery of mitochodria formed in anoxic cells is functional.

Prolongation imbibition of wheat and pea seeds under anoxia brought about a gradual loss of mitochondrial capacity for oxidative phosphorylation. In the case of pea seeds, the embryos were unable to perform oxidative phosphorylation after 72 hours of anaerobic imbibition. The formation of mitochondrial ultrastructure in the embryos of anaerobically imbibed seeds could not be blocked by inhibiting protein synthesis using cycloheximide or chloramphenicol (Vartapetian *et al.*, 1987). However, seed imbibition was not accompanied by the formation of normal mitochondrial ultrastructure if the seeds had been killed by prior exposure to high temperature (100 °C) or to chloroform or diethyl ether vapour (Vartapetian, 1991). Based on these results, it is concluded that anoxic formation of mitochondrial ultrastructure in seed embryos occurs by membrane assemblage from pre-existing protein lipids using glycolytic ATP.

CONCLUSIONS

During the last twenty years our understanding of how flood-tolerant and flood-sensitive plants adapt or are injured by anaerobic conditions has increased considerably. It is evident that the avoidance of anaerobiosis by translocation of oxygen from shoots to roots and metabolic adaptation involving substantial shifts in cell biochemistry are two principal strategies of plant adaptation to oxygen-free environments. Both mechanisms are much less apparent in flood-sensitive plants. There are grounds for believing that energy imbalances within plant cells, accompanied by considerable perturbations to protein and lipid metabolism, are the principal mechanisms causing plant injury and death under anoxia in flooding-sensitive species. Nevertheless, in spite of the marked progress to date, elucidation of whole plant and molecular mechanisms of plant adaptation and damage in anaerobic environments remains an outstanding problem for plant ecological physiology and biochemistry in the future.

REFERENCES

Aleshin EP, Yakovlev BV, Gulak VJ, Kolesnikov GP (1971) The pathways of energy transfer in rice coleoptiles (in Russian). Doklady VASKHNIL No 4: 23-25

Andreev VY, Generozova IP, Vartapetian BB (1991) Energy status and mitochondrial ultrastructure of excised pea root at anoxia and post-anoxia. Plant Physiology and Biochemistry 29: 171-176

ApRees T, Jenkin LET, Smith AM, Wilson PM (1987) The metabolism of flood tolerant plants. In: Crawford RMM (ed) Plant life in aquatic and amphibious habitats. Blackwell Scientific Publications, Oxford, pp 227-238

ApRees T, Wilson P (1984) Effects of reduced supply of oxygen on the metabolism of roots of *Glyceria maxima* and *Pisum sativum.* Zeitschrift Pflanzenphysiologie 114: 493-503

Armstrong W (1979) Aeration in higher plants. In: Woolhouse CO (ed) Advances in botanical research. Academic Press, London, pp 226-332

Armstrong W (1980) Root aeration in the wetland condition. In: Hook DD, Crawford RMM (eds) Plant life in anaerobic environments, 2nd printing. Ann Arbor Science, Michigan, pp 269-297

Armstrong W, Beckett PM, Justin SIIFW, Lythe S (1991) Modelling, and other aspects of root aeration by diffusion. In: Jackson MB, Davies DD, Lambers H (eds) Plant life under oxygen deprivation. SPB Academic Publishing, The Hague, pp 267-282

Aspart L, Got A, Delseny M, Mocquot B, Pradet A (1983) Adaptation of ribonucleic acid metabolism to anoxia in rice embryos. Plant Physiology 72: 115-121

Atwell BJ, ApRees T (1986) Distribution of protein synthesized by seedlings of *Oryza sativa* grown in anoxia. Journal of Plant Physiology 123: 401-408

Chirkova TV, Khazova IV, Astafurova TP (1974) On the metabolic regulation of plant adaptation to transient anaerobiosis. Soviet Plant Physiology 21: 102-107

Costes C, Vartapetian BB (1978) Plants grown in vacuum: the ultrastructure and functions of mitochondria. Plant Science Letters 11: 115-119

Crawford RMM (1980) Metabolic adaptations to anoxia. In: Hook DD, Crawford RMM (eds) Plant life in anaerobic environments. 2nd printing. Ann Arbor Science, Michigan, pp 119-136

Crawford RMM, Vartapetian BB (1984) Effect of ethanol and anoxia on ultrastructure of plant mitochondria (in Russian). Doklady Akademii nauk SSSR 275: 1279-1280

Drew MC, Jackson MB, Giffard SC, Campbell R (1981) Inhibition by silver ions of gas space (aerenchyma) formation in adventitious roots of ***Zea mays*** L. subjected to exogenous ethylene or to oxygen deficiency. Planta 153: 217-224

Jackson MB, Fenning TM, Drew MC, Saker LR (1985) Stimulation of ethylene production and gas-space (aerenchyma) formation in adventitious roots of *Zea mays* L. by small partial pressures of oxygen. Planta 165: 486-492

Jackson MB, Herman B, Goodenough A (1982) An examination of the importance of ethanol in causing injury to flooded plants. Plant, Cell and Environment 5: 163-172

Kennedy RA, Rumpho ME, VanderZee DV (1983) Germination of *Echinocloa crus-galli* (L.) (barnyard grass) seeds under anaerobic conditions. Plant Physiology 72: 787-794

Konings H (1982) Ethylene-promoted formation of aerenchyma in seedling roots of *Zea mays* L. under aerated and non-aerated conditions. Physiologia Plantarum 54: 119-124

Maslova IP, Tchernyadeva IF, Vartapetian BB (1975) Soluble proteins and alcohol dehydrogenase of rice seedlings in anoxia. In: Twelfth International Botanical Congress Abstracts Vol. 2. Nauka, Leningrad, p 365

McManmon M, Crawford RMM (1971) A metabolic theory of flooding tolerance: The significance of enzyme distribution and behaviour. The New Phytologist 70: 299-306

Mocquot B, Pradet A, Litvak S (1977) DNA synthesis and anoxia in rice coleoptiles. Plant Science Letters 9: 365-371

Mocquot B, Prat Ch, Mouches C, Pradet A (1981) Effect of anoxia on energy charge and protein synthesis in rice embryos. Journal of Plant Physiology 68: 636-640

Neue HU, Becker-Heidmann P, Seharpenseel HW (1990) Organic matter dynamics, soil properties, and cultural practices in rice lands and their relationship to methane production. In: Bouwman AF (ed) Soils and the greenhouse effect, John Wiley and Sons, Chichester, pp 457-466

Neunylov BA (1979) Biological basis for rice production in the Far East (in Russian). Journal of General Biology 40: 485-496

Nuritdinov N, Vartapetian BB (1981) A quantitative assay of O_2 transport in cotton plants at different temperatures. Physiologie Vègetale 19: 211-217

Pradet A, Prat C (1976) Metabolisme energetique au cours de la germination du Riz en anoxie. In: Jacques R (ed) Etudes de biologie vegetale CNRS, Paris, pp 561-574

Raskin I, Kende H (1985) Mechanism of aeration in rice. Science 228: 327-329

Saglio PH, Raymond P, Pradet A (1980) Metabolic activity and energy charge of excised maize root tips under anoxia. Plant Physiology 66: 1053-1057

Soldatenkov SV, Chirkova TV (1963) The role of leaves in root respiration under anaerobic conditions (in Russian). Soviet Plant Physiology 10: 452-458

Vartapetian BB (1973) Aeration of roots in relation to molecular oxygen transport in plants. In: Slater RO (ed) Proceedings of the Uppsala Symposium 1970, UNESCO, Paris, pp 259-265

Vartapetian BB (1978) Structure and function of mitochondria from rice coleoptiles grown under strictly anaerobic conditions. In: Ducet G, Lance C (eds) Plant mitochondria. Elsevier, Amsterdam, pp 411-418

Vartapetian BB (1982) Pasteur effect visualization by electron microscopy. Naturwissenschaften 69: 99

Vartapetian BB (1991) Flood sensitive plants under primary and secondary anoxia: ultrastructural and metabolic responses. In: Jackson MB, Davies DD, Lambers H (eds) Plant life under oxygen deprivation, SPB Academic Publishing, The Hague, pp 201-216

Vartapetian BB, Agapova LP, Averianov AA, Veselovsky VA (1974a) New approach to study of oxygen transport in plants using a chemiluminescent method. Nature (London) 249: 269

Vartapetian BB, Andreeva IN (1986) Mitochondrial ultrastructure of three hygrophyte species at anoxia and in anoxic glucose-supplemented medium. Journal of Experimental Botany 37: 685-692

Vartapetian BB, Andreeva IN, Kursanov AL (1974b) Appearance of unusual mitochondria in rice coleoptiles at conditions of secondary anoxia. Nature (London) 248: 258-259

Vartapetian BB, Andreeva IN, Kursanov AL (1974c) Erratum. Nature (London) 250: 84

Vartapetian BB, Andreeva IN, Kozlova GI (1976) The resistance to anoxia and the mitochondrial fine structure of rice seedlings. Protoplasma 88: 215-224

Vartapetian BB, Andreeva IN, Kozlova GI, Agapova LP (1977) Mitochondrial ultrastructure in roots of mesophyte and hydrophyte at anoxia and after glucose feeding. Protoplasma 91: 243-256

Vartapetian BB, Andreeva IN, Nuritdinov N (1980) Plant cells under oxygen stress. In: Hook DD, Crawford RMM (eds) Plant life in anaerobic environments. 2nd printing. Ann Arbor Science, Michigan, pp 13-88

Vartapetian BB, Mazliak P, Lance C (1978) Lipid biosynthesis in rice coleoptiles grown in the presence or in the absence of oxygen. Plant Science Letters 13: 321-328

Vartapetian BB, Nuritdinov N (1976) Molecular oxygen transport in plants. Naturwissenschaften 63: 246

Vartapetian BB, Snkhchian HH, Generozova IP (1987). In: Crawford RMM (ed) Plant life in aquatic and amphibious habitats. Blackwell Scientific Publications, Oxford, pp 205-223

Vartapetian BB, Zakhmilova NA (1990) Ultrastructure of wheat seedling mitochondria under anoxia and post-anoxia. Protoplasma 156: 39-44

Webb T, Armstrong W (1983) The effects of anoxia and carbohydrates on the growth viability of rice, pea and pumpkin roots. Journal of Experimental Botany 34: 579-603

RESPONSES TO TOTAL SUBMERGENCE IN TOLERANT AND INTOLERANT RIVERSIDE SPECIES

C.W.P.M. Blom, L.A.C.J. Voesenek and A.J.M. van der Sman
Department of Ecology
Faculty of Sciences
University of Nijmegen
Toernooiveld
6525 ED Nijmegen
The Netherlands

ABSTRACT

Submergence during the growing season often occurs in river areas throughout the world. In order not to succumb, plants have developed mechanisms to overcome periods of flooding that may differ greatly in duration, depth and frequency. To study these mechanisms, a group of related species, mainly from the genus *Rumex*, has been investigated as a model. *Rumex* occurs in regularly flooded to non-flooded sites in river areas. Adaptive behaviour ranges from changes in life-history strategies to responses at the level of hormones. Submergence may occur at several stages of the life cycle and these should all be included in studies concerning flooding resistance. In this life-history approach, adaptive responses at stages from seed bank to reproduction are described and experimentally proved. *Rumex* species from frequently flooded areas have persistent seed banks and are characterised by germination flushes after flooding in the spring. Species from higher elevations have transient seed banks and the seeds germinate in the autumn. A trade-off between survival of submergence and early reproduction was found in species occurring in the most frequently flooded sites. Submergence of the wetland species causes leaf re-orientation. This change, from prostrate to vertical, is accompanied by a stimulation of the extension growth of both petioles and internodes. Petiole elongation is due to increased cell expansion and is functionally related to the restoration of contact between shoots and air. This re-establishes oxygen-dependent metabolic processes. Both petiole and stem elongation are correlated with increased internal concentrations of ethylene. Measurements with a laser-driven photoacoustic detection technique reveal variations in endogenous ethylene levels among the various *Rumex* species. The differences in elongation reactions between the species are the result of differences in the internal concentration of ethylene and, predominantly, in sensitivity to this hormone. In wetland species, protrusion of a substantial part of the shoot above the water results in the formation of new aerenchymatous roots. Their high porosities enhance internal aeration, resulting in restored aerobic respiratory activities. Underwater photosynthesis is another adaptive response leading to the maintenance of basic physiological and biochemical processes during submergence. In some

NATO ASI Series, Vol. I 16
Interacting Stresses on Plants in a Changing Climate
Edited by M. B. Jackson, C. R. Black

tolerant species the older plants survive the adverse underwater conditions through the slow consumption of carbohydrates stored in their tap roots. Thus, a broad spectrum of adaptive responses at the level of avoidance or tolerance strategies is involved in the survival of resistant riverside plants during and after total submergence.

INTRODUCTION

Flooding occurs in many river areas throughout the world. In some regions flooding leads to uncontrolled submergence of cultivated land and forms one of the main problems for food supply. In other regions, large-scale flooding is artificially induced to stimulate the growth of, for example, deep-water rice. In densely populated areas in Western Europe flooding of river areas has been regulated by a system of dykes, canals and polders. However, in spite of these protective barriers, large changes in river-water discharge have brought about unpredictable floods that can occur during the growing season. The outdyke areas in the Netherlands provide such highly dynamic conditions. These river forelands are characterised by highly fluctuating water levels (Blom, 1990; Blom *et al.*, 1990) due to dramatic peaks in water discharge by the river Rhine (Fig. 1). Precipitation peaks in the catchment areas are superimposed on the more predictable yearly pattern of relatively high water levels in winter and spring and low levels in summer and autumn. In order to survive, plants must be adapted to the adverse conditions caused by flooding. Responses depend on the severity of floods that can range from waterlogging of the soil to submergence of the whole plant. The ability of plants to recover after a flood differs between species, but is strongly related to the duration of the flooding period.

Our research aims to reveal adaptive reactions in plants exposed to waterlogging and submergence. A group of related species from the genus *Rumex* has been chosen as a model system. The flooding characteristics of these species are compared with species from other genera that occur in similar habitats. This approach enables us to obtain an impression of the variation in adaptive responses. The research approaches are at two levels; first, investigations of survival strategies in life histories and second, the detection of morphological and physiological components of the adaptation reaction. Life-history studies are carried out in the field and in experimental outdoor plots to detect the behaviour of populations under natural conditions. Morphological alterations in plants often occur in response to changing environmental circumstances and are mostly initiated by a change in physiological processes. We aim to reveal these relationships in form and function under alternating drained, waterlogged and submerged conditions. Metabolic and hormonal reactions are studied to gain insight into the basis of physiological and morphological adaptation during and after flooding. The ultimate aim of our research is to combine results from field and laboratory studies to elucidate the phenotypic variation and genetic differentiation in the adaptive responses of plants to flooding. Furthermore, model species are also used to generate realistic hypotheses on the behaviour of co-occurring species. This fundamental knowledge of development at the individual and community level is not only important for scientific reasons but can also be helpful in the management of nature reserves that occur abundantly in the rapidly changing river foreland.

THE SPECIES UNDER STUDY

Irregular and unpredictable floods, together with differences in elevation and variation in chemical and physical soil characteristics have their impact on plant species distribution in river forelands. A typical plant zonation can often be distinguished (Van de Steeg, 1984; Blom *et al.*, 1990). *Rumex* species are abundantly represented in river areas and belong to the few genera with species clearly zoned along a flooding gradient (Fig. 1). *Rumex acetosa* and *R. thyrsiflorus* occur in higher lying habitats that are seldom flooded and then only for very short periods. *Rumex crispus*, *R. obtusifolius* and *R. conglomeratus* inhabit the more frequently but not very severely flooded zones. *Rumex palustris* and *R. maritimus* are restricted to the banks of the river and the low lying former river beds characterised by frequent and prolonged flooding. *Chenopodium rubrum* and *Plantago major* ssp. *pleiosperma* are also situated directly along the river and can be found adjacent to both *Rumex* species from the wettest environments. The

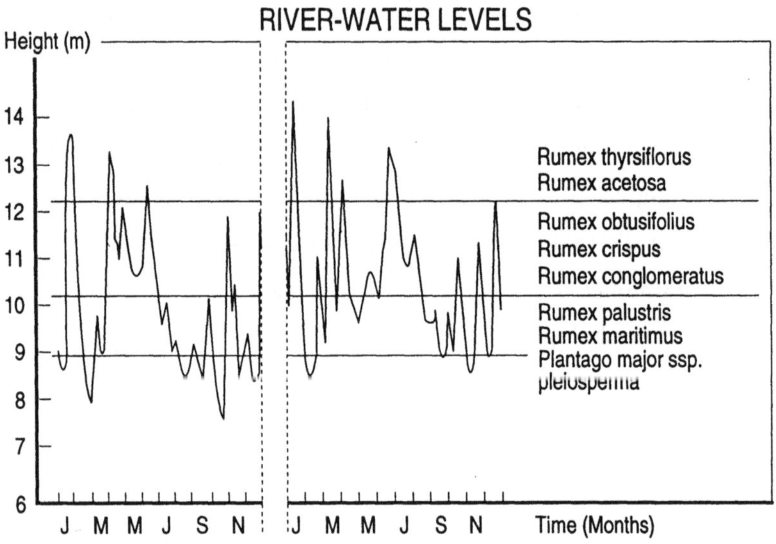

FIGURE 1. Water levels of the river Rhine near Nijmegen, The Netherlands, during two successive years in relation to the zonation of some species in the river foreland. Height is expressed as distance above sea-level. From Blom et al. *(1990).*

distribution of different *Rumex* species along a flooding gradient was a principal reason for choosing *Rumex* as a model for comparative studies. A second argument is the relative easy experimental handling of these species. Large amounts of seeds can be collected in the field which possess a high germinability, and the seeds as well as full-grown plants are of manageable size. These plants have a high relative growth rate and grow fast in experimental systems. The behaviour of the *Rumex* species has been compared with both *Chenopodium rubrum* and *Plantago major* ssp. *pleiosperma.* These species can also be easily handled in experiments.

LIFE-HISTORY STRATEGIES

The variety of habitats, i.e., higher and lower dikes, former river beds, sand banks, and man-made depressions, provides a small-scale mosaic structure of the river foreland. This diversity in types of landscape is intensified by the effects of the transient floods that differ in intensity, duration and frequency. Plant populations cope with these highly dynamic conditions through different life-history strategies. To study adaptive responses at life-history levels at least two approaches can be chosen. The first is the 'one genus-more habitats' approach and the second the 'more species-one habitat' approach. The 'one genus-more habitats' approach is based on the idea that the properties of phylogenetically related species are nearly similar and the differences among the individual species reflect adaptations to their specific environments. River forelands in the Netherlands are characterised by clearly distinguishable zones (Fig. 1). Within the genus *Rumex* the different species apparently match the particular conditions found in their respective zones. An excellent example in which this approach has been applied is the multidisciplinary study on *Plantago* (Kuiper and Bos, 1992). In this research, the demographic, physiological and genetic properties of species from the genus *Plantago* were compared in relation to their location in grasslands. The 'one genus-more habitats' approach can also be applied at the population level. Lotz and Blom (1986) studied the plasticity in life-histories of *Plantago major* ssp. *pleiosperma* by means of demographic descriptions of natural populations on the bank of the river Rhine and performed reciprocal transplant experiments. In combination with greenhouse tests on the effects of nutrient supply, plants from different populations were compared. Representatives from river-side populations showed a high degree of plasticity in reproductive effort. In contrast, plants from populations occurring in less dynamic sites showed higher plasticity in shoot-root ratio. These results demonstrated varying degrees of phenotypic plasticity in life-history characteristics in *P. major* spp. *pleiosperma* populations from different sites. Moreover, the phenotypic plasticity expressed in either vegetative or generative development turns out to be as important a response to selective forces as interspecific competition or temporal variability.

The 'more species-one habitat' approach originates from the observation that in one zone a variety of species from different genera are able to cope with similar environmental circumstances. Field studies have revealed that co-occurring species may possess more than one survival strategy. Thus, adjacent plant species in one habitat have developed plural solutions to the problems imposed by their common environment. How plant species occurring in one habitat deal with germination, establishment, growth, reproduction and longevity under conditions of highly fluctuating water levels was studied by Van der Sman (1992). She compared the life histories of *Rumex palustris, R. maritimus* and *Chenopodium rubrum* in the lowest sites of river forelands. Field observations, greenhouse studies and experiments performed in outdoor plots clearly revealed differences between the life histories of these species. *Rumex palustris* can be considered as a short-lived perennial, *R. maritimus* as mainly biennial, whereas *C. rubrum* is strictly annual. *Chenopodium rubrum* flowers and reproduces either early in the season or, when germinating as late as September, before the onset of the winter (Van der Sman *et al.*, 1988, 1992, 1993). This species relies on the production of many seeds in a short period. Reproductive capacity depends on the time of emergence and establishment, events that are determined by the subsidence of floods. Flowering is induced earlier in early germinating cohorts than in later ones, but later established plants have a diminished size requirement for flowering and complete their life cycles in a shorter time (Table 1). In spite of a reduced seed output, reproductive effort and seed size increase in later cohorts. However, the plants fail to set seed when they emerge very late in the season or are severely submerged during their vegetative or reproductive stages. *Rumex palustris* and, to a lesser degree, *R. maritimus* rely primarily on adaptive responses upon flooding during their established phase. *R. palustris* especially is able to postpone flowering to the following year. Under favourable conditions, *Rumex maritimus* may complete its life cycle in one season. Both species have a size-requirement before flowering is initiated and the prevailing flooding-conditions largely determine whether this is achieved in the first season. Flowering induction is also controlled by photoperiod. *Rumex palustris* needs a longer critical photoperiod before the initiation of reproductive development than *R. maritimus*. Flooding during the vegetative phase also affects the survival of these species to a large extent. During the established phase, *Chenopodium rubrum* is the most sensitive followed by *R. maritimus*, whereas *R. palustris* and *Plantago major* ssp. *pleiosperma* are the most resistant species (Lotz, 1989; Van der Sman, 1992).

TABLE 1. *Reproductive development of four cohorts of* Chenopodium rubrum *in outdoor experimental plots.*

Planting date (1987)	% flowering	Onset of flowering	Duration of life cycle (d)	Total no. of main leaves	Final plant height (cm)	Seed weight per plant (g)	Seed weight to stem weight ratio*
18 May	100	7 August	143	61.8 ± 1.1 a	70.4 ± 1.2 a	8.6 ± 0.8 a	0.89 ± 0.15 b
29 June	100	18 August	112	46.2 ± 1.6 b	53.4 ± 1.6 b	5.9 ± 0.5 a	1.28 ± 0.20 b
10 August	100	9 September	97	24.0 ± 0.9 c	21.8 ± 0.6 c	4.6 ± 0.3 a	3.87 ± 0.44 a
21 September	50	28 October	**	–	–	–	–

Values are median values per cohort or mean ± 1 SE (n=20).

* The ratio of seed weight to stem weight is calculated for 10 plants per cohort; stem weight to tap root weight ratio remained constant in all cohorts.

** Plants died in December due to frost before the fruits ripened.

Means followed by the same letter are not significantly different (Bonferronit-test, $p < 0.05$), (Van der Sman, Blom and Van de Steeg, 1991).

MORPHOLOGICAL ADAPTATIONS

It is obvious that the morphology and anatomy of terrestrial plants are not normally suited for conditions in over-wet environments. Consequently, the root and shoot systems of plants from riparian habitats must be very plastic to cope with alternating dry and wet circumstances. Riverside species, whether tolerant or intolerant of submergence, only germinate and establish in relatively dry periods. Thus, both groups of species develop shoots and roots initially adapted to drained conditions. Tolerant plants resist submergence through their ability to change their morphological characteristics within a limited period of time. The most adverse factor induced by submergence is the rapid decrease of oxygen in the soil, caused partly by hindrance of gas exchange between the atmosphere and soil and partly by the respiration of roots and micro-organisms (Armstrong, 1979; Kozlowski, 1984; Laanbroek, 1990). In anoxic soils, the formation of aerenchymatous roots is beneficial to the survival of plants (Smirnoff and Crawford, 1983; Laan *et al.*, 1989b, 1991; Armstrong *et al.*, 1991). Most *Rumex* species have a tap root and, as a response to flooding, new laterals with a high porosity are formed in the tolerant species. Root-porosity values in the higher level species, *R. thyrsiflorus* and *R. acetosa*, are less than 10%, while the lower level species, *R. crispus* and *R. maritimus*, have values higher than 10%. Important determinants of the degree of flooding tolerance of *Rumex* species appear to be the number and formation rate of new aerenchymatous roots (Fig. 2). Positive correlations were found between the relative growth rate of plants, the development of new roots and the degree of flooding tolerance in the *Rumex* species (Voesenek *et al.*, 1989). Not only do the roots of flood-tolerant plants have large quantities of aerenchyma, but the shoot systems also appear to be very porous. This system of channels and open spaces provides a pathway for the longitudinal and radial transport of oxygen to restore the respiration and growth of submerged plants (Armstrong, 1979; Laan *et al.*, 1990). High porosities lead to radial oxygen loss from roots to the rhizosphere, and this occurs more often in tolerant than in intolerant *Rumex* species (Fig. 3). Advantages of this phenomenon include prevention of iron and manganese toxicity

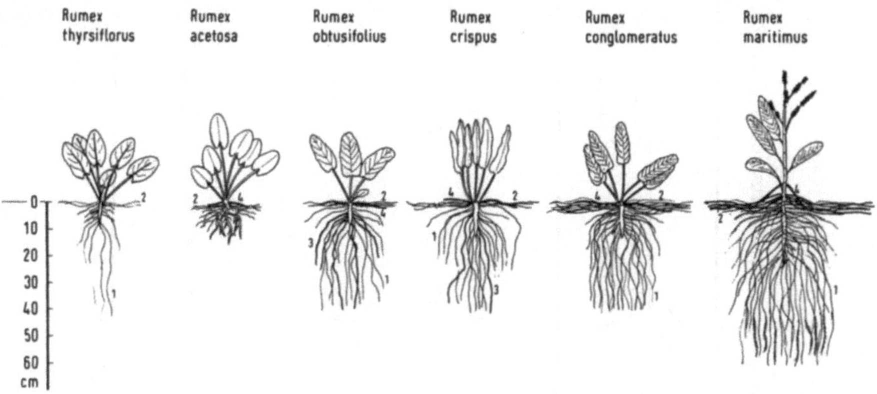

FIGURE 2. *Root morphology and rooting depth of* Rumex *species after three weeks of flooding in river sand (age of the plants at the start of the experiment six weeks; original lateral root system not shown). Root type: 1, downward-growing laterals; 2, horizontally-growing laterals; 3, thick downward-growing laterals; 4, adventitious roots. From Laan* et al. *(1989b).*

(e.g., Ernst, 1990; Gambrell *et al.*, 1991) and stimulation of the uptake of nutrients (Table 2). A comparison of nitrifying capacity in the rhizosphere of the flooding intolerant *R. thyrsiflorus* with that of the tolerant species *R. palustris* (Table 3) indicated a nitrifying population that remains active for longer after waterlogging of soils in which *R. palustris* grows (see Engelaar *et al.*, 1991). The aerenchymatous tissues of *R. palustris* increased oxygen concentrations in the soil, which in turn stimulated the growth and activity of ammonium-oxidising bacteria. In the presence of *R. thyrsiflorus*, the activity and growth of these bacteria were inhibited under flooded, but not under drained conditions.

A plant suddenly facing total submergence must renew contact with the open air to obtain sufficient oxygen in its tissue to avoid anoxia and initiate the development of new aerenchymatous laterals. Voesenek and Blom (1989), Laan and Blom (1990) and Voesenek *et al.* (1990b) clearly demonstrated that *Rumex* species normally occurring on lower sites are more successful in attempting to extend their leaves above the water surface than species from higher elevations (Fig. 4). Van der Sman *et al.* (1991) demonstrated in *R. maritimus* not only a

TABLE 2. *Effect of flooding on biomass production and the nutrient content of leaves of* Rumex *species. From Laan* et al. *(1989a).*

Species/ treatment	Shoot dry weight (g)	Leaf nutrient content (μmol g^{-1} dry wt)				
		Nitrogen	Phosphorus	Potassium	Calcium	Magnesium
R. thyrsiflorus						
Drained	15.2 ±1.4	1878 ± 45	143 ± 10	1238 ± 37	323 ± 11	435 ± 22
Flooded	7.2 ±1.4	1006 ± 64	44 ± 2	355 ± 15	190 ± 2	218 ± 3
R. crispus						
Drained	13.6 ±1.4	1372 ± 54	90 ± 5	857 ± 39	532 ± 27	311 ± 7
Flooded	12.7 ±2.5	702 ± 35	58 ± 10	347 ± 23	315 ± 13	160 ± 8
R. maritimus						
Drained	24.8 ±2.6	1018 ± 91	59 ± 2	478 ± 11	615 ± 19	401 ± 5
Flooded	25.4 ±3.9	1052 ± 52	59 ± 2	272 ± 10	761 ± 40	398 ± 11

Means of five replicates plus SD; plants grown in a clay/sand mixture for eight weeks.

TABLE 3. *Mean nitrifying capacity (nmol NH_4^+ g^{-1} dry soil h^{-1}) of drained and waterlogged sand/clay mixture with either* Rumex thyrsiflorus *or* Rumex palustris *(± 1 SD, n=4-5). Nitrifying capacity was determined as the amount of NH_4 converted to NO_2^- and NO_3^- after incubating fresh soil and medium containing NH_4^+, for six hours. Waterlogging started after three weeks. Different letters indicate significant differences in nitrifying capacity ($p < 0.05$).*

	Rumex thyrsiflorus		*Rumex palustris*	
Week	Drained	Waterlogged	Drained	Waterlogged
4	29.1 ± 3.89 a		50.6 ± 19.3 a	
6	159.0 ± 88.10 b	57.60 ± 84.80 ab	314.0 ± 154.0 b	109.0 ± 27.00 c
10	95.3 ± 42.80 b	3.48 ± 0.63 c	154.0 ± 56.0 bc	55.6 ± 7.48 a

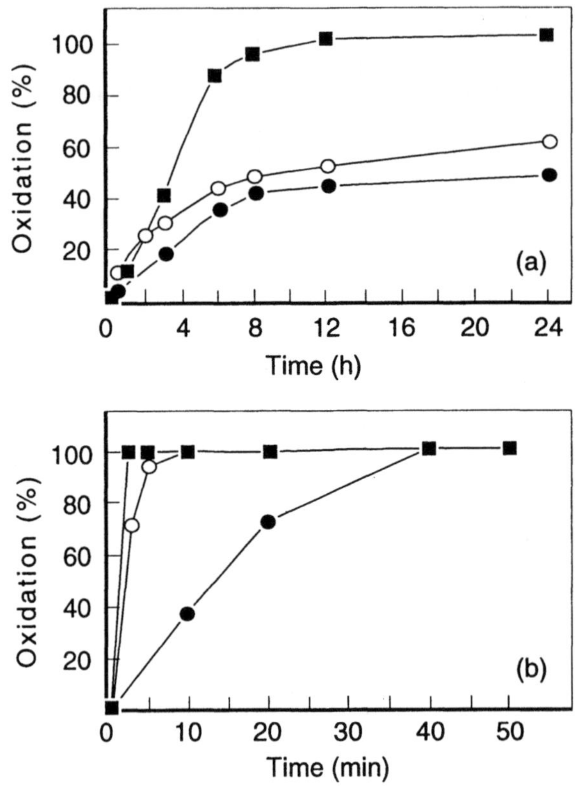

***FIGURE 3.** Timecourse of readily perceived oxidation of leuco-methylene blue as blue colouration rate of primary (a) and newly formed (b) lateral roots of* Rumex thyrsiflorus (●), R. crispus (○) *and* R. maritimus (■). *The data are expressed as percentage oxidation of total root length. From Laan* et al. *(1989a).*

positive relationship between protrusion and survival, but also a positive correlation between the height of the shoot above the water level and the amount of seed produced (Table 4).

METABOLIC RESPONSES

Anaerobic conditions due to flooding have a negative affect on the growth and survival of nearly all terrestrial plants. The duration of anaerobiosis mainly determines the degree of resistance to flooding in plants. Flood-tolerant species have developed morphological and physiological adaptations to overcome shorter

TABLE 4. *Mean seed production (g plant^{-1} ± 1 SE) per stem length class (primary stem length in cm) and the number of plants in this class (n) for three series of* Rumex maritimus *plants differing in planting dates. From Van der Sman* et al. *(1991).*

Stem length	Series I, 2 May 1988		Series II, 16 May 1988		Series III, 30 May 1988	
	Drained seed (n)	Flooded seed (n)	Drained seed (n)	Flooded seed (n)	Drained seed (n)	Flooded seed (n)
<50	18.6 ± 1.5 a (4)	3.7 ± 0.2 c (3)	18.9 ± 0.3 a (5)	2.8 ± 1.1 c (4)	17.6 ± 0.9 a (4)	
50-54	18.3 ± 1.2 a (4)	7.3 ± 0.4 d (4)	19.4 ± 0.5 a (3)	6.3 (1)	18.6 ± 2.1 a (3)	
>54	19.2 ± 0.2 a (4)	11.4 ± 1.0 bc (5)	20.0 ± 0.7 a (4)	13.7 ± 0.7 b (7)	18.9 ± 1.6 a (5)	10.1 ± 0.8 c (12)

Means of the same letter are not significantly different.

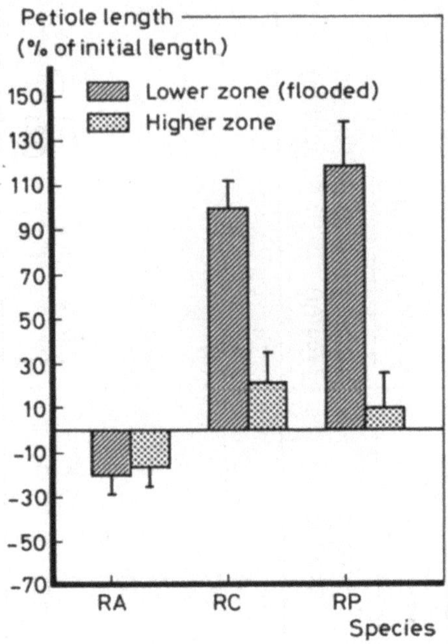

FIGURE 4. The length of the longest petiole (± 1 SE; n = 108) of three Rumex *species (RA =* Rumex acetosa; *RC =* Rumex crispus; *RP =* Rumex palustris*) in a lower and higher zone in the river area. During the study period (three weeks) the lower zone was flooded for four days. From Voesenek and Blom (1989).*

or longer periods of flooding. Shoot emergence and the development of new roots take considerable time and can thus be considered as long-term adaptive responses. In the period between the onset of submergence and the formation of the new aerenchymatous roots other mechanisms must operate to maintain, at least partially, the basic metabolic functions. Two of these mechanisms are well known: first, underwater photosynthesis restores sugar and oxygen supplies to the roots (Bowes, 1987; Gaynard and Armstrong, 1987; Laan and Blom, 1990; Voesenek *et al.*, 1992), and second, anaerobic respiration, which demands a high sugar supply (Crawford, 1982; Jackson and Drew, 1984; Brändle, 1990). A relationship between the flooding resistance of *Rumex* species and their ability to photosynthesise under water, or to respire anaerobically, appears evident. *Rumex acetosa* and *R. thyrsiflorus*, occupants of the higher-lying habitats, succumb rapidly to submergence. These species cannot cope with sustained periods of root anaerobiosis

(Laan, 1990; Voesenek, 1990). In contrast, *R. crispus, R. palustris* and to a lesser extent *R. maritimus* survive prolonged submergence and may use underwater photosynthesis to satisfy needs for oxygen and sugars (Laan and Blom, 1990; Voesenek *et al.*, 1992). In general, *R. maritimus* relies more than *R. crispus* and *R. palustris* on fast shoot growth and allocation of biomass to the lateral roots rather than the tap root. *Rumex maritimus* must quickly react to adverse conditions; its resistance to submergence strongly depends on avoidance of anaerobiosis and also on the availability and generation of sugars. Therefore, all resistance responses to submergence are directed towards gaining oxygen, either by shoot elongation to restore the free diffusion pathway between the aerial environment and the root system, or by underwater photosynthesis (Laan and Blom, 1990; Van der Sman *et al.*, 1991). Compared with *R. maritimus* and *R. crispus, R. palustris* seems to be intermediate in its reactions. For a perennial species such as *R. crispus*, the ready availability of oxygen to resist the early phase of submergence seems to be least important. This species shows decreased metabolic rates and uses the starch reserves in its tap-root to overcome periods of anaerobiosis (Laan and Blom, 1990).

A further metabolic acclimation to resist the adverse conditions that higher plants meet in the anaerobic environments of submerged areas is acceleration of glycolysis. This 'Pasteur effect' provides the formation of more ATP per unit time and is mostly accompanied by the induction of pyruvate decarboxylase and alcohol dehydrogenase, two key enzymes catalysing the final steps in anaerobic fermentation. However, flood-tolerant plants cannot rely entirely on this form of adaptative mechanism. Due to the low efficiency of sugar utilisation, the necessary ATP level can only be maintained for a short time. Over longer periods a shortage of respirable sugars is likely to occur (Jackson and Drew, 1984). So, the fermentation reaction is a short-term response that enables the plant to bridge the period between the beginning of anaerobiosis and the functioning of the more long-term adaptive characteristics. However, there are indications that flood-intolerant species may have a relatively high level of ADH-induction and/or 'Pasteur-effect' (Crawford, 1978, 1982).

In an attempt to relate the metabolic responses of ***Rumex*** species to their resistance to submergence and zonation in the flooding gradient, experiments were conducted to establish if any differences in the ethanolic fermentation in the primary root tips exist between *R. acetosa, R. crispus* and *R. palustris* (Voesenek *et al.*, 1992). Under hypoxic conditions, the period for which the root tips of the flood-intolerant species survived was approximately 37% less than that of tolerant species. Distinct differences in ethanol production under hypoxic conditions were also found. Ethanol concentrations after 10 hours did not exceed 50 μmol g^{-1} dry weight in *R. acetosa*, whereas *R. palustris* and *R. crispus* produced *c.* 350 and 300 μmol g^{-1} dry weight, respectively (Voesenek *et al.*, 1992). These results prove a positive relationship between tolerance of hypoxia and fermentation rate expressed as ethanol production. In addition, the tolerance of hypoxia *in vitro* of primary roots of *Rumex* species appears to be positively correlated with the flooding resistance of plants in the field.

HORMONAL REGULATION

Upon total submergence, shoot elongation and the subsequent formation of a new root system are adaptive responses often observed in plants located in riparian habitats. Questions about how these responses are initiated and what kind of internal plant substances are involved have been the subject of much recent physiological and ecophysiological research (e.g. Jackson, 1990). Submergence of terrestrial plants can be considered as a form of environmental stress that triggers the action of rapidly acting plant substances. Such fast-acting substances are the hormones which play an important role in stress reactions. In plants facing submergence, ethylene is well-known as a substance responsible for root and shoot extension. Diffusion of this gas from the plant to the air is strongly reduced in submerged environments. Endogenously produced ethylene (Musgrave *et al.*, 1972) and the ingress of exogenous gas of micro-biological origin cause the accumulation of this gas in plant tissues (Smith and Restall, 1971; Konings and Jackson, 1979; Konings, 1982; Voesenek *et al.*, 1992). Plants respond differently to the ethylene enrichment; rice is very sensitive and elongates more rapidly at levels below 0.1 Pa, while barley is not able to respond in this manner (Smith and Robertson, 1971). In *Rumex*, variation between species in their reaction to ethylene is also apparent.

Rumex acetosa, *R. crispus* and *R. palustris* show increased internal ethylene concentrations after submergence, but only the two latter species are able to increase extension growth (Voesenek and Blom, 1989; Voesenek *et al.*, 1990); the sensitivity of *R. acetosa* to this hormone is clearly lower than that of both other species (Fig. 5). By means of a highly sensitive laser-driven photoacoustic detector assembled in line with a flow-through system (Harren *et al.*, 1990; Voesenek *et al.*, 1990b), the ethylene production of intact plants can be monitored. Less than two hours after the onset of flooding, ethylene concentration increased in *R. crispus* and *R. palustris*, species from the lower areas. After approximately five hours, a gas peak ten times larger than that from plants growing under drained conditions was found. This increase was followed by a fall in production and, after another 24 hours, by a second increase up to twenty times the control level (Voesenek *et al.*, 1990). *Rumex acetosa*, occurring on the higher sites in the field, responded rather differently, with only a slightly enhanced ethylene production being observed. The functionality of ethylene-mediated shoot elongation upon submergence was reported by Laan and Blom (1990), Voesenek (1990) and Van der Sman *et al.* (1991). A positive relationship between stem elongation and seed output in *R. maritimus* was observed by Van der Sman *et al.* (1991) in an experiment in which plants were successively submerged and de-submerged. During submergence, increased production as well as entrapment caused a high gas concentration that was registrated by the photoacoustic laser equipment (Fig. 6). Upon de-submergence, two peaks of ethylene release were found. The first peak was explained as immediate release of entrapped ethylene and the second by the conversion of ACC to ethylene directly after the fall of the water level. The function of the extra ethylene synthesis could be to allow the continuation of fast shoot growth even after restoration of shoot-atmosphere contact. Moreover, the extent to which the shoots emerged from the water level was positively correlated not only with

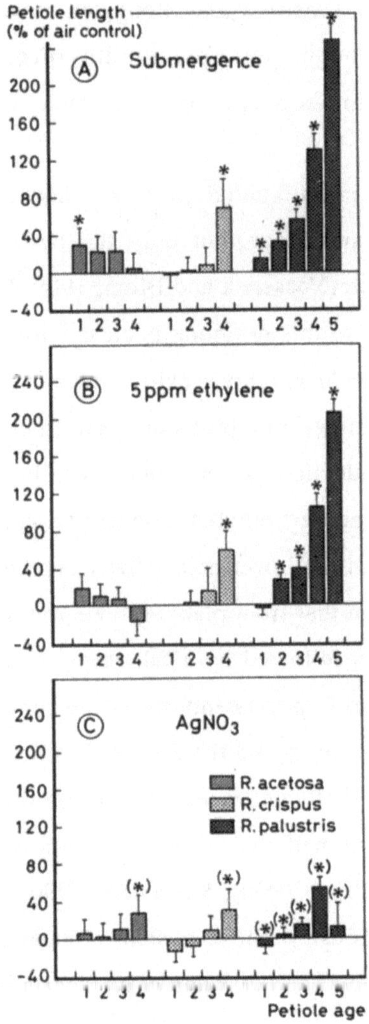

FIGURE 5. The length of Rumex *petioles (± 1 SE; n = 12) of different ages after submergence (A), exposure to an ethylene-air gas mixture (0.5 Pa) (B), and submergence in a solution with the ethylene action inhibitor $AgNO_3$ (C) under laboratory conditions, expressed as percentages of air controls (1 = oldest petiole; 4/5 = youngest petiole); * significantly different from control level [LSD; $P < 0.05$]. From Voesenek and Blom (1989).*

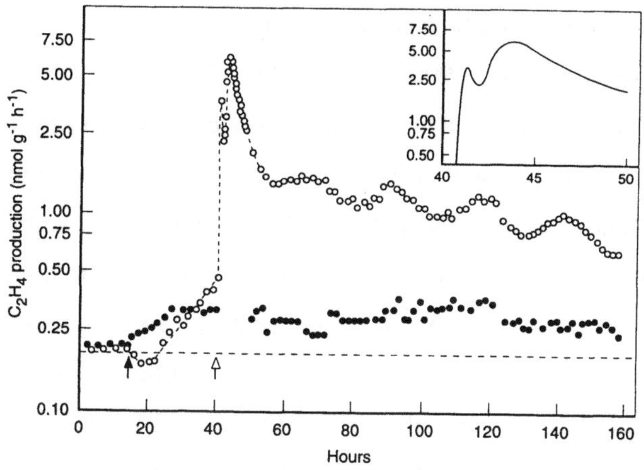

FIGURE 6. *Ethylene production (logarithmically presented) of* Rumex maritimus *plants at the bolting stage monitored using a laser-driven photoacoustic detection system. --, mean ethylene production of drained plants;* ●, *plant waterlogged from* ↑ *onwards;* ○, *plant submerged from* ↑ *until* ⇧ *and thereafter waterlogged. The insert represents the two peaks in ethylene production occurring directly after submergence in more detail. Results are expressed in terms of the above-ground dry weight, determined at the end of the measurement. From Van der Sman* et al. *(1991).*

survival but also with the amount of seed produced (Table 4), which is undeniably of ecological advantage. Until recently, studies of the relationship between ethylene and the behaviour of *Rumex* upon submergence have been primarily directed towards the shoot reaction, but in the present research root responses to ethylene and other hormones are also involved. It is well known, for example in tomato, that ethylene may influence root growth depending on the production level, the surface area/volume ratio, the permeability of surface layers, root porosity, and the depth of the water layer (Smith and Robertson, 1971; Konings and Jackson, 1979; Jackson, 1985). Reaction depends on the concentration; lower concentrations of the gas stimulate root growth, while higher concentrations inhibit extension. Furthermore, the kind of response will also depend on the sensitivity of the root tissue to ethylene (Jackson and Pearce, 1991). The porosity of the root and shoot systems and the related internal aeration determine, to a large extent, the effects of ethylene in the plant.

In *Rumex*, aerenchyma formation is promoted by oxygen shortage caused by submergence (Blom *et al.,* 1990; Laan, 1990; Voesenek, 1990). Aerenchyma promotion is generally mediated by increased ethylene (Jackson *et al.*, 1985). Other hormones may be also involved in the adaptive responses to flooding (Jackson and Pearce, 1991; Karssen *et al.*, 1992). There are indications that some hormones act together with ethylene (Smulders and Horton, 1991). Auxin, for example, is a necessary adjunct for ethylene synthesis and stimulates ethylene biosynthesis, but also promotes growth independently of ethylene. Other hormones involved in submergence reactions may be gibberellins that act with ethylene in stimulating growth (Musgrave *et al.*, 1972; Raskin and Kende, 1984).
Although little seems to be known on the possible roles of abscisic acid and cytokinins, the conclusion seems to be justified that ethylene, and to a much lesser degree auxin and gibberellin, are involved in the submergence responses of plants. The presence of some available oxygen, however, remains a requirement for the activities of these hormones.

CONCLUSIONS

Irregular and unpredictable flooding during the growing season forces plants to a rapid mobilisation of resistance mechanisms. Due to the suddenness of the stress, flood-resistant plants either possess avoidance mechanisms to escape the unfavourable conditions or have sufficient physiological and morphological plasticity to tolerate the enmity of the outside world. Avoidance mechanisms are generally expressed in different life-history strategies and are mostly genetically determined. The 'more species-one habitat approach' is applied to compare the adaptive responses of different species in one location that is exposed to severe submergence stress. This approach reveals that several solutions at the level of the life-history strategy have been evolved to overcome a common environmental problem. The 'one genus-more habitats approach' demonstrates that the characteristics of specific environments, for example frequency and duration of submergence, strongly determine differences in responses among species, even when they are phylogenetically related.

Tolerance mechanisms seem to be an expression of phenotypic plasticity acting at the level of metabolic, hormonal and morphological responses. The degree of flooding tolerance in plants depends on their ability to exhibit both a short-term, metabolic response and long-term, morphological and physiological adaptation. Covered by a layer of water, tolerant riparian plants react with a more upright orientation of their leaves, faster extension of petioles or stems, a decrease in the mechanical strength of the upper parts, a new superficially growing root system with adventitious laterals and the formation of aerenchyma in shoots and roots. The changed morphological characteristics are initiated by the action of hormones and certain metabolic or biochemical reactions that are triggered by submergence and lower amounts of oxygen in the submerged tissues. The adaptive changes in morphology and anatomy restore physiological processes that are indispensable for the maintenance and survival of the plant. This gives an excellent illustration of the strong relationship between form and function in plants.

REFERENCES

Armstrong W (1979) Aeration in higher plants. Advances in Botanical Research 7: 225-332

Armstrong W, Justin SHFW, Beckett PM, Lythe S (1991) Root adaptation to soil waterlogging. Aquatic Botany 39: 57-73

Blom CWPM (1990) Responses to flooding in weeds from river areas. In: Kawano S (ed) Biological approaches and evolutionary trends in plants. Academic Press, London, pp 81-94

Blom CWPM, Bögemann GM, Laan P, Sman AJM Van der, Steeg HM Van de, Voesenek LACJ (1990) Adaptations to flooding in plants from river areas. Aquatic Botany 38: 29-47

Bowes G (1987) Aquatic plant photosynthesis: Strategies that enhance carbon gain. In: Crawford RMM (ed) Plant life in aquatic and amphibious habitats. British Ecological Society, Special Symposium 5. Blackwell Scientific, Oxford, UK, pp 79-98

Brändle RA (1990) Flooding resistance of rhizomatous amphibious plants. In: Jackson MB, Davies DD, Lambers H (eds) Plant life under oxygen deprivation. SPB Academic, The Hague, The Netherlands, pp 35-46

Crawford RMM (1978) Metabolic adaptation to anoxia. In: Hook DD, Crawford RMM (eds) Plant life in anaerobic environments. Ann Arbor Scientific, Ann Arbor, Michigan, USA, pp 119-136

Crawford RMM (1982) Physiological responses to flooding. In: Lange O, Nobel PS, Osmond CB, Ziegler H (eds) Encyclopedia of plant physiology, New Series, Volume 12B, Physiological plant ecology. II. Water relations and carbon assimilation. Springer-Verlag, Berlin, Germany, pp 453-477

Engelaar WMHG, Bodelier PLE, Laanbroek HJ, Blom CWPM (1991) Nitrification in the rhizosphere of a flooding-resistant and a flooding-non-resistant *Rumex* species under drained and waterlogged conditions. FEMS Microbiology Ecology 86: 33-42

Ernst WHO (1990) Ecophysiology of plants in waterlogged and flooded environments. Aquatic Botany 38: 73-90

Gambrell RP, DeLaune RD, Patrick WH (1991) Redox processes in soils following oxygen depletion. In: Jackson MB, Davies DD, Lambers H (eds) Plant life under oxygen deprivation. SPB Academic, The Hague, The Netherlands, pp 101-117

Gaynard TJ, Armstrong W (1987) Some aspects of internal aeration in amphibious species. In: Crawford RMM (ed) Plant life in aquatic and amphibious habitats. British Ecological Society, Special Symposium 5. Blackwell Scientific, Oxford, UK, pp 303-320

Harren FJM, Bijnen FGC, Reuss J, Voesenek LACJ, Blom CWPM (1990) Sensitive intracavity photoacoustic measurements with a CO_2 waveguide laser. Applied Physics B 50: 137-144

Jackson MB (1985) Ethylene and the response of plants to excess water in their environment - a review. In: Roberts JA, Tucker GA (eds) Ethylene and plant development. Butterworths, London, UK, pp 241-265

Jackson MB (1990) Hormones and developmental change in plants subjected to submergence or soil waterlogging. Aquatic Botany 38: 49-72

Jackson MB, Drew MC (1984) Effect of flooding on herbaceous plants. In: Kozlowski TT (ed) Flooding and plant growth. Academic Press, London, UK, pp 47-128

Jackson MB, Davies DD, Lambers H (eds) (1990) Plant life under oxygen deprivation. SPB Academic, The Hague, The Netherlands

Jackson MB, Fenning TM, Drew MC, Saker LR (1985) Stimulation of ethylene production and gas space (aerenchyma) formation in adventitious roots of *Zea mays* L. by small partial pressures of oxygen. Planta 165: 486-492

Jackson MB, Pearce DME (1991) Hormones and morphological adaptation to aeration stress in rice. In: Jackson MB, Davies DD, Lambers H (eds) Plant life under oxygen deprivation. SPB Academic, The Hague, The Netherlands, pp 47-67

Karssen CM, Van Loon LC, Vreugdenhil D (eds) (1992) Progress in plant growth regulation. Current plant science and biotechnology in agriculture. Kluwer Academic Publishers, Dordrecht, Boston, London

Konings H (1982) Ethylene-promoted formation of aerenchyma in seedling roots of *Zea mays* L. under aerated and non-aerated conditions. Physiologia Plantarum 54: 119-124

Konings H, Jackson MB (1979) A relationship between rates of ethylene production by roots and the promoting or inhibiting effects of exogenous ethylene and water on root elongation. Zeitschrift für Pflanzenphysiologie 92: 385-397

Kozlowski TT (ed) (1984) Flooding and plant growth. Academic Press, London

Kuiper PJC, Bos M (eds) (1991) Plantago: A multidisciplinary study. Ecological Studies 89. Springer-Verlag, Berlin, Heidelberg, New York

Laan P (1990) Mechanisms of flood-tolerance in *Rumex* species. PhD Thesis, Katholieke Universiteit Nijmegen, The Netherlands

Laan P, Berrevoets MJ, Lythe S, Armstrong W, Blom CWPM (1989b) Root morphology and aerenchyma formation as indicators for the flood-tolerance of *Rumex* species. Journal of Ecology 77: 693-703

Laan P, Blom CWPM (1990) Growth and survival responses of *Rumex* species to flooded and submerged conditions: the importance of shoot elongation, underwater photosynthesis and reserve carbohydrates. Journal of Experimental Botany 41: 775-783

Laan P, Clement JMAM, Blom CWPM (1991) Growth and development of *Rumex* roots as affected by hypoxic and anoxic conditions. Plant and Soil 136: 145-151

Laan P, Smolders A, Blom CWPM, Armstrong W (1989a) The relative roles of internal aeration, radial oxygen losses, iron exclusion and nutrient balance in flood-tolerance of *Rumex* species. Acta Botanica Neerlandica 38: 131-145

Laan P, Tosserams M, Blom CWPM, Veen B (1990) Internal oxygen transport in *Rumex* species and its significance for respiration under hypoxic conditions. Plant and Soil 122: 39-46

Laanbroek HJ (1990) Bacterial cycling of minerals that affect plant growth in waterlogged soils: a review. Aquatic Botany 38: 109-125

Lotz LAP (1989) Variation in life-history characteristics between and within populations of *Plantago major* L. PhD Thesis, Katholieke Universiteit Nijmegen, The Netherlands

Lotz LAP, Blom CWPM (1986) Plasticity in life-history traits of *Plantago major* L. ssp. *pleiosperma* Pilger. Oecologia 69: 25-30

Musgrave A, Jackson MB, Ling E (1972) *Callitriche* stem elongation is controlled by ethylene and gibberellin. Nature New Biology 238: 93-96

Raskin I, Kende H (1984) Role of gibberellin in the growth response of submerged deep water rice. Plant Physiology 76: 947-950

Smirnoff N, Crawford RMM (1983) Variation in the structure and response to flooding of root aerenchyma in some wetland plants. Annals of Botany 51: 237-249

Smith KA, Restall SWF (1971) The occurrence of ethylene in anaerobic soil. Journal of Soil Science 22: 430-443

Smith KA, Robertson PD (1971) Effect of ethylene on root extension of cereals. Nature (London) 234: 148-149

Smulders MJM, Horton RF (1991) Ethylene promotes elongation growth and auxin promotes radial growth in *Ranunculus sceleratus* petioles. Plant Physiology 96: 806-811

Van der Sman AJM (1992) Flooding resistance and life histories of short-lived floodplain herbs. PhD Thesis, Katholieke Universiteit Nijmegen, The Netherlands

Van der Sman AJM, Blom CWPM, Steeg HM Van de (1992) Phenology and seed production in *Chenopodium rubrum, Rumex maritimus* and *Rumex palustris* as related to photoperiod in river forelands. Canadian Journal of Botany 70: 392-400

Van der Sman AJM, Joosten NN, Blom CWPM (1993) Flooding regimes and life-history characteristics of short-lived species in river forelands. Journal of Ecology (in press)

Van der Sman AJM, Tongeren OFR Van, Blom CWPM (1988) Growth and reproduction of *Rumex maritimus* and *Chenopodium rubrum* under different waterlogging regimes. Acta Botanica Neerlandica 37, 439-450

Van der Sman AJM, Voesenek LACJ, Blom CWPM, Harren FJM, Reuss J (1991) The role of ethylene in shoot elongation with respect to survival and seed output of flooded *Rumex maritimus* L. plants. Functional Ecology 5: 304-313

Van de Steeg HM (1984) Effects of summer inundation on flora and vegetation of river foreland in the Rhine area. Acta Botanica Neerlandica 33: 365-366

Voesenek LACJ (1990) Adaptations of *Rumex* in flooding gradients. PhD Thesis, Katholieke Universiteit Nijmegen, The Netherlands

Voesenek LACJ, Blom CWPM (1989) Growth responses of *Rumex* species in relation to submergence and ethylene. Plant, Cell and Environment 12: 433-439

Voesenek LACJ, Blom CWPM, Pouwels RHW (1989) Root and shoot development of *Rumex* species under waterlogged conditions. Canadian Journal of Botany 67: 1865-1869

Voesenek LACJ, Harren FJM, Bögemann GM, Blom CWPM, Reuss J (1990a) Ethylene production and petiole growth in *Rumex* plants induced by soil waterlogging. Plant Physiology 94: 1071-1077

Voesenek LACJ, Perik PJM, Blom CWPM, Sassen MMA (1990b) Petiole elongation in *Rumex* species during submergence and ethylene exposure: The relative contributions of cell division and cell expansion. Plant Growth Regulation 9: 13-17

Voesenek LACJ, Sman AJM Van der, Harren FJM, Blom CWPM (1992) An amalgamation between hormone physiology and plant ecology: a review on flooding resistence and ethylene. Plant Growth Regulation 11: 171-188

INORGANIC CARBON ASSIMILATION AND GROWTH OF AQUATIC MACROPHYTES

Tom Vindbæk Madsen
Department of Plant Ecology
Aarhus University
Nordlandsvej 68
DK-8240 Risskov
Denmark

ABSTRACT

A major constraint associated with photosynthesis underwater is the rate of supply of dissolved inorganic carbon. The potential limitation of photosynthesis by the rate of supply of carbon is a result of low diffusion rates in water combined with a thick diffusive boundary layer. Aquatic macrophytes possess several features regarded as adaptations which alleviate carbon constraints. These include use of the rich sediment and atmospheric CO_2 pools in conjunction with physiological and biochemical mechanisms which enhance the carbon uptake efficiency (use of bicarbonate and C4 acids). Physiological and biochemical features are particularly plastic, and only operate in plants under carbon stress. Inorganic carbon availability affects the growth of submerged macrophytes at both sub-saturating and saturating light levels and is potentially limiting to growth under natural conditions.

INTRODUCTION

The aquatic environment is inhabited by a variety of photosynthetic organisms which encompass small unicellular bacteria, unicellular and multicellular algae and macroalgae, and rooted angiosperms of considerable dimensions. The success of these plants in growing and reproducing in water shows that the aquatic environment is a feasible habitat for photosynthetic organisms. However, there are numerous constraints on photosynthesis and growth in water, and the aquatic environment is usually considered to be less suitable for plant growth than the terrestrial environment. The most important external factors limiting photosynthesis in water are probably inorganic carbon and light, but other physical and chemical constraints may also be of significance. Among the advantages of growing submerged, the more significant are the unlimited supply of water and the support

NATO ASI Series, Vol. I 16
Interacting Stresses on Plants in a Changing Climate
Edited by M. B. Jackson, C. R. Black

provided by water which reduces the requirements for supporting tissue in macroscopic water plants.

This chapter describes the most pronounced differences between the terrestrial and aquatic environments and discusses the constraints associated with growth under water, with emphasis on the conditions important for the growth of submerged freshwater macrophytes. The efficiency and restrictions imposed by the structural and physiological features of submerged macrophytes which are believed to alleviate these constraints will also be discussed. In particular, features important for inorganic carbon uptake and photosynthesis will be considered.

AIR AND WATER AS ENVIRONMENTS FOR PHOTOSYNTHESIS

The photosynthesis and growth of terrestrial plants are often limited by the availability of water, but this is obviously not the case for water plants. The water potential of fresh water is close to zero, which effectively eliminates the need for stomata or other protective features against transpiration. The need for extensive water transport systems is also reduced. Consequently, vascular tissues are reduced in most submerged macrophytes, though they are still operative in many species, presumably for the uptake of mineral salts from the sediment. The sediment is the major source for phosphorus, nitrogen, iron and most micronutrients for rooted submerged macrophytes, whereas calcium, magnesium and a number of other salts abundant in the water are taken up through the leaves (Barko and Gunnison, 1991). Other advantages of growing in water include the high density of water relative to air and its high heat capacity. The former provides mechanical support for the plants and reduces the need for non-photosynthetic supporting tissue, which in terrestrial plants makes up a substantial fraction of the total shoot biomass. The high heat capacity of water means that diurnal and seasonal temperature fluctuations are reduced compared to air. The likelihood that submerged macrophytes will be exposed to extreme and potentially harmful temperatures is therefore reduced.

The main constraints associated with life in water are the reduced light availability, and presumably of more importance, the low exchange rate of gases and solutes between the open water and the plant surfaces. Other significant constraints include the lack of a transpiration stream to assist nutrient uptake from the sediment, the frequently anoxic sediments, mechanical stress imposed by moving water and sediment instability. The latter two features are particularly important in streams and on exposed shores.

Except for plants positioned very close to or on the water surface, water is a shade environment. Light is absorbed and scattered by dissolved inorganic and organic matter and suspended particles (e.g., phytoplankton, algae), resulting in a logarithmic attenuation of light with depth. The attenuation can be described by $I_z=I_0e^{-kz}$, which gives the remaining light intensity I_z after passing through a water column of thickness z (in metres) with a specific attenuation constant k. I_0 is the light intensity at the water surface. The attenuation constant varies from less than 0.3 in clear water lakes to more than 0.9 in eutrophic and highly productive lakes. Consequently, the depth at which light is reduced to 1% of the surface irradiance varies from more than 15 to less than 0.05 m.

The low exchange rate of gases and solutes results from a combination of the reduced diffusion rates in water and a substantial boundary layer. All exchange across the boundary layer is by diffusion, which is about 10^4 times slower in water than in air. The thickness of the boundary layer surrounding submerged macrophytes varies from 100-500 μm and changes with the morphology of the plants and the flow rate of water over the leaves (Wheeler, 1980; Smith and Walker, 1980; Madsen, 1984). For terrestrial plants, the boundary layer is only a few micrometres thick (Raven, 1984). The exchange rate across the boundary layer can be described by Fick's First Law of Diffusion, $F = D(C_e-C_i)\delta^{-1}$, where F is the flux, D is the diffusion coefficient, C_e and C_i the concentration of the diffusing substance in the external medium and at the leaf surface respectively, and δ is the length of the diffusion path. Thus, from a photosynthetic point of view, to obtain the same flux of CO_2 across the boundary layer in air and water and hence the same rate of photosynthesis, the CO_2 concentration in water must be more than 10^4 higher than in air. However, the concentration of CO_2 in water in equilibrium with air is about the same as in air. Higher CO_2 concentrations may be encountered, particularly in groundwater-fed streams and in stagnant water close to the sediment surface, but even in these habitats the CO_2 concentration seldom exceeds 350 μM (Maberly, 1985; Rebsdorf *et al.*, 1991). CO_2 is not the only carbon species in water, however, since two other forms are present, bicarbonate (HCO_3^-) and carbonate (CO_3^{2-}). The three forms interact with water in a series of chemical equilibria controlled by pH following the scheme:

$$CO_2 + H_2O \rightleftharpoons H_2CO_3$$

$$H_2CO_3 \rightleftharpoons HCO_3^- + H^+$$

$$HCO_3^- \rightleftharpoons CO_3^{2-} + H^+$$

$$CO_2 + OH^- \rightleftharpoons HCO_3^-$$

In fresh water, the bicarbonate concentration varies typically from close to zero in acid lakes and streams to over 5 mM in hardwater systems. In seawater the concentration is less variable at about 2.3 mM. All submerged macrophytes can use CO_2 in photosynthesis; in addition, some (about 50% of the species tested) can use HCO_3^- whereas carbonate seems to be inaccessible (Spence and Maberly, 1985; Maberly, 1992). The ability to use bicarbonate for photosynthesis allows exploitation of a substantially larger carbon pool than that available to species restricted to CO_2 use alone. However, even for HCO_3^- users the availability of carbon in water is greatly restricted as compared to air.

The exchange rate of inorganic carbon across the air-water interface is a slow diffusive process and may be insufficient to counterbalance carbon uptake by the plants, causing the carbon content of the water to decline. Such photosynthetic carbon depletion is particularly common in productive lakes in the summer, where the concentration of CO_2 and HCO_3^- may be reduced almost to zero and pH may rise to above 10 (Van *et al.*, 1976; C. Howard-Williams, personal communication, 1992), which further restricts the availability of inorganic carbon in water as compared to air, where diurnal changes in CO_2 are minimal. The low availability of inorganic carbon in water is reflected by the low inorganic carbon affinities and high carbon saturation needs of submerged macrophytes. The CO_2 half-saturation constants for submerged macrophytes range typically from 60 to 300 μM and the CO_2 concentrations needed to saturate photosynthesis are about 350-650 μM. For bicarbonate-users, the HCO_3^- half-saturation constants are about 0.5-8 mM and the saturation needs are commonly above 1.5-2.0 mM (Allen and Spence, 1981; Sand-Jensen, 1983; Madsen, 1991; Sand-Jensen *et al.*, 1992; Madsen, personal observation). Comparison of these kinetic figures with the naturally occurring concentrations of CO_2 (0-350 μM) and HCO_3^- (0-5 mM) emphasises the potential of inorganic carbon to limit photosynthesis.

CARBON ASSIMILATION IN AQUATIC MACROPHYTES

Plants growing under water show a number of features which can be seen as adaptations to submergence. The most common, and nearly universal among submerged freshwater macrophytes, are reduced leaf thickness, thin cuticle and epidermal chloroplasts. All these features reduce the transport resistance to CO_2 (and other solutes) and enhance the carboxylation efficiency at low external concentrations of inorganic carbon. In addition, a number of other, less universal, adaptations to submergence are found. These characteristics, which may be restricted to few species or common to larger groups, tend to alleviate the carbon constraints either by allowing the plants to explore alternative carbon sources or by enhancing the carbon use efficiency through physiological or biochemical mechanisms. The former are here referred to as exploitation strategies and the latter as carbon-concentrating mechanisms, acknowledging that these mechanisms may enhance the internal concentration of inorganic carbon in the cells above the level that can be obtained and maintained by passive uptake of CO_2.

Exploitation strategies. As a result of heterotrophic activity, the CO_2 concentration in lake and stream sediments is often orders of magnitude higher than in the overlying water. A few submerged plant species restricted to oligotrophic, softwater lakes, including *Littorella uniflora, Lobelia dortmanna*, and species belonging to the genus *Isoetes* possess features which allow them to exploit this CO_2 source (Wium-Andersen, 1971; Søndergaard and Sand-Jensen, 1979;

Richardson *et al.*, 1984; Madsen, 1987a). These features include low stature, rosette growth form, short subulate leaves, large root surface area, high root : shoot ratio and a well-developed longitudinal air-channel system running from the roots to the shoots. These structures provide an efficient short-distance transport route for sediment-CO_2, a large CO_2 assimilating interface with the environment and a low resistance diffusion pathway for CO_2. However, since transport is by diffusion and therefore inherently slow, constraints are imposed on the morphology of plants that make use of sediment-CO_2 for photosynthesis. In particular, these limits set the maximum leaf length and minimum cross-sectional area of the lacunal system in roots, stems and leaves. Consequently, plants with long thin stems, less developed lacunae and leaves that reach to the water surface, which is the predominant growth form for many submerged macrophytes, are unable to acquire sufficient CO_2 from the sediment to cover their needs (Loczy *et al.*, 1983; Madsen and Sand-Jensen, 1991).

The amount of carbon acquired through the roots compared with that absorbed through the leaves varies with the concentration of sediment-CO_2 around the roots relative to that around the leaves. In *Littorella uniflora* (L.) Aschers, 25-60% of the total quantity of carbon fixed is acquired through the roots at a sediment : water CO_2 ratio of 1, as compared to >95% at a sediment : water ratio of 40 (Boston *et al.*, 1987a,b). Under natural conditions, the sediment : water CO_2 ratio varies from 10 to 50. Despite high sediment-CO_2 concentrations, the carbon supply may still limit photosynthesis in species relying on sediment-CO_2, although the availability may be as good as or even better than in air. Terrestrial forms of *Littorella uniflora* maintain their utilisation of sediment-CO_2, which constitutes more than 80% of the total carbon uptake, at rhizopheric CO_2 concentrations above 0.5 mM (Nielsen *et al.*, 1991).

Aerial leaves. Many aquatic macrophytes have the ability to develop aerial leaves at some stage during their life cycle. From a photosynthetic point of view, access to the atmosphere is beneficial because: 1) the transport resistance is lower than in water, which enhances the availability of inorganic carbon and; 2) the transpiration stream may enhance nutrient uptake from the sediment (Bowes, 1987). The aerial leaves may either be similar to submerged leaves (homophyllous) or be of a different morphological type (heterophyllous).

The restriction imposed on photosynthesis by the high transport resistance in water can be evaluated by measuring photosynthesis for the same leaf in air and in water. For submerged leaves of *Myriophyllum brasilience*, the half-saturation constant for CO_2 was 706 μM in water but only 122 μM in water vapour saturated air (Salvucci and Bowes, 1982). In accordance with this, Nielsen (1992) found that the initial slope of photosynthesis versus CO_2 concentration was 5-8 times steeper in air than in water for a number of submerged amphibious species.

Aerial leaves generally have higher carbon and light saturated rates of photosynthesis than submerged leaves (Salvucci and Bowes, 1982; Nielsen, 1992), a difference which is accompanied by a higher ribulose bisphosphate carboxylase-oxygenase (Rubisco) content in aerial than in submerged leaves (Salvucci and Bowes, 1982; Spencer and Bowes, 1985; Farmer *et al.*, 1986). The carboxylation efficiency is also usually higher in aerial than in submerged leaves, at least when measured in air (Salvucci and Bowes, 1982; Nielsen, 1992). When measured in water, the difference may be negligible, as was found in a comparative study of submerged, floating and aerial leaves of *Batrachium peltatum* (Nielsen, 1992). The apparently higher metabolic capacity of aerial as compared to submerged leaves makes the former well suited to take advantage of the higher CO_2 supply rate in air and may reflect a physiological response to the enhanced CO_2 supply in air.

The higher photosynthetic efficiency in air than in water is also reflected by differences in the growth rates of plants with and without access to the atmosphere. *Callitriche cophocarpa*, a stream macrophyte which forms both submerged leaves and apical rosettes of floating leaves, grew about twice as fast in air as in water at 50 μM CO_2 (a common concentration in streams inhabited by *Callitriche*) (Fig. 1). However, the difference in growth rate between air and water could be eliminated by increasing the CO_2 concentration of the water and thereby enhancing the supply of CO_2 to the submerged plants. At 300 μM CO_2 in the water, the growth rate was the same as that for plants in air, while submerged plants exposed to dissolved CO_2 above 300 μM performed better than those which had access to the atmosphere.

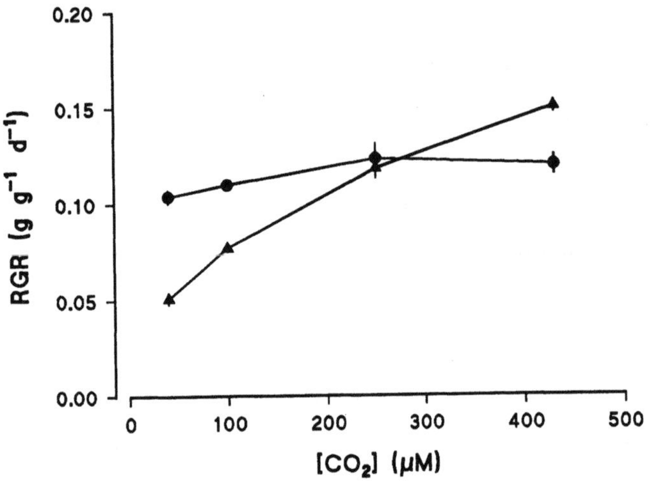

FIGURE 1. The relative growth rate (g g^{-1} d^{-1}) of Callitriche cophocarpa *in relation to the CO_2 concentration of the water. The growth rates were determined for plants grown at 15 °C and 200 μmol PAR m^{-2} s^{-1} with (●) or without (▲) access to the atmosphere. The CO_2 concentration in the atmosphere was kept at 350 ppm. Mean±S.E. Redrawn from Breinholt (1989).*

Concentrating mechanisms. Many submerged aquatic macrophytes do not follow the scheme predicted by CO_2 uptake and C3 photosynthesis. The photorespiratory activity is low (as indicated by low CO_2 compensation points), the oxygen sensitivity of photosynthesis is low, and the inorganic carbon compensation point in alkaline media is lower than can be explained by the effects of CO_2 uptake alone. It is therefore evident that submerged macrophytes must possess carbon uptake systems that supplement CO_2 use and C3 photosynthesis and modify their photosynthetic characteristics. These systems, which have been described in detail during the last decade or so, include utilisation of HCO_3^- ions and some form of C4 photosynthesis. The most widespread of these is bicarbonate use, which has been found in approximately 50% of the species tested (Spence and Maberly, 1985). C4 photosynthesis and the use of C4 acids have, on the other hand, been described only in a few species (Keeley, 1981; Madsen, 1985; Bowes and Salvucci, 1989).

The ability to utilise HCO_3^- for photosynthesis substantially enhances the carbon pool available to the plants, since the HCO_3^- content in water is 10 to 50 times higher than the CO_2 content. For example, in a medium with a dissolved inorganic carbon concentration of 1 mM at air-equilibrium of CO_2 (11.9 μM at 25 °C), approximately 3% of the carbon is available to plants restricted to CO_2 use, whereas bicarbonate users can extract about 80% (assuming a CO_2 compensation point of 4 μM and an HCO_3^- compensation point of 50 μM, and allowing for pH changes associated with photosynthesis). Furthermore, since the CO_2 affinity of CO_2-users and HCO_3^--users is about the same, the higher concentration of carbon available to the bicarbonate-users results in higher photosynthetic rates than in CO_2-users at any given sub-saturating CO_2 concentration (Madsen and Maberly, 1991). Whether this always translates into higher growth rates in bicarbonate-users is not known, but bicarbonate-users and non-users often grow next to each other in lakes and streams. The synthesis and maintenance of photosynthetic machinery with the ability to use bicarbonate is associated with higher carbon costs, which may counterbalance the benefit of higher carbon assimilation rates.

The information that has accumulated over the last decade shows that several mechanisms for exogenous HCO_3^- use may have evolved in aquatic macrophytes and several models to explain HCO_3^- uptake have been suggested. Two models, proposed for submerged angiosperms and *Chara*, suggest that the driving force for HCO_3^- use is an ATPase-mediated extrusion of H^+ from the cells to the boundary layer, but the actual carbon species suggested to cross the plasmalemma differs. The first model, and the most likely mechanism in *Chara* (Lucas, 1985), suggests that HCO_3^- is taken up *via* a H^+/HCO_3^- co-transport system in the plasmalemma. The other model, which operates in *Elodea canadensis* and in *Potamogeton* species (Prins *et al.*, 1979; Staal *et al.*, 1989), suggests that the periplasmic space and the boundary layer immediately adjacent to the leaf surface are acidified, which causes the equilibrium between HCO_3^- and CO_2 to shift towards CO_2. The CO_2 concentration in the boundary layer is thereby enhanced which creates a driving force for the diffusive entry of CO_2 into the cells. To avoid or reduce alkalinisation of the cytosol, H^+ extrusion has to be compensated for by OH^- extrusion (or H^+ uptake) and the two processes have to be spatially separated to avoid futile cycling of H^+ and OH^-. In *Chara*, the two processes are separated in distinct bands along the giant cells, a pattern which is seen as alternating acid and alkaline bands. In *Elodea canadensis* and *Potamogeton*, acidification is restricted to the abaxial leaf surfaces and alkalinisation to the adaxial. The pH difference between the two surfaces of the polarised leaves may be greater than 5 units (Elzenga and Prins, 1989). A consequence of the spatial separation of acidification and alkalinisation is that it limits the region of the leaf surface actually available for HCO_3^- uptake to only 50% in polarised leaves, for example. In accordance with this, the affinity and uptake capacity for HCO_3^- is often lower than for CO_2.

In species with polarised leaves or alkaline banding, the alkalinisation might be accompanied by precipitation of $CaCO_3$ and thick layers of marl are often found in these species, particularly in alkaline waters. Recently, calcification has been suggested to be more directly involved in HCO_3^- uptake rather than just being a inevitable consequence of the alkalinisation process. The model presented by McConnaughey (1991) suggests that Ca^{2+} extruded from the cells via a Ca^{2+} ATPase located within the alkaline bands precipitates with CO_2 which leaks from the cell. However, the loss of CO_2 is compensated for by the production of two CO_2 molecules from HCO_3^- ions using the protons generated in the calcification process. Evidence supporting this model has so far been obtained for *Chara corallina* (McConnaughey, 1991; McConnaughey and Falk, 1991).

The use of C4 acids for photosynthesis appears to have evolved along two different lines in submerged macrophytes. In *Littorella uniflora*, *Isoetes lacustris* and other isoetids a CAM-like (Crassulacean Acid Metabolism) system operates (Keeley, 1981; Richardson *et al.*, 1984; Madsen, 1985), while *Hydrilla verticillata* and other elodeids and some marine macroalgae have a photosynthetic system which physiologically and biochemically resembles that in terrestrial C4 plants (Bowes and Salvucci, 1984; Reiskind *et al.*, 1988).

The isoetid CAM plants show a net influx of inorganic carbon throughout most of the light and dark periods, and the internal supply of CO_2 due to decarboxylation of malate has only a small effect on the acquisition of external CO_2 in the light. Conversely, decarboxylation is inhibited when external CO_2 is high and saturating for photosynthesis (Madsen, 1987b). Thus, the CAM-like systems can improve the diurnal carbon balance in isoetids possessing the system. The isoetids all grow in oligotrophic, softwater lakes where carbon may be limiting for photosynthesis, even though the plants can acquire CO_2 from the sediment-CO_2 pool. In *Hydrilla*, which has a photosynthetic system resembling C4 photosynthesis, the enhanced internal concentration of inorganic carbon (Reiskind, personal communication, 1993) improves the carboxylation efficiency of Rubisco by inhibiting oxygenation and thereby enhances the carbon uptake efficiency. It can be calculated that aquatics with a C4-like system, but without the ability to use bicarbonate will have a carbon extraction capacity comparable to bicarbonate-users (Prins and Elzenga, 1989). It must also be expected that the carboxylation efficiency may increase above that of plants restricted to CO_2 use as a result of reduced oxygen inhibition, thus enhancing the rate of photosynthesis under conditions of limited carbon supply. However, in an environment where HCO_3^- is the quantitatively dominant form of inorganic carbon and the supply of inorganic carbon is limiting for photosynthesis, possession of a C4 system will probably not be sufficient to obtain *in situ* photosynthetic rates as rapid as those maintained by HCO_3^--users. Thus, the C4 system in aquatic plants may not be regarded as a primary mechanism for improving the inorganic carbon uptake capacity of the plants. Rather, the C4 system may be an important mechanism for ensuring efficient recycling of respiratory and photorespiratory CO_2. This could protect the plants from photoinhibitory damage and reduce carbon loss when the external carbon concentration is low, for example, under extreme conditions of photosynthetic carbon depletion. In this context, it is interesting to note that *Hydrilla* possesses an efficient HCO_3^- uptake system in addition to the C4 system (Madsen, personal observation), and can grow in dense mats where near zero CO_2 and HCO_3^- concentrations and high O_2 concentrations and high light often are encountered.

Aquatic plants with C4-like physiology lack Kranz anatomy. However, for the successful operation of a C4 system, spatial separation of the C4 and C3 carboxylation processes are crucial since this eliminates competition for CO_2. Using an immunocytochemical gold-labelling technique, Reiskind *et al.* (1989) found that in *Hydrilla* Rubisco was located in the chloroplast whereas phospho-enolpyruvate carboxylase was in the cytosol. This observation indicates that the C3 and C4 carboxylation processes take place in different cell compartments and that the successful operation of the C4 system depends on intracellular rather than intercellular separation. Whether this pattern is representative of other aquatic C4-like species is unknown.

The operation of HCO_3^- uptake systems, C4 systems or CAM requires investment in additional photosynthetic apparatus and may increase the running costs of inorganic carbon uptake above those associated with passive diffusive entry of CO_2 and C3 photosynthesis. These investments are very difficult to quantify and will be beneficial only if they lead to a higher net carbon gain, increased N-use efficiency or in other ways improve the photosynthetic efficiency of the plants. This implies that they are competitively advantageous only under a confined range of environmental conditions. Intuitively, it seem obvious that the ability to use bicarbonate, for example, will not be an important feature for the competitive outcome among species in waters with high CO_2 content and low bicarbonate. However, aquatic macrophytes show a striking intraspecific plasticity with respect to biochemistry and physiology, and tend to suppress the synthesis of accessory carbon uptake systems when, from a teleological viewpoint, they are not needed (Salvucci and Bowes, 1981; Sand-Jensen and Gordon, 1986). This can be exemplified by a series of experiments conducted with *Ranunculus peltatum*, a stream macrophyte which has the potential to use bicarbonate. The plants were grown in water with 0.5 mM HCO_3^- and CO_2 concentrations varying from 20 µM to 600 µM. After two weeks of growth, the CO_2 and HCO_3^- affinity (defined as the initial slope of photosynthesis versus CO_2 or HCO_3^-) was measured (Fig. 2). The affinity for both CO_2 and HCO_3^- decreased with increasing CO_2; however, the decrease in HCO_3^- affinity was more pronounced and at 600 µM CO_2 the ability

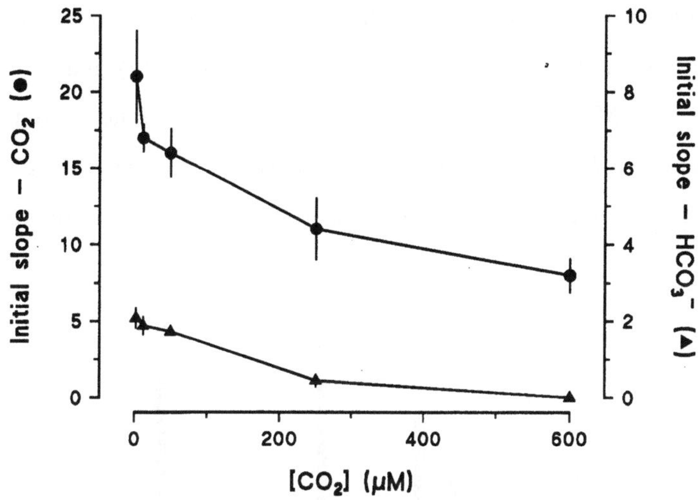

FIGURE 2. Initial slope of the CO_2 (●) and HCO_3^- (▲) response curves (μmol O_2 g^{-1} DW h^{-1}/mM CO_2 or HCO_3^-) for Ranunculus peltatum *grown at varying CO_2 concentrations at 15 °C and 400 μmol m^{-2} s^{-1}. Mean±S.D.*

to use HCO_3^- appeared to be totally suppressed. The light- and carbon-saturated rates of photosynthesis were not affected by increasing the CO_2 used during the growth period (data not shown), indicating that the observed response to enhanced CO_2 probably was not related to any toxic effect of CO_2. The capacity for acclimation in response to environmental variables has also been observed in the C4-like *Hydrilla* as low temperatures and short days suppress its ability to operate the C4 system (Salvucci and Bowes, 1981). In aquatics without accessory carbon uptake systems, CO_2 affinity has also been observed to change in response to CO_2 availability, although the change was small (Madsen, 1991). Thus, besides being the substrate for photosynthesis, inorganic carbon has other growth-regulating effects important for the physiology of submerged macrophytes.

GROWTH OF SUBMERGED MACROPHYTES

The constraints imposed by inorganic carbon on the photosynthesis of submerged macrophytes are also reflected in constraints on growth, although the number of studies conducted is small (Madsen and Sand-Jensen, 1987; Titus, 1992). In outdoor flow-through tanks exposed to full sun, *Elodea canadensis* grew more slowly at low inorganic carbon concentrations (0.1 mM) than at high concentrations (1 mM) at air-levels of CO_2 (Madsen and Sand-Jensen, 1987). In *Ranunculus aquatilis* grown at varying CO_2 and HCO_3^- concentrations, both carbon species had a stimulatory effect on growth (Fig. 3). The affinity for CO_2 was

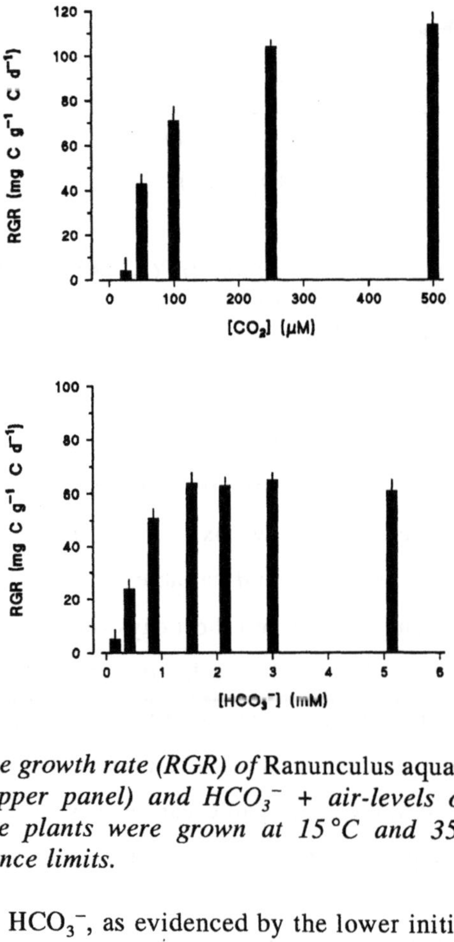

FIGURE 3. Relative growth rate (RGR) of Ranunculus aquatilis *(mg g^{-1} d^{-1}) grown at varying CO_2 (upper panel) and HCO_3^- + air-levels of CO_2 (lower panel) concentrations. The plants were grown at 15 °C and 350 µmol m^{-2} s^{-1} PAR. Mean±95% confidence limits.*

higher than that for HCO_3^-, as evidenced by the lower initial slope of the plot of growth against HCO_3^- (70 mg C g^{-1} C d^{-1} mM^{-1}) as compared to growth against CO_2 (1500 mg C g^{-1} C d^{-1} mM^{-1}) and the lower half-saturation constant for HCO_3^- (0.8 mM) than for CO_2 (0.05 mM). Lower maximum growth rates were also obtained when HCO_3^- was used as the carbon source; 0.065 mg C g^{-1} d^{-1} compared to 0.12 at CO_2 saturation. The different growth capacity of *Ranunculus* in response to CO_2 and HCO_3^- parallels the photosynthetic responses to CO_2 and HCO_3^- generally observed in bicarbonate-users (Sand-Jensen, 1983) and probably reflects the different uptake mechanisms for the two carbon species.

Enhanced inorganic carbon levels not only increase the carbon supply to the plants but also reduce inefficiencies in carbon uptake by suppressing photo-respiration. Inorganic carbon availability is therefore likely to affect plant growth not only at saturating but also at sub-saturating light levels. Experiments with *Elodea canadensis* have demonstrated that the light-use efficiency at low light is strongly affected by carbon availability, as is the maximum growth rate under conditions of high light (Fig. 4). Concomitant with the enhanced light use

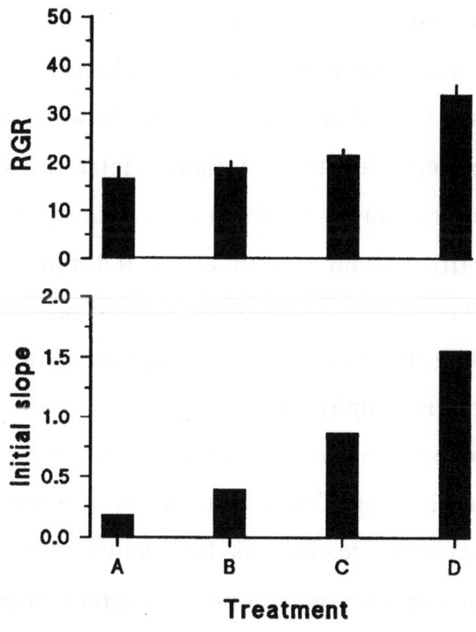

FIGURE 4. Upper panel, maximum relative growth rate (RGR, mg C g^{-1} C d^{-1}) and lower panel, the initial slope of RGR versus light (mg C g^{-1} C d^{-1}/μmol PAR m^{-2} s^{-1}) for Elodea canadensis *grown at 7°C at varying light intensities under four different carbon regimes providing increasing levels of carbon availability: A, 0.2 mM HCO_3^- and air-levels of CO_2; B, 1 mM HCO_3^- and air-levels of CO_2; C, 1 mM HCO_3^- and 150 μM CO_2; D, 1 mM HCO_3^- and 600 μM CO_2. Means±95% confidence limits for RGR.*

efficiency, the compensation point declined from 18 μmol m^{-2} s^{-1} at low carbon availability to 5 μmol m^{-2} s^{-1} at high carbon availability (data not shown). The light compensation point and the light use efficiency at low light usually set the lower depth limits for biomass production and survival of submerged macrophytes. Inorganic carbon availability can therefore be predicted to affect not only growth close to the water surface, but also growth at greater depths and the depth to which submerged macrophytes will penetrate.

The experiments referred to above have all been performed in tanks under laboratory or semi-natural conditions. Whether the results can be extrapolated to natural conditions is therefore open to question. In a first attempt to address this question, a recent study examined the effects of CO_2 and HCO_3^- enrichment on the *in situ* growth of *Elodea canadensis* (Vadstrup, 1992). The experiments were carried out in a meso-trophic softwater lake ([DIC]: 0.3-0.4 mM and CO_2 concentrations close to air-levels were maintained during the experimental period) during which the CO_2 and HCO_3^- concentrations of the water immediately adjacent to the plants were enhanced. Elevated CO_2 (approximately 500 μM) increased the relative growth rate of the plants by about 60% relative to the control and an increase in [HCO_3^-] from about 0.3 to about 0.4 mM enhanced growth by approximately 25%. This result resembled the pattern observed in the laboratory experiments, emphasising the importance of inorganic carbon as a limiting factor for growth in submerged macrophytes.

The range of inorganic carbon levels applied in the experiments referred to above are comparable to the range found in natural waters inhabited by submerged macrophytes, at least in temperate climates. Inorganic carbon limitation on growth may therefore be a common phenomenon under natural conditions. However, rate limitation should be distinguished from yield limitation (O'Brien, 1972), and so it is not possible to conclude that inorganic carbon will be the main limiting factor for biomass. Other factors such as light and nutrient availability and mechanical factors that affect loss of biomass (for example, grazing, water flow and other physical disturbances) may be more important. Inorganic carbon will only be limiting if the growth rates set by carbon are so low that the limits set by other factors are not reached during the growing season. It should be noted, however, that the maximum biomass level set by light may be modified by inorganic carbon availability through its effects on minimum light requirements for growth.

ACKNOWLEDGEMENTS

I thank Dr George Bowes for comments on the manuscript.

REFERENCES

Allen ED, Spence DHN (1981) The differential ability of aquatic plants to utilize the inorganic carbon supply in freshwaters. The New Phytologist 87: 269-283

Barko JW, Gunnison D (1991) Sediment interactions with submersed macrophyte growth and community dynamics. Aquatic Botany 41: 41-65

Boston HL, Adams MS, Pienkowski TP (1987a) Utilization of sediment CO_2 by selected North American isoetids. Annals of Botany 60: 485-494

Boston HL, Adams MS, Pienkowski TP (1987b) Models of the use of root-CO_2 by selected North American isoetids. Annals of Botany 60: 495-503

Bowes G (1987) Aquatic plant photosynthesis: Strategies that enhance carbon gain. In: Crawford RMM (ed) Inorganic carbon uptake by aquatic photosynthetic organisms. American Society of Plant Physiologists pp 187-210

Bowes G, Salvucci ME (1984) *Hydrilla*: inducible C_4-type photosynthesis without Kranz anatomy. In: Sybesma C (ed) Advances in photosynthesis research. Vol. III. Martinus Nijhoff, The Hague, The Netherlands, pp 829-832

Bowes G, Salvucci ME (1989) Plasticity in the photosynthetic carbon metabolism of submersed aquatic macrophytes. Aquatic Botany 34: 233-266

Breinholt M (1989) Vækst hos flydebladsplanter og undervandsplanter af *Callitriche stagnalis* Sendtner. MS Thesis, Aarhus University, Denmark

Elzenga JTM, Prins HBA (1989) Light-induced polar pH changes in leaves of *Elodea canadensis*. I. Effects of carbon concentration and light intensity. Plant Physiology 91: 62-67

Farmer AM, Maberly SC, Bowes G (1986) Activities of carboxylation enzymes in freshwater macrophytes. Journal of Experimental Botany 37: 1568-1573

Keeley JE (1981) *Isoetes howellii*: a submerged aquatic CAM plant? American Journal of Botany 68: 420-424

Loczy C, Carignan R, Planas, D (1983) The role of roots in carbon uptake by the submerged macrophyte *Myriophyllum spicatum*, *Vallisneria americana* and *Heteranthera dubia*. Hydrobiologia 98: 3-7

Lucas WJ (1985) Bicarbonate utilization by *Chara*: a re-analysis. In: Lucas WJ, Berry JA (eds) Inorganic carbon uptake by aquatic photosynthetic organisms. American Society of Plant Physiologists, Rockville MD, USA, pp 229-254

Maberly SC (1985) Photosynthesis by *Fontinalis antipyretica* II. Assessment of environmental factors limiting photosynthesis and production. The New Phytologist 100: 141-155

Maberly SC (1992) Carbonate ions appear to neither inhibit nor stimulate use of bicarbonate ions in photosynthesis by *Ulva lactuca.* Plant, Cell and Environment 15: 255-260

Madsen TV (1984) Resistance to CO_2 fixation in the submerged aquatic macrophyte *Callitriche stagnalis* Scop. Journal of Experimental Botany 35: 338-347

Madsen TV (1985) A community of submerged aquatic CAM plants in Lake Kalgaard, Denmark. Aquatic Botany 23: 97-108

Madsen TV (1987a) Sources of inorganic carbon acquired through CAM in *Littorella uniflora* (L.) Aschers. Journal of Experimental Botany 38: 367-377

Madsen TV (1987b) Interactions between internal and external CO_2 pools in the photosynthesis of the aquatic CAM plants *Littorella uniflora* (L.) Aschers. and *Isoetes lacustris* L. The New Phytologist 106: 35-50

Madsen TV (1991) Inorganic carbon uptake kinetics of the stream macrophyte *Callitriche cophocarpa* sendt. Aquatic Botany 40: 321-332

Madsen TV, Maberly SC (1991) Diurnal variation in light and carbon limitation of photosynthesis by two species of submerged freshwater macrophytes with a differential ability to use bicarbonate. Freshwater Biology 26: 175-187

Madsen TV, Sand-Jensen K (1987) Photosynthetic capacity, bicarbonate affinity and growth of *Elodea canadensis* exposed to different concentrations of inorganic carbon. Oikos 50: 176-182

Madsen TV, Sand-Jensen K (1991) Photosynthetic carbon assimilation in aquatic macrophytes. Aquatic Botany 41: 5-40

McConnaughey TA (1991) Calcification in *Chara corallina*: CO_2 hydroxylation generates protons for bicarbonate assimilation. Limnology and Oceanography 36: 619-628

McConnaughey TA, Falk RH (1991) Calcium-proton exchange during algal calcification. Biological Bulletin 180: 185-195

Nielsen SL (1992) Vækst og fotosyntesetilpasninger hos amphibiske planter. Ph.D. Dissertation, Aarhus University, Denmark

Nielsen SL, Gacia E, Sand-Jensen K (1991) Land plants of amphibious *Littorella uniflora* (L.) Aschers. maintain utilization of CO_2 from the sediment. Oecologia 88: 258-262

O'Brien WJ (1972) Limiting factors in phytoplankton algae: Their meaning and measurement. Science 178: 616-617

Prins HBA, Elzenga JTM (1989) Bicarbonate utilization: function and mechanism. Aquatic Botany 34: 59-83

Prins HBA, Snel JFH, Helder RJ, Zanstra PE (1979) Photosynthetic bicarbonate utilization in the aquatic angiosperm *Potamogeton* and *Elodea.* Hydrobiological Bulletin 13: 106-111

Raven JA (1984) Energetics and transport in aquatic plants. M.B.L. Lectures in Biology 4, AR Liss, New York, USA

Rebsdorf A, Thyssen N, Erlandsen M (1991) Regional and temporal variation in pH, alkalinity and carbon dioxide in Danish streams in relation to land use. Freshwater Biology 25: 419-436

Reiskind JB, Seamon PT, Bowes G (1988) Alternative methods of photosynthetic carbon assimilation in marine macroalgae. Plant Physiology 87: 686-692

Reiskind JB, Berg, RH, Salvucci ME, Bowes G (1989) Immunogold localization of primary carboxylase in leaves of aquatic macrophytes and a C_3-C_4 intermediate species. Plant Science 61: 43-52

Richardson K, Griffiths H, Reed JA, Griffiths NM (1984) Inorganic carbon assimilation in the isoetids *Isoetes lacustris* L. and *Lobelia dortmanna* L. Oecologia 61: 115-121

Salvucci ME, Bowes G (1981) Induction of reduced photorespiratory activity in submersed and amphibious macrophytes. Plant Physiology 67: 335-340

Salvucci ME, Bowes G (1982) Photosynthetic and photorespiratory responses of the aerial and submerged leaves of *Myriophyllum brasiliense*. Aquatic Botany 13: 147-164

Sand-Jensen K (1983) Photosynthetic carbon sources of stream macrophytes. Journal of Experimental Botany 34: 198-210

Sand-Jensen K, Gordon DM (1986) Variable affinity of *Elodea canadensis* Michaux in response to different HCO_3^- and CO_2 concentrations during growth. Oecologia 70: 426-432

Sand-Jensen K, Pedersen MF, Nielsen SL (1992) Photosynthetic use of inorganic carbon among primary and secondary water plants in streams. Freshwater Biology 27: 283-293

Smith FA, Walker NA (1980) Photosynthesis of aquatic plants: effects of unstirred layers in relation to assimilation of CO_2 and HCO_3^- and to isotope discrimination. The New Phytologist 86: 245-259

Spence DHN, Maberly SC (1985) Occurrence and ecological importance of HCO_3^- use among aquatic higher plants. In: Lucas WJ, Berry JA (eds) Inorganic carbon uptake by aquatic photosynthetic organisms. American Society of Plant Physiologists, Rockville MD, USA, pp 125-143

Spencer W, Bowes G (1985) *Limnophila* and *Hygrophila*: a review and physiological assessment of their weed potential in Florida. Journal of Aquatic Plant Management 23: 7-16

Staal M, Elzenga JTM, Prins HBA (1989) ^{14}C fixation by leaves and leaf cell protoplasts of the submerged aquatic angiosperm *Potamogeton lucens* L.: carbon dioxide or bicarbonate? Plant Physiology 90: 1035-1040

Søndergaard M, Sand-Jensen K (1979) Carbon uptake by leaves and roots of *Littorella uniflora* (L.) Aschers. Aquatic Botany 6: 1-12

Titus JE (1992) Submerged macrophyte growth at low pH. II. CO_2 x sediment interactions. Oecologia 92: 391-398

Vadstrup M (1992) Effekter af kulstofberigelse på vækst hos to submergente makrofyter *Elodea canadensis* og *Callitriche cophocarpa*. MS Thesis, Aarhus University, Denmark

Van TK, Haller WT, Bowes G (1976) Comparison of the photosynthetic characteristics of three submerged aquatic plants. Plant Physiology 58: 761-768

Wheeler WH (1980) Effect of boundary layer transport on the fixation of carbon by the giant kelp *Macrocystis pyrifera*. Marine Biology 56: 103-110

Wium-Andersen S (1971) Photosynthetic uptake of free CO_2 by the roots of *Lobelia dortmanna*. Physiologia Plantarum 25: 245-248

A MICROELECTRODE STUDY OF OXYGEN DISTRIBUTION IN THE ROOTS OF INTACT MAIZE SEEDLINGS

W. Armstrong, S. Cringle[1], M. Brown[1] and H. Greenway[2]
Department of Applied Biology
University of Hull
UK

ABSTRACT

A preliminary platinum microelectrode study of radial oxygen distribution in roots is described. Roots of intact maize seedlings were embedded in solid agar to impede the radial oxygen supply and ensure that there was a substantial dependence upon cortical gas-phase diffusion from the shoot. Radial oxygen profiles were measured at various positions along the roots using platinum microelectrodes driven through the agar and into the root, mostly in 5 μm steps, by a servo-operated piezoelectric microdriver. Any damage caused by the electrodes appeared to be very localised and there was no evidence that it greatly affected the oxygen regimes. In general, the profiles were in good agreement with earlier mathematical modelling predictions of oxygen distribution in roots and, by adjustment of shoot oxygen concentrations, some anoxia could be induced readily within the stele, while the cortex remained aerobic. Interestingly, whilst the meristematic region of the growing root appeared to be aerobic throughout, much of the root cap and some of the root cap material ensheathing the meristematic region were found to be anoxic.

INTRODUCTION

Oxygen enters roots in one or both of two ways, the balance between the two depending upon the relative magnitudes and distribution of diffusive resistances, tissue respiratory demands and oxygen source concentrations (Armstrong, 1979, 1988; Armstrong *et al.*, 1990). In drained soils or vigorously aerated culture solutions, inward radial diffusion through the root surface is likely to dominate; in waterlogged soil or stagnant culture solution, cortical gas-phase diffusion from

---

[1] Lions' Eye Institute, Queen Elizabeth II Hospital and
[2] Department of Crop and Pasture Science, University of Western Australia

NATO ASI Series, Vol. I 16
Interacting Stresses on Plants in a Changing Climate
Edited by M. B. Jackson, C. R. Black

shoot to root may be the only significant means of aeration and there may even be a substantial radial loss of oxygen from root to soil.

The first very clear demonstration of the limitations of radial aeration in roots was that of Berry and Norris (1949), who found that the respiratory critical oxygen pressure (COPR) of isolated onion root segments increased with temperature and respiratory demand. The most likely explanation for these findings was thought to be the development of a central stelar core of anoxic tissue at oxygen pressures below the COPR for the segments.

The mathematical modelling of diffusive aeration in roots has since demonstrated that there may be a considerable potential in roots for the development of some stelar anoxia whether oxygen is supplied by radial liquid-phase diffusion from the soil (or solution culture) (Armstrong and Beckett, 1985) *or* by cortical longitudinal gas-phase transport from the shoot (Armstrong and Beckett, 1987). It was suggested that relatively shallow radial oxygen gradients might be expected in the cortex, with relatively steep gradients in the stele, particularly in sub-apical zones; also that in growing roots, stelar anoxia in sub-apical regions might often precede meristem anoxia. A critical oxygen pressure for extension growth (COPE) in rice which was less than the COPR (Armstrong and Webb, 1985) has been interpreted as evidence of this, and strong support has recently emerged from comparative studies of ethanol production and PDC and ADH activities in the stele and cortex of growing maize roots (Thompson and Greenway, 1991). Again in growing maize roots, a peak in ACC and ethene production at cortical oxygen concentrations of approximately 3-5% has been thought to indicate stelar anoxia (Jackson *et al.*, 1985).

Direct oxygen measurements within roots have rarely been reported: almost the whole of our knowledge concerning the concentrations and distribution of oxygen in intact roots comes either from measurements of radial oxygen loss (Armstrong, 1979), mathematical modelling, or from metabolic evidence such as ethanol production and enzymatic activities (Thomson and Greenway, 1991) and energy charge values (Drew *et al.*, 1985). Apart from one radial profile study with mini-electrodes which shows a sharp drop in oxygen concentration within the endodermal boundary of mangrove pneumatophores (Andersen and Kristensen, 1988), we are unaware of any detailed radial oxygen distribution data extending to the stele of roots, although previous attempts have been made to use micro-electrodes for this purpose (Bowling, 1973; Hook and McKevlin, 1988).

In this chapter we describe the results of an exploratory microelectrode study of radial and longitudinal oxygen distribution in the primary roots of maize seedlings dependent for their oxygen supply chiefly upon gas-phase oxygen transport from the shoot. The data, which include radial oxygen profiles at several points along the root, confirm the profile pattern of the previous mathematical modelling and confirm that stelar anoxia can readily be induced. In addition, they provide a first insight into the oxygen distribution patterns in the meristem and root cap.

MATERIALS AND METHODS

Plant material. Seeds of maize (*Zea mays* hybrid GH5010) were imbibed for 14 hours in aerated de-ionised water and then germinated at 25 °C in the dark on plastic mesh suspended over aerated 0.5 mol m^{-3} calcium sulphate solution. Plants were transferred to the experimental system 4-5 days after germination when the shoots were approaching or at the first leaf stage and the roots were 60-90 mm long. Only plants with relatively straight roots were chosen for study.

Apparatus and procedures. Agar (1% w:v) containing KCl at a strength of 5 mol m^{-3} to act as a background electrolyte was poured while still hot into a cylinder containing the glass and Perspex assembly illustrated in Fig. 1. At this stage a glass rod of approximately the same diameter as the maize roots occupied the position in Fig. 1 shown as a hole in the agar. The rod was held in a position about 4 mm from the facing edge of the assembly by two plastic end caps (not shown). The sliding Perspex plates shown on this facing edge were held in position by a number of elastic bands sleeving the assembly. Once the agar had cooled and set, the whole assembly was withdrawn from the cylinder and any excess agar adhering to its outer surface was removed. The assembly was then submerged in

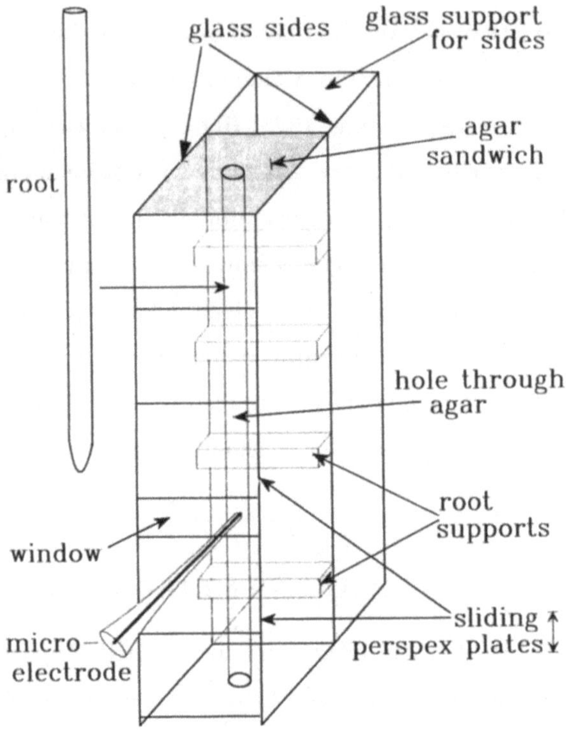

FIGURE 1. Diagrammatic representation (not to scale) of the assembly used to position roots in 1% agar to restrict the radial oxygen supply and provide ready access for microelectrode penetration.

water, the glass rod withdrawn to create a hole, and the end caps removed. A maize seedling was then gently manipulated so that its primary root slid into the hole in the agar as far as the root-shoot junction.

The assembly which now supported the plant with the root embedded in the agar was removed from the water with a finger retained over the lower end of the hole, and the base was made to stand in a reservoir of dilute electrolyte (5 mol m-3 KCl). The assembly was now secured firmly in the upright position, a small Perspex hood with gas-entry and exit ports was sleeved over the top of the assembly and sealed around it with Terostat to enclose the shoot, and the fibre junction of a Ag:AgCl anode was inserted into the electrolyte reservoir.

A glass-sheathed, bare-tipped platinum microelectrode (tip diameter *c.* 5 μm: Alder *et al.*, 1990) was secured on a piezoelectric driver and connected, together with the anode, to a polarographic circuit set at -0.6 V and the oxygen-reduction current output was measured to an accuracy of 1 pA. The electrode was viewed through a pivoting binocular microscope for positioning and during driving. The piezoelectric driver was computer-controlled (Cringle *et al.*, 1990) and set to move in steps of from 1-20 μm depending upon the requirements and at a frequency of one step every two seconds. The oxygen concentration was recorded at each step. The data were displayed on a computer screen and stored on disk before transferring to Sigmaplot software for graph plotting.

The gas composition around the shoot, supplied from O_2 and N_2 cylinders *via* a gas-mixer, exerted primary control on the oxygen concentration in the root since the 'radial' impedance to oxygen diffusion in the agar and the Perspex plates ensured that most of the oxygen reaching the root travelled via the cortical gas-space from the shoot. To allow the electrode entry to the agar and root, 'windows' were created as required by sliding apart the Perspex plates to a sufficient degree (see Fig. 1). The electrode was then driven in a direction normal to the root surface, and positioned close to the root surface so that it would subsequently follow a radial track through the root. Most profiles were begun with the electrode at or a few micrometres away from the root, and were usually programmed to extend for 1200 μm. This usually took the electrode through the root and a short distance beyond - the *in-track*. When set in reverse the electrode was drawn back again at the same step frequency - the *out-track*. One difficulty experienced, which was only partially overcome by constructing the root supports (Fig. 1), was that at various stages, particularly during in-tracks, the electrode moved the root for a few steps rather than penetrating further into the root. Consequently in-track and out-track data were often thrown out of phase and the radial dimensions of tissue zones can appear greater than their true values. One solution is to observe the root continuously during the trackings and note the extent of any root movements; superficial data can subsequently be omitted. A possible example of the problem can be seen in Fig. 3 at M.

Since the electrode response in the agar and root decayed to about one-sixth of that in air-saturated Ringers solution, the equilibrium current at the agar window-surface was calibrated to read air-saturation.

RESULTS AND DISCUSSION

Results at different positions along a single root are shown in Figs. 2-6, 8 and 9. The first example (Fig. 2) is of an oxygen profile across the root shown in perspective with the oxygen gradient developed in the agar on the 'window' side

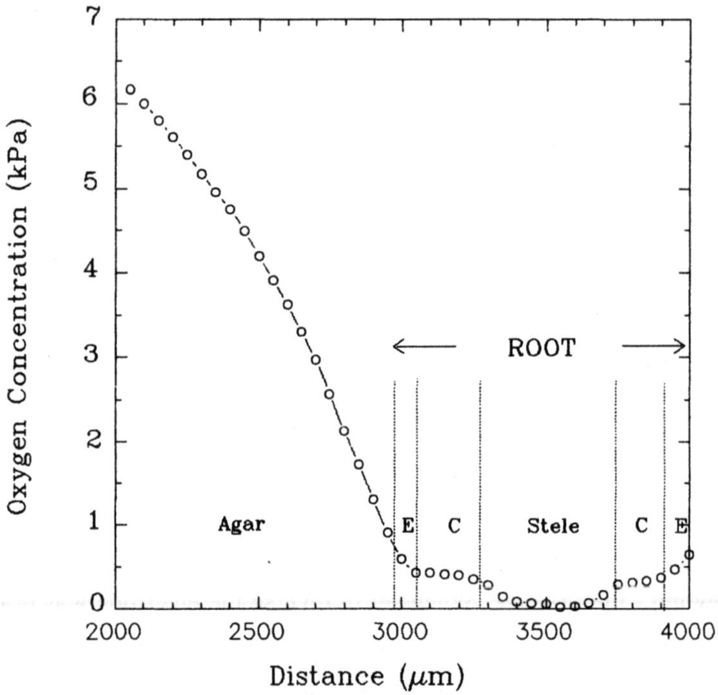

FIGURE 2. Microelectrode 'radial' oxygen profile across a primary root of maize and embedding agar. Maize shoot in 10% oxygen. Profile position 70 mm from base of 89 mm root. E = epidermal and hypodermal layers; C = cortex.

of the root. In this instance, since the measurements were 70 mm from the root-shoot junction and the shoot was in an atmosphere of 10% oxygen, oxygen concentrations in the root were very low and some oxygen flux from agar to root is apparent from the gradient through the epidermal/hypodermal cylinder; shallower gradients can be seen in the cortex with steeper ones within the stele, as anticipated by mathematical modelling.

A higher resolution out-track profile 35 mm from the base of the root is shown in Fig. 3. Out-track profiles were generally less 'noisy' than in-track profiles, a property attributed to what was later found to be very localised flooding

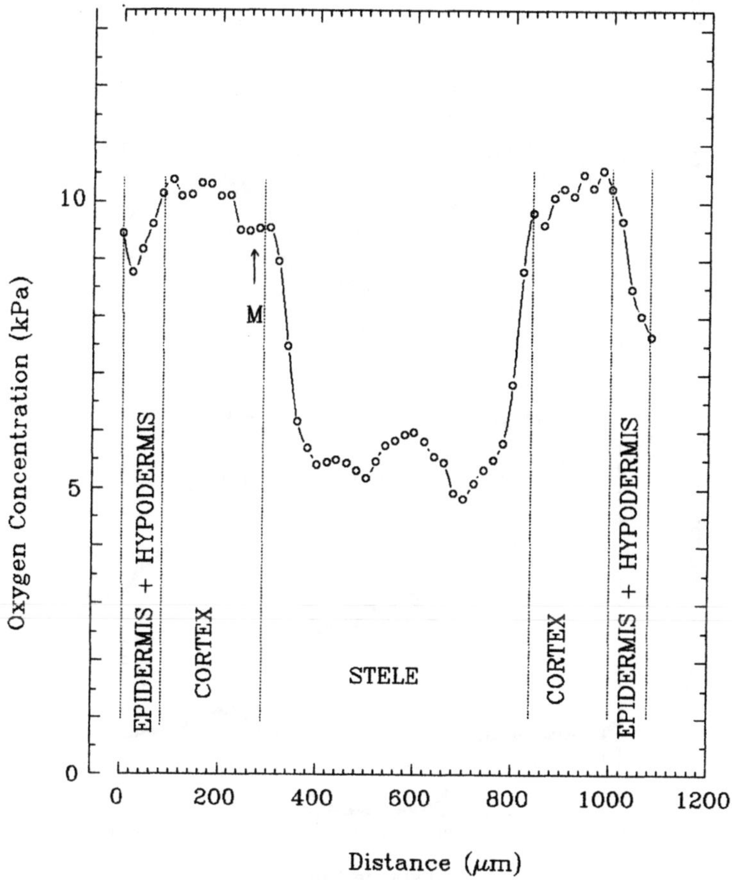

FIGURE 3. *Microelectrode polarographic 'radial' oxygen profile across a primary root of maize (L = 85 mm) at a position 35 mm from base. Root embedded in agar (1% w/v); oxygen supply chiefly from the shoot through the non-aerenchymatous cortex. M indicates the position at which the electrode has probably moved the root for a few steps.*

of the intercellular gas-spaces. The proximity of the electrode to these spaces is thought to account for the spikiness ('noise') in the cortex in the inflow plot (Fig. 4-1). In Fig. 3 it may be noted that a radial inflow of oxygen from the agar

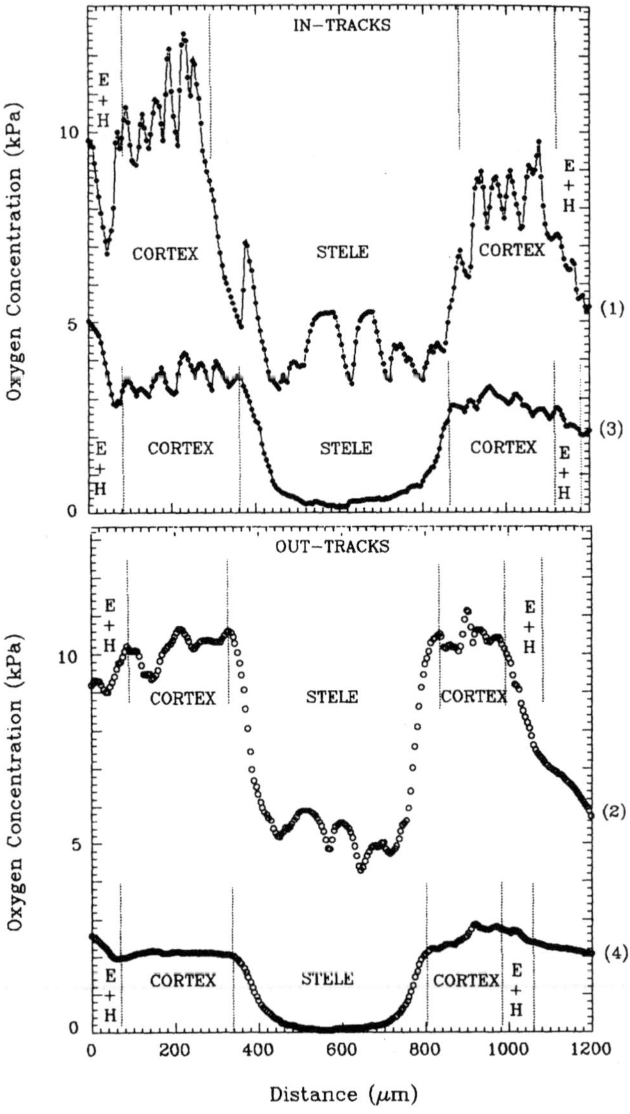

FIGURE 4. *Microelectrode polarographic 'radial' oxygen profiles across a primary root of maize (L = 85 mm) at a position 35 mm from base. Graphs (1) and (2), shoot in 21% oxygen; graphs (3) and (4) shoot in 10% oxygen. E + H = epidermis plus hypodermal layers.*

is only evident on the side of the root nearest to the agar window, and this oxygen is apparently consumed within the epidermal/hypodermal cylinder. On the far side of the root the gradient indicates a very definite radial outflow to the agar. These patterns are replicated in Fig. 4 in a second pair of profiles (in-track 1 and out-track 2) obtained at approximately the same level on the same root. The cortical oxygen concentration at 35 mm is approximately that which is mathematically predictable for non-aerenchymatous maize seedling roots from the respiration and porosity data of Thomson and Greenway (1991) and Thomson *et al.* (1990). Unfortunately, as transverse sections of the appropriate regions later demonstrated, it was not always possible to ensure a strictly radial track across the roots, and the tracks in Figs. 3 and 4 (1 and 2), although traversing the stele, were somewhat tangential. This probably accounts for the spikiness as well as the relative truncation of the curves in what might have been expected to be the centre of the pith region. The electrode was in fact passing across the inner fringe of the metaxylem elements and in-filling pith- or xylem-parenchyma. Nevertheless the steepness of the oxygen gradient within the outer part of the stele is very striking and indicative of high respiratory demand and/or much higher radial diffusive resistance than in the cortex. In the profiles shown, the oxygen concentration halved from 10 to 5% between the cortex and the apparent stelar centre and, had the profiles been strictly radial, an even greater decline could have been anticipated.

The profiles shown in Figs. 4-3 and 4-4 were obtained when the oxygen concentration around the shoot was lowered from 21 to 10%. As expected, the oxygen concentration in the cortex dropped by more than half, but unless respiratory demand across the transect had been reduced (e.g., in the stele), a simple displacement downwards of Fig. 4 (1 and 2) would have been expected (Armstrong and Beckett, 1985, 1987). Such a decline in stelar respiration could be indicated by the almost flat curve in the mid-stelar zone, and conditions here are clearly approaching anoxia. Again, however, this could have been a function of a tangential track across the stele, and the need to ascertain later the exact position taken by tracks by sectioning the roots is further emphasised. Subsequent measurements with the technique should resolve the question.

As expected, at 60 mm from the root-shoot junction the profiles showed much lower cortical and stelar oxygen concentrations (Figs. 5-1 and 5-2). Here,

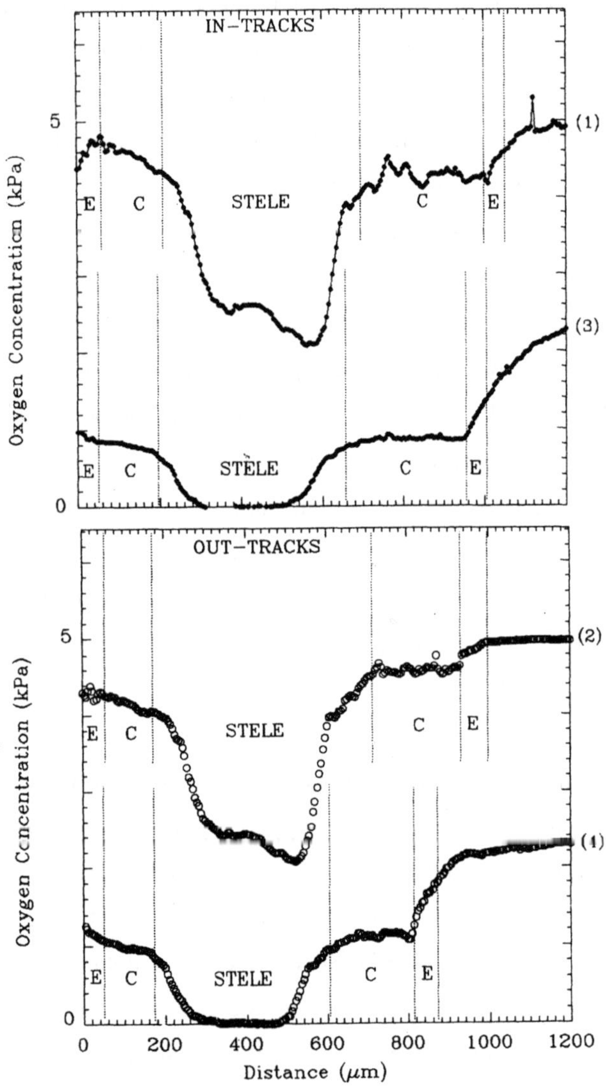

FIGURE 5. Microelectrode polarographic 'radial' oxygen profiles across primary root pf maize (L = 89 mm) at position 60 mm from base. Graphs (1) and (2), shoot in 21% oxygen; graphs (3) and (4), shoot in 10% oxygen. Dashed lines mark the approximate boundaries of the stele, cortex (C) and epidermal + hypodermal layers (E).

however, there appeared to be a shallow gradient across the cortex which might indicate a lack of substantial radial gas-space connection in the cortex. Again there was a steep decline in stelar oxygen, but the tracks were somewhat tangential and hence the full extent of the decline cannot be judged. However, at this level, the change to 10% oxygen around the shoot produced most definitely a core of stelar anoxia. The profile bottomed-out to zero oxygen near the pith boundary, the anoxic core diameter being at least 200 μm. So far as we are aware, this is the first direct measurement of stelar anoxia in roots.

At 70 mm from the base (Fig. 6), oxygen concentrations in the cortex were lower still, and some oxygen flux from the agar to the epidermis at both sides of the root was apparent even with the shoot in air. The gradients from cortex to epidermis and cortex to stele indicate radial flows from the cortex in both directions and further suggest some appreciable radial resistance to diffusion in the cortex at this level, as well as showing that the cortex is still the major oxygen source. The large changes effected by changing the shoot atmosphere to 10% oxygen also confirm the dominance of the cortex and longitudinal transport in the aeration of the root. The stele did not appear to develop an anoxic core even with 10% oxygen around the shoot but it came very close: the plot may have been too tangential to detect it. Plots 1 and 2, however, were very close to radial and, since the stelar concentration, although less than half that in the cortex, was only about 1.8 kPa lower in what must have been a very respiratorily active region, we tentatively conclude that the stelar diffusive resistance was significantly less here than at the 35 mm level (Figs. 3 and 4). This latter point is again suggested by the data in Fig. 8. The position of these transects is shown in Fig. 7 (upper line), and was across the meristematic region of cortex and stele. Again the stelar centre did not become anoxic, and in the cortex there were radial gradients in both directions. A most interesting feature of these profiles is the anoxia detected in the root cap material sheathing the tip at this level, and the gradients show oxygen being supplied to the cap material from both the agar and root.

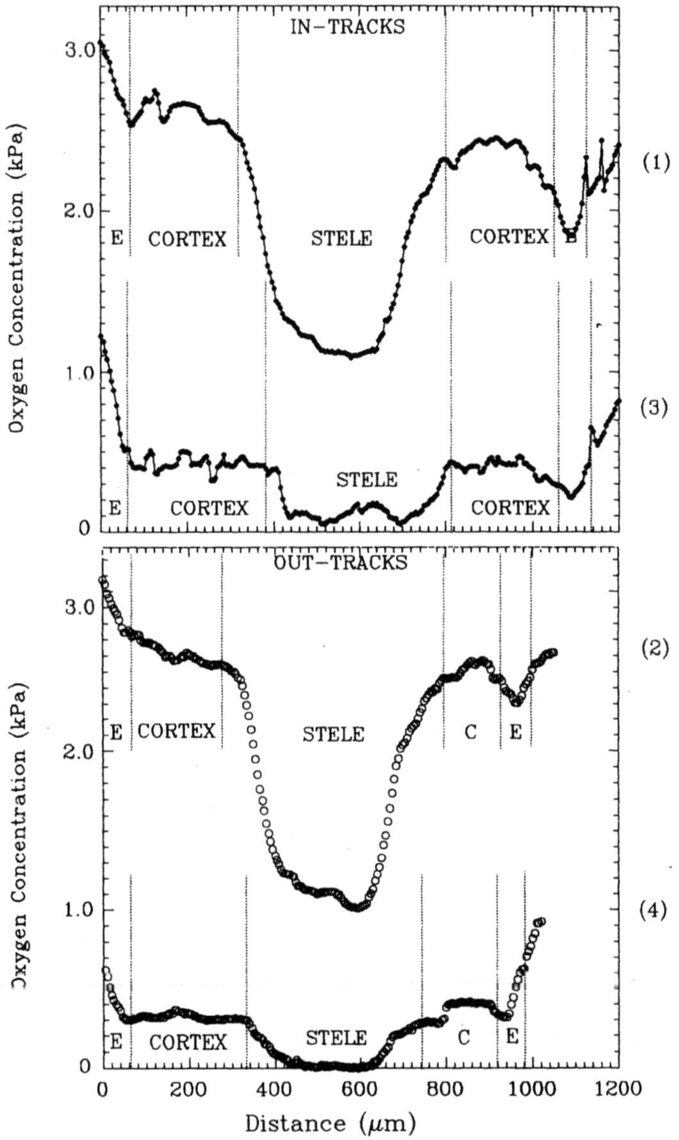

FIGURE 6. *Microelectrode polarographic 'radial' oxygen profiles across a primary root of maize (L = 90 mm) at a position 70 mm from base. Tracks (1) and (2), shoot in 21% oxygen; tracks (3) and (4), shoot in 10% oxygen. Dashed lines mark the approximate outer boundaries of the stele, cortex (C) and epidermal + hypodermal layers (E).*

The final transect (Fig. 9) passed through the root cap proper (see Fig. 7, lower line). At this level, the root cap was bathed in the electrolyte near the bottom of the hole and the oxygen concentration in the electrolyte was relatively high. Despite this, it can be seen that a considerable part of the track within the cap is anoxic. Since the root and cap continued to grow, we conclude that oxygen is not necessary throughout the cap at this level either for the growth of the root or cap. The very different lengths of the in-track and out-tracks were due to the considerable movement of the root by the electrode, particularly on the in-track.

Finally it should be noted that the damage caused by the electrode was sufficiently localised that it did not obviously interfere with growth (during the 12 hours of the experiment the root grew by *c.* 5 mm, and during the following 48 hours it grew by *c.* 30 mm).

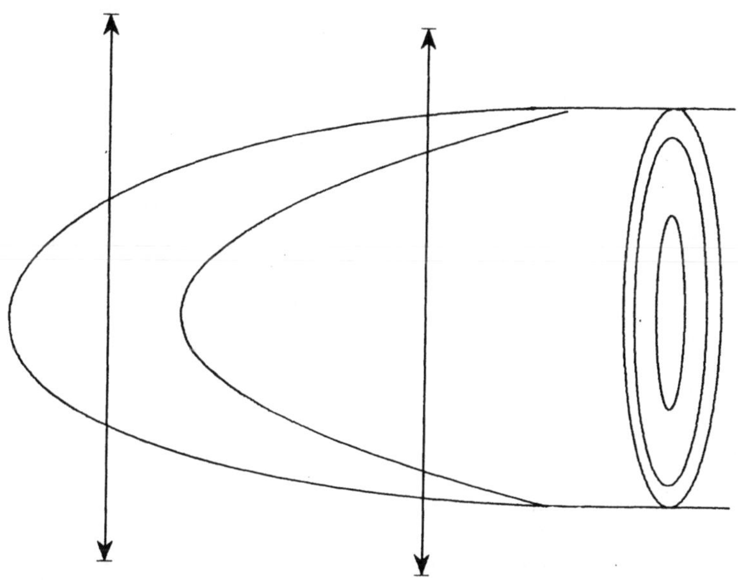

FIGURE 7. *Approximate positions of the electrode tracks in Figs. 8 and 9. Upper track (data shown in Fig. 8) passes through sheathing root cap material and through the cortical and stelar meristem; the lower track (data in Fig. 9) is through the root cap proper.*

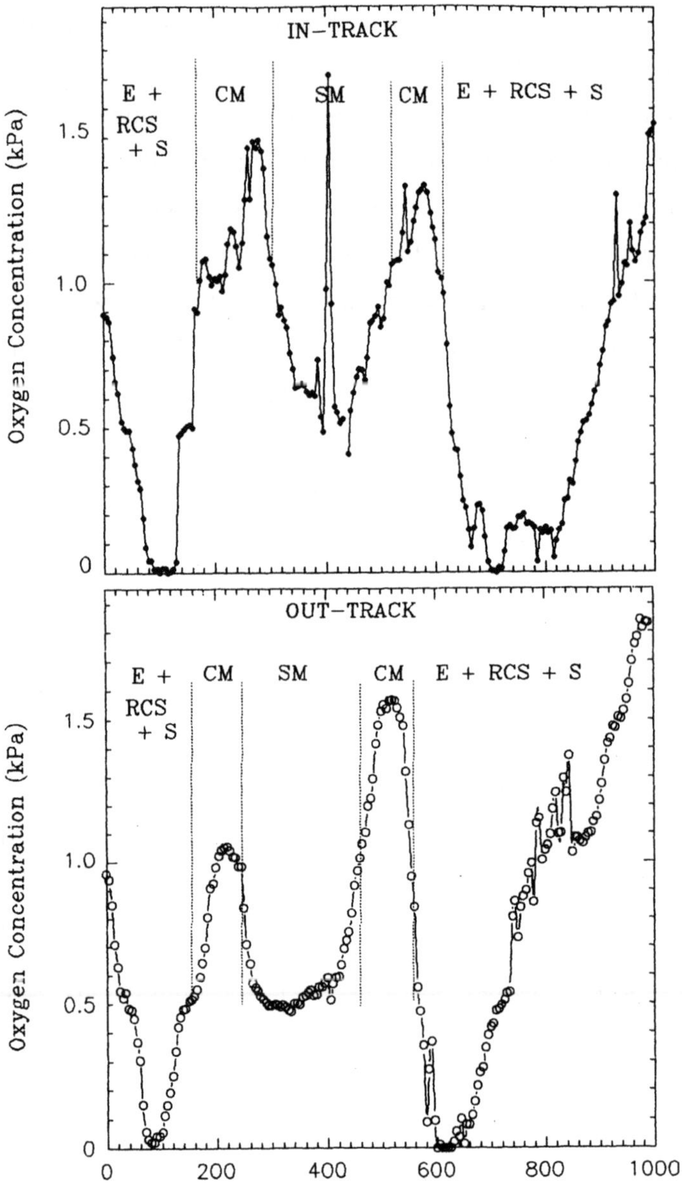

FIGURE 8. *Microelectrode polarographic 'radial' oxygen profiles across a primary root of maize (L = 90 mm) at an apical position within the zone of root cap sheathing the meristem. E = epidermis plus hypodermis, RCS = root cap sheath, CM = cortical meristem, SM = stelar meristem, S = sloughing cells from root cap sheath.*

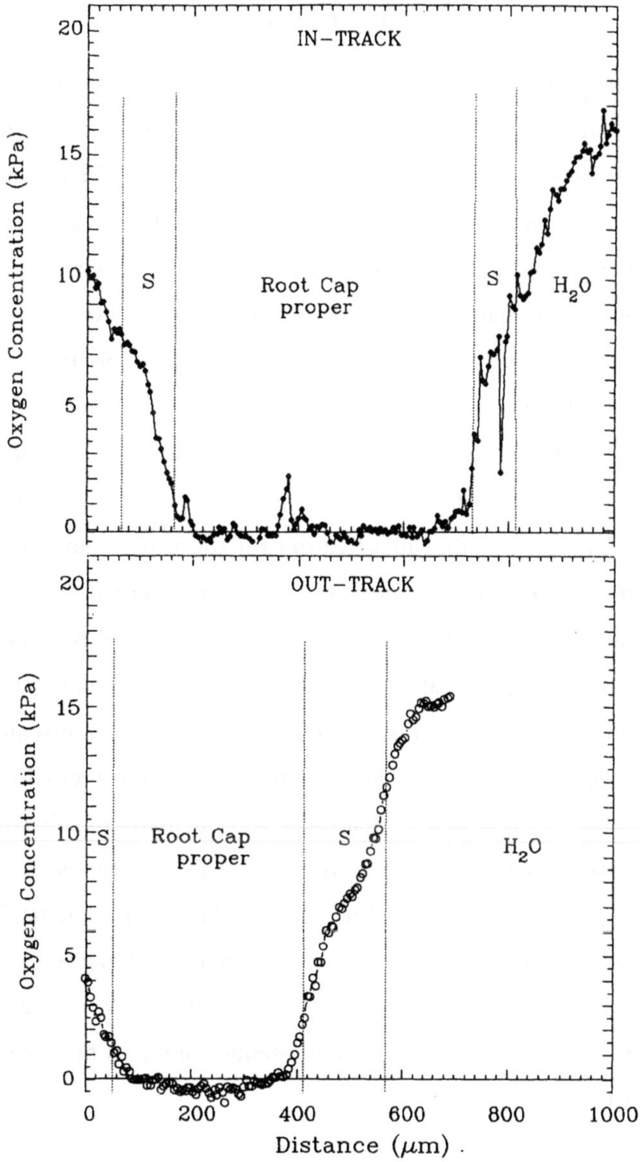

***FIGURE 9**. Microelectrode polarographic 'radial' oxygen profiles across root cap of a maize primary root. Root embedded in agar (1% w/v). Root length = 90 mm and oxygen supply to the root was chiefly from the shoot through a non-aerenchymatous cortex. The root cap was projecting into a water-filled channel in the agar. S = sloughing cap cells.*

CONCLUSIONS

In the examples shown, although stelar anoxia was not found in sub-apical parts, it was readily induced, and moreover the concentration difference between the cortex and stele was more substantial towards the base of the root than nearer to the apex. If the root had been a little older and longer, or had been subject to a radial oxygen sink and not the rather weak oxygen source which often prevailed here, it is conceivable that some sub-apical stelar anoxia would have been found even though the shoot was in 21% oxygen. It is even possible that there was some stelar anoxia which went undetected because of tangential tracking.

It would be foolish to overstress the relevance of this study to climatic change. However, it confirms much that has been predicted from mathematical modelling and provides further justification for the use of modelling to forecast effects of climatic change on root aeration. It should be noted that the likely effect of increased temperature on the aeration of roots such as these would be steeper oxygen gradients both axially along the root and radially, the increased respiratory demand outweighing the reduction in diffusive resistance consequent upon the raised temperature-dependent diffusion coefficients. The principal result is likely to be poorer aeration and hence shallower rooting patterns. Similar effects on coastal vegetation may be anticipated from any sea level rise as a consequence of global warming: in this case the diffusion path through the plants from shoot to root will be extended by any increased level of shoot submergence – the greater diffusive resistance that this would impose would again cause a deterioration in the potential for root aeration. The ultimate consequence might be a narrowing of the vegetated coastal strip.

ACKNOWLEDGEMENTS

WA gratefully acknowledges the financial support of the Royal Society of London which made this study possible. Also we are indebted to Dr Valerie Alder for the use of the facilities of her laboratory at the Lions' Eye Institute, Perth, Western Australia.

REFERENCES

Alder VA, Yu D-Y, Cringle SJ, Su E-N (1990) Changes in vitreal oxygen tension distribution in the streptozotocin diabetic rat. Diabetologia 34: 469

Anderson FO, Kristensen E (1988) Oxygen microgradients in the rhizosphere of the mangrove *Axicennia marina.* Marine Ecology 44: 201-204

Armstrong W (1979) Aeration in higher plants In: Advances in botanical research Vol. 7. Academic Press, pp 225-332

Armstrong W (1988) Aeration in roots. In: Cherry T (ed) Environmental stress physiology, NATO ASI series, Vol. G19. Springer-Verlag, Berlin, Germany, pp 197-206

Armstrong W, Beckett PM (1985) Root aeration in unsaturated soil: a multi-shelled model of oxygen distribution and diffusion with and without sectoral blocking of the diffusion path. The New Phytologist 100: 293-311

Armstrong W, Beckett PM (1987) Internal aeration and the development of stelar anoxia in submerged roots: A multishelled mathematical model combining axial diffusion of oxygen in the cortex with radial losses to the stele, the wall layers and the rhizosphere. The New Phytologist 105: 221-245

Armstrong W, Beckett PM, Justin SHFW, Lythe S (1990) Modelling and other aspects of root aeration. In: Jackson MB, Davies DD, Lambers H (eds) Plant life under oxygen stress, SPB Academic, The Hague, The Netherlands, pp 267-282

Armstrong W, Webb T (1985) A critical oxygen pressure for root extension in rice. Journal of Experimental Botany 36: 1573-1582

Berry LJ, Norris WE (1949) Studies of onion root respiration. I. Velocity of oxygen consumption in different segments of root at different temperatures as a function of partial pressure of oxygen. Biochimica et Biophysica Acta 3: 593-606

Bowling DJF (1973) Measurement of gradient of oxygen partial pressure across the intact root. Planta 111: 323-328

Cringle SJ, Yu D-Y, Alder VA (1990) Intraretinal and intravitreal oxygen distribution in the rat. Advances in Experimental Medical Biology - Oxygen Transport to Tissue XIII

Drew MC, Saglio PH, Pradet A (1985) Larger adenylate energy charge and ATP/ADP ratios in aerenchymatous roots of *Zea mays* in anaerobic media as a consequence of improved internal oxygen transport. Planta 165: 51-58

Hook DD, McKevlin MR (1988) Use of oxygen microelectrodes to measure aeration in the roots of intact tree seedlings. In: Hook DD, McKee WH Jr, Smith HK, Gregory J, Burrell VF Jr, Devoe MR, Sojka RE, Gilbert S, Banks R, Stolzy LH, Brooks C, Matthews TD, Shea TH (eds) The ecology and management of wetlands, Timber Press, Portland, Oregon, USA, pp 467-476

Jackson MB, Fenning TM, Drew MC, Saker LR (1985) Stimulation of ethylene production and gas-space (aerenchyma) formation in adventitious roots of *Zea mays* L. by small partial pressures of oxygen. Planta 165: 486-492

Thomson CJ, Greenway H (1991) Metabolic evidence for stelar anoxia in maize roots exposed to low oxygen concentrations. Plant Physiology 96: 1294-1301

Thomson CJ, Armstrong W, Waters I, Greenway H (1990) Aerenchyma formation and associated oxygen movement in seminal and nodal roots of wheat. Plant, Cell and Environment 13: 395-403

LONG TERM SURVIVAL OF RHIZOMATOUS SPECIES UNDER OXYGEN DEPRIVATION

Thomas Henzi and Roland Brändle
Institute of Plant Physiology
University of Berne
Altenbergrain 21
3013 Berne
Switzerland

ABSTRACT

The underground organs of geophytes, such as rhizomes and tubers, are usually not the most sensitive to flooding and anoxia stress. However, they do show considerable variability. Prolonged flooding resistance is dependent on efficient oxygen transport and release through porous tissues, as well as metabolic adaptations. The mechanism for maintaining a sufficient oxygen supply and its underlying components are well understood. Similarly, details of anaerobic energy metabolism are well known. Some very general information is available concerning the synthesis of essential gene products, proteins and other macromolecules. On the other hand, our knowledge of the extent of lipid and membrane integrity in wetland and dryland rhizomes under oxygen deprivation and re-aeration is rather limited. The lipid composition of rhizomes from *Acorus calamus*, *Iris germanica* and *Schoenoplectus lacustris*, kept under strict anoxia but retaining 25-50% regeneration capacity, shows a saturation of the linolenic acid (C18:3) component of most individual polar lipids. The saturation gradually increases in the order *A. calamus* < *S. lacustris* < *I. germanica.* The total quantity of polar lipids decreases only slightly in *A. calamus* and *S. lacustris*, whereas in *I. germanica* a rapid loss of glyco- and phospholipids can be observed under prolonged oxygen deprivation. Free fatty acids increase markedly in *I. germanica*, but only to a limited extent in the other species. Re-exposure to air induces a marked formation of peroxidation products (malondialdehyde and ethane) in *I. germanica*, less so in *S. lacustris* and only to a limited extent in *A. calamus*, which is able to re-constitute the previous lipid desaturation state under these conditions. *A. calamus* proved to be the most anoxia-tolerant species, followed by the other wetland species, *S. lacustris*. The dryland species *I. germanica*, with its unstable lipids, is the most anoxia-intolerant species of the three.

NATO ASI Series, Vol. I 16
Interacting Stresses on Plants in a Changing Climate
Edited by M. B. Jackson, C. R. Black

INTRODUCTION

Underground perennating shoots are typified by ample energy reserves and the presence of buds for regeneration and vegetative propagation. With regard to flooding resistance and oxygen deprivation, remarkable differences exist between organs and species. Wetland plant rhizomes are usually much more resistant than dryland plant rhizomes or similar kinds of overwintering shoots. As far as we know, the rhizome of the sweet flag (*Acorus calamus*) is one of the most tolerant of prolonged oxygen shortage. It can survive up to three months in an anaerobic incubator at room temperature and still recover, whereas potato tubers are killed in less than three days. By comparison, the wetland species *Schoenoplectus lacustris* survives in the anaerobic incubator for about two months, while *Iris germanica* rhizomes remain viable for less than three weeks. Through evolution, wetland plant rhizomes have presumably become adapted to oxygen deprivation. This will have taken place in response to selective pressure arising from the annual imposition of oxygen shortage in winter and during the very early phases of growth through sediment and water in late winter and early spring. Morphological as well as physiological adaptations have evolved. The morphological adaptations consist mainly of the formation of a porous internal structure allowing oxygen transport from the shoot to the underground organs and radial oxygen release into the immediate surroundings (Grosse *et al.*, 1991; Armstrong *et al.*, 1992). Shootless rhizomes, submerged or buried in the mud, need additional metabolic strategies for survival of flooding and re-aeration (Summers and Jackson, this volume). Some important aspects, such as energy metabolism and the preservation of membrane integrity, are briefly discussed in this chapter.

METABOLIC ASPECTS

The energy requirement of flooded or oxygen-deprived underground organs is provided by fermentation processes (Brändle, 1991). However, so far, no major differences in metabolic pathways have been discovered between adapted and non-adapted rhizomes (Monk *et al.*, 1984). The difference between an organ that is resistant and one that is non-resistant is a question of metabolic regulation rather than the evolution of new pathways (Brändle, 1991). It seems that the main achievement is not just to save energy but to dispose of fermentative end products such as ethanol, which can be released easily (Studer and Brändle, 1987). Non-tolerant tissues, for example potato tubers, produce and accumulate considerable amounts of lactic acid (Sieber and Brändle, 1991). Lactic acid is one of the main causes of cytoplasmic acidosis and eventual cell death (Roberts *et al.*, 1985; Saglio, this volume). Besides ethanol and lactic acid, some other NADH re-oxidising pathways and accessory end products are known, of which alanine and malate are of some importance (Crawford, 1978). However, in flood-tolerant rhizomes we cannot find a pronounced 'Pasteur effect' under artificially induced long-term anoxia. This is only seen at the onset of anoxia. Adenylate energy charge (AEC) values remain high after an initial drop, and the total adenylate pools are stable (Steinmann and Brändle, 1981; Sieber and Brändle, 1991). This indicates an equilibrated metabolism in adapted tissue, albeit probably at a lower overall level of activity (Raymond and Pradet, 1980), but it seems to be at least sufficient for maintenance and the synthesis of the most important macromolecules, such as proteins (Bucher, 1988). In non-tolerant potato tubers, AEC drops to a very low level and the adenylate pool is depleted very quickly (Sieber and Brändle, 1991).

Severe hypoxic conditions in the natural habitat do not appear to impede the development and regeneration of wetland plant rhizomes (Brändle, 1990). Activities of glycolytic enzymes, for example alcohol dehydrogenase (ADH), lactate dehydrogenase (LDH), pyruvate decarboxylase (PDC), and the presence of their mRNAs are correlated with season and developmental stage. They increase during winter to a maximum in early spring when the submersed buds start to grow (Haldemann and Brändle, 1986; Bucher and Kuhlemeier, unpublished data). Simultaneously, a sharp decrease in carbohydrate level and the synthesis of nitrogen-rich amino acids such as arginine, asparagine and glutamine takes place (Haldemann and Brändle, 1988). Of these, arginine especially could help to counteract cytoplasmic acidosis.

The preservation of a regulated and adapted metabolism in wetland plant rhizomes and the fact that rhizomes kept under strict anoxia can restore their full respiration capacity when returned to air requires high membrane and lipid stability (Monk *et al.*, 1984). Long-term survival is thus fully dependent on membrane integrity and function.

STABILITY OF MEMBRANE LIPIDS

Table 1 shows the changes in fatty acid composition of rhizomes with about 50% regeneration capacity retained following prolonged anoxic treatment. In *A. calamus*

***TABLE 1**. Statistically significant changes in fatty acid composition of various polar lipids in rhizomes: + = increase, − = decrease. The rhizomes were incubated in an anaerobic incubator in an atmosphere consisting of 85% N_2, 10% H_2 and 5% CO_2 in the presence of a Pd catalyst (Forma Scientific, Marietta, USA). The lipids were extracted according to Bligh and Dyer (1959). The fatty acid composition was determined according to Christie (1982). The separation of the fatty acid methyl esters was performed by a Sigma 300 Dual FID GC (oven 190 °C, injector 220 °C, detector 230 °C) equipped with packed steel columns (10% DEGS on Chromsorb W/AW). MGDG = monogalactosyldiacylglycerol, DGDG = digalactosyldi-acylglycerol, DPG = diphosphatidylglycerol, PA = phosphatidic acid, PC = phosphatidylcholine, PE = phosphatidylethanol-amine, PG = phosphatidylglycerol, PI = phosphatidylinositol.*

	A. calamus 70 d anoxia					*S. lacustris* 35 d anoxia					*I. germanica* 14 d anoxia				
	16:0	18:0	18:1	18:2	18:3	16:0	18:0	18:1	18:2	18:3	16:0	18:0	18:1	18:2	18:3
MGDG				+	−				+	−	+	+			−
DGDG				+	−		+	+		−	+	+			−
DPG		+	−		−		+	+		−		+			−
PA				+	−			+		−		+			
PC					−			+		−		+			
PE								+		−		+			
PG		+			−		+	+		−		+	−	+	−
PI				+	−			+		−		+	−		

a decrease in linolenic acid (C18:3) is found in most of the lipids analysed, concomitant with an increase in linoleic acid (C18:2) in many lipids. *S. lacustris* also loses linolenic acid, but small changes in the content of linoleic acid occur.

Here, an increase takes place in the monounsaturated oleic acid (C18:1) and, in some lipids, in the saturated stearic acid (C18:0). Loss of linolenic acid is observed in many lipids of *I. germanica*. Even oleic acid decreases in some lipids. The saturated fatty acid species, especially palmitic acid, are found in higher amounts. Thus, oxygen deprivation leads to a loss of linolenic acid in all three species. In *A. calamus*, there is a compensatory rise in the amount of linoleic acid and the overall desaturation level therefore remains high. A higher saturation level of fatty acids occurs in *S. lacustris*, but the most pronounced saturated fatty acid content is observed in *I. germanica*. Shorter periods of anoxia induce only minor changes in all three species (Henzi, 1991). Re-exposure of rhizomes to air has no effect on the fatty acid composition of *S. lacustris* and *I. germanica*, whereas *A. calamus* is able to reconstitute the former fatty acid composition within a few days (Henzi, 1991). The sum of the polar lipids decreases only slightly in *A. calamus* (Fig. 1). Even after 70 days no decrease in polar lipids can be seen and the free fatty acids show hardly any change. A similar response is exhibited by *S. lacustris*. In contrast, *I. germanica* rhizomes show a rapid loss of phospho- and glycolipids under anoxia. The decline is detectable after seven days of anoxia and by 21 days only traces of polar lipids remain. The observed loss is not lipid-specific, and all individual phospho- and glycolipids decrease comparably (Henzi, 1991). On the other hand, a rapid accumulation of free fatty acids occurs in stressed rhizomes of *I. germanica*. After 21 days of anoxic treatment, the content is about ten-fold higher than in unstressed plants. The increase is due to the hydrolysis of the ester bond between the fatty acyl moiety and the glycerol backbone of the lipid.

After re-exposure of anaerobically incubated rhizomes to air, the lipid peroxidation products, malondialdehyde and ethane (Figs. 2 and 3), occur in all species (Pryor *et al.*, 1976; Sandmann and Böger, 1983). However, in *A. calamus* peroxidation products appear only after 70 days of anaerobic pre-incubation. In *S. lacustris*, oxidative damage is visible only after 35 days of preincubation and increases with prolonged anoxic preincubation. In *I. germanica*, by contrast, a few days of anoxia are sufficient to abolish the anti-oxidative defence mechanisms. As a consequence, anoxic pre-incubation for only seven days results in the production of considerable amounts of the peroxidation products, malondialdehyde and ethane. Longer periods of anoxia lead to considerably higher production of malondialdehyde and ethane during the first hours of post-anoxia. Ethane formation never occurs during normoxia and anoxia in any of the three species investigated.

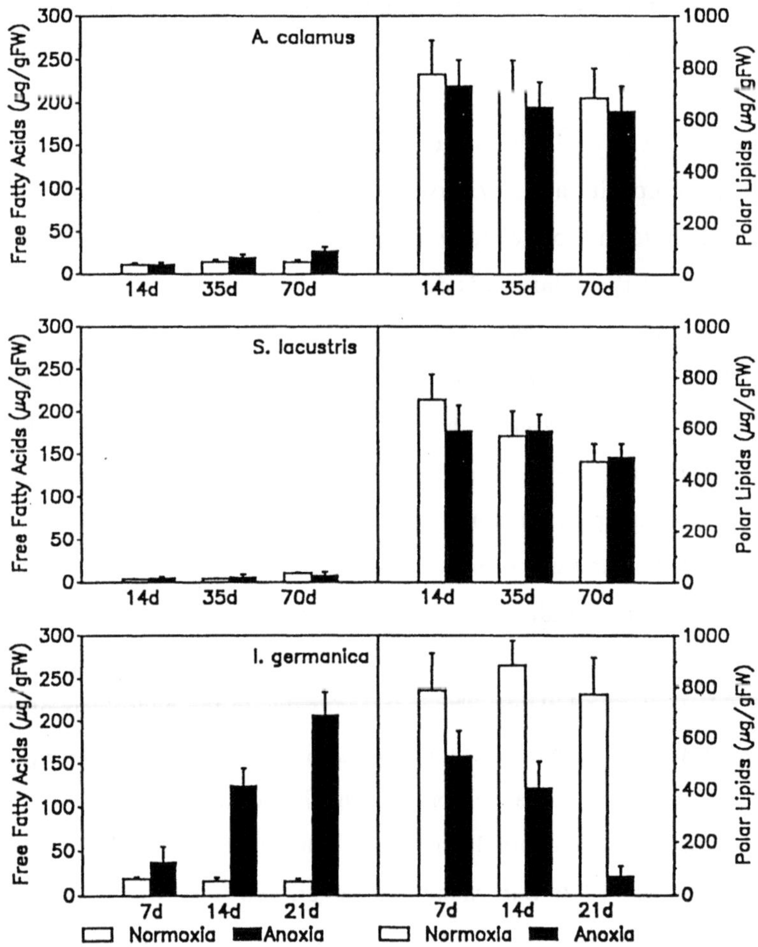

FIGURE 1. Content of total polar lipids and free fatty acids in rhizomes of A. calamus, S. lacustris *and* I. germanica *after time-dependent anoxic treatments. The internal standard, heptadecanoic acid methyl ester, was added for quantification to the samples prior to gas chromatographic analysis.*

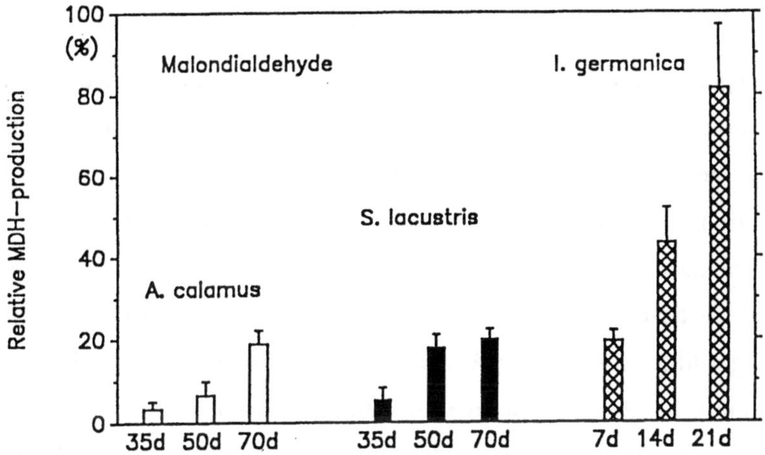

FIGURE 2. Accumulation of malondialdehyde in rhizomes of A. calamus, S. lacustris *and* I. germanica *during the first six hours of post-anoxia. The thiobarbituric acid assay was performed to detect post-anoxic MDH (Hunter* et al., *1983). The evaluation of the MDH measurements followed that described by Heath and Packer (1965). A first measurement took place immediately after the anaerobic incubation and a second after re-oxygenation. Because of the large differences in the background TBA values of the three rhizomes, a percentage presentation of colour intensity has been chosen. Rhizomes which were either kept under normoxia or anoxia showed no significant changes of the coloured complex.*

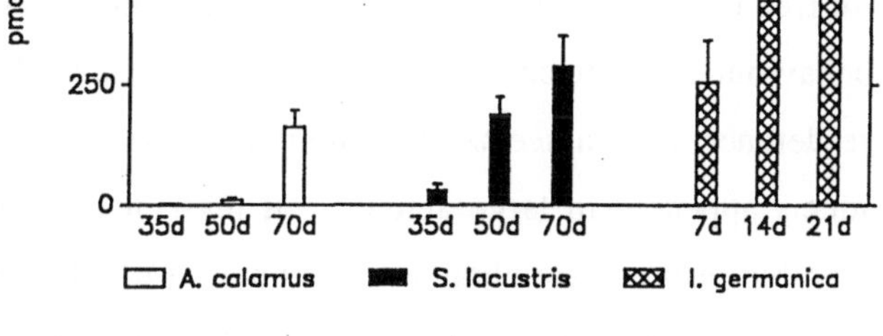

FIGURE 3. Ethane production in rhizomes of A. calamus, S. lacustris *and* I. germanica *during the first 24 hours of post-anoxia. Ethane was analysed in the headspace of rhizomes according to Kunert and Böger (1984).*

DISCUSSION

Under oxygen deprivation, the three rhizomatous species examined showed only minor alterations and differences with regard to energy metabolism. The main difference between the species was an unlimited increase in tissue ethanol concentration in *I. germanica* (Monk *et al.*, 1984). This ethanol may be broken down by catalase during post-anoxia to cause a surge of toxic acetaldehyde that may result in tissue damage. Marked differences were found between the three species regarding lipid stability. *A. calamus* proved to be the most tolerant species in this regard; its desaturation level of membrane lipids decreased only slightly, with no lipid breakdown occurring during 70 days of strict anoxia. Moreover, no lipid peroxidation took place upon re-aeration. The anti-oxidative defence mechanisms, presumably depending on glutathione, ascorbate, α-tocopherol or SOD, seemed hardly to be impaired (Halliwell, 1991). *S. lacustris* was able to endure about 50 days without oxygen before the first signs of lipid damage appeared. The least anoxia-tolerant species we have examined is the dryland plant, *I. germanica*, in which 14 days without oxygen were sufficient to increase the saturation of membrane lipids drastically. Only if the anoxic periods were very short was there no effect on the desaturation level in *I. germanica*, as has also been shown in isolated membranes of the anoxia-intolerant *Triticum aestivum* (Hetherington *et al.*, 1987; Henzi, 1991). In addition, marked lipid breakdown occurred in *I. germanica* leaving only traces of lipids but a marked accumulation of free fatty acids. Wheat plants subjected to anoxia for seven days showed a similar increase in free fatty acids and a decrease in polar lipids (Hetherington *et al.*, 1988). This behaviour could be common to all sensitive species. In these cases the anti-oxidative defence system seems to be inhibited, since post-anoxic lipid peroxidation products appeared in large amounts, at least in the rhizomes of *I. germanica*. Comparable results were obtained by Hunter *et al.* (1983) who found increased concentrations of malondialdehyde in the rhizomes of *I. germanica*, while the wetland species *I. pseudacorus* showed no rise. Thus, the less flood-tolerant *I. germanica* and other sensitive rhizomatous species suffer not only from lipid degradation during anoxia, but also from lipid peroxidation during post-anoxia. However, potato tubers seem to be even more sensitive since less than 24 hours of anoxia allows the formation of post-anoxic peroxidation products (M. Sieber, University of Berne, personal communication).

REFERENCES

Armstrong J, Armstrong W, Beckett PM (1992) Venturi- and humidity-induced pressure flows enhance rhizome aeration and rhizosphere oxidation. The New Phytologist 120: 197-207

Bligh EG, Dyer WJ (1959) A rapid method of total lipid extraction and purification. Canadian Journal of Biochemistry and Physiology 37: 911-917

Brändle R (1990) Überlebensstrategie der Rhizome von Sumpf- und Röhrichtpflanzen. In: Sukopp H, Krauss M (eds) Oekologie, Gefährdung und Schutz von Röhrichtpflanzen, Landschaftsentwicklung und Umweltforschung Nr. 71. Technische Universität, Berlin, Germany, pp 103-120

Brändle R (1991) Flooding resistance of rhizomatous amphibious plants. In: Jackson MB, Davies DD, Lambers H (eds) Plant life under oxygen stress: ecology, physiology and biochemistry. SPB Academic Publishing, The Hague, The Netherlands, pp 35-46

Bucher M (1988) Anaerobe Proteine von *Acorus calamus* und anderen Sumpf- und Röhrichtpflanzen. Master of Science Thesis, University of Bern, Switzerland

Christie WW (1982) Lipid analysis, 2nd edn. Pergammon International Library, London

Crawford RMM (1978) Biochemical and ecological similarities in marsh plants and diving animals. Naturwissenschaften 65: 194-201

Grosse W, Büchel HB, Tiebel H (1991) Pressurized ventilation in wetland plants. Aquatic Botany 39: 89-98

Haldemann C, Brändle R (1986) Jahreszeitliche Unterschiede im Reservestoffgehalt und von Gärungsprozessen in Rhizomen von Sumpf- und Röhrichtpflanzen aus dem Freiland. Flora 178: 307-313

Haldemann C, Brändle R (1988) Amino acid composition in rhizomes of wetland species in their natural habitat and under anoxia. Flora 180: 407-411

Halliwell B (1991) The biological toxicity of free radicals and other reactive oxygen species. In: Aruoma OI, Halliwell B (eds) Free radicals and food additives. Taylor & Francis, London New York Philadelphia, pp 37-57

Heath RL, Packer L (1965) Effect of light on lipid peroxidation in chloroplasts. Biochemical and Biophysical Research Communications 19: 716-720

Henzi T (1991) Auswirkungen von Anoxia und Postanoxia auf die Lipide und Fettsäuren in Rhizomen von *Acorus calamus* (L.), *Schoenoplectus lacustris* (L.) Palla und *Iris germanica* (L.). PhD Thesis, University of Bern, Switzerland

Hetherington PR, Broughton HL, McKersie BD (1988) Ice-encasement injury to microsomal membranes isolated from winter wheat crowns II. Changes in membrane lipids during ice encasement. Plant Physiology 86: 740-743

Hetherington PR, McKersie BD, Borochov A (1987) Ice-encasement injury to microsomal membranes from winter wheat crowns I. Comparison of membrane properties after lethal ice encasement and during a post-thaw period. Plant Physiology 85: 1068-1072

Hunter MIS, Hetherington AM, Crawford RMM (1983) Lipid peroxidation - a factor in anoxia intolerance in Iris species? Phytochemistry 22: 1145-1147

Kunert KJ, Böger P (1984) The diphenyl ether herbicide oxyfluorfen: action of antioxidants. Journal of Agricultural and Food Chemistry 32: 725-728

Monk LS, Crawford RMM, Brändle R (1984) Fermentation rates and ethanol accumulation in relation to flooding tolerance in rhizomes of monocotyledonous species. Journal of Experimental Botany 35: 738-745

Pryor WA, Stanley JP, Blair E (1976) Autoxidation of polyunsaturated fatty acids: II. A suggested mechanism for the formation of TBA-reactive materials from prostaglandin-like endoperoxides. Lipids 11: 370-379

Raymond P, Pradet A (1980) Stabilization of adenine nucleotide ratios at various values by an oxygen limitation of respiration in germinating lettuce (*Lactuca sativa*) seeds. Biochemical Journal 190: 39-44

Roberts JKM, Andrade FH, Anderson IC (1985) Further evidence that cytoplasmic acidosis is a determinant of flooding intolerance in plants. Plant Physiology 77: 492-494

Sandmann G, Böger P (1983) Peroxidative formation of C3-hydrocarbons from an ω-4 polyunsaturated fatty acid (16:3ω4) in the alga *Bumilleriopsis*. Lipids 18: 37-41

Sieber M, Brändle R (1991) Energy metabolism in rhizomes of *Acorus calamus* (L.) and in tubers of *Solanum tuberosum* (L.) with regard to their anoxia tolerance. Acta Botanica 104: 279-282

Steinmann F, Brändle R (1981) Die Überflutungstoleranz der Teichbinse (*Schoenoplectus lacustris* L. Palla): Beziehungen zwischen der Sauerstoffversorgung und der 'Adenylate Energy Charge' der Rhizome in Abhängigkeit von der Sauerstoffkonzentration in der Umgebung. Flora 171: 307-314

Studer C, Brändle R (1987) Ethanol, acetaldehyde, ethylene release and ACC concentration of rhizomes from marsh plants under normoxia, hypoxia and anoxia. In: Crawford RMM (ed) Plant life in aquatic and amphibious habitats. Blackwell, Oxford, UK, pp 293-301

PROMOTION OF STEM EXTENSION IN AN AQUATIC MONOCOT (*POTAMOGETON PECTINATUS* L.) BY THE COMPLETE ABSENCE OF OXYGEN, AND BY PARTIAL OXYGEN SHORTAGE

Jacky E. Summers and Michael B. Jackson
Department of Agricultural Sciences
University of Bristol
AFRC Institute of Arable Crops Research
Long Ashton Research Station
Bristol BS18 9AF
UK

ABSTRACT

Germinating tubers of the aquatic monocot *P. pectinatus* are shown to possess several unusual characteristics. Their rate of stem extension can be promoted by strictly anaerobic solutions, an effect sustained for several days. This leads to a stem several centimetres longer than that formed under aerated conditions. This appears to be the first report of such a phenomenon. Some cell production continues under anaerobic conditions as judged by cell counts of epidermal strips, and by the relative increases in length by cells and internodes. Stem extension is also promoted by solutions that are partially deficient in oxygen (5-8 kPa oxygen). The effect cannot be attributed to ethylene action, since applications of the gas fail to promote stem elongation, although leaf extension is stimulated. Ethylene production by germinating tubers is extremely small, even in the presence of oxygen. Production of ethylene is affected only slightly by treatment with aminoethoxyvinylglycine (AVG), an inhibitor of ethylene biosynthesis in many higher plants, or by additions of 1-aminocyclopropane-1-carboxylic acid (ACC), the putative precursor of ethylene. The low activity of ACC and AVG suggests that the conventional pathway of ethylene production from methionine, via *S*-adenosylmethionine and ACC, is suppressed in this species.

INTRODUCTION

Potamogeton pectinatus L., commonly called sago-leaved [US] or fennel-leaved [UK] pondweed, is a submersed, aquatic monocot found in many parts of the world. It is a hardy perennial and grows in fresh, saline, still or fast-flowing water of different trophic levels, at altitudes extending from sea level in Europe to over 5000 m in Tibet and Venezuela (Yeo, 1965). Inevitably, this considerable

NATO ASI Series, Vol. I 16
Interacting Stresses on Plants in a Changing Climate
Edited by M. B. Jackson, C. R. Black

ecological amplitude is associated with a wide range of strategies for survival in different environments. In common with other aquatic species, *P. pectinatus* is anatomically, morphologically and physiologically diverse (Wiegleb and Brux, 1991). Variation in morphology extends to a variable chromosome number, although no close correlation between this and other mutable characteristics has been shown (Hettiarachchi and Triest, 1986). Morphology is also influenced strongly by the environment, and this can lead to more than one form of a characteristic being present on a single plant and can present problems for correct identification in the field. However, despite this highly plastic morphology, it is possible to outline some general and easily recognisable features of the species. Thus, *P. pectinatus* is a monocot, with rhizomes that terminate in tubers during late summer and autumn. These tubers overwinter in mud, sand or gravel of the pond or stream bed, and consist of two swollen internodes and a short terminal bud comprising two compressed internodes and an apical meristem surrounded by miniature leaves. Tubers are produced singly or in chains, each tuber being separated by a rhizomatous internode. Smaller, axillary tubers can also form on the shoots. In spring, the tubers germinate by elongation of the stem of the apical bud which forms more nodes and internodes as it lengthens. Each new node produces a thin lanceolate leaf and associated basal sheath. However, the two original nodes each produce only a leaf sheath. Roots are often produced at the nodes. Many tubers are produced in the autumn and become detached from the mother plant. This explains why it is difficult to eradicate *P. pectinatus* from waterways and drainage channels as pulling up mature plants (Devlin and Karczmarczyk, 1985) leaves the tubers to re-colonise the area in the following year.

A notable feature of *P. pectinatus* is its well-developed network of inter-connected intercellular spaces (aerenchyma) that permeate the rhizomes, stems and leaves. This adaptation is thought to help aerate the roots and shoot bases by acting as a conduit for molecular oxygen diffusing from better aerated organs near the surface of the water. However, germinating tubers cannot be aerated *via* aerenchyma since they commence growth from the bed of the stream, below the soil/water interface. Here, oxygen entry and its internal distribution is completely precluded by deep submergence, a covering of mud or similar material, the absence of light, and the strong likelihood of the immediate environment being anaerobic. This chapter describes the capabilities of the shoot of overwintered tubers to grow under these highly stressful conditions, which would be fatal to most plant species. The research has concentrated on shoot extension, respiration and the ability of germinating tubers to synthesise and respond to the plant hormone ethylene (ethene) when subjected to various degrees of oxygen shortage, including the complete absence of oxygen.

EFFECT OF OXYGEN SUPPLY ON SHOOT GROWTH

Tubers were grown in 0.5 mM calcium sulphate in anaerobic or aerobic environments. An anaerobic workbench (Forma Scientific Inc., Marietta, Ohio, USA) imposed oxygen-free conditions on the tubers by circulating a mixture of hydrogen and nitrogen gases (1 : 9) over a palladium catalyst which reduced any traces of oxygen to water. Aerobic conditions were provided using a workbench with similar dimensions to the anaerobic workbench but vented to the atmosphere. In each experiment, six or more tubers of similar weight for each treatment were grown individually in darkness in 15 cm tall, 20.4 ml test-tubes, weighted down by approximately 6 g of 5 mm diameter glass ballotini. Solutions used in the oxygen-free treatments were outgassed in the anaerobic workbench for 24 hours prior to use. The experimental period was usually five days.

Comparison of aerobic and anaerobic conditions. To create an aerobic environment, the solution in which the tubers were grown was sparged gently with air (2.5 ml min^{-1}) to ensure an equilibrium partial pressure of dissolved oxygen of 20.8 kPa. Oxygen in solutions within the anaerobic work bench was undetectable using a Radiometer oxygen electrode linked to a Strathkelvin Model 781 oxygen meter (Strathkelvin Instruments, Glasgow, UK). Vigorous and sustained shoot elongation was observed under anaerobic conditions. Within two days, stems had grown longer in anaerobic than in aerobic solutions, and the difference increased up to at least the fifth day (Fig. 1a). Longer-term tests showed that growth in anaerobic conditions could be sustained for up to 14 days. Extension growth was located in both preformed internodes. Leaves gave a different response to stems, with anaerobic conditions inhibiting rather than promoting their extension. However, the anaerobic stimulation of stem elongation was proportionately greater

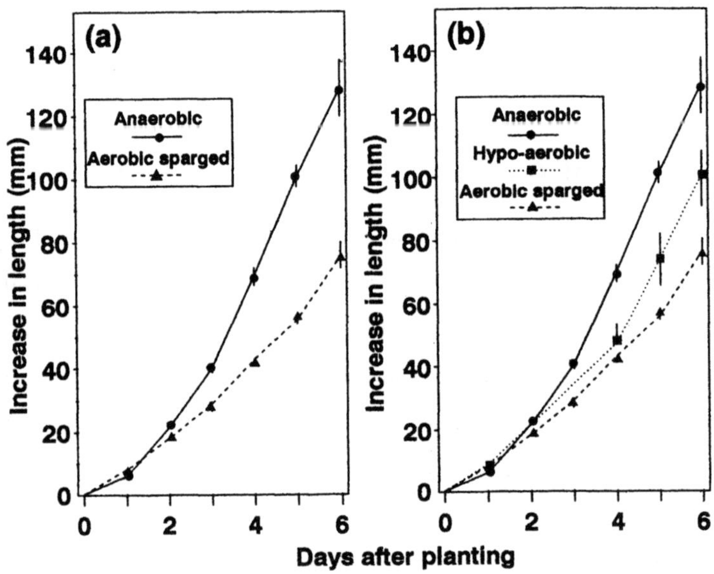

FIGURE 1. Effects of oxygen shortage on elongation by the apical shoot (stem + leaf) of germinating tubers of Potamogeton pectinatus: *(a) comparison between anaerobic and aerobic [air-sparged] solutions; (b) comparison between aerobic [air-sparged], anaerobic and hypo-aerobic solutions [unsparged, open to the air]. Means are of six replicates. Vertical lines are double standard errors.*

than the inhibition of leaf growth. This led, overall, to a longer shoot in the absence of oxygen. To assess the relative contributions of cell expansion and production to the growth of the stems, epidermis was removed from the first and second internodes using fine forceps and stained using 0.5% thionine before cells were counted or their size measured using a microscope linked to a video camera, a Calcomp 200 digitising tablet and a computer running VIDS II software (Analytical Measuring Systems Ltd., Saffron Walden, Essex, UK). Cell numbers and the lengths and widths of cells along the entire length of each of the two pre-formed internodes were compared with those of epidermes from ungerminated shoots. After five days in anaerobic conditions, cell lengths in both internodes were longer than those in aerobic conditions. There was no difference in cell width, implying a promotion of total cell volume under anaerobiosis. There was no difference between the dimensions of cells of the first and second internodes within each treatment. Comparing the amount of stem extension with cell extension on a proportional basis showed that cell extension accounted fully for stem elongation of the first (i.e., older) internode in both treatments (Fig. 2). In the younger, second internode, stem length increased more than cell length, suggesting that some cell production was occurring, even in anaerobic plants. Counts of the total

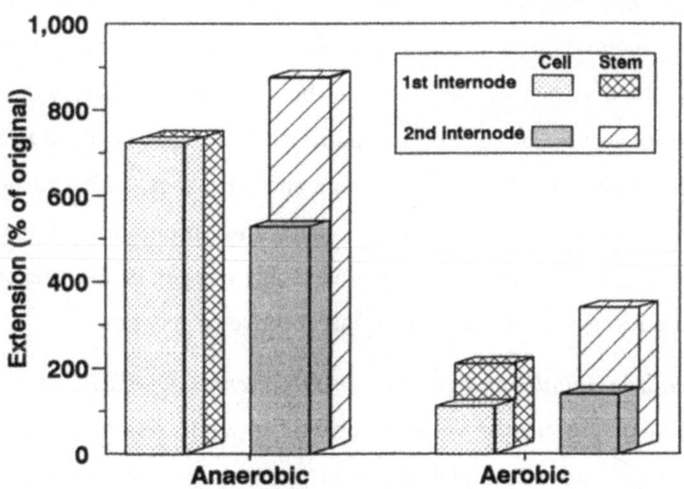

FIGURE 2. Effect of five days aerobic or anaerobic conditions on extension growth by stem internodes and internodal epidermal cells of the apical shoot of Potamogeton pectinatus *tubers. Results show that, in either treatment, the extent of elongation by the second (younger) internode cannot be fully accounted for by cell elongation. Thus, some cell production must have taken place, even in the absence of oxygen.*

number of cells in rows running from one node to the next, compared with an estimate of the original number of cells (using epidermes from ungerminated tubers), confirmed that cell production was continuing in the absence of oxygen. In aerobically growing stems, the number of cells approximately doubled in five days, while in anaerobic internodes, final cell number increased by approximately 50% in the same time.

Effect of sparging anaerobic solutions. One possible explanation for the faster stem elongation under anaerobic conditions is that one or more growth promoters accumulated because losses by radial diffusion were minimised by the unstirred solution. This idea was supported by the finding that gentle sparging with anaerobic gas mixture (2.5 ml min^{-1}) decreased, but did not eliminate the promotion of stem extension induced by anaerobiosis. The effect of sparging was localised in the stem, with the leaf showing no significant response. Sparging did not change the oxygen supply, and thus may have acted by thinning the boundary layers and thereby maximising the concentration gradients driving the outward diffusion of potential growth promoters. Carbon dioxide could be one growth promoter involved as this gas was found to be evolved by *P. pectinatus* at similar rates under both aerobic or anaerobic conditions. In rice (Raskin and Kende, 1984; Pearce *et al.*, 1992) and *P. distinctus* (Suge and Kusanagi, 1975) carbon dioxide can promote shoot extension. However, until now, promotion of extension by carbon dioxide has been observed only when some oxygen is also present. Further work is needed to assess its activity in the absence of oxygen.

Stem extension in aquatic and gaseous environments. One method of increasing the radial loss of gaseous or volatile substances while retaining fully any non-volatile solutes is to allow the shoot to grow in a water-saturated gaseous environment rather than under water. As expected from the sparging experiment, anaerobic growth of the shoot was slower out of water than under water, with the effect once again being localised in the stem. This encourages the view that a gas or volatile substance promotes extension in the absence of oxygen. An alternative explanation, that buoyant tension generated by submergence is responsible for the promoting influence of submergence (Musgrave and Walters, 1974), remains to be tested.

The effect of partial oxygen shortage. To induce hypoxia rather than anoxia or normoxia, plants were grown in unsparged solutions (i.e., hypo-aerobic) in the aerobic workbench. These stagnant conditions generated oxygen partial pressures of 5-8 kPa when measured with an oxygen electrode after five days. The amount of shoot elongation in these hypo-aerobic solutions over this period was intermediate between that of plants grown in anaerobic and aerobic conditions (Fig. 1b). The promoting effect was located in the stem rather than the leaves, although the effect was less marked than that in anaerobic plants. The explanation for faster extension by hypoxic stems may be similar to that given for other species showing this response. In the coleoptiles of rice (Pearce *et al.*, 1992) and the stems of deep-water rice (Raskin and Kende, 1984), small oxygen partial pressures, and increases in carbon dioxide and ethylene each contribute positively to extension growth by these organs when submerged. The promoting effect of ethylene in most aquatic and amphibious plants when some oxygen is present is well known (Musgrave *et al.*, 1971; Jackson, 1990), and, in certain tissues, ethylene biosynthesis has been reported to be increased by partial but not complete oxygen shortage (Jackson, 1982; Metraux and Kende, 1983; Jackson *et al.*, 1984). The possibility that ethylene might be involved in the stimulation of stem elongation by hypoxia in the shoots of *P. pectinatus* is described in the following section.

ETHYLENE PRODUCTION AND ACTION

Effects of ACC and AVG on ethylene production. Plants were grown for five days in the aerobic workbench in unstirred solutions (5 - 8 kPa oxygen) containing 0, 0.1, 1 or 10 mM 1-aminocyclopropane-1-carboxylic acid (ACC, the precursor of ethylene), or 0, 0.1, 1, or 10 μM aminoethoxyvinylglycine (AVG, an inhibitor of ACC synthesis). After two and a half days, the test-tubes were drained and capped with 'Subaseal' stoppers to allow ethylene to accumulate for four hours before measurements were taken using FID gas chromatography. After making these measurements, the tubes were flushed with air and resealed. A second set of measurements was then taken 20 hours later. Plants were then re-submerged in fresh ACC or AVG solutions for a further two days when shoot lengths and ethylene were measured for a third time. Carbon dioxide and oxygen were also assayed. The latter indicated that aerobic respiration was taking place during the periods of ethylene collection (RQ > 0.65; CO_2 production rate 6.8 x 10^{-5} mol kg^{-1} s^{-1}). Ethylene production by control plants was extremely small. Surprisingly, neither AVG nor ACC changed substantially this slow rate of ethylene production, and shoot extension remained unaffected. The absence of a strong enhancement of ethylene production by the ethylene precursor ACC was particularly unexpected since most species release copious amounts of the gas when supplied with this compound (Adams and Yang, 1979). One possible explanation is that all the exogenous ACC was oxidised to ethylene prior to the analyses two and a half and five days after treatment. To eliminate this possibility, solutions of 0, 1, 10 or 25 mM ACC were supplied to tubers with large well-developed shoots for only four hours before sealing. Ethylene, carbon dioxide and oxygen were then measured four hours later, prior to a second treatment with ACC (for 15 hours) and further analyses of the gases. For comparison with a conventional land plant, pea seedlings (*Pisum sativum* L., cv. Meteor) similar in size to germinating *Potamogeton* tubers were also included. The results confirmed that the basal rate of ethylene production by *P. pectinatus* shoots was extremely slow compared to that of pea shoots. A very small stimulation in production by *P. pectinatus* was seen in response to increasing amounts of ACC. However, even in 25 mM ACC, the strongest concentration used, *P. pectinatus* produced less ethylene than peas without exogenous ACC. These results indicate that *P. pectinatus* produces almost no ethylene, suggesting that the well-known pathway for ethylene biosynthesis, involving the conversion of methionine to ACC via *S*-adenosylmethionine, and the oxidation of ACC to ethylene, either does not operate, or operates only very weakly.

Effects of ethylene on shoot growth. The majority of aquatic and semi-aquatic species so far examined respond to ethylene with enhanced elongation rates by leaves or stems when oxygen is also available (see Voesenek *et al.* (1992) for a recent review), although there are exceptions (e.g., *Echinochloa oryzoides*; Pearce *et al.* (1992)). To test the responsiveness of *P. pectinatus*, ethylene gas (5-10 μl l^{-1}) was supplied to tubers for five days in a humid gaseous environment provided by a sealed 300 x 300 x 300 mm glass tank. Controls were enclosed in a similar atmosphere together with the granular ethylene absorbent 'Ethysorb' (Stayfresh Ltd., London, UK), which has alkaline potassium permanganate as the active ingredient. Plants treated with ethylene were taller than controls. However, almost all of the growth stimulation was located in the leaf, with the stem elongating very little either in the presence or absence of ethylene. Since it was *stem* rather than leaf elongation which was enhanced by hypoxia, the positive growth response to solutions that are only partially deficient in oxygen cannot be attributed to the action of ethylene.

CONCLUSIONS

An outstanding feature of shoot growth by germinating tubers of *P. pectinatus* is the stimulation of internode extension in response to decreasing oxygen content, that can be sustained for several days. The most vigorous response was seen in strictly oxygen-free conditions. The effect was the result of greater cell elongation. However, even in anaerobic conditions, cell production appeared to continue, although its rate was reduced by 50%. This unexpected finding now requires further confirmation by cytological examination of anaerobic meristems. Leaf extension was slightly inhibited by oxygen shortage, but the overall effect of oxygen deprivation was to lengthen the shoot, through a positive effect on stem length. This is presumed to be of adaptive significance through increasing the likelihood of the shoot reaching better illuminated and aerated zones closer to the surface of the water. Once this is achieved, an extensive aerenchyma would facilitate internal transport of oxygen to basal parts, thus supporting aerobic growth, especially by roots which fail to emerge in the absence of oxygen. Although there are several species where shoot growth under anaerobic conditions has been demonstrated e.g., rhizomatous shoots of *Schoenoplectus, Scirpus, Typha* and *Potamogeton filiformis* (Barclay and Crawford, 1982), or the seedling shoots of germinating rice (Pearce *et al.*, 1992) or *Echinochloa* (VanderZee and Kennedy, 1981), all examples appear to involve leaf growth rather than stem growth, and only in the case of the rice coleoptile is elongation actually enhanced. Menegus *et al.* (1992) recently reported that the root and cotyledon of geminating seeds of the water chestnut (*Trapa natans*) can be stimulated when made anoxic, but stem growth was arrested by such conditions. Thus, *Potamogeton pectinatus* appears to be the first species for which stem extension has been reported to be enhanced by the complete absence of oxygen. We know of no other plant where cell production has been shown to take place without oxygen. The most vigorous anaerobic growth occurred in unstirred solutions. Sparging or water saturated atmospheres attenuated the effect slightly. We speculate that a growth promoter, possibly carbon dioxide, may be active in anoxic shoots and be depleted by sparging or by providing a gaseous rather than aqueous environment. Some stimulation of stem growth was also seen when tubers were germinated in solutions containing small but not extinguished amounts of oxygen (5 - 8 kPa). In many aquatic plants ethylene is an active promoter of extension growth under such circumstances. However, this does not apply to *P. pectinatus* for two reasons. Firstly, this species produces almost undetectable amounts of ethylene and does not appear to utilise the methionine/ACC/ethylene pathway common to most plants. Secondly, when ethylene is administered, no stimulation of growth by the stem takes place.

REFERENCES

Adams DO, Yang SF (1979) Ethylene biosynthesis: identification of 1-aminocyclopropane-1-carboxylic acid as an intermediate in the conversion of methionine to ethylene. Proceedings of the National Academy of Sciences USA 76: 170-174

Barclay AM, Crawford RMM (1982) Plant growth and survival under strict anaerobiosis. Journal of Experimental Botany 33: 541-549

Devlin RM, Karczmarczyk SJ (1975) Responses of *Potamogeton pectinatus* L. to norflurazon. Aquatic Botany 1: 263-268

Hettiarachchi PL, Triest L (1986) Isozyme polymorphism of *Potamogeton pectinatus* L. Proceedings of the 7th International Symposium on Aquatic Weeds. European Weed Research Society/Association of Applied Biologists, Loughborough, Leicestershire, pp 163-167

Jackson MB (1982) Ethylene as a growth promoting hormone under flooded conditions. In: Wareing PF (ed) Plant growth regulators 1982. Academic Press, London, pp 291-301

Jackson MB (1990) Hormones and developmental change in plants subjected to submergence or soil waterlogging. Aquatic Botany 39: 49-72

Jackson MB, Dobson CM, Herman B, Merryweather A (1984) Modification of 3,5-diiodo-4-benzoic acid (DIHB) activity and stimulation of ethylene production by small concentrations of oxygen in the root environment. Plant Growth Regulation 2: 251-262

Menegus F, Cattaruzza L, Scaglioni L, Ragg E (1992) Effects of oxygen level on metabolism and development of seedlings of *Trapa natans* and two ecologically related species. Physiologia Plantarum 86: 168-172

Metraux J-P, Kende H (1983) The role of ethylene in the growth response of submerged deepwater rice. Plant Physiology 72: 441-446

Musgrave A, Walters J (1974) Ethylene and buoyancy control rachis elongation of the semi-aquatic fern *Regnellidium diphyllum*. Planta 121: 51-56

Musgrave A, Jackson MB, Ling E (1972) Callitriche stem elongation is controlled by ethylene and gibberellin. Nature New Biology 81: 93-96

Pearce DME, Hall KC, Jackson MB (1992) The effects of oxygen, carbon dioxide and ethylene on ethylene biosynthesis in relation to shoot extension in seedlings of rice (*Oryza sativa*) and barnyard grass (*Echinochloa oryzoides*). Annals of Botany 69: 441-447

Raskin I, Kende H (1984) Regulation of growth in stem sections of deep-water rice. Planta 160: 66-72.

Suge H, Kusanagi T (1975) Ethylene and carbon dioxide: Regulation of growth in two perennial aquatic plants, arrowhead and pond weed. Plant and Cell Physiology 16: 65-72

VanderZee D, Kennedy RA (1981) Germination and seedling growth in *Echinochloa crus-galli* var. Oryzicola under anoxic conditions: structural aspects. American Journal of Botany 68: 1269-1277

Voesenek LACJ, van der Sman AJM, Harren FJM, Blom CWPM (1992) An amalgamation between hormone physiology and plant ecology: a review on flooding resistance and ethylene. Journal of Plant Growth Regulation 11: 171-188

Wiegleb G, Brux H (1991) Comparison of life history characters of broadleaved species of the genus *Potamogeton* L. I. General characterization of morphology and reproductive strategies. Aquatic Botany 39: 131-146

Yeo RR (1965) Life history of sago pondweed. Weeds 13: 314-321

METABOLIC ACCLIMATION TO ANOXIC CONDITIONS AND BIOCHEMICAL BASIS OF DEATH

P. Saglio
Station de Physiologie Végétale
INRA, Centre de Recherches de Bordeaux
B.P. 81, 33883 Villenave d'Ornon Cedex
France

ABSTRACT

Exposure to hypoxic conditions induces a metabolic acclimation which improves the survival of maize root systems under subsequent extreme anoxic conditions. The regulation of cytoplasmic pH is considered to be the major factor in plant tissue survival under anoxia. The rapid acidification of the cytosol following transfer to anoxic conditions has been attributed to lactic acid accumulation. However, the pH drop and accumulation rate of lactic acid have different kinetics. This implies that lactic acid alone cannot account for the initial cytoplasmic acidification in anoxic maize root tips. Nevertheless, lactic acid might play a decisive role in the susceptibility of plant tissues to anoxia by producing an overacidification of the cytoplasm during the first hour of the anaerobic incubation after the main drop of pH. Hypoxically pretreated (acclimated) maize root tips accumulate less lactic acid and have higher cytoplasmic pH values under anoxia than non-acclimated ones. This lower accumulation is the result of an increased efflux of lactic acid and reduced production. The higher efflux of lactate from acclimated root tips is abolished by the presence of cycloheximide during the hypoxic pretreatment, suggesting that protein synthesis might be involved in this excretion. Similar strategies are encountered in some cereal organs reported to be very resistant to anoxic conditions; for example, rice embryos produce only trace amounts of lactic acid while barley aleurone tissues produce large amounts of lactic acid which is efficiently excreted into the medium. The mechanism of lactic acid excretion induced in maize root tips by the hypoxic pretreatment, either carrier-mediated or triggered by the pH gradient, may be an important determinant of the tolerance of plant tissues to anoxic conditions.

The root systems of most dryland species sustain injury or death when suddenly exposed to anaerobic conditions (Jackson and Drew, 1984). However, a 'training' in hypoxic conditions induces acclimation mechanisms which, in maize roots, improve their survival capacity under subsequent extreme anoxic conditions (Drew *et al.*, 1985; Johnson *et al.*, 1989). These mechanisms lead to two kinds of survival

NATO ASI Series, Vol. I 16
Interacting Stresses on Plants in a Changing Climate
Edited by M. B. Jackson, C. R. Black

strategy. One involves dramatic morphological modification of the root cortex with the formation of lysigenous aerenchyma (Drew *et al.*, 1979). The formation of aerenchyma is slow (days) and is an irreversible process stimulated by ethylene accumulation in the tissues under poorly aerated conditions (Drew *et al.*, 1979; Campbell and Drew, 1983). In the absence of aerenchyma, root porosity is low (Drew *et al.*, 1985; Saglio and Belgrand, 1986) and the contribution of internally transported oxygen to the respiration of the tissues, as quantified by the adenylate energy charge or the ATP/ADP ratio (Pradet and Raymond, 1983), decreases rapidly along the root from the stem base to the tip (Saglio *et al.*, 1983). The formation of an aerenchymatous structure, which resembles that of wetland species with a large internal gas-filled volume, clearly improves the internal movement of oxygen to the root apex (Drew *et al.*, 1985).

A second strategy consists of a metabolic acclimation that improves the energy status of anoxic tissues, and contributes to their more prolonged survival in the absence of oxygen (Saglio *et al.*, 1988; Johnson *et al.*, 1989). This response is rapid (hours) and leads to the induction of a limited number of 'anaerobic polypeptides' (Sachs *et al.*, 1980, and see this volume). Most of these polypeptides are different from heat shock proteins (Mocquot *et al.*, 1987) and those which have been identified are enzymes involved in the glycolytic and fermentative pathways (Sachs and Ho, 1986; Rivoal *et al.*, 1989; Ricard *et al.*, 1991). Abscisic acid, which is known to trigger a number of plant responses to stress (Creelman, 1989), has recently been reported to induce tolerance of anoxia in corn, probably via the mediation of the synthesis of some proteins including alcohol dehydrogenase (ADH) (Hwang and Vantoai, 1991). However, the significance of these proteins for the mechanisms of tolerance to anoxia has rarely been examined and remains unclear. In hypoxically acclimated maize root tips, lactate dehydrogenase (LDH) activity increases at the same time that the rate of anoxic lactate production declines (Xia and Saglio, 1992). The results obtained with the hypoxically-inducible LDH in barley roots and aleurone tissues suggest some unrecognised function for this enzyme (Hoffman *et al.*, 1986; Hanson and Jacobsen, 1986). It is interesting to note that in mammalian cancer cells, anoxia has been reported to induce synthesis of a distinct LDH isoenzyme interacting with the regulation of a number of systems including certain nucleic acids (Anderson and Farkas, 1988). Furthermore, the role of ADH, which is known to increase dramatically during hypoxic acclimation, is controversial (Johnson *et al.*, 1989), and evidence has been provided showing that ADH level is not closely correlated with increased survival of maize root tips under anoxia (Roberts *et al.*, 1989).

Metabolic studies of the survival of plant tissues in the absence of oxygen have focussed mainly on the possible toxicity of the end-products of fermentation. In the 1970s, Crawford (1977) proposed that sensitivity to anoxia was proportional to the rate of ethanol production. This idea is now discounted because of the high diffusion rate of ethanol in the external medium and its low toxicity under anoxic conditions for plant tissues (Jackson *et al.*, 1982). However, ethanol can be indirectly toxic to some tissues by its metabolisation to acetaldehyde following a return to aerobic conditions, as shown for carrot cells (Perata and Alpi, 1991) and several anoxia-intolerant species (Monk *et al.*, 1987).

Regulation of cytoplasmic pH is an important factor involved in the survival of plant tissues under anoxia. Roberts *et al.* (1984b) used ^{31}P-NMR to study the cytoplasmic acidosis of maize root tips during the transition from normoxia to anoxia, and concluded that the initial acidosis is attributable to a transient accumulation of lactic acid which provides the signal triggering ethanol production (Davies *et al.*, 1974). The death of the root tips in the absence of oxygen is attributed to a poor regulation of cytoplasmic pH controlled by the balance between lactic acid and ethanol fermentation (Roberts *et al.*, 1984a). However, recent studies also using ^{31}P-NMR show that the fast initial cytoplasmic acidification and lactate accumulation do not have similar timecourses (Menegus *et al.*, 1991; Saint-Gès *et al.*, 1991). Therefore, lactic acid alone does not account for the initial cytoplasmic acidification in anoxic maize root tips. However, a recent study (Xia and Saglio, 1992) suggests that lactic acid might indeed play a decisive role in the susceptibility of plant tissues to anoxia by overacidifying the cytoplasm during the first hour of anaerobic incubation, after the main drop of pH. In this study it was shown that hypoxically pretreated (acclimated) maize root tips accumulate less lactic acid and have higher cytoplasmic pH values under anoxia than non-acclimated ones. This lower accumulation is the result of an increased efflux of lactic acid which is eventually associated with reduced production. This higher efflux of lactate from acclimated root tips is abolished by incubation in the presence of cycloheximide during the hypoxic pretreatment, suggesting that protein synthesis might be involved in this excretion. Similar strategies are encountered in some cereal organs reported to be very resistant to anoxic conditions. Rice embryos produce only trace quantities of lactic acid (Ricard *et al.*, 1991), while barley aleurone tissues (Hanson and Jacobsen, 1984) produce large amounts of lactic acid, which does not accumulate in the tissues because it is efficiently excreted into the medium.

REFERENCES

Anderson GR, Fakas BK (1988) The major anoxic stress response protein p34 is a distinct lactate dehydrogenase. Biochemistry 27: 2187-2193

Campbell R, Drew MC (1983) Electron microscopy of gas space (aerenchyma) formation in adventitious roots of *Zea mays* L. subjected to oxygen shortage. Planta 157: 350-357

Crawford RMM (1977) Tolerance to anoxia and ethanol metabolism in germinating seeds. The New Phytologist 79: 511-517

Creelman RA (1989) Abscisic acid physiology and biosynthesis in higher plants. Physiologia Plantarum 75: 131-136

Davies DD, Greco S, Kenworthy P (1974) The control of the production of lactate and ethanol by higher plants. Planta 118: 297-310

Drew MC, Jackson MB, Giffard S (1979) Ethylene promoted adventitious rooting and development of cortical air spaces(aerenchyma) in roots may be adaptive responses to flooding in *Zea mays* L. Planta 147: 83-88

Drew MC, Saglio PH, Pradet A (1985) Adenylate energy charge and ATP/ADP ratios in aerenchymatous roots of *Zea mays* in anaerobic media as a consequence of improved internal oxygen transport. Planta 165: 51-58

Hanson AD, Jacobsen JV (1984) Control of lactate dehydrogenase, lactate glycolysis, and α-amylase by O_2 in barley aleurone layers. Plant Physiology 75: 566-572

Hoffman NE, Bent AF, Hanson AD (1986) Induction of lactate dehydrogenase isoenzymes by oxygen deficit in barley root tissue. Plant Physiology 82: 658-663

Hwang SY, Vantoai TT (1991) Abscisic acid induces anaerobiosis tolerance in corn. Plant Physiology 97: 593-597

Jackson MB, Drew MC (1984) Effect of flooding on growth and metabolism of herbaceous plants. In: TT Kozlowski (ed) Flooding and plant growth. Academic Press, Orlando, pp 47-128

Jackson MB, Herman B, Goodenough A (1982) An examination of the importance of ethanol in causing injury to flooded plants. Plant, Cell and Environment 5: 163-172

Johnson J, Cobb BG, Drew MC (1989) Hypoxic induction of anoxia tolerance in root tips of *Zea mays*. Plant Physiology 91: 837-841

Menegus F, Cattaruzza L, Mattana M, Beffagna N, Ragg E (1991) Response to anoxia in rice and wheat seedlings. Change in the pH of intracellular compartments, glucose-6-phosphate, and metabolic rate. Plant Physiology 95: 760-767

Mocquot B, Ricard B, Pradet A (1987) Rice embryos can express heat-shock genes under anoxia. Biochimie 69: 677-681

Monk LS, Brändle R, Crawford RMM (1987) Catalase activity and post-anoxic injury in monocotyledonous species. Journal of Experimental Botany 38: 233-246

Perata P, Alpi A (1991) Ethanol-induced injuries to carrot cells. The role of acetaldehyde. Plant Physiology 95: 748-752

Pradet A, Raymond P (1983) Adenine nucleotide ratios and adenylate energy charge in energy metabolism. Annual Review of Plant Physiology 34: 199-224

Ricard B, Rivoal J, Spiteri A, Pradet A (1991) Anaerobic stress induces the transcription of sucrose synthase in rice. Plant Physiology 95: 669-674

Rivoal J, Ricard B, Pradet A (1989) Glycolytic and fermentative enzyme induction during anaerobiosis in rice seedlings. Plant Physiology and Biochemistry 27: 43-52

Roberts JKM, Callis J, Jardetsky O, Walbot V, Freeling M (1984a) Cytoplasmic acidosis as a determinant of flooding intolerance in plants. Proceedings of the National Academy of Sciences, USA 81: 6029-6033

Roberts JKM, Callis J, Wemmer D, Walbot W, Jardetsky O (1984b) Mechanism of cytoplasmic pH regulation in hypoxic maize root tips and its role in survival under hypoxia. Proceedings of the National Academy of Sciences, USA 81: 3379-3383

Roberts JKM, Chang K, Webster C, Callis J, Wemmer D, Walbot V (1989) Dependence of ethanolic fermentation, cytoplasmic pH regulation and viability on the activity of alcohol dehydrogenase in hypoxic maize root tips. Plant Physiology 89: 1275-1278

Sachs MM, Ho THD (1986) Alteration of gene expression during environmental stress in plants. Annual Review of Plant Physiology 37: 363-376

Sachs MM, Freeling M, Okimoto R (1980) The anaerobic proteins of maize. Cell 20: 761-767

Saglio PH, Belgrand M (1986) Determination of root porosity by gas chromatography of helium released by preloaded tissues. Physiologie Vègetale 24: 117-122

Saglio PH, Raymond P, Pradet A (1983) Oxygen transport and root respiration of maize seedlings. A quantitative approach using the correlation between ATP/ADP and the respiration rate controlled by oxygen tension. Plant Physiology 72: 1035-1039

Saglio PH, Drew MC, Pradet A (1988) Metabolic acclimation to anoxia induced by low (2-4 kPa) partial pressure oxygen pretreatment (hypoxia) in root tips of *Zea mays*. Plant Physiology 86: 61-66

Saint-Gès V, Roby C, Bligny R, Pradet A, Douce R (1991) Kinetic studies of the variation of cytoplasmic pH, nucleotide-triphosphates (^{31}P-NMR) and lactate during normoxic and anoxic transitions in maize root tips. European Journal of Biochemistry 200: 477-482

Xia JH, Saglio PH (1992) Lactic acid efflux as a mechanism of hypoxic acclimation of maize root tips to anoxia. Plant Physiology 100: 40-46

ENERGY AND FERMENTATION METABOLISM IN HYPOXIC RICE COLEOPTILES – A MULTINUCLEAR NMR APPROACH

Teresa W-M. Fan, Andrew N. Lane[1] and Richard M. Higashi[2]
Department of Land, Air and Water Resources
University of California
Davis CA 95616
USA

ABSTRACT

In vivo ^{31}P NMR was employed to follow changes in intracellular pH, concentrations of phosphate metabolites, and the unidirectional rate of ATP synthesis in rice shoot tips during hypoxia. Under extended hypoxia (16-22 hours), rice shoot tips exhibited a small and transient decrease in cytoplasmic pH, while the ATP concentration and unidirectional rate constant of ATP synthesis were maintained at 90 and 75% of the control level, respectively. This is in contrast to actively growing flood-sensitive tissues such as maize root tips which showed a two-fold decrease in ATP concentration and a ten-fold decline in the rate of ATP synthesis. In addition, carbon flow through biochemical pathways including fermentation was followed using ^{13}C-enriched acetate and bicarbonate precursors, and ^{1}H NMR, ^{13}C NMR and GC-MS techniques. Ethanol, alanine, succinate, γ-aminobutyrate and lactate were major fermentation products that accumulated in hypoxic rice shoot tips; succinate and γ-aminobutyrate (*via* glutamate) were major carbon sinks for acetate metabolism through the Krebs cycle. Moreover, both the wet weight and length of rice shoot tips increased comparably under control and hypoxic conditions. These results indicated that hypoxic rice shoot tips possessed active fermentation metabolism and continued to operate the Krebs cycle, both of which contributed to maintenance of ATP synthesis, reduced cytoplasmic acidification, and ultimately flood tolerance. However, the lack of unique fermentation products in rice shoot tips suggested that a quantitative instead of a qualitative difference in fermentation metabolism may account for this tolerance. In addition, rice shoot tips may utilise other oxidative pathways such as nitrate respiration to couple with the Krebs cycle to maintain their high ATP turnover under hypoxia.

[1] National Institute for Medical Research, Mill Hill, London NW7 1AA, UK
[2] University of California Bodega Marine Laboratory, Bodega Bay, CA 94923, USA

NATO ASI Series, Vol. I 16
Interacting Stresses on Plants in a Changing Climate
Edited by M. B. Jackson, C. R. Black

INTRODUCTION

When aerobic tissues are subjected to hypoxia, a typical response is a decline in ATP concentration ([ATP][1]) due to a reduction in mitochondrial oxidative phosphorylation, which is accompanied by an excess production of NADH and cytoplasmic acidosis as a result of glycolytic activity. To cope with these problems, different tissues possess various fermentation pathways to remove excess H^+ and regenerate NAD^+ so that glycolysis can continue. In animal tissues, lactate is the major fermentation product, whereas ethanol is its counterpart in hypoxic plant tissues (Davies, 1980). Since synthesis of ethanol removes twice as many H^+ as synthesis of lactate per mole of pyruvate, plant tissues are generally considered to be better at regulating hypoxia-induced intracellular acidosis than animal tissues. In spite of this, hypoxic plant tissues still exhibit a cytoplasmic pH drop ranging from 0.3 to 0.8 units (Roberts *et al.*, 1984; Fan *et al.*, 1988; Menegus *et al.*, 1991). In the case of rice shoots, a decrease of 0.4 units in cytoplasmic pH was reported recently (Menegus *et al.*, 1991). The combination of reduced ATP synthesis and cytoplasmic acidosis is considered to be the main cause of cell injury (Garlick *et al.*, 1979; Hochachka and Mommsen, 1983).

For actively growing tissues such as maize root tips (Roberts *et al.*, 1984), the [ATP] decrease and cytoplasmic acidosis are greater than in the corresponding mature roots (Fan *et al.*, 1988) during hypoxia, perhaps due to a greater metabolic demand of the tips. Consequently, growing tips are less tolerant of hypoxia than mature parts of the plant. However, metabolically active rice coleoptiles are

---

[1] *Abbreviations*: [x], concentration of x; NMR, nuclear magnetic resonance; ATP_{γ}, $_{\alpha}$ & $_{\beta}$, γ, α and β-resonances of ATP; T_1, spin-lattice relaxation time; FIRFT, fast inversion recovery Fourier transformation; k_{app}, apparent first-order rate constant; GC-MS, gas chromatography-mass spectrometry; TOCSY, total correlation spectroscopy; HETCORR, heteronuclear shift correlation spectroscopy; GAB, γ-aminobutyrate; PEP, phosphoenolpyruvate; MES, 2-[N-morpholino]-ethanesulphonic acid; MTBSTFA, N-methyl-N-(*tert*-butyldimethylsilyl)trifluoroacetamide; NaOAc, sodium acetate; α-KG, α-ketoglutarate; TCA, tricarboxylic acid.

known to maintain growth or elongation even under extremely low oxygen tension (Öpik, 1973). This raises the basic question of whether rice shoot tips under hypoxia somehow maintain a high level of ATP synthesis or are metabolically more efficient. To address this question, we have employed a steady-state ^{31}P nuclear magnetic resonance (NMR) saturation transfer technique (Gadian, 1982) to measure *in vivo* the unidirectional rate constant of ATP synthesis, along with the changes in cytoplasmic pH and phosphate metabolites in excised rice shoot tips under both hypoxic and control conditions (Fan *et al.*, 1992). More recently, we have utilised ^{13}C-enriched acetate and carbonate precursors in conjunction with measurements by ^{13}C NMR, ^{1}H NMR and gas chromatography-mass spectrometry (GC-MS) to examine the carbon flow from precursors to various fermentation products. In both studies, the tip portion (first 3-5 mm) of the shoot was chosen for its high metabolic demand. This chapter describes experiments and results from both the *in vivo* NMR (Fan *et al.*, 1992) and the ^{13}C isotope studies.

MATERIALS AND METHODS

Rice (*Oryza sativa* L. Calrose M201) seeds, obtained from University of California Rice Research Station, Biggs, CA, USA, were germinated and grown under etiolating conditions according to our previous procedure (Fan *et al.*, 1992). For *in vivo* ^{31}P NMR experiments, tips 3-5 mm in length were excised with a single cut and incubated at room temperature with constant aeration in 10 mM sodium acetate, 0.5 mM $CaSO_4$, 50 mM glucose at pH 6.0 (acetate buffer) until ready for packing into the NMR tube.

Growth of rice shoot tips was measured under 'hypoxic' and 'extreme hypoxic' conditions. The former treatment was implemented in open flasks sparged with O_2-free N_2 using the same acetate buffer as in the *in vivo* ^{31}P NMR experiments (Fan *et al.*, 1992); a trace amount of O_2 exchange was expected to occur during the treatment. Growth measurement under 'extreme hypoxia' was carried out in all buffers (see below) employed in this study using a sealed sparging flask under a constant positive pressure of N_2 to eliminate O_2 exchange.

The system for *in vivo* ^{31}P NMR measurements has been described previously (Fan *et al.*, 1986a; Fan *et al.*, 1992). Briefly, rice shoot tips were superfused in acetate buffer (750 ml) sparged either with air or O_2-free N_2 at 25 °C with a flow rate of >15 ml min^{-1}. The O_2 level fell below detectable limits (100 μg l^{-1}) by a Clark-type O_2 electrode in under 15 minutes. *In vivo* ^{31}P NMR spectra were recorded with either a 12 mm probe at 145.7 MHz on a Nicolet NT-360 or a 10 mm probe at 161.9 MHz on a wide-bore Bruker AM-400 NMR spectrometer (Fan *et al.*, 1992). In the former case, spectral parameters used a 70° excitation pulse, an interpulse delay of 1 s, 2048 data points, a spectral width of 8 kHz, and 2048 acquisitions per file. In the latter case, a 90° excitation pulse, 4096 points over a spectral width of 8 kHz (acquisition time of 0.256 s), an interpulse delay of 0.744 s, and 1800 acquisitions per file were used. Free induction decays were zero-filled to 4096 or 8192 points and apodised with a line-broadening factor of 30 Hz before Fourier transformation. The spectra were referenced to an external capillary containing methylene diphosphonate in D_2O (Roberts *et al.*, 1984), which also provided a lock signal. Saturation transfer measurements were made on the AM-400 instrument using a selective excitation unit whose calibrated output was attenuated to give a square pulse of B_1 = 108 Hz. Thirty-two scans with irradiation for 3.3 s on the γ-resonance of ATP (ATP_γ) and off resonance (equidistant from the carrier opposite to ATP_γ) were acquired alternately in separate files, and cycled until sufficient signal had been acquired (typically 16-20 hours). The apparent spin-lattice relaxation times (T_1) were measured with saturation of the ATP_γ signal using FIRFT (Gupta *et al.*, 1980; Gadian, 1982). Eight relaxation delays from 1 ms to 5 s were used and the interpulse delay was 2 s. The value of T_1 was determined from the magnetisation recovery curves using a nonlinear regression fit routine (Fan *et al.*, 1992). The apparent first-order rate constant for the synthesis of ATP (k_{app}) was determined from the extent of saturation of the cytoplasmic phosphate resonance and the measured T_1 value according to Gadian (1982).

For the carbonate treatment, the buffer used included 10 mM MES, 0.5 mM $CaSO_4$, 50 mM glucose and 5 mM either unlabelled or $NaH^{13}CO_3$ at pH 6.0. For the acetate treatment, the buffer contained 10 mM MES, 0.5 mM $CaSO_4$, 50 mM glucose, 5 mM $NaHCO_3$ and 10 mM either unlabelled or sodium ^{13}C-2-acetate (Cambridge Isotope Laboratories, Woburn, MA, USA). Approximately 1.9-3.1 g (wet weight) of shoot tips were incubated in the dark in the unlabelled or ^{13}C-enriched buffer (125-200 ml) at 25 °C for 22 hours with either CO_2-free air or N_2-sparging at a flow rate of 70-80 ml min^{-1}. Ethanol from the medium was trapped in a vacuum trap (Ace Glass, Vineland, NJ, USA) immersed in a dry ice-methanol bath. At the end of the treatment, the tips were rinsed with either aerated or N_2-sparged deionised water and lyophilised to trap tissue ethanol in a vacuum trap immersed in a dry ice-methanol bath. The lyophilised tissue was then extracted and derivatised as described previously (Fan *et al.*, 1986b). Briefly, tissue was extracted with 5% perchloric acid (PCA), titrated to pH 3.5 with K_2CO_3, lyophilised again and derivatised with MTBSTFA:pyridine (1 : 1) before GC analysis for non-volatile metabolites. The GC analysis was modified from our previous procedure (Fan *et al.*, 1986b) using a 0.18 mm i.d. x 40 m DB-1 column (J&W Scientific, Folsom, CA, USA) with the temperature of the column held at 60 °C for two minutes, followed by ramping to 150 °C at 20 °C min^{-1}, then to 300 °C at 6 °C min^{-1}, with injector and flame ionisation detector temperatures at 280 and 320 °C, respectively, and with a H_2 carrier gas velocity of 60 cm s^{-1}. GC-MS used the same GC conditions, except that He carrier gas was used at 30 cm s^{-1}, and detection was by a Trios-2 quadrupole mass spectrometer (VG Instruments, Altrincham, UK) with electron energy at 70 eV, source temperature at 200 °C, and transfer line at 300 °C. Ethanol was measured on the same column using an initial temperature of 40 °C for two minutes, followed by a 20 °C min^{-1} increase to 150 °C, injector temperature of 100 °C, and flame ionisation detector temperature of 320 °C. For analysis by NMR, the extracts were further titrated to neutral pH with KOH before lyophilisation, followed by passage through Chelex 100 resin (200-400 mesh, BioRad, Richmond, CA, USA), lyophilisation, and redissolution in 0.5 ml D_2O.

All NMR measurements on the extracts were performed on a Bruker AM-500 NMR spectrometer (Karlsruhe, Germany) except for ^{1}H TOCSY (Bax and Davis, 1985) which was carried out on a Bruker AM-400 NMR spectrometer. The spectral conditions for the 1-D ^{1}H NMR spectra included 90° pulse, 6 kHz spectral width, 1.5 s relaxation delay, 16 k data points, presaturation of solvent, 128-256 scans and 0.5 Hz line-broadening; for the 1-D ^{13}C NMR spectra, 90° pulse, 26.3 kHz spectral width, 1 s relaxation delay, 32 k data points, Waltz-16 ^{1}H decoupling, 2048-8192 scans and 3 Hz line-broadening; for the 2-D ^{1}H-^{13}C HETCORR spectrum, 26.3 kHz spectral width in F2 with 4096 points and 5 kHz in F1 with 300 time increments at 128 scans per increment, zerofilling of F2 to 8192 and F1 to 1024 points, 3 Hz line-broadening in F2, and apodising with 60°-shifted sine-squared bell in F1; for the 2-D ^{1}H TOCSY spectrum, 4.5 kHz spectral width, 8.3 kHz spin-lock strength and 45 ms spin-lock time using MLEV-17, 1.5 s relaxation delay, 2048 (F2) by 384 (F1) points which were zerofilled to 4096 by 2048 complex points, 96 scans per time increment, and processing with 36° and 60°-shifted sine-squared function in F2 and F1, respectively.

RESULTS AND DISCUSSION

Effect of hypoxia on energy metabolism. As shown previously (Fan *et al.*, 1992), the excised shoot tips continued to grow in the presence of glucose under both control and hypoxic conditions (for both growth and *in vivo* ^{31}P NMR experiments). The wet weight increased by about 29% after 15 hours under both conditions, while the tip length increased by 43 and 49% for control and hypoxic shoot tips, respectively. This suggests that some of the size increase was due to elongation. There was no statistically significant difference between the values for the two treatments. However, under extreme hypoxia, the shoot tips showed less of an increase (0-35%) than the controls (64-113%) in wet weight after 22 hours of treatment. This indicates that rice shoot tips are capable of sustaining growth under hypoxia and also under extreme hypoxia although at a much reduced rate. It should be noted that the extent of hypoxia for *in vivo* NMR experiments approximated to that for hypoxic growth experiments, where a trace amount of O_2 exchange might be occurring along the long superfusion line. This is supported by the similar growth response acquired from the two setups.

The *in vivo* ^{31}P NMR spectra of rice shoot tips under both control and hypoxic conditions are shown in Fig. 1 (adapted from Fan *et al.*, 1992). Resonances were assigned to sugar phosphates (between -12 and -13 ppm), cytoplasmic phosphate (-14.05 in air, -14.25 ppm in N_2), vacuolar phosphate (-16.1 ppm), and γ (-21.9 ppm), α (-26.9 ppm) and β (-35.5 ppm) phosphate peaks of ATP (Gerasimowicz *et al.*, 1986), assuming that ATP is the dominant nucleoside triphosphate. The relative areas of the ATP_γ and ATP_β peaks were similar, indicating that little ADP was present in either case. Using an appropriate calibration curve (Gadian, 1982), cytoplasmic pH values were estimated from the chemical shifts of the cytoplasmic phosphate peak.

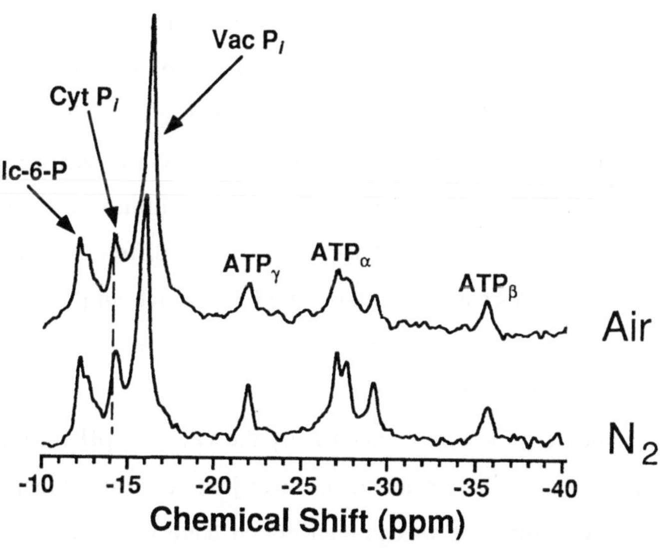

FIGURE 1. In vivo *^{31}P NMR spectra of control and hypoxic rice shoot tips. These spectra were acquired using the Bruker AM-400 spectrometer with conditions as described in the text. Upper spectrum: control (air); lower spectrum: hypoxia (N_2). Peaks assigned are glucose-6-phosphate (Glc-6-P), cytoplasmic phosphate (Cyt P_i), vacuolar phosphate (Vac P_i), and γ, α, and β phosphates of adenosine triphosphate (ATP_γ, ATP_α and ATP_β).*

After eight hours of treatment, the cytoplasmic pH was 7.50±0.07 in air and 7.30±0.05 in N_2 (see Table 1). The timecourses of cytoplasmic pH changes under both control and hypoxic conditions were also measured. Relative to the control, a slight acidification occurred after eight hours of hypoxia but returned to normal after 16 hours (data not shown). Hence, the cytoplasm exhibited only a transient acidification by about 0.2 pH units during extended hypoxia, in contrast to mature maize roots (Fan *et al.*, 1988) and root tips (Roberts *et al.*, 1984) where permanent pH decreases of 0.3 and 0.6 units, respectively were observed. An acid shift of 0.4 units and partial recovery of the cytoplasmic pH in rice shoots under severe hypoxia was reported recently (Menegus *et al.*, 1991).

These different findings may be attributed to the type of tissues and experimental protocols used. We have used the 'Calrose' variety of *Oryza sativa* and a single excision of the apical 3-5 mm part of the shoot (shoot tips), whereas 'Arborio' variety and a larger portion (*c.*17 mm) of the shoot with multiple excisions were used by Menegus *et al.* (1991). Differential tolerance of submergence is known to occur for different cultivars of rice and the number of excisions would lead to a different extent of 'wounding', which could complicate the hypoxic response. In addition, the incubation medium was circulated at all times during our NMR measurements, while no circulation was employed during hypoxia by Menegus *et al.* (1991). This could lead to differences in the accumulation of metabolic wastes such as ethanol and CO_2, which might in turn affect the hypoxic response. Finally, it is possible that different O_2 levels were present in our system and that used by Menegus *et al.* (1991).

We were unable to determine accurately the vacuolar pH from the chemical shift of the vacuolar phosphate peak since it was just out of the phosphate titration range. However, vacuolar pH was judged not to have increased significantly in either case, based on the constant vacuolar phosphate chemical shift (Fig. 1). This again differs from the vacuolar alkalinisation reported for larger rice shoot sections (Menegus *et al.*, 1991).

The changes in relative phosphate metabolite concentrations after 8 hours of hypoxia are listed in Table 1. The decrease in [ATP] reported here was the maximal change observed; in two out of six cases, no change in [ATP] was observed. The change in [cytoplasmic phosphate] was also small, with an increase of about 20%. This increase was again a transient event (data not shown), similar to the cytoplasmic pH. In contrast, in growing maize root tips [cytoplasmic phosphate] increased nearly threefold, while [ATP] decreased about twofold (Roberts *et al.*, 1984). The mature maize roots again showed intermediate behaviour, with [cytoplasmic phosphate] increasing and [ATP] decreasing by about 40% (Fan *et al.*, 1988).

Since rice shoot tips continued to grow and their [ATP] was maintained during hypoxia (Table 1), this raises the question of whether the rate of ATP synthesis was also maintained. We have therefore measured this rate *in vivo* by ^{31}P NMR saturation transfer. Fig. 2 (adapted from Fan *et al.*, 1992) shows a typical set of saturation transfer spectra under control conditions with the on-resonance irradiation on ATP_γ for 3.3 s. The fractional magnetisation transfer from ATP_γ to the cytoplasmic phosphate peak was determined from the area of the cytoplasmic phosphate peak in the difference spectrum normalised against that in the off-resonance spectrum. The mean fractional transfer was 0.23±0.05 in air and 0.28±0.05 in N_2 (Table 1).

To determine the pseudo first-order rate constant for ATP synthesis, the T_1 of the phosphate resonance was required. This was determined under both control and hypoxic conditions with saturation of ATP_γ as described above. The T_1 values under these two conditions were similar, 1.8±0.3 s in air and 2.0±0.3 s in N_2 (Table 1). Thus, the k_{app} for ATP synthesis was 0.13±0.04 s^{-1} for control and 0.1±0.03 s^{-1} for hypoxia. Based on this and the observation that [ATP] and [cytoplasmic phosphate] did not change significantly, we conclude that the rates of ATP synthesis and utilisation both decreased by less than 25% in rice shoot tips between control and hypoxia. This is in contrast with maize root tips, where the rate of ATP synthesis decreased 10-fold under hypoxia (Roberts *et al.*, 1985).

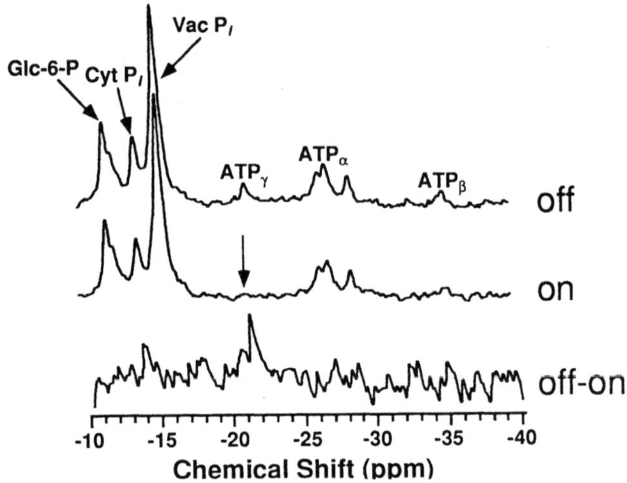

FIGURE 2. In vivo *^{13}P NMR saturation transfer spectra of rice shoot tips. The magnetisation transfer experiment was performed as described in the text. The peak at -22 ppm (ATP_γ) was irradiated as indicated by the arrow. The difference spectrum (off-on) was generated by subtracting on-resonance (on) from off-resonance (off) free induction decays and multiplying by a 40 Hz line-broadening prior to Fourier transformation. Magnetisation transfer from ATP_γ to Cyt P_i is evident in the difference spectrum which is shown at 4x scale.*

TABLE 1. *Response of phosphate metabolism of rice shoot tips to hypoxia. Cyt P_i, ATP_γ and ATP_β are peak areas of the cytoplasmic phosphate resonance and γ and β-resonances of ATP normalised to that of the vacuolar phosphate peak, which did not change significantly during hypoxia. $\Delta M/M^0$ is the fractional change in peak area of the cytoplasmic phosphate resonance upon saturation of the ATP_γ resonance. Standard deviations represent the statistical error in measuring peak properties due to random noise and variation from replicate samples (n=4). k_{app} is the apparent pseudo first-order rate constant for synthesis of ATP as determined from saturation transfer experiments. Data from Fan* et al. *(1992).*

Property	Condition		Ratio (N_2/Air)
	Air	N_2	
pH_i	7.50 ± 0.07	7.30 ± 0.05	-
Cyt P_i	0.23 ± 0.05	0.28 ± 0.05	1.2
ATP_γ	0.08 ± 0.01	0.08 ± 0.01	1.0
ATP_β	0.09 ± 0.01	0.08 ± 0.01	0.89
$\Delta M/M^0$	0.23 ± 0.04	0.20 ± 0.04	0.87
T_1(s)	1.80 ± 0.20	2.00 ± 0.20	1.1
k_{app} (s^{-1})	0.13 ± 0.04	0.10 ± 0.03	0.77

Effect of hypoxia on fermentation patterns. During short-term hypoxia, the primary fermentation end products in rice shoots have been shown to be ethanol, lactate and Ala (Bertani *et al.*, 1980; Menegus *et al.*, 1989; Fan *et al.*, 1992). To determine the major fermentation products under the more extended hypoxic conditions employed in this report, both volatile (ethanol) and non-volatile products produced by control and hypoxic shoot tips were measured by high-resolution GC (Fan *et al.*, 1992). The results are shown in Table 2. It is clear that all hypoxic treatments resulted in the accumulation of ethanol, GAB, succinate, Ala, lactate and α-glycerol-3-phosphate in rice shoot tips, with the first four being the major products. The former four products have been observed previously by us (Fan *et al.*, 1986a; Fan *et al.*, 1992) and others (Menegus *et al.*, 1988, 1989). The synthesis of these compounds involves H^+ and/or NADH consumption, which presumably helps alleviate cytoplasmic acidosis and allows glycolysis to continue. Small increases in the concentration of several amino acids (Val, Gly, Ile, Pro, Ser, Phe, Lys and His; Table 2 and results not shown) were also observed, which could be due to a breakdown of proteins and/or accumulation of Glu (a primary amino donor for the synthesis of these amino acids). Moreover, [Leu], [malate], [citrate] and [Asp] decreased, which could result from consumption of these compounds for synthesis of fermentation products and/or competition for precursors with fermentation reactions.

There were also differences in the hypoxia-induced changes in metabolite pattern between the bicarbonate and bicarbonate+acetate treatments (Table 2). The bicarbonate/N_2 treatment led to a higher accumulation of Ala but less of GAB, while the reverse occurred for the bicarbonate+acetate/N_2 treatment. These differences could result from the different biochemical pathways for acetate and bicarbonate incorporation. Acetate presumably stimulates carbon flow through the first half of the TCA cycle with GAB as the carbon sink, while bicarbonate would be incorporated into malate through PEP carboxylation, which in turn could lead to Ala synthesis via malate dehydrogenase and transaminase activities. A third difference was the large decrease in [malate] in the bicarbonate+acetate/N_2

TABLE 2. [a]Concentrations of metabolites in rice shoot tips under various treatments.

	Treatment							
	$NaHCO_3$/Air	$NaHCO_3/N_2$	$NaH^{13}CO_3$/Air	$NaH^{13}CO_3/N_2$	NaOAc/Air	$NaOAc/N_2$	^{13}C-$NaOAc/N_2$	^{13}C-$NaOAc/N_2$
Ala	3.63	48.20	3.81	66.31	5.22	15.36	4.56	19.85
Ascorbate	61.34	52.12	40.49	67.25	71.77	104.30	150.22	178.09
Asp	3.74	2.93	5.58	3.94	6.91	2.24	8.91	3.12
Citrate	40.72	22.61	31.94	16.14	75.55	46.04	78.14	48.85
[b]Ethanol	0.00	75.87	0.37	3.46	3.33	149.84	0.06	112.70
[c]Ethanol	-	1575.32	90.53	2885.99	154.74	5050.63	194.17	2034.87
GAB	0.87	14.85	1.69	23.22	1.75	83.30	2.12	89.25
Glu	4.37	5.94	34.47	16.27	9.21	29.92	10.04	44.82
α-Glycerol-3-P	0.00	7.90	4.49	10.86	11.91	22.61	10.63	19.81
Lactate	6.83	20.17	8.21	13.25	9.54	15.24	6.78	12.58
Leu	31.51	26.74	36.63	22.50	25.61	19.69	35.19	20.57
Malate	156.58	150.76	163.37	99.50	449.26	67.40	433.87	79.39
Succinate	0.81	34.38	1.02	38.28	3.83	46.97	2.77	42.18
Val	0.68	3.40	1.08	4.82	1.30	3.50	1.41	4.18
[d]Unknown	1989.26	1503.09	1961.60	1065.03	2872.63	1044.74	3125.37	1918.82

[a] Concentrations (in μmol g^{-1} dry wt) of all metabolites except ethanol were determined using a minibore DB-1 column and MTBSTFA-derivatives; ethanol was quantified on the DB-1 column without derivatisation; other metabolites determined but not reported include *n*-acetylaspartate, *t*-aconitate, Arg, Asn, Cys, fumarate, Gln, Gly, His, β-hydroxybutyrate, hydroxyproline, Ile, α-ketoglutarate, Lys, Met, ornithine, Phe, Pro, pyruvate, Ser, taurine, Thr, Trp and Tyr, the unknown compound had a similar retention time to that of α-ketoglutarate.

[b] Ethanol from tissues.

[c] Ethanol from treatment media.

[d] Concentration calculated based on the GC response of α-ketoglutarate standard.

treatment (also noted previously by Fan *et al.*, 1992) and a smaller reduction in the bicarbonate/N_2 treatment. This difference was caused largely by the different [malate] maintained by control shoot tips. Under aerated conditions, acetate treatment could cause additional accumulation of malate via the glyoxylate cycle (Zubay, 1983).

To trace the incorporation of ^{13}C label into various metabolites, both ^{13}C NMR and GC-MS were employed to analyse the ethanol trap and PCA extracts used for GC analysis. The assignment of ^{13}C NMR resonances was based on the following criteria: chemical shift, ^{13}C-^{1}H heteronuclear spin correlations, ^{1}H-^{1}H spin correlations, and knowledge of the metabolite concentration. An example of the assignment strategy and detection of selective enrichment is illustrated in Fig. 3. Using GAB as an example, the initial assignment of its C_2 and C_4 ^{13}C NMR resonances was according to chemical shift. This assignment was confirmed by the correlation from the C_2 carbon to α proton and the C_4 carbon to γ proton resonances of GAB (Fig. 3a), and by total correlations among α, β and γ proton resonances (Fig. 3b). It should also be noted that a pair of satellite peaks was present in the ^{1}H TOCSY spectrum (Fig. 3b), which could be accounted for by the ^{13}C-^{1}H coupling resulting from ^{13}C enrichment at the C-2 position. Moreover, the intensity of the C_2-GAB peak was much higher than that of the C_4-GAB peak in the ^{13}C NMR spectrum (Fig. 3a), which confirmed this selective enrichment.

Using ^{13}C-2-NaOAc as the labelled precursor, several metabolites associated with the TCA cycle were significantly enriched with ^{13}C under both control and hypoxic conditions, as shown in Fig. 4. Comparing the labelled with unlabelled spectra, C_2-acetate, C_4-Glu, C_2 or C_3-succinate, C_2 or C_4-citrate and $C_{2,3}$-malate were enriched with ^{13}C under both conditions but to a different extent. Malate was enriched most heavily under control conditions (Fig. 4a), while acetate, Glu, GAB, citrate and succinate were most significantly enriched under hypoxic conditions (Fig. 4b). These results were generally confirmed by the GC-MS data (data not shown). In addition, C_2-GAB appeared to be enriched only under hypoxic and $C_{1,4}$-malate only under control conditions (Fig. 4 and results not shown). Moreover, ethanol was enriched at the C-2 position based on ^{13}C NMR, ^{1}H NMR and GC-MS

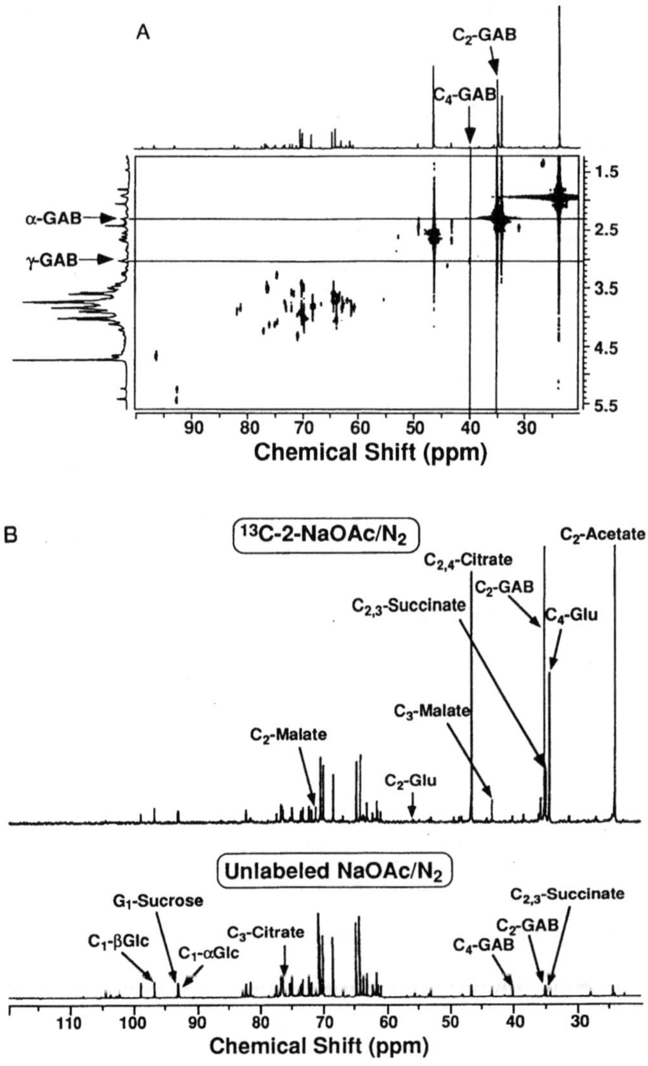

FIGURE 3. *^{13}C-1H HETCORR and 1H TOCSY spectra of the PCA extracts of rice shoot tips subjected to hypoxia/Na$^{13}C_2$-acetate treatment. These spectra were recorded as described in the Materials and Methods. The HETCORR spectrum (A) was displayed along with the corresponding 1-D high-resolution ^1H and ^{13}C NMR spectra. The spin correlations from C_2 carbon to α protons and C_4 carbon to γ protons of GAB are indicated by the cross lines. The ^1H TOCSY spectrum (B) is also presented with its 1-D high-resolution ^1H spectrum. α-GAB/Sat denotes the two ^{13}C satellite peaks for the α protons of GAB. Square boxes trace the spin correlations among the three methylene protons of GAB, including those arising from the satellites.*

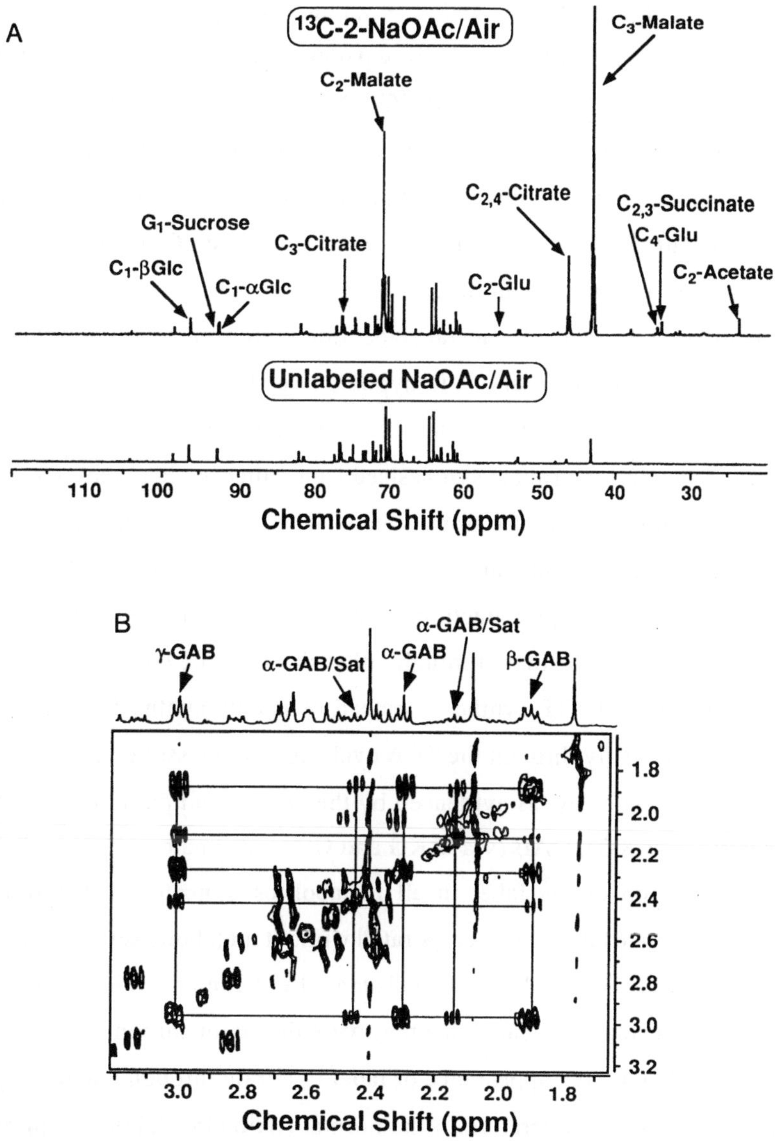

FIGURE 4. ^{13}C NMR Spectra of the PCA extracts of rice shoot tips subjected to acetate treatment. Spectral parameters were described in the Materials and Methods. The spectra acquired from unlabelled and ^{13}C-labelled acetate (^{13}C-2-NaOAc) treatments were compared for both control (A) and hypoxic (B) rice shoot tips. C_x denotes the position of the carbon, while G_1 represents the C_1 carbon in the glucose moiety of the sucrose molecule.

data (results not shown). Sugars and other amino and organic acids did not appear to be significantly enriched. The labelling pattern was different with $NaH^{13}CO_3$ as the labelled precursor (^{13}C NMR and GC-MS data not shown). The overall label incorporation was much reduced compared with the acetate treatment. The control shoot tips exhibited little selective enrichment, while a small amount of enrichment seemed to occur at C_1 or C_4 and C_2 or C_3-succinate, C_2-GAB, C_4-Glu, and C_1-malate under hypoxic conditions.

The incorporation of $^{13}C_2$-acetate label into various metabolites coupled with the changes in their concentrations in hypoxic ricc shoot tips allow the following suggestions to be made. First, under extended hypoxia, ethanol was the most significant fermentation product synthesised from glucose via glycolysis. Ethanol could also be produced directly from acetate (when present) without scrambling, as evidenced by the enrichment of C_2-ethanol with $^{13}C_2$-NaOAc as the labelled precursor. Second, the large reduction in [malate] could be due to a diversion of carbon into GAB and succinate through either direction of the TCA cycle and of PEP to substrate-level ATP synthesis instead of malate synthesis. Third, acetate was metabolised mostly through the TCA cycle and, to a lesser extent, through the ethanol synthesis pathway, as evidenced by the enrichment pattern in citrate, Glu (presumably via α-KG), GAB (via α-KG and Glu), succinate, malate and ethanol. Fourth, the position of the label in all metabolites indicated that a significant amount of the acetate label was incorporated without metabolic scrambling (Fig. 5, adapted from Zubay, 1983). Fifth, the substantial enrichment and accumulation of GAB, Glu and succinate indicated that they were the major carbon sinks for acetate under hypoxia. Lastly, the appearance of label in citrate through malate suggested that the TCA cycle was operative in the normal (oxidative) direction in hypoxic rice shoot tips, although the possibility of simultaneous operation in the reverse (reductive) direction cannot be eliminated. Evidence for a partial functioning of the TCA cycle has also been obtained for an anaerobic phytoplankton using $H^{14}CO_3^-$ (Vanlerberghe *et al.*, 1990) and barnyard grass using [^{14}C]acetate (Rumpho and Kennedy, 1983).

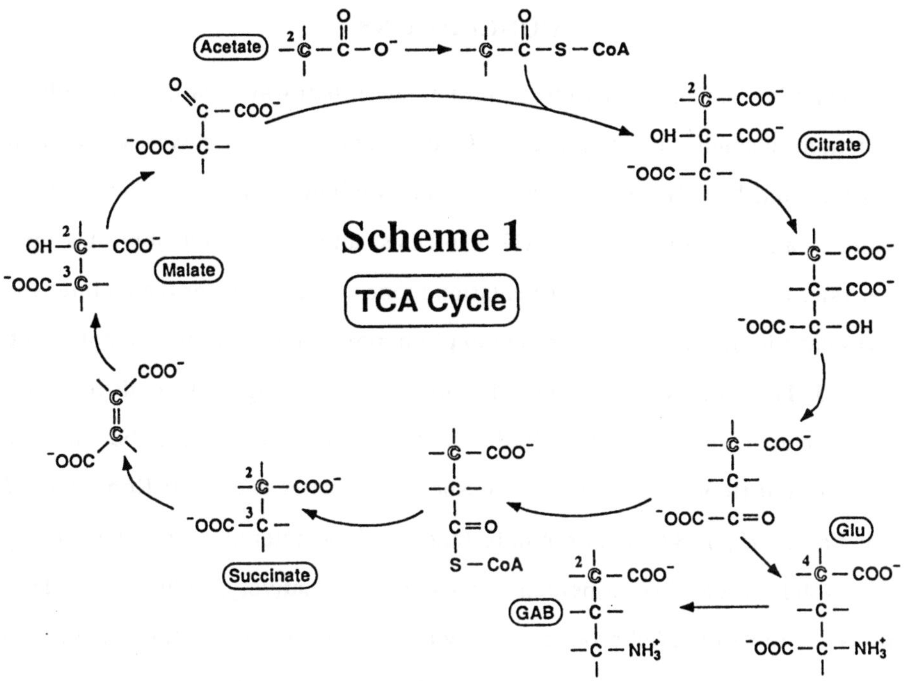

FIGURE 5. This scheme illustrates the incorporation pathways of ^{13}C label from ^{13}C-2-NaOAC into various intermediates in the TCA cycle and connecting reactions in hypoxic rice coleoptiles. The position of the carbon selectively enriched with ^{13}C label in various metabolites is noted by the outlined C.

As for the bicarbonate treatment, it is assumed that label incorporation was via carboxylation reactions catalysed largely by PEP carboxylase, pyruvate carboxylase or malate dehydrogenase. Therefore, it is to be expected that malate would be the main vehicle of enrichment and that the extent of enrichment would be less since label could be lost easily through decarboxylation. The observation of little selective enrichment under control conditions is consistent with the loss through decarboxylation and/or extensive metabolic scrambling. However, selective enrichment at the non-carboxyl positions of GAB, Glu and succinate under hypoxic conditions suggested that the carbon initially fixed could be scrambled into other positions instead of being lost through decarboxylation, which may be related to the different biochemical pathways activated during hypoxia.

CONCLUSIONS

It is clear that rice shoot tips utilise fermentation pathways, particularly ethanol production, to ameliorate the impact of O_2 deficiency, including excess production of protons and NADH resulting in part from continued glycolysis. Fermentation reactions can also contribute to ATP synthesis, although presumably at a much lower rate. However, whether fermentation alone can account for the maintenance of cytoplasmic pH and energy production in hypoxic rice shoot tips remains to be answered. This question is particularly intriguing in the light of the difference in energy metabolism and yet the similarity in fermentation products between rice shoot tips and flood-sensitive maize tissues (Roberts *et al.*, 1984; Roberts *et al.*, 1985; Fan *et al.*, 1988). It is possible that the flood tolerance of rice shoot tips resides solely in a higher fermentation capacity. Alternatively, additional oxidative pathways coupled to ATP production may also contribute to this tolerance in rice. The present finding that the oxidative TCA cycle was partly functioning in hypoxic rice shoot tips provided evidence for the latter. If so, how do rice shoot tips keep the TCA cycle going when O_2 is depleted? Is there any alternative electron acceptor such as nitrate to take the place of O_2 in hypoxic rice shoot tips? Answers to these questions would require a careful account of carbon fluxes through fermentation pathways and correlation of this information to rates of ATP turnover and N metabolism under hypoxic conditions. A combination of *in vivo* and *in vitro* approaches with isotope enrichment studies utilising ^{13}C NMR, ^{15}N NMR, ^{31}P NMR and GC-MS should help to resolve these questions.

ACKNOWLEDGEMENTS

The authors wish to thank Steven Scardaci and William Brannon for their generous gift of rice seeds and Tim Colmer for his assistance on the ^{13}C label studies.

REFERENCES

Bax A, Davis DG (1985) MLEV-17-based two-dimensional homonuclear magnetization transfer spectroscopy. Journal of Magnetic Resonance 65: 355-360

Bertani A, Brambilla I, Menegus F (1980) Effect of anaerobiosis on rice seedlings: growth, metabolic rate and fate of fermentation products. Journal of Experimental Botany 31: 325-331

Davies DD (1980) Anaerobic metabolism and the production of organic acids. In: Davies DD (ed) The biochemistry of plants, Vol. 2. Academic Press, New York, pp 581-607

Fan TW-M, Higashi RM, Lane AN (1986a) Monitoring of hypoxic metabolism in superfused plant tissues by *in vivo* ^{1}H NMR assignments. Archives of Biochemistry and Biophysics 251: 674-687

Fan TW-M, Higashi RM, Lane AN (1988) An *in vivo* ^{1}H and ^{31}P NMR investigation of the effect of nitrate on hypoxic metabolism in maize roots. Archives of Biochemistry and Biophysics 266: 592-606

Fan TW-M, Higashi RM, Lane AN, Jardetzky O (1986b) Combined use of ^{1}H NMR and GC-MS for metabolite monitoring and *in vivo* ^{1}H-NMR assignments. Biochimica et Biophysica Acta 882: 154-167

Fan TW-M, Lane AN, Higashi RM (1992) Hypoxia does not affect rate of ATP synthesis and energy metabolism in rice shoot tips as measured by ^{31}P NMR *in vivo*. Archives of Biochemistry and Biophysics 294: 314-318

Gadian DG (1982) NMR and its applications to living systems. Oxford University Press, London

Garlick PB, Radda GK, Scilla PO (1979) Studies of acidosis in the ischaemic heart by phosphorus nuclear magnetic resonance. Biochemical Journal 184: 547-554

Gerasimowicz V, Tu S-I, Pfeffer PE (1986) Energy facilitated Na^+ uptake in excised corn roots via ^{31}P and ^{23}Na NMR. Plant Physiology 81: 925-928

Gupta RK, Ferretti JA, Becker ED, Weiss GH (1980) A modified fast inversion-recovery technique for spin-lattice relaxation measurements. Journal of Magnetic Resonance 38: 447-452

Hochachka PW, Mommsen TP (1983) Protons and anaerobiosis. Science 219: 1391-1397

Menegus F, Cattaruzza L, Chersi A, Fronza G (1989) Differences in the anaerobic lactate-succinate production and in the changes of cell sap for plants with high and low resistance to anoxia. Plant Physiology 90: 29-32

Menegus F, Cattaruzza L, Chersi A, Serva A, Fronza G (1988) Production and organ distribution of succinate in rice seedlings during anoxia. Physiologia Plantarum 74: 444-449

Menegus F, Cattaruzza L, Mattana M, Beffagna N, Ragg E (1991) Response to anoxia in rice and wheat seedlings. Changes in the pH of intracellular compartments, glucose-6-phosphate level, and metabolic rate. Plant Physiology 95: 760-767

Öpik H (1973) Effect of anaerobiosis on respiratory rate, cytochrome oxidase activity and mitochondrial structures in coleoptiles of rice (*Oryza sativa* L.). Journal of Cell Science 12: 725-759

Roberts JKM, Callis J, Wemmer DE, Walbot V, Jardetzky O (1984) Mechanism of cytoplasmic pH regulation in hypoxic maize root tips and its role in survival under hypoxia. Proceedings of the National Academy of Sciences 81: 3379-3383

Roberts JKM, Lane AN, Clark RA, Nieman RH (1985) Relationship between the rate of synthesis of ATP and the concentrations of reactants and products of ATP hydrolysis in maize root tips, determined by ^{31}P nuclear magnetic resonance. Archives of Biochemistry and Biophysics 240: 712-722

Rumpho ME, Kennedy RA (1983) Anaerobiosis in *Echinochloa crus-galli* (barnyard grass) seedlings. Intermediary metabolism and ethanol tolerance. Plant Physiology 72: 44-49

Vanlerberghe GC, Horsey AK, Weger HG, Turpin DH (1989) Anaerobic carbon metabolism by the tricarboxylic acid cycle. Evidence for partial oxidative and reductive pathways during dark ammonium assimilation. Plant Physiology 91: 1551-1557

Zubay, G.L (1983) Biochemistry. Addison-Wesley Publishing Company, Reading, UK, pp 336

ENDOGENOUS PHYTOHORMONES AND GERMINATION OF RICE UNDER ANOXIA: INDOLEACETIC ACID AND ABSCISIC ACID

Sergio Mapelli and Alcide Bertani
Consiglio Nazionale delle Ricerche
Istituto Biosintesi Vegetali
via Bassini 15
20133 Milano
Italy

ABSTRACT

Indole acetic acid (IAA) and abscisic acid (ABA) are present in cereal seeds, and it is believed they play roles in seed maturation and germination. Modification of environmental conditions can inhibit or modify germination and seedling elongation. Our aim is to analyse the role played by IAA and ABA during the anoxic germination of rice (*Oryza sativa*), the only crop plant able to germinate in the absence of oxygen. Examination of the metabolism of IAA and ABA in rice seeds during the first week of anoxic germination shows that IAA esters are not hydrolysed, while ABA is released rapidly into the growth medium. Comparison of flooding tolerant and intolerant rice cultivars (Japonica and Indica ecotypes respectively) indicates that IAA levels are not closely related to the elongation capacity of the anaerobic coleoptile. However, an ability to excrete ABA may be an adaptative mechanism that enhances the elongation of anaerobic coleoptiles.

INTRODUCTION

The involvement of growth regulators in plant cell extension growth has been widely reported (Zeroni and Hall, 1980; Cleland, 1981). The biological material most often used for these studies has been cereal coleoptiles (Went and Thimann, 1937). Among cereal crops, only rice can extend its coleoptile vigorously (by cell elongation) when seeds are germinated under water, in hypoxia, or when oxygen is completely lacking (Taylor, 1942; Yamada, 1954; Bertani *et al.*, 1980). When rice seeds are sown in paddy fields, the coleoptile elongates to reach the water surface, allowing oxygen from the air to enter the plant and diffuse internally to the root, thus permitting root growth to begin. The lowland Japonica varieties, normally sown in flooded soils, grow longer coleoptiles than Indica varieties in

NATO ASI Series, Vol. I 16
Interacting Stresses on Plants in a Changing Climate
Edited by M. B. Jackson, C. R. Black

hypoxic conditions. Thus, Indica rice is not usually flooded at germination (Takahashi, 1978; Atwell *et al.*, 1982). On the other hand, some Indica rice varieties, for example deepwater rices, elongate more vigorously than Japonica rice during flooding of whole plants, giving faster and greater internode extension (Vergara *et al.*, 1976).

In order to obtain further information on the physiological role of endogenous IAA and ABA during germination and coleoptile elongation of rice in anoxic environments, we have analysed some aspects of their metabolism in different Indica and Japonica ecotypes.

INDOLEACETIC ACID

Some aspects related to the positive effects induced by auxin and anaerobiosis on coleoptile cell elongation have been reported previously. Treatment with IAA can decrease cytoplasmic pH, which may be the primary action of the hormone (Brummer and Parish, 1983; Schopfer, 1989), and low oxygen tension also induces acidification of the cytoplasm (Roberts *et al.*, 1984; Menegus *et al.*, 1990). At the level of metabolism of cell wall polysaccharides and proteins, wall hexosamine and hydroxyproline-containing glycoproteins were found to be controlled by auxin (Kato and Fujii, 1982; Hoson, 1987) and anoxia (Hoson and Wada, 1980; Hoson and Masuda, 1992). Despite these similarities between the effects of auxin and anoxia, the absence of synergistic, additive or inhibitory interactions between IAA and anaerobiosis on rice coleoptile elongation is notable (Pegoraro *et al.*, 1988) for Arborio rice, a Japonica cultivar. In accordance with these observations, we have also demonstrated the absence of membrane auxin-binding proteins (putative auxin receptors) in rice coleoptiles subject to anoxia (Zaina *et al.*, 1989). Thus, the physiological significance of free IAA in anoxic rice coleoptiles remains uncertain. The origin of free IAA in aerobically and anerobically germinated rice seedlings is the endosperm (Mapelli *et al.*, 1986). An endosperm origin for IAA in cereal seedlings has also been reported for oat and maize (Momonaki *et al.*, 1983; Jackson and McWha, 1984; Kamisaka *et al.*, 1988).

Bandurski *et al.* (1986) proposed a working theory according to which IAA is oxidised during its action as a growth-promoting hormone. This theory may also be applicable to aerobic Arborio rice coleoptiles (Pegoraro *et al.*, 1988). Accordingly, we have measured IAA in free and conjugated forms, and the extent of its metabolism during germination in Japonica and Indica rice ecotypes showing different tolerances to flooding at germination (Takahashi, 1978). Fig. 1 shows IAA levels in both free and conjugated forms during six days of germination of Arborio Japonica rice. Dry seeds contained both free and bound forms of methanol extractable IAA. During germination in aerobic conditions, fast hydrolysis of the conjugated IAA was evident and free IAA levels also decreased. On the sixth day of germination only 13% of the total IAA formerly present in the dry seeds was still measurable. Under anoxia, the decrease of IAA was less, and on the sixth day 64% of the total IAA and 80% of the bound form was still present. In Fig. 2 the IAA contents of different cultivars of rice, Roma (a second Japonica cultivar), Lemont and California Belle (two Indica ecotypes), are compared. The patterns of IAA hydrolysis and utilisation were very similar to those shown by the Arborio cultivar (Fig. 1). After five days of aerobic germination, Roma, Lemont and California Belle showed 14, 20 and 30% respectively of the original IAA content of the dry seeds. In anoxia, the methanol extractable IAA remained above 80% of the dry seed content for all cultivars after five days of germination. These data for the IAA content of dry seeds and young seedlings strengthen the hypothesis (Pegoraro *et al.*, 1988) that IAA is not strongly involved in the promotion of elongation of rice coleoptiles by anaerobic conditions since the patterns of IAA mobilisation and utilisation were similar for the Japonica and Indica cultivars, which have contrasting levels of tolerance to flooded conditions.

ABSCISIC ACID

In addition to IAA, the effects of atmospheric gas composition (CO_2, ethylene) on rice coleoptile growth have been much studied, either in aerobic environments or under oxygen-deficient and flooded conditions (Raskin and Kende, 1983; Pearce

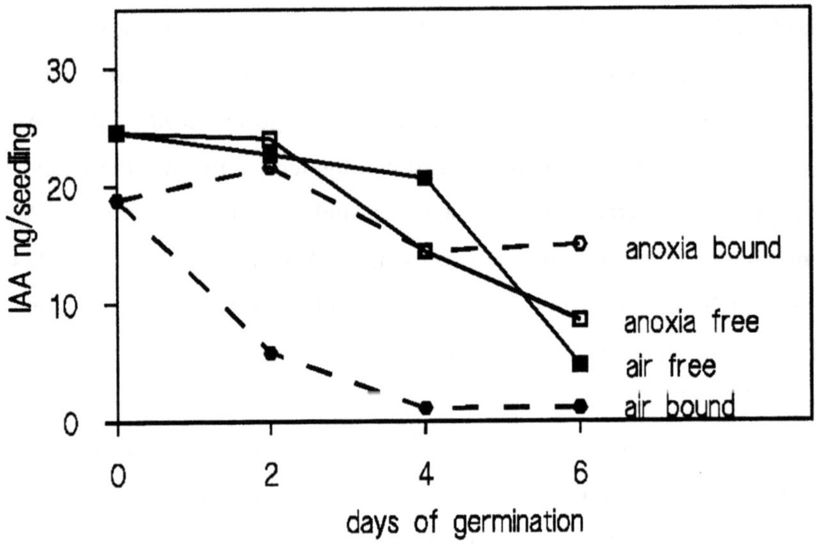

FIGURE 1. Changes in free and ester-bound IAA during the germination of Arborio Japonica rice in air and anaerobic environments.

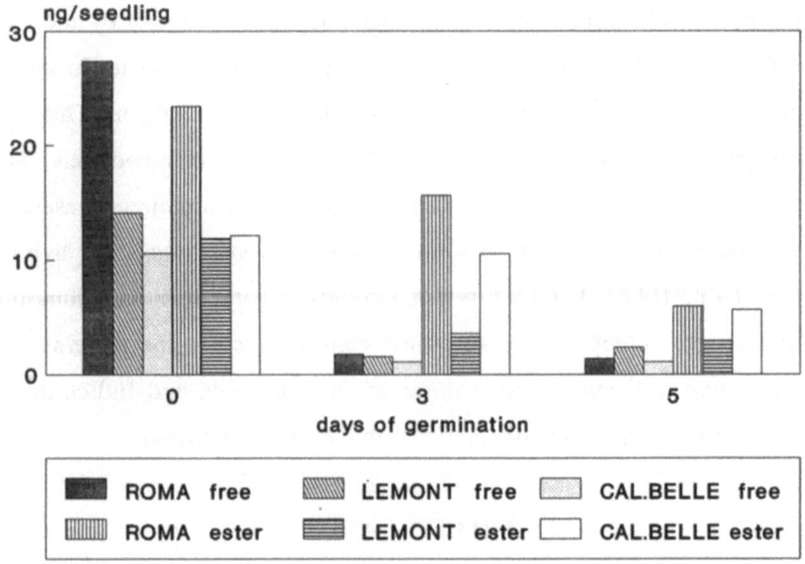

FIGURE 2. Changes in free and ester-bound IAA in Japonica and Indica rice cultivars during aerobic germination.

and Jackson, 1991). The action of other growth regulators has received little or no attention. Zeevaart (1983) found that endogenous ABA in the leaves of *Xanthium strumarium* was reduced by ethylene application. Abscisic acid is known also to be involved in grain filling and to inhibit germination and shoot elongation (Walker-Simmons, 1987; Oishi and Bewley, 1990; Saab *et al.*, 1990). Thus, it is of interest to investigate if ABA influences seed germination and coleoptile elongation in rice under anaerobic conditions.

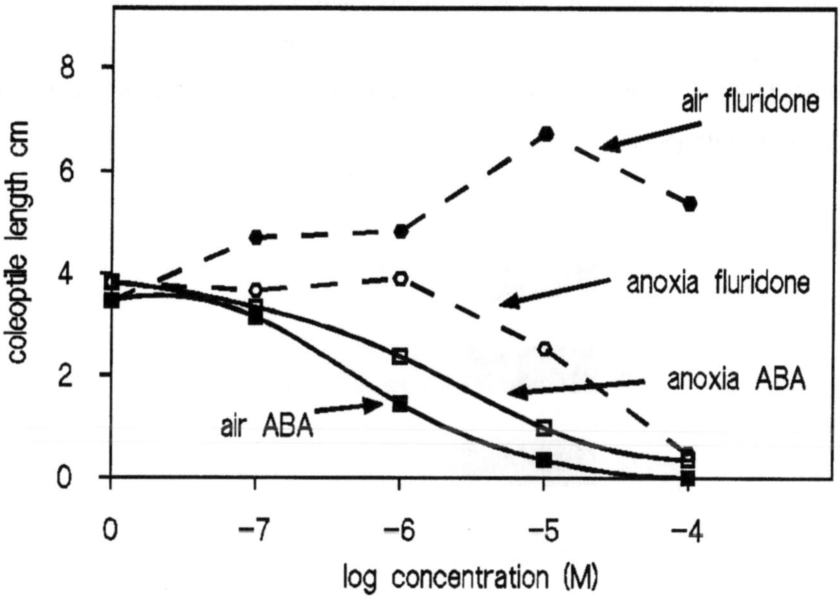

FIGURE 3. *Effects of ABA and fluridone on coleoptile elongation of rice in aerobic and anaerobic conditions.*

Fig. 3 shows that ABA inhibits rice coleoptile elongation both in air and under anoxia. Our experiments with other growth substances (cytokinins, gibberellins, auxins, brassinolide) indicate that ABA appears to be the only phytohormone able to modify anaerobic rice coleoptile elongation (Mapelli *et al.*, 1992). We therefore suppose that the mechanism of ABA inhibition of extension is similar in air and under anaerobiosis. The use of fluridone, an inhibitor of carotenoid biosynthesis and consequently of ABA biosynthesis (Gamble and Mullet, 1986), stimulates coleoptile elongation in air but not when oxygen is absent (Fig. 3). The analyses of endogenous ABA in rice seedlings indicate that fluridone treatment reduced the ABA level in air-grown coleoptiles, while having no effect on anoxic coleoptiles (Fig. 4). Surprisingly, the anoxically grown seedlings released into the growth medium 5-8 times the quantity of ABA within the seedling, irrespective of fluridone treatment. The amount of ABA released was not significant in air grown seedlings (Table 1). The data concerning the effect of fluridone on aerobic rice coleoptile elongation and ABA content agree with the behaviour reported for maize seedlings in which it appears that the ABA content, as modified by low water potential or fluridone treatment, is negatively correlated with shoot growth (Saab *et al.*, 1990).

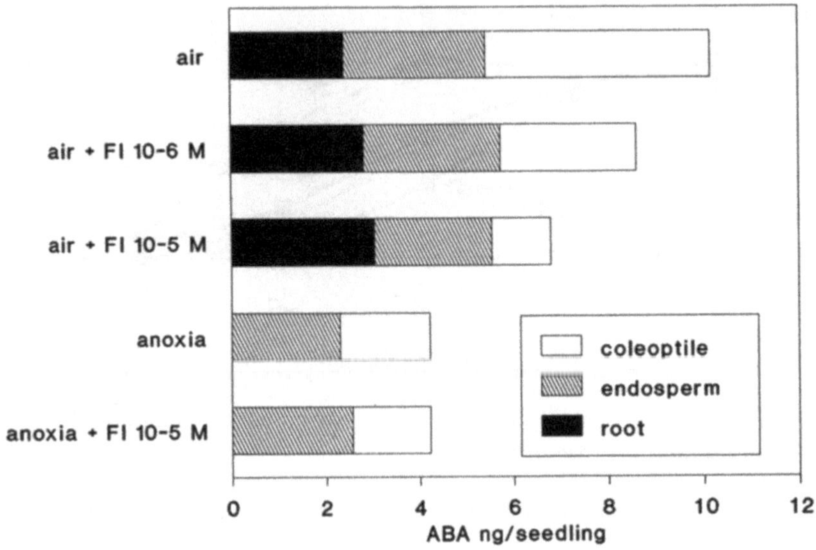

FIGURE 4. *Endogenous ABA levels of Arborio rice seedlings germinated for five days in air or under anoxia and under anoxia in presence of fluridone.*

TABLE 1. ***Abscisic acid released into the growth medium from cv. Arborio rice seedlings grown for five days in air or anoxia and in the presence or absence of fluridone (FI).***

Growth conditions	ABA (ng seedling^{-1})
Air	1.27
+ FI 10^{-6} M	1.68
+ FI 10^{-5} M	2.05
Anoxia	17.25
+ FI 10^{-6} M	30.73
+ FI 10^{-5} M	25.40

However, we do not consider reduced ABA to be the only factor influencing the elongation of aerobic rice coleoptiles treated with fluridone since fluridone is not a specific ABA biosynthesis inhibitor. It also inhibits pigment biosynthesis and probably that of other compounds, for example coumarins (Tan *et al.*, 1991). The decrease in ABA, pigments and coumarins may favour higher cell wall and membrane extensibility. These conditions are naturally present in anoxic rice coleoptiles and are artificially induced by IAA and fluridone application in air-grown seedlings. The absence of effect of fluridone on ABA levels in anaerobic rice coleoptiles is compatible with the observation that ABA biosynthesis requires oxygen (Zeevaart *et al.*, 1989). If this hypothesis is valid for rice, the ABA content of the anaerobic seedling and that released into the growth medium could originate from the endosperm, where analyses showed 7.51 and 153 ng per seed of methanol extractable ABA and peptide-bound ABA respectively. The increase in the total amount of free ABA (i.e., endogenous plus excreted) under anoxia confirms that this phytohormone is stress-inducible under conditions of oxygen shortage. The data in Fig. 4 and Table 1 are drawn from experiments with the rice cultivar Arborio but we have obtained similar results with other Japonica and Indica types.

CONCLUSIONS

The data extend our knowledge of the possible roles and involvement of IAA and ABA in regulating coleoptile elongation in rice under aerobic or anaerobic conditions. Both substances are present in considerable quantities in dry rice seeds in both free and conjugated forms. Thus, new synthesis during seedling emergence could be superfluous. ABA and IAA have antagonistic effects on rice coleoptile growth in air but are not the primary substances involved in controlling the extension of anoxic coleoptiles. Anoxia, in common with other environmental stresses, also provokes an increase in free ABA. However, its ready release into the medium will prevent a marked endogenous accumulation and thus minimise its effect on coleoptile elongation.

ACKNOWLEDGEMENTS

Generous gifts of rice caryopses by Dr G. Baldi of the Ente Nazionale Risi, Mortara, Italy and Dr T-T Chang of the International Rice Research Institute, Los Banos, Philippines are gratefully acknowledged. We thank Dr F Locatelli and Dr L Zanichelli for technical co-operation. The research work was supported by the National Research Council of Italy, Special Project RAISA - Sub-project 2, paper 772.

REFERENCES

Atwell BJ, Waters I, Greenway H (1982) The effect of oxygen and turbulence on elongation of coleoptiles of submergence-tolerant and -intolerant rice cultivars. Journal of Experimental Botany 33: 1030-1044

Bandurski S, Schulze A, Reinecke DM (1986) Biosynthetic and metabolism aspects of auxins. In: Bopp M (ed) Plant growth substances 1985. Springer-Verlag, Berlin, pp 83-91

Bertani A, Brambilla I, Menegus F (1980) Effect of anaerobiosis on rice seedlings: growth, metabolic rate, and fate of fermentation products. Journal of Experimental Botany 31: 325-331

Bialek K, Cohen JD (1986) Isolation and partial characterization of the major amide-linked conjugate of indole-3-acetic acid from *Phaseolus vulgaris* L. Plant Physiology 80: 99-104

Brummer B, Parish RW (1983) Mechanisms of auxin-induced plant cell elongation. FEBS Letters 161: 9-13

Cleland RE (1981) Wall extensibility: hormones and wall extension. In: Tanner W, Loewus FA (eds) Encyclopaedia of plant physiology, Vol. 13B. Springer-Verlag, Berlin, pp 255-276

Gamble PE, Mullet JE (1986) Inhibition of carotenoid accumulation and abscisic acid biosynthesis in fluridone-treated dark grown barley. European Journal of Biochemistry 160: 117-121

Hoson T (1987) Inhibiting effect of auxin on glucosamine incorporation into cell walls of rice coleoptiles. Plant Cell Physiology 28: 301-308

Hoson T, Masuda Y (1992) Relationship between polysaccharide synthesis and cell wall loosening in auxin-induced elongation of rice coleoptile segments. Plant Science 83: 149-154

Hoson T, Wada S (1980) Role of hydroxyproline-rich cell wall protein in growth regulation of rice coleoptile grown on or under water. Plant Cell Physiology 21: 511-524

Jackson DL, McWha JA (1984) Grain or coleoptile tip as the source of IAA in cereal shoots? Plant, Cell and Environment 7: 15-21

Kamisaka S, Herrera MT, Zarra I, Inouhe M, Miyamoto K, Masuda Y (1988) Inhibition of the coleoptile growth in *Avena sativa* by endosperm removal - Changes in auxin, osmotica and cell wall. Plant Cell Physiology 29: 467-474

Kato R, Fujii T (1982) Effect of IAA on the transport of hydroxyproline-protein(s) in *Zea* coleoptiles. Plant Cell Physiology 23: 509-513

Leverone LA, Kossenjans W, Jayasimihulu K, Caruso JL (1991) Evidence of zein-bound indoleacetic acid using gas chromatography-selected ion monitoring-mass spectrometry analysis and immunogold labelling. Plant Physiology 96: 1070-1075

Mapelli S, Lombardi L, Bertani A (1993) The effect of growth substances on rice germination and growth under anoxia. Life Science Advances (in press)

Mapelli S, Rocchi P, Bertani A (1986) ABA and IAA in rice seedlings under anaerobic conditions. Biologia Plantarum 28: 57-61

Menegus F, Cattaruzza L, Mattana M, Beffagna N, Ragg E (1991) Response to anoxia in rice and wheat seedlings: changes in the pH of intracellular compartments, glucose-6-phosphate level, and metabolic rate. Plant Physiology 95: 760-767

Momonoki YS, Schulze A, Bandurski RS (1983) Effect of deseeding on the indole-3-acetic acid content of shoots and roots of *Zea mays* seedlings. Plant Physiology 72: 526-529

Oishi MY, Bewley JD (1990) Distinction between the response of developing maize kernels to fluridone and desiccation in relation to germinability, α-amylase activity, and abscisic acid content. Plant Physiology 94: 592-598

Pearce DME, Jackson MB (1991) Comparison of growth responses of Barnyard grass (*Echinocloa oryzoides*) and rice (*Oryza sativa*) to submergence, ethylene, carbon dioxide and oxygen shortage. Annals of Botany 68: 201-209

Pegoraro R, Mapelli S, Torti G, Bertani A (1988) Indole-3-acetic acid in rice coleoptile elongation under anoxia. Journal of Plant Growth Regulation 7: 85-94

Percival FW, Bandurski RS (1976) Esters of indole-3-acetic acid from *Avena* seeds. Plant Physiology 58: 60-67

Raskin I, Kende H (1983) Regulation of growth in rice seedlings. Journal of Plant Growth Regulation 2: 193-203

Roberts JK, Callis J, Wemmer D, Walbot V, Jardetzky O (1984) Mechanism of cytoplasmic pH regulation in hypoxic maize root tips and its role in survival under hypoxia. Proceedings of the National Academy of Sciences USA 81: 3379-3383

Saab IN, Sharp RE, Pritchard J, Voetberg GS (1990) Increased endogenous abscisic acid maintains primary root growth and inhibits shoot growth of maize seedlings at low water potentials. Plant Physiology 93: 1329-1336

Schopfer P (1989) pH-dependence of extension growth in *Avena* coleoptiles and its implications for the mechanism of auxin action. Plant Physiology 90: 202-207

Takahashi N (1978) Adaptive importance of mesocotyl and coleoptile growth in rice under different moisture regimes. Australian Journal of Plant Physiology 5: 511-517

Tan K-S, Hoson T, Masuda Y, Kamisaka S (1991) Correlation between cell wall extensibility and the content of diferulic and ferulic acids in cell walls of *Oryza sativa* coleoptiles grown under water and in air. Physiologia Plantarum 83: 397-403

Taylor L (1942) Influence of oxygen tension on respiration, fermentation, and growth in wheat and rice. American Journal of Botany 29: 721-738

Tsunoda S, Takahashi N (eds) (1984) Biology of rice. Japan Scientific Societies Press, Tokyo

Vergara BS, Jackson B, DeDatta SK (1976) Deep-water rice and its response to deep-water stress. In: Climate and rice. International Rice Research Institute, Los Banos, Philippines, p 301

Walker-Simmons N (1987) ABA levels and sensitivity in developing whcat embryos of sprouting resistant and susceptible cultivars. Plant Physiology 84: 61-66

Went FW, Thimann KV (1937) Phytohormones. Macmillan, London

Xu N, Coulter KM, Bewley JD (1990) Abscisic acid and osmoticum prevent germination of developing alfalfa embryos, but only osmoticum maintains the synthesis of developmental proteins. Planta 182: 382-390

Yamada N (1954) Auxin relationship of the rice coleoptile. Plant Physiology 29: 92-96

Zaina S, Bertani A, Lombardi L, Mapelli S, Torti G (1989) Membrane-associated binding sites for indoleacetic acid in rice coleoptile. Planta 179: 222-227

Zeevaart JAD (1983) Metabolism of abscisic acid and its regulation in *Xanthium* leaves during and after water stress. Plant Physiology 71: 477-481

Zeevaart JAD, Heath TG, Gage DA (1989) Evidence for a universal pathway of abscisic acid biosynthesis in higher plants from ^{18}O incorporation patterns. Plant Physiology 91: 1594-1601

Zeroni M, Hall MA (1980) Molecular effects of hormone treatment on tissue. In: MacMillan J (ed) Encyclopaedia of plant physiology, Vol. 9. Springer-Verlag, Berlin, p 511

PRELIMINARY EVIDENCE OF NITRATE ASSIMILATION DURING THE ANAEROBIC GERMINATION OF RICE

M. Mattana, A. Bertani, N. Aurisano and R. Reggiani
Istituto Biosintesi Vegetali
Consiglio Nazionale Ricerche
20133 Milano
Italy

ABSTRACT

The utilisation of nitrate stored in the seed was studied during the anaerobic germination of rice (*Oryza sativa* L.). After 11 days of germination, 74% of the initial nitrate content had been utilised. At all times during anaerobic germination, nitrate reductase and nitrite reductase activities were undetectable in the caryopsis, but present in the coleoptile. Their activities increased, as did the activities of glutamine synthetase and Fd-glutamate synthase. Nitrate reductase and nitrite reductase activities were shown to be induced by exogenous nitrate and strongly depressed by cycloheximide. Feeding experiments with ^{15}N nitrate demonstrated that the coleoptile is able to reduce nitrate under anaerobic conditions and to incorporate nitrogen into amino acids. The role of nitrate utilisation during the anaerobic germination of rice is discussed.

INTRODUCTION

Rice seeds germinating in flooded soils may be subjected to anaerobiosis. Under flooding, gaseous diffusion of oxygen is very slow and the coefficient of diffusion is 10000-fold smaller than in air. Unlike most higher plants, rice is able to germinate in a strictly anaerobic environment (Opik, 1973). Under these conditions rice seeds emit only a white coleoptile, while root growth is blocked. Indeed, meristematic root cells require oxygen to divide and elongate (Kordan, 1976). In anoxia, coleoptiles elongate until they reach an aerobic zone from which oxygen can be transferred internally to the seed, thus leading to normal growth of the radicle and leaf (Kordan, 1976; Takahashi, 1978). Coleoptile elongation under anaerobic conditions depends on cell enlargement, with cell division being restricted to the first hours of germination (Opik, 1973). Turgor pressure and cell wall plasticity are central factors in controlling cell enlargement. Zarra and Masuda

NATO ASI Series, Vol. I 16
Interacting Stresses on Plants in a Changing Climate
Edited by M. B. Jackson, C. R. Black

(1979) have shown that, in hypoxic conditions, cell wall plasticity is elevated and rice coleoptile elongation is controlled by turgor pressure. On the other hand, wall plasticity is low under aerobic conditions and coleoptile elongation is inhibited accordingly.

During anaerobic germination, the rice coleoptile assimilates nutrients from the endosperm in order to sustain growth and maintain high cell turgor pressure (Atwell *et al.*, 1982). Major contributions to the osmotic potential were attributed to K^+, sugars and probably carboxylic acids (Menegus *et al.*, 1984). Furthermore, it was shown that free amino acids also contribute to maintaining osmotic potential in anaerobic rice coleoptiles. In a timecourse experiment, we have shown that, as the anoxic germination of rice seeds (*Oryza sativa* L. cv. Arborio) proceeded, the fresh weight of the coleoptile increased following a trend similar to that of the amino acid pool (Fig. 1). The fastest rate of coleoptile growth occurred between the fourth and eighth days of germination.

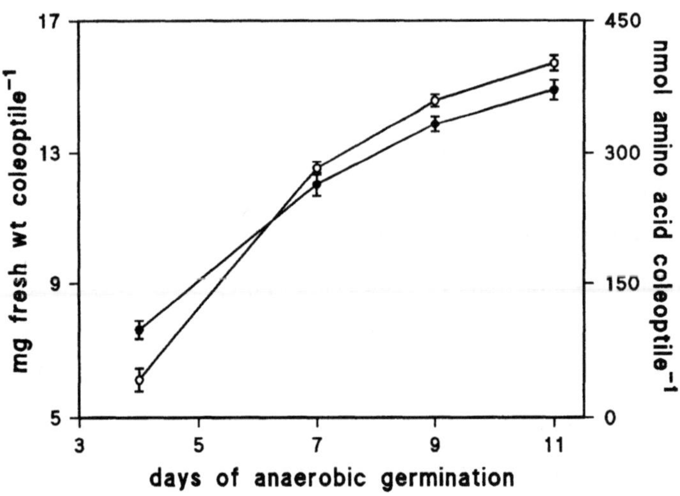

FIGURE 1. *Changes in fresh weight (open circles) and amino acid level (closed circles) in rice coleoptiles during anaerobic germination.*

Among the nutrients coming from the endosperm, nitrate is of particular interest since it may be metabolised by the cell. In well-oxidised soils, this ion is stable (at a redox potential higher than 220 mV) but when soils are waterlogged, the redox potential falls below 0 mV and the nitrate disappears because of bacterial reduction and denitrification (Gambrell and Patrick, 1978). In view of this, we aimed to address the following questions:

1. is nitrate present in the rice seed?
2. is nitrate reduced during anaerobic growth of the coleoptile?

NITRATE IN ANAEROBIC RICE SEEDLINGS

Information in the literature concerning the presence of nitrate in rice seeds and its translocation, reduction and assimilation under anoxia is lacking. To obtain information on the presence of nitrate in anaerobic rice tissues, a timecourse experiment was set up to determine the concentration of the ion. Nitrate was assayed by both colorimetric (Reggiani *et al.*, 1985a) and ion-selective electrode methods (Ingold, Switzerland). The concentration of nitrate at different times of germination was determined in the caryopsis, embryo and coleoptile. Nitrate was found in both the caryopsis and embryo of dry seeds (Fig. 2).

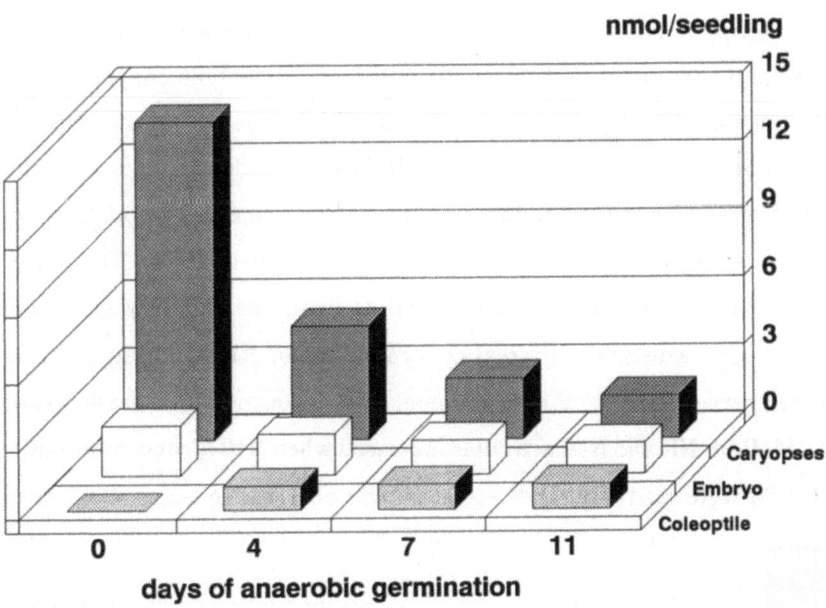

***FIGURE 2.** Nitrate content of the coleoptile, embryo and caryopsis of rice seedlings during 11 days of anaerobic germination.*

During anaerobic germination, the nitrate content of the caryopsis decreased, with the highest rate of disappearance being observed during the first four days of germination. The level of nitrate decreased slightly in the embryo but remained low and fairly constant in the coleoptile. However, at all times during anaerobic germination, the ion was present only in trace amounts in the growth medium (<<1 nmol seedling^{-1}), while nitrite was always undetectable in the growth medium and in the three tissues analysed. This indicated that nitrate was metabolised by anaerobic rice seedlings. From this experiment, we estimated that after 11 days of anaerobic germination, the level of nitrate had decreased by 74% of the initial content (about 12 nmol per caryopsis).

NITRATE REDUCTASE (NR) AND NITRITE REDUCTASE (NIR) ACTIVITIES

Since the total nitrate content of the seedling decreased so markedly during 11 days of anaerobic germination, we measured NR and NiR activities in the different tissues. In the caryopsis, the two activities were undetectable throughout 11 days of germination, while the embryo exhibited only a low activity of NiR (14.8 pmol h^{-1} per embryo). In the coleoptile, activity by both enzymes was found and this increased during the 11 days (Fig. 3). However, in this tissue, NiR activity was 600-850 times higher than that of NR. Nitrate reductase is one of the best examples of a plant enzyme induced by its substrate, and this response is coupled to an increase in NiR activity (Redinbaugh and Campbell, 1991). Thus, effects of exogenous nitrate on NR and NiR activities in the coleoptile were studied by supplementing seeds with different concentrations of KNO_3 during eight days of anaerobic germination. In this experiment nitrate concentration ranged from 10^{-6} to 10^{-2} M. Both NR and NiR activities increased when KNO_3 concentrations higher than 10^{-5} M were used (Fig. 4). No increase was observed in the nitrate-free sample (control).

It was of interest to compare the absolute amount of nitrate supplied during germination (shown at the bottom of Fig. 4) with the amount of endogenous nitrate that had disappeared from the caryopsis after eight days of incubation. Induction of NR and NiR activities occurred when the amount of nitrate supplied was several times higher than the amount of nitrate provided by the seed (about 11 nmol per seed).

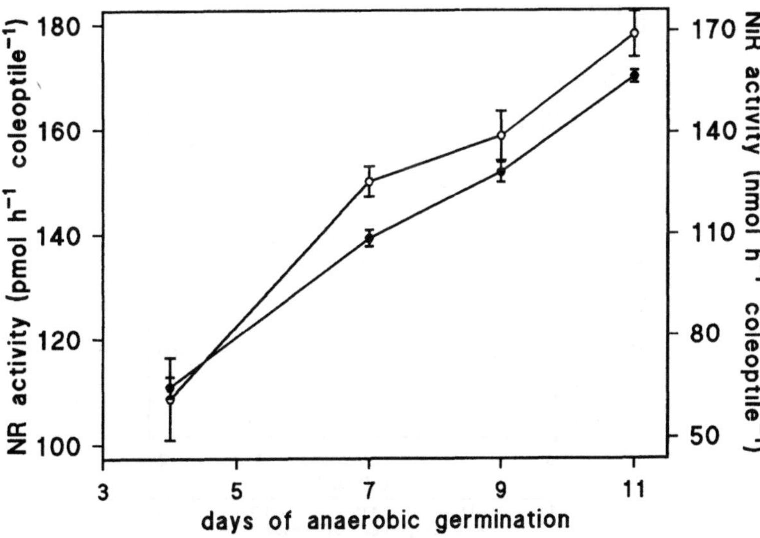

FIGURE 3. NR (open circles) and NiR (closed circles) activities in rice coleoptiles during anaerobic germination.

We investigated the possibility that protein synthesis may maintain NR and NiR activities in the coleoptile. For this purpose, seven day old anaerobic rice seedlings were supplied with 0.5 μg ml^{-1} cycloheximide (CH, a protein synthesis inhibitor) for one additional day. NR and NiR activities in the absence of CH were 169.2 pmol h^{-1} per tissue and 84.2 nmol h^{-1} per tissue, respectively. In the presence of CH, a strong decrease, of about 66% for NR and 80% for NiR, was observed.

ASSIMILATION OF NITRATE

To verify if the nitrogen of the nitrate source could be assimilated into amino acids, six day old anaerobic rice seedlings were supplied with 1 mM $K^{15}NO_3$ and incubated for two additional days. The abundance of ^{15}N was determined in the

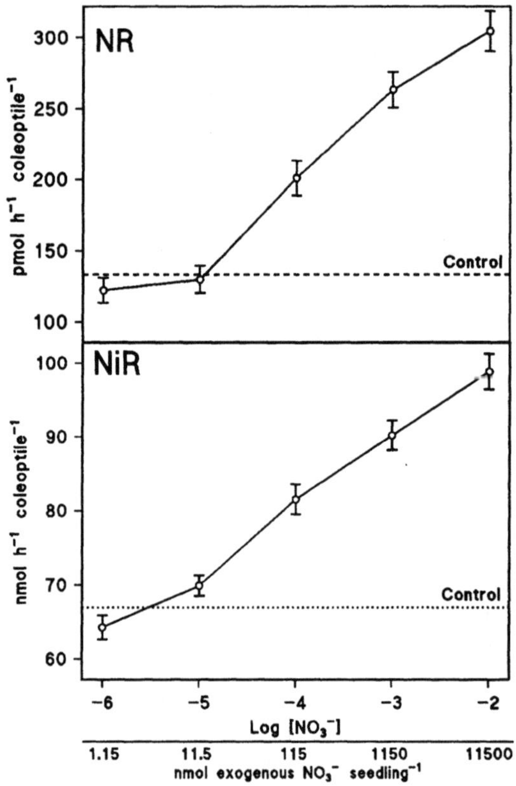

FIGURE 4. NR and NiR activities in rice coleoptiles after eight days of anaerobic germination in the absence (control) or in the presence of different KNO_3 concentrations. The absolute amount of nitrate supplied to each seedling is indicated below the lower graph.

acid-soluble (including also unreduced labelled nitrate) and amino acid (assimilated ^{15}N) fractions. In the absence of labelled nitrate (Fig. 5) we found in both fractions values of ^{15}N abundance similar to those naturally occurring for this isotope (0.37%). However, when $K^{15}NO_3$ was added, there was a similar enrichment for the ^{15}N isotope in the two fractions (Fig. 5). This result suggests that rapid assimilation of nitrogen nitrate into amino acids occurs under anoxia. In fact, a higher enrichment in the acid-soluble fraction would be expected if the reduction rate was lower than nitrate uptake. Considering that assimilation of nitrogen nitrate ultimately involves the glutamine synthetase-glutamate synthase pathway (GS-GOGAT), the activities of GS, NADH-GOGAT and Fd-GOGAT in the anaerobic coleoptile were also measured. The activity of NADH-GOGAT was always absent during germination while the GS and Fd-GOGAT activities increased with the duration of anaerobic germination (Fig. 6).

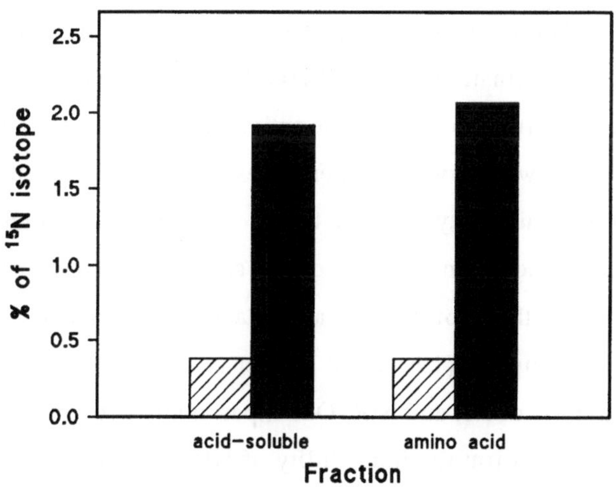

FIGURE 5. Abundance of ^{15}N isotope in the acid-soluble and amino acid fractions of rice coleoptiles fed between six and eight days of anaerobic germination with 1 mM $K^{15}NO_3$.

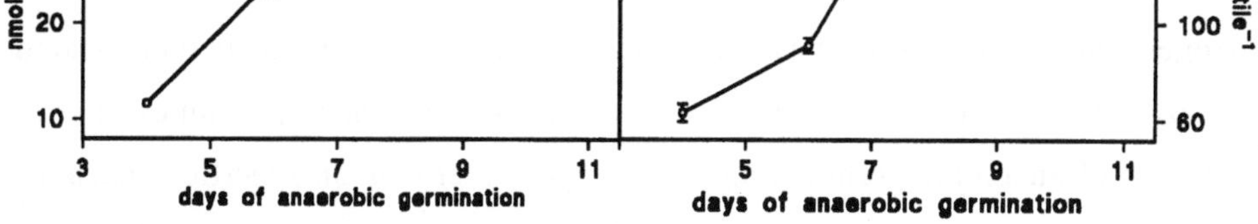

FIGURE 6. GS and Fd-GOGAT activities in rice coleoptiles during anaerobic germination.

CONCLUSIONS

The level of nitrate in rice caryopses decreased during anaerobic germination (Fig. 2), while both NR and NiR activities were undetectable in this tissue. On the contrary, all the enzymes of nitrate reduction and assimilation were present in the coleoptile and their activities increased during germination (Figs. 3 and 6). $^{15}NO_3^-$ was reduced by the coleoptile and nitrogen incorporated into amino acids (Fig. 5). It is conceivable that the rate-limiting step in the assimilation of nitrate during the first four days of germination is the NR reaction, since the level of activity of this enzyme was several times lower than the activity of the other assimilating enzymes (Figs. 3 and 6). However, with longer times, the level of seed storage nitrate decreased and its availability is likely to have become increasingly limiting with time. As far as the induction of NR and NiR activities is concerned, our findings are consistent with those of Ogawa and Ida (1987), Hayakawa *et al.* (1990), Redinbaugh and Campbell (1991) and Melzer *et al.* (1989), showing that nitrate induces the synthesis of NR and NiR. Our experiments with nitrate supplied to the seedlings (Fig. 4), confirm the inducibility of NR and NiR activities and suggest that the basal activity observed in the coleoptile is induced in the absence of exogenous nitrate by ions translocated from the caryopsis. Moreover, the strong inhibition of the two activities observed when CH was used indicates that continuous synthesis of NR and NiR proteins is occurring during anaerobic germination of rice.

What is the physiological role of nitrate during anaerobic germination of rice? From our results we estimate that, after four days of anaerobic germination, nitrate nitrogen can account for about 10% of the nitrogen present in the amino acid pool of the coleoptile. At later stages, nitrate becomes less important as a nitrogen source and most of the amino acid present in the coleoptile is translocated directly from the seed. Nevertheless, nitrate reduction can be important even after four days of anaerobic germination for its reported effects on specific aspects of anaerobic metabolism. Previous research (Reggiani *et al.*, 1985a,b; Brambilla *et al.*, 1986; Reggiani *et al.*, 1986) indicated that anaerobic nitrate reduction, affecting pH and the redox state of the cell, produced beneficial effects on carbohydrate metabolism. A better understanding of the role of nitrate during anoxic germination will result from the study and characterisation of rice varieties with different contents of nitrate in their seeds. Preliminary data suggest that the amount of nitrate in seeds of various rice cultivars can differ substantially (Table 1).

TABLE 1*. Nitrate content in the dry seed of different cultivars of rice.*

Rice cultivar	Nitrate content[a]	Rice cultivar	Nitrate content[a]
Lemont	140.0	Arborio	457.6
Roma	108.7	Argo	290.1
Tanginbozu	330.3	IR36	424.7

[a] nmol g^{-1} dry weight

REFERENCES

Atwell BJ, Waters I, Greenway H (1982) The effect of oxygen and turbulence on elongation of coleoptiles of submergence-tolerant and intolerant rice cultivars. Journal of Experimental Botany 33: 1030-1044

Brambilla I, Bertani A, Reggiani R (1986) Effect of inorganic nitrogen nutrition (ammonium and nitrate) on aerobic and anaerobic metabolism in excised rice roots. Journal of Plant Physiology 123: 419-428

Hayakawa T, Kamachi K, Oikawa M, Ojima K, Yamaya T (1990) Response of glutamine synthetase and glutamate synthase isoforms to nitrogen sources in rice cell cultures. Plant and Cell Physiology 31: 1071-1077

Gambrell RP, Patrick WH (1978) Chemical and microbiological properties of anaerobic soils and sediments. In: Hook DD, Crawford RMM (eds) Plant life in anaerobic environments. Ann Arbor Science, Ann Arbor, USA, pp 375-423

Kordan HA (1976) Oxygen as an environmental factor in influencing normal morphogenetic development in germinating rice seedlings. Journal of Experimental Botany 27: 947-952

Melzer JM, Kleinhofs A, Warner RL (1989) Nitrate reductase regulation: effects of nitrate and light on nitrate reductase mRNA accumulation. Molecular and General Genetics 217: 341-346

Menegus F, Brambilla I, Bertani A (1984) Nutrient translocation pattern and accumulation of free amino acids in rice coleoptile elongation under anoxia. Physiologia Plantarum 61: 203-208

Ogawa M, Ida S (1987) Biosynthesis of ferredoxin-nitrite reductase in rice seedlings. Plant and Cell Physiology 28: 1501-1508

Opik H (1973) Effect of anaerobiosis on respiratory rate, cytochrome oxidase activity and mitochondrial structures in coleoptile of rice (*Oryza sativa* L.). Journal of Cell Science 12: 725-739

Redinbaugh MG, Campbell WH (1991) Higher plant responses to environmental nitrate. Physiologia Plantarum 82: 640-650

Reggiani R, Brambilla I, Bertani A (1985a) Effect of exogenous nitrate on anaerobic metabolism in excised rice roots. I. Nitrate reduction and pyridine nucleotide pools. Journal of Experimental Botany 36: 1193-1199

Reggiani R, Brambilla I, Bertani A (1985b) Effect of exogenous nitrate on anaerobic metabolism in excised roots. II. Fermentative activity and adenylic energy charge. Journal of Experimental Botany 36: 1698-1704

Reggiani R, Brambilla I, Bertani A (1986) Effect of exogenous nitrate on anaerobic metabolism in excised roots. III. Glycolytic intermediates and enzymatic activities. Journal of Experimental Botany 37: 1472-1478

Takahashi N (1978) Adaptive importance of mesocotyl and coleoptile growth in rice under different moisture regimes. Australian Journal of Plant Physiology 5: 511-517

Zarra J, Masuda Y (1979) Growth and cell wall changes in rice coleoptiles growing under different conditions. I. Changes in turgor pressure and cell wall polysaccharides during intact growth. Plant and Cell Physiology 20: 1117-1124

MOLECULAR GENETIC BASIS OF METABOLIC ADAPTATION TO ANOXIA IN MAIZE AND ITS POSSIBLE UTILITY FOR IMPROVING TOLERANCE OF CROPS TO SOIL WATERLOGGING

Martin M. Sachs
USDA/ARS/MWA Plant Physiology and Genetics Research Unit and
Department of Agronomy
University of Illinois at Urbana/Champaign
S108 Turner Hall
1102 S. Goodwin Avenue
Urbana IL 61801
USA

ABSTRACT

Anaerobic treatment dramatically alters the pattern of gene expression in maize seedlings. During anaerobiosis there is an immediate repression of pre-existing protein synthesis and the simultaneous selective synthesis of four polypeptides with a molecular weight of ~33 kDa (the transition polypeptides). After 90 minutes, this is followed by specific transcription and mRNA accumulation as well as the selective synthesis of approximately twenty anaerobic-specific proteins. Among these anaerobic proteins (ANPs) are enzymes involved in glycolysis and related processes, such as alcohol dehydrogenase (ADH), aldolase, anolase, glucose phosphate isomerase, glyceraldehyde-3-phosphate dehydrogenase, pyruvate decarboxylase and sucrose synthase. This is most likely a natural response of plants to flooding. Similar responses have been observed in every plant species examined to date. We have analysed the anaerobic response and flooding tolerance of maize using both molecular and genetic approaches, and have identified 15 different anaerobic-specific cDNA clones. In timecourse experiments, most of these have been shown to hybridise to mRNAs that have kinetics of anaerobic induction very similar to those previously reported for *Adh1* and *Adh2* (i.e., accumulation is first apparent after 90 minutes and reaches its highest level after approximately five hours). One cDNA encodes an RNA that accumulates a little earlier, while a few others encode mRNAs that accumulate later than the *Adh* mRNAs. One of these does not begin accumulating until after six hours of anaerobiosis and peaks after 48 hours. Among the cDNAs analysed are those that encode cytosolic glyceraldehyde-3-phosphate dehydrogenase (GAPC; E.C. 1.2.1.12). We have found that GAPC is encoded by a small multigene family consisting of four genes (*Gpc1* - *Gpc4*). There are apparently two sub-families with two genes each. One sub-family is anaerobically inducible and shows organ-specific expression, while the other shows essentially constitutive expression and is neither anaerobically inducible nor does it show any organ-specificity. On the genetic side, we have analysed a maize line that is ADH1 null but

NATO ASI Series, Vol. I 16
Interacting Stresses on Plants in a Changing Climate
Edited by M. B. Jackson, C. R. Black

whose seedlings can survive 24 hours of anoxic treatment (all previously tested ADH1 null lines survived for only a few hours, while most ADH1$^+$ lines can survive 72 hours). We have found that the factor involved behaves as a single recessive locus that segregates independently of *Adh1* and *Adh2*. We have also analysed maize germplasm for variation in flood tolerance and have found a few accessions that can survive anoxia for up to six days. Genetic analysis indicates that this may be due to a single dominant locus.

INTRODUCTION

When heavy rains shortly after planting leave standing water in the field, this can lead to unacceptable crop losses and may necessitate replanting. Waterlogged soils are rapidly depleted of oxygen by micro-organisms, leaving the germinating seedlings in an anoxic environment. Measures to enhance the tolerance of crop plants to this type of flooding would be very beneficial to farmers. The response of plants (especially maize) to anoxia has been the subject of extensive genetic, physiological, biochemical and molecular genetic analysis that may eventually lead to such improvements.

Molecular studies of the maize anaerobic response began with the extensive analysis of the maize alcohol dehydrogenase (ADH) system by Drew Schwartz and coworkers (reviewed by Freeling and Birchler, 1981; Freeling and Bennett, 1985; Gerlach *et al.*, 1986). Hageman and Flesher (1960) were the first to show that ADH activity increases as a result of flooding maize seedlings. Freeling (1973) later reported that ADH activity increased at a zero order rate between 5 and 72 hours of anaerobic treatment, reflecting a simultaneous expression of two unlinked genes, *Adh1* and *Adh2*. By producing ADH null mutations, Schwartz (1969) showed that ADH activity is essential for the survival of maize seeds and seedlings during flooding. Normal seedlings survive ~3 days of anoxia (Schwartz, 1969; Lemke-Keyes and Sachs, 1989b), while ADH nulls survive only a few hours of flooding. ADH is the major terminal enzyme of fermentation in plants and is responsible for recycling NAD$^+$ during anoxia. It has been suggested that ethanolic fermentation permits tight cytoplasmic pH regulation, thus preventing acidosis from competing with lactic fermentation (Roberts *et al.*, 1984a, 1984b, 1985). Alternatively, a plant that is ADH null may accumulate toxic levels of acetaldehyde (the substrate for ADH produced by pyruvate decarboxylase; PDC). While some minimum level of ADH activity appears to be required for any flooding tolerance, natural variation in long term seedling survival of anaerobic stress does not appear to be correlated with variation in the levels of ADH activity (Lemke-Keyes and Sachs, 1989b). In this study, more than 400 ADH^+ maize lines were examined, and variation in flooding tolerance among these lines was from two to six days. Selected lines, showing variation in flooding tolerance, also varied in induced ADH activity levels over a five-fold range, but there was no apparent relationship between these two traits (Lemke-Keyes and Sachs, 1989b). A mutation (*atn*; Lemke-Keyes and Sachs, 1989a) that allows ADH null seedlings to survive 24 hours of anoxia may allow further understanding of the role of ADH in flooding tolerance.

THE ANAEROBIC RESPONSE

The maize anaerobic response is analogous to the heat-shock response observed in many organisms, including plants. There is a repression of pre-existing protein synthesis and *de novo* synthesis of a new set of proteins. Except for one possible overlap, anaerobiosis induces a different set of proteins in maize from that induced by heat shock (Sachs *et al.*, 1980; Kelley and Freeling, 1982; Cooper and Ho, 1983). Regulation of both transcription and translation appear to be involved in the anaerobic stress response (Sachs *et al.*, 1980; Dennis *et al.*, 1988b). This is most likely a natural response of plants to flooding.

In an early study (Lin and Key, 1967), anaerobic treatment was shown to cause a near complete dissociation of polysomes and a rapid repression of protein synthesis in soybean roots. Anaerobic treatment of maize seedlings also disrupts polysomes (E.S. Dennis and A.J. Pryor, personal communication; Russell and Sachs, unpublished). This phenomenon was shown to result from a redirection of protein synthesis (Sachs *et al.*, 1980; Russell and Sachs, 1992) in maize as well as soybean (Russell *et al.*, 1990). As is the case in many heat-shock systems (see Schlesinger *et al.*, 1982), maize seedlings subjected to anoxia (e.g., an argon atmosphere) display an immediate repression of pre-existing (aerobic) protein synthesis along with the induction of a new set of proteins (Sachs *et al.*, 1980). In the first five hours of anaerobic treatment there is a transition period during which there is a rapid increase in the synthesis of a class of polypeptides with an approximate molecular weight of 33 kDa (see Fig. 1). These have been referred to as the transition polypeptides (TPs). After approximately 90 minutes of anoxia, the synthesis of an additional group of ~20 polypeptides is induced (Fig. 1). This group of anaerobic polypeptides (ANPs) represents more than 70% of the total labelled amino acid incorporation after five hours of anaerobiosis. By this time, the synthesis of the TPs is at a minimal level; however, these polypeptides accumulate to a high level during early anaerobiosis and have been shown by pulse-chase experiments to be very stable. The synthesis of the ANPs continues at a constant rate for up to ~72 hours of anaerobic treatment, depending on which maize line is being examined. After 72 hours, protein synthesis decreases concurrently with the start of cell death (Sachs *et al.*, 1980). In similar experiments, it was shown that the rate of amino acid uptake in the anoxic root was about half that of the aerobic control, while incorporation into polypeptides was reduced to 1% (Russell, 1990). Similar effects of anoxia on amino acid uptake and incorporation were also observed in rice roots (Mocquot *et al.*, 1981).

Hypoxia (low oxygen) has been found to induce ANPs in maize primary roots, but does not cause complete repression of pre-existing protein synthesis (Sachs, unpublished). Significant levels of 'aerobic' protein synthesis are still observed at O_2 concentrations as low as 0.2% (Kelley and Freeling, 1982; Russell and Sachs, unpublished observations). In addition, a novel set of polypeptides, not normally observed under either aerobic or anaerobic conditions, is synthesised

***FIGURE 1** (opposite). Protein synthesis in a maize primary root during anaerobic treatment. Fluorographs of native-SDS 2-D polyacrylamide gels loaded with extracts from maize primary roots after the following treatments: A) one hour pulse-labelling with [^{3}H]-Leucine under aerobic conditions; B-E) pulse-labelling with [^{3}H]-Leucine for the specified times under anaerobic conditions. The arrow labelled 'TPs' indicates the position of the transition polypeptides. The unlabelled arrow indicates the position of alcohol dehydrogenase-1 (ADH1). From Sachs* et al. *(1980).*

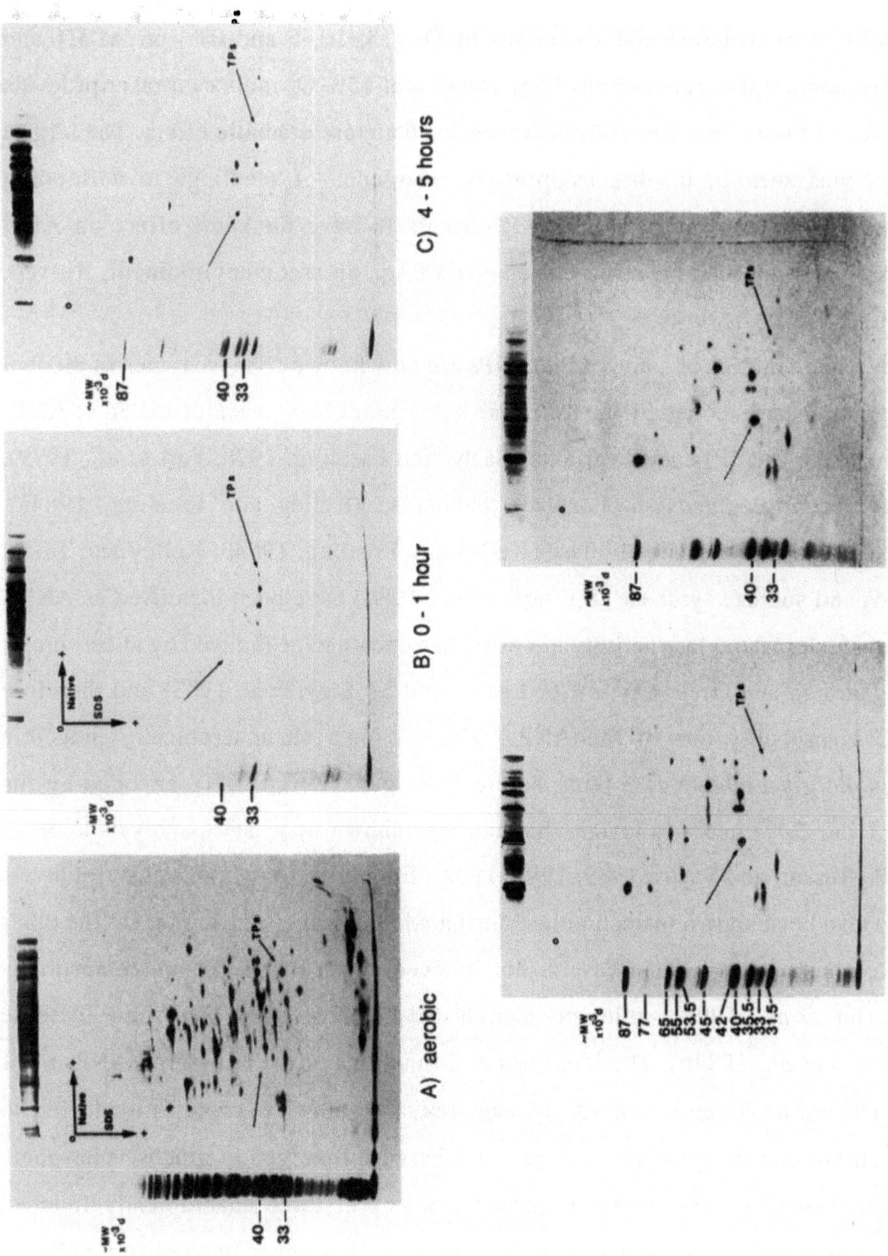

~MW x10³d
40
33
Native
SDS
TPs
A) aerobic
B) 0 - 1 hour
87
C) 4 - 5 hours
77
65
55
53.5
45
42
35.5
31.5
D) 12 - 17 hours
E) 50 - 55 hours

under hypoxic conditions (Kelley and Freeling, 1982). Paul and Ferl (1991a) tested the effects of different concentrations of O_2 (15, 10, 5 and 0%) on ADH1 and ADH2 transcript accumulation. They found that 15% O_2 induces transcript levels by ~four-fold while lower concentrations have a more dramatic effect. The largest effect was seen in the 0% samples. Submergence of seedlings in a flooding solution, or treatment with 0.2% O_2 appears to have the same effect on ADH activity and transcript levels as true anoxia (e.g., an argon environment; Russell, 1990).

The identities of some of the ANPs are now known. The isozymes of alcohol dehydrogenase encoded by the *Adh1* and *Adh2* genes have been identified as ANPs through the use of genetic variants (Sachs and Freeling, 1978; Ferl *et al.*, 1979). More recently, glucose-6-phosphate isomerase (Kelley and Freeling, 1984a), fructose-1,6-diphosphate aldolase (Kelley and Freeling, 1984b; Kelley and Tolan, 1986) and sucrose synthase (Springer *et al.*, 1986) have been identified as ANPs. Pyruvate decarboxylase activity has also been shown to be induced by anaerobiosis (Wignarajah and Greenway, 1976; Laszlo and St. Lawrence, 1983) and therefore PDC is probably one of the ANPs. There is also an anaerobically inducible cytosolic glyceraldehyde-3-phosphate dehydrogenase (GAPDH), encoded by the *Gpc3* and *Gpc4* genes in maize, that has been shown by western analysis to be an ANP (Russell and Sachs, 1989, 1991, 1992). Enolase activity and transcript levels have also been shown to accumulate during anoxia (Lal *et al.*, 1991). On the other hand, alanine aminotransferase is not induced anaerobically in maize seedlings (Watson *et al.*, 1992), although alanine levels have been shown to increase (Roberts *et al.*, 1984b). The identities and functions of the remaining ANPs or of the TPs are as yet unknown. Of the anaerobically inducible proteins identified so far, all are either glycolytic enzymes or otherwise involved in glucose-phosphate metabolism. However, recent sequence analysis of three anaerobically induced transcripts, with no sequence homology to any other genes entered in GenBank, indicates that some of the ANPs may be novel proteins (Peschke and Sachs, 1993a; Vogel and Freeling, personal communication). In the light of the inability of maize seedlings to survive even six hours of flooding in the absence of an active *Adh1* gene (Schwartz, 1969), it appears that at least one function of the anaerobic response is to enable the plant to produce as much ATP as possible when there is an oxygen deficit, as would occur in waterlogged soils.

In the presence of air, the roots, coleoptile, mesocotyl, endosperm, scutellum and the anther wall of maize all synthesise a tissue-specific spectrum of polypeptides. Most organs (e.g., the scutellum and endosperm of the developing kernel) actually synthesise many or all of the ANPs constitutively, along with many other proteins under aerobic conditions (Okimoto *et al.*, 1980). Only the root tissue of mature plants and seedlings and other pre-emergent seedling organs (e.g., the coleoptile) shows a dramatic increase in the synthesis of ANPs by anaerobic induction. Under anaerobic conditions, however, all of the above organs selectively synthesise predominantly the ANPs. Moreover, except for a few characteristic qualitative and quantitative differences, the patterns of anaerobic protein synthesis in these diverse organs are remarkably similar (Okimoto *et al.*, 1980). This has also been observed in maize tissue culture (Woodman, personal communication) and, while most culture strains tested have high constitutive levels of the ANPs under aerobic conditions, one particular culture strain (P3377) actually displays anaerobic induction (Duncan *et al.*, 1985; Paul and Ferl, 1991a, b). On the other hand, maize leaves that have emerged from the coleoptile do not incorporate labelled amino acids under anaerobic conditions, and do not survive even a brief exposure to anaerobiosis (Okimoto *et al.*, 1980).

The maize primary root tip is also very sensitive to anoxic treatment, surviving for only a few hours in the absence of oxygen. However, it has been shown (Saglio *et al.*, 1988) that a hypoxic pretreatment allows longer term survival of the root tip under anoxia. In contrast, the more proximal part of the primary root and the pre-emergent shoot (i.e., those with an intact coleoptile) can survive anoxia for a few days without a hypoxic pretreatment (Sachs *et al.*, 1980).

A shift in the pattern of protein synthesis due to anaerobiosis has been observed in the root tissue of many other plant species including rice (Bertani, 1981; Pradet, 1985), soybean (Russell *et al.*, 1990), sorghum, barley, pea and carrot (M. Freeling and coworkers, unpublished). In anaerobically treated barley aleurone cells, lactate dehydrogenase (LDH) activity increases (Hanson and Jacobsen, 1984) as do ADH mRNA and activity levels (Hanson *et al.*, 1984). In the barley system, the gibberellic acid (GA_3) induction of α-amylase synthesis and its mRNA accumulation are suppressed by anaerobiosis. The 'anaerobic' pattern of protein synthesis is perturbed only slightly by GA_3 when aleurone cells incorporate labelled amino acids in a nitrogen atmosphere (Hanson and Jacobsen, 1984).

The rapid repression of pre-existing protein synthesis caused by anaerobic treatment is correlated with a near complete dissociation of polysomes in primary roots of soybeans (Lin and Key, 1967) and maize (E.S. Dennis and A.J. Pryor, personal communication; Sachs and Russell, unpublished). This does not result from degradation of 'aerobic' mRNA, because the mRNAs encoding the pre-existing proteins remain translatable in an *in vitro* system for at least five hours after the anaerobic treatment is initiated (Sachs *et al.*, 1980; Russell and Sachs, 1992). This is in agreement with the observation that the polysomes, dissociated by anaerobiosis, rapidly reform up to 80 - 90% of their pre-treatment levels even in the absence of new RNA synthesis when soybean seedlings are returned to air (Lin and Key, 1967).

The induction of anaerobic polypeptide synthesis occurs in two phases: the immediate synthesis of the TPs and the appearance after a 90 minute lag of approximately 20 ANPs. In heat shock, the appearance of the heat-shock polypeptides (HSPs) is immediate with kinetics of induction very similar to the rapid appearance of the TPs. Both the HSPs and the TPs exhibit a decreased rate of synthesis after five hours of stress treatment (Sachs *et al.*, 1980; Cooper and Ho, 1983). It is interesting that the only potential overlap in protein synthesis observed in maize in response to heat shock or anaerobiosis is that of a 33 kDa TP and a similar sized HSP (Kelley and Freeling, 1982). In soybeans, there appears to be a ~27-28 kDa HSP that is induced by a number of different environmental stresses, including both heat shock and anaerobiosis (Czarnecka *et al.*, 1984). The anaerobically inducible GAPDH was also found to be heat inducible in maize seedling roots and shoots at the level of transcription (Russell and Sachs, 1989), but not at the level of translation (Russell and Sachs, 1992).

MOLECULAR ANALYSIS OF THE GENES INVOLVED IN THE ANAEROBIC RESPONSE

The molecular analysis of the maize anaerobic response was initiated by synthesising cDNA clones from high molecular weight poly(A)$^+$ RNA of anaerobically treated maize seedling roots (Gerlach *et al.*, 1982). Cloned anaerobic-specific cDNAs were identified by differential hybridisation to recombinant bacterial colonies using radio-labelled cDNAs of mRNA from anaerobic and aerobic roots. Colonies were selected that hybridised specifically with 'anaerobic' cDNAs. The anaerobic-specific cDNA clones were grouped into families on the basis of cross hybridisation to each other, and several of these families were analysed by hybrid-selected translation and RNA gel blot (Northern) hybridisation. The *Adh1* and *Adh2* cloned cDNA families were subsequently identified by immunological analysis and amino acid/nucleotide sequence comparisons (Gerlach *et al.*, 1982; Dennis *et al.*, 1985). These cDNA clone families and the genes encoding them were then analysed extensively (Dennis *et al.*, 1984, 1985; Sachs *et al.*, 1986; Peschke and Sachs, 1993a).

The anaerobic-specific cDNA clones were used as probes to measure gene expression in maize seedling roots and shoots. In both tissues, the levels of mRNA hybridisable to these cDNAs increase during anaerobic treatment (Peschke and Sachs, 1993a; Russell and Sachs, 1989). This has been quantified rigorously in the case of *Adh1* and *Adh2*. The kinetics of mRNA increase are the same for *Adh1* and *Adh2*. In both cases, the mRNA level first appears to increase after 90 minutes of anaerobic treatment. The mRNA level continues to increase until it reaches a plateau ~50-fold above the aerobic level after five hours of anaerobiosis. This level is maintained until after 48 hours in the case of *Adh1*, but starts declining after 10 hours in the case of *Adh2* (Dennis *et al.*, 1985). This pattern of increasing and decreasing mRNA levels is reflected in the previously described rates of *in vivo* anaerobic protein synthesis for *Adh1* and *Adh2* (Sachs *et al.*, 1980). PDC1 and

PDC2 also appear to have similar kinetics of mRNA level increase (Peschke and Sachs, 1993b) as do the inducible cytosolic GAPDH (Russell and Sachs, 1989) and two unidentified, anaerobically inducible mRNAs (Sachs *et al.*, unpublished). However, aldolase (Hake *et al.*, 1985; Kelley and Tolan, 1986; Dennis *et al.*, 1988a), sucrose synthase (ANP87; Hake *et al.*, 1985; Springer *et al.*, 1986), PDC3 (Peschke and Sachs, 1993b) and two other anaerobically inducible mRNAs (Peschke and Sachs, 1993a) show different kinetics of induction. *In vitro* run-off transcription experiments show an increase in transcription rate of the *Adh1* and *Adh2* genes in root cells (Rowland and Strommer, 1985, 1986; Dennis *et al.*, 1988b) and P3377 tissue culture (Paul and Ferl, 1991a) during anoxia, strongly indicating that the increase in the levels of anaerobic-specific mRNAs resulting from anaerobiosis is due to induced transcription of the anaerobic-specific genes.

A comparison of the regions of the *Adh1* and *Adh2* genes upstream from the site of transcription initiation reveals only a few islands of homology. One of these is an 11 base pair (bp) homologous region (Dennis *et al.*, 1985) that includes the 'TATA box' (Goldberg, 1979), a sequence that is found upstream from the vast majority of higher eukaryote genes that have been analysed (Maniatis *et al.*, 1987). The two genes also have three additional 8 bp regions of homology. It was postulated that one or more of these sequences might account for the anaerobic control of these genes (Dennis *et al.*, 1985). This is reminiscent of the heat-shock response element, a 10 bp consensus sequence determined in the heat-shock system that appears to be responsible for the high temperature induction of *Hsp* gene transcription (Pelham and Bienz, 1982). This sequence is conserved in the promoters of *Hsp* genes in many organisms ranging from *Drosophila* to maize. *In vitro* mutagenesis coupled with transformation and gene expression studies, as well as the analysis of the 5′ regions of the other anaerobic genes will be necessary to determine which, if any, of these homologous regions might be important in regulating the induced expression of the anaerobic response genes. The leader sequences of anaerobic-specific mRNAs must also be compared to determine if there are common structural features that allow their selective translation during anoxia.

Some groups have used electroporation to introduce foreign genes into maize protoplasts for transient expression assays (Fromm *et al.*, 1985, 1986; Howard *et al.*, 1987). Using this method, in association with *in vitro* mutagenesis of the *Adh1* promoter, Walker *et al.* (1987a) found two regions upstream from the site of transcription initiation that appeared to control the anaerobic induction of mRNA accumulation. A particular nucleotide sequence found in both of these regions has been called the 'anaerobic response element' (ARE). Its core nucleotide sequence is GGTTT (Walker *et al.*, 1987b). Upstream sequences homologous to the 'ARE' detected by Walker *et al.* (1987a) are found in both the *Adh1* and *Adh2* genes of maize, an *Adh1*-like gene from pea (Llewellyn *et al.*, 1987), the *Adh* gene of *Arabidopsis thaliana* (Chang and Meyerowitz, 1986), and also in the anaerobically inducible genes encoding sucrose synthase and aldolase in maize (reviewed by Walker *et al.*, 1987b). However, the region containing the 'ARE' from the maize *Adh2* gene does not appear to drive anaerobic induction in similar transformation experiments (C. Franche-Bogusz and W.J. Peacock, personal communication), and a region that is homologous to the 'ARE' has been found upstream from a maize gene encoding a constitutive GAPDH (*Gpc1*), whose expression actually decreases during anaerobic treatment (Martinez *et al.*, 1989; Russell and Sachs, 1989). Additionally, several other genes, that are not at all related to the anaerobic response, have been noted to have similar 'ARE' sequences in similar positions (Russell and Sachs, 1991). Reports by Olive *et al.* (1990, 1991) present data indicating that the 'ARE', and protein factors binding to it, may be necessary, but are not sufficient for, hypoxic activation of gene expression. It may be that the region called the 'ARE' is actually a general enhancer element and not the anaerobic analogue of the heat-shock response element.

Chromatin structure in the promoter regions of the *Adh1* and *Adh2* genes under aerobic and anaerobic conditions was analysed by Paul and Ferl (1991a). Using dimethyl sulphate footprinting, the positions of DNA binding proteins were determined. In the 5′ region flanking the *Adh1* gene there are two types of DNA-binding factors. One set that is constitutively associated with this region includes the previously identified 'ARE' sequence (Walker *et al.*, 1987a) and another set of factors that bind only during anaerobic or hypoxic stress. In contrast, all of the factors binding the 5′ region flanking *Adh2* are constitutively present and unchanged by anoxia. While *Adh2* has an 'ARE'-like sequence, this region does not appear to bind any protein factors and the only sequence in common between the two genes that does bind a protein factor has been shown not to be involved in anoxic induction (Walker *et al.*, 1987a; Paul and Ferl, 1991a).

These data indicate that the molecular mechanisms involved in transcriptional activation by anaerobic induction will not be as straightforward as those observed in the heat-shock response. Analysis of translational regulation in the anaerobic response also requires additional studies (Bailey-Serres and Freeling, 1990; Russell and Sachs, 1992). It is therefore important to analyse additional genes involved in the maize anaerobic response. In recent years several other cDNA and genomic clones encoding ANPs have been isolated and are being characterised. These include the cDNAs and genes encoding sucrose synthase (Springer *et al.*, 1986) aldolase (Kelley and Tolen, 1986; Dennis *et al.*, 1988a), inducible cytosolic GAPDH (Russell and Sachs, 1989; R. Cerff, personal communication), and PDC1 (Kelley, 1989; Kelley *et al.*, 1991). In addition, cDNAs encoding enolase (Lal *et al.*, 1991), PDC2 and PDC3 (Peschke and Sachs, 1993b), plus three additional cDNAs representing anaerobically induced transcripts with unknown functions (Peschke and Sachs, 1993a; Vogel and Freeling, personal communication) are being studied. A maize genomic sequence with homology to a barley cDNA encoding LDH has also been described recently (Hondred and Hanson, 1990; Good and Paetkau, 1992).

We are also genetically characterising genes involved in flooding tolerance (Lemke-Keyes and Sachs, 1989b; Peschke and Sachs, unpublished). It appears that genetic factors found in a few exotic maize accessions allow seedlings to survive flooding twice as long as their alleles in standard North American inbreds and hybrids. In crosses and backcrosses, these factors behave, in each case, as a simple dominant trait. Whether the gene(s) involved in flooding tolerance encode ANP(s) or are involved in the anaerobic response remains to be determined.

Molecular, genetic and physiological analysis of the genes involved in the anaerobic response and flooding tolerance needs to be performed either in intact seedlings (Okimoto *et al.*, 1980) or a tissue culture system similar to P3377 (Duncan *et al.*, 1985; Paul and Ferl, 1991a, b). This is necessary to test the expression of anaerobically stimulated promoters in tissues where they are normally inducible by anoxic treatment (e.g., the seedling root and shoot). Such a system is also important for analysis of selective translation of anaerobic mRNAs. Previous work analysing the effects of *in vitro* mutagenesis on *Adh1* promoter function (e.g., Walker *et al.*, 1987a) was done using transient expression assays in protoplasts derived from the Black Mexican Sweet maize cell culture line. This line has high constitutive expression of the ANPs under aerobic conditions, and is thus not ideal for studying transcriptional regulation. Additionally, protoplasts derived from this line do not survive hypoxic treatments where O_2 levels fall below 5%, making studies on translational regulation impossible. Fortunately, methods of introducing foreign and engineered genes into maize plants have been developed and this technology will soon be routine (Rhodes *et al.*, 1988; Fromm *et al.*, 1990; Gordon-Kamm *et al.*, 1990; Gould *et al.*, 1991; Walters *et al.*, 1992; Spencer *et al.*, 1992).

CONCLUSIONS

The anaerobic stress response is a model system for the analysis of coordinate gene expression in higher organisms. Studies presently under way will determine the molecular mechanisms involved in: a) the anaerobic induction of transcription of the 'family' of genes encoding the ANPs and; b) the selective translation during anoxia of mRNAs encoding both the transition and anaerobic polypeptides. Understanding the response to anaerobic stress is also important for the future breeding of flood tolerant or resistant crop plants. In addition to molecular, physiological and biochemical approaches, genetic analysis has been essential in the establishment of causal relationships between the induction of the stress proteins and the establishment of resistance to the stress condition. In most stress responses, it is not difficult to detect the induction of new proteins during stress. However, the induction of new proteins does not necessarily establish a relationship with stress resistance. Novel protein synthesis may well be the consequence of damage caused by a stress condition. Thus, genetic mutants are necessary to test the physiological role of a stress protein. Perhaps the best example of this is the demonstration that ADH null mutants in maize fail to survive more than a few hours of anaerobic stress (Schwartz, 1969) but grow well in nonflooding conditions. On the other hand, their ADH^+ siblings will survive approximately three days of flooding (Schwartz, 1969; Lemke-Keyes and Sachs, 1989b).

With the advances in genetic transformation techniques, it is likely that genes encoding stress proteins from tolerant plants will be introduced into plants that are normally sensitive to the stress condition. This approach is not only important for the molecular analysis of DNA sequences that are responsive to an anaerobic stimulus, but will also pave the way for constructing 'tailor-made' flood resistant crop plants.

ACKNOWLEDGMENTS

I would like to thank O. Yefet, D. Hirsch, M. Watters, L. Small, M. Lidell and J.-K. Zhang for technical assistance, M. Veith for photographic assistance, and V.M. Peschke and C.C. Subbaiah for their critical reading of this manuscript. The author's research is supported by the US Public Health Service, NIH Grant R01 GM34740.

REFERENCES

Bailey-Serres J, Freeling M (1990) Hypoxic stress-induced changes in ribosomes of maize seedling roots. Plant Physiology 94: 1237-1243

Bertani A, Menegus F, Bollini R (1981) Some effects of anaerobiosis on protein metabolism in rice roots. Zeitschrift für Pflanzenphysiologie 103: 37-43

Chang C, Meyerowitz EM (1986) Molecular cloning and DNA sequence of the *Arabidopsis thaliania Alcohol dehydrogenase* gene. Proceedings of the National Academy of Sciences USA 83: 1408-1412

Cooper P, Ho T-HD (1983) Heat shock proteins in maize. Plant Physiology 71: 215-222

Czarnecka E, Edelman L, Schöffl F, Key JL (1984) Comparative analysis of physical stress responses in soybean seedlings using cloned heat shock cDNAs. Plant Molecular Biology 3: 45-58

Dennis ES, Gerlach WL, Pryor AJ, Bennetzen JL, Ingles A, Llewellyn D, Sachs MM, Ferl RJ, Peacock WJ (1984) Molecular analysis of the *Alcohol dehydrogenase* (*Adh1*) gene of maize. Nucleic Acids Research 12: 3983-4000

Dennis ES, Gerlach WL, Walker JC, Lavin M, Peacock WJ (1988a) Anaerobically regulated aldolase gene of maize: A chimaeric origin? Journal of Molecular Biology 202: 759-767

Dennis ES, Sachs MM, Gerlach WL, Beach L, Peacock WJ (1988b) The *Ds1* transposable element acts as an intron in the mutant allele *Adh1-Fm335* and is spliced from the message. Nucleic Acids Research 16: 3815-3828

Dennis ES, Sachs MM, Gerlach WL, Finnegan EJ, Peacock WJ (1985) Molecular analysis of the *Alcohol dehydrogenase 2* (*Adh2*) gene of maize. Nucleic Acids Research 13: 727-743

Duncan DR, Williams ME, Zehr BE, Widholm JM (1985) The production of callus capable of plant regeneration from immature embryos of numerous *Zea mays* genotypes. Planta 165: 322-332

Ferl RJ, Dlouhy SR, Schwartz D (1979) Analysis of maize alcohol dehydrogenase by native-SDS two dimensional electrophoresis and autoradiography. Molecular and General Genetics 169: 7-12

Freeling M (1973) Simultaneous induction by anaerobiosis or 2,4-D of multiple enzymes specified by two unlinked genes: Differential *Adh1-Adh2* expression in maize. Molecular and General Genetics 127: 215-227

Freeling M, Bennett DC (1985) Maize *Adh1*. Annual Review of Genetics 19: 297-323

Freeling M, Birchler JA (1981) Mutants and variants of the Alcohol dehydrogenase - 1 gene in maize. In: Setlow JK, Hollaender A (eds) Genetic engineering principles and methods, Vol 3. Plenum Press, pp 223-264

Fromm ME, Morrish F, Armstrong C, Williams R, Thomas J, Klein TM (1990) Inheritance and expression of chimeric genes in the progeny of transgenic maize plants. Biotechnology 8: 833-839

Fromm M, Taylor LP, Walbot V (1985) Expression of genes transferred into monocot and dicot plant cells by electroporation. Proceedings of the National Academy of Sciences USA 82: 5824-5828

Gerlach WL, Pryor AJ, Dennis ES, Ferl RJ, Sachs MM, Peacock WJ (1982) cDNA cloning and induction of the *Alcohol dehydrogenase* gene (*Adh1*) of maize. Proceedings of the National Academy of Sciences USA 79: 2981-2985

Gerlach WL, Sachs MM, Llewellyn D, Finnegan EJ, Dennis ES (1986) Maize *Alcohol dehydrogenase*: A molecular perspective. In: Blonstein AD, King PJ (eds) Plant gene research, Vol 3, A genetic approach to plant biochemistry. Springer-Verlag Wein, New York, pp 73-100

Goldberg ML (1979) Sequence analysis of *Drosophila* histone genes. PhD dissertation, Stanford University, USA

Good AG, Paetkau DH (1992) Identification and characterization of a hypoxically induced maize lactate dehydrogenase gene. Plant Molecular Biology 19: 693-697

Gordon-Kamm WJ, Spencer TM, Mangano ML, Adams TR, Daines RJ, Start WG, O'Brien JV, Chambers SA, Adams Jr WR, Willetts NG, Rice TB, Mackey CJ, Kruger RW, Kausch AP, Lemaux PG (1990) Transformation of maize cells and regeneration of fertile transgenic plants. The Plant Cell 2: 603-618

Gould J, Devey M, Hasegawa O, Ulian EC, Peterson G, Smith RH (1991) Transformation of *Zea mays* L. using *Agrobacterium tumefaciens* and the shoot apex. Plant Physiology 95: 426-434

Hageman RH, Flesher D (1960) The effect of anaerobic environment on the activity of alcohol dehydrogenase and other enzymes of corn seedlings. Archives of Biochemistry and Biophysics 87: 203-209

Hake S, Kelley PM, Taylor WC, Freeling M (1985) Coordinate induction of alcohol dehydrogenase 1, aldolase, and other anaerobic RNAs in maize. Journal of Biological Chemistry 260: 5050-5054

Hanson AD, Jacobsen JV (1984) Control of lactate dehydrogenase, lactate glycolysis, and α-amylase by O_2 deficit in barley aleurone layers. Plant Physiology 75: 566-572

Hanson AD, Jacobsen JV, Zwar JA (1984) Regulated expression of three *Alcohol dehydrogenase* genes in barley aleurone layers. Plant Physiology 75: 573-581

Hondred D, Hanson AD (1990) Hypoxically inducible barley lactate dehydrogenase: cDNA cloning and molecular analysis. Proceedings of the National Academy of Sciences USA 87: 7300-7304

Howard EA, Walker JC, Dennis ES, Peacock WJ (1987) Regulated expression of an alcohol dehydrogenase 1 chimeric gene introduced into maize protoplasts. Planta 170: 535-540

Kelley PM (1989) Maize pyruvate decarboxylase mRNA is induced anaerobically. Plant Molecular Biology 13: 213-222

Kelley PM, Freeling M (1982) A preliminary comparison of maize anaerobic and heat-shock proteins. In: Schlesinger MJ, Ashburner M, Tissières A (eds) Heat shock from bacteria to man. Cold Spring Harbor Laboratory Press, pp 315-319

Kelley PM, Freeling M (1984a) Anaerobic expression of maize glucose phosphate isomerase I. Journal of Biological Chemistry 259: 673-677

Kelley PM, Freeling M (1984b) Anaerobic expression of maize fructose-1,6-diphosphate aldolase. Journal of Biological Chemistry 259: 14180-14183

Kelley PM, Godfrey K, Lal SK, Alleman M (1991) Characterization of the maize pyruvate decarboxylase gene. Plant Molecular Biology 17: 1259-1261

Kelley PM, Tolan DR (1986) The complete amino acid sequence for the anaerobically induced aldolase from maize derived from cDNA clones. Plant Physiology 82: 1076-1080

Lal SK, Johnson S, Conway T, Kelley PM (1991) Characterization of a maize cDNA that complements an enolase-deficient mutant of *Escherichia coli*. Plant Molecular Biology 16: 787-795

Laszlo A, St Lawrence P (1983) Parallel induction and synthesis of PDC and ADH in anoxic maize roots. Molecular and General Genetics 192: 110-117

Lemke-Keyes CA, Sachs MM (1989a) Anaerobic tolerant null: A mutant that allows *Adh1* nulls to survive anaerobic treatment. Journal of Heredity 80: 316-319

Lemke-Keyes CA, Sachs MM (1989b) Genetic variation for seedling tolerance to anaerobic stress in maize germplasm. Maydica 34: 329-337

Lin CY, Key JL (1967) Dissociation and reassembly of polyribosomes in relation to protein synthesis in the soybean root. Journal of Molecular Biology 26: 237-247

Llewellyn DJ, Finnegan EJ, Ellis JG, Dennis ES, Peacock WJ (1987) Structure and expression of an *Alcohol dehydrogenase 1* gene from *Pisum sativum* (cv 'Greenfeast'). Journal of Molecular Biology 195: 115-123

Maniatis T, Goodbourn S, Fischer JA (1987) Regulation of inducible and tissue-specific gene expression. Science 236: 1237-1245

Martinez P, Martin W, Cerff R (1989) Structure, evolution and anaerobic expression of the nuclear gene for cytosolic glyceraldehyde-3-phosphate dehydrogenase from maize. Journal of Molecular Biology 208: 551-565

Mocquot B, Prat C, Mouches C, Pradet A (1981) Effect of anoxia on energy charge and protein synthesis in rice embryo. Plant Physiology 68: 636-640

Okimoto R, Sachs MM, Porter EK, Freeling M (1980) Patterns of polypeptide synthesis in various maize organs under anaerobiosis. Planta 150: 89-94

Olive MR, Peacock WJ, Dennis ES (1991) The anaerobic responsive element contains 2 GC-rich sequences essential for binding a nuclear-protein and hypoxic activation of the maize *Adh1* promoter. Nucleic Acids Research 19: 7053-7060

Olive MR, Walker JC, Singh K, Dennis ES, Peacock WJ (1990) Functional properties of the anaerobic responsive element of the maize *Adh1* gene. Plant Molecular Biology 15: 593-604

Paul A-L, Ferl RJ (1991a) *Adh1* and *Adh2* regulation. Maydica 36: 129-134

Paul A-L, Ferl RJ (1991b) *In vivo* footprinting reveals unique cis-elements and different modes of hypoxic induction in maize *Adh1* and *Adh2*. The Plant Cell 3: 159-168

Pelham HRB, Bienz M (1982) A synthetic heat-shock promoter element confers heat-inducibility on the herpes simplex virus thymidine kinase gene. European Molecular Biology Journal 1: 1473-1477

Peschke VM, Sachs MM (1993a) Characterization and expression of anaerobically induced maize genes. Plant Molecular Biology (submitted)

Peschke VM, Sachs MM (1993b) Multiple maize pyruvate decarboxylase genes are induced by hypoxia. Plant Molecular Biology (submitted)

Pradet A, Mocquot B, Raymond P, Morisset C, Aspart L, Delseny M (1985) Energy metabolism and synthesis of nucleic acids and proteins under anoxic stress. In: Key JL, Kosuge T (eds) Cellular and molecular biology of plant stress. UCLA symposia on molecular and cellular biology, new series Vol 22. Alan R Liss, New York, pp 227-245

Rhodes CA, Pierce DA, Mettler IJ, Mascarenhas D, Detmer JJ (1988) Genetically transformed maize plants from protoplasts. Science 240: 204-207

Roberts JKM, Andrade FH, Anderson IC (1985) Further evidence that cytoplasmic acidosis is a determinant of flooding intolerance in plants. Plant Physiology 77: 492-494

Roberts JKM, Callis J, Jardetzky O, Walbot V, Freeling M (1984a) Cytoplasmic acidosis as a determinant of flooding intolerance in plants. Proceedings of the National Academy of Sciences USA 81: 6029-6033

Roberts JKM, Callis J, Wemmer D, Walbot V, Jardetzky O (1984b) Mechanism of cytoplasmic pH regulation in hypoxic maize root tips and its role in survival under hypoxia. Proceedings of the National Academy of Sciences USA 81: 3379-3383

Rowland LJ, Strommer JN (1985) Insertion of an unstable element in an intervening sequence of maize *Adh1* affects transcription but not processing. Proceedings of the National Academy of Sciences USA 82: 2875-2879

Rowland LJ, Strommer JN (1986) Anaerobic treatment of maize roots affects transcription of *Adh1* and transcript stability. Molecular Cell Biology 6: 3368-3372

Russell DA (1990) The genetics and expression of cytosolic glyceraldehyde-3-phosphate dehydrogenase. PhD dissertation, Washington University, USA

Russell DA, Sachs MM (1989) Differential expression and sequence analysis of the maize glyceraldehyde-3-phosphate dehydrogenase gene family. The Plant Cell 1: 793-803

Russell DA, Sachs MM (1991) The maize cytosolic glyceraldehyde-3-phosphate dehydrogenase gene family: Organ-specific expression and genetic analysis. Molecular and General Genetics 229: 219-228

Russell DA, Sachs MM (1992) Protein synthesis in maize during anaerobic and heat stress. Plant Physiology 99: 615-620

Russell DA, Wong DM-L, Sachs MM (1990) The anaerobic response of soybean. Plant Physiology 92: 401-407

Sachs MM, Dennis ES, Gerlach WL, Peacock WJ (1986) Two alleles of maize *Alcohol dehydrogenase 1* have 3′ structural and poly(A) addition polymorphisms. Genetics 113: 449-467

Sachs MM, Freeling M (1978) Selective synthesis of alcohol dehydrogenase during anaerobic treatment of maize. Molecular and General Genetics 161: 111-115

Sachs MM, Freeling M, Okimoto R (1980) The anaerobic proteins of maize. Cell 20: 761-767

Saglio PH, Drew MC, Pradet A (1988) Metabolic acclimation to anoxia induced by low (2-4 kPa partial pressure) oxygen pretreatment (hypoxia) in root tips of *Zea mays*. Plant Physiology 86: 61-66

Schlesinger M, Ashburner M, Tissières A (1982) Heat shock from bacteria to man. Cold Spring Harbor Laboratory Press

Schwartz D (1969) An example of gene fixation resulting from selective advantage in suboptimal conditions. American Naturalist 103: 479-481

Spencer TM, O'Brien JV, Start WG, Adams TR, Gordon-Kamm WJ, Lemaux PG (1992) Segregation of transgenes in maize. Plant Molecular Biology 18: 201-210

Springer B, Werr W, Starlinger P, Bennett DC, Zokolica M, Freeling M (1986) The *Shrunken* gene on chromosome 9 of *Zea mays* L. is expressed in various plant tissues and encodes an anaerobic protein. Molecular and General Genetics 205: 461-468

Walker JC, Howard EA, Dennis ES, Peacock WJ (1987a) DNA sequences required for anaerobic expression of the maize *Alcohol dehydrogenase-1* gene. Proceedings of the National Academy of Sciences USA 84: 6624-6628

Walker JC, Llewellyn DJ, Mitchell LE, Dennis ES (1987b) Anaerobically regulated gene expression; molecular adaptations of plants for survival under flooded conditions. In: Miflin BJ, Miflin HF (eds) Oxford surveys of plant molecular and cell biology, Vol 4. Oxford University Press, pp 71-93

Walters DA, Vetch CS, Potts DE, Lundquist RC (1992) Transformation and inheritance of a hygromycin phosphotransferase gene in maize plants. Plant Molecular Biology 18: 189-200

Watson NR, Peschke VM, Russell DA, Sachs MM (1992) Analysis of the L-alanine:2-oxoglutarate aminotransferase isozymes in maize. Biochemical Genetics 30: 371-383

Wignarajah K, Greenway H (1976) Effect of anaerobiosis on activities of alcohol dehydrogenase and pyruvate decarboxylase in roots in *Zea mays*. The New Phytologist 77: 575-584

METABOLIC ACCLIMATION IN WINTER CEREALS BY INTERACTING LOW TEMPERATURE STRESSES

C.J. Andrews
Plant Research Centre
Agriculture Canada
Ottawa K1A OC6
Canada

ABSTRACT

Overwintering crop plants are subject to an array of stresses before and during winter. This chapter is concerned with the effects of these stresses and their interactions on the survival of winter cereal plants and the ability of the crop to provide an adequate economic yield. Interactions influencing survival at low temperature have been studied in controlled environment simulations. Low temperature flooding (2 °C) markedly reduces the cold hardiness of winter cereal seedlings, but simultaneously increases their tolerance to ice encasement at 1 °C. This constitutes a metabolic acclimation to severe hypoxia, and provides a valuable means to search for increased tolerance of ice encasement. Pre-ice flooding induces higher levels of ethanol and CO_2 production in plant crowns during subsequent ice encasement, together with higher activity of alcohol dehydrogenase and higher total adenylates and adenylate energy charge. We have been unable to demonstrate increased activity of the glycolytic kinases (PFK, PK), but the accumulated evidence indicates an increased flux through glycolysis during ice encasement as a result of pre-ice flooding. Several polypeptides unique to the flood-ice treatments of winter wheat have been revealed by ^{35}S-methionine labelling and image-analysis of 2-D electrophoretograms. Current work is examining the nature of the unique polypeptides to determine if they may be exploited as markers for ice tolerance.

INTRODUCTION

The stresses associated with winter markedly limit the distribution and economic viability of autumn-sown crops. When these stresses are very severe on a regional basis, autumn-sown crops may be replaced by annual crops which totally avoid winter stresses. These generally yield less than equivalent autumn-sown crops after full survival, but a more stable yield can be realised.

The need to use lower yielding annual crops can be avoided by developing winter crops with greater stress tolerance resulting in a more consistent winter

NATO ASI Series, Vol. I 16
Interacting Stresses on Plants in a Changing Climate
Edited by M. B. Jackson, C. R. Black

survival. Even with the currently predicted warming of climate and the potential redistribution of agricultural zones, the requirement to use winter crops close to the limits of their survival will continue, in order to maximise yields and economic return. An additional advantage of autumn-sown and perennial crops over annuals is that the soil surface is covered for a longer period, thus reducing soil erosion and removing the need for further field operations to anchor the surface soil layers. In moving towards this goal of increased production of winter crops, we have been concerned with the limitations to survival for winter wheat in eastern Canada, and a search for the means to increase this survival.

THE COMPONENTS OF WINTER STRESS

Winter air temperatures frequently fall to potentially damaging levels, even for fully acclimated winter cereals. Actual damage is dependent upon the presence or absence of insulating snow layers, since soil temperature at the crown level of wheat plants increases markedly shortly after the onset of permanent snow cover each winter (Andrews and Pomeroy, 1977) to a normally stable level of approximately -1 °C. The insulating snow layer may be removed by wind and temperature at the crown level then decreases to potentially damaging levels, together with the additional stress of desiccation. The insulating snow layer may also be reduced or removed by thawing or rain, which accumulates as flooding at the soil surface. Subsequently, this freezes over the plants when the temperature again decreases below freezing. The resulting ice encasement can be severely damaging to cereal plants, and is the main subject of this chapter. A number of biotic stresses may also influence wheat plants during the winter period (snow moulds, Barley Yellow Dwarf Virus and others) and interact with the above stresses, but will not be discussed further here.

Ice encasement damage. Cold hardened winter wheat plants are injured during ice encasement at -1 °C as a consequence of severe hypoxia due to the almost total impermeability of ice to the respiratory gases (Rakitina, 1965). The accumulation of ethanol, CO_2 and lactic acid has been documented previously (Andrews and

Pomeroy, 1979), with the toxicity of CO_2 being of greatest significance. There is a breakdown of the K^+ uptake system in cells (Pomeroy and Andrews, 1985) and tissues (Gao *et al.*, 1983) during ice encasement, and evidence that the ATPase activity of plasma membrane enriched fractions is reduced by exposure to ice encasement. A summary of these investigations has been given elsewhere (Andrews and Pomeroy, 1991).

The flood-ice interaction. Under field conditions, an ice encasement period is frequently preceded by low temperature flooding, either in the autumn or in winter during a thaw period. These events are difficult to separate experimentally in the field, but can be conveniently arranged in controlled environments. Such experiments have shown that exposure to flooding decreases the subsequent cold hardiness of wheat plants, but increases their tolerance to ice encasement at -1 °C. These changes are also seen when a moderate freezing stress follows flooding; cold hardiness and ice tolerance are reduced but the interaction with the flooding remains (Table 1). The decrease in cold hardiness after flooding may be explained by the associated increase in crown water content according to the close correlation established by Fowler *et al.* (1981). The increase in ice tolerance is not easily explained, particularly as Rakitina (1977) showed that flooding and ice encasement acted synergistically to increase damage when the latter stress was applied at -5 °C.

The promotive effect of low temperature flooding on ice tolerance at -1 °C is an example of a hypoxic acclimation, where exposure of plants to mild hypoxia (flooding) increases tolerance to severe hypoxia or anoxia (ice encasement). Indeed, the flooding effect may be mimicked by a nitrogen atmosphere. The promotive effect of flooding on ice tolerance is demonstrable in wheat and rye, but not in barley (Andrews and Pomeroy, 1981), and has also been shown in winter Brassica (*B. napus*) but not in timothy (*Phleum pratense*) (Andrews, unpublished data). A similar response has been shown at warm temperatures in maize roots, where exposure to a lowered oxygen concentration (2-4 kPa) increased root survival during anoxia (Saglio *et al.*, 1988; Hole *et al.*, 1992).

TABLE 1. Effects of low temperature flooding and freezing on the cold tolerance of cold-hardened Fredrick winter wheat. From Gao et al. *(1983).*

Treatment	Survival (%)	Cold hardiness (°C)	Ice tolerance (days)
None	100	-17.1	5.4
-6 °C 1 week	100	-16.4	4.3
-8 °C 1 week	88	-14.6	2.2
Flood 1 week (FL)	100	-13.8	7.5
Flood -6 °C 1 week	86	-11.1	5.8
Flood -8 °C 1 week	75	- 9.8	5.8

THE INCREASE OF ICE TOLERANCE BY FLOODING

The increased ice tolerance of cold hardened winter wheat plants induced by pre-ice flooding is associated with a substantially increased production of ethanol (Andrews and Pomeroy, 1983) and a smaller increase in CO_2 in the plants during ice encasement (Table 2). Because both metabolites are potentially toxic to cold hardened cereal plants, their increased accumulation must be linked with a separate process beneficial to subsequent plant survival. This process is proposed to be an acceleration of glycolysis as a result of the flooding exposure, prior to the formation of encasing ice.

A significant indicator of glycolytic acceleration is the higher level of total adenylates (AdN) in ice-encased winter wheat crowns that have previously been exposed to low temperature flooding than in those that have not (Fig. 1) (Andrews and Pomeroy, 1989). This occurs despite the fact that low temperature flooding itself induces a loss of AdN. However, this loss is similar after several days of flooding to that found after only 1.5 hours of flooding (flooding shock), which even 'non-flood' plants experience during the formation of the ice layer. It is noteworthy that winter barley plants do not show higher AdN in ice after flooding, nor do they show greater survival in ice after this treatment, as discussed above.

TABLE 2. Ethanol and CO_2 concentrations in the crowns of winter wheat plants at the LD_{50} kill point in ice. Plants were previously flooded at low temperature (FL) or not (NFL).

	LD_{50} (days)		Ethanol ($nmol\ g^{-1}$ dry wt.)		CO_2 ($nmol\ g^{-1}$ dry wt.)	
	NFL	FL	NFL	FL	NFL	FL
Fredrick	8.8	14.4	0.55	2.45	0.58	1.01
Norstar	11.2	13.5	1.05	1.95	0.72	1.00

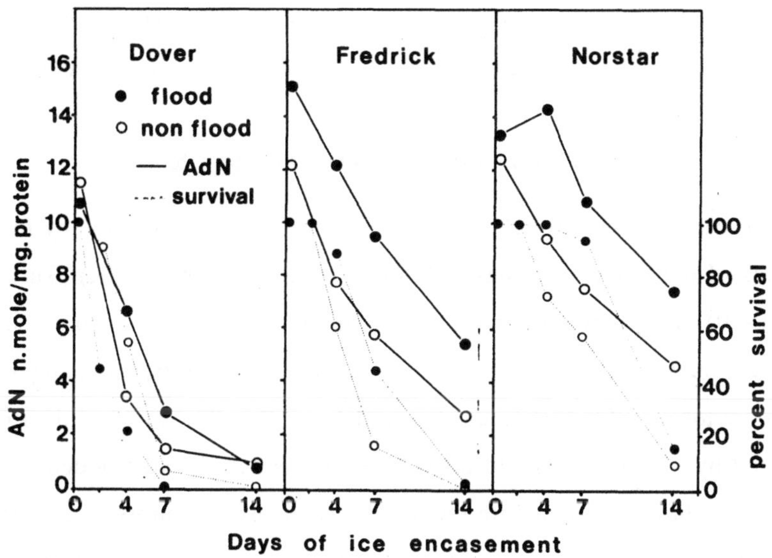

FIGURE 1. Changes in total adenylates (AdN) and survival of crowns of cold-hardened winter wheat and barley during ice encasement with and without a previous two week exposure to low temperature flooding. From Andrews and Pomeroy (1989).

Pre-ice flooding substantially increases ATP levels in wheat crowns after a brief two day period of ice encasement, which does not reduce crown regrowth in survival tests. ADP and AMP are less affected, such that the adenylate energy charge is higher in iced plants after a flooding episode than without it. The ATP generation is believed to be a substrate level phosphorylation from two sites of glycolysis, linked with the generation of ethanol and CO_2. The ATP contributes to the increased survival of the tissues by providing increased substrate for plasma membrane and vacular ATPase membrane pumps which maintain ionic balance. The operation of various biosynthetic pathways is dependent upon ATP, and ATP is also required continuously to prime the glycolytic sequence itself. Ice encasement reduced the activity of an ATPase of plasma membrane enriched fractions of cold-hardened wheat seedlings (Andrews and Pomeroy, 1991). An increased level of cellular ATP generated in response to a pre-ice flooding treatment could protect ATPases against disruption, or allow more efficient functioning of the disrupted enzyme. Either process could explain the increased survival in ice, but the relevant experimentation has not yet been undertaken.

CONTROL OF GLYCOLYTIC FLUX

The rate of carbon flow through glycolysis is known to be controlled by feedback inhibitions of phosphofructokinase (PFK) and pyruvate kinase (PK) (Turner and Turner, 1980), though major complexities exist in the regulation of both enzymes (Duff *et al.*, 1989). In cold-hardened wheat crowns, no significant changes in the activity of cytosolic PFK have been observed due to flooding, or when the flooding is followed by ice (Table 3). No differences in the activity of flooded and non-flooded plants have emerged with the use of proteinase inhibitors or a wide array of modifications in extraction or assay procedures. The conclusion is thus in accord with the reports of Bailey-Serres *et al.* (1988) and Bertani *et al.* (1991) that measurable changes in the glycolytic kinases are not observed in response to flooding. Neither were there major differences in the activity of phosphofructo-phosphotransferase (PFP) observed in flooded as compared to non-flooded wheat. This enzyme uses pyrophosphate rather than ATP as a phosphate donor to maintain the glycolytic flux and might be expected to conserve cellular reserves of adenylate when its regeneration is markedly reduced during anoxia.

***TABLE 3**. Activity of glycolytic enzymes from crowns of cold-hardened Fredrick winter wheat after flooding and ice encasement treatments.*

	PFK	PFP	PK	ALDO
		(nmol mg^{-1} protein min^{-1})		
Non Flood	39	41	144	75
Flood	46	46	146	63
Non Flood-Ice	35	46	154	99
Flood-Ice	33	33	213	91

PFK = ATP-dependent phosphofructokinase; PFP = PPi-dependent phosphofructophosphotransferase; PK = pyruvate kinase; ALDO = aldolase.

Small increases in the activity of pyruvate kinase have been recorded for flooded as compared to non-flooded cold-hardened wheat, but no changes were observed at an early stage of flooding (four days) which had already induced an increase in plant survival during a subsequent ice treatment. An alternative method for measuring the activity of PK is by determining the relative levels of its substrate end-product. When arrested in liquid nitrogen, pyruvate : phosphoenol pyruvate (PEP) ratios were slightly increased by flooding, indicating that the enzyme from the flooded plants utilised PEP more rapidly. Further enzyme assays have indicated that no detectable increases in the activity of aldolase are induced by flooding (Table 3).

The last enzymatic step in hypoxic glycolysis is catalysed by alcohol dehydrogenase (Adh), which reduces acetaldehyde to ethanol with the concurrent regeneration of NAD. This contributes to the maintenance of the glycolytic pathway, allowing the continued generation of substrate level phosphorylations. Adh is significantly active in cold-hardened, aerobic winter wheat, but its activity increases markedly when seedlings are flooded at low temperature (Table 4). After a subsequent brief period of ice encasement, activity increases to a higher level in seedlings that had been flooded than those that had not. Only a minor increase in activity occurs after ice alone. These activities are correlated with increased ethanol production and increased adenylate levels in plants in ice following previous exposure to flooding. Preliminary investigations have shown that the activity of pyruvate decarboxylase (PDC) doubles in response to flooding. While Adh is not generally considered as a regulatory enzyme in glycolysis, evidence has recently been presented for an adaptive role in flooding tolerance for PDC (Chan and Burton, 1992).

TABLE 4. *Activity of alcohol dehydrogenase in crowns of cold-hardened winter wheat and barley after low temperature flooding (FL) and ice encasement (Ice).*

	Non-FL	FL	Non-FL-Ice	FL-Ice
		(nmol mg^{-1} protein min^{-1})		
Dover barley	36	116	73	189
Fredrick wheat	132	247	224	518
Norstar wheat	159	345	231	510

Assayed with acetaldehyde as substrate, pH 8.1.

POLYPEPTIDE SYNTHESIS

When cold-hardened wheat crowns are incubated in the presence of ^{35}S-methionine, they incorporate the label in varying amounts and in different distributions depending on the stress imposed immediately following the incubation period (Hope and Andrews, unpublished). The total quantity of label incorporated is reduced by flooding and severely reduced by ice encasement, but is little changed from control levels when flooding is followed by ice encasement. This indicates an enhanced incorporation by plants in ice when they have been previously exposed to flooding. Separation of polypeptides on 2-D polyacrylamide gels and quantification by digital image analysis have shown that only three new polypeptides are formed during ice encasement, 16 during flood only, but that many are formed in ice preceded by flooding. The nature of these polypeptides is unknown, but the observations indicate an enhanced synthetic activity in crown tissue as a result of the interactive stresses which, in conjunction with increased enzyme activity and total adenylates, is presumed to be associated with increased survival of the plants.

ISOZYME EXPRESSION

The possibility exists that the new polypeptides revealed by the 2-D gel analysis are isozymes of glycolytic enzymes. To investigate this, enzyme proteins were separated by starch gel electrophoresis and visualised with a series of specific enzyme-linked stains (Wendel and Weeden, 1989). Although a large number of enzymes have been visualised by these techniques, only Adh has as yet demonstrated any substantial and consistent variation due to treatment. This enzyme appears in extracts from cold-hardened wheat tissues as three bands, representing (Hart, 1983) the expression of a gene from each of the three genomes of wheat (ABD), sorting into a homodimer and heterodimers with different mobilities (Fig. 2). Flooding induces expression of a second locus, again with a homodimer and two sets of heterodimers, but with lower intensity of the more mobile band.

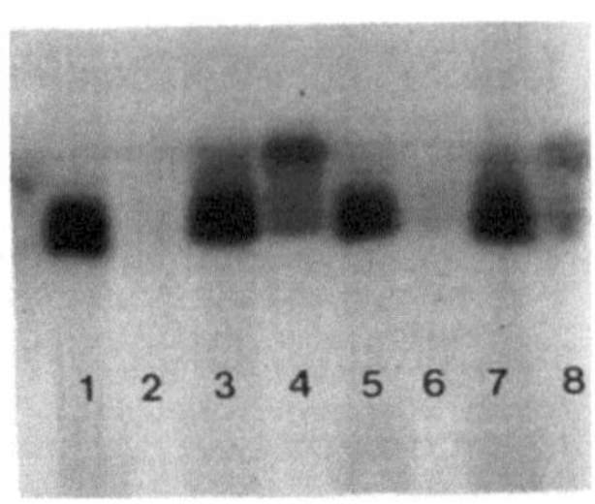

FIGURE 2. *Alcohol dehydrogenase (Adh) isozymes of cold-hardened wheat on starch gel: Lane 1, Non-flood (NF) crowns; 2, NF roots; 3, Flooded (F) crowns; 4, F roots; 5, NF-Iced crowns; 6, NF-Iced roots; 7, F-Iced crowns; 8, F-Iced roots.*

Root expression is generally lower for the original locus, but for the flood-induced upper locus, the faster dimer is predominant. The second locus is not induced by exposure to ice, but is present at an increased intensity when ice is preceded by flooding. It is probable that the extra flood-induced isozymes are responsible for the increased Adh activity (Table 4) after flooding, and after ice preceded by flooding. No other enzyme tested showed similar consistent changes after flooding, nor did any of the enzyme systems show a loss of isozyme bands after one week of ice. Shikimate dehydrogenase showed a major loss of activity in a fast allele

after ice exposure, which potentially could influence the synthesis of aromatic amino acids. It is also notable that superoxide dismutase (SOD) activity was found in both roots and crowns after flooding and after ice, but no consistent changes after treatment were seen, and no additional expression of isozymes or increased activity was recorded, as reported by Monk *et al.* (1989) for flood-tolerant species. The presence of SOD within tissues has the potential for protection against peroxidation during the influx of oxygen during the post-anoxic period.

CONCLUSIONS

The goal has been to locate a marker associated with flooding which confers increased tolerance of wheat plants to ice encasement. Such a marker has not yet been found, but further analysis of the Adh isozymes may possibly yield a candidate. Also, a closer understanding of the fine control of the glycolytic enzymes in this system may lead to the identification of a valuable selectable marker. It is clear that changes in gene expression are induced by these interacting low temperature stresses. Further genetic analysis may lead to the isolation of flood-regulated genes which may, or may not, be associated with the glycolytic sequence. The acclimative interaction between flooding and ice encasement demonstrated here in winter wheat has given an opportunity to determine limiting components of ice tolerance, and to move towards a directed improvement of this tolerance. Meanwhile, selections under applied ice encasement in the field and in controlled environments have resulted in advances in winter survival and in subsequent grain yields.

ACKNOWLEDGEMENTS

I wish to thank Dr S.I. Warwick of the Centre for Land and Biological Resources Research of Agriculture Canada for providing facilities and guidance for the isozyme analyses.

REFERENCES

Andrews CJ, Pomeroy MK (1977) Changes in survival of winter cereals due to ice cover and other simulated winter conditions. Canadian Journal of Plant Science 57: 1141-1149

Andrews CJ, Pomeroy MK (1979) Toxicity of anaerobic metabolites accumulating in winter wheat seedlings during ice encasement. Plant Physiology 64: 120-125

Andrews CJ, Pomeroy MK (1981) The effect of flooding pretreatment on cold hardiness and survival of winter cereals in ice encasement. Canadian Journal of Plant Science 61: 507-513

Andrews CJ, Pomeroy MK (1983) The influence of flooding pretreatment on metabolic changes in winter cereal seedlings during ice encasement. Canadian Journal of Botany 61: 142-147

Andrews CJ, Pomeroy MK (1989) Metabolic acclimation to hypoxia in winter cereals: low temperature flooding increases adenylates and survival in ice encasement. Plant Physiology 91: 1063-1068

Andrews CJ, Pomeroy MK (1991) Low temperature anaerobiosis in ice encasement damage to winter cereals. In: Jackson MB, Davies DD, Lambers H (eds) Plant life under oxygen deprivation. SPB Academic Publishing, The Hague, The Netherlands, pp 85-89

Bailey-Serres J, Kloeckner-Gruissem B, Freeling M (1988) Genetic and molecular approaches to the study of the anaerobic response and tissue specific gene expression in maize. Plant, Cell and Environment 11: 351-357

Bertani A, Reggiani R (1991) Anaerobic metabolism in rice roots. In: Jackson MB, Davies DD, Lambers H (eds) Plant life under oxygen deprivation. SPB Academic Publishing, The Hague, The Netherlands

Chan JWY, Burton RS (1992) Variation in alcohol dehydrogenase and flood tolerance in white clover, *Trifolium repens*. Evolution 46: 721-734

Duff SME, Moorhead GBG, Lefebvre DD, Plaxton WG (1989) Phosphate starvation inducible bypasses of adenylate and phosphate dependent glycolytic enzymes in *Brassica nigra* suspension cells. Plant Physiology 90: 1275-1278

Fowler DB, Gusta LV, Tyler JJ (1981) Selection for winter hardiness in wheat. III. Screening methods. Crop Science 21: 896-901

Gao J-Y, Andrews CJ, Pomeroy MK (1983) Interactions among flooding, freezing, and ice encasement in winter wheat. Plant Physiology 72: 303-307

Hart GE (1983) Hexaploid wheat (*Triticum aestivum* L. em Thell). In: Tanksley SD, Orton TJ (eds) Isozymes in plant genetics and breeding Part B. Elsevier Scientific Publishers, Amsterdam, The Netherlands, pp 35-56

Hole DJ, Cobb BG, Hole PS, Drew MC (1992) Enhancement of anaerobic respiration in root tips of *Zea mays* following low oxygen (hypoxic) acclimation. Plant Physiology 99: 213-218

Monk LS, Fagerstedt KV, Crawford RMM (1989) Superoxide dismutase as an anaerobic polypeptide - A key factor in recovery from oxygen deprivation in *Iris pseudacorus*. Plant Physiology 85: 1016-1020

Pomeroy MK, Andrews CJ (1985) Effect of low temperature and calcium on survival and membrane properties of isolated winter wheat cells. Plant Physiology 78: 484-488

Rakitina ZG (1965) The permeability of ice for O_2 and CO_2 in connection with the study of the reasons for winter cereal mortality under ice crust. Soviet Plant Physiology 12: 795-803

Rakitina ZG (1977) Effect of a surrounding ice crust on winter wheat plants as a function of their flooding before freezing. Soviet Plant Physiology 254: 317-324

Saglio PH, Drew MC, Pradet A (1988) Metabolic acclimation to anoxia induced by low (2-4 kPa partial pressure) oxygen pretreatment (hypoxia) in root tips of *Zea mays*. Plant Physiology 86: 61-66

Turner JF, Turner DH (1980) The regulation of glycolysis. In: Stumpf PK, Conn EE (eds) The biochemistry of plants 2. Academic Press, New York, pp 279-316

Wendel JF, Weeden NF (1989) Visualization and interpretation of plant isozymes. In: Soltis DE, Soltis PM (eds) Isozymes in plant biology. Advances in plant biology 4. Dioscoroides Press, Portland, Oregon, pp 5-45

METABOLIC AND CELLULAR IMPACT OF ICE ENCASEMENT ON HERBAGE PLANTS

B. E. Gudleifsson
Agricultural Research Institute
Mödruvellir Experimental Station
Oseyri 2
IS-603 Akureyri
ICELAND

ABSTRACT

In northern maritime areas, with high winter precipitation and unstable winter temperatures fluctuating around zero, thaw water and precipitation freeze to form an ice layer covering the ground, which is highly impermeable to gases. In this chapter the literature on the impact of ice encasement on the metabolism and cellular structure of herbaceous plants is reviewed. Plants encased in ice become anaerobic and accumulate metabolites such as carbon dioxide, ethanol, lactate and malate. These substances are potentially injurious to ice encased plants, especially the plasma membrane. The primary site of injury is the ion transport system, while the semipermeable properties of the lipid bilayer are also damaged at higher stress levels.

INTRODUCTION

Research activity on winter stresses has concentrated mostly on freezing injury to plants, indicating that freezing damage is often considered as being more important and having a greater impact on agriculture than other winter stresses. However, in some areas, ice encasement damage is commonly experienced by herbaceous plants. This kind of damage has repeatedly been reported to cause large economic losses and to be a practical problem to local farmers (Brink *et al.*, 1939; Rakitina, 1965; Andersen, 1971; Gudleifsson, 1975, 1989). Ice crusts may form over fields at different times during autumn, winter or spring. Most frequently, ice cover occurs when snow melts and thaw water accumulates on the surface, especially in depressions in the landscape (Gudleifsson, 1975; Andersen, 1976). At the end of the thaw, the water refreezes to form an ice layer which is highly impermeable to gases (Hemmingsen, 1959). These weather conditions are frequently experienced

NATO ASI Series, Vol. I 16
Interacting Stresses on Plants in a Changing Climate
Edited by M. B. Jackson, C. R. Black

in northern maritime areas with high winter precipitation and unstable winter temperatures fluctuating around zero (Gudleifsson and Larsen, 1992). Large or small areas may be covered for different durations with ice of varying thickness, with resulting ice encasement damage to the plants beneath (Jónsson, 1938; Friðriksson, 1954; Andersen, 1963; Woehrle, 1963; Beard, 1964; Gudleifsson, 1971; Årsvoll, 1973; Andrews and Pomeroy, 1975; Hakamata *et al.*, 1978; Andrews *et al.*, 1986). The expected global warming may reduce the probability of ice encasement damage at locations where it is common at present. However, the expected temperature rise may also increase the probability of winter thaws and subsequent ice encasement at locations currently dominated by stable snow cover.

Plants encased in ice are readily killed by freezing temperatures that do not affect the survival of unencased plants. Nevertheless, freezing is not believed to be the main cause of plant death under ice (Gudleifsson and Larsen, 1992). The effects of ice encasement are similar in many respects to those of flooding since both can impose anaerobic stress. Flooding induces the accumulation of metabolites in winter cereals similar to those observed during icing. Although metabolic rates are slowed at the depressed temperatures of ice encasement, the development of anoxia is faster than in flooding, because of the high impermeability of ice to gases and the absence of the continuous leaching of metabolites that occurs during flooding (Andrews, 1977; Pomeroy and Andrews, 1979; McKersie *et al.*, 1982). Ice encasement is therefore a more severe type of stress than flooding.

METABOLIC IMPACT OF ICE ENCASEMENT

Plants encased in ice are exposed to either partial or complete oxygen deprivation (Rakitina, 1970). As a result, their normal pathways of respiration are restricted or blocked and pathways of anaerobic respiration predominate. This results in the production of potentially toxic metabolites and a decrease in the production of metabolic energy (ATP). Early studies indicated that plants encased in ice become depleted of oxygen and accumulate carbon dioxide (Sprague and Graber, 1940,

1943; Freyman and Brink, 1967; Rakitina, 1970) and that this is coupled with reduced survival (Rakitina, 1965). Later it was demonstrated that ethanol (Andrews, 1977; McKersie *et al.*, 1982), lactate (Pomeroy and Andrews, 1978a) and malate (Andrews and Pomeroy, 1983) also accumulate during ice encasement (Fig. 1).

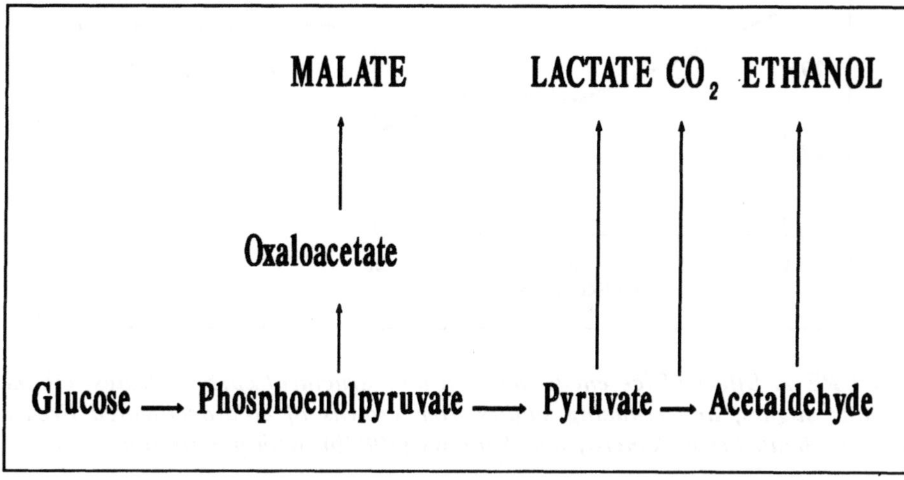

FIGURE 1. Summary diagram illustrating the major end products in anaerobic respiration. Based on Crawford (1978).

During ice encasement, herbage plants utilise storage carbohydrates accumulated during prior hardening (McKersie *et al.*, 1982; Andrews *et al.*, 1984). McKersie *et al.* (1982) and Pomeroy and Andrews (1983b) found that the levels of non-structural carbohydrates and ethanol-soluble carbohydrates declined during ice encasement of winter cereals, while the plant content of reducing sugars increased (Fig. 2). In winter cereals, the increase in reducing sugars results primarily from an accumulation of fructose (Pomeroy and Andrews, 1983b). The loss of plant viability was not associated with depletion of carbohydrate reserves. McKersie *et al.* (1982) and Gao *et al.* (1983) obtained no evidence to indicate that survival, cold hardiness and ice encasement tolerance were directly related to carbohydrate levels.

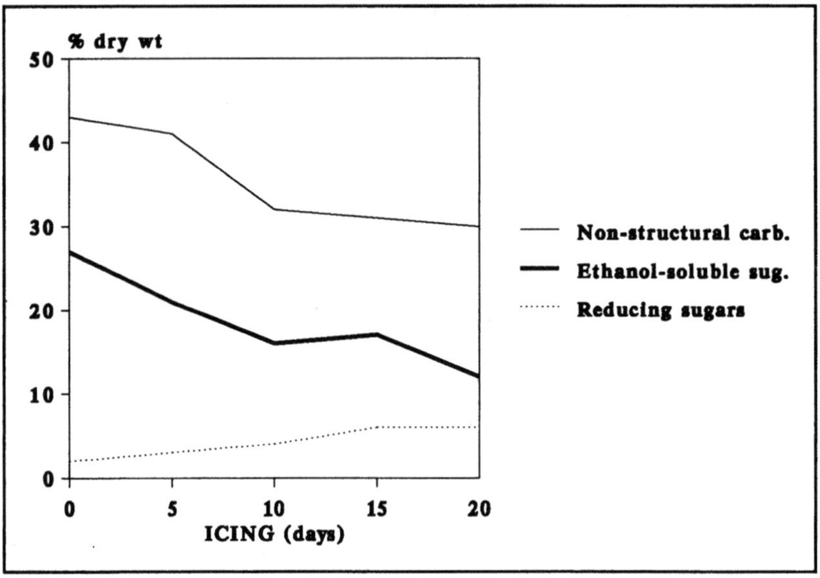

FIGURE 2. Effect of ice encasement on non-structural carbohydrates, ethanol-soluble sugars, and reducing sugars from crowns of cold-acclimated Fredrick winter wheat. From Pomeroy and Andrews (1983b), with permission.

Tissues of plants encased in ice are carbon dioxide-enriched and oxygen-deficient, and Rakitina (1970) demonstrated that carbon dioxide accumulates more rapidly than oxygen is utilised. This was apparent on the first day of ice encasement at -5 °C. After five days, the carbon dioxide content increased by a factor of approximately 100 while oxygen decreased by three to six-fold. Freyman and Brink (1967) demonstrated that flushing the soil of ice encased plants or plants in tubes with carbon dioxide was injurious. Sprague and Graber (1943), working with alfalfa, found a direct relationship between carbon dioxide concentration and the injury sustained. Freyman and Brink (1967) and Rakitina (1970) concluded that injury to plants under an ice crust occurs not as a result of oxygen deficiency but because of the injurious action of carbon dioxide and other products of anaerobic metabolism.

Ice encasement of field plants reduces respiration and increases ethanol levels, although the latter decrease again during the spring thaw (Pomeroy and Andrews, 1978a). Andrews (1977) measured tissue and leachate ethanol after ice encasement of winter cereals. Concentrations expressed on a fresh weight basis were higher in crowns than in roots or leaves and at LD_{50} the ethanol concentration in crowns was highest in the hardiest cultivars.

Non-iced field plants did not accumulate ethanol but accumulated lactate (Fig. 3). The lactate concentration in field plants increased in early autumn, prior to the occurrence of significant environmental stress, indicating that lactate accumulation is not associated with stress damage (Pomeroy and Andrews, 1978a). Lactate does not accumulate to sufficiently high levels in ice to account for plant death (Andrews and Pomeroy, 1979).

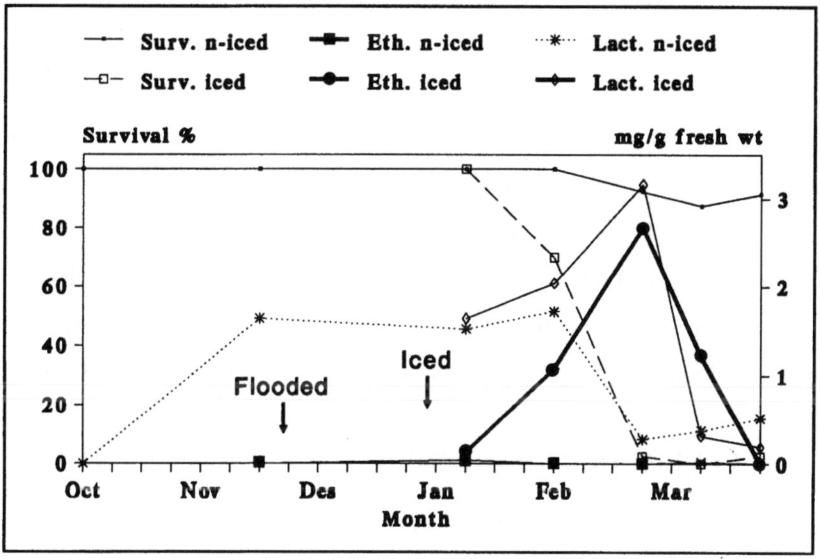

FIGURE 3. Effect of flooding and ice encasement under field conditions on survival and ethanol and lactic acid levels in the crowns of Fredrick winter wheat. Redrawn from Pomeroy and Andrews (1983a); data from Pomeroy and Andrews (1978a), with permission.

Andrews and Pomeroy (1983) found some malate accumulation in ice-encased winter cereal plants. This malate did not substitute for ethanol and sterilisation of plants had no effect, indicating that the malate was not of bacterial origin. On the other hand, McKersie *et al.* (1982) demonstrated that winter cereal crowns deplete malate under icing condition. Gudleifsson (1986) demonstrated that timothy plants produce carbon dioxide, ethanol, lactate and malate during ice encasement. Only malate accumulation was higher in timothy than in winter cereals.

Comparison of the balance between carbohydrate depletion and metabolite accumulation (ethanol, carbon dioxide, lactate) did not indicate accumulation of any major unidentified metabolite in winter wheat (Pomeroy and Andrews, 1983b). On the other hand, McKersie *et al.* (1982) calculated the balance between carbohydrates broken down and metabolites produced and concluded that the hardiest winter wheat cultivars must produce unidentified metabolites in addition to carbon dioxide and ethanol. Lactate might be one of them, and based on work with flood tolerant plants, others might include succinate, σ-aminobutyrate and alanine (Crawford, 1978; Zemlianukhin and Ivanov, 1978). Accumulation of these products in herbaceous plants during ice encasement has not been examined. In some cases, a strong odour emitted by grasslands has been observed immediately after the thaw of ice (Woehrle, 1963; Gudleifsson, 1977; Gudleifsson *et al.*, 1986). This odour might originate from some of the substances produced by plants or soil when anaerobic under the ice sheet. Butyric acid has been isolated from ice-encased timothy plants, but this is probably of bacterial origin (Gudleifsson, 1986).

CELLULAR DAMAGE DURING ICE ENCASEMENT

Plants killed by freezing lose their structure, and their tissues become flaccid, discoloured and water-soaked (Palta *et al.*, 1978; Steponkus, 1984). Microscopic studies during and after cooling to lethal, subfreezing temperatures show cell collapse and destruction (Pomeroy and Andrews, 1978b; Levitt, 1980; Singh and Andrews, 1981). On the other hand, plants killed by ice-encasement often appear healthy when the ice thaws (Gudleifsson, 1979, 1986) but gradually wilt and die during regeneration (Tanino and McKersie, 1985).

Tanino and McKersie (1985) examined tissue viability in wheat crowns and demonstrated that most crown cells appeared to survive icing stress that was lethal to the rest of the plant. After lethal icing stress, the crown was unable to regrow and cell viability, measured as the ability to reduce triphenyltetrazolium chloride (TTC), was gradually lost during the regrowth period. Thus, in winter cereals, no specific regions within the crown were more susceptible to injury than others (Tanino and McKersie, 1985), but in timothy (Gudleifsson and Larsen, 1992) the extreme apex of the crown is the first part to be killed during ice encasement, while the basal region survives for some time after the plant has lost its ability to regenerate. The last part of the crown tissue to die is the intermediate region. The death of the apex could explain why plants fail to regrow after ice encasement, even though most crown cells survive.

Singh and Andrews (1981) observed no major structural changes in winter wheat cells during lethal ice encasement at -1 °C, while during lethal freezing at colder temperatures cell collapse and ultrastructural dissolution of membranes were observed. This difference between freezing and ice encasement damage demonstrates that cell death during ice encasement does not result from disruptive mechanical and dehydrative stresses produced by ice formation during freezing (Pomeroy and Andrews, 1978b; Singh and Andrews, 1981). After ice encasement of field grown winter wheat plants and plants grown in controlled environments, Pomeroy and Andrews (1978a,b) and Andrews and Pomeroy (1979) observed characteristic ultrastructural cellular changes during the early stages of ice encasement. The changes included proliferation of the ER membrane system, which often resulted in formation of concentric whorls of parallel membranes. They also observed membrane-bound electron-dense areas within the cytoplasm which appeared to be associated with the expanded endoplasmic reticulum. The structural integrity of many cellular organelles was relatively unaffected by ice encasement, but after seven days of ice encasement of plants grown in controlled environments disorganisation of internal mitochondrial membranes was observed and after 14 days, when the plants were killed, separation of the inner and outer membranes of the nuclear envelope and formation of large pores were observed. These ultrastructural changes disappeared rapidly in surviving plants during recovery (Pomeroy and Andrews, 1977).

By using a freeze-fixation technique, Singh and Andrews (1981) were able to study cells during ice encasement. They observed neither dramatic disruption of the protoplasm nor destruction of membrane ultrastructure during the stress. During longer periods of ice encasement, numerous vesicles appeared in the cytoplasm, but there was no reorganisation of the endoplasmic membranes to form concentric rings. The changes observed by Pomeroy and Andrews (1978a,b) and Andrews and Pomeroy (1979) therefore seem to have occurred during the post-thaw period. Singh and Andrews (1981) concluded that neither mechanical effects of encasing ice nor dehydration are factors in lethal injury during ice encasement. This indicates that death caused by ice encasement is not a result of mechanical stress but rather a result of biochemical stress, related to an accumulation of metabolites.

To test the toxicity of the metabolites accumulated during anaerobic respiration, Andrews (1977) and Andrews and Pomeroy (1977b) exposed winter cereal plants to various concentrations of metabolites at -1 °C. Plant survival was reduced by separate exposure to ethanol, lactate and carbon dioxide, and when all three were supplied together, plants were killed at rather low concentrations. Comparison was made between the concentrations needed to kill plants when administered externally, and the endogenous concentrations measured in iced plants. Winter cereal crowns tolerated about three or four times higher external ethanol and lactate concentrations than those produced in ice, indicating that these metabolites, although potentially toxic, are not the only damaging factors in ice encasement. On the other hand, the crowns tolerated much lower concentrations of carbon dioxide during experimental exposure than when encased in ice (Table 1). This indicates a more significant role for carbon dioxide in ice encasement damage than for ethanol or lactate (Andrews and Pomeroy, 1979, 1990).

Exposure to external lactate alone (0.5%) had little effect on cell ultra-structure. Ethanol (10%) caused proliferation of membrane whorls and nuclear rearrangement. Carbon dioxide (100%) did not induce proliferation of membranes but induced a condensed form of the nucleus (Andrews and Pomeroy, 1979). Exposure to carbon dioxide and ethanol simultaneously resulted in a proliferation of membranes and nuclear condensation similar to that seen in ice-encased plants.

Membrane permeability was also increased markedly. It was concluded that cell damage caused by the combined effects of accumulated ethanol and carbon dioxide occurs at the membrane level and changes permeability and destroys membrane functions.

TABLE 1. *Comparison of endogenous metabolic concentrations at the 50% kill point in cold-hardened seedlings of Fredrick winter wheat induced by ice encasement and the concentrations tolerated by seedlings exposed to the exogenous metabolites. From Andrews and Pomeroy (1979), with permission.*

Metabolite	LD_{50} ice exposure (mg g^{-1} fresh wt.)	LD_{50} metabolite exposure (mg g^{-1} fresh wt.)
Ethanol	2.94	14.01
Lactate	0.67	2.17
CO_2	2.45	1.24

Hetherington *et al.* (1987, 1988) studied the impact of ice encasement on microsomal membranes isolated from the crowns of winter wheat plants and found that the protein and lipid components were both profoundly altered. During early icing there were marked qualitative alterations in the composition of the membrane proteins, with the loss of several distinct polypeptides and the appearance of a number of new ones. After longer durations of icing, plant mortality was associated with an increased leakage of cellular electrolytes from the crown, suggesting membrane injury. Membrane physical properties and chemical composition were adversely affected and degradation was evident. This injury included a loss of membrane proteins and de-esterification of phospholipids, with the loss of the polar head group from the membrane. Free fatty acids, another de-esterification product, remained in the bilayer, reducing membrane fluidity and introducing a physical phase separation. Further injury, including lipid peroxidation, occurred during a post-thaw period when crowns were removed from ice encasement. Lipolytic enzymes may be responsible for this, while free radicals possibly perturbate enzymes involved in electron transfer reactions.

Gao *et al.* (1983), working with winter wheat subjected to various low temperature stresses, obtained indications that stress-induced metabolic changes occur before the applied stress was severe enough to reduce survival. By applying ice encasement or ethanol treatments to isolated winter wheat cells, Pomeroy *et al.* (1983) demonstrated that the passive efflux of amino acids increased gradually but the uptake of ^{86}Rb, an analogue of potassium, declined much more rapidly. Electron-spin resonance studies revealed no major changes in molecular ordering within the cell membranes following these treatments. Hetherington *et al.* (1988) demonstrated that membrane damage, detected as an increase in microviscosity and electrolyte leakage or a loss of semipermeability, began at an early stage of ice encasement. On the other hand, Pomeroy *et al.* (1983) and Pomeroy and Andrews (1985), working with cell suspensions at -1 °C, demonstrated that ice encasement resulted in a decline in ion uptake before any increase in ion efflux or decrease in cell viability could be detected. Thus, during ice encasement, the decline in membrane transport capacity approximates to the decline in survival of intact plants or crowns, while reduction in cell viability and ion efflux occur at higher stress levels. This demonstrates that the ion transport system of these cells is the primary site of injury due to ice encasement and that damage to the ion pump system precedes damage to the bilayer.

Theoretically, the ATPase enzymes of the plasma membrane could be rapidly inhibited by metabolites accumulating within tissues in the early stages of ice encasement. Working with mitochondria, Andrews and Pomeroy (1977a) found a slow decline in respiratory properties during ice encasement and the activity was little impaired even when 50% of the plants were killed. Tanino and McKersie (1985) also observed that mitochondrial activity, measured as TTC reduction, was high after ice encasement lethal to plants, but was gradually reduced during regeneration. Although oxygen consumption potential is high in ice-encased plants, the ATP supply could be sufficiently low to inhibit the activity of energy-dependent ATPase membrane ion pumps, thus explaining damage to the ion transport system. During ice encasement there is a general decline in adenylate energy charge (AEC) and total adenylates (Andrews and Pomeroy, 1989b).

Pomeroy and Andrews (1986) noted that the levels of the three nucleotides (ATP, ADP, AMP) decreased gradually in approximate relation to a decline in cell viability, but that AEC remained high even when cell viability was severely reduced (Andrews and Pomeroy, 1989b). Therefore, the reduction in ion uptake in ice is due neither to decreased levels of ATP nor to a decline in the AEC system of the cell, as is the case for anoxia. Furthermore, it cannot be attributed to a shortage of carbohydrate energy substrates (Gao *et al.*, 1983), since fermentation processes provide sufficient substrate level ATP to maintain a high energy charge and synthetic functions in surviving cells (Andrews and Pomeroy, 1990). Inhibition of enzyme activity or denaturation by rapidly accumulating anaerobic products remains as a possibility.

TABLE 2. *Survival (%) of isolated cells of winter wheat in various treatments at -1 °C. Gas and ice treatments in 30 mM Tris Mes buffer, pH 6.5, Ca^{++} at 10 mM. Bicarbonate and carbonate treatments in 100 mM Tris Mes buffer. Survival is expressed as a percentage of freshly isolated cells. From Andrews and Pomeroy (1989a), with permission.*

Treatment	7 days	14 days
Air	93	87
Air + Ca	92	85
N_2	86	74
N_2 + Ca	86	77
CO_2	61	21
CO_2 + Ca	72	42
Iced	81	60
Iced + Ca	92	83
$KHCO_3$ 100 mM	77	51
$KHCO_3$ 500 mM	58	16
K_2CO_3 100 mM	68	27
K_2CO_3 500 mM	0	0

The damaging effect of carbon dioxide is supported by the fact that carbonate and bicarbonate ions reduce cell survival and promote amino acid leakage from cells more rapidly than anoxia alone (Table 2). The membrane ATPase activity is strongly inhibited by carbonate and bicarbonate ions, but much less so by ethanol and lactic acid (Andrews and Pomeroy, 1989a, 1990). This indicates that carbon dioxide might be the main cause of damage to membrane ion pumps during ice encasement stress. A 10 mM suspension of Ca^{++} protects isolated cells against damage (Table 2) and also reverses the inhibitory effect of ice exposure on the ^{86}Rb uptake system (Pomeroy and Andrews, 1985; Pomeroy and Andrews, 1986), while other cations such as Li^{++} and Mg^{++} do not. Calcium seems to have a stabilising effect on the plasma membranes (Andrews and Pomeroy, 1989a).

CONCLUSIONS

These results indicate that the plasma membrane is an early target and a primary site of injury due to ice encasement, and that damage to the ion transport system is the earliest manifestation of this injury. This happens prior to injury to the semipermeable properties of the cell and precedes significant losses of cell viability or changes in other physiological or metabolic properties of the cell, and is partially reversible by calcium.

REFERENCES

Andersen IL (1963) Investigation on the wintering of meadow plants in Northern Norway II. Some investigations on damages caused by ice and water choking on meadows (Overvintringsundersökelser i eng i Nord-Norge. II. Noen undersökelser over is- og vannskader i eng). Forskning og forsök i landbruket 14: 639-669

Andersen IL (1971) Experiments on winter damages on different grass species (Overvintringsforsök med ulike grasarter). Forskning og forsök i landbruket 22: 121-134

Andersen IL (1976) Winter damages on grasslands (Overvintringsskader på eng). Aktuelt fra Landbruksdepartementets opplysningstjeneste Nr. 2 1976: 118-126

Andrews CJ (1977) Accumulation of ethanol in ice-encased winter cereals. Crop Science 17: 157-161

Andrews CJ, Pomeroy MK (1975) Survival and cold hardiness of winter wheats during partial and total ice immersion. Crop Science 15: 561-566

Andrews CJ, Pomeroy MK (1977a) Mitochondrial activity and ethanol accumulation in ice-encased winter cereal seedlings. Plant Physiology 59: 1174-1178

Andrews CJ, Pomeroy MK (1977b) The effect of anaerobic metabolites on survival and ultrastructure of winter wheat in relation to ice encasement. Plant Physiology 61 Supplement 17

Andrews CJ, Pomeroy MK (1979) Toxicity of anaerobic metabolites accumulating in winter wheat seedlings during ice encasement. Plant Physiology 64: 120-125

Andrews CJ, Pomeroy MK (1983) The influence of flooding pre-treatment on metabolic changes in winter cereal seedlings during ice encasement. Canadian Journal of Botany 61: 142-147

Andrews CJ, Pomeroy MK (1989a) Ice-encasement injury at the cellular and membrane level. Icelandic Agricultural Sciences 2: 57-61

Andrews CJ, Pomeroy MK (1989b) Metabolic acclimation to hypoxia in winter cereals. Low temperature flooding increases adenylates and survival in ice encasement. Plant Physiology 91: 1063-1068

Andrews CJ, Pomeroy MK (1990) Low temperature anaerobiosis in ice encasement damage to winter cereals. In: Jackson MB, Davies DD, Lambers H (eds) Plant life under oxygen deprivation. SPS Academic Publishing, The Hague, The Netherlands, pp 85-99

Andrews CJ, Pomeroy MK, Seaman WL (1986) The response of fall-sown cereals to winter stress in eastern Ontario. Canadian Journal of Plant Science 66: 25-37

Andrews CJ, Seaman WL, Pomeroy MK (1984) Changes in cold hardiness, ice tolerance and total carbohydrates of winter wheat under various cutting regimes. Canadian Journal of Plant Science 64: 547-558

Årsvoll K (1973) Winter damage in Norwegian grasslands, 1968-1971. Meldinger fra Norges landbrukshögskole 52, pp 21

Beard JB (1964) Effects of ice, snow and water covers on Kentucky bluegrass, annual bluegrass and creeping bentgrass. Crop Science 4: 638-640

Brink RA, Keller W, Eisenhart C (1939) Differential survival of alfalfa strains under an ice sheet. Journal of Agricultural Research 59: 59-71

Crawford RMM (1978) Metabolic adaptation to anoxia. In: Hook DD, Crawford RMM (eds) Plant life in anaerobic environments. Ann Arbor Science, Ann Arbor, USA, pp 119-136

Freyman S, Brink VC (1967) Nature of ice-sheet injury to alfalfa. Agronomy Journal 59: 557-560

Fridriksson S (1954) Investigations on winter injury of plants in Icelandic hayfields 1951-1952 (Rannsóknir á kali túna árin 1951 og 1952). Department of Agriculture, Report Series B No 7, 72 pp

Gao J-Y, Andrews CJ, Pomeroy MK (1983) Interactions among flooding, freezing, and ice encasement in winter wheat. Plant Physiology 72: 303-307

Gudleifsson BE (1971) Winter damages on grasslands in Iceland, extent and causes (Overvintringsskadar i grasmark på Island, omfang og årsaker). Lisensiatoppgave, Norges landbrukshögskole, pp 130
Gudleifsson BE (1975) Winter damages in grasslands (Over vintringsskadar i grasmark). Nordisk Jordbruksforskning 58: 498-504
Gudleifsson BE (1977) Ice encasement damages and smell from damaged fields (Svellkal og kallykt). Ársrit Ræktunarfélags Norðurlands 74: 70-76
Gudleifsson BE (1979) Descriptions of grass diseases No 11. Ice and water damage. Weibulls Gräs-tips 22
Gudleifsson BE (1986) Ice encasement damages on grasses and winter cereals. NJF seminar Nr. 84 Lantbruksväxternas övervintring 59-65
Gudleifsson BE (1989) Extent and importance of ice-encasement damages on graminaceous plants in the Nordic countries. Icelandic Agricultural Sciences 2: 7-14
Gudleifsson BE, Andrews CJ, Björnsson H (1986) Cold hardiness and ice tolerance of pasture grasses grown and tested in controlled environments. Canadian Journal of Plant Science 66: 601-608
Gudleifsson BE, Larsen A (1992). Ice encasement as a component of winterkill of herbage plants. In: Li PH, Christersson L (eds) Advances in plant cold hardiness. CRC Press, Boca Raton, Ann Arbor, London, Tokyo, pp 229-249
Hakamata T, Noshiro M, Hirashima T, Nose I (1978) Investigation of actual condition on the winter killing of pasture species in the Nemuro-Kushiro district - Exploration of factors by the quantification No 1. Journal of Japanese Grassland Science 23: 280-288
Hemmingsen E (1959) Permeation of gases through ice. Tellus 11: 355-359
Hetherington PR, McKersie BD, Borochov A (1987) Ice encasement injury to microsomal membranes from winter wheat crowns. I. Comparison of membrane properties after lethal ice encasement and during a post-thaw period. Plant Physiology 85: 1068-1072
Hetherington PR, Broughton HL, McKersie BD (1988) Ice-encasement injury to microsomal membranes isolated from winter wheat crowns. II. Changes in membrane lipids during ice encasement. Plant Physiology 86: 740-743
Jónsson Ó (1938) Kal. Ársrit Ræktunarfélags Norðurlands 34: 19-47
Levitt J (1980) Responses of plants to environmental stresses. I. Chilling, freezing, and high temperature stresses. Academic Press, New York, pp 497
McKersie BD, McDermott BM, Hunt LA, Poysa V (1982) Changes in carbohydrate levels during ice encasement and flooding of winter cereals. Canadian Journal of Botany 60: 1822-1826
Palta JP, Levitt J, Stadelmann EJ (1978) Plant viability assay. Cryobiology 15: 249-255
Pomeroy MK, Andrews CJ (1978a) Metabolic and ultrastructural changes in winter wheat during ice encasement under field conditions. Plant Physiology 61: 806-811
Pomeroy MK, Andrews CJ (1978b) Ultrastructural changes in shoot apex cells of winter wheat seedlings during ice encasement. Canadian Journal of Botany 56: 786-794

Pomeroy MK, Andrews CJ (1979) Metabolic and ultrastructural changes associated with flooding at low temperature in winter wheat and barley. Plant Physiology 64: 635-639

Pomeroy MK, Andrews CJ (1983a) Flooding and ice encasement damage to winter wheat. In: Fowler DB, Gusta LV, Slinkard AE, Hobin BA (eds) New frontiers in winter wheat production. Western Canada Winter Wheat Conference 1983, pp 39-56

Pomeroy MK, Andrews CJ (1983b) Responses of winter cereals to various low temperature stresses. In: Randall DD, Blewins DG, Larson RL, Rapp BJ (eds) Current topics in plant biochemistry and physiology, Vol 2. University of Missouri, Columbia, USA, pp 96-106

Pomeroy MK, Andrews CJ (1985) Effect of low temperature and calcium on survival and membrane properties of isolated winter wheat cells. Plant Physiology 78: 484-488

Pomeroy MK, Andrews CJ (1986) Changes in adenine nucleotides and energy charge in isolated winter wheat cells during low temperature stress. Plant Physiology 81: 361-366

Pomeroy MK, Pihakaski SJ, Andrews CJ (1983) Membrane properties of isolated winter wheat cells in relation to icing stress. Plant Physiology 72: 535-539

Rakitina ZG (1965) The permeability of ice for O_2 and CO_2 in connection with a study of the reasons for winter cereal mortality under the ice crust. Soviet Plant Physiology 12: 795-803

Rakitina ZG (1970) Effect of an ice crust on gas composition of the internal atmosphere in winter wheat. Soviet Plant Physiology 17: 755-759

Singh J, Andrews CJ (1981) Comparisons of ultrastructural changes during extracellular freezing at -10 °C and ice-encasement at -1 °C in winter wheat crown cells by the method of freeze-fixation. Cryo-Letters 2: 117-124

Sprague VG, Graber LF (1940) Physiological factors operative in ice-sheet injury of alfalfa. Plant Physiology 15: 661-673

Sprague MA, Graber LF (1943) Ice sheet injury to alfalfa. Journal of American Society of Agronomy 35: 881-894

Steponkus PL (1984) Role of the plasma membrane in freezing injury and cold acclimation. Annual Review of Plant Physiology 35: 543- 584

Tanino KK, McKersie BD (1985) Injury within the crown of winter wheat seedlings after freezing and icing stress. Canadian Journal of Botany 63: 432-436

Woehrle T (1963) Ice sheet damage. The golf course reporter 31: 22-24

Zemlianukhin AA, Ivanov BF (1978) Metabolism of organic acids of plants in the condition of hypoxia. In: Hook DD, Crawford RMM (eds) Plant life in anaerobic environments. Ann Arbor Science, Ann Arbor, USA, pp 203-268

RESPONSES OF NODULATED LEGUMES TO OXYGEN DEFICIENCY

C. Arrese-Igor, M. Royuela and P.M. Aparicio-Tejo
Departamento de Producción Agraria
Universidad Pública de Navarra
Campus de Arrosadía
E-31106-Pamplona
Spain

ABSTRACT

Legumes are intolerant of flooding and symbiotic nitrogen fixation is particularly sensitive to waterlogging. Morphologically, the most obvious features of nodules subjected to flooding are the enhancement of lenticel-like structures on their surfaces and increased nodule cortication. These anatomical features increase the supply of oxygen to the bacteroid-containing tissue. Nevertheless, nitrogen fixation is severely impaired under waterlogged conditions. Exposure to subambient oxygen concentrations decreases the nitrogenase activity of detached root nodules. However, some recovery may occur in intact nodulated legumes after exposure to short-term moderate anoxygenic conditions. Tolerance of long-term moderate oxygen deficiency by grain legumes with determinate nodules is achieved by several structural adaptive mechanisms (e.g., the nodule surface is covered with loosely packed cells or the nodules have an increased number of unoccluded intercellular spaces, enhancing gaseous permeability) and biochemical adaptations (increased expression of ureide-forming enzymes and leghemoglobin content). Nevertheless, the tolerance mechanisms displayed by these plants under conditions of moderate hypoxia fail to provide an adequate environment for nitrogen fixation under severe hypoxia. This adaptation involves changes in photosynthate allocation to underground organs, aerenchyma formation in both roots and nodules to help maintain high rates of respiration (and hence mineral uptake), a less compact cortex with an increased nodule gas permeability, increased recycling of CO_2, a less demanding nitrogenase activity and a more efficient nodule structure.

INTRODUCTION

Oxygen permeability is necessary for normal plant development. Direct effects of soil oxygen on plants have been reviewed elsewhere (Glinski and Stepniewski, 1985; Drew and Stolzy, 1991). These effects also apply to nodulated legumes, but in this particular case some singular features should be taken into account. In legume root nodules there is a paradox between oxygen exclusion to avoid

NATO ASI Series, Vol. I 16
Interacting Stresses on Plants in a Changing Climate
Edited by M. B. Jackson, C. R. Black

nitrogenase damage and the high oxygen demand from the highly active respiring bacteroids. The equilibrium between both processes is achieved by means of two different mechanisms: a fine control of the gas permeability, the so-called oxygen diffusion barrier, located in the inner cortex, and the occurrence of leghemoglobin, which allows a rapid oxygen flux despite a very low oxygen concentration (~20 nM) in the infected region.

Recent research regarding oxygen and nodule functioning has focussed on supra-ambient oxygen concentrations. However, these conditions are unlikely to occur in soils, whilst oxygen shortage can be brought about by a range of natural circumstances. This chapter reviews the available information on nodule performance under hypoxia as well as recent progress in this field.

EFFECTS OF WATERLOGGING AND SHORT-TERM OXYGEN DEFICIENCY ON NITROGEN FIXATION

Waterlogging is the most common naturally-occurring circumstance causing oxygen shortage in soils. In general, legumes are intolerant of flooding (Justin and Armstrong, 1987), although there are marked differences between legumes (Heinrichs, 1970). There is some evidence in the literature that symbiotic nitrogen fixation is particularly sensitive to waterlogging (Minchin and Pate, 1975 and references therein).

The effect of waterlogging varies with the duration of flooding and plant age: the earlier the stage of development, the greater the effect (Minchin *et al.*, 1978). Morphologically, the most obvious feature of nodules subjected to flooding is the enhancement of lenticel-like structures on the nodule surface of soybean (Pankhurst and Sprent, 1975) or increased nodule cortication if these structures are absent or very reduced, as in cowpea (Minchin and Summerfield, 1976). These anatomical features are thought to be involved in increasing the supply of oxygen to the bacteroid-containing tissue. Indeed, it has been reported that nodules displaying these anatomical adaptations have a prolonged bacteroid survival (Sprent, 1971; Minchin and Summerfield, 1976) as compared to legumes not having these features, such as pea (Minchin and Pate, 1975). Nevertheless, despite these structural adaptations, nitrogen fixation is severely impaired under waterlogged conditions expressed either on a plant basis or a nodule dry weight basis (Minchin and Summerfield, 1976), reducing both vegetative growth and seed yield (Minchin *et al.*, 1978).

However, waterlogging is more than simply anaerobiosis, as removal of toxic products may be impaired and metabolically generated gases can be entrapped because of their lower diffusibility in water (Kawase, 1981). An early study showed that exposure to subambient oxygen concentrations decreased the nitrogenase activity of detached root nodules (Bergersen, 1962). In short-term studies using intact plants, acetylene reduction was also rapidly suppressed in soybean and french bean subjected to 1 kPa oxygen (Wasfi and Prioul, 1986), although no experiments were performed to assess the possibility of recovery.

However, Criswell *et al.* (1976) described for the first time experimental evidence for adaptation of nodulated legumes to anoxygenic conditions in short-term studies (up to four days). They showed that, despite a dramatic decline in nitrogenase activity in plants subjected to 6 kPa oxygen within the first four hours, there was complete recovery within 24 hours. This recovery was stable for at least 96 hours. When plants were subjected to 2 kPa oxygen, they only partially recovered their nitrogenase activity, although a trend to recover the original activity could be observed after four days. Similar conclusions were reached by Weisz and Sinclair (1987). These observations have considerable agricultural significance but, surprisingly, until very recently no attempt has been made to verify these observations in the long term and ascertain the mechanisms by which this adaptation is achieved. On the other hand, this result could also be criticised in terms of the analytical methods used to estimate nitrogenase activity. The widely used standard acetylene reduction assay has received severe criticism, as it has been shown that addition of acetylene causes a rapid decline in nitrogenase activity (Minchin *et al.*, 1983, 1986). Therefore the abovementioned results could reflect shortcomings in the methods rather than a true adaptive response.

TOLERANCE OF LONG-TERM MODERATE OXYGEN DEFICIENCY

In this context, some significant contributions have been published recently dealing with adaptation by nodulated legumes to long-term microaerobiosis. Interestingly, soybeans subjected to 10 kPa oxygen for one month did not show any significant difference in vegetative yield (James *et al.*, 1991). Even exposure to 4.7 kPa oxygen for 15 days had no effect on soybean growth parameters (Parsons and Day, 1990). Unfortunately, no data are available on grain yield in either study.

Tolerance of moderate levels of hypoxia involved several structural adaptation mechanisms. The nodule surface in low oxygen-grown plants was covered with loosely packed cells (Parsons and Day, 1990), probably representing an enhanced expression of lenticel development as previously described for nodules subjected to waterlogging (Pankhurst and Sprent, 1975). Transverse sections of the same nodules also showed an increased number of unoccluded intercellular spaces, increasing the area of intercellular spaces and allowing more efficient transport of oxygen into the bacteroid-containing tissue. Occlusion in normal air-grown nodules is provided by the presence of a glycoprotein, which can be recognised by a monoclonal antibody to a carbohydrate epitope of a glycoprotein found in the infection thread matrix. The presence of this glycoprotein is significantly reduced in hypoxic nodules (James *et al.*, 1991). From the lack of significant differences in growth parameters, both authors concluded that these structural adaptations are sufficient to compensate for the difference in external oxygen supply, enhancing the gaseous diffusive properties of these nodules so that internal oxygen concentration may not be significantly affected. No significant changes in leghemoglobin or bacteroid content, nor in other physiological parameters were reported.

Atkins and co-workers (Dakora and Atkins, 1990a, b, c; Atkins *et al.*, 1990; Dakora *et al.*, 1991) produced a series of papers dealing with the performance of the tropical grain legume, cowpea, under different external oxygen concentrations. Cowpea plants grown under 10 kPa behaved as normal air-grown plants, whilst plants subjected to 5 kPa had a slightly poorer performance compared to control plants. This was more evident during the first stages of nodule development, but thereafter they showed a trend towards recovery, yielding similar amounts of nitrogen fixed as controls. These results were further confirmed by measurements of acetylene reduction (Dakora and Atkins, 1990a). Nodules showed similar morphological adaptations to those mentioned above for soybean (Dakora and Atkins, 1990c). This strong performance during moderate oxygen deficiency is likely to have occurred because of an increased gaseous permeability within these nodules (Dakora and Atkins, 1990b). This result also provides experimental support for the theoretical proposals of Parsons and Day (1990) and James *et al.* (1991). However, these results from Atkins' group also suggest more subtle changes than previously envisaged. Thus, a lower pO_2 may also induce biochemical adaptations. In this way, hypoxic nodules may come to contain more ureide-forming enzymes, allowing a fairly stable ureide composition in the xylem sap (Atkins *et al.*, 1990). Also, these nodules have an increased leghemoglobin content when expressed either on an infected cell or on a bacteroid basis (Dakora *et al.*, 1991).

Overall, these observations strongly suggest that during moderate oxygen deficiency, determinate nodules of legumes show a full set of morphological and biochemical adaptations that provide an adequate supply of oxygen to the bacteroids.

TOLERANCE TO LONG-TERM SEVERE OXYGEN DEFICIENCY

Despite these promising results during moderate oxygen deficiency, little is known concerning conditions of severe hypoxia. However, in natural circumstances such as soil flooding, the latter is more likely to occur (Glinski and Stepniewski, 1985). Dakora and Atkins (1990a) showed clearly that the adaptation mechanisms displayed by cowpea plants during moderate hypoxia fail to provide an adequate environment for nitrogen fixation below 5 kPa pO_2, and consequently that plant growth was dramatically impaired. In contrast, nodulated lucerne plants subjected to 1 kPa oxygen did not show any significant decrease in shoot dry weight (Arrese-Igor and Aparicio-Tejo, 1990). This occurred despite a dramatic reduction in both nodule and root dry weight. This ability to maintain shoot growth seems to be related to a smaller assimilate allocation to the roots, resulting in a substantial readjustment of the shoot/root ratio. This fact is relevant because of the agronomic use of this species as a forage crop.

The underground organs of oxygen-deprived lucerne were observed to have a 'hairy' look because of the occurrence of outgrowths that appear under the light microscope as patches of loosely packed cells. The presence of loosely packed parenchyma cells in nodules following waterlogging (Pankhurst and Sprent, 1975) or exposure to low oxygen pressure (Parsons and Day, 1990) has also been described in determinate nodules. However, these structures have been related to lenticels, but lenticels do not occur in lucerne nodules (Corby *et al.*, 1983). In fact, these loosely packed cells present similarities to the aerenchyma described in some species following waterlogging (Kawase, 1981). Aerenchyma enhances oxygen diffusion (Armstrong, 1971) and increases root energetic charge (Drew *et al.*, 1985). The abundance of aerenchyma in roots from plants grown under 1 kPa pO_2 might have a detrimental effect on root strength. This possibility is met by an increase in schlerenchyma fibres in the cortex which presumably provide increased mechanical strength.

Mineral uptake did not show any significant reduction when external oxygen supply was decreased. Also, root respiration was virtually unaffected by growth conditions. The insensitivity of root respiration to external oxygen concentration agrees with previous short-term studies (Wasfi and Prioul, 1986). Our results (Arrese-Igor, 1990) could also support the suggestion that the reported critical oxygen concentration based on excised root measurements may be greatly overestimated because of artifacts following excision, such as flooding of the intercellular air spaces (Armstrong and Gaynard, 1976).

Nitrogenase activity, measured by the acetylene reduction assay, showed no difference on a nodule dry weight basis between nodules which had developed under 1 kPa oxygen and control nodules. This result was further supported by measurements of plant organic nitrogen content. This suggests that internal oxygen concentrations were similar to those in control nodules. However, hypoxic nodules showed an increased heme synthesis on a bacteroid basis and an enhanced dissimilatory nitrate reductase activity. Thus, internal oxygen concentration was indeed lower in these nodules.

Apparent CO_2 evolution was five times higher in normal nodules than in nodules under 1 kPa oxygen. Conversely, when nodule respiration was monitored as oxygen uptake, we failed to detect any significant difference. This could be due to the recycling of CO_2 because phosphoenolpyruvate carboxylase is more efficient at low oxygen concentrations. It has been reported that PEPC activity can recycle up to 25% of gross respiration (Anderson *et al.*, 1987).

Microaerobic-grown nodules showed a less compact cortex, with an increase in the unoccluded intercellular space from 7.6 to 48.1%. This was correlated with an increased gas permeability, as shown by their increased sensitivity to repeated acetylene reduction assays and because they reached steady-state respiration rates more rapidly than control nodules. The protein content of nodules was differentially affected by oxygen availability; the plant fraction was largely insensitive, whilst the bacteroid protein decreased by 50% under 1 kPa oxygen. It has been shown that, in determinate nodules such as soybean, degradative effects related to senescence markedly affect cytosol protein content (Pfeiffer *et al.*, 1983). However, in determinate nodules, bacteroids seemed to be more sensitive to stress (Arrese-Igor *et al.*, 1990), although because of the meristematic properties of these nodules, the generation of newly infected tissue is possible upon recovery. This decrease in bacteroid protein is not due to diminished bacteroid size, but to the occurrence of a central zone of infected tissue depleted of bacteroids. Therefore, specific nitrogenase activity on a bacteroid basis is even higher in 1 kPa oxygen, based both on the acetylene reduction assay and on measurements of nitrogen content, compared to normal air-grown nodules. The altered morphology of these 1 kPa oxygen-grown nodules with an empty core seems to provide a more efficient environment for nitrogenase activity and suggests that in normal nodules the performance of different infected cell layers could vary considerably due to oxygen gradients. This observation that a bacteroid-depleted core in the infected region allows a more efficient nodule performance opens up the possibility of manipulating nodule development in order to improve nitrogen fixation.

CONCLUSIONS

Determinate nodules of grain legumes show structural adaptations that help to maintain adequate rates of nitrogen fixation during moderate oxygen deficiency, but fail under severe hypoxia. Adaptation of lucerne, a forage legume with indeterminate nodules, to 1 kPa pO_2 involves changes in photosynthate allocation to underground organs, aerenchyma formation in both root and nodules that helps to maintain high rates of respiration and mineral uptake, a less compact cortex with an increased nodule gas permeability, increased recycling of CO_2, a less demanding nitrogenase activity and a more efficient nodule structure.

ACKNOWLEDGEMENTS

We are grateful to Drs C de Lorenzo and MR de Felipe for excellent technical assistance with microscopy studies and Dr M Rodriguez Juliá and Miss C Bastida for NPK analyses. Experimental work discussed in this chapter was financed by Dirección General de Investigación Científica y Técnica (PA86-0086) of Spain.

REFERENCES

Anderson MP, Heichel GH, Vance CP (1987) Nonphotosynthetic CO_2 fixation by alfalfa (*Medicago sativa* L.) roots and nodules. Plant Physiology 85: 283-289

Armstrong W (1971) Radial oxygen losses from intact rice roots as affected by distance from the apex, respiration and waterlogging. Physiologia Plantarum 25: 192-197

Armstrong W, Gaynard TJ (1976) The critical oxygen pressures for respiration in higher plants. Physiologia Plantarum 37: 200-205

Arrese-Igor C (1990) Effects of nitrate and oxygen on the symbiotic nitrogen fixation of lucerne. PhD Thesis, University of the Basque Country, Bilbao

Arrese-Igor C, Aparicio-Tejo PM (1990) Lucerne nitrogen fixation in low rhizosphere oxygen. In: Gresshoff PM, Roth LE, Stacey G, Newton WE (eds) Nitrogen fixation: achievements and objectives. Chapman and Hall, New York

Arrese-Igor C, García-Plazaola JI, Hernández A, Aparicio-Tejo PM (1990) Effect of low nitrate supply to nodulated lucerne on time course of activities of enzymes involved in inorganic nitrogen metabolism. Physiologia Plantarum 80: 185-190

Atkins CA, Dakora FD, Storer PJ (1990) Effect of oxygen pressure on synthesis and export of nitrogenous solutes by nodules of cowpea. Planta 182: 565-571

Bergersen FJ (1962) The effects of partial pressure of oxygen upon respiration and nitrogen fixation by soybean root nodules. Journal of General Microbiology 29: 113-125

Corby HDL, Polhill RM, Sprent JI (1983) Taxonomy. In: Broughton WJ (ed) Nitrogen fixation, Vol 3: Legumes. Oxford University Press, Oxford, pp 1-35

Criswell JG, Havelka UD, Quebedaux B, Hardy RWF (1976) Adaptation of nitrogen fixation by intact soybean nodules to altered rhizosphere pO_2. Plant Physiology 58: 622-625

Dakora FD, Appleby CA, Atkins CA (1991) Effect of pO_2 on the formation and status of leghemoglobin in nodules of cowpea and soybean. Plant Physiology 95: 723-730

Dakora FD, Atkins CA (1990a) Effect of pO_2 on growth and nodule functioning of symbiotic cowpea (*Vigna unguiculata* L Walp.). Plant Physiology 93: 948-955

Dakora FD, Atkins CA (1990b) Effect of pO_2 during growth on the gaseous diffusional properties of nodules of cowpea (*Vigna unguiculata* L Walp.). Plant Physiology 93: 956-961

Dakora FD, Atkins CA (1990c) Morphological and structural adaptation of nodules of cowpea to functioning under sub- and supra-ambient oxygen pressure. Planta 182: 572-582

Drew MC, Saglio PH, Pradet A (1985) Larger adenylate energy charge and ATP/ADP ratios in aerenchymatous roots of *Zea mays* in anaerobic media as a consequence of improved internal oxygen transport. Planta 165: 51-58

Drew MC, Stolzy LH (1991) Growth under oxygen stress. In: Waisel Y, Eshel A, Kafkafi U (eds) Plant roots: The hidden half. Marcel Dekker, New York, pp 331-350

Glinski J, Stepniewski W (1985) Soil aeration and its role for plants. CRC Press, Boca Raton, Florida, USA

Heinrichs DH (1970) Flooding tolerance of legumes. Canadian Journal of Plant Science 50: 435-438

James EK, Sprent JI, Minchin FR, Brewin NJ (1991) Intercellular location of glycoprotein in soybean nodules: effect of altered rhizosphere oxygen concentration. Plant, Cell and Environment 14: 467-476

Justin SHFW, Armstrong W (1987) The anatomical characteristics of roots and plant response to flooding. The New Phytologist 106: 465-495

Kawase M (1981) Anatomical and morphological adaptation of plants to waterlogging. HortScience 16: 30-34

Minchin FR, Pate JS (1975) Effects of water, aeration and salt regime on nitrogen fixation in a nodulated legume. Definition of an optimum root environment. Journal of Experimental Botany 90: 60-69

Minchin FR, Summerfield RJ (1976) Symbiotic nitrogen fixation and vegetative growth of cowpea (*Vigna unguiculata* L Walp.) in waterlogged conditions. Plant and Soil 45: 113-127

Minchin FR, Summerfield RJ, Eaglesham ARJ, Stewart KA (1978) Effects of short-term waterlogging on growth and yield of cowpea (*Vigna unguiculata*). Journal of Agricultural Science 90: 355-366

Minchin FR, Witty JF, Sheehy JE (1986) Further errors in the acetylene reduction assay: effects of plant disturbance. Journal of Experimental Botany 37: 1581-1591

Minchin FR, Witty JF, Sheehy JE, Muller M (1983) A major error in the acetylene reduction assay: decreases in nodular nitrogenase activity under assay conditions. Journal of Experimental Botany 34: 641-649

Pankhurst CE, Sprent JI (1975) Surface features of soybean root nodules. Protoplasma 85: 85-98

Parsons R, Day DA (1990) Mechanism of soybean nodule adaptation to different oxygen pressures. Plant, Cell and Environment 13: 501-512

Pfeiffer NE, Torres CM, Wagner FW (1983) Proteolytic activity in soybean root nodules. Plant Physiology 71: 797-802

Sprent JI (1971) Effects of water stress on nitrogen fixation in root nodules. Plant and Soil, Special Volume, 225-228

Wasfi M, Prioul JL (1986) A comparison of inhibition of French-bean and soybean nitrogen fixation by nitrate, 1% oxygen or direct assimilate deprivation. Physiologia Plantarum 66: 481-490

Weisz PR, Sinclair TR (1987) Regulation of soybean nitrogen fixation in response to rhizosphere oxygen. I. Role of nodule respiration. Plant Physiology 84: 900-905

SECTION IV

Drought

EFFECTS OF DROUGHT AND ELEVATED CO_2 ON PLANT WATER USE EFFICIENCY AND PRODUCTIVITY

Theodore C. Hsiao[1]
Instituto Agronomico Mediterraneo-Bari
70010 Valenzano
Bari
Italy

ABSTRACT

The biomass water use efficiency of specific crops is usually almost constant at any given ambient CO_2 concentration when normalised for the evaporative demand of the environment, regardless of the severity of water stress. This conservative behaviour is rooted in two basic tenets of plant productivity, the capture of radiation and assimilation of CO_2 in exchange for water lost. The dominant factor is probably radiation capture, which supplies the energy for both transpiration and photosynthesis. Radiation capture depends on the extent of canopy cover, and hence on leaf growth, a process most sensitive to water stress. When canopy cover is incomplete, even mild water stress may reduce leaf growth and radiation interception in a way that compounds with time, leading to closely linked reductions in assimilation and transpiration. If canopy cover is complete, water stress, when severe enough, may reduce photosynthesis and stomatal opening. The intercellular CO_2 concentration (C_i), however, remains constant in many instances, and decreases in others. Using the equations for CO_2 transport only in the gaseous phase and the equation for transpiration, and taking energy balance into account, it is shown that the photosynthetic WUE of single leaves decreases in the case of constant C_i, and remains about the same or increases in the case of decreased C_i, depending on the magnitude of the changes in C_i and leaf temperature. Elevated CO_2 reduces stomatal aperture, steepens the CO_2 gradient for assimilation and accelerates the development of leaf area, particularly in C3 plants. The first effect lowers transpiration per unit effective leaf area. The reduction, however, may not be large due to energy balance requirements, especially for closed canopies under low wind conditions. The second effect raises C_i sufficiently despite the lower stomatal conductance to effect a net increase in photosynthesis per unit effective leaf area. The combined impact of the first two effects on WUE, again evaluated using the equations for assimilation and transpiration, is that the percentage increase in photosynthetic WUE due to elevated CO_2 is almost proportional to the ratio of the new to the original concentration

---

[1] Permanent address: Department of Land, Air and Water Resources, University of California, Davis, CA 95616, USA.

NATO ASI Series, Vol. I 16
Interacting Stresses on Plants in a Changing Climate
Edited by M. B. Jackson, C. R. Black

of CO_2, in agreement with some published data. The third effect of enhancing leaf area development would be to increase both transpiration and photosynthesis per unit land area prior to the completion of canopy cover, but this would have little effect on photosynthetic WUE. In contrast, consumptive WUE may improve as a consequence of soil shading and reduced soil evaporation. The compounding effect with time of elevated CO_2 when the canopy is incomplete provides an explanation for the phenomenon of a disproportionately larger enhancement of biomass compared to the enhancement in photosynthesis per unit leaf area effected by elevated CO_2. Effects of water stress on photosynthetic WUE under high CO_2 should be minimal and similar to those observed under normal CO_2 concentrations. The conceptual framework provided by the aforementioned equations offers a rational basis for the systematic evaluation of WUE, although its application at the canopy level remains to be tested.

INTRODUCTION

Atmospheric CO_2 will rise for decades or centuries to come. Whether this will result in a hotter and drier climate is not yet certain, although the dominant opinion points in that direction. Regardless of resultant changes in climate, elevated CO_2 will have a direct impact on plant productivity and water use, by enhancing photosynthesis and effecting partial stomatal closure. Numerous studies, although nearly all conducted under conditions different from the open field, have shown that plant growth (Kimball and Idso, 1983; Lawlor and Mitchell, 1991) and water use efficiency (Kimball and Idso, 1983; Morison, 1985; Eamus, 1991) are enhanced under elevated levels of CO_2. The extent of the enhancement, however, is highly variable; for the same species, some studies may show little response to elevated CO_2 whereas others show marked responses. Some of the discrepancies may be attributed to different growth conditions, or to factors other than CO_2, such as mineral nutrients, being limiting. Many discrepancies, however, cannot be easily evaluated without a more systematic framework. An initial framework is proposed in this paper, which highlights some of the key parameters which are often ignored but should be measured in studies of plant productivity and water use efficiency under elevated CO_2 and accentuated stresses.

CROP PRODUCTIVITY IN RELATION TO LIGHT CAPTURE AND CO_2 ASSIMILATION

Some basic aspects of plant productivity need to be outlined before considering the effects of elevated CO_2 and stresses on productivity and water use efficiency. The basic framework is the carbon economy of the plant – the acquisition and expenditure of carbon as photosynthetic assimilates, and the partitioning of the carbon gained to different organs. The acquisition of carbon in turn depends on the capture and use of solar radiation for photosynthesis.

For crop species, the desired product or yield consists of the carbon-based products in the harvested part of the plant. The accretion of this product is intimately tied to CO_2 assimilation and the partitioning of the assimilates over the growing season, and can be summarised as:

$$Y = HI \cdot c \left(\int_e^h A\,dt - \int_e^h R\,dt \right) \tag{1}$$

where Y is the yield per plant or per unit area of land at harvest and HI represents the harvest index or proportion of the biomass in the harvested part of the plant. The rate of net CO_2 assimilation per plant or per unit of land area, A, is for the day-time period only, and is integrated over the period between the time of plant emergence, e, and harvest, h. The loss of carbon during the night is represented by the rate of night respiration, R, also integrated between emergence and harvest. The factor c accounts for the difference in mass per mole of carbon between the total biomass and the assimilated CO_2 (Penning de Vries, 1975). The inclusion of day-time respiration in the net assimilation term and the assignment of a separate term for night-time respiration are arbitrary, for the convenience of considering water use efficiency and interpreting experimental data. Although written for crop species, this equation is also useful for many wild plants. The only change in view-point is that Y should be taken to represent reproductive organs such as seeds or tubers, on which the survival of the species depends.

Eq. 1 expresses the total carbon gained as an integral of time. Implicit is the need to integrate over space to account for variations in the photosynthetic ability of the leaves making up the canopy, as well as variations in the local environment of the leaves. Fortunately a simpler approach is available by considering canopy CO_2 assimilation or biomass accumulation in terms of the quantity of photosynthetically active radiation (PAR) intercepted by the canopy and the efficiency of use of the intercepted PAR for CO_2 assimilation or biomass production. The efficiency of use of the intercepted PAR for canopy photosynthesis is fundamental but data are limited. Many more studies have been conducted on biomass production in relation to radiation interception by the canopy. The amount of biomass produced per unit of PAR intercepted – here termed the radiation use efficiency (RUE) – is a reflection of PAR use efficiency for photosynthesis, for a given pattern of respiration and chemical composition of the biomass. It turned out that during the life cycle of a given herbaceous crop, RUE averaged over a number of days is relatively constant when other factors are not limiting until the beginning of maturation and senescence (Williams *et al.*, 1965; Monteith, 1977; Gallagher and Biscoe, 1978; Fischer, 1983). Thus, the major variable determining how much CO_2 is assimilated and how much biomass is produced by a crop is the cumulative quantity of PAR intercepted by its canopy. Surprisingly, RUE appears to be similar in C3 and C4 species, at least within the limited number of comparisons available (Charles-Edwards *et al.*, 1986; Held and Hsiao, in preparation).

The near constancy of RUE, or the close to linear relationship between biomass production and PAR interception, depends on the following factors:

1. most of the intercepted PAR is absorbed by the canopy and the proportion absorbed is relatively constant, ranging between 95 and 97%;
2. canopy photosynthesis responds to increasing PAR up to the level encountered at noon on clear summer days (2.0 mmol $m^{-2} s^{-1}$ or higher), either in a linear (Hesketh and Baker, 1967; Sale, 1977) or curvilinear

(Puckridge and Ratkowsky, 1971; Sale, 1977; Connor *et al.*, 1985; Campbell *et al.*, 1990) fashion;

3. photosynthesis by a canopy is proportional to its PAR interception, with a near-constant light use efficiency for different growth stages prior to maturation and senescence (Hesketh and Baker, 1967; Puckridge and Ratkowsky, 1971; Puech-Suanzes *et al.*, 1989);
4. after the canopy is complete and intercepts most of the radiation, the rate of canopy photosynthesis at midday on sunny days varies only slightly over a period of several weeks when other factors are not limiting (Puckridge and Ratkowsky, 1971; Jones *et al.*, 1986; Puech-Suanzes *et al.*, 1989);
5. photosynthesis (Björkman, 1981), particularly that of canopies (Sale, 1977), is relatively insensitive to temperature within the normal temperature range for the growth of a plant;
6. short-term fluctuations in RUE tend to average out over longer time intervals.

Examples of the near linear light response over a daily timecourse are given in Fig. 1 for cotton, a C3 crop, and sorghum, a C4 crop. The lower rate of assimilation at given PAR levels in the afternoon than in the morning seen in Fig. 1 has been frequently reported (for example, Sale, 1977; Connor *et al.*, 1985; Puech-Suanzes *et al.*, 1989) and attributed to various causes, including saturation of sinks by assimilates, water stress or high temperature. An example where canopy photosynthesis was linearly related to radiation interception when data collected at different times and at different canopy sizes were plotted together is shown in Fig. 2. The corollary is that biomass accumulation is maximal and relatively constant during the period just after canopy closure and before senescence begins, in agreement with experimental observations (Gallagher and Biscoe 1978; Fischer, 1983).

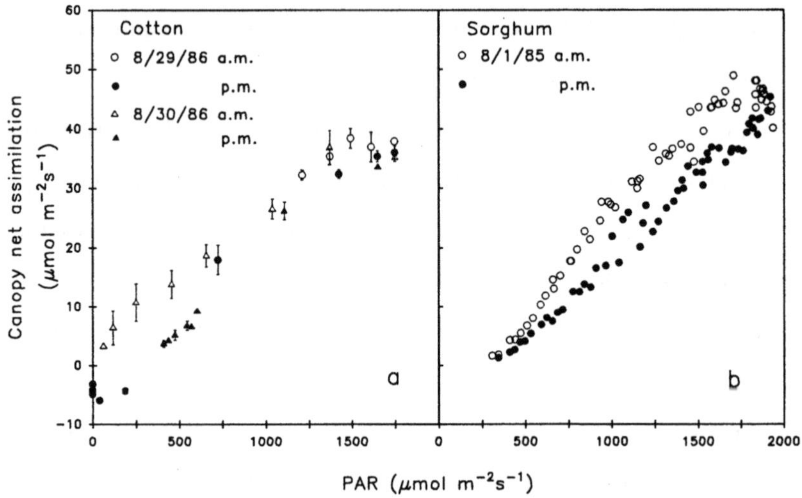

FIGURE 1. Net photosynthesis per unit land area by complete canopies of (a) cotton and (b) sorghum in the field in relation to incident PAR flux. Variation in PAR flux was the result of the daily change in solar angle, with the associated changes in temperature. The data for cotton (Puech-Suanzes et al., *1989) were taken with a canopy chamber, with the soil covered to exclude soil and root respiration. The data for sorghum (Held and Hsiao, 1993) were obtained using the micrometeorological method of Bowen ratio/energy balance/CO_2 gradient (Held* et al., *1990) and are uncorrected for soil and root respiration, which was estimated to be 2-3 μmol m^{-2} s^{-1}.*

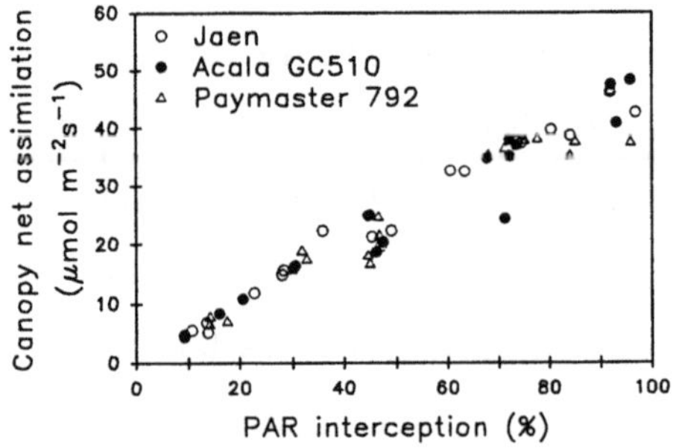

FIGURE 2. Canopy net photosynthesis per unit land area of three cotton cultivars at midday in relation to PAR interception. Data were obtained over the major part of the season as the canopy developed. From Puech-Suanzes et al. *(1989).*

Recently Demetriades-Shah *et al.* (1992), on the basis of statistical analysis, questioned the validity of the linear relationship between biomass production and radiation interception, and argued that light is only one of the factors that determines the rate of photosynthesis and cannot be considered as the dominant limiting factor. They failed to consider, however, the physiological basis and the photosynthetic data for the causal relationship between light interception and biomass production. Their point is well taken in as much as light use efficiency does decline when stresses of various kinds are sufficiently severe. For example, it is known that water stress may reduce the photosynthetic light use efficiency of canopies (Jones *et al.*, 1986; Puech-Suanzes *et al.*, 1989), and that other stresses can also be inhibitory (Fischer, 1983). The evidence suggests, however, that the reductions in RUE are often very small and that RUE is much more resistant to stress than leaf area per plant and canopy size. This point will be elaborated on below.

Since RUE is a relatively conservative parameter, the net carbon gain of a plant over the season is readily affected by the amount of PAR it captures. In this analysis, the growing season may be divided into three phases (Hsiao, 1982), as depicted in Fig. 3. Early in the season, the canopy is incomplete and much of the incident radiation is not intercepted. Thus the major limiting factor for productivity is the degree of canopy cover. Any enhancing or retarding effect on leaf growth will be amplified with time (Hsiao, 1982; Bradford and Hsiao, 1982) as the production of additional photosynthetic area is a function of existing area. As the canopy enlarges and reaches full size or canopy closure, the effective radiation capture surface reaches a maximum and the amount of radiation captured becomes dependent on the level of insolation. Late in the season, senescence sets in and the light capturing surface declines. More importantly, there is also a decline in RUE through the decline in the photosynthetic capacity of old and senescent leaves.

The expansive growth of leaves making up the canopy is highly sensitive to environmental variables, including water supply, salinity, temperature and nutritional status. Thus in the canopy-cover limiting phase, even mild stresses can slow the development of the light capturing surface and hence biomass accumulation. This aspect will be elaborated on later. Once the canopy is full, mild stresses sufficient to reduce leaf expansive growth are probably not sufficient to inhibit photosynthesis per unit leaf area or per unit of absorbed PAR, and in such cases biomass accumulation would not be affected. However, during the maturation

FIGURE 3. Conceptual depiction of the seasonal pattern of biomass accumulation for annual crops in the absence of significant stress periods. The early phase is approximately exponential and occurs when the canopy is incomplete and PAR interception is determined by canopy size. The middle phase is approximately linear and occurs when the canopy is complete and intercepts almost all the incident PAR. The final phase covers the period of maturation and leaf senescence when net assimilation declines and finally becomes zero.

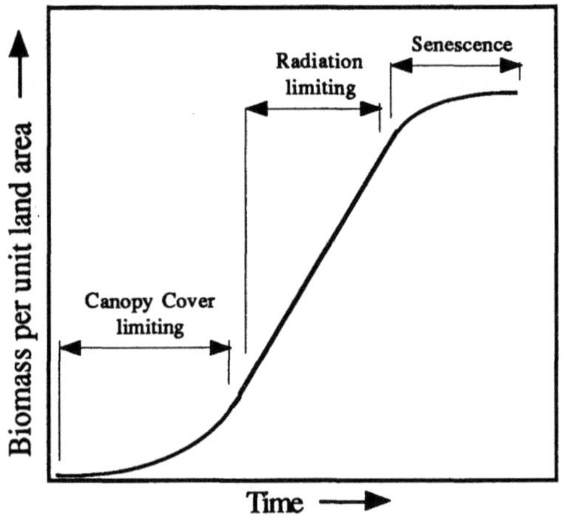

and senescence phase, sufficiently severe stress accelerates the senescence of older leaves and reduces their photosynthetic capacity. The loss of green leaf area is quantified in the literature in terms of reduced green leaf area duration (green leaf area index integrated over time). Grain yield is often well correlated with green leaf area duration because most of the assimilates for grain filling come from photosynthesis late in the season.

The negative side of the carbon balance is respiration. For the purpose of evaluating productivity, respiration is often divided, somewhat simplistically, into that necessary for growth and that required for the maintenance of existing tissue (McKree, 1970). The more active the growth, the higher the rate of respiration, which supplies the energy and building blocks necessary for the synthetic processes associated with growth. Maintenance respiration, however, is more closely proportional to the size of the plant. Over the life cycle of a crop, growth respiration constitutes a major portion of the total respiration during the early stages, but maintenance respiration becomes increasingly important as the plant gains biomass until senescence begins, when it becomes the dominant component of total respiration. In crops grown for grain or fruits, respiration by the fruiting bodies can be a major component (Bolaños and Hsiao, 1991). Respiration is highly important, consuming one fourth to two thirds of the total assimilates over the plant life cycle, according to the highly variable estimates summarised by Amthor (1989, Table 6.1). A major uncertainty is the lack of good quantitative data on daily canopy photosynthesis and night-time respiration spanning the season.

In contrast to photosynthesis, the rate of which does not vary greatly with temperature around the optimum (Björkman, 1981), respiration is highly sensitive to temperature. A value of approximately 2 is common for the Q_{10} of respiration, the ratio of the rate at a temperature 10 °C higher to the rate at the basal temperature. This geometric response does not always hold, however, and Q_{10} can be substantially higher or lower than 2 (Amthor, 1989). For example, for leaves of field grown tomato, the respiratory response to temperature was approximately geometrical with a Q_{10} of 2.4. Respiration of fruits from the same plants, however, showed a nearly linear dependence on temperature and the slope of the response varied according to the ontogenetic stage of the fruit (Bolaños and Hsiao, 1991). Because of the strong effect of temperature, much work remains to be done on quantifying cumulative respiration under field conditions which are characterised by strong diurnal temperature changes. Diurnal temperature patterns can also have a significant impact on net canopy photosynthesis rates through effects on the respiration of organs within the canopy.

A word should be said about the well known difference in photorespiration between C3 and C4 species, which is apparent in the values for net assimilation. Photorespiration results from the dual functions of the photosynthetic enzyme, RuBP carboxylase/oxygenase. Higher temperatures favour the oxygenating over the carboxylating activity of the enzyme and increase the release of assimilated CO_2 by photorespiration. This contributes to the flattening of the A vs. temperature curves of C3 species. High CO_2 concentration has the opposite effect of suppressing oxygenation while enhancing carboxylation, and thus ameliorates the effect of high temperature (Björkman, 1981). These interactions may be crucial in keeping C3 species competitive in the future world of elevated CO_2 and warmer temperatures.

Environmental conditions can also affect yield by altering the proportion of biomass allocated to the harvestable organs. Severe water deficiency at anthesis can prevent pollination in grain crops. The reduction in the number of potential grains limits the amount of assimilates in the harvestable organ and HI is lowered. Hence, the reduction in yield would be more than proportional to the reduction in biomass. Heat stress at the time of microspore formation in cereal crops can cause the pollen formed to be largely sterile. Again HI and grain number are reduced, although there may be only very minor effects on the amount of total biomass produced. Less dramatic is the change in the partitioning of carbon to the roots. It is well known that root growth is favoured over shoot growth by mild to moderate water stress (Sharp and Davies, 1979). This change, though of adaptive advantage (Bradford and Hsiao, 1982), would reduce HI slightly.

EVAPOTRANSPIRATION IN RELATION TO ENERGY SUPPLY

It is necessary to outline the basic aspects of the evapotranspiration (ET) process before considering water use efficiency. ET is the sum of evaporation from the soil (E) and transpiration from the plant (T). Water used by ET is referred to as consumptive use. The pivotal consideration is energy supply, as ET can only be sustained by the continuous input of energy corresponding to the rate of ET in terms of the latent heat of vaporisation. The energy comes from the sun as net absorbed radiation (net radiation flux) of any wavelength, and also from the surrounding air when the foliage or soil is cooler and hence there is a net transfer of sensible heat to the leaves or soil surface. Not all of the net radiation is necessarily used to evaporate water, as part of it can be dissipated as sensible heat when the temperature of the leaves or soil is higher than the surrounding air. The fraction of the net radiation used for photosynthesis is usually negligible (less than 3-4%).

The net radiation of the plant canopy is generally dominant in determining transpiration rates provided the stomata are substantially open. In many situations, the daily rate of transpiration, when expressed in energy units as latent heat flux, is closely approximated by the net radiation flux to the plant canopy. Deviations are caused by significant fluxes of sensible heat, which can supply appreciable quantities of energy for transpiration when the canopy is sparse and the exposed soil surface is relatively dry. The increase in soil surface temperature resulting from radiative heating then warms the air and enhances sensible heat transfer to the canopy. In dry areas, extra energy is also supplied to the canopy, regardless of whether it is sparse or complete, by large scale advection (horizonal flow of energy and associated materials, in this case warm and relatively dry air), especially in crop fields surrounded by fallow areas under windy conditions. In areas of high humidity, leaf temperature is often higher than air temperature and transpiration in terms of energy equivalent tends to be less than net radiation due to the sensible heat flux away from the foliage. Sensible heat also dissipates a considerable portion of the energy from net radiation if the stomata are substantially closed by stresses or leaf senescence, restricting water vapour loss and transpirational cooling.

In the field it is nearly impossible to separate transpiration and soil evaporation experimentally without altering the natural energy supply for T or for E. Hence, it is often necessary to consider consumptive use of water instead of transpirational use in calculating water use efficiency. In general, soil evaporation follows the same principles as transpiration, except there is no stomatal control. Instead, the water vapour concentration at the soil surface is reduced by soil drying. The main source of energy for E is also net radiation, in this case for the soil. E from exposed soil is much higher than from soil shaded by plants at the same level of soil surface wetness. When soil is wetted frequently, E constitutes the major part of ET early in the season when the canopy is sparse, but declines with time as the soil is increasingly shaded by the developing canopy. When the canopy is complete, E is only a small fraction of the ET even when the soil surface is wet (Ritchie and Burnett, 1971).

DEFINITIONS OF WATER USE EFFICIENCY

The above descriptions of plant productivity and ET serve as a basis for considerations of the efficiency of water use by plants. Water use efficiency (WUE) can be viewed in a number of ways (Fischer and Turner, 1978). The most basic is in terms of the carbon gained through photosynthesis relative to the water lost through transpiration. The ratio of net assimilation to transpiration, sometimes known as transpiration ratio, is here termed photosynthetic WUE. Photosynthetic WUE can be used to describe the behaviour of single leaves, plants, or a canopy made up of a population of plants, and can refer to near instantaneous values (A/T), or to the overall outcome when the rates of A and T are integrated over periods ranging from minutes to a day or longer ($\int Adt/\int Tdt$). When referring to whole plants, the respiration of the non-photosynthetic parts must be accounted for in arriving at values for A. Unfortunately, many of the data reported in the literature are of a near instantaneous or short term nature, usually only for individual leaves, and are of very limited value in deducing WUE in the field over time intervals of significant duration. Because of the paucity of data for longer spans of weeks and months due to the difficulties involved in the continuous operation of instruments to monitor A, it is usually necessary to consider instead biomass transpirational WUE, the amount of biomass produced per unit of water transpired, or biomass consumptive WUE, which includes the water evaporated from the soil in addition to that transpired. Biomass consumptive WUE is closely linked to photosynthetic WUE, but differs from the latter by the factor c of Eq. 1, by the night respiration integral, and by soil evaporation E, as follows:

$$\text{Biomass consumptive WUE} = \frac{c(\int Adt - \int Rdt)}{\int ETdt} \tag{2}$$

Most of the available data on long-term WUE are based on consumptive use. A few studies are based on measurements of transpiration obtained by covering the soil (which increases sensible heat flux from the covered soil to the leaves) or by deducting E estimated by various methods.

The overall WUE of a crop at harvest may bear little relation to the photosynthetic WUE at any particular stage of growth, or even to its biomass transpirational WUE. Yield water use efficiency (harvested yield per unit of water used) would be more relevant. In addition to being a function of biomass WUE integrated over the life of the crop, yield WUE is also dependent on the partitioning of carbon between the plant organs, i.e., on HI, such that:

$$\text{Yield WUE} = \text{HI (biomass WUE)} \tag{3}$$

Eq. 3 is applicable to either consumptive or transpirational water use. The yield WUE of most crops has been greatly improved with the development of modern high yielding cultivars as an inadvertent result of breeding for higher yields, which has increased HI. Although important in the context of the effects of stresses (Fischer and Turner, 1978) and high CO_2 (Lawlor and Mitchell, 1991), yield WUE is not considered further here due to a space limitation.

VARIATION AND CONSTANCY IN WATER USE EFFICIENCY

At ambient CO_2 concentration, the instantaneous photosynthetic WUE varies markedly with time of the day, as exemplified by data collected at Davis, California (Fig. 4). The main cause of the variation lies in the discordance between

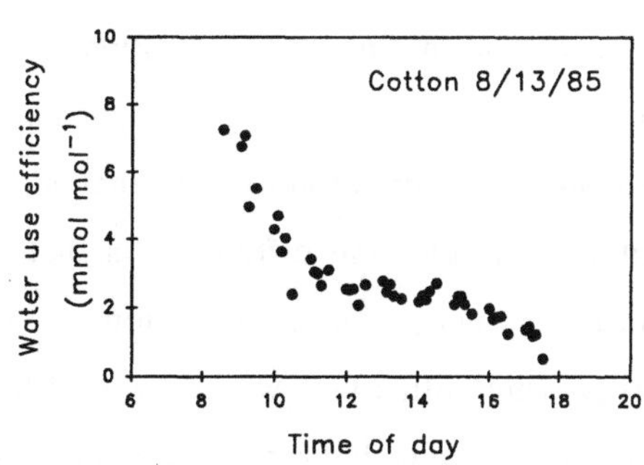

FIGURE 4. *Diurnal timecourse of photosynthetic water use efficiency of a full cotton canopy in the open field. The day was basically sunny with 1.79 mmol m^{-2} s^{-1} of incident PAR and 2.44 kPa vapour saturation pressure deficit at midday. Measurements were made with the micro-meteorological method of Fig. 1 and no correction was made for soil and root respiration (see Fig. 1) or soil evaporation, which should be minimal due to the canopy cover. From Held and Hsiao (1993).*

the daily pattern of PAR flux and canopy temperature. Air and hence canopy temperature are lower in the morning than in the afternoon. As a consequence, the water vapour concentration within the leaf is higher in the afternoon than in the morning at the same PAR. Thus, T is higher in the afternoon while A remains similar, resulting in a lower photosynthetic WUE. Biomass WUE, a time integral, however, tends to fall within a narrow range of values for a given species, probably because the diurnal trends in weather are similar for most days and the variations in weather from day to day tend to average out over periods of weeks. In fact, when normalised for the evaporative demand of the atmosphere (de Wit, 1958; Arkley, 1963; Tanner and Sinclair, 1983), biomass WUE at ambient CO_2 concentration has been found to be nearly constant for many crop species under different weather and regimes of water supply, i.e., the relationship between biomass produced and water consumed is linear. Several hundred such near-linear relationships can be found in the literature (see Hanks, 1983).

What are the reasons for this conservative behaviour of biomass WUE? Although not widely discussed in the literature, it appears that the behaviour is rooted in two basic tenets of plant productivity – the influence of intercepted radiation on both A and T, and the sharing of the transport pathway by CO_2 and water vapour as they pass between the atmosphere and the intercellular space of leaves. The dominant factor is probably radiation capture. As shown in Fig. 5, the extent of radiation capture depends on the effective leaf area or canopy size, which in turn is dependent on the actual leaf area (or leaf area index) as modulated by leaf distribution and geometry. Plant distribution over the land area and factors which are highly species specific, such as leaf distribution and foliar angles within the canopy, are important determinants. A and T share mostly the same source of captured energy. One difference is that the sensible heat flux can either supply additional energy for transpiration or dissipate some of the absorbed radiative energy which otherwise would be used for transpiration, as discussed earlier. The other difference is that assimilation utilises only radiation in the photosynthetically active band (approximately 330-730 nm, see McCree, 1981) whereas transpiration

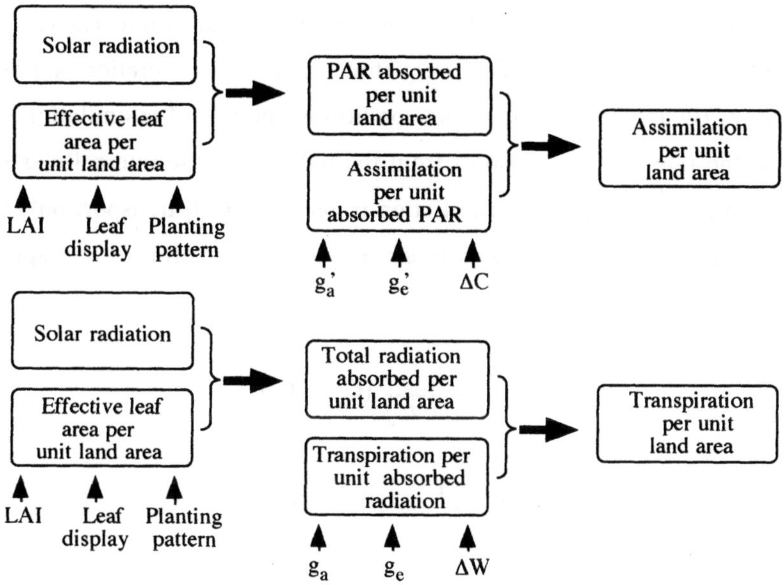

FIGURE 5. *Similarities and differences in factors affecting assimilation and transpiration per unit land area. Arrows indicate causal relations. Leaf area index = LAI;* g'_a *and* g'_e *represent respectively the conductances of the air boundary layer and leaf epidermis to* CO_2*;* g_a *and* g_e *have analogous meaning but are for water vapour;* ΔC *and* ΔW *are respectively the difference in* CO_2 *and water vapour concentration between the bulk air and the leaf intercellular space. Modified from Hsiao and Bradford (1983).*

uses radiation of any wavelength. At the same time, PAR is a fairly constant portion of the incident solar radiation (Varlet-Grancher *et al.*, 1989), and the ratio of absorptance for PAR and non-PAR is relatively constant for many leaves. Consequently, any change in the amount of radiation captured by the canopy, whether the result of changes in weather or in effective leaf area, affects A (or biomass production) and T similarly and WUE is kept nearly constant. This commonality (Hsiao and Bradford, 1983) has not been emphasised in the literature, which has often attributed the tight relationship between biomass produced and water consumed to the shared pathway for CO_2 and water vapour transport (Kanemasu, 1983; Tanner and Sinclair, 1983) shown in the central portion of Fig. 5.

The shared pathway for CO_2 and water vapour for single leaves can be formalised with two simple transport equations, one for assimilation and one for transpiration. By assuming steady state conditions, the rate of assimilation can be equated to the rate of CO_2 transport from the bulk air to the intercellular space, making it possible to consider assimilation in terms of CO_2 transport only in the gaseous phase. The resultant equation, along with the equation for transpiration, is:

$$A = \frac{1}{r'_a + r'_e} (C_a - C_i) \tag{4a}$$

$$T = \frac{1}{r_a + r_e} (W_i - W_a) \tag{4b}$$

where the resistances to gas transport through the air boundary layer and leaf epidermis are denoted respectively by r_a and r_e, with r_e being the resistance made up by the two parallel resistances of the stomata and cuticle. The resistances for CO_2 are indicated by a prime, in contrast to the resistances for water vapour which have no prime. Due to its heavier molecular mass, CO_2 diffuses more slowly than water vapour and $r' = 1.6r$, or $r = 0.625r'$, for both the boundary layer and epidermal segments of the pathway (Farquhar and Sharkey, 1982). The driving forces for CO_2 and water vapour transport are respectively the difference in the concentration of CO_2 (ΔC) between the bulk air (C_a) and the intercellular space (C_i), and the difference in the water vapour concentration (ΔW) between the intercellular space (W_i) and the bulk air (W_a). The advantage of not going beyond the gaseous phase for CO_2 is that complex metabolic changes in the liquid phase associated with changes in environmental conditions need not be dealt with directly. Instead, they are reflected in the absolute value of C_i and the relative value of C_i to C_a. Since only physical processes are involved in the gaseous phase, all terms in the equation for A are well defined and can be experimentally determined.

Fig. 5 shows that, for a given amount of radiation absorbed, A and T are dependent additionally on their respective resistances r'_a and r'_e, and r_a and r_e. Since the resistances for CO_2 and water vapour differ by a constant factor, any change in the resistance of either the boundary layer or the stomata would have a similar impact on A and T. On the other hand, Fig. 5 and Eq. 4a and 4b also demonstrate that the driving forces for A (ΔC) and T (ΔW) are totally different and could lead to changes in photosynthetic WUE following changes in certain conditions. The most obvious is the increase in ΔC under conditions of elevated CO_2, leading to increased WUE, as will be discussed later. Also obvious is why the WUE of a given species differs under different climates unless the quantity of water consumed is normalised for the evaporative demand of the atmosphere. Often the normalisation takes the form of dividing the total ET by the mean vapour pressure deficit of the air (Tanner and Sinclair, 1983). The latter is an approximation of the mean ΔW, as ΔW is directly proportional to the vapour pressure deficit of the air if leaf temperature is identical to air temperature.

Based on Eq. 4a and 4b, and recognising that $r = 0.625r'$, the photosynthetic WUE (hereafter referred to simply as WUE unless otherwise stated) of single leaves can be written (Farquhar *et al.*, 1989) as:

$$WUE = \frac{(r_a + r_e)}{(r'_a + r'_e)} \frac{\Delta C}{\Delta W} = 0.625 \frac{\Delta C}{\Delta W} \quad (5)$$

regardless of whether the plant is well watered and growing under optimal conditions or is affected by stress. The equation highlights the importance of knowing ΔC and ΔW when evaluating WUE.

Eq. 4 and 5 are written for single leaves but provide a conceptual basis for the examination of canopy WUE. The problems of scaling up from processes at the leaf level to those at the canopy are complex and receiving much current attention (e.g., Kim and Verma, 1991). On the other hand, the indications are that the 'big leaf' model for canopy ET (Penman-Monteith combination equation, Monteith, 1973) and canopy A are reasonably valid for short canopies such as herbaceous crops which completely or almost completely cover the ground (Jarvis and McNaughton, 1986; McNaughton and Jarvis, 1991). Applications to the canopies of trees and forests require much more caution (Jarvis *et al.*, 1976). In any event, the basic elements in many scaling-up models are single leaves or their equivalent. Therefore information developed in terms of the concept represented by Eq. 4 and 5 and their derivatives would be valuable.

FURTHER CONSIDERATIONS OF STRESS EFFECTS ON WATER USE EFFICIENCY

Detailed discussion of WUE as affected by water supply needs to take into account the timing of water stress relative to the growth stage and the severity of the water stress. Only the first two stages of the life cycle of an annual crop as depicted in Fig. 4 are considered here because of space limitations.

*FIGURE 6. Interception of PAR by bean (*Phaseolus vulgaris*) canopies in the field at midday in relation to time after planting and as influenced by irrigation. Water was applied weekly to replace ET losses in the irrigated treatment. The unirrigated treatment grew solely on stored water in the soil, which was at field capacity at planting. Leaf water potential measured on three dates was on average 0.14 MPa lower in the unirrigated plants and leaf epidermal resistances did not differ significantly between the treatments. k is a proportionality constant representing the fractional daily increase in effective PAR interception area per unit of existing interception area, and r is the regression coefficient for exponential functions of PAR interception vs. time represented by the lines. Modified from Bradford and Hsiao (1982).*

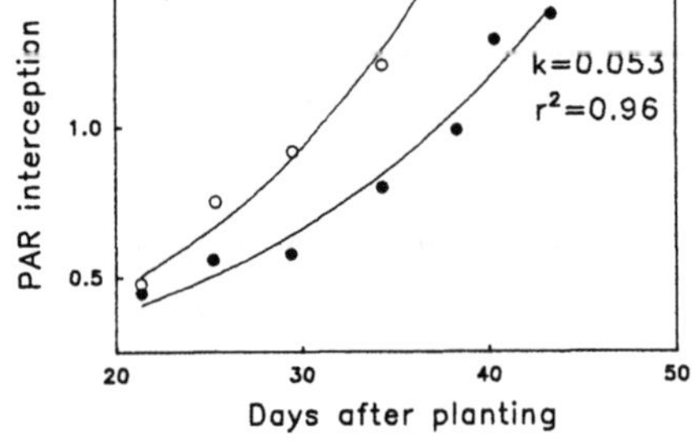

Mild water stress. Of all known plant processes, the expansive growth of leaves is the most sensitive to water stress and is the first to be inhibited as water stress develops (Boyer, 1970; Acevedo *et al.*, 1971; Hsiao, 1973; Bradford and Hsiao, 1982). When canopy cover is incomplete and the capture of incident radiation only partial (first phase of Fig. 3), continued enlargement of the canopy is crucial for the plants to reach their maximal rates of radiation capture and biomass production. Even mild water stress can reduce the rate of leaf area development and hence PAR capture by the canopy. As pointed out previously (Hsiao, 1982; Bradford and Hsiao, 1982), this effect compounds with time and a small percentage reduction in the daily leaf growth rate can result in much larger reductions in biomass or the radiation intercepting canopy area. In Fig. 6, the effective canopy area of a young bean crop, as reflected by PAR interception at midday, increased exponentially with time. The daily rate of this increase relative to the existing canopy area is represented by k. For the unirrigated and irrigated treatments respectively, the values of k were 0.053 and 0.068. That is, the daily relative increase in effective canopy area was 5.3% for the unirrigated compared to 6.8% for the irrigated treatment, a reduction of 22% due to water stress. The total reduction in PAR interception on day 40 after about 20 days of stress was, however, larger at 36% because of the compounding effect of time. When the above ground biomass sampled at different times was plotted against cumulative intercepted PAR, the data for the irrigated and unirrigated treatments fell on the same line (Fig. 7).

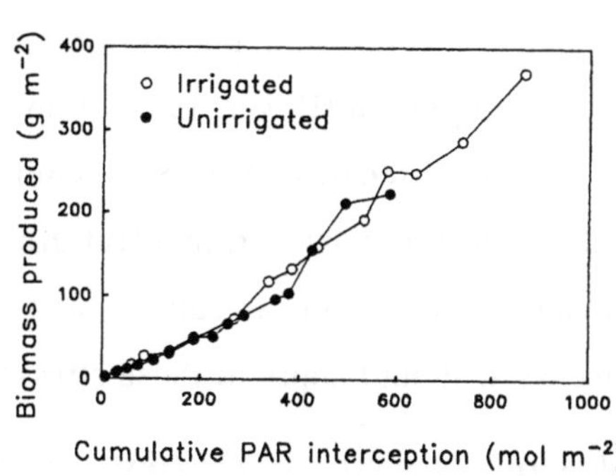

FIGURE 7. *Above-ground biomass production in relation to cumulative PAR interception by irrigated and unirrigated bean (*Phaseolus vulgaris*) in the field. The crops and treatments were as in Fig. 6. Biomass was sampled at regular intervals shortly after plant emergence. The daily time-course of PAR interception was measured at regular intervals and interpolated for intervening days, then summed to the appropriate date. Unpublished data of F. Arruda and T.C. Hsiao.*

Thus, for any sampling date, the biomass produced by the unirrigated treatment was lower, in proportion to its smaller quantity of intercepted PAR. These results indicate that RUE was not altered by the mild water stress and that canopy size and PAR interception accounted for all the difference in biomass production. This interpretation does rely on the assumption that the root biomass is a small portion of total biomass (as in many field crops) and that changes in root to shoot ratio were minor.

As assimilation and transpiration are both dependent on radiation capture, when biomass is reduced only by restricted canopy development, biomass and transpirational water use are reduced in a closely linked way, leading to very little change in biomass transpirational WUE during this stage. However, a slight decrease is possible due to small increases in leaf temperature resulting from the sensible heat flux from the less shaded and hence hotter soil surface if that surface is often dry. Biomass consumptive WUE could be reduced more severely if the soil is frequently wetted by rain or irrigation; because of the slower canopy development, more soil would remain unshaded and E would be higher. The reduction in WUE, however, is not likely to be marked, as E constitutes only a minor part of the ET integral under most situations. The exception is when the canopy is sparse for most of the season and the soil is frequently wetted. If the plant canopy covers the ground completely and is already intercepting nearly all the incident radiation (phase 2 of Fig. 3), or vegetative growth is complete, mild water stress should have little or no effect on productivity and WUE since leaf area is already sufficient and stomatal opening and photosynthesis are less sensitive to water stress than expansive growth (Boyer, 1970; Acevedo *et al.*, 1971; Hsiao, 1973).

This discussion has been confined to the effects of water stress. The indications are, however, that leaf growth is also very sensitive to other environmental stresses such as salinity and low temperature, and that what has been discussed should apply to those stresses as well.

Moderate to severe water stress. Under more severe water stress, in addition to inhibition of leaf growth, CO_2 assimilation and stomatal opening would also be reduced, frequently along with the photosynthetic capacity of the leaf, and WUE during any of the three phases (Fig. 3) may change either due to changes in ΔC or in ΔW. Designating WUE for well watered plants as WUE_O and for water stressed plants as WUE_S and utilising Eq. 5, the ratio of WUE under stress to that under well watered conditions would be:

$$\frac{WUE_S}{WUE_O} = \frac{\Delta C_S \, \Delta W_O}{\Delta C_O \, \Delta W_S} \tag{6}$$

For many species C_i tends to remain constant over a range of environmental conditions, including water stress (Wong *et al.*, 1979; reviewed by Morison, 1987), as long as the atmospheric CO_2 concentration is constant. The available data indicate that there may even be a near constancy in C_i as temperature is varied over a 15 °C range around the optimum for photosynthesis (Björkman, 1981). In some cases C_i remains constant under water stress, but in others is reduced. In plants with stomata which respond to ΔW, C_i tends to decrease linearly with increasing ΔW (Ball and Berry, 1982; Morison, 1987). Examples of constancy and changes in C_i are given in Fig. 8 for tomato leaves. The C_i of these leaves remained almost constant over a wide range of PAR fluxes and increased only when PAR was low. Of the three leaves shown in Fig. 8b, C_i for two remained essentially constant as water potential decreased over the range tested, while C_i of the third showed some decline.

In cases where C_i remains constant, ΔC also remains constant provided C_a does not change. Hence, according to Eq. 6 the ratio of WUE with and without stress would simply be $\Delta W_O/\Delta W_S$. For the same atmospheric humidity (same W_a), ΔW would depend on W_i, which in turn is determined by leaf temperature as the intercellular space is virtually saturated with water vapour and saturation vapour concentration is a function of temperature. The stomata are less open and transpiration is reduced by many stresses. Leaf temperature then rises, as dictated by the principle of energy balance, leading to a higher W_i. Consequently $\Delta W_O/\Delta W_S < 1$ as long as C_i remains constant, and $WUE_S < WUE_O$. The extent to which WUE

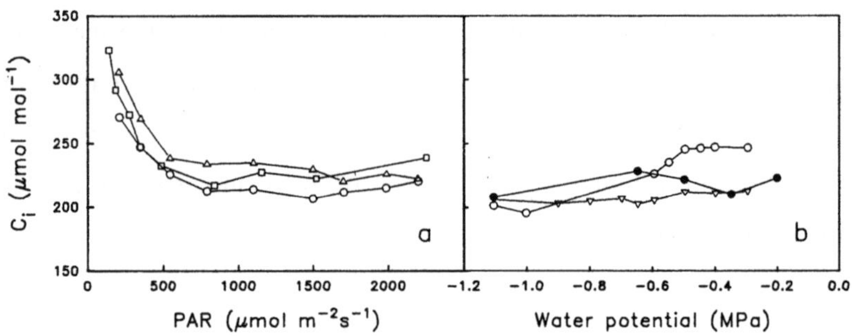

FIGURE 8. *Intercellular CO_2 concentration (C_i) of the youngest fully expanded and fully exposed leaves of field grown tomato plants in relation to (a) PAR flux and (b) leaf water potential. Each line represents a leaf from a different plant. Atmospheric CO_2 concentration (C_a) was ambient. In (b), PAR was approximately 2 mmol m^{-2} s^{-1}. The apical leaflet of a leaf was monitored in a gas exchange chamber while the remaining leaflets were covered by plastic film and aluminium foil to stop transpiration. One covered leaflet was excised at intervals for the measurement of leaf water potential in a pressure chamber as the plant dried out. The measured water potential was assumed to be that of the exposed and assimilating leaflet. Data in (a) are from Bolaños and Hsiao (1991) and in (b) are unpublished results of the same authors.*

is reduced would depend on the degree of stomatal closure, the thermal and radiation environment, air humidity, and boundary layer conditions affecting sensible heat transfer. The impact of these factors shows up in the ratio of ΔW. Other things being equal, the more severe the stress and the greater the stomatal closure, the more WUE would be reduced. It should be pointed out that the situations described are not those occurring in the gas exchange chambers typically used to determine the WUE of single leaves. There, by keeping the chamber or leaf temperature constant, the principle of energy balance as it operates in the open field is circumvented and W_i is kept constant. Hence, WUE would remain unchanged provided C_i remains the same. An interesting point made clear by Eq. 6 is that the impact on WUE of a given increase in leaf temperature effected by water stress will increase as relative humidity rises. This is significant when comparing estimates of changes in WUE made in the field with those in controlled environments, where relative humidity is often higher.

The conclusion that WUE is reduced by stresses which cause stomatal closure in plants whose C_i remains constant is not surprising, except for the case of water stress. The tendency is to view the plant as a well adapted organism with capabilities for improving the efficiency of utilisation of the resource that is the most limiting. In the case of water stress, this appears not to be true for plants which maintain their C_i nearly constant. For plants which exhibit a reduced C_i, either directly due to the stress or indirectly through the increase in ΔW, the direction and extent of the change in WUE would depend on the increase in ΔC due to the reduction in C_i relative to the increase in ΔW. In any event, if the linear relationship between biomass production and crop water use is any indication, it would be unusual for WUE to be changed markedly by water stress. What is probable is that in many cases the main limiting factor for the production of most of the biomass is a lack of adequate leaf area and radiation interception, which would be associated with an essentially constant transpirational biomass WUE. The remaining portion of the biomass is produced when water stress is sufficiently severe to cause stomatal closure and possibly alter C_i, with consequent changes in WUE. The overall biomass WUE, however, would be nearly constant because it reflects more closely the WUE of the major portion of the biomass produced.

IMPACT OF HIGH CO_2 ON WATER USE EFFICIENCY AND PRODUCTIVITY

It is well known that assimilation is increased at high CO_2 concentrations, especially in C3 species, even though partial stomatal closure and consequent reduction in transpiration are generally induced. Morison (1987) concluded from a literature review of 80 observations in 25 species that stomatal conductance at a CO_2 concentration double the present ambient level is approximately 60% of that under ambient CO_2, regardless of whether the species is C3 or C4. With the increase in A and reduction in T, WUE is generally enhanced, but to a highly variable degree (Enoch, this volume). Although the effect of elevated CO_2 on WUE has been reviewed and evaluated by several authors (Kimball and Idso, 1983; Morison, 1985; Eamus, 1991), there is still no reference framework to analyse and reconcile these highly variable results. It is hoped that the analysis proposed below, though simplistic to a certain degree, will serve such a purpose.

Using an approach similar to that adopted for water stress effects and denoting the new CO_2 level with the subscript N, the ratio of WUE under the new [CO_2] to that under present day [CO_2] in the absence of water stress can be expressed as follows:

$$\frac{WUE_N}{WUE_O} = \frac{(C_{e,N} - C_{i,N})}{(C_{e,O} - C_{i,O})} \frac{\Delta W_{ei,O}}{\Delta W_{ei,N}} \tag{7}$$

Eq. 7 is based on transport over the epidermal path segment between the air immediately adjacent to the leaf epidermis, denoted by subscript *e* to describe its location, and the intercellular space (*i*). ΔW_{ei} is the difference in W between the two locations. For steady state transport, the rates of A and T are the same for the epidermal and the boundary layer segment as they are connected in series; hence, Eq 7 is an expression of the ratio of the new to the original WUE.

A major question is how C_i is affected by C_a or C_e. Starting with the work of Wong *et al.* (1979), analyses (Ball and Berry, 1982; Morison, 1987) show that the ratio of C_i to C_e, designated by α ($\alpha = C_i/C_e$), at various values of C_e is generally a conservative quantity, but decreases linearly with increases in ΔW. Substituting αC_e for C_i in Eq. 7, the expression becomes:

$$\frac{WUE_N}{WUE_O} = \frac{(1 - \alpha_N)}{(1 - \alpha_O)} \frac{C_{e,N}}{C_{e,N}} \frac{\Delta W_{ei,O}}{\Delta W_{ei,N}} \tag{8a}$$

For situations where the air is highly turbulent or stomatal opening is limited, $C_a \cong C_e$ and $\Delta W \cong \Delta W_{ei}$; the ratio of WUE can then be approximated as:

$$\frac{WUE_N}{WUE_O} \cong \frac{(1 - \alpha_N)\, C_{a,N}\, \Delta W_O}{(1 - \alpha_O)\, C_{a,N}\, \Delta W_N} \tag{8b}$$

α is conservative in that it generally remains nearly constant as air CO_2 concentration or PAR flux varies over short periods (Ball and Berry, 1982; Morison, 1987). Very few data are available on the long-term effects of different concentrations of CO_2 on α. If α remains essentially unchanged when plants are grown at high CO_2 (Radoglou *et al.*, 1992; Ryle *et al.*, 1992) then $\alpha_N = \alpha_O$ and Eq. 8b becomes:

$$\frac{WUE_N}{WUE_O} \cong \frac{C_{a,N}\, \Delta W_O}{C_{a,O}\, \Delta W_N} \tag{9}$$

Accordingly, as a first approximation, the change in WUE would be less than proportional to the change in CO_2 concentration, regardless of whether the species is C3 or C4. How much less is determined by the ratio of ΔW_O to ΔW_N, which in turn is determined by the extent to which leaf temperature is increased in the high CO_2 environment by partial stomatal closure. In apparent agreement with the prediction of Eq. 9, the data of Allen and coworkers (Allen, 1990) show that the WUE of soybean was increased by 120-130% when the CO_2 concentration at which the plants were grown was increased from 330 to 800 μmol mol^{-1}, an increase of 142%. Some studies have shown that WUE increases linearly with CO_2 concentration (Eamus and Jarvis, 1989).

Returning to the question of constancy of α, given the likelihood that the climate is going to become hotter under elevated CO_2 (larger ΔW), α may decline slightly. However, for plants grown for long periods under high CO_2, α may increase slightly at a given ΔW due to the often observed reduction in the photosynthetic capacity of the leaves of plants grown in high CO_2. These counteracting effects may minimise changes in α, causing the increase in WUE to be somewhat less than proportional to that in CO_2, as pointed out above. As for yield WUE, in many cases the HI of plants grown in elevated CO_2 did not change significantly, although there were more harvestable assimilate sinks per plant (Lawlor and Mitchell, 1991). Hence, improvement in yield WUE would be the sole result of improvements in photosynthetic and biomass WUE. Only in root crops was HI enhanced under high CO_2 (Idso *et al.*, 1988).

The elevated CO_2 level in the future world is likely to be accompanied by more frequent atmospheric droughts. However, plants, especially C4 species, may not necessarily experience more severe water stress because of the improvement in WUE under high CO_2. In any event, the combined effect of water stress and high CO_2 can be examined using the same approach. Denoting the situation of water stress under high CO_2 by the subscript $N+S$, $\Delta C_{N+S}/\Delta C_N = 1$ if C_i is not altered by the stress. Utilising Eq. 6 and 9:

$$\frac{WUE_{N+S}}{WUE_O} = \frac{WUE_{N+S}}{WUE_N}\,\frac{WUE_N}{WUE_O} \cong \frac{C_{a,N}}{C_{a,O}}\,\frac{\Delta W_O}{\Delta W_{N+S}} \tag{10}$$

Eq. 10 is basically the same as Eq. 9, except that ΔW must be evaluated for the stressed treatment under the new level of CO_2.

What about productivity under high CO_2, regardless of WUE? The ratio of assimilation rates under the original and at elevated CO_2 concentration, denoted by A_N and A_O respectively, can be expressed as:

$$\frac{A_N}{A_O} \cong \frac{(r'_{a,O} + r'_{e,O})}{(r'_{a,N} + r'_{e,N})}\,\frac{(1 - \alpha_N)}{(1 - \alpha_O)}\,\frac{C_{a,N}}{C_{a,O}} = \frac{g'_{ae,N}}{g'_{ae,O}}\,\frac{(1 - \alpha_N)}{(1 - \alpha_O)}\,\frac{C_{a,N}}{C_{a,O}} \tag{11}$$

Here it is convenient to express the path transmission characteristics in terms of conductances, g', with the subscript *ae* denoting the overall conductance for the boundary layer and epidermal segments. For the simplest case of constant α and low boundary layer resistance, the equation shows that the enhancing effect of higher C_a on assimilation and primary productivity is reduced in proportion to the reduction in gas phase conductance caused by stomatal closure. For example, if C_a is doubled but gas phase conductance is reduced by 25%, the new assimilation rate would be 50% higher than the original. Again, information on the constancy or variability of α is crucial to evaluate the applicability of this relationship. Eq. 11 is consistent with published results for cotton and maize (Wong, 1979), giving good estimates of the observed difference in A between leaves on plants grown at ambient and double ambient CO_2 with a good nitrogen supply.

The enhanced rate of assimilation per unit leaf area under high CO_2 may have a variable effect on biomass production, depending on the growth phase. The more rapid assimilation under high CO_2 usually leads to faster leaf area development (Cure *et al.*, 1989). In the canopy cover limiting phase (Fig. 3), this effect compounds with time, resulting in a larger percentage increase in biomass than in assimilation per unit leaf area. This is consistent with the results of Wong (1979). The photosynthetic rate of cotton leaves was enhanced by 50% under high CO_2, but biomass was doubled. In contrast, during the radiation-limiting phase, leaf area is already adequate and there would be no compounding effect of adding leaf area. The increase in biomass production would then be more closely proportional to the increase in photosynthetic rate. In fact, excessive leaf area may be a respiratory drain on the carbon pool in such cases. There is much conflict in the literature concerning the extent of the enhancing effect of high CO_2 on biomass production within individual plant species. However, much of the difference might have been explained if a distinction had been made between the canopy-limiting and the radiation-limiting phases.

ACKNOWLEDGMENT

This work was partly supported by USDA Competitive Grant 88-37/30-3849 and by Regional Research Project W-154.

REFERENCES

Acevedo E, Hsiao TC, Henderson DW (1971) Immediate and subsequent growth responses of maize leaves to changes in water status. Plant Physiology 48: 631-636

Allen LH Jr. (1990) Plant responses to rising carbon dioxide and potential interactions with air pollutants. Journal of Environmental Quality 19: 15-34

Amthor JS (1989) Respiration and crop productivity. Springer, New York Berlin Heidelberg London Paris Tokyo

Arkley RJ (1963) Relationships between plant growth and transpiration. Hilgardia 34: 559-584

Ball JT, Berry JA (1982) The C_i/C_s ratio: a basis for predicting stomatal control of photosynthesis. Carnegie Institution of Washington Yearbook 1981: 88-92

Björkman O (1981) The response of photosynthesis to temperature. In: Grace J, Ford ED, Jarvis PG (eds) Plants and their atmospheric environment. Blackwell, Oxford London Edinburgh Boston Melbourne, pp 273-301

Bolaños JA, Hsiao TC (1991) Photosynthesis and respiratory characterization of field grown tomato. Photosynthesis Research 28: 21-32

Boyer JS (1970) Leaf enlargement and metabolic rates in corn, soybean, and sunflower at various leaf water potentials. Plant Physiology 46: 233-235

Bradford KJ, Hsiao TC (1982) Physiological responses to moderate water stress. In: Lange OL, Nobel PS, Osmond CB, Ziegler H (eds) Physiological plant ecology II Water relations and carbon assimilation. Encyclopedia of plant physiology, Vol. 12B. Springer, Berlin Heidelberg New York, pp 263-324

Campbell WJ, Allen LH Jr, Bowes G (1990) Response of soybean canopy photosynthesis to CO_2 concentration, light, and temperature. Journal of Experimental Botany 41: 427-433

Charles-Edwards DA, Doley D, Rimmington GM (1986) Modelling plant growth and development. Academic Press, Sidney Orlando San Diego New York Austin London Montreal Tokyo Toronto

Connor DJ, Palta JA, Jones TR (1985) Response of sunflower to strategies of irrigation. II. Crop photosynthesis and transpiration. Field Crop Research 12: 281-293

Cure JD, Rufty TW Jr, Israel DW (1989) Alterations in soybean leaf development and photosynthesis in a CO_2-enriched atmosphere. Botanical Gazette 150: 337-345

de Wit CT (1958) Transpiration and crop yields. Versl. Landbouwk. Onderz. 64.6. Institute of Biological and Chemical Research on Field Crops and Herbage, Wageningen, The Netherlands

Demetriades-Shah TH, Fuchs M, Kanemasu ET, Flitcroft I (1992) A note of caution concerning the relationship between cumulated intercepted solar radiation and crop growth. Agricultural and Forest Meteorology 58: 93-207

Eamus D (1991) The interaction of rising CO_2 and temperatures with water use efficiency. Plant, Cell and Environment 14: 843-852

Eamus D, Jarvis PG (1989) The direct effects of increase in the global atmospheric CO_2 concentration on natural and commercial temperate trees and forests. Advances in Ecological Research 19: 1-55

Farquhar GD, Ehleringer JR, Hubick KT (1989) Carbon isotope discrimination and photosynthesis. Annual Review of Plant Physiology and Plant Molecular Biology 40: 503-537

Farquhar GD, Sharkey TD (1982) Stomatal conductance and photosynthesis. Annual Review of Plant Physiology 33: 317-345

Fischer RA (1983) Wheat. In: Productivity of field crops. International Institute of Rice Research, Los Baños, Philippines, pp 129-154

Fischer RA, Turner NC (1978) Plant productivity in the arid and semiarid zones. Annual Review of Plant Physiology 29: 277-317

Gallagher JN, Biscoe PV (1978) Radiation absorption, growth and yield of cereals. Journal of Agricultural Science, Cambridge 91: 47-60

Hanks JR (1983) Yield and water-use relationships: an overview. In: Taylor HM, Jordan WR, Sinclair TR (eds) Limitations to efficient water use in crop production. Agronomy Society of America, Madison, Wisconsin, USA, pp 393-411

Held AA, Hsiao TC (1993) Canopy assimilation and evapotranspiration of well watered crops in the field: I. Closed canopies and role of stomata (in preparation)

Held AA, Steduto P, Orgaz F, Matista A, Hsiao TC (1990) Bowen ratio/energy balance technique for estimating crop net CO_2 assimilation, and comparison with a canopy chamber. Theoretical and Applied Climatology 42: 203-213

Hesketh J, Baker D (1967) Light and carbon assimilation by plant communities. Crop Science 7: 285-293

Hsiao TC (1973) Plant responses to water stress. Annual Review of Plant Physiology 24: 519-570

Hsiao TC (1982) The soil-plant-atmosphere continuum in relation to drought and crop production. In: Drought resistance in crops, with emphasis on rice. International Rice Research Institute, Los Baños, Philippines, pp 39-52

Hsiao TC, Bradford KJ (1983) Physiological consequences of cellular water deficits. In: Taylor HM, Jordan WR, Sinclair TR (eds) Limitations to efficient water use in crop production. Agronomy Society of America, Madison, Wisconsin, USA, pp 227-265

Idso SB, Kimball BA, Mauney JR (1988) Effects of atmospheric CO_2 enrichment on root : shoot ratios of carrot, radish, cotton and soybean. Agriculture, Ecosystems and Environment 21: 293-299

Jarvis PG, James GB, Landsberg JJ (1976) Coniferous forest. In: Monteith JL (ed) Vegetation and the atmosphere. Case studies. Vol. 2. Academic Press, London, UK, pp 171-240

Jarvis PG, McNaughton KG (1986) Stomatal control of transpiration: scaling up from leaf to region. Advances in Ecological Research 15: 1-49

Jones JW, Zur B, Bennett JM (1986) Interactive effects of water and nitrogen stresses on carbon and water vapor exchange of corn canopies. Agricultural and Forest Meteorology 38: 113-126

Jones P, Jones JW, Allen LH (1985) Seasonal carbon and water balances of soybeans grown under stress treatments in sunlit chambers. Transactions of the American Society of Agricultural Engineers 28: 2021-2028

Kanemasu ET (1983) Yield and water-use relationships: Some problems of relating grain yield to transpiration. In: Taylor HM, Jordan WR, Sinclair TR (eds) Limitations to efficient water use in crop production. Agronomy Society of America, Madison, Wisconsin, USA, pp 413-417

Kim J, Verma SB (1991) Modeling canopy photosynthesis: scaling up from a leaf to canopy in a temperate grassland ecosystem. Agricultural and Forest Meteorology 57: 187-208

Kimball BA, Idso SB (1983) Increasing atmospheric CO_2: effects on crop yield, water use and climate. Agricultural Water Management 7: 55-72

Lawlor DW, Mitchell RAC (1991) The effects of increasing CO_2 on crop photosynthesis and productivity: a review of field studies. Plant, Cell and Environment 14: 807-818

McKree KJ (1970) An equation for the rate of respiration of white clover plants grown under controlled conditions. In: Setlik I (ed) Prediction and measurement of photosynthetic productivity. Centre for Agricultural Publishing and Documentation, Wageningen, The Netherlands, pp 221-229

McKree KJ (1981) Photosynthetically active radiation. In: Lange OL, Nobel PS, Osmond CB, Ziegler H (eds) Physiological plant ecology I. Responses to the physical environment. Encyclopedia of plant physiology, Vol. 12A. Springer, Berlin Heidelberg New York, pp 41-55

McNaughton KG, Jarvis PG (1991) Effects of spatial scale on stomatal control of transpiration. Agricultural and Forest Meteorology 54: 279-301

Monteith JL (1973) Principles of environmental physics. Edward Arnold, London, UK

Monteith JL (1977) Climate and the efficiency of crop production in Britain. Philosophical Transactions of the Royal Society London B 281: 277-294

Morison JIL (1985) Sensitivity of stomata and water use efficiency to high CO_2. Plant, Cell and Environment 8: 467-474

Morison JIL (1987) Intercellular CO_2 concentration and stomatal response to CO_2. In: Zeiger E, Farquhar GD, Cowan IR (eds) Stomatal function. Stanford University Press, Stanford, California, USA, pp 229-251

Penning de Vries FWT (1975) Use of assimilates in higher plants. In: Cooper JP (ed) Photosynthesis and productivity in different environments. Cambridge University Press, Cambridge, UK, pp 459-480

Puckridge DW, Ratkowsky DA (1971) Photosynthesis of wheat under field conditions. IV The influence of density of leaf area index on the response to radiation. Australian Journal of Agricultural Research 22: 11-20

Puech-Suanzes I, Hsiao TC, Fereres E, Henderson DW (1989) Water stress effects on the carbon exchange rates of three upland cotton (*Gossipium hirsutum*) cultivars in the field. Field Crop Research 21: 239-255

Radoglou KM, Aphalo P, Jarvis PG (1992) Response of photosynthesis, stomatal conductance and water use efficiency to elevated CO_2 and nutrient supply in acclimated seedlings of *Phaseolus vulgaris* L. Annals of Botany 70: 257-264

Ritchie JT, Burnett E (1971) Dryland evaporative flux in a subhumid climate. II. Plant influences. Agronomy Journal 63: 56-62

Ryle GJA, Woledge J, Tewson V, Powell CE (1992) Influence of elevated CO_2 and temperature on the photosynthesis and respiration of white clover dependent on N_2 fixation. Annals of Botany 70: 213-220

Sale PJM (1977) Net carbon exchange rates of field-grown crops in relation to irradiance and dry weight accumulation. Australian Journal of Plant Physiology 4: 555-569
Sharp RE, Davies WJ (1979) Solute regulation and growth by roots and shoots of water-stressed maize plants. Planta 146: 319-326
Tanner CB, Sinclair TR (1983) Efficient water use in crop production: research or re-search? In: Taylor HM, Jordan WR, Sinclair TR (eds) Limitations to efficient water use in crop production. Agronomy Society of America, Madison, Wisconsin, USA, pp 1-27
Varlet-Grancher C, Gosse G, Chartier M, Sinoquet H, Bonhomme R, Allirand JM (1989) Mise au point: rayonnement solaire absorbé ou intercepté par un couvert végétal. Agronomie 9: 419-439
Williams WA, Loomis RS, Lepley CR (1965) Vegetative growth of corn as affected by population density. I. Productivity in relation to interception of solar radiation. Crop Science 5: 211-215
Wong SC (1979) Elevated atmospheric partial pressure of CO_2 and plant growth. I. Interactions of nitrogen nutrition and photosynthetic capacity in C3 and C4 plants. Oecologia 44: 68-74
Wong SC, Cowan IR, Farquhar GD (1979) Stomatal conductance correlates with photosynthetic capacity. Nature (London) 282: 424-426

WATER RELATIONS IN CONTROLLED ENVIRONMENTS AND THE FIELD

John W. Radin
US Department of Agriculture
Agricultural Research Service
Western Cotton Research Laboratory
Phoenix
Arizona 85040
USA

ABSTRACT

Water availability is often a primary limitation to biomass production. With the aim of explaining and predicting water use in the field, plant physiologists frequently study the fundamental biology of water uptake, transport and evaporation in controlled environments (glasshouses, growth chambers). However, despite the accumulation of knowledge, an understanding of the fundamental biological principles operating in controlled environments rarely leads to improved predictions of plant water relations in the field. This paper discusses some of the factors contributing to these apparently irreconcilable differences. Of the many possible sources of error in controlled environment experiments, a few predominate. Common problems are: (1) rooting restrictions resulting from using plants grown in pots; (2) the use of plants at an inappropriate developmental stage, usually very early in their life cycle and; (3) environmental conditions which fail to simulate the field environment adequately, and are critical because they fundamentally alter plant water status or responses to water status. In many cases, resistance to water stress in the field varies with developmental changes in the water transport capacity of the root system as the plant matures. To study such effects in controlled environments requires mature plants, grown in large pots that do not alter normal root development, and an evaporative demand sufficient to reveal hydraulic limitations imposed by the root system. Studies in controlled environments can never completely match those in the field, but the gap between the two can be narrowed. The sources of error in controlled environments are often easily addressed. Focussing on such critical factors will remove obstacles to interpreting results, improve relevance to the field, and speed the application of basic principles to predicting performance under field conditions.

NATO ASI Series, Vol. I 16
Interacting Stresses on Plants in a Changing Climate
Edited by M. B. Jackson, C. R. Black

INTRODUCTION

In green plants, as in almost all living organisms, water constitutes by far the majority of tissue mass. Terrestrial plants survive and grow only because they can transfer water from the soil to their aerial organs. The physical principles governing water transport require that aerial tissues be subjected to water deficits (i.e., water potentials below that of the source of water in the soil), which are generated by transpiration (Davies, 1986). Small deficits are harmless, but when water is scarce, the balance between uptake and transpiration is disturbed, and tissue deficits become more severe (plants become stressed). Numerous mechanisms have evolved which promote adaptation to water scarcity, and these aspects of stress physiology have been reviewed over the years (Hsiao, 1973; Turner, 1979; Bradford and Hsiao, 1982; Passioura 1983; Taylor *et al.*, 1983; Boyer, 1985; Passioura, 1988a; Zeevaart and Creelman, 1988; Cornish and Radin, 1990; Davies and Zhang, 1991). The object of this chapter is not to re-review the literature, but to interpret and discuss the state of our knowledge about the basic biology of water stress. The specific issue is how to make such knowledge acquired in artificial controlled environments applicable to the field.

Historically, plant physiology has been occupied with the pursuit of mechanisms by which plants grow and develop. Plant physiologists assume that detailed knowledge of mechanisms will eventually lead to better plant material or management schemes for farmers. For various reasons to be explained below, the pursuit of mechanisms has largely involved experiments in controlled environments which minimise variability. The environment is often optimal except for the stressful factor being tested. As a result, our knowledge of environmental responses is very narrowly based in terms of the range of possible environments. None-theless, eventual application of the knowledge requires that observed responses to stress must be similar across all environments likely to occur in the field. This is crucial because, even under the best of circumstances, we are predicting in advance that a genotype or management scheme will be beneficial, without knowing what next year's environment will be like. Especially in the context of unpredictable global climate change, establishing the generality of any observed plant responses must be a paramount objective of all experiments.

Applicability also requires quantification. This need is visualised by a simplified general relationship adapted from Passioura (1983) that expresses the dependence of yield, Y, on the capture and conversion of a limiting resource (in this case water) to generate biomass, M:

$$Y = \begin{array}{c}\text{Resource}\\ \text{capture}\\ \text{rate}\end{array} \times \begin{array}{c}\text{Resource}\\ \text{utilisation}\\ \text{efficiency}\end{array} \times \begin{array}{c}\text{Harvest}\\ \text{index}\end{array} \times \text{Time} \qquad [1]$$

$$Y\,(g) = g\,H_2O\,d^{-1} \times g\,M\,g^{-1}\,H_2O \times g\,Y\,g^{-1}\,M \times d$$

If the various terms of this equation are independent they may be treated in isolation; such an approach frequently underlies physiological studies in which production is broken into its component processes for separate study. A simple analysis, however, reveals that the terms are not independent. For example, osmotic adjustment can increase the water potential gradient from the soil to the leaves and thereby maintain water uptake rate. However, the accumulation of osmotica in leaves sequesters carbon from assimilate pools, and harvest index is decreased. Theoretically, should osmotic adjustment increase or decrease growth and yield under stress? The answer to this quantitative question is not obvious (Munns, 1988). Despite the attention the subject has received over the years, it is difficult to demonstrate any net benefit associated with osmotic adjustment (McCree and Richardson, 1987).

If the premise stated above is accepted, that studies of isolated physiological processes in a narrow range of environments are seldom applicable to growth limitations in the field, then we must learn how to overcome the obstacles to applicability. This chapter highlights some important factors underlying water-limited yields in the field, and discusses why experiments in controlled environments seldom address these factors. Fortunately, some of the obstacles to applicability can be removed.

APPROPRIATE USE OF CONTROLLED ENVIRONMENTS

There are several valid reasons for using controlled environment facilities (growth chambers, glasshouses):

1. To minimise uncontrolled environmental variation. Every variation from a standard environment leads to unintended, unknown consequences and irreproducible experiments. Without standardised conditions, deductions about cause and effect may be beyond reach. Nonetheless, this argument in favour of controlled conditions restates a theme of the Introduction, that plant responses to most stimuli vary with environment, and plant physiologists often choose to work with a narrow information base.

2. To separate covarying environmental factors. In the field, many variables are linked, such as temperature and irradiance on a daily time scale, or temperature and photoperiod on a yearly time scale. Growth chambers or other artificial environments offer the means to separate such factors and to test their effects independently.

3. To allow continuous experimentation. In most cropping areas, periods of crop growth and production usually alternate with periods of dormancy or fallow. Controlled environment facilities allow year-long work with actively growing plants. Thus, from the standpoint of the scientist's available time, controlled environment facilities are efficient.

Despite these reasons for using them, controlled environments impose experimental constraints. Environmental chambers are expensive to build and operate, especially if realistically high irradiances are maintained. As a result, available space is limited, and experiments tend to be small with respect to both the number and size of plants. Furthermore, as plants grow, variability between individuals increases, and this problem is exacerbated if the plants shade each other and compete for available light (especially in low-light environments). Together, these problems strongly discourage the use of mature plants, especially in dense stands. Experiments also tend to be very short-term. Articles in a randomly chosen issue of the journal *Plant Physiology* (vol. 98, number 4, April 1992) illustrate the typical use of controlled environments. Twenty-nine articles reported experiments with higher plants grown in controlled environments, with an average plant age of

14 days. Six studies utilised field-grown material but were not field experiments (e.g., experiments using vegetables purchased at the local market). In addition, six other papers relied solely upon cultured cells. There were no papers reporting experiments in realistic environments (i.e., in the field, or in carefully simulated field-like artificial environments).

Fundamental studies of plant water relations and water stress often follow the same pattern. Detailed knowledge of mechanisms arises from very young plants, grown in isolation under environmental conditions that are prescribed by mechanical limitations to cooling or lighting capability. Experiments are short-term, which dictates the use of small pots so that water can be depleted relatively rapidly to provide stress. Years of such experiments have produced a well-established set of responses to stress including declining water potential, stomatal closure, net synthesis of the phytohormone abscisic acid (ABA) which mediates stomatal closure, slowing of leaf expansion via effects on cell division and enlargement, acceleration of leaf senescence, osmotic adjustment in leaves and roots, changes in cell wall properties, decreased photosynthetic capacity, and alteration of ion uptake and transport (Bradford and Hsiao, 1982). Some of these cell-level responses are considered to confer fitness for a stressful environment and thus are worthy of transfer to crops at risk from stress injury. As a result, they have been studied intensively at the level of gene action (e.g., Bray, 1988; Mason *et al.*, 1988). The long-term goal is to engineer these traits into stress-sensitive crop plants.

The order of these events seems to be reversed. Focussing on fundamental mechanisms as a starting point for problem-solving presumes that the narrow basis for those fundamentals is adequate to predict behaviour in a broader set of environments. Indeed, the very need for controlled environments to obtain interpretable results (see point 1 above) implicitly argues that the narrow basis may not be adequate. The isolation of fundamental studies from the 'real world' has led to divergences such that even the thermodynamic framework for describing stress has been questioned as a useful approach in the field (Sinclair and Ludlow, 1985). Dracup (1991) has discussed similar problems with salt tolerance work.

An alternative strategy is to start with field-based observations, with subsequent controlled environment work designed specifically to isolate and improve critical factors that have already been carefully identified. Fig. 1 shows a flow diagram depicting this strategy. After identifying a problem and formulating a hypothesis, some preliminary tests are required to test the feasibility of the hypothesis. If the tests fail, a new hypothesis can be developed with minimal waste of effort. To bypass these steps and begin fundamental studies directly is to risk being diverted by negative outcomes into the 'loops' near the bottom of the diagram. With a hypothesis that is not applicable to the field, there is no emerging from these loops and no hope of attaining the goal, i.e., escaping the bottom of the diagram with a tangible product to alleviate the problem.

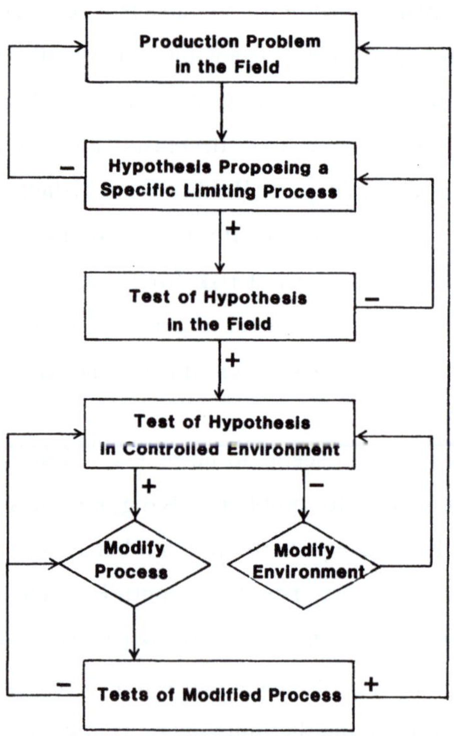

FIGURE 1. Flow diagram for the chain of events leading to problem-solving through research. The + and - symbols indicate pathways to be followed after positive and negative outcomes of tests.

The strategy is illustrated by studies of the water relations of soybean cultivars. Even under well-watered conditions in the field, low-yielding cultivars become water-stressed during hot afternoons, when transpiration is most rapid and the stress inhibits photosynthesis (Boyer *et al.*, 1980). The water transport capacity of these plants is inadequate to meet peak demand. By selecting for increased yield, plant breeders have unknowingly improved water transport. Improved transport is associated with greater root proliferation throughout the soil, especially at depths greater than 1 m (Boyer *et al.*, 1980). This discovery of an important source of yield improvement would not have occurred in a standard controlled environment, as there would have been no basis for differentiating between genotypes. First, evaporative demand in controlled environments is less than in the field, and stress levels that inhibit photosynthesis would not have been reached. Second, observation of differences in rooting density below 1 m would have been precluded in small pots, which may be only 10-20 cm deep. Third, the critical characteristics are those of fully grown plants, and could not have been seen in the small plants typically used in controlled environments.

In the light of this example, how can controlled environments be used to maximise the likelihood that studies will be applicable to the field? One approach is to simulate realistic growth conditions regardless of inconvenience, for example, by growing plants in containers in the glasshouse with soil volumes approximating those available in the field (Passioura, 1977). A second approach is to study traits under unrealistic conditions, but only after the expression of the traits under such conditions has been carefully validated by comparison to field behaviour. For example, Kaspar *et al.* (1984) grew several soybean genotypes in the glasshouse in long tubes of soil which allowed the roots to grow downward without hindrance. Across genotypes, rapid early taproot elongation in the glasshouse was associated with deeper root proliferation and deeper water extraction in the field. Thus a simple modification of normal cultural practice in the controlled environment produced data that were readily applicable to the field. Having reached the 'Modify

Process' step shown in Fig. 1, detailed fundamental and molecular studies of the factors contributing to root length under those conditions are now appropriate. In other words, the mechanisms of root growth can be studied with confidence that improvements will increase water extraction in the field and thus solve a problem of poor water uptake.

WATER RELATIONS OF PLANTS GROWN IN SMALL POTS

Studies of the water relations of soybean in the field (Boyer *et al.*, 1980) illustrate the three major problems involved in the routine application of results from controlled environments to the field: the growth of plants in small pots, use of plants at an inappropriate developmental stage, and important differences in environment that alter critical plant characteristics. I will not dwell upon the effects of small pots, except to point out that they may have multiple effects. Small pots limit the amount of available water, leading to possible water stress even when the pot is fully charged with water (sometimes labelled 'hydraulic growth restriction'), and the rapid onset of severe stress when water is withheld from the soil. The rapid onset of stress precludes the occurrence of slower stress responses that would normally occur in the field (Jones and Rawson, 1979). Small pots also physically restrict root growth, with negative consequences for shoot growth. In cotton, for example, rooting restriction is associated with a non-hydraulic (i.e., not caused by water) restriction of top growth and photosynthesis (Thomas and Strain, 1991). Growth rate was proportional to pot volume over a range from 2.5 to at least 12 l in one study (Ben-Porath and Baker, 1990) and 2 to 25 l in another (Carmi and Shalhevet, 1983). In the latter case, plant water potentials did not indicate a significant difference in the water status of the tops. These non-hydraulic effects remain an active but controversial area of research (Kramer, 1988; Passioura, 1988b; Ludlow *et al.*, 1989). Root restriction due to the strength of dry soil has been proposed to explain some of the effects of drought on plants (Masle and Passioura, 1987; Davies and Zhang, 1991).

STAGES OF PLANT DEVELOPMENT AND WATER RELATIONS

Phenological adaptation, in which a plant's development is matched to the demands of its environment, may be the most important single stress adaptation factor in the field. This principle is illustrated by the difference between maize, a determinate crop, and cotton, an indeterminate crop. With many crops, the flowering process is especially sensitive to water stress. In maize, stress at anthesis leads to failure of pollination or poor development of the fertilised embryos (Herrero and Johnson, 1983; Westgate and Boyer, 1985). Because flowering occurs over a defined interval, yield responds strongly to stress at anthesis (but see Sinclair *et al.*, 1990).

In contrast, because of its indeterminate flowering habit, cotton flowers and fruits more slowly and over a more extended period than maize. As a result, its overall reproductive productivity is less subject to short-term environmental perturbations. For example, in cotton grown for several years under stress of unpredictable timing, the most indeterminate line was also the highest-yielding (Quisenberry and Roark, 1976). Its developmental plasticity enabled it to take advantage of rains whenever they occurred.

In cotton, as in maize, severe or prolonged water stress disrupts flowering and fruiting. More importantly though, flowering begins while the plants are relatively small and before the crop has produced a full canopy. Vegetative and reproductive growth occur simultaneously over an extended interval. Stress limits vegetative growth, and because flowers are produced only in the axils of leaves, the smaller plant has fewer flowering sites. The substantial overlap of vegetative and reproductive growth periods necessarily leads to some linkage between the two processes, so that a stressed plant tends to carry only the number of fruits that it can mature. At the same time, the smaller plants conserve more water for later use when the fruits are maturing. Delaying water use until after anthesis is very effective in protecting yield against water stress (Passioura, 1983). In contrast, when vegetative and reproductive growth are temporally separated, as in maize, the balance between the two depends upon the randomly fluctuating environment to a much greater degree.

ROOT DYNAMICS AND WATER RELATIONS

The importance of root development for water relations in the field has already been mentioned. Again using cotton as an example, total root length generally reaches a maximum at the beginning of flowering (Taylor and Klepper, 1974) and then declines slowly, perhaps because fruits are strong sinks that divert assimilates from the roots (Jurgens *et al.*, 1978). The root decline can make fruiting plants sensitive to conditions of high evaporative demand even when the soil is wet, as the degenerated root system may no longer be able to meet the demands of the shoot for water. Chronic mild water stress is the result, even in wet soil.

Water stress during heavy fruiting of cotton has a particularly strong effect on dieback of the root system (Taylor and Klepper, 1974) and, as a result, the plants become extremely sensitive to soil water stress. Presumably because fruits alter the whole-plant carbon balance, regrowth after rewatering is also much slower than during vegetative growth (Taylor and Klepper, 1974). Thus the effect of stress on the root system is not only more severe during fruiting than during vegetative growth, but recovery after rewatering is more prolonged.

Radin *et al.* (1989) monitored root degeneration in cotton by determining the leaf water potentials and apparent hydraulic conductance of plants in the field. They found little evidence of hydraulic restriction during vegetative growth, but the restriction greatly increased as fruiting progressed. Root degeneration during fruiting was much less on high-frequency (daily drip) irrigation, as compared to application of water at the standard 14 day intervals. The difference was ascribed to incipient stress at the end of the long irrigation intervals (Radin *et al.*, 1989), which had relatively little effect during vegetative growth but became injurious during heavy fruiting. In a later study, Radin *et al.* (1992) halved the interval between furrow-applied irrigations for one month during heavy fruiting, but with the total amount of applied water almost unchanged. This single change caused sustained increases in water uptake capacity and increased yield by 25%. These results indicate that the aspects of the water relations of cotton most critical to stress resistance are those observed only during fruiting. As with soybeans, the critical traits appear to be related to the root system.

With modern high-yielding genotypes that partition a large fraction of assimilates to fruits, root degeneration may be more widespread than with slower-fruiting, lower-yielding genotypes. Tests of this proposition using cotton have so far been inconclusive (unpublished results of the author). Nonetheless, a recent study of indeterminate peanut genotypes (Rao *et al.*, 1989) is consistent with this analysis. When drought was imposed during seed fill, percentage yield loss was strongly positively correlated with yield potential (Fig. 2). In other words, high yield potential engendered high sensitivity to stress during seed fill. When drought occurred before seed fill, however, there was no such correlation (Rao *et al.*, 1989). The experiment was structured so that partitioning to fruits was the major genetic factor affecting pod yield. The results strongly indicate that the severity of drought injury during fruiting is proportional to the intensity of partitioning to yield. It may be inferred that the site of injury is the root system, but since Rao *et al.* (1989) did not report on either the water status or rooting densities of the genotypes involved, this inference cannot be confirmed. Similar deleterious effects of fruits on rooting are seen in grapes (McLean *et al.*, 1992) and poplar trees, but generally not in cereal grains or dry beans (A.J.M. Smucker, Michigan State University, personal communication). Smucker inferred that perennials, or other plants with strong vegetative sinks as well as reproductive sinks for storage of assimilates, are most likely to have root systems so severely limited by assimilate availability that water uptake is restricted.

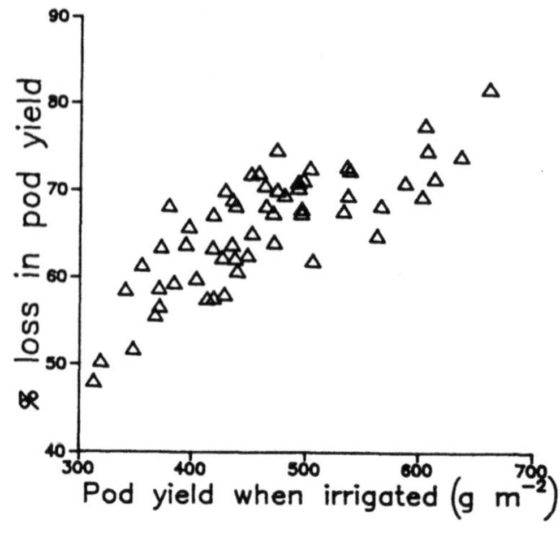

FIGURE 2. Sensitivity to water stress of peanut genotypes differing in yield potential. Stress was limited to the seed-filling period. Data from Rao et al. *(1989). Reproduced with permission of the publisher.*

ATMOSPHERIC ENVIRONMENT AND WATER RELATIONS

Assume for the moment that pots are large enough not to restrict root growth, and that developmental stage, determinacy and the number of competing sinks of the plants have all been taken into account. One must still ask whether controlled environments effectively simulate critical aspects of the field environment. There are usually many differences between the field and controlled environments, so identification of the most critical parameters is intellectually challenging. Nonetheless, simplification is often possible. Again drawing from the example of the water relations of soybeans (Boyer *et al.*, 1980), a critical factor in the controlled environment is the evaporative demand (a function of ambient temperature and humidity), which must approach that in the field for the steady-state plant water balance to be realistic. Similarly, with cotton, of the many differences between the two environments, one factor explains much of the divergence in plant responses: ambient temperature (Radin, 1992). Controlled environments are often maintained at an ambient temperature approximating to the optimum growth temperature, at first glance a logical choice. In most cotton-growing areas of the world, however, temperatures exceed that optimum for much of the season. In this and in some other heat-adapted species (Drake *et al.*, 1970), high temperatures increase stomatal conductance. The effect of this is to enhance the cooling of leaves by evaporation of water. Burke *et al.* (1990) developed these concepts into a 'thermal stress index' for cotton based upon leaf temperatures rather than air temperatures. They found an excellent relationship between the index and crop performance, concluding that stomatal regulation of leaf temperature is an important component of thermal adaptation.

Radin (1992) described numerous lines of evidence that indicate an important role of ABA in the temperature responses of stomata. Both in the field and in controlled environments, ABA concentrations in leaves are very low at high temperatures, and some other manifestations of ABA action on stomata disappear. In both environments, reintroduction of exogenous ABA restores stomatal behaviour to that seen at cooler temperatures. High temperature accelerates the conversion of ABA to its metabolites, primarily phaseic acid, in leaf discs. Thus it appears that an important aspect of plant adaptation to the field environment may be reduced to one or a few metabolic steps in a single biochemical pathway, the ABA oxidation pathway. Along with the rooting-limited soybean cultivars, this is another system that has been successfully transferred from field to controlled environments. As a result, fundamental studies of this system are likely to have direct applications for crop productivity.

CONCLUSIONS

Stress-resistant plants can balance incoming and outgoing fluxes of water better than their stress-sensitive counterparts. Predicting stress resistance in the field requires a quantitative understanding of the factors affecting those fluxes. Many parameters that affect water balance and water fluxes in the field are often systematically excluded from study in physiological experiments under controlled environments. Root distribution and the partitioning of assimilates between roots and fruits are among those factors; much information implicates partitioning in the determination of stress resistance during reproduction, but few controlled environment studies allow unrestricted root growth in large containers. Fewer still carry the observations all the way through to the reproductive stage of growth. Evaporative demand is also critical to a quantitative understanding of water status in the field, but is seldom simulated accurately in controlled environments.

Physiological relevance is certainly not achieved by working only in the field. To the contrary, rigorous studies require the use of controlled environments. Nevertheless, deductions about yield-limiting processes from a narrowly based set of observations are often misleading. Variable plant responses to environmental factors mean that basic studies of yield limiting factors must be designed around field-based observations, rather than *vice versa*. When this protocol is followed, experiments in controlled environments are easily designed to produce results applicable to the field.

Not all aspects of the environment need match field conditions for an experiment to simulate field responses and for an environment to be 'realistic'. For example, stomatal regulation of transpiration by cotton is largely determined by ambient temperature independently of most other factors. The examples described here indicate that closing the gap between controlled environments and the field can be rather straightforward. Only after this occurs can fundamental biological principles elucidated in controlled environments be applied to predicting or modifying performance in the field.

REFERENCES

Ben-Porath A, Baker DN (1990) Taproot restriction effects on growth, earliness, and dry weight partitioning of cotton. Crop Science 30: 809-814

Boyer JS (1985) Water transport. Annual Review of Plant Physiology 36: 473-516

Boyer JS, Johnson RR, Saupe SG (1980) Afternoon water deficits and grain yields in old and new soybean cultivars. Agronomy Journal 72: 981-986

Bradford KJ, Hsiao TC (1982) Physiological responses to moderate water stress. In: Lange OL, Nobel PS, Osmond CB, Zeigler H (eds) Physiological plant ecology II. Water relations and carbon assimilation. Encyclopedia of plant physiology, Vol. 12B. Springer-Verlag, Berlin Heidelberg New York, pp 264-324

Bray EA (1988) Drought- and ABA-induced changes in polypeptide and mRNA accumulation in tomato leaves. Plant Physiology 88: 1210-1214

Burke JJ, Hatfield JL, Wanjura DF (1990) A thermal stress index for cotton. Agronomy Journal 82: 526-530

Carmi A, Shalhevet J (1983) Root effects on cotton growth and yield. Crop Science 23: 875-878

Cornish K, Radin JW (1990) From metabolism to organism: an integrative view of water stress emphasizing abscisic acid. In: Katterman FJ (ed) Environmental injury to plants. Academic Press, Orlando, pp 89-112

Davies WJ (1986) Transpiration and the water balance of plants. In: Steward FC, Sutcliffe JF, Dale JE (eds) Plant physiology: a treatise, Vol. IX. Water and solutes in plants. Academic Press, Orlando, USA, pp 49-137

Davies WJ, Zhang J (1991) Root signals and the regulation of growth and development of plants in drying soil. Annual Review of Plant Physiology and Plant Molecular Biology 42: 55-76

Dracup M (1991) Increasing salt tolerance of plants through cell culture requires greater understanding of tolerance mechanisms. Australian Journal of Plant Physiology 18: 1-15

Drake B, Raschke K, Salisbury FB (1970) Temperatures and transpiration resistances of *Xanthium* leaves as affected by air temperature, humidity, and wind speed. Plant Physiology 46: 324-330

Herrero MP, Johnson RR (1983) Drought stress and its effects on maize reproductive systems. Crop Science 21: 105-110

Hsiao TC (1973) Plant responses to water stress. Annual Review of Plant Physiology 24: 519-570

Jones MM, Rawson HM (1979) Influence of rate of development of leaf water deficits upon photosynthesis, leaf conductance, water use efficiency, and osmotic potential in sorghum. Physiologia Plantarum 45: 103-111

Jurgens SK, Johnson RR, Boyer JS (1978) Dry matter production and translocation in maize subjected to drought during grain fill. Agronomy Journal 70: 678-682

Kaspar TC, Taylor HM, Shibles RM (1984) Taproot-elongation rates of soybean cultivars in the glasshouse and their relation to field rooting depth. Crop Science 24: 916-920

Kramer PJ (1988) Changing concepts regarding plant water relations. Plant, Cell and Environment 11: 565-568

Ludlow MM, Sommer KJ, Flower DJ, Ferraris R, So HB (1989) Influence of root signals resulting from soil dehydration and high soil strength on the growth of crop plants. In: Randall DD, Blevins DG (eds) Current topics in plant biochemistry and physiology, Vol. 8. University of Missouri, Columbia, USA, pp 81-99

Masle J, Passioura JB (1987) Effect of soil strength on the growth of young wheat plants. Australian Journal of Plant Physiology 14: 643-656

Mason HS, Mullet JE, Boyer JS (1988) Polysomes, messenger RNA, and growth in soybean stems during development and water deficit. Plant Physiology 86: 725-733

McCree KJ, Richardson SG (1987) Stomatal closure vs. osmotic adjustment: a comparison of stress responses. Crop Science 27: 539-543

McLean M, Howell GS, Smucker AJM (1992) A minirhizotron system for *in situ* root observation studies of Seyval grapevines. American Journal of Enology and Viticulture 43: 87-89

Munns R (1988) Why measure osmotic adjustment? Australian Journal of Plant Physiology 15: 717-726

Passioura JB (1977) Grain yield, harvest index, and water use of wheat. Journal of the Australian Institute of Agricultural Science 43: 117-120

Passioura JB (1983) Roots and drought resistance. Agricultural Water Management 7: 265-280

Passioura JB (1988a) Water transport in and to roots. Annual Review of Plant Physiology and Plant Molecular Biology 39: 245-265

Passioura JB (1988b) Response to Dr. P.J. Kramer's article, 'Changing concepts regarding plant water relations.' Plant, Cell and Environment 11: 569-571

Quisenberry JE, Roark B (1976) Influence of indeterminate growth habit on yield and irrigation water-use efficiency in upland cotton. Crop Science 16: 762-765

Radin JW (1992) Reconciling water-use efficiencies of cotton in field and laboratory. Crop Science 32: 13-18

Radin JW, Mauney JR, Kerridge PC (1989) Water uptake by cotton roots during fruit filling in relation to irrigation frequency. Crop Science 29: 1000-1005

Radin JW, Reaves LL, Mauney JR, French OF (1992) Yield enhancement in cotton by frequent irrigations during fruiting. Agronomy Journal 84: 551-557

Rao RCN, Williams JH, Singh M (1989) Genotypic sensitivity to drought and yield potential of peanut. Agronomy Journal 81: 887-893

Sinclair TR, Bennett JM, Muchow RC (1990) Relative sensitivity of grain yield and biomass accumulation to drought in field-grown maize. Crop Science 30: 690-693

Sinclair TR, Ludlow MM (1985) Who taught plants thermodynamics? The unfulfilled potential of plant water potential. Australian Journal of Plant Physiology 12: 213-217

Taylor HM, Jordan WR, Sinclair TR (eds)(1983) Limitations to efficient water use in crop production. American Society of Agronomy, Madison, Wisconsin, USA

Taylor HM, Klepper B (1974) Water relations of cotton. I. Root growth and water use as related to top growth and soil water content. Agronomy Journal 66: 584-588

Thomas RB, Strain BR (1991) Root restriction as a factor in photosynthetic acclimation of cotton seedlings grown in elevated carbon dioxide. Plant Physiology 96: 627-634

Turner NC (1979) Drought resistance and adaptation to water deficits in crop plants. In: Mussell H, Staples RC (eds) Stress physiology in crop plants. Wiley-Interscience, New York, USA, pp 343-372

Westgate ME, Boyer JS (1985) Carbohydrate reserves and reproductive development at low leaf water potentials in maize. Crop Science 25: 762-769

Zeevaart JAD, Creelman RA (1988) Metabolism and physiology of abscisic acid. Annual Review of Plant Physiology and Plant Molecular Biology 39: 439-473

THE PHYSIOLOGY OF SURVIVAL AT THE LIMITS OF FARMING IN THE DRY TROPICS

G.R. Squire
Scientific Adviser's Unit
Scottish Office Agriculture and Fisheries Department
47 Robb's Loan
Edinburgh EH14 1TW
UK

ABSTRACT

The dominant form of agriculture changes systematically along any spatial gradient of increasing dryness in the tropics. Between humid and arid regions, dense stands of perennial trees give way to annual or biennial cereals, legumes and root crops, and these in turn to mixed arable and animal husbandry, and eventually to extensive rangeland. Such transitions reflect changes in two main attributes of the commonest plants. One is the 'return' of resources intercepted by dry matter invested in the resource-capturing structures (τ), a value that decreases as the resources become scarcer. For example, if expressed as total radiation intercepted per unit of vegetative dry matter in shoots, τ is typically 5 MJ g^{-1} for perennial palms and 1 MJ g^{-1} for short-season dryland cereals. The other major attribute is the minimum dry mass (W_v) that plants must accumulate before any economic yield is produced. W_v ranges from 10^5 g $plant^{-1}$ for some tree crops to 10-20 g $plant^{-1}$ for large annuals, 1-4 g $plant^{-1}$ for small cereals and legumes, and a small fraction of a gram for rangeland grasses. In contrast to these two attributes, the conversion ratios change much less systematically in relation to aridity. Transitions – whether from maize to millet, or arable to rangeland – increase the chance that the returns of dry matter will exceed W_v, and therefore that surplus dry matter will be available for economically important structures. For example, the transition from arable cropping to rangeland seems to occur when the total dry matter per plant is similar to W_v for the most adaptable dryland cereals. τ and W_v are inherently determined by physiological attributes of the plant. Particularly important are the longevity of leaves and roots, their expansion rates and area/mass or length/mass relations, the determinacy of growth habit, and the sensitivity of reproduction to drought and high temperature. Generally, longevity is the most important attribute and the one that can perhaps be most influenced by husbandry, water-harvesting practices or by reducing stand density. This paper examines the way these physiological attributes discriminate both between species and among stands of the same species, using examples from different tropical environments.

NATO ASI Series, Vol. I 16
Interacting Stresses on Plants in a Changing Climate
Edited by M. B. Jackson, C. R. Black

INTRODUCTION

In his account of travels across the Empty Quarter in Arabia, Thesiger (1959) tells how his companions among the bedu knew of seeds that can lie for years in the hot dry sand, then germinate when wetted, and of grazing that can remain green for years following a single fall of rain lasting perhaps a day and a night. These and many other dryland plants are adapted to drought in both their dormant and expanding states. In settled agriculture, however, the farmer has some control over the dormant state by planting seeds and tubers in anticipation of rain, whereas the weather and soil still strongly influence the expanding state. Accordingly, it is mainly attributes of the expanding state of agricultural plants that change systematically along any spatial gradient of dryness in the tropics.

In humid lowland or moist upland areas, the dominant agriculture consists of continuously expanding perennial shrubs and trees, yielding fruits, leaves and wood. Along with these perennials are grown annual cereals, legumes and root crops, that grow for only a few months. In seasonally humid regions, these annuals dominate, with the perennials remaining as individuals or small copses, giving a variety of produce. In dry regions, the changes occur more subtly: with decreasing moisture, large, stress intolerant cereals, taking half a year to reach maturity, give way to smaller, more tolerant short-season sorghums and millets. In areas drier still, plants grown mainly for their reproductive products are replaced by rangeland shrubs and grasses, whose leaves and stems are grazed by animals.

Over the range from the humid to the arid tropics, the physiological attributes that change to give rise to these differences in the dominant agricultural species are those defining the size, plasticity and longevity of the plant and its parts. Size and plasticity are determined by many attributes of the plant's architecture and the arrangement of its meristems. Nevertheless, broad comparisons between species can be made in terms of the minimum vegetative dry mass (W_v) that the plant must accumulate before any of its dry matter can be allocated to reproductive or other economic yield. Whether a plant in a given environment can achieve and surpass this mass, and so produce economic yield, is influenced by another set of attributes which determine the 'returns' of dry matter that a plant makes from 'investment' of dry matter in structures that intercept or take up the environmental resources. The longevity of these structures has a strong influence on this 'investment efficiency'.

This paper first examines the changes in W_v and investment efficiency that occur among some plants typically dominant or frequent in zones of different dryness in the tropics; and second, shows how these attributes are influenced by population density and nutrition. The account concentrates on plants and environments near the border between arable cropping and grazed rangeland.

SIZE, PLASTICITY AND THE MINIMUM VEGETATIVE MASS

It is difficult to measure the minimum vegetative mass, W_v, and few studies have attempted specifically to do so. W_v can be best obtained from experiments in which the resources available to each individual are varied over a wide range, for example by altering the availability of solar radiation, water or nutrients.

The value of W_v has been measured for oil and coconut palms where the limiting factor is the solar radiation available to each plant (Squire and Corley, 1987; Moss, 1992), and for some dryland plants where the limiting factor is rainfall (for example, Rees, 1986; Jones, Nyamudesa and Busangavanye, interpreted by Squire, 1990a). For most crops, however, W_v can be estimated from the results of population density experiments by extrapolation to the y-axis of a plot of reproductive, or economic, dry matter at harvest against total dry matter. The value so obtained is therefore an apparent value, and most of the values cited here are such. The oil palm growing in the humid tropics is one of the largest agricultural plants, and when grown in humid, fertile conditions has a W_v value around 80 kg plant^{-1}. For some large cassavas and pigeon peas W_v is around 1 kg plant^{-1}, and for some types of rice and small pigeon pea about 2 g plant^{-1}. Squire (1990b) gave examples of the analysis and corresponding values of W_v for a range of tropical crops.

In population density experiments, the forcing variable constraining growth is usually intercepted solar radiation acting on the mean dry matter available per individual. If drought or nutrient stress are the forcing variables, they sometimes act directly on expanding meristems rather than through dry matter itself. Any constraint on expansion in these circumstances tends to make W_v a smaller value than would be obtained in a corresponding plot from an experiment on population density. The value of W_v is therefore influenced by the nature of the limiting environmental factor (Squire, 1990b). Despite this, it usually remains a good enough discriminant of species.

SIZE, LONGEVITY AND INVESTMENT EFFICIENCY

The quantity of dry matter produced from a given investment in dry matter depends on two groups of attributes (in addition to the resources themselves). One group determines the resources that are captured by dry matter invested in vegetative structures, while the other determines the conversion of these resources to new dry matter. The following analysis concentrates on interception of resources, since conversion has been investigated in several dryland species (Natarajan and Willey, 1980; Reddy and Willey, 1980; Squire *et al.*, 1984; Rees, 1986), and is generally more stable than interception in the face of changing environmental factors. The return of resources per unit investment is here represented by the symbol τ, with subscripts 's', 'w' and 'n' denoting respectively solar radiation, water or nutrients. For the analysis described here, the investment is taken to be all the dry matter in the appropriate vegetative structures – the whole canopy or the whole root system, not just the leaves or fine roots. So for example, the units MJ g^{-1} for τ_s refer to total solar radiation intercepted per unit of above-ground vegetative dry matter. The values are cumulative over a defined period, which for annuals usually extends from sowing to the time of maximum mass.

The value of τ for a stand depends on two main factors, the quantity of resources intercepted per unit of intercepting surface (resource/surface) and the surface area or length supported per unit of vegetative mass (surface area/mass ratio). The value of τ is obtained by integrating resource/mass over time during the growth period. A typical resource/surface area relation for the canopy is the asymptotic relation between fractional radiation interception (f) and leaf area index, L (Kasanga and Monsi, 1954). As L increases, f also increases but τ_s decreases. This can be shown by taking any set of values for f and L (see references cited) and plotting f/L against L. Attributes such as leaf orientation and the degree of clumping of the foliage influence the relation between f and L, but do not differ systematically between the agricultural plants in different zones of dryness. For example, the relations are similar for oil and coconut palms and some dryland cereals (Squire, 1990b). There is less information on the shape of the corresponding curve for extraction of water in relation to the size of the root system, but some evidence indicates that the shape is similar to the relation between f and L (for example, Cooper *et al.*, 1987). Below ground, therefore, the quantity of water extracted per unit root length might decrease as the size of the root system increases.

The relation between surface area and vegetative mass is much more stable than that between resource capture and surface area. For a given canopy or group of canopies of the same species, the dry matter required to support a unit leaf area changes relatively little as L varies; leaves and their supports do not become much thinner as the canopy becomes denser. The ratio may also be similar for canopies of very different size and structure. Among the species referred to in this chapter, the leaf area per unit mass of above-ground vegetative tissue ranged mostly from 3-4 $m^2\ kg^{-1}$, for example, in maize in Zimbabwe (Allison, 1969), cassava in Sierra Leone (Enyi, 1972) and oil palm in Malaysia (Corley, 1973), to 6-8 $m^2\ kg^{-1}$, for example, in small pigeon pea in Australia (Rowden *et al.*, 1981) and rice in Japan (Akita, 1982). Limited evidence below ground suggests the length/mass relation for fine roots is similar at about 100 $m\ g^{-1}$ among several dryland annuals such as pearl millet and groundnut (Squire *et al.*, 1984; Simmonds and Azam Ali, 1989). Because of this conservatism in the surface/mass ratio, τ responds to L much as f/L does, decreasing as the size and mass of the canopy increase.

For a canopy or root system of given size, τ increases with time (simply as more radiation is intercepted or water abstracted), at least until the tissue senesces. In several of the experiments cited, this effect of time usually counteracts the effect in the opposite direction of the size of the intercepting or abstracting structures. Therefore, τ usually increases with time throughout growth to maximum mass. For example, τ_s for a determinate pearl millet canopy in India was 0.48 MJ g^{-1} at 30 d, 0.75 at 50 d and 1.2 at 65 d (original data in Squire *et al.*, 1984); for an indeterminate cassava in Colombia the corresponding values were 2.1 MJ g^{-1} at six months and 3.1 at 12 months (original data in Veltkamp, 1985). Again, limited data for roots suggest the same effect of time. For example, τ_n (expressed as P accumulated into shoots per unit root length) for pigeon pea in India doubled from 0.025 to 0.044 g km^{-1} between 30 and 60 d; changes of similar magnitude occurred for N and K in pigeon pea, and for all three nutrients in sorghum in the same experiment (Natarajan and Willey, 1980). There is, however, some evidence to suggest that τ_n levels off after a certain period of growth (unlike τ_s and τ_w), especially when nutrients are recycled in the plant (for example, as shown by data in Gregory, 1979 and Squire *et al.*, 1984).

This effect of time tends to be the main discriminant between species, such that τ increases with increasing longevity of the canopy. Among the examples shown in Table 1, the massive palms were two to four times more efficient at making returns from investment in canopy structures than small short-season dryland cereals. (The values of τ cited here are only a general indicator for a species since τ may vary substantially with population density and soil conditions, as described later). Superimposed on these differences in τ are further differences in the photosynthetic conversion efficiencies. While different species of a group such as C3 crops have a generally similar conversion efficiency for intercepted solar radiation, the increase in atmospheric dryness between the humid and dry tropics tends to reduce the conversion efficiency, and hence the investment ratio, of dryland plants independently of any effect on τ (Squire, 1990b).

TABLE 1. Differences between species in canopy duration, cumulative intercepted total radiation (S_i), and τ_s expressed as cumulative intercepted radiation per unit vegetative dry mass above ground. Sources of original data: oil palm, Corley (1973), Squire (1990b); cassava, Veltkamp (1985); pigeon pea, Natarajan and Willey (1980); groundnut, Reddy and Willey (1980); pearl millet, Squire et al. *(1984).*

Species	Site	Duration (d)	S_i (GJ m^{-2})	τ_s (MJ g^{-1})
Oil palm	Malaysia, poor soil	365	3.0	5.6
Oil palm	Malaysia, rich soil	365	5.5	3.0
Cassava	Colombia	365	4.0	3.1
Pigeon pea	Hyderabad, India	140	1.1	2.2
Groundnut	Hyderabad, India	103	0.81	2.1
Pearl millet	Hyderabad, India	68	0.45	1.4

IMPLICATIONS FOR SURVIVAL AND HUSBANDRY

At the humid end of the range of environments considered here, the perennial, indeterminate habit of the dominant species allows them to intercept a large fraction of the available resources – commonly 80% of the radiation incident on the canopy from planting to felling. The structures they make to achieve this high interception require much dry matter (a large W_v), but the availability of ample resources and the plants' high investment and conversion efficiencies enable them to produce much more dry matter than W_v.

As resources become scarcer, farmers adapt by growing plants with smaller W_v values. At the dry limit of arable agriculture, the scarce resources and small τ of the plants result in the investment efficiency being near unity in many seasons, so there is little dry matter surplus to that required for vegetative growth. Here, there are several options available to the farmer to increase his chance of success. One is to grow very drought tolerant species that probably have a very small W_v. There seems, however, to be a widespread reluctance to part with maize, despite its relatively high W_v, for the better adapted, though small-seeded cereals, whose grain is less palatable and easier for birds to steal or spoil. Other options are to attempt to manipulate the resources or the investment ratio.

The most common ways of increasing the resources available to plants are by some form of 'water-harvesting' to increase the amount of water infiltrating into the soil, and by manuring. Water-harvesting often has the intended effect, especially in seasons when a large part of the rain is received in intense falls. It has not been adopted more widely because the extra yields are thought not always to justify the effort needed to form the surface of fields into troughs or pits to catch the rainwater.

Fertiliser, as dung or minerals, potentially has the effect of increasing total production by stimulating the plants to intercept more of the other resources, radiation and water. However, there is evidence that, by increasing leaf area and root length, fertiliser reduces τ both for nutrients and the other resources. This is consistent with the general relation between size of structure and investment efficiency. For example, in stands of barley in Syria, fertiliser increased root length and the seasonal transpired water, but reduced τ_w (Table 2). Such an effect would be tolerable if the total uptake of resources is consistently increased, but in the dry tropics the uptake of additional water in response to added fertiliser is never assured. There are also much longer term effects of organic manures on soil structure, but these are difficult to quantify.

TABLE 2. *Effect of fertiliser on τ_w, expressed as cumulative water uptake per unit dry matter in roots, for barley in Syria. Original data of Cooper* et al. *(1987) for the cultivar Arabic Abiad at Jindiress.*

Fertiliser	Transpired water ($kg\ m^{-2}$)	Root mass ($g\ m^{-2}$)	τ_w ($kg\ g^{-1}$)
Without	190	630	0.30
With	230	1020	0.23

Reducing population density is the most common and often the most effective means of manipulating τ. Whenever population is reduced, τ increases, and this is as true for perennials in the humid tropics as for the annuals in dryland agriculture. The example in Table 3 shows that τ_s of sorghum increased substantially as the sown population decreased from 21 to 1 plant m^{-2}. The result was that more of the total dry matter was available to be allocated to grain at low populations. In some dry years at this site, there was no yield at the highest populations, for which τ must have been close to unity. The data of Simmonds and Azam Ali (1989) give an example of a growing stand of groundnut which also seems to have had an investment efficiency near unity.

TABLE 3. *Effect of plant population density on τ_s and other attributes of dryland sorghum in Botswana; τ_s is expressed as total solar radiation intercepted during the season per unit vegetative dry matter above ground. Mean insolation was 23.2 MJ m^{-2} d^{-1}. Analysis using original data of Rees (1986) and additional material supplied to author.*

Intended population (m^{-2})	1.3	5.3	21.3
Actual population (m^{-2})	1.6	5.1	13.5
Total stand dry mass (g m^{-2})	322	453	503
Grain dry mass (g m^{-2})	116	149	96
Intercepted radiation (MJ m^{-2})	1020	1100	1155
τ_s (MJ g^{-1})	5.0	3.6	2.8

As arable cropping spreads to drier areas, ultimately the returns of dry matter from even the most drought tolerant plants such as finger millet, proso millet and bambara groundnut will become habitually so low that reproductive growth is rare. At this point, settled arable agriculture gives way to pastoralism. The W_v of the plants grazed by cattle is very small, but economic yield is still possible because the vegetation is foraged over a wide area and its products concentrated in the bodies of the animals (Tessema and Emojong, 1984). However, the reduction in the number of livestock that occurs in drought years (Mbugua, 1986) depends not just on the attributes of the vegetation, but also on the availablity and dispersion of sources of drinking water.

CONCLUSIONS

If climate changes to give reduced or more unreliable rainfall, or soils degrade so that they hold less water, agriculture will have to continue to adapt. Research has assisted, and may still be able to assist farming in these areas in several ways. One way is to devise more reliable criteria for manipulating investment efficiency through plant population density, as in 'response farming' (Stewart and Kashasha, 1984). A second is for plant breeding to produce plants with a higher investment efficiency if the physiological factors controlling this attribute were better understood (Robinson, 1986).

Ultimately, however, arable farmers will be forced to grow species or cultivars with a smaller W_v than the main crops presently grown. Perhaps the greatest service breeding could provide for arid regions is to widen the choice of plants whose W_v is small enough to produce yield when the existing staple crops fail. Breeders have already made much progress in this respect, for example in providing early-flowering maize cultivars. But in many dry regions, there are already species such as bambara groundnut (Linnemann and Azam Ali, 1992) and proso millet (Penninkhoff, 1984) which have a small W_v, but also certain characteristics that make them less preferred as food than other less adapted plants such as maize. Breeding would contribute to the more widespread use of such species, for example by making the grain of some of the cereals less attractive to pests, more palatable to humans or livestock and more easily stored.

REFERENCES

Akita K (1982) Studies on competition and compensation of crop plants. IX. Effect of planting density of the characters of rice plant. Scientific report of the Faculty of Agriculture, Kobe University, Japan 15: 5-10 (in Japanese)

Allison JCS (1969) Effect of plant population on the production and distribution of dry matter in maize. Annals of Applied Biology 63: 135-144

Cooper PJM, Gregory PJ, Keatinge JDH, Brown SC (1987) Effects of fertilizer, variety and location on barley production under rainfed conditions in northern Syria. 2. Soil water dynamics and crop water use. Field Crops Research 16: 67-84

Corley RHV (1973) Effects of plant density on growth and yield of oil palm. Experimental Agriculture 9: 169-180

Enyi BAC (1972) Effects of spacing on growth, development and yield of single and multi-shoot plants of cassava (*Manihot esculenta* Crantz). 2. Physiological factors. East African Agricultural and Forestry Journal 38: 27-34

Gregory PJ (1979) Uptake of N, P and K in irrigated and unirrigated pearl millet (*Pennisetum typhoides*). Experimental Agriculture 15: 217-223

Kasanga H, Monsi M (1954) On the light transmission of leaves, and its meaning for the production of dry matter in plant communities. Japanese Journal of Botany 14: 304-324

Linnemann AR, Azam Ali SN (1993) Bambara groundnut (*Vigna subterranea* (L) Verdc). In: Underutilised crops, series II. Vegetables and pulses. Chapman and Hall, London (in press)

Mbugua SW (1986) Monitoring livestock and wildlife in Kenya. In: Hansen R M, Woie B M, Child R D (eds) Range development and research in Kenya. Winrock International Institute for Agricultural Development, Morrilton, Arizona, pp 413-430

Moss JRJ (1992) Measuring light interception and the efficiency of light utilisation by the coconut palm (*Cocos nucifera*). Experimental Agriculture 28: 273-85

Natarajan M, Willey RW (1980) Sorghum-pigeonpea intercropping and the effects of plant population density. 2. Resource use. Journal of Agricultural Science, Cambridge 95: 59-65

Penninkhoff P (1984) The prospects of growing proso millet in arid and semi-arid areas of Kenya. East African Agricultural and Forestry Journal 44: 298-305

Reddy MS, Willey RB (1980) Growth and resource-use studies in an intercrop of pearl millet/groundnut. Field Crops Research 4: 13-24

Rees DJ (1986) The effects of population density, row spacing and intercropping on the interception and utilisation of solar radiation by *Sorghum bicolor* and *Vigna unguiculata* in semi-arid conditions in Botswana. Journal of Applied Ecology 23: 917-928

Robinson D (1986) Compensatory changes in the partitioning of dry matter in relation to nitrogen uptake and optimal variations in growth. Annals of Botany 58: 841-848

Rowden R, Gardiner D, Whiteman P, Wallis ES (1981) Effects of planting density on growth, light interception and yield of a photoperiod-insensitive pigeon pea (*Cajanus cajan*). Field Crops Research 4: 210-213

Simmonds LP, Azam Ali SN (1989) Population, growth and water use of groundnut maintained on stored water. 4. The influence of population on water supply and demand. Experimental Agriculture 25: 87-98

Squire GR (1990a) Effects of changes in climate and physiology around the dry limits of agriculture in the tropics. In: Jackson M T, Ford-Lloyd B, Parry M (eds) Climatic change and plant genetic resources. Belhaven Press, London New York, pp 116-147

Squire GR (1990b) The physiology of tropical crop production. CAB International, Wallingford

Squire GR, Corley RHV (1987) Oil palm. In: Sethuraj M R, Raghavendra A S (eds) Tree crop physiology. Elsevier Science Publishers BV, Amsterdam, pp 141-167

Squire GR, Gregory PJ, Monteith JL, Russell MB, Piara Singh (1984) Control of water use by pearl millet (*Pennisetum typhoides* S & H) Experimental Agriculture 20: 135-139

Stewart JI, Kashasha DAR (1984) Rainfall criteria to enable response farming through crop-based climate analysis. East African Agricultural and Forestry Journal 44: 58-79

Tessema S, Emojong EE (1984) The influence of stocking rates and grazing management on liveweight changes in cattle, sheep and goats grazing natural pastures. East African Agricultural and Forestry Journal 44: 383-399

Thesiger, W. (1959) Arabian sands. Penguin Books, London

Veltkamp HJ (1985) Physiological causes of yield variation in cassava (*Manihot esculenta* Crantz). Agricultural University Wageningen Papers 85-6. Agricultural University, Wageningen

PLANT WATER STRESS UNDER LOW RAINFALL, HIGH TEMPERATURE SUMMER CONDITIONS IN PORTUGAL

Manuel T Oliveira
Department of Crop Science
UTAD
Vila Real
Portugal

ABSTRACT

Cima Corgo is a sub-region of the Douro Demarcated Area where agriculture is the main activity and vines are the single most important crop, and in certain areas the only crop. Steep slopes, shallow soils, high summer temperatures and little or no summer rainfall combine to create a stressful environment for crop production in most years. Agricultural production has been made possible by using drought resistant crops such as vines and appropriate cultivation techniques. Indeed, the physiography of the entire region has been profoundly changed to create a soil environment capable of storing sufficient water from winter rainfall to last throughout the summer. Effective crop management and cultivation techniques are crucial in preventing dramatic production losses during the drier seasons which have become more frequent during the last decade.

INTRODUCTION

Vine plants *(Vitis vinifera* L.*)* respond to high evaporative demand, exceeding the capacity of the soil to supply water sufficiently rapidly, by using various morphological and physiological responses. The extensive root system explores large soil volumes which are built up through cultivation practices. The plants have stomata only on the lower leaf surface, where they are protected from direct sunlight by the trelling system used to support the plant. In mid-morning, as air temperature rises and atmospheric humidity decreases sharply, the plants can close their stomata within a few minutes and are able to withstand the consequent increase in canopy temperature to as much as 7-9 °C above air temperature. As a last resort, the plants withdraw water from their fruits and shed their leaves.

No amount of soil water can prevent the daily formation of water stress between 9 am and 6 pm from mid-June to mid-August because the development

NATO ASI Series, Vol. I 16
Interacting Stresses on Plants in a Changing Climate
Edited by M. B. Jackson, C. R. Black

of stress is mainly an atmosphere-controlled process. However, the presence of sufficient soil water makes recovery possible during the night and ultimately avoids the last resort plant response of leaf-shedding. With soil volume-building techniques, it may not be possible to provide further increases in storage capacity, which is already approximately 46% by volume. Infiltration rate is probably also close to its maximum. In normal years, soil water storage is usually sufficient for the plants to produce an economic yield, but in drier years irrigation is essential. Because of the difficult terrain and lack of easily accessible water sources, irrigation must be applied carefully. When and how much water to apply are questions still to be answered.

THE DOURO VALLEY

The Douro Valley lies within the Demarcated Area and is a v-shaped valley with no flood plain. The river's course is rather irregular but flows in a generally east-west direction. Along both river banks the area below the 400 m contour has a characteristic and well-defined climate, morphology and vegetation. The geomorphology of the region has been greatly affected by human activity, which has created a peculiar appearance resulting mainly from the plantation of vineyards on terraces (Real *et al.*, 1978). The Douro Valley is composed of pre-cambric clayey schists with several levels of metamorphic alteration and some granite outcrops. Soils are generally shallow, acidic and of low fertility (Castro, 1967). The climate has Mediterranean characteristics, but with a continental influence. Rainfall decreases from west to east, while the average summer temperature follows the inverse pattern (Portugal Serviço Meteorológico Nacional, 1965).

The Douro Valley is an agricultural region where the main crops are vines, olives, almonds and peaches. The vineyards predominate and, in some areas, are the only crop (Real *et al.*, 1978). The predominance of such crops is due to climatic and soil conditions which create a water stressed environment, mainly during summer, where only crops with drought-resistant characteristics can thrive. Crop management and cultivation techniques have been developed to increase soil water storage and minimise plant water use. The work reported here is concerned with soil water storage, plant water use and plant responses to water deficits in vineyards.

TEMPERATURE AND RAINFALL

Rainfall in the Douro Valley is distributed very irregularly throughout the year (Fig. 1). Total precipitation is higher at the western end of the Demarcated Area, and most rain falls during the winter months and very little between May and August. The mean annual temperature varies little between different locations within the Demarcated Area, although the average summer temperature tends to be higher in the most easterly areas (Portugal Serviço Meteorológico Nacional, 1965).

Weather records have been kept for many years at three weather stations within the Douro Valley: Régua is the most westerly station, Almendra the easternmost and Adorigo is situated closest to our field site and is intermediate between Régua and Almendra. Table 1 shows mean temperatures and precipitation for the period 1931-1960, while seasonal averages for selected years during the last decade are shown in Figs. 2-4.

***TABLE 1**. Average temperature (T, °C) and rainfall (R, mm). Source: ADVID (1992).*

Year	Period	Régua		Adorigo		Almendra	
		T	R	T	R	T	R
1931-1960	APR-AUG	15.5	856.7	15.7	604.6	15.2	437.1
		19.9	166.9	20.4	142.5	20.7	121.4
1982	APR-AUG	15.9	568.0				
		20.2	89.2				
1985	APR-AUG	16.0	960.4	14.8	473.7	15.6	460.3
		20.1	193.5	18.8	99.6	20.2	150.1
1988	APR-AUG	15.7	749.0	15.0	502.4	15.4	447.9
		18.7	351.1	19.5	295.9	19.2	293.3
1991	APR-AUG	16.1	669.4	15.5	412.0	15.3	271.4
		21.1	53.1	20.7	38.1	21.2	9.1

In general during the period since 1981, the Demarcated Area has experienced a decline in precipitation relative to the 1931-60 average (Table 1). Although the annula precipitation recorded in specific years at some stations did not differ greatly from the 30 year average, the quantity of rainfall received between April and August has given some cause for concern since precipitation during this period is critical for crop production as the water supplies are marginal.

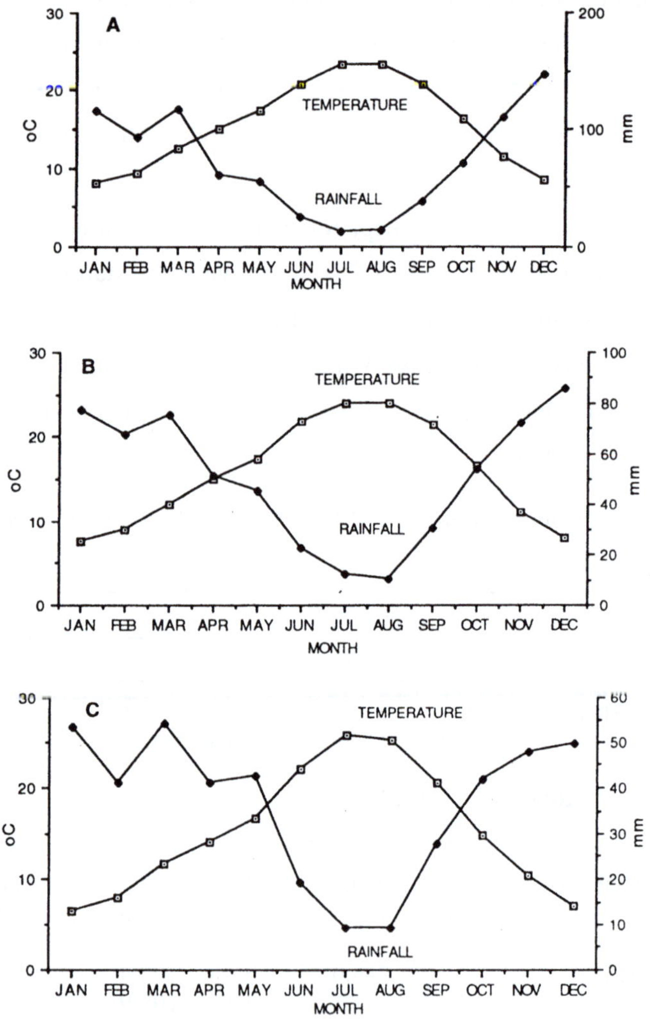

***FIGURE 1.** Average temperature and rainfall, 1931-60. A, Régua Station; B, Adorigo Station; and C, Almendra Station.*

In 1981-82 there was a drought and in 1991-92 a more severe one. However, 1988 brought a very humid summer with excessive rainfall which induced soil waterlogging, higher incidence of diseases and crop production losses. It is thus clear that water deficits in summer are an important factor for agricultural production since water supplies for irrigation are not easily obtainable.

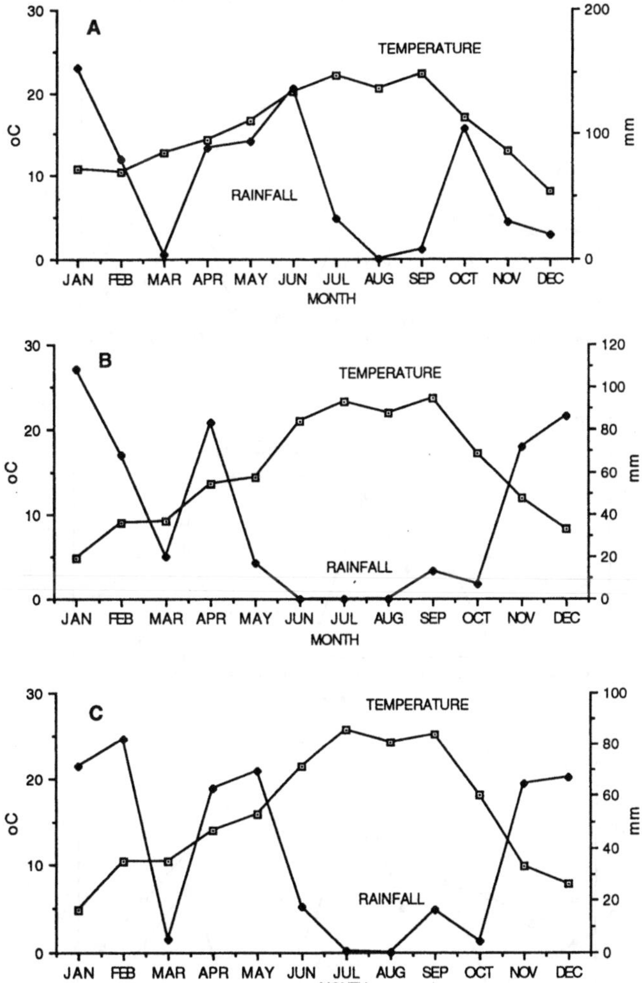

FIGURE 2. Monthly average temperature and rainfall, 1985. A, Régua Station; B, Adorigo Station; and C, Almendra Station.

The geomorphology of the region presents additional difficulties since the slope of the land is steep, usually above 10%, and slopes of 100% or greater are not unusual. Unmodified soils are generally about 10 cm deep, overlying consolidated rocky material. Such conditions are unsuitable for crops. To create an increased depth of soil for root growth and water storage, the normal procedure is to rip the substrate to a depth of 1.2 m and then rebuild the land either to the

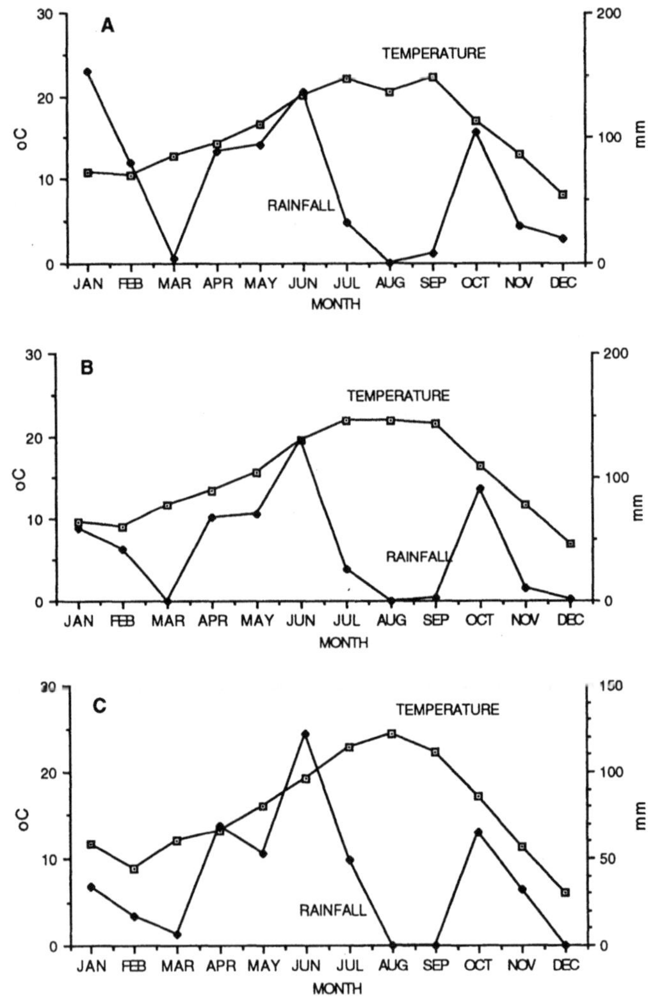

FIGURE 3. Monthly average temperature and rainfall, 1988. A, Régua Station; B, Adorigo Station; and C, Almendra Station.

original inclination or as a terraced slope. This approach creates a porous mixture of fine material and gravel with a large water holding capacity and a high infiltration rate. Typically, porosity is about 46% by volume and infiltration rates exceed 8 mm min^{-1}.

The air temperatures recorded by weather stations provide little information about the temperatures experienced by crops growing in the field. Terracing and

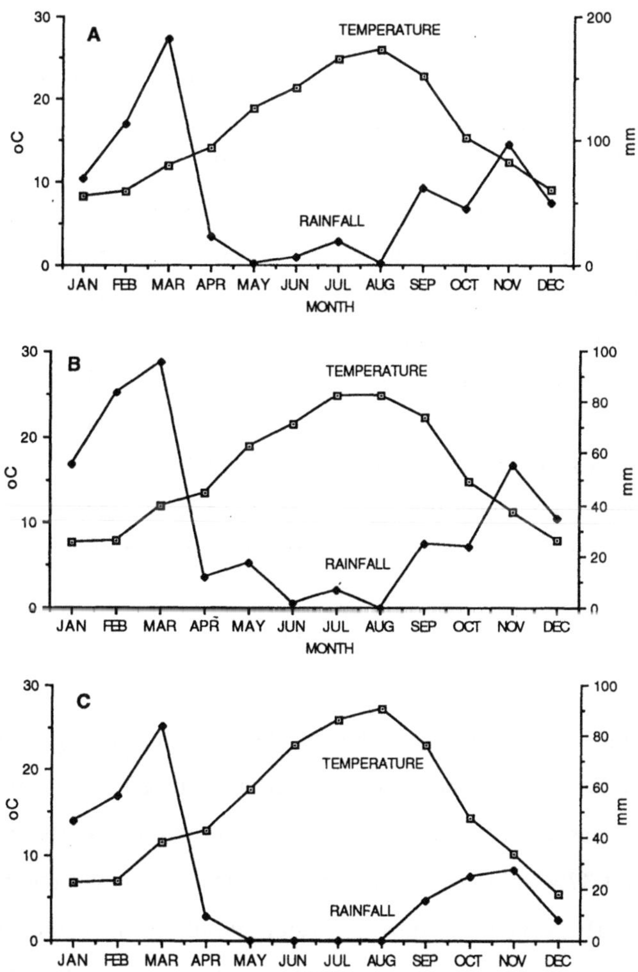

FIGURE 4. Monthly average temperature and rainfall, 1991. A, Régua Station; B, Adorigo Station; and C, Almendra Station.

the cultivation practices used in vineyards (or other orchards for that matter) create their own individual microenvironment. For example, terraced land has an increased exposed area, cultivation and herbicide applications are used to keep the surface free of vegetation other than the crop, and the trelling system keeps soil cover to a minimum. Air temperatures recorded adjacent to the crop canopy are higher than those recorded at the weather stations. For instance, the temperatures shown in Table 2 were recorded between 15 July and 29 July 1992 and on 27 July 1992 at 1500 h local time.

TABLE 2. *Soil (5 cm depth) and air temperatures (°C) between July 15-29, 1992.*

Location	MIN	MAX	July 27 (1500 h)
SOIL – non-irrigated terrace, outer row	21.4	41.4	38.0
SOIL – non-irrigated terrace, inner row	21.1	33.7	31.5
SOIL – irrigated terrace, outer row	18.0	36.0	32.0
SOIL – irrigated terrace, inner row	19.4	32.5	30.0
AIR – 2 m above canopy	13.1	41.8	41.5
AIR – inside canopy, outer row	13.3	42.6	41.9
AIR – inside canopy, inner row	13.5	43.3	42.9

SOIL WATER

The soil volume created by the cultivation practices described above is located on the original consolidated rock. Although the soil has a large water holding capacity, the underlying rock prevents deep percolation before the soil reaches field capacity. Our measurements indicate that deep percolation starts at a soil water content of about 37% volume, but such high values were never recorded between late May and early November; thus the water stored during winter is not lost by internal drainage during the period of most intensive transpiration by the crop.

Nevertheless, soil water is stored at depths below 45 - 50 cm, whereas the surface horizons contain little or no moisture for most of the summer. Indeed, no adventitious roots are found above 45 cm. Thus quantities of precipitation that do not moisten the soil sufficiently to ensure deep infiltration do not contribute to the crop water supply. Fortunately, as stated above, these soils have very rapid infiltration rates. After the unusually dry winter of 1991-92, very little water was stored in the soil by late March 1992 but 92 mm of rain were then received over a 15 day period without any surface runoff, as estimated indirectly by monitoring soil water storage and internal drainage (Oliveira, unpublished data). In an irrigation trial in June 1992, an average infiltration rate of 8.3 mm min^{-1} was measured over a 30 minute time period (Oliveira, unpublished data).

PLANT WATER STRESS

The grapevine is the predominant crop in the Douro Valley largely because it can withstand water stress for considerable periods. The response of grapevine to water deficits in the Douro Valley, as evaluated from measurements of leaf water potential, shows a remarkable resemblance to data reported by several authors working in different environments. The diurnal fluctuation in mean leaf water potentials in grapevine at mid-growing season 1987, when soil water storage and weather conditions were little different from the long-term averages, is shown in Table 3.

TABLE 3. Diurnal timecourse of leaf water potential in grapevine (Oliveira, 1987).

Time (h)	Leaf water potential (MPa)
0630	-0.60
0930	-1.02
1230	-1.40
1530	-1.39
1830	-1.33

In the 1991 growing season, the severity of crop water stress was evaluated using an index (CWSI) based on infra-red thermography, and a progressive increase in stress was observed as soil water reserves were depleted (Oliveira, 1992). Over longer time periods, CWSI was correlated with soil water content, but on specific days this index was affected most strongly by the quantity of incoming radiation; thus vine plants growing close to the terrace wall showed higher CWSI values than those further from the wall. The CWSI data for 1991 suggest that there is a positive correlation between CWSI and sugar accumulation in the fruit expressed in terms of both absolute and relative concentrations. Thus, since a certain content of sugar is essential to the quality of the produce, crop water stress should not be permitted to decrease below a certain threshold level which has yet to be determined. However, excessive levels of stress cause the plant to adopt defence mechanisms which decrease both the quality and quantity of fruit produced.

Because rainfall was below average in the summer of 1992 (Oliveira, unpublished), the vine plants were forced to resort to one of their last resort defences against extreme deficits by withdrawing water from their fruit, while some of the more sensitive varieties started shedding leaves, a more dramatic defensive form. Several reports show indisputably that vines also control the size and number of clusters of fruit produced according to water availability and, because grapevines are a perennial crop, exposure to severe stress during one growing season will have effects on production in the following season.

It is impossible, and certainly undesirable, to eliminate water stress entirely in grapevines since this is a weather-dependent process, and the prevailing high temperature and radiation in the Douro Valley will maintain high atmospheric demand for water, in excess of the ability of the soil to supply water no matter how moist it might be. Once potential transpiration exceeds the hydraulic conductivity of the soil, the plants must curtail their water losses by closing their stomata. This process will raise canopy temperature well above the prevailing air temperature. Vine plants are well suited to resist these stresses since they have stomata only on the underside of their leaves and can withstand elevated canopy temperatures, which we have found to reach as much as 9 °C above ambient air temperature (Oliveira, 1992). Crop management also has a role since the plants are trelled in a way that creates a two-faced row of plants, each consisting basically of two planes of leaves, one well illuminated and the other shaded, with the undersides of the leaves turned inwards.

At night, the plants can recover their turgidity, providing there is sufficient available soil water. Otherwise, water stress will build up during the next day from an already partially dessicated condition to reach an even higher level than before; eventually, recovery becomes impossible as the soil dries.

We have derived estimates of the evapotranspiration of terraced vineyards from measurements of soil water depletion and internal drainage. Results showing the seasonal timecourse of evaporation in the summer of 1991 are shown in Fig. 5 (Oliveira, 1992). Soil water storage for the same period is shown in Fig. 6. As can be seen, evapotranspiration was closely related to soil water storage, suggesting that soil water availability prevails over evaporative demand, and that vine plants adjust their transpiration to the available water supplies.

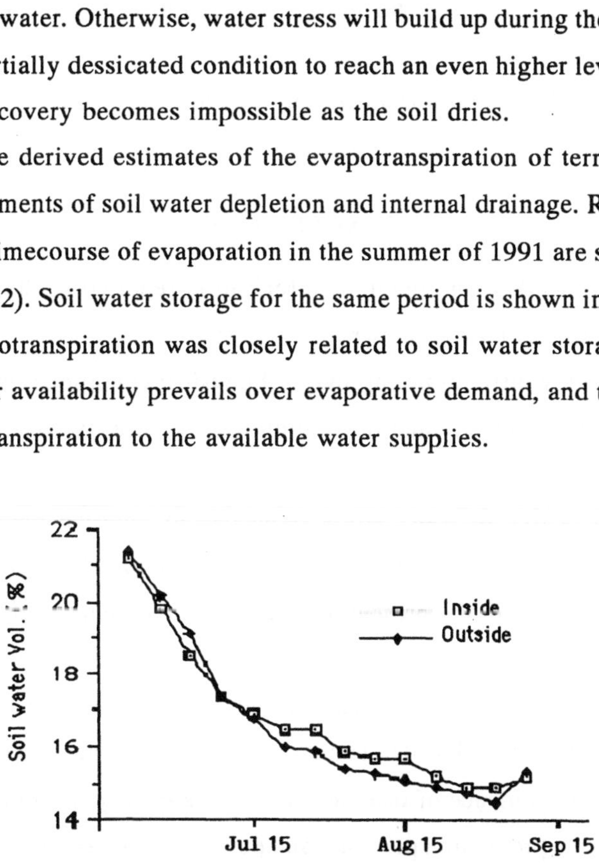

FIGURE 5. *Weekly variation in soil water content between June 14 and September 12, 1991.*

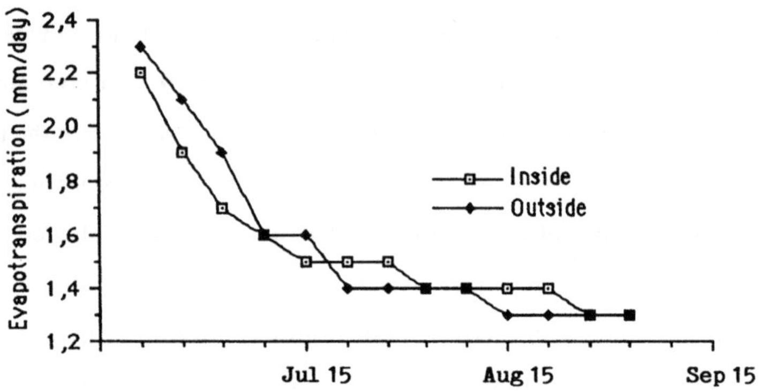

FIGURE 6. Weekly variation in evapotranspiration between June 14 and September 7, 1991.

CONCLUSIONS

On the basis of our data and previous information reported in the literature, the following conclusions may be drawn:

1. It will be very difficult to improve soil physical conditions further since the water holding capacity and infiltration rate are already close to their upper limits.
2. A certain degree of water stress is needed to improve the quality of the produce, but extreme water stress must be avoided since production is jeopardised in both the present and subsequent seasons. Although the daily level of water stress cannot be controlled, the overnight recovery of turgidity may be facilitated by supplying water to the soil if needed.
3. Close control of soil water content is important since excess water may increase the incidence of diseases and reduce quality, while too little water will decrease production per plant and per unit land area. Since dry years are apparently becoming more frequent in a region where summer precipitation is already marginal, it may be essential in future to apply irrigation. However, further data are needed to evaluate when and how much water to apply.

4. Irrigation might increase production per unit land area. Thus, in order to maintain overall production constant, it may be necessary to reduce the total area under vineyards, making it possible to utilise only the most suitable soils.

REFERENCES

Castro LMF (1967) Carta litológica de Portugal. Notícia explicativa. Agronomia Lusit. 29: 289-374

Oliveira M (1987) Economia da água na vinha. Caso da vinha do Douro - Cima Corgo. PhD Thesis. UTAD, Vila Real, Portugal, pp 147

Oliveira M (1992) Influence of planting system upon vineyard water stress and berry sugar accumulation. 4th International Symposium on Grapevine Physiology, Trento and Turin, Italy, pp 10

Portugal Serviço Meteorológico Nacional (1965) O Clima de Portugal. Fasc. 13. Lisboa, Portugal, pp 207

Real F, Dionisio G, Martins A, Mendes J (1978) Aspectos climáticos, geológicos, fisiográficos e de aptidão agro-silvo-pastoril da área do Douro. IUTAD, Vila Real, Portugal, pp 55

IDENTIFICATION OF STRESS TOLERANCE TRAITS IN SUGAR BEET

N. Clarke, H. Hetschkun, C. Jones, E. Boswell and H. Marfaing
Broom's Barn Experimental Station
Higham
Bury St Edmunds
Suffolk IP28 6NP
United Kingdom

ABSTRACT

Drought stress is probably the largest single cause of lost potential yield in UK sugar beet production and global warming seems certain to exacerbate this effect. Irrigation, the obvious corrective measure for drought stress, is only used to a limited extent, and this is unlikely to change. Modern varieties of sugar beet are derived from seashore beet, which is accustomed to a saline environment and has specific mechanisms to combat water stress. Drought-tolerance has, however, been a low priority in the selection and breeding of European sugar beet varieties in the past. We are trying to identify traits responsible for stress tolerance and develop suitable screening methods. During drought, beet leaves are subjected to both heat and water stress. We are using measurements of photosynthetic efficiency (principally chlorophyll fluorescence) to measure the effects of both of these stresses upon the plant. A portable plant efficiency analyser (Hansatech Instruments Ltd) is being used to analyse the chlorophyll-fluorescence induction kinetics (Kautsky effect) of dark-adapted leaves. A number of parameters are recorded, but the principal area of study is the use of the ratio of the variable fluorescence (F_v) to the maximal fluorescence (F_m) as a measure of photosynthetic efficiency. During water stress, sugar beet accumulates amino acids (mainly glutamine) and glycine betaine. Unfortunately these, together with sodium and potassium, are the principal impurities which reduce beet quality for processing. An HPLC system is being used to separate and measure sucrose, invert sugars and glycine betaine in stressed plants. The content of these components, other quality parameters (amino acids, sodium and potassium) and total sap osmotic concentration are being correlated with the estimated varietal stress resistance with the aim of identifying the characteristic(s) responsible for stress tolerance. Similarly, plants are being screened for the presence and quantity of the major classes of desiccation stress and heat shock proteins.

NATO ASI Series, Vol. I 16
Interacting Stresses on Plants in a Changing Climate
Edited by M. B. Jackson, C. R. Black

INTRODUCTION

In the recent years of drought (1989-91), soil moisture deficits have been well above average in the UK, and yield losses in sugar beet crops due to stress have been as great as 50% (Dunham and Clarke, 1992). Global warming will exacerbate the situation; even if the average summer rainfall stays the same, drought stress will become more common as higher temperatures increase evapotranspiration. The main deterrents to irrigation are its cost and doubts about the availability of water in years of greatest need. The main object of our work is to assess the prospects of improving yields and quality by altering the stress tolerance of sugar beet.

Sugar beet has a deep root system which enables it to utilise extensive volumes of soil water (Dunham, 1993). However, the plant shows apparently poor control of its internal water supply because wilting of leaves occurs frequently under conditions of high evaporative demand, even when water is readily available to the roots (Milford and Lawlor, 1975). The ability of the plant to compensate for decreasing soil water potential is limited and leaf water potential tends to decrease faster than soil water potential as the soil dries (Lawlor and Milford, 1975).

Modern cultivated varieties of sugar beet (*Beta vulgaris*) are so closely related to wild maritime forms (*Beta maritima*) that they can be regarded as the same species (Ford-Lloyd, 1986). These wild forms colonise saline environments and have evolved specialised mechanisms to combat water deficiency stress (WDS). *Beta* species can therefore be expected to have a high level of tolerance to WDS. However, conventional breeding objectives in Western Europe have been root shape, higher yields and sugar content, lower impurities and disease resistance. A number of these legitimate goals may have inadvertently reduced WDS tolerance.

European hybrid sugar beet varieties have rarely been based solely on highly inbred lines. The most common system has been to use an F_1 hybrid between an inbred male sterile line and an unrelated inbred O-type as the female parent, and an open-pollinated line or population as the pollinator parent (Bosemark, 1993). Whilst hybrids from a broad-based pollinator are likely to have a more stable performance in a range of environments, it does mean that a variety has to be treated as a fairly disparate population.

THE EFFECT OF DROUGHT UPON BEET QUALITY

A number of soluble components of sugar beet storage roots reduce beet quality by decreasing crystallisation during processing. These components (potassium, K^+; sodium, Na^+; amino acids, termed α-amino-N compounds; glycine betaine) are referred to as impurities and their effect upon quality can be estimated using the equation of Carruthers *et al.* (1962):

$$\text{Juice Purity} = 97 - 0.8\ [2.5\ (K^+) + 3.5\ (Na^+) + 10\ (\alpha\text{-amino-N}) + \text{glycine betaine}]$$

where the impurities are expressed as g 100 g^{-1} sugar. Fig. 1 shows the effect of the accumulated summer potential deficit on α-amino-N compounds and measured impurities (excluding glycine betaine) in beet at final harvest in a long-term field experiment at Broom's Barn. The correlation of impurity levels with drought stress is clear. Evidence from the National Institute of Agricultural Botany (NIAB) suggests that, when seasonal effects are eliminated, plant breeding has reduced impurities by over 7% between 1979 and 1989, compared with only a 1% increase in sugar content (Kimber, 1990). However, this improvement in quality has probably significantly affected the sensitivity of the crop to WDS, as is discussed below.

DROUGHT PHYSIOLOGY

Loss of water to the atmosphere can be rapidly reduced by stomatal closure (Kaiser, 1987). If the water potential of the leaf cells decreases in line with falling soil water potential, then water uptake, photosynthesis and growth can continue. This is achieved by reducing the solute potential of both cytoplasm and vacuole. Whilst the vacuole can accumulate those solutes present in abundance (Na^+, K^+ and organic anions), high concentrations of inorganic ions are toxic to cytosolic functions. It is now widely accepted that plants accumulate 'compatible osmotic solutes' in the cytoplasm when the osmotic adjustment required is high (Yancey *et al.*, 1982), and this is the major role attributed to the glycine betaine accumulated in sugar beet in response to WDS (Hanson and Wyse, 1982). There is now strong evidence that betaine also has a role as a stabilising compound or osmoprotectant in stressed cells (Rhodes and Rich, 1988). As well as quaternary

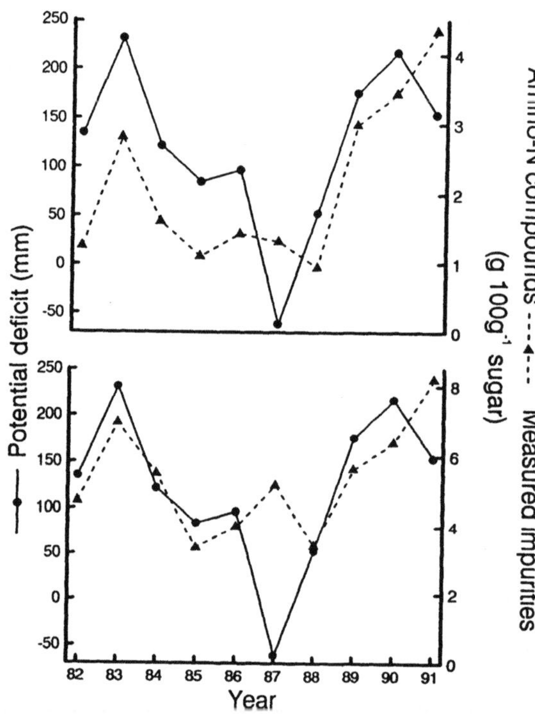

FIGURE 1. Correlation between accumulated soil moisture deficits (June-August) and beet quality in a long-term fertiliser experiment at Broom's Barn.

ammonium compounds such as betaine, amino acids are also believed to have a role as compatible solutes (Rhodes and Handa, 1989). It has not been demonstrated whether the accumulated α-amino-N compounds, principally the amino acid glutamine and to a lesser extent asparagine (Burba *et al.*, 1984), are acting as compatible solutes in stressed beet. Asparagine has been implicated as a compatible solute (Rhodes and Handa, 1989), and both glutamine and asparagine are amides of acidic amino acids with many of the features associated with known organic compatible solutes. Both glycine betaine and α-amino-N compounds accumulate in the roots following stress relief, but whilst the α-amino-N compounds can be utilised during regrowth following stress relief, glycine betaine is not readily degraded (Hanson and Wyse, 1982). Both components may have a role to play in sugar accumulation in the root. Thus, the high sucrose content of the vacuoles of storage cells is counterbalanced by the compatible solutes in the cytoplasm. Bell *et al.* (1992) suggested that the strong correlation between α-amino-N and both glycine betaine and the osmotic potential of the root sap provides circumstantial evidence to support this view. Hecker *et al.* (1979) found that, in general, selection for low α-amino-N content was deleterious to sucrose production.

The ability of sugar beet plants to combat WDS by osmotic adjustment is limited and wilting commences in the oldest leaves of the rosette (Lawlor and Milford, 1975). In our own experiments, WDS was imposed by withholding water from young sugar beet plants (10-12 leaf stage) growing in a 50:50 sand:peat mixture. This method was employed as it was found that polyethylene glycol (PEG 8000) accumulated in the leaves when used to control soil water potential. The plants were kept in a controlled environment room (16 hour day, 20 °C, 200 μmol m^{-2} s^{-1} of PAR, ~ 50% relative humidity). Fig. 2 shows how the relative water content (RWC) of the leaves decreased during a period of drought. The onset of this reduction coincided with the first signs of wilting and a decline in the rate of net photosynthesis and stomatal conductance to about 40% of the control. By this time 70% of the water capacity of the soil had been lost.

Chlorophyll fluorescence is known to be a sensitive indicator of irreversible damage to photosynthesis and the physiology of the plant in general resulting from a variety of stresses including WDS (Bolhar-Nordenkampf *et al.*, 1989). The ratio of the variable fluorescence (F_v) to the maximal fluorescence (F_m) quantifies damage to the primary electron acceptors of photosystem II (PSII). In our experiments, the RWC of the leaf tissue had fallen below 40% before a significant reduction in F_v/F_m was detected (Fig. 2). At this stage the soil contained 20% of its maximum water capacity. Further exposure to stress induced an almost exponential reduction in F_v/F_m whilst soil moisture content remained constant. These changes in photosynthetic activity and efficiency are similar to those observed in other species (Kaiser, 1987).

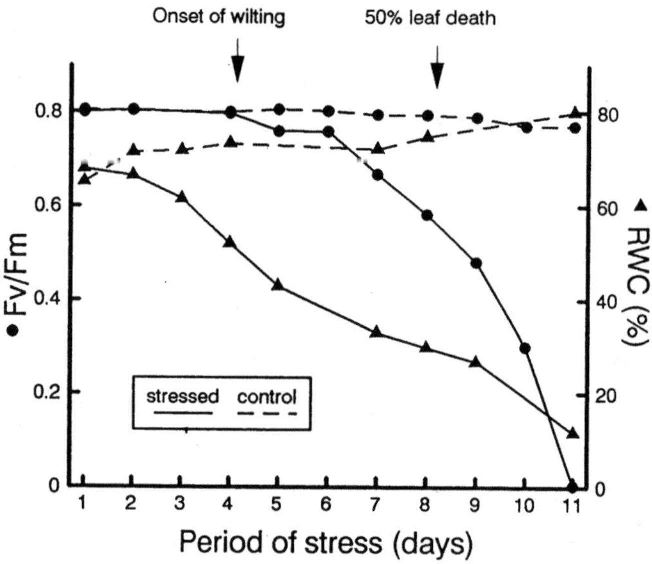

FIGURE 2. The effect of water stress on the relative water content and chlorophyll fluorescence of the leaves of young sugar beet plants.

The age of the leaf had a large influence upon its physiology. Thus, in the absence of stress, the net photosynthetic rate and stomatal conductance of the oldest leaves of young plants were less than 50% of those of the youngest leaves (Figs. 3A and B). Furthermore, the reduced F_v/F_m values for these oldest leaves suggest that they were senescing. After four days of WDS, the relative effect upon the oldest leaves was far greater; for example, the net photosynthetic rate of the youngest leaves was halved, whilst that of the oldest leaves was effectively eliminated. Stomatal closure appeared to account for virtually all of the reduction in photosynthetic activity except in the oldest leaves (Figs. 3A and B). Fig. 3C demonstrates that PSII of the oldest leaves was severely damaged by WDS, whereas that of the youngest was intact, and thus a significant loss of photosynthetic activity was due to reduced efficiency.

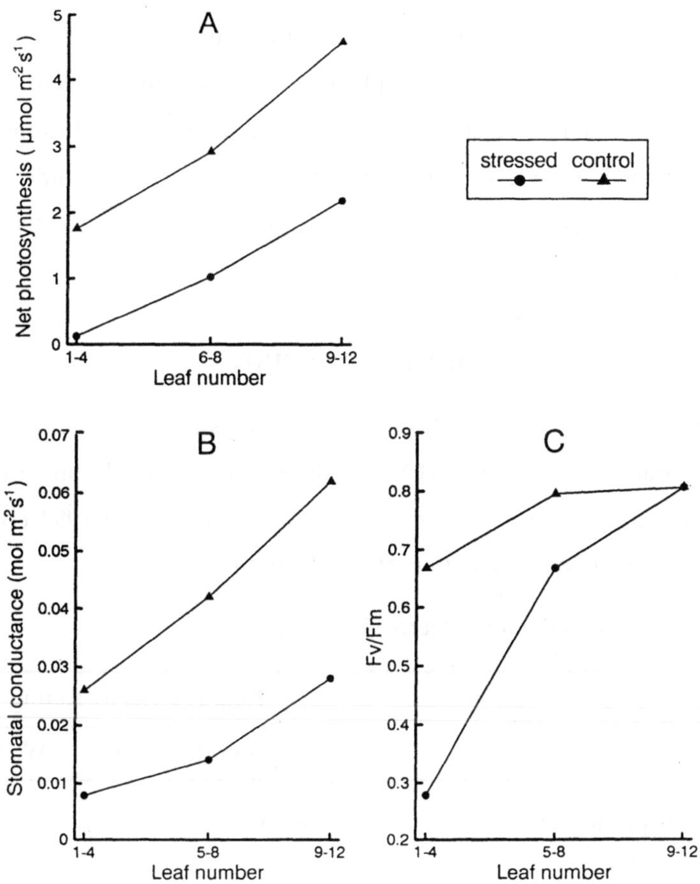

FIGURE 3*. The effect of leaf age upon selected physiological components of young sugar beet plants.*

These results agree with those of Lawlor and Milford (1975), who also demonstrated that the greater stomatal conductance of the young leaves was not due to differing leaf water potentials since their stomata closed at potentials 0.2-0.3 MPa lower than those of mature leaves. Furthermore, an increase in the humidity of the air increased stomatal conductance without altering leaf water potential. This effect was greatest for mature leaves but decreased as the leaves wilted. These authors suggested that the increased susceptibility of the more mature leaves to wilting may be due to their lower mechanical strength. They also proposed that the reduced sensitivity of the stomata of young leaves to changes in leaf water potential may be due to a lower osmotic potential, which would increase turgor, or because of differences in the plant growth regulator status of the leaves. This differential behaviour related to leaf age is obviously central to the survival of the plant under high WDS. The plant appears to sacrifice the older leaves as water is preferentially drawn towards younger leaves due to their higher transpiration rates.

VARIETAL TESTING

The varietal response to WDS has been investigated using five varieties (Table 1). In addition to our standard variety (E), the others chosen were deemed to be stress susceptible and tolerant by NIAB in the drought years of 1989-1991 (Dixon and McCullagh, personal communication of NIAB data).

***TABLE 1**. Details of varieties selected for screening.*

Variety	Ploidy	Breeder	Relative NIAB Rating	
			Regression analysis	Wilting
A	triploid	1	tolerant	tolerant
B	diploid	2	tolerant	intermediate
C	triploid	3	sensitive	sensitive
D	triploid	4	sensitive	intermediate
E	triploid	2	intermediate	tolerant

The NIAB estimates are based upon two factors: relative susceptibility to wilting in the field and a regression analysis of relative sugar yield with the coefficient of variation for the experiment. The latter calculation does not distinguish between types of stress, but in the years in question the major stress was WDS. In addition, varieties from different seed companies were selected to give an indication of the range of tolerance available. In particular, the selection of breeding lines may have been influenced by the relative susceptibility to drought at breeder's trial sites. To give an indication of the effect of ploidy, a diploid variety was selected to compare with the triploids. In order to reduce other effects, the variety selected was from the same breeder as one of the triploid varieties. When European sugar beet breeders had to decide upon the method of production of genetic monogerm varieties, most chose to develop monogerm triploid hybrids based on diploid monogerm male steriles and multigerm tetraploid pollinators (Bosemark, 1993). However, as a result of selective breeding, the agronomic quality of the male steriles has improved to such an extent that diploid hybrids are now appearing on the European market which compare favourably with the triploids, and are cheaper to produce.

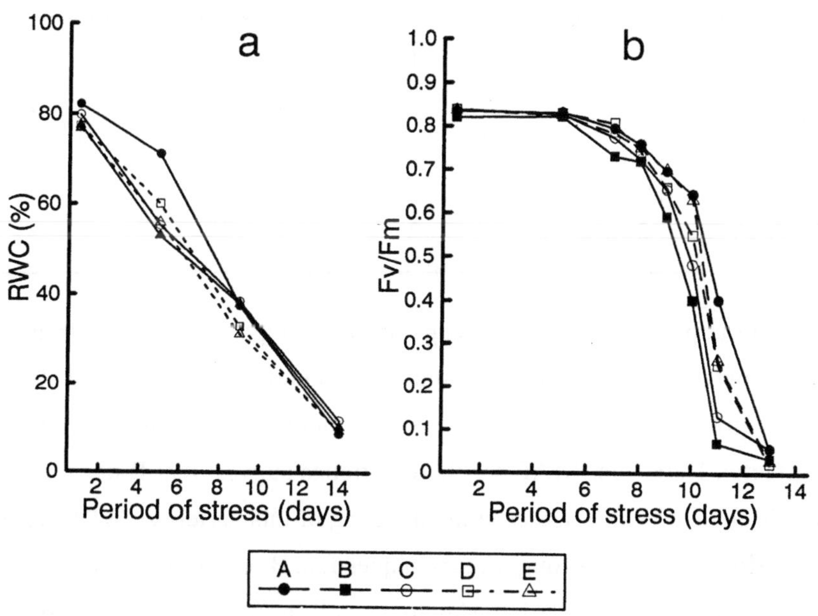

FIGURE 4*. The use of relative water content (a) and chlorophyll fluorescence (b) to screen for stress tolerance in five varieties of sugar beet (A-E).*

Variety A stood out from all the others in having a significantly higher RWC after five days of WDS (Fig. 4a). The chlorophyll fluorescence studies also indicated that A was the most tolerant variety, followed by E, D, C and B (Fig. 4b and Table 2). This ranking was determined largely from measurements taken after 10 and 11 days of stress. The rankings were significant in terms of the standard errors of the means, except between varieties C and B. This order agrees well with the order of A, B and E, D and C derived from the NIAB data. The one variety out of order is B, which these results suggest to be the least tolerant.

TABLE 2. *Relative performance of varieties in NIAB field trials (1989-1991) and chlorophyll fluorescence stress tests. The lower numbers indicate greater stress tolerance, α-amino-N content and sugar yield.*

Variety	Overall NIAB stress tolerance	Chlorophyll fluorescence stress tolerance	NIAB α-amino-N	NIAB sugar yield
A	1	1	2	2
B	2=	4=	5	1
C	5	4=	3	5
D	4	3	1	4
E	2=	2	4	3

We are at present assessing a range of physiological parameters from these varieties to obtain information on the source of tolerance. Their relative ability to accumulate glycine betaine and α-amino-N compounds in response to stress is an obvious factor. Data from NIAB for these varieties suggest that B has the lowest α-amino-N content, which agrees with our rating of this variety as WDS sensitive (Table 2). However, in the drought years in question, B produced the greatest sugar yields. The reason why this variety should do well in field tests and poorly in our stress tests may be related to growth rate. The stress test depends on all varieties being approximately the same size when watering is withheld, since larger plants using more water would experience stress sooner. Indeed, all plants were at the 9-

10 leaf stage except for varieties B (some plants were at the 11-12 leaf stage) and C (7-8). However, the leaves of variety B were noticeably narrower, with a smaller surface area. Furthermore, B and C were the least tolerant even though they were at opposite extremes in terms of developmental stage. It is difficult to draw firm conclusions from the limited number of varieties examined in the present study, but one obvious difference is that B is a diploid variety. NIAB data show that the only diploid in the list of varieties currently recommended to growers has the greatest sugar yield; however, in the provisional list, two triploids are ranked top and outproduce this diploid. When variety B is ignored, both stress ratings agree with those of NIAB based on sugar yield data in the drought years of 1989-91 (Table 2). However, whilst the α-amino-N ratings do not agree with the stress rating, in the absence of data for glycine betaine we are only considering part of the total compatible solutes. Furthermore, the NIAB data are for final yield and we cannot be sure that α-amino-N contents at this time accurately reflect levels during periods of WDS.

HEAT SHOCK

In the field, the damage caused by drought will inevitably be exacerbated by high radiation fluxes and heat damage. We have investigated the effect of a four hour heat shock upon the survival of seedlings and the chlorophyll fluorescence of young plants. In both cases the lethal upper temperature limit (LT_{50}) in the absence of other stresses appears to be 42 °C. This temperature is at the higher end of the range (35-45 °C) reported for other species (Gounaris *et al.*, 1984). Since Loman (1986) has reported that the upper surface of turgid sugar beet leaves was 7.3 °C warmer than air temperature (21.8 °C), it is possible that wilted sugar beet leaves suffering from photoinhibition induced by WDS and high radiation may occasionally experience temperatures approaching 42 °C. We hope to make such measurements in the next growing season. Experiments to determine whether there are differences between varieties in their response to heat shock are under way. To date, we have not studied either the combined effects of multiple stresses, or prolonged periods of heat shock.

THE FUTURE OF SELECTION FOR WDS TOLERANCE

The yield of sugar beet is directly related to intercepted radiation (Brown *et al.*, 1980), and so the premature loss of leaf cover will directly affect final yield. Chlorophyll fluorescence is therefore a particularly useful tool for assessing the response of sugar beet plants to stress as it provides a sensitive method for detecting irreversible leaf damage. Furthermore, as individual measurements require illumination for only one second, numerous plants can be screened, thereby reducing the influence of the diverse genetic composition of specific varieties. The data collected so far indicate that the technique may offer a useful rapid screening method for WDS as, except for the diploid variety, rankings agreed with NIAB data from field trials in the drought years of 1989-91. It seems likely that, in the absence of a method for controlling soil water potential, the differences in relative growth rate between varieties may pose a problem which increases with the diversity of the material under examination.

We are presently investigating the physiological basis of the varietal responses to stress. An HPLC system is being used to separate and measure sucrose, raffinose, invert sugars and glycine betaine in stressed plants. The content of these components, other quality parameters (amino acids, sodium and potassium), plant growth regulator levels and sap osmotic potential are being correlated with the estimated varietal stress tolerance with the aim of identifying characteristic(s) responsible for stress tolerance. Similarly, plants are being screened for the presence and quantity of the major classes of proteins produced during desiccation stress and heat shock.

We are particularly interested in the ability of varieties to minimise the decrease in RWC as available soil moisture declines and reduce leaf damage as RWC falls. In addition to conventional breeding approaches, it may be possible to alter the stress response and/or root quality through genetic manipulation, for example, by introducing genes which alter the ability of the plant to produce glycine betaine. Unlike many other stresses, it is unlikely that any one factor will greatly alter WDS tolerance, particularly in a tolerant crop such as sugar beet. However, the crop can only benefit as more components of stress tolerance are identified and improved upon.

REFERENCES

Bell CI, Jones J, Milford GJF, Leigh RA (1992) The effects of crop nutrition on sugar beet quality. Aspects of Applied Biology 32: 19-26

Bosemark NO (1993) Genetics and breeding. In: Cooke DA, Scott RK (eds) The sugar beet crop: science into practice. Chapman and Hall, London, pp 67-119

Bolhar-Nordenkampf HA, Long SP, Baker NR, Oquist G, Schreibers U, Lechner EG (1989) Chlorophyll fluorescence as a probe of the photosynthetic competence of leaves in the field: a review of current instrumentation. Functional Ecology 3: 497-514

Brown KF, Messen AB, Dunham RJ, Biscoe PV (1987) Effect of drought on growth and water use of sugar beet. Journal of Agricultural Science Cambridge 109: 421-435

Burba M, Nitzschke U, Ritterbusch R (1984) Die N-assimilation der planze unter besonder Berucksichtigung der zuckerrube (*Beta vulgaris* L.). Zucker-industrie 109: 613-627

Carruthers A, Oldfield JFT, Teague HJ (1962) Assessment of beet quality. Proceedings 15th Annual Technical Conference, British Sugar Corporation Ltd

Dunham RJ (1993) Water use and irrigation. In: Cooke DA, Scott RK (eds) The sugar beet crop: science into practice. Chapman and Hall, London, pp 279-309

Dunham RJ, Clarke NA (1992) Coping with stress. British Sugar Beet Review 60: 10-13

Ford-Lloyd BV (1986) Infraspecific variation in wild and cultivated beets and its effect upon infraspecific classification. In: Styles BT (ed) Infraspecific classification of wild and cultivated plants. The Systematics Association, special vol. 29, pp 331-344

Gounaris K, Brain ARR, Quinn PJ, Williams WP (1984) Structural reorganisation of chloroplast thylakoid membranes in response to heat-stress. Biochimica et Biophysica Acta 766: 198-208

Hanson AD, Wyse R (1982) Biosynthesis, translocation, and accumulation of betaine in sugar beet and its progenitors in relation to salinity. Plant Physiology 70: 1191-1198

Hecker RJ, Martin SS, Smith GA (1979) Effects of mass selection for amino nitrogen content. Sugar Beet Research c: 25-27

Kaiser WM (1987) Effects of water deficit on photosynthetic capacity. Physiologia Plantarum 71: 142-149

Kimber D (1990) Improvements in sugar beet varieties and seed. British Sugar Beet Review 58: 9-11

Lawlor DW, Milford GFJ (1975) The control of water and carbon dioxide flux in water-stressed sugar beet. Journal of Experimental Botany 26: 657-665

Loman G (1986) The climate of a sugar beet stand: dynamics, impact on the crop and possibilities of improvement. The Royal University of Lund, Sweden

Milford GFJ, Lawlor DW (1975) Effects of varying air and soil moisture on the water relations and growth of sugar beet. Annals of Applied Biology 80: 93-102

Rhodes D, Handa S (1989) Amino acid metabolism in relation to osmotic adjustment in plant cells. In: Cherry JH (ed) Environmental stress in plants. Springer, Berlin Heidelberg New York, pp 41-62

Rhodes D, Rich PJ (1988) Preliminary genetic studies of the phenotype of betaine deficiency in *Zea mays* L. Plant Physiology 88: 102-108

Yancy PH, Clark ME, Hand SC, Bowlus RD, Somero GN (1982) Living with water stress: evolution of osmolyte systems. Science 217: 1214-1222

STOMATAL RESPONSES TO ABSCISIC ACID IN NATURAL ENVIRONMENTS

W. Hartung and H. Heilmeier[1]
Julius-von-Sachs-Institut für Biowissenschaften
Lehrstuhl Botanik I
Universität Würzburg
Mittlerer Dallenbergweg 64
D 8700 Würzburg
Germany

ABSTRACT

The role of abscisic acid (ABA) as a stress hormone is now well established. Its synthesis and transport to the sites of action are stimulated under stress conditions. In leaves, ABA induces stomatal closure, thereby reducing transpiration and improving water relations. Ion- and water-uptake and root growth are also affected by ABA. These phenomena have all been studied intensively under laboratory and glasshouse conditions, but confirmation of the findings was not possible until recently because of analytical problems. With the availability of sensitive ABA assays we have investigated the role of ABA under natural conditions in almond trees and several species from the Negev desert (Israel), *Ceanothus* shrubs from the Chaparall in South California and *Prunus* and *Cornus* trees from a steep, dry and sun exposed habitat in the Main valley north of Würzburg. In all cases we investigated the relations between xylem sap ABA concentration, water potential and leaf conductance (g). ABA rises slightly with decreasing water potential until a threshold is reached, but thereafter a dramatic increase in xylem sap ABA can be observed. The threshold depends on both species and conditions. In some cases clear relations can only be observed between predawn water potential (a measure of soil water potential) and xylem sap ABA. When xylem sap ABA is plotted against g, in most cases there is a highly sensitive phase where tiny increases in ABA are sufficient to decrease g dramatically. ABA can be increased above this range in many cases, except in the *Prunus* and *Cornus* trees from the Main valley. Similar relations have been observed in stressed field-grown maize plants. The physiological significance of this excess ABA is not yet understood. Elevated CO_2 is an important environmental factor causing stomatal closure, and in laboratory experiments CO_2 has been shown to enhance guard cell sensitivity to ABA and *vice versa*. We injected ABA into the veins of cotton leaves grown in open-top chambers at high CO_2 concentration (650 $\mu l\ l^{-1}$) under highly stressed field conditions in Arizona and measured the stomatal conductance of the treated leaves. The results confirmed the earlier observations from laboratory experiments with excised leaves and epidermal

---

[1] Lehrstuhl Pflanzenökologie, Universität Bayreuth, Postfach 10 12 51, D 8580 Bayreuth, Germany

NATO ASI Series, Vol. I 16
Interacting Stresses on Plants in a Changing Climate
Edited by M. B. Jackson, C. R. Black

strips of interactions between CO_2 and ABA. The importance of varying guard cell sensitivity to ABA under field conditions has also been reported for *Arbutus* grown in a natural Portuguese macchia, nutrient deficient cotton and stressed field-grown maize plants. Our investigations indicate that ABA also has an important stress-physiological role as a root-to-shoot signal influencing stomatal responses to environmental perturbations under natural conditions.

INTRODUCTION

Abscisic acid (ABA) is now well established as an effective stress hormone in plants. Its biosynthesis and release from sites of synthesis and transport to target cells is stimulated under stress conditions. ABA plays its role in ameliorating water stress through its effects on processes such as root growth (stimulation), water uptake (increased) and stomatal opening (decreased). After arriving at the guard cells, ABA may induce stomatal closure, which in turn reduces water loss and improves the water relations of the leaf. Environmental factors which may induce stomatal movements and influence ABA relations include soil water deficits, heat, frost and salt stress, which all reduce the water potential of root and leaf tissues, illumination and increased cellular CO_2 concentration.

The role of ABA as a hormonal stress mediator dependent on the water status of plants is thus well established. Light/dark-dependent stomatal movements are, however, not likely to be ABA-mediated, since the darkening of leaves causes only small increases of ABA concentration in the apoplastic fluid, as postulated by a mathematical model and our own unpublished measurements (Slovik and Hartung, 1992). These increases seem to be much too small to induce stomatal movements. Interactions between the action of CO_2 and ABA have been described by Raschke (1975), Mansfield (1976), Snaith and Mansfield (1982) and Blackman and Davies (1984), who reported that ABA and CO_2 act simultaneously, with each increasing the sensitivity of the guard cells to the other.

Most of the experiments described above were performed with isolated systems (excised leaves, leaf disks or epidermal strips), and only a few attempts have been made to confirm these findings with results obtained from experiments using plants grown under naturally fluctuating conditions. We have therefore investigated the role of ABA in plants growing under the natural conditions in a desert and other extreme environments. ABA in the xylem sap of apoplastic leaf fractions (Hartung *et al.*, 1988) was assayed because this fluid originates from the only compartment in direct contact with the primary site of ABA action, the apoplast surrounding the guard cells (Hartung, 1983). The objectives were to examine under field conditions the relation between:

- water shortage and the ABA concentration of xylem sap
- leaf conductivity and xylem sap ABA concentration
- the role of elevated CO_2 concentration and ABA action

Experiments were performed at the following locations:

- in the Negev Desert (Israel) using almond trees grown in large lysimeters and several wild desert plants
- in the Chaparral formation of South California using *Ceanothus* shrubs grown in pots in the field
- on a steep, dry and sun-exposed slope of the Main valley near Gambach, north of Würzburg (Germany) using shrubs.

RESULTS AND DISCUSSION

Daily courses of xylem sap ABA concentration and stomatal conductance. Many attempts have been made to correlate daily fluctuations in bulk leaf ABA with those for stomatal conductance or CO_2 assimilation, all with little success. In some instances there were negligible fluctuations in leaf ABA, and in others no clear relationship between bulk leaf ABA and leaf conductance could be observed. This was probably due to the use of leaf extracts containing several types of tissue and many different compartments (Burschka *et al.*, 1983; and references therein). With the development of extremely sensitive immunoassays, the analysis of large numbers of small samples became feasible and we therefore decided to re-investigate this problem using xylem sap. The use of xylem sap has two advantages: 1) it is in direct contact with the leaf apoplast, and thus with the primary site of action, the outer surface of the guard cell plasma membrane; and

2) it provides an answer to the hypothesis of a root-sourced stress signal to the shoot, which has become very convincing since the pioneering work of Davies and coworkers (see references in Davies and Zhang, 1991).

In the case of almond trees grown under arid conditions in the Negev Desert we found no consistent daily courses of xylem sap ABA concentration ($[ABA]_{xyl}$). There were daily fluctuations in $[ABA]_{xyl}$ which in some cases were correlated with leaf conductance (Fig. 1a), but in other instances were not. In many cases $[ABA]_{xyl}$ remained constant throughout the day, although there was a pronounced diurnal variation in leaf conductance (Figs. 1b and c).

In most cases we found no daily fluctuations of $[ABA]_{xyl}$ in well watered, unstressed plants of *Ceanothus thyrsiflorus* grown in pots in the Chaparral formation under natural conditions (Fig. 2a). In a few examples, ABA concentration was slightly higher before dawn and gradually declined afterwards, a result which agrees well with predictions of model calculations performed by Tardieu and Davies (1993). However, when the soil began to dry out, $[ABA]_{xyl}$ increased dramatically and daily ABA fluctuations with a peak during the late morning were observed (Fig. 2b). These plants had greatly reduced leaf conductances and xylem water potentials. The limited soil volume accessible to these potted plants should have caused large diurnal fluctuations of water potential in the rhizosphere. Since xylem sap ABA comes from the roots, it is considered to transmit information concerning water status within the root environment to the shoot.

Environmental factors affecting ABA concentration in the leaf apoplast. Cowan *et al.* (1982) postulated that light-dependent changes in the pH gradients between chloroplasts and extraplastidic compartments control apoplastic ABA concentrations. Since xylem sap ABA comes from the roots, direct light-dependent changes in this source of ABA cannot be expected. When considering the leaf alone, light-dependent redistribution has been shown to increase apoplastic ABA only by a factor of 1.6 (Slovik and Hartung, 1992). This observation has been confirmed by ABA assays of the apoplastic sap of darkened and illuminated sunflower leaves (Hartung, unpublished). Thus light plays no significant direct role

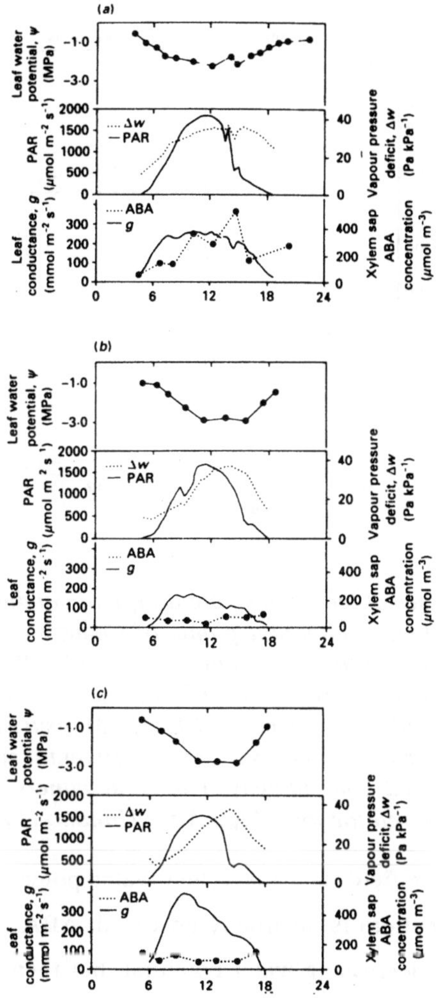

FIGURE 1. Daily courses of leaf water potential, photosynthetically active radiation, vapour pressure deficit between leaf and air, leaf conductance and xylem sap ABA concentration of Prunus dulcis *at Avdat (Israel), 1987: (a), 2 year old tree in a 3 m deep lysimeter, 16 June; (b), 2 year old tree in a 3 m deep lysimeter, 5 September; (c), 1 year old tree in a 3 m deep lysimeter, 15 September. After Wartinger* et al. *(1990), with permission.*

in determining ABA concentration in the leaf apoplast. However, light may affect ABA delivery to the guard cells indirectly through light/dark-dependent changes in the rate of transpiration.

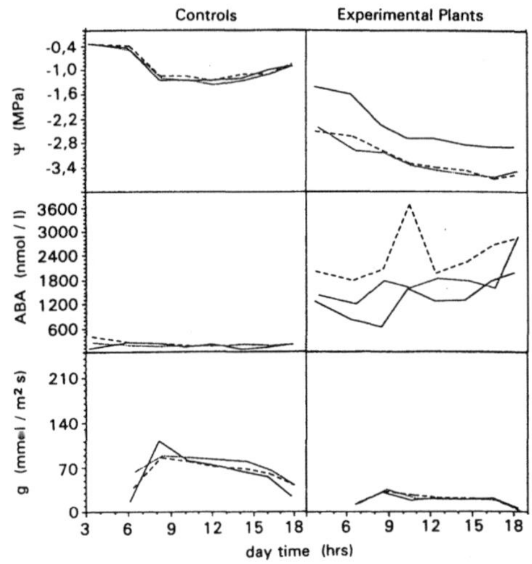

FIGURE 2. Daily courses of leaf water potential, leaf conductance and xylem sap ABA concentration of three individual plants of Ceanothus thyrsiflorus *(150 cm tall) grown at Santa Margarita Ecological Reserve (SMER), Southern California (27 May, 1991). Plants were grown in 19 l pots in soil with a maximum water holding capacity of 73% of soil dry weight (105 °C). Measurements were performed four weeks after transferring the plants to the experimental site. The control plants (left panels) were watered continuously. The right panels show data for plants which had not been watered for four days. After Hanano (1992), with permission.*

When we plot our field data for xylem water potential against xylem sap ABA, a characteristic pattern is invariably observed. ABA concentration initially increases slightly with decreasing water potential, but then rises dramatically by 10-30 fold once a threshold level of stress is reached, depending on the species. For example, in *Anastatica hiërochuntica*, a winter annual of the Negev Desert, the threshold level of stress corresponds to a xylem water potential of -0.8 MPa (Fig. 3). Other species show very different relationships, including the woody desert shrubs *Hammada scoparia* and *Zygophyllum dumosum* (threshold water potential of -5 to -6 MPa, measured at the end of the vegetative period), *Ceanothus thyrsiflorus* (-1.5 MPa; Fig. 4) and *Chilopsis linearis* (-1.0 MPa, grown in the Chaparral in Southern California). In laboratory experiments, similar relationships were observed by Zabadal (1974) for leaf strips (threshold of -1.1 MPa) and Rivier *et al.* (1983) for roots (threshold of -0.3 MPa). However, in almond trees grown

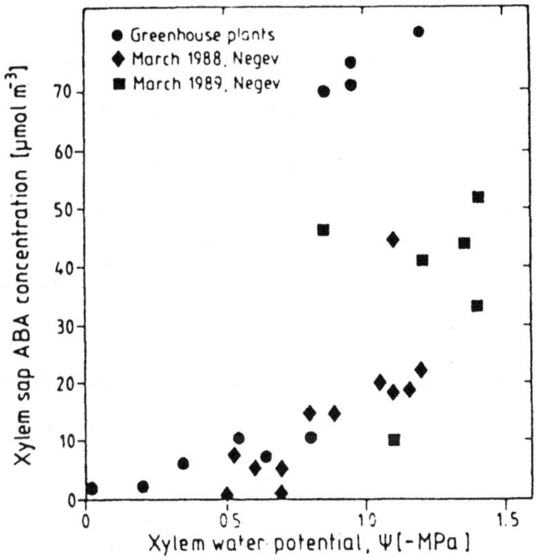

FIGURE 3. Relationship between xylem sap ABA concentration and xylem water potential in Anastatica hiërochuntica *plants.* ● *, 4 week old greenhouse plants grown in Würzburg during the winter, 1988-1989;* ◆ *, plants from Makhtesh Ramon (Negev Desert, Israel), March, 1988;* ■ *, plants from Makhtesh Ramon, March 1989. From Hartung* et al. *(1990), with permission.*

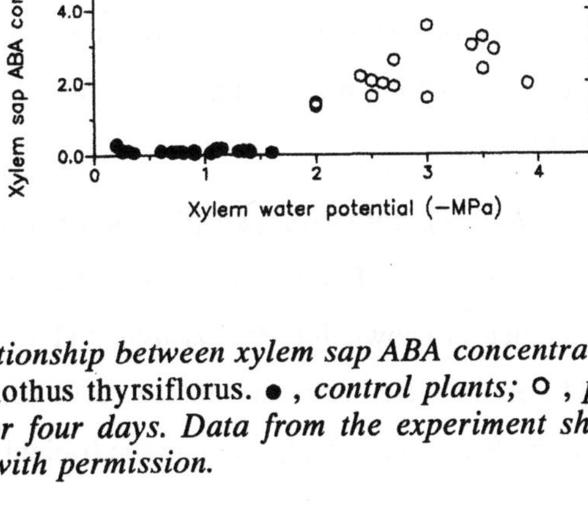

FIGURE 4. Relationship between xylem sap ABA concentration and xylem water potential in Ceanothus thyrsiflorus. ● *, control plants;* ○ *, plants which had not been irrigated for four days. Data from the experiment shown in Fig. 2. After Hanano (1992), with permission.*

in large lysimeters in the Negev Desert, it was the predawn water potential (which is assumed to provide an estimate of soil water potential adjacent to the roots) that exhibited a significant relationship with $[ABA]_{xyl}$ (Wartinger *et al.*, 1990). Similarly, in field-grown maize plants, the predawn value for $[ABA]_{xyl}$ was closely related to predawn water potential and transpirable soil water (Tardieu *et al.*, 1992).

Seasonal courses of xylem sap ABA concentration in Prunus dulcis ***and*** Ceanothus thyrsiflorus. Whereas no consistent diurnal relationships existed between xylem sap ABA and leaf conductance in almond trees grown in large lysimeters in the Negev desert, clear seasonal fluctuations of $[ABA]_{xyl}$ could be observed. With increasing soil drying from April to July, there was a marked increase of $[ABA]_{xyl}$ (Fig. 5) which seemed to be influenced by the relative amount of water given in spring and was reflected by the relationship between predawn water potential and $[ABA]_{xyl}$ (see above). The incrase in $[ABA]_{xyl}$ paralleled the decline in leaf conductance. At the end of the drying cycle $[ABA]_{xyl}$ decreased, probably because of the drought-dependent decay of the fine root system. We conclude that under these conditions the roots might act as sensors for soil water status during the first part of soil drying.

Unstressed *Ceanothus* plants, grown in much smaller pots under the naturally fluctuating climate of the Chaparral, showed a convincing relationship between $[ABA]_{xyl}$, leaf conductance and xylem water potential over a three month period (Fig. 6). The increase in leaf temperature in April paralleled the decreases in daily mean water potential (but not predawn water potential) and leaf conductance and the increases in xylem sap ABA concentration and xylem sap pH. From these long term experimental data we conclude that factors other than root water status may induce increased $[ABA]_{xyl}$.

Relationship between xylem sap ABA concentration and leaf conductance. The relationship between xylem sap ABA concentration and leaf conductance or maximum leaf conductance is shown for *Prunus dulcis* (Fig. 7) and *Ceanothus thyrsiflorus* (Fig. 8). In both species two parts to the curve can be distinguished. In the first, relatively small changes of $[ABA]_{xyl}$ are sufficient to cause dramatic changes in leaf conductance. Thereafter a relatively wide range of $[ABA]_{xyl}$ has little effect on stomatal conductance. The range of high stomatal sensitivity depends on the species involved. In *Prunus dulcis* it is up to 100 nM, in *Ceanothus*

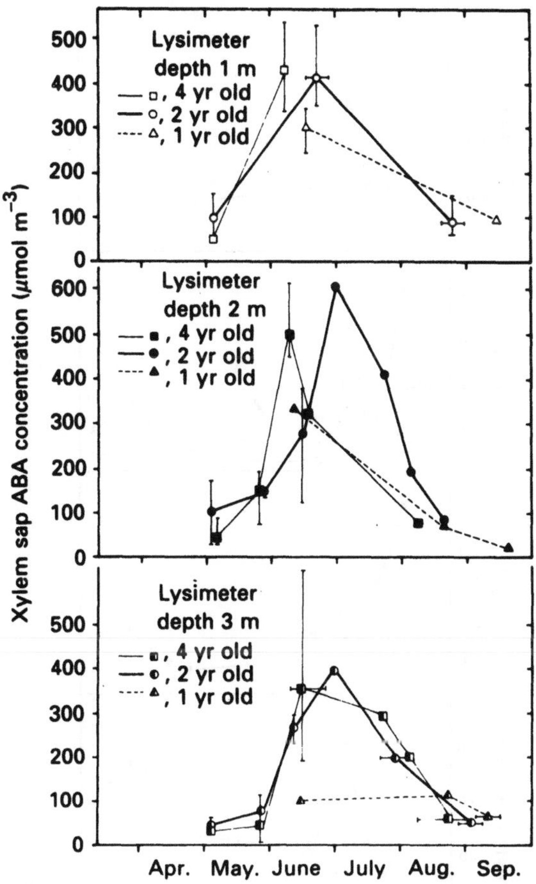

FIGURE 5. Seasonal courses of xylem sap ABA concentration at the time of maximum leaf conductance in Prunus dulcis *at Avdat (Israel), 1987. Data points represent individual trees or mean values of two or three replicates per treatment. The bars indicate the range of the values. From Wartinger* et al. *(1990), with permission.*

up to 400 nM and in *Zea mays* up to 200 nM (Tardieu and Davies, 1992). The second range of $[ABA]_{xyl}$ can only be observed in stressed systems, as in *Ceanothus* one week after the onset of soil drying (Fig. 8). In other cases, such as

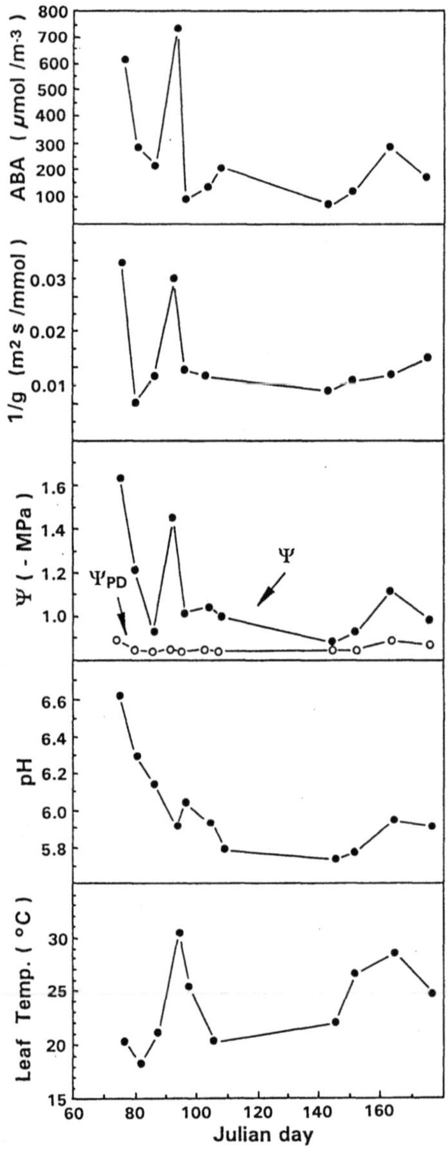

FIGURE 6. Seasonal courses of daily average values of xylem sap ABA concentration, leaf conductance, leaf water potential (●), predawn water potential (○), xylem sap pH and leaf temperature in well watered Ceanothus thyrsiflorus *plants. For further experimental details see legend for Fig. 2. From Hanano (1992), with permission.*

highly stressed *Prunus avium* (Fig. 9) and *Cornus sanguineum* shrubs from the sun-exposed slopes of the Main valley, the second range of $[ABA]_{xyl}$ is missing. The role of this second phase is not yet understood, but we may speculate that these high $[ABA]_{xyl}$ levels are necessary to maintain elevated concentrations of ABA in the guard cell apoplast during stress, when the transpiration stream is strongly suppressed by the reduced leaf conductance. A non-stomatal role for ABA by increasing stress resistance might also be proposed, as has been suggested by Werner *et al.* (1991) for bryophytes and Bartels (1990) for resurrection plants.

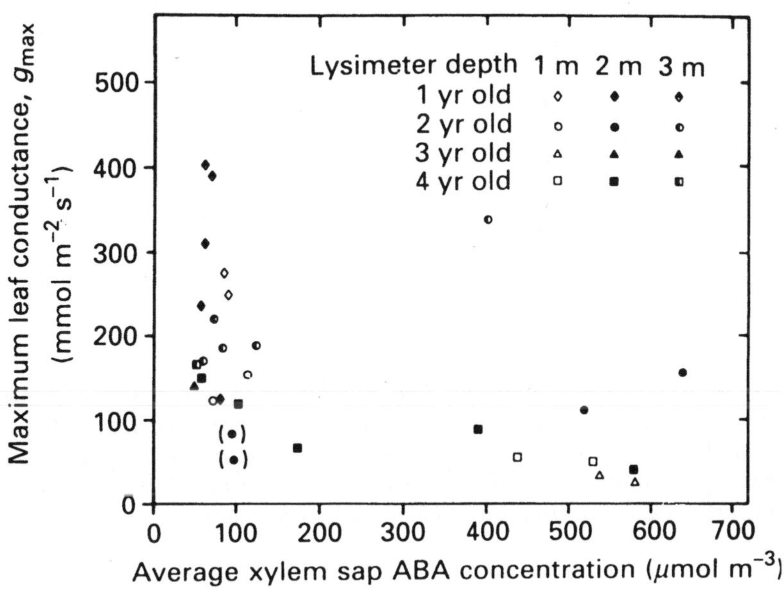

FIGURE 7. Relationship between daily average xylem sap ABA concentration and maximum leaf conductance of Prunus dulcis ***at Avdat 1987. Data in parentheses represent measurements in autumn. From Wartinger*** et al. ***(1990), with permission.***

The role of CO_2. The influence of CO_2 is of particular interest, especially because of the continuing increase in atmospheric CO_2 concentration over recent decades and its possible interaction with rising temperature and drought stress. It is known that CO_2 sensitises the guard cells to ABA and *vice versa*, and that ABA

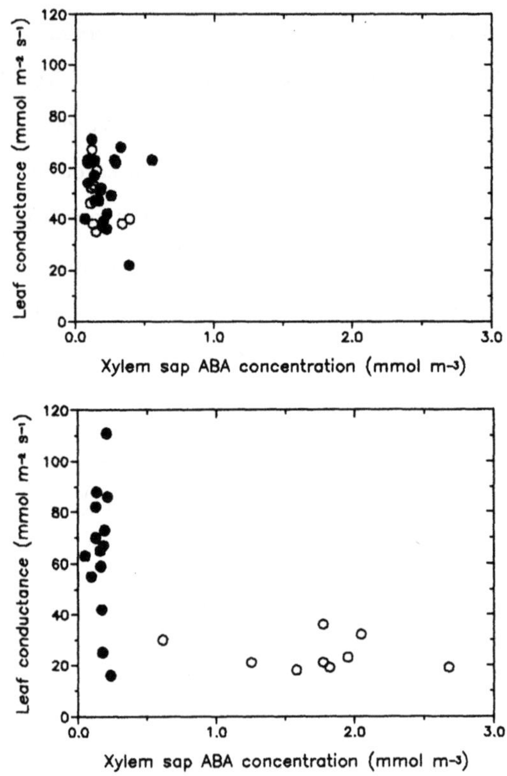

FIGURE 8. Relationship between leaf conductance and xylem sap ABA concentration in Ceanothus thyrsiflorus, *grown under the conditions shown in the legend for Fig. 2. Closed symbols, control treatment; open symbols, stressed treatment. Upper panel, start of the experiment; lower panel, 9 days after the start of the experiment. After Hanano (1992), with permission.*

concentrations which are in themselves ineffective can exert effects on stomata when intercellular CO_2 concentration is increased. Together with J.W. Radin, this phenomenon was reinvestigated under natural conditions using field-grown cotton plants in Arizona. Plants were grown in open-top chambers where they could be exposed to increased CO_2 concentrations. Small quantities of ABA were injected into a main vein on one half of the leaf, and the other half was used as a control. When plants were grown under conditions of elevated CO_2 (650 $\mu l\ l^{-1}$), the stomata reacted more rapidly and to a greater extent than in control plants grown under ambient CO_2 (Fig. 10). Auxin and cytokinins may be important factors in modulating ABA effects under conditions of elevated CO_2 (Blackman and Davies, 1984).

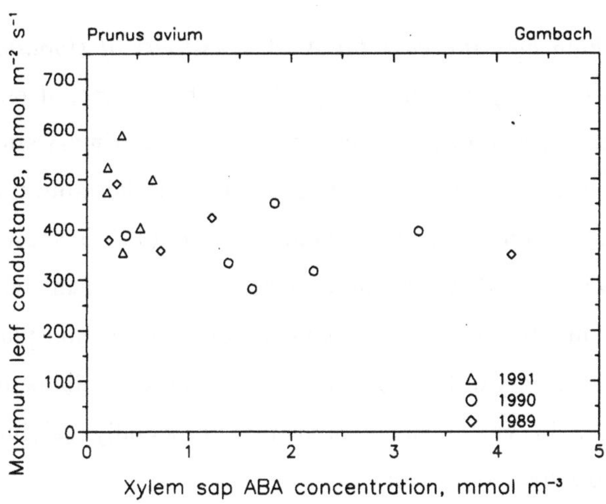

FIGURE 9. Relationship between leaf conductance and xylem sap ABA concentration in Prunus avium *trees growing on a steep, dry and sun-exposed slope of the Main valley near Gambach, Bavaria.*

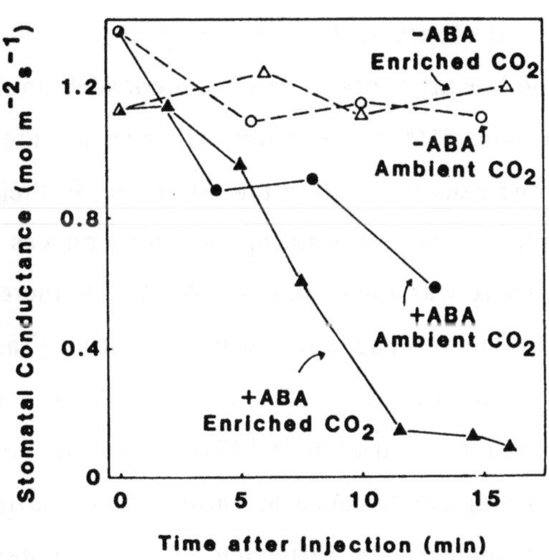

FIGURE 10. Effect of ABA on the stomatal conductances of leaves in ambient and CO_2-enriched atmospheres. ABA was injected into a lateral vein at time zero. Conductance was determined separately on the injected half of the leaf (+ABA) and the uninjected half, which served as a control (-ABA). From Radin et al. *(1988), with permission.*

Other factors which may increase the ABA sensitivity of stomata. It has been repeatedly pointed out by Trewavas (e.g., 1981) that hormonal reactions do not depend only on changes of hormone concentration at the primary site of action, but also on changes in the sensitivity of the guard cells to hormones. For sunflower plants grown in a climate chamber Schurr *et al.* (1992) showed significant correlations between the ABA concentration of the xylem sap and leaf conductance for individual plants, but not for the whole set of experimental plants. The great variation in the sensitivity of leaf conductance of individual plants to $[ABA]_{xyl}$ was due to differences in xylem sap composition, especially calcium and nitrate concentration and xylem sap or leaf pH.

There is increasing evidence that the concept of variable stomatal sensitivity to ABA also applies to stomatal reactions under natural conditions. For example, Burschka *et al.* (1983) introduced small quantities of ABA into the transpiration stream of intact *Arbutus unedo* plants grown under natural conditions in a Portuguese macchia, and observed that sensitivity to injected ABA was highest during the morning and late afternoon. The effect depended on the irrigation of the plants. Tardieu and Davies (1992) presented evidence that the stomata of maize plants also became more sensitive during the afternoon. In their experiments, the decreased water potential of the plants during the afternoon was responsible for the increased sensitivity. There was a tendency for $[ABA]_{xyl}$ to increase in these plants during the day, perhaps due to reduced root water potentials (Tardieu *et al.*, 1991). In the case of the almond trees grown in large lysimeters in the Negev Desert, there was no consistent diurnal variation in $[ABA]_{xyl}$ (see above). The relationship between $[ABA]_{xyl}$ and leaf conductance at various times during several diurnal timecourses does not indicate an influence of water potential on stomatal sensitivity, provided leaf water potential is above a certain threshold (Fig. 11). Thus we conclude that, in a system where large diurnal fluctuations in the water potential of the rhizosphere are not to be expected and the flow of water through the soil to the roots is not limiting, there will be no great diurnal change in either $[ABA]_{xyl}$ or in stomatal sensitivity to ABA caused by changes in leaf water potential.

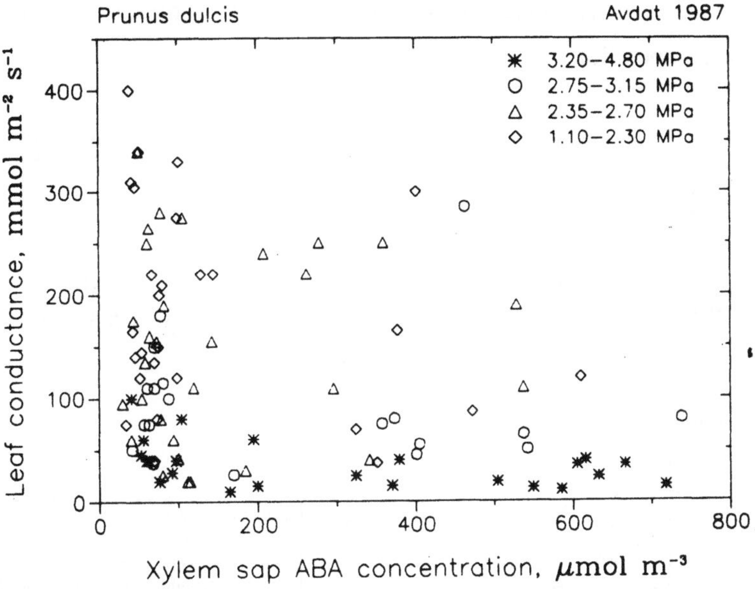

FIGURE 11. Relationship between leaf conductance and xylem sap ABA concentration as a function of leaf water potential for Prunus dulcis ***at Avdat in 1987 at several times during diurnal courses taken over the whole drying cycle. For experimental details see Wartinger*** et al. ***(1990).***

As shown by Schurr *et al.* (1992), nutritional status may affect the sensitivity of guard cells to ABA. Many careful field investigations by Radin and coworkers have shown that nitrogen and phosphorus deficiency may both increase the guard cell sensitivity of field grown cotton plants to ABA (Radin *et al.*, 1982; Radin, 1984). In such experiments the balance between ABA and cytokinins may be important.

The hypothesis that the ratio of ABA to cytokinins in xylem sap is important in determining stomatal aperture was tested for *Prunus dulcis* in the Negev Desert by measuring the daily courses of the xylem sap concentrations of ABA and several cytokinin fractions (nucleotides and free bases plus ribosides of the zeatin and N^6-(D^2-isopentenyl)adenine type). When ABA was not limiting for stomatal opening, there was a positive correlation between the ratio of the xylem sap concentrations of free bases and ribosides of zeatin-type cytokinins to ABA and leaf conductance (Fig. 12). Thus we conclude that the increased sensitivity of stomata to ABA in lysimeter-grown almond trees was primarily a function of xylem sap composition.

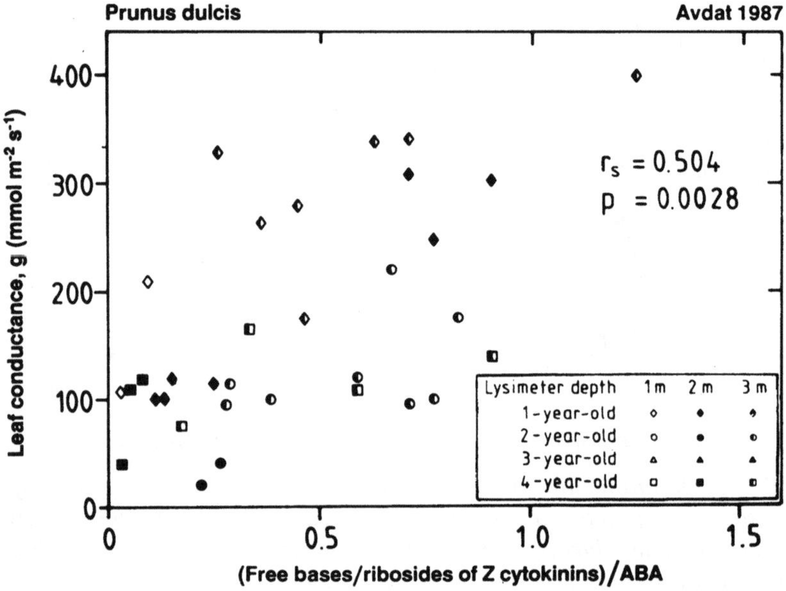

FIGURE 12. Relationship between the ratio of free bases and ribosides of zeatin-type cytokinins to ABA in the xylem sap and leaf conductance of Prunus dulcis *at Avdat, 1987. The Spearman rank correlation r_s was highly significant. From Wartinger (1991), with permission.*

CONCLUSIONS

From the experiments discussed above, we conclude that drying roots may act as sensors for soil water status and affect the concentration of abscisic acid in the xylem sap of plants under naturally fluctuating conditions. Increased ABA concentrations will reduce stomatal opening, which may improve water use during a drying cycle. However, there appear to be other factors which also influence xylem sap ABA concentration. Moreover, the sensitivity of stomata to ABA depends on a number of different factors, including plant water and nutritional status and balances with other plant growth substances in the xylem sap, especially cytokinins. Thus we have to deal with a complex system for sensing environmental stress and intraplant signalling involving thresholds, feedbacks and amplifications at different levels of integration ranging from the single cell to the whole plant.

ACKNOWLEDGEMENTS

Most of the work presented here was supported by the Deutsche Forschungsgemeinschaft (SFB 137, TP B1 [H.H.] and SFB 251, TP 3 [W.H.]).

REFERENCES

Bartels D, Schneider K, Terstappen G, Piatkowski D, Salamini F (1990) Molecular cloning of abscisic acid-modulated genes which are induced during dessication of the resurrection plant *Craterostigma plantagineum*. Planta 181: 27-34

Blackman PG, Davies WJ (1984) Modification of the CO_2 responses of maize stomata by abscisic acid and by naturally-occurring and synthetic cytokinins. Journal of Experimental Botany 35: 174-179

Burschka C, Tenhunen JD, Hartung W (1983) Diurnal variations in abscisic acid content and stomatal response to applied abscisic acid in leaves of irrigated and non irrigated *Arbutus unedo* plants under naturally fluctuating environmental conditions. Oecologia 58: 128-131

Cowan IR, Raven JA, Hartung W, Farquhar GD (1982) A possible role for abscisic acid in coupling stomatal conductance and photosynthetic carbon metabolism in leaves. Australian Journal of Plant Physiology 9: 489-498

Davies WJ, Zhang J (1991) Root signals and the regulation of growth and development of plants in drying soil. Annual Review of Plant Physiology and Molecular Biology 42: 55-76

Hanano R (1992) *Ceanothus thyrsiflorus* under water stress in its natural environment: The role of xylem sap abscisic acid. Diploma Thesis, University of Würzburg, Germany

Hartung W (1983) The site of action of abscisic acid at the guard cell plasmalemma of *Valerianella locusta*. Plant, Cell and Environment 6: 427-428

Hartung W, Heilmeier H, Wartinger A, Kettemann I, Schulze E-D (1990) Ionic and abscisic acid relationships of *Anastatica hiërochuntica* under arid conditions. Israel Journal of Botany 39: 373-382

Hartung W, Radin JW, Hendrix DL (1988) Abscisic acid movement into the apoplastic solution of water stressed cotton leaves. Role of apoplastic pH. Plant Physiology 86: 908-913

Mansfield TA (1976) Delay in the response of stomata to abscisic acid in CO_2-free air. Journal of Experimental Botany 27: 559-564

Radin JW (1984) Stomatal responses to water stress and ABA in phosphorus deficient cotton plants. Plant Physiology 76: 392-394

Radin JW, Hartung W, Kimball BA, Mauney JR (1988) Correlation of stomatal conductance with photosynthetic capacity of cotton only in a CO_2-enriched atmosphere: Mediation by abscisic acid? Plant Physiology 88: 1058-1062

Radin JW, Parker LL, Guinn G (1982) Water relations of cotton plants under nitrogen deficiency. V. Environmental control of ABA-accumulation and stomatal sensitivity to ABA. Plant Physiology 70: 1066-1070

Raschke K (1975) Simultaneous requirement of carbon dioxide and abscisic acid for stomatal closing in *Xanthium strumarium*. Planta 125: 243-259

Rivier L, Léonard JF, Cottier JP (1983) Rapid effect of osmotic stress on the content and exodiffusion of abscisic acid in *Zea mays* roots. Plant Science Letters 31: 133-137

Schurr U, Gollan T, Schulze E-D (1992) Stomatal response to drying soil in relation to changes in the xylem sap composition of *Helianthus annuus*. II. Stomatal sensitivity to abscisic acid imported from the xylem sap. Plant, Cell and Environment 15: 561-567

Slovik S, Hartung W (1992) Compartmental distribution and redistribution of abscisic acid in intact leaves. II. Model analysis. Planta 187: 26-36

Snaith PJ, Mansfield TA (1982) Control of the CO_2 responses of stomata by indol-3yl-acetic acid and abscisic acid. Journal of Experimental Botany 33: 360-365

Tardieu F, Davies WJ (1992) Stomatal response to abscisic acid is a function of current plant water status. Plant Physiology 98: 540-545

Tardieu F, Davies WJ (1993) Root-shoot communication and whole plant regulation of water flux. In: Environmental plant biology, Vol. II. Bios Scientific Publishers, Oxford (in press)

Tardieu F, Katerji N, Bethenod O, Zhang J, Davies WJ (1991) Maize stomatal conductance in the field: its relationship with soil and plant water potentials, mechanical constraints and ABA concentration in the xylem sap. Plant, Cell and Environment 14: 121-126

Tardieu F, Zhang J, Davies WJ (1992) What information is conveyed by an ABA signal from maize roots in drying field soil? Plant, Cell and Environment 15: 185-191

Trewavas A (1981) How do plant growth substances work? Plant, Cell and Environment 4: 203-228

Wartinger A (1991) Der Einfluß von Austrocknungszyklen auf Blattleitfähigkeit, CO_2-Assimilation, Wachstum und Wassernutzung von *Prunus dulcis* (Miller) D.A.Webb. Inaugural Dissertation, University Bayreuth, Germany

Wartinger A, Heilmeier H, Hartung W, Schulze E-D (1990) Daily and seasonal courses of leaf conductance and abscisic acid in the xylem sap of almond trees (*Prunus dulcis*) under desert conditions. The New Phytologist 116: 581-587

Werner O, Ros Espin RM, Bopp M, Atzorn R (1991) Abscisic acid induced drought tolerance in *Funaria hygrometrica* Hedw. Planta 186: 99-103

Zabadal TJ (1974) A water potential threshold for the increase of abscisic acid in leaves. Plant Physiology 53: 125-127

ABA AND THE CONTROL OF GROWTH AND PHYSIOLOGY OF STRESSED PLANTS

W.J. Davies, C.L. Trejo and S.J. Palmer
Institute of Environmental and Biological Sciences
University of Lancaster
Bailrigg
Lancaster
LA1 4YQ
UK

ABSTRACT

Plants which experience perturbation of their immediate climatic and edaphic environment often contain increased amounts of the plant hormone, abscisic acid (ABA). In this regard, the most studied environmental stress is the effect of soil drying, which can result in increases in ABA concentration of several orders of magnitude. In addition, flooding of the soil, temperature stress and nitrogen stress also induce ABA accumulation. ABA can be a potent regulator of stomatal behaviour and may also regulate the growth of roots and leaves. In this chapter we describe how ABA concentration in the xylem stream of plants growing in drying soil may reflect the water status of the soil surrounding their roots. This may mean that the physiology, growth and development of stressed plants can be regulated by ABA acting as a measure of the degree of stress. We consider the nature of the ABA signal under different circumstances and show that along with variation in the supply of ABA, there may be variation in the sensitivity of the response to the signal. The concept of sensitivity is discussed. The importance of variation in the flux of the hormone and in the rate of ABA metabolism in the leaf are considered.

INTRODUCTION

In his review of half a century of progress in water relations research, Kramer (1988) cited as one of our most important advances the realisation that measurement of the water relations variables of plants, as opposed to the soil, could help explain the effects of drought on plant growth and development. Other authors have noted that many of the effects of other stresses (for example, salinity, nutrient imbalance, flooding, air pollution and low temperatures) can also be explained by a perturbation of plant water relations. It is therefore not surprising that measurement of shoot water potential, solute potential and turgor has become

NATO ASI Series, Vol. I 16
Interacting Stresses on Plants in a Changing Climate
Edited by M. B. Jackson, C. R. Black

a recommended part of many studies in stress biology. If this symposium had been held in the early 1980s, it seems likely that the interacting effects of many stresses on plants would have been largely attributed to their effects on plant water status.

Over the last thirty years, hormone physiologists have become increasingly interested in the involvement of plant growth regulators in the control of the physiology, growth and development of plants under stress. In the years following the identification of the five major classes of growth regulator, an enormous number of papers reported the effects of hormones applied externally to a wide range of plant species. During this period it became fashionable to compare effects of hormones with effects of the environment on developmental and physiological responses and suggest a regulatory role for the hormone(s) in the processes under consideration.

Many authors have published comprehensive surveys of the effects of growth regulators applied externally to different plant species in many different situations. Nevertheless, largely because of technical limitations in many laboratories, we still have very little information on the natural synthesis and distribution of many growth regulators in plants. Still more rare are good quality measurements of this type, combined with adequate measurements of the environmental variable of interest and other aspects of the stress biology of the plant.

In 1981, Trewavas criticised the types of conclusion drawn from many application-type experiments. One of the grounds for criticism was that the endogenous concentrations of naturally-occurring regulators were often several orders of magnitude lower than those apparently promoting the same effects when applied externally. We will argue here that we now know enough about natural variation in the concentration and distribution of the plant growth regulator, abscisic acid (ABA), to begin to ascribe an ecophysiological role to this compound. These arguments are not based on extrapolation of effects of random concentrations of the compound sprayed onto plants held under arbitrary conditions, but rather on the need to explain particular physiological and developmental responses to environmental variation.

CAN WE EXPLAIN EFFECTS OF SOIL DRYING ON SHOOT GROWTH AND PHYSIOLOGY?

It is well known that positive leaf turgor is a prerequisite for leaf growth and stomatal opening. It is not surprising then that drought-induced limitations in leaf growth and stomatal opening have commonly been ascribed to effects of soil drying on the uptake of water by the plant and the water relations of the shoots. Careful observations over the last 10-15 years have suggested, however, that plants growing in drying soil do not always show reductions in leaf turgor and that, in many cases, stomatal closure may act to sustain turgor at values which are comparable to those shown by well-watered plants.

In an attempt to understand why stomata remain closed under these circumstances, various techniques have been used to destroy the link between drying soil and reduced water uptake. Passioura and colleagues (e.g., Gollan *et al.*, 1986; Passioura, 1988) have used a device that applies pressure to the roots of plants in drying soil, such that the applied pressure exactly counteracts the increase in soil suction as the soil dries. This means that the shoot water relations of plants in drying soil can be sustained artificially at values close to those shown by well-watered plants. It is crucial to note that, despite this, some interaction between roots and drying soil generates an influence that restricts leaf growth rate and leaf conductance.

Gowing *et al.* (1991) described an experiment that suggests that this influence is a chemical inhibitor which moves from roots to shoots, probably in the transpiration stream. These authors used plants with roots split between two separate rooting containers. The soil in one container was allowed to dry while the other was watered regularly. This treatment resulted in a limitation of leaf growth and stomatal opening, even though the shoot water relations of these plants were not perturbed compared to plants watered on both halves of the root system. Most importantly, when the roots in contact with drying soil were severed from the plants, the leaf growth rates of these plants increased to the rates shown by the well-watered plants, even though no more water was made available to the roots. This result suggests that this treatment removed the source of an inhibitor of leaf growth which was generated as a result of the association between roots and drying soil.

WHAT IS THE INHIBITOR OF LEAF GROWTH AND STOMATAL OPENING?

There now seems little doubt that, in some plants, roots in drying soil produce increased amounts of abscisic acid (ABA) which move in the transpiration stream to the shoots (see Davies and Zhang, 1991). This compound is a potent regulator of stomatal behaviour, even when there is no clear relation between conductance and shoot water relations (Fig. 1). ABA also influences the growth of both leaves

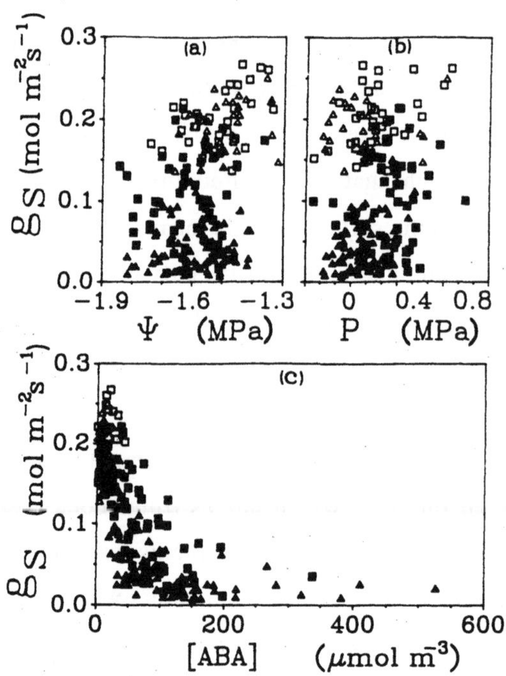

FIGURE 1. Stomatal conductance of leaves of maize plants growing in the field as a function of leaf water potential (ψ), leaf turgor (P) and ABA concentration in the xylem of the leaves. Values are paired measurements taken on individual leaves of plants that were either well watered or droughted on plots of well mixed soil, and plants subjected to the same watering treatment on plots of compacted soil. From Tardieu et al. *(1992a).*

and roots (Sharp, this volume). Abscisic acid is also produced in leaves that lose turgor as the soil dries and in many situations leaf-sourced ABA must also contribute to the ABA message in the xylem. This need not be the case, however, since in some circumstances leaf water relations are regulated and there is no reason to argue that extra ABA is produced here.

Some argument in this area has centred around the question of whether plants produce enough ABA to explain the responses that are seen. Zhang and Davies (1991) explained all the limitation in stomatal opening that they saw in maize in terms of the extra ABA produced by soil drying. However, Munns and King (1988) argued that in wheat the extra ABA produced by plants in drying soil is much less than is required to produce the observed degree of stomatal closure. Furthermore, removal of ABA from the transpiration stream did not remove all the antitranspirant activity. Further work by Munns' group revealed that this activity stems from a polymerisation reaction which takes place while the sap is stored, although it seems possible that this process could also occur in the plant (Munns, personal communication).

Claims and counterclaims about the adequacy or otherwise of different ABA doses must all be viewed with some scepticism until we can actually measure the local concentration of the hormone at the site of action in the leaf. At least for stomatal action, this is the concentration in the apoplast close to the guard cells. This concentration must be a function of the concentration of ABA in the xylem sap arriving in the leaf, but Gowing *et al.* (1993) have shown that stomatal behaviour is also affected by the total amount of ABA arriving, which is a function of transpiration rate. Tardieu *et al.* (1992a), working in the field, have shown clearly that ABA concentration in the xylem sap is also greatly affected by the transpiration flux. Any bioassay designed to quantify the magnitude of a root signal should clearly take into account the important influence of the transpiration flux.

The measurement of local concentrations of ABA in the leaf is particularly important because of the way that the hormone distributes between different compartments of the leaf. The work of Hartung and colleagues (see this volume) has shown that very large amounts of ABA can be sequestered away from sites of action on the guard cells so that there is not always a clear relationship between the quantity of ABA in the leaf and the apparent stomatal response. Stomatal behaviour is commonly more closely linked to ABA arriving in the transpiration stream. We should not, however, ignore the possibility that a root signal could be any influence which allows previously-sequestered ABA access to sites of action for control of stomatal behaviour or leaf growth. This might be a change in the pH or ionic balance of the xylem sap.

One controlling influence on the local ABA balance that has received little attention is the metabolism of ABA within the leaf. Trejo *et al.* (1993) have been able to attribute apparent differences in stomatal sensitivity to ABA to the rate at which the leaf metabolises the ABA supplied. Stomata in detached epidermis are apparently highly sensitive to applied ABA because epidermal tissue has a very restricted capacity to break down ABA. When ABA was applied to pieces of whole leaves, ABA metabolism kept pace with the rate at which the hormone entered the leaf. This resulted in only a very restricted accumulation of ABA and a very low apparent stomatal sensitivity (Fig. 2). We know that ABA metabolism is highly sensitive to changes in the physiological status of the leaf (Zeevaart and Creelman, 1988) and it therefore seems likely that any root signalling system will be highly dependent upon the balance between ABA arriving at the sites of action and the rates of sequestration and metabolism.

There is clearly much that we do not understand about the regulation of leaf functioning by root-sourced hormones. We may know enough, however, to suggest a framework of analysis to help agronomists and ecologists understand how plants growing in drying soil regulate their physiology, growth and development in the field. It may be that we will conclude that there is little point in trying to quantify growth regulator concentrations, fluxes and distributions in plants in certain kinds of experiment, but if we do need to do this, we need to be sure that we are asking the correct kinds of questions.

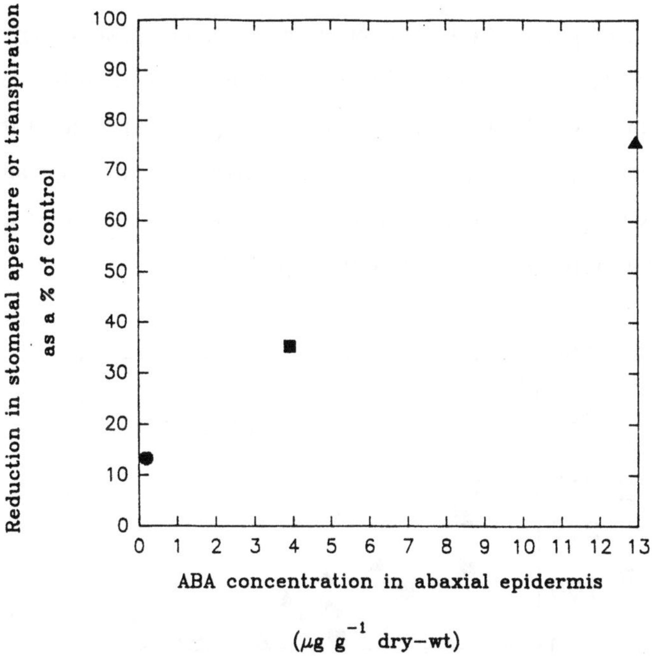

FIGURE 2. Reduction in stomatal aperture or transpiration as a function of the ABA concentration in the abaxial epidermis of Commelina communis *L. after incubation in, or feeding of ABA at 10^{-3} mol m^{-3} for three hours. The treatments were: a) (circles) leaf pieces floated on ABA solutions and stomatal apertures measured using a microscope; b) (squares) shoots placed in ABA solutions for a gravimetric transpiration bioassay; and c) (triangles) epidermis floated on ABA solutions and stomatal apertures measured using a microscope. From Trejo* et al. *(1993).*

ABA AND RESPONSES OF PLANTS UNDER NATURAL CONDITIONS

In our experiments, unwatered maize plants growing in the field in France showed no clear reduction in water potential when compared to plants growing on plots that were watered regularly. Despite this, the stomatal conductance of unwatered plants was substantially reduced (Tardieu *et al.*, 1992a). The combination of these two responses meant that there was no clear relationship between shoot water relations and stomatal conductance. The restriction of stomatal opening was apparently a result of small increases in ABA concentration in the xylem (Fig. 1).

In non-transpiring plants, ABA concentrations provided the shoot with a 'measure' of the availability of soil water and it is important to note that the data for plants from all treatments fell on the same relationship (Fig. 3a). During the day, there were two clear relationships between xylem ABA concentration and soil water availability, with plants growing on compacted soil with restricted access to soil water showing higher ABA concentrations (Fig. 3b).

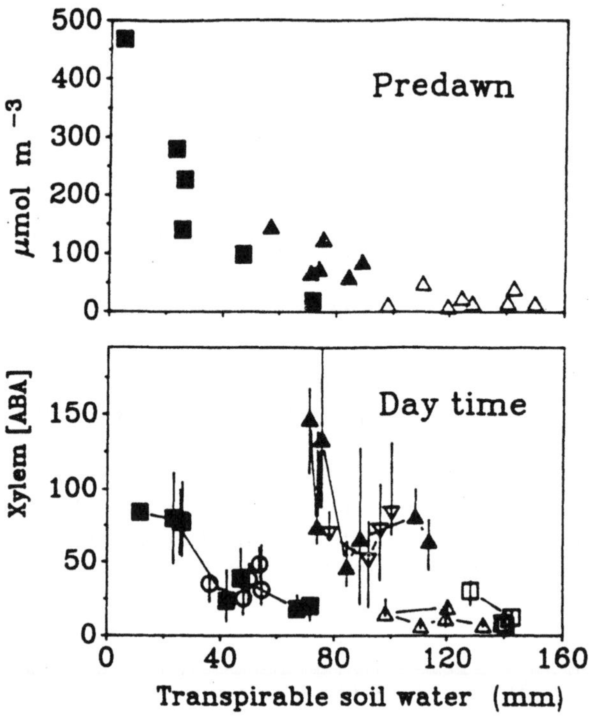

FIGURE 3. Relation between ABA concentration in the xylem sap of maize leaves from plants growing in the field and transpirable soil water before dawn and during the day. Daytime values correspond to the median; bars show quartiles (non-symmetrical with respect to the mean value). From Tardieu et al. *(1992b).*

These data show that the ABA relations of the shoot reflect not only the quantity of ABA synthesised in response to drying soil but also the extent to which that ABA is diluted by the transpiration flux. High transpiration rates will be expected to result in relatively low ABA concentrations. Indeed, we have shown that the concentration of ABA in the xylem declines during the day as the vapour pressure deficit increases. Without this effect, we might expect that increased water loss would result in increased ABA delivery later in the day. The implications of ABA dilution with increased transpiration flux are that on hot dry days, the signalling system will be least effective unless the shoot processes are responding to something other than a simple modulation of concentration.

One possibility suggested by Gowing *et al.* (1993) is that the stomata respond to an *amount* of ABA arriving in the transpiration stream, with high transpiration rates producing high localised concentrations around the guard cells. Tardieu *et al.* (1993) examined this possibility for the maize crop growing in the field but failed to define any clear relationship between the flux of ABA and the physiological response. This is not entirely surprising, since we have no information on the partitioning and metabolism of ABA in the leaves of these field-grown plants.

If we are to argue that ABA has an important controlling influence on stomatal behaviour on a day-to-day and even an hour-to-hour basis, it is necessary to explain the common observation that droughted plants show reduced conductances during the afternoon, even though xylem ABA concentrations are comparatively low. Detailed analysis of the data presented in Fig. 1 showed that there is an apparent change in the sensitivity of the stomata to the ABA signal, with an increased response developing later in the day (Tardieu and Davies, 1992). Further analysis suggests that this is a function of the changing water status of the leaf and that stomata are sensitised to ABA as leaf water potential falls. Using epidermal strips incubated at different water potentials, we have shown that there is a fundamental variation in the sensitivity of the guard cells to ABA. There is also the possibility that our field observations reflect an effect of leaf water potential on the rate at which ABA in the leaf is broken down, and therefore on the local concentration around the guard cell (Trejo *et al.*, 1993).

Tardieu and Davies (1993) used the relationships described above to produce a simple model describing the control of stomatal behaviour by chemical and hydraulic signals. Despite a number of multiple feedback effects, the model is accessible to intuition. Two important results of the modelling exercise are that, as well as regulation of stomatal behaviour, the system also results in the regulation of shoot water relations and ABA balance. Both of these results are expected from our field observations. It is interesting to speculate that the regulation of ABA concentration results in a relatively stable signal for the long-term regulation of development as a function of the availability of soil water.

In the context of the subject of this meeting, we will now consider the possibility that ABA may be a common factor in the developmental responses of plants to several environmental stresses.

ABA AND THE EFFECTS OF INTERACTING STRESSES

As noted above, many environmental stresses can reduce plant turgor, with a consequent synthesis of extra ABA. We know that ABA is a potent molecule but in most cases we still need to determine to what extent ABA action is responsible for the physiological and developmental responses observed. We note here three examples of different stress responses in which ABA seems to be involved and highlight limitations in our understanding.

Flooded plants commonly contain very large amounts of ABA (see for example, Jackson, 1991). It is clear that this ABA acts to limit stomatal opening and restrict water loss. This response may be necessary to sustain shoot turgor under circumstances where, paradoxically, water supply through the roots may be restricted. Despite this problem, there is no detectable water deficit in many of the leaves of many flooded plants. There is still some controversy over where this extra ABA is coming from. The possibilities that have received attention are the roots, older leaves that have lost some turgor or younger turgid leaves that are exporting or metabolising reduced amounts of ABA. It may well be that all three sources contribute to an altered hormone balance.

Another example of an interaction between edaphic and climatic stresses where there may be a central role for ABA is in plants growing in cold soil. Here, water supply is also restricted by the hydraulic resistance of the root system and the problem of turgor maintenance may be exacerbated by high insolation or evaporative demand. Under these circumstances, high ABA concentrations are commonly found in plants and it has been suggested that the hormone may increase the water flux through roots at low temperatures (Davies *et al.*, 1982). ABA should also close the stomata to restrict water loss, but there are now several reports that ABA has little or no effect on stomata at chilling temperatures (Cornic and Ghashghaie, 1991) and may even increase stomatal conductance (Rodriguez and Davies, 1982). We have little understanding of how temperature may affect signal transduction at the guard cell membrane.

In plants which receive only limited supplies of nitrogen, ABA concentrations are often enhanced (for example, Chapin *et al.*, 1988), but is not always clear why this is so. In a careful piece of work with plants growing in solution culture, Palmer *et al.* (1993) could find no clear difference between the water relations of plants growing at two different nitrogen supply rates. Furthermore, the concentrations of nutrients in the solution bathing the roots were low and similar in both treatments (amounts of nutrient supplied were different). It seems therefore that the increased ABA concentrations found in plants grown with a limited nitrogen supply were not the result of a signal such as a decrease in the external concentration of nitrate, but rather an internal change resulting from a lower uptake of nitrate. Whatever the specific signal, changes in the ABA balance as a result of variation in nutrient availability may contribute to the modification of physiology, growth and development under these conditions. The extent to which this is the case has yet to be determined.

CONCLUSIONS

There is now a strong case for arguing that we cannot understand the drought responses of plants without detailed measurements of their ABA balance. No study of the effects of soil drying would be complete without a complete suite of water relations measurements made at appropriate times and on appropriate parts of the plant. We would argue that measurements of ABA concentrations should also form a central part of many drought studies. It remains to be seen whether ABA plays a similar role in the modulation of other environmental effects. The potent effects of the molecule on the regulation of gene expression suggests that this may indeed be the case.

REFERENCES

Chapin FS, Walter CHS, Clarkson DT (1988) Growth response of barley and tomato to nitrogen stress and its control by abscisic acid, water relations and photosynthesis. Planta 173: 352-366

Cornic G, Ghashghaie J (1991) Effect of temperature on net CO_2 assimilation and photosystem II Quantum yield of electron transfer of French bean (*Phaseolus vulgaris* L.) leaves. Plant Physiology 185: 255-260

Davies WJ, Rodriguez JL, Fiscus EL (1982) Stomatal behaviour and water movement through roots of wheat plants treated with abscisic acid. Plant, Cell and Environment 5: 485-493

Davies W, Zhang J (1991) Root signals and the regulation of growth and development of plants in drying soil. Annual Review of Plant Physiology and Plant Molecular Biology 42: 55-76

Gollan T, Passioura JB, Munns R (1986) Soil water status affects the stomatal conductance of fully turgid wheat and sunflower plants. Australian Journal of Plant Physiology 13: 459-464

Gowing DJ, Davies WJ, Jones HG (1990) A positive root-sourced signal as an indicator of soil drying in apple, Malus x domestica Borkh. Journal of Experimental Botany 41: 1535-1540

Gowing DJ, Davies WJ, Jones HG (1993) Stomatal control as a function of concentration or amount of ABA? Plant, Cell and Environment (in press)

Jackson MB (1991) Regulation of water relationships in flooded plants by ABA from leaves, roots and xylem sap. In: Davies WJ, Jones HG (eds) Abscisic acid: physiology and biochemistry. Bios Scientific Publishers, Oxford, pp 217-226

Kramer PJ (1988) Changing concepts regarding plant water relationships. Plant, Cell and Environment 11: 565-568

Munns R, King RW (1988) Abscisic acid is not the only stomatal inhibitor in the transpiration stream. Plant Physiology 88: 703-708

Palmer SJ, McDonald AJS, Davies WJ (1993) Elevated abscisic acid concentrations in sunflower as a result of limitation in nitrate supply. Journal of Experimental Botany (in press)

Passioura JB (1988) Root signals control leaf expansion in wheat seedlings growing in drying soil. Australian Journal of Plant Physiology 15: 687-693

Rodriguez JL, Davies WJ (1982) The effects of temperature and ABA on stomata of *Zea mays* L. Journal of Experimental Botany 33: 977-987

Tardieu F, Davies WJ (1992) Stomatal response to abscisic acid is a function of current plant water status. Plant Physiology 98: 540-545

Tardieu F, Davies WJ (1993) Integration of hydraulic and chemical signalling in the control of stomatal conductance and water status of droughted plants. Plant, Cell and Environment (in press)

Tardieu F, Katerji N, Bethenod O, Zhang J, Davies WJ (1992a) Maize stomatal conductance in the field: its relationship with soil and plant water potentials, mechanical constraints and ABA concentration in the xylem sap. Plant, Cell and Environment 15: 185-191

Tardieu F, Zhang J, Davies WJ (1992b) What information is conveyed by an ABA signal from maize roots in drying field soil? Plant, Cell and Environment 15: 185-191

Tardieu F, Zhang J, Gowing DJ (1993) A model of stomatal control by both ABA concentration in the xylem sap and leaf water status: test of the model and of alternative mechanisms for droughted and ABA-fed field-grown maize. Plant, Cell and Environment (in press)

Trejo CL, Davies WJ, Ruiz LMP (1993) Sensitivity of stomata to ABA: an effect of the mesophyll. Plant Physiology (in press)

Trewavas AJ (1981) How do plant growth substances work? Plant, Cell and Environment 4: 203-228

Zeevaart JAD, Creelman RA (1988) Metabolism and physiology of abscisic acid. Annual Review of Plant Physiology and Plant Molecular Biology 39: 439-473

Zhang J, Davies WJ (1991) Antitranspirant activity in the xylem sap of maize plants. Journal of Experimental Botany 42: 317-321

REGULATION OF ROOT GROWTH AT LOW WATER POTENTIALS

Robert E. Sharp, Eric S. Ober and Yajun Wu
Department of Agronomy
University of Missouri
Columbia
MO 65211
USA

ABSTRACT

In many species, roots continue to grow at low water potentials that cause complete inhibition of shoot growth. Recent studies show that the differential sensitivity of root and shoot growth to low water potentials is regulated, at least in part, by the hormone abscisic acid (ABA). When ABA accumulation in maize seedlings at low water potential is inhibited using either fluridone or the *vp5* mutant to block carotenoid (and ABA) biosynthesis, primary root elongation is severely inhibited, while shoot growth is promoted as compared with untreated or wild-type seedlings growing at the same water potential. Fluridone-induced inhibition of root elongation at low water potential may be reversed by applying exogenous ABA, confirming the requirement for increased endogenous levels of ABA for the maintenance of root growth under dry conditions. In fluridone-treated seedlings, ABA levels at low water potential decreased to a similar extent throughout the root growth zone, while inhibition of elongation increased with distance from the apex. These results suggest a developmental gradient in tissue responsiveness to endogenous ABA. Current studies suggest that ABA may have a number of roles in maintaining root cell elongation at low water potentials. Firstly, proline is a major contributor to osmotic adjustment in the growth zone of maize primary roots, and its accumulation is partially prevented by treatment with fluridone. This effect can be reversed by adding exogenous ABA. Secondly, activity of the enzyme xyloglucan endotransglycosylase (XET), which is believed to be important for cell wall yielding, is markedly increased in the root growth zone at low water potential, and this response is also inhibited by fluridone, suggesting regulation by ABA.

NATO ASI Series, Vol. I 16
Interacting Stresses on Plants in a Changing Climate
Edited by M. B. Jackson, C. R. Black

INTRODUCTION

It is well known that water-limited conditions can induce more extensive root development. Plants in drying soil may develop a broader and more deeply-penetrating root system (Weaver, 1926), exhibit a proliferation of lateral roots (Jupp and Newman, 1987), and produce a greater root biomass than plants growing under well-watered conditions (Sharp and Davies, 1979). These responses contrast with the generally high sensitivity of growth of the aerial plant parts to the development of water deficits (Hsiao, 1973). There is considerable genetic variability in the ability to develop more extensive root systems under water-limited conditions, and this response can be positively associated with yield in dry years (e.g., Sponchiado *et al.*, 1989). However, despite the importance and long-standing recognition of the effects of soil drying on root system morphology, the physiological mechanisms that regulate root growth under water-limited conditions have received little attention and remain poorly understood.

Westgate and Boyer (1985) showed that the growth of maize nodal roots was intrinsically less sensitive than that of the aerial plant parts to low water potentials in the growing tissues. In addition, root growth continued at low water potentials that were completely inhibitory to shoot growth, as is also found in maize seedlings (Fig. 1; Sharp *et al.*, 1988). These results indicate that the differential sensitivity of root and shoot growth to water-limited conditions cannot be explained simply by the exposure of the shoot to the desiccating effects of the atmosphere, but must involve an internal control system. This chapter reviews recent advances in our understanding of the mechanisms involved in maintaining root growth at low water potentials, and focusses on the role of the hormone, abscisic acid (ABA).

ROLE OF ABA IN THE REGULATION OF ROOT GROWTH AT LOW WATER POTENTIALS

It is well established that ABA accumulates to high concentrations in plant tissues experiencing water deficits. In addition to its role in regulating stomatal behaviour,

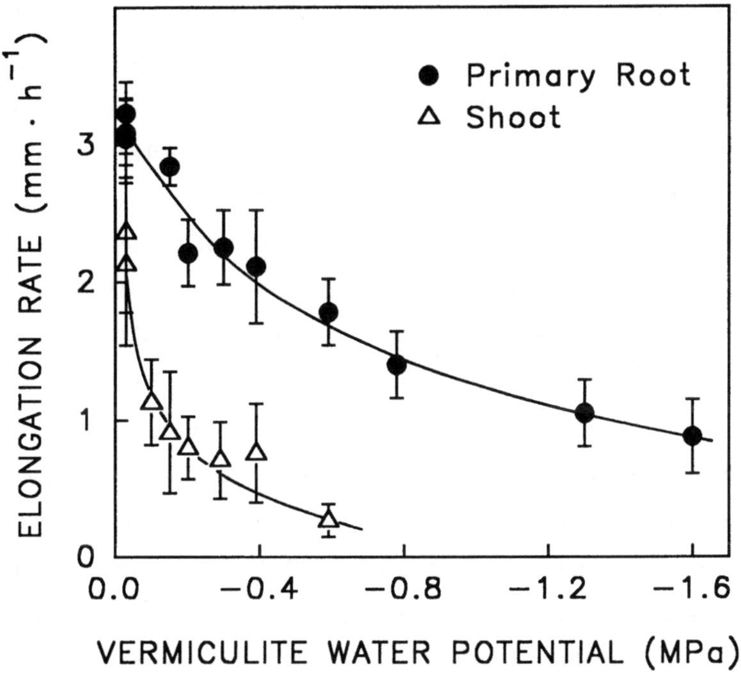

FIGURE 1. Elongation rate of the primary root and shoot of maize seedlings (cv. FR27 x FRMo17) grown at various vermiculite water potentials. The seedlings were transplanted to the various water potential treatments 36 hours after initial planting, and were grown in the dark at 29 °C at near saturation humidity. For roots, measurements were made at root lengths of 5 cm, when elongation rates were approximately constant. The data for shoots represent maximum elongation rates after transplanting. Data are means ± 1 SD. Modified from Sharp (1990).

ABA may play an important role in the differential sensitivity of root and shoot growth to low water potentials (Watts *et al.*, 1981; Hsiao and Bradford, 1983; Zeevaart and Creelman, 1988). Speculation has largely relied on studies in which ABA has been applied to well-watered plants. Such treatments have almost invariably resulted in inhibition of shoot growth, whereas a number of studies have reported a promotion of root growth (e.g., Gaither *et al.*, 1975; Watts *et al.*, 1981). Even in roots, however, high levels of applied and endogenous ABA have generally been associated with growth inhibition (Jones *et al.*, 1987; Saugy *et al.*, 1989).

In most studies of the effects of applied ABA on root growth at high water potentials, measurements were not made to determine whether the resulting internal levels of ABA were similar to those found in roots growing at low water potentials. Furthermore, even when applied ABA is shown to accumulate to realistic concentrations (Creelman *et al.*, 1990), the resulting growth responses might not adequately explain those which occur at low water potentials because of possible differences in ABA compartmentation (Hartung *et al.*, 1988), sensitivity (Trewavas, 1981) and response (Rodriguez and Davies, 1982) under different environmental conditions. Manipulation of endogenous ABA levels in plants growing at low water potentials by chemical or genetic means is more likely to produce unambiguous results, and these approaches were adopted in a recent study in this laboratory of the role of ABA in the maintenance of primary root growth in maize at low water potentials (Saab *et al.*, 1990, 1992). Accumulation of ABA in seedlings transplanted shortly after germination to vermiculite maintained at a water potential of -1.6 MPa was reduced by about two-thirds when either fluridone, an inhibitor of carotenoid (and ABA) synthesis, or the mutant *vp5*, which carries a lesion at the same step as that blocked by fluridone, was used. Both treatments were associated with severe inhibition of root elongation rate as compared to untreated and wild-type roots growing at the same water potential (Fig. 2). There was little effect on either ABA levels or root elongation rate at high water potential. These results indicate that high levels of ABA are required for the maintenance of primary root growth at low water potentials. This has been confirmed in subsequent experiments in which fluridone-induced inhibition of root elongation rate at low water potential was reversed by applying ABA to raise the internal ABA level in the root growth zone to that of untreated roots (RE Sharp, GS Voetberg and IN Saab, unpublished data from an experiment conducted at a water potential of -0.3 MPa to facilitate uptake of ABA from the vermiculite). In contrast, raising the ABA concentration in well-watered roots to the usual level in roots at low water potential resulted in growth inhibition, showing that the response of root growth to ABA varies with tissue water status.

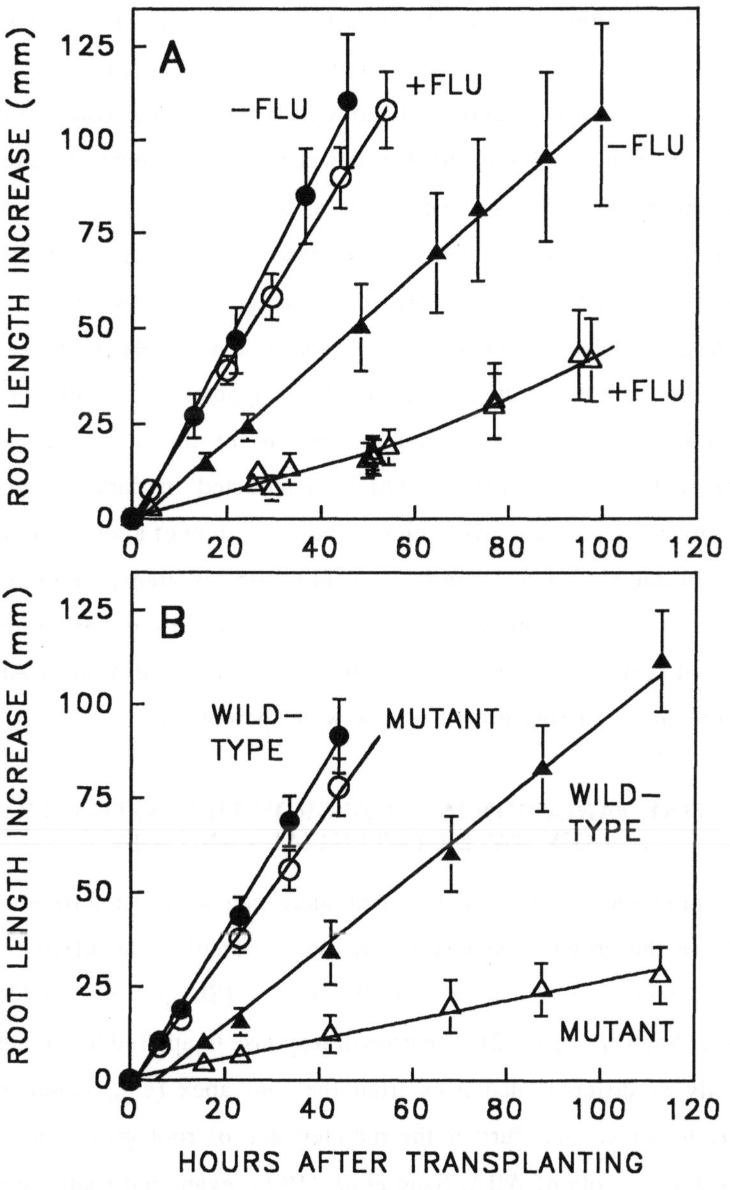

***FIGURE 2.** Elongation of the primary root of: (A) fluridone-treated (+FLU); and (B)* vp5 *maize seedlings compared to untreated (-FLU) and wild-type seedlings after transplanting to vermiculite maintained at high (-0.03 MPa, circles) or low (-1.6 MPa, triangles) water potential. Fluridone (10 μM) was mixed with the vermiculite in which the seeds were germinated and into which the seedlings were transplanted. Data points are means ± 1 SD. Modified from Saab* et al. *(1990).*

Similar experiments yielded contrasting results for shoot growth: in both fluridone-treated and *vp5* seedlings shoot growth was promoted at low water potential as compared with untreated and wild-type seedlings (Saab *et al.*, 1990). Additional experiments indicated that the contrasting effects of reduced ABA levels on the growth of roots and shoots at low water potentials could not be explained by competition for water or seed reserves (Saab *et al.*, 1992).

The results of these studies provide evidence that ABA plays a major role in regulating the balance between root and shoot development in maize seedlings at low water potentials through independent and opposite effects on the growth of the two organs. The use of a seedling system for this investigation had the advantage that the experiments could be conducted in darkness under near-saturating humidity, thereby avoiding any indirect effect of altered ABA levels on plant growth due to stomatal control of photosynthesis, transpiration and overall plant water status. Further, as described below, the relative simplicity of the seedling root system offers hope of identifying the mechanism(s) of ABA action in the maintenance of root growth at low water potentials.

SPATIAL VARIATION IN ROOT GROWTH SENSITIVITY TO LOW WATER POTENTIALS AND ABA

It is becoming increasingly evident that there can be large differences in the sensitivity of the growth response to low water potentials at different locations within the growing zone of roots and other organs (Sharp *et al.*, 1988; Pritchard *et al.*, 1991; Saab *et al.*, 1992). The growth response to applied hormones can also vary greatly at different distances from the root apex (e.g., Goodwin, 1972). Therefore, to understand further the maintenance of root growth at low water potentials and the role of ABA, Saab *et al.* (1992) examined spatial gradients in ABA levels and the responsiveness of growth to the hormone in the maize primary root system described above.

In roots growing in vermiculite maintained at a water potential of -1.6 MPa, longitudinal expansion was completely inhibited in the basal region of the growing zone but was unaffected near the apex (Fig. 3A; Sharp *et al.*, 1988). ABA

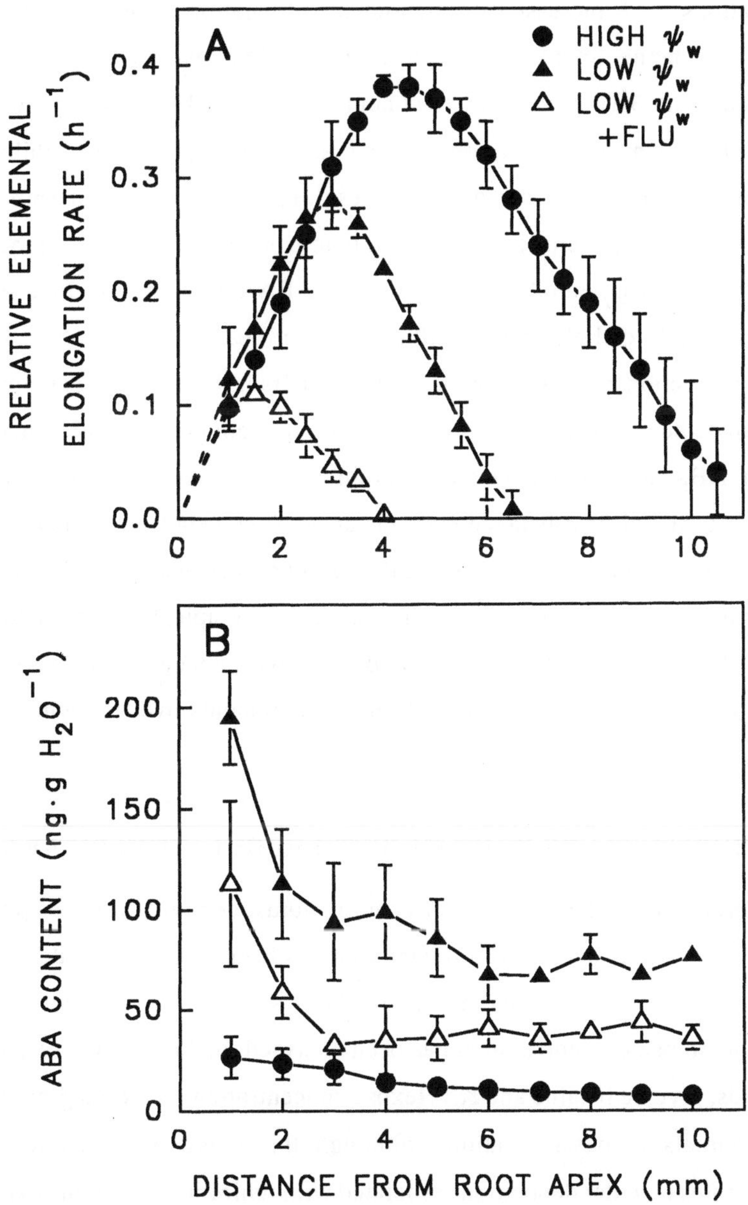

FIGURE 3. *Spatial distribution of: (A) relative elemental elongation rate; and (B) ABA content in the apical 10 mm of maize primary roots growing in vermiculite maintained at high water potential (-0.03 MPa), low water potential (-1.6 MPa), or low water potential plus fluridone (FLU, 10 μM). Data points are means ± 1 SD. Modified from Saab* et al. *(1992).*

concentrations were increased throughout the elongation zone, with the highest concentration being found in the apical few millimetres where elongation was maintained (Fig. 3B). Fluridone treatment reduced ABA concentration at all locations to a similar extent, but caused an increasingly large inhibition of elongation with distance from the apex, resulting in a further shortening of the growth zone. Measurements of cortical cell length profiles indicated that the effects of both low water potential and reduction of ABA levels by fluridone involved large decreases in cell elongation, although in both cases the data also suggested that cell production rates were reduced. The results indicate that a developmental gradient in responsiveness to ABA exists within the root growth zone, whereby the ability of ABA to maintain cell elongation at low water potentials appears to decrease with increasing distance from the apex. Such detailed characterisation is an essential step towards understanding the cellular mechanisms involved. Clearly, mechanisms that act either to maintain growth in the apical region or to inhibit growth in the basal region need to be addressed with adequate spatial resolution since the composite picture obtained from measurements over the whole growth zone may well be misleading.

MECHANISM OF ROOT GROWTH REGULATION BY ABA

Osmotic adjustment. The process of osmotic adjustment at low water potentials in the growth zone of the maize primary root has been examined in detail (Sharp *et al.*, 1988, 1990; Voetberg and Sharp, 1991). A spectrum of solutes as potential contributors to osmotic adjustment has been analysed, including hexoses, sucrose, amino acids, glycine betaine and K^+. Hexose concentration increased greatly at low water potentials in basal locations, although this could be accounted for by decreased rates of water deposition associated with inhibition of volume expansion. K^+ concentration increased only slightly, although as discussed later, even small changes in cytoplasmic K^+ concentration may be important in other respects. Glycine betaine accumulation was not detected and changes in most amino acids were minor, except for proline. Proline concentration increased greatly throughout the elongation zone, particularly in the apical region where elongation was maintained. At a water potential of -1.6 MPa, proline concentration (expressed on a bulk tissue basis) was 120 mmol in the apical millimetre (Fig. 4; Voetberg and Sharp, 1991), which accounted for almost half of the decrease in osmotic potential. Moreover, the contribution of proline to the osmotic potential in this region was approximately equal to the magnitude of turgor at the same water potential, as measured with a pressure probe (Spollen and Sharp, 1991). Therefore, the accumulation of proline is probably essential to the maintenance of cell elongation in this region.

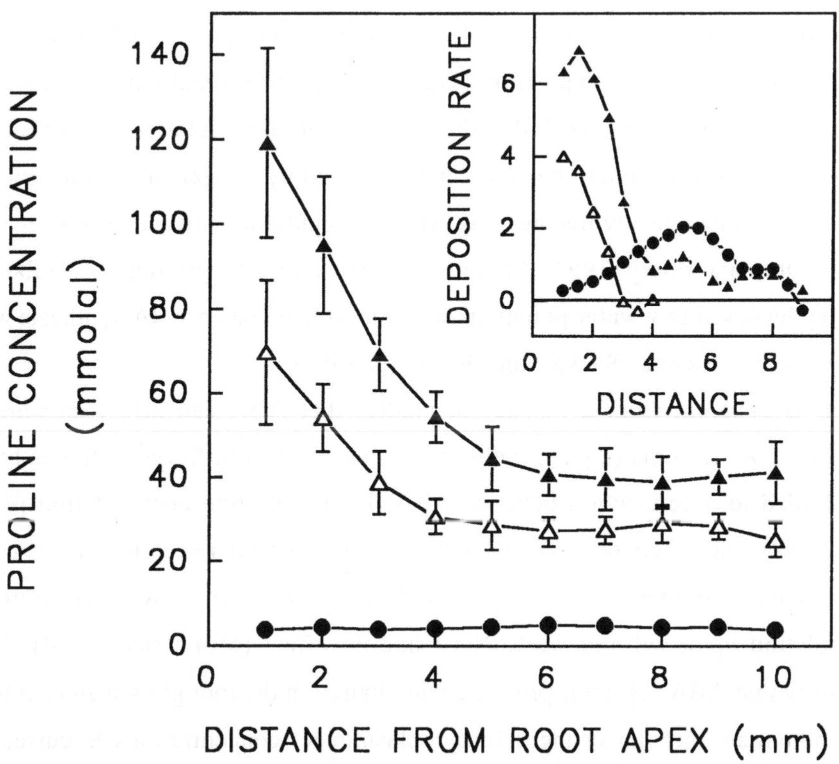

FIGURE 4. *Spatial distribution of proline concentration in the apical 10 mm of maize primary roots growing in vermiculite maintained at high water potential (-0.03 MPa) (●), low water potential (-1.6 MPa) (▲), or low water potential plus fluridone (10 μM) (Δ). Data points are means ± 1SD. The inset shows the spatial distribution of net proline deposition rate (nmoles mm^{-1} length h^{-1}) for the same treatments, calculated from proline contents and growth velocity distributions (from Saab* et al., *1992) using the continuity equation (Silk* et al., *1984). Modified from Ober and Sharp (1993).*

The dramatic increase in proline concentration in the root growth zone at low water potentials resulted from substantially increased rates of net proline deposition in conjunction with decreased volume expansion (Fig. 4, inset; Voetberg and Sharp, 1991). These results provided the first clear demonstration of a major role for increased rates of solute deposition in the osmotic adjustment of growing regions in higher plants exposed to low water potentials. It is therefore important to understand the control system involved, and recent evidence indicates that ABA plays a key role. In roots growing at a water potential of -1.6 MPa, treatment with fluridone to reduce the accumulation of ABA also reduced proline accumulation throughout the elongation zone (Fig. 4). This was associated with a 75% decrease in the net rate of proline deposition (Fig. 4, inset). Additional experiments at a higher water potential (-0.28 MPa) showed that the accumulation of proline in fluridone-treated roots could be completely restored by exogenous application of ABA, confirming the involvement of ABA in regulating proline levels in this system (Ober and Sharp, 1993). In contrast to these results, proline accumulation in barley leaves at low water potentials was not affected when ABA synthesis was blocked with fluridone (Stewart and Voetberg, 1987).

It is apparent from a number of studies that ABA can affect membrane transport of K^+ in roots (e.g., Erlandson and Jensen, 1990). Several studies have also alluded to a connection between ABA, K^+ and proline accumulation (e.g., Pesci, 1989), and others reinforce an apparent interaction between K^+ and proline. For example, pretreatment of maize leaf disks with KCl increased proline accumulation upon wilting, while NaCl had no effect (Mukherjee, 1974). The possibility that ABA regulates proline accumulation in the root growth zone at low water potentials through changes in cytoplasmic K^+ concentrations is currently under investigation. The metabolic basis by which proline deposition increases in the root growth zone at low water potentials is uncertain, but increases in the rate of proline synthesis and decreases in proline oxidation have both been reported in response to low water potential treatments in other systems.

Cell wall yielding. An important factor influencing the maintenance of root growth at low water potentials appears to be the ability to adjust the cell wall yielding properties such that greater longitudinal expansion can occur per unit of turgor. Several earlier investigations showed that root elongation recovered despite decreased turgor in the root tips following immersion in solutions of low water potential (Kuzmanoff and Evans, 1981; Hsiao and Jing, 1987; Itoh *et al.*, 1987). More recently, Spollen and Sharp (1991) showed that the complete maintenance of elongation in the apical region of maize primary roots at a water potential of -1.6 MPa (Fig. 3A) occurred even though turgor was less than half the value in well-watered roots.

Cell wall yielding properties are the result of physical and metabolic processes, either of which could be altered to allow maintenance of root growth at reduced turgor. Fry *et al.* (1992) recently characterised an enzyme, xyloglucan endotransglycosylase (XET), that has the property of cleaving and reforming bonds within and between xyloglucan polymers that link cellulose microfibrils, thereby contributing to wall yielding while conserving xyloglucan. In recent work in this laboratory, the spatial distribution of XET activity in the growth zone of maize primary roots was determined to establish whether it might be associated with the maintenance of elongation at low water potentials. Comparison of Figs. 5 and 3A shows that XET activity (expressed on a cell wall dry weight basis) was well correlated with the distribution of elongation at high water potential and was greatly enhanced in the apical 6 mm at low water potential, in correspondence with the maintenance of elongation in this region. The increase in activity at low water potential was also evident when expressed on a unit length, fresh weight or soluble protein basis (it should be noted, however, that the fraction of XET activity in the cell walls is unknown). These results suggest that an increase in XET activity may be important in increasing cell wall extensibility and allowing continued root elongation at low water potentials despite incomplete turgor maintenance.

The enhancement of XET activity in the root growth zone at low water potentials was inhibited when ABA accumulation was decreased with fluridone

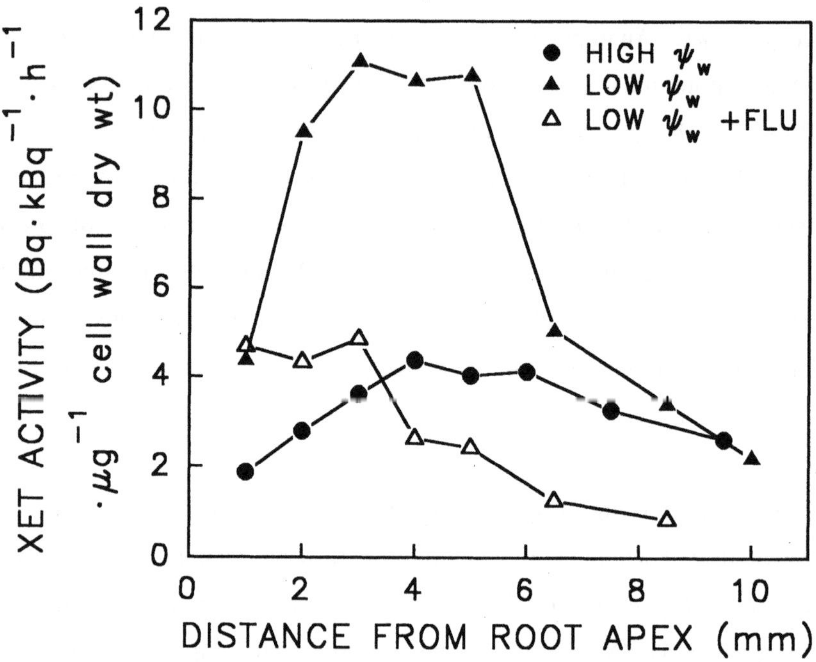

FIGURE 5. *Spatial distribution of xyloglucan endotransglycosylase (XET) activity per µg cell wall dry weight in the apical 10 mm of maize primary roots growing in vermiculite maintained at high water potential (-0.03 MPa), low water potential (-1.6 MPa), or low water potential plus fluridone (FLU, 10 µM). Modified from Wu* et al. *(1993).*

(Fig. 5). Comparison of these data with the effect of fluridone on the spatial growth distribution (Fig. 3A) indicates a close correlation between the two responses. Although cause and effect in this relationship remain to be established, it is noteworthy that a decrease in XET activity did not accompany cessation of growth in the basal region of untreated roots at low water potential. This suggests that growth inhibition in the latter case is due to a different, as yet unidentified factor, and that decreased wall extensibility due to inhibition of XET activity is a likely component of growth inhibition in the apical region when ABA levels are decreased. While several earlier studies have provided evidence that applied ABA can reduce cell wall extensibility in both shoot and root tissues at high water potential (Van Volkenburgh and Davies, 1983; Jones *et al.*, 1987; Kutschera *et al.*, 1987), a role for endogenous ABA in increasing wall extensibility in roots at low water potentials has not been previously identified.

CONCLUSIONS

A role has been established for ABA accumulation in the maintenance of primary root growth in maize seedlings at low water potentials, and evidence has been obtained that the hormone is involved in the regulation of both osmotic adjustment and cell wall yielding. Further work is needed to address the mechanism(s) of action of ABA in these processes. Locations within the zone of elongation have been identified where cell expansion is differentially responsive to low water potentials and endogenous ABA, and recent experiments have demonstrated qualitative and quantitative changes in polypeptide patterns associated with these responses (IN Saab, DT-H Ho, RE Sharp, unpublished data). The regulation of genes encoding these proteins, and their possible function in growth regulation are exciting areas for future study. The approaches taken in the study of ABA should also apply in assessing the roles of other hormones in the regulation of root growth at low water potentials, about which very little is known.

Regulation of root growth relative to shoot development at the seedling stage is of obvious practical significance for successful crop establishment under dry conditions. However, as emphasised by Radin elsewhere in this volume, regulation of important characteristics of root system development in mature plants under water-limited conditions may not be adequately predicted from studies using plants at very early stages of their life cycle. To investigate whether ABA has a similar role in mature plants, it would be of particular interest to examine the effects of soil drying on root development in genotypes selected for contrasting drought-induced ABA accumulation (Quarrie, 1991).

REFERENCES

Creelman RA, Mason HS, Bensen RJ, Boyer JS, Mullet JE (1990) Water deficit and abscisic acid cause differential inhibition of shoot *versus* root growth in soybean seedlings. Analysis of growth, sugar concentration, and gene expression. Plant Physiology 95: 205-214

Erlandson AGI, Jensen P (1990) Stimulation or inhibition of K^+ ($^{86}Rb^+$) influx in wheat roots depending on different ABA treatments. Physiologia Plantarum 78: 331-334

Fry SC, Smith RC, Renwick KJ, Martin DJ, Hodge SK, Matthews KJ (1992) Xyloglucan endotransglycosylase, a new wall-loosening enzyme activity from plants. Biochemistry Journal 282: 821-828

Gaither DH, Lutz DH, Forrence LE (1975) Abscisic acid stimulates elongation of excised pea root tips. Plant Physiology 55: 948-949

Goodwin RH (1972) Studies on roots. V. Effects of indoleacetic acid on the standard root growth pattern in *Phleum pratense*. Botanical Gazette 133: 224-229

Hartung W, Radin JW, Hendrix DL (1988) Abscisic acid movement into the apoplastic solution of water-stressed cotton leaves. Role of apoplastic pH. Plant Physiology 86: 908-913

Hsiao TC (1973) Plant responses to water stress. Annual Review of Plant Physiology 24: 519-570

Hsiao TC, Bradford KC (1983) Physiological consequences of cellular water deficits. In: Taylor HM, Jordan WR, Sinclair TR (eds) Limitations to efficient water use in crop production. American Society of Agronomy, Madison, USA, pp 227-267

Hsiao TC, Jing J (1987) Leaf and root expansive growth in response to water deficits. In: Cosgrove DJ, Knievel DP (eds) Physiology of cell expansion during plant growth. American Society of Plant Physiologists, Rockville, USA, pp 180-193

Itoh K, Nakamura Y, Kawata H, Yamada T, Ohta E, Sakata M (1987) Effect of osmotic stress on turgor pressure in mung bean root cells. Plant and Cell Physiology 28: 987-994

Jones H, Leigh RA, Tomos AD, Wyn Jones RG (1987) The effect of abscisic acid on cell turgor pressures, solute content and growth of wheat roots. Planta 170: 190-197

Jupp AP, Newman EI (1987) Morphological and anatomical effects of severe drought on the roots of *Lolium perenne* L. The New Phytologist 105: 393-402

Kutschera U, Bergfeld R, Schopfer P (1987) Cooperation of epidermis and inner tissues in auxin-mediated growth of maize coleoptiles. Planta 170: 168-170

Kuzmanoff KM, Evans ML (1981) Kinetics of adaptation to osmotic stress in lentil (*Lens culinaris* Med.) roots. Plant Physiology 68: 244-247

Mukherjee I (1974) Effect of potassium on proline accumulation in maize during wilting. Physiologia Plantarum 31: 288-291

Ober ES, Sharp RE (1993) Regulation of proline accumulation by endogenous abscisic acid in maize primary roots at low water potentials. Plant Physiology (in press)

Pesci P (1989) Involvement of Cl^- in the increase in proline induced by ABA and stimulated by potassium chloride in barley leaf segments. Plant Physiology 89: 1226-1230

Pritchard J, Wyn Jones RG, Tomos AD (1991) Turgor, growth and rheological gradients of wheat roots following osmotic stress. Journal of Experimental Botany 42: 1043-1049

Quarrie SA (1991) Implications of genetic differences in ABA accumulation for crop production. In: Davies WJ, Jones HG (eds) Abscisic acid: physiology and biochemistry. BIOS Scientific Publishers, Oxford, pp 227-244

Rodriguez JL, Davies WJ (1982) The effects of temperature and ABA on stomata of *Zea mays* L. Journal of Experimental Botany 33: 977-987

Saab IN, Sharp RE, Pritchard J, Voetberg GS (1990) Increased endogenous abscisic acid maintains primary root growth and inhibits shoot growth of maize seedlings at low water potentials. Plant Physiology 93: 1329-1336

Saab IN, Sharp RE, Pritchard J (1992) Effect of inhibition of abscisic acid accumulation on the spatial distribution of elongation in the primary root and mesocotyl of maize at low water potentials. Plant Physiology 99: 26-33

Saugy M, Mayor G, Pilet P-E (1989) Endogenous ABA in growing maize roots: light effects. Plant Physiology 89: 622-627

Sharp RE (1990) Comparative sensitivity of root and shoot growth and physiology to low water potentials. In: Davies WJ, Jeffcoat B (eds) Importance of root to shoot communication in the responses to environmental stress. Monograph 21. British Society for Plant Growth Regulation, Bristol, pp 29-44

Sharp RE, Davies WJ (1979) Solute regulation and growth by roots and shoots of water-stressed maize plants. Planta 147: 43-49

Sharp RE, Hsiao TC, Silk WK (1990) Growth of the maize primary root at low water potentials. II. Role of growth and deposition of hexose and potassium in osmotic adjustment. Plant Physiology 93: 1337-1346

Sharp RE, Silk WK, Hsiao TC (1988) Growth of the maize primary root at low water potentials. I. Spatial distribution of expansive growth. Plant Physiology 87: 50-57

Silk WK, Walker RC, Labavitch J (1984) Uronide deposition rates in the primary root of *Zea mays*. Plant Physiology 74: 721-726

Spollen WG, Sharp RE (1991) Spatial distribution of turgor and root growth at low water potentials. Plant Physiology 96: 438-443

Sponchiado BN, White JW, Castillo JA, Jones PG (1989) Root growth of four common bean cultivars in relation to drought tolerance in environments with contrasting soil types. Experimental Agriculture 25: 249-257

Stewart CR, Voetberg G (1987) Abscisic acid accumulation is not required for proline accumulation in wilted leaves. Plant Physiology 83: 747-749

Trewavas A (1981) How do plant growth substances work? Plant, Cell and Environment 4: 203-228

Van Volkenburgh E, Davies WJ (1983) Inhibition of light-stimulated leaf expansion by abscisic acid. Journal of Experimental Botany 34: 835-845

Voetberg GS, Sharp RE (1991) Growth of the maize primary root at low water potentials. III. Role of increased proline deposition in osmotic adjustment. Plant Physiology 96: 1125-1130

Watts S, Rodriguez JL, Evans S, Davies WJ (1981) Root and shoot growth of plants treated with abscisic acid. Annals of Botany 47: 595-602

Weaver JE (1926) Root development of field crops. McGraw-Hill, New York

Westgate ME, Boyer JS (1985) Osmotic adjustment and the inhibition of leaf, root, stem and silk growth at low water potentials in maize. Planta 164: 540-549

Wu Y, Spollen WG, Sharp RE, Fry SC, Hetherington PR (1993) Root growth maintenance at low water potentials: Increased activity of xyloglucan endo-transglycosylase. Plant Physiology (submitted)

Zeevaart JAD, Creelman RA (1988) Metabolism and physiology of abscisic acid. Annual Review of Plant Physiology and Plant Molecular Biology 39: 439-473

STRESS-ENHANCED METABOLISM OF ABSCISIC ACID

Jan A.D. Zeevaart
MSU-DOE Plant Research Laboratory
Michigan State University
East Lansing
MI 48824-1312
USA

ABSTRACT

During water stress, abscisic acid (ABA) accumulates rapidly to high levels in leaves and roots, and various approaches have established that this increase is due to *de novo* synthesis rather than to liberation of ABA from a conjugate. Biosynthesis takes place via the indirect pathway in which ABA is derived from 9-*cis* xanthophylls, such as violaxanthin and neoxanthin. The first cleavage product is xanthoxin. Current evidence suggests that the rate-limiting step for ABA biosynthesis is the cleavage reaction. ^{18}O from molecular oxygen is readily incorporated into the side chain of ABA and more slowly into the ring of the ABA molecule. The ^{18}O labelling patterns of ABA from different sources vary due to differences in the size of the precursor pools (xanthophylls) and the turnover rates of intermediates which can result in exchange of ^{18}O with the medium. Thus, although there are differences in the ^{18}O labelling patterns of ABA from different materials, it is clear that in all systems investigated (stress or developmentally regulated) ABA is an oxidative cleavage product of larger precursor molecules, the xanthophylls. In *Xanthium* leaves, phaseic acid (PA) is the major catabolite of ABA, whereas the glucosyl ester of ABA (ABA-GE) accumulates slowly during prolonged stress. Upon rehydration of wilted leaves only PA is formed. Thus leaf water status is the determining factor for ABA metabolism since water deficits cause rapid ABA biosynthesis, and restoration of turgor results in rapid conversion of ABA to PA until the pre-stress ABA level is restored.

INTRODUCTION

Abscisic acid (ABA) is the stress hormone *par excellence*: its biosynthesis is greatly stimulated by water deficits (Zeevaart and Creelman, 1988) and it plays a major role in ameliorating various stresses (Davies and Jones, 1991; Hetherington

NATO ASI Series, Vol. I 16
Interacting Stresses on Plants in a Changing Climate
Edited by M. B. Jackson, C. R. Black

and Quatrano, 1991). This chapter deals with the biosynthesis and catabolism of ABA and how these processes are regulated by the water status of the plant.

In order to respond to a stimulus, an organism must first detect the stimulus, and then transduce the signal into the response, in this case ABA biosynthesis. We discuss here how the water stress stimulus is translated into biochemistry, and which step(s) in the ABA biosynthetic pathway is (are) stimulated by water deficits.

PERCEPTION OF THE STIMULUS

In early work on ABA accumulation, emphasis was placed on the decline of leaf water potential below a threshold value (e.g., Zabadal, 1974), but later work has shown that loss of turgor is more closely correlated with the initiation of ABA biosynthesis. Using the pressure bomb technique, Pierce and Raschke (1980) showed that ABA levels rose sharply at different water potentials for different leaves depending on the water potential at which turgor became zero. Furthermore, non-penetrating solutes induced accumulation of ABA, whereas rapidly penetrating solutes that caused a decrease in the osmotic potential of the tissue and only a transient loss of turgor did not (Creelman and Zeevaart, 1985). These results suggest that relaxation of the plasmalemma and associated conformational changes may be the signal for ABA biosynthesis.

TRANSMISSION OF THE PERCEIVED STRESS SIGNAL

The results of experiments in which inhibitors of RNA or protein synthesis were applied prior to dehydration suggest that gene transcription and protein synthesis are required before an increase in ABA biosynthesis can occur (Guerrero and Mullet, 1986). Thus, if the stress stimulus is perceived by the plasmalemma, there must be transmission of a signal to the nucleus. At present, there is no evidence as to the mechanism(s) involved. In animal systems it has been found that a signal perceived by the membrane can induce phosphorylation of transcription factor proteins. As a result, these factors become associated with and then enter the nucleus where they stimulate transcription (Schindler *et al.*, 1992). It remains to be determined whether similar transduction pathways also operate in plants, but the animal system may serve as a model for future work in plants.

BIOSYNTHESIS OF ABSCISIC ACID

There is now conclusive evidence that ABA is a breakdown product of larger molecules, the xanthophylls. This evidence is based on three different experimental approaches.

First, inhibitors of carotenoid biosynthesis (fluridone, norflurazon) or mutants (*vp2*, *vp5* in maize) that block the conversion of phytoene to phytofluene also inhibit accumulation of ABA. However, these inhibitors and mutations cause photobleaching and chloroplast degradation in light, so that their effects can be considered as non-specific with respect to ABA biosynthesis. More conclusive evidence has been obtained from the ABA-deficient mutant *aba* of *Arabidopsis*. This mutant has normal green leaves and is impaired in the epoxidation of zeaxanthin to antheraxanthin and violaxanthin. Consequently, the levels of violaxanthin and neoxanthin, the precursors of ABA, are very low in the *aba* mutant (Rock and Zeevaart, 1991).

Second, by incubating tissue that was synthesizing ABA rapidly in an atmosphere containing $^{18}O_2$, it was shown that three ^{18}O atoms from $^{18}O_2$ are incorporated into the ABA molecule (Zeevaart *et al.*, 1989). One ^{18}O atom is rapidly incorporated into the carboxyl group, whereas the other two atoms are slowly incorporated as oxygens on the ring (Fig. 1). The ^{18}O enrichment on the ring was found to vary between tissues, depending on their xanthophyll content (Zeevaart *et al.*, 1991). The unequal labelling with ^{18}O at the various positions in the ABA molecule indicates that it is a cleavage product of a larger precursor molecule, a xanthophyll.

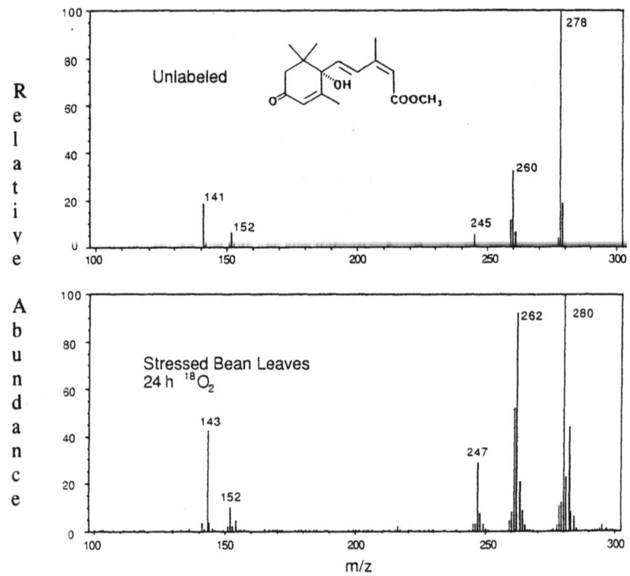

FIGURE 1. Negative chemical ionisation mass spectra of methyl abscisate (MeABA). Top: Unlabelled MeABA; molecular ion M^- at m/z 278; m/z 141 contains the side chain and m/z 152 includes the ring of the MeABA molecule. Bottom: MeABA from wilted bean leaves incubated in $^{18}O_2$ for 24 hours. M^- has shifted from m/z 278 to m/z 280 with one ^{18}O atom in the side chain, since m/z 141 has shifted to m/z 143. ^{18}O is also present on the ring, as is evident from the presence of the m/z 154 ion. From Zeevaart et al. *(1989).*

Third, a 1 : 1 stoichiometry has been established between the disappearance of violaxanthin and neoxanthin and the appearance of ABA and its catabolites, phaseic acid (PA) and dihydrophaseic acid (DPA). In light-grown leaves, the levels of carotenoids are many times higher than the amount of ABA formed upon dehydration, so that in such material correlations between carotenoid losses and ABA accumulation cannot be made. However, etiolated bean leaves (Li and Walton, 1990; Parry *et al.*, 1990) and roots (Parry *et al.*, 1992) have low carotenoid levels as well as the ability to synthesise ABA in response to water stress. These materials were therefore instrumental in observing decreases in the levels of violaxanthin and neoxanthin and a concomitant rise in ABA levels.

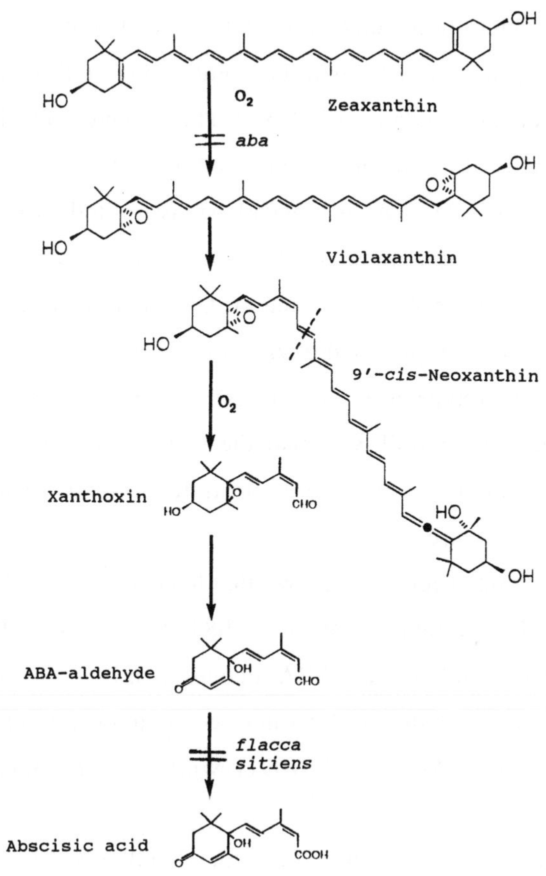

FIGURE 2*. Proposed pathway for the biosynthesis of ABA in higher plants. The biochemical lesions of the* aba *mutant in* Arabidopsis *and of the* flacca *and* sitiens *mutants in tomato are indicated.*

Feeding studies with *cis*- and *trans*-xanthoxin indicate that only the *cis* form is converted to ABA (Sindhu and Walton, 1988). This implies that the xanthophyll precursor must have a *cis* bond that corresponds to the one in xanthoxin and ABA (Fig. 2). In higher plants, the only two such xanthophylls that have been identified are 9-*cis*-violaxanthin and 9'-*cis*-neoxanthin (Li and Walton, 1990; Parry *et al.*, 1990). The biosynthetic route of these xanthophylls, as well as their relationship to their all-*trans* forms has not been worked out, but all-*trans*-violaxanthin functions as a precursor of 9′-*cis*-neoxanthin (Parry *et al.*, 1992). This conversion involves isomerisation (all-*trans*→9-*cis*) and allene formation (violaxanthin→neoxanthin), but the exact order in which these conversions take place is not known.

The cleavage product, xanthoxin, is rapidly converted to ABA with ABA-aldehyde as intermediate (Sindhu *et al.*, 1990). The enzyme activities for oxidation of xanthoxin and ABA-aldehyde are not affected by water stress and appear to be constitutive, so that the rate of ABA biosynthesis is limited by the cleavage step. Thus, the formation of xanthoxin rather than the conversion of xanthoxin to ABA is the rate-limiting step for ABA production. Cleavage to produce xanthoxin is presumably carried out by a specific dioxygenase, since carbon monoxide, a strong inhibitor of P-450 monooxygenases, does not inhibit ABA accumulation (Creelman *et al.*, 1992). A further possibility is that the formation of 9′-*cis*-neoxanthin is enhanced by dehydration, so that more substrate becomes available for ABA production during water stress.

Mutants in several species are known that have a lesion in the conversion of ABA-aldehyde to ABA (Sindhu and Walton, 1988; Taylor *et al.*, 1988). These mutants are leaky and accumulate *trans* ABA-alcohol. A minor 'shunt' pathway from ABA-aldehyde via ABA-alcohol to ABA has been demonstrated in the *flacca* and *sitiens* mutants of tomato (Rock *et al.*, 1991), which may account for their leaky phenotype.

Carotenoids are not only constituents of the thylakoids in the chloroplast, but are also present in the chloroplast envelope (Douce and Joyard, 1990). Considering that gene transcription and protein synthesis are required for ABA biosynthesis (see above), the cleavage enzyme may be cytosolic, with the cleavage of 9′-*cis*-neoxanthin taking place at the surface of the chloroplast.

CATABOLISM OF ABA

The conversion of ABA to PA is greatly accelerated when turgor is restored in dehydrated leaves, so that within approximately four hours the ABA level returns to that of turgid leaves (Fig. 3), with a concomitant increase in PA (Zeevaart, 1980; Pierce and Raschke, 1981). The biochemical basis for this increased capacity to remove ABA is not understood, but could be due to increased activity of ABA-hydroxylase, *de novo* synthesis of the enzyme, or increased accessibility of the substrate to the enzyme.

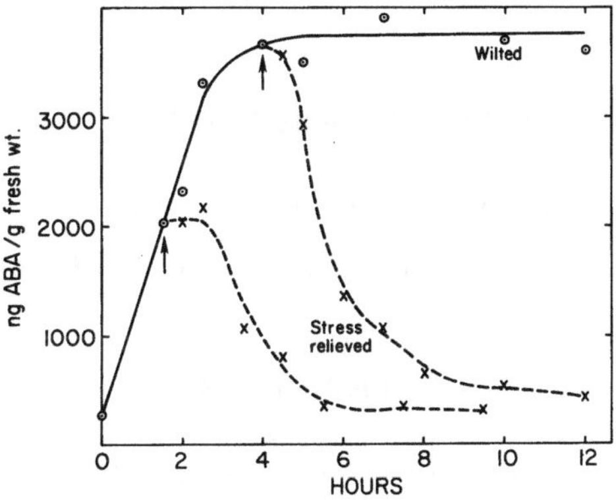

FIGURE 3. *Accumulation of ABA in dehydrated* Xanthium *leaves, and the decline in ABA content following rehydration. Arrows indicate when leaves were rehydrated. From Zeevaart (1980).*

ABA can also be conjugated to glucose to form the glucosyl ester of ABA (ABA-GE). This conjugate accumulates slowly during stress but, in contrast to ABA, does not decline following relief of stress (Pierce and Raschke, 1981; Zeevaart, 1983). The level of ABA-GE may therefore serve as an indicator of the cumulative effects of stress to which a particular leaf has been exposed (Zeevaart, 1983).

CONCLUSIONS

The long-standing dispute as to whether ABA is synthesised via the direct or indirect route (Zeevaart and Creelman, 1988) has finally been resolved in favour of the latter alternative. Nevertheless, several details of the ABA biosynthetic pathway remain to be established, such as the isomerisation of *trans*-xanthophylls to 9-*cis*-xanthophylls, the cellular location of the cleavage step, the nature of the substrate for the cleavage reaction, and the cleavage step itself.

Biosynthesis of ABA is greatly stimulated by loss of turgor. Conversely, when turgor is restored, the rate of ABA biosynthesis decreases and the conversion of ABA to PA is much accelerated. How the sensing of turgor in plant cells is transduced into large changes in the rate of ABA metabolism remains a challenging question.

ACKNOWLEDGEMENTS

The work reported here was supported by the US Department of Energy Grant DE-FG02-91ER20021 and the National Science Foundation Grant DMB-8703847.

REFERENCES

Creelman RA, Bell E, Mullet JE (1992) Involvement of a lipoxygenase-like enzyme in abscisic acid biosynthesis. Plant Physiology 99: 1258-1260

Creelman RA, Zeevaart JAD (1985) Abscisic acid accumulation in spinach leaf slices in the presence of penetrating and non-penetrating solutes. Plant Physiology 77: 25-28

Davies WJ, Jones HG (eds) (1991) Abscisic acid: Physiology and biochemistry. BIOS Scientific, Oxford

Douce R, Joyard J (1990) Biochemistry and function of the plastid envelope. Annual Review of Cell Biology 6: 173-216

Guerrero F, Mullet JE (1986) Increased abscisic acid biosynthesis during plant dehydration requires transcription. Plant Physiology 80: 588-591

Hetherington AM, Quatrano RS (1991) Mechanisms of action of abscisic acid at the cellular level. The New Phytologist 119: 9-32

Li Y, Walton DC (1990) Violaxanthin is an abscisic acid precursor in water-stressed dark-grown bean leaves. Plant Physiology 92: 551-559

Parry AD, Babiano MJ, Horgan R (1990) The role of *cis*-carotenoids in abscisic acid biosynthesis. Planta 182: 118-128

Parry AD, Griffiths A, Horgan R (1992) Abscisic acid in roots II. The effects of water-stress in wild-type and abscisic-acid-deficient mutant (*notabilis*) plants of *Lycopersicon esculentum* Mill. Planta 187: 192-197

Pierce M, Raschke K (1980) Correlation between loss of turgor and accumulation of abscisic acid in detached leaves. Planta 148: 174-182

Pierce M, Raschke K (1981) Synthesis and metabolism of abscisic acid in detached leaves of *Phaseolus vulgaris* L. after loss and recovery of turgor. Planta 153: 156-165

Rock CD, Heath TG, Gage DA, Zeevaart JAD (1991) Abscisic alcohol is an intermediate in abscisic acid biosynthesis in a shunt pathway from abscisic aldehyde. Plant Physiology 97: 670-676

Rock CD, Zeevaart JAD (1991) The *aba* mutant of *Arabidopsis thaliana* is impaired in epoxy-carotenoid biosynthesis. Proceedings of the National Academy of Science, USA 88: 7496-7499

Schindler C, Shuai K, Prezioso VR, Darnell Jr JE (1992) Interferon-dependent tyrosine phosphorylation of a latent cytoplasmic transcription factor. Science 257: 809-813

Sindhu RK, Griffin DH, Walton DC (1990) Abscisic aldehyde is an intermediate in the enzymatic conversion of xanthoxin to abscisic acid in *Phaseolus vulgaris* L. leaves. Plant Physiology 93: 689-694

Sindhu RK, Walton DC (1988) Xanthoxin metabolism in cell-free preparations from wild type and wilty mutants of tomato. Plant Physiology 88: 178-182

Taylor IB, Linforth RST, Al-Naieb RJ, Bowman WR, Marples BA (1988) The wilty tomato mutants *flacca* and *sitiens* are impaired in the oxidation of ABA-aldehyde to ABA. Plant, Cell and Environment 11: 739-745

Zabadal TJ (1974) A water potential threshold for the increase of abscisic acid in leaves. Plant Physiology 53: 125-127

Zeevaart JAD (1980) Changes in the levels of abscisic acid and its metabolites in excised leaf blades of *Xanthium strumarium* during and after water stress. Plant Physiology 66: 672-678

Zeevaart JAD (1983) Metabolism of abscisic acid and its regulation in *Xanthium* leaves during and after water stress. Plant Physiology 71: 477-481

Zeevaart JAD, Creelman RA (1988) Metabolism and physiology of abscisic acid. Annual Review of Plant Physiology and Plant Molecular Biology 39: 439-473

Zeevaart JAD, Heath TG, Gage DA (1989) Evidence for a universal pathway of abscisic acid biosynthesis in higher plants from ^{18}O incorporation patterns. Plant Physiology 91: 1594-1601

Zeevaart JAD, Rock CD, Fantauzzo F, Heath TG, Gage DA (1991) Metabolism of abscisic acid and its physiological implications. In: Davies WJ, Jones HG (eds) Abscisic acid: Physiology and biochemistry. BIOS Scientific, Oxford, pp 39-52

MOLECULAR BIOLOGICAL RESPONSES TO DROUGHT IN MAIZE

Montserrat Pagès, Josep Vilardell, Anders B. Jensen, Mª Mar Albà, Margarita Torrent and Adela Goday
Departamento de Genética Molecular
Centro de Investigación y Desarrollo (CSIC)
Jorge Girona Salgado 18-26
08034 Barcelona
Spain

ABSTRACT

Almost all plants can withstand extremes of osmotic stress at some stages of their life cycle. This is the case in cereal seed whose embryos can tolerate reductions in water content of about 80%. Such severe desiccation kills cells from other parts of the plant. During growth, plants also endure mild water deficits (10-15% reduction of their water content) in dry periods. However, the molecular mechanisms contributing to tolerance of dehydration are largely unresolved. The hormone, abscisic acid (ABA), appears to modulate the responses of plants under conditions of water deficit. Developmental studies of seeds and physiological studies of the effects of water stress indicate that ABA controls quantitative and qualitative aspects of the accumulation of specific mRNAs and proteins. Despite considerable research in this field, the function of the ABA-responsive proteins remains elusive and the mode of action of the hormone is unknown. The *rab* genes of maize are expressed in different plant tissues in response to ABA and osmotic stress. Current evidence indicates that more than one mechanism determines the level of expression of the *rab* genes. One level of control involves transcription of the *rab* genes. Specific sequence elements directing hormone responsiveness are localised in discrete regions of the *rab* gene promoters. *rab* mRNAs are induced in all tissues by ABA. In contrast, the level of *rab* proteins differs between embryo and vegetative tissues since they are strongly accumulated in dry embryos but poorly represented in the overall protein pattern of ABA-treated or water-stressed vegetative tissues. These data indicate that different levels of control are involved in the regulation of *rab* mRNA stability and protein accumulation during ABA-induced responses.

INTRODUCTION

Cereal seeds can withstand reductions in water content of about 80% of their fresh weight, whereas other parts of the plant can only endure mild water deficits. Long periods without significant rainfall are a common feature of arid and semi-arid

NATO ASI Series, Vol. I 16
Interacting Stresses on Plants in a Changing Climate
Edited by M. B. Jackson, C. R. Black

lands and water conservation is therefore a major concern for agronomic productivity. In spite of considerable research on the desiccation resistance of plants, the factors contributing to the tolerance of cells to dehydration are largely unresolved. Current evidence (reviewed by Skriver and Mundy, 1990) indicates that plants subjected to osmotic stress respond with a series of physiological, morphological and biochemical changes which appear to be elicited by the reduction of cell turgor. Among these, closure of the stomata, changes in ionic composition, and accumulation of organic osmolytes and specific proteins have been described.

The plant hormone abscisic acid (ABA) appears to modulate the responses of plants under conditions of water deficit (reviewed by Davies and Mansfield, 1983; Zeevaart and Creelman, 1988). Developmental studies in embryos have shown that ABA accumulates during the stages preceding desiccation (King, 1976; Jones and Brenner, 1987), altering the rate of transcription of a specific set of genes (Quatrano, 1986; Sánchez-Martínez *et al.*, 1986). ABA levels increase in water-stressed vegetative tissues (Wright and Hiron, 1969) and alter the expression of genes, as can also be induced by exogenous ABA in non-stressed tissues (Mundy and Chua, 1988; Gómez *et al.*, 1988; Pla *et al.*, 1989).

IDENTIFICATION OF THE *rab*17 GENE IN MAIZE

Differential screening of a maize dry embryo cDNA library using first strand cDNA from young embryos incubated in the presence or absence of ABA has led to the cloning and characterisation of a specific set of genes from maize, which are involved in general plant ABA responses (Goday *et al.*, 1988). One of these genes, which specifically hybridises with a size-selected mRNA is the maize *rab*17 gene (rab: responsive to ABA). The *rab*17 gene has been identified as ABA-inducible in embryo and vegetative tissues (Goday *et al.*, 1988) and its mRNA is also induced by water-stress in leaves (Pla *et al.*, 1989; Vilardell *et al.*, 1990).

Sequence analysis of a *rab*17 partial cDNA and a corresponding genomic clone showed that of the three possible different open reading frames of the protein sequence, only one was plausible due to the absence of stop codons. S1 mapping indicates that the transcription start site is located 96 nucleotides upstream of the first ATG codon which contains the consensus translation initiation site (A at -3 and C at -1 and -4 from the ATG). Confirmation of the predicted protein sequence was obtained by amino acid analysis of the protein purified from acrylamide gels, and the peptide sequence following cyanogen bromide digestion (Plana *et al.*, 1991).

Screening of a dry embryo cDNA library with the partial cDNA clone resulted in the identification of three full-length cDNAs that shared a virtually identical open reading frame but differed in the size of the 3′ untranslated region (A.B. Jensen, unpublished). This result is consistent with the S1 analysis in which no different transcription start sites were detected in the 5′ region of the gene *rab*17.

The map position of the *rab*17 gene in maize chromosomes was determined by RFLP mapping (Benjamin Burr, Brookhaven National Laboratory, New York). In maize only one *rab*17 locus was identified in the long arm of chromosome 6, position 6L074. However, the exact number of copies of the *rab*17 gene in maize is still unknown.

THE *rab*17 PROTEIN

The deduced amino acid sequence of the *rab*17 protein indicates that it is a basic glycine-rich protein containing a cluster of serine residues and conserved tracts of amino acids showing extensive homology with *rab* proteins from other plant species such as rice *rab*21 (Mundy and Chua, 1988), cotton *Lea*-D11 (Baker *et al.*, 1988), barley dehydrins (Close *et al.*, 1989) and tomato TAS14 (Godoy *et al.*, 1990). The high conservation of *rab* genes, protein structure and regulation in monocots and dicots suggest that they may play an important role in desiccation tolerance.

The study of *rab*17 protein structure and its post-translational modifications may be a way to elucidate the function of the protein and the mechanism(s) by which ABA regulates protein function. Three dimensional structure algorithms suggest that the *rab*17 protein may have three different protein domains. The initial 60 amino acids form a random coil, although a β-turn may be present at position 2-5 (Fig. 1). Next, there is a cluster of serine residues which is probably flanked by two β-turns; this special configuration would make the serines accessible to phosphorylation. Previous work has shown that *rab*17 protein is highly phosphorylated (Goday *et al.*, 1988) and the phosphorylation has been found to be

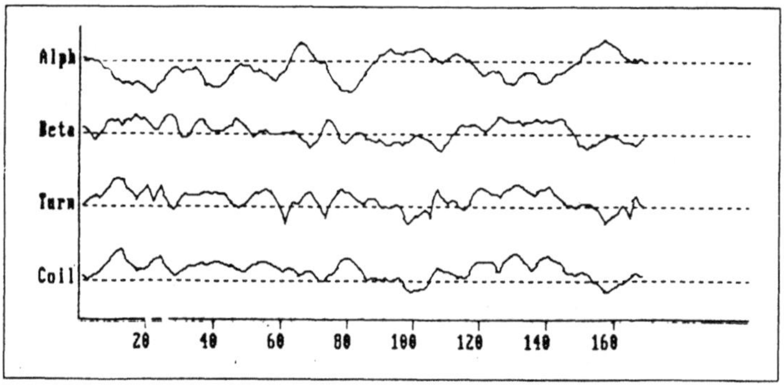

***FIGURE 1**. Secondary structure prediction algorithm of the* rab*17 maize protein.*

exclusively located on serine residues by means of phosphoamino acid analysis (Vilardell *et al.*, 1990) and peptide HPLC (Plana *et al.*, 1991). In the third domain two α-helices are present flanking an unstructured region that could adopt a β-sheet or turn conformation. These α-helical regions correspond to a 14 amino acid repeat which is very basic and contains the sequence KIKEKLPG. It appears twice in the C terminal domain of the maize *rab*17 protein, and it is a consensus repeat in most of the *rab* protein identified so far in other plants. The α-helical regions are expected to be amphiphilic and this may be significant for the functional role of these proteins. Based on this feature and because *rab* proteins are highly hydrophillic (Fig. 2), hypothetical roles in desiccation survival have been proposed (Dure *et al.*, 1989). Extreme heat stability of ABA-induced polypeptides has been described previously (Jacobsen and Shaw, 1989) and is a common feature with hydrophilicity, high content of Glx, abundance, timecourse accumulation and

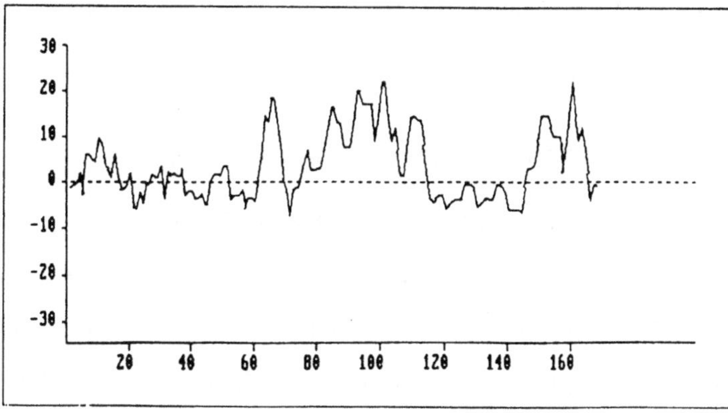

FIGURE 2. Hydrophilicity plot of the predicted amino acid sequence of the rab17 protein. Hydrophilicity is plotted as a function of the residue number.

hormonal control. *rab*17 protein from mature maize embryos is readily extractable and remains in solution after heating at 100 °C for 10 minutes (Fig. 3). On the other hand, the existence of protein specific structural domains may permit the interaction of the *rab* protein with other molecules. For instance, the C-terminal

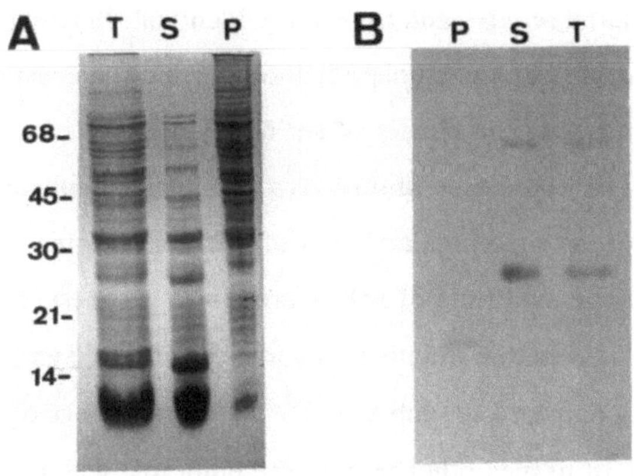

*FIGURE 3. Extraction of heat-stable proteins of maize mature embryo: A, Coomassie blue staining of total (T), soluble (S) and pelleted (P) proteins after heat treatment; equivalent amounts of proteins were loaded on each lane; B, immunoblotting of the same extracts incubated with rabbit anti-*rab17 *serum (1/2000 dilution) and stained with alkaline phosphatase. Molecular weight markers are shown on the left in kDa.*

end contains a possible α-helix : β-turn : α-helix structure characteristic of DNA-binding proteins.

EXPRESSION OF *rab*17 PROTEIN

The pattern of expression of *rab*17 protein indicates that it is synthesised during normal embryogenesis at the time when desiccation begins (Sanchez-Martinez *et al.*, 1986). The *rab*17 protein accumulates in mature and dry embryos as a highly phosphorylated protein (Goday *et al.*, 1988), disappearing during the first hours of germination.

In previous work we have shown that the *rab*17 protein is phosphorylated *in vitro* by casein kinase 2 and that phosphorylation is restricted to the serine cluster region both *in vivo* and *in vitro* (Plana *et al.*, 1991). The amino acid environment of the serine residues contains the consensus pseudo-substrate site of casein kinase 2 (Vilardell *et al.*, 1990). Phosphoamino acid analysis of *rab*17 protein phosphorylated *in vivo* or by casein kinase 2 *in vitro* shows that ^{32}P-incorporation is exclusively in serines (Plana *et al.*, 1991). Phosphopeptide maps derived from *rab*17 phosphorylated *in vitro* and *in vivo* are identical. Purification by HPLC and partial sequencing of the phosphopeptide indicate that it corresponds to the region of the protein containing the cluster of serine residues.

The pattern of expression and *in vivo* phosphorylation of *rab*17 protein in maize embryos, young seedlings and calli of different origins indicates that ABA treatment induces the synthesis of *rab*17 protein in all these tissues, although in calli and young leaves the relative abundance of the synthesised protein is extremely low as compared to embryos. Furthermore, induction by water-deficits of the *rab*17 gene in tissues other than maize embryos, where it occurs naturally, shows that mRNA accumulation is easily achieved, whereas accumulation of *rab*17 protein is dependent upon the physiological stress conditions. Thus, severe and short term water stress in maize leaves do not allow protein accumulation, whereas a mild, long-term physiological water deficit imposed under controlled conditions (Close *et al.*, 1989) effectively induces protein accumulation in the tissue (Fig. 4).

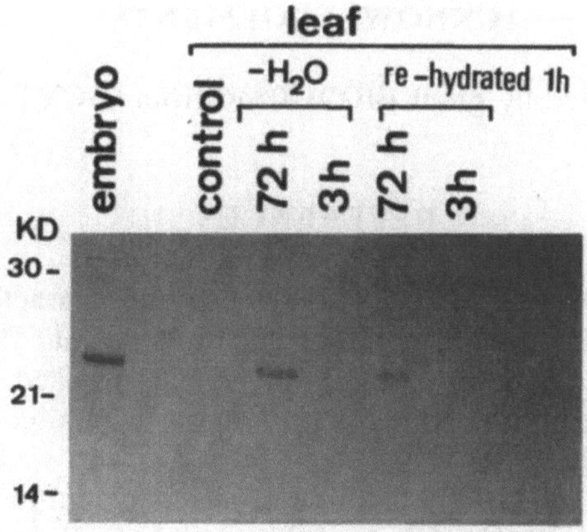

FIGURE 4. ***Immunodetection of*** rab*17* ***protein in maize leaves subjected to mild and severe water stress. Seven day old leaves were dehydrated for 72 hours according to Close*** et al. ***(1989) under a controlled relative humidity atmosphere (mild stress) or for three hours under a constant stream of air with no other precautions (severe stress). Following desiccation, seedlings were placed in water and rehydrated for one hour.*** rab*17* ***was detected in mature embryos and in leaves subjected to mild stress for 72 hours; lower levels of*** rab*17* ***protein were still present after the one hour rehydration period.*** rab*17* ***protein was not detectable in severely dehydrated leaves.***

The degree of *rab*17 protein phosphorylation varies between the different tissues studied, with embryogenic tissues showing relatively large quantities of accumulated phosphorylated protein compared to calli or water-stressed leaves (Vilardell *et al.*, 1990). These differences could be due to variation in the casein kinase 2 activity in the tissues examined, inaccessibility of *rab*17 as substrate in these tissues, or the absence of specific cellular storage structures. These observations indicate that phosphorylation itself may play a major role in regulating *rab*17 functionality, but that mechanisms other than phosphate binding may also reversibly regulate the protein activity. To better understand the function(s) and regulation of the *rab*17 protein, it is important to know its cellular distribution, tissue-specific phosphorylation, and its relevance in stress tolerance by using transgenic plants carrying the *rab*17 gene. The availability of these data will help to unravel the function(s) of this ubiquitous but enigmatic protein.

ACKNOWLEDGEMENTS

This work was supported by grant BIO91-0546 from CICYT.

REFERENCES

Baker JC, Steele C, Dure L III (1988) Sequence and characterization of 6 LEA proteins and their genes from cotton. Plant Molecular Biology 11: 277-291

Close TJ, Kortt AA, Chandler PM (1989) A cDNA-based comparison of dehydration-induced proteins (dehydrins) in barley and corn. Plant Molecular Biology 13: 95-108

Davies MJ, Mansfield TA (1983) The role of abscisic acid in drought avoidance. In: Addicott FT (ed) Abscisic acid. Praeger, New York, pp 237-268

Dure L, Crouch M, Harada, J, Ho THD, Mundy J, Quatrano RS, Thomas T, Sung ZR (1989) Common amino acid sequence domains among the LEA proteins of higher plants. Plant Molecular Biology 12: 475-486

Goday A, Sánchez-Martínez D, Gómez J, Puigdomènech P, Pagès M (1988) Gene expression in developing *Zea mays* embryos: Regulation by abscisic acid of a highly phosphorylated 23- to 25-kD group of proteins. Plant Physiology 88: 564-569

Godoy JA, Pardo JM, Pintor-Toro JA (1990) A tomato cDNA inducible by salt stress and abscisic acid: nucleotide sequence and expression pattern. Plant Molecular Biology 15: 695-705

Gómez J, Sánchez-Martínez D, Stiefel V, Rigau J, Puigdomènech P, Pagès M (1988) A gene induced by the plant hormone abscisic acid in response to water stress encodes a glycine-rich protein. Nature (London) 334: 262-264

Jacobsen JV, Shaw DC (1989) Heat-stable proteins and abscisic acid action in barley aleurone cells. Plant Physiology 91: 1520-1526

Jones RJ, Brenner ML (1987). Distribution of abscisic acid in maize kernel during grain filling. Plant Physiology 83: 905-909

King RW (1976) Abscisic acid in developing wheat grains and its relationship to grain growth and maturation. Planta 133: 43-61

Mundy J, Chua NH (1988) Abscisic acid and water-stress induce the expression of a novel rice gene. EMBO Journal 7: 2279-2286

Pla M, Goday A, Vilardell J, Gómez J, Pagès M (1989) Differential regulation of the ABA induced 23-25 kD proteins in embryos and vegetative tissues of the viviparous mutants of maize. Plant Molecular Biology 13: 385-394

Plana M, Itarte E, Eritja R, Goday A, Pagès M, Martínez MC (1991) Phosphorylation of maize RAB-17 protein by casein kinase 2. Journal of Biological Chemistry 266: 22510-22514

Quatrano RS (1986) Regulation of gene expression by abscisic acid during angiosperm embryo development. Oxford Surveys of Plant Molecular and Cell Biology 3: 745-760

Sánchez-Martínez D, Puigdomênech P, Pagès M (1986) Regulation of gene expression in developing *Zea mays* embryos: Protein synthesis during embryogenesis and early germination of maize. Plant Physiology 82: 543-549

Skriver K, Mundy J. (1990) Gene expression in response to abscisic acid and osmotic stress. Plant Cell 2: 503-512

Vilardell J, Goday A, Freire MA, Torrent M, Martínez MC, Torné JA, Pagès M (1990) Gene sequence, developmental expression and protein phosphorylation of RAB-17 in maize. Plant Molecular Biology 14: 423-432

Wright ST, Hiron RWP (1969) (+)-Abscisic acid, the growth inhibitor induced in detached wheat leaves by a period of wilting. Nature (London) 224: 719-720

Zeevaart JAD, Creelman RA (1988) Metabolism and physiology of abscisic acid. Annual Review of Plant Physiology and Plant Molecular Biology 39: 439-473

REPLACEMENT OF GLYCINE BETAINE BY ß-ALANINE BETAINE, CHOLINE-O-SULPHATE OR DIMETHYLSULPHONIOPROPIONATE IN PLANTS ADAPTED TO INTERACTING STRESSES

Andrew D. Hanson and Bala Rathinasabapathi
IRBV, Université de Montréal
4101 Rue Sherbrooke Est
Montreal
PQ HIX 2B2
Canada

ABSTRACT

Glycine betaine has been shown to act as a cytoplasmic osmolyte in many organisms. In flowering plants, the capacity to accumulate glycine betaine in response to drought or salt stress is widespread, and biochemical and immunological evidence indicates that the glycine betaine biosynthesis pathway is of ancient origin. Despite the efficacy and probable evolutionary age of glycine betaine, it appears to have been replaced by similar compounds in a few angiosperm taxa. Some of these taxa have been very successful in harsh environments. Examples are choline-O-sulphate and ß-alanine betaine in the Plumbaginaceae, and dimethylsulphoniopropionate in *Spartina* (Gramineae) and *Wedelia* (Compositae). There is little evidence that these compounds are more effective than glycine betaine in a physico-chemical sense. However, they may be superior metabolically in environments where drought or salinity is compounded by a second stress. In the case of choline-O-sulphate, this may be high internal SO_4^{2-} levels. For ß-alanine betaine, the second stress may be hypoxia; unlike glycine betaine, ß-alanine betaine synthesis does not require O_2. For dimethylsulphoniopropionate, the interacting stress may be low N availability.

INTRODUCTION

Various zwitterionic quaternary ammonium compounds, and their sulphonium analogues, can act as non-toxic ('compatible') osmolytes or 'osmoprotectants' in the cytoplasm of osmotically-stressed cells of bacteria (Csonka and Hanson, 1991), marine algae (Blunden and Gordon, 1986), animals (Somero, 1986) and higher plants (Rhodes and Hanson, 1993). The biosynthetic pathways of such compounds are potential targets for the genetic engineering of osmotic stress tolerance in

NATO ASI Series, Vol. I 16
Interacting Stresses on Plants in a Changing Climate
Edited by M. B. Jackson, C. R. Black

microorganisms and crop plants (LeRudulier *et al.*, 1984; McCue and Hanson, 1990). Among higher plants, the commonest of these compounds is glycine betaine (I). However, certain taxa accumulate other osmoprotectants, which include ß-alanine betaine (II), choline-O-sulphate (III), and dimethylsulphoniopropionate (DMSP, IV).

$(CH_3)_3{}^+N\text{-}CH_2\text{-}COO^-$
I

$(CH_3)_3{}^+N\text{-}CH_2\text{-}CH_2\text{-}COO^-$
II

$(CH_3)_3{}^+N\ CH_2\ CH_2\ OSO_3^-$
III

$(CH_3)_2{}^+S\ CH_2\ CH_2\ COO^-$
IV

Compounds II-IV occur in only a few taxa which tend to be found in particularly harsh environments with multiple stresses. Because glycine betaine is widespread among higher plants, and because comparative biochemical and immunological evidence suggests that the glycine betaine biosynthesis pathway appeared early in angiosperm evolution (Weretilnyk *et al.*, 1989), it can be viewed as the archetypal angiosperm betaine. If this is so, then compounds II-IV can be seen as evolutionarily novel substitutes which presumably confer some advantage(s) that glycine betaine does not. Such advantages are important to explore in relation to the genetic engineering of crop stress resistance. Accordingly, this chapter summarises recent evidence on the occurrence and biosynthetic pathways of compounds I-IV, considers possible advantages of II-IV over glycine betaine and briefly discusses metabolic engineering of the accumulation of these compounds.

TAXONOMIC DISTRIBUTION

Glycine betaine has been shown unequivocally to accumulate in members of the following diverse families: Chenopodiaceae, Amaranthaceae, Avicenniaceae, Compositae, Convolvulaceae, Plumbaginaceae, Solanaceae, Leguminosae and Gramineae (Rhodes and Hanson, 1993). Typical levels in salt- or water-stressed plants are 20-200 μmol g^{-1} dry weight, concentrations which would be osmotically significant if the glycine betaine is mainly in cytoplasmic compartments, as in the Chenopodiaceae spinach (Robinson and Jones, 1986) and *Atriplex* (Matoh *et al.*, 1987).

Unlike the wide distribution of glycine betaine, choline-O-sulphate and ß-alanine betaine are both known to accumulate only in the Plumbaginaceae. This family comprises many species adapted to semi-arid, saline, high-latitude or high-altitude environments. Choline-O-sulphate has been found in all genera of Plumbaginaceae so far tested (seven of a total of eleven), whereas ß-alanine betaine appears to be confined to certain sections of *Limonium* and related genera, for example, *Armeria* (Hanson and Gage, 1991; Hanson *et al.*, 1991; Hanson and Gage, unpublished). The known distribution of DMSP accumulation is even narrower, being restricted to certain species of the salt marsh grass *Spartina* (Dacey *et al.*, 1987) and to the Pacific coastal strand species *Wedelia biflora* (Compositae) (Gorham, 1986).

Before leaving the topic of distribution, it should be noted that the current understanding of the distribution of compounds I-IV is in fact highly incomplete. The available data are fragmentary and biased towards herbaceous plants of saline or dry habitats in temperate regions; the majority of families have not been examined.

BIOSYNTHETIC PATHWAYS

Glycine betaine. For plants from several families, *in vivo* isotope tracer studies (e.g., Weretilnyk *et al.*, 1989; Lerma *et al.*, 1991) have confirmed that glycine betaine is synthesised by a two step oxidation of choline, via the intermediate betaine aldehyde. The enzymes mediating both steps have been studied in Chenopodiaceae, and that mediating the second step in Gramineae.

The first step is catalysed by a stromal, ferredoxin-dependent choline monooxygenase (CMO) which introduces an atom of oxygen from O_2 into choline, giving the hydrated form of betaine aldehyde (Lerma *et al.*, 1988; Brouquisse *et al.*, 1989). The reduced ferredoxin is photosynthetically generated, and choline oxidation is promoted by light, both *in vivo* (Lerma *et al.*, 1988) and in isolated chloroplasts (Weigel *et al.*, 1988). Spinach CMO has been partially purified; it has a native M_r of 98000, a pH optimum of approximately 8 and is stimulated by Mg^{2+} (Brouquisse *et al.*, 1989).

The second step in glycine betaine synthesis is catalysed by a pyridine nucleotide-dependent betaine aldehyde dehydrogenase (BADH) which has a strong preference for NAD^+ (Weretilnyk and Hanson, 1989; Arakawa *et al.*, 1990). In spinach leaves the majority of the BADH activity is in the chloroplast stroma; the remainder apparently resides in a cytosolic isozyme (Weigel *et al.*, 1986). Inheritance studies of BADH isozyme variants indicate that the stromal enzyme is a homodimer of a nuclear-encoded protein (Weretilnyk and Hanson, 1988). BADH has been purified to homogeneity from spinach (Arakawa *et al.*, 1987; Weretilnyk and Hanson, 1989) and, in agreement with the genetic data, the purified enzyme is a dimer with subunits of M_r ~60000. The spinach enzyme has a broad pH optimum around 8.5 and is substrate-specific. BADH cDNA clones containing the entire protein coding sequence have been isolated from spinach (Weretilnyk and Hanson, 1990) and sugar beet (McCue and Hanson, 1992). Surprisingly for a chloroplastic enzyme, the deduced amino acid sequences for these clones do not include typical transit peptides.

Choline-O-sulphate. Choline-O-sulphate has been shown by *in vivo* radiotracer studies to be formed from choline by leaves and roots of *Limonium* species (Hanson *et al.*, 1991) and by leaves of *Aegialitis annulata* (Benson and Atkinson, 1967). The enzyme responsible has not been identified in higher plants, although choline sulphotransferases which use 3′-phosphoadenosine-5′-phosphosulphate as sulphate donor are known from microorganisms.

ß-alanine betaine. *In vivo* radiotracer studies of *Limonium* species have shown that ß-alanine betaine is synthesised in both leaves and roots by methylation of ß-alanine, with the methyl groups coming from methionine (Larher, 1976; Hanson *et al.*, 1991). The *N*-methyltransferase(s) responsible has not yet been investigated, and the origin of the ß-alanine moiety is not known. The simplest possibility, aspartate decarboxylation, is unlikely because [^{14}C]aspartate does not act as a precursor of ß-alanine or ß-alanine betaine in *Limonium* (Hanson, unpublished).

Dimethylsulphoniopropionate. DMSP biosynthesis has been studied in higher plants and marine algae by *in vivo* radiotracer methods. In the green alga, *Ulva lactuca*, the carbon skeleton and the sulphur and methyl groups of DMSP are derived from methionine (Greene, 1962; Kahn, 1964); this is probably also so in the red alga, *Chondria coerulescens* (Chillemi *et al.*, 1990), in brown algae (Pokorny *et al.*, 1970), and in *Wedelia biflora* (Hanson, unpublished). DMSP therefore presumably originates from methionine via deamination, decarboxylation, oxidation and methylation, although the order of these steps is not known. The enzymes involved have not been identified. Methylation seems not to be the first step, because while S-methylmethionine is a very common metabolite of methionine (Pokorny *et al.*, 1970; Mudd and Datko, 1990), [^{14}C]*S*-methylmethionine was not converted to DMSP in *Ulva* or *Chondria* (Greene, 1962; Chillemi *et al.*, 1990). Methylation also seems unlikely to be the final step, because ß-methylthiopropionate was not a detectable metabolite of [^{14}C]methionine in DMSP-accumulating algae (Pokorny *et al.*, 1970), and because in *Wedelia* ß-[^{14}C]methylthiopropionate was a poor precursor of DMSP (Hanson, unpublished).

POSSIBLE ADVANTAGES OF COMPOUNDS II-IV OVER GLYCINE BETAINE

One possibility is that compounds II-IV are superior to glycine betaine in a physicochemical sense, in that they may be even less disruptive to cellular functions when present at high concentration. There is as yet little evidence that this is the case: in bacteria, glycine betaine is a more potent alleviator of osmotic stress than most other osmoprotectants (Csonka and Hanson, 1991), and generally also ranks as most effective in studies with isolated enzymes (Wyn Jones, 1984; Mason and Blunden, 1989). A second possibility is that compounds II-IV offer some *metabolic* advantage over glycine betaine, in that their synthesis better fits the particular metabolic context of the species in which they occur. Plausible arguments can be made for this for all three compounds, with the metabolic context in each case being set by a second stress.

Choline-O-sulphate. In this case the second stress may be a high level of SO_4^{2-} within the plant. The Plumbaginaceae, the family in which choline-O-sulphate accumulation occurs, possess salt glands in their leaves (Faraday and Thompson, 1986), which can excrete large amounts of Na^+ and Cl^-, but not SO_4^{2-} (Hanson and Gage, 1991). This can produce internal accumulation of SO_4^{2-} during growth on saline soils, which often contain at least seawater levels of SO_4^{2-} (30 mM). Because SO_4^{2-} ions are inhibitory to metabolism, conjugation with choline may represent a means of detoxifying part of the excess SO_4^{2-}. Clearly, glycine betaine could not fulfil any such detoxification function.

ß-alanine betaine. Here, the second stress may be O_2 deficiency, which frequently accompanies salinity in both agricultural and natural soils. The oxygenase (CMO) in the biosynthetic pathway of glycine betaine has an absolute requirement for O_2. There is no such direct O_2 requirement for ß-alanine betaine biosynthesis from ß-alanine (Hanson *et al.*, 1991). Thus, ß-alanine betaine synthesis could proceed in anoxic roots, whereas glycine betaine synthesis could not. The habitats of glycine betaine- and ß-alanine betaine-accumulating species of *Limonium* are consistent with this: the former tend to occur in dry, sandy or rocky soils, while many of the latter colonise salt-marshes. Another possible factor favouring ß-alanine betaine over glycine betaine in the Plumbaginaceae is elimination of the competition between choline-O-sulphate and glycine betaine for choline (Hanson *et al.*, 1991).

Dimethylsulphoniopropionate. For DMSP, the second stress appears to be low nitrogen availability. In both *Spartina* and *Wedelia*, DMSP co-occurs with glycine betaine, and in *Spartina* field experiments have shown that N-fertilisation increases the level of glycine betaine and decreases that of DMSP (Dacey *et al.*, 1987). Because DMSP contains S in place of N, and because glycine betaine can represent up to about 20% of total reduced plant N, substituting DMSP for glycine betaine can have a major sparing effect on the N budget, and hence make significantly more N available for growth. The same rationale has been proposed for the widespread occurrence of DMSP in marine algae (Grone and Kirst, 1992).

METABOLIC ENGINEERING CONSIDERATIONS

Compounds I-IV are all related to primary metabolites by short (one- to three-step) biosynthetic pathways. Assuming that only these steps would need to be inserted into target species, then the genes for one to three enzymes would be involved. It is quite possible in principle to engineer biosynthetic sequences of this size, once the enzymes are known and the corresponding cDNA or genomic clones are available. The limitation to doing this lies at the level of biochemistry and cDNA cloning. As outlined above, only one gene (BADH) for glycine betaine biosynthesis has been cloned, and the enzymes for the synthesis of compounds II-IV have not yet been identified. For ß-alanine betaine, these enzymes are presumably one to three *N*-methyltransferases, for choline-O-sulphate, a sulphotransferase, and for DMSP, the enzymes mediating the deamination, oxidative decarboxylation and methylation steps in its synthesis from methionine.

It has been emphasised for some time that metabolic engineering of glycine betaine accumulation may have potential for increasing crop resistance to salt- and drought stress (LeRudulier *et al.*, 1984; McCue and Hanson, 1990). The value of engineering compounds such as II-IV has had less attention. Nevertheless, the preceding discussion of their possible advantages suggests that they could be more appropriate than glycine betaine for certain environments. The kinds of agricultural environments in which compounds I-IV might be beneficial can be summarised as follows:

Compound		Features	Environments
I	Glycine betaine	O_2 required	Dry and/or saline
II	ß-alanine betaine	O_2 not required	Hypoxic saline
III	Choline-O-sulphate	Detoxifies SO_4^{2-}	High-SO_4^{2-} salinity
IV	DMSP	S replaces N	Low N

REFERENCES

Arakawa K, Katayama M, Takabe T (1990) Levels of betaine and betaine aldehyde dehydrogenase in the green leaves, and etiolated leaves and roots of barley. Plant and Cell Physiology 31: 797-803

Arakawa K, Takabe T, Sugiyama T, Akazawa T (1987) Purification of betaine-aldehyde dehydrogenase from spinach leaves and preparation of its antibody. Journal of Biochemistry 101: 1485-1488

Benson AA, Atkinson MR (1967) Choline sulfate and phosphate in salt excreting plants. Federation Proceedings 26: 394

Blunden G, Gordon SM (1986) Betaines and their sulphonio analogues in marine algae. Progress in Phycological Research 4: 39-79

Brouquisse R, Weigel P, Rhodes D, Yocum CF, Hanson AD (1989) Evidence for a ferredoxin-dependent choline monooxygenase from spinach chloroplast stroma. Plant Physiology 90: 322-329

Chillemi R, Patti A, Morrone R, Piattelli M, Sciuto S (1990) The role of methylsulfonium compounds in the biosynthesis of *N*-methylated metabolites in *Chondria coerulescens*. Journal of Natural Products 53: 87-93

Csonka LN, Hanson AD (1991) Prokaryotic osmoregulation: genetics and physiology. Annual Review of Microbiology 45: 569-606

Dacey JWH, King GM, Wakeham SG (1987) Factors controlling emission of dimethylsulphide from salt marshes. Nature (London) 330: 643-645

Faraday CD, Thomson WW (1986) Structural aspects of the salt glands of the Plumbaginaceae. Journal of Experimental Botany 37: 461-470

Gorham J (1986) Separation and quantitative estimation of betaine esters by high-performance liquid chromatography. Journal of Chromatography 361: 301-310

Greene RC (1962) Biosynthesis of dimethyl-ß-propiothetin. Journal of Biological Chemistry 237: 2251-2254

Grone T, Kirst GO (1992) The effect of nitrogen deficiency, methionine and inhibitors of methionine metabolism on the DMSP contents of *Tetraselmis subcordiformis* (Stein). Marine Biology 112: 497-503

Hanson AD, Gage D (1991) Identification and determination by fast atom bombardment mass spectrometry of the compatible solute choline-O-sulfate in *Limonium* species and other halophytes. Australian Journal of Plant Physiology 18: 317-327

Hanson AD, Rathinasabapathi B, Chamberlin B, Gage DA (1991) Comparative physiological evidence that ß-alanine betaine and choline-O-sulfate act as compatible osmolytes in halophytic *Limonium* species. Plant Physiology 97: 1199-1205

Kahn V (1964) Glycine as a methyl donor in dimethyl-ß-propiothetin synthesis. Journal of Experimental Botany 15: 225-231

Larher F (1976) Sur quelques particularités du métabolisme azoté d'une halophyte: *Limonium vulgare* Mill. DSc Thesis, Université de Rennes, France

Lerma C, Hanson AD, Rhodes D (1988) Oxygen-18 and deuterium labeling studies of choline oxidation by spinach and sugar beet. Plant Physiology 88: 695-702

Lerma C, Rich PJ, Ju GC, Yang W-J, Hanson AD, Rhodes D (1991) Betaine deficiency in maize: complementation tests and metabolic basis. Plant Physiology 95: 1113-1119

LeRudulier D, Strom AR, Dandekar AM, Smith LT, Valentine RC (1984) Molecular biology of osmoregulation. Science 224: 1064-1068

Mason TG, Blunden G (1989) Quaternary ammonium and tertiary sulfonium compounds of algal origin as alleviators of osmotic stress. Botanica Marina 32: 313-316

Matoh T, Watanabe J, Takahashi E (1987) Sodium, potassium, chloride, and betaine concentrations in isolated vacuoles from salt-grown *Atriplex gmelini* leaves. Plant Physiology 84: 173-177

McCue KF, Hanson AD (1990) Drought and salt tolerance: towards understanding and application. Trends in Biotechnology 8: 358-362

McCue KF, Hanson AD (1992) Salt-inducible betaine aldehyde dehydrogenase from sugar beet: cDNA cloning and expression. Plant Molecular Biology 18: 1-11

Mudd SH, Datko AH (1990) The *S*-methylmethionine cycle in *Lemna paucicostata.* Plant Physiology 93: 623-630

Pokorny M, Marcenko E, Keglevic D (1970) Comparative studies of L- and D-methionine metabolism in lower and higher plants. Phytochemistry 9: 2175-2188

Rhodes D, Hanson AD (1993) Quaternary ammonium and tertiary sulfonium compounds in higher plants. Annual Review of Plant Physiology and Plant Molecular Biology 44 (in press)

Robinson SP, Jones GP (1986) Accumulation of glycine betaine in chloroplasts provides osmotic adjustment during salt stress. Australian Journal of Plant Physiology 13: 659-668

Somero GN (1986) Protons, osmolytes, and fitness of internal milieu for protein function. American Journal of Physiology 251: R197-R213

Weigel P, Lerma C, Hanson AD (1988) Choline oxidation by intact spinach chloroplasts. Plant Physiology 86: 54-60

Weigel P, Weretilnyk EA, Hanson AD (1986) Betaine aldehyde oxidation by spinach chloroplasts. Plant Physiology 82: 753-759

Weretilnyk EA, Bednarek S, McCue KF, Rhodes D, Hanson AD (1989) Comparative biochemical and immunological studies of the glycine betaine synthesis pathway in diverse families of dicotyledons. Planta 178: 342-352

Weretilnyk EA, Hanson AD (1988) Betaine aldehyde dehydrogenase polymorphism in spinach: genetic and biochemical characterization. Biochemical Genetics 26: 143-151

Weretilnyk EA, Hanson AD (1989) Betaine aldehyde dehydrogenase from spinach leaves: purification, *in vitro* translation of the mRNA, and regulation by salinity. Archives of Biochemistry and Biophysics 271: 56-63

Weretilnyk EA, Hanson AD (1990) Molecular cloning of a plant betaine-aldehyde dehydrogenase, an enzyme implicated in adaptation to salinity and drought. Proceedings of the National Academy of Sciences USA 87: 2745-2749

Wyn Jones RG (1984) Phytochemical aspects of osmotic adaptation. Recent Advances in Phytochemistry 18: 55-78

WALL EXTENSIBILITY AND THE GROWTH OF SALT STRESSED LEAVES

Peter M. Neumann
Plant Physiology Laboratory
Faculty of Agricultural Engineering
Technion IIT
Haifa 32000
Israel

ABSTRACT

An understanding of the physical restraints underlying the inhibition by environmental stresses of leaf expansion and the consequent reductions in photosynthetic area and crop yield may facilitate the selection or engineering of more resistant varieties. Physical restraints to leaf expansion may involve reductions in turgor and/or the mechanical extensibility of growing cell walls. Several recent reports suggest that reductions in wall extensibility rather than turgor may underlie growth inhibition induced by salinity and water stress. We have developed a rapid method for assaying the plastic, elastic and total extensibilities of growing leaves *in vivo*. Extensibility was assayed by applying and removing a 2 g weight to and from the leaf tip and assaying the resultant changes in position with a sensitive displacement transducer (LVDT). The technique was used to investigate the degree to which inhibition of primary leaf elongation in salinised maize seedlings is associated with reductions in mechanical extensibility. Segmental analysis of expanding epidermal tissues revealed that the capacity for cell elongation was inhibited throughout the basal growing zone of the leaf, but that this could be reversed within hours of transferring plants to non-saline nutrient solutions. Leaf elongation rates in plants salinised for three days were 9 $\mu m\ min^{-1}$ as compared with 16 $\mu m\ min^{-1}$ in control plants, a reduction of 44%. The elastic extensibility of the leaves was only slightly reduced by salinity, but plastic extensibility was reduced by 32%. This reduction occurred in the growing tissues since the plastic extensibility of mature fully expanded leaf tissues was comparatively low and unaffected by salinisation. Moreover, the reduction was not caused by hydraulic or turgor differences between the growing leaf tissues since similar reductions were measured after freeze-thaw treatment of the growing tissues to dissipate turgor and hydraulic restraints. Similar results were also obtained after short-term (20 minute) exposure to NaCl or PEG. We conclude that sustained reductions in the extensibility of expanding cell walls make an important contribution to the observed inhibition of leaf growth by salinity stress.

NATO ASI Series, Vol. I 16
Interacting Stresses on Plants in a Changing Climate
Edited by M. B. Jackson, C. R. Black

INTRODUCTION

The decreasing availability and increasing salinity of irrigation water, together with the progressive salinisation of agricultural soils is inhibiting agricultural productivity in many semi-arid and arid regions of the world. A better understanding of the physiological and molecular mechanisms by which salinity and water deficits inhibit the growth of crop plants could assist in future selection or engineering of more tolerant varieties (reviewed by Blum, 1988). It is well known that one of the first plant responses to salinity stress is a reduction of leaf growth rate, with consequent reductions in the leaf area available for photosynthesis. Excessive accumulation of salts can lead to the death of tissues, organs and whole plants (reviewed by Munns and Termaat, 1986). This chapter deals with investigations of possible mechanisms underlying the early inhibition of leaf growth by excessive salinity in the root environment.

BIOPHYSICAL RESTRAINTS TO GROWTH UNDER STRESS CONDITIONS

Plant growth results from the massive and irreversible expansion of young cells produced by ongoing meristematic divisions. Salinisation can inhibit both cell division and cell expansion in growing tissues of roots and leaves (Kurth *et al.*, 1986; Neumann *et al.*, 1988; Zidan *et al.*, 1990). However, it is cell expansion which produces the ongoing volume increases associated with growth. Thus, the biophysical model of cell expansion introduced by Lockhart (1965) provides a useful theoretical framework for considering mechanisms by which environmental stresses could inhibit plant growth. This model (reviewed by Tomos *et al.*, 1989) relates rates of cell or tissue expansion to rates of water uptake for volume expansion and associated rates of cell wall yielding as follows:

$$RGR = \frac{ml.}{m+L}(wo\text{-}w\pi\text{-}Y) \qquad (1)$$

According to Eq. 1, relative growth rate, RGR, can be defined by four physical parameters, where m reflects wall extensibility, L the water conductance, wo-wπ

the water potential gradient between the external water source and intracellular water (which in turn defines the maximum turgor pressure) and Y represents the yield threshold turgor pressure which must be exceeded to initiate wall expansion. The cell parameters in Eq. 1 (m, L, $w\pi$, Y) are defined by ongoing biochemical activities which may change in response to environmental changes. The equation is however limited to describing conditions of steady state growth.

A simple explanation for the early inhibition of leaf elongation rate by salinity is that accumulation of salts in the external root medium reduces the external water potential, wo, thus reducing $wo\text{-}w\pi$ and hence growth rate. This type of response, with associated turgor loss, has been observed in growing bean leaves after 24 hours of salinisation (Neumann *et al.*, 1988). In contrast, rapid stress-induced osmotic adjustment in growing tissues of barley leaves decreased cellular water potentials within hours of stress exposure, thus balancing the reduction in external water potential and restoring turgor; however, the leaf growth rates of salinised barley plants were still reduced, even after the apparent recovery of bulk turgor (Matsuda and Riazi, 1981). Micro pressure probe measurements in the elongating cells of barley leaf tissues similarly revealed that salinity-induced reductions in elongation rate were maintained after rapid recovery (within one hour) of turgor pressure (Thiel *et al.*, 1988). Finally, long term leaf growth of salinised wheat and barley plants was not restored by pressurising the salinised nutrient solution to restore control levels of external water potential (Munns and Termaat, 1986). Thus, additional factors (other than reductions in external water potential and cell turgor pressure) appear to be involved in the inhibition of cereal leaf growth by salinity. According to Eq. 1, these factors could be either hydraulic (changes in L) or wall mechanical (changes in m or Y). Several recent reports have shown that salinity can reduce root hydraulic conductivity (e.g., Evlagon *et al.*, 1990; Azaizeh and Steudle, 1991 and references therein); such reductions might contribute to long term reductions in leaf growth. There is relatively little direct experimental evidence concerning the possibility that salinity inhibits elongation by reducing the mechanical extensibility of the growing cell walls (i.e., wall hardening). Iraki *et al.* (1989) found that salinity reduced the long term growth of tissue-cultured tobacco cells primarily by altering the *in vivo* mechanical characteristics of the cell walls. However, growth-inhibitory levels of salinity had little inhibitory effect on the *in vitro* extensibility of cell wall preparations from growing bean and barley leaves, as assayed with the Instron technique (Neumann *et al.*, 1988; Lynch *et al.*, 1988). This may be due to the fact that sample excision and preparation can alter wall properties (Cosgrove 1987; Mattysek *et al.*, 1988); thus wall extensibility determined *in vitro* by the Instron technique may not always reflect a direct proportional relationship to growth rate *in vivo* (Taiz, 1984; Cleland, 1984; Kutshera and Schopfer, 1986a, b). Several recent reports have described the use of creep extensiometers for the *in vivo* assay of wall extensibility characteristics. This approach involves the application and removal of a unidirectional force in the direction of growth of excised coleoptile tissues (Kutschera and Schopfer, 1986 a, b), or intact stems (Nonami and Boyer, 1990) and leaves (Cramer and Bowman, 1992; Neumann and Etzion, 1993). The increases and decreases in length induced by the application and removal of the force were used to determine the plastic (irreversible) and elastic (reversible) extensibility of the growing material. Each of these reports suggested a relationship between the rate of growth (irreversible size increase) and *in vivo* plastic extensibility of the cell walls.

LONG TERM EFFECTS OF SALINITY ON WALL EXTENSIBILITY AND GROWTH OF MAIZE SEEDLING PRIMARY LEAVES

In recent years we have used relatively salt sensitive maize plants to investigate the biophysical and biochemical mechanisms involved in the inhibition of root and leaf elongation by salinity. These investigations did not support hypotheses suggesting that salinity inhibits the growth of maize seedlings via effects on the capacity for wall acidification (Zidan *et al.*, 1990), competitive displacement of membrane-bound calcium (Zidan *et al.*, 1991), or reductions in the tissue concentrations of essential mineral nutrients (Zidan *et al.*, 1992). Salinity did reduce the hydraulic conductivity of primary roots but this effect was partially inhibited when calcium levels in the nutrient solution were raised to prevent salinity-induced deficiencies (Evlagon *et al.*, 1990). Some preliminary results concerning the use of a creep extensiometer to investigate short and long term effects of root salinisation on the *in vivo* extensibility and growth of the primary leaves of intact maize seedlings are now presented.

Maize seedlings (*Zea mays* L. cv. Halamish) were germinated and grown on dilute nutrient solution with additional calcium ±100 mM NaCl, as previously described (Zidan *et al.*, 1990). The long term inhibitory effects of salinity on the elongation of primary leaves are shown in Fig. 1.

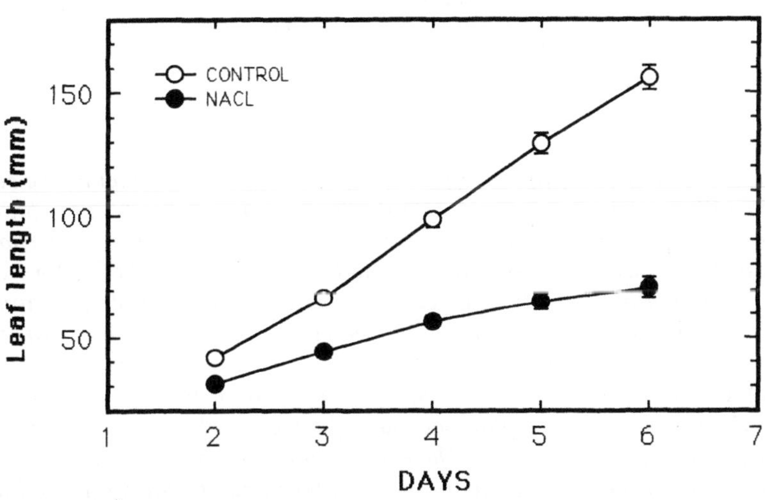

FIGURE 1*. Effect of salinity on leaf elongation. Germinated maize seedlings were transferred to aerated nutrient solutions ±100 mM NaCl and the length of the primary leaf was measured daily after a two day adjustment period. Bars indicate SE larger than symbols, n≥10.*

The *in vivo* plastic and elastic extensibilities of the primary leaves of intact two day old control plants and three day old salinised plants of equivalent size were assayed *in vivo*, as shown schematically in Fig. 2. Table 1 indicates that reductions in the leaf growth rate of salinised plants were accompanied by reductions in *in vivo* plastic extensibility. Salinity had relatively little effect on elastic extensibility. Assay of *in vivo* extensibility of mature (fully expanded) leaf sections excised from growing leaves revealed comparatively small plastic

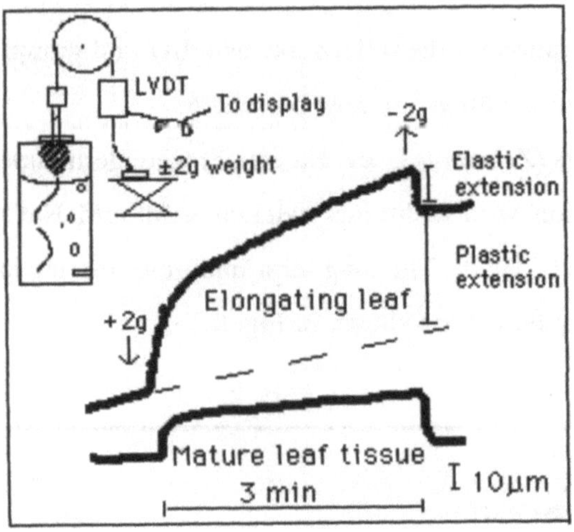

FIGURE 2. In vivo *assay of elastic and plastic extension in intact maize seedling leaves. Seedlings were firmly fixed with the roots in aerated nutrient solution and the primary leaf attached via a pulley to a position transducer. The change in position of the transducer core following addition and removal of a 2 g weight is shown schematically by the bold traces. The distance between the position of the leaf tip after three minutes with a 2 g weight attached and the extrapolated position without additional force (shown by dashed line) are used to determine total leaf extension. The elastic component of total extension is given by the distance the leaf contracts within seconds of removal of the weight. The plastic component is the distance between the extrapolated and actual position of leaf tip after removing the weight. Extension was proportional to the weight added and repeat cycles of weight addition and removal gave similar results (data not shown).*

extensibility values, which were unaffected by salinity. This indicated that the salinity-induced changes in the plastic extensibility of intact leaves occurred in the elongating basal tissues, rather than in the mature fully elongated tissues. Freeze-thaw treatment of the elongating leaf tissues disrupted turgor and resulted in large increases in elastic extensibility, as also noted by others (Kutschera and Schopfer, 1986b; Nonami and Boyer, 1990). More importantly, this treatment did

TABLE 1. *Inhibitory effects of salinisation on* in vivo *extensibility in intact maize seedlings. Primary leaves of maize plants were assayed after two or three days respectively in control or salinised nutrient solutions. Mean leaf elongation rates were 16 μm min^{-1} for the control and 9 μm min^{-1} for the salinity treatment. Freeze-thaw treatment of the growing tissues eliminated elongation growth. Plastic (P), elastic (E) and total (T) extensibilities are means ±SE for ≥20 separate plants for each* in vivo *assay, 11 plants for each freeze-thaw assay and three plants for each mature leaf assay.*

Leaf		P	E	T
			(μm (3 min.2 g)$^{-1}$)	
Control		32±2	12±1	44±2
Salinised		22±1	11±1	32±2
Control	(mature)	2±1	8±1	11±1
Salinised	(mature)	4±1	8±2	13±2
Control	(freeze-thawed)	29±2	79±7	108±8
Salinised	(freeze-thawed)	20±2	59±2	79±7

not eliminate the salinity-induced differences in plastic extensibility found by *in vivo* assay. These *in vitro* results therefore indicate that salinity-induced changes in plastic extensibility were caused by changes in wall mechanical properties, and were not artefactual consequences of differences in water availability or the pressure exerted on the cell walls during the *in vivo* assay.

The changes in extensibility induced by salinisation for three days were accompanied by inhibition of segmental elongation and an apparent shortening of the elongation zone from 7 to 5 mm (Fig. 3). However, the growth inhibition in the apical 2 mm was reversed within 17 hours of transferring NaCl-treated plants to non-saline control media. This reversal was not established one hour after salt removal (data not shown), and therefore appeared to be caused by a controlled

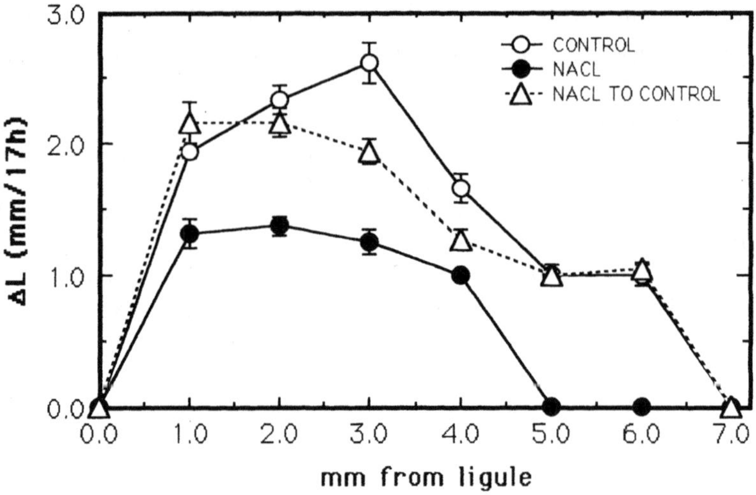

FIGURE 3. Segmental elongation along growing basal tissues of maize primary leaves. The elongating tissues situated above the ligule in intact primary leaves from two day control and three day salinised seedlings were marked by pin pricks at 1 mm intervals. One set of salinised plants was transferred to non-saline nutrient solution immediately after marking. ΔL represents changes in distances between the pin holes, measured 17 hours later. Each point is a mean value for ≥8 plants. Bars show SE.

acceleration of elongation rate rather than the rapid swelling of tissues due to sudden rehydration. The reversal of the inhibition of cell elongation in the growing zone by salt removal suggests that the associated inhibition of wall extensibility should also be a reversible process. This was investigated by assaying rapid leaf responses to the removal of NaCl.

SHORT TERM EFFECTS OF CHANGING SALINITY LEVELS ON WALL EXTENSIBILITY AND GROWTH OF MAIZE SEEDLING PRIMARY LEAVES

The upper trace in Fig. 4 is a typical timecourse showing primary leaf elongation rate (GR) and plastic extensibility (P) before and after removal of 100 mM NaCl from the root medium of a maize seedling which had been salinised for three days.

The elongation rate increased from 12 to 15 μm min^{-1} and during the same period the plastic extensibility increased from 37 to 52 μm (3 min.2 g)$^{-1}$; the treatment had relatively little effect on elastic extensibility. The lower trace shows that the reverse response was obtained when the roots of non-salinised control plants were

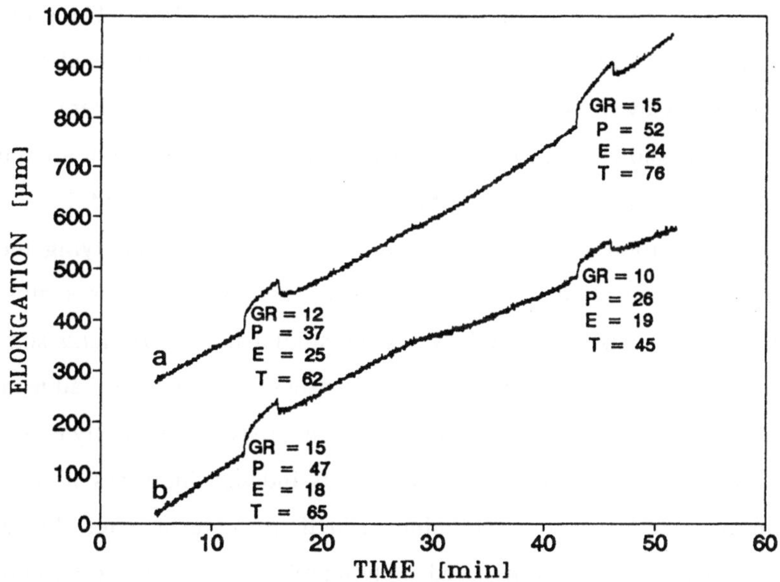

FIGURE 4. Kinetics of: a) enhancement of leaf elongation following removal of 100 mM NaCl from the rooting medium of a three day salinised plant; and b) inhibition of leaf elongation following addition of 100 mM NaCl to the rooting medium of a non-salinised plant. Solutions were changed at 26 minutes GR denotes leaf elongation rate (μm min^{-1}); P, E and T are respectively the plastic, elastic and total in vivo *extensibility values (μm (3 min.2 g)$^{-1}$).*

exposed to 100 mM NaCl. Presumably the rapid changes in leaf elongation rate induced by sudden changes in the osmolality of the root medium initially reflected the increase or decrease in the availability of xylem water to expanding leaf cells (by affecting wo-wπ in Eq. 1). However, plastic extensibility also changed over the same period; the changes in wall properties were also detectable *in vitro* (i.e., after freeze-thaw treatment) and were located in the growing tissues (data not shown).

Numerous previous investigations of the growth responses of plants to sudden environmental changes have provided indirect indications that wall mechanical properties can change within minutes of stressing lower plants (Green *et al.*, 1971) or the roots (Kuzmanoff and Evans, 1981; Hsaio and Jing, 1987; Neumann and David, 1992), stems (Cosgrove, 1987; Matyssek *et al.*, 1988) and leaves (Hsaio and Jing, 1987) of higher plants. Rapid changes in wall properties can complicate experimental approaches for determining wall yield thresholds (see Cosgrove, 1987; Hsaio and Jing, 1987; Matyssek *et al.*, 1988). The problem of adjustment of growth rate in response to the application of weights also occurred during *in vivo* assay of wall extensibility in intact soybeans; accelerated rates of stem extension showed further increases about 10 minutes after applying the weight. As this suggested the onset of cellular adjustment to the additional weight, the assays were routinely completed within 3-6 minutes of applying the force (Nonami and Boyer, 1990). We completed our assays within three minutes of applying force to maize primary leaves since similar rate adjustments sometimes occurred with longer extension periods. The onset of the secondary adjustment of wall properties during the *in vivo* assay of growing tissues may partly explain why Cramer and Bowman (1992) observed salinity-induced increases in wall yield threshold in maize leaves, but only minor reductions in plastic extensibility. Their assay of plastic extensibility involved applying relatively large forces (up to 61 g) for 45 minutes, and the final extensibility values could therefore reflect additional weight-induced feedback changes in wall extensibility.

MECHANISMS REGULATING RAPID AND LONG TERM CHANGES IN WALL MECHANICAL PROPERTIES

The onset of changes in the mechanical properties of leaf cell walls within 20 minutes of changing the root environment makes it unlikely that the walls were simply responding to the changes in NaCl concentrations. This conclusion is strengthened by the observation that the reductions in plastic extensibility associated with salinisation of the roots were also obtained when non-ionic osmoticum (PEG) was used in place of NaCl (data not shown). Presumably, close metabolic interactions between hydraulic and wall mechanical factors are involved in the rapid leaf responses to root-applied stresses. The controls underlying the maintenance of lower wall extensibility in the growing leaf tissues of salinised plants after osmotic adjustment has occurred are unclear. Root to leaf signals, involving altered fluxes of nutrient ions or growth regulators such as cytokinins or abscisic acid, may have a role (Neumann and Stein, 1983; Davies and Zhang, 1991).

CONCLUSIONS

Our data provide evidence that the inhibition of cell expansion and leaf growth in maize under saline conditions is readily reversible and involves both rapid and long term adjustments of wall mechanical properties. Future investigations should reveal more about the mechanisms by which roots signal environmental changes to the shoot and the molecular transduction of such signals into biophysical changes in wall properties. A better understanding of these events may eventually facilitate selection or engineering of more salt tolerant plants.

ACKNOWLEDGEMENTS

This work was supported by a grant from the Fund for Basic Research administered by the Israel Academy of Sciences and Humanities.

REFERENCES

Azaizeh H, Steudle E (1991) Effects of salinity on water transport in excised maize (*Zea mays* L.) roots. Plant Physiology 97: 1136-1145

Blum A (1988) Plant breeding for stress environments. CRC Press, Boca Raton, Florida, USA

Cleland RE (1984) The Instron technique as a measure of immediate-past wall extensibility. Planta 160: 514-120

Cosgrove DJ (1987) Wall relaxation in growing stems: comparison of four species and assessment of measurement techniques. Planta 171: 266-278

Cramer GR, Bowman DC (1992) Kinetics of maize leaf elongation II. Responses of a Na-excluding and a Na-including cultivar to varying Na/Ca salinities. Journal of Experimental Botany 43: 857-864

Davies WJ, Zhang J (1991) Root signals and the regulation of growth and development of plants in drying soil. Annual Review of Plant Physiology and Plant Molecular Biology 42: 55-76

Evlagon D, Ravina I, Neumann PM (1990) Interactive effects of salinity and calcium on hydraulic conductivity, osmotic adjustment and growth in primary roots of maize seedlings. Israel Journal of Botany 39: 239-247

Green PB, Erikson RO, Buggy J (1971) Metabolic and physical control of cell elongation rate: *In vivo* studies in Nitella. Plant Physiology 47: 423-430

Hsiao TC, Jing J (1987) Leaf and root expansive growth in response to water deficits. In: Cosgrove DJ, Knievel DP (eds) Physiology of cell expansion during plant growth. American Society for Plant Physiology, Rockville, Maryland, USA, pp 180-192

Iraki NM, Bressan RA, Hasegawa PM, Carpita NC (1989) Alteration of the physical and chemical structure of growth limited plant cells adapted to osmotic stress. Plant Physiology 91: 39-47

Kurth E, Cramer ER, Lèuchli A, Epstein E (1986) Effects of NaCl and $CaCl_2$ on cell enlargement and cell production in cotton roots. Plant Physiology 82: 1102-1106

Kutschera U, Schopfer P (1986a) Effect of auxin and abscisic acid on cell wall extensibility in maize coleoptiles. Planta 167: 527-535

Kutschera U, Schopfer P (1986b) *In vivo* measurement of cell wall extensibility in maize coleoptiles: Effects of auxin and abscisic acid. Planta 169: 437-442

Kuzmanoff KM, Evans ML (1981) Kinetics of adaption to osmotic stress on Lentil (*Lens culinaris* Med.) roots. Plant Physiology 68: 244-247

Lockhart JA (1965) An analysis of irreversible plant cell elongation. Journal of Theoretical Biology 8: 264-275

Lynch J, Thiel G, Lauchli A (1988) Effects of salinity on the extensibility and Ca availability in the expanding region of growing barley leaves. Botanica Acta 101: 355-361

Matsuda K, Riazi A (1981) Stress-induced osmotic adjustment in growing regions of barley leaves. Plant Physiology 68: 571-576

Matyssek R, Maruyama S, Boyer JS (1988) Rapid wall relaxation in elongating tissues. Plant Physiology 86: 1163-1167

Munns R, Termaat A (1986) Whole plant responses to salinity. Australian Journal of Plant Physiology 13: 143-160

Neumann PM, David L (1992) Physical restraints underlying short term inhibition by auxin of root elongation in intact maize seedlings. Journal of Plant Growth Regulation 11: 119-123

Neumann PM, Etzion O (1993) Salinity reduces *in vivo* extensibility of expanding maize leaf tissues. Current Topics in Plant Biochemistry and Physiology 11 (In press)

Neumann PM, Stein Z (1983) Xylem transport and the regulation of leaf metabolism. What's New in Plant Physiology 14: 33-36

Neumann PM, Van Volkenburgh E, Cleland RE (1988) Salinity stress inhibits bean leaf expansion by reducing turgor, not wall extensibility. Plant Physiology 88: 233-237

Nonami H, Boyer JS (1990) Wall extensibility and cell hydraulic conductivity decrease in enlarging stem tissues at low water potentials. Plant Physiology 93: 1610-1619

Taiz L (1984) Plant cell expansion: Regulation of cell wall mechanical properties. Annual Review of Plant Physiology 35: 585-657

Thiel G, Lynch J, Lèuchli A (1988) Short term effects of salinity stress on the turgor and elongation of barley leaves. Journal of Plant Physiology 132: 38-44

Tomos AD, Malone M, Pritchard J (1989) The biophysics of differential growth. Environmental and Experimental Botany 29: 7-23

Zidan I, Azaizeh H, Neumann PM (1990) Does salinity reduce growth in maize root epidermal cells by inhibiting their capacity for cell wall acidification? Plant Physiology 93: 7-11

Zidan I, Jacoby B, Ravina I, Neumann PM (1991) Sodium does not compete with calcium in saturating plasmamembrane sites regulating ^{22}Na influx in salinized maize roots. Plant Physiology 96: 331-334

Zidan I, Shaviv A, Ravina I, Neumann PM (1992) Does salinity inhibit maize leaf growth by reducing tissue concentrations of essential mineral nutrients? Journal of Plant Nutrition 15: 1407-1419

GROWTH AND COMPOSITION OF NITROGEN AND WATER STRESSED PEPPER PLANTS, THEIR ASSESSMENT BY REMOTE SENSING AND THEIR HERBIVORY EFFECTS

J. Peñuelas, C. Biel, M. Estiarte, I. Filella, R. Matamala, L. Serrano and R. Savé
Institut de Recerca i Tecnologia Agroalimentàries
Carretera de Cabrils s/n
08348 Cabrils
Barcelona
Spain

ABSTRACT

Nitrogen and water deficiencies are expected to become more common as a result of global climate change. To test their effects, severe nitrogen (N) limitation and mild water stresses were imposed that reduced leaf, stem, root and fruit biomass of peppers (*Capsicum annuum*). Root/shoot ratio and specific leaf weight were increased in N-limited plants. Low water and N availability decreased stomatal conductance and resulted in higher leaf temperatures. Except in fruits, there was a decrease in the nitrogen content of all organs in N-limited plants that resulted in higher C/N ratios. Lower leaf N content and higher leaf phenolic concentration may have been responsible for the lower relative growth rate (RGR) of insects (*Heliothis armigera*) feeding on the foliage. The red edge of the reflectance spectra of leaves clearly showed a shift to shorter wavelengths (blue shift). This was a useful index of low chlorophyll levels in N-limited plants.

INTRODUCTION

The expected promotion of plant growth by higher atmospheric CO_2 concentrations may increase the demand for mineral nutrients. Thus, nitrogen (N) could become one of the first limiting factors (Lemon, 1983). The increased carbon gain associated with higher atmospheric CO_2 levels should increase the amount of carbon relative to other essential elements in plant tissues (Curtis *et al.*, 1989). Peñuelas and Matamala (1990) found decreases of 31% in leaf nitrogen content from AD 1750 (278 ppm of CO_2 in the atmosphere) to today (350 ppm) in herbarium specimens of 14 plants collected in Catalonia. However, changes in

NATO ASI Series, Vol. I 16
Interacting Stresses on Plants in a Changing Climate
Edited by M. B. Jackson, C. R. Black

climate originating from the greenhouse effect are expected to influence water availability to vegetation (Hayes, 1991). We here discuss the preliminary results of studies on the interacting effects of nitrogen and water shortage on plant growth, plant nitrogen, chlorophyll and phenolic contents, and the implications for herbivores. The influence of both nitrogen and water deficiencies on plant production and composition is well documented (Hanks and Rasmussen, 1982; Van Keulen *et al.*, 1989) but the effects on the basic metabolic processes are far less studied. Phenolic compounds have been implicated in a large number of ecological processes such as plant-herbivore interactions, decomposition of litter and plant/micro-organism interactions. Therefore, possible changes in their concentration have important potential effects on ecosystems in a changing climate.

We also describe reflectance changes in stressed plants that can be useful for following ecophysiological changes by remote sensing. We have focussed on changes in the red edge, the point of maximum slope in the reflectance spectra of vegetation. This occurs between wavelengths of 680-750 nm where the reflectance changes from very low in the red absorption region of chlorophyll to very high in the near infrared because of leaf scattering. Some authors have already demonstrated the relationship between the red edge and chlorophyll concentration at the leaf level (Thomas and Gausman, 1977; Horler *et al.*, 1980). However, the relationship is not clear at the plant and canopy levels. Demetriades-Shah and Steven (1988) found no correlation at all between the canopy red edge and the chlorophyll content of its component leaves. In contrast, Curran (1990) found the red edge to be sensitive to total canopy chlorophyll. We here present results that also demonstrate a strong correlation at the plant and canopy level, thus showing the promise of remote sensing for the assessment of N-limitation. We also raise the possibility of remote sensing for water-stress by measuring leaf and canopy temperature and near infrared reflectance.

MATERIALS AND METHODS

Experimental system. Seeds of peppers (*Capsicum annuum*) were planted in pots in a controlled environment chamber with a 14 hour photoperiod (300 $\mu mol\ m^{-2}\ s^{-1}$)

and 10 hour dark period, 70% relative humidity and 25 °C day/20 °C night and grown for four months. Plants were raised in 2.5 l pots containing a substrate consisting of 1:1 (v:v) peat moss and sand. A 2 x 2 factorial experiment established two levels of irrigation and two levels of nitrogen supply. Water-stressed plants were irrigated with liquid fertiliser only when soil matric potential reached -0.04 MPa ('low water' treatment -LW) or -0.01 MPa ('high water' control treatment -HW). Two levels of nitrogen supply were established by irrigating with 'high nitrogen' solution (-HN) containing 4.88 meq l^{-1} NO_3, 0.63 meq l^{-1} H_2PO_4, 4.41 meq l^{-1} SO_4, 2.76 meq l^{-1} K, 8.71 meq l^{-1} Ca, 2.01 meq l^{-1} Mg, 1.36 meq l^{-1} NH_4, 17.8 μeq l^{-1} Fe, 9.09 μeq l^{-1} Mn, 1.82 μeq l^{-1} Zn, 7.87 μeq l^{-1} Cu, 4.62 μeq l^{-1} B, 1.04 μeq l^{-1} Mo, 25 μeq l^{-1} Mg. The low nitrogen (LN) solution contained only 1.2 meq l^{-1} NO_3 but was otherwise unchanged.

Biomass measurements. Leaf area was estimated every two weeks from leaf lengths using second order regressions established experimentally for each treatment. Fresh and dry weights of roots, stems, leaves and fruits of three plants from each treatment were measured every month. Four plants were harvested at the end of each experiment. Leaf areas were measured with a leaf area meter (LI-COR 3000). Plant components were then dried at 60 °C to constant weight. Specific leaf weight (SLW) was also determined. Transpiration rate, stomatal conductance and leaf temperature were measured with a LI-COR Model LI-1600 steady-state porometer and an IRGA-porometer LI-COR 6000 gas exchange system (LI-COR Inc., Lincoln, NE, USA) in fully-developed leaves of each treatment every three or four days.

Chemical analyses. Roots, stems, leaves and fruits were oven-dried for two days at 60 °C and analysed for total nitrogen in a Carlo-Erba NA 1500 automatic analyser. A SPAD chlorophyll meter (Minolta Model 502) was used to measure chlorophyll concentration (Monje and Bugbee, 1992). The relative units were previously calibrated by measuring chlorophyll concentration by shaking fresh leaf discs in N,N'-dimethylformamide (Porra *et al.*, 1989). For the analysis of total phenolic compounds 0.25 g of dry leaves were extracted with 70% aqueous methanol for 45 minutes. Total phenolics were analysed using the Folin-Ciocalteau method modified according to Marigo (1973). Different phenolic compounds were analysed by reverse phase HPLC. Oven-dried leaves (0.5 g) were extracted with 25 ml ethyl acetate, 12.5 ml ammonium sulphate (40%) and 1.2 ml metaphosphoric (20%) for 20 minutes. After separation, re-extraction with ethyl acetate was carried out for 10 minutes. The two superior phases were concentrated in a rotary evaporator, dissolved in 4 ml of methanol, and filtered with 0.45 μm Teknokroma pro-x filters (modified from Amiott, 1989). HPLC analyses were performed after injection onto a Nucleosil C18 column (25 x 0.4 cm, 10 μm) with a photodiode array detector (Shimadzu SPD-M6A) reading at 278-282 nm. The mobile phase consisted of a linear gradient from H_2O-MeOH 100:0 to 0:100 in 33 minutes at 2 ml min^{-1}.

Herbivore feeding. *Heliothis armigera*, a polyphagous noctuid insect pest was used to evaluate the nutritional value of pepper leaves. Caterpillars were obtained from the countryside and maintained for several generations in a chamber at 22 °C with a 16 hour light : 8 hour dark cycle. They were fed on an artificial diet (Poitout and Bues, 1970). Caterpillars weighing 0.04-0.12 g were evenly distributed in six replicates per treatment, after being isolated for four hours without food prior to the test. They were weighed and distributed in Petri dishes containing a single leaf. Caterpillars were allowed to feed for 24 hours.

To determine the effect of the nutritional quality of leaves we examined the relative growth rate (RGR) of the caterpillars and their efficiency of conversion of ingested biomass (ECI) measured as biomass gained (fresh weight) per unit of ingested leaf biomass (dry weight). To distinguish the effect of nitrogen availability from that of other compounds such as phenolics we also measured the nitrogen use efficiency (NUE) of caterpillars, defined as biomass gained per unit of nitrogen ingested.

Spectral reflectance. Canopy radiance was detected with a narrow-bandwidth visible/near-infrared spectroradiometer fitted with 15° field of view optics (model SE590 with detector model CE390WB-R, Spectron Engineering Inc., Denver, Colorado, USA), and expressed as spectral reflectance after standardisation by radiance of a level white standard (Spectralon Labsphere Inc. panel). The instrument detects 252 approximately evenly spaced spectral bands between 390 and 1100 nm. All spectral measurements were made in a nadir orientation of the radiometer held at arm's length.

Data analysis. Spectral reflectance was analysed using the Igor program from Wavemetrics (Oregon, USA). Statistical analyses was carried out using SSPS version 4.0 and Systat version 5.1.

RESULTS

Plants responded to nitrogen and water deprivation by changing the assimilation and partitioning of carbon and other resources such as nitrogen. All organs of stressed plants had a smaller biomass (roots, stems, leaves and fruits). The decrease in biomass produced by N-limitation was larger than that produced by water limitation (Table 1). The seasonal course of plant growth showed a clear negative effect of N-deficiency. Low nitrogen treatments produced very small plants regardless of the water level (Fig. 1). In N-limited plants, root biomass decreased less than above-ground biomass and so N-limited plants possessed higher root/shoot ratios (Table 1). As expected, leaf nitrogen content was greater in the high nitrogen (HN) treatment than in the low nitrogen (LN) treatment (Table 1). No clear differences were found between the two watering levels. There were no significant differences in the leaf carbon concentration, which ranged between 33.7 and 37.7%. Thus, the C/N ratio was higher in N-limited leaves. Similar trends were found in stems and roots but not in fruits. There was no decrease in the N-content of the fruit of N-limited plants, showing that this variable was largely independent of nitrogen availability. Nevertheless, fruit yield decreased with both nitrogen and water shortage. Chlorophyll concentration was related to leaf nitrogen content ($r=0.96$, $p<0.01$), with larger concentrations in plants from HN treatments (Fig. 2). Lower water availability produced smaller stomatal conductances and higher leaf temperatures (Table 1). Similar effects, although weaker, were produced by nitrogen shortage (Table 1).

***TABLE 1.** Biomass variables (total, leaf, stem and root dry biomass, leaf area, specific leaf weight and root/shoot ratio), nitrogen content and C/N ratio of leaves, stems, roots and fruits, and stomatal conductance and leaf minus air temperature of pepper plants growing in low water-low nitrogen (LW-LN), high water-low nitrogen (HW-LN), low water-high nitrogen (LW-HN) and high water-high nitrogen (HW-HN) treatments (see methods). Values are means of six plants ± SE (different letters show significant differences at the $p<0.05$ level) measured 126 days after germination.*

	LW - LN	HW - LN	LW - HN	HW - HN
Total dry weight (g plant^{-1})	3.47 ± 0.23c	4.56 ± 0.60c	15.89 ± 0.90b	25.00 ± 1.34a
Leaf dry weight (g plant^{-1})	1.14 ± 0.16c	1.44 ± 0.10c	7.55 ± 0.33b	11.27 ± 0.63a
Stem dry weight (g plant^{-1})	1.02 ± 0.50c	1.70 ± 0.04c	5.20 ± 0.45b	9.90 ± 0.59a
Root dry weight (g plant^{-1})	1.32 ± 0.07d	1.99 ± 0.09c	3.13 ± 0.23b	4.59 ± 0.28a
Yield (g plant^{-1})	1.28 ± 0.30c	2.76 ± 0.79c	12.72 ± 1.08b	19.70 ± 1.71a
Total leaf area (cm^{-2} plant^{-1})	181.00 ± 61.00c	183.00 ± 47.00c	1805.0(± 371.00b	2701.00 ± 380.00a
SLW (g m^{-2})	63.38 ± 5.64a	61.84 ± 12.26ab	42.65 ± 1.85b	41.87 ± 0.93b
Root/shoot (%)	63.20 ± 5.53a	63.70 ± 3.47a	24.58 ± 1.30b	21.38 ± 1.16b
Leaf nitrogen (%)	1.43 ± 0.10b	1.45 ± 0.04b	3.69 ± 0.17a	3.30 ± 0.03a
Stem nitrogen (%)	0.76 ± 0.00b	1.09 ± 0.16a	1.86 ± 0.34a	1.59 ± 0.42ab
Root nitrogen (%)	1.38 ± 0.08b	1.27 ± 0.00b	2.50 ± 0.05a	2.29 ± 0.04a
Fruit nitrogen (%)	2.03 ± 0.04c	3.18 ± 0.01a	2.51 ± 0.30abc	2.38 ± 0.11b
Leaf C/N	23.60 ± 0.10b	24.90 ± 0.14a	10.20 ± 0.29c	10.40 ± 0.21c
Stem C/N	50.47 ± 1.14a	36.29 ± 0.58b	19.97 ± 1.74c	22.22 ± 1.38c
Root C/N	27.91 ± 0.05b	29.89 ± 0.14a	14.90 ± 1.31c	16.49 ± 0.86c
Stomatal conductance (mmol m^{-2} s^{-1})	41.23 ± 6.35c	94.93 ± 16.66b	40.50 ± 4.60c	203.87 ± 14.12a
Temperature leaf-air (°C)	0.46 ± 0.06a	0.50 ± 0.06a	0.63 ± 0.05a	0.08 ± 0.04b

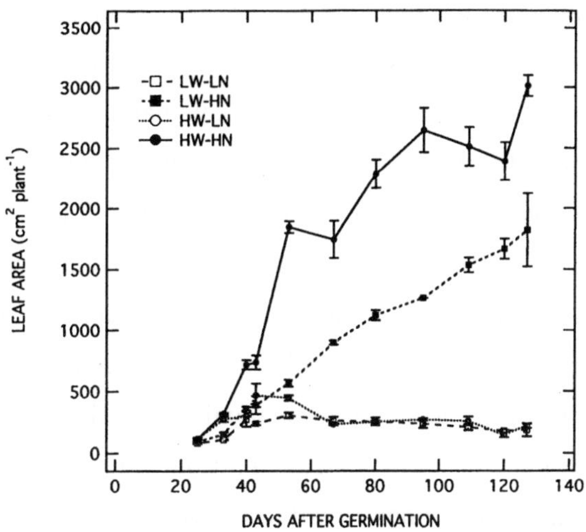

FIGURE 1. Timecourse of changes in leaf area of pepper plants growing in low water-low nitrogen (LW-LN), high water-low nitrogen (HW-LN), low water-high nitrogen (LW-HN) and high water-high nitrogen (HW-HN) treatments for up to 126 days. Means ± SE (n = 4).

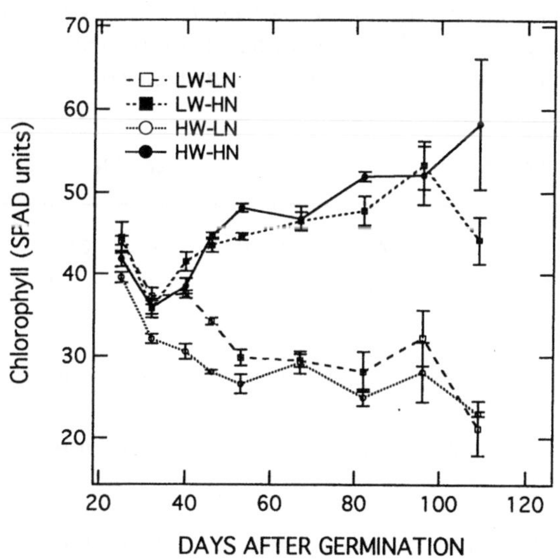

FIGURE 2. Timecourse of changes in chlorophyll content of leaves of pepper plants expressed as SPAD relative units. Total chlorophyll content in mg m^{-2} can be estimated by the regression 14.3 x SPAD - 249.7 (r= 0.93, n=20). Means ± SE (n = 5). Abbreviations as for Fig. 1.

Concentrations of total phenolic compounds in leaves were higher in plants from LN treatments than in those from HN treatments, i.e., higher in plants with a larger C/N ratio. The different water treatments had no significant influence on the total phenolics, although plants from LW treatments tended to contain slightly higher concentrations than those from HW treatments (Table 2). The chromatograms of leaves from different treatments showed the same pattern of decreased content of phenolic compounds in the less stressed treatments. Fig. 3 shows concentrations of different phenolics identified by their relative retention times in HPLC as derivatives of benzoic acid, cinnamic acid, flavonols and flavanones. Concentrations were increased by water and nitrogen deficiencies, with the largest values and the maximum number of different compounds being found in LW-LN plants. Only three of the 10 phenolics found were detected in all four treatments (Fig. 3).

TABLE 2. *Leaf carbon/nitrogen ratios and total phenolic concentration of pepper plants (mg gallic acid g^{-1} leaf dry weight), and relative growth rate (RGR), efficiency of conversion of biomass ingested (ECI) and nitrogen use efficiency of the insect* Heliothis arnigera *feeding on leaves from pepper plants provided with low water-low nitrogen (LW-LN), high water-low nitrogen (HW-LN), low water-high nitrogen (LW-HN) and high water-high nitrogen (HW-HN).*

	LW - LN	HW - LN	LW - HN	HW - HN
Leaf C/N	23.60 ± 0.10b	24.90 ± 0.14a	10.20 ± 0.29c	10.40 ± 0.21c
Total phenolic compounds (mg gallic acid g^{-1} DW)	71.53 ± 5.62a	65.09 ± 4.80a	29.16 ± 1.12b	24.44 ± 1.36b
RGR of insects ($g\ g^{-1}\ d^{-1}$)	0.13 ± 0.10b	0.12 ± 0.02b	0.39 ± 0.01a	0.37 ± 0.06a
Biomass gained/ food ingested ($g\ g^{-1}$ DW)	0.22 ± 0.07b	0.21 ± 0.02b	0.72 ± 0.08a	0.77 ± 0.13a
Biomass gained/ nitrogen ingested ($g\ g^{-1}$ N)	0.16 ± 0.02b	0.15 ± 0.02b	0.20 ± 0.02a	0.24 ± 0.04a

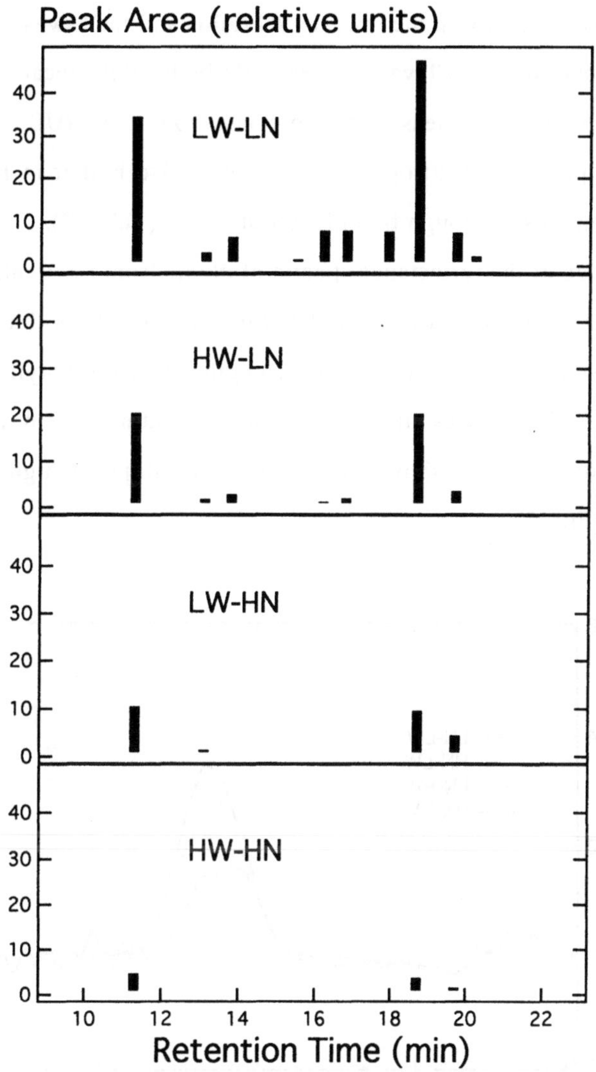

FIGURE 3. *Relative amounts of simple phenolic compounds, expressed as peak area from HPLC chromatograms (relative units), from leaves of pepper growing in low water-low nitrogen (LW-LN), high water-low nitrogen (HW-LN), low water-high nitrogen (LW-HN) and high water-high nitrogen (HW-HN) treatments.*

The RGR of *Heliothis armigera* was about 300% higher when fed with leaves from the HN treatments compared with those fed with LN leaves (Table 2). The efficiency of conversion of ingested biomass (ECI) was also more than 300%

higher in HN treatments (with the lowest C/N ratios and the lowest concentrations of phenolic compounds). ECI values were relatively high because the biomass gained by the insects was measured before egestion. The NUE, defined as the biomass gained per unit of nitrogen ingested, was also higher in HN treatments than in LN treatments, although by only about 150% (Table 2).

The red edge of the reflectance spectra of plants from LN treatments showed a shift to shorter wavelengths (blue shift) (Fig. 4). There was thus a significant correlation between canopy red edge wavelength and chlorophyll concentration ($r=0.92$, $p<0.01$). There was also a tendency to saturation in the red edge-chlorophyll relationship, as shown by a change in slope at higher chlorophyll concentrations (Fig. 5).

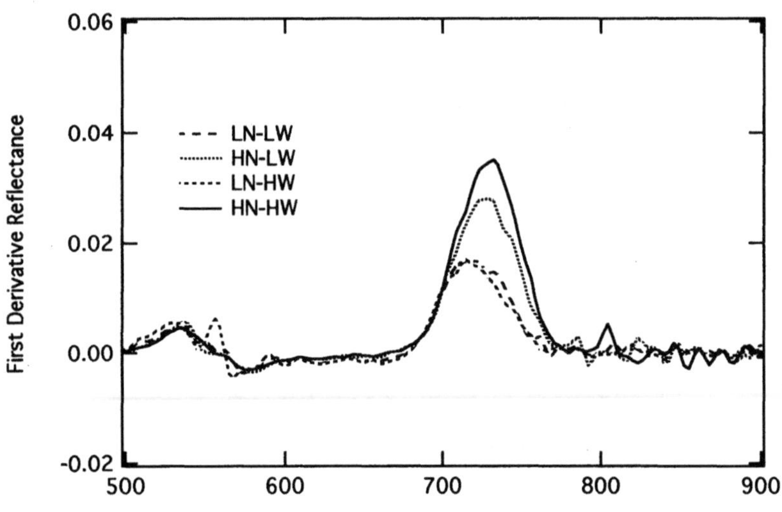

FIGURE 4. First derivative of reflectance spectra of pepper plants from the low water-low nitrogen (LW-LN), high water-low nitrogen (HW-LN), low water-high nitrogen (LW-HN) and high water-high nitrogen (HW-HN) treatments. Each spectrum is a mean of five leaf spectra. Notice the blue shift of the 'red edge' in the N-limited plants.

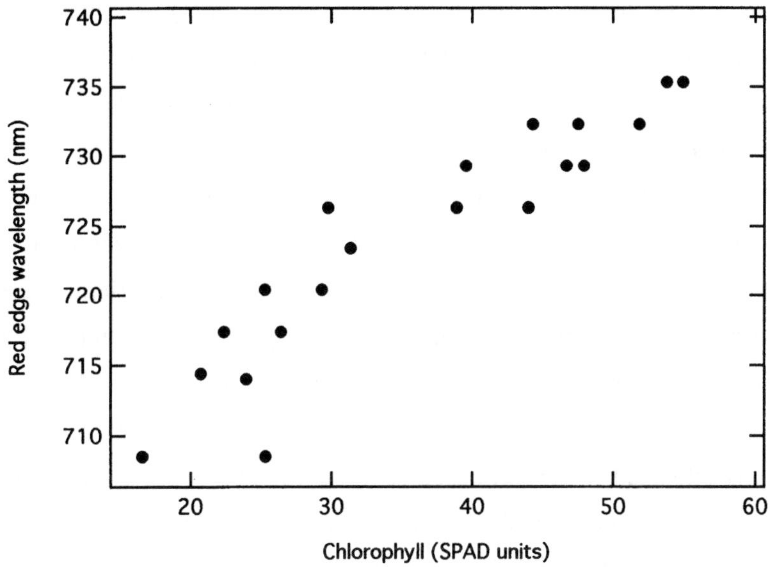

FIGURE 5. Red edge versus leaf chlorophyll content in pepper plants exposed to the stresses listed in Fig. 4. Total chlorophyll content in mg m^{-2} can be estimated by the regression 14.3 x SPAD - 249.7 (r= 0.93, n=20).

DISCUSSION

Our experimental design produced severe nitrogen limitation in LN plants and slight water limitation in LW plants. Plant growth and carbon partitioning changed in response to the stresses. In N-limited plants, and also in water-limited plants to a lesser extent, growth slowed thus decreasing sink strength (MacAdam *et al.*, 1989). Carbon accumulated in the leaves, as shown by their higher specific leaf weight (McDonald, 1990; Dijkstra, 1990) and was redirected to roots, as shown by the higher root/shoot ratios. The initial effect of water stress was to lower stomatal conductance, and this was translated into higher leaf temperatures. These higher temperatures have been used to assess plant water status (Peñuelas *et al.*, 1992). LW plants seemed to avoid water deficiency *per se* by reducing water loss through the lowering of stomatal conductance. Nitrogen deficiency also induced stomatal closure, as has been found by Radin and Ackerson (1981). All these changes led to a loss of productivity in biomass (Table 1). The lower N concentration per unit dry weight in the leaves, stems and roots of N-stressed plants might have been caused in part by increased starch (Lambers *et al.*, 1981) but phenolics, other carbon compounds that are influenced by N availability, also increased.

As the plants studied were genetically homogeneous, the differences in phenolic content were dependent on nitrogen and water stress. The intraspecific variation in the content of carbon-based secondary compounds, like phenolics and terpenes, has been explained by the carbon/nutrient balance hypothesis (Bryant *et al.*, 1983). This assumes that resource availability (mineral nutrients) determines the individual carbon/nutrient balance in a plant species. Under mineral nutrient deficiency, the decline in growth is generally greater than that in photosynthesis. This implies a nutrient dilution in the plant, and the carbon excess relative to nutrients is allocated to carbon-based compounds. The changes we found in the phenolic composition are consistent with this carbon/nutrient balance hypothesis.

The higher RGR and ECI of insects fed with leaves from HN treatments suggest the N content of leaves was an important determinant of insect growth. This agrees with the available evidence in the literature (Mattson, 1980). However, the NUE of caterpillars (biomass gained by nitrogen ingested) was higher in HN treatments, indicating that some other factor in addition to nitrogen availability was determining the higher efficiency in biomass transformation. One possibility is that the lower contents of certain phenolic compounds, considered quantitative defences against herbivores, were less inhibiting for larval growth (Feeny, 1976). Phenolics can accumulate as deterrents, making the foliage unpalatable to mammals and birds and can reduce its digestibility (Coley, 1983). These effects on herbivores are an important issue in a changing climate where the availability of nitrogen and water may change sufficiently to influence the levels of phenolics and other related secondary carbon-based compounds.

The ground-level remote sensing measurements indicate that the red edge could provide a useful assessment of vegetation chlorophyll concentration. As chlorophyll concentration was found to be closely correlated with nitrogen level (see also Everitt *et al.*, 1985), measurement of the red edge may give indirect information about the nitrogen status of vegetation (Peñuelas *et al.*, 1993). Leaf temperature together with other spectral signals in the near infrared region may also provide a good assessment of the water status of vegetation (Peñuelas *et al.*, 1993).

ACKNOWLEDGEMENTS

We thank M. Bosch and X. Carbonell for technical help. The research was made possible by a grant AGR90-0458 from the CICYT (Spain). We gratefully acknowledge FPI (Spain) and CIRIT (Catalonia) fellowships to M.E., I.F. and R.M.

REFERENCES

Amiott MJ, Aubert S, Gonnet M, Tacchini M (1989) Les composés phénoliques des miels: étude préliminaire sur l'identification et la quantification par familles. Apidologie 20: 115-125

Bryant JP, Chapin III FS, Klein DR (1983) Carbon/nutrient balance of boreal plants in relation to vertebrate herbivory. Oikos 40: 357-368

Coley PD (1983) Herbivory and defensive characteristics of tree species in a lowland tropical forest. Ecological Monographs 53: 209-233

Curran PJ, Dungan JL, Golz HL (1990) Exploring the relationship between reflectance red edge and chlorophyll content in slash pine. Tree Physiology 7: 33-48

Curtis PS, Drake BG, Whigham DF (1989) Nitrogen and carbon dynamics in C_3 and C_4 estuarine marsh plants grown under elevated CO_2 *in situ*. Oecologia 78: 297-301

Demetriades-Shah TH, Steven MD (1988) High resolution spectral indices for monitoring crop growth and chlorosis. In: Proceedings 4th International Colloquium on 'Spectral signatures of objects in remote sensing', Aussois, France 18-22 January 1988, ESA SP-287, European Space Agency, Paris

Dijkstra P (1990) Cause and effect of differences in specific leaf area. In: Lambers H, Cambridge ML, Konings H, Pons TL (eds) Causes and consequences of variation in growth rate and productivity of higher plants. SPB Academic Publishing, The Hague, The Netherlands, pp 125-140

Everitt JH, Richardson AJ, Gausman HW (1985) Leaf reflectance-chlorophyll relations in buffelgrass. Photogrammetric Engineering and Remote Sensing 51: 463-466

Feeny PP (1976) Plant apparency and chemical defence. In: Wallace J, Mansell RL (eds) Biochemical interactions between plants and insects. Recent Advances in Phytochemistry 10, Plenum Press, New York, pp 1-40

Hanks RJ, Rasmussen VP (1982) Predicting crop production as related to plant stress. Advances in Agronomy 35: 193-215

Hayes JT (1991) Global climate change and water resources. In: Wyman (ed) Global climate change and life on earth. Chapman and Hall, New York, pp 18-42

Horler DNH, Barber J, Barringer AR (1980) Effects of heavy metals on the absorbance and reflectance spectra of plants. International Journal of Remote Sensing 1: 121-136

Keulen H van, Goudriean J, Seligman NG (1989) Modelling the effects of nitrogen on canopy development and crop growth. In: Russell G, Marshall B, Jarvis PG (eds) Plant canopies: their growth, form and function. Cambridge University Press, pp 83-104

Lambers H, Posthumus FS, Stulen I, Lanting L, Van de Dijk SJ, Hofstra R (1981) Energy metabolism of *Plantago lanceolata* as dependent on the supply of mineral nutrients. Physiologia Plantarum 51: 85-92

Lemon ED (ed) (1983) CO_2 and plants. The response of plants to rising levels of atmospheric carbon dioxide. AASS Selected Symposium 84, Westview Press, Boulder, Colorado, USA

MacAdam JW, Volenec JJ, Nelson CJ (1989) Effects of nitrogen on mesophyll cell division and epidermal cell elongation in tall fescue leaf blades. Plant Physiology 89: 549-556

Marigo G (1973) Sur une méthode de fractionnement et déstimation des composés .phénoliques chez les vegetaux. Analisis 2: 106-110

Mattson WJ (1980) Herbivory in relation to plant nitrogen content. Annual Review of Ecology Systems 11: 119-161

McDonald AJS (1990) Phenotypic variation in growth rate as affected by N-supply: its effects on net assimilation rate (NAR), leaf weight ratio (LWR) and specific leaf area (SLA). In: Lambers H, Cambridge ML, Konings H, Pons TL (eds) Causes and consequences of variation in growth rate and productivity of higher plants. SPB Academic Publishing, The Hague, The Netherlands, pp 35-44

Monje OA, Bugbee B (1992) Inherent limitations of nondestructive chlorophyll meters. A comparision of two types of meter. HortScience 27: 69-71

Peñuelas J, Filella I, Biel C, Serrano L, Savé R (1993) The reflectance of the 950-970 nm region as an indicator of plant water status. International Journal of Remote Sensing (in press)

Peñuelas J, Matamala R (1990) Changes in N and S content, stomatal density and specific leaf area of 14 plant species during the last three centuries of CO_2 increase. Journal of Experimental Botany 41: 1119-1124

Peñuelas J, Savé R, Marfà O, Serrano L (1992) Remotely measured canopy temperature of greenhouse strawberries as an indicator of water status and yield under mild and very mild water stress conditions. Agricultural and Forest Meteorology 58: 63-77

Poitout S, Bues R (1970) Elevage de plusieurs espècies de lépidoptères *Noctuidae* sur milieu artificiel riche et sur milieu artificiel simplifié. Annual Zoology Ecology Animal 1: 245-264

Porra RJ, Thompson WA, Kriemann PE (1989) Determination of accurate extinction coefficients and simultaneous equations for assaying chlorophylls a and b extracted with four different solvents: verification of the concentration of chlorophyll standards by atomic absorption spectroscopy. Biochimica et Biophysica Acta 975: 384-394

Radin JW, Ackerson RC (1981) Water relations of cotton plants under nitrogen deficiency. III Stomatal conductance, photosynthesis and abscisic acid accumulation during drought. Plant Physiology 67: 115-119

Thomas JR, Gausman HW (1977) Leaf reflectance v. leaf chlorophyll and carotenoid concentrations for eight crops. Agronomy Journal 69: 799

SCREENING METHODS FOR DETERMINING DROUGHT RESISTANCE IN BARLEY

S. Nogués[1] and L. Alegre
University of Barcelona
Department of Plant Biology
Unit of Plant Physiology
645 Avenue Diagonal
E-08028 Barcelona
Spain

ABSTRACT

The effect of drought on the photosynthetic functioning of twenty genotypes of barley (*Hordeum vulgare* L.) was examined using a modulated fluorescence technique coupled to measurements of simple fluorescence, net photosynthesis and leaf water potential. While any of these parameters can be used to score genotypes with extreme responses to water stress, all fail when there is only slight variability. For genotypes growing under mild water stress, it is necessary to use several screening tests simultaneously to assess their ability to score genotypes for resistance to water stress. Nevertheless, when stressed and control plants were compared, drought-tolerant genotypes showed smaller decreases than sensitive genotypes in simple fluorescence, measured as the (P-S)/S ratio, photochemical quenching (q_Q) and net photosynthesis.

INTRODUCTION

In recent years increased attention has been given to the effects on plants of adverse environmental conditions. The influence of drought on photosynthesis is well documented and has been the subject of several critical reviews (Bradford and Hsiao, 1982; Jones, 1983; Blum, 1989; Farquhar *et al.*, 1989). Early experiments indicated that the decrease in net leaf CO_2 uptake was mainly the result of stomatal closure. However, it was shown that non-stomatal effects on photosynthetic processes were also induced by mild drought stress (Sharkey and Seeman, 1989).

[1] Present address: University of Essex, Department of Biology, Colchester CO4 3SQ, UK

NATO ASI Series, Vol. I 16
Interacting Stresses on Plants in a Changing Climate
Edited by M. B. Jackson, C. R. Black

Non-intrusive techniques have an important role in the detection of stress effects on photosynthetic activity. A well-established technique is the measurement of chlorophyll fluorescence (Schreiber and Bilger, 1987). When leaves are exposed to unfavourable environmental conditions, inhibition of some of the photochemical processes controlling fluorescence yield *in vivo* causes substantial changes in the characteristics of the chlorophyll fluorescence curves, which can therefore be used to provide simple and rapid indicators of stress conditions in plants (Renger and Schreiber, 1986; Lichtenthaler *et al.*, 1986. These fluorescence indicators can be used to quantify the responses of different genotypes to environmental stress and hence to rank them according to their relative stress tolerance (Havaux and Lannoye, 1985a).

In this study, we have used simple and modulated chlorophyll fluorescence and measurements of net photosynthesis and leaf water potential to assess the suitability of these methods for screening studies of drought tolerance in twenty barley genotypes.

MATERIALS AND METHODS

Twenty genotypes of barley (*Hordeum vulgare* L.) were used; details of the genotypes used and their vegetative cycle and source are given in Table 1. Plants were grown in a glasshouse at the Free University of Brussels, Belgium. Maximum daily photosynthetic photon flux density was approximately 1200 μmol m^{-2} s^{-1} (natural illumination), with a 16 h photoperiod, 27/14 °C day/night air temperature and 78% relative humidity. Seeds were sown in plastic pots (13 cm diameter) filled with a Terofor compost-sand mixture (1 : 1). The compost contained all required nutrients in amounts sufficient to allow maximum growth. Plants were watered daily until the third leaf emerged.

Water stress treatment. Water stress was imposed by withholding water, and its development was monitored by regular measurements of leaf ψ_W using a hydraulic press (Havaux *et al.*, 1988b). The third fully expanded non-senescing leaf was used in all studies. Experiments began when leaf ψ_W in the morning was lower than

TABLE 1. *List of barley genotypes used, their vegetative cycle and source; CIMMYT, International Maize and Wheat Improvement Centre; ICARDA, International Centre for Agricultural Research in the Dry Areas.*

Name	Order number	Vegetative cycle	Source
Albada	1	Alternative	CIMMYT
Alpha	2	Winter	France
Mogador	3	Winter	France
Alexis	4	Spring	Germany
Beka	5	Spring	Spain
EVI-89-213	6	Spring	Spain
EVI-89-215	7	Spring	Spain
Zaida	8	Spring	Spain
Kym	9	Spring	Great Britain
Cameo	10	Spring	Great Britain
Dobla	11	Spring	Spain
Plaisant	12	Winter	France
Barbarrosa	13	Winter	France
ID-4016	14	Alternative	CIMMYT
Hatif	15	Winter	France
Albacete	16	Winter	Spain
Dacil	17	Winter	Spain
Pane	18	Winter	Spain
Roxana	19	Alternative	ICARDA
AD-8	20	Winter	Spain

-0.4±0.1 MPa in control plants and -1.2±0.2 MPa in stressed plants, about 12 days after starting the water stress treatment.

Gas exchange. Gas exchange measurements were made on the third fully expanded leaf using an infrared gas analyser (ADC MK2, Analytical Development

Co. Ltd, Hoddesdon, Herts, UK; Parkinson and Legg, 1971; Parkinson, 1985). Net photosynthesis was calculated from concomitant measurements of CO_2 uptake, transpirational water loss and temperature using equations developed by von Caemmerer and Farquhar (1981). Three Philips lamps provided a PAR flux density of approximately 1000 μmol m^{-2} s^{-1}. A water-filled cooling circuit placed between the leaves being measured and the lamps controlled the temperature of the system.

Simple chlorophyll fluorescence. Simple chlorophyll fluorescence induction curves were measured at room temperature in intact third leaves using a portable SF-10 fluorometer (Richard Brancker Research Ltd, Ottawa, Canada) (Havaux *et al.*, 1988b). Fluorescence was excited by red light (670 nm, 15 μmol m^{-2} s^{-1}) produced by a light-emitting diode, and was detected at wavelengths above 700 nm using a photodiode. The fluorescence signal was digitised by a RS-232 digital card (Datatech Enterprises Co. Ltd, Taiwan) and analysed using a microcomputer. In accordance with Havaux *et al.* (1988b), the simple fluorescence [(P-S)/S] ratio was used as an indicator of drought stress. The decrease in fluorescence quenching (q) relative to control plants was calculated (C, expressed as % of control) as:

$$q_{control} = [(P\text{-}S)/S]_{control} \qquad q_{stressed} = [(P\text{-}S)/S]_{stressed}$$

$$C\ (\%\ of\ control) = (q_{stressed}/q_{control}) \times 100$$

Modulated chlorophyll fluorescence. Measurements of the yield of modulated chlorophyll fluorescence emission were made using a pulse amplitude modulation fluorimeter (PAM 101-102-103, Heinz Walz, Mess-und Regeltechnik, Effeltrich, Germany; Schreiber *et al.*, 1986). Leaves were dark-adapted for 30 minutes prior to measurement of fluorescence induction. The initial level of chlorophyll fluorescence (F_o) was elicited with a very dim 650 nm light source modulated at 1.6 kHz. Chlorophyll fluorescence was detected at wavelengths above 700 nm by a PIN photodiode shielded with a long-pass filter. Variable modulated fluorescence (F_v) was induced by nonmodulated white light (1000 μmol m^{-2} s^{-1}) supplied by an Osram xenoplot HLX light source. The superimposition of short pulses (1 s) of intense saturating light (4000 μmol m^{-2} s^{-1}) applied at 10 second intervals allowed the determination of F_m. The initial fluorescence level (F_o), the variable fluorescence level (F_v), the maximal level of fluorescence reached during the flash (F_m) and the photochemical quenching at the steady state (q_Q) were calculated as described previously (Schreiber *et al.*, 1986). Following Screiber *et al.* (1986) and Havaux *et al.* (1988a), we examined the variation in photochemical quenching (q_Q): $q_Q = (F_m - F_v)/(F_m - F_o)$ and its percentage decrease relative to the control C (% of control). During the fluorescence measurements, leaves were exposed to normal air.

RESULTS

Table 2 shows the values of C (% control) for the different parameters used to score the twenty barley genotypes according to their resistance to water stress. Water stress caused substantial inhibition of the slow chlorophyll fluorescence quenching from P to S (measured as the (P-S)/S ratio). The differences in C (% control) for the P-S/S ratio indicate that the genotypes examined show differing responses in terms of simple fluorescence, and that there was a gradation from genotypes such as Dobla (11) to Alexis (4). According to Havaux and Lannoye (1988b), this parameter may be a good indicator of the influence of drought on plants.

Water stress influenced the modulated fluorescence levels. A gradual reduction in q_Q, expressed as a percentage of the control, was observed in plants subjected to water stress. A range of sensitivity between genotypes from approximately 90% in Plaisant (12) and Dobla (11) to 57% in EV1-89-213 (6) was observed (Table 2). The differences in net photosynthesis between control and stressed plants were greater in the genotypes EV1-89-213 (6), EV1-89-215 (7) and Alexis (4) than in Plaisant (12) and Dobla (11). There was no correlation between the criteria tested and leaf water potential (Table 3).

A screening of the twenty genotypes for water stress resistance calculated on the basis of the scores shown in Table 2 is summarised in Fig. 1. The results indicate that Dobla and Plaisant were the genotypes most tolerant to water stress, while EV1-89-213, EV1-89-215 and Alexis were the most sensitive.

TABLE 2. *Relative values of C (% of control) for simple fluorescence* $[(q_{stressed}/q_{control}) \times 100]$*, net photosynthesis and leaf water potential for the twenty genotypes of barley examined. Genotypes with same number of stars were not significantly different. S = Score; G = Genotype Order Number; HG = Homogenous Groups from Multiple Range Analysis of Tukey's method (p = 0.05). Values of C are the means of three observations.*

	Simple fluorescence			Modulated fluorescence			Net photosynthesis			Water potential		
S	G	C	HG	G	C	HG	G	C	HG	G	C	HG
20	9	131	*	12	90	*	11	45	*	2	125	*
19	5	129	**	11	87	**	2	28	**	14	150	**
18	11	108	***	2	87	**	12	25	**	3	175	***
17	13	101	***	20	86	**	9	22	***	11	200	****
16	1	98	***	16	84	***	3	22	***	12	200	****
15	10	97	***	17	83	***	20	20	****	20	200	****
14	20	92	***	3	82	***	19	15	*****	6	200	****
13	14	90	***	19	82	***	5	15	*****	18	225	****
12	12	82	***	10	82	***	16	13	*****	19	225	****
11	16	73	***	18	81	***	8	10	*****	5	250	****
10	8	71	***	7	81	***	1	9	****	8	250	****
9	15	68	***	15	81	***	18	8	****	10	275	***
8	6	63	***	4	80	***	15	8	****	15	275	***
7	17	55	***	13	80	***	13	5	***	16	275	***
6	18	45	***	5	79	***	17	5	***	4	275	***
5	3	29	***	8	79	***	14	5	***	7	275	***
4	19	29	***	1	75	***	6	4	***	9	275	***
3	2	21	***	9	67	***	7	3	**	1	300	**
2	7	14	**	14	62	**	10	2	**	17	300	**
1	4	12	*	6	57	*	4	1	*	13	325	*

TABLE 3. Correlation matrix between four tests: Simple fluorescence (SF), modulated fluorescence (MF), net photosynthesis (A) and leaf water potential (ψ_W). Critical value (p = 0.05) = ± 0.379. n = 20.

	SF	MF	A	ψ_W
SF	1.000			
MF	-0.221	1.000		
A	0.233	0.402	1.000	
ψ_W	0.215	0.038	-0.503	1.000

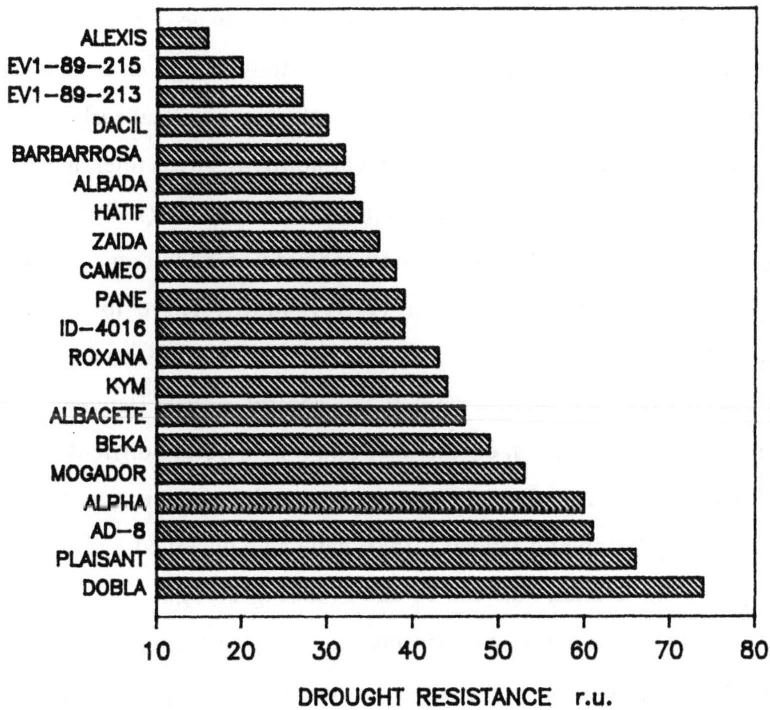

***FIGURE 1.** Screening of twenty barley genotypes according to their drought resistance calculated using the scores shown in Table 2.*

DISCUSSION

All the parameters used in this work [(P-S)/S ratio, q_Q, A and ψ_W] are influenced by water stress and could be used as rapid screening tests for drought tolerance in barley although, applied alone, they are more appropriate as indicators of extreme responses to water stress rather than of mild variations in response to drought. It is thus difficult to choose a single technique for screening genotypes for their resistance to water stress and it is much better to use several parameters simultaneously.

Havaux and Lannoye (1998b) first reported the use of the ratio (P-S)/P as a rapid indicator for comparing variability in drought tolerance between large numbers of wheat genotypes, suggesting that this screening method could also be used to rank other cereals for drought stress. The rapidity and sensitivity of the *in vivo* chlorophyll fluorescence methods described here have permitted comparison of 20 genotypes of barley, and this comparative study has given a measure of the range of variability in drought tolerance in cereals. Our results clearly indicate that barley genotypes are characterised by high variability in stress tolerance (Table 2). In response to water stress there was a rapid decrease in the (P-S)/P ratio. At the same value of ψ_W = -1.2 MPa, those genotypes considered tolerant of water stress showed only a slight decrease in the (P-S)/P ratio relative to control plants (C % control ∼ 100), whereas those genotypes considered to be sensitive to drought showed a sharp decrease in the ratio (C % control ∼ 30). Between these two extremes there was a gradation of tolerance and it is therefore difficult to establish the boundaries between tolerant, intermediate and sensitive genotypes. Other authors including Epron and Dreyer (1990) have also observed a reduction in the initial slope of the relation between F_p and F_t during drought.

In order to obtain additional information on the influence of water stress on chlorophyll *a* fluorescence, we studied the decrease in q_Q in relation to water stress. Havaux *et al.* (1988b) showed that leaf desiccation caused drastic changes in the different modulated fluorescence levels, resulting in a strong reduction in q_Q, and that the percentage decrease in this value was a good index of the level of drought resistance.

As shown in Table 2, there is a similarity between the data obtained using the (P-S)/P ratio and q_Q in those genotypes which are either more resistant or more sensitive to water stress. Drought reduced q_Q in all the genotypes tested, but the percentage inhibition was highly dependent on the degree of drought resistance of the genotype concerned. In drought-tolerant genotypes, the decrease in q_Q was limited (C % control ~ 87 for Dobla), whereas in drought-sensitive genotypes q_Q was strongly affected (C % control ~ 57 for EV1-89-213). The value of q_Q can thus be used as a rapid indicator of the degree of drought tolerance of barley genotypes. However, as for simple fluorescence, although q_Q reduction is a good method for distinguishing between tolerant and sensitive genotypes, it fails for intermediate genotypes.

Water stress has been shown to influence photosynthesis in many ways (Schulze, 1986; Turner, 1986; Nilsen, 1992). In the early stages of water stress, photosynthesis is inhibited by stomatal closure as a result of either an atmospheric stimulus (e.g., vapour pressure) or a root signal. In our experiments, water stress decreased net photosynthesis, but the net photosynthesis rates of the sensitive varieties hardly differed from those of the tolerant genotypes.

REFERENCES

Blum A (1989) Breeding methods for drought resistance. In: Jones HG, Flowers TJ, Jones MB (eds) Plants under stress: Biochemistry, physiology and ecology and their application to plant improvement. Society for Experimental Biology, Vol. 39. Cambridge University Press, UK, pp 197-215

Bradford KJ, Hsiao TC (1982) Physiological responses to moderate water-stress. In: Lange OL, Nobel PS, Osmond CB, Ziegler H (eds) Encyclopaedia of Plant Physiology, New Series, Vol. 2A: Physiological plant ecology II. Springer Verlag, Berlin, Germany, pp 264-324

Epron D, Dreyer E (1990) Stomatal and nonstomatal limitation of photosynthesis by leaf water deficits in three oak species: a comparison of gas exchange and chlorophyll a fluorescence data. Annals of Scientific Forestry 47: 435-450

Farquhar GD, Wong SC, Evans JR, Hubick KT (1989) Photosynthesis and gas exchange. In: Jones HG, Flowers TJ, Jones MB (eds) Plants under stress: Biochemistry, physiology and ecology and their application to plant improvement. Society for Experimental Biology, Vol. 39. Cambridge University Press, UK, pp 47-70

Havaux M, Ernez M, Lannoye R (1988a) Correlation between heat tolerance and drought tolerance in cereals demonstrated by rapid chlorophyll fluorescence. Journal of Plant Physiology 133: 555-560

Havaux M, Ernez M, Lannoye R (1988b) Sélection de variétés de blé dur (*Triticum durum* Desf.) et de blé tendre (*Triticum aestivum* L.) adaptées à la sécheresse par la mesure de l'extinction de la fluorescence de la chlorophylle *in vivo*. Agronomie 8: 193-199

Havaux M, Lannoye R (1985a) *In vivo* chlorophyll fluorescence and delayed light emission as a rapid screening technique for stress tolerance in crop plants. Zeitschrift Pflanzenzchtung 95: 1-13

Havaux M, Lannoye R (1985b) Drought tolerance of hard wheat cultivars measured by a rapid chlorophyll fluorescence test. Journal of Agricultural Science (Cambridge) 104: 501-504

Jones HG (1983) Drought and drought tolerance. In: Jones HG (ed) Plants and microclimate: A quantitative approach to environmental plant physiology. Cambridge University Press, UK, pp 212-226

Lichtenthaler HK, Buschmann C, Rinderle U, Schmuck G (1986) Application of chlorophyll fluorescence in ecophysiology. Radiation and Environmental Biophysics 25: 297-308

Parkinson KJ (1985) Porometry. In: Marshall B, Woodward FI (eds) Instrumentation for environmental physiology. Cambridge University Press, UK, pp 171-191

Parkinson KJ, Legg BJ (1971) A new method for calibrating infrared gas analysers. Journal of Physics E: Scientific Instruments 4: 598-600

Nilsen ET (1992) The influence of water-stress on leaf and stem photosynthesis in *Spartium junceum* L. Plant, Cell and Environment 15: 455-461

Renger G, Schreiber U (1986) Practical applications of fluorimetric methods to algae and higher plant research. In: Govindjee A, Amesz J, Fork DC (eds) Light emission by plants and bacteria. Academic Press, New York, USA, pp 587-619

Schulze ED (1986) Whole plant responses to drought. Australian Journal of Plant Physiology 13: 127-141

Schreiber U, Schliwa U, Bilger W (1986) Continuous recording of photochemical and non-photochemical quenching with a new type of modulation fluorometer. Photosynthesis Research 10: 51-61

Schreiber U, Bilger W (1987) Rapid assessment of stress effects on plant leaves by chlorophyll fluorescence measurements. In: Tenhunen JD, Catarino FM, Lange OL, Oechel WC (eds) Plant response to stress: Functional analysis in Mediterranean ecosystems. Springer Verlag, Berlin, Germany, pp 27-53

Sharkey TD, Seemann JR (1989) Mild water stress on carbon-reduction-cycle intermediates, ribulose biphosphate carboxylase activity, and spatial homogeneity of photosynthesis in intact leaves. Plant Physiology 89: 1060-1065

Turner NC (1986) Adaptation to water deficits: a changing perspective. Australian Journal of Plant Physiology 13: 175-189

von Caemerer S, Farquhar GD (1981) Some relationships between the biochemistry of photosynthesis and the gas exchange of leaves. Planta 153: 376-387

SECTION V

Cold

LOW TEMPERATURE INDUCED BIOCHEMICAL MECHANISMS: IMPLICATIONS FOR COLD ACCLIMATION AND DE-ACCLIMATION

C. Stushnoff, R.L. Remmele Jr.[1], V. Essensee[1] and M. McNeil[2]
Departments of Horticulture and Biochemistry, [1]Horticulture and [2]Microbiology
Colorado State University
Fort Collins
Colorado 80523
USA

ABSTRACT

Acclimation and de-acclimation of woody plants is correlated with endogenous levels of raffinose family oligosaccharides (RFO), but not to total sugars or other oligosaccharides. Synthesis begins at the onset of cold acclimation and reaches a maximum in mid-winter at peak hardiness. RFO levels decline as cortical stem tissues de-acclimate. Exposure for 24-48 hours at 25°C induces substantial dehardening. Arrhenius plots depict the rate of raffinose accumulation during acclimation, and the rate of degradation during de-acclimation.

INTRODUCTION

Low temperature injury has been a concern to plant scientists and producers of food crops for more than a century, resulting in numerous reports that describe differences in genotypic responses and injury symptoms following stress induced by exposure to low temperature episodes. Injury can occur over a very broad range of temperatures. It is necessary to distinguish between events associated with chilling sensitivity, which occur above ice nucleation temperatures, and injury induced when ice is present in plant tissues. Furthermore, events that stabilise tissues in over-wintering herbaceous perennials are likely to differ from those that permit survival of tissues in deciduous and evergreen woody plants. It has been known for many years that changes in temperature and photoperiod associated with seasonal variation in temperate climates are the primary environmental stimuli that trigger transitions from the cold tender state to the cold hardy state, and *vice versa* (Sakai and Larcher, 1987). Perennials must respond to annual seasonal cycles that

NATO ASI Series, Vol. I 16
Interacting Stresses on Plants in a Changing Climate
Edited by M. B. Jackson, C. R. Black

trigger metabolic events leading to acclimation and must have the genetic potential to tolerate climatic extremes for their entire life cycle if they are to survive.

Membrane stability is a prerequisite during low temperature stress of thylakoid membranes (Steffen *et al.*, 1989; Hallgren and Oquist, 1990; Hincha, 1990) and plasma membranes during cold acclimation and freezing (Steponkus, 1984) and during cryopreservation (Steponkus *et al.*, 1988). There is ample evidence that numerous endogenous metabolic changes accompany the transition to the cold hardy state (Weiser, 1970). Soluble endogenous cryoprotectants such as sugars and proteins have been shown to protect photosynthetic membranes during freezing (Hallgren and Oquist, 1990). At this time, however, neither the molecular mechanisms responsible for cryoprotection nor the nature of the disturbances in macromolecular structure and function that cause lethal injury are well understood.

Elucidation of these mechanisms is not simple because many of the metabolic changes that occur during acclimation are difficult to distinguish from those that occur because of maturation and cessation of growth. The latter may have little to do with a specific cryoprotective function. Our search for endogenous cryoprotectants has been targeted at buds and cortical tissues of woody plant species. Several requirements can be set forth as prerequisites to qualify a metabolite as a specific endogenous cryoprotectant.

- The metabolite must be inducible at the first stage of development, permitting induction of cold hardiness, and preferably not before.
- Changes in endogenous cryoprotectants may be qualitative, but must also be quantitative and highly correlated to changes in cold hardiness.
- Conditions that result in a loss of hardiness should also result in a corresponding quantitative decrease of the cryoprotectant. It seems improbable that products requiring high expenditure of carbon resources would be allowed to disturb the carbon balance under non-stress conditions (Amthor and McCree, 1990).
- If the mechanism has a broad base, similar trends should be detectable in different families, genera, species and cultivars.

- The response should be inducible under both natural and controlled environments.
- The product should have biochemical and biophysical properties that modify the freezing behaviour of aqueous plant solutions and/or play a role in stabilising membranes and biomolecules in the presence of tissue ice and at low temperatures.

MATERIALS AND METHODS

Cortical tissues and dormant buds from current-season's growth were analysed monthly for soluble oligosaccharides by gas chromatography and for cold hardiness status by controlled freezing tests, conductivity and recovery ratings. Samples were taken from plants grown at Fort Collins, Colorado, USA for the following species: 'Altaglow' Saskatoon berry, (*Amelanchier alnifolia* Nutt.), 'Dolgo' crabapple (*Malus baccata x domestica*), 'Red Delicious' apple (*Malus domestica* Borkh.), 'Red Lake' currant (*Ribes rubrum* L.), Western sandcherry (*Prunus besseyi* Bailey), 'Valiant' grape (*Vitis riparia x labrusca*), Red-Osier dogwood (*Cornus sericea* L.), Schubert chokecherry (*Prunus virginiana* L.) and Douglas fir (*Pseudotsuga menziesii, var. glauca* Beissn.). Samples of several apple cultivars ('Anna', 'Dolgo', 'Golden Delicious', 'K-14' and 'Mutsu') were shipped by air express from the National Clonal Apple Repository, Geneva, New York, USA. Dehardening was induced by exposing the samples to 25 °C at 100% relative humidity (RH) for 0, 12, 24, 36, 48 and 72 hours. Tissues were plunged into liquid nitrogen, lyophilised, ground and passed through a 100 mesh sieve before derivatisation by trimethylsilylation for gas chromatographic analysis (Sweeley *et al.*, 1963). Experiments were conducted to determine the following: (1) seasonal changes under an outdoor environment for a complete growth cycle; (2) induction of the hardening response at the onset of acclimation under controlled conditions, and (3) associated changes during dehardening under controlled conditions.

RESULTS AND DISCUSSION

Relationship of endogenous oligosaccharides to season-long changes in hardiness. The raffinose family oligosaccharides (RFO) were the only sugars studied that correlated significantly with hardiness for all seven taxa sampled at Fort Collins over the entire season (Table 1). Pronounced quantitative seasonal changes, consistent with changes in hardiness and environmental temperatures (Fig. 1), can be seen in data normalised by expressing a ratio of raffinose + stachyose/sucrose (mol g^{-1} dry wt.) (Chen and Burris, 1990). The polyol, sorbitol, correlated significantly with hardiness in 'Red Delicious' but not in any of the other taxa (Table 1, Fig. 2). Total sugars were correlated to season-long hardiness only for 'Red Delicious' apple where sorbitol is the predominant oligosaccharide (Table 1). This supports the contention that cold hardiness cannot be explained on the basis of colligative properties alone. Our data implicate RFO biosynthesis and degradation, but the specific molecular mechanism responsible for stabilisation at low temperatures is presently unknown.

TABLE 1. *Linear correlation of endogenous oligosaccharides (mol g^{-1} dry wt.) to season-long (12 months) cold-hardiness, based on leakage analyses and tissue browning (Lowest Survival Temperature for 50% of the samples) for seven woody plant taxa. Data are r values. Significance levels: * ≤.05, ** ≤.01, *** ≤.001; no notation = not significant.*

	Malus species		*Prunus species*		*Cornus*	*Ribes*	*Vitis*
	Dolgo Crabapple	Red Delicious Apple	Choke Cherry	Western Sand Cherry	Red-osier Dogwood	Red Lake Currant	Valiant Grape
Glucose	-0.61	-0.24	0.29	0.11	0.14	-0.10	0.28
Sucrose	-0.35	-0.68	-0.86*	-0.38	-0.14	-0.11	-0.54
Raffinose	-0.85**	-0.84**	-0.85*	-0.85**	-0.73*	-0.86***	-0.74*
Stachyose	-0.83**	-0.86**	-0.99***	-0.86**	-0.70*	-0.83**	-0.75*
Sorbitol	-0.04	-0.86**	0.48	0.22	-0.02	0	-0.16
Raffinose/ sucrose	-0.88***	-0.79*	-0.81*	-0.81**	-0.77*	-0.92***	-0.78*
Raffinose + stachyose/ sucrose	-0.88***	-0.78*	-0.82*	-0.81**	-0.78**	-0.93***	-0.76*
Total sugars	-0.34	-0.92***	0.18	-0.03	-0.18	-0.50	-0.39

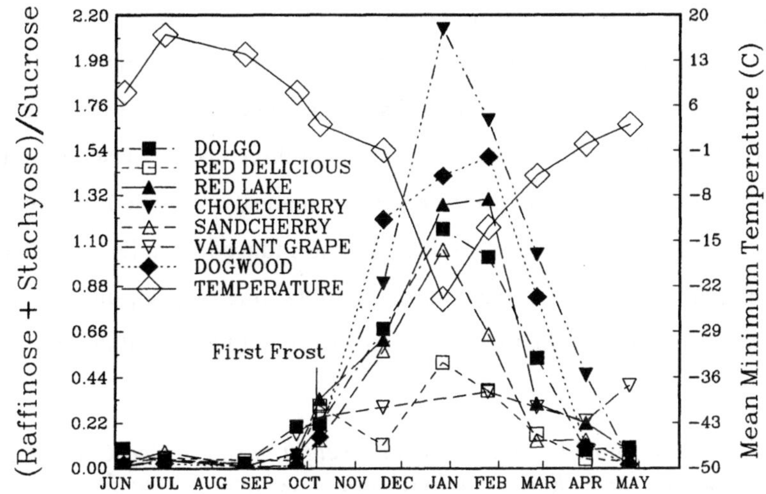

***FIGURE 1.** Seasonal changes in raffinose + stachyose/sucrose ratio compared to the mean minimum temperature for seven days preceding sampling.*

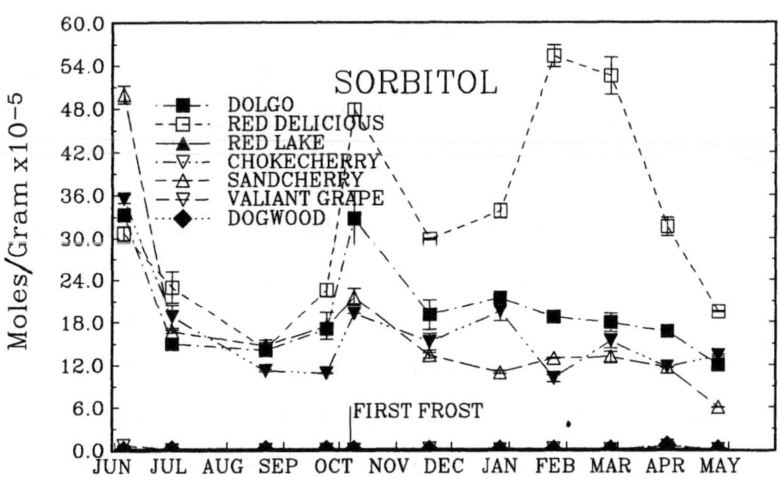

***FIGURE 2.** Seasonal changes in endogenous sorbitol. Bars indicate standard errors.*

The first stage of cold hardening. Woody plant tissues can be induced to cold harden by exposing them to a suitable low temperature regime after their vegetative maturity stage has passed (Nissila and Fuchigami, 1978; Friesen and Stushnoff, 1989). Analyses of oligosaccharides in three taxa, exposed to inductive conditions in growth cabinets after vegetative maturity, revealed an increase in glucose, raffinose and stachyose (Table 2). The induced samples were hardy to -20 °C, compared to -7 °C for the controls. At this time of year, outdoor samples were not yet exposed to inductive conditions, were not hardy and did not increase in RFO.

TABLE 2. Oligosaccharide content (%, dry weight basis) of Red-Osier dogwood, 'Dolgo' crabapple and 'Red Delicious' apple cortical tissue for control and hardiness-induced stems. Data are means ± sd.

Taxon	Treatment	Glucose	Sucrose	Raffinose	Stachyose	Sorbitol
Red-Osier dogwood	Control[y]	0.38±.01	2.40±.04	<.02	<.02	1.9±.08
	Induced[z]	1.62±.56	2.25±.16	0.95±.05	0.17±.01	<.02
Dolgo crabapple	Control	0.14±.01	1.47±.11	0.36±.18	0.11±.01	3.1±.17
	Induced	0.32±.07	2.18±.49	1.52±.18	0.42±.02	3.3±.66
Red Delicious apple	Control	0.20±.06	1.32±.01	0.05±.00	<.02	4.1±.22
	Induced	0.62±.09	2.35±.03	0.86±.08	0.18±.01	2.9±.01

[y] 10 days at 21 °C (12 hour day)/10 °C night.
[z] 10 days at 10 °C (12 hour day)/4 °C night + 24 hours at -4 °C on day 7.

Changes in oligosaccharides during controlled de-acclimation. Cold hardy, dormant bud twigs from five *Malus* cultivars, *Amelanchier alnifolia*, 'Altaglow' and *Pseudotsuga menziesii*, Douglas fir, were held at 25 °C and 100% RH to induce de-acclimation. A substantial loss in cold hardiness occurred in all taxa after 24-48 hours, and at the same time RFO levels dropped in cortical and bud tissues (Table 3, Figs. 3 and 4). These results are consistent with data for outdoor tissues undergoing dehardening with the onset of spring (Fig. 1). The rapid response to warm temperatures for both loss of cold hardiness (Table 3) and RFO degradation was observed for several taxa collected from different locations. Extreme fluctuation in temperature, especially towards the end of the winter, followed by freezing conditions has long been recognised as one of the most damaging

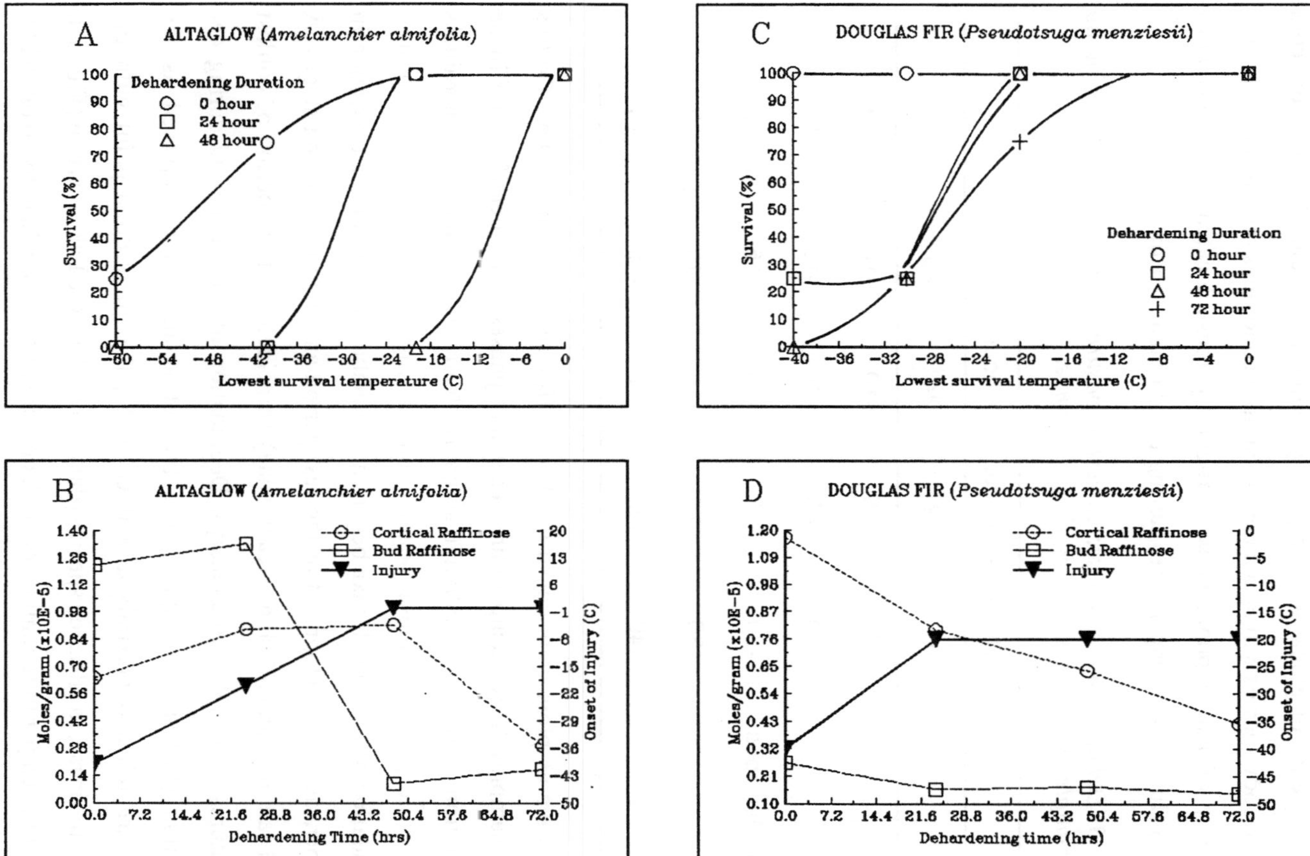

***FIGURE 3**. (A) De-acclimation and (B) raffinose degradation in* A. alnifolia, *'Altaglow'; (C) de-acclimation and (D) raffinose degradation in* P. menziesii, *Douglas fir.*

conditions facing growers of fruit trees, landscape plants and forest species (Hanninen, 1991). If temperature fluctuation during the dormant season increases as a consequence of global warming, the potential for greater damage from cold injury to woody species will also increase. Similar conditions may also threaten the biodiversity of native populations, particularly those located in regions most subject to fluctuating temperatures during the dormant season.

***TABLE* 3.** *Dehardening response; lowest exposure temperature (°C) with no injury. Samples from Geneva, New York, December 1991.*

	Dehardening duration (hours at 25 °C)				
Taxon	0	12	24	48	72
Anna	-30	-15	-10	-10	-10
Dolgo	-60	-40	-36	-15	-15
Golden Delicious	-30	-26	-20	-15	-10
K-14	-30	-20	-10	-10	-10
Mutsu	-40	-20	-10	-10	-10

Arrhenius plots were constructed for raffinose levels for the Fort Collins outdoor samples. Three temperature-dependent phases were identified: raffinose accumulation (during acclimation in autumn), steady state (at maximum hardiness in mid-winter) and raffinose degradation (during de-acclimation in spring). Accumulation of RFO in all taxa began in September, and attained a maximum in December, during development of the rest period and cold acclimation stages. A reduced rate of accumulation (steady-state) coincided with the period of coldest temperatures (November to January). Degradation of RFO occurred during de-acclimation (February to April). The rate of accumulation increased as the temperature fell and maximised at 0 °C (36.6), the intersection point with the steady state phase, in all seven species [data shown for 'Red Lake' currant (Fig. 5)]. Perhaps low temperature inhibits degradation or shifts metabolism to favour RFO

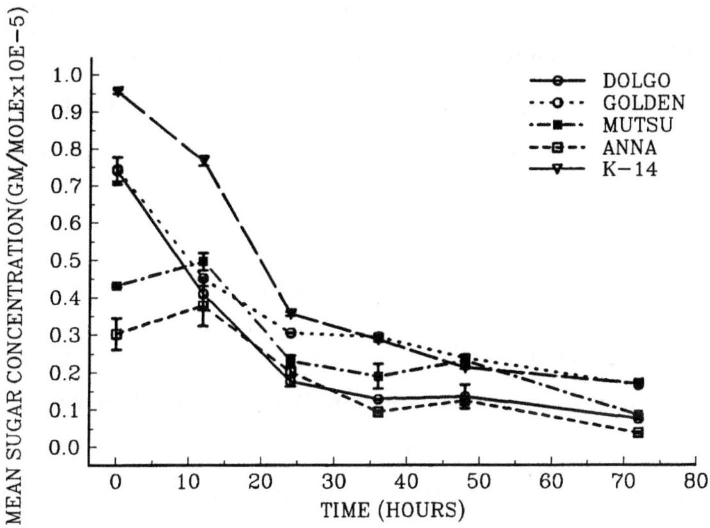

FIGURE 4. Raffinose + stachyose/sucrose ratios for five apple cultivars during dehardening at 25 °C and 100% RH for periods of 0-72 hours.

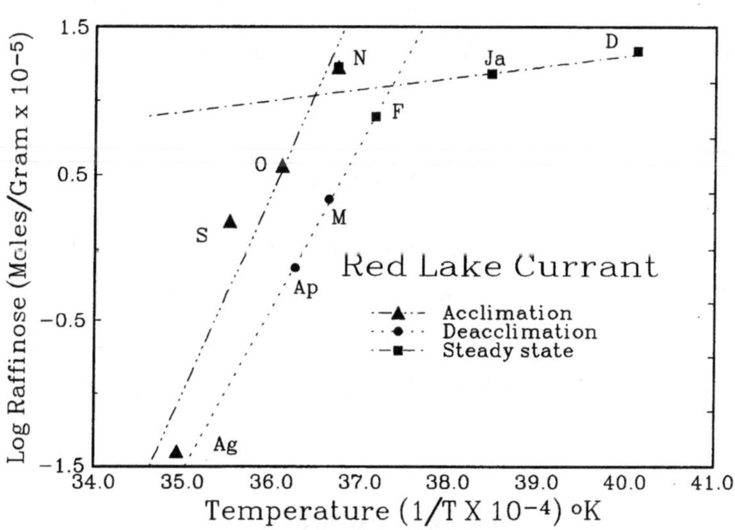

FIGURE 5. Arrhenius plot of raffinose accumulation by cortical tissues of Ribes rubrum, *'RedLake' currant. Monthly samples are abbreviated as shown.*

accumulation. In either case, the linear Arrhenius relation suggests a temperature-dependent process. The onset of degradation, at the intersection of the de-acclimation stage with steady state (Fig. 5), was very close to -3 °C for all seven taxa studied (data not shown). The steady state level was taxon-specific, being higher for the hardiest taxa and lower for the least hardy, but the maximum rates of accumulation and the onset of degradation were remarkably consistent for all taxa studied.

CONCLUSIONS

Certain soluble sugars and polyols can be good endogenous cryoprotectants because of their well known capacity to modify the freezing behaviour of aqueous solutions and stabilise membranes. Several are compatible osmolytes, in the non-structural carbon pool. Their synthesis and degradation can be rapidly induced by an environmental stimulus such as temperature. The raffinose family oligosaccharides are the only sugars we have studied that meet the criteria set forth for endogenous cryoprotectants. Their induction and degradation are temperature-dependent, endogenous levels are quantitatively related to hardiness and the response has been detected in taxa representing several families, including a coniferous species.

ACKNOWLEDGEMENTS

The authors gratefully acknowledge provision of plant material by Phil Forsline, USDA-ARS, Geneva, New York and graphics and technical assistance by Timothy Walsh, Jim Coleman and Don Archer. This work was supported by USDA-CSRS-CRGO #59002419, USDA-ARS-NSSL, Fort Collins and Colorado Experiment Station (project 690).

REFERENCES

Amthor JS, McCree KJ (1990) Carbon balance of stressed plants: A conceptual model for integrating research results. In: Alscher RG, Cumming JR (eds) Stress responses in plants: adaptation and acclimation mechanisms. Wiley-Liss, New York, USA, pp 1-15

Chen Y, Burris JS (1990) Role of carbohydrates in desiccation tolerance and membrane behavior in maturing maize seed. Crop Science 30: 971-975

Friesen LJ, Stushnoff C (1989) Vegetative maturity of *Amelanchier alnifolia* Nutt. compared to red-osier dogwood and Rescue crabapple. Canadian Journal of Plant Science 69: 955-960

Hallgren JE, Oquist G (1990) Adaptations to low temperatures. In: Alscher RG, Cumming JR (eds) Stress responses in plants: adaptation and acclimation mechanisms. Wiley-Liss, New York, USA, pp 265-293

Hanninen H (1991) Does climatic warming increase the risk of frost damage in northern trees? Plant, Cell and Environment 14: 449-454

Hincha DK (1990) Differential effects of galactose containing saccharides on mechanical freeze-thaw damage to isolated thylakoid membranes. Cryo-Letters 11: 437-444

Nissila PC, Fuchigami LH (1978) The relationship between vegetative maturity and the first stage of cold acclimation. Journal of the American Society for Horticultural Science 103: 710-711

Sakai A, Larcher W (1987) Frost survival of plants, responses and adaptations to freezing stress. Ecological Studies 62, Springer Verlag, Berlin, Germany

Steffen KL, Arora R, Palta JP (1989) Relative sensitivity of photosynthesis and respiration to freeze-thaw stress in herbaceous species: importance of realistic freeze-thaw protocols. Plant Physiology 89: 1372-1379

Steponkus PL (1984) Role of the plasma membrane in freezing injury and cold acclimation. Annual Review of Plant Physiology 35: 543-584

Steponkus PL, Uemura M, Balsamo RA, Arvinte T, Lynch DV (1988) Transformation of the cryobehavior of rye protoplasts by modification of plasma membrane lipid composition. Proceedings of the National Academy of Sciences, USA 85: 9026-9030

Sweeley CC, Bentley R, Makita M, Wells WW (1963) Gas-liquid chromatography of trimethylsilyl derivatives of sugars and related substances. Journal of the American Chemical Society 85: 2497

Weiser CJ (1970) Cold resistance and injury in woody plants. Science 169: 1269-1278

MOLECULAR MECHANISMS OF FREEZE-THAW INJURY AND COLD ACCLIMATION IN HERBACEOUS PLANTS: MERGING PHYSIOLOGICAL AND GENETIC APPROACHES

Jiwan P. Palta, Laurie S. Weiss, James F. Harbage, John B. Bamberg[1] and Julie M. Stone[2]
Department of Horticulture
University of Wisconsin
Madison
WI 73706
USA

ABSTRACT

In nature several factors including ice nucleation, temperature, freezing (cooling) rate, duration of exposure to ice, thawing rate and post-thaw conditions contribute to the degree of injury caused by frost episodes. Our results show that an increase in cooling rate from 1 to 4 °C h^{-1} made the difference between survival and death. We have also found that in the critical temperature range where injury occurs, the thaw rate influenced the degree and type of injury. The plasma membrane is a key site of alteration by freeze-thaw stress and cold acclimation. Important properties in this respect include membrane lipids and proteins (plasma membrane ATPase, kinase, desaturase), and the concentration of membrane and cytosolic calcium. Plasma membrane ATPase appears to be an important site of cellular response to freeze-thaw stress and an alteration in the function of this enzyme is one of the earliest manifestations of stress. Our results provide evidence that these alterations could be mediated by perturbation of cellular Ca^{2+} and/or changes in membrane lipid composition. These results provide an insight into the mechanisms of incipient injury and recovery following injury. To understand the genetics of freezing stress resistance, we performed an interspecific hybridisation of two diploid potato species that vary in freezing tolerance and cold acclimation ability. The species were *Solanum commersonii*, which is freezing tolerant and able to cold-acclimate (double its freezing tolerance in 10 days at chilling temperatures) and *Solanum cardiophyllum*, which is freezing sensitive and unable to cold-acclimate. Analysis of the backcross progenies shows that non-acclimated freezing tolerance and acclimation ability are genetically

---

[1] USDA, ARS Vegetable Research Unit, Potato Introduction Station, Sturgeon Bay, WI 54235, USA.
[2] Present Address: Department of Biochemistry, University of Missouri, Columbia, MO 65211, USA.

NATO ASI Series, Vol. I 16
Interacting Stresses on Plants in a Changing Climate
Edited by M. B. Jackson, C. R. Black

distinct traits that segregate independently. Generation mean analysis revealed that cold-acclimation ability can be explained by a simple additive-dominance model. Our results indicate that the ability to cold-acclimate is genetically relatively simple and should be amenable to selection at the diploid level. We have recently performed lipid analysis of purified plasma membrane preparations obtained from the parents, F_1 and the backcross progenies. Analysis shows that the relative increase in linoleic acid (18:2) in the plasma membrane is highly correlated to the cold acclimation ability. An increase in 18:2 co-segregated with the capacity to acclimate. Our results suggest that specific membrane lipids play a role in the genetic ability of the plant material to cold-acclimate.

INTRODUCTION

Resistance to freezing stress enables plants to maintain their function and/or survive following a frost episode. Herbaceous plants survive freeze-thaw stress because of their tolerance of extracellular ice. Many herbaceous species possess cold acclimation ability, and in such plants freezing tolerance is increased following a brief exposure to chilling temperatures. The plasma membrane is recognised as a key site of cellular responses to freeze-thaw stress. It follows that this membrane would be expected to change to impart a greater degree of tolerance to freeze-thaw stress following cold acclimation.

Physiologists and plant breeders/geneticists have independently investigated the physiological and genetic mechanisms of resistance to freezing stress. These two approaches have rarely been combined in the same study, but we have recently used both approaches to investigate the mode of inheritance of freezing stress resistance in the progenies of two diploid potato species. Our results show that progress in the improvement of freezing stress can be made by individually selecting for various components of this resistance and recombining them to obtain the desired plants. The use of genetic variants within a narrow population allows meaningful relationships between physiological properties and freezing stress resistance to be determined. By investigating the membrane lipid composition of the segregating progenies, we have been able to study membrane lipids that co-segregate with various components of freezing stress resistance. A brief account of this research is given here.

The freezing tolerance of individual plant species ultimately determines their ability to survive frost episodes in the natural environment. The degree of injury depends upon several factors including cooling rate, warming (thawing) rate, duration of freezing, the lowest temperature achieved and post-thaw conditions. A brief discussion of the impact of some of these factors on the degree of injury under laboratory conditions is given to demonstrate the importance of using realistic protocols in laboratory studies designed to investigate the mechanisms of freeze-thaw injury.

ICE IS EXTRACELLULAR DURING MOST FROST EPISODES IN NATURE

Ice formation depends upon ice nucleation and cooling rates (Steffen *et al.*, 1989; Palta and Weiss, 1993). During most spring and autumn frosts, ice is formed in the extracellular space (Pearce, 1988) and, as air temperature drops, ice grows in the extracellular space, withdrawing water from the cell. The dehydration and mechanical stresses the cell experiences from extracellular ice contribute largely to the freezing injury (Palta *et al.*, 1977d; Levitt, 1980; Sakai and Larcher, 1987). Several other factors including the duration of exposure to ice, cooling rates, thawing rates and post-thaw conditions can also influence the injury (Palta and Weiss, 1993). Therefore, it is extremely important that experimental studies of the mechanism of injury by freeze-thaw stress accurately simulate the type and magnitude of stresses associated with injury. Our results show that the outcome depends upon the freeze-thaw protocol used.

Ice nucleation. The temperature at which ice is initiated can greatly influence plant survival of a frost episode. If ice formation is initiated at relatively warm temperature (e.g., -1 or -2 °C), ice formation begins in the extracellular space because of the low concentration of solutes as compared to intracellular water and the high prevalence of ice nucleators in the extracellular space. As the nucleation temperature is lowered, the incidence of intracellular ice formation (which is lethal to the cell) increases. In support of this, a dramatic increase in injury was found in potato leaves when ice was initiated at about -4 °C as compared to similar leaf tissue frozen to the same temperature but with ice initiated at about -1 °C (Steffen *et al.*, 1989). Therefore, it should be possible to increase survival in species tolerant of extracellular ice by promoting ice nucleation at warmer temperatures. Ice nucleation temperature depends on the freezing temperature, the size of the tissue and the prevalence of ice nucleators such as dust, ice nucleating bacteria, wind and agitation (Levitt, 1980; Lindow *et al.*, 1982; Ashworth *et al.*, 1985; Palta and Weiss, 1993). We have found that the leaf and stem tissue of potato nucleates at -1 to -3 °C under field conditions (Steffen and Palta, 1987; Weiss and Palta, 1991a; Palta and Weiss, 1993).

Cooling rates. The maximal cooling rates below 0 °C are usually less than 2 °C h^{-1} during typical spring and autumn frosts (Steffen *et al.*, 1989). This conclusion was reached from examination of climatological data (Steffen *et al.*, 1989) and records of air and leaf temperatures in the field during frost episodes (Palta and Weiss, 1993). Our results show that simply increasing the cooling rate from 1 to 2.9 °C h^{-1} made the difference between survival and death of potato leaf tissue (Steffen *et al.*, 1989). Thus, laboratory tests should employ realistic cooling rates of 1-2 °C h^{-1}. Unfortunately, many studies investigating the mechanism of freezing injury continue to use relatively fast freezing rates (see Palta and Weiss, 1993). The likelihood of supercooling and rapid intracellular ice formation following nucleation increases when plants are subjected to very rapid cooling rates. As discussed above, this type of intracellular ice formation always results in lethal injury.

THAW RATE INFLUENCES INJURY

During thawing, as extracellular ice melts, water re-enters the cell and the cell expands. The speed of rehydration is directly influenced by the thaw rate. To maintain viability, the membrane and cytoplasm must expand along with the cell wall. During a very rapid thaw, the membrane and cytoplasm may not be able to keep pace with the expanding cell wall, resulting in mechanical tearing of the membrane from the cell wall. In standard freeze-thaw protocols, the thaw rate is not controlled. Usually frozen tissue is transferred to ice to thaw, assuming that this type of thaw is non-injurious. We have recently shown that thawing in ice is relatively rapid and that the extent of injury to membrane functions is less if the thaw rate is reduced (Weiss and Palta, 1991b; Palta and Weiss, 1993). In this study, the impact of thaw rate on ion leakage, photosynthesis and respiration was studied in non-acclimated and cold-acclimated potato leaf tissue. We found that the sensitivity of photosynthesis to thaw rate following cold acclimation was altered. In non-acclimated leaf tissue, slow thawing resulted in greater reduction of photosynthesis. However, ion leakage was consistently greater at fast thaw rates than at slow thaw rates irrespective of the state of cold acclimation. These results suggest that cold acclimation induces tolerance to the duration of freezing stress in the chloroplast but does not improve tolerance to the expansion stresses encountered during rehydration.

PLASMA MEMBRANE H^+-ATPase IS AN EARLY SITE OF PERTURBATION BY FREEZE-THAW STRESS

Plasma membrane ATPase as a site of cellular response to freeze-thaw stress was first proposed by Palta *et al.* (1977a,b). We have provided several kinds of evidence in support of this hypothesis:

1. Enhanced ion leakage from cells following freeze-thaw injury occurs from viable cells and is not necessarily the result of membrane rupture or complete loss of membrane semipermeability (Palta *et al.*, 1977a).
2. In spite of enhanced ion leakage and a water-soaked appearance following freeze-thaw injury, the cells were able to plasmolyse, exhibited protoplasmic streaming and bright fluorescence when treated with fluorescein diacetate, an indicator of cell viability (Palta *et al.*, 1977a; Palta and Li, 1980; Arora and Palta, 1988), and showed no significant change in membrane permeability to water and non-electrolytes (Palta and Li, 1980).

3. The major cation leaking from the cell is K^+ (Palta *et al.*, 1977a). There is a specific increase in the passive K^+ permeability of the injured cell membranes (Palta and Li, 1982). Since transport of K^+ across the cell membrane is coupled with the activity of the H^+-ATPase, these results suggest an alteration in the functions of H^+-ATPase by freeze-thaw stress.
4. Injury is reversible and full recovery of tissue from injury is possible (Palta *et al.*, 1977c). Since recovery requires that all leaked ions (primarily K^+) be pumped back into the cell, these results further suggest involvement of H^+ ATPase in recovery.
5. At the incipient level of injury (recoverable), there is a selective loss of plasma membrane ATPase activity (Table 1; Iswari and Palta, 1989a,b).
6. The loss of the water-soaked appearance (recovery) is accompanied by a reduction (return to normal) in ion leakage rate from the cells and plasma membrane ATPase activity (Table 2; Arora and Palta, 1991).

TABLE 1. *Effect of degree of freeze-thaw stress in* Solanum commersonii *leaflets on the specific activity of plasma membrane ATPase, CCO and CCR. Data calculated from Iswari and Palta (1989b).*

Category of damage	Visual water soaking (% area)	Ion leakage (% of total)	Specific activity of ATPase (% of control)		Specific activity (% of control)	
			Mic	PM	CCR	CCO
Control	0	9 ± 2	100	100	100	100
Slight	10-20	14 ± 6	216	189	105	102
	10-20	27 ± 6	200	216	-	-
Substantial	70-90	76 ± 9	26	24	98	108
	70-90	-	45	68	-	-
Total	100	85 ± 5	ND	ND	62	127
	100	-	19	27	-	-

Specific activity of ATPase in control was about 0.06 μmol mg^{-1} protein min^{-1} in microsomes and 0.3 μmol mg^{-1} protein min^{-1} in the plasma membrane fraction. Cytochrome c oxidase (CCO) activity in control was 3.7 nmol mg^{-1} protein min^{-1}. Cytochrome C reductase (CCR) activity in control was 30 nmol mg^{-1} protein min^{-1}. ND = not detectable. Mic = Microsomal membranes. PM = plasma membrane.

TABLE 2. *Vanadate sensitive ATPase activity in the microsomal membrane fraction and ion leakage rate of onion scale tissue following reversible freeze-thaw injury and recovery. Standard assays were used to measure the activity of vanadate sensitive ATPase (in the presence of 100 μM vanadate) in the microsomes isolated from the same tissue which was used for ion leakage rate measurements. Data from Arora and Palta (1991).*

Expt. No.	Treatment	Ion leakage rate measured* (% h^{-1})		Vanadate-sensitive ATPase activity	
		After thaw	After recovery	After thaw	After recovery
1	Control				
	(unfrozen)	25.5 ± 0.4	22.4 ± 1.4	49.2 (65.2)**	44.2 (78.3)
	Freeze-thaw	31.3 ± 2.1	21.3 ± 4.0	25.0 (41.1)	35.5 (61.1)
2	Control				
	(unfrozen)	24.1 ± 1.1	20.0 ± 1.5	50.2 (73.1)	51.3 (86.0)
	Freeze-thaw	30.7 ± 1.7	22.3 ± 0.3	38.9 (53.5)	56.2 (84.8)

* Ion leakage rate (efflux h^{-1}) was expressed as percentage of total internal ions.

** Values in parentheses are total activity in the absence of vanadate. The values for ion leakage rate and ATPase activity (expressed as μmol P_i 50 g^{-1} fresh weight 35 min^{-1}) are from two representative experiments and are averages of triplicates and duplicates, respectively.

7. Recovery of injury can be prevented by treating the tissue with 100 μM sodium orthovanadate, an inhibitor of the plasma membrane ATPase (Arora and Palta, 1991).

In addition to our own research, several independent studies have provided evidence in support of the ATPase hypothesis (Jian *et al.*, 1982; Pomeroy *et al.*, 1983; Hellergren *et al.*, 1985; Uemura and Yoshida, 1984). We have pursued two lines of research to understand mechanisms by which plasma membrane ATPase can be altered by freeze-thaw stress. These alterations in the functions of plasma membrane ATPase can result from (1) changes in membrane/cytosolic Ca^{2+} concentration and; (2) changes in membrane lipids associated with these proteins.

PERTURBATION OF MEMBRANE/CYTOSOLIC CALCIUM CAN TRIGGER/MEDIATE CELLULAR RESPONSE TO FREEZE-THAW STRESS AND COLD ACCLIMATION.

Recent evidence demonstrates the role of Ca^{2+} as a metabolic and developmental controller in plants (Hepler and Wayne, 1985; Pooviah and Reddy, 1987; Leonard and Hepler, 1990). Calcium concentration in the cytoplasm is kept very low, usually within the range 10^{-6} to 10^{-8} M. Small changes in the absolute quantity of Ca^{2+} can create 10- to 100-fold changes in Ca^{2+} concentration without disturbing the ionic balance. This feature makes Ca^{2+} an excellent candidate for triggering or mediating cellular responses to stresses such as freeze-thaw stress. This response can be brought about by activating or deactivating Ca^{2+} and/or Ca^{2+}-calmodulin dependent protein kinases (Arora and Palta, 1989; Palta, 1989). From the analysis of ion effluxes, Palta *et al.* (1977a,b) found a significant leakage of Ca^{2+} from freeze-injured cells. More recently, Arora and Palta (1988) found a decrease in membrane-associated calcium in both reversibly and irreversibly injured cells. These results suggest a role of membrane/cytosolic calcium in the initiation and progression of freeze-thaw injury and recovery from injury.

Membrane functions can be impaired if a stress results in the displacement of calcium from the membrane. Following freeze-thaw stress, membrane Ca^{2+} could be displaced by K^+ which leaks out of the cell and accumulates in the extracellular space (Palta and Li, 1978; Arora and Palta, 1986). These studies suggest that tighter binding of Ca^{2+} with the membrane may provide some tolerance to freeze-thaw stress which would result in increased membrane-associated Ca^{2+}. The results of Arora and Palta (1989) support this suggestion.

Membrane and/or cytosolic calcium may also play a role in triggering and mediating cold acclimation. Metabolic changes leading to cold acclimation could result from the activation or deactivation of Ca^{2+} and/or of Ca^{2+}-calmodulin dependent protein kinases. In support of this suggestion, changes in the activity of plasma membrane-associated, Ca^{2+} dependent protein kinase during cold acclimation were reported by Iswari *et al.* (1990).

PLASMA MEMBRANE LIPID IS A SITE OF CELLULAR RESPONSE DURING FREEZE-THAW STRESS AND COLD ACCLIMATION

Many species cold-acclimate, thereby increasing their freezing resistance (Levitt, 1980). In perennial woody species this acclimation is gradual (over a period of weeks or months) in response to low temperatures and long nights (short days) in autumn. Many herbaceous plant species also cold-acclimate when exposed to chilling temperatures. Since the plasma membrane is a key site of alteration by freeze-thaw stress, it follows that this membrane must undergo changes to be able to tolerate a higher degree of freeze-thaw stress following acclimation. Changes in the physical properties and composition of membrane lipids have been investigated for the last 20 years. Several investigators reported an increase in the unsaturation of fatty acids following cold acclimation (Grenier *et al.*, 1972; de la Roche *et al.*, 1972; Willemot, 1975), while others failed to detect such changes (de la Roche *et al.*, 1975; Vigh *et al.*, 1985). These contradictory results can be explained in part by the fact that these researchers analysed the lipid composition of whole tissues, organelle preparations or crude membrane preparations rather than that of the purified membrane of interest. Our recent results support this view (Palta and Meade, 1987; Palta and Weiss, 1993). We compared the fatty acid composition of whole leaf tissue and purified plasma membranes and found no significant changes following cold acclimation in the fatty acid composition when leaf tissue was analysed (Table 3). However, in purified plasma membrane preparations following cold acclimation, a significant increase in linoleic acid (18:2) and a decrease in palmitic acid (16:0) were observed. A similar consistent increase in 18:2 (as freezing tolerance increased with cold acclimation) has been found in cranberry (*Vaccinium macrocarpon* Ait.) leaves and pine (*Pinus* sp.) needles (Abdallah and Palta, 1989; Sutinen and Palta, 1989, respectively). A specific increase in 18:2 (with a concomitant decrease in 16:0) will profoundly influence the biophysical properties of the membrane. As the melting point of 18:2 is much lower than that of 16:0, an increase in 18:2 would reduce the melting point of membrane lipids.

Therefore, the plasma membrane of cold-acclimated plants would be expected to remain 'more fluid' at freezing temperatures.

Phase transition events, for example a change from a liquid-crystalline state to a gel state, may occur in certain lipids and at certain lipid domains as the temperature is lowered. Such a phase change could impart a damaging effect on the membrane. An increased proportion of 18:2 may help prevent these injurious phase transitions in certain key domains in the membrane.

TABLE 3. *Relative changes after cold acclimation in key fatty acids and in the degree of unsaturation of these fatty acids in the polar lipid fractions obtained from leaf and plasma membranes of* Solanum commersonii.

	Relative proportion unsaturated/ saturated fatty acids		Relative changes in the proportion of various fatty acids after acclimation (% change as compared to non-acclimated)	
Material used	Non-acclimated	Cold-acclimated	16:0	18:2
Leaf tissue	4.35	4.35	−1.6	+0.8
Plasma membrane	1.22	1.34	−10.1	+28.5

COMPARATIVE STUDIES OF RELATED POTATO SPECIES PROVIDE AN EXCELLENT TOOL FOR INVESTIGATING PLASMA MEMBRANE LIPID CHANGES RELATED TO COLD ACCLIMATION

In order to understand the relationship between plasma membrane lipid composition and cold acclimation, most researchers have followed changes in lipid composition following cold acclimation in a single clone (Uemura and Yoshida, 1984; Yoshida and Uemura, 1984; Lynch and Steponkus, 1987). In these studies it is impossible to distinguish lipid changes related to cold acclimation *per se* (i.e., the development of freezing tolerance) from those resulting from low temperature adjustment of plant metabolism. In addition, some lipid changes during cold acclimation may be associated with aging, especially when relatively long periods (over four weeks) of acclimation are employed.

Many comparative studies aimed at establishing the relationship between resistance to freezing stress and membrane lipid composition have used unrelated plant species. A major drawback of this approach is the difficulty in distinguishing lipid differences that are related to freezing resistance or cold acclimation capacity from those that are unrelated to freezing resistance and are the result of differences in plant materials.

We have avoided such problems by making comparative studies of lipid changes in genetically related cold-acclimating and non-acclimating potato species. In these studies we compared the plasma membrane lipid composition of *Solanum commersonii* (freezing tolerant and able to cold-acclimate) with that of *Solanum tuberosum* (freezing sensitive and unable to cold-acclimate), both before and after cold acclimation. A specific increase in the fatty acid 18:2 and a decrease in fatty acid 16:0 were seen only in *S. commersonii* following acclimation treatment (Palta and Weiss, 1993; Table 4). These studies suggest that an increase in 18:2 may be related to the development of freezing tolerance during acclimation.

TABLE 4. *Plasma membrane ATPase activity and key lipid changes in purified plasma membrane, after cold acclimation in acclimating (*Solanum commersonii*) and non-acclimating (*S. tuberosum*) potato species.*

Species	Treatment	Freezing tolerance* (°C)	SA** of PM ATPase	Key phospholipid fatty acids		Ratio
				16:0	18:2	18:2/16:0
S. commersonii	Control	−4	0.637	34.4	31.1	0.90
	Cold	−9	1.042	30.5	37.2	1.22
S. tuberosum	Control	−2.5	0.550	31.7	30.6	0.97
	Cold	−2.5	0.570	32.1	31.6	0.98

* Temperature at which irreversible injury occurred.

** SA – Specific activity of plasma membrane ATPase (μmol P_i mg^{-1} min^{-1}).

INCREASES IN LINOLEIC ACID (18:2) CAN HELP EXPLAIN INCREASED ACTIVITY OF PLASMA MEMBRANE ATPase FOLLOWING COLD ACCLIMATION

As discussed earlier, we have provided evidence that plasma membrane H^+-ATPase is a sensitive and an early site of cellular response to freeze-thaw stress. It follows that plasma membrane H^+-ATPase should respond in some manner as freezing tolerance increases during cold acclimation. We have found an increase in the specific activity of plasma membrane ATPase in parallel with an increase in freezing tolerance following cold acclimation of leaf tissue of *Solanum commersonii*, a potato species capable of cold acclimation (Palta and Meade, 1989). In this species, an increase in 18:2 and a decrease in 16:0 coincided with an increase in plasma membrane ATPase activity (Table 4). These studies suggest an increase in plasma membrane ATPase can be brought about *via* changes in lipids associated with this protein. Studies by Palmgren *et al.* (1988) support this suggestion, since they observed an *in vitro* increase in plasma membrane ATPase activity following addition of 18:2 fatty acid. We have found similar changes in 18:2 and plasma membrane ATPase activity in pine needles during cold acclimation and de-acclimation under natural conditions (Sutinen *et al.*, 1990). Our studies provide evidence that changes in 18:2 could modulate plasma membrane ATPase during cold acclimation.

NON-ACCLIMATED FREEZING TOLERANCE AND COLD ACCLIMATION CAPACITY ARE UNDER INDEPENDENT GENETIC CONTROL

The mode of inheritance of freezing stress resistance has been investigated in several species. Most research has been conducted with wheat (Gullord, 1975; Gullord *et al.* 1975; Stutka, 1981, 1984; Brule-Babel and Fowler, 1988; Stutka and Veisz, 1988; Limin and Fowler, 1989) and other cereals (Amirshashi and Patterson, 1956; Pfahler, 1966; Brule-Babel and Fowler, 1989), and it has been generally concluded that inheritance of freezing stress resistance in cereals involves multiple genes with both additive and non-additive effects (Marshall, 1982; Stushnoff *et al.*,

1984; Blum, 1988; Thomashow, 1990). In some studies, freezing resistance has been shown to be partially dominant, whereas in others freezing sensitivity has been shown to be partially dominant. These apparently conflicting results have been, in part, explained in terms of the freezing protocols used. Under severe freezing stress, freezing sensitivity was dominant, whereas at moderate freezing stress, freezing resistance was dominant (Stutka and Veisz, 1988).

No systematic attempt has been made to separate various components of freezing stress in winter cereals. This has been largely due to heavy reliance on field selection. Our recent studies with potato species show that it should be possible to screen separately for freezing tolerance (in the non-acclimated state) and the capacity to cold-acclimate (Stone *et al.*, 1993). In these studies, two wild potato species, *Solanum commersonii* (freezing tolerance of -4.5 °C and able to cold-acclimate down to -9.6 °C in 14 days at 4/2 °C day/night temperature) and *Solanum cardiophyllum* (freezing tolerance of -1.6 °C and unable to cold-acclimate), were used as diploid parents. The progeny, developed from the backcrosses of the F_1 products with the parents, were screened for freezing tolerance before and after cold acclimation. Freezing tolerance (non-acclimated state) and the capacity to cold-acclimate segregated independently. In relatively few of the individuals examined, it was possible to nearly recover both parental types in the backcross progeny for cold acclimation capacity and freezing tolerance. From these studies (Stone *et al.*, 1993), it appears that:

1. Freezing tolerance (non-acclimated state) and cold acclimation must have separate genetic mechanisms.
2. Freezing tolerance (non-acclimated state) and cold acclimation appear to be quantitatively inherited.
3. Capacity to acclimate can be explained by a simple additive dominance model and lack of tolerance and lack of acclimation appear to be somewhat dominant.
4. Few major genes must control freezing tolerance and cold acclimation in potato species. Thus, selection for these two traits in breeding programmes should not be difficult, at least at the diploid level.

From the evaluation of the backcross populations of potato species, it was suggested that a few major genes are responsible for freezing tolerance and cold acclimation capacity (Stone *et al.*, 1993). Interestingly, the results of a recent study of winter field pea (*Pisum sativum* subsp. *arvense*) also suggest as few as three or four genes or linkage groups could be responsible for 'winter hardiness' (Liesenfeld *et al.*, 1986). These results suggest that inheritance of freezing stress resistance is less complex than originally thought. They also show that it should be possible to select separately for improved freezing tolerance (in the non-acclimated state) and the ability to cold-acclimate.

CONTENTS OF SPECIFIC PLASMA MEMBRANE FATTY ACIDS CO-SEGREGATE WITH NON-ACCLIMATED FREEZING TOLERANCE AND COLD ACCLIMATION CAPACITY IN *SOLANUM* PROGENIES

As stated above, we have recently demonstrated independent genetic control for two components of freezing tolerance, namely non-acclimated freezing tolerance and acclimation capacity (Stone *et al.*, 1993). This was established by investigating the freezing tolerance of a segregating backcross population derived from freezing-sensitive, non-acclimating *Solanum cardiophyllum*, and freezing-tolerant, acclimating *S. commersonii*. These *Solanum* progenies provide ideal material to investigate the association of specific plasma membrane lipids with different components of freezing stress resistance. Using this material, we have found that the relative proportion of specific fatty acids is correlated with specific traits of freezing stress resistance (Harbage *et al.*, 1992). The two traits, non-acclimated freezing tolerance and acclimation capacity, correlated best with different fatty acids (Table 5; Fig. 1). In these studies we have also found that these correlations are highly significant only when backcross populations of the two parents are analysed separately (Table 5). A brief account of these results is given below.

The relative proportion of 18:2 fatty acid in the plasma membrane was correlated positively with non-acclimated freezing tolerance, acclimated freezing tolerance and acclimation capacity (Table 5; Fig. 1) only in the *S. commersonii* population. Furthermore, the relative proportion of 18:3 correlated negatively with non-acclimated and acclimated freezing tolerance. However, in the *S. cardiophyllum* population, the relative proportion of 16:0 fatty acid correlated positively with non-acclimated tolerance and negatively with acclimation capacity (Table 5; Fig. 1). These results suggest that the fatty acids associated with non-acclimated freezing tolerance are different from those associated with acclimation capacity.

***TABLE 5.** Correlation coefficients of plasma membrane fatty acid contents purified from non-acclimated and cold-acclimated leaves (generated from* S. commersonii *and* S. cardiophyllum, *their F_1 hybrid, and backcrosses to each parent) with relative freezing tolerance (RFT) in non-acclimated and acclimated states and of changes in the fatty acid contents with changes in freezing tolerance following cold acclimation (cold acclimation capacity = acclimated RFT - non-acclimated RFT).*

Population	Trait	Correlation coefficients		
		Fatty acid		
		16:0	18:2	18:3
Total	Non-acclimated RFT	0.65**	0.29^{NS}	−0.65**
	Acclimated RFT	0.20^{NS}	0.54**	−0.68**
	Cold acclimation capacity	−0.54**	0.45*	−0.01^{NS}
S. commersonii	Non-acclimated RFT	0.07^{NS}	0.58*	−0.79**
	Acclimated RFT	−0.20^{NS}	0.83**	−0.86**
	Cold acclimation capacity	−0.24^{NS}	0.65**	−0.06^{NS}
S. cardiophyllum	Non-acclimated RFT	0.70*	−0.21^{NS}	−0.03^{NS}
	Acclimated RFT	−0.36^{NS}	0.24^{NS}	−0.31^{NS}
	Cold acclimation capacity	−0.64*	0.33^{NS}	−0.17^{NS}

^{NS} Non-significant
*, ** Significant at the 5% and 1% levels, respectively.

In the backcross population derived from the cross of the F_1 with the tolerant, acclimating parent, *S. commersonii*, the non-acclimated freezing tolerance varied from -2.3 °C (similar to the F_1) to -4.5 °C (similar to the parent) and the acclimation capacity varied from 0.3 °C (similar to the F_1) to 5.7 °C (similar to the parent). Our results show that in this population, the relative proportion of 18:2

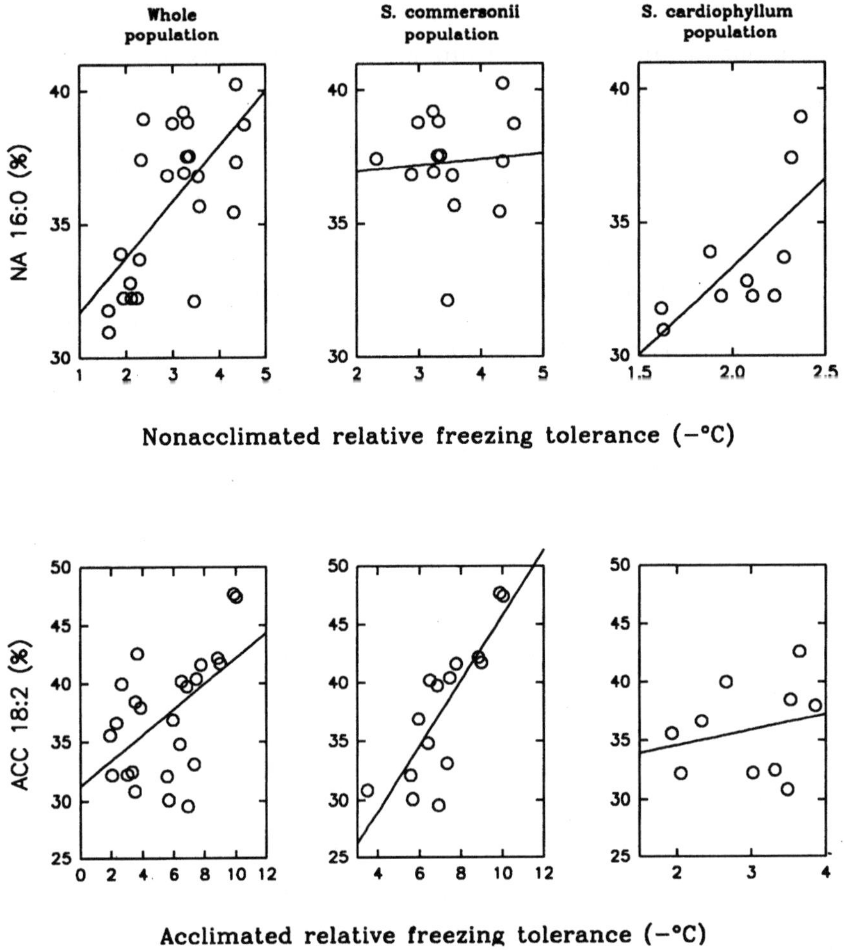

FIGURE 1. Correlation of relative freezing tolerance (RFT) with plasma membrane palmitate (16:0) content before acclimation and of RFT with plasma membrane linoleate (18:2) content after acclimation in Solanum *progeny segregating for freezing tolerance and acclimation capacity.*

was most closely correlated with the freezing stress resistance traits. These results suggest a specific role for 18:2 in conferring freezing tolerance in potato species. Furthermore, since the F_1 was closer to the sensitive parent for 18:2 content, our results suggest that the mode of inheritance of this lipid is largely recessive.

Clearly separate examination of closely related populations was necessary to establish this relationship between fatty acids and freezing tolerance.

AN INCREASED PROPORTION OF LINOLEIC ACID (18:2) DURING COLD ACCLIMATION IS IMPORTANT FOR THE DEVELOPMENT OF FREEZING TOLERANCE

A change in the physical state of membrane lipids from a flexible liquid crystalline state to a solid gel state has been proposed as one of the primary responses leading to temperature-induced injury. In the liquid crystalline or fluid phase the membrane bilayer is comparatively pliable (fluid), since the fatty acid chains possess rotational freedom. In the gel phase, the phospholipid fatty acid tails completely lose their motional freedom. A preponderance of unsaturated fatty acids such as linoleic and linolenic acid enhances the fluid phase. The reason for this is the presence of 'kinks' or bends at the location of the unsaturated double bond which increases the pliability. Naturally occurring lipids show broad phase transitions due to heterogeneity of lipid species and the wide variety of molecular species with differing acyl chain fatty acids (Gurr and Harwood, 1991). Recent evidence suggests that the membrane may exist as a patchwork of areas, each with varying lipid and protein composition, allowing the possibility that domains of lipids more prone to changes to gel phase could occur within the membrane. Membrane functions including membrane permeability and membrane bound enzyme activity can be impaired if phase change results.

A significant increase in the proportion of 18:2 would clearly help maintain membrane fluidity at low temperature. For example, the phase transition temperature of 18:0 (sn1), 18:2 (sn2) phosphatidylcholine (18:0/18:2 PC) is -16 °C as compared to 1 °C for 16:0/18:1 PC and 41 °C for 16:0/16:0 PC (Gurr and Harwood, 1991). A possible specific interaction of 18:2 with other membrane enzymes could be another significant aspect of changes in 18:2 during cold acclimation. For example, as discussed earlier, there is evidence that 18:2 may modulate the activity of plasma membrane ATPases (Palmgren *et al.*, 1988). The reason for this type of specific role of 18:2 could be the fact that 18:2 has a larger molecular area (almost two- and four-fold greater than 18:1 and 18:0) and a smaller length because of the two 'kinks' at the sites of unsaturated bonds (Cook, 1985). (For a detailed discussion see Leshem, 1992.) The results of our studies suggest an interesting potential role of 18:2 in cold acclimation and freezing tolerance.

CONCLUSIONS

During natural frost episodes, several factors including ice nucleation, cooling rate, duration of exposure to ice, thawing rate and post-thaw conditions contribute to the degree of injury. Our results demonstrate the importance of using realistic freeze-thaw protocols during laboratory experiments designed to understand the mechanism of injury.

The plasma membrane is a key site of freeze-thaw injury and cold acclimation. Using related potato species and onion bulb tissue, we have shown that plasma membrane H^+-ATPase is a sensitive and early site of cellular response to freeze-thaw stress. Perturbation of the plasma membrane functions can be mediated by stress-induced changes in the membrane lipids and changes in the membrane/cytosolic calcium. We have suggested a sequence of events that can lead to perturbation of the functions of plasma membrane H^+-ATPase following freeze-thaw stress (see Palta, 1989) and may explain injury (enhanced ion leakage) as well as recovery from injury.

By merging physiological and genetic approaches, we have gained an important insight into the mode of inheritance of two major components of freezing stress resistance, namely non-acclimated freezing tolerance and acclimation capacity, and into the specific plasma membrane lipids associated with these components. Our results show independent genetic control of non-acclimated freezing tolerance and acclimation capacity. By analysing plasma membrane lipid fatty acid composition in segregating backcross progenies, we have found that linoleic acid (18:2) has an important role in determining freezing tolerance and acclimation capacity. Our studies demonstrate that by merging genetic and physiological approaches, we can gain important knowledge which would not be possible with either approach alone.

ACKNOWLEDGMENTS

This research was supported by the USDA National Research Initiative Grant (Agreement No. 91-37100-6636) and the College of Agriculture and Life Sciences, University of Wisconsin-Madison.

REFERENCES

Abdallah AY, Palta JP (1989) Specific changes in membrane polar lipid fatty acid composition coincide with initiation of fruit ripening and changes in freezing stress resistance of leaves of cranberry. Plant Physiology 89: 100

Amirshashi MC, Patterson FL (1956) Cold resistance of parent varieties, F_2 populations and F_3 lines of 20 oat crosses. Agronomy Journal 48: 184-188

Arora R, Palta JP (1986) Protoplasmic swelling as a symptom of freezing injury in onion bulb cells. Its simulation in extracellular KCl and prevention by calcium. Plant Physiology 82: 625-629

Arora R, Palta JP (1988) *In vivo* perturbation of membrane-associated calcium by freeze-thaw stress in onion bulb cells. Simulation of this perturbation in extracellular KCl and its alleviation by calcium. Plant Physiology 87: 622-628

Arora R, Palta JP (1989) Perturbation of membrane calcium as a molecular mechanism of freezing injury. In: Cherry JH (ed) Environmental stress in plants: Biochemical and biophysical mechanisms. NATO ASI Series Vol. 19. Springer-Verlag, New York, USA, pp 281-290

Arora R, Palta JP (1991) A loss in plasma membrane ATPase activity and its recovery coincides with incipient freeze-thaw injury and post-thaw recovery in onion bulb scale tissue. Plant Physiology 95: 845-852

Ashworth EN, Davis GA, Anderson JA (1985) Factors affecting ice nucleation in plant tissues. Plant Physiology 79: 1033-1037

Blum A (1988) Plant breeding for stress environments. CRC Press, Boca Raton, USA, pp 79-127

Brule-Babel AL, Fowler DB (1988) Genetic control of cold hardiness and vernalization requirement in winter wheat. Crop Science 28: 879-884

Brule-Babel AL, Fowler DB (1989) Use of controlled environments for winter cereal cold hardiness evaluation: controlled freeze tests and tissue water content as prediction tests. Canadian Journal of Plant Science 69: 355-366

Cook HW (1985) Fatty acid desaturation and chain elongation. In: Vance DE, Vance JE (eds) Biochemistry of lipids and membranes. Benjamin Cummings Publishing, California, USA, pp 181-212

Grenier G, Tremolieres A, Therrien HP, Willemot C (1972) Changements dans les lipides de la luzerne en conditions menant a l'endurcissement au froid. Canadian Journal of Botany 50: 1681-1689

Gullord M (1975) Genetics of freezing hardiness in winter wheat (*Triticum aestivum* L.) PhD Thesis, Michigan State University, East Lansing, USA

Gullord M, Olien CR, Everson EH (1975) Evaluation of freezing hardiness in winter wheat. Crop Science 15: 153-157

Gurr MI, Harwood JL (1991) Lipid biochemistry: An introduction. Chapman and Hall, London, UK, pp 338-344

Harbage JF, Weiss LS, Stone JM, Bamberg JB, Palta JP (1992) Increased plasma membrane linoleic acid is correlated with acclimation capacity in a *Solanum* population segregating for freezing tolerance and acclimation capacity. Plant Physiology 99: 10

Hellergren J, Widell S, Lundborg T (1985) ATPase in relation to freezing in purified plasma membrane vesicles from nonacclimated seedings of *Pinus sylvestris*. Acta Horticulturae 168: 161-166

Hepler PK, Wayne RO (1985) Calcium and plant development. Annual Review of Plant Physiology 36: 397-439

Iswari S, Palta JP (1989a) Plasma membrane ATPase as a site of functional alteration during cold acclimation and freezing injury. In: Li PH (ed) Low temperature stress physiology in crops. CRC Press, Boca Raton, USA, pp 123-137

Iswari S, Palta JP (1989b) Plasma membrane ATPase activity following reversible and irreversible freezing injury. Plant Physiology 90: 1088-1095

Iswari S, Weiss LS, Bertics PJ, Palta JP (1990) Plasma membrane bound protein kinase in cold acclimating and nonacclimating potato species. Plant Physiology 93: 84

Jian CL, Sun LH, Dong HZ, Sun DL (1982) Changes in ATPase activity during freezing injury and cold hardening. In: Li PH, Sakai A (eds) Plant cold hardiness and freezing stress: Mechanisms and crop implications. Vol. 2. Academic Press, New York, USA, pp 243-259

Leshem YY (1992) Plant membranes: A biophysical approach to structure, development and senescence. Kluwer Academic Publishers, Dordrecht, The Netherlands

Leonard RT, Hepler PK (eds) (1990) Calcium in plant growth and development. American Society of Plant Physiologists, Rockville, USA

Levitt J (1980) Responses of plants to environmental stresses. Vol. 1: Chilling, freezing, and high temperature stresses. Academic Press, New York, USA

Liesenfeld DR, Auld DL, Murray GA, Swensen JB (1986) Transmittance of winter hardiness in segregating populations of peas. Crop Science 26: 49-54

Limin AE, Fowler DB (1989) The influence of cell size and chromosome dosage on cold hardiness expression in the Triticeae. Genome 32: 667-671

Lindow SE, Arny DC, Upper CD (1982) Bacterial ice nucleation: a factor in frost injury to plants. Plant Physiology 79: 1084-1089

Lynch DV, Steponkus PL (1987) Plasma membrane lipid alterations associated with cold acclimation of winter rye seedlings (*Secale cereale* L. cv Puma). Plant Physiology 83: 761-767

Marshall HG (1982) Breeding for tolerance to heat and cold. In: Christiansen MN, Lewis CF (eds) Breeding plants for less favorable environments. John Wiley and Sons, New York, USA, pp 47-69

Palmgren MG, Sommarin M, Ulvskov P, Jorgensen PL (1988) Modulation of plasma membrane H^+-ATPase from oat roots by lysophosphatidylcholine, free fatty acids and phospholipase A_2. Physiologia Plantarum 74: 11-19

Palta, JP (1989) Plasma membrane ATPase as a key site of perturbation in response to freeze-thaw stress. In: Randall D, Blevins D, Campbell W (eds) Current topics in plant biochemistry and physiology Vol. 8. University of Missouri, Columbia, USA, pp 841-868

Palta JP, Levitt J, Stadelmann EJ (1977a) Freezing injury in onion bulb cells. I. Evaluation of the conductivity method for an analysis of ion and sugar efflux from injured cells. Plant Physiology 60: 393-397

Palta JP, Levitt J, Stadelmann EJ (1977b) Freezing injury in onion bulb cells. II. Post thawing injury or recovery. Plant Physiology 60: 398-401

Palta JP, Levitt J, Stadelmann EJ (1977c) Freezing tolerance of onion bulbs and significance of freeze-induced tissue infiltration. Cryobiology 14: 614-619

Palta JP, Levitt J, Stadelmann EJ, Burke MJ (1977d) Dehydration of onion cells: A comparison of freezing vs. desiccation and living vs. dead cells. Physiologia Plantarum 41: 273-279

Palta JP, Li PH (1978) Cell membrane properties in relation to freezing injury. In: Li PH, Sakai A (eds) Plant cold hardiness and freezing stress: Mechanisms and crop implications. Academic Press, New York, USA, pp 93-115

Palta JP, Li PH (1980) Alterations in membrane transport properties by freezing injury in herbaceous plants. Evidence against the rupture theory. Physiologia Plantarum 50: 169-175

Palta JP, Meade LS (1987) Fatty acid composition changes after cold acclimation: comparison of leaf, callus and purified membranes. Plant Physiology 83: 71

Palta JP, Meade LS (1989) During cold acclimation of potato species, an increase in 18:2 and decrease in 16:0 in plasma membrane phospholipid coincides with an increase in freezing stress resistance. Plant Physiology 89: 89

Palta JP, Weiss LS (1993) Ice formation and freezing injury: An overview on the survival mechanisms and molecular aspects of injury and cold acclimation in herbaceous plants. In: Li PH, Christersson L (eds) Advances in plant cold hardiness. CRC Press, Boca Raton, USA, pp 143-176

Pearce RS (1988) Extracellular ice and cell shape in frost-stressed cereal leaves: A low temperature scanning electron microscopy study. Planta 175: 313-324

Pfahler PL (1966) Small grain improvement by breeding and selection. Report of the Florida Agriculture Experiment Station, pp 52

Pomeroy MK, Pihakaski JS, Andrews CJ (1983) Membrane properties of isolated winter wheat cells in relation to icing stress. Plant Physiology 72: 535-539

Poovaiah BW, Reddy ASN (1987) Calcium messenger systems in plants. CRC Critical Reviews in Plant Sciences 6: 47-102

de la Roche AI, Andrews CJ, Pomeroy MK, Weinberger PA, Katers M (1972) Lipid changes in winter wheat seedlings (*Triticum aestivum*) at temperatures inducing cold hardiness. Canadian Journal of Botany 50: 1681-1689

de la Roche AI, Pomeroy MK, Andrews CJ (1975) Changes in fatty acid composition in wheat cultivars of contrasting hardiness. Cryobiology 12: 506-512

Sakai A, Larcher W (1987) Frost survival in plants: Responses and adaptations to freezing stress. Springer-Verlag, New York, USA

Steffen KL, Arora R, Palta JP (1989) Sensitivity of photosynthesis and respiration to a freeze-thaw stress: Role of realistic freeze-thaw protocol. Plant Physiology 89: 1372-1379

Steffen KL, Palta JP (1987) Photosynthesis as a key process in plant response to low temperature: Alteration during low temperature acclimation and impairment during incipient freeze-thaw injury. In: Li PH (ed) Plant cold hardiness. Alan R. Liss, New York, USA, pp 67-99

Stone JM, Palta JM, Bamberg JB, Weiss LS, Harbage JF (1993) Inheritance of freezing resistance in tuber-bearing *Solanum* species: evidence for the independent control of nonacclimated freezing tolerance and cold acclimation capacity. Proceedings of the National Academy of Sciences (In press)

Stushnoff C, Fowler DB, Brule-Babel A (1984) Breeding and selection for resistance to low temperature. In: Voss PB (ed) Plant breeding - A contemporary basis. Pergamon Press, Elmsford, NY, USA, pp 115-136

Stutka J (1981) Genetic studies of frost resistance in wheat. Theoretical and Applied Genetics 59: 145-152

Stutka J (1984) A ten-parental diallele analysis of frost resistance in winter wheat. Zeitschrift fuer Pflanzenzuechtung 93: 147-157

Stutka J, Veisz O (1988) Reversal of dominance in a gene on chromosome 5A controlling frost resistance in wheat. Genome 30: 313-317

Sutinen MK, Palta JP (1989) Seasonal changes in phospholipid fatty acid composition coincide with changes in freezing stress resistance in pine needles. Plant Physiology 89: 28

Sutinen MK, Rybarczyk SJ, Palta JP (1990) Simultaneous changes in the plasma membrane polar lipids' fatty acid composition and the vanadate-sensitive ATPase activity during cold acclimation of pine needles. Plant Physiology 93: 85

Thomashow MF (1990) Molecular genetics of cold acclimation in higher plants. In: Scandalios JG (ed) Advances in genetics, Vol. 28, Genomic responses to environmental stress. Academic Press, New York, USA, pp 99-125

Uemura M, Yoshida S (1984) Involvement of plasma membrane alterations in cold acclimation of winter rye seedlings (*Secale cereale* L. cv Puma). Plant Physiology 75: 818-826

Vigh L, Horvath I, van Hasselt PR, Kuiper PJC (1985) Effect of frost hardening on lipid and fatty acid composition of chloroplast thylakoid membranes in two wheat varieties of contrasting hardiness. Plant Physiology 79: 756-759

Weiss LS, Palta JP (1991a) Ice nucleation in potato leaves and the influence of thaw rates on the level of injury by freeze-thaw stress. 5th International Conference on Biological Ice Nucleation. Madison, Wisconsin, USA

Weiss LS, Palta JP (1991b) Influence of thaw rates on the level of injury by freeze-thaw stress in a cold hardy potato species. HortScience 26: 732

Willemot C (1975) Stimulation of phospholipid biosynthesis during frost hardening of winter wheat. Plant Physiology 55: 356-359

Yoshida S, Uemura M (1984) Protein and lipid compositions of isolated plasma membranes from orchid grass (*Dactylis glomerata* L.) and changes during cold acclimation. Plant Physiology 75: 31-37

MOLECULAR BIOLOGY OF COLD TOLERANCE

Roger S. Pearce[1], M. Alison Dunn[2], Monica A. Hughes[2]
Departments of [1]Biological and Nutritional Science and
[2]Biochemistry and Genetics
University of Newcastle upon Tyne
NE1 7RU
UK

ABSTRACT

We have studied genes expressed in the shoot meristem during acclimation to low temperature in barley, a highly adaptable crop species. The shoot meristem region is important for acclimation of cereals to low temperature, since it is the centre for perception of cold, is critical for frost survival and is central to vernalisation changes. Differential screening of a cDNA library was used to obtain clones for genes up-regulated in cv. Igri plants grown at day/night temperatures of 6/2 °C as compared to plants grown at 20/15 °C (10 hour photoperiod). Clones for twelve cold up-regulated genes were selected for further study. Transcript sizes varied from 500 to 1400 bases. Run-on transcription analysis indicated that up-regulation of several of the genes (including BLT4 and BLT101) was transcriptionally controlled, whereas up-regulation of others was controlled post-transcriptionally (including BLT14 and BLT801). Based on the amino-acid sequence inferred from the base sequence, BLT4 has strong homology with a non-specific phospholipid transfer protein, BLT63 has strong homology with protein synthesis elongation factor 1α, BLT411 has homology with catalase and BLT801 has homology with maize glycine-rich protein. BLT14 and BLT101 have no homology with published sequences of known function. Some of the genes are members of multigene families (BLT4, BLT14, BLT63). Two genes homologous to BLT14 were studied in rye. Both were more strongly expressed in the more frost-hardy cv. Puma than in cv. Rhayader. However, one was more strongly up-regulated in mature leaves, whereas the other was most strongly up-regulated in the root. The barley homologue BLT14 was most strongly expressed in the shoot meristem. This indicates that one possible significance of multigene families may be that they code for organ-specific isoforms. Drought hardens cv. Igri and drought and ABA strongly up-regulate BLT4, but have little or no effect on the mRNA levels of BLT14 and BLT101. Nutrient stress also induces these genes, and again BLT4 was up-regulated most strongly. Thus non-specific lipid transfer proteins, apparently coded for by BLT4 and other members of this multigene family, may have a function when barley is exposed to a range of environmental stresses. Moving cv. Igri plants between 20/15 and 6/2 °C resulted in changes in hardiness and levels of gene expression within one day. Young plants of cv. Igri were also grown in a range of temperature environments: 18/10 °C, constant 12, 9 and 6 °C, and 6/2 °C (all 10 hour

NATO ASI Series, Vol. I 16
Interacting Stresses on Plants in a Changing Climate
Edited by M. B. Jackson, C. R. Black

photoperiod). Vernalisation was similar in all treatments. Hardiness increased progressively with reduced temperature below 12 °C and so too did the level of mRNA of BLT4 and BLT14, indicating some correlation between the level of up-regulation of at least these genes and frost hardiness. In general, the experiments indicated that hardening to frost and the level of gene expression are very sensitive to environmental change.

INTRODUCTION

Adaptation to the prevailing temperature environment is a central factor influencing the distribution and performance of plants. Plant response to low temperature is important because survival, growth rate, and development can all be affected. Temperature changes impinge directly on almost every step in plant function, and there will be indirect effects. Thus, responses to temperature are likely to be complex, and it may be difficult to separate important from unimportant changes.

Molecular analysis of adaptation to cold has several attractions. It is now clear that the expression of some genes varies with growth temperature (see below). Investigations at the molecular level will be crucial in discovering how low temperature is sensed and the responses mediated in plants. Molecular analysis also initiates a route, through genetic engineering, for distinguishing features which are critical to adaptation from ones which are fortuitous and unimportant.

Low temperature adaptation is likely to prove important in relation to climate change. Milder winters may lead to earlier dehardening, laying plants open to damage from later sharp frosts, which will still occur. Plants which vernalise differ in the degree of cold exposure required, and so a milder climate could affect the ability of some to complete their life cycle. There is also an approximate inverse relationship between low-temperature growth and frost hardiness, and thus the balance between the ability to exploit milder winter weather for growth and the ability to survive frosts could be crucial for some species; the optimum balance point could be altered by climate change. Finally, there could be effects of higher CO_2 concentrations on acclimation, particularly as carbohydrate metabolism has a critical role in hardening. Interaction with other stresses will also be important. Sometimes this will be because several stresses can occur together, as with cold and waterlogging (Andrews, this volume), but in addition, other stresses may have similar effects to cold. Thus, freezing of temperate species dehydrates the cells because ice forms extracellularly, drawing water from within the cell (Pearce and Ashworth, 1992). Frost-tolerance may be induced by drought as well as by cold.

Our research has been mainly with barley, concentrating particularly on the shoot meristem. Barley has the attraction of having a smaller genome than other cereals and a wealth of differently adapted types. Barley and many other temperate Gramineae show three major adaptations to low temperature: hardening to frost, vernalisation and growth adjustment. All are expressed in the shoot meristem region: the crown is critical for frost survival (Olien, 1967, 1974; Pearce, 1980; Gusta *et al.*, 1982), the growth and developmental responses are centred there, and it is the site of temperature perception for growth acclimation (Watts, 1972; Peacock, 1975; Thomas and Stoddart, 1984).

DIVERSITY OF COLD UP-REGULATED GENES

We chose to investigate genes which are up-regulated during growth at low temperature. Alternatively, genes expressed over a range of temperatures might be involved in acclimation if the way in which the gene product functions is altered by temperature. It would be difficult to identify these without specific mutations. A cDNA library was made from shoot meristems of barley (*Hordeum vulgare* L.) cv. Igri grown for several weeks in a 6/2 °C day/night environment. The library was screened with labelled cDNA synthesised from mRNA isolated from shoot meristems of Igri grown at 20/15 °C or 6/2 °C (Dunn *et al.*, 1990). This identified a large number of clones for cold up-regulated genes, from which clones for twelve different genes were selected for further analysis. This indicates an extensive response of barley to low temperature at the level of gene expression. Sequences homologous to the cold-induced barley genes are also present in wheat (Fig. 1), and at least some homologous genes are cold up-regulated in oats (BLT4; Dunn *et al.*, 1991) and rye (BLT14; Zhang *et al.*, 1993).

In cold-grown barley, mRNA transcripts for the cold-induced genes range in size from 550 to 2400 bases. The open reading frame of one of the smallest, BLT14 (Zhang *et al.*, 1993), corresponds to a polypeptide with a relative molecular mass of about 8000. *In vitro* mRNA translation experiments indicate that the steady-state level of mRNA coding for a polypeptide of about 77000 M_r increases dramatically in the shoot meristems of cold-grown barley (Hughes and Pearce, 1988); the larger mRNA transcripts could theoretically code for polypeptides of about this size. Similarly, in a variety of other species, cold induces substantial up-regulation of genes coding for small (down to 16000 M_r), medium (from 40000-90000 M_r) and large (up to 180000 M_r) polypeptides (Guy *et al.*, 1985; Meza-Basso *et al.*, 1986; Mohapatra *et al.*, 1987; Kurkela *et al.*, 1988; Gilmour *et al.*, 1988; Durham *et al.*, 1991).

A variety of polypeptides is also made in response to heat shock. However, unlike heat shock, cold does not induce a general down-regulation of the genes previously expressed at more moderate temperatures (Guy, 1990), though in some species some genes coding for proteins involved in photosynthesis are down-regulated (Meza-Basso *et al.*, 1986; Hahn and Walbot, 1989). With a few important exceptions (Anderson *et al.*, 1992), the genes up-regulated in response to cold do not correspond to genes for heat-shock proteins. Usually, the pattern in cold-acclimation (so far based on herbaceous species) is consistent with a response which continues the normal metabolic functions necessary for growth (albeit slow) while adding functions necessary for acclimation.

Southern blot analysis of barley chromosome addition lines in wheat shows the chromosomal location of the cold up-regulated genes; BLT14 is on chromosome 2 (Dunn *et al.*, 1990), BLT4 on chromosome 3 (Fig 1; Dunn *et al.*, 1991) and BLT101 on chromosome arm 4L (Goddard *et al.*, 1992). Traditional genetic analysis has identified groups of genes determining the spring habit in barley (*Sh*-) and wheat (*Vrn*). The barley genes are located on chromosomes 5L (*Sh*3), 7L (*Sh*2) and 7 (*Shs*) (Takahashi and Yasuda, 1970). Several of the wheat genes are located on chromosome 5L (*Vrn*1 on 5A, *Vrn*3 on 5D, *Vrn*4 on 5B), and one, *Vrn*5, on the short arm of chromosome 7B (Worland *et al.*, 1987). The chromosomal locations of BLT14, BLT4 and BLT101 indicate that they are not specifically associated with the major vernalisation genes in barley (though this does not necessarily exclude a role in vernalisation). On the other hand, traditional genetic analysis indicates a complex control of frost hardiness (Nilan, 1964; Worland *et al.*, 1987). The location across several chromosomes of cold up-regulated genes identified by the molecular analysis is consistent with this.

FUNCTION OF THE COLD UP-REGULATED GENES

Several of the cDNA clones have been completely sequenced, and in some cases the amino acid sequence inferred indicates homology with other genes of known function. There is strong homology between BLT4 and plant non-specific lipid transfer proteins (nsLTPs) (Hughes *et al.*, 1992). The function of nsLTPs in plants is unclear: their ability to transfer lipids between membrane systems is a function expressed *in vitro*; signal sequences in nascent plant nsLTPs indicate they would enter the secretory pathway, and other evidence confirms an extra-protoplasmic location for some of them (Chasan, 1991; Sterk *et al.*, 1991). Nevertheless, the ability to transfer lipids indicates that nsLTPs can bind lipids, and this could be relevant to the changes in membrane lipid quantity and composition which occur during acclimation (Sakai and Larcher, 1987; Steponkus, 1990). A role in the control of development is also possible (Sterk *et al.*, 1991).

Only one of our cold up-regulated genes (BLT14) has homology with a gene previously reported (in barley coleoptile) to be up-regulated in response to cold, and its function is unknown (Cattivelli and Bartels, 1990; Zhang *et al.*, 1993). This lack of homology may partly reflect our choice of shoot meristem tissue as the main object of study, since most other groups have investigated genes up-regulated in leaves. Given the complexity of the meristem and its central role in acclimation, this may give particular interest to the genes we have found to be up-regulated. Other genes up-regulated in the shoot meristem of barley grown at low temperature have homology with protein synthesis elongation factor 1α (BLT63), with catalase (BLT411) or with maize glycine-rich protein (BLT801). The capacity for protein synthesis increases during acclimation (Guy, 1990), perhaps to compensate for the effect of low temperature on rate, and an increase in elongation factor 1α could be part of this increased capacity.

MULTIGENE FAMILIES

Multigene families may have a role in adaptation. It is suggested that different environmental conditions may elicit expression of different multigene family members, resulting in the synthesis of different isoforms of the protein coded for (Smith, 1990). Two of the three genes studied in most detail, BLT14 and BLT4 (Dunn *et al.*, 1990, 1991), are members of small multigene families, both on single chromosomes. This was inferred from Southern blots washed at moderate stringency, together with gene copy number reconstruction experiments. The latter show that both BLT14 and BLT4 occur in single copies per haploid genome (Dunn *et al.*, 1990, and unpublished, respectively). Fig. 1 shows restriction fragments from the genome of barley cv. Betzes (B) and from the genome of wheat cv. Chinese Spring (W), revealed by probing with BLT4 (Dunn *et al.*, 1991). In both cases several bands are evident, and in barley the most strongly hybridising band is the smallest.

Two genes, RLT1412 and RLT1421, with considerable sequence homology to BLT14 were obtained by probing a cDNA library from rye cv. Puma with BLT14 (Zhang *et al.*, 1993). Both genes were strongly up-regulated in response to cold in two winter cultivars of rye, Puma (a very hardy cultivar) and Rhayader (a Welsh cultivar used for forage). Rhayader hardened much less than Puma when grown at 6/2 °C as compared with 20/15 °C, as judged by a leakage test for the leaf blades and a regrowth test for the crown. The up-regulation of RLT1412 and RLT1421 in both leaf blades and shoot meristems was higher in Puma than in Rhayader, indicating a possible role in hardiness for these two homologous genes.

The two rye genes were up-regulated in different parts of the rye plant, with a similar pattern being observed in both cultivars. Gene RLT1421 gave the highest steady state mRNA level in mature leaf tissue, whereas RLT1412 gave the highest

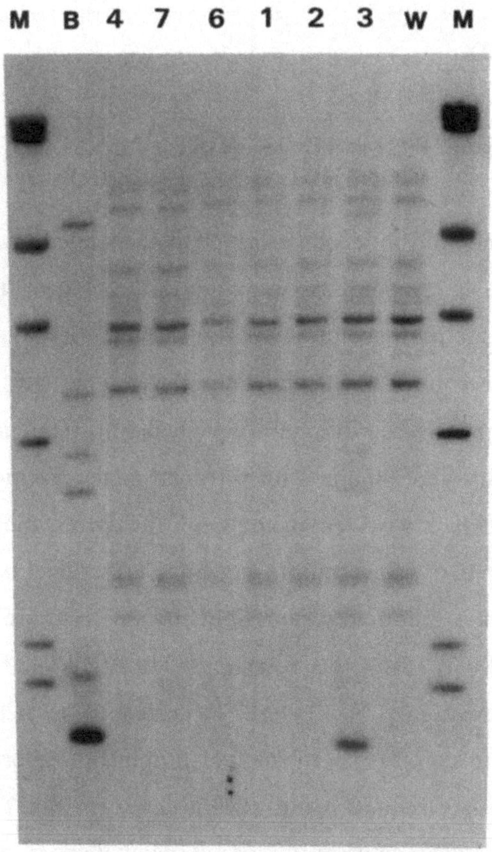

FIGURE 1. Southern blot analysis of BLT4. DNA was digested with DraI. *Barley chromosome addition lines (chromosomes 4, 7, 6, 1, 2 and 3), barley parent (B) and background wheat parent cv. Chinese Spring (W) are indicated. Lambda* Hind*III digest size markers in lanes 1 and 10 (23.1, 9.4, 6.6, 4.4, 2.3, 2.0 kb). Washed at moderate stringency (0.6xSSPE, 0.1% SDS at 65 °C). Only the chromosome 3 addition line includes fragments corresponding to the bands in the parent barley cultivar (B). The wheat parent (W) also includes fragments hybridising with the probe. The smallest barley fragment is 1.6 kb (transcript size is 750 bases) and there are five less prominent bands (2.0, 3.9, 4.4, 5.4, 9.7 kb). Reproduced with permission from Dunn* et al. *(1991).*

level in root tissues and a much lower level in the mature leaf. Although recovered from a shoot meristem library, both genes were least strongly up-regulated in this tissue. In contrast, the parent clone, BLT14, was most strongly up-regulated in the meristem. Thus, we suggest that one possible significance of multigene families in environmental adaptation is that the genes could code for organ-specific isoforms (Zhang *et al.*, 1993).

TRANSCRIPTIONAL OR POST-TRANSCRIPTIONAL CONTROL OF EXPRESSION

Identification of the cold-induced barley genes depended upon increased steady-state mRNA levels in the cold, as compared to the control. Run-on transcription analysis measures the presence of newly transcribed mRNA, and thus can be used to test whether the control of steady-state mRNA level acts through the rate of transcription or is post-transcriptional. Eleven cold up-regulated genes were tested and different results were obtained for different genes (Dunn, unpublished data). The level of BLT4/9 mRNA (a member of the BLT4 multigene family) was controlled by the rate of transcription. In contrast, the level of BLT14 mRNA was controlled post-transcriptionally, possibly by mRNA stability; BLT14 was transcribed at relatively low levels as compared to BLT4/9 in both cold and control environments. Interestingly, BLT14 mRNA increased by at least ten-fold during cold acclimation, to comprise a substantial proportion (about 0.2%) of the total mRNA in the cold-acclimated plants (Dunn *et al.*, 1990). Thus control through mRNA stability can achieve quantitatively large changes in mRNA level. Steady-state mRNA levels of three cold up-regulated genes from *Arabidopsis thaliana* are post-transcriptionally controlled and a fourth is controlled by the rate of transcription (Hajela *et al.*, 1990). Thus there are at least two pathways of cold up-regulation of mRNA levels.

Control of expression of these genes is yet more complex, because at least some are differentially up-regulated in different parts of the plant (see above), indicating that there will also be organ-specific controls. In addition, some are responsive to drought as well as cold (see below). We have no information on the control of the quantities of the proteins coded for by the genes.

RESPONSE TO OTHER STRESSES

Water deficiency induces frost-resistance in many plants, including barley. All the cold up-regulated genes reported for *Arabidopsis thaliana* are also up-regulated by water stress or ABA (Hajela *et al.*, 1990; Kurkela and Franck, 1990; Nordin *et al.*, 1991), and in alfalfa ABA up-regulates some cold up-regulated genes (Mohapatra *et al.*, 1988). The question arises of whether this is the norm. Some alfalfa genes are up-regulated in response to cold but not drought (Mohapatra *et al.*, 1989). Amongst barley genes, BLT4 and BLT4/9 (another member of the BLT4 multigene family), are drought and ABA up-regulated (Dunn *et al.*, 1991), but the expression of others, including BLT14 (Hughes and Dunn, 1990), is less affected. Nutrient stress, which does not induce substantial hardening to frost, substantially up-regulates BLT4 and to some extent BLT14 and BLT101 (Pearce *et al.*, unpublished data). Other stresses have not been investigated extensively. Since BLT4 may code for a lipid transfer protein (Hughes *et al.*, 1992), understanding the normal role of such proteins could clarify the mechanisms underlying adaptation to both cold and drought.

RESPONSE TO DIFFERENT TEMPERATURE ENVIRONMENTS

The rate at which plants respond to changes in the temperature environment may be important for surviving and exploiting the winter period. Dehardening can be faster than initial hardening (Gusta and Fowler, 1976; Chen and Li, 1980). Several studies show that up-regulation of cold-induced genes occurs rapidly, often within hours of transfer to cold, and that down-regulation on return to higher temperatures is also rapid (Hajela *et al.*, 1990; Kurkela and Franck, 1990; Houde *et al.*, 1991; Nordin *et al.*, 1991; Orr *et al.*, 1992). We investigated whether changes in hardiness paralleled changes in the levels of mRNA. BLT4 and BLT14 were both detectably up-regulated during the first day of transfer from 20/15 to 6/2 °C and substantially down-regulated during the first day of transfer back to 20/15 °C (Dunn, unpublished data). Similarly, hardiness increased to a small but significant extent during the first day at 6/2 °C and fell substantially on the first day of return to 20/15 °C (Pearce, unpublished data).

Experiments were also carried out to investigate the effect on the levels of mRNA and frost hardiness of small differences in growth temperature. Barley cv. Igri was grown to the third leaf stage at 18/10 °C (10 hour photoperiod). Some plants were maintained at 18/10 °C, while others were transferred to constant temperatures of 12, 9 or 6 °C, or to a 6/2 °C regime for 7 or 14 days (all with a 10 hour photoperiod). Hardiness was assessed by a regrowth test. All treatments vernalised the plants, whereas the usual control of 20/15 °C does not. The extent of hardening was inversely related to growth temperature (Fig. 2). Comparison of constant 12, 9 or 6 °C treatments shows that a three degree difference in growth conditions conferred differences in hardiness of several degrees. There was a slight

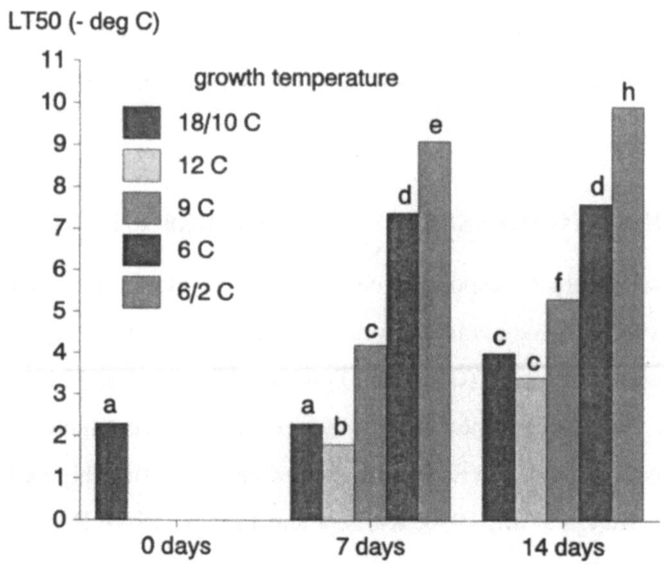

FIGURE 2. Hardiness as indicated by LT_{50} (regrowth test) of barley cv. Igri grown in different temperature environments for 7 or 14 days (all previously in 18/10 °C). Treatments with different letters are significantly different at $p < 0.05$, based on a Chi-squared test of the original survival data ($n = 16$).

but significant fall in hardiness following transfer from 18/10 °C to a constant 12 °C for seven days. Hardiness tended to increase with time in all treatments.

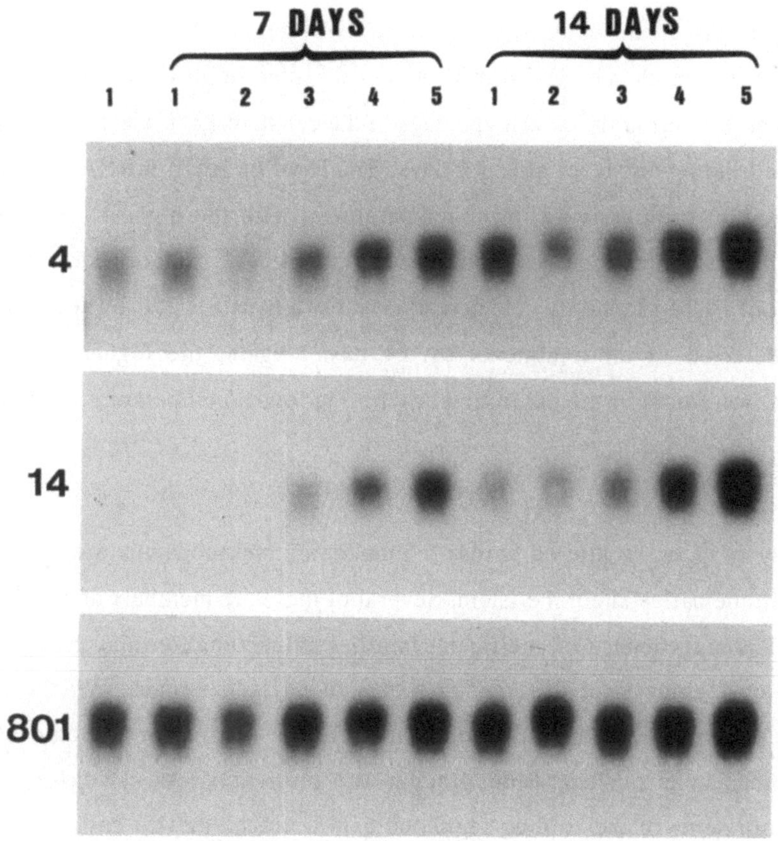

FIGURE 3. *Northern blot analysis of BLT14, BLT4 and BLT801 in the shoot meristem of barley cv. Igri grown for 7 or 14 days in different temperature environments (previously in 18/10 °C). The same filter was used for each blot.*

Lane	*Treatment*	*Lane*	*Treatment*
1	*time 0 (18/10 °C)*	*7*	*18/10 °C, 14 days*
2	*18/10 °C, 7 days*	*8*	*12 °C, 14 days*
3	*12 °C, 7 days*	*9*	*9 °C, 14 days*
4	*9 °C, 7 days*	*10*	*6 °C, 14 days*
5	*6 °C, 7 days*	*11*	*6/2 °C, 14 days*
6	*6/2 °C, 7 days*		

Fig. 3 shows mRNA levels for three genes, BLT14, BLT4 and BLT801. Gene BLT801 is presented for comparison with the others because its expression was constitutive in plants grown in this range of environments. Differences between growth environments in its mRNA level were mainly within the range of experimental error. In contrast, the other two genes showed marked low temperature responses. The BLT4 mRNA was detected in all environments. The mRNA level fell during the first seven days in the constant 12 °C treatment, which also showed the lowest level after 14 days. The level of BLT4 mRNA increased progressively at progressively lower temperatures, with the highest level being found in the 6/2 °C treatment after both seven and 14 days. Unlike BLT4 mRNA, the mRNA of BLT14 was only detected at or below a constant 9 °C by seven days, but was detected in all treatments by 14 days. Again, the highest level of expression was found in the environment with the lowest temperature.

CONCLUSIONS

A diversity of genes scattered across a number of chromosomes are cold up-regulated in the barley shoot meristem. Several of these are members of multigene families. The significance of multigene families in acclimation may be at least partly the up-regulation of different members in different organs. Whether this relates to differing hardening potentials in different organs, or to some other functional aspect is unclear. Both drought and nutrient stress can up-regulate expression of some of the cold up-regulated genes, indicating that environmental interactions exist at the molecular level.

The following conclusions can be drawn from the observed responses to different temperature regimes:

1. The uppermost temperature at which up-regulation is detectable differs between different cold up-regulated genes.
2. At least for BLT4 and BLT14, cold up-regulation is not 'all or nothing', since there was a progressive (but apparently non-linear) increase in mRNA level as growth temperature decreased (starting at the highest inductive temperature) over the temperature range investigated.

3. Although temperature has a major effect on the level of expression and hardiness, both tended to increase with time, indicating they are influenced by factors other than temperature.
4. Over the range of temperatures giving detectable induction, changes in the levels of BLT4 and BLT14 mRNAs parallel changes in the level of hardiness. It remains to be proved whether this reflects a causal link.
5. Differences in temperature of a few degrees may have large effects on mRNA levels and hardiness. Thus the implication for the effects of climatic warming is that levels of hardiness and of up-regulation of genes induced during cold acclimation will be quantitatively altered by changes in growth temperature.

ACKNOWLEDGEMENTS

The research reported here was supported by the AFRC and the University of Newcastle upon Tyne Research Committee.

REFERENCES

Anderson JV, Neven LG, Hofig A, Guy CL (1992) Identification of a spinach BIP which shows upregulation during cold acclimation. Plant Physiology 99: S-146

Cattivelli L, Bartels D (1990) Molecular cloning and characterisation of cold-regulated genes in barley. Plant Physiology 93: 1504-1510

Chasan R (1991) Lipid transfer proteins: Moving molecules. The Plant Cell 3: 842-3

Chen HH, Li PH (1980) Characteristics of cold acclimation and deacclimation in tuber-bearing Solanum species. Plant Physiology 65: 1146-1148

Dunn MA, Hughes MA, Pearce RS, Jack PL (1990) Molecular characterisation of a barley gene induced by cold treatment. Journal of Experimental Botany 41: 1405-1413

Dunn MA, Hughes MA, Zhang L, Pearce RS, Quigley AS, Jack PL (1991) Nucleotide sequence and molecular analysis of the low temperature induced cereal gene, BLT4. Molecular and General Genetics 229: 389-394

Durham RE, Moore GA, Haskell D, Guy CL (1991) Cold acclimation changes in freezing tolerance and translatable RNA content in *Citrus grandis* and *Poncirus trifoliata*. Physiologia Plantarum 82: 519-522

Gilmour SJ, Hajela RK, Thomashow MF (1988) Cold acclimation in *Arabidopsis thaliana*. Plant Physiology 87: 745-750

Goddard NJ, Dunn MA, Hughes MA (1992) Molecular analysis of the barley low temperature induced gene, BLT101. Journal of Experimental Botany 43: 57

Gusta LV, Fowler DB (1976) Dehardening and rehardening of spring-collected winter wheats and a winter rye. Canadian Journal of Plant Science 56: 775-779

Gusta LV, Fowler DB, Tyler NJ (1982) Factors influencing hardiness and survival in winter wheat. In: Li PH, Sakai A (eds) Plant cold hardiness and freezing stress, Vol 2. Academic Press, New York, pp 23-40

Guy CL (1990) Cold acclimation and freezing stress tolerance: Role of protein metabolism. Annual Review of Plant Physiology and Molecular Biology 41: 187-223

Guy CL, Nieme KJ, Bramble R (1985) Altered gene expression during cold acclimation of spinach. Proceedings of the National Academy of Science, USA 82: 3673-3677

Hahn M, Walbot V (1989) Effects of cold-treatment on protein synthesis and mRNA levels in rice leaves. Plant Physiology 91: 930-938

Hajela RK, Horvath DP, Gilmour SJ, Thomashow MF (1990) Molecular cloning and expression of *cor* (cold-regulated) genes in *Arabidopsis thaliana*. Plant Physiology 93: 1246-1252

Houde H, Danyluk J, Lalberte J-F, Dhindsa RS, Sarhan F (1991) Cloning and characterization of four cDNAs specifically induced by low temperature in wheat. Plant Physiology 96: S-161

Hughes MA, Dunn MA (1990) The effect of temperature on plant growth and development. Biotechnology and Genetic Engineering Reviews 8: 161-187

Hughes MA, Dunn MA, Pearce RS, White AJ, Zhang L (1992) An abscisic acid responsive low temperature barley gene has homology with a maize phospholipid transfer protein. Plant, Cell and Environment 15: 861-865

Hughes MA, Pearce RS (1988) Low temperature treatment of barley plants causes altered gene expression in shoot meristems. Journal of Experimental Botany 39: 1461-1467

Kurkela S, Franck M (1990) Cloning and characterisation of a cold- and ABA-inducible *Arabidopsis* gene. Plant Molecular Biology 15: 137-144

Kurkela S, Franck M, Heino P, Lang V, Palva T (1988) Cold induced gene expression in *Arabidopsis thaliana* L. Plant Cell Reports 7: 495-498

Meza-Basso L, Alberdi M, Raynal M, Fererro-Cadinanos M-L, Delseny M (1986) Changes in protein synthesis in rapeseed (*Brassica napus*) seedlings during a low temperature treatment. Plant Physiology 82: 733-738

Mohapatra SS, Poole RJ, Dhindsa RS (1987) Cold acclimation, freezing resistance and protein synthesis in alfalfa (*Medicago sativa* L. cv. Saranac). Journal of Experimental Botany 38: 1697-1703

Mohapatra SS, Poole RJ, Dhindsa RS (1988) Abscisic acid-regulated gene expression in relation to freezing tolerance in alfalfa. Plant Physiology 87: 468-473

Mohapatra SS, Wolfram L, Poole RJ, Dhindsa RS (1989) Molecular cloning and relationship to freezing tolerance of cold-acclimation-specific genes of alfalfa. Plant Physiology 89: 375-380

Nilan RA (1964) The cytological genetics of barley 1951-1962. Monographic Supplement No. 3, Research Studies, Washington State University Press, USA

Nordin K, Heino P, Palva ET (1991) Separate signal pathways regulate the expression of a low-temperature-induced gene in *Arabidopsis thaliana* (L.) Heynh. Plant Molecular Biology 16: 1061-1071

Olien CR (1967) Freezing processes in the crown of 'Hudson' barley (*Hordeum vulgare* L.). Crop Science 4: 91-95

Olien CR (1974) Research Report number 247, Michigan State University Agricultural Experimental Station, East Lansing, USA

Orr W, In B, White TC (1992) Complementary DNA sequence of a low temperature-induced *Brassica napus* gene with homology to the *Arabidopsis thaliana* kin1 gene. Plant Physiology 98: 1532-1534

Peacock JM (1975) Temperature and leaf growth of *Lolium perenne* II. The site of temperature perception. Annals of Applied Ecology 12: 115-123

Pearce RS (1980) Relative hardiness to freezing of laminae, roots and tillers of tall fescue. The New Phytologist 84: 449-463

Pearce RS, Ashworth EN (1992) Cell shape and localisation of ice in leaves of overwintering wheat during frost stress in the field. Planta 188: 324-331

Sakai A, Larcher W (1987) Frost survival of plants. Springer, Berlin, Germany

Smith H (1990) Signal perception, differential expression within multigene families and the molecular basis of phenotypic plasticity. Plant, Cell and Environment 13: 585-594

Steponkus PL, Lynch DV, Uemura M (1990) The influence of cold acclimation on the lipid composition and cryobehaviour of the plasma membrane of isolated rye protoplasts. Philosophical Transactions of the Royal Society London B 326: 571-583

Sterk P, Booij H, Schellekens GA, Van Kammen A, De Vries SC (1991) Cell-specific expression of the carrot EP2 lipid transfer protein gene. The Plant Cell 3: 907-921

Takahashi R, Yasuda S (1970) Genetics of earliness and growth in barley. In: Barley genetics II. Washington State University Press, USA, pp 388-408

Thomas H, Stoddart JL (1984) Kinetics of leaf growth in *Lolium temulentum* at optimal and chilling temperatures. Annals of Botany 53: 341-347

Watts WR (1972) Leaf extension of *Zea mays*. II. Leaf extension in response to independent variation of the temperature of the apical meristem, and of the air around the leaves and the root zone. Journal of Experimental Botany 23: 713-721

Worland AJ, Gale MD, Law CN (1987) Wheat genetics. In: Lupton F G H (ed) Wheat breeding. Chapman and Hall, London, UK, pp 129-171

Zhang L, Dunn MA, Pearce RS, Hughes MA (1993) Analysis of organ specificity of a low temperature responsive gene family in rye (*Secale cereale* L.). Plant Physiology (in press)

REDESIGNING CROPS FOR INCREASED TOLERANCE TO FREEZING STRESS

Peter L. Steponkus, Matsuo Uemura and Murray S. Webb
Department of Soil, Crop and Atmospheric Sciences
Cornell University
Ithaca
New York 14853
USA

ABSTRACT

The process of cold acclimation is a complex developmental phenomenon that invokes the orchestration of many different processes, including hormonal responses to environmental cues, altered gene activity, the synthesis of new gene products, and alterations that influence virtually every facet of cellular metabolism. Although seemingly disparate, many of these metabolic changes ultimately contribute to the increased cryostability of cellular membranes, destabilisation of which is the primary cause of freezing injury. Therefore, to provide a rational integration of the many different facets of the cold acclimation process, it is necessary to first identify the freeze/thaw-induced lesions in cellular membranes and to elucidate the biophysical mechanisms and biochemical alterations responsible for their occurrence. Although all cellular membranes are vulnerable to freeze-induced destabilisation, maintenance of the structural integrity of the plasma membrane is a prerequisite for survival because of the central role that it plays during a freeze-thaw cycle. Cryomicroscopy studies of isolated protoplasts complemented with electron microscopy studies of freeze-induced ultrastructural changes are yielding a comprehensive analysis of the phenomenology of freezing injury and the identification of specific 'lesions' in the plasma membrane, which vary depending on the stage of acclimation and the nadir temperature to which the protoplasts are cooled. Of the lesions identified to date (expansion-induced lysis, lamellar-to-hexagonal II phase transitions, and the fracture-jump lesion), all are a consequence of freeze-induced dehydration; however, whereas expansion-induced lysis is the result of cellular dehydration and the large osmotic excursions incurred during a freeze-thaw cycle, lamellar-to-hexagonal II phase transitions and the fracture-jump lesion are consequences of the removal of water that is closely associated with cellular membranes. These studies, together with a detailed analysis of the plasma membrane lipid composition at the molecular species level and procedures to alter selectively the lipid composition of the plasma membrane have provided a foundation for mechanistic studies to establish directly the relationships between alterations in the plasma membrane lipid composition and its increased cryostability after cold acclimation. In order to determine if genotypic diversity in freezing tolerance is associated with genotypic differences in membrane lipid

NATO ASI Series, Vol. I 16
Interacting Stresses on Plants in a Changing Climate
Edited by M. B. Jackson, C. R. Black

composition, we are extending these studies to other cereals (e.g., oat and barley), which are less freezing tolerant than winter rye. These studies reveal that the freeze-induced membrane lesions in a spring oat cultivar (Ogle) are qualitatively similar to those observed in rye, but occur at substantially higher temperatures. Moreover, the lipid composition of the plasma membrane isolated from leaves of the spring oat is vastly different from that of winter rye leaves. These studies suggest that strategies for improving the freezing tolerance of cereals should include techniques to modify membrane lipid composition.

INTRODUCTION

Environmental constraints are the primary determinants of the geographical distribution and productivity of native and cultivated plant species. Even in areas considered suitable for the cultivation of a given species, unpredictable and aberrant fluctuations in temperature result in decreased yield or crop failure. For instance, more than 70% of the shortfall in the average annual yield below the yield potential of the major field-crops grown in the USA is attributable to temperature and water stress (Boyer, 1982). In contrast, only 10% of the shortfall is attributable to the collective influence of diseases, insects and weeds. High-value horticultural crops are especially vulnerable to freezing temperatures, with substantial economic losses occurring annually. As a result, over $100 million are spent annually in attempts to minimise damage to frost-sensitive crops. The majority of these expenditures are for energy-costly practices to modify the ambient microclimate. In spite of these practices, annual losses resulting from freezing injury are estimated to be $1.5 billion in the United States and $14 billion worldwide. In January 1991, a single night of subzero temperatures in Southern California resulted in a $730 million loss of agricultural crops.

Only modest increases in freezing tolerance would have a dramatic impact on agricultural productivity and profitability. The development of genotypes with increased freezing tolerance would greatly diminish the use of energy-costly practices to modify the microclimate and would provide a more reliable means to minimise crop losses resulting from freezing stresses. As a result, the improvement of freezing tolerance of agronomic, horticultural and arboreal crops is a major concern. Nevertheless, in spite of considerable effort by plant breeders to increase the 'cold hardiness' of agricultural crop species, progress has been minimal. In part, this is attributable to a lack of a fundamental understanding of plant responses to freezing stress and attributes responsible for adaptation (long-term phytogenetic evolution) or acclimation (metastable changes elicited within a single life cycle in response to the prevailing environmental conditions) to freezing temperatures. Without such information, the selection of appropriate genetic stocks and the development of inheritance information to use them effectively is severely impeded.

GENETIC ENGINEERING APPROACHES

Currently, there is much interest in using techniques of genetic engineering and molecular biology for improving the freezing tolerance of crop species. Several different approaches are under investigation including transformation of extremely frost-sensitive species with genes for bacterial ice-nucleating proteins and fish antifreeze proteins to alter the freezing process, transformation of less freezing-tolerant cultivars with genes from more freezing-tolerant cultivars, construction of genetic maps of freezing tolerance using RFLP (Restriction Fragment Length Polymorphism) mapping, and studies of alterations in gene expression during cold acclimation of freezing-tolerant species.

Some investigators have suggested that transformation of freezing-sensitive species with the gene for the antifreeze protein from winter flounder to inhibit ice formation will lead to increased freezing tolerance (Davies, 1987; Cutler *et al.*, 1989). Although there has been some success in effecting expression of the flounder antifreeze gene in plants (Georges *et al.*, 1990; Hightower *et al.*, 1991), significant increases in freezing tolerance have yet to be reported. Alternatively, it has been suggested that an increase in the freezing tolerance of freezing-sensitive species may be achieved by incorporating the gene for ice-nucleating proteins from bacteria. The rationale for this approach is based on the assumption that the extreme freezing sensitivity of species such as potato (*Solanum tuberosum*) is a consequence of intracellular ice formation, which presumably occurs because the cytosol becomes excessively supercooled before extracellular ice is nucleated. Therefore it is inferred that, if extracellular ice nucleation occurs at a higher temperature, the cells will dehydrate and the extent of supercooling and the probability of intracellular ice formation will be minimised (T.H.H. Chen, personal communication). In the studies to date, the bacterial ice-nucleating protein has been expressed in plants (Baertlein *et al.*, 1992), but the influence on freezing tolerance has not been published.

Whether either of the two approaches will succeed remains to be demonstrated. The studies do, however, underline the need for a greater understanding of the mechanism of freezing injury in extremely sensitive species. For example, if freezing-sensitive species succumb to freezing because of extreme sensitivity to freeze-induced dehydration rather than intracellular ice formation, transformation with the bacterial gene for ice-nucleating proteins will do little to improve their freezing tolerance. Therefore, studies to determine if freezing injury in extremely freezing-sensitive species is a consequence of either intracellular ice formation or extreme sensitivity to freeze-induced dehydration will have a large influence on the choice of genetic engineering approaches to be taken.

It is often suggested that isolation of a gene associated with cold acclimation of freezing-tolerant species or cultivars may be used to transform less freezing-tolerant species. However, such suggestions overlook the fact that freezing tolerance is a quantitative trait that is controlled by several genes. For instance, in reviewing studies on the inheritance of freezing tolerance in wheat, Thomashow (1990) noted that 11 of the 21 chromosomes of hexaploid wheat are reported to be associated with freezing tolerance. Furthermore, cold acclimation undoubtedly involves both the expression of new gene products and more subtle modifications in metabolism that result in an increase in freezing tolerance. Given the multigenic nature of freezing tolerance, construction of genetic maps of freezing tolerance using RFLP mapping should be useful in breeding programmes for improvement of freezing tolerance.

MOLECULAR BIOLOGY

Because of the potential to improve freezing tolerance by genetic engineering approaches there is considerable interest in molecular biology studies of the cold acclimation process. Since the report of Guy *et al.* (1985) that new gene products (mRNA and polypeptides) are synthesised during cold acclimation, there have been many studies of the molecular biology of cold acclimation in a diverse array of species (Mohapatra *et al.*, 1987a,b, 1988; Guy, 1990; Lin *et al.*, 1990; Thomashow, 1990; Hajela *et al.*, 1990; Gilmour and Thomashow, 1991; Gilmour *et al.*, 1992). The appearance of 'cold-regulated' (COR) polypeptides closely parallels the development of freezing tolerance (Guy and Haskell, 1987; Mohaptra *et al.*, 1987a,b) and the level of expression of the *cor* genes is positively correlated with the freezing tolerance of several alfalfa cultivars (Mohaptra *et al.*, 1989). Nevertheless, the identity and function of the COR polypeptides remain to be elucidated.

There is a prevailing attitude that the COR proteins have a direct cryo-protective function. For instance, Kurkela and Franck (1990) have suggested that KIN1, which is a cold-regulated polypeptide in *Arabidopsis thaliana*, may function as an antifreeze protein. This is consistent with recent reports of the occurrence in plants of proteins with antifreeze activity (Urrutia *et al.*, 1992; Griffith *et al.*, 1992). However, the COR6.6 polypeptide, which was also isolated from *A. thaliana* and is almost identical to KIN1 (Gilmour and Thomashow, 1992), does not have antifreeze activity (M.F. Thomashow, personal communication). Alternatively, others (Thomashow, 1990; Houde *et al.*, 1992) suggest that COR polypeptides may act as cryoprotectants to minimise the impact of freeze-induced dehydration. Lin and Thomashow (1992) have reported that COR15, which was also isolated from *A. thaliana*, exhibits cryoprotection for the cold-labile enzyme lactate de-hydrogenase. They have suggested that the COR polypeptides may be homologues of the cryoprotective polypeptides, which were originally described by Volger and Heber (1975) and more recently studied by Hincha (Hincha *et al.*, 1989, 1990).

Although it is possible that the COR polypeptides serve a direct cryoprotective function, it must be acknowledged that cold acclimation is a complex developmental phenomenon that involves the orchestration of many different processes and alterations in metabolic regulation that influence virtually every facet of metabolism. Therefore, there is the possibility that some COR polypeptides function to alter metabolic pathways, the end products of which result in cryoprotection. For example, there is substantial evidence that the accumulation of cytosolic sugars and alterations in membrane lipid composition are important facets of the cold acclimation process. Alternatively, it is possible that metabolic regulation of carbohydrate and lipid metabolism occurs by more subtle regulatory mechanisms that do not require the intervention of COR polypeptides.

FREEZE-INDUCED LESIONS IN THE PLASMA MEMBRANE

Clearly, redesigning crop species for increased tolerance to freezing stress by either traditional breeding or genetic engineering approaches requires a better understanding of the cold acclimation process. One approach is to consider that the many, seemingly disparate, metabolic changes that occur during cold acclimation ultimately contribute to the increased cryostability of cellular membranes, destabilisation of which is the primary cause of freezing injury. Therefore, to provide a rational integration of the many different facets of the cold acclimation process, it is first necessary to identify the freeze-induced lesions in cellular membranes and to elucidate the biophysical mechanisms and biochemical alterations responsible for their occurrence. Although all cellular membranes are vulnerable to freeze-induced destabilisation, maintenance of the structural integrity of the plasma membrane is a prerequisite for survival because of the central role that it plays during a freeze/thaw cycle (see Steponkus, 1984).

Our approach has been to determine the cryobehaviour of the plasma membrane *in situ* by cryomicroscopic studies of protoplasts isolated from winter rye leaves (*Secale cereale* L. cv. Puma). These studies, complemented with electron microscopy studies of ultrastructural changes in the plasma membrane, are yielding a comprehensive analysis of the phenomenology of freezing injury and the identification of specific 'lesions' in the plasma membrane, which vary depending on the stage of cold acclimation and the nadir temperature to which the protoplasts are cooled. They also provide a focus for elucidating the biophysical and biochemical events that are responsible for the various lesions. These studies, together with a molecular species analysis of the plasma membrane lipid composition (Lynch and Steponkus, 1987) and procedures to modify selectively the lipid composition of the plasma membrane (Arvinte and Steponkus, 1988), have provided a foundation for mechanistic studies to establish the relationships between alterations in the plasma membrane lipid composition and its increased cryostability during cold acclimation (Steponkus *et al.*, 1988). Comprehensive reviews of these studies have been presented by Steponkus and Lynch (1989a), Steponkus *et al.* (1990) and Steponkus and Webb (1992); for this discussion, only the highlights will be presented.

Under conditions that minimise the probability of intracellular ice formation (cooling rates ≤1 °C min^{-1}), freezing injury in protoplasts is a consequence of freeze-induced osmotic stresses and cell dehydration, with destabilisation of the plasma membrane a primary manifestation of injury. However, the mechanism of injury and the associated changes in the ultrastructure of the plasma membrane that result from freeze-induced dehydration depend on the stage of cold acclimation and the sub-freezing temperature, which determines the magnitude of the osmotic stress and the extent of cell dehydration. In some instances, injury is the result of cellular dehydration and the large osmotic excursions incurred during a freeze/thaw cycle; in other instances, membrane destabilisation is a consequence of the removal of water that is closely associated with cellular membranes. Most important, the cryostability of the plasma membrane increases during cold acclimation – in large part, because of changes in membrane lipid composition.

In protoplasts that are isolated from non-acclimated rye leaves (NA protoplasts) and cooled to temperatures of -2 to -5 °C, destabilisation of the plasma membrane is a consequence of osmotic excursions that occur during a freeze/thaw cycle. Although the protoplasts exhibit normal osmometric behaviour during freezing and thawing of the suspending medium, lysis occurs during osmotic expansion before the initial size is regained. This form of injury, referred to as *expansion-induced lysis*, occurs because the plasma membrane undergoes endocytotic vesiculation during osmotic contraction (Dowgert and Steponkus, 1984; Gordon-Kamm and Steponkus, 1984a) and large area reductions are irreversible. Because intrinsic elastic expansion of the plasma membrane is limited to 2 to 3% (Wolfe and Steponkus, 1983), osmotic expansion requires that additional material be reincorporated into the plasma membrane during osmotic expansion. However, the rate of incorporation is not sufficiently rapid to preclude the attainment of lytic tensions (Wolfe *et al.*, 1985, 1986).

Expansion-induced lysis does not occur in protoplasts isolated from cold-acclimated rye leaves (ACC protoplasts) (Dowgert and Steponkus, 1984). In ACC protoplasts, osmotic contraction results in the formation of exocytotic extrusions of the plasma membrane (Gordon-Kamm and Steponkus, 1984b). Because the surface area of the plasma membrane is conserved during osmotic contraction, lytic tensions do not occur during subsequent osmotic expansion (Dowgert *et al.*, 1987). That the differential behaviour of NA and ACC protoplasts during osmotic contraction (formation of endocytotic vesicles *vs.* exocytotic extrusions) is a consequence of alterations in the lipid composition that occur during cold acclimation has been directly demonstrated: (1) liposomes prepared from the plasma membrane lipid extracts of non-acclimated and cold acclimated rye leaves also exhibit the differential behaviour (Steponkus and Lynch, 1989b); and (2) fusion of NA protoplasts with liposomes composed of phospholipids isolated from the plasma membrane of cold-acclimated leaves results in a transformation in the cryobehaviour of the plasma membrane such that exocytotic extrusions are formed during osmotic contraction and >90% of the protoplasts survive over the range of -2 to -5 °C (Steponkus *et al.*, 1988; Uemura and Steponkus, 1989). This transformation in the behaviour of the plasma membrane and the associated increase in freezing tolerance can also be effected by artificial enrichment of the plasma membrane with mono- or di-unsaturated species of phosphatidylcholine.

A second form of injury occurs in NA protoplasts at temperatures below -5 °C. The protoplasts respond osmotically during freezing of the suspending medium and attain a volume predicted from the Boyle-van't Hoff relationship; however, they are osmotically unresponsive following thawing of the suspending medium. This form of injury, referred to as *loss of osmotic responsiveness*, is associated with several changes in membrane ultrastructure, including the formation of large aparticulate domains in the plasma membrane, the appearance of aparticulate lamellae sub-tending the plasma membrane, and lamellar-to-hexagonal II phase transitions in the plasma membrane and subtending lamellae (Gordon-Kamm and Steponkus, 1984c). The H_{II} phase is also observed in regions of the chloroplast envelope and endoplasmic reticulum (Sugawara and Steponkus, 1990). These changes in membrane ultrastructure are a consequence of freeze-induced dehydration rather than exposure to subzero temperatures *per se* and can be induced by osmotic dehydration. Their occurrence is attributed to bilayer-bilayer interactions that occur when bilayers are brought into close approach by osmotic pressures sufficient to overcome the hydration forces associated with the hydrophilic surfaces of bilayers (see Steponkus and Webb, 1992).

Neither large, aparticulate domains nor the H_{II} phase are observed in ACC protoplasts frozen to temperatures as low as -40 °C (Gordon-Kamm and Steponkus, 1984c), with the decreased propensity for H_{II} phase formation occurring within the first 7-10 days of the cold acclimation period (Sugawara and Steponkus, 1990). The decreased propensity for H_{II} phase formation is a consequence of alterations in membrane lipid composition that occur during this period (see Steponkus *et al.*, 1990): (1) liposomes of plasma membrane lipids from non-acclimated leaves undergo dehydration-induced lamellar-to-H_{II} phase transitions, whereas liposomes prepared from plasma membrane lipids of cold-acclimated leaves do not (Cudd and Steponkus, 1988); and (2) H_{II} phase formation can be minimised by enrichment of the plasma membrane with dilinoleoylphosphatidylcholine (Sugawara and Steponkus, 1990). The decreased propensity for H_{II} phase formation is, in part, ascribed to an alteration in the hydration characteristics of the plasma membrane because of increases in highly hydrated phospholipid species (e.g., phosphatidylcholine) and decreases in less hydrated species (cerebrosides) during cold acclimation (Steponkus and Lynch, 1989a). In addition, alterations in lipid composition influence the propensity for H_{II} phase formation at a given water content (Steponkus and Webb, 1992).

Expansion-induced lysis and loss of osmotic responsiveness that is associated with lamellar-to-hexagonal II phase transitions are the two forms of injury that limit the freezing tolerance of NA protoplasts. As such, they represent the first stress barriers that must be overcome during the process of cold acclimation. For both, this occurs during the first 7-10 days of the cold acclimation period, and the freezing tolerance (LT_{50}) of the protoplasts increases from -5 to -12 °C; however, an additional three weeks of acclimation are required to attain the maximum freezing tolerance (LT_{50} = -27 °C). With protoplasts that are isolated from cold-acclimated leaves (ACC protoplasts) and frozen to lethal temperatures (-12 °C for one week ACC protoplasts, -27 °C for four week ACC protoplasts), injury is manifested as a loss of osmotic responsiveness.

Freeze-fracture electron microscopy (FFEM) studies reveal that the predominant change in the ultrastructure of the plasma membrane that occurs at the lethal temperature is a deviation of the fracture plane in localised regions of the plasma membrane, i.e., the fracture plane 'jumps' from the plasma membrane to closely appressed aparticulate lamellae or to aparticulate regions of various endomembranes (predominantly chloroplast envelopes and the endoplasmic reticulum, and at a lower frequency, the tonoplast) that are in close apposition with the plasma membrane (Fujikawa and Steponkus, 1990). In protoplasmic fracture faces (PF), the fracture jump results in a discontinuity in the plasma membrane with the subtending lamellae visible in this region. In exoplasmic fracture faces (EF) of the plasma membrane, aparticulate lamellae are apparent on the inner surface of the plasma membrane in the region of the fracture jump. In both PF and EF fracture faces, the region of the plasma membrane that is involved in the fracture jump ranges from 0.1 to 2.5 μm in diameter. We use the phrase *fracture-jump lesion* to refer to the occurrence of a localised deviation in the fracture plane in aparticulate regions of membranes that are in close apposition and to distinguish it from a deviation in the fracture plane that results from a simple cross fracture.

There is a close correspondence between the occurrence of the fracture-jump lesion and injury following freezing to lethal temperatures in protoplasts isolated from leaves of rye seedlings at various stages of cold acclimation. The fracture-jump lesion occurs in one week ACC protoplasts frozen to -10 °C and increases in frequency over the range of -10 to -20 °C; in four week ACC protoplasts, it is first observed in protoplasts frozen to -20 °C and increases in frequency over the range of -20 to -40 °C; its occurrence in two and three week ACC protoplasts is intermediate between these two extremes. For all, the temperature range over which the frequency of the fracture-jump lesion increases coincides with the temperature range over which survival decreases from >80 to <20% (see Uemura and Steponkus, 1989).

We are currently investigating what freeze-induced alterations in the plasma membrane and subtending lamellae are responsible for the fracture-jump lesion. Two possibilities are under investigation: (1) the formation of interlamellar attachments and localised fusion of the plasma membrane and subtending lamellae; and (2) interdigitation of lipids that have undergone an $L_\alpha \rightarrow L_\beta$ phase transition in domains of the plasma membrane. The occurrence of either would be strongly influenced by alterations in membrane lipid composition. It is also important to note that the occurrence of freeze-induced lamellar-to-hexagonal II phase transitions in NA protoplasts and the fracture-jump lesion in ACC protoplasts are not unique to isolated protoplasts; these lesions are also observed in non-acclimated and cold acclimated rye leaves, respectively (Webb and Steponkus, 1992a).

Having established the involvement of plasma membrane lipid alterations in the cold acclimation of winter rye, we have recently initiated studies to determine if genotypic differences in freezing tolerance are associated with differences in membrane lipid composition. For this, we have chosen first to contrast rye, which is the most freezing-tolerant of the winter cereals, with oat (*Avena sativa*), which is the least freezing-tolerant of the winter cereals. For these studies, both winter

(Kanota) and spring (Ogle) cultivars of oat were used. A detailed analysis of the freezing tolerance at various stages of cold acclimation (Uemura and Steponkus, 1992a) revealed that injury in non-acclimated seedlings of both oat cultivars occurred after freezing to -2 °C, which is significantly higher than the temperature (-5 °C) at which injury occurs in leaves of non-acclimated rye seedlings. In both the spring and winter cultivars of oat, the freezing tolerance increases progressively during the four week period of cold acclimation; however, both the rate and extent to which the freezing tolerance increase are less in the spring cultivar. With the winter cultivar, survival increases to >75% over the range of -2 to -6 °C after the first week of cold acclimation, with progressive increases in survival over the range of -6 to -10 °C during the next three weeks of cold acclimation. In contrast, the freezing tolerance of the spring cultivar increases only over the range of -2 to -4 °C after the first week of cold acclimation, with further increases in survival during the next three weeks limited to the range of -4 to -8 °C.

To determine what forms of injury limited the freezing tolerance of the spring cultivar, the freezing tolerance of protoplasts isolated from leaves of non-acclimated and four week acclimated seedlings was determined. For both, the freezing tolerance of the protoplasts is similar to that of the leaves from which they were isolated. With NA protoplasts, expansion-induced lysis is the predominant form of injury at temperatures between -2 and -4 °C and can be precluded by either thawing in a hypertonic medium to minimise osmotic expansion or artificially enriching the plasma membrane with dilinoleoylphosphatidylcholine prior to the freeze/thaw cycle. The incidence of loss of osmotic responsiveness (LOR) increases from ~20% at -4 °C to >80% at -6 °C. In ACC protoplasts, LOR is the predominant form of injury and occurs in 50% of the ACC protoplasts frozen to -8 °C. These results demonstrate that the spring cultivar of oat undergoes the initial stage of cold acclimation to preclude expansion-induced lysis, but further increases in freezing tolerance are limited because of dehydration-induced LOR at temperatures of -6 °C or below – even after four weeks of cold acclimation.

To characterise the ultrastructural manifestations of freezing injury that accompany LOR, protoplasts isolated from leaves of non-acclimated and cold-acclimated spring oat seedlings were freeze-dehydrated at temperatures between -2 and -10 °C and then examined by freeze-fracture electron microscopy (Webb and Steponkus, 1992b). The inverted hexagonal (H_{II}) phase first occurred in regions of the plasma membrane and chloroplast envelopes of NA protoplasts at -3 °C. The frequency of occurrence of the H_{II} phase increased at -4 °C and was observed in >90% of the protoplasts at -10 °C. In contrast, the H_{II} phase is not observed in ACC protoplasts at either -3 or -10 °C. Instead, injury in ACC protoplasts at -10 °C was associated with the frequent occurrence of the fracture-jump lesion in the plasma membrane and subtending membranes, aparticulate lamellae in close apposition to the plasma membrane and, in some cases, extensive aggregation of plasma membrane IMP. These observations demonstrate that the ultrastructural manifestations of freezing injury in oat are qualitatively very similar to those described previously for the more freezing-tolerant rye, but occur at substantially higher temperatures.

A lipid molecular species analysis of the plasma membrane revealed that there are vast differences in the plasma membrane lipid composition of spring oat (Ogle) and winter rye (Puma) leaves (Uemura and Steponkus, 1992b). The plasma membrane of non-acclimated leaves of oat consists of 8 mol% free sterols, 6 mol% sterylglucosides, 27 mol% acylated sterylglucosides, 27 mol% glucocerebrosides, and 29 mol% phospholipids. In contrast, the plasma membrane of rye leaves contains 38 mol% free sterols, 6 mol% sterylglucosides, 3 mol% acylated sterylglucosides, 16 mol% glucocerebrosides, and 37 mol% phospholipids. Furthermore, there are significant differences in (1) the predominant sterol moieties in the free sterols, sterylglucosides and acylated sterylglucosides, (2) the predominant cerebroside molecular species, and (3) the proportions of di-unsaturated species of phosphatidylcholine and phosphatidylethanolamine, which are lower in oat than in rye.

During cold acclimation, the proportion of phospholipids increases significantly in both oat and rye as a result of increases in the proportions of phosphatidylcholine and phosphatidylethanolamine. The relative proportions of di-unsaturated species in both phosphatidylcholine and phosphatidylethanolamine increase during cold acclimation, but the increase is higher in rye than in oat. As with the rye plasma membrane, the major increase in the proportion of phospholipids in oat occurs during the first week of cold acclimation – when expansion-induced lysis of the protoplasts isolated from the leaves was precluded. In rye, the proportion of glucocerebrosides decreases significantly during cold acclimation; in oat, there is only a small decrease in glucocerebrosides such that after cold acclimation the mol% of glucocerebrosides in the plasma membrane is 2.5-fold greater in oat than in rye. Also, during cold acclimation of rye, the proportion of free sterols increases and the proportions of sterylglucosides and acylated sterylglucosides decrease; in oat, there is no change in either the proportion of free sterols or sterylglucosides. Although the mol% of acylated sterylglucosides decreases slightly during cold acclimation of spring oat, it is 20-fold greater than that which is present in the plasma membrane of winter rye. Studies are in progress to determine which of these differences in lipid composition are associated with the increased sensitivity of spring oat to freeze-induced dehydration.

CONCLUSIONS

Collectively, these studies demonstrate that modifications in membrane lipid composition are causally related to the increase in freezing tolerance of rye and oat and suggest that differences in membrane lipid composition are, in part, responsible for the genotypic differences in freezing tolerance between winter rye and spring oat. Therefore, strategies to modify membrane lipid composition warrant consideration in attempts to redesign crops for increased freezing tolerance. However, it should be emphasised that freezing tolerance is a multifaceted trait and lipid alterations are but one facet. There are good reasons to be equally concerned about the role of endogenous cryoprotectants, such as sugars, that influence the cryostability of cellular membranes. Similarly, the influence of cryoprotective proteins, though less well-documented, also needs to be elucidated. Comprehensive studies will be required for a complete understanding of the cold acclimation process and elucidation of the basis for genotypic differences in freezing tolerance, which will contribute to the development of rational strategies for redesigning crop species for improved freezing tolerance.

ACKNOWLEDGEMENTS

Portions of these studies were supported by grants from the USDA Competitive Grants Program (Grant No. 88-37264-3988) and the United States Department of Energy (Grant DE-FG03-84ER13214).

REFERENCES

Arvinte T, Steponkus PL (1988) Characterization of the pH-induced fusion of liposomes with the plasma membrane of rye protoplasts. Biochemistry 27: 5671-5677

Baertlein DA, Lindow SE, Panopoulos NJ, Lee SP, Mindrinos MN, Chen THH (1992) Expression of a bacterial ice nucleation gene in plants. Plant Physiology 1730-1736

Boyer JS (1982) Plant productivity and environment. Science 218: 443-448

Cudd' A, Steponkus PL (1988) Lamellar-to-hexagonal$_{II}$ phase transitions in liposomes of rye plasma membrane lipids after osmotic dehydration. Biochimica et Biophysica Acta 941: 278-286

Cutler AJ, Saleem M, Kendall E, Gusta LV, Georges F, Fletcher GL (1989) Winter flounder antifreeze protein improves the cold hardiness of plant tissues. Journal of Plant Physiology 135: 351-354

Davies PL (1987) Antifreeze proteins: prospects for transferring freeze resistance. New Biotechnology 1: 11-16

Dowgert MF, Steponkus PL (1984) Behavior of the plasma membrane of isolated protoplasts during a freeze-thaw cycle. Plant Physiology 75: 1139-1151

Dowgert MF, Wolfe J, Steponkus PL (1987) The mechanics of injury to isolated protoplasts following osmotic contraction and expansion. Plant Physiology 83: 1001-1007

Fujikawa S, Steponkus PL (1990) Freeze-induced alterations in the ultrastructure of the plasma membrane of rye protoplasts isolated from cold-acclimated leaves. Cryobiology 27: 665-666

Georges F, Saleem M, Cutler AJ (1990) Design and cloning of a synthetic gene for the flounder antifreeze protein and its expression in plant cells. Gene 91: 159-165

Gilmour SJ, Artus NN, Thomashow MF (1992) cDNA sequence analysis and expression of two cold-regulated genes of *Arabidopsis thaliana.* Plant Molecular Biology 18: 13-21

Gilmour SJ, Thomashow MF (1991) Cold acclimation and cold-regulated gene expression in ABA mutants of *Arabidopsis thaliana*. Plant Molecular Biology 17: 1233-1240

Gilmour SJ, Thomashow MF (1992) Characterization of COR 6.6, a polypeptide encoded by a cold-regulated gene from *Arabidopsis thaliana*. Plant Physiology 99: S-11

Gordon-Kamm WJ, Steponkus PL (1984a) The behavior of the plasma membrane following osmotic contraction of isolated protoplasts: implications in freezing injury. Protoplasma 123: 83-94

Gordon-Kamm WJ, Steponkus PL (1984b) The influence of cold acclimation on the behavior of the plasma membrane following osmotic contraction of isolated protoplasts. Protoplasma 123: 161-173

Gordon-Kamm WJ, Steponkus PL (1984c) Lamellar-to-hexagonal$_{II}$ phase transitions in the plasma membrane of isolated protoplasts after freeze-induced dehydration. Proceedings of the National Academy of Sciences USA 81: 6373-6377

Griffith M, Ala P, Yang DSC, Hon WC, Moffat BA (1992) Antifreeze protein produced endogenously by winter rye leaves. Plant Physiology 100: 593-596

Guy CL (1990) Cold acclimation and freezing stress tolerance: Role of protein metabolism. Annual Review of Plant Physiology and Plant Molecular Biology 41: 187-223

Guy CL, Haskell DL (1987) Induction of freezing tolerance in spinach is associated with the synthesis of cold acclimation induced proteins. Plant Physiology 84: 872-878

Guy CL, Niemi KJ, Brambl R (1985) Altered gene expression during cold acclimation of spinach. Proceedings of the National Academy of Sciences USA 82: 3673-3677

Hajela RK, Horvath DP, Gilmour SJ, Thomashow MF (1990) Molecular cloning and expression of *cor* (cold-regulated) genes in *Arabidopsis thaliana*. Plant Physiology 93: 1246-1252

Hightower R, Baden C, Penzes, Lund P, Dunsmuir P (1991) Expression of antifreeze proteins in transgenic plants. Plant Molecular Biology 17: 1013-1021

Hincha DK, Heber U, Schmitt JM (1989) Freezing ruptures thylakoid membranes in leaves, and rupture can be prevented *in vitro* by cryoprotective proteins. Plant Physiology and Biochemistry 27: 795-801

Hincha DK, Heber U, Schmitt JM (1990) Proteins from frost-hardy leaves protect thylakoids against mechanical freeze-thaw damage *in vitro*. Planta 180: 416-419

Houde M, Danyluk J, Laliberté JF, Rassart E, Dhindsa RS, Sarhan F (1992) Cloning, characterization, and expression of a cDNA encoding a 50-kilodalton protein specifically induced by cold acclimation in wheat. Plant Physiology 99: 1381-1387

Kurkela S, Franck M (1990) Cloning and characterization of a cold- and ABA-inducible *Arabidopsis* gene. Plant Molecular Biology 15: 137-144

Lin C, Guo WW, Everson E, Thomashow MF (1990) Cold acclimation in *Arabidopsis* and wheat. A response associated with expression of related genes encoding 'boiling-stable' polypeptides. Plant Physiology 94: 1078-1083

Lin C, Thomashow MF (1992) A cold-regulated *Arabidopsis* gene encodes a polypeptide having potent cryoprotective activity. Biochemical and Biophysical Research Communications 183: 1103-1108

Lynch DV, Steponkus PL (1987) Plasma membrane lipid alterations associated with cold acclimation of winter rye seedlings (*Secale cereale* L. cv Puma). Plant Physiology 83: 761-767

Mohapatra SS, Poole RJ, Dhindsa RS (1987a) Changes in protein patterns and translatable messenger RNA populations during cold acclimation of alfalfa. Plant Physiology 84: 1172-1176

Mohapatra SS, Poole RJ, Dhindsa RS (1987b) Cold acclimation, freezing resistance and protein synthesis in alfalfa (*Medicago sativa* L. cv. Saranac). Journal of Experimental Botany 38: 1697-1703

Mohapatra SS, Poole RJ, Dhindsa RS (1988) Detection of two membrane polypeptides induced by abscisic acid and cold acclimation - possible role in freezing tolerance. Plant and Cell Physiology 29: 727-730

Mohapatra SS, Wolfraim L, Poole RJ, Dhindsa RS (1989) Molecular cloning and relationship to freezing tolerance of cold-acclimation-specific genes of alfalfa. Plant Physiology 89: 375-380

Steponkus PL (1984) Role of the plasma membrane in freezing injury and cold acclimation. Annual Review of Plant Physiology 35: 543-584

Steponkus PL, Lynch DV (1989a) Freeze/thaw-induced destabilization of the plasma membrane and the effects of cold acclimation. Journal of Bioenergetics and Biomembranes 21: 21-41

Steponkus PL, Lynch DV (1989b) The behavior of large unilamellar vesicles of rye plasma membrane lipids during freeze/thaw-induced osmotic excursions. Cryo-Letters 10: 43-50

Steponkus PL, Lynch DV, Uemura M (1990) The influence of cold acclimation on the lipid composition and cryobehaviour of the plasma membrane of isolated rye protoplasts. Philosophical Transactions of the Royal Society, London B 326: 571-583

Steponkus PL, Uemura M, Balsamo RA, Arvinte T, Lynch DV (1988) Transformation of the cryobehavior of rye protoplasts by modification of the plasma membrane lipid composition. Proceedings of the National Academy of Sciences, USA 85: 9026-9030

Steponkus PL, Webb MS (1992) Freeze-induced dehydration and membrane destabilization in plants. In: Somero GN, Osmond CB, Bolis CL (eds) Water and life: Comparative analysis of water relationships at the organismic, cellular and molecular level, Springer-Verlag, Berlin, pp 338-362

Sugawara Y, Steponkus PL (1990) Effect of cold acclimation and modification of the plasma membrane lipid composition on lamellar-to-hexagonal$_{II}$ phase transitions in rye protoplasts. Cryobiology 27: 667

Thomashow MF (1990) Molecular genetics of cold acclimation in higher plants. Advances in Genetics 28: 99-131

Uemura M, Steponkus PL (1989) Effect of cold acclimation on the incidence of two forms of freezing injury in protoplasts isolated from rye leaves. Plant Physiology 91: 1131-1137

Uemura M, Steponkus PL (1992a) Cold acclimation and freezing injury of oat leaves. Cryobiology 29: 710

Uemura M, Steponkus PL (1992b) A contrast of the plasma membrane lipid composition of spring oat and winter rye leaves. Cryobiology 29: 711

Urrutia ME, Duman JG, Knight CA (1992) Plant thermal hysteresis proteins. Biochimica et Biophysica Acta Protein Structure and Molecular Enzymology 1121: 199-206

Volger HG, Heber U (1975) Cryoprotective leaf proteins. Biochimica et Biophysica Acta 412: 335-349

Webb MS, Steponkus PL (1992a) Freeze-induced membrane ultrastructural alterations in leaves of non-acclimated rye seedlings. Cryobiology 29: 709-710

Webb MS, Steponkus PL (1992b) Ultrastructural manifestations of freezing injury in protoplasts isolated from cold-acclimated and non-acclimated oat leaves. Plant Physiology 99: S-127

Wolfe J, Dowgert MF, Steponkus PL (1985) Dynamics of membrane exchange of the plasma membrane and the lysis of isolated protoplasts during rapid expansions in area. Journal of Membrane Biology 86: 127-138

Wolfe J, Dowgert MF, Steponkus PL (1986) Mechanical study of the deformation and rupture of the plasma membranes of protoplasts during osmotic expansion. Journal of Membrane Biology 93: 63-74

Wolfe J, Steponkus PL (1983) Mechanical properties of the plasma membrane of isolated protoplasts - mechanism of hyperosmotic and extracellular freezing injury. Plant Physiology 71: 276-285

THE MOLECULAR MECHANISM OF THE LOW-TEMPERATURE TOLERANCE OF PLANTS STUDIED BY GENE TECHNOLOGY OF MEMBRANE LIPIDS

N. Murata, H. Wada, Z. Gombos and I. Nishida
Department of Regulation Biology
National Institute for Basic Biology
Myodaiji
Okazaki 444
Japan

ABSTRACT

It has been stated that the fatty-acid unsaturation of membrane lipids is related to the tolerance (or sensitivity) of plants to high- and low-temperature stresses. This claim has not been supported directly in previous studies because the level of unsaturation has always been altered indirectly by changing the growth temperature. The change in growth temperature may be expected to alter not only the lipids, but also a number of other cellular components and enzymes. Therefore, to study the relationship between temperature tolerance and the lipid, it is necessary to establish an experimental system in which *only* the fatty-acid unsaturation of membrane lipids is altered, while the growth temperature remains constant. Such a system is now available through gene technology that modifies membrane lipids using genes for the desaturation of fatty acids. Recently, we have isolated genes termed *desA* and *desB*) for fatty-acid desaturation at the Δ^{12} and Δ^{15} positions from the cyanobacterium *Synechocystis* PCC6803. By mutagenesis and transformation with desaturase genes, we obtained strains of this cyanobacterium that had discretely different levels of fatty-acid unsaturation of their membrane lipids. The characteristic features of temperature tolerance of these strains suggest that the fatty-acid unsaturation of membrane lipids is related to the tolerance against low-temperature stress, but not to that against high-temperature stress. A second line of research has been the gene-technological modification of fatty-acid unsaturation of phosphatidylglycerol, one of the membrane lipids of higher-plant chloroplasts. In these studies, we cloned genes for acyl-ACP:glycerol-3-phosphate acyltransferase from *Arabidopsis thaliana* and *Cucurbita moschata* (squash) and introduced them into *Nicotiana tabacum* (tobacco). Leaves of the resultant tobacco revealed distinctly different levels of unsaturation of the phosphatidylglycerol. By comparing the tolerance of the transformants against low-temperature stress, we have obtained evidence that low-temperature tolerance is enhanced by unsaturation of the membrane lipid.

NATO ASI Series, Vol. I 16
Interacting Stresses on Plants in a Changing Climate
Edited by M. B. Jackson, C. R. Black

INTRODUCTION

It has long been argued that the tolerance (or sensitivity) of plants to high- and low-temperature stresses is related to the fatty-acid unsaturation of membrane lipids (Raison *et al.*, 1982; Murata and Nishida 1989). This statement has not been supported experimentally because in previous studies the level of unsaturation has been altered by changing growth temperature. For example, when plants are grown at two different temperatures, the level of unsaturation of membrane lipids is greater in 'low-temperature plants' than in 'high-temperature plants' and heat stability is enhanced in the low-temperature plants. However, cold tolerance is also enhanced in the low-temperature plants when compared with high-temperature plants. It has been suggested that the increases in heat stability and low-temperature sensitivity in high-temperature plants are related to an increased saturation of membrane lipids (Raison *et al.*, 1982). However, no concrete evidence has been provided for the contribution of saturated fatty acids to heat stability and low-temperature sensitivity. Since the change in growth temperature is accompanied by changes not only in the unsaturation of fatty acids but also in a number of other metabolic components and factors, it is possible that such metabolic changes, and not the unsaturation of fatty acids in membrane lipids, allow plants to be tolerant of low- and high-temperature stresses. To verify whether the unsaturation of fatty acids contributes to temperature tolerance, it is first necessary to establish a system in which only the unsaturation of fatty acids can be altered without any effect on other metabolic factors.

Recently, we have isolated a desaturase gene, *desA*, for desaturation at the Δ^{12} position of fatty acids of membrane lipids from the cyanobacterium *Synechocystis* PCC6803 (Wada *et al.*, 1990). It is now possible to obtain strains of *Synechocystis* PCC6803 that contain membrane lipids with defined levels of unsaturation of fatty acids generated by mutation and transformation with *desA* (Wada *et al.*, 1992). Our experimental results using this system suggest that the unsaturation of fatty acids contributes to protection against low-temperature damage, but not to high-temperature inactivation of photosynthesis.

We have found that the chilling sensitivity of higher plants is well correlated with the level of cis-unsaturated fatty acids (18:1+18:2+18:3) in phosphatidylglycerol from chloroplast membranes (Murata *et al.*, 1982; Murata, 1983; Roughan, 1985). The level of cis-unsaturated fatty acids in phosphatidylglycerol is highly dependent on the substrate selectivity of acyl-ACP: glycerol-3-phosphate acyltransferase (hereafter glycerol-P acyltransferase) which transfers the acyl group from acyl-ACP to the C-1 position of glycerol 3-phosphate (Frentzen *et al.*, 1983, 1987). Therefore, it was predicted that gene engineering for this enzyme could convert plants from chilling-sensitive to chilling-resistant and *vice versa.*

TEMPERATURE ACCLIMATION IN *Synechocystis*

The cyanobacterium *Synechocystis* PCC6803 contains mono-galactosyl diacylglycerol, digalactosyl diacylglycerol, sulpho-quinovosyl diacylglycerol, and phosphatidylglycerol as major components of membrane lipids. Major components of fatty acids are 16:0, 18:1(9), 18:2(9,12), 18:3(6,9,12) or 18:3γ, 18:3 (9,12,15) or 18:3α, and 18:4(6,9,12,15) (Wada and Murata, 1989).

When cells of *Synechocystis* grown at 22 °C were compared with cells grown at 34 °C, the lipid class composition was not significantly changed. On the other hand, the molecular species composition, which can be specified by the composition of fatty acids at the *sn*-1 and *sn*-2 positions of the glycerol moiety, was markedly affected by growth temperature (Table 1). When wild-type cells grown at 22 °C were compared with wild-type cells grown at 34 °C, the proportions of 16:0/16:0, 18:1/16:0 and 18:2/16:0 were lowered whereas 18:3α/16:0 and 18:4/16:0 newly appeared. This suggests that the lipids were more desaturated at low temperature.

The heat stabilities of these cells were compared. In this experiment, cells grown either at 22 or 34 °C were exposed to various high temperatures for 20 minutes, and then photosynthetic activity was assayed at 30 °C. Temperatures for 50% inactivation were 46 and 49 °C in cells grown at 22 and 34 °C. This result clearly demonstrates that cells grown at 34 °C are more resistant to heat than cells grown at 22 °C. We further compared the two kinds of cell with respect to their low-temperature stability (Fig. 1). In this experiment, cells grown at 22 or 34 °C were exposed to various low temperatures in the light for 60 minutes, and then photosynthetic activity was assayed at 30 °C. Although there was no marked difference between the two kinds of cell in the inactivation at 10 and 20 °C, the cells grown at 34 °C suffered much more damage than those grown at 22 °C when exposed to 0 °C. This result demonstrates that cells grown at 22 °C are more tolerant of low temperature than cells grown at 34 °C.

TABLE 1. *Effect of growth temperature and mutation on major molecular species of membrane lipids in* Synechocystis *PCC6803.*

Strain	Growth temp. (°C)	Molecular species (% of total molecular species)					
		16:0 / 16:0 / X	18:1 / 16:0 / X	18:2 / 16:0 / X	18:3γ / 16:0 / X	18:3α / 16:0 / X	18:4 / 16:0 / X
WT	22	2	4	16	42	16	16
WT	34	16	16	24	34	0	0
Fad6	34	14	22	50	0	0	0
Fad6/*desA*::Km^r	34	12	80	0	0	0	0

This series of experiments demonstrates that when the growth temperature is increased the lipids become more saturated and the heat stability is increased. The next step was to establish whether the pronounced heat stability of cells grown at 34 °C was caused by the unsaturation of membrane lipids and whether the low-temperature stability of cells grown at 22 °C also resulted from the unsaturation of membrane lipids. To answer these questions it is necessary to establish a system in which the unsaturation of membrane lipids could be modified while the growth temperature was kept constant.

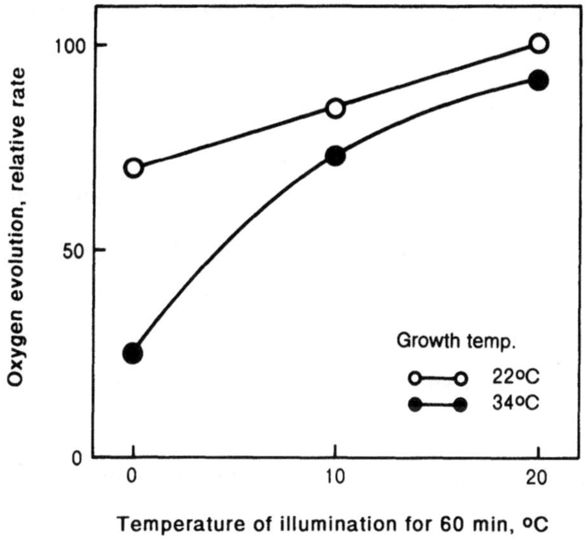

FIGURE 1. The effect of growth temperature on the low-temperature sensitivity of Synechocystis *PCC6803. Cells grown at 22 and 34 °C were exposed to designated temperatures for 60 minutes in the light. After prompt warming, photosynthetic activity was assayed by oxygen evolution at 30 °C.*

GENE-TECHNOLOGICAL MODIFICATION OF FATTY-ACID UNSATURATION IN *SYNECHOCYSTIS*

Synechocystis PCC6803 belongs to Group 4 of the cyanobacteria in accordance with the mode of fatty-acid desaturation of membrane lipids (Murata *et al.*, 1992a). In this group, four steps of fatty-acid desaturation occur at Δ^6, Δ^9, Δ^{12} and Δ^{15} positions of C18 fatty acids esterified at the sn-1 position. We have isolated mutant strains, Fad6 and Fad12, which are defective in the desaturation at Δ^6 and Δ^{12} positions, respectively (Wada and Murata, 1989). We have subsequently cloned desaturase genes for the Δ^{12} (Wada *et al.*, 1990) and Δ^{15} positions (unpublished), which are termed *desA* and *desB*, respectively. By transformation of wild type and Fad6 with the *desA* gene which had been disrupted with a kanamycin-resistant gene cartridge, we could disrupt the *desA* gene, producing an artificial mutant for the Δ^{12} desaturation.

Fatty acids of total glycerolipids were analysed for the wild type, Fad6 and the transformant of *Synechocystis* PCC6803, each grown at 34 °C. The most abundant fatty acids in the wild type were 16:0, 18:1, 18:2 and 18:3γ. The major fatty acids in Fad6 were 16:0, 18:1 and 18:2, indicative of a defect in desaturation of fatty acids at the Δ^6 position (Wada and Murata, 1989). The transformant, Fad6/*desA*::Km^r, was defective in desaturation at the Δ^6 and Δ^{12} positions (Wada *et al.*, 1992). Consequently, this strain did not contain polyunsaturated fatty acids and possessed only 16:0 and 18:1(9) as the most abundant fatty acids. Table 1 shows the major molecular species of membrane lipids in the wild type, Fad6 and Fad6/*desA*::Km^r. Using these three strains, we studied the effects of unsaturation of fatty acids on low-temperature and heat stabilities with minimum interference by other metabolic factors.

Heat stability was studied in all three strains. In this experiment, cells of wild type, Fad6 and Fad6/*desA*::Km^r grown at 34 °C were exposed to various temperatures in the range from 40 to 55 °C for 20 minutes in darkness, and photosynthetic activity at 34 °C was measured by oxygen evolution after immediate cooling. The temperature for 50% inactivation was 48 °C in all strains. A similar type of experiment was carried out with supplementation of 1,4-benzoquinone to investigate the heat stability of the oxygen-evolving activity. The temperature for 50% inactivation was 49 °C in all strains. These findings suggest that the elimination of unsaturation, step by step, with mutation of the Δ^6 desaturase gene and disruption of the Δ^{12} desaturase gene has no significant effect on the heat stability of photosynthesis and oxygen evolution.

The low-temperature stabilities of the three strains were compared. In this experiment, cells of wild type, Fad6 and Fad6/*desA*::Km^r grown at 34 °C were exposed to 10, 20 and 30 °C for 60 minutes in the light, and photosynthetic activity at 34 °C was measured after immediate warming. The experimental results indicate that low-temperature treatment in the light produced partial damage to the cells of wild type and Fad6. In Fad6/*desA*::Km^r, on the other hand, the activity declined to 15% upon treatment at 20 °C and totally disappeared upon treatment at 10 °C. In contrast, when the three strains were exposed to low temperature in darkness, no damage to photosynthesis was observed. These observations suggest that Fad6/*desA*::Km^r cells are much more susceptible to low temperature in the light than wild type and Fad6 cells. Although the photoinhibition of all strains of *Synechocystis* PCC6803 was more pronounced at low temperature than at growth temperatures, the photoinhibition at low temperatures is accelerated further by elimination of polyunsaturated fatty acids from membrane lipids. Therefore, we conclude that the fatty-acid unsaturation enables the cyanobacterium to tolerate low temperatures but not high temperatures.

GENE-TECHNOLOGICAL MODIFICATION OF FATTY-ACID UNSATURATION IN TOBACCO

We have cloned cDNA for glycerol-P acyltransferase from a chilling-sensitive plant, *Cucurbita moschata* (squash), and a chilling-resistant plant, *Arabidopsis thaliana*. These cDNAs were inserted into a vector plasmid, pBI121, and introduced into tobacco. The plasmids constructed with cDNAs from squash and *Arabidopsis* were termed pSQ and pARA, respectively. The introduced genes were incorporated into the genome of tobacco in the transformed plants.

Analysis of lipids from leaves of transgenic tobacco plants revealed no discernible change in overall fatty-acid composition between transformed and non-transformed plants. Analysis of individual lipid classes demonstrated that only phosphatidylglycerol was significantly altered in its fatty-acid composition by transformation with cDNAs for glycerol-P acyltransferases (Table 2). When the squash cDNA was introduced with pSQ, the level of *cis*-unsaturated fatty acids in phosphatidylglycerol was significantly lowered. In contrast, introduction of the *Arabidopsis* cDNA with pARA caused a small but significant increase in these fatty acids.

A chilling sensitivity experiment was performed by exposing a leaf disk of transformants to 1 °C for four hours under illumination, returning it to 27 °C in darkness and monitoring photosynthetic activity (Murata *et al.*, 1992b). The transformant with pBI-121 and wild-type tobacco showed some chilling sensitivity. Upon introduction of the squash cDNA for glycerol-P acyltransferase this sensitivity was markedly increased. However, introduction of the *Arabidopsis* cDNA for glycerol-P acyltransferase increased chilling tolerance.

***TABLE 2.** The proportion of* cis-*unsaturated molecules of phosphatidylglycerol from tobacco plants transformed with cDNA for glycerol-P acyltransferase from* Arabidopsis *and squash.*

Plant		*cis*-unsaturated	Saturated
		(% of total PG)	
Squash		36	64
Arabidopsis		80	20
Tobacco	Wild type (Control)	64	36
	pBI-121	64	36
	pARA	72	28
	pSQ	38	62

Exposure of whole plants to 1 °C for ten days clearly induced an altered leaf appearance, including chlorosis and total deterioration, among transformants with different constructs (Murata *et al.*, 1992b). Leaf damage was more obvious when the plants were kept at 25 °C for two days after the low-temperature treatment. Leaves of wild type and transformants with pBI-121 suffered partial chlorosis following the low-temperature treatment. Transformants with pSQ were severely damaged and some leaves were completely dead. In contrast, transformants with pARA were much more resistant to low temperatures than the wild type and transformants with pBI-121. These differences in low-temperature sensitivity were well correlated with the extent of fatty-acid unsaturation of phosphatidylglycerol.

CONCLUSIONS

The present study demonstrates that the fatty-acid unsaturation of membrane lipids can be modified by genetic manipulation of fatty-acid desaturases and glycerol-P acyltransferases in cyanobacteria and higher plants. In accordance with the increase in fatty-acid unsaturation of membrane lipids, low-temperature stability can be enhanced. Heat stability is not correlated with fatty-acid unsaturation of membrane lipids.

REFERENCES

Frentzen M, Heinz E, McKeon TA, Stumpf PK (1983) Specificities and selectivities of glycerol-3-phosphate acyltransferase from pea and spinach chloroplasts. European Journal of Biochemistry 125: 629-636

Frentzen M, Nishida I, Murata N (1987) Properties of the plastidial acyl-(acyl-carrier protein): glycerol-3-phosphate acyltransferase from the chilling-sensitive plant squash (*Cucurbita moschata*). Plant and Cell Physiology 28: 1195-1201

Murata N (1983) Molecular species composition of phosphatidylglycerols from chilling-sensitive and chilling-resistant plants. Plant and Cell Physiology 24: 81-86

Murata N, Nishida I (1989) Lipid in relation to chilling sensitivity of plants. Wang CY (ed) Chilling injury of horticultural crops. CRC Press, Boca Raton, USA, pp 181-199

Murata N, Ishizaki-Nishizawa O, Tasaka Y, Hayashi H, Higashi S, Nishida I (1992b) Genetically engineered alteration in chilling sensitivity of plants. Nature (London) 356: 710-713

Murata N, Sato N, Takahashi N, Hamazaki Y (1982) Compositions and positional distributions of fatty acids in phospholipids from leaves of chilling-sensitive and chilling-resistant plants. Plant and Cell Physiology 23: 1071-1079

Murata N, Wada H, Gombos Z (1992a) Modes of fatty acid desaturation in cyanobacteria. Plant and Cell Physiology 33: 933-941

Raison JK, Roberts JK, Berry JA (1982) Correlations between the thermal stability of chloroplast (thylakoid) membranes and the composition and fluidity of their polar lipids upon acclimation of the higher plant, *Nerium oleander*, to growth temperature. Biochimica et Biophysica Acta 688: 218-228

Roughan PG (1985) Phosphatidylglycerol and chilling sensitivity in plants. Plant Physiology 77: 740-746

Wada H, Murata N (1989) *Synechocystis* PCC6803 mutants defective in desaturation of fatty acids. Plant and Cell Physiology 30: 971-978

Wada H, Gombos Z, Murata N (1990) Enhancement of chilling tolerance of a cyanobacterium by genetic manipulation of fatty acid desaturation. Nature (London) 347: 200-203

Wada H, Gombos Z, Sakamoto T, Murata N (1992) Genetic manipulation of the extent of desaturation of fatty acids in membrane lipids in the cyanobacterium *Synechocystis* PCC6803. Plant and Cell Physiology 33: 535-540

ASPECTS OF THE CELLULAR AND MOLECULAR BASIS OF COLD TOLERANCE IN PLANTS

Alain M. Boudet, Mireille Cabané, Nathalie Leborgne and Chantal Teulières
Centre de Biologie et Physiologie Végétales
URA CNRS 1457
Université Paul Sabatier
118 route de Narbonne
31062 Toulouse Cedex
France

ABSTRACT

The diversity of the reactions of plants to low temperatures is illustrated by the classical distinction between chilling-sensitive and freezing-sensitive species. In an attempt to improve our understanding of the cellular and molecular basis underlying this diversity, we have investigated the fluidity of specific membranes in relation to cold resistance in *Eucalyptus gunnii*, a freezing-sensitive tree, and the changes in protein synthesis induced by cold acclimation in soybean (*Glycine max*), a chilling-sensitive species. We have been able to obtain isolated vacuoles and to purify, by free-flow electrophoresis, specific membranes (tonoplast and plasmalemma) from two cell lines of *Eucalyptus gunnii* which exhibit differential frost tolerance. The lateral and rotational mobilities of lipids in these different systems were studied by two biophysical techniques: fluorescence recovery after photobleaching (FRAP) and fluorescence polarisation. Our results show, for the first time, that intrinsically the tonoplast exhibits a higher fluidity than the plasma membrane. In addition, the membranes from the frost tolerant line or from acclimated lines were always more fluid than those from the sensitive line. These data strongly suggest a correlation between membrane fluidity and freezing tolerance in *Eucalyptus*. Through a progressive exposure to low temperatures, soybean plants were acclimated to a temperature of 8 °C. As is the case with freezing-sensitive plants, acclimation improved the cold tolerance of chilling-sensitive plants. In order to assess changes in protein synthesis related to cold acclimation, proteins were labelled *in vivo* with ^{35}S methionine, separated by two dimensional gel electrophoresis and the derived autoradiograms were subjected to computer analysis. The comparison of soluble proteins stimulated during acclimation and at low temperatures during the post-acclimation phase revealed that no new polypeptides were synthesised. The effects of acclimation on protein synthesis are essentially quantitative. Of the soluble proteins, the synthesis of a polypeptide homologous to the heat shock protein 70 family is stimulated. This polypeptide could protect soybean proteins from denaturation and aggregation at low, non-freezing temperatures.

NATO ASI Series, Vol. I 16
Interacting Stresses on Plants in a Changing Climate
Edited by M. B. Jackson, C. R. Black

INTRODUCTION

Plants may be divided into two classes depending upon their responses to low temperature; chilling-sensitive species which are injured by low temperatures (above zero) and freezing-sensitive species which are injured by sub-zero temperatures. Due to the diversity of individual plant responses to low temperatures and the complexity of environmental conditions, the cellular and molecular mechanisms directly associated with cold tolerance have not been fully characterised. Cold acclimation is the process whereby freezing-sensitive plants adapt themselves naturally during late summer and autumn to low temperatures. This phenomenon has been extensively studied since it allows comparisons to be made between frost-sensitive and frost-tolerant plants in one and the same genotype.

Experimental results over the last twenty years have shown that three main factors are frequently correlated with induced cold resistance:

- changes in membrane properties including increased fluidity;
- accumulation of solutes with potential cryoprotective properties;
- synthesis of new proteins resulting from new gene expression.

However, in most cases, although correlations can be made, it is difficult to distinguish between the cause and effect of cold tolerance. Furthermore, for most of the time, the above parameters are considered independently and it is difficult to establish whether any hierarchy or interrelationship exists between them. For example, despite the significant investment of several groups in the characterisation of cold-regulated genes, it has not yet been demonstrated that the changes in gene expression occurring in many plants during cold acclimation are causally related to increased resistance. The problem seems to be more complex than is the case with genes regulated by other stresses (e.g., heat shock, anaerobiosis, pathogen attack) since most cold-regulated genes differ between species. Although such a lack of homogeneity in response does not support the theory that cold stimulates a common protective mechanism, the results of recent manipulations of membrane lipids, through protoplast/liposome fusion procedures (Steponkus *et al.*, 1989), or through genetic alteration of fatty acid unsaturation (Murata *et al.*, 1992) present convincing evidence that plant membrane lipids at least play a common role in cold tolerance. In addition, cryoprotective solutes and/or specific proteins might act to improve frost tolerance. These different compounds are likely to be under the control of key enzymes which would represent potential targets for the future engineering of plants with increased resistance. However, to identify the regulatory genes involved in these processes, more information is needed on the cellular sites of cold injury and the molecular mechanisms of cold tolerance. In this chapter we report new data on cold tolerance in two species of economic interest: *Eucalyptus gunnii*, a freezing-sensitive tree, and *Glycine max*, a chilling-sensitive crop.

LATERAL AND ROTATIONAL MOBILITIES OF LIPIDS IN SPECIFIC CELLULAR MEMBRANES OF *EUCALYPTUS GUNNII*

Two cell lines of *Eucalyptus gunnii* have been obtained from two trees exhibiting differential frost sensitivity *in natura* (Teulières *et al.*, 1989). At the cellular level, these two lines (832 (frost-tolerant) and 867 (frost-sensitive)) exhibit a differential response in cell viability after freezing in controlled conditions. In both lines, cells held at 4 °C (control) were stable and no decrease in viability was observed during storage in the cryostat. After inducing freezing of the medium at -1 °C, the viability of cells decreased progressively, but at different rates depending on the cell line. 50% of the cells from cell line 832 (frost tolerant) were still viable at -8 °C, whereas all cells from cell line 867 (frost sensitive) died at -5 °C. These results demonstrate a good correlation between the viability of specific cell lines and the frost tolerance observed in nature of the *Eucalyptus* trees from which these cell lines originated. They confirm the results previously observed by Teulières *et al.* (1989) on the same two cell lines more than three years ago, and demonstrate the ability of these cells to retain differential freezing-sensitivity in long-term culture. Moreover, these observations suggest that frost tolerance can be ascribed to factors occurring at the cellular level and that these *Eucalyptus* cell lines represent a convenient model system for investigations of the molecular basis of cold tolerance.

Previous work on the isolation of pure vacuoles (Boudet *et al.*, 1981) and the purification of membrane fractions (plasmalemma and tonoplast) (Canut *et al.*, 1991) from different plant tissues persuaded us to investigate the fluidity of different specific membranes (plasmalemma and tonoplast) in these two *Eucalyptus* cell lines. Although the plasmalemma is generally considered to be the prime site of cold injury, the results concerning the tonoplast are still very limited and controversial. For example, cytological observations of vacuole integrity during cold treatment suggest either a lower sensitivity of the tonoplast to cold injury in comparison with the plasma membrane (Singh, 1979, Pukacki and Puckacka, 1987) or, in contrast, that the vacuolar membrane plays a crucial role in cold adaptation (Ziegler and Kandler, 1980).

Lateral diffusion of fluorescent probes in vacuolar membranes of frost-tolerant and -sensitive cell lines estimated by fluorescence recovery after photobleaching (FRAP). Vacuoles were prepared from protoplasts and their membranes labelled with the fluorescent lipid probe NBD-PC, 1-oleoyl-2-(7nitro-2,1,3-benzoxadiazol-4-yl) aminocaproyl phosphatidylcholine, according to the procedures described by Leborgne *et al.* (1992). After labelling, a diffuse and homogeneous fluorescence was observed at the vacuole surface under the fluorescence microscope but NBD-PC was not internalised in the aqueous compartment of the vacuoles. These labelled vacuoles were then used for fluorescence recovery after photobleaching (FRAP) experiments (Leborgne *et al.*, 1992) to study the lateral mobility of lipids in the tonoplast.

Two sets of diffusion coefficients (D) were obtained for NBD-PC diffusing laterally in the tonoplast of frost-tolerant and frost-sensitive cells respectively. Despite the large variability observed in the individual measurements, it was noted that the lateral mobility of lipids determined at different temperatures was generally lower in the tonoplast of sensitive cells than in the tonoplast of tolerant cells. For example, mean D (10^{-9} cm^2 s^{-1}) values for sensitive and tolerant lines were respectively 1.65 and 2.14 at 7 °C. For each cell line, the lateral diffusion coefficient D of the probe decreased with decreasing temperature.

Despite several attempts under different experimental conditions, it was impossible to obtain significant values of D for the plasmalemma after labelling intact protoplasts with different probes. First of all it should be stressed that for technical reasons (depth of focus of the objective used or probe internalisation), it is extremely difficult to determine if the fluorescence signal which is observed originates only from the plasma membrane. Moreover, the abnormally high photobleaching rate of the HEDAF (5-(N-hexadecanoyl-aminofluorescein)) probe (probably due to an accelerated photo-oxidation process resulting from the release of activated forms of oxygen by freshly isolated protoplasts) and the very rapid fluorescence recoveries for the NBD-PC probe (two orders of magnitude faster than those usually encountered in biological membranes) suggested that FRAP experiments were extremely difficult to carry out under these conditions, in contradiction to previous reports (Metcalf *et al.*, 1986).

Rotational mobility of 1,6-diphenylhexatriene (DPH) in membrane fractions of frost-tolerant and -sensitive cell lines estimated by fluorescence polarisation. Purified tonoplast and plasmalemma fractions were obtained from the two cell lines by free-flow electrophoresis as described by Canut *et al.* (1991). These purified membranes were labelled with the fluorescent probe diphenylhexatriene (DPH) for fluorescence polarisation experiments under the conditions described by Leborgne *et al.* (1992) to evaluate the rotational mobility of lipids. The polarisation rates (p) were determined at different temperatures (4 to 26 °C) for the two membranes. For both tonoplast and plasma membrane fractions, no phase transition could be detected over the temperature range studied and p progressively decreased as the temperature increased, suggesting an increase in membrane fluidity. At all temperatures, DPH exhibited lower polarisation rates in the tonoplast and plasma membrane from frost- tolerant cells than in frost-sensitive cells. In addition, at any temperature the p of DPH was lower in the tonoplast than in the plasma membrane, suggesting a lower fluidity of the latter membrane.

Experiments were also performed on frost-sensitive cells *after acclimation*. In this case, cell viability in cold conditions was increased after progressive exposure of the cell suspensions to decreasing culture temperatures (18, 13 and 8 °C). In all the experiments concerning differently acclimated cells from the same genotype a positive correlation was found between cell viability at low temperature and an increase in the 'fluidity' of the microsomal membranes as determined by fluorescence polarisation.

***Membrane fluidity and cold resistance in* Eucalyptus gunnii.** FRAP experiments carried out with the probe NBD-PC at different temperatures resulted in nearly 100% fluorescence recovery. Since this probe has been shown to have most of the attributes of a phospholipid molecule, the results suggest both the absence of lipid domains and the onset of lipid phase transitions in the tonoplast. Significant differences in D values were observed between the two cell lines at both normal and low temperatures, strongly suggesting that the vacuolar membranes of frost-sensitive and -tolerant cells have different lipid fluidity. The polarisation rate, p, of DPH was lower in the tonoplast of frost-resistant cells than in the tonoplast of frost-sensitive cells and, in both cases, p decreased progressively and regularly with increasing temperature, excluding the lipid phase transition in these membranes – at least over the temperature range investigated. This result is in good agreement with the existence of mobile fractions close to 100%, which also strongly suggests the absence of an immobilised lipid fraction in these membranes. The fluorescence lifetime of DPH in these membranes was not measured and, under these conditions, changes in the polarisation rate of the probe cannot definitely be related to changes in its rotational mobility and hence membrane fluidity. However, in many membrane systems, a decrease in the polarisation rate of DPH has been shown to be correlated with an increase in membrane fluidity as measured by other biophysical or biochemical techniques (van Blitterswijk, 1984). It is then interesting to note that, when comparing the vacuolar membranes from frost-tolerant and -sensitive cells, a positive correlation exists between the lateral motion of the phospholipid probe, NBD-PC, and polarisation data: at a given temperature, the higher the lateral diffusion rate of NBD-PC, the greater the fluidity of the membrane and the lower the p of DPH. Such a correlation between fluorescence recovery and polarisation data strongly suggests that the tonoplast may be considered as a 'bulk' in terms of lipid fluidity and provides further support for the idea that the lipid phase is homogeneous in this membrane.

The polarisation data for purified plasma membranes of the two cell lines and the microsomal membranes of the acclimated sensitive cell line always demonstrated a lower p of DPH and 'higher fluidity' in the more cold-tolerant material. In addition, in all conditions tested, the tonoplast appeared to be more fluid than the plasmalemma, as suggested by work on etiolated mung bean (Yoshida and Uemura, 1986) in which a higher lipid to sterol ratio was observed in the tonoplast than in the plasma membrane.

In conclusion, this work supports the correlation between increased membrane fluidity and improved freezing tolerance. Our data provide functional evidence of the higher fluidity of specific membranes (plasmalemma and tonoplast) in more cold-tolerant cells. Moreover, as proposed by Steponkus and Lynch (1989), our work supports the hypothesis that the plasma membrane is a major target for freezing damage because of its lower fluidity as compared with the tonoplast. Indeed, the cryostability of the vacuolar membrane in *Eucalyptus* cells seems to be higher than that of the plasmalemma, since additional experiments performed in our laboratory on *Eucalyptus* protoplasts submitted to a freezing programme clearly demonstrated that, at temperatures which induce disruption of the plasma membrane and rupturing of protoplasts, most vacuoles remain intact.

It would therefore now be of interest to determine the specific biochemical changes correlated with an increase in membrane fluidity and freezing tolerance occurring in the plasma membrane of *Eucalyptus* cells.

COLD ACCLIMATION AND PROTEIN SYNTHESIS IN SOYBEAN

Cold acclimation of soybean - a chilling-sensitive species. Soybean (*Glycine max* L. Merr), a plant of sub-tropical origin, is an example of a chilling-sensitive crop injured by low, non-freezing temperatures. For several freezing-sensitive species of temperate origin, cold acclimation, which increases freezing tolerance, has been extensively exploited in order to understand tolerance mechanisms at the cellular and molecular levels. Among the biochemical changes associated with acclimation, increasing interest during recent years has focussed on the modifications of protein synthesis and genome expression (Mohapatra *et al.*, 1989; Guy, 1990; Kurkela and Franck, 1990; Gilmour *et al.*, 1992). From these results it is possible to draw some preliminary conclusions concerning the role of protein synthesis in cold acclimation:

- after transfer to low temperature the synthesis of the majority of proteins produced at the control temperature is maintained;
- low temperatures induce (or stimulate) the synthesis of a limited number of proteins;
- these induced (stimulated) proteins do not seem to be the same in different species.

Different genes corresponding to proteins induced during cold acclimation have been cloned, particularly in *Arabidopsis thaliana.* They can exhibit some homologies with genes induced by water deficits (Grossi *et al.*, 1992) and their products also display similarities with type I fish antifreeze proteins (Kurkela and Franck, 1990; Kurkela and Borg-Franck, 1992), group II LEA (late embryogenesis abundant proteins) (Gilmour *et al.*, 1992; Sutton *et al.*, 1992) and, in some cases, remain soluble during boiling in aqueous solution (Houde *et al.*, 1992). The high hydrophilicity of this last class of proteins may be important in trapping enough water inside the cell to prevent the local dehydration that may occur during freezing stress.

Chilling-sensitive species of tropical or subtropical origin are not naturally submitted to cold acclimation. While a few studies have focussed on alterations in protein synthesis in response to low temperatures in these species (Ort *et al.*, 1989; Hahn and Walbot, 1989) the possibility of cold acclimation was not taken into account. In order to study the proteins related to cold tolerance in one and the same genotype of soybean we attempted to acclimate this species and examine the changes in protein synthesis related to cold acclimation.

Sixteen day old plants grown at 25 °C day/23 °C night were transferred to 14 °C day/8 °C night for five days in order to induce acclimation. After this period, the plants were transferred to 8 °C to evaluate the cold acclimation, and at the same time 16 day old plants were directly transferred from 25 °C/23 °C to 8 °C. Two cultivars, Verdon and Maple Arrow, were tested for their ability to acclimate to low temperatures. Following the transfer to 14 °C/8 °C, plant growth was greatly limited, as previously shown (Cabané *et al.*, 1992), and both non-acclimated and acclimated plants showed similar growth patterns when transferred to 8 °C. At 8 °C, control plants were strongly affected (cessation of growth and rapid yellowing). Nevertheless, yellowing was more rapid in non-acclimated plants than in acclimated ones. After 22 days at 8 °C, non-acclimated plants showed chlorosis whereas acclimated plants appeared normal. After 27 days at 8 °C, the quantity of chlorophyll was two-fold higher in acclimated plants than in non-acclimated plants in both cultivars. No differences were noticed between the two cultivars.

After previous exposure to 14 °C/8 °C, soybean plants showed a better tolerance to 8 °C than plants that had not been previously exposed, thus indicating that soybean is able to acclimate to low temperatures. From these results we conclude that cold acclimation appears to be a general phenomenon in plants and also in other living organisms, since it has also been observed in some insects (Czajka and Lee, 1990).

Changes in protein synthesis related to cold acclimation in soybean. Leaf proteins of the cultivar Verdon were labelled at different stages: non-acclimated plants at the end of acclimation at 14 °C, and two and seven days after transfer to 8 °C for both non-acclimated and acclimated plants. The first trifoliate leaf which had completed its growth at these stages was used for labelling experiments with methionine ^{35}S to avoid any changes in protein synthesis resulting from differential growth. Proteins were resolved by two-dimensional gel electrophoresis and detected by direct autoradiography to allow the quantification of the spots. For each labelling condition, four replicate gels, derived from independent experiments, were selected for analysis on the basis of the quality of the separation, and the autoradiograms were computer-analysed according to Tarroux *et al.* (1987) and Cabané *et al.* (1993). The synthesis of a large number of polypeptides was either repressed (106 spots) or stimulated (91 spots) during the acclimation period. No new polypeptides were characterised and thus the effects of acclimation on soybean protein synthesis are essentially quantitative.

In plants transferred for two days to 8 °C, 32 polypeptides were more abundantly synthesised in acclimated plants as compared to non-acclimated plants. These proteins are essentially different from those stimulated during acclimation (only six spots are common). The common spots could represent proteins that are important for adaptation to cold conditions, whereas the other spots might represent proteins of basic metabolism whose synthesis is enhanced as a result of acclimation. Of the proteins stimulated both during acclimation and after transfer to 8 °C, polypeptide n° 54 was selected because of its relative abundance in the protein pattern. This polypeptide has a Mr of 78.000 and a pI of 5. It is synthesised in plants grown under normal conditions but its synthesis is stimulated during the acclimation period. At the end of the five days of acclimation at 14 °C this labelled peptide was found to be three-fold more abundant in acclimated plants than in non acclimated plants. Polypeptide n° 54 was concentrated from 50 two-dimensional electrophoretic gels and digested with trypsin. The tryptic peptides were separated by reverse phase HPLC and the protein and tryptic peptides sequenced (Cabané *et al.*, 1993).

Comparison of microsequence data for polypeptide n° 54 with the protein database (NBRF) revealed that this polypeptide shares homology with members of the heat shock proteins of the 70 kDa (HSP 70) family (Fig. 1). Twenty to 23 residues out of 29 were identical when the peptide sequence was compared to HSP 70s from plants and micro-organisms. Moreover, the Mr and pI of this polypeptide are in keeping with the properties of this family of proteins. In agreement with these results, increased RNA levels for HSPs have also been found in chilled soybean (Kuznetsov *et al.*, 1987).

	N-terminal	peak n°1	peak n°2
spot n°54	E K V V G I D L G	T T P S V V A Y T	Y V V L V P A Y F N D
Hsc 70 tomato	(7) G P A I G I D L G (15)	(40) T T P S Y V G F T (48)	(146) A V V T V P A Y F N D (156)
Hsp 70 soybean	(6) G K A I G I D L G (14)	(39) T T P S Y V A F T (47)	(145) A V V T V P A Y F N D (155)
Hsp 70 maize	(6) G P A I G I D L G (14)	(39) T T P S Y V G F T (47)	(144) A V V T V P A Y F N D (154)
Hsp 70 petunia	(7) G P A I G I D L G (15)	(40) T T P S Y V G F T (48)	(146) A V V T V P A Y F N D (156)
SSC1 yeast	(30) G S V I G I D L G (38)	(63) T T P S V V A F T (71)	(166) A V V T V P A Y F N D (176)
Dna K *E.coli*	(2) G K I I G I D L G (8)	(40) T T P S I I A Y T (48)	(147) A V I T V P A Y F N D (157)

***FIGURE 1.** Comparison of amino-acid sequences of NH_2 terminal and two tryptic digest peptides from polypeptide n° 54 with the sequences of HSP 70 from tomato, soybean, maize, petunia, yeast and* E. coli. *Identical amino acid residues are boxed.*

Some HSP 70 related proteins known as heat shock cognates are also found in different organisms under normal conditions. Since the protein characterised in soybean is relatively abundant in normal conditions it could also be a heat shock cognate protein. It has been proposed that under stress conditions these proteins interact with damaged proteins to avoid their aggregation (Pelham, 1986). Moreover, it is known that low temperatures induce changes in protein folding that may result in aggregation (Jaenicke, 1990). Thus the HSP 70 induced by acclimation in soybean could play a protective role by avoiding the aggregation of proteins in cold conditions, thereby protecting cells from resulting damage. Other adaptative advantages could be related to the potential involvement of heat shock cognate proteins in protein oligomeric assembly and transport. In this context it is interesting to point out that in yeast, two heat shock cognate proteins of 70 kDa are stimulated under cold conditions and are necessary for growth at low temperatures (Craig and Jacobsen, 1985). Furthermore, the induction of this HSP 70 polypeptide at 14 °C in soybean is in good agreement with the results obtained recently by Neven *et al.* (1992). These authors have shown that exposure of young spinach seedlings and other freezing-sensitive species to 5 °C leads to an increase in the synthesis of a 70 kDa heat shock protein. This induced protein has been characterised by both sequence analysis and immunological detection in spinach. Furthermore, a similar induction was also observed at 5 °C for chilling-sensitive species such as tomato and pepper. These results are particularly interesting since, in combination with ours and those of other workers (Yacoob and Filion, 1987), they demonstrate, for the first time, the increased synthesis of the same family of proteins in different chilling- and freezing-sensitive plants in response to low, non-freezing temperatures. Thus the accumulation of HSP 70 cognate proteins during acclimation might have an important adaptative value.

It should be noted that the HSP 70 polypeptide induced in several species at 5 °C is already stimulated in soybean at 14 °C. When soybean plants are directly transferred to 8 °C, the rate of synthesis (estimated by the intensity of labelling) after two days of this HSP 70 is two-fold higher in acclimated plants as compared to non-acclimated plants, whereas at seven days after transfer the rate of synthesis is almost equivalent in both types. If this protein plays a role in cold tolerance, it may be that its synthesis is not stimulated rapidly enough by a direct transfer to 8 °C, and that it is the acclimation process which stimulates its synthesis to a level which allows the plant better to tolerate conditions at 8 °C.

CONCLUSIONS

Our results for *Eucalyptus* are in agreement with the correlation generally found between freezing resistance and membrane fluidity. In addition, they strongly support the idea that the tonoplast is more fluid than the plasmalemma and that it is this latter membrane which could be a prime target for cold injury.

Chilling acclimation in soybean involves subtle modifications of the synthesis of a large number of proteins. Among these, the characterisation of an HSP 70 homologue suggests that proteins with a general protective function could play a role in the adaptative process. The recent demonstration of this common response in different plant species exposed to low, non-freezing temperatures supports this hypothesis. In this way the molecular mechanisms of cold resistance, which are generally considered to be different in chilling-sensitive and freezing-sensitive species, might in fact be similar in the two types of plant. However, as cold resistance is likely to be under multigenic control, cumulative biochemical changes involving membrane properties, protection of proteins at low temperatures and cryoprotective effects could all be necessary for plants to adapt fully to a wide range of temperatures.

ACKNOWLEDGEMENTS

Part of this work was performed with the kind collaboration of Dr Tocanne. University Paul Sabatier, CNRS, AFOCEL and ASEDIS-SO are gratefully acknowledged for their support. We also thank Dr Simon Hawkins for checking the manuscript.

REFERENCES

Boudet AM, Canut H, Alibert G (1981) Isolation and characterization of vacuoles from *Melilotus alba* mesophyll. Plant Physiology 68: 1354-1358

Cabané M, Calvet P, Vincens P, Boudet AM (1993) Characterization of cold-acclimation-related proteins in soybean and identification of one as a member of heat shock protein (HSP 70) family. Planta (in press)

Cabané M, Vincens P, Boudet AM (1992) Protein synthesis at low temperatures in two soybean cultivars differing by their cold sensitivity. Physiologia Plantarum 85: 573-580

Canut H, Baudracco S, Cabané M, Boudet AM, Marigo G (1991) Preparation of sealed tonoplast and plasma membrane vesicles from *Catharantus roseus* (L.) G. Don. cells by free-flow electrophoresis. Planta 184: 448-456

Connor J, Gillum K, Schroit AJ (1990) Maintenance of lipid asymmetry in red blood cells and ghosts: effect of divalent cations and serum albumin on the transbilayer distribution of phosphatidylserine. Biochimica et Biophysica Acta 1025: 82-86

Craig EA, Jacobsen K (1985) Mutations in cognate gene of *Saccharomyces cerevisiae* result in reduced growth at low temperatures. Molecular and Cellular Biology 5: 3517-3524

Czajka MC, Lee RE (1990) A rapid cold-hardening response protecting against cold shock injury in *Drosophila melanogaster*. Journal of Experimental Biology 148: 245-254

Gilmour SJ, Artus NN, Thomashow MF (1992) cDNA sequence analysis and expression of two cold-regulated genes of *Arabidopsis thaliana*. Plant Molecular Biology 18: 13-21

Grossi M, Cattivelli L, Terzi V, Stanca AM (1992) Modification of gene expression induced by ABA, in relation to drought and cold stress in barley shoots. Plant Physiology and Biochemistry 30: 97-103

Guy CL (1990) Cold acclimation and freezing stress tolerance: role of protein metabolism. Annual Review of Plant Physiology and Plant Molecular Biology 41: 187-223

Hahn M, Walbot V (1989) Effects of cold-treatment on protein synthesis and mRNA levels in rice leaves. Plant Physiology 91: 930-938

Houde M, Danyluk J, Laliberté JF, Rassart E, Dhindsa RS, Sarhan F (1992) Cloning, characterization and expression of a cDNA encoding a 50-kilodalton protein specifically induced by cold acclimation in wheat. Plant Physiology 99: 1381-1387

Jaenicke R (1990) Protein structure and function at low temperatures. Philosophical Transactions of the Royal Society of London, Series B 326: 535-553

Kurkela S, Borg-Franck M (1992) Structure and expression of kin 2, one of two cold- and ABA-induced genes of *Arabidopsis thaliana*. Plant Molecular Biology 19: 689-692

Kurkela S, Franck M (1990) Cloning and characterization of a cold- and ABA-inducible *Arabidopsis* gene. Plant Molecular Biology 15: 137-144

Kuznetsov VV, Kimpel JA, Goekjian G, Key JL (1987) Elements of nonspecificity in responses of the plant genome to chilling and heat stress. Soviet Plant Physiology 34: 685-693

Leborgne N, Dupou-Cézanne L, Teulières C, Canut H, Tocanne JF, Boudet AM (1992) Lateral and rotational mobilities of lipids in specific cellular membranes of *Eucalyptus gunnii* cultivars exhibiting different freezing tolerance. Plant Physiology 100: 246-254

Metcalf TN, Wang JL, Schindler M (1986) Lateral diffusion of phospholipids in the plasma membrane of soybean protoplasts: evidence for membrane lipid domains. Proceedings of the National Academy of Science, USA 83: 95-99

Mohapatra SS, Wolfraim L, Poole RJ, Dhindsa RS (1989) Molecular cloning and relationship to freezing tolerance of cold-acclimation-specific genes of alfalfa. Plant Physiology 89: 375-380

Murata N, Ishizaki-Nishizawa O, Higashi S, Nayashi H, Tasaka Y, Nishida I (1992) Genetically engineered alteration in the chilling sensitivity of plants. Nature (London) 356: 710-713

Neven LG, Haskell DW, Guy CL, Denslow N, Klein PA, Green LG, Silverman A (1992) Association of 70 kilodalton heat-shock cognate proteins with acclimation to cold. Plant Physiology 99: 1362-1369

Ort DR, Martino S, Wise RR, Kent J, Cooper P (1989) Changes in protein synthesis induced by chilling and their influence on the chilling sensitivity of photosynthesis. Plant Physiology and Biochemistry 27: 785-793

Pelham HRB (1986) Speculations on the functions of the major heat shock and glucose regulated proteins. Cell 46: 959-961

Pukacki P, Puckacka S (1987) Freezing stress and membrane injury of norway spruce tissues. Physiologia Plantarum 69: 156-160

Singh J (1979) Freezing of protoplasts isolated from cold hardened and non hardened winter rye. Plant Science Letters 16: 195-201

Steponkus PL, Lynch DV (1989) Freeze/thaw-induced destabilization of the plasma membrane and the effects of cold acclimation. Journal of Bioenergetic and Biomembrane 21: 21-41

Sutton F, Ding X, Kenefick, DG (1992) Group 3 LEA gene HVA_1 regulation by cold acclimation and deacclimation in two barley cultivars with varying freeze resistance. Plant Physiology 99: 338-340

Tarroux P, Vincens P, Rabilloud T (1987) HERMeS. A second generation approach to the automatic analysis of two-dimensional electrophoresis gels. Part V: Data analysis. Electrophoresis 8: 187-199

Teulières C, Feuillet C, Boudet AM (1989) Differential characteristics of cell suspension cultures initiated from *Eucalyptus gunnii* clones differing by their frost tolerance. Plant Cell Reports 8: 407-410

van Blitterswijk WJ (1984) Alterations in lipid fluidity in the plasma membrane of tumor cells. In: M Shinitzky (ed) Physiology of membrane fluidity, Vol II. CRC Press, Boca Raton, Florida, pp 53-83

Yacoob RK, Filion WG (1987) The effects of cold-temperature stress on gene expression in maize. Biochemistry and Cell Biology 65: 112-119

Yoshida S, Uemura M (1986) Lipid composition of plasma membranes and tonoplasts isolated from etiolated seedings of mung bean (*Vigna radiata*). .Plant Physiology 82: 807-812

Ziegler P, Kandler O (1980) Tonoplast stability as a critical factor in frost injury and hardening of spruce (*Picea abies* L. Karst) needles. Zeitschrift für Pflanzenphysiologie 99: 393-410

SECTION VI

Reports of Working Groups

REPORT OF ATMOSPHERIC POLLUTION WORKING GROUP

Compiled by JD Barnes

Assisted by C Biel, CR Black, VJ Black, L van der Eerden, HZ Enoch, AM Farmer, BR Jordan, TA Mansfield, S Nogues, JJ van Oosten, G Schenone and G Terry

STATUS

1. Concerns about global warming and the role of rising atmospheric carbon dioxide (CO_2) content seem to have relegated studies of the effects of air pollutants to a position of secondary importance. There are very strong arguments for a continuing high profile for research on the impacts of nitrogenous air pollutants (NO_x, NH_3), ozone (O_3) and non-methane hydrocarbons (NMHCs). Atmospheric concentrations of these pollutants are increasing steadily and it is highly probable that they will continue to do so. Furthermore, these changes in atmospheric composition will be accompanied by increasing levels of UV-B radiation.

2. High profile research on the impact of air pollution and increasing levels of UV-B radiation is needed to: (i) set air quality guidelines, particularly for O_3 and nitrogenous air pollutants; (ii) enable critical levels/loads of pollutants which will form the basis for future abatement strategies to be established on sound scientific principles; (iii) further establish the impact of air pollution and UV-B on natural vegetation; (iv) determine the fundamental physiological and biochemical disturbances that underlie the impact of pollutants and UV-B on plant growth and yield; (v) understand how plants will respond to rising levels of CO_2, UV-B and climate change against a shifting background of air pollution.

3. Gaseous air pollutants and increasing levels of UV-B may modify plant responses to rising atmospheric CO_2. At this meeting it has been shown that NO_x, NH_3, O_3 and UV-B may counteract (at least in part) the stimulation of

NATO ASI Series, Vol. I 16
Interacting Stresses on Plants in a Changing Climate
Edited by M. B. Jackson, C. R. Black

photosynthesis induced by elevated CO_2. Consequently, in polluted regions the beneficial impact of elevated CO_2 on plant growth and yield may be less than current model predictions suggest.

4. There is growing recognition of the importance and complexity of interactions between air pollutants, increasing levels of UV-B and a number of biotic and abiotic stresses. Air pollutants can predispose plants to physical stress, but stress itself may modify the impact of pollutants.

5. Current legislation aimed at reducing NH_3 emissions may inadvertently enhance the effects of rising CO_2 and O_3.

6. Growth stimulation induced by CO_2 and nitrogenous air pollutants must not necessarily be considered as a positive effect since this may be accompanied by increased sensitivity to environmental stress. Furthermore, in both natural and managed ecosystems, differential growth responses will affect the future competitive fitness of plants. The relative importance of various weed species in agroecosystems may change, but selection of adapted crop varieties and management methods may minimise negative effects. In natural ecosystems, extinction of species may well increase.

7. To predict changes in photosynthesis occurring in response to rising atmospheric CO_2, its effects on leaf biochemistry and stomatal resistance must be addressed. Furthermore, it is important to distinguish between short-term effects of elevated CO_2 on leaf performance and longer term effects resulting from growth in an elevated CO_2 atmosphere.

FUTURE WORK

1. The vast majority of studies on the responses of plants to elevated atmospheric CO_2 have been conducted under favourable controlled environment conditions. Consequently, there is little information on which to base predictions of how plants will respond to rising CO_2 in the field, where they are subjected to various environmental stresses. Future research must

address this gap in our knowledge, paying particular attention to: (i) the impact of CO_2 on plant growth under sub-optimal conditions and; (ii) the way in which CO_2 may influence plant responses to both biotic and abiotic stress. These criticisms and recommendations apply equally to studies of the effects of gaseous air pollutants, climate change and increasing UV-B radiation.

2. Further research is needed to predict the impact of rising CO_2 on natural vegetation.

3. In studies of the combined effects of elevated CO_2 and air pollutants, plants should be exposed to more CO_2 x pollutant combinations than simply double the present ambient CO_2 concentration in order to establish response surfaces, since dose/response relationships may well not be linear.

4. Little is known about the impact of rising CO_2, increasing UV-B radiation and gaseous pollutants on crops or native species in arid or semi-arid environments. There is an urgent need to establish their impact on mediterranean, sub-tropical and tropical ecosystems.

5. There is a need for an improved understanding of the molecular events underlying plant responses to environmental stress. This will allow common defence strategies to be identified and enable some general predictions to be made regarding plant responses. The technology is available to make substantial progress, but there is a strong need for a multidisciplinary approach. It is recommended that a committee be established with the primary aim of developing links between research scientists with complementary skills. Means by which modern molecular biological and immunological techniques can be disseminated need to be addressed.

6. What should be the distribution of funding between research on climate change and biological effects? At present, there is huge investment in climatological research, the rationale being that biological problems cannot be properly targeted until future climatic changes can be predicted. This

policy is misguided because biological responses will have a major influence on the nature of climatic changes, and these effects are likely to become apparent before statistically significant changes in climate can be detected. For example, if atmospheric concentrations of O_3 and NO_x continue to rise, the capacity of plants to assimilate CO_2 and thus sequester the additional carbon available from the increasing atmospheric CO_2 is likely to decrease. Consequently, there will be a tendency for atmospheric CO_2 concentrations to rise even higher than present predictions, thereby intensifying global warming. In future, we need to stimulate a better dialogue between climatologists and biologists, and this will not be achieved whilst the present disproportionate research funding is allowed to continue.

7. Reliable, mechanistically-based models which simulate the impact of elevated CO_2 and air pollutants on plant growth, yield and survival are urgently needed. Current mechanistic simulation models need to be refined to encompass several key factors, including interactions with gaseous air pollutants, differential growth responses and increased sensitivity to environmental stresses. To achieve this, it will be necessary to promote greater dialogue between different scientific disciplines; one way to achieve this may be through the introduction of specific scientific databases.

8. Experimental approaches:

   (i) On a field scale, most research on the impacts of air pollutants (particularly O_3) on crops and young trees has been conducted using open-top chambers (OTCs). These provide a useful tool for air pollution research, providing a compromise between the totally artificial environment of laboratory-based chambers and the natural conditions achieved in free-air exposure systems. However, the limitations of this approach must be recognised; in particular, OTCs create an environment which differs significantly from that in an unenclosed field. Modifications in microclimate and micrometeorology may influence pollutant uptake, plant development and the way in which other stresses interactively affect plant growth and physiology. Attention should be given to achieving standardised chamber design and operation, improving light quality by the use of alternative cladding materials, minimising differences in environmental conditions between the inside and outside of chambers, and increasing the efficiency of filtration.

(ii) Field release systems for elevated CO_2 studies require further development. Whilst such systems allow realistic estimation of the effects of elevated CO_2, existing systems have the potentially serious drawback that there is no control over air quality other than with respect to CO_2. As pollutants characteristic of the season and location are always present, the data obtained are site- and season-specific. The challenge will be to develop cost-effective exposure systems that enable plants to be exposed in their field environment, but also permit control of pollutants, and possibly UV-B. The feasibility of using natural sources of CO_2 for experimental purposes should be considered further.

(iii) The growth of plants in small pots should be avoided wherever possible since restricted rooting capacity can alter assimilate partitioning and strongly influence plant responses to CO_2 and gaseous air pollutants. Moreover, rapid soil drying during the imposition of drought treatments provides little opportunity for plants to undergo adaptive responses.

(iv) Laboratory-based studies are essential so that the fundamental mechanisms underlying plant responses to CO_2 and gaseous air pollutants can be elucidated, but it is important that such studies are carefully focussed to ensure that the results obtained are relevant to the natural environment.

REPORT OF FLOODING AND SUBMERGENCE WORKING GROUP

Compiled by MB Jackson
Assisted by CJ Andrews, RM Brändle, RMM Crawford, TW-M Fan, S Mappelli, M Mattana, WH Patrick Jr, MM Sachs, PHM Saglio, JE Summers, BB Vartapetian and LACJ Voesenek

STATUS

The expectation that the warming effects of carbon dioxide (CO_2) and other greenhouse gases will impose drier growing conditions in some areas has overshadowed attendant problems of flooding and submergence elsewhere, as the climate changes. Increased flooding is predicted as a result of changing patterns of rainfall and rising sea levels caused by melting ice and the thermal expansion of the warmer seas. Agriculturally important coastal plains are expected to suffer increasingly from decreased crop yields and manageability of the soil as water levels rise. In drier zones, greater reliance on irrigation will accelerate the rise in groundwater levels, leading to flooding and associated salinity. In addition, the long-term pressures on food production, resulting from a predicted doubling of the world's population to 10 billion by 2060, and the warming and drying of southerly latitudes will force increased cropping in more northerly regions. In these regions, prolonged spring flooding together with freezing of flood water can be expected to cause problems of ice-encasement in the marginal zones. In more temperate latitudes, an increasing prevalence of storms in spring or early summer could prove especially damaging to young crop plants. Changes in the balance of species in the natural flora are anticipated since only a minority are equipped biochemically or morphologically to tolerate over-wet growing conditions. Even species well-adapted to flooding will come under increasing pressure if intervening periods of lowered water tables become shorter. Such damaging effects on estuarine species have already been recorded in the subsiding Mississippi Deltaic Plain.

NATO ASI Series, Vol. I 16
Interacting Stresses on Plants in a Changing Climate
Edited by M. B. Jackson, C. R. Black

Research has been concentrated in four main areas:

1. Biochemical lesions that lead to cell death, particularly in anoxic roots and rhizomes.

2. Biochemical and molecular biological characteristics associated with tolerance of oxygen shortage.

3. Morphological adaptations that enhance resilience, especially: (a) the formation of replacement root systems with internal pathways for oxygen transport; (b) faster shoot extension that helps raise the apex to, or above, the water level to allow entry of oxygen and carbon dioxide; (c) stomatal closure and related changes that minimise injury to the shoot as a result of impaired root functioning.

4. Oxidative injury, especially to rhizomes, caused by oxygen upon re-exposure to the air after prolonged periods of flooding.

5. Mechanisms of underwater photosynthesis.

FUTURE WORK

1. The shoots, or in one case the roots, of a very small number of species are capable of extending when anaerobic, sometimes more rapidly in anaerobic than in aerated conditions, e.g., the rice coleoptile or the stem of *Potamogeton pectinatus*. A priority task is to identify the genes, enzymes and metabolic pathways that sustain or even enhance growth in conditions that prove fatal to most plant cells. Differential screening using cDNA libraries, reverse genetics and other molecular technologies, and non-invasive analyses of cell metabolites by nuclear magnetic resonance are expected to contribute to achieving these aims. Further evidence for a role for nitrate as an alternative acceptor of respiratory electrons in anoxia-tolerant tissues, and as a nutrient supporting amino acid production anaerobically, should be actively sought.

2. There is a need to identify patterns of gene expression that underpin processes such as the formation of internal gas-filled space (aerenchyma) and enhanced stem extension that increase tolerance and can be promoted by ethylene. The greater ethylene production in tissues that are partially deficient in oxygen also requires a metabolic and molecular biological explanation.

3. Widening the number of species studied may reveal novel mechanisms for survival of anoxia. This is exemplified by the recent discovery that several Arctic species can survive without oxygen under warm conditions for several weeks as whole green plants.

4. With the prospect of greater reliance on cereal production in more northerly latitudes, studies of the causes of injury induced by ice-encasement and identifying more tolerant genotypes could usefully be expanded. Similarly, problems of salinity associated with rising water tables should prompt renewed investigation of the interacting effects of salinity and flooding.

5. Research into links between membrane integrity, the degree of fatty acid unsaturation and tolerance of anaerobiosis and post-anoxic injury should be expanded. Prospects for genetically engineered improvements in flood-tolerance using genes for fatty acid desaturation may follow the example already set by studies of cold tolerance.

REPORT OF DROUGHT STRESS WORKING GROUP

Compiled by JAD Zeevaart
Assisted by WJ Davies, W Hartung, TC Hsiao, AD Hanson, T Jesko, JW Radin, M Pagès, RE Sharp and GR Squire

STATUS AND FUTURE WORK

Lack of water is the single most important environmental factor affecting the productivity of agricultural and natural ecosystems. Moreover, almost all other environmental stresses interact with, or exert their effects through plant water relations. This is true of thermal stresses, salinity, gaseous pollutants and the rising level of atmospheric CO_2. It is therefore particularly important to have a sound understanding of water stress injury and resistance.

More than for other physicochemical environmental stresses, the effects of water deficits depend on levels of organisation above that of the organelle or single cell. It is difficult at present to integrate the results of very exciting studies at the molecular level with those at the whole plant or community level. This inability is a reflection of the limited work on the physiology and biochemistry of water stress, fields which alone are capable of connecting the potential of molecular biology with the behaviour of plants and crops.

Relative to other stresses, there is far less understanding of the physicochemical basis of water stress injury and resistance. Specifically, most effects of agricultural importance involve water losses in the 10-40% range. Although it is highly likely that losses of this magnitude are connected with profound changes in solvent-solute-macromolecule-membrane interactions, there is essentially no work in this area.

The majority of our understanding of injury by and adaptation to stress comes from a few crop species. Very limited work on ecological physiology and biochemistry suggests that many adaptations to drought have been evolved by native species and that some of these may be controlled by a sufficiently small number of genes to be manipulated by current recombinant-DNA techniques. There

NATO ASI Series, Vol. I 16
Interacting Stresses on Plants in a Changing Climate
Edited by M. B. Jackson, C. R. Black

is almost certainly potential for exploiting such adaptations for crop improvement, but realising this potential will rest on an improved understanding of the physiological and biochemical mechanisms involved.

REPORT OF COLD STRESS WORKING GROUP

Compiled by C Stushnoff

Assisted by A Boudet, BE Gudleifsson, Y Leshem, N Murata, JP Palta, RS Pearce and PL Steponkus

STATUS

Low temperature and frost are among the most important environmental constraints limiting the productivity and distribution of plants on the earth's surface. Although future climatic changes are difficult to predict with certainty, it is clear that, if global warming increases the frequency and severity of mid- and late-winter fluctuations, especially warming episodes, there will be an important damaging impact on plant survival in temperate regions. Loss of cold hardiness is driven by warm temperatures, especially after endogenous rest is satisfied, and perennial plants become increasingly susceptible to freeze-damage under such circumstances.

During the last decade, new information has emerged on mechanisms controlling low temperature tolerance and injury in plants. It is known that cell membranes and cryoprotectants are involved in the process of acclimation, with the membranes being the primary sites of injury. On the other hand, at the molecular level neither the mechanism of stabilisation against freezing stress nor the events resulting in death are well known. Accordingly, there has been minimal progress in our ability to control injury, especially in breeding winter cereals.

FUTURE WORK

It is essential to define key objectives in this field of research. We still need more information on mechanisms in order to manipulate resistance to low temperature stress. Gene technology provides new opportunities to test critical hypotheses. The mechanisms controlling perception of environmental signals and the processes that mediate the induction of biochemical changes which bring about resistance to stress must be identified. More integration between disciplines is needed to establish mechanisms of acclimation and the processes involved in injury. There

NATO ASI Series, Vol. I 16
Interacting Stresses on Plants in a Changing Climate
Edited by M. B. Jackson, C. R. Black

is a need to integrate physiology, biochemistry, molecular biology and genetics to a greater extent. Global interactions drawing upon the skills of personnel within these disciplines can be facilitated through collaborative programmes, such as those organised by NATO, and would be very welcome.

SUBJECT INDEX

Printing: Druckhaus Beltz, Hemsbach
Binding: Buchbinderei Schäffer, Grünstadt

Springer-Verlag and the Environment

We at Springer-Verlag firmly believe that an international science publisher has a special obligation to the environment, and our corporate policies consistently reflect this conviction.

We also expect our business partners – paper mills, printers, packaging manufacturers, etc. – to commit themselves to using environmentally friendly materials and production processes.

The paper in this book is made from low- or no-chlorine pulp and is acid free, in conformance with international standards for paper permanency.

MIX
Papier aus verantwortungsvollen Quellen
Paper from responsible sources
FSC® C105338
FSC
www.fsc.org

If you have any concerns about our products,
you can contact us on
ProductSafety@springernature.com

In case Publisher is established outside the EU,
the EU authorized representative is:
Springer Nature Customer Service Center GmbH
Europaplatz 3, 69115 Heidelberg, Germany

Printed by Libri Plureos GmbH
in Hamburg, Germany

Signals and Communication Technology

The series "Signals and Communications Technology" is devoted to fundamentals and applications of modern methods of signal processing and cutting-edge communication technologies. The main topics are information and signal theory, acoustical signal processing, image processing and multimedia systems, mobile and wireless communications, and computer and communication networks. Volumes in the series address researchers in academia and industrial R&D departments. The series is application-oriented. The level of presentation of each individual volume, however, depends on the subject and can range from practical to scientific.

More information about this series at http://www.springer.com/series/4748

Frank Nielsen
Editor

Geometric Structures of Information

Springer

Editor
Frank Nielsen
Sony Computer Science Laboratories, Inc.
Tokyo, Japan

and

Ecole Polytechnique
Palaiseau, France

ISSN 1860-4862 ISSN 1860-4870 (electronic)
Signals and Communication Technology
ISBN 978-3-030-02519-9 ISBN 978-3-030-02520-5 (eBook)
https://doi.org/10.1007/978-3-030-02520-5

Library of Congress Control Number: 2018960356

This Springer imprint is published by the registered company Springer Nature Switzerland AG
The registered company address is: Gewerbestrasse 11, 6330 Cham, Switzerland

Preface

This collective book is devoted to the recent advancements of the *Geometric Structures of Information* (GSI) and *Information Geometry* (IG). The book consists of twelve contributed chapters that have been carefully peer reviewed and revised. Each chapter has been assigned to two to five reviewers. I am very thankful for the reviewers' expertise, feedback, and insight and would like to acknowledge them in alphabetical order:

Ayan Basu, Charles Casimiro Cavalcante, Shinto Eguchi, Gaëtan Hadjeres, Tomoyuki Ichiba, Peter Jupp, Amor Keziou, Takashi Kurose, Bertrand Lods, Hiroshi Matsuzoe, Subrahamanian Moosath K. S., Cyrus Mostajeran, Jan Naudts, Nigel Newton, Frank Nielsen, Richard Nock, Atsumi Ohara, Xavier Pennec, Giovanni Pistone, Hector Roman Quiceno-Echavarría, Johannes Ruf, George Ruppeiner, Salem Said, Aida Toma, Barbara Trivellato, Pavan Turaga, Paul Vos, and Konstantinos Zografos. I apologize for any potential omission.

I list below the collection of chapters in order of appearance in this book:

1. Naudts and Zhang present in their paper entitled "Rho-Tau Embedding of Statistical Models" the (ρ, τ) monotone embeddings of statistical models (for any two increasing functions ρ and τ) and discuss the gauge freedom and its induced geometry.
2. Montrucchio and Pistone studied a class of nonparametric deformed statistical models, its associated statistical divergence, and the underlying geometry in their paper "A Class of Non-parametric Deformed Exponential Statistical Models."
3. Henmi and Matsuzoe survey and report recent results on statistical manifolds admitting torsion (SMAT) and precontrast functions and show how to get this SMAT structure from an estimating function in their paper entitled "Statistical Manifolds Admitting Torsion and Partially Flat Spaces."
4. Ohara studies some transformation on the probability simplex from the viewpoint of affine differential geometry and provides some applications in his paper entitled "Conformal Flattening on the Probability Simplex and Its Applications to Voronoi Partitions and Centroids."

5. Nielsen and Hadjeres introduce a series of computationally friendly information-geometric dualistic manifolds approximating a computationally untractable statistical manifold in their paper "Monte Carlo Information-Geometric Structures."
6. Wong in his paper "Information Geometry in Portfolio Theory" applies the principle of information geometry to financial problems.
7. Maroufy and Marriott describe the use of information geometry for Cox regression in survival analysis in their paper entitled "Generalising Frailty Assumptions in Survival Analysis: A Geometric Approach."
8. Broniatowski and Stummer present a unifying view of dissimilarities in their detailed paper "Some Universal Insights on Divergences for Statistics, Machine Learning and Artificial Intelligence."
9. Chirikjian studies information theory in interaction with Lie groups in his paper called "Information-Theoretic Matrix Inequalities and Diffusion Processes on Unimodular Lie Groups."
10. Said, Bombrun, and Berthoumieu proved that the Fisher-Rao information metric of any location-scale model is a warped Riemannian metric provided that the model is invariant under the action of some Lie group in their paper "Warped Riemannian Metrics for Location-Scale Models."
11. Nielsen and Sun propose to use Hilbert's projective geometry for modeling the probability simplex and the elliptope of correlation matrices in their paper entitled "Clustering in Hilbert's Projective Geometry: The Case Studies of the Probability Simplex and the Elliptope of Correlation Matrices."
12. Finally, Barbaresco presents an introduction on Koszul's pioneering work on homogeneous bounded domains that has revealed itself as the elementary structures of information geometry in his paper "Jean-Louis Koszul and the Elementary Structures of Information Geometry."

I would also like to thank Christoph Baumann and his support staff at Springer Nature that made the publishing process very smooth. Finally, I heartily thank Prof. Hiroaki Kitano (President of Sony Computer Science Laboratories, Inc.) for his kind support and encouragements over the years.

Tokyo, Japan
August 2018

Frank Nielsen
Sony Computer Science Laboratories, Inc.

Contents

Rho-Tau Embedding of Statistical Models

Jan Naudts and Jun Zhang

Abstract Two strictly increasing functions ρ and τ determine the rho-tau embedding of a statistical model. The Riemannian metric tensor is derived from the rho-tau divergence. It depends only on the product $\rho'\tau'$ of the derivatives of ρ and τ. Hence, once the metric tensor is fixed still some freedom is left to manipulate the geometry. We call this the *gauge freedom*. A sufficient condition for the existence of a dually flat geometry is established. It is shown that, if the coordinates of a parametric model are affine then the rho-tau metric tensor is Hessian and the dual coordinates are affine as well. We illustrate our approach using models belonging to deformed exponential families, and give a simple and precise characterization for the rho-tau metric to become Hessian.

1 Introduction

A *statistical manifold* [1, 2, 7] is an abstract manifold $\mathbb{M}$ equipped with a Riemannian metric g and an Amari—Chentsov tensor T. If the manifold is a smooth differentiable manifold then it can be realized [8] as a *statistical model.*

Most studies of statistical models are based on the widely used logarithmic embedding of probability density functions. Here, more generally embeddings are considered. Recent work [11, 12, 23] unifies the formalism of rho-tau embeddings [19] with statistical models belonging to deformed exponential families [10]. The present exposition continues this investigation.

The notion of a statistical manifold has been generalized in the non-parametric setting [14, 15, 20, 21] to include Banach manifolds. The corresponding terminology is used here, although up to now only a few papers have combined non-parametric manifolds with deformed exponential families [9, 13, 16, 18].

J. Naudts (✉)
Universiteit Antwerpen, Antwerpen, Belgium
e-mail: jan.naudts@uantwerpen.be

J. Zhang
University of Michigan, Ann Arbor, MI, USA
e-mail: junz@umich.edu

F. Nielsen (ed.), *Geometric Structures of Information*, Signals and Communication Technology, https://doi.org/10.1007/978-3-030-02520-5_1

The rho-tau divergence is discussed in the next section. Eguchi [4, 5] proved under rather general conditions that, given a differentiable manifold, a divergence function defines a metric tensor and a pair of connections. These are derived in Sect. 4, respectively Sect. 6. Parametric statistical models are discussed in Sect. 7, which discusses Hessian geometry, and Sect. 8, which deals with deformed exponential families.

2 The Statistical Manifold

The points of a given statistical manifold $\mathbb{M}$ are assumed to be random variables over some measure space $(\mathscr{X}, \mu)$. A random variable X is defined as any measurable real function. The expectation, if it exists, is denoted $\mathbb{E}_\mu X$. Throughout the text it is assumed that the manifold is differentiable and that for each X in $\mathbb{M}$ the tangent plane $T_X\mathbb{M}$ is well-defined.

The derivative of a random variable is again a random variable. Therefore one can expect that the tangent vectors at a point X of $\mathbb{M}$ are random variables with vanishing expectation value. Let us assume that these tangent vectors can be used as a local chart in the vicinity of the point X and that they belong to some Banach space $\mathscr{B}$. Then $\mathbb{M}$ is a Banach manifold, provided a number of technical conditions are satisfied.

In the simplest case the manifold $\mathbb{M}$ consists of all strictly positive probability distributions on a discrete set $\mathscr{X}$. These probability distributions can be considered as positive-valued random variables with expectation equal to 1. The space $\mathscr{B}$ of all random variables is a Banach space for instance for the L^1 norm. The manifold $\mathbb{M}$ is a Banach manifold. Our approach here is the same as that adopted in [21], where random variables are called χ-functions, and functions of random variables are called χ-functionals.

In the more general situation the choice of an appropriate norm for the tangent vectors is not so simple. See the work of Pistone et al. [14–16].

3 Rho-Tau Divergence

Given a strictly convex differentiable function h and a pair of real-valued random variables P and Q the Bregman divergence [3] is given by

$$\mathscr{D}(P, Q) = \mathbb{E}_\mu \left[h(P) - h(Q) - (P - Q)h'(Q)\right], \tag{1}$$

where h' denotes the derivative of h. A generalization involving two strictly increasing real functions $\rho(u)$ and $\tau(u)$ is proposed in [19]. For the sake of completeness the definition is repeated here. Throughout the text these functions ρ and τ are assumed to be at least once, sometimes twice differentiable.

There exists a strictly convex function f with the property that $f' \circ \rho = \tau$. It is given by

$$f(u) = \int^{\rho^{-1}(u)} \tau(v)\mathrm{d}\rho(v). \tag{2}$$

The convex conjugate function f^* is therefore given by

$$f^*(u) = \int^{\tau^{-1}(u)} \rho(v)\mathrm{d}\tau(v), \tag{3}$$

provided the lower boundary of the integrals is chosen appropriately.

The original definition [19] of the rho-tau divergence can be written as

$$\mathscr{D}_{\rho,\tau}(P, Q) = \mathbb{E}_\mu \left[f(\rho(P)) + f^*(\tau(Q)) - \rho(P)\tau(Q) \right] \tag{4}$$

which is assumed to be $\leq +\infty$. The reformulation given below simplifies the proof of some of its properties.

Definition 1 Let be given two strictly increasing differentiable functions ρ and τ, defined on a common open interval D in $\mathbb{R}$. The rho-tau divergence of two random variables P and Q with values in D is given by

$$\mathscr{D}_{\rho,\tau}(P, Q) = \mathbb{E}_\mu \left(\int_Q^P [\tau(v) - \tau(Q)]\, \mathrm{d}\rho(v) \right). \tag{5}$$

This definition is equivalent to (4). To see this, split (5) into two parts. Use (2) to write the former contribution as $\mathbb{E}_\mu f \circ \rho(P) - \mathbb{E}_\mu f \circ \rho(Q)$ and the latter as $-\mathbb{E}_\mu \tau(Q)[\rho(P) - \rho(Q)]$. Use partial integration to prove that $f \circ \rho + f^* \circ \tau = \rho\tau$. This definition also generalizes (1). To see this take $I = f$, $\rho = \mathrm{id}$, and $\tau = I'$.

Note that the integral in (5) is a Stieltjes integral, which is well-defined because ρ and τ are strictly increasing functions. The result is non-negative. Hence, the μ-expectation is either convergent or it diverges to $+\infty$.

Let P and Q be two random variables with joint probability distribution $p(\zeta, \eta)$. Then (5) can be written as

$$\begin{aligned} \mathscr{D}_{\rho,\tau}(P, Q) &= \int p(\zeta, \eta)\mathrm{d}\zeta\mathrm{d}\eta \left(\int_\eta^\zeta [\tau(v) - \tau(\eta)]\, \mathrm{d}\rho(v) \right) \\ &\leq \int p(\zeta, \eta)\mathrm{d}\zeta\mathrm{d}\eta\, |\tau(\zeta) - \tau(\eta)|\, |\rho(\zeta) - \rho(\eta)| \\ &\leq \left\{ \mathbb{E}_\mu |\tau(P) - \tau(Q)|^2 \mathbb{E}_\mu |\rho(P) - \rho(Q)|^2 \right\}^{1/2}. \end{aligned} \tag{6}$$

To obtain the latter the Cauchy–Schwarz inequality is used.

Theorem 1 *$\mathscr{D}_{\rho,\tau}(P, Q) \geq 0$ with equality if $P = Q$. If μ is faithful, i.e. $\mathbb{E}_\mu P = 0$ implies $P = 0$ for any non-negative P, then $\mathscr{D}_{\rho,\tau}(P, Q) = 0$ implies $P = Q$.*

Proof From (5) it is immediately clear that $\mathscr{D}_{\rho,\tau}(P, Q) \geq 0$ and $\mathscr{D}_{\rho,\tau}(P, P) = 0$. Assume now that $\mathscr{D}_{\rho,\tau}(P, Q) = 0$. By assumption this implies that

$$\int_Q^P [\tau(v) - \tau(Q)]\, \mathrm{d}\rho(v) = 0 \quad \mu\text{-almost everywhere.}$$

However, because τ and ρ are strictly increasing the integral is strictly positive unless $P = Q$, μ-almost everywhere. □

It can be easily verified that the rho-tau divergence satisfies the following generalized Pythagorean equality for any three points P, Q, R

$$\mathscr{D}_{\rho,\tau}(P, Q) + \mathscr{D}_{\rho,\tau}(Q, R) - \mathscr{D}_{\rho,\tau}(P, R) = \mathbb{E}_\mu \{[\rho(P) - \rho(Q)][\tau(R) - \tau(Q)]\}.$$

The general expression for the rho-tau entropy is

$$S_{\rho,\tau}(P) = -\mathbb{E}_\mu f(\rho(P)) + \text{ constant} = -\mathbb{E}_\mu \int^P \tau(u)\mathrm{d}\rho(u). \tag{7}$$

See for instance Section 2.6 of [23]. The function f is a strictly convex function which, given ρ, can still be chosen arbitrarily and then determines τ. The following identity holds

$$\mathscr{D}_{\rho,\tau}(P, Q) = -S_{\rho,\tau}(P) + S_{\rho,\tau}(Q) - \mathbb{E}_\mu [\rho(P) - \rho(Q)]\, \tau(Q). \tag{8}$$

In [12, 23], we also discuss rho-tau cross-entropy, as well as the notion of “dual entropy” arising out of rho-tau embedding.

Rho-tau divergence $\mathscr{D}_{\rho,\tau}(P, Q)$ is a special form of the more general divergence function $\mathscr{D}^{(\alpha)}_{f,\rho}(P, Q)$ arising out of convex analysis, see [19, 20]:

$$\mathscr{D}^{(\alpha)}_{f,\rho}(P, Q) = \frac{4}{1-\alpha^2} \times \mathbb{E}_\mu \left\{\frac{1-\alpha}{2} f(\rho(P)) + \frac{1+\alpha}{2} f(\rho(Q)) - f\left(\frac{1-\alpha}{2}\rho(P) + \frac{1+\alpha}{2}\rho(Q)\right)\right\}. \tag{9}$$

Clearly

$$\lim_{\alpha\to 1} \mathscr{D}^{(\alpha)}_{f,\rho}(P, Q) = \mathscr{D}_{\rho,\tau}(P, Q) = \mathscr{D}_{\tau,\rho}(Q, P);$$
$$\lim_{\alpha\to -1} \mathscr{D}^{(\alpha)}_{f,\rho}(P, Q) = \mathscr{D}_{\rho,\tau}(Q, P) = \mathscr{D}_{\tau,\rho}(P, Q);$$

with $f' \circ \rho = \tau$ (and equivalent $(f^*)' \circ \tau = \rho$, with f^* denoting convex conjugate of f). Though in $\mathscr{D}^{(\alpha)}_{f,\rho}(P, Q)$ the two free functions are f (a strictly convex function) and ρ (a strictly monotone increasing function), as reflected in its subscripts, there is only notational difference from the ρ, τ specification of two function's choice. This is because for f, f^*, ρ, τ, a choice of any two functions (one of which would have to be either ρ or τ) would specify the remaining two. See [19, 22].

4 Tangent Vectors

The rho-tau divergence introduced above can be used to fix a Riemannian metric on the tangent planes of the statistical manifold $\mathbb{M}$.

In the standard situation of the Fisher-Rao metric the point P is a probability density function p^θ, parametric with $\theta \in \mathbb{R}^n$. A short calculation gives

$$\partial_j \mathbb{E}_\mu p^\theta Y = \left\langle \partial_j \log p^\theta, Y \right\rangle_\theta, \tag{10}$$

with $\langle X, Y \rangle_\theta = \mathbb{E}_\mu p^\theta XY$, and where ∂_j is an abbreviation for $\partial/\partial\theta^j$. The metric tensor is then given by

$$g_{ij}(\theta) = \langle \partial_i \log p^\theta, \partial_j \log p^\theta \rangle_\theta.$$

The score variables $\partial_j \log p^\theta$ have vanishing expectation and span the tangent plane at the point p^θ.

These expressions are now generalized. Fix P in $\mathbb{M}$. Make the assumption that there exists some open neighborhood U of P in $\mathbb{M}$ and a one-to-one correspondence χ_P between elements Q of U and tangent vectors $X = \chi_P(Q)$ of $T_P\mathbb{M}$, satisfying $\chi_P(P) = 0$. This map χ_P is used as a local chart centered at the point P. The directional derivative d_X is then defined as

$$d_X P := \lim_{\varepsilon \to 0} \frac{\chi_P^{-1}(\varepsilon X) - \chi^{-1}(0)}{\varepsilon},$$

and is assumed to exist for all $X \in T_P\mathbb{M}$. Here, we leave the topology unspecified.

Now we take one of the two increasing functions ρ and τ, say ρ, to define a two-point correlation function $\mathbb{E}_\mu \rho(P) Y$, and the other function, τ, to act as a deformed logarithmic function replacing the logarithmic function which appears in the definition of the standard scores. The expression analogue to (10) now involves derivatives of $\mathbb{E}_\mu \rho(P) Y$ and of $\tau(P)$. It becomes

$$d_X \mathbb{E}_\mu \rho(P) Y = \left\langle d_X \tau(P), Y \right\rangle_P, \tag{11}$$

with

$$\langle X, Y \rangle_P = \mathbb{E}_\mu \frac{\rho'(P)}{\tau'(P)} XY.$$

This relation should hold for any P in $\mathbb{M}$ and X in $T_P\mathbb{M}$, and for any random variable Y. The metric tensor $g_{XY} \equiv g(X, Y)$ becomes

$$\begin{aligned} g_{XY}(P) &= \langle d_X\tau(P), d_Y\tau(P)\rangle_P \\ &= \mathbb{E}_\mu \rho'(P)\tau'(P) d_X P d_Y P. \end{aligned} \tag{12}$$

This metric tensor is related to the divergence function introduced in the previous section by

$$d_Y^P d_X^Q \mathscr{D}_{\rho,\tau}(P, Q)\bigg|_{P=Q} = -g_{XY}(P),$$

where d^P is the derivative acting only on P and d^Q acts only on Q. See [21] for the derivation of the metric tensor in the form of (12) for the non-parametric setting.

In the case of a model p^θ which belongs to the exponential family the tangent plane can be identified with the coordinate space. The chart becomes $\chi_{p^\theta}(p^\zeta) = \zeta - \theta$ so that

$$d_\zeta p^\theta := \lim_{\varepsilon \to 0} \frac{1}{\varepsilon}\left(p^{\theta+\varepsilon(\zeta-\theta)} - p^\theta\right).$$

If $(\zeta - \theta)_i = \delta_{i,j}$ then $d_\zeta p^\theta = \partial_j p^\theta$ follows and (11) reduces to (10).

5 Gauge Freedom

From (12) it is clear that the metric tensor depends only on the product $\rho'\tau'$ and not on ρ and τ separately. This implies that once the metric tensor is fixed there remains one function to be chosen freely, either the embedding ρ or the deformed logarithm τ, keeping $\rho'\tau'$ fixed. This is what we call the gauge freedom of the rho-tau formalism.

The notion of gauge freedom is common in Physics to mark the introduction of additional degrees of freedom which do not modify the model but control some of its appearances. Here, the Riemannian metric of the manifold is considered to be an essential feature while the different geometries such as the Riemannian geometry or Amari's dually flat geometries are attributes which give a further characterization.

It is known for long that distinct choices of the divergence function can lead to the same metric tensor. The present formalism offers the opportunity to profit from this freedom. Quantities such as the divergence function, the entropy or the alpha-family of connections depend on the specific choice of both ρ and τ. This is illustrated further on. Some examples are found in Table 1.

The simplest choice to fix the gauge is $\rho = \text{id}$. Several classes generalizing Bregman divergences found in the literature, e.g. [6, 10], belong to this case. The phi-divergence of [10] is obtained by choosing τ equal to the deformed logarithm $\log_\phi$ (see Sect. 8), the derivative of which is $1/\phi$. This implies $\rho'\tau' = 1/\phi$, which is also the condition for the deformed metric tensor of [10] to be conformally equivalent with

Table 1 Examples of ρ, τ combinations

$\rho(u)$	$\tau(u)$	$(\rho'\tau')(u)$	$f(u)$	$f^*(u)$
u	$\log u$	$\frac{1}{u}$	$u[\log u - 1]$	e^u
$2\sqrt{u}$	$2\sqrt{u}$	$\frac{1}{u}$	$\frac{1}{2}u^2$	$\frac{1}{2}u^2$
u	$\log_q(u)$	$\frac{1}{u^q}$	$\frac{u}{2-q}\left[\log_q(u) - 1\right]$	$\frac{1}{2-q}\left[\exp_q(u)\right]^{2-q}$
$\rho(u)$	$\log_\rho(u)$	$\frac{\rho'}{\rho}(u)$	$u[\log u - 1]$	e^u
u	$\log_\phi(u)$	$\frac{1}{\phi(u)}$	$u\log_\phi(u) - \int_1^u \frac{v}{\phi(v)}dv$	$\int_1^{\exp_\phi(u)} \frac{v}{\phi(v)}dv$

(12). The U-divergence of [6] is obtained by taking τ equal to the inverse function of U'. These were discussed in detail in [11, 12, 23].

Also of interest is the gauge defined by $\rho(u) = 1/\tau'(u)$. Let $\log_\rho$ be the corresponding deformed logarithm (see (21) below). It satisfies $\log_\rho(u) = \tau(u) - \tau(1)$. Hence, the entropy becomes

$$S_{\rho,\tau}(P) = -\mathbb{E}_\mu \rho(P)\tau(P) + \mathbb{E}_\mu P + \text{ constant.}$$

The divergence becomes

$$\mathscr{D}_{\rho,\tau}(P,Q) = \mathbb{E}_\mu \rho(P)\left[\log_\rho(P) - \log_\rho(Q)\right] - \mathbb{E}_\mu\left[P - Q\right].$$

This expression is an obvious generalization of the Kullback–Leibler divergence.

6 Induced Geometry

A divergence function not only fixes a metric tensor by taking two derivatives, it also fixes a pair of torsion-free connections by taking an extra derivative w.r.t. the first argument [4, 5]. In particular, the rho-tau-divergence (5) determines an alpha-family of connections [11, 19, 21].

A covariant derivative ∇_Z with respect to a vector field Z is defined by

$$\langle \nabla_Z d_X \tau(P), d_Y \tau(P)\rangle_P = -d_Z^P d_Y^P d_X^Q \mathscr{D}_{\rho,\tau}(P,Q)\Big|_{Q=P}.$$

A short calculation of the righthand side, with $\mathscr{D}_{\rho,\tau}$ defined by (4), gives

$$\langle \nabla_Z d_X \tau(P), d_Y \tau(P)\rangle_P = \mathbb{E}_\mu\left[d_X \tau(P)\right] d_Z d_Y \rho(P).$$

Let $\nabla_Z^{(1)} = \nabla_Z$ and let $\nabla_Z^{(-1)}$ be the operator obtained by interchanging ρ and τ. This is

$$\begin{aligned}\langle \nabla_Z^{(-1)} d_X \tau(P), d_Y \tau(P) \rangle_P &= \mathbb{E}_\mu \left[d_X \rho(P) \right] d_Z d_Y \tau(P) \\ &= \langle d_X \tau(P), d_Z d_Y \tau(P) \rangle_P. \end{aligned} \tag{13}$$

This shows that $\nabla_Z^{(-1)}$ is the adjoint of d_Z with respect to g. In addition one has

$$\langle \nabla_Z^{(1)} d_X \tau(P), d_Y \tau(P) \rangle_P + \langle d_X \tau(P), \nabla_Z^{(-1)} d_Y \tau(P) \rangle_P = d_Z \, g_{XY}(P). \tag{14}$$

The latter expression shows that the connections $\nabla^{(1)}$ and $\nabla^{(-1)}$ are the dual of each other with respect to g. The alpha-family of connections is then obtained by linear interpolation with $\alpha \in [-1, 1]$

$$\nabla_Z^{(\alpha)} = \frac{1+\alpha}{2} \nabla_Z^{(1)} + \frac{1-\alpha}{2} \nabla_Z^{(-1)}, \tag{15}$$

such that the covariant derivatives $\nabla^{(\alpha)}$ and $\nabla^{(-\alpha)}$ are mutually dual. In particular, $\nabla^{(0)}$ is self-dual and therefore coincides with the Levi-Civita connection. The family of α-connections (15) is induced by the divergence function $\mathscr{D}_{f,\rho}^{(\alpha)}(P, Q)$ given by (9), with corresponding α-values. Furthermore, upon switching $\rho \leftrightarrow \tau$ in the divergence function, the designation of 1-connection vs (-1)-connection also switches.

From (13) it is clear that the covariant derivative $\nabla_Z^{(-1)}$ vanishes on the tangent plane when

$$\langle d_X \tau(P), d_Z d_Y \tau(P) \rangle_P = 0 \quad \text{for all } X, Y \in T_P \mathbb{M}. \tag{16}$$

If this holds for all P in $\mathbb{M}$ then the $\nabla^{(-1)}$-geometry is flat. This implies that the dual geometry $\nabla^{(1)}$ is also flat — see Theorem 3.3 of [1]. The interpretation of (16) is that all second derivatives $d_Z d_Y \tau(P)$ are orthogonal to the tangent plane.

7 Parametric Models

The previous sections deal with the geometry of arbitrary manifolds consisting of random variables, without caring whether they possess special properties. Now parametric models with a Hessian metric g are considered.

From here on the random variables of the manifold $\mathbb{M}$ are probability distribution functions p^θ, labeled with coordinates θ belonging to some open convex subset U of $\mathbb{R}^n$. The manifold is assumed to be differentiable. In particular, the θ^i are covariant coordinates and the assumption holds that the derivatives $\partial_i p^\theta \equiv \partial p^\theta / \partial \theta^i$ form a basis for the tangent plane $T_\theta \mathbb{M} \equiv T_{p^\theta} \mathbb{M}$. The simplifications induced by this setting are that the tangent planes are finite-dimensional and that the dual coordinates belong again to $\mathbb{R}^n$. For general Banach manifolds both properties need not to hold. The assumptions imply that the metric tensor

$$g_{ij}(\theta) = \langle \partial_i \tau(p^\theta), \partial_j \tau(p^\theta) \rangle_\theta$$

is a strictly positive-definite matrix.

The metric g of the manifold $\mathbb{M}$ is said to be Hessian if there exists a strictly convex function $\Phi(\theta)$ with the property that $g_{ij}(\theta) = \partial_i \partial_j \Phi(\theta)$. See for instance [17]. Let $\Psi(\eta)$ denote the convex dual of $\Phi(\eta)$. This is

$$\Psi(\eta) = \sup_\theta \{ \langle \eta, \theta \rangle - \Phi(\theta) : \theta \in U \}.$$

Let U^* denote the subset of $\mathbb{R}^n$ of η for which the maximum is reached at some θ in U. This θ is unique and defines a bijection $\theta \mapsto \eta$ between U and U^*. These η are dual coordinates for the manifold $\mathbb{M}$. Conversely [11], if there exist coordinates η_i for which $g_{ij}(\theta) = \partial_j \eta_i$ then the rho-tau metric tensor g is Hessian.

The condition (16) for $\nabla^{(-1)}$ to vanish can now be written as

$$\langle \partial_i \tau(p^\theta), \partial_k \partial_j \tau(p^\theta) \rangle_\theta = 0, \qquad \text{for all } \theta \in U \text{ and for all } i, j, k. \tag{17}$$

Theorem 2 *Assume that the θ^i are affine coordinates such that $\nabla^{(-1)} = 0$. Then*

(1) the metric tensor g is Hessian;
(2) the η_i are affine coordinates for the $\nabla^{(1)}$-geometry.

Proof (1) The metric tensor (12) becomes

$$g_{ij}(p^\theta) = \langle \partial_i \tau(p^\theta), \partial_j \tau(p^\theta) \rangle_\theta = \mathbb{E}_\mu \left(\partial_i \tau(p^\theta) \right) \partial_j \rho(p^\theta) = \mathbb{E}_\mu \left(\partial_j \tau(p^\theta) \right) \partial_i \rho(p^\theta).$$

This implies

$$\partial_k g_{ij}(p^\theta) = \mathbb{E}_\mu \left(\partial_k \partial_i \tau(p^\theta) \right) \partial_j \rho(p^\theta) + \mathbb{E}_\mu \left(\partial_i \tau(p^\theta) \right) \partial_k \partial_j \rho(p^\theta),$$

but also

$$\partial_k g_{ij}(p^\theta) = \mathbb{E}_\mu \left(\partial_k \partial_j \tau(p^\theta) \right) \partial_i \rho(p^\theta) + \mathbb{E}_\mu \left(\partial_j \tau(p^\theta) \right) \partial_k \partial_i \rho(p^\theta).$$

These equations simplify by means of (17). The result is

$$\partial_k g_{ij}(p^\theta) = \mathbb{E}_\mu \left(\partial_i \tau(p^\theta) \right) \partial_k \partial_j \rho(p^\theta) = \mathbb{E}_\mu \left(\partial_j \tau(p^\theta) \right) \partial_k \partial_i \rho(p^\theta).$$

This implies that $\partial_k g_{ij}(\theta) = \partial_i g_{kj}(\theta)$. Hence there exist functions $\eta_j(\theta)$ such that $g_{ij}(\theta) = \partial_i \eta_j(\theta)$. As remarked above, it is proved in [11] that this suffices to conclude that the metric g is Hessian.

(2) Let us show that

$$\eta(t) = (1-t)\eta^{(1)} + t\eta^{(2)}. \tag{18}$$

is a solution of the Euler-Lagrange equations

$$\frac{\mathrm{d}^2}{\mathrm{d}t^2}\theta^i + \Gamma^i_{km}\left(\frac{\mathrm{d}}{\mathrm{d}t}\theta^k\right)\left(\frac{\mathrm{d}}{\mathrm{d}t}\theta^m\right) = 0. \tag{19}$$

Here, the Γ^i_{km} are the coefficients of the connection $\Gamma^{(1)}$ induced by the $\nabla^{(1)}$-geometry. They follow from

$$\Gamma_{ij,k} = \partial_i g_{jk}(\theta). \tag{20}$$

One has

$$\frac{\mathrm{d}}{\mathrm{d}t}\theta^i = \frac{\partial\theta^i}{\partial\eta_j}\frac{\mathrm{d}\eta_j}{\mathrm{d}t} = g^{ij}(\theta)\left[\eta_j^{(2)} - \eta_j^{(1)}\right]$$

and

$$\begin{aligned}\frac{\mathrm{d}^2}{\mathrm{d}t^2}\theta^i &= \frac{\mathrm{d}}{\mathrm{d}t}g^{ij}(\theta)\left[\eta_j^{(2)} - \eta_j^{(1)}\right]\\ &= \left[\partial_k g^{ij}(\theta)\right]\frac{\mathrm{d}\theta^k}{\mathrm{d}t}\left[\eta_j^{(2)} - \eta_j^{(1)}\right]\\ &= \left[\partial_k g^{ij}(\theta)\right]g^{kl}(\theta)\left[\eta_l^{(2)} - \eta_l^{(1)}\right]\left[\eta_j^{(2)} - \eta_j^{(1)}\right]\\ &= \left[\partial_k g^{ij}(\theta)\right]g_{jm}(\theta)\left(\frac{\mathrm{d}}{\mathrm{d}t}\theta^k\right)\left(\frac{\mathrm{d}}{\mathrm{d}t}\theta^m\right).\end{aligned}$$

The l.h.s. of (19) becomes

$$\text{l.h.s.} = \left\{\left[\partial_k g^{ij}(\theta)\right]g_{jm}(\theta) + \Gamma^i_{km}\right\}\left(\frac{\mathrm{d}}{\mathrm{d}t}\theta^k\right)\left(\frac{\mathrm{d}}{\mathrm{d}t}\theta^m\right).$$

This vanishes because (20) implies

$$\Gamma^i_{km} = -\left[\partial_k g^{ij}(\theta)\right]g_{jm}(\theta).$$

□

It is important to realize that the discussion in this section is generic for parametric models, without assuming particular parametric families.

8 The Deformed Exponential Family

A repeated measurement of n independent random variables $F_1, \ldots, F_n$ results in a joint probability distribution $\pi(\zeta_1, \ldots, \zeta_n)$, which describes the probability that the true value of the measured data equals ζ. More generally, the model can be taken to be a deformed exponential family, obtained by using a deformed exponential function $\exp_\phi$. Following [10], a deformed logarithm $\log_\phi$ is defined by

$$\log_\phi(u) = \int_1^u \mathrm{d}v \, \frac{1}{\phi(v)}, \tag{21}$$

where $\phi(v)$ is strictly positive and integrable on the open interval $(0, +\infty)$. The deformed exponential function $\exp_\phi(u)$ is the inverse function of $\log_\phi(u)$. It is defined on the range $\mathscr{R}$ of $\log_\phi(u)$, but is eventually extended with the value 0 if $u < \mathscr{R}$ and with the value $+\infty$ if $u > \mathscr{R}$.

The expression for the probability density function then becomes

$$p^\theta(x) = \exp_\phi\left(\sum_{k=1}^n \theta^k F_k(x) - \alpha(\theta)\right). \tag{22}$$

The function $\alpha(\theta)$ serves to normalize p^θ and is assumed to exist within the open convex domain $U \subset \mathbb{R}^n$ in which the model is defined. One can show [10] that it is a convex function. However, in general it does not coincide with the potential $\Phi(\theta)$ of the previous section. The explanation is that *escort probabilities* come into play. Indeed, from

$$0 = \partial_i \mathbb{E}_\mu p^\theta = \mathbb{E}_\mu \phi(p^\theta)\left[F_i - \partial_i \alpha\right]$$

follows that

$$\partial_i \alpha = \tilde{\mathbb{E}}_\theta F_i,$$

with the escort expectation $\tilde{\mathbb{E}}_\theta$ defined by

$$\tilde{\mathbb{E}}_\theta Y = \frac{\mathbb{E}_\mu \phi(p^\theta) Y}{\mathbb{E}_\mu \phi(p^\theta)}.$$

Only in the non-deformed case, when $\phi(u) = u$, the escort $\tilde{\mathbb{E}}_\theta$ coincides with the model expectation $\mathbb{E}_\theta$. Then the dual coordinates η_i satisfy $\eta_i = \mathbb{E}_\theta F_i = \partial_i \alpha(\theta)$.

In general, the rho-tau metric tensor g of the deformed exponential model is *not* Hessian.We have the following Theorem (see [12])

Theorem 3 *With respect to the (deformed) ϕ-exponential family p^θ obeying (22), the rho-tau metric tensor g is*

(a) conformal to Hessian if

$$\rho' \tau' \phi = \phi';$$

(b) Hessian if

$$\rho'\tau'\phi = id.$$

In case (a), the rho-tau metric tensor is conformally equivalent with the metric tensor obtained by taking the Hessian of the normalization function α; in case (b) the potential $\Phi(\theta)$ is constructed in [10]. However, there still leaves a gauge freedom. The question is then whether one can choose ρ and τ so that condition (16) for the dually flat geometry is satisfied. A sufficient condition is that $\rho = \text{id}$ and $\tau = \log_\phi$. This is the rho-affine gauge. In this gauge both the θ^i and the η_i coordinates are affine and the model has a dually flat structure.

Acknowledgements The research reported here is supported by DARPA/ARO Grant W911NF-16-1-0383 (PI: Jun Zhang).

References

1. Amari, S., Nagaoka, H.: Methods of Information Geometry. AMS Monograph. Oxford University Press, Oxford (2000). (Originally published in Japanese by Iwanami Shoten, Tokyo, Japan, 1993.)
2. Ay, N., Jost, J., LÊ, H.V., Schwachhöfer, L.: Information Geometry. Springer, Berlin (2017)
3. Bregman, L.M.: The relaxation method of finding the common point of convex sets and its application to the solution of problems in convex programming. USSR Comput. Math. Phys. **70**, 200–217 (1967)
4. Eguchi, S.: Second order efficiency of minimum contrast estimators in a curved exponential family. Ann. Stat. **11**, 793–803 (1983)
5. Eguchi, S.: A differential geometric approach to statistical inference on the basis of contrast functionals. Hiroshima Math. J. **15**, 341–391 (1985)
6. Eguchi, S.: Information geometry and statistical pattern recognition. Sugaku Expositions (Amer. Math. Soc.) **19**, 197–216 (2006). (originally Sūgaku 56 (2004) 380 in Japanese)
7. Lauritzen, S.: Statistical manifolds. In: Amari, S., Barndorff-Nielsen, O., Kass, R., Lauritzen, S., Rao, C.R. (eds.) Differential Geometry in Statistical Inference. Lecture Notes, vol. 10, pp. 163–216. IMS, Hayward (1987)
8. Lê, H.V.: Statistical manifolds are statistical models. J. Geom. **84**, 83–93 (2005)
9. Montrucchio, L., Pistone, G.: Deformed exponential bundle: the linear growth case. In: Nielsen, F., Barbaresco, F. (eds.) GSI 2017 LNCS Proceedings of Geometric Science of Information, pp. 239–246. Springer, Berlin (2017)
10. Naudts, J.: Estimators, escort probabilities, and phi-exponential families in statistical physics. J. Ineq. Pure Appl. Math. **5**, 102 (2004)
11. Naudts, J., Zhang, J.: Information geometry under monotone embedding. Part II: Geometry. In: Nielsen, F., Barbaresco, F. (eds.) GSI 2017 LNCS Proceedings of Geometric Science of Information, pp. 215–222. Springer, Berlin (2017)
12. Naudts, J., Zhang J.: Information geometry under monotone embedding. Inf. Geom. (under review)
13. Newton, N.J.: An infinite-dimensional statistical manifold modeled on Hilbert space. J. Funct. Anal. **263**, 1661–1681 (2012)
14. Pistone, G., Sempi, C.: An infinite dimensional geometric structure on the space of all the probability measures equivalent to a given one. Ann. Stat. **33**, 1543–1561 (1995)
15. Pistone, G., Rogantin, M.P.: The exponential statistical manifold: mean parameters, orthogonality and space transformations. Bernoulli **5**, 721–760 (1999)

16. Pistone, G.: κ-exponential models from the geometrical viewpoint. Eur. Phys. J. B **70**, 29–37 (2009)
17. Shima, H.: The Geometry of Hessian Structures. World Scientific, Singapore (2007)
18. Vigelis, R.F., Cavalcante, C.C.: On ϕ-families of probability distributions. J. Theor. Probab. **26**, 870–884 (2013)
19. Zhang, J.: Divergence function, duality, and convex analysis. Neural Comput. **16**, 159–195 (2004)
20. Zhang, J.: Referential duality and representational duality on statistical manifolds. In: Proceedings of the Second International Symposium on Information Geometry and Its Applications, Tokyo, Japan, pp. 58–67 (2005)
21. Zhang, J.: Nonparametric information geometry: from divergence function to referential-representational biduality on statistical manifolds. Entropy **15**, 1 (2013)
22. Zhang, J.: On monotone embedding in information geometry. Entropy **17**, 4485–4499 (2015)
23. Zhang, J., Naudts, J.: Information geometry under monotone embedding. Part I: Divergence functions. In: Nielsen, F., Barbaresco, F. (eds.) GSI 2017 LNCS Proceedings of Geometric Science of Information, pp. 205–214. Springer, Berlin (2017)

A Class of Non-parametric Deformed Exponential Statistical Models

Luigi Montrucchio and Giovanni Pistone

Abstract We study the class on non-parametric deformed statistical models where the deformed exponential has linear growth at infinity and is sub-exponential at zero. This class generalizes the class introduced by N.J. Newton. We discuss the convexity and regularity of the normalization operator, the form of the deformed statistical divergences and their convex duality, the properties of the escort densities, and the affine manifold structure of the statistical bundle

1 Introduction

In this paper we study a geometry on the set $\mathscr{P}$ of strictly positive probability densities on a probability space $(\mathbb{X}, \mathscr{X}, \mu)$. In some cases one is led to consider the set $\overline{\mathscr{P}}$ of probability densities i.e., without the restriction of strict positivity. There is a considerable literature on the Information Geometry in the sense defined in the Amari and Nagaoka monograph [2] on $\mathscr{P}$. There is also a non-parametric approach i.e., we are not considering the geometry induced on the parameter set of a given statistical model but on the full set of densities. This was done in [21, 23] by using logarithmic chart to represent densities.

A different approach, that leads to the construction of an Hilbert manifold on $\mathscr{P}$, has been proposed by N.J. Newton in [18, 19]. It is based on the use of the chart $p \mapsto p - 1 - \log p$ instead of a purely logarithmic chart. This paper presents a variation on the same theme by enlarging the class of permitted charts.

Let $\mathscr{M} \subset \mathscr{P}$. At each $p \in \mathscr{M}$, the Hilbert space of square-integrable random variables $L^2(p)$ provides a fiber that sits at $p \in \mathscr{M}$, so we can define the *Hilbert bundle* with base $\mathscr{M}$. The Hilbert bundle, or similar bundles with fibers which are

L. Montrucchio
Collegio Carlo Alberto, Piazza Vincenzo Arbarello 8, 10122 Turin, Italy
e-mail: luigi.montrucchio@unito.it

G. Pistone (✉)
de Castro Statistics, Collegio Carlo Alberto, Piazza Vincenzo Arbarello 8, 10122 Turin, Italy
e-mail: giovanni.pistone@carloalberto.org

F. Nielsen (ed.), *Geometric Structures of Information*, Signals and Communication Technology, https://doi.org/10.1007/978-3-030-02520-5_2

vector spaces of random variables, provides a convenient framework for Information Geometry, cf. [1, 12, 21].

If $\mathscr{M}$ is an exponential manifold in the sense of [23], there exists a splitting of each fiber $L^2(p) = \mathscr{H}_p \oplus \mathscr{H}_p^{\perp}$, such that each $\mathscr{H}_p$ contains a dense vector sub-space which is an expression of the tangent space $T_p\mathscr{M}$ of the manifold. Moreover, the manifold on $\mathscr{M}$ is an affine manifold (it can be defined by an atlas whose transition mapping are affine) and it is also an Hessian manifold (the inner product on each fiber is the second derivative of a potential function, [24]).

When the sample space is finite and $\mathscr{M}$ is the full set $\mathscr{P}$ of positive probability densities, then $\mathscr{H}_p$ is the space of centered square integrable random variables $L_0^2(p)$ and moreover there is an identification of the fiber with the tangent space $\mathscr{H}_p \simeq T_p\mathscr{P}$. A similar situation occurs even when $\mathscr{M}$ is a finite-dimensional exponential family. It is difficult to devise set-ups other than those mentioned above, where the identification of the Hilbert fiber with the tangent space holds true. In fact, a necessary condition would be the topological linear isomorphism among fibers. One possible option would be to take as fibers the spaces of bounded functions $L_0^{\infty}(p)$, see G. Loaiza and H.R. Quiceno [14].

This difficulty is overcome in the N.J. Newton's setting. On a probability space $(X, \mathscr{X}, \mu)$, he considers the "balanced chart" $\mathscr{M} \ni p \mapsto \log p + p - 1 \in L_0^2(\mu)$. In this chart, all the tangent spaces are identified with the fixed Hilbert space $L_0^2(\mu)$ so that the statistical Hilbert bundle is trivialized.

N.J. Newton balanced chart falls in a larger class of "deformation" of the usual logarithmic representation. It is in fact an instance of the class of "deformed logarithm" as defined by J. Naudts [17]. It is defined as $\log_A(x) = \int_1^x dt/A(t)$, where A is a suitable increasing function. If A is bounded, then a special class of deformed logarithms results. It includes N.J. Newton balanced chart as well as other deformed logarithms, notably the G. Kaniadakis logarithm [10, 11, 20].

In this paper, we try a mixture of the various approaches by considering deformed logarithms with linear growth as established by N.J. Newton, but we do not look for a trivialization of the Hilbert bundle. Instead we construct an affine atlas of charts, each one centered at a $p \in \mathscr{M}$. This is obtained by adapting the construction of the exponential manifold of [21] to the deformed exponential models as defined by J. Naudts [17]. Moreover, we allow for a form of general reference measure by using an idea introduced by R.F. Vigelis and C.C. Cavalcante [26]. That is, each density has the form $q = \exp_A(u - K_p(u) + \log_A p)$, where $\exp_A = \log_A^{-1}$ is an exponential-like function which has a linear growth at $+\infty$ and is dominated by an exponential at $-\infty$.

The formalism of deformed exponentials is discussed in Sect. 2. This section is intended to be self-contained and contains material from the references discussed above without an explicit mention. The following Sect. 3 is devoted to the study of non-parametric deformed exponential families. In Sect. 4 we introduce the formulation of the divergence, in accordance with our approach. In Sect. 5 the construction of the Hilbert statistical bundle is outlined.

A first version of this piece of research has been presented at the GSI 2017 Conference [16] and we refer to that paper for some of the proofs.

2 Deformed Exponential

Let us introduce a class of the deformed exponential, according to the formalism introduced by [17]. Assume to be given a function A from $]0, +\infty[$ onto $]0, a[$, strictly increasing, continuously differentiable and such that $\left\|A'\right\|_\infty < \infty$. This implies $a = \|A\|_\infty$ and $A(x) \leq \left\|A'\right\|_\infty x$, so that $\int_0^1 d\xi/A(\xi) = +\infty$.

The A-logarithm is the function

$$\log_A(x) = \int_1^x \frac{d\xi}{A(\xi)}, \quad x \in]0, +\infty[.$$

The A-logarithm is strictly increasing from $-\infty$ to $+\infty$, its derivative $\log_A'(x) = 1/A(x)$ is positive and strictly decreasing for all $x > 0$, hence $\log_A$ is strictly concave.

By inverting the A-logarithm, one obtains the A-exponential, $\exp_A = \log_A^{-1}$. The function $\exp_A :]-\infty, +\infty[\to]0, +\infty[$ is strictly increasing, strictly convex, and is the solution to the Cauchy problem

$$\exp_A'(y) = A(\exp_A(y)), \quad \exp_A(0) = 1 . \tag{1}$$

As a consequence, we have the linear bound

$$\left|\exp_A(y_1) - \exp_A(y_2)\right| \leq \|A\|_\infty |y_1 - y_2| . \tag{2}$$

The behavior of the A-logarithm is linear for large arguments and super-logarithmic for small arguments. To derive explicit bounds, set

$$\alpha_1 = \min_{x \leq 1} \frac{A(x)}{x}, \quad \alpha_2 = \max_{x \leq 1} \frac{A(x)}{x},$$

namely, they are the best constants such that $\alpha_1 x \leq A(x) \leq \alpha_2 x$ for $0 < x \leq 1$. Note that $\alpha_1 \geq 0$ while $\alpha_2 > 0$. If in addition also $\alpha_1 > 0$, then

$$\frac{1}{\alpha_2} \log x \leq \log_A x \leq \frac{1}{\alpha_1} \log x , \quad 0 < x \leq 1 . \tag{3}$$

If otherwise $\alpha_1 = 0$, the left inequality is true only.

For $x \geq 1$ we have $A(1) \leq A(x) < \|A\|_\infty$, hence

$$\frac{1}{\|A\|_\infty}(x - 1) < \log_A x \leq \frac{1}{A(1)}(x - 1) , \quad x \geq 1 . \tag{4}$$

Under the assumptions made on the function A, the coefficient $\alpha_1 > 0$, if and only if $A'(0+) > 0$.

2.1 Examples

The main example of A-logarithm is the N.J. Newton A-logarithm [18], with

$$A(\xi) = 1 - \frac{1}{1+\xi} = \frac{\xi}{1+\xi} ,$$

so that

$$\log_A(x) = \log x + x - 1 .$$

There is a simple algebraic expression for the product,

$$\log_A(x_1x_2) = \log_A(x_1) + \log_A(x_2) + (x_1 - 1)(x_2 - 1) .$$

Other similar examples are available in the literature. One is a special case of the G. Kaniadakis' exponential of [9], generated by

$$A(\xi) = \frac{2\xi^2}{1+\xi^2} .$$

It turns out

$$\log_A x = \frac{x - x^{-1}}{2} ,$$

whose inverse provides

$$\exp_A(y) = y + \sqrt{1+y^2} .$$

A remarkable feature of the G. Kaniadakis' exponential is the relation

$$\exp_A(y)\exp_A(-y) = \left(y + \sqrt{1+y^2}\right)\left(-y + \sqrt{1+y^2}\right) = 1$$

Notice that the A function for N.J. Newton exponential is concave, while the A function of G. Kaniadakis exponential is not.

Another example is $A(\xi) = 1 - 2^{-\xi}$, which gives $\log_A(x) = \log_2(1 - 2^{-x})$ and $\exp_A(y) = \log_2(1 + 2^y)$.

Notable examples of deformed exponentials that do not fit into our set of assumptions are Tsallis q-logarithms, see [25]. For instance, for $q = 1/2$,

$$\log_{1/2} x = 2\left(\sqrt{x} - 1\right) = \int_1^x \frac{d\xi}{\sqrt{\xi}}.$$

In this case, $\log_{1/2}(0+) = -\int_0^1 d\xi/\sqrt{\xi} = -2$, so that the inverse is not defined for all real numbers. Tsallis logarithms provide models having heavy tails, which is not the case in our setting.

2.2 Superposition Operator

The deformed exponential will be employed to represent positive probability densities in the type $p(x) = \exp_A[u(x)]$, where u is a random variable on a probability space $(\mathbb{X}, \mathscr{X}, \mu)$. For this reason, we are interested in the properties of the *superposition operator*

$$S_A : u \mapsto \exp_A \circ u \tag{5}$$

defined in some convenient functional setting. About superposition operators, see e.g. [3, Ch. 1] and [4, Ch. 3].

It is clear from the Lipschitz condition (2) that $\exp_A(u) \leq 1 + \|A\|_\infty |u|$, which in turn implies that the superposition operator S_A maps $L^\alpha(\mu)$ into itself for all $\alpha \in [1, +\infty]$ and the mapping is uniformly Lipschitz with constant $\|A\|_\infty$. Notice that we are assuming that μ is a finite measure.

The superposition operator $S_A : L^\alpha(\mu) \to L^\alpha(\mu)$ is 1-to-1 and its image consists of all positive random variables f such that $\log_A f \in L^\alpha(\mu)$. The following proposition intercepts a more general result [19]. We give a direct proof here for sake of completeness and because our setting includes deformed logarithms other than the case treated there.

Proposition 1 *1. For all $\alpha \in [1, \infty]$, the superposition operator S_A of Eq. (5) is Gateaux-differentiable with derivative*

$$dS_A(u)[h] = A(\exp_A(u))h \; . \tag{6}$$

2. S_A is Fréchet-differentiable from $L^\alpha(\mu)$ to $L^\beta(\mu)$, for all $\alpha > \beta \geq 1$.

Proof 1. Equation (1) implies that for each couple of random variables $u, h \in L^\alpha(\mu)$

$$\lim_{t \to 0} t^{-1} \left(\exp_A(u + th) - \exp_A(u)\right) - A(\exp_A(u))h = 0$$

holds point-wise. Moreover, if each $\alpha \in [1, \infty[$, by Jensen inequality we infer that if $t > 0$ then

$$\left|t^{-1} \left(\exp_A(u + th) - \exp_A(u)\right) - A(\exp_A(u))h\right|^\alpha \leq$$
$$t^{-1} |h|^\alpha \int_0^t \left|A(\exp_A(u + rh)) - A(\exp_A(u))\right|^\alpha \, dr \leq (2 \|A\|_\infty)^\alpha |h|^\alpha \, .$$

Now, dominated convergence forces the limit to hold in $L^\alpha(\mu)$. If $t < 0$, it sufficies to replace h with $-h$.

Whenever $\alpha = \infty$, we can use the second-order bound

$$\left|t^{-1}\left(\exp_A(u+th)-\exp_A(u)\right)-A(\exp_A(u))h\right| = \\ |t|^{-1}h^2\left|\int_0^t (t-r)\frac{d}{dr}A(\exp_A(u+rh))\,dr\right| \leq \frac{t}{2}\|h\|_\infty^2\left\|A'\right\|_\infty\|A\|_\infty .$$

As $\left\|A'\cdot A\right\|_\infty<\infty$, the RHS goes to 0 as $t\to 0$ uniformly for each $h\in L^\infty(\mu)$.

2. Given $u, h\in L^\alpha(\mu)$, thanks again to Taylor formula,

$$\int\left|\exp_A(u+h)-\exp_A(u)-A(\exp_A(u))h\right|^\beta d\mu \leq \\ \int|h|^\beta\int_0^1\left|A(\exp_A(u+rh))-A(\exp_A(u))\right|^\beta dr\,d\mu .$$

By means of Hölder inequality, with conjugate exponents α/β and $\alpha/(\alpha-\beta)$, the RHS is bounded by

$$\left(\int|h|^\alpha\,d\mu\right)^{\frac{\beta}{\alpha}}\left(\iint\left|A(\exp_A(u+rh))-A(\exp_A(u))\right|^{\frac{\alpha\beta}{\alpha-\beta}}dr\,d\mu\right)^{\frac{\alpha-\beta}{\alpha}} .$$

Consequently,

$$\|h\|_{L^\alpha(\mu)}^{-1}\left\|\exp_A(u+h)-\exp_A(u)-A(\exp_A(u))h\right\|_{L^\beta(\mu)} \leq \\ \left(\iint\left|A(\exp_A(u+rh))-A(\exp_A(u))\right|^{\frac{\alpha\beta}{\alpha-\beta}}dr\,d\mu\right)^{\frac{\alpha-\beta}{\alpha\beta}} .$$

In order to show that the RHS vanishes as $\|h\|_{L^\alpha(\mu)}\to 0$, observe that for all $\delta>0$ we have

$$\left|A(\exp_A(u+rh))-A(\exp_A(u))\right| \leq \begin{cases} 2\|A\|_\infty & \text{always,}\\ \left\|A'\right\|_\infty\|A\|_\infty\delta & \text{if } |h|\leq\delta, \end{cases}$$

so that, decomposing the double integral as $\iint=\iint_{|h|\leq\delta}+\iint_{|h|>\delta}$, we obtain

$$\iint\left|A(\exp_A(u+rh))-A(\exp_A(u))\right|^\gamma dr\,d\mu \leq \\ (2\|A\|_\infty)^\gamma\,\mu\{|h|>\delta\}+\left(\left\|A'\right\|_\infty\|A\|_\infty\delta\right)^\gamma \leq \\ (2\|A\|_\infty)^\gamma\,\delta^{-\alpha}\int|h|^\alpha\,d\mu+\left(\left\|A'\right\|_\infty\|A\|_\infty\delta\right)^\gamma ,$$

where $\gamma=\alpha\beta/(\alpha-\beta)$ and we have used Cebičev inequality. Now it is clear that the last bound implies the conclusion for each $\alpha<\infty$. The case $\alpha=\infty$ follows a fortiori. □

Remark 1 It is not generally true that the superposition operator S_A be Fréchet differentiable for $\alpha = \beta$, cf. [3, §1.2]. We repeat here the well known counter-example.

Assume μ is a non-atomic probability measure. For each $\lambda \in \mathbb{R}$ and $\delta > 0$ define the simple function

$$h_{\lambda,\delta}(x) = \begin{cases} \lambda & \text{if } |x| \leq \delta, \\ 0 & \text{otherwise.} \end{cases}$$

For each $\alpha \in [1, +\infty[$ we have

$$\lim_{\delta \to 0} \left\| h_{\lambda,\delta} \right\|_{L^\alpha(\mu)} = \lim_{\delta \to 0} |\lambda| \, \mu \left\{ |x| \leq \delta \right\}^{1/\alpha} = 0 \, .$$

Differentiability at 0 in $L^\alpha(\mu)$ would imply for all λ

$$0 = \lim_{\delta \to 0} \frac{\left\| \exp_A(h_{\lambda,\delta}) - 1 - A(1) h_{\lambda,\delta} \right\|_{L^\alpha(\mu)}}{\left\| h_{\lambda,\delta} \right\|_{L^\alpha(\mu)}} =$$
$$\lim_{\delta \to 0} \frac{\left| \exp_A(\lambda) - 1 - A(1)\lambda \right| \mu \left\{ x \middle| \, |x| \leq \delta \right\}^{1/\alpha}}{|\lambda| \, \mu \left\{ x \middle| \, |x| \leq \delta \right\}^{1/\alpha}} = \left| \frac{\exp_A(\lambda) - 1}{\lambda} - A(1) \right| ,$$

which is a contradiction.

Remark 2 Theorems about the differentiability of the deformed exponential are important because of computations like $\frac{d}{d\theta} \exp_A(v(\theta)) = \exp'_A(v(\theta))\dot{v}(\theta)$ are essential for the geometrical theory of statistical models. Several variations in the choice of the combination domain space - image space are possible. Also, one could look at a weaker differentiability property than Frechét differentiability. Our choice is motivated by the results of the following sections. A large class of cases is discussed in [19].

Remark 3 It would also be worth to study the action of the superposition operator on spaces of differentiable functions, for example Gauss-Sobolev spaces of P. Malliavin [15]. If μ is the standard Gaussian measure on $\mathbb{R}^n$, and u is a differentiable function such that u, $\frac{\partial}{\partial x_i} u \in L^2(\mu)$, $i = 1, \dots, n$, then it follows that $\exp_A(u) \in L^2(\mu)$ as well as $\frac{\partial}{\partial x_i} \exp_A(u) \in L^2(\mu)$, since

$$\frac{\partial}{\partial x_i} \exp_A(u(x)) = A(\exp_A(u(x)) \frac{\partial}{\partial x_i} u(x) \, .$$

We do not pursue this line of investigation here.

3 Deformed Exponential Family Based on $\exp_A$

According to [5, 26], let us define the deformed exponential curve in the space of positive measures on $(\mathbb{X}, \mathscr{X})$ as follows

$$t \mapsto \mu_t = \exp_A(tu + \log_A p) \cdot \mu \,, \quad u \in L^1(\mu) \,.$$

We have the following inequality:

$$\exp_A(x + y) \leq \|A\|_\infty x^+ + \exp_A(y).$$

Actually, it is true for $x \leq 0$, as being $\exp_A$ increasing. For $x = x^+ > 0$ the inequality follows from Eq. (2). As a consequence, each μ_t is a finite measure, $\mu_t(\mathbb{X}) \leq t \|A\|_\infty \int u^+ \, d\mu + 1$, with $\mu_0 = p \cdot \mu$. The curve is actually continuous and differentiable in $L^1(\mu)$ because the point-wise derivative of the density $p_t = \exp_A(tu + \log_A(p))$ is $\dot{p}_t = A(p_t)u$ so that $|\dot{p}_t| \leq \|A\|_\infty |u|$. In conclusion $\mu_0 = p \cdot \mu$ and $\dot{\mu}_0 = A\,(p)\,u \cdot \mu$.

There are two ways to normalize the density p_t to total mass 1, either dividing by a normalizing constant $Z(t)$ to get the statistical model $t \mapsto \exp_A(tu + \log_A p)/Z(t)$ or, subtracting a constant $\psi(t)$ from the argument to get the model $t \mapsto \exp_A(tu - \psi(t) + \log_A(p))$. Unlike the standard exponential case, where these two methods lead to the same result, this is not the case for deformed exponentials where $\exp_A(\alpha + \beta) \neq \exp_A(\alpha)\exp_A(\beta)$. We choose in the present paper the latter option.

Here we use the ideas of [5, 17, 26] to construct deformed non-parametric exponential families. Recall that we are given: the probability space $(\mathbb{X}, \mathscr{X}, \mu)$; the set $\mathscr{P}$ of the positive probability densities and the function A satisfying the conditions set out in Sect. 2. Throughout this section, the density $p \in \mathscr{P}$ will be fixed.

The following proposition is taken from [16] where a detailed proof is given.

Proposition 2 *1. The mapping $L^1(\mu) \ni u \mapsto \exp_A(u + \log_A p) \in L^1(\mu)$ has full domain and is $\|A\|_\infty$-Lipschitz. Consequently, the mapping*

$$u \mapsto \int g \exp_A(u + \log_A p)\, d\mu$$

is $\|g\|_\infty \cdot \|A\|_\infty$-Lipschitz for each bounded function g.

2. *For each $u \in L^1(\mu)$ there exists a unique constant $K_p(u) \in \mathbb{R}$ such that $\exp_A(u - K_p(u) + \log_A p) \cdot \mu$ is a probability.*
3. *$K_p(u) = u$ if, and only if, u is constant. In such a case,*

$$\exp_A(u - K_p(u) + \log_A p) \cdot \mu = p \cdot \mu \,.$$

Otherwise, $\exp_A(u - K_p(u) + \log_A p) \cdot \mu \neq p \cdot \mu$.

4. *A density q is of the form $q = \exp_A(u - K_p(u) + \log_A p)$, with $u \in L^1(\mu)$ if, and only if, $\log_A q - \log_A p \in L^1(\mu)$.*

5. *If*
$$\exp_A(u - K_p(u) + \log_A p) = \exp_A(v - K_p(v) + \log_A p) \;,$$
with $u, v \in L^1(\mu)$, *then* $u - v$ *is constant.*
6. *The functional* $K_p \colon L^1(\mu) \to \mathbb{R}$ *is translation invariant. More specifically,*
$$K_p(u + c) = K_p(u) + cK_p(1)$$
holds for all $c \in \mathbb{R}$.
7. $K_p : L^1(\mu) \to \mathbb{R}$ *is continuous and convex.*

3.1 Escort Density

For each positive density $q \in \overline{\mathscr{P}}$, its *escort density* is defined as
$$\text{escort}\,(q) = \frac{A(q)}{\int A(q)\,d\mu}\,,$$
see [17]. Notice that $0 \leq A(q) \leq \|A\|_\infty$. In particular, $\widetilde{q} = \text{escort}\,(q)$ is a bounded positive density. Hence, $\text{escort}\left(\overline{\mathscr{P}}\right) \subseteq \overline{\mathscr{P}} \cap L^\infty(\mu)$. Clearly, the inclusion $\text{escort}\,(\mathscr{P}) \subseteq \mathscr{P} \cap L^\infty(\mu)$ is true as well.

Proposition 3 *1. The mapping* $\text{escort}\colon \overline{\mathscr{P}} \to \overline{\mathscr{P}} \cap L^\infty(\mu)$ *is a.s. injective.*
2. *A bounded positive density* $\widetilde{q}$ *is an escort density, i.e.,* $\widetilde{q} \in \text{escort}\left(\overline{\mathscr{P}}\right)$ *if, and only if,*
$$\lim_{\alpha \uparrow \|A\|_\infty} \int A^{-1}\left(\alpha \frac{\widetilde{q}}{\|\widetilde{q}\|_\infty}\right) d\mu \geq 1\,. \tag{7}$$
3. *Condition* (7) *is fulfilled if* $\mu\left\{\widetilde{q} = \|\widetilde{q}\|_\infty\right\} > 0$. *In particular, every density taking a finite number of different values, i.e., a simple density, is an escort density.*
4. *If* $\widetilde{q}_1 = \text{escort}\,(q_1)$ *is an escort density, and* q_2 *is a bounded positive density such that*
$$\mu\left\{\widetilde{q}_1 > t\,\|\widetilde{q}_1\|_\infty\right\} \leq \mu\left\{q_2 > t\,\|q_2\|_\infty\right\}, \quad t > 0\,,$$
then q_2 *is an escort density as well.*

Proof 1. Let $\text{escort}\,(q_1) = \text{escort}\,(q_2)$ for μ-almost all x. Say, $\int A \circ q_1\,d\mu \geq \int A \circ q_2\,d\mu$. Then $A(q_2(x)) \leq A(q_1(x))$, for μ-almost all x. Since A is strictly increasing, it follows $q_2(x) \leq q_1(x)$ for μ-almost all x, which, in turn, implies $q_1 = q_2$ μ-a.s. because both μ-integrals are equal to 1. Thus the escort mapping is a.s. injective.
2. Fix a $\widetilde{q} \in \overline{\mathscr{P}} \cap L^\infty(\mu)$, and define the function

$$f(\alpha) = \int A^{-1}\left(\alpha \frac{\widetilde{q}}{\|\widetilde{q}\|_\infty}\right) d\mu, \quad \alpha \in [0, \|A\|_\infty[\,.$$

It is finite, increasing, continuous and $f(0) = 0$. It is clear that the range condition (7) is necessary because $\widetilde{q} = \text{escort}\,(q)$ implies $q = A^{-1}\left(\left(\int A(q)\, d\mu\right) \widetilde{q}\right)$ and, in turn, $1 = \int A^{-1}\left(\left(\int A(q)\, d\mu\right) \widetilde{q}\right)\, d\mu$, given that q is a probability density. If we take $\alpha = \int A(q)\, d\mu\ \|\widetilde{q}\|_\infty \leq \|A\|_\infty$, the range condition is satisfied. Conversely, if the range condition holds, there exists $\alpha \leq \|A\|_\infty$ such that $q = A^{-1}\left(\alpha \frac{\widetilde{q}}{\|\widetilde{q}\|_\infty}\right)$ is a positive probability density whose escort is $\widetilde{q}$.

3. This is a special case of Item 2, in that

$$f(\alpha) = \int A^{-1}\left(\alpha \frac{\widetilde{q}}{\|\widetilde{q}\|_\infty}\right) d\mu \geq A^{-1}(\alpha) \mu \left\{\widetilde{q} = \|\widetilde{q}\|_\infty\right\}\,.$$

Therefore, $f(\alpha) \uparrow +\infty$, as $\alpha \uparrow \|A\|_\infty$.

4. For each bounded positive density q we have

$$\int A^{-1}\left(\frac{q}{\|q\|_\infty}\right) d\mu = \int_0^{+\infty} \mu\left\{\frac{q}{\|q\|_\infty} > A(t)\right\} dt = \int_0^{\|A\|_\infty} \mu\left\{\frac{q}{\|q\|_\infty} > s\right\} \frac{1}{A'\left(A^{-1}(s)\right)}\, ds\,.$$

Now the necessary condition of Item 3. Follows from Item 1. and our assumptions. □

The previous proposition shows that the range of the escort mapping is uniformly dense as it contains all simple densities. Moreover, in the partial order induced by the rearrangement of the normalized density (that is for each q the mapping $t \mapsto \mu\left\{\frac{q}{\|q\|_\infty} > t\right\}$), it contains the full right interval of each element. But the range of the escort mapping is not the full set of bounded positive densities, unless the σ-algebra $\mathscr{X}$ is generated by a finite partition. To provide an example, consider on the Lebesgue unit interval the densities $q_\delta(x) \propto (1 - x^{1/\delta})$, $\delta > 0$, and $A(x) = x/(1+x)$. The density q_δ turns out to be an escort if, and only if, $\delta \leq 1$.

3.2 *Gradient of the Normalization Operator K_p*

Proposition 2 shows that the functional K_p is a global solution of an equation. We now study its local properties by the implicit function theorem as well as the related subgradients of the convex function K_p. We refer to [7, Part I] for the general theory of convex functions in infinite dimension.

For every $u \in L^1(\mu)$, let us write

$$q(u) = \exp_A(u - K_p(u) + \log_A p) \tag{8}$$

while $\widetilde{q}(u) = \text{escort}\,(q(u))$ denotes its escort density.

Proposition 4 *1. The functional $K_p \colon L^1(\mu) \to \mathbb{R}$ is Gateaux-differentiable with derivative*

$$\left.\frac{d}{dt} K_p(u + tv)\right|_{t=0} = \int v\widetilde{q}(u)\, d\mu\, .$$

It follows that $K_p \colon L^1(\mu) \to \mathbb{R}$ is monotone and globally Lipschitz.

2. *For every $u, v \in L^1(\mu)$, the inequality*

$$K_p(u + v) - K_p(u) \geq \int v\widetilde{q}(u)\, d\mu$$

holds, i.e., the density $\widetilde{q}(u) \in L^\infty(\mu)$ is the unique subgradient of K_p at u.

Proof 1. Consider the equation

$$F(t, \kappa) = \int \exp_A(u + tv - \kappa + \log_A p)\, d\mu - 1 = 0, \quad t, \kappa \in \mathbb{R}\, ,$$

so that $\kappa = K_p(u + tv)$. Derivations under the integral hold by virtue of the bounds

$$\left|\frac{\partial}{\partial t} \exp_A(u + tv - \kappa + \log_A p)\right| = \\ \left|A(\exp_A(u + tv - \kappa + \log_A p))v\right| \leq \|A\|_\infty\, |v|$$

and

$$\left|\frac{\partial}{p} \partial\kappa \exp_A(u + tv - \kappa + \log_A p)\right| = \left|A(\exp_A(u + tv - \kappa + \log_A p))\right| \leq \|A\|_\infty\, .$$

Furthermore, the partial derivative with respect to κ is never zero. Thanks to the implicit function theorem, there exists the derivative $(d\kappa/dt)_{t=0}$ which is the desired Gateaux derivative. Since $\widetilde{q}(u)$ is positive and bounded, K_p is monotone and globally Lipschitz.

2. Thanks to the convexity of $\exp_A$ and the derivation formula, we have

$$\exp_A(u + v - K_p(u + v) + \log_A p) \geq q + A(q)(v - (K_p(u + v) - K_p(v)))\, ,$$

where $q = \exp_A(u - K_p(u) + \log_A p)$. If we take μ-integral of both sides,

$$0 \geq \int vA(q)\, d\mu - (K_p(u + v) - K_p(v)) \int A(q)\, d\mu\, .$$

Isolating the increment $K_p(u+v) - K_p(v)$, the desired inequality obtains. Therefore, $\widetilde{q}(u)$ is a subgradient of K_p at u. From Item 1. we deduce that $\widetilde{q}(u)$ is the unique subgradient and further $\widetilde{q}(u)$ is the Gateaux differential of K_p at u. □

We can also establish Fréchet-differentiability of the functional, under more stringent assumptions.

Proposition 5 *Let* $\alpha \geq 2$.

1. *The superposition operator*

$$L^\alpha(\mu) \ni v \mapsto \exp_A(v + \log_A p) \in L^1(\mu)$$

is continuously Fréchet differentiable with derivative

$$d\exp_A(v) = (h \mapsto A(\exp_A(v + \log_A p))h) \in \mathscr{L}(L^\alpha(\mu), L^1(\mu)) .$$

2. *The functional* $K_p : L^\alpha(\mu) \to \mathbb{R}$, *implicitly defined by the equation*

$$\int \exp_A(v - K_p(v) + \log_A p)\, d\mu = 1, \quad v \in L^\alpha(\mu)$$

is continuously Fréchet differentiable with derivative

$$dK_p(v) = (h \mapsto \int h\widetilde{q}(v)\, d\mu) ,$$

where $\widetilde{q}(u) = \text{escort}\,(q(u))$.

Proof 1. Setting $\beta = 1$ in Proposition 1, we get easily the assertion. It remains just to check that the Fréchet derivative is continuous, i.e., that the Fréchet derivative is a continuous map $L^\alpha(\mu) \to \mathscr{L}(L^\alpha(\mu), L^1(\mu))$. If $\|h\|_{L^\alpha(\mu)} \leq 1$ and $v, w \in L^\alpha(\mu)$ we have

$$\int \left|(A[\exp_A(v + \log_A p)] - A[\exp_A(w + \log_A p)])h\right|\, d\mu$$
$$\leq \|A[\exp_A(v + \log_A p) - A[\exp_A(w + \log_A p)]\|_{L^\sigma(\mu)} ,$$

where $\sigma = \alpha/(\alpha - 1)$ is the conjugate exponent ofα. On the other hand,

$$\|A[\exp_A(v + \log_A p) - A[\exp_A(w + \log_A p)]\|_{L^\sigma(\mu)}$$
$$\leq \left\|A'\right\|_\infty \|A\|_\infty \|v - w\|_{L^\sigma(\mu)}$$

and so the map $L^\alpha(\mu) \to \mathscr{L}(L^\alpha(\mu), L^1(\mu))$ is continuous whenever $\alpha \geq \sigma$, i.e., $\alpha \geq 2$.

2. Fréchet differentiability of K_p is a consequence of the Implicit Function Theorem in Banach spaces, see [6], applied to the C^1-mapping

$$L^\alpha(\mu) \times \mathbb{R} \ni (v, \kappa) \mapsto \int \exp_A(v - \kappa + \log_A p)\, d\mu \, .$$

The value of the derivative is given by Proposition 4. □

4 Deformed Divergence

In analogy with the standard exponential case, define the A-divergence between probability densities as

$$D_A(q \| p) = \int \left(\log_A q - \log_A p\right) \operatorname{escort}(q)\, d\mu, \quad \text{for } q, p \in \mathscr{P} \, .$$

Since $\log_A$ is strictly concave with derivative $1/A$, we have

$$\log_A(x) \leq \log_A(y) + \frac{1}{A(y)}(x - y)$$

for all $x, y > 0$ and with equality if, and only if, $x = y$. Hence

$$A(y)\left(\log_A(y) - \log_A(x)\right) \geq y - x \, . \tag{9}$$

It follows in particular that $D_A(\cdot \| \cdot)$ is a well defined, possibly extended valued, function.

Observe further that by Proposition 2, $\log_A q - \log_A p \in L^1(\mu)$, and so $D_A(q \| p) < \infty$, whenever $q = q(u)$.

The binary relation D_A is a faithful divergence in that it satisfies the following Gibbs' inequality.

Proposition 6 *It holds $D_A(q \| p) \geq 0$ and $D_A(q \| p) = 0$ if and only if $p = q$.*

Proof From inequality (9) it follows

$$\begin{aligned} D_A(q \| p) &= \frac{1}{\int A(q)\, d\mu} \int \left(\log_A q - \log_A p\right) A(q)\, d\mu \\ &\geq \frac{1}{\int A(q)\, d\mu} \int (q - p)\, d\mu = 0. \end{aligned}$$

Moreover, equality holds if and only if $p = q$ μ-a.e. □

There are other alternative definitions that may fully candidate to be a divergence measure. For instance:

$$I_A(q\|p) = -\int \log_A(p/q)q\, d\mu.$$

or also

$$\widetilde{D}_A(q\|p) = \int A(q/p)\log_A(p/q)p\, d\mu.$$

By means of the concavity of $\log_A$, it is not difficult to check that both satisfy Gibbs' condition of Proposition 6, as well as they equal the Kullback–Leibner functional in the non-deformed case. Observe further that the functional $I_A(q\|p)$ is closely related to Tallis' divergence (see [25] and also [14]). In fact, if one replaces $\log_A$ with the q-logarithm, one gets just Tallis' q-divergence.

However our formulation for the divergence is motivated by the structure of the deformed exponential representation. As it will be now seen, our definition of divergence is more adapted to the present setting and it turns out be closely related to the normalizing operator.

In the equation

$$q = \exp_A(u - K_p(u) + \log_A p), \quad u \in L^1(\mu)\,,\ q \in \mathscr{P}\,, \tag{10}$$

the random variable u is identified up to an additive constant for any fixed density q. There are at least two options for selecting an interesting representative member in the equivalence class.

One option is to impose the further condition $\int u\widetilde{p}\, d\mu = 0$, where $\widetilde{p} = \text{escort}\,(p)$, the integral being well defined, given that the escort density is bounded. This restriction provides a unique element u_q. On the other hand, if we solve Eq. (10) with respect to $u - K(u)$, we get the desired relation:

$$K_p(u_q) = E_{\widetilde{p}}\left[\log_A p - \log_A q\right] = D_A(p\|q), \tag{11}$$

where $u = u_q$ is uniquely characterized by the two equations: $E_{\widetilde{p}}[u] = 0$ and $q = \exp_A(u - K_p(u) + \log_A p)$.

Observe further that Eq. (11) entails the relation

$$K_p(u) = D_A(p\|q(u)) \quad \forall u \in L^1(\mu)\,.$$

The previous choice is that followed in the construction of the non-parametric exponential manifold, see [22, 23].

With regard to the non-deformed case, Eq. (11) yields the Kulback-Leibler divergence with p and q exchanged, with respect to what is considered more natural in Statistical Physics, see for example the comments [13].

For this purpose, we undertake another choice for the random variable in the equivalence class. More specifically, in Eq. (10) the random variable u will be now centered with respect to $\widetilde{q} = \text{escort}\,(q)$, i.e., $E_{\widetilde{q}}[u] = 0$.

To avoid confusion let us rewrite Eq. (10) as follows and where for convenience the function K_p is replaced with $H_p = -K_p$:

$$q = \exp_A(v + H_p(v) + \log_A p), \quad v \in L^1(\mu), \quad E_{\tilde{q}}[v] = 0, \tag{12}$$

so that

$$H_p(v_q) = E_{\tilde{q}}\left[\log_A q - \log_A p\right] = D_A(q \| p),$$

where $v = v_q$ is the solution to the two equations $E_{\tilde{q}}[v] = 0$ and $q = \exp_A(v + H_p(v) + \log_A p)$. There are hence two notable representations of the same probability density q:

$$q = \exp_A(u - K_p(u) + \log_A p) = \exp_A(v + H_p(v) + \log_A p)$$

which implies $u_q - v_q = K_p(u_q) + H_p(v_q)$. This, in turn, leads to

$$-E_{\tilde{p}}\left[v_q\right] = E_{\tilde{q}}\left[u_q\right] = K_p(u_q) + H_p(v_q) = K_p(u_q) - K_p(v_q).$$

This provides the following remarkable relation

$$H_p(v_q) = E_{\tilde{q}}\left[u_q\right] - K_p(u_q). \tag{13}$$

4.1 Variational Formula

We now present a variational formula in the spirit of the classical one by Donsker-Varadhan. Next proposition provides the convex conjugate of K_p, in the duality $L^\infty(\mu) \times L^1(\mu)$.

In what follows, the operator $\eta \mapsto \hat{\eta}$ denotes the inverse of the escort operator, i.e., $\eta = \text{escort}\left(\hat{\eta}\right)$. In the light of the results established in Sect. 3.1, this operator maps a dense subset of $\overline{\mathscr{P}} \cap L^\infty(\mu)$ onto $\overline{\mathscr{P}}$.

Proposition 7 *1. The convex conjugate function of K_p:*

$$K_p^*(w) = \sup_{u \in L^1(\mu)} \left(\int wu \, d\mu - K_p(u) \right), \quad w \in L^\infty(\mu) \tag{14}$$

has domain contained into $\overline{\mathscr{P}} \cap L^\infty(\mu)$. More precisely,

$$\text{escort}(\mathscr{P}) \subseteq dom K_p^* \subseteq \overline{\mathscr{P}} \cap L^\infty(\mu).$$

2. *$K_p^*(w) \geq 0$ for all $w \in L^\infty(\mu)$. For any $\eta \in$ escort $(\mathscr{P})$, the conjugate $K_p^*(\eta)$ is given by the* Legendre transform*:*

$$K_p^*(\eta) = \int \eta\, u_{\hat{\eta}}\, d\mu - K_p(u_{\hat{\eta}})\ .$$

So that $K_p^*(\eta) = H_p(v_{\hat{\eta}}) = D_A(\hat{\eta}\|p)$; *equivalently:*

$$K_p^*(\text{escort}\,(q)) = D_A(q\|p) \quad \forall p, q \in L^1(\mu).$$

3. *It holds the inversion formula*

$$K_p\,(u) = \max_{\eta \in \text{escort}(\mathscr{P})} \left(\int \eta u\, d\mu - D_A(\hat{\eta}\|p) \right)$$

$$= \max_{q \in \mathscr{P}} \left(\int \text{escort}\,(q)\, u\, d\mu - D_A(q\|p) \right), \quad \forall u \in L^1(\mu).$$

Proof 1. It follows from the fact that K_p is monotone and translation invariant. Let us first suppose $w \notin L_+^\infty(\mu)$. That means that

$$\int w\chi_C\, d\mu < 0$$

is true for some indicator function χ_C. If we consider the cone generated by the function $-\chi_C$, we can write

$$K_p^*\,(w) \geq \sup_{u \in\, cone(-\chi_C)} \left(\int wu\, d\mu - K_p\,(u) \right) \geq \sup_{u \in\, cone(-\chi_C)} \int wu\, d\mu = +\infty,$$

since $K_p\,(u) \leq 0$ when $u \in cone(-\chi_C)$. Now consider the case in which $w \geq 0$. If we set $u = \lambda \in \mathbb{R}$, we have $K_p(\lambda) = \lambda$ and consequently

$$K_p^*(w) \geq \sup_{\lambda \in \mathbb{R}} \left(\lambda \int w\, d\mu - \lambda \right)\ . \tag{15}$$

This sup is $+\infty$, unless $\int w\, d\mu = 1$. Hence, $K_p^*\,(w) < \infty$ implies $w \in \overline{\mathscr{P}}$. Summarizing, the domain of K_p^* is contained into $\overline{\mathscr{P}} \cap L^\infty(\mu)$, and this proves one of the two claimed inclusions. The other one will be a direct consequence of the next point.

2. Equation (15) implies $K_p^* \geq 0$. By Proposition 4 the concave and Gateaux differentiable function $u \mapsto \int \eta u\, d\mu - K_p(u)$ has derivative at u given by $\eta - dK_p(u) = \eta - \text{escort}\,(q(u))$, where $q(u) = \exp_A(u - K_p(u) + \log_A p)$. Under our assumptions, the derivative vanishes at $u = u_{\hat{\eta}}$ and the sup in the definition of K_p^* is attained at that point. The maximum value is $K_p^*(\eta) = \int \eta u\, d\mu - K_p(u)$, by setting $u = u_{\hat{\eta}}$.
   The last formula follows straightforward from Eq. (13).
3. For a well-known property of Fenchel–Moreau duality theory, we have:

$$K_p(u) \geq \int wu \, d\mu - K_p^*(w) \quad \forall u \in L^1(\mu), \quad \forall w \in L^\infty(\mu)$$

$$K_p(u) = \int wu \, d\mu - K_p^*(w) \iff w \in \partial K_p(u).$$

Clearly in our case $\partial K_p(u)$ is a singleton and the image of ∂K_p is the set escort $(\mathscr{P})$. Therefore

$$K_p(u) = \max_{w \in \operatorname{escort}(\mathscr{P})} \left(\int wu \, d\mu - K_p^*(w) \right).$$

By Item 2 the desired inversion formula obtains. □

5 Hilbert Bundle Based on $\exp_A$

We shall introduce the Hilbert manifold of probability densities as defined in [18, 19]. A slightly more general set-up than the one used in that reference will be introduced. By means of a general A function, we provide an atlas of charts, and define a linear bundle as an expression of the tangent space.

Let $\mathscr{P}(\mu)$ denote the set of all μ-densities on the probability space $(\mathbb{X}, \mathscr{X}, \mu)$ of the kind

$$q = \exp_A(u - K_1(u)), \quad u \in L^2(\mu), \quad E_\mu[u] = 0. \tag{16}$$

Notice that $1 \in \mathscr{P}(\mu)$ because we can take $u = 0$.

Proposition 8 *1. $\mathscr{P}(\mu)$ is the set of all densities q such that $\log_A q \in L^2(\mu)$, in which case $u = \log_A q - E_\mu\left[\log_A q\right]$.*

2. *If in addition $A'(0+) > 0$, then $\mathscr{P}(\mu)$ is the set of all densities q such that both q and $\log q$ are in $L^2(\mu)$.*
3. *Let $A'(0+) > 0$. On a product space with reference probability measures μ_1 and μ_2, and densities respectively q_1 and q_2, we have $q_1 \in \mathscr{P}(\mu_1)$ and $q_2 \in \mathscr{P}(\mu_2)$ if, and only if, $q_1 \otimes q_2 \in \mathscr{P}(\mu_1 \otimes \mu_2)$.*

Proof 1. From Eq. (16), it follows $\log_A q = u - K_1(u) \in L^2(\mu)$, provided $u \in L^2(\mu)$. Conversely, let $\log_A q \in L^2(\mu)$. Equation (16) yields

$$u = \log_A q - K_1(u) \quad and \quad K_1(u) = -\log_A q.$$

Therefore $u = \log_A q - E_\mu\left[\log_A q\right]$ and $u \in L^2(\mu)$.

2. Write

$$\left|\log_A q\right|^2 = \left|\log_A q\right|^2 (q < 1) + \left|\log_A q\right|^2 (q \geq 1)$$

and use the bounds of Eqs. (3) and (4) to get

$$E_\mu\left[\left|\log_A q\right|^2\right] \le \frac{1}{\alpha_2^2} E_\mu\left[|\log q|^2 (q<1)\right] + \frac{1}{A(1)^2} E_\mu\left[|q-1|^2 (q \ge 1)\right] \le \frac{1}{\alpha_2^2} E_\mu\left[|\log q|^2\right] + \frac{1}{A(1)^2} E_\mu\left[q^2\right] .$$

We deduce that the two conditions q and $\log q$ in $L^2(\mu)$ imply $\log_A q \in L^2(\mu)$. Conversely, let $\log_A q \in L^2(\mu)$. By means of the other two bounds (recall that $\alpha_1 > 0$) we have too

$$E_\mu\left[\left|\log_A q\right|^2\right] \ge \frac{1}{\alpha_1^2} E_\mu\left[|\log q|^2 (q<1)\right] + \frac{1}{\|A\|_\infty^2} E_\mu\left[(q-1)^2 (q \ge 1)\right] .$$

Consequently, $E_\mu\left[(q-1)^2(q \ge 1)\right] < +\infty$. This in turn gives $E_\mu\left[(q-1)^2\right] < +\infty$, and so $q \in L^2(\mu)$.
Once again, the previous inequality provides the condition $E_\mu\left[|\log q|^2 (q<1)\right] < +\infty$. On the other hand, $E_\mu\left[|\log q|^2 (q \ge 1)\right] < +\infty$ since $|\log q|^2 (q \ge 1) \le (q-1)^2(q \ge 1)$. Therefore, $\log q \in L^2(\mu)$.

3. We deduce by the previous item that: $q_1 \otimes q_2 \in \mathscr{P}(\mu_1 \otimes \mu_2)$ if and only if both $q_1 \otimes q_2$ and $\log(q_1 \otimes q_2)$ are in $L^2(\mu_1 \otimes \mu_2)$.

The first condition is equivalent to both $q_1 \in L^2(\mu_1)$ and $q_2 \in L^2(\mu_2)$. The second one is equivalent to $\log q_1 + \log q_2 \in L^2(\mu_1 \otimes \mu_2)$. On the other hand, we have

$$E_{\mu_1 \otimes \mu_2}\left[(\log q_1 + \log q_2)^2\right] = E_{\mu_1}\left[\log^2 q_1\right] + E_{\mu_2}\left[\log^2 q_2\right] + 2\, E_{\mu_1}\left[\log q_1\right] E_{\mu_2}\left[\log q_2\right]. \tag{17}$$

By Eq. (17), $q_1 \in \mathscr{P}(\mu_1)$ and $q_2 \in \mathscr{P}(\mu_2)$ imply $q_1 \otimes q_2 \in \mathscr{P}(\mu_1 \otimes \mu_2)$.
Conversely, assume $q_1 \otimes q_2 \in \mathscr{P}(\mu_1 \otimes \mu_2)$. This implies that it holds, $E_{\mu_1 \otimes \mu_2}\left[(\log q_1 + \log q_2)^2\right] < +\infty$. Since $E_{\mu_i}\left[\log q_i\right] \le E_{\mu_1}\left[q_i - 1\right] = 0$. We have $E_{\mu_1}\left[\log q_1\right] E_{\mu_2}\left[\log q_2\right] \ge 0$. In view of Eq. (17), we can infer that $q_1 \in \mathscr{P}(\mu_1)$ and $q_2 \in \mathscr{P}(\mu_2)$ □

We proceed now to define an Hilbert bundle with base $\mathscr{P}(\mu)$. The notion of Hilbert bundle has been introduced in Information Geometry by [1]. We are here using an adaptation to the A-exponential of arguments elaborated by [8, 21]. Notice that the construction depends in a essential way on the specific conditions we are assuming for the present class of deformed exponential.

At each $q \in \mathscr{P}(\mu)$ the escort density $\widetilde{q}$ is bounded, so that we can define the fiber given by the Hilbert spaces

$$\mathscr{H}_q = \left\{u \in L^2(\mu) | E_{\widetilde{q}}\left[u\right] = 0\right\}$$

with scalar product $\langle u, v\rangle_q = \int uv \, d\mu$. The Hilbert bundle is

$$H\mathscr{P}(\mu) = \left\{(q, u) | q \in \mathscr{P}(\mu), u \in \mathscr{H}_q\right\} .$$

For each $p, q \in \mathscr{P}(\mu)$ the mapping $\mathbb{U}_p^q u = u - E_{\tilde{q}}[u]$ is a continuous linear mapping from $\mathscr{H}_p$ to $\mathscr{H}_q$. Moreover, $\mathbb{U}_q^r \mathbb{U}_p^q = \mathbb{U}_p^r$. In particular, $\mathbb{U}_q^p \mathbb{U}_p^q$ is the identity on $\mathscr{H}_p$ and so $\mathbb{U}_p^q$ is an isomorphism of $\mathscr{H}_p$ onto $\mathscr{H}_q$.

In the next proposition an affine atlas of charts is constructed in order to define our Hilbert bundle which is an expression of the tangent bundle. The velocity of a curve $t \mapsto p(t) \in \mathscr{P}(\mu)$ is given in the Hilbert bundle by the so called A-score that, in our case, takes the form $A(p(t))^{-1}\dot{p}(t)$, where $\dot{p}(t)$ is computed in $L^1(\mu)$.

The following proposition is taken from [16] where a detailed proof is presented.

Proposition 9 *1. Fix $p \in \mathscr{P}(\mu)$. A positive density $q \in \mathscr{P}(\mu)$ if and only if*

$$q = \exp_A(u - K_p(u) + \log_A p), \text{ with } u \in L^2(\mu) \text{ and } E_{\tilde{p}}[u] = 0.$$

2. *For any fixed $p \in \mathscr{P}(\mu)$ the mapping $s_p \colon \mathscr{P}(\mu) \to \mathscr{H}_p$ defined by*

$$q \mapsto \log_A q - \log_A p + D_A(p \| q)$$

is injective and surjective, with inverse $e_p(u) = \exp_A(u - K_p(u) + \log_A p)$.

3. *The atlas $\left\{s_p | p \in \mathscr{P}(\mu)\right\}$ is affine with transitions*

$$s_q \circ e_p(u) = \mathbb{U}_p^q u + s_p(q) .$$

4. *The velocity of the differentiable curve $t \mapsto p(t) \in \mathscr{P}(\mu)$ in the chart s_p is $ds_p(p(t))/dt \in \mathscr{H}_p$. Conversely, given any $u \in \mathscr{H}_p$, the curve*

$$p \colon t \mapsto \exp_A(tu - K_p(tu) + \log_A p)$$

satisfies $p(0) = p$ and has velocity u at $t = 0$, expressed in the chart s_p. If the velocity of a curve is $t \mapsto \dot{u}(t)$, in a chart s_p, then $\mathbb{U}_p^q \dot{u}(t)$ is its velocity in the chart s_q.

5. *If $t \mapsto p(t) \in \mathscr{P}(\mu)$ is differentiable with respect to the atlas then it is differentiable as a mapping in $L^1(\mu)$. It follows that the A-score is well-defined and is the expression of the velocity of the curve $t \mapsto p(t)$ in the moving chart $t \mapsto s_{p(t)}$.*

We end here our discussion of the geometry of the Hilbert bundle, because our aim is limited to show the applicability of the analytic results obtained in the previous section. A detailed discussion of the relevant geometric objects e.g., the affine covariant derivative, is not attempted here.

6 Final Remarks

A non-parametric Hilbert manifold based on a deformed exponential representation of positive densities has been firstly introduced by N.J. Newton [18, 19]. We have derived regularity properties of the normalizing functional K_p and discussed the relevant Fenchel conjugation. In particular, we have discussed some properties of the escort mapping and a form of the divergence that appears to be especially adapted to our set-up. We have taken a path different from that of N.J. Newton original presentation. We allow for a manifold defined by an atlas containing charts centered at each density in the model. In conclusion, we have discussed explicitly a version of the Hilbert bundle as a family of codimension 1 sub-vector spaces of the basic Hilbert space.

Acknowledgements The Authors wish to thank the anonymous referees whose comments have led to a considerable improvement of the paper. L. Montrucchio is Honorary Fellow of the Collegio Carlo Alberto Foundation. G. Pistone is a member of GNAMPA-INdAM and acknowledges the support of de Castro Statistics and Collegio Carlo Alberto.

References

1. Amari, S.: Dual connections on the Hilbert bundles of statistical models. In: Geometrization of Statistical Theory (Lancaster, 1987), pp. 123–151. ULDM Publ. (1987)
2. Amari, S., Nagaoka, H.: Methods of Information Geometry. American Mathematical Society, Providence (2000). Translated from the 1993 Japanese original by Daishi Harada
3. Ambrosetti, A., Prodi, G.: A Primer of Nonlinear Analysis. Cambridge Studies in Advanced Mathematics, vol. 34. Cambridge University Press, Cambridge (1993)
4. Appell, J., Zabrejko, P.P.: Nonlinear Superposition Operators. Cambridge Tracts in Mathematics, vol. 95. Cambridge University Press, Cambridge (1990). https://doi.org/10.1017/CBO9780511897450
5. Ay, N., Jost, J., Lê, H.V., Schwachhöfer, L.: Information Geometry. Springer, Berlin (2017)
6. Dieudonné, J.: Foundations of Modern Analysis. Academic Press, New York (1960)
7. Ekeland, I., Témam, R.: Convex Analysis and Variational Problems. Classics in Applied Mathematics, vol. 28, English edn. Society for Industrial and Applied Mathematics (SIAM) (1999). https://doi.org/10.1137/1.9781611971088. Translated from the French
8. Gibilisco, P., Pistone, G.: Connections on non-parametric statistical manifolds by Orlicz space geometry. IDAQP **1**(2), 325–347 (1998)
9. Kaniadakis, G.: Non-linear kinetics underlying generalized statistics. Phys. A **296**(3–4), 405–425 (2001)
10. Kaniadakis, G.: Statistical mechanics in the context of special relativity. Phys. Rev. E **66**, 056, 125 1–17 (2002)
11. Kaniadakis, G.: Statistical mechanics in the context of special relativity. ii. Phys. Rev. E **72**(3), 036,108 (2005). https://doi.org/10.1103/PhysRevE.72.036108
12. Kass, R.E., Vos, P.W.: Geometrical Foundations of Asymptotic Inference. Wiley Series in Probability and Statistics: Probability and Statistics. Wiley, New York (1997). https://doi.org/10.1002/9781118165980. A Wiley-Interscience Publication
13. Landau, L.D., Lifshits, E.M.: Course of Theoretical Physics. Statistical Physics, vol. V, 3rd edn. Butterworth-Heinemann, Oxford (1980)

14. Loaiza, G., Quiceno, H.R.: A q-exponential statistical Banach manifold. J. Math. Anal. Appl. **398**(2), 466–476 (2013). https://doi.org/10.1016/j.jmaa.2012.08.046
15. Malliavin, P.: Integration and Probability. Graduate Texts in Mathematics, vol. 157. Springer-Verlag (1995). With the collaboration of Hlne Airault, Leslie Kay and Grard Letac. Edited and translated from the French by Kay, With a foreword by Mark Pinsky
16. Montrucchio, L., Pistone, G.: Deformed exponential bundle: the linear growth case. In: Nielsen, F., Barbaresco, F. (eds.) Geometric Science of Information. LNCS, vol. 10589, pp. 239–246. Springer (2017). Proceedings of the Third International Conference, GSI 2017, Paris, France, November 7–9
17. Naudts, J.: Generalised Thermostatistics. Springer-Verlag London Ltd. (2011). https://doi.org/10.1007/978-0-85729-355-8
18. Newton, N.J.: An infinite-dimensional statistical manifold modelled on Hilbert space. J. Funct. Anal. **263**(6), 1661–1681 (2012). https://doi.org/10.1016/j.jfa.2012.06.007
19. Newton, N.J.: Infinite-dimensional statistical manifolds based on a balanced chart. Bernoulli **22**(2), 711–731 (2016). https://doi.org/10.3150/14-BEJ673
20. Pistone, G.: κ-exponential models from the geometrical viewpoint. Eur. Phys. J. B Condens. Matter Phys. **71**(1), 29–37 (2009). https://doi.org/10.1140/epjb/e2009-00154-y
21. Pistone, G.: Nonparametric information geometry. In: Nielsen, F., Barbaresco, F. (eds.) Geometric Science of Information. Lecture Notes in Computer Science, vol. 8085, pp. 5–36. Springer, Heidelberg (2013). Proceedings First International Conference, GSI 2013 Paris, France, August 28–30
22. Pistone, G., Rogantin, M.: The exponential statistical manifold: mean parameters, orthogonality and space transformations. Bernoulli **5**(4), 721–760 (1999)
23. Pistone, G., Sempi, C.: An infinite-dimensional geometric structure on the space of all the probability measures equivalent to a given one. Ann. Stat. **23**(5), 1543–1561 (1995)
24. Shima, H.: The Geometry of Hessian Structures. World Scientific Publishing Co. Pte. Ltd., Hackensack, NJ (2007). https://doi.org/10.1142/9789812707536
25. Tsallis, C.: Possible generalization of Boltzmann-Gibbs statistics. J. Stat. Phys. **52**(1–2), 479–487 (1988)
26. Vigelis, R.F., Cavalcante, C.C.: On ϕ-families of probability distributions. J. Theor. Probab. **26**, 870–884 (2013)

Statistical Manifolds Admitting Torsion and Partially Flat Spaces

Masayuki Henmi and Hiroshi Matsuzoe

Abstract It is well-known that a contrast function defined on a product manifold $M \times M$ induces a Riemannian metric and a pair of dual torsion-free affine connections on the manifold M. This geometrical structure is called a statistical manifold and plays a central role in information geometry. Recently, the notion of pre-contrast function has been introduced and shown to induce a similar differential geometrical structure on M, but one of the two dual affine connections is not necessarily torsion-free. This structure is called a statistical manifold admitting torsion. The notion of statistical manifolds admitting torsion has been originally introduced to study a geometrical structure which appears in a quantum statistical model. However, it has been shown that an estimating function which is used in "classical" statistics also induces a statistical manifold admitting torsion through its associated pre-contrast function. The aim of this paper is to summarize such previous results. In particular, we focus on a partially flat space, which is a statistical manifold admitting torsion where one of its dual connections is flat. In this space, it is possible to discuss some properties similar to those in a dually flat space, such as a canonical pre-contrast function and a generalized projection theorem.

1 Introduction

A statistical manifold is a Riemannian manifold with a pair of dual torsion-free affine connections and it plays a central role in information geometry. This geometrical structure is induced from an asymmetric (squared) distance-like smooth function called a contrast function by taking its second and third derivatives [1, 2]. The Kullback–Leibler divergence on a regular parametric statistical model is a typical example of contrast functions and its induced geometrical objects are the Fisher met-

M. Henmi (✉)
The Institute of Statistical Mathematics, 10-3 Midori-cho, Tachikawa, Tokyo 190-8562, Japan
e-mail: henmi@ism.ac.jp

H. Matsuzoe
Nagoya Institute of Technology, Gokiso-cho, Showa-ku, Nagoya, Aichi 466-8555, Japan
e-mail: matsuzoe@nitech.ac.jp

F. Nielsen (ed.), *Geometric Structures of Information*, Signals and Communication Technology, https://doi.org/10.1007/978-3-030-02520-5_3

ric, the exponential and mixture connections. The geometrical structure determined by these objects plays an important role in the geometry of statistical inference, as is widely known [3, 4].

A statistical manifold admitting torsion (SMAT) is a Riemannian manifold with a pair of dual affine connections, where only one of them must be torsion-free but the other is *not* necessarily so. This geometrical structure naturally appears in a quantum statistical model (i.e. a set of density matrices representing quantum states) [3] and the notion of SMAT was originally introduced to study such a geometrical structure from a mathematical point of view [5]. A pre-contrast function was subsequently introduced as a generalization for the first derivative of a contrast function and it was shown that an pre-contrast function induces a SMAT by taking its first and second derivatives [6].

In statistics, an estimating function is a function defined on a direct product of parameter and sample spaces, and it is used to obtain an estimator by solving its corresponding estimating equation. Henmi and Matsuzoe [7] showed that a SMAT also appears in "classical" statistics through an estimating function. More precisely, an estimating function naturally defines a pre-contrast function on a parametric statistical model and a SMAT is induced from it.

This paper summarizes such previous results, focusing on a SMAT where one of its dual connections is flat. We call this geometrical structure a partially flat space. Although this space is different from a dually flat space in general since one of the dual connections in a SMAT possibly has torsion, some similar properties hold. For example, the canonical pre-contrast function can be naturally defined on a partially flat space, which is an analog of the canonical contrast function (or canonical divergence) in a dually flat space. In addition, a generalized projection theorem holds with respect to the canonical pre-contrast function. This theorem can be seen as a generalization of the projection theorem in a dually flat space. This paper is an extended version of the conference proceedings [8]. We consider a statistical problem to see an example of statistical manifolds admitting torsion induced from estimating functions and discuss some future problems, neither of which were included in [8].

2 Statistical Manifolds and Contrast Functions

Through this paper, we assume that all geometrical objects on differentiable manifolds are smooth and restrict our attention to Riemannian manifolds, although the most of the concepts can be defined for semi-Riemannian manifolds.

Let (M, g) be a Riemannian manifold and ∇ be an affine connection on M. The *dual connection* ∇^* of ∇ with respect to g is defined by

$$Xg(Y, Z) = g(\nabla_X Y, Z) + g(Y, \nabla^*_X Z) \quad (\forall X, \forall Y, \forall Z \in \mathscr{X}(M)),$$

where $\mathscr{X}(M)$ is the set of all vector fields on M.

For an affine connection ∇ on M, its curvature tensor field R and torsion tensor field T are defined by the following equations as usual:

$$R(X, Y)Z := \nabla_X \nabla_Y Z - \nabla_Y \nabla_X Z - \nabla_{[X,Y]} Z,$$
$$T(X, Y) := \nabla_X Y - \nabla_Y X - [X, Y]$$

$(\forall X, \forall Y, \forall Z \in \mathscr{X}(M))$. It is said that an affine connection ∇ is *torsion-free* if $T = 0$. Note that for a torsion-free affine connection ∇, $\nabla^* = \nabla$ implies that ∇ is the Levi-Civita connection with respect to g. Let R^* and T^* be the curvature and torsion tensor fields of ∇^*, respectively. It is easy to see that $R = 0$ always implies $R^* = 0$, but $T = 0$ does not necessarily imply $T^* = 0$.

Let ∇ be a torsion-free affine connection on a Riemannian manifold (M, g). Following [9], we say that (M, g, ∇) is a *statistical manifold* if and only if ∇g is a symmetric (0, 3)-tensor field, that is

$$(\nabla_X g)(Y, Z) = (\nabla_Y g)(X, Z) \quad (\forall X, \forall Y, \forall Z \in \mathscr{X}(M)). \tag{1}$$

This condition is equivalent to $T^* = 0$ under the condition that ∇ is a torsion-free. If (M, g, ∇) is a statistical manifold, so is (M, g, ∇^*) and it is called the *dual statistical manifold* of (M, g, ∇). Since ∇ and ∇^* are both torsion-free for a statistical manifold (M, g, ∇), $R = 0$ implies that ∇ and ∇^* are both flat. In this case, (M, g, ∇, ∇^*) is called a *dually flat space* [3].

Let ϕ be a real-valued function on the direct product $M \times M$ of a manifold M and $X_1, \ldots, X_i, Y_1, \ldots, Y_j$ be vector fields on M. The functions $\phi[X_1, \ldots, X_i | Y_1, \ldots, Y_j]$, $\phi[X_1, \ldots, X_i | \,]$ and $\phi[\, | Y_1, \ldots, Y_j]$ on M are defined by the equations

$$\phi[X_1, \ldots, X_i | Y_1, \ldots, Y_j](r) := (X_1)_p \cdots (X_i)_p (Y_1)_q \cdots (Y_j)_q \phi(p, q)|_{p=r, q=r}, \tag{2}$$

$$\phi[X_1, \ldots, X_i | \,](r) := (X_1)_p \cdots (X_i)_p \phi(p, r)|_{p=r}, \tag{3}$$
$$\phi[\, | Y_1, \ldots, Y_j](r) := (Y_1)_q \cdots (Y_j)_q \phi(r, q)|_{q=r} \tag{4}$$

for any $r \in M$, respectively [1]. Using these notations, a *contrast function* ϕ on M is defined to be a real-valued function on $M \times M$ which satisfies the following conditions [1, 2]:

(*a*) $\phi(p, p) = 0 \quad (\forall p \in M)$,
(*b*) $\phi[X | \,] = \phi[\, | X] = 0 \quad (\forall X \in \mathscr{X}(M))$,
(*c*) $g(X, Y) := -\phi[X | Y] \quad (\forall X, \forall Y \in \mathscr{X}(M))$ is a Riemannian metric on M.

Note that these conditions imply that

$$\phi(p, q) \geq 0, \quad \phi(p, q) = 0 \iff p = q$$

in some neighborhood of the diagonal set $\{(r, r)|r \in M\}$ in $M \times M$. Although a contrast function is not necessarily symmetric, this property means that a contrast function measures some discrepancy between two points on M (at least locally). For a given contrast function ϕ, the two affine connections ∇ and ∇^* are defined by

$$g(\nabla_X Y, Z) = -\phi[XY|Z], \quad g(Y, \nabla^*_X Z) = -\phi[Y|XZ]$$

$(\forall X, \forall Y, \forall Z \in \mathscr{X}(M))$. In this case, ∇ and ∇^* are both torsion-free and dual to each other with respect to g. This means that both of (M, g, ∇) and (M, g, ∇^*) are statistical manifolds. In particular, (M, g, ∇) is called the statistical manifold induced from the contrast function ϕ.

A typical example of contrast functions is the Kullback–Leibler divergence on a statistical model. Let $S = \{p(\boldsymbol{x}; \boldsymbol{\theta}) \mid \boldsymbol{\theta} = (\theta^1, \dots, \theta^d) \in \Theta \subset \boldsymbol{R}^d\}$ be a regular parametric statistical model, which is a set of probability density functions with respect to a dominating measure ν on a sample space Ω. Each element is indexed by a parameter (vector) $\boldsymbol{\theta}$ in an open subset Θ of $\boldsymbol{R}^d$ and the set S satisfies some regularity conditions, under which S can be seen as a differentiable manifold. The Kullback–Leibler divergence of the two density functions $p_1(\boldsymbol{x}) = p(\boldsymbol{x}; \boldsymbol{\theta}_1)$ and $p_2(\boldsymbol{x}) = p(\boldsymbol{x}; \boldsymbol{\theta}_2)$ in S is defined to be

$$\phi_{KL}(p_1, p_2) := \int_\Omega p_2(\boldsymbol{x}) \log \frac{p_2(\boldsymbol{x})}{p_1(\boldsymbol{x})} \nu(d\boldsymbol{x}).$$

It is easy to see that the Kullback–Leibler divergence satisfies the conditions (a), (b) and (c), and so it is a contrast function on S. Its induced Riemannian metric and dual connections are Fisher metric g^F, the exponential connection $\nabla^{(e)}$ and mixture connection $\nabla^{(m)}$, respectively. They are given as follows:

$$g^F_{jk}(\boldsymbol{\theta}) := g^F(\partial_j, \partial_k) = E_{\boldsymbol{\theta}}\{s^j(\boldsymbol{x}, \boldsymbol{\theta}) s^k(\boldsymbol{x}, \boldsymbol{\theta})\},$$

$$\begin{cases} \Gamma^{(e)}_{ij,k}(\boldsymbol{\theta}) := g^F(\nabla^{(e)}_{\partial_i} \partial_j, \partial_k) = E_{\boldsymbol{\theta}}[\{\partial_i s^j(\boldsymbol{x}, \boldsymbol{\theta})\} s^k(\boldsymbol{x}, \boldsymbol{\theta})] \\ \Gamma^{(m)}_{ik,j}(\boldsymbol{\theta}) := g^F(\partial_j, \nabla^{(m)}_{\partial_i} \partial_k) = \int_\Omega s^j(\boldsymbol{x}, \boldsymbol{\theta}) \partial_i \partial_k p(\boldsymbol{x}; \boldsymbol{\theta}) \nu(d\boldsymbol{x}) \end{cases},$$

where $E_{\boldsymbol{\theta}}$ indicates that the expectation is taken with respect to $p(\boldsymbol{x}; \boldsymbol{\theta})$, $\partial_i = \frac{\partial}{\partial \theta^i}$ and $s^i(\boldsymbol{x}; \boldsymbol{\theta}) = \partial_i \log p(\boldsymbol{x}; \boldsymbol{\theta})$ $(i = 1, \dots, d)$. As is widely known, this geometrical structure plays the most fundamental and important role in the differential geometry of statistical inference [3, 4].

3 Statistical Manifolds Admitting Torsion and Pre-contrast Functions

A statistical manifold admitting torsion is an abstract notion for the geometrical structure where only one of the dual connections is allow to have torsion, which naturally appears in a quantum statistical model [3]. The definition is obtained by generalizing (1) in the definition of statistical manifold as follows [5].

Let (M, g) be a Riemannian manifold and ∇ be an affine connection on M. We say that (M, g, ∇) is a *statistical manifold admitting torsion* (SMAT for short) if and only if

$$(\nabla_X g)(Y, Z) - (\nabla_Y g)(X, Z) = -g(T(X, Y), Z) \quad (\forall X, \forall Y, \forall Z \in \mathscr{X}(M)). \quad (5)$$

This condition is equivalent to $T^* = 0$ in the case where ∇ possibly has torsion, and it reduces to (1) if ∇ is torsion-free. Note that (M, g, ∇^*) is not necessarily a statistical manifold although ∇^* is torsion-free. It should be also noted that (M, g, ∇^*) is a SMAT whenever a torsion-free affine connection ∇ is given on a Riemannian manifold (M, g).

For a SMAT (M, g, ∇), $R = 0$ does not necessarily imply that ∇ is flat, but it implies that ∇^* is flat since $R^* = 0$ and $T^* = 0$. In this case, we call (M, g, ∇, ∇^*) a *partially flat space*.

Let ρ be a real-valued function on the direct product $TM \times M$ of a manifold M and its tangent bundle TM, and $X_1, \ldots, X_i, Y_1, \ldots, Y_j, Z$ be vector fields on M. The function $\rho[X_1, \ldots, X_i Z | Y_1, \ldots, Y_j]$ on M is defined by

$$\rho[X_1, \ldots, X_i Z | Y_1, \ldots, Y_j](r) := (X_1)_p \cdots (X_i)_p (Y_1)_q \cdots (Y_j)_q \rho(Z_p, q)|_{p=r, q=r}$$

for any $r \in M$. Note that the role of Z is different from those of the vector fields in the notation of (2). The functions $\rho[X_1, \ldots, X_i Z | \,]$ and $\rho[\, | Y_1, \ldots, Y_j]$ are also defined in the similar way to (3) and (4).

We say that ρ is a *pre-contrast function* on M if and only if the following conditions are satisfied [6, 7]:

(*a*) $\rho(f_1 X_1 + f_2 X_2, q) = f_1 \rho(X_1, q) + f_2 \rho(X_2, q)$
$(\forall f_1, \forall f_2 \in C^\infty(M),\ \forall X_1, \forall X_2 \in \mathscr{X}(M),\ \forall q \in M)$.

(*b*) $\rho[X | \,] = 0 \ (\forall X \in \mathscr{X}(M)) \ \big(i.e.\ \rho(X_p, p) = 0 \ (\forall p \in M)\big)$.

(*c*) $g(X, Y) := -\rho[X | Y] \ (\forall X, \forall Y \in \mathscr{X}(M))$ is a Riemannian metric on M.

Note that for any contrast function ϕ on M, the function ρ_ϕ which is defined by

$$\rho_\phi(X_p, q) := X_p \phi(p, q) \quad (\forall p, \forall q \in M,\ \forall X_p \in T_p(M))$$

is a pre-contrast function on M. The notion of pre-contrast function is obtained by taking the fundamental properties of the first derivative of a contrast function as axioms. For a given pre-contrast function ρ, two affine connections ∇ and ∇^* are defined by

$$g(\nabla_X Y, Z) = -\rho[XY | Z], \quad g(Y, \nabla^*_X Z) = -\rho[Y | XZ]$$

$(\forall X, \forall Y, \forall Z \in \mathscr{X}(M))$ in the same way as for a contrast function. In this case, ∇ and ∇^* are dual to each other with respect to g and ∇^* is torsion-free. However, the affine connection ∇ possibly has torsion. This means that (M, g, ∇) is a SMAT and it is called the SMAT induced from the pre-contrast function ρ.

4 Canonical Pre-contrast Functions in Partially Flat Spaces

In a dually flat space (M, g, ∇, ∇^*), it is well-known that the canonical contrast functions (called ∇ and ∇^*- divergences) are naturally defined, and the Pythagorean theorem and the projection theorem are stated in terms of the ∇ and ∇^*- geodesics and the canonical contrast functions [3, 4]. In a partially flat space (M, g, ∇, ∇^*), where $R = R^* = 0$ and $T^* = 0$, it is possible to define a pre-contrast function which can be seen as canonical, and a projection theorem holds with respect to the "canonical" pre-contrast function and the ∇^*-geodesic.

Proposition 1 (Canonical Pre-contrast Functions) *Let (M, g, ∇, ∇^*) be a partially flat space (i.e. (M, g, ∇) is a SMAT with $R = R^* = 0$ and $T^* = 0$) and (U, η_i) be an affine coordinate neighborhood with respect to ∇^* in M. The function ρ on $TU \times U$ defined by*

$$\rho(Z_p, q) := -g_p(Z_p, \dot{\gamma}^*(0)) \ \ (\forall p, \forall q \in U, \forall Z_p \in T_p(U)), \tag{6}$$

is a pre-contrast function on U, where $\gamma^* : [0, 1] \to U$ is the ∇^*-geodesic such that $\gamma^*(0) = p, \gamma^*(1) = q$ and $\dot{\gamma}^*(0)$ is the tangent vector of γ^* at p. Furthermore, the pre-contrast function ρ induces the original Riemannian metric g and the dual connections ∇ and ∇^* on U.

Proof For the function ρ defined as (6), the condition (a) in the definition of pre-contrast functions follows from the bilinearity of the inner product g_p. The condition (b) immediately follows from $\dot{\gamma}^*(0) = 0$ when $p = q$. By calculating the derivatives of ρ with the affine coordinate system (η_i), it can be shown that the condition (c) holds and that the induced Riemannian metric and dual affine connections coincide with the original g, ∇ and ∇^*. □

In particular, if (U, g, ∇, ∇^*) is a dually flat space, the pre-contrast function ρ defined in (6) coincides with the directional derivative $Z_p\phi^*(\cdot, q)$ of ∇^*-divergence $\phi^*(\cdot, q)$ with respect to Z_p (cf. [10, 11]). Hence, the definition of (6) seems to be natural one and we call the function ρ in (6) the *canonical pre-contrast function* in a partially flat space (U, g, ∇, ∇^*).

From the definition of the canonical pre-contrast function, we can immediately obtain the following theorem.

Corollary 1 (Generalized Projection Theorem) *Let (U, η_i) be an affine coordinate neighborhood in a partially flat space (M, g, ∇, ∇^*) and ρ be the canonical pre-contrast function on U. For any submanifold N in U, the following conditions are* *equivalent:*

> (i) *The* ∇^*- *geodesic starting at* $q \in U$ *is perpendicular to* N *at* $p \in N$.
> (ii) $\rho(Z_p, q) = 0$ *for any* Z_p *in* $T_p(N)$.

If (U, g, ∇, ∇^*) is a dually flat space, this theorem reduces to the projection theorem with respect to the ∇^*-divergence ϕ^*, since $\rho(Z_p, q) = Z_p\phi^*(p, q)$. In this sense, it can be seen as a generalized version of the projection theorem in dually flat spaces, and this is also one of the reasons why we consider the pre-contrast function ρ defined in (6) as canonical.

5 Statistical Manifolds Admitting Torsion Induced from Estimating Functions

As we mentioned in Introduction, a SMAT naturally appears through an estimating function in a "classical" statistical model as well as in a quantum statistical model. In this section, we briefly explain how a SMAT is induced on a parametric statistical model from an estimating function. See [7] for more details.

Let $S = \{p(\boldsymbol{x}; \boldsymbol{\theta}) \mid \boldsymbol{\theta} = (\theta^1, \ldots, \theta^d) \in \Theta \subset \boldsymbol{R}^d\}$ be a regular parametric statistical model. An estimating function on S, which we consider here, is a $\boldsymbol{R}^d$-valued function $\boldsymbol{u}(\boldsymbol{x}, \boldsymbol{\theta})$ satisfying the following conditions:

$$E_{\boldsymbol{\theta}}\{\boldsymbol{u}(\boldsymbol{x}, \boldsymbol{\theta})\} = \boldsymbol{0}, \quad E_{\boldsymbol{\theta}}\{\|\boldsymbol{u}(\boldsymbol{x}, \boldsymbol{\theta})\|^2\} < \infty, \quad \det\left[E_{\boldsymbol{\theta}}\left\{\frac{\partial \boldsymbol{u}}{\partial \boldsymbol{\theta}}(\boldsymbol{x}, \boldsymbol{\theta})\right\}\right] \neq 0 \;\; (\forall \boldsymbol{\theta} \in \Theta).$$

The first condition is called the unbiasedness of estimating functions, which is important to ensure the consistency of the estimator obtained from an estimating function. Let $X_1, \ldots, X_n$ be a random sample from an unknown probability distribution $p(\boldsymbol{x}; \boldsymbol{\theta}_0)$ in S. The estimator $\hat{\boldsymbol{\theta}}$ for $\boldsymbol{\theta}_0$ is called an M-estimator if it is obtained as a solution to the estimating equation

$$\sum_{i=1}^{n} \boldsymbol{u}(\boldsymbol{X}_i, \boldsymbol{\theta}) = \boldsymbol{0}. \tag{7}$$

The M-estimator $\hat{\boldsymbol{\theta}}$ has the consistency

$$\hat{\boldsymbol{\theta}} \longrightarrow \boldsymbol{\theta}_0 \;\; \text{(in probability)}$$

as $n \to \infty$ and the asymptotic normality

$$\sqrt{n}(\hat{\boldsymbol{\theta}} - \boldsymbol{\theta}_0) \longrightarrow N\left(\boldsymbol{0}, \mathrm{Avar}\left(\hat{\boldsymbol{\theta}}\right)\right) \;\; \text{(in distribution)}$$

as $n \to \infty$ under some additional regularity conditions [12], which are also assumed in the following discussion. The matrix $\mathrm{Avar}(\hat{\boldsymbol{\theta}})$ is the asymptotic variance-covariance matrix of $\hat{\boldsymbol{\theta}}$ and is given by

$$\mathrm{Avar}(\hat{\boldsymbol{\theta}}) = \{A(\boldsymbol{\theta}_0)\}^{-1} B(\boldsymbol{\theta}_0) \{A(\boldsymbol{\theta}_0)\}^{-T}, \tag{8}$$

where $A(\boldsymbol{\theta}) := E_{\boldsymbol{\theta}} \{(\partial \boldsymbol{u}/\partial \boldsymbol{\theta})(\boldsymbol{x}, \boldsymbol{\theta})\}$, $B(\boldsymbol{\theta}) := E_{\boldsymbol{\theta}} \left\{\boldsymbol{u}(\boldsymbol{x}, \boldsymbol{\theta})\boldsymbol{u}(\boldsymbol{x}, \boldsymbol{\theta})^T\right\}$ and $-T$ means transposing an inverse matrix (or inverting a transposed matrix).

In order to induce the structure of SMAT on S from an estimating function, we consider the notion of *standardization* of estimating functions. For an estimating function $\boldsymbol{u}(\boldsymbol{x}, \boldsymbol{\theta})$, its standardization (or *standardized estimating function*) is defined by

$$\boldsymbol{u}_*(\boldsymbol{x}, \boldsymbol{\theta}) := E_{\boldsymbol{\theta}} \left\{\boldsymbol{s}(\boldsymbol{x}, \boldsymbol{\theta})\boldsymbol{u}(\boldsymbol{x}, \boldsymbol{\theta})^T\right\} \left[E_{\boldsymbol{\theta}} \left\{\boldsymbol{u}(\boldsymbol{x}, \boldsymbol{\theta})\boldsymbol{u}(\boldsymbol{x}, \boldsymbol{\theta})^T\right\}\right]^{-1} \boldsymbol{u}(\boldsymbol{x}, \boldsymbol{\theta}),$$

where $\boldsymbol{s}(\boldsymbol{x}, \boldsymbol{\theta}) = (\partial/\partial\boldsymbol{\theta}) \log p(\boldsymbol{x}; \boldsymbol{\theta})$ is the score function for $\boldsymbol{\theta}$ [13]. Geometrically, the i-th component of the standardized estimating function $\boldsymbol{u}_*(\boldsymbol{x}, \boldsymbol{\theta})$ is the orthogonal projection of the i-th component of the score function $\boldsymbol{s}(\boldsymbol{x}, \boldsymbol{\theta})$ onto the linear space spanned by all components of the estimating function $\boldsymbol{u}(\boldsymbol{x}, \boldsymbol{\theta})$ in the Hilbert space

$$\mathscr{H}_{\boldsymbol{\theta}} := \{a(\boldsymbol{x}) \mid E_{\boldsymbol{\theta}}\{a(\boldsymbol{x})\} = 0,\ E_{\boldsymbol{\theta}}\{a(\boldsymbol{x})^2\} < \infty\}$$

with the inner product $< a(\boldsymbol{x}), b(\boldsymbol{x}) >_{\boldsymbol{\theta}} := E_{\boldsymbol{\theta}}\{a(\boldsymbol{x})b(\boldsymbol{x})\}$ $(\forall a(\boldsymbol{x}), \forall b(\boldsymbol{x}) \in \mathscr{H}_{\boldsymbol{\theta}})$. The standardization $\boldsymbol{u}_*(\boldsymbol{x}, \boldsymbol{\theta})$ of $\boldsymbol{u}(\boldsymbol{x}, \boldsymbol{\theta})$ does not change the estimator since the estimating equation obtained from $\boldsymbol{u}_*(\boldsymbol{x}, \boldsymbol{\theta})$ is equivalent to the original estimating equation (7). In terms of the standardization, the asymptotic variance-covariance matrix (8) can be rewritten as

$$\mathrm{Avar}(\hat{\boldsymbol{\theta}}) = \{G(\boldsymbol{\theta}_0)\}^{-1},$$

where $G(\boldsymbol{\theta}) := E_{\boldsymbol{\theta}} \left\{\boldsymbol{u}_*(\boldsymbol{x}, \boldsymbol{\theta})\boldsymbol{u}_*(\boldsymbol{x}, \boldsymbol{\theta})^T\right\}$. The matrix $G(\boldsymbol{\theta})$ is called a Godambe information matrix [14], which can be seen as a generalization of the Fisher information matrix.

As we have seen in Sect. 2, the Kullback–Leibler divergence ϕ_{KL} is a contrast function on S. Hence, the first derivative of ϕ_{KL} is a pre-contrast function on S and given by

$$\rho_{KL}((\partial_j)_{p_1}, p_2) := (\partial_j)_{p_1}\phi_{KL}(p_1, p_2) = -\int_{\Omega} s^j(\boldsymbol{x}, \boldsymbol{\theta}_1)p(\boldsymbol{x}; \boldsymbol{\theta}_2)\nu(d\boldsymbol{x})$$

for any two probability distributions $p_1(\boldsymbol{x}) = p(\boldsymbol{x}; \boldsymbol{\theta}_1)$, $p_2(\boldsymbol{x}) = p(\boldsymbol{x}; \boldsymbol{\theta}_2)$ in S and $j = 1, \ldots, d$. This observation leads to the following proposition [7].

Proposition 2 (Pre-contrast Functions from Estimating Functions) *For an estimating function $\boldsymbol{u}(\boldsymbol{x}, \boldsymbol{\theta})$ on the parametric model S, a pre-contrast function $\rho_{\boldsymbol{u}}$:*

$TS \times S \to \boldsymbol{R}$ is defined by

$$\rho_{\boldsymbol{u}}((\partial_j)_{p_1}, p_2) := -\int_{\Omega} u_*^j(\boldsymbol{x}, \boldsymbol{\theta}_1) p(\boldsymbol{x}; \boldsymbol{\theta}_2) \nu(d\boldsymbol{x}) \tag{9}$$

for any two probability distributions $p_1(\boldsymbol{x}) = p(\boldsymbol{x}; \boldsymbol{\theta}_1)$, $p_2(\boldsymbol{x}) = p(\boldsymbol{x}; \boldsymbol{\theta}_2)$ in S and $j = 1, \ldots, d$, where $u_*^j(\boldsymbol{x}, \boldsymbol{\theta})$ is the j-th component of the standardization $\boldsymbol{u}_*(\boldsymbol{x}, \boldsymbol{\theta})$ of $\boldsymbol{u}(\boldsymbol{x}, \boldsymbol{\theta})$.

The use of the standardization $\boldsymbol{u}_*(\boldsymbol{x}, \boldsymbol{\theta})$ instead of $\boldsymbol{u}(\boldsymbol{x}, \boldsymbol{\theta})$ ensures that the definition of the function $\rho_{\boldsymbol{u}}$ does not depend on the choice of coordinate system (parameter) of S. In fact, for a coordinate transformation (parameter transformation) $\boldsymbol{\eta} = \Phi(\boldsymbol{\theta})$, the estimating function $\boldsymbol{u}(\boldsymbol{x}, \boldsymbol{\theta})$ is changed into $\boldsymbol{v}(\boldsymbol{x}, \boldsymbol{\eta}) = \boldsymbol{u}(\boldsymbol{x}, \Phi^{-1}(\boldsymbol{\eta}))$ and we have

$$\boldsymbol{v}_*(\boldsymbol{x}, \boldsymbol{\eta}) = \left(\frac{\partial \boldsymbol{\theta}}{\partial \boldsymbol{\eta}}\right)^T \boldsymbol{u}_*(\boldsymbol{x}, \boldsymbol{\theta}).$$

This is the same as the transformation rule of coordinate bases on a tangent space of a manifold. The set of all components of the standardized estimating function $\boldsymbol{u}_*(\boldsymbol{x}, \boldsymbol{\theta})$ can be seen as a representation of the coordinate basis $\{(\partial_1)_p, \ldots, (\partial_d)_p\}$ on the tangent space $T_p(S)$ of S, where $p(\boldsymbol{x}) = p(\boldsymbol{x}; \boldsymbol{\theta})$.

The proof of Proposition 2 is straightforward. In particular, the condition (b) in the definition of pre-contrast function follows from the unbiasedness of the (standardized) estimating function. The Riemannian metric g, dual connections ∇ and ∇^* induced from the pre-contrast function $\rho_{\boldsymbol{u}}$ are given as follows:

$$g_{jk}(\boldsymbol{\theta}) := g(\partial_j, \partial_k) = E_{\boldsymbol{\theta}}\{u_*^j(\boldsymbol{x}, \boldsymbol{\theta}) u_*^k(\boldsymbol{x}, \boldsymbol{\theta})\} = G(\boldsymbol{\theta})_{jk},$$

$$\begin{cases} \Gamma_{ij,k}(\boldsymbol{\theta}) := g(\nabla_{\partial_i}\partial_j, \partial_k) = E_{\boldsymbol{\theta}}[\{\partial_i u_*^j(\boldsymbol{x}, \boldsymbol{\theta})\} s^k(\boldsymbol{x}, \boldsymbol{\theta})] \\ \Gamma^*_{ik,j}(\boldsymbol{\theta}) := g(\partial_j, \nabla^*_{\partial_i}\partial_k) = \int_{\Omega} u_*^j(\boldsymbol{x}, \boldsymbol{\theta}) \partial_i \partial_k p(\boldsymbol{x}; \boldsymbol{\theta}) \nu(d\boldsymbol{x}) \end{cases},$$

where $G(\boldsymbol{\theta})_{jk}$ is the (j, k) component of the Godambe information matrix $G(\boldsymbol{\theta})$. Note that ∇^* is always torsion-free since $\Gamma^*_{ik,j} = \Gamma^*_{ki,j}$, whereas ∇ is not necessarily torsion-free unless $\boldsymbol{u}_*(\boldsymbol{x}, \boldsymbol{\theta})$ is integrable with respect to $\boldsymbol{\theta}$ (*i.e.* there exists a function $\psi(\boldsymbol{x}, \boldsymbol{\theta})$ satisfying $\partial_j \psi(\boldsymbol{x}, \boldsymbol{\theta}) = u_*^j(\boldsymbol{x}, \boldsymbol{\theta})$ $(j = 1, \ldots, d)$).

If it is integrable and ∇ is torsion-free, it is possible to construct a contrast function on S, from which the pre-contrast function $\rho_{\boldsymbol{u}}$ in (9) is obtained by taking its first derivative, as follows:

$$\phi_{\boldsymbol{u}}(p_1, p_2) = \int_{\Omega} \{\psi(\boldsymbol{x}, \boldsymbol{\theta}_1) - \psi(\boldsymbol{x}, \boldsymbol{\theta}_2)\} p(\boldsymbol{x}; \boldsymbol{\theta}_2) \nu(d\boldsymbol{x}),$$

where $\partial_j \psi(\boldsymbol{x}, \boldsymbol{\theta}) = u_*^j(\boldsymbol{x}, \boldsymbol{\theta})$ $(j = 1, \ldots, d)$ and $p_l(\boldsymbol{x}) = p(\boldsymbol{x}; \boldsymbol{\theta}_l)$ $(l = 1, 2)$.

Table 1 Votes cast in the n-th constituency ($n = 1, \dots, N$)

Party	C	L	Total
C	X_{1n}	$m_{1n} - X_{1n}$	m_{1n}
L	X_{2n}	$m_{2n} - X_{2n}$	m_{2n}
Total	X_n	$m_n - X_n$	m_n

6 Example

In this section, we consider the estimation problem of voter transition probabilities described in [15] to see an example of statistical manifolds admitting torsion (SMAT) induced from estimation functions.

Suppose that we had two successive elections which were carried out in N constituencies, and that the two political parties C and L contended in each election. The table below summarizes the numbers of voters in the n-th constituency for the respective elections. It is assumed that we can observe only the marginal totals m_{1n}, m_{2n}, X_n and $m_n - X_n$, where X_n is a random variable and the others are treated as fixed constants. Let θ^1 and θ^2 be the probabilities that a voter who votes for the parties C and L in Election 1 votes for C in Election 2, respectively. They are the parameters of interest here. Then, the random variables X_{1n} and X_{2n} in Table 1 are assumed to independently follow the binomial distributions $B(m_{1n}, \theta^1)$ and $B(m_{2n}, \theta^2)$, respectively.

In the n-th constituency, the probability function of the observation $X_n = X_{1n} + X_{2n}$ is given by

$$p_n(x_n; \boldsymbol{\theta}) = \sum_{x_{1n}=0}^{m_{1n}} \binom{m_{1n}}{x_{1n}} \binom{m_{2n}}{x_n - x_{1n}} \left(\theta^1\right)^{x_{1n}} \left(1 - \theta^1\right)^{m_{1n} - x_{1n}} \left(\theta^2\right)^{x_n - x_{1n}} \left(1 - \theta^2\right)^{m_{2n} - x_n + x_{1n}},$$

where $\boldsymbol{\theta} = \left(\theta^1, \theta^2\right)$. The statistical model S in this problem consists of all possible probability functions of the observed data $\boldsymbol{X} = (X_1, \dots, X_N)$ as follows:

$$S = \left\{ p(\boldsymbol{x}; \boldsymbol{\theta}) \,\middle|\, \boldsymbol{\theta} = \left(\theta^1, \theta^2\right) \in (0, 1) \times (0, 1) \right\},$$

where $p(\boldsymbol{x}; \boldsymbol{\theta}) = \prod_{n=1}^{N} p_n(x_n; \boldsymbol{\theta})$ $(\boldsymbol{x} = (x_1, \dots, x_N))$ since $X_1, \dots, X_N$ are independent.

Although the maximum likelihood estimation for $\boldsymbol{\theta}$ is possible based on the likelihood function $L(\boldsymbol{\theta}) = p(\boldsymbol{X}; \boldsymbol{\theta})$, it is a little complicated since X_{1n} and X_{2n} are not observed in each n-th constituency. An alternative approach for estimating $\boldsymbol{\theta}$ is to use the quasi-score function $\boldsymbol{q}(\boldsymbol{x}, \boldsymbol{\theta}) = (q^1(\boldsymbol{x}, \boldsymbol{\theta}), q^2(\boldsymbol{x}, \boldsymbol{\theta}))^T$ [15] as an estimating function, where

$$q^1(\boldsymbol{x}, \boldsymbol{\theta}) = \sum_{n=1}^{N} \frac{m_{1n}\{x_n - \mu_n(\boldsymbol{\theta})\}}{V_n(\boldsymbol{\theta})}, \quad q^2(\boldsymbol{x}, \boldsymbol{\theta}) = \sum_{n=1}^{N} \frac{m_{2n}\{x_n - \mu_n(\boldsymbol{\theta})\}}{V_n(\boldsymbol{\theta})}.$$

Here, $\mu_n(\boldsymbol{\theta})$ and $V_n(\boldsymbol{\theta})$ are the mean and variance of X_n, respectively, i.e.

$$\mu_n(\boldsymbol{\theta}) = E(X_n) = m_{1n}\theta^1 + m_{2n}\theta^2 \tag{10}$$
$$V_n(\boldsymbol{\theta}) = V(X_n) = m_{1n}\theta^1\left(1-\theta^1\right) + m_{2n}\theta^2\left(1-\theta^2\right).$$

In this example, the random variables $X_1, \ldots, X_N$ in the observed data are independent, but not identically distributed. However, it is possible to apply the results in Sect. 5 by considering the whole of the left hand side of (7) as an estimating function and modifying the results in this case. Note that the estimating function $\boldsymbol{q}(\boldsymbol{x}, \boldsymbol{\theta})$ is already standardized since the i-th component $q^i(\boldsymbol{x}, \boldsymbol{\theta})$ of $\boldsymbol{q}(\boldsymbol{x}, \boldsymbol{\theta})$ is obtained by the orthogonal projection of the i-th component of the score function $\boldsymbol{s}(\boldsymbol{x}, \boldsymbol{\theta})$ for $\boldsymbol{\theta}$ onto the linear space spanned by $\{x_1 - \mu_1(\boldsymbol{\theta}), \ldots, x_N - \mu_N(\boldsymbol{\theta})\}$. In fact, the orthogonal projection is calculated as follows:

$$\begin{aligned}
&E_\theta\left\{\boldsymbol{s}(\boldsymbol{x},\boldsymbol{\theta})(\boldsymbol{x}-\boldsymbol{\mu}(\boldsymbol{\theta}))^T\right\}\left[E_\theta\left\{(\boldsymbol{x}-\boldsymbol{\mu}(\boldsymbol{\theta}))(\boldsymbol{x}-\boldsymbol{\mu}(\boldsymbol{\theta}))^T\right\}\right]^{-1}(\boldsymbol{x}-\boldsymbol{\mu}(\boldsymbol{\theta}))\\
&= -E_\theta\left\{\frac{\partial}{\partial\boldsymbol{\theta}^T}(\boldsymbol{x}-\boldsymbol{\mu}(\boldsymbol{\theta}))\right\}\left[E_\theta\left\{(\boldsymbol{x}-\boldsymbol{\mu}(\boldsymbol{\theta}))(\boldsymbol{x}-\boldsymbol{\mu}(\boldsymbol{\theta}))^T\right\}\right]^{-1}(\boldsymbol{x}-\boldsymbol{\mu}(\boldsymbol{\theta}))\\
&= \begin{pmatrix} m_{11} & \cdots & m_{1N} \\ m_{21} & \cdots & m_{2N}\end{pmatrix}\begin{pmatrix} V_1(\boldsymbol{\theta}) & \cdots & 0 \\ \vdots & \ddots & \vdots \\ 0 & \cdots & V_n(\boldsymbol{\theta})\end{pmatrix}^{-1}\begin{pmatrix} x_1-\mu_1(\boldsymbol{\theta}) \\ \vdots \\ x_N-\mu_N(\boldsymbol{\theta})\end{pmatrix} = \begin{pmatrix} q^1(\boldsymbol{x},\boldsymbol{\theta}) \\ q^2(\boldsymbol{x},\boldsymbol{\theta})\end{pmatrix},
\end{aligned}$$

where $\boldsymbol{x} = (x_1, \ldots, x_N)^T$ and $\boldsymbol{\mu}(\boldsymbol{\theta}) = (\mu_1(\boldsymbol{\theta}), \ldots, \mu_N(\boldsymbol{\theta}))^T$. In addition, the estimating function $\boldsymbol{q}(\boldsymbol{x}, \boldsymbol{\theta})$ is not integrable with respect to $\boldsymbol{\theta}$ since $\partial q^1/\partial\theta^2 \neq \partial q^2/\partial\theta^1$. From Proposition 2 and the fact that $\boldsymbol{q}(\boldsymbol{x}, \boldsymbol{\theta})$ itself is a standardized estimating function, we immediately obtain the pre-contrast function $\rho_q : TS \times S \to \boldsymbol{R}$ defined by $\boldsymbol{q}(\boldsymbol{x}, \boldsymbol{\theta})$, where

$$\rho_q((\partial_i)_{p_1}, p_2) = -\sum_{\boldsymbol{x}} q^i(\boldsymbol{x}, \boldsymbol{\theta}_1)p(\boldsymbol{x}; \boldsymbol{\theta}_2) = \sum_{n=1}^{N}\frac{m_{in}\{\mu_n(\boldsymbol{\theta}_1) - \mu_n(\boldsymbol{\theta}_2)\}}{V_n(\boldsymbol{\theta}_1)}$$

with $p_l(\boldsymbol{x}) = p(\boldsymbol{x}; \boldsymbol{\theta}_l) \in S$ $(l = 1, 2)$. The pre-contrast function ρ_q induces the statistical manifold admitting torsion as follows.
Riemannian metric g:

$$g_{ij}(\boldsymbol{\theta}) = \sum_{\boldsymbol{x}} q^i(\boldsymbol{x}, \boldsymbol{\theta})q^j(\boldsymbol{x}, \boldsymbol{\theta})p(\boldsymbol{x}; \boldsymbol{\theta}) = \sum_{n=1}^{N}\frac{1}{V_n(\boldsymbol{\theta})}m_{in}m_{jn}.$$

Dual affine connections ∇^* and ∇:

$$\begin{aligned}
\Gamma^*_{ij,k}(\boldsymbol{\theta}) &= \sum_{\boldsymbol{x}}\{\partial_i\partial_j p(\boldsymbol{x}; \boldsymbol{\theta})\}q^k(\boldsymbol{x}, \boldsymbol{\theta})\\
&= \sum_{n=1}^{N}\frac{m_{kn}}{V_n(\boldsymbol{\theta})}\left[\sum_{\boldsymbol{x}} x_n\{\partial_i\partial_j p(\boldsymbol{x}; \boldsymbol{\theta})\} - \mu_n(\boldsymbol{\theta})\sum_{\boldsymbol{x}}\{\partial_i\partial_j p(\boldsymbol{x}; \boldsymbol{\theta})\}\right]
\end{aligned}$$

$$
\begin{aligned}
&= \sum_{n=1}^{N} \frac{m_{kn}}{V_n(\boldsymbol{\theta})} \left[\partial_i \partial_j \sum_{\boldsymbol{x}} x_n p(\boldsymbol{x}; \boldsymbol{\theta}) - \mu_n(\boldsymbol{\theta}) \partial_i \partial_j \sum_{\boldsymbol{x}} p(\boldsymbol{x}; \boldsymbol{\theta}) \right] \\
&= \sum_{n=1}^{N} \frac{m_{kn}}{V_n(\boldsymbol{\theta})} \partial_i \partial_j \mu_n(\boldsymbol{\theta}) = 0 \ \ \text{(from (10))} \\
\Gamma_{ij,k}(\boldsymbol{\theta}) &= \Gamma^*_{ij,k}(\boldsymbol{\theta}) - \partial_i g_{jk}(\boldsymbol{\theta}) \ \ \text{(from the duality between } \nabla \text{ and } \nabla^*\text{)} \\
&= \sum_{n=1}^{N} \frac{1 - 2\theta^i}{V_n(\boldsymbol{\theta})^2} m_{in} m_{jn} m_{kn}.
\end{aligned}
$$

In this example, the statistical model S is ∇^*-flat since the coefficient of ∇^* with respect to the parameter $\boldsymbol{\theta}$ is equal to zero. Furthermore, this shows that $\boldsymbol{\theta}$ provides an affine coordinate system for ∇^*. Although the curvature tensor of ∇ vanishes because the curvature tensor of ∇^* vanishes and ∇ is dual to ∇^*, the statistical model S is not ∇-flat because ∇ is not torsion-free, which comes from the non-integrability of the estimating function $\boldsymbol{q}(\boldsymbol{x}, \boldsymbol{\theta})$. Hence, this geometrical structure provides an example of partially flat spaces, which was discussed in Sect. 4.

7 Future Problems

In this paper, we have summarized existing results on statistical manifolds admitting torsion, especially focusing on partially flat spaces. Although some results that are not seen in the standard theory of information geometry have been obtained, including a generalized projection theorem in partially flat spaces and statistical manifolds admitting torsion induced from estimating functions in statistics, a lot of (essential) problems have been unsolved. We discuss some of them to conclude this paper.

(1) The canonical pre-contrast function and the generalized projection theorem in a partially flat spase (M, g, ∇, ∇^*) are described only in terms of the flat connection ∇^*. In this sense, it can be said that these are a concept and a theorem for the Riemannian manifold (M, g) with the flat connection ∇^*. What is the role of the affine connection ∇ in the partially flat space (M, g, ∇, ∇^*), especially when ∇ is not torsion-free?
(2) The canonical pre-contrast function is defined in terms of the Riemannian metric g and the ∇^*-geodesic in a partially flat space (U, g, ∇, ∇^*) *without* using the affine coordinate system (η_i) on U. Hence, this function can be defined in a general statistical manifold admitting torsion (M, g, ∇) as long as the ∇^*-geodesic uniquely exists. What is the condition under which this function is a pre-contrast function that induces the original Riemannian metric g, dual affine connections ∇ and ∇^*? What properties does the (canonical) pre-contrast function have in this case? These problems are closely related to the works by [10, 11], who try to define a canonical divergence (canonical contrast function) on a general statistical manifold beyond a dually flat space.

(3) The definition of pre-contrast functions from estimating functions is obtained by replacing the score function which appears in the pre-contrast function as the derivative of Kullback–Leibler divergence with the standardized estimating functions. However, this is not the unique way to obtain a pre-contrast function from an estimating function. For example, if we consider the β-divergence [16] (or density power divergence [17]) as a contrast function, its first derivative is also a pre-contrast function and takes the same form as (9) in Proposition 2. However, the estimating function which appears in the pre-contrast function is not standardized. Although the standardization seems to be natural, further consideration is necessary on how to define a pre-contrast function from a given estimating function.
(4) For the example considered in Sect. 6, we can show that the pre-contrast function ρ_q coincides with the canonical pre-contrast function in the partially flat space (S, g, ∇, ∇^*) and the generalized projection theorem (Corollary 1 in Sect. 4) can be applied. However, its statistical meaning has not been clarified yet. Although it is expected that the SMAT induced from an estimating function has something to do with statistical inference based on the estimating function, the clarification on it is a future problem.

Acknowledgements This work was supported by JSPS KAKENHI Grant Numbers JP15K00064, JP15K04842.

References

1. Eguchi, S.: Geometry of minimum contrast. Hiroshima Math. J. **22**, 631–647 (1992)
2. Matsuzoe, H.: Geometry of contrast functions and conformal geometry. Hiroshima Math. J. **29**, 175–191 (1999)
3. Amari, S., Nagaoka, H.: Methods of Information Geometry. American Mathematical Society, Providence; Oxford University Press, Oxford (2000)
4. Amari, S.: Information Geometry and Its Applications. Springer, Berlin (2016)
5. Kurose, T.: Statistical manifolds admitting torsion. Geometry and Something, Fukuoka University (2007)
6. Matsuzoe, H.: Statistical manifolds admitting torsion and pre-contrast functions. Information Geometry and Its Related Fields, Osaka City University (2010)
7. Henmi, M., Matsuzoe, H.: Geometry of pre-contrast functions and non-conservative estimating functions. AIP Conf. Proc. **1340**, 32–41 (2011)
8. Henmi, M.: Statistical manifolds admitting torsion, pre-contrast functions and estimating functions. Lect. Notes Comput. Sci. **10589**, 153–161 (2017)
9. Kurose, T.: On the divergences of 1-conformally flat statistical manifolds. Tohoku Math. J. **46**, 427–433 (1994)
10. Henmi, M., Kobayashi, R.: Hooke's law in statistical manifolds and divergences. Nagoya Math. J. **159**, 1–24 (2000)
11. Ay, N., Amari, S.: A novel approach to canonical divergences within information geometry. Entropy **17**, 8111–8129 (2015)
12. van der Vaart, A.W.: Asymptotic Statistics. Cambridge University Press, Cambridge (2000)
13. Heyde, C.C.: Quasi-Likelihood and Its Application. Springer, Berlin (1997)
14. Godambe, V.: An optimum property of regular maximum likelihood estimation. Ann. Math. Stat. **31**, 1208–1211 (1960)

15. McCullagh, P., Nelder, J.A.: Generalized Linear Models, 2nd edn. Chapman and Hall, Boca Raton (1989)
16. Eguchi, S., Kano, Y.: Robustifying maximum likelihood estimation. In: Research Memorandum of the Institute of Statistical Mathematics, vol. 802 (2001)
17. Basu, A., Harris, I.R., Hjort, N.L., Jones, M.C.: Robust and efficient estimation by minimizing a density power divergence. Biometrika **85**, 549–559 (1998)

Conformal Flattening on the Probability Simplex and Its Applications to Voronoi Partitions and Centroids

Atsumi Ohara

Abstract A certain class of information geometric structure can be conformally transformed to dually flat one. This paper studies the transformation on the probability simplex from a viewpoint of *affine differential geometry* and provides its applications. By restricting affine immersions with certain conditions, the probability simplex is realized to be 1-conformally flat statistical manifolds immersed in $\mathbf{R}^{n+1}$. Using this fact, we introduce a concept of *conformal flattening* for such manifolds in order to obtain the corresponding dually flat statistical (Hessian) ones with conformal divergences, and show explicit forms of potential functions and affine coordinates. Finally, we demonstrate applications of the flattening to nonextensive statistical physics, Voronoi partitions and weighted centroids on the probability simplex with respect to *geometric divergences*, which are not necessarily of Bregman type.

Keywords Conformal flattening · Affine differential geometry · Escort probability · Geometric divergence · Conformal divergence

1 Introduction

In the theory of information geometry for statistical models, the logarithmic function is crucially significant to give a standard information geometric structure for exponential family [1, 2]. By changing the logarithmic function to another one we can deform the standard structure to a new one while keeping its basic property as a statistical manifold, which consists of a pair of mutually dual affine connections (∇, ∇^*) with respect to Riemannian metric g. There exist several ways [3–6] to introduce functions to deform a statistical manifold structure and these functions are sometimes called *embedding* or *representing functions*.

A. Ohara (✉)
Department of Electrical and Electronics Engineering,
University of Fukui, Bunkyo 3-9-1, Fukui 910-8507, Japan
e-mail: ohara@fuee.u-fukui.ac.jp

F. Nielsen (ed.), *Geometric Structures of Information*, Signals and Communication Technology, https://doi.org/10.1007/978-3-030-02520-5_4

Affine immersion [7] can be regarded as one of possible ways. Further, Kurose [8] has proved that *1-conformally flat* statistical manifolds (See Appendix) realized by a certain class of affine immersions can be conformally transformed to dually flat ones, which are the most fruitful information geometric structures.

In this paper we call the transformation *conformal flattening* and give its explicit formula in order to elucidate the relations between representing functions and realized information geometric structures. We also discuss its applicability to computational geometric topics. These are interpreted as generalizations of the results in [9, 10], where the arguments are limited to conformal flattening of the alpha-geometry [1, 2] (See also Sect. 2.4).

The paper is organized as follows: In Sect. 2 we first discuss the affine immersion of the probability simplex and its geometric structure realized by the associated *geometric divergence*. Next, the conformally flattening transformation is given and the obtained dually flat structure with the associated *conformal divergence* is investigated. Section 3 describes applications of the conformal flattening. We consider a Voronoi partition and a weighted centroid with respect to the geometric divergence on the probability simplex. While geometric divergences are not of Bregman type in general, geometric properties such as conformality and projectivity are well utilized in these topics. We also see that *escort probabilities*, which are interpreted as the dual affine coordinates for the flattened geometry, play important roles. Section 4 includes concluding remarks. Finally, a short review on statistical manifolds and affine differential geometry is given in Appendix.

2 Affine Immersion of the Probability Simplex

Let $\mathcal{S}^n$ be the relative interior of the probability simplex, i.e.,

$$\mathcal{S}^n := \left\{ p = (p_i) \,\middle|\, p_i \in \mathbf{R}_+, \ \sum_{i=1}^{n+1} p_i = 1 \right\},$$

where $\mathbf{R}_+$ denotes the set of positive numbers.

Consider an affine immersion [7] (f, ξ) of the simplex $\mathcal{S}^n$ (see also Appendix). Let D be the canonical flat affine connection on $\mathbf{R}^{n+1}$. Further, let f be an immersion from $\mathcal{S}^n$ into $\mathbf{R}^{n+1}$ and ξ be a transversal vector field on $\mathcal{S}^n$ (cf. Fig. 1). For a given *affine immersion* (f, ξ) of $\mathcal{S}^n$, the induced torsion-free connection ∇ and the affine fundamental form h are defined from the Gauss formula by

$$D_X f_*(Y) = f_*(\nabla_X Y) + h(X, Y)\xi, \quad X, Y \in \mathcal{X}(\mathcal{S}^n), \tag{1}$$

where f_* is the differential of f and $\mathcal{X}(\mathcal{S}^n)$ is the set of vector fields on $\mathcal{S}^n$.

It is well known [7, 8] that the realized geometric structure $(\mathcal{S}^n, \nabla, h)$ is a statistical manifold if and only if (f, ξ) is nondegenerate and equiaffine, i.e., h is nondegenerate

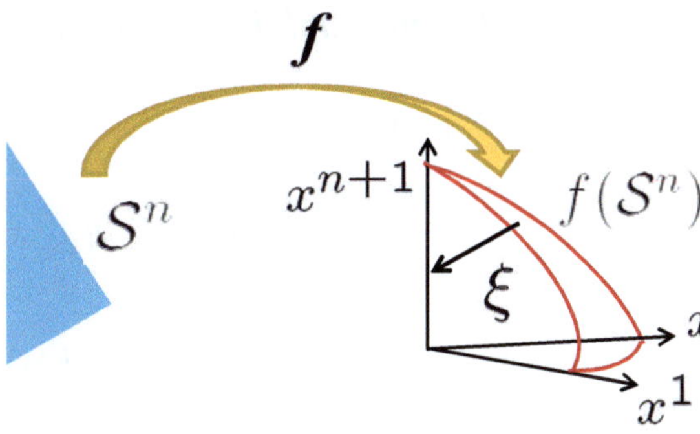

Fig. 1 An affine immersion (f, ξ) from $\mathcal{S}^n$ to $\mathbf{R}^{n+1}$

and $D_X\xi$ is tangent to $\mathcal{S}^n$ for any $X \in \mathcal{X}(\mathcal{S}^n)$. Furthermore, the statistical manifold $(\mathcal{S}^n, \nabla, h)$ is 1-conformally flat [8] (but not necessarily dually flat nor of constant curvature).

Now we consider the affine immersion with the following assumptions.

Assumptions 1. The affine immersion (f, ξ) is nondegenerate and equiaffine,

2. Let $\{x^i\}$ be an affine coordinate system for D on $\mathbf{R}^{n+1}$. The immersion f is given by the component-by-component and a common representing function L, i.e.,

$$f : \mathcal{S}^n \ni p = (p_i) \mapsto x = (x^i) \in \mathbf{R}^{n+1}, \quad x^i = L(p_i), \ i = 1, \dots, n+1,$$

3. The representing function $L : (0, \ 1) \to \mathbf{R}$ is sign-definite (or non-zero), concave with $L'' < 0$ and strictly increasing, i.e., $L' > 0$, Hence, the inverse of L denoted by E exists, i.e., $E \circ L = \mathrm{id}$.
4. Each component of ξ satisfies $\xi^i < 0, \ i = 1, \dots, n+1$ on $\mathcal{S}^n$.

Remark 1 From the assumption 3, it follows that $L'E' = 1$, $E' > 0$ and $E'' > 0$. Regarding sign-definiteness of L, note that we can adjust $L(u)$ to $L(u) + c$ by a suitable constant c without loss of generality since the resultant geometric structure is unchanged (See Proposition 1) by the adjustment. For a fixed L satisfying the assumption 3, we can choose ξ that meets the assumptions 1 and 4. For example, if we take $\xi^i = -|L(p_i)|$ then (f, ξ) is called *centro-affine*, which is known to be equiaffine [7]. The assumptions 3 and 4 also assure positive definiteness of h (The details are described in the proof of Proposition 1). Hence, (f, ξ) is non-degenerate and we can regard h as a Riemannian metric on $\mathcal{S}^n$.

2.1 Conormal Vector and the Geometric Divergence

Define a function Ψ on $\mathbf{R}^{n+1}$ by

$$\Psi(x) := \sum_{i=1}^{n+1} E(x^i),$$

then $f(\mathcal{S}^n)$ immersed in $\mathbf{R}^{n+1}$ is expressed as a level surface of $\Psi(x) = 1$. Denote by $\mathbf{R}_{n+1}$ the dual space of $\mathbf{R}^{n+1}$ and by $\langle \nu, x \rangle$ the pairing of $x \in \mathbf{R}^{n+1}$ and $\nu \in \mathbf{R}_{n+1}$. The conormal vector [7] $\nu : \mathcal{S}^n \to \mathbf{R}_{n+1}$ for the affine immersion (f, ξ) is defined by

$$\langle \nu(p), f_*(X) \rangle = 0, \ \forall X \in T_p\mathcal{S}^n, \qquad \langle \nu(p), \xi(p) \rangle = 1 \tag{2}$$

for $p \in \mathcal{S}^n$. Using the assumptions and noting the relations:

$$\frac{\partial \Psi}{\partial x^i} = E'(x^i) = \frac{1}{L'(p_i)} > 0, \quad i = 1, \ldots, n+1,$$

we have

$$\nu_i(p) := \frac{1}{\Lambda} \frac{\partial \Psi}{\partial x^i} = \frac{1}{\Lambda(p)} E'(x^i) = \frac{1}{\Lambda(p)} \frac{1}{L'(p_i)}, \quad i = 1, \ldots, n+1, \tag{3}$$

where Λ is a normalizing factor defined by

$$\Lambda(p) := \sum_{i=1}^{n+1} \frac{\partial \Psi}{\partial x^i} \xi^i = \sum_{i=1}^{n+1} \frac{1}{L'(p_i)} \xi^i(p). \tag{4}$$

Then we can confirm (2) using the relation $\sum_{i=1}^{n+1} X^i = 0$ for $X = (X^i) \in \mathcal{X}(\mathcal{S}^n)$. Note that $\upsilon : \mathcal{S}^n \to \mathbf{R}_{n+1}$ defined by

$$\upsilon_i(p) := \Lambda(p)\nu_i(p) = \frac{1}{L'(p_i)}, \quad i = 1, \ldots, n+1,$$

also satisfies

$$\langle \upsilon(p), f_*(X) \rangle = 0, \ \forall X \in T_p\mathcal{S}^n. \tag{5}$$

Further, it follows, from (3), (4) and the assumption 4, that

$$\Lambda(p) < 0, \quad \nu_i(p) < 0, \quad i = 1, \ldots, n+1,$$

for all $p \in \mathcal{S}^n$.

It is known [7] that the affine fundamental form h can be represented by

$$h(X,Y) = -\langle \nu_*(X), f_*(Y)\rangle, \quad X, Y \in T_p\mathcal{S}^n.$$

In our case, it is calculated via (5) as

$$\begin{aligned} h(X,Y) &= -\Lambda^{-1}\langle v_*(X), f_*(Y)\rangle - X(\Lambda^{-1})\langle v, f_*(Y)\rangle \\ &= -\frac{1}{\Lambda}\sum_{i=1}^{n+1}\left(\frac{1}{L'(p_i)}\right)' L'(p_i)X^iY^i = \frac{1}{\Lambda}\sum_{i=1}^{n+1}\frac{L''(p_i)}{L'(p_i)}X^iY^i. \end{aligned}$$

Since h is positive definite from the assumptions 3 and 4, we can regard it as a Riemannian metric.

Utilizing these notions from affine differential geometry, we can introduce the function ρ on $\mathcal{S}^n \times \mathcal{S}^n$, which is called a *geometric divergence* [8], as follows:

$$\begin{aligned} \rho(p,r) &= \langle \nu(r), f(p) - f(r)\rangle = \sum_{i=1}^{n+1}\nu_i(r)(L(p_i) - L(r_i)) \\ &= \frac{1}{\Lambda(r)}\sum_{i=1}^{n+1}\frac{L(p_i) - L(r_i)}{L'(r_i)}, \qquad p, r \in \mathcal{S}^n. \end{aligned} \tag{6}$$

We can easily see that ρ is a contrast function [2, 11] of the geometric structure $(\mathcal{S}^n, \nabla, h)$ because it holds that

$$\rho[X|] = 0, \quad h(X,Y) = -\rho[X|Y], \tag{7}$$

$$h(\nabla_X Y, Z) = -\rho[XY|Z], \quad h(Y, \nabla^*_X Z) = -\rho[Y|XZ], \tag{8}$$

where $\rho[X_1 \ldots X_k | Y_1 \ldots Y_l]$ stands for

$$\rho[X_1 \ldots X_k|Y_1 \ldots Y_l](p) := (X_1)_p \ldots (X_k)_p (Y_1)_r \ldots (Y_l)_r \rho(p,r)|_{p=r}$$

for $p, r \in \mathcal{S}^n$ and $X_i, Y_j \in \mathcal{X}(\mathcal{S}^n)$.

2.2 Conformal Divergence and 1-Conformal Transformation

Let σ be a positive function on $\mathcal{S}^n$. Associated with the geometric divergence ρ, the *conformal divergence* [8] of ρ with respect to a conformal factor $\sigma(r)$ is defined by

$$\tilde{\rho}(p,r) = \sigma(r)\rho(p,r), \qquad p, r \in \mathcal{S}^n. \tag{9}$$

The divergence $\tilde{\rho}$ can be proved to be a contrast function for $(\mathcal{S}^n, \tilde{\nabla}, \tilde{h})$, which is 1-conformally transformed geometric structure from $(\mathcal{S}^n, \nabla, h)$, where $\tilde{h}$ and $\tilde{\nabla}$ are given by

$$\tilde{h} = \sigma h, \tag{10}$$
$$h(\tilde{\nabla}_X Y, Z) = h(\nabla_X Y, Z) - d(\ln \sigma)(Z)h(X, Y). \tag{11}$$

When there exists such a positive function σ that relates $(\mathcal{S}^n, \nabla, h)$ with $(\mathcal{S}^n, \tilde{\nabla}, \tilde{h})$ as in (10) and (11), they are called 1-conformally equivalent and $(\mathcal{S}^n, \tilde{\nabla}, \tilde{h})$ is also a statistical manifold [8].

2.3 *Main Result*

Generally, the induced structure $(\mathcal{S}^n, \tilde{\nabla}, \tilde{h})$ from the conformal divergence $\tilde{\rho}$ is not also dually flat, which is the most abundant structure in information geometry. However, by choosing the conformal factor σ carefully, we can demonstrate that $(\mathcal{S}^n, \tilde{\nabla}, \tilde{h})$ is dually flat. Hereafter, we call such a transformation as *conformal flattening*.

Define

$$Z(p) := \sum_{i=1}^{n+1} \nu_i(p) = \frac{1}{\Lambda(p)} \sum_{i=1}^{n+1} \frac{1}{L'(p_i)},$$

then it is negative because each $\nu_i(p)$ is. The conformal divergence of ρ with respect to the conformal factor $\sigma(r) := -1/Z(r)$ is

$$\tilde{\rho}(p, r) = -\frac{1}{Z(r)}\rho(p, r).$$

Proposition 1 *If the conformal factor is given by $\sigma = -1/Z$, then the statistical manifold $(\mathcal{S}^n, \tilde{\nabla}, \tilde{h})$ that is 1-conformally transformed from $(\mathcal{S}^n, \nabla, h)$ via (10) and (11) is dually flat. Further, $\tilde{\rho}$ is the canonical divergence where mutually dual pair of affine coordinates (θ^i, η_i) and a pair of potential functions (ψ, φ) are explicitly given by*

$$\theta^i(p) = x^i(p) - x^{n+1}(p) = L(p_i) - L(p_{n+1}), \quad i = 1, \dots, n \tag{12}$$
$$\eta_i(p) = \frac{\nu_i(p)}{Z(p)} =: P_i(p), \quad i = 1, \dots, n, \tag{13}$$
$$\psi(p) = -x_{n+1}(p) = -L(p_{n+1}), \tag{14}$$
$$\varphi(p) = \frac{1}{Z(p)} \sum_{i=1}^{n+1} \nu_i(p) x^i(p) = \sum_{i=1}^{n+1} P_i(p) L(p_i). \tag{15}$$

Proof Using given relations, we first show that the conformal divergence $\tilde{\rho}$ is the canonical divergence [2] for $(\mathcal{S}^n, \tilde{\nabla}, \tilde{h})$:

$$\begin{aligned}\tilde{\rho}(p, r) &= -\frac{1}{Z(r)}\langle \nu(r), f(p) - f(r)\rangle = \langle P(r), f(r) - f(p)\rangle \\ &= \sum_{i=1}^{n+1} P_i(r)(x^i(r) - x^i(p)) \\ &= \sum_{i=1}^{n+1} P_i(r)x^i(r) - \sum_{i=1}^{n} P_i(r)(x^i(p) - x^{n+1}(p)) - \left(\sum_{i=1}^{n+1} P_i(r)\right) x^{n+1}(p) \\ &= \varphi(r) - \sum_{i=1}^{n} \eta_i(r)\theta^i(p) + \psi(p). \end{aligned} \tag{16}$$

Next, let us confirm that $\partial\psi/\partial\theta^i = \eta_i$. Since $\theta^i(p) = L(p_i) + \psi(p),\ i = 1, \ldots, n$, we have

$$p_i = E(\theta^i - \psi), \quad i = 1, \ldots, n+1,$$

by setting $\theta^{n+1} := 0$. Hence, we have

$$1 = \sum_{i=1}^{n+1} E(\theta^i - \psi).$$

Differentiating by θ^j, we have

$$\begin{aligned} 0 &= \frac{\partial}{\partial\theta^j} \sum_{i=1}^{n+1} E(\theta^i - \psi) = \sum_{i=1}^{n+1} E'(\theta^i - \psi)\left(\delta^i_j - \frac{\partial\psi}{\partial\theta^j}\right) \\ &= E'(x^j) - \left(\sum_{i=1}^{n+1} E'(x^i)\right)\frac{\partial\psi}{\partial\theta^j}. \end{aligned}$$

This implies that

$$\frac{\partial\psi}{\partial\theta^j} = \frac{E'(x^j)}{\sum_{i=1}^{n+1} E'(x^i)} = P_j = \eta_j.$$

Together with (16) and this relation, φ is confirmed to be the Legendre transform of ψ.

The dual relation $\partial\varphi/\partial\eta_i = \theta^i$ follows automatically from the property of the Legendre transform. □

Remark 2 Since the conformal metric is $\tilde{h} = -h/Z$, it is also positive definite. The dual affine connections ∇^* and $\tilde{\nabla}^*$ are known to be projectively equivalent [8]. Hence, ∇^* is projectively (or -1-conformally) flat. Further, the following corollary implies that the realized affine connection ∇ is also projectively equivalent to the flat connection $\tilde{\nabla}$ if we use the centro-affine immersion, i.e., $\xi^i = -L(p_i)$ [7, 8] (See also Appendix). Note that the expressions of the dual coordinates $\eta_i(p) = P_i(p)$ can be interpreted as a generalization of the *escort probability* [12] because it is a normalization of deformed probabilities $1/L'(p_i)$ (see the following subsection).

Corollary 1 *The choice of ξ does not affect the obtained dually flat structure $(\mathcal{S}^n, \tilde{\nabla}, \tilde{h})$.*

Proof We have the following alternative expressions of $\eta_i = P_i$ with respect to L and E:

$$P_i(p) = \frac{1/L'(p_i)}{\sum_{k=1}^{n+1} 1/L'(p_k)} = \frac{E'(x_i)}{\sum_{i=1}^{n+1} E'(x_i)} > 0, \quad i = 1, \ldots, n.$$

Hence, all the expressions in Proposition 1 does not depend on ξ, and the statement follows. □

2.4 Examples

Ex.(1) If we take L to be the logarithmic function $L(t) = \ln(t)$, the conformally flattened geometry immediately defines the standard dually flat structure $(g^{\mathrm{F}}, \nabla^{(1)}, \nabla^{(-1)})$ on the simplex $\mathcal{S}^n$, where g^F denotes the Fisher metric. We see that $-\varphi(p)$ is the entropy, i.e., $\varphi(p) = \sum_{i=1}^{n+1} p_i \ln p_i$ and the conformal divergence is the KL divergence (relative entropy), i.e., $\tilde{\rho}(p, r) = D^{(\mathrm{KL})}(r||p) = \sum_{i=1}^{n+1} r_i(\ln r_i - \ln p_i)$.

Ex.(2) Next let the affine immersion (f, ξ) be defined by the following L and ξ:

$$L(t) := \frac{1}{1-q} t^{1-q}, \quad x^i(p) = \frac{1}{1-q}(p_i)^{1-q},$$

and

$$\xi^i(p) = -q(1-q)x^i(p),$$

with $0 < q$ and $q \neq 1$, then it realizes the alpha-geometry [2] $(\mathcal{S}^n, \nabla^{(\alpha)}, g^F)$ with $q = (1+\alpha)/2$. Since the immersion (f, ξ) is centro-affine and the length of ξ is suitably scaled, $(\mathcal{S}^n, \nabla^{(\alpha)}, g^F)$ is of constant curvature $\kappa = (1-\alpha^2)/4$. The associated geometric divergence is the alpha-divergence, i.e.,

$$\rho(p, r) = D^{(\alpha)}(p, r) = \frac{4}{1-\alpha^2}\left(1 - \sum_{i=1}^{n+1}(p_i)^{(1-\alpha)/2}(r_i)^{(1+\alpha)/2}\right). \tag{17}$$

Following the procedure of conformal flattening described in the above, we have [9]

$$\Psi(x) = \sum_{i=1}^{n+1}((1-q)x^i)^{1/1-q}, \quad \Lambda(p) = -q, \ (constant)$$

$$\nu_i(p) = -\frac{1}{q}(p_i)^q, \quad -\frac{1}{Z(p)} = \frac{q}{\sum_{k=1}^{n+1}(p_i)^q},$$

and obtain dually flat structure $(\tilde{h}, \tilde{\nabla}, \tilde{\nabla}^*)$ via the formulas in Proposition 1:

$$\eta_i = \frac{(p_i)^q}{\sum_{k=1}^{n+1}(p_k)^q}, \quad \theta^i = \frac{1}{1-q}(p_i)^{1-q} - \frac{1}{1-q}(p_{n+1})^{1-q} = \ln_q(p_i) - \psi(p),$$

$$\psi(p) = -\ln_q(p_{n+1}), \quad \varphi(p) = \ln_q\left(\frac{1}{\exp_q(S_q(p))}\right), \quad \tilde{h}(p) = -\frac{1}{Z(p)}g^F(p).$$

Here, $\ln_q$ and $S_q(p)$ are the *q-logarithmic function* and the *Tsallis entropy* [12], respectively defined by

$$\ln_q(t) = \frac{t^{1-q}-1}{1-q}, \quad S_q(p) = \frac{\sum_{i=1}^{n+1}(p_i)^q - 1}{1-q}.$$

3 Construction of Voronoi Partitions and Centroids with Respect to Geometric Divergences

In the previous section we have seen that various geometric divergences ρ can be constructed on the statistical manifold $\mathcal{S}^n$ by changing the representing function L and the transversal vector field ξ.

We demonstrate interesting applications of the conformal flattening to topics related with computational geometry, which are Voronoi partitions and centroids for the geometric divergence on a 1-conformally flat statistical manifold. We find that escort probabilities P_i (dual coordinates η_i) play important roles.

In this section, subscripts by Greek letters such as p_λ are used to denote the λ-th point in $\mathcal{S}^n$ among given ones while subscripts by Roman letters such as p_i denote the ith coordinate of a point $p = (p_i) \in \mathcal{S}^n$.

3.1 *Voronoi Partitions*

Let ρ be a geometric divergence defined in (6) on a 1-conformal statistical manifold $(\mathcal{S}^n, \nabla, h)$. For given m points p_λ, $\lambda = 1, \ldots, m$ on $\mathcal{S}^n$ we define *Voronoi regions* on $\mathcal{S}^n$ with respect to the geometric divergence ρ as follows:

$$\mathrm{Vor}^{(\rho)}(p_\lambda) := \bigcap_{\mu \neq \lambda} \{ r \in \mathcal{S}^n | \rho(p_\lambda, r) < \rho(p_\mu, r) \}, \quad \lambda = 1, \ldots, m.$$

An *Voronoi partition (diagram)* on $\mathcal{S}^n$ is a collection of the Voronoi regions and their boundaries. For example, if we take $L(t) = t^{1-q}/(1-q)$ as in Sect. 2.4, the corresponding Voronoi partition is the one with respect to the alpha-divergence $D^{(\alpha)}$ in (17) on $(\mathcal{S}^n, \nabla^{(\alpha)}, g^F)$ (Cf. the figures in [10]). Note that $D^{(\alpha)}$ approaches the Kullback–Leibler (KL) divergence if $\alpha \to -1$, and $D^{(0)}$ is called the Hellinger distance. Further, the partition is also equivalent to that with respect to *Rényi divergence* [13] defined by

$$D_\alpha(p, r) := \frac{1}{\alpha - 1} \ln \sum_{i=1}^{n+1} (p_i)^\alpha (r_i)^{1-\alpha}$$

because of their one-to-one functional relationship.

The acclaimed algorithm using projection of a convex polyhedron [14, 15] has been known to commonly work well to construct Voronoi partitions for the KL divergence [16–18] as well as the Euclidean distance. Furthermore, the algorithm is generally applicable if a divergence function δ is of *Bregman type* [19], which is represented by the remainder of the first order Taylor expansion of a convex potential function in a suitable coordinate system. Geometrically speaking, this implies that

(i) the divergence δ is a *canonical divergence* [2] associated with a dually flat structure, i.e, it is of Bregman type:

$$\begin{aligned}
\delta(p, r) &= \psi(\theta(r)) + \varphi(\eta(r)) - \sum_{i=1}^{n} \theta^i(p) \eta_i(r) \\
&= \varphi(\eta(r)) - \left\{ \varphi(\eta(p)) + \sum_{i=1}^{n} \theta^i(p) \left(\eta_i(r) - \eta_i(p)\right) \right\}, \qquad (18) \\
\theta^i &= \frac{\partial \varphi(\eta)}{\partial \eta_i}, \quad i = 1, \ldots, n,
\end{aligned}$$

(ii) its affine coordinate system $\eta = (\eta_i)$ is chosen to realize the corresponding Voronoi partitions. In this coordinate system with one extra complementary coordinate the polyhedron is expressed as the upper envelope of m hyperplanes tangent to the potential function $\varphi(\eta)$ at $\eta(p_\lambda)$, $\lambda = 1, \ldots, m$.

Unfortunately a problem for the case of our Voronoi partition is that the geometric divergences ρ on $\mathcal{S}^n$ is *not* of Bregman type generally, i.e., they *cannot* be represented as a remainder of any convex potentials as in (18).

The following theorem, however, claims that the problem is resolved via Proposition 1. In other words, we can still apply the projection algorithm by conformally flattening a statistical manifold $(\mathcal{S}, \nabla, h)$ to a dually flat structure $(\mathcal{S}, \tilde{\nabla}, \tilde{h})$ and by invoking the conformal divergence $\tilde{\rho}$, which is always of Bregman type, and escort probabilities $\eta_i(p) = P_i(p)$ as a coordinate system.

The similar result is proved in [10] for the case of the α-divergence $D^{(\alpha)}$. However, the proof there was based on the fact that $(\mathcal{S}^n, \nabla^{(\alpha)}, g^F)$ is a statistical manifold *of constant curvature* in order to use the *modified Pythagorean relation* (See Appendix). In the following theorem, the assumption is relaxed to a 1-conformally flat statistical manifold $(\mathcal{S}, \nabla, h)$ and we prove with the usual Pythagorean relation on dually flat space.

Here, we denote the space of escort distributions by $\mathcal{E}^n$ and represent the point on $\mathcal{E}^n$ by $P = (P_1, \ldots, P_n)$ because $P_{n+1} = 1 - \sum_{i=1}^n P_i$ and $\mathcal{E}^n$ is also the probability simplex.

Theorem 1 *(i) The bisector of two points p_λ and p_μ defined by $\{r | \rho(p_\lambda, r) = \rho(p_\mu, r)\}$ is a simultaneously ∇^*- and $\tilde{\nabla}^*$-autoparallel hypersurface on $\mathcal{S}^n$.*

(ii) Let $\mathcal{H}_\lambda$, $\lambda = 1, \ldots, m$ be the hyperplane in $\mathcal{E}^n \times \mathbf{R}$ which is respectively tangent at $(P(p_\lambda), \varphi(p_\lambda))$ to the hypersurface $\{(P, y) = (P(p), \varphi(p)) | p \in \mathcal{S}^n\}$. The Voronoi partition with respect to ρ can be constructed on $\mathcal{E}^n$ by projecting the upper envelope of all $\mathcal{H}_\lambda$'s along the y-axis.

Proof (i) We construct a bisector for points p_λ and p_μ. Consider the $\tilde{\nabla}$-geodesic $\tilde{\gamma}$ connecting p_λ and p_μ, and let $\bar{p}$ be the midpoint on $\tilde{\gamma}$ satisfying $\tilde{\rho}(p_\lambda, \bar{p}) = \tilde{\rho}(p_\mu, \bar{p})$. Note that the point $\bar{p}$ satisfies $\rho(p_\lambda, \bar{p}) = \rho(p_\mu, \bar{p})$ by the conformal relation (9). Denote by $\mathcal{B}$ the $\tilde{\nabla}^*$-autoparallel hypersurface that is orthogonal to $\tilde{\gamma}$ at $\bar{p}$ with respect to the conformal metric $\tilde{h}$. Note that $\mathcal{B}$ is simultaneously ∇^*-autoparallel because of the projective equivalence of ∇^* and $\tilde{\nabla}^*$ as is mentioned in Remark 2.

Using these setup and the fact that $(\mathcal{S}^n, \tilde{\nabla}, \tilde{h})$ is dually flat, we have the following relation from the Pythagorean theorem [2]

$$\tilde{\rho}(p_\lambda, r) = \tilde{\rho}(p_\lambda, \bar{p}) + \tilde{\rho}(\bar{p}, r) = \tilde{\rho}(p_\mu, \bar{p}) + \tilde{\rho}(\bar{p}, r) = \tilde{\rho}(p_\mu, r),$$

for all $r \in \mathcal{B}$. Using the conformal relation (9) again, we have $\rho(p_\lambda, r) = \rho(p_\mu, r)$ for all $r \in \mathcal{B}$. Hence, $\mathcal{B}$ is a bisector of p_λ and p_μ.

(ii) Recall the conformal relation (9) between ρ and $\tilde{\rho}$, then we see that $\mathrm{Vor}^{(\rho)}(p_\lambda) = \mathrm{Vor}^{(\tilde{\rho})}(p_\lambda)$ holds on $\mathcal{S}^n$, where

$$\mathrm{Vor}^{(\tilde{\rho})}(p_\lambda) := \bigcap_{\mu \neq \lambda} \{r \in \mathcal{S}^n | \tilde{\rho}(p_\lambda, r) < \tilde{\rho}(p_\mu, r)\}.$$

Proposition 1 and the Legendre relations (16) imply that $\tilde{\rho}(p_\lambda, r)$ is represented with the escort probabilities, i.e., the dual coordinates $(P_i) = (\eta_i)$ by

$$\tilde{\rho}(p_\lambda, r) = \varphi(P(r)) - \left(\varphi(P(p_\lambda)) + \sum_{i=1}^n \frac{\partial \varphi}{\partial P_i}(p_\lambda)\{P_i(r) - P_i(p_\lambda)\} \right),$$

By definition the hyperplane $\mathcal{H}_\lambda$ is expressed by

$$\mathcal{H}_\lambda = \left\{ (P(r), y(r)) \;\middle|\; y(r) = \varphi(P(p_\lambda)) + \sum_{i=1}^{n} \frac{\partial \psi^*}{\partial P_i}(p_\lambda)\{P_i(r) - P_i(p_\lambda)\},\ r \in \mathcal{S}^n \right\}.$$

Hence, we have $\tilde{\rho}(p_\lambda, r) = \varphi(P(r)) - y(r)$. Thus, we see, for example, that the bisector on $\mathcal{E}^n$ for p_λ and p_μ is represented as a projection of $\mathcal{H}_\lambda \cap \mathcal{H}_\mu$. Thus, the statement follows. □

As a special case of the above theorem for $\rho = D^{(\alpha)}$, examples of Voronoi partitions with respect to $D^{(\alpha)}$ on usual probability simplex $\mathcal{S}^n$ and escort probability simplex $\mathcal{E}^n$ are given with their figures in [10].

Remark 3 Voronoi partitions for broader class of divergences that are not necessarily associated with any convex potentials are theoretically studied [20] from more general affine differential geometric points of views.

On the other hand, if the domain is extended from $\mathcal{S}^n$ to the positive orthant $\mathbf{R}_+^{n+1}$, then the α-divergence there can be expressed as a Bregman divergence [1, 2, 21]. Hence, the α-geometry on $\mathbf{R}_+^{n+1}$ is dually flat. Using this property, α-Voronoi partitions on $\mathbf{R}_+^{n+1}$ is discussed in [22].

However, while both of the above mentioned methods require constructions of the convex polyhedrons in the space of dimension $d = n + 2$, the new one proposed in this paper does in the space of dimension $d = n + 1$. Since it is known [23] that the optimal computational time of polyhedrons depends on the dimension d by $O(m \log m + m^{\lfloor d/2 \rfloor})$, the new one is slightly better when n is even and m is large.

3.2 Weighted Centroids

Let $p_\lambda,\ \lambda = 1, \ldots, m$ be given m points on $\mathcal{S}^n$ and $w_\lambda > 0,\ \lambda = 1, \ldots, m$ be their weights. Define the *weighted ρ-centroid* $c^{(\rho)} \in \mathcal{S}^n$ by the minimizer of the following problem:

$$\min_{p \in \mathcal{S}^n} \sum_{\lambda=1}^{m} w_\lambda \rho(p, p_\lambda).$$

Theorem 2 *The weighted ρ-centroid $c^{(\rho)}$ for given m points $p_1, \ldots, p_m$ on $\mathcal{S}^n$ is expressed by*

$$P_i(c^{(\rho)}) = \frac{1}{\sum_{\lambda=1}^{m} w_\lambda Z(p_\lambda)} \sum_{\lambda=1}^{m} w_\lambda Z(p_\lambda) P_i(p_\lambda), \quad i = 1, \ldots, n+1,$$

with weights w_λ, escort probabilities $P(p_\lambda)$ and the conformal factors $\sigma(p_\lambda) = -1/Z(p_\lambda) > 0$ for $p_\lambda,\ \lambda = 1, \ldots, m$.

Proof Denote $\theta^i(p)$ by θ^i simply. Using (9), we have

$$\sum_{\lambda=1}^{m} w_\lambda \rho(p, p_\lambda) = -\sum_{\lambda=1}^{m} w_\lambda Z(p_\lambda)\tilde{\rho}(p, p_\lambda)$$
$$= -\sum_{\lambda=1}^{m} w_\lambda Z(p_\lambda) \left\{ \psi(\theta) + \psi^*(\eta(p_\lambda)) - \sum_{i=1}^{n} \theta^i \eta_i(p_\lambda) \right\}.$$

Then the optimality condition is

$$\frac{\partial}{\partial \theta^i} \sum_{\lambda=1}^{m} w_\lambda \rho(p, p_\lambda) = -\sum_{\lambda=1}^{m} w_\lambda Z(p_\lambda)\{\eta_i - \eta_i(p_\lambda)\} = 0, \quad i = 1, \ldots, n,$$

where $\eta_i = \eta_i(p)$. Thus, the statements for $i = 1, \ldots, n$ hold from $\eta_i = P_i$ in Proposition 1. For $i = n + 1$, we have as follows:

$$P_{n+1}(c^{(\rho)}) = 1 - \sum_{i=1}^{n} P_i(c^{(\rho)})$$
$$= \frac{1}{\sum_{\lambda=1}^{m} w_\lambda Z(p_\lambda)} \sum_{\lambda=1}^{m} w_\lambda Z(p_\lambda) \left\{ 1 - \sum_{i=1}^{n} P_i(p_\lambda) \right\}$$
$$= \frac{1}{\sum_{\lambda=1}^{m} w_\lambda Z(p_\lambda)} \sum_{\lambda=1}^{m} w_\lambda Z(p_\lambda) P_{n+1}(p_\lambda).$$

□

4 Concluding Remarks

We have realized 1-conformally flat structures $(\mathcal{S}^n, \nabla, h)$ by changing affine immersions (f, ξ) or representing functions L, considered their conformal flattening and explicitly derived the corresponding dually flat structure, i.e., mutually dual potentials and affine coordinate systems.

Applications of the conformal flattening to topics in computational geometry are also demonstrated. As a result the geometric divergence, which is not generally of Bregman type, can be easily treated via the traditional computation algorithm. Recently, conformal divergences for Bregman-type divergences are proposed from different viewpoints and their properties are exploited [24, 25].

Extensions of the conformal flattening to other non-flat statistical manifolds or families of continuous probability distributions are left in the future work. While relations with the gradient flows (replicator flows, in a special case) on $(\mathcal{S}^n, \nabla, h)$ or

$(\mathcal{S}^n, \tilde{\nabla}, \tilde{h})$ can be found in [26], searching for the other applications of the technique would be also of interest.

Acknowledgements A. O. is partially supported by JSPS Grant-in-Aid (C) 15K04997.

Appendix: A Short Review of Statistical Manifolds and Affine Differential Geometry

We shortly summarize the basic notions and results in information geometry [1, 2], Hessian domain [27] and affine differential geometry [7, 8], which are used in this paper. See for the details and proofs in the literature.

Statistical Manifolds

For a torsion-free affine connection ∇ and a pseudo-Riemannian metric g on a manifold $\mathcal{M}$, the triple $(\mathcal{M}, \nabla, g)$ is called a *statistical (Codazzi) manifold* if it admits another torsion-free affine connection ∇^* satisfying

$$Xg(Y, Z) = g(\nabla_X Y, Z) + g(Y, \nabla^*_X Z) \tag{19}$$

for arbitrary X, Y and Z in $\mathcal{X}(\mathcal{M})$, where $\mathcal{X}(\mathcal{M})$ is the set of all tangent vector fields on $\mathcal{M}$. We say that ∇ and ∇^* *duals* of each other with respect to g, and (g, ∇, ∇^*) is called *dualistic structure* on $\mathcal{M}$.

A statistical manifold $(\mathcal{M}, \nabla, g)$ is said to be of *constant curvature* $\kappa \in \mathbf{R}$ if the curvature tensor R of ∇ satisfies

$$R(X, Y)Z = \kappa\{g(Y, Z)X - g(X, Z)Y\}. \tag{20}$$

When the constant κ is zero, the statistical manifold is called *flat*, or *dually flat*, because the dual curvature tensor R^* of ∇^* also vanishes automatically [2, 27].

For $\alpha \in \mathbf{R}$, statistical manifolds $(\mathcal{M}, \nabla, g)$ and $(\mathcal{M}, \tilde{\nabla}, \tilde{g})$ are said to be α-*conformally equivalent* [8] if there exists a positive function σ on $\mathcal{M}$ satisfying

$$\begin{aligned}\tilde{g}(X, Y) &= \sigma g(X, Y)\\ g(\tilde{\nabla}_X Y, Z) &= g(\nabla_X Y, Z) - \frac{1+\alpha}{2}(d\ln\sigma)(Z)g(X, Y)\\ &\quad + \frac{1-\alpha}{2}\{(d\ln\sigma)(X)g(Y, Z) + (d\ln\sigma)(Y)g(X, Z)\}.\end{aligned}$$

Statistical manifolds $(\mathcal{M}, \nabla, g)$ and $(\mathcal{M}, \tilde{\nabla}, \tilde{g})$ are α-conformally equivalent if and only if $(\mathcal{M}, \nabla^*, g)$ and $(\mathcal{M}, \tilde{\nabla}^*, \tilde{g})$ are $-\alpha$-conformally equivalent. In particular, -1-conformal equivalence means *projective equivalence* of ∇ and $\tilde{\nabla}$, which implies that a ∇-pregeodesic curve is simultaneously $\tilde{\nabla}$-pregeodesic [7]. A statistical manifold $(\mathcal{M}, \nabla, g)$ is called *α-conformally flat* if it is locally α-conformally equivalent to a flat statistical manifold. It is known that a statistical manifold is of constant curvature if and only if it is ± 1-conformally flat, when $\dim \mathcal{M} \geq 3$ [8].

Affine Differential Geometry

Let $\mathcal{M}$ be an n-dimensional manifold and consider an *affine immersion* [7] (f, ξ), which is the pair of an immersion f from $\mathcal{M}$ into $\mathbf{R}^{n+1}$ and a transversal vector field ξ along $f(\mathcal{M})$. By a given affine immersion (f, ξ) of $\mathcal{M}$ and the usual flat affine connection D of $\mathbf{R}^{n+1}$, the Gauss and Weingarten formulas are respectively obtained as follows:

$$D_X f_*(Y) = f_*(\nabla_X Y) + h(X, Y)\xi,$$
$$D_X \xi = -f_*(SX) + \tau(X)\xi.$$

Here, ∇, h, S and τ are called, respectively, *induced connection, affine fundamental form, affine shape operator* and *transversal connection form*. In this case, we say that the affine immersion realizes $(\mathcal{M}, \nabla, h)$ in $\mathbf{R}^{n+1}$. If h is non-degenerate (resp. $\tau = 0$ on $\mathcal{M}$), the affine immersion (f, ξ) is called *non-degenerate* (resp. *equiaffine*). It is known that non-degenerate and equiaffine (f, ξ) realizes a statistical manifold $(\mathcal{M}, \nabla, h)$ by regarding h as a pseudo-Riemannian metric g.

Such a statistical manifold is characterized as follows:

Proposition 2 *[8] A simply connected statistical manifold $(\mathcal{M}, \nabla, g)$ can be realized by a non-degenerate and equiaffine immersion if and only if it is 1-conformally flat.*

Let a point o in $\mathbf{R}^{n+1}$ be chosen as origin and consider an immersion f from $\mathcal{M}$ to $\mathbf{R}^{n+1}\backslash\{o\}$ so that $\xi = -\overrightarrow{of(p)}$ is transversal to $f(\mathcal{M})$ for $p \in \mathcal{M}$. Such an affine immersion (f, ξ) is called *centro-affine*, where the Weingarten formula is $D_X\xi = -f_*(X)$, or $S = I$ and $\tau = 0$. This implies that a centro-affine immersion, if it is non-degenerate, realizes a statistical manifold of constant curvature because of the Gauss equation:

$$R(X, Y)Z = h(X, Z)SX - h(X, Z)SY.$$

Further, the realized affine connection ∇ is projectively flat [7].

Denote the dual space of $\mathbf{R}^{n+1}$ by $\mathbf{R}_{n+1}$ and the pairing of $x \in \mathbf{R}^{n+1}$ and $y \in \mathbf{R}_{n+1}$ by $\langle y, x \rangle$. Define a map $\nu : \mathcal{M} \to \mathbf{R}_{n+1} \backslash \{o\}$ as follows:

$$\langle \nu_p, \xi_p \rangle = 1, \quad \langle \nu_p, f_*(X) \rangle = 0 \quad (\forall X \in T_p\mathcal{M}).$$

Such ν_p is uniquely defined and is called the *conormal vector*.

The pair $(\nu, -\nu)$ can be regarded as a centro-affine immersion from $\mathcal{M}$ into the dual space $\mathbf{R}_{n+1}$ equipped with the usual flat connection D^*. The formulas are

$$\begin{aligned} D^*_X(\nu_* Y) &= \nu(\nabla^*_X Y) + h^*(X, Y)(-\nu), \\ D^*_X(-\nu) &= -\nu_*(X), \end{aligned}$$

where $h^*(X, Y) = h(SX, Y)$, and ∇^* is dual of ∇ with respect to h. Hence, when (f, ξ) realizes a statistical manifold $(\mathcal{M}.\nabla, h)$ with $S = I$, then $(\nu, -\nu)$ realizes its dual statistical manifold $(\mathcal{M}, \nabla^*, h)$ [7]. Both manifolds are of constant curvature.

For a statistical manifold $(\mathcal{M}, \nabla, h)$ realized by a non-degenerate and equiaffine immersion (f, ξ), we can define a *contrast function* ρ that induces the structure $(\mathcal{M}, \nabla, h)$

$$\rho(p, q) = \langle \nu(q), f(p) - f(q) \rangle, \quad (p, q \in \mathcal{M}).$$

The function ρ is called the *geometric divergence* of $(\mathcal{M}, \nabla, h)$ [8]. For a statistical manifold $(\mathcal{M}, \tilde{\nabla}, \tilde{h})$ that is 1-conformally equivalent to $(\mathcal{M}, \nabla, h)$, one of its contrast function is given by $\tilde{\rho}(p, q) = \sigma(q)\rho(p, q)$ for a certain positive function σ. The contrast function $\tilde{\rho}$ is called the *conformal divergence* [8].

A statistical manifold $(\mathcal{M}, \nabla, g)$ of constant curvature κ is studied from a view-point of affine differential geometry [8]. It is known that $(\mathcal{M}, \nabla, g)$ realized in $\mathbf{R}^{n+1}$ has the following geometric properties:

P1 For three points p, q and r in $\mathcal{M}$ let the ∇-geodesic connecting p and q and the ∇^*-geodesic connecting q and r are orthogonal at q. Then the following modified Pythagorean relation holds:

$$\rho(p, r) = \rho(p, q) + \rho(q, r) - \kappa\rho(p, q)\rho(q, r),$$

P2 An arbitrary ∇-geodesic on $\mathcal{M}$ is the intersection of a two-dimensional subspace in $\mathbf{R}^{n+1}$ and $\mathcal{M}$,

P3 The volume element θ on $\mathcal{M}$ induced from $\mathbf{R}^{n+1}$ satisfies $\nabla\theta = 0$,

and so on. A typical example of the statistical manifold of non-zero constant curvature is the alpha-geometry $(\mathcal{S}^n, \nabla^{(\alpha)}, g^F)$, where $\kappa = (1 - \alpha^2)/4$. In this case, the modified Pythagorean relation induces the widely-known nonextensivity relation of Tsallis entropy [21, Remark 2].

References

1. Amari, S.-I.: Differential-Geometrical Methods in Statistics. Lecture Notes in Statistics, vol. 28. Springer, New York (1985)
2. Amari, S-I., Nagaoka, H.: Methods of Information Geometry. Translations of Mathematical Monographs, vol. 191. American Mathematical Society and Oxford University Press, Oxford (2000)
3. Zhang, J.: Divergence function, duality, and convex analysis. Neural Comput. **16**, 159–195 (2004)
4. Eguchi, S.: Information geometry and statistical pattern recognition. Sugaku Expos. **19**, 197–216 (2006). (originally 2004 Sugaku 56 380–399 in Japanese)
5. Naudts, J.: Continuity of a class of entropies and relative entropies. Rev. Math. Phys. **16**(6), 809–822 (2004)
6. Harsha, K.V., Subrahamanian Moosath, K.S.: F-geometry and Amari's α-geometry on a statistical manifold. Entropy **16**(5), 2472–2487 (2014)
7. Nomizu, K., Sasaki, T.: Affine Differential Geometry. Cambridge University Press, Cambridge (1993)
8. Kurose, T.: On the divergences of 1-conformally flat statistical manifolds. Tohoku Math. J. **46**, 427–433 (1994)
9. Ohara, A., Matsuzoe, H., Amari, S.-I.: A dually flat structure on the space of escort distributions. J. Phys. Conf. Ser. **201**, 012012 (2010)
10. Ohara, A., Matsuzoe, H., Amari, S.: Conformal geometry of escort probability and its applications. Mod. Phys. Lett. B **26**(10), 1250063-1–1250063-14 (2012)
11. Eguchi, S.: Geometry of minimum contrast. Hiroshima Math. J. 631–647 (1992)
12. Tsallis, C.: Introduction to Nonextensive Statistical Mechanics: Approaching a Complex World. Springer, New York (2009)
13. Rényi, A.: On measures of entropy and information. In: Proceedings of the 4th Berkeley Symposium on Mathematical Statistics and Probability, vol. 1, pp. 547–561. University of California Press, Berkeley (1961)
14. Edelsbrunner, H., Seidel, R.: Voronoi diagrams and arrangements. Discrete Comput. Geom. **1**, 25–44 (1986)
15. Edelsbrunner, H.: Algorithms in Combinatorial Geometry. Springer, Berlin (1987)
16. Onishi, K., Takayama, N.: Construction of Voronoi diagram on the upper half-plane. IEICE Trans. Fundam. **E79-A**, 533–539 (1996)
17. Onishi, K., Imai, H.: Voronoi diagram in statistical parametric space by Kullback-Leibler divergence. In: Proceedings of the 13th ACM-SIAM Symposium on Computational Geometry, pp. 463–465 (1997)
18. Imai, H., Inaba, M.: Divergence-based geometric clustering and its underlying discrete proximity structures. IEICE Trans. Inf. Syst. **E83-D**, 27–35 (2000)
19. Boissonnat, J.-D., Nielsen, F., Nock, N.: Bregman Voronoi diagram. Discrete Comput. Geom. **44**, 281–307 (2010)
20. Matsuzoe, H.: Computational Geometry from the Viewpoint of Affine Differential Geometry. In: Nielsen, F. (eds.) Emerging Trends in Visual Computing, pp. 103–123. Springer, Berlin (2009)
21. Ohara, A.: Geometry of distributions associated with Tsallis statistics and properties of relative entropy minimization. Phys. Lett. A **370**, 184–193 (2007)
22. Nielsen, F., Nock, R.: The dual Voronoi diagrams with respect to representational Bregman divergences. In: International Symposium on Voronoi Diagrams (ISVD), DTU Lyngby, Denmark. IEEE Press, New York (2009)
23. Chazelle, B.: An optimal convex hull algorithm in any fixed dimension. Discrete Comput. Geom. **10**, 377–409 (1993)
24. Nielsen, F., Nock, R.: Total Jensen divergences: definition, properties and clustering. In: IEEE International Conference on Acoustics, Speech and Signal Processing (ICASSP) (2015)

25. Nock, R., Nielsen, F., Amari, S.-I.: On conformal divergences and their population minimizers. IEEE Trans. IT **62**(1), 527–538 (2016)
26. Ohara, A.: Conformal flattening for deformed information geometries on the probability simplex. Entropy **20**, 186 (2018)
27. Shima, H.: The Geometry of Hessian Structures. World Scientific, Singapore (2007)

Monte Carlo Information-Geometric Structures

Frank Nielsen and Gaëtan Hadjeres

Abstract Exponential families and mixture families are parametric probability models that can be geometrically studied as smooth statistical manifolds with respect to any statistical divergence like the Kullback–Leibler (KL) divergence or the Hellinger divergence. When equipping a statistical manifold with the KL divergence, the induced manifold structure is dually flat, and the KL divergence between distributions amounts to an equivalent Bregman divergence on their corresponding parameters. In practice, the corresponding Bregman generators of mixture/exponential families require to perform definite integral calculus that can either be too time-consuming (for exponentially large discrete support case) or even do not admit closed-form formula (for continuous support case). In these cases, the dually flat construction remains theoretical and cannot be used by information-geometric algorithms. To bypass this problem, we consider performing stochastic Monte Carlo (MC) estimation of those integral-based mixture/exponential family Bregman generators. We show that, under natural assumptions, these MC generators are almost surely Bregman generators. We define a series of dually flat information geometries, termed Monte Carlo Information Geometries, that increasingly-finely approximate the untractable geometry. The advantage of this MCIG is that it allows a practical use of the Bregman algorithmic toolbox on a wide range of probability distribution families. We demonstrate our approach with a clustering task on a mixture family manifold. We then show how to generate MCIG for arbitrary separable statistical divergence between distributions belonging to a same parametric family of distributions.

F. Nielsen (✉)
Sony Computer Science Laboratories, Tokyo, Japan
e-mail: Frank.Nielsen@acm.org

G. Hadjeres
Sony Computer Science Laboratory, Paris, France
e-mail: Gaetan.Hadjeres@sony.com

F. Nielsen (ed.), *Geometric Structures of Information*, Signals and Communication Technology, https://doi.org/10.1007/978-3-030-02520-5_5

Keywords Computational information geometry · Statistical manifold
Dually flat information geometry · Bregman generator
Stochastic Monte Carlo integration · Mixture family · Exponential family
Clustering

1 Introduction

We concisely describe the construction and properties of dually flat spaces [1, 8] in Sect. 1.1, define the statistical manifolds of exponential families and mixture families in Sect. 1.2, and discuss about the computational tractability of Bregman algorithms in dually flat spaces in Sect. 1.3.

1.1 Dually Flat Space: Bregman Geometry

A smooth (potentially asymmetric) distance $D(\cdot,\cdot)$ is called a *divergence* in information geometry [1, 8], and induces a differential-geometric dualistic structure [1, 2, 8, 17]. In particular, a strictly convex and twice continuously differentiable D-dimensional real-valued function F, termed a *Bregman generator*, induces a dually connection-flat structure via a corresponding Bregman Divergence (BD) [4] $B_F(\cdot,\cdot)$ given by:

$$B_F(\theta_1 : \theta_2) := F(\theta_1) - F(\theta_2) - \langle \theta_1 - \theta_2, \nabla F(\theta_2)\rangle, \tag{1}$$

where $\langle y, x\rangle := y^\top x$ denotes the inner product, and $\nabla F(\theta) := (\partial_i F(\theta))_i$ denotes the gradient vector of partial first-order derivatives with respect to vector parameter θ. We use the standard notational convention of information geometry [1, 8]: $\partial_i :=: \frac{\partial}{\partial \theta^i}$ to indicate[1] a contravariant vector [18] $\theta = (\theta^i)_i$.

The Legendre–Fenchel transformation [30] :

$$F^*(\eta) = \sup_\theta\{\langle \theta, \eta\rangle - F(\theta)\}, \tag{2}$$

is at the heart of the duality of flat structures by defining two global affine coordinate systems: The *primal affine θ-coordinate system* and the *dual affine η-coordinate system*, so that any point P of the manifold $\mathcal{M}$ can either be accessed by its *primal* $\theta(P)$ coordinates or equivalently by its *dual* $\eta(P)$ coordinates. We can switch between these two dual coordinates as follows:

[1]The :=: symbol means it is a notational convention equality, like $\sum_{i=1}^k x_i :=: x_1 + \ldots x_k$. It differs from $a := b$ which denotes the symbol of a quantity equality by definition.

$$\eta = \eta(\theta) = \nabla F(\theta) = (\partial_i F(\theta))_i, \tag{3}$$
$$\theta = \theta(\eta) = \nabla F^*(\eta) = (\partial^i F^*(\eta))_i, \tag{4}$$

with reciprocal gradients $\nabla F^* := (\nabla F)^{-1}$. We used the notational convention $\partial^i :=: \frac{\partial}{\partial \eta_i}$ which indicates the covariant vector [18] $\eta = (\eta_i)_i$.

The metric tensor g of the dually flat structure $(\mathcal{M}, F)$ can either be expressed using the θ- or η-coordinates using the Hessians of the potential functions [53]:

$$G(\theta) = \nabla^2 F(\theta), \tag{5}$$
$$G^*(\eta) = \nabla^2 F^*(\eta). \tag{6}$$

It defines a smooth bilinear form $\langle v, v' \rangle_g$ on $\mathcal{M}$ so that for two vectors v, w of a tangent plane T_P:

$$\langle v, v' \rangle_g = \theta(v)^\top G(\theta)\theta(w), \tag{7}$$
$$= \eta(v)^\top G^*(\eta)\eta(w), \tag{8}$$

where $\theta(v) = (v^i)_i$ and $\eta(v) = (v_i)_i$ denote the contravariant coefficients and covariant coefficients of a vector v, respectively. That is, any vector $v \in T_P$ can be written either as $v = \sum_i v^i e_i$ or as $\sum_i v_i e^{*i}$, where $\{e_i\}_i$ and $\{e^{*i}\}_i$ are the dual basis [18] of the vector space structure of T_P.

Matrices $G(\theta)$ and $G^*(\eta)$ are symmetric positive definite (SPD, denoted by $G(\theta) \succ 0$ and $G^*(\eta) \succ 0$), and they satisfy the Crouzeix identity [13]:

$$G(\theta)G^*(\eta) = I, \tag{9}$$

where I stands for the $D \times D$ identity matrix. This indicates that at each tangent plane T_P, the dual coordinate systems are biorthogonal [57] (with $\{e_i\}_i$ and $\{e^{*i}\}_i$ forming a dual basis [18] of the vector space structure of T_P):

$$\langle e_i, e^{*j} \rangle = \delta_i^j, \tag{10}$$

with δ_i^j the Krönecker symbol: $\delta_i^j = 1$ if and only if (iff) $i = j$, and 0 otherwise. We have:

$$\frac{\partial \eta_i}{\partial \theta^j} = g_{ij}(\theta) = \langle e_i, e_j \rangle, \tag{11}$$
$$\frac{\partial \theta^i}{\partial \eta_j} = g^{ij}(\eta) = \langle e^{*i}, e^{*j} \rangle. \tag{12}$$

The convex conjugate functions $F(\theta)$ and $F^*(\eta)$ are called *dual potential functions*, and define the global metric [53].

Table 1 Overview of the dually differential-geometric structure $(\mathcal{M}, F)$ induced by a Bregman generator F. Notice that if F and ∇F^* are available in closed-form then so are ∇F and F^*

Manifold $(\mathcal{M}, F)$	Primal structure	Dual structure
Affine coordinate system	$\theta(\cdot)$	$\eta(\cdot)$
Conversion $\theta \leftrightarrow \eta$	$\theta(\eta) = \nabla F^*(\eta)$	$\eta(\theta) = \nabla F(\theta)$
Potential function	$F(\theta) = \langle\theta, \nabla F(\theta)\rangle - F^*(\nabla F(\theta))$	$F^*(\eta) = \langle\eta, \nabla F^*(\eta)\rangle - F(\nabla F^*(\eta))$
Metric tensor g	$G(\theta) = \nabla^2 F(\theta)$	$G^*(\eta) = \nabla^2 F^*(\eta)$
	$g_{ij} = \partial_i \partial_j F(\theta)$	$g^{ij} = \partial^i \partial^j F^*(\eta)$
Geodesic ($\lambda \in [0, 1]$)	$\gamma(P, Q) := \{(PQ)_\lambda = (1-\lambda)\theta(P) + \lambda\theta(Q)\}_\lambda$	$\gamma^*(P, Q) := \{(PQ)^*_\lambda = (1-\lambda)\eta(P) + \lambda\eta(Q)\}_\lambda$

Table 1 summarizes the differential-geometric structures of dually flat spaces. Since Bregman divergences are *canonical divergences*[2] of dually flat spaces [1], the geometry of dually flat spaces is also referred to the *Bregman geometry* [15] in the literature.

Definition 1 (*Bregman generator*) A Bregman generator is a strictly convex and twice continuously differentiable real-valued function $F : \mathbb{R}^D \to \mathbb{R}$.

Let us cite the following well-known properties of Bregman generators [4]:

Property 1 *(Bregman generators are equivalent up to modulo affine terms) The Bregman generator $F_2(\theta) = F_1(\theta) + \langle a, \theta\rangle + b$ (with $a \in \mathbb{R}^D$ and $b \in \mathbb{R}$) yields the same Bregman divergence as the Bregman divergence induced by F_1, $B_{F_2}(\theta_1 : \theta_2) = B_{F_1}(\theta_1 : \theta_2)$, and therefore the same dually flat space $(\mathcal{M}, F_2) \cong (\mathcal{M}, F_1)$.*

Property 2 *(Linearity rule of Bregman generators) Let F_1, F_2 be two Bregman generators and $\lambda_1, \lambda_2 > 0$. Then $B_{\lambda_1 F_1 + \lambda_2 F_2}(\theta : \theta') = \lambda_1 B_{F_1}(\theta : \theta') + \lambda_2 B_{F_2}(\theta : \theta')$.*

In practice, the algorithmic toolbox in dually flat spaces (e.g., clustering [4], minimum enclosing balls [39], hypothesis testing [31] and Chernoff information [32], Voronoi diagrams [6, 34], proximity data-structures [45, 46], etc.) can be used whenever the dual Legendre convex conjugates F and F^* are both available in closed-form (see Type 1 of Table 4). In that case, both the primal $\gamma(P, Q) := \{(PQ)_\lambda\}_\lambda$ and dual $\gamma^*(P, Q) := \{(PQ)^*_\lambda\}_\lambda$ geodesics are available in closed form. These dual geodesics can either be expressed using the θ or η-coordinate systems as follows:

$$(PQ)_\lambda = \begin{cases} \theta((PQ)_\lambda) = \theta(P) + \lambda(\theta(Q) - \theta(P)), \\ \eta((PQ)_\lambda) = \nabla F(\theta((PQ)_\lambda)) = \nabla F(\nabla F^*(\eta(P)) + \lambda(\nabla F^*(\eta(Q)) - \nabla F^*(\eta(P)))), \end{cases} \tag{13}$$

[2]That is, we can associate to any dually flat manifold a divergence that amounts to a Bregman divergence [1].

Table 2 Some fundamental Bregman clustering algorithms [4, 22, 41] (of the Bregman algorithmic toolbox) that illustrate which closed-form are required to be run in practice

Algorithm	$F(\theta)$	$\eta(\theta) = \nabla F(\theta)$	$\theta(\eta) = \nabla F^*(\eta)$	$F^*(\eta)$
Right-sided Bregman clustering	✓	✓	×	×
Left-sided Bregman clustering	×	×	✓	✓
Symmetrized Bregman centroid	✓	✓	✓	✓
Mixed Bregman clustering	✓	✓	✓	✓
Maximum Likelihood Estimator for EFs	×	×	✓	×
Bregman soft clustering (≡ EM)	×	✓	✓	✓

$$(PQ)^*_\lambda = \begin{cases} \eta((PQ)^*_\lambda) = \eta(P) + \lambda(\eta(Q) - \eta(P)), \\ \theta((PQ)^*_\lambda) = \nabla F^*(\eta((PQ)^*_\lambda)) = \nabla F^*(\nabla F(\theta(P)) + \lambda(\nabla F(\theta(Q)) - \nabla F(\theta(P)))) \end{cases} \tag{14}$$

That is, the primal geodesic corresponds to a straight line in the primal coordinate system while the dual geodesic is a straight line in the dual coordinate system. However, in many interesting cases, the convex generator F or its dual F^* (or both) are not available in closed form or are computationally intractable, and the above Bregman toolbox cannot be used. Table 2 summarizes the closed-form formulas required to execute some fundamental clustering algorithms [4, 22, 41] in a Bregman geometry.

Let us notice that so far the points $P \in \mathcal{M}$ in the dually flat manifold have no particular meaning, and that the dually flat space structure is generic, not necessarily related to a statistical flat manifold. We shall now quickly review the dualistic structure of statistical manifolds [24].

1.2 Geometry of Statistical Manifolds

Let $I_1(x; y)$ denote a *scalar divergence*. A *statistical divergence* between two probability distributions P and Q, with Radon-Nikodym derivatives $p(x)$ and $q(x)$ with respect to (w.r.t.) a base measure μ defined on the support $\mathcal{X}$, is defined as:

$$I(P : Q) = \int_{x\in\mathcal{X}} I_1\,(p(x) : q(x))\,\mathrm{d}\mu(x). \tag{15}$$

A statistical divergence is a measure of dissimilarity/discrimination that satisfies $I(P : Q) \geq 0$ with equality iff. $P = Q$ (a.e., reflexivity property) . For example, the Kullback–Leibler divergence is a statistical divergence:

$$\mathrm{KL}(P : Q) := \int_{x\in\mathcal{X}} \mathrm{kl}(p(x) : q(x))\mathrm{d}\mu(x), \tag{16}$$

with corresponding scalar divergence:

$$\mathrm{kl}(x : y) := x \log \frac{x}{y}. \tag{17}$$

The KL divergence between P and Q is also called the *relative entropy* [11] because it is the difference of the *cross-entropy* $h^\times(P : Q)$ between P and Q with the Shannon entropy $h(P)$ of P:

$$\mathrm{KL}(P : Q) = h^\times(P : Q) - h(P), \tag{18}$$

$$h^\times(P : Q) := \int_{x\in\mathcal{X}} p(x) \log \frac{1}{q(x)} \mathrm{d}\mu(x), \tag{19}$$

$$h(P) := \int_{x\in\mathcal{X}} p(x) \log \frac{1}{p(x)} \mathrm{d}\mu(x) = h^\times(P : P). \tag{20}$$

Thus we distinguish a statistical divergence from a parameter divergence by stating that a statistical divergence is a separable divergence that is the definite integral on the support of a scalar divergence.

In information geometry [1, 8], we equip a probability manifold $\mathcal{M} = \{p(x;\theta) : \theta \in \Theta\}$ with a *metric tensor* g (for measuring angles between vectors and lengths of vectors in tangent planes) and a *pair of dual torsion-free connections* ∇ and ∇^* (for defining parallel transports and geodesics) that are defined by their Christoffel symbols Γ_{ijk} and Γ^*_{ijk}. These geometric structures $(\mathcal{M}, D) := (\mathcal{M}, g_D, \nabla_D, \nabla^*_D)$ can be induced by *any smooth* C^∞ divergence $D(\cdot : \cdot)$ [1, 2, 8, 17] as follows:

$$g_{ij}(x) = \left.\frac{\partial^2}{\partial x_i \partial x_j} D(x : y)\right|_{y=x}, \tag{21}$$

$$\Gamma_{ijk}(x) = -\left.\frac{\partial^3}{\partial x_i \partial x_j \partial y_k} D(x : y)\right|_{y=x}. \tag{22}$$

The *dual divergence* $D^*(p : q) := D(q : p)$ highlights the *reference duality* [57], and the dual connection ∇^* is induced by the dual divergence $D^*(\cdot : \cdot)$ (∇^* is defined by $\Gamma^*_{ijk}(x) = -\left.\frac{\partial^3}{\partial x_i \partial x_j \partial y_k} D^*(x : y)\right|_{y=x}$). Observe that the metric tensor is self-dual: $g^* = g$.

Let us give some examples of parametric probability families and their statistical manifolds induced by the Kullback–Leibler divergence.

1.2.1 Exponential Family Manifold (EFM)

We start by a definition:

Definition 2 (*Exponential family*) Let μ be a prescribed base measure and $t(x)$ a sufficient statistic vector. We can build a corresponding exponential family:

$$\mathcal{E}_{t,\mu} := \{p(x;\theta) \propto \exp(\langle t(x), \theta \rangle)\}_\theta, \tag{23}$$

where $p(x;\theta) := \frac{\mathrm{d}P(\theta)}{\mathrm{d}\mu}(x)$.

The densities are normalized by the cumulant function F:

$$F(\theta) := \log \left(\int_{x \in \mathcal{X}} \exp(\langle t(x), \theta \rangle) \mathrm{d}\mu(x) \right), \tag{24}$$

so that:

$$p(x;\theta) = \exp(\langle t(x), \theta \rangle - F(\theta)). \tag{25}$$

The function F is a Bregman generator on the natural parameter space:

$$\Theta := \left\{ \theta : \int_{x \in \mathcal{X}} \exp(\langle t(x), \theta \rangle) \mathrm{d}\mu(x) < \infty \right\}. \tag{26}$$

If we add an extra carrier term $k(x)$ and consider the measure $\nu(x) := \frac{\mu(x)}{\exp(k(x))}$, we get the generic form of an exponential family [36]:

$$\mathcal{E}_{t,k,\nu} := \{p(x;\theta) \propto \exp(\langle t(x), \theta \rangle + k(x)) : \theta \in \Theta\}. \tag{27}$$

We call the function F the *Exponential Family Bregman Generator*, or EFBG for short in the remainder.

It turns out that $(\mathcal{E}_{t,\mu}, \mathrm{KL}, \nabla_{\mathrm{KL}}, \nabla^*_{\mathrm{KL}}) \cong (\mathcal{M}, F)$ (meaning the information-geometric structure of the statistical manifold is isomorphic to the information-geometry of a dually flat manifold) so that:

$$\mathrm{KL}(p(x;\theta_1) : p(x;\theta_2) = B_F(\theta_2 : \theta_1), \tag{28}$$

$$= B_{F^*}(\eta_1 : \eta_2), \tag{29}$$

with $\eta = E_{p(x;\theta)}[t(x)]$ the dual parameter called the expectation parameter or moment parameter.

1.2.2 Mixture Family Manifold (MFM)

Another important type of families of probability distributions are the mixture families:

Definition 3 (*Mixture family*) Given a set of k prescribed statistical distributions $p_0(x), \ldots, p_{k-1}(x)$, all sharing the same support $\mathcal{X}$ (say, $\mathbb{R}$), a *mixture family* $\mathcal{M}$ of order $D = k - 1$ consists of all *strictly convex combinations* of these component distributions [43, 44]:

$$\mathcal{M} := \left\{ m(x;\eta) = \sum_{i=1}^{k-1} \eta_i p_i(x) + \left(1 - \sum_{i=1}^{k-1} \eta_i\right) p_0(x) : \eta_i > 0, \sum_{i=1}^{k-1} \eta_i < 1 \right\}. \tag{30}$$

It shall be understood from the context that $\mathcal{M}$ is a shorthand for $\mathcal{M}_{p_0(x),\ldots,p_D}$.

It turns out that $(\mathcal{M}, \mathrm{KL}, \nabla_{\mathrm{KL}}, \nabla^*_{\mathrm{KL}}) \cong (\mathcal{M}, G)$ so that:

$$\mathrm{KL}(m(x;\eta) : m(x;\eta')) = B_G(\eta : \eta'), \tag{31}$$

for the Bregman generator being the Shannon negative entropy (also called Shannon information):

$$G(\eta) = -h(m(x;\eta)) = \int_{x\in\mathcal{X}} m(x;\eta) \log m(x;\eta) \mathrm{d}\mu(x). \tag{32}$$

We call function G the *Mixture Family Bregman Generator*, or MFBG for short in the remainder.

For a mixture family, we prefer to use the notation η instead of θ for indexing the distribution parameters as it is customary in textbooks of information geometry [1, 8]. One reason comes from the fact that the KL divergence between two mixtures amounts to a BD on their respective parameters (Eq. 31) while the KL divergence between exponential family distributions is equivalent to a BD on the swapped order of their respective parameters (Eq. 28), see [3, 19]. Thus in order to get the same order of arguments for the KL between two exponential family distributions, we need to use the dual Bregman divergence on the dual η parameter, see Eq. 29.

1.2.3 Cauchy Family Manifold (CFM)

This example is only given to emphasize that probability families may neither be exponential nor mixture families [28].

A Cauchy distribution has probability density defined on the support $\mathcal{X} = \mathbb{R}$ by:

$$p(x;\mu,\sigma) = \frac{1}{\pi\sigma\left(1 + \left(\frac{x-\mu}{\sigma}\right)^2\right)}. \tag{33}$$

The space of all Cauchy distributions:

$$\mathcal{C} = \{p(x;\mu,\sigma) : \mu \in \mathbb{R}, \sigma > 0\}. \tag{34}$$

is a location-scale family [23]. It is not an exponential family nor a mixture family.

Table 3 compares the dually flat structures of mixture families with exponential families. In information geometry, $(\mathcal{E}_{t,k,\mu}, \mathrm{KL}, \nabla_{\mathrm{KL}}, \nabla^*_{\mathrm{KL}}) = (\mathcal{E}_{t,k,\mu}, g, \nabla^e, \nabla^m)$ and $(\mathcal{M}, \mathrm{KL}, \nabla_{\mathrm{KL}}, \nabla^*_{\mathrm{KL}}) = (\mathcal{M}, g, \nabla^m, \nabla^e)$ where g is the *Fisher information metric*

Table 3 Characteristics of the dually flat geometries of Exponential Families (EFs) and Mixture Families (MFs)

	Exponential Family	Mixture Family
Density	$p(x;\theta) = \exp(\langle\theta, x\rangle - F(\theta))$	$m(x;\eta) = \sum_{i=1}^{k-1} \eta_i f_i(x) + c(x)$
		$f_i(x) = p_i(x) - p_0(x)$
Family/Manifold	$\mathcal{M} = \{p(x;\theta) : \theta \in \Theta^\circ\}$	$\mathcal{M} = \{m(x;\eta) : \eta \in H^\circ\}$
Convex function ($\equiv ax + b$)	F: cumulant	F^*: negative entropy
Dual coordinates	moment $\eta = E[t(x)]$	$\theta^i = h^\times(p_0 : m) - h^\times(p_i : m)$
Fisher Information $g = (g_{ij})_{ij}$	$g_{ij}(\theta) = \partial_i\partial_j F(\theta)$ $g = \mathrm{Var}[t(X)]$	$g_{ij}(\eta) = \int_{\mathcal{X}} \frac{f_i(x)f_j(x)}{m(x;\eta)} \mathrm{d}\mu(x)$
		$g_{ij}(\eta) = -\partial_i\partial_j h(\eta)$
Christoffel symbol	$\Gamma_{ij,k} = \frac{1}{2}\partial_i\partial_j\partial_k F(\theta)$	$\Gamma_{ij,k} = -\frac{1}{2}\int_{\mathcal{X}} \frac{f_i(x)f_j(x)f_k(x)}{m^2(x;\eta)} \mathrm{d}\mu(x)$
Entropy	$-F^*(\eta)$	$-F^*(\eta)$
Kullback–Leibler divergence	$B_F(\theta_2 : \theta_1)$	$B_{F^*}(\eta_1 : \eta_2)$
	$= B_{F^*}(\eta_1 : \eta_2)$	$= B_F(\theta_2 : \theta_1)$

tensor and ∇^e and ∇^m are the exponential and mixture connections, respectively. These connections are dual to each others, see [8].

1.3 Computational Tractability of Dually Flat Statistical Manifolds

The previous section explained the dually flat structures (i.e., Bregman geometry) of the exponential family manifold and of the mixture family manifold. However these geometries may be purely theoretical as the Bregman generator F may not be available in closed form so that the Bregman toolbox cannot be used in practice. This work tackles this problem faced in exponential and mixture family manifolds by proposing the novel framework of *Monte Carlo Information Geometry* (MCIG). MCIG approximates the untractable Bregman geometry by considering the Monte Carlo stochastic integration of the definite integral-based ideal Bregman generator.

But first, let us quickly review the five types of tractability of Bregman geometry in the context of statistical manifolds by giving an illustrating family example for each type:

Type 1. F and ∇F^* are both available in closed-form, and so are ∇F and F^*. For example, this is the case of the *the Gaussian exponential family*. The normal distribution [36] has sufficient statistic vector $t(x) = (x, x^2)$ so that its log-normalizer is

$$F(\theta) = \log\left(\int_{-\infty}^{+\infty} \exp(\theta_1 x + \theta_2 x^2)\mathrm{d}x\right). \tag{35}$$

Since $\int_{-\infty}^{\infty} \exp(\theta_1 x + \theta_2 x^2) = \sqrt{\frac{\pi}{-\theta_2}} \exp(-\frac{\theta_1^2}{4\theta_2})$ for $\theta_2 < 0$, we find:

$$F(\theta) = \log\left(\int \exp(\theta_1 x + \theta_2 x^2)\mathrm{d}x\right) = -\frac{\theta_1^2}{4\theta_2} + \frac{1}{2}\log\frac{\pi}{-\theta_2}. \tag{36}$$

This is in accordance with the direct canonical decomposition [36] of the density $p(x;\theta) = \exp(\langle t(x), \theta\rangle - F(\theta))$ of the normal density $p(x;\mu,\sigma) = \frac{1}{\sqrt{2\pi}\sigma} \exp(-\frac{(x-\mu)^2}{2\sigma^2})$.

Remark 1 When $F(\theta)$ can be expressed using the canonical decomposition of exponential families, this means that the definite integral $\log(\int \exp(\langle t(x), \theta\rangle + k(x))\mathrm{d}x)$ is available in closed form, and vice-versa.

Type 2. F is available in closed form (and so is ∇F) but ∇F^* is not available in closed form (and therefore F^* is not available too). This is for example the *Beta exponential family*. A Beta distribution $\mathrm{Be}(\alpha, \beta)$ has density on support $x \in (0, 1)$:

$$p(x;\alpha,\beta) = \frac{1}{B(\alpha,\beta)} x^{\alpha-1}(1-x)^{\beta-1}, \tag{37}$$

where $B(\alpha,\beta) = \frac{\Gamma(\alpha)\Gamma(\beta)}{\Gamma(\alpha+\beta)}$, and $(\alpha > 0, \beta > 0)$ are the shape parameters. The Beta family of distributions is an exponential family with $\theta = (\alpha, \beta)$, $t(x) = (\log(x), \log(1-x))$, $k(x) = -\log(x) - \log(1-x)$ and $F(\theta) = \log B(\theta_1, \theta_2) = \log\Gamma(\theta_1) + \log\Gamma(\theta_2) - \log\Gamma(\theta_1+\theta_2)$. Note that we could also have chosen $\theta = (\alpha-1, \beta-1)$ and $k(x) = 0$. Thus $\nabla F(\theta) = (\psi(\theta_1) - \psi(\theta_1+\theta_2), \psi(\theta_2) - \psi(\theta_1+\theta_2))$ where $\psi(x) = \frac{\Gamma'(x)}{\Gamma(x)}$ is the digamma function. Inverting the gradient $\nabla F(\theta) = \eta$ to get $\eta = \nabla F^*(\theta)$ is not available in closed-form.[3]
Type 3. This type of families has discrete support $\mathcal{X}$ and thus requires an exponential time to compute the log-normalizer. For example, consider the Ising models [5, 9, 21]: Let $G = (V, E)$ be an undirected graph of $|V|$ nodes and $|E|$ edges. Each node $v \in V$ is associated with a binary random variable $x_v \in \{0, 1\}$. The probability of an Ising model is defined as follows:

$$p(x;\theta) = \exp\left(\sum_{v\in V}\theta_v x_v + \sum_{(v,w)\in E}\theta_{vw}x_v x_w - F(\theta)\right). \tag{38}$$

[3]To see this, consider the digamma difference property: $f_\Delta(\theta) = \psi(\theta) - \psi(\theta+\Delta) = -\sum_{i=0}^{\Delta-1}\frac{1}{x+i}$ for $\Delta \in \mathbb{N}$. We cannot invert $f_\Delta(\theta)$ since it involves solving the root of a high-degree polynomial.

The vector $t(x) = (\ldots, x_v, \ldots, x_{vw}, \ldots)$ of sufficient statistics is D-dimensional with $D = |V| + |E|$. The log-normalizer is:

$$F(\theta) = \log\left(\sum_{(x_v)_v \in \{0,1\}^{|V|}} \left(\exp \sum_{v \in V} \theta_v x_v + \sum_{(v,w) \in E} \theta_{vw} x_v x_w\right)\right). \quad (39)$$

It requires to sum up $2^{|V|}$ terms.

Type 4. This type of families has a Bregman generator which is not available in closed-form. For example, this is the case of the *Polynomial Exponential Family* [10, 42] (PEF) which are helpful to model a multimodal distribution (instead of using a statistical mixture). Consider the following vector of sufficient statistics $t(x) = (x, x^2, \ldots, x^D)$ for defining an exponential family:

$$\mathcal{E}_{t(x),\mu} = \left\{ p(x;\theta) = \exp\left(\sum_{i=1}^{D} \theta_i x^i - F(\theta)\right) : \theta \in \Theta \right\}. \quad (40)$$

(Beware that here, $x^i = \mathrm{Pow}(x, i) := \underbrace{x \times \cdots \times x}_{i \text{ times}}$ denotes the ith power of x (monomial of degree i), and not a contravariant coefficient of a vector x.)

In general, the definite integral of the cumulant function (the Exponential Family Bregman Generator, EFBG) of Eq. 24 does not admit a closed form, but is analytic. For example, choosing $t(x) = x^8$, we have:

$$F(\theta) = \log \int_{-\infty}^{\infty} \exp(\theta x^8) \mathrm{d}x = \log 2 + \log \Gamma(9/8) - \frac{1}{8} \log(-\theta), \quad (41)$$

for $\theta < 0$. But $\int_{-\infty}^{\infty} \exp(-x^8 - x^4 - x^2) \mathrm{d}x \simeq 1.295$ is not available in closed form.

Type 5. This last category is even more challenging from a computational point of view because of log-sum terms. For example, the *mixture family*. As already stated, the negative Shannon entropy (i.e., the Mixture Family Bregman Generator, MFBG) is not available in closed form for statistical mixture models [43]. It is in fact even worse, as the Shannon entropy of mixtures is not analytic [56].

This paper considers approximating the computationally untractable generators of statistical exponential/mixture families (type 4 and type 5) using stochastic Monte Carlo approximations.

In [12], Critchley et al. take a different approach of the computational tractability by discretizing the support $\mathcal{X}$ into a finite number of bins, and considering the corresponding discrete distribution. However, this approach does not scale well with the dimension of the support. Our Monte Carlo Information Geometry scales to arbitrary high dimensions because it relies on the fact that the Monte Carlo stochastic estimator is independent of the dimension [52].

1.4 Paper Organization

In Sect. 2, we consider the MCIG structure of mixture families: Namely, Sect. 2.1 considers first the uni-order families to illustrate the basic principle. It is followed by the general case in Sect. 2.2. Similarly, Sect. 3 handles the exponential family case by first explaining the uni-order case in Sect. 3.1 before tackling the general case in Sect. 3.2. Sect. 4 presents an application of the computationally-friendly MCIG structures for clustering distributions in dually flat statistical mixture manifolds. In Sect. 5, we show how to construct non-flat MCIG structures of a parametric family of distributions given by a statistical separable divergence. Finally, we conclude and discuss several perspectives in Sect. 6.

2 Monte Carlo Information Geometry of Mixture Families

Recall the definition of a statistical mixture model (Definition 3): Given a set of k prescribed statistical distributions $p_0(x), \ldots, p_{k-1}(x)$, all sharing the same support $\mathcal{X}$, a *mixture family* $\mathcal{M}$ of order $D = k - 1$ consists in all *strictly convex combinations* of the $p_i(x)$'s [43]:

$$\mathcal{M} := \left\{ m(x;\eta) = \sum_{i=1}^{k-1} \eta_i p_i(x) + \left(1 - \sum_{i=1}^{k-1} \eta_i\right) p_0(x) : \eta_i > 0, \sum_{i=1}^{k-1} \eta_i < 1 \right\}. \tag{42}$$

The differential-geometric structure of $\mathcal{M}$ is well studied in information geometry [1, 8] (although much less than for the exponential families), where it is known that:

$$\mathrm{KL}(m(x;\eta) : m(x;\eta')) = B_G(\eta : \eta'), \tag{43}$$

for the Bregman generator being the Shannon negative entropy (MFBG):

$$G(\eta) = -h(m(x;\eta)) = \int_{x \in \mathcal{X}} m(x;\eta) \log m(x;\eta) \mathrm{d}\mu(x). \tag{44}$$

The negative entropy $G(\eta) = \int_{x\in\mathcal{X}} m(x;\eta) \log m(x;\eta) \mathrm{d}\mu(x)$ is a smooth and strictly convex function which induces a dually flat structure with Legendre convex conjugate:

$$F(\theta) = G^*(\theta) = -\int_{x \in \mathcal{X}} p_0(x) \log m(x;\eta) \mathrm{d}\mu(x) = h^\times(p_0(x) : m(x;\eta)), \tag{45}$$

interpretable as the cross-entropy of $p_0(x)$ with the mixture $m(x;\eta)$ [43].

Notice that the component distributions may be heterogeneous like $p_0(x)$ being a fixed Cauchy distribution, $p_1(x)$ being a fixed Gaussian distribution, $p_2(x)$ a Laplace distribution, etc. Except for the case of the finite categorical distributions (that are both interpretable as either a mixture family and an exponential family, see [1]), $G(\eta)$ provably does not admit a closed form [56] (i.e., meaning that the definite integral of Eq. 32 does not admit a simple formula using common standard functions). Thus the dually-flat geometry $(\mathcal{M}, G)$ is a theoretical construction which cannot be explicitly used by Bregman algorithms.

One way to tackle the lack of closed form in Eq. 32, is to approximate the definite integrals whenever they are used by using Monte Carlo stochastic integration. However, this is computationally very expensive, and, even worse, it cannot guarantee that the overall computation is consistent.

Let us briefly explain the meaning of *consistency*: We can estimate the KL between two distributions p and q by drawing m variates $x_1, \ldots, x_m \sim p(x)$, and use the following MC KL estimator:

$$\widehat{\mathrm{KL}}_m(p : q) := \frac{1}{m} \sum_{i=1}^{m} \log \frac{p(x_i)}{q(x_i)}. \tag{46}$$

Now, suppose we have $\mathrm{KL}(p : q) \leq \mathrm{KL}(q : r)$, then their MC estimates may not satisfy $\widehat{\mathrm{KL}}_m(p : q) < \widehat{\mathrm{KL}}_m(q : r)$ (since each time we evaluate a $\widehat{\mathrm{KL}}_m$ we draw different samples). Thus when running a KL/Bregman algorithm, the more MC stochastic approximations of integrals are performed in the algorithm, the less likely is the output consistent. For example, consider computing the Bregman Voronoi diagram [34] of a set of n mixtures belonging to a mixture family manifold (say, with $D = 2$) using the algorithm explained in [34]: Since we use for each BD calculation or predicate evaluation relying on F or F^* stochastic Monte Carlo integral approximations, this MC algorithm may likely not deliver a proper combinatorial structure of the Voronoi diagram: The Voronoi structure is likely to be inconsistent.

Let us now show how Monte Carlo Information Geometry (MCIG) approximates this computationally untractable $(\mathcal{M}, G)$ geometric structure by defining a consistent and computationally-friendly dually-flat information geometry $(\mathcal{M}, \tilde{G}_{\mathcal{S}})$ for a finite number m of identically and independently distributed (iid) random samples $\mathcal{S}$.

2.1 MCIG of Order-1 Mixture Family

In order to highlight the principle of MCIGs, let us first consider a mixture family of order $D = 1$. That is, we consider a set of mixtures of $k = 2$ components with density:

$$m(x; \eta) = \eta p_1(x) + (1 - \eta) p_0(x) = p_0(x) + \eta(p_1(x) - p_0(x)), \tag{47}$$

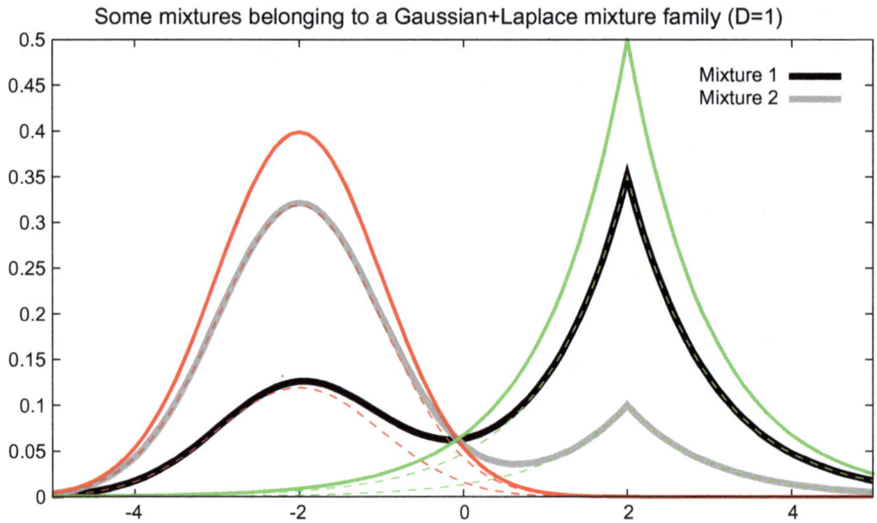

Fig. 1 Example of a mixture family of order $D = 1$ ($k = 2$): $p_0(x) \sim \text{Gaussian}(-2, 1)$ (red) and $p_1(x) \sim \text{Laplace}(2, 1)$ (green). The two mixtures are $m_1(x) = m(x; \eta_1)$ (black) with $\eta_1 = 0.7$ and $m_2(x) = m(x; \eta_2)$ (grey) with $\eta_2 = 0.2$. Weighted component distributions are displayed in dashed

with parameter η ranging in $(0, 1)$. The two prescribed component densities $p_0(x)$ and $p_1(x)$ (with respect to a base measure μ, say the Lebesgue measure) are defined on a common support $\mathcal{X}$. Densities $p_0(x)$ and $p_1(x)$ are assumed to be linearly independent [8].

Figure 1 displays an example of uni-order mixture family with heterogeneous components: $p_0(x)$ is chosen as a Gaussian distribution while $p_1(x)$ is taken as a Laplace distribution. A mixture $m(x; \eta)$ of $\mathcal{M}$ is visualized as a point P (here, one-dimensional) with $\eta(P) = \eta$.

Let $\mathcal{S} = \{x_1, \ldots, x_m\}$ denote a iid sample from a fixed *proposal distribution* $q(x)$ (with $q(x) > 0$ for $x \in \mathcal{X}$, and $q(x)$ independent of η). We approximate the Bregman generator $G(\eta)$ using Monte Carlo stochastic integration with importance sampling as follows:

$$G(\eta) \simeq \tilde{G}_{\mathcal{S}}(\eta) := \frac{1}{m} \sum_{i=1}^{m} \frac{1}{q(x_i)} m(x_i; \eta) \log m(x_i; \eta). \tag{48}$$

Let us prove that the Monte Carlo function $\tilde{G}_{\mathcal{S}}(\eta)$ is a proper Bregman generator. That is, that $\tilde{G}_{\mathcal{S}}(\eta)$ is strictly convex and twice continuously differentiable (Definition 1).

Write for short $m_x(\eta) := m(x;\eta)$ so that $G(\eta) = \int_{x\in\mathcal{X}} m_x(\eta)\log m_x(\eta)\mathrm{d}\mu(x)$ is approximated by $\frac{1}{m}\sum_{i=1}^{m}\frac{1}{q(x_i)}m_{x_i}(\eta)\log m_{x_i}(\eta)$. Since $\frac{1}{m}\frac{1}{q(x_i)} > 0$, it suffices to prove that the function $g_x(\eta) = m_x(\eta)\log m_x(\eta)$ is strictly convex wrt parameter η. Then we shall conclude that $\tilde{G}_{\mathcal{S}}(\eta)$ is strictly convex because it is a finite positively weighted sum of strictly convex functions.

Let us write the first and second derivatives of $g_x(\eta)$ as follows:

$$g_x(\eta)' = m_x(\eta)'(\log m_x(\eta) + 1), \tag{49}$$

$$g_x(\eta)'' = m_x(\eta)''(\log m_x(\eta) + 1) + \frac{(m_x(\eta)')^2}{m_x(\eta)}. \tag{50}$$

Since $m_x'(\eta) = p_1(x) - p_0(x)$ and $m_x''(\eta) = 0$, we get:

$$g_x(\eta)'' = \frac{(p_1(x) - p_0(x))^2}{m_x(\eta)}. \tag{51}$$

Thus it follows that:

$$\tilde{G}_{\mathcal{S}}''(\eta) = \frac{1}{m}\sum_{i=1}^{m}\frac{1}{q(x_i)}\frac{(p_1(x_i) - p_0(x_i))^2}{m(x_i;\eta)} \geq 0. \tag{52}$$

It is strictly convex provided that there exists at least one x_i such that $p_1(x_i) \neq p_0(x_i)$.

Let $\mathcal{D} \subset \mathcal{X}$ denote the degenerate set $\mathcal{D} = \{x \in \mathcal{X} : p_1(x) = p_0(x)\}$. For example, if $p_0(x)$ and $p_1(x)$ are two distinct univariate normal distributions, then $|\mathcal{D}| = 2$ (roots of a quadratic equation), and

$$\mu_q(\mathcal{D}) := \int_{x\in\mathcal{X}} 1_{[p_0(x)=p_1(x)]}q(x)\mathrm{d}\mu(x) = 0. \tag{53}$$

Assumption 1 *(AMF1D)* We assume that $p_0(x)$ and $p_1(x)$ are linearly independent (non-singular statistical model, see [8]), and that $\mu_q(\mathcal{D}) = 0$.

Lemma 1 (Monte Carlo Mixture Family Function is a Bregman generator) *The Monte Carlo Mixture Family Function (MCMFF) $\tilde{F}_{\mathcal{S}}(\theta)$ is a Bregman generator almost surely.*

Proof When there exists a sample $x \in \mathcal{S}$ with two distinct densities $p_0(x)$ and $p_1(x)$, we have $(p_1(x_i) - p_0(x_i))^2 > 0$ and therefore $\tilde{G}_{\mathcal{S}}''(\eta) > 0$. The probability to get a degenerate sample is almost zero.

To recap, the MCMFF of the MCIG of uni-order family has the following characteristics:

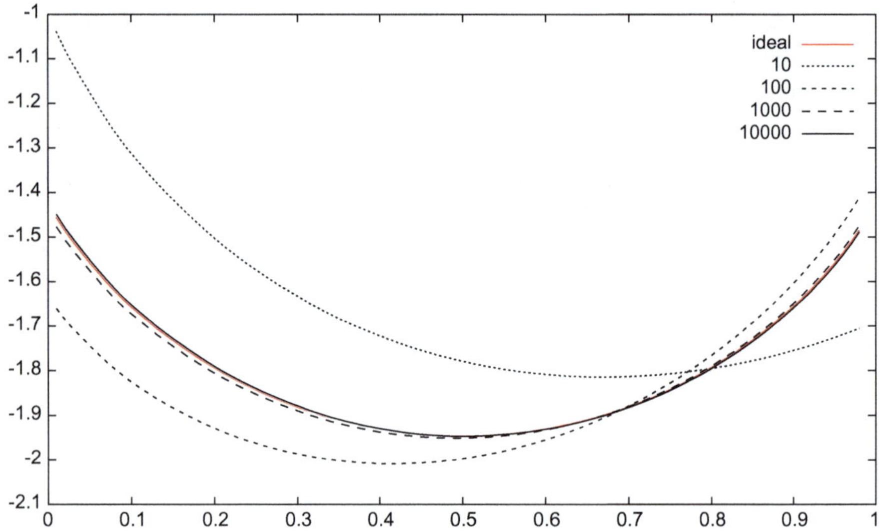

Fig. 2 A series $G_{\mathcal{S}}(\eta)$ of Bregman Monte Carlo Mixture Family generators (for $m = |\mathcal{S}| \in \{10, 100, 1000, 10000\}$) approximating the untractable ideal negentropy generator $G(\eta) = -h(m(x;\eta))$ (red) of a mixture family with prescribed Gaussian distributions $m(x;\eta) = (1-\eta)p(x;0,3) + \eta p(x;2,1)$ for the proposal distribution $q(x) = m(x;\frac{1}{2})$

Monte Carlo Mixture Family Generator 1D:

$$\tilde{G}_{\mathcal{S}}(\eta) = \frac{1}{m}\sum_{i=1}^{m}\frac{1}{q(x_i)}m(x_i;\eta)\log m(x_i;\eta), \tag{54}$$

$$\tilde{G}'_{\mathcal{S}}(\eta) = \theta = \frac{1}{m}\sum_{i=1}^{m}\frac{1}{q(x_i)}(p_1(x_i) - p_0(x_i))(1+\log m(x_i;\eta)), \tag{55}$$

$$\tilde{G}''_{\mathcal{S}}(\eta) = \frac{1}{m}\sum_{i=1}^{m}\frac{1}{q(x_i)}\frac{(p_1(x_i)-p_0(x_i))^2}{m(x_i;\eta)}. \tag{56}$$

Note that $(G^*)'$ and G^* may be calculated numerically but not in closed-form. We may also MC approximate ∇G^* since $\theta = (h^\times(p_0 : m) - h^\times(p_i : m))_i$.

Thus we change from type 5 to type 2 the computational tractability of mixtures by adopting the MCIG approximation.

Figure 2 displays a series of Bregman mixture family MC generators for a mixture family for different values of $|\mathcal{S}| = m$.

As we increase the sample size of $\mathcal{S}$, the MCMFF Bregman generator tends to the ideal mixture family Bregman generator.

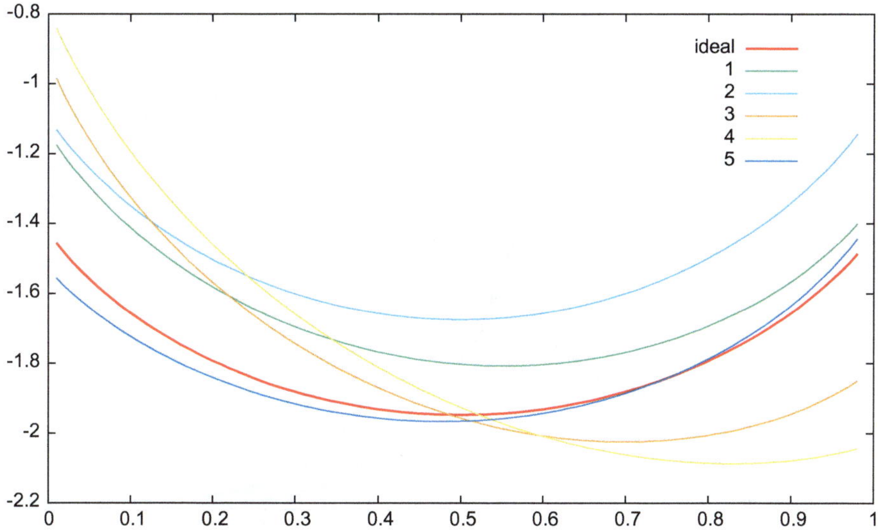

Fig. 3 The Monte Carlo Mixture Family Generator $\hat{G}_{10}$ (MCMFG) considered as a random variable: Here, we show five realizations (i.e., $\mathcal{S}_1, \ldots, \mathcal{S}_5$) of the randomized generator for $m = 5$. The ideal generator is plot in thick red

Theorem 1 (Consistency of MCIG) *Almost surely,* $\lim_{m\to\infty}(\mathcal{M}, \tilde{G}_{\mathcal{S}}) = (\mathcal{M}, G)$ *when* $\mu_q(\mathcal{D}) = 0$.

Proof It suffices to prove that $\lim_{m\to\infty} \tilde{G}_{\mathcal{S}}(\eta) = G(\eta)$. The general theory of Monte Carlo stochastic integration yields a consistent estimator provided that the following variance is bounded

$$\mathrm{Var}_q\left[\frac{m(x;\eta)\log m(x;\eta)}{q(x)}\right] < \infty. \tag{57}$$

For example, when $m(x;\eta)$ is a mixture of prescribed isotropic gaussians (say, from a KDE), and $q(x)$ is also an isotropic Gaussian, the variance is bounded. Note that q is the proposal density wrt the base measure μ.

In practice, the proposal distribution $q(x)$ can be chosen as the uniform mixture of the fixed component distributions:

$$q(x) = \frac{1}{m}\sum_{i=0}^{D} p_i(x). \tag{58}$$

Notice that the Monte Carlo Mixture Family Function is a random variable (r.v. for short) estimator itself by considering a vector of iid variables instead of a sample variate: $\hat{G}_m(\eta)$. Figure 3 displays five realizations of the random variable $\hat{G}_m(\eta)$ for $m = 10$.

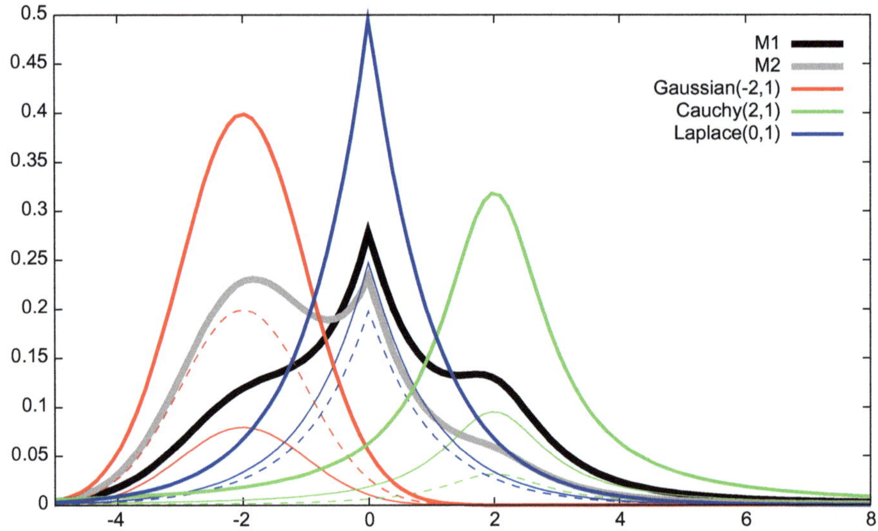

Fig. 4 Example of a mixture family of order $D = 2$ ($k = 3$): $p_0(x) \sim$ Gaussian$(-2, 1)$ (red), $p_1(x) \sim$ Laplace$(0, 1)$ (blue) and $p_2(x) \sim$ Cauchy$(2, 1)$ (green). The two mixtures are $m_1(x) = m(x; \eta_1)$ (black) with $\eta_1 = (0.3, 0.5)$ and $m_2(x) = m(x; \eta)$ (gray) with $\eta = (0.1, 0.4)$

2.2 *General D-Order Mixture Case*

Here, we consider statistical mixtures with $k = D + 1 > 2$ prescribed distributions $p_0(x), \dots, p_D(x)$. The component distributions are linearly independent so that they define a non-singular statistical model [8].

We further strengthen conditions on the prescribed distributions as follows:

Assumption 2 *(AMF)* We assume that the linearly independent prescribed distributions further satisfy:

$$\sup_{B\in\mathcal{B}} \left\{ \mu_q(B) : \exists\lambda \neq (0), \sum_{i\neq j} \lambda_i \left(p_i|_B - p_j|_B \right) = 0 \right\} = 0, \quad \forall j, \tag{59}$$

where the supremum is over all subsets B of the σ-algebra $\mathcal{B}$ of the probability space with support $\mathcal{X}$ and measure μ, with $p_i|_B$ denoting the restriction of p_i to subset B. In other words, we impose that the components $(p_i)_i$ still constitute an affinely independent family when restricted to any subset of positive measure.

For example, Figure 4 displays two mixture distributions belonging to a 2D mixture family with Gaussian, Laplace and Cauchy component distributions.

Recall that the mixture family Monte Carlo generator is:

$$\tilde{G}_{\mathcal{S}}(\eta) = \frac{1}{m}\sum_{i=1}^{m} \frac{1}{q(x_i)} m(x_i;\eta) \log m(x_i;\eta). \tag{60}$$

In order to prove that G is strictly convex, we shall prove that $\nabla^2 \tilde{G}_{\mathcal{S}}(\eta) \succ 0$ almost surely. It suffices to consider the basic Hessian matrix $\nabla^2 g_x = (\partial^i \partial^j g_x(\eta))_{ij}$ of $g_x(\eta) = m_x(\eta) \log m_x(\eta)$. We have the partial first derivatives:

$$\partial^i g_x(\eta) = (p_i(x) - p_0(x))(1 + \log m(x;\eta)), \tag{61}$$

and the partial second derivatives:

$$\partial^i \partial^j g_x(\eta) = \frac{(p_i(x) - p_0(x))(p_j(x) - p_0(x))}{m(x;\eta)}, \tag{62}$$

so that

$$\partial^i \partial^j \tilde{G}_{\mathcal{S}}(\eta) = \frac{1}{m}\sum_{l=1}^{m} \frac{1}{q(x_l)} \frac{(p_i(x_l) - p_0(x_l))(p_j(x_l) - p_0(x_l))}{m(x_l;\eta)}. \tag{63}$$

Theorem 2 (Monte Carlo Mixture Family Function is a Bregman generator) *The Monte Carlo multivariate function $\tilde{G}_{\mathcal{S}}(\eta)$ is always convex and twice continuously differentiable, and strictly convex almost surely.*

Proof Consider the D-dimensional vector:

$$v_l = \begin{bmatrix} \frac{p_1(x_l)-p_0(x_l)}{\sqrt{q(x_l)m(x_l;\eta)}} \\ \vdots \\ \frac{p_D(x_l)-p_0(x_l)}{\sqrt{q(x_l)m(x_l;\eta)}} \end{bmatrix}. \tag{64}$$

Then we rewrite the Monte Carlo generator $\tilde{G}_{\mathcal{S}}(\eta)$ as:

$$\partial^i \partial^j \tilde{G}_{\mathcal{S}}(\eta) = \frac{1}{m}\sum_{l=1}^{m} v_l v_l^\top. \tag{65}$$

Since $v_l v_l^\top$ is always a symmetric positive semidefinite matrix of rank one, we conclude that $\tilde{G}_{\mathcal{S}}(\eta)$ is a symmetric positive semidefinite matrix when $m < D$ (rank deficient) and a symmetric positive definite matrix (full rank) almost surely when $m \geq D$.

3 Monte Carlo Information Geometry of Exponential Families

We follow the same outline as for mixture familes: Sect. 3.1 first describes the univariate case. It is then followed by the general multivariate case in Sect. 3.1.

3.1 MCIG of Order-1 Exponential Family

We consider the order-1 exponential family of parametric densities with respect to a base measure μ:

$$\mathcal{E} := \{p(x;\theta) = \exp(t(x)\theta - F(\theta) + k(x)) : \theta \in \Theta\}, \tag{66}$$

where Θ is the natural parameter space, such that the log-normalizer/cumulant function [1] is

$$F(\theta) = \log\left(\int \exp(t(x)\theta + k(x))\mathrm{d}\mu(x)\right). \tag{67}$$

The sufficient statistic function $t(x)$ and 1 are linearly independent [8].

We perform Monte Carlo stochastic integration by sampling a set $\mathcal{S} = \{x_1, \ldots, x_m\}$ of m iid variates from a proposal distribution $q(x)$ to get:

$$F(\theta) \simeq \tilde{F}^{\dagger}_{\mathcal{S}}(\theta) := \log\left(\frac{1}{m}\sum_{i=1}^{m}\frac{1}{q(x_i)}\exp(t(x_i)\theta + k(x_i))\right). \tag{68}$$

Without loss of generality, assume that x_1 is the element that minimizes the sufficient statistic $t(x)$ among the elements of $\mathcal{S}$, so that $a_i = t(x_i) - t(x_1) \geq 0$ for all $x_i \in \mathcal{S}$.

Let us factorize $\frac{1}{q(x_1)}\exp(t(x_1)\theta + k(x_1))$ in Eq. 68 and remove an affine term from the generator $\tilde{F}_{\mathcal{S}}(\theta)$ to get the equivalent generator (see Property 1):

$$\begin{aligned}
\tilde{F}^{\dagger}_{\mathcal{S}}(\theta) &\equiv \tilde{F}_{\mathcal{S}}(\theta), && (69)\\
\tilde{F}_{\mathcal{S}}(\theta) &= \log\left(1 + \sum_{i=2}^{m}\exp((t(x_i) - t(x_1))\theta + k(x_i) - k(x_1) - \log q(x_i) + \log q(x_1))\right), && (70)\\
&= \log\left(1 + \sum_{i=2}^{m}\exp(a_i\theta + b_i)\right), && (71)\\
&:= \mathrm{lse}_0^+(a_2\theta + b_2, \ldots, a_m\theta + b_m), && (72)
\end{aligned}$$

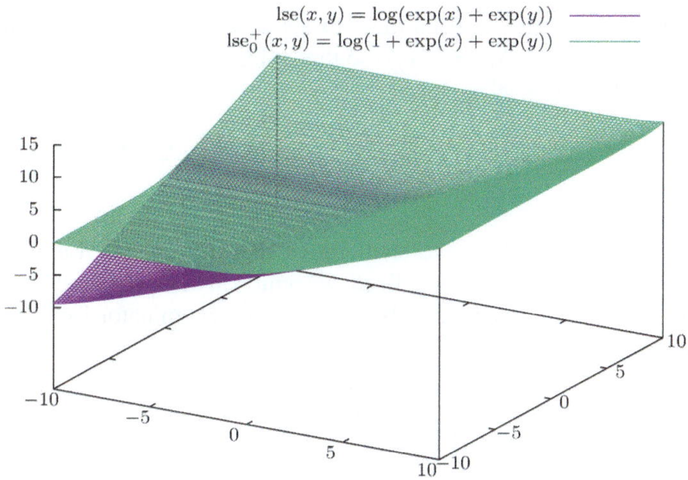

Fig. 5 Graph plots of the lse and lse_0^+ functions: The lse function (violet) is only convex while the lse_0^+ function (green) is always guaranteed to be strictly convex

with $a_2, \ldots, a_m > 0$ and $b_i = k(x_i) - k(x_1) - \log q(x_i) + \log q(x_1)$. Function $\mathrm{lse}_0^+(x_1, \ldots, x_m) = \mathrm{lse}(0, x_1, \ldots, x_m)$ is the log-sum-exp function [20, 47] $\mathrm{lse}(x_1, \ldots, x_m) = \log \sum_{i=1}^n \exp(x_i)$ with an additional argument set to zero.

Let us notice that the lse_0^+ function is always *strictly convex* while the lse function is only convex[4] [7], p. 74. Figure 5 displays the graph plots of the lse and lse_0^+ functions. Let us clarify this point with a usual exponential family: The binomial family. The binomial distribution is a categorical distribution with $D = 1$ (and 2 bins). We have $F(\theta) = \log(1 + \exp(\theta)) = \mathrm{lse}(0, \theta) := \mathrm{lse}_0^+(\theta)$. We check the strict convexity of $F(\theta)$: $F'(\theta) = \frac{e^\theta}{1+e^\theta}$ and $F''(\theta) = \frac{e^\theta}{(1+e^\theta)^2} > 0$.

We write for short $\mathrm{lse}_0^+(x) = \mathrm{lse}_0^+(x_1, \ldots, x_d)$ for a d-dimensional vector x.

Theorem 3 (lse_0^+ is a Bregman generator) *Multivariate function* $\mathrm{lse}_0^+(x)$ *is a Bregman generator.*

Proof is deferred to Appendix 7.

Lemma 2 (Univariate Monte Carlo Exponential Family Function is a Bregman generator) *Almost surely, the univariate function $\tilde{F}_{\mathcal{S}}(\theta)$ is a Bregman generator.*

Proof The first derivative is:

$$\eta = \tilde{F}_{\mathcal{S}}'(\theta) = \frac{\sum_{i=2}^m a_i \exp(a_i\theta + b_i)}{1 + \sum_{i=2}^m \exp(a_i\theta + b_i)} \geq 0, \tag{73}$$

[4]Function lse can be interpreted as a vector function, and is C^2, convex but not strictly convex on $\mathbb{R}^m$. For example, lse is affine on lines since $\mathrm{lse}(x + \lambda 1) = \mathrm{lse}(x) + \lambda$ (or equivalently $\mathrm{lse}(x_1, \ldots, x_m) = \lambda + \mathrm{lse}(x_1 - \lambda, \ldots, x_m - \lambda)$). It is affine only on lines passing through the origin.

and is strictly greater than 0 when there exists at least two elements with distinct sufficient statistics (i.e., $t(x_i) \neq t(x_j)$) so that at least one $a_i > 0$.

The second derivative is:

$$\tilde{F}''_{\mathcal{S}}(\theta) = \frac{\left(\sum_{i=2}^m a_i^2 \exp(a_i\theta + b_i)\right)\left(1 + \sum_{i=2}^m \exp(a_i\theta + b_i)\right) - \left(\sum_{i=2}^m a_i \exp(a_i\theta + b_i)\right)^2}{(1 + \sum_{i=2}^m \exp(a_i\theta + b_i))^2} =: \frac{\text{Num}}{\text{Den}} \tag{74}$$

For each value of $\theta \in \Theta$, we shall prove that $\tilde{F}''_{\mathcal{S}}(\theta) > 0$. Let $c_i = c_i(\theta) = \exp(a_i\theta + b_i) > 0$ for short (θ being fixed, we omit it in the c_i notation in the calculus derivation). Consider the numerator Num since the denominator Den is a non-zero square, hence strictly positive. We have:

$$\text{Num} > \left(\sum_{i=2}^m a_i^2 c_i\right)\left(\sum_{i=2}^m c_i\right) - \left(\sum_{i=2}^m a_i c_i\right)^2, \tag{75}$$

$$\text{Num} > \sum_{ij} a_i^2 c_i c_j - \sum_i a_i^2 c_i^2 - 2\sum_{i<j} a_i a_j c_i c_j, \tag{76}$$

$$\text{Num} > \sum_{i=j} a_i^2 c_i^2 + \sum_{i \neq j} a_i^2 c_i c_j - \sum_i a_i^2 c_i^2 - 2\sum_{i<j} a_i a_j c_i c_j, \tag{77}$$

$$\text{Num} > \sum_{i<j} a_i^2 c_i c_j + \sum_{i>j} a_i^2 c_i c_j - 2\sum_{i<j} a_i a_j c_i c_j, \tag{78}$$

$$\text{Num} > \sum_{i<j} a_i^2 c_i c_j + \sum_{i<j} a_j^2 c_i c_j - 2\sum_{i<j} a_i a_j c_i c_j, \tag{79}$$

$$\text{Num} > \sum_{i<j} (a_i^2 + a_j^2 - 2a_i a_j) c_i c_j, \tag{80}$$

$$\text{Num} > \sum_{i<j} (a_i - a_j)^2 c_i c_j > 0. \tag{81}$$

Therefore the numerator is strictly positive if at least two a_i's are distinct.

Thus we add the following assumption:

Assumption 3 *(AEF1D)* For all $y \in \text{dom}(t)$, $E_q[1_{t(x)=y}] = 0$.

To recap, the MCEFF of the MCIG of uni-order family has the following characteristics:

Monte Carlo Mixture Family Generator 1D:

$$\tilde{F}_{\mathcal{S}}(\theta) = \mathrm{lse}_0^+(a_2\theta + b_2, \ldots, a_m\theta + b_m), \tag{82}$$

$$a_i = t(x_i) - t(x_1), \tag{83}$$

$$b_i = k(x_i) - k(x_1) - \log q(x_i) + \log q(x_1), \tag{84}$$

$$\tilde{F}'_{\mathcal{S}}(\theta) = \frac{\sum_{i=2}^m a_i \exp(a_i\theta + b_i)}{1 + \sum_{i=2}^m \exp(a_i\theta + b_i)} =: \eta, \tag{85}$$

$$\tilde{F}''_{\mathcal{S}}(\theta) = \frac{\left(\sum_{i=2}^m a_i^2 \exp(a_i\theta + b_i)\right)\left(1 + \sum_{i=2}^m \exp(a_i\theta + b_i)\right) - \left(\sum_{i=2}^m a_i \exp(a_i\theta + b_i)\right)^2}{(1 + \sum_{i=2}^m \exp(a_i\theta + b_i))^2} \tag{86}$$

3.2 The general D-Order Case

The difference of sufficient statistics $a_i = t(x_i) - t(x_1)$ is now a vector of dimension D:

$$a_i = \begin{bmatrix} a_i^1 \\ \vdots \\ a_i^D \end{bmatrix}. \tag{87}$$

We replace the scalar multiplication $a_i\theta$ by an inner product $\langle a_i, \theta\rangle$ in Eq. 72, and let $c_i(\theta) = \exp(\langle a_i, \theta\rangle + b_i)$ with $b_i = k(x_i) - k(x_1) - \log q(x_i) + \log q(x_1)$. Then the Monte Carlo Exponential Family Function (MCEFF) writes concisely as:

$$\tilde{F}_{\mathcal{S}}(\theta) = \log\left(1 + \sum_{l=2}^m c_l(\theta)\right), \tag{88}$$

$$:= \mathrm{lse}_0^+(c_2(\theta), \ldots, c_m(\theta)), \tag{89}$$

Theorem 4 (Monte Carlo Exponential Family Function is a Bregman Generator) *Almost surely, the function $\tilde{F}_{\mathcal{S}}(\theta)$ is a proper Bregman generator.*

Proof We have the gradient of first-order partial derivatives:

$$\eta_i = \partial_i \tilde{F}_{\mathcal{S}}(\theta) = \frac{\sum_{l=2}^m a_l^i c_l(\theta)}{1 + \sum_{l=2}^m c_l(\theta)}, \tag{90}$$

and the Hessian matrix of second-order partial derivatives:

$$\partial_i \partial_j \tilde{F}_{\mathcal{S}}(\theta) = \frac{(\sum_{l=2}^m a_l^i a_l^j c_l(\theta))(1 + \sum_{l=2}^m c_l(\theta)) - (\sum_{l=2}^m a_l^i c_l(\theta))(\sum_{l=2}^m a_l^j c_l(\theta))}{(1 + \sum_{l=2}^m c_l(\theta))^2} =: \frac{\text{Num}}{\text{Den}}. \tag{91}$$

Let us prove that the Hessian matrix $\nabla^2 \tilde{F}_{\mathcal{S}}(\theta) = (\partial_i \partial_j \tilde{F}_{\mathcal{S}}(\theta))_{ij}$ is always symmetric positive semi-definite, and symmetric positive definite almost surely.

Indeed, we have:

$$\text{Num} = \underbrace{\sum_k a_k^i a_k^j c_k}_{:=D} + \underbrace{\sum_{k,l} a_k^i a_k^j c_k c_l - \sum_{k,l} a_k^i c_k a_l^j c_l}_{:=E}. \tag{92}$$

Let us rewrite D as $D = CA^\top A$ with $C = \text{diag}(c_1, \ldots, c_D)$. It follows that matrix D is symmetric positive definite. Let us prove that matrix E is also SPD:

$$\begin{aligned} E &\overset{\star}{=} \sum_{k<l} a_k^i a_k^j c_k c_l + \sum_{l<k} a_k^i z_k^j c_k c_l - \sum_{k<l} a_k^i a_l^j c_k c_l - \sum_{l<k} a_k^i a_l^j c_k c_l, & (93)\\ &\overset{\star\star}{=} \sum_{k<l} \left(a_k^i a_k^j + a_l^i a_l^j - a_k^i a_l^j - a_l^i a_k^j \right) c_k c_l, & (94)\\ &= \sum_{k<l} (a_k^i - a_l^i)(a_k^j - a_l^j) c_k c_l. & (95) \end{aligned}$$

$\star$: The terms $l = k$ vanish
$\star\star$: After a change of variable $l \leftrightarrow k$ in the second and fourth sums of Eq. 93.

Thus Eq. 95 can be rewritten as $(a_k - a_l)(a_k - a_l)^\top c_k c_l$ where $a_k = \begin{bmatrix} a_k^1 \\ \vdots \\ a_k^D \end{bmatrix}$. It follows that E is a positively weighted sum of rank-1 symmetric positive semi-definite matrices, and is therefore symmetric positive semi-definite.

We want $y^T E y > 0$ for all $y \neq 0 \in \mathbb{R}^D$. Suppose that there exists $y \neq 0 \in \mathbb{R}^D$ such that $y^T E y = 0$. Noting that $a_k^i - a_l^i = t_i(x_k) - t_i(x_l)$, we can write this as

$$\sum_{k<l} \left(\sum_i y_i c_i (t_i(x_k) - t_i(x_l)) \sum_j y_j c_j (t_j(x_k) - t_j(x_l)) \right) = 0, \tag{96}$$

which implies

$$\sum_i y_i c_i \left(t_i(x_k) - t_i(x_l) \right) \sum_j y_j c_j \left(t_j(x_k) - t_j(x_l) \right) = 0, \quad \forall k < l, \tag{97}$$

since each of these terms is non negative. In particular, we have the existence of a $y \neq 0 \in \mathbb{R}^D$ such that

$$\sum_i y_i t_i(x_k) = \sum_i y_i t_i(x_l), \quad \forall y \neq 0, \quad \forall k < l. \tag{98}$$

To get almost surely a Monte Carlo Bregman generator, we introduce the following assumption:

Assumption 4 (*AEF*) The sufficient statistics (t_i) verify that for all $\lambda \neq 0$ and all $y \in dom(\sum_i \lambda_i t_i)$:

$$E_q \left[1_{\sum_i \lambda_i t_i(x) = y} \right] = 0.$$

4 Application to Clustering

In this section, we demonstrate the practical use of MCIG to cluster a set of mixtures in Sect. 4.1, and consider in Sect. 4.2 parallel calculations/aggregations of Monte Carlo Exponential/Mixture Functions.

4.1 *Clustering Mixtures on the Mixture Family Manifold*

Consider clustering a set of n mixtures $m(x; \eta_1), \ldots, m(x; \eta_n)$ of the mixture family manifold. Prior work considered clustering the mixture components (e.g., Gaussian components) to simplify mixtures by using the Bregman k-means [14, 37]. This prior work can be interpreted as a Gaussian component quantization procedure.

Here, we address the different problem of clustering the mixtures themselves, not their components.

Since $\mathrm{KL}(m(x; \eta_i) : m(x; \eta_j)) = B_G(\eta_i : \eta_j)$ for $G(\eta) = -h(m(x; \eta))$ (Shannon information), we may approximate the KL divergence from the MC Bregman Divergence (MCBD) $\tilde{G}_S$ as follows:

$$\mathrm{KL}(m(x; \eta_i) : m(x; \eta_j)) = B_G(\eta_i : \eta_j), \tag{99}$$

$$\simeq B_{\tilde{G}_S}(\eta_i : \eta_j). \tag{100}$$

One advantage of using a MCIG is that all divergence computations $B_{\tilde{G}_S}$ performed during the execution of a Bregman algorithm are consistent by reusing the same variates of $\mathcal{S}$. In particular, this also guarantees to always have nonnegative estimated KL divergences.

The traditional way to MC estimate the KL divergence is to consider the MC stochastic integration of the extended Kullback–Leibler divergence [4]:

$$\widehat{\mathrm{eKL}}_m(p:q) := \frac{1}{m}\sum_{i=1}^{m}\left(\log\frac{p(x_i)}{q(x_i)} + \frac{q(x_i)}{p(x_i)} - 1\right), \tag{101}$$

for $x_1, \ldots, x_m \sim p(x)$. Indeed, if we just used the MC KL estimator:

$$\widehat{\mathrm{KL}}_m(p:q) := \frac{1}{m}\sum_{i=1}^{m}\log\frac{p(x_i)}{q(x_i)}, \tag{102}$$

we may endup with negative values to our estimated KL, depending on the sample variates! This never happens for eKL which is a statistical divergence for the scalar divergence $\mathrm{ekl}(p:q) = p\log\frac{p}{q} + q - p \geq 0$.

Bregman k-means [4, 22] can be applied using either the sided or their symmetrized centroid [40]: The right-sided centroid is always the center of mass of the parameters. The left-sided centroid requires to compute $F'(\theta)$ and its reciprocal inverse function $(F'(\theta))^{-1}$ (wlog, assuming $D = 1$ for simplicity[5]). Although $F'(\theta)$ is available in closed form (and define the dual parameter θ):

$$\tilde{G}'_S(\eta) = \frac{1}{m}\sum_{i=1}^{m}\frac{1}{q(x_i)}\left(p_1(x_i) - p_0(x)\right)\left(1 + \log m(x;\eta)\right) = \theta, \tag{103}$$

the dual parameter of $(\mathcal{M}, G)$ cannot be written as a simple function $\eta = F^{*\prime}(\eta)$. Notice that $\theta = \tilde{G}'_S(\eta)$ is an increasing function of η and that the inverting operation can be performed numerically. Indeed, we can compute $\eta = (\tilde{G}'_S)^{-1}(\theta) = \tilde{G}^*_S(\theta)$ using a numerical scheme (e.g., bisection search).

The symmetric Jeffreys divergence is:

$$\begin{aligned} J(m(x;\eta_i):m(x;\eta_j)) &= \mathrm{KL}(m(x;\eta_i):m(x;\eta_j)) + \mathrm{KL}(m(x;\eta_j):m(x;\eta_i)), &(104)\\ &= B_G(\eta_i:\eta_j) + B_G(\eta_j:\eta_i), &(105)\\ &= B_G(\eta_i:\eta_j) + B_{G^*}(\theta_i:\theta_j), &(106)\\ &= \langle \Delta\theta_{ij}, \Delta\eta_{ij}\rangle, &(107)\end{aligned}$$

where $\Delta\theta_{ij} = \theta_i - \theta_j$ and $\Delta\eta_{ij} = \eta_i - \eta_j$.

We may approximate the J divergence by considering the Monte Carlo Bregman generator in Eq. 105:

$$J(m(x;\eta_i):m(x;\eta_j)) \simeq B_{\tilde{G}_S}(\eta_i:\eta_j) + B_{\tilde{G}_S}(\eta_j:\eta_i). \tag{108}$$

We can then apply the technique of mixed Bregman clustering [49] that considers two centers per cluster. Moreover a fast probabilistic initialization, called *mixed Bregman k-means++* [49], allows one to guarantee a good initialization with high probability (without computing centroids but requiring to compute divergences).

---

[5]Otherwise, we need to consider monotone operator theory [25] to invert $\nabla F(\theta)$.

Another technique to bypass the computation of the gradient $\nabla\tilde{G}_{\mathcal{S}}$ in the BD consists in taking the scaled skew α-Jensen divergence [35] for an infinitesimal value of α. Indeed, we have the α-Jensen divergence defined by:

$$J_F^\alpha(p:q) = (1-\alpha)F(p) + \alpha F(q) - F((1-\alpha)p + \alpha q), \tag{109}$$

and asymptotically this skewed Jensen divergences yield the sided Bregman divergences [35] as follows:

$$\lim_{\alpha\to 0^+} \frac{J_F^\alpha(p:q)}{\alpha} = B_F(q:p), \tag{110}$$

$$\lim_{\alpha\to 1^-} \frac{J_F^\alpha(p:q)}{1-\alpha} = B_F(p:q), \tag{111}$$

Thus we have for small values of $\alpha > 0$ (say, $\alpha = 0.001$):

$$J(m(x;\eta_i) : m(x;\eta_j)) = B_G(\eta_i:\eta_j) + B_G(\eta_j:\eta_i), \tag{112}$$

$$\simeq \frac{1}{\alpha} J_{\tilde{G}_{\mathcal{S}}}^\alpha(\eta_i:\eta_j) + \frac{1}{1-\alpha} J_{\tilde{G}_{\mathcal{S}}}^{1-\alpha}(\eta_i:\eta_j). \tag{113}$$

The last equation Eq.113 is the symmetrized skew Jensen divergence studied in [29].

Figure 6 plots the result of a 2-cluster clustering wrt the Jeffreys' divergence for a set of $n = 8$ mixtures.

4.2 *Parallelizing Information Geometry*

We can distribute the Monte Carlo information geometry either on a multicore machine with l cores with shared memory or on a cluster of l machines with distributed memory, or even consider hybrid architectures.

Let $(M, \tilde{F}_{\mathcal{S}_1}), \ldots, (M, \tilde{F}_{\mathcal{S}_l})$ be a set of l information-geometric manifolds obtained from iid sample sets $\mathcal{S}_1, \ldots, \mathcal{S}_l$. Let $\oplus_{i=1}^s \mathcal{S}_i$ be a partition of $\mathcal{S}$.

4.2.1 Multicore Architectures

On a multicore architecture, we may evaluate the mixture family Bregman divergence $B_{\tilde{G}_{\mathcal{S}}}(\eta:\eta')$ by evaluating $B_{\tilde{G}_{\mathcal{S}_i}}(\theta:\theta')$, and using the compositionality rule of Bregman generators in BDs (Property 2) with:

$$\tilde{G}_{\mathcal{S}}(\theta) = \sum_{i=1}^{l} \frac{|\mathcal{S}_i|}{|\mathcal{S}|} \tilde{G}_{\mathcal{S}_i}(\eta). \tag{114}$$

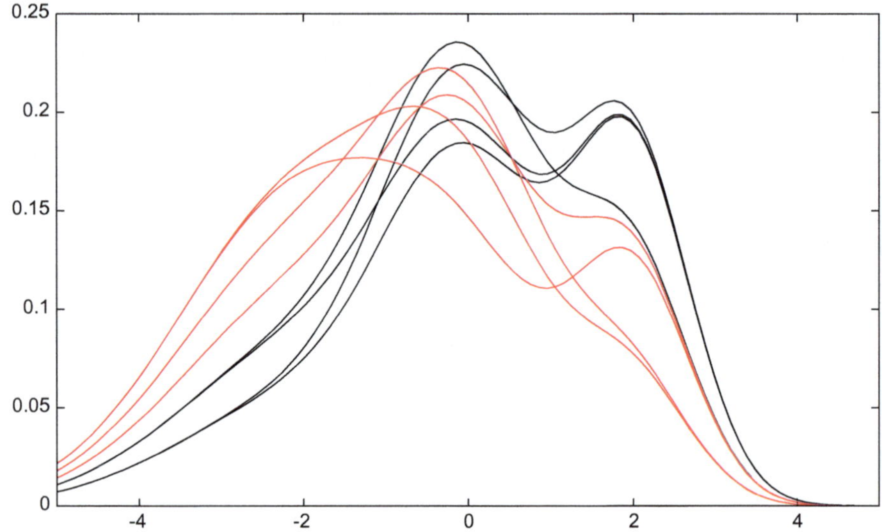

Fig. 6 Clustering a set of $n = 8$ statistical mixtures of order $D = 2$ with $K = 2$ clusters: Each mixture is represented by a 2D point on the mixture family manifold. The Kullback–Leibler divergence is equivalent to an integral-based Bregman divergence that is computationally untractable: The Bregman generator is stochastically approximated by Monte Carlo sampling

That is, $\tilde{G}_{\mathcal{S}}(\eta)$ is the *arithmetic weighted mean* of the mixture sub-generators.

For the exponential families, recall that we have:

$$\tilde{F}_{\mathcal{S}}(\theta) = \log\left(\sum_{i=1}^{s} \frac{|\mathcal{S}_i|}{|\mathcal{S}|} \exp(\tilde{F}_{\mathcal{S}_i})\right). \tag{115}$$

That is, $\tilde{F}_{\mathcal{S}}(\theta)$ can be interpreted as an *exponential mean* (quasi-arithmetic mean, called f-mean [35] for the monotonically increasing function $f(x) = \exp(x)$) of the sub-generators. Thus we can perform the computation of the MC Bregman generators on multi-core architectures easily with a MapReduce strategy [33].

Fact 1 (MapReduce evaluation of MC Bregman generators) *The MCMF or MCEF functions can be computed in parallel using a quasi-arithmetic mean MapReduce operation.*

4.2.2 Cluster Architectures

Since the MC Bregman generators can be interpreted as random variables $\tilde{G}_m(\theta)$ and $\tilde{F}_m(\theta)$, we may obtain robust estimate [51] by carrying the calculations on l MCIGs on a cluster architecture, and then integrate those l geometries.

Given a sequence of matching parameters $\theta_1 \in (M, \tilde{F}_{s_1}), \ldots, \theta_l \in (M, \tilde{F}_{s_l})$, we aggregate these parameters by doing the *KL-averaging* method [26]. This amounts to compute a sided centroid for θ.

5 Information-Geometric Structures Induced by Statistical Separable Divergences

In this section, we consider Monte Carlo sampling to define a (tractable) statistical divergence that approximates another (untractable) statistical divergence, and uses this MC statistical divergence to define an information-geometric manifold.

The core structure of information geometry [1] is a manifold M equipped with a pair of dual connections, ∇ and ∇^* coupled to the metric tensor g: (M, g, ∇, ∇^*). In terms of differential geometry, the definition of this coupling is expressed as

$$X\langle Y, Z\rangle_g = \langle \nabla_X Y, Z\rangle_g + \langle Y, \nabla_X^* Z\rangle_g, \tag{116}$$

where X, Y and Z are smooth vector fields on M. The coupling of connections to the metric tensor means that the dual parallel transport is compatible with the metric:

$$\langle u, v\rangle_{c(0)} = \left\langle \prod_{c(0)\to c(t)}^{\nabla} u, \prod_{c(0)\to c(t)}^{\nabla^*} v \right\rangle_{c(t)}, \tag{117}$$

where c is a smooth curve (parallel transport is path dependent, except for dually flat connections). The notation $\prod_{c(0)\to c(t)}^{\nabla} u$ means that vector $u \in T_{c(0)} = T_p$ is parallel transported along smooth curve c to tangent plane $T_{c(t)}$ with respect to the affine connection ∇. From this (M, g, ∇, ∇^*) structure, a statistical manifold [24] (M, g, C) can be defined, where $C(X, Y, Z) = \langle \nabla_X Y - \nabla_X^* Y, Z\rangle$ is a totally symmetric cubic tensor, termed the Amari–Chentsov cubic tensor. It follows a one-parameter family of dual connections [1] (with ∇^0 being the Levi-Civita metric connection): $(M, g, \nabla^{-\alpha}, \nabla^{\alpha})$ so that if connection ∇^{α} has constant curvature κ then its dual connection has also the same curvature. Furthermore, one can build [1, 16, 17] a pair of dual connections coupled to a metric from any smooth divergence D: $(M, {}^D g, {}^D\nabla, {}^D\nabla^*)$. Figure 7 summarizes the fundamental structures of parametric information geometry and their relationships.

Let us consider a separable statistical divergence:

$$D[p : q] := \int d(p(x) : q(x))\mathrm{d}\mu(x), \tag{118}$$

where $d(x : y)$ is a scalar divergence. For example, the f-divergences [1] are obtained for $i_f(x : y) = x f(x/y)$:

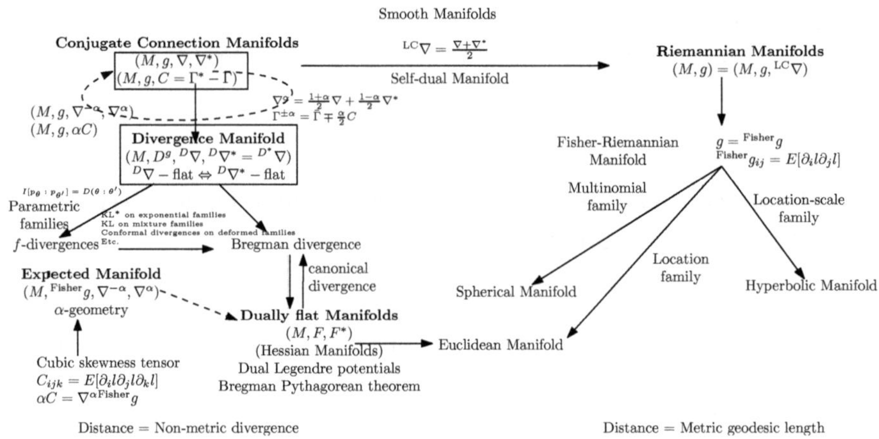

Fig. 7 The web of fundamental information-geometric structures. An arrow $a \to b$ means that geometric structure b is a special case of the (meta-)structure a

$$I_f[p : q] := \int p(x) f\left(\frac{p(x)}{q(x)}\right) \mathrm{d}\mu(x) = \int i_f(p(x) : q(x)) \mathrm{d}\mu(x), \tag{119}$$

The f-divergences are the only statistical separable divergences that satisfy the information monotonicity property [1]. On a parametric family of distributions $\{p_\theta\}$, the statistical f-divergences amount to equivalent parameter divergences:

$$D_f(\theta_1 : \theta_2) := I_f[p_{\theta_1} : p_{\theta_2}] \tag{120}$$

The information-geometric structure induced by this (parameter) divergence is $(M, {}^{D_f}g, {}^{D_f}\nabla, {}^{D_f}\nabla^*)$, and the dual connections correspond to the expected α-connections[1] for f-divergences.

It may happen that D_f, although well-defined, may not be available in closed form. In that case, we approximate the divergence by Monte Carlo stochastic integration by drawing a set $\mathcal{S}_m = \{x_1, \ldots, x_m\}$ of m iid variates from p_{θ_1}:

$$\tilde{D}_{\mathcal{S}_m}(\theta_1 : \theta_2) := \frac{1}{m} \sum_{i=1}^{m} \frac{1}{p_{\theta_1}(x_i)} d(p_{\theta_1}(x_i) : p_{\theta_2}(x_i)). \tag{121}$$

We need to assert that $\tilde{D}_{\mathcal{S}_m}$ is a smooth divergence: The smoothness of the divergence $\tilde{D}_{\mathcal{S}_m}$ follows from the smoothness divergence of the corresponding scalar divergence d. Then we need to guarantee that $\tilde{D}_{\mathcal{S}_m}(\theta_1 : \theta_2) = 0$ iff $\theta_1 = \theta_2$. Since $d(p_{\theta_1}(x) : p_{\theta_2}(x)) = 0$ if and only if $p_{\theta_1}(x) = p_{\theta_2}(x)$, we need to assert that with high probability $p_{\theta_1}(x) \neq p_{\theta_2}(x)$ when $\theta_1 \neq \theta_2$. Let $I = \max_{\theta_1, \theta_2} \mu(\{p_{\theta_1}(x) = p_{\theta_2}(x), x \in \mathcal{X}\})$. When $I = 0$, then almost surely $\tilde{D}_{\mathcal{S}}$ is a divergence. This condition holds when the probability densities intersect in at most a finite number of points. It follows the corresponding information-geometric structure $(M, {}^{\tilde{D}_{S_m}}g, {}^{\tilde{D}_{S_m}}\nabla, {}^{\tilde{D}_{S_m}}\nabla^*)$ (with its associated one-family of α-connections) such that asymptotically, we have:

$$\lim_{m\to\infty}(M, {}^{\tilde{D}_{S_m}}g, {}^{\tilde{D}_{S_m}}\nabla, {}^{\tilde{D}_{S_m}}\nabla^*) = (M, {}^{D}g, {}^{D}\nabla, {}^{D}\nabla^*), \tag{122}$$

as desired.

Let us quickly report two examples to illustrate these divergence-based sequences of information-geometric structures:

- Polynomial Exponential Families (PEFs) of order D with the γ-divergences [50]: Let us notice that we do not need to normalize the PEF distributions in order to sample variates, and that the γ-divergence D_γ is a projective divergence [48] (invariant by positive rescaling of the distributions) which tends to the KL divergence when $\gamma \to 0$. Since the densities of any two distinct PEF distributions of order D intersect in at most $D+1$ points, we check that $I=0$. Thus for $\gamma \to 0$ and $m \to \infty$, we tend to a dually flat manifold. As an application, we can consider clustering these PEFs on a MCIG manifold.
- Consider a mixture family $\{m_\eta(x) = (1-\eta)p_1(x) + \eta p_2(x), \eta \in (0,1)\}$ of order $D=1$ for the two mixture component distributions p_1 and p_2, linearly independent. We have $m_{\eta_1}(x) = m_{\eta_2}(x)$ iff $p_1(x) = p_2(x)$ (holds only for this particular case of $D=1$). Assume $I=0$ for the component distributions, then we obtain a sequence of Monte Carlo information-geometric structures that tend asymptotically to the dually flat mixture manifold.

In the later case, we consider the MCIG manifold for a 1D mixture manifold with respect to an arbitrary divergence. Notice that the divergence-based MCIG for the exponential/mixture manifold may not be flat for KL. In Sect. 2.2, we took the different approach of approximating the negative differential entropy via Monte-Carlo, ensuring that all sequence of MCIG manifolds are dually flat.

6 Conclusion and Perspectives

In this work, we proposed a new type of *randomized information-geometric structure* to cope with computationally untractable information-geometric structures (types 4 and 5 in the classification of Table 4): Namely, the Monte Carlo Information Geometry [38] (MCIG). MCIG performs stochastic integration of the ideal but computationally intractable definite integral-based Bregman generator (e.g. Eq. 32 for mixture family) for mixture family and Eq. 24 for exponential family). We proved that the MC Bregman generators for the mixture family and the exponential family are almost surely strictly convex and differentiable (Theorem 2 and Theorem 4, respectively), and therefore yield a computationally tractable information-geometric structure (Type 2 in the classification of Table 4). Thus we can get a series of *consistent* and *computationally-friendly* information-geometric structures that tend asymptotically

Table 4 A smooth and strictly convex function F induces a dually flat structure: We classify those structures according to their computational tractability properties

Type	F	∇F^*	Example
Type 1	Closed-form	Closed-form	Gaussian (exponential) family
Type 2	Closed-form	Not closed-form	Beta (exponential) family
Type 3	Comp. intractable	Not closed-form	Ising family [54]
Type 4	Not closed-form	Not closed-form	Polynomial exponential family [42]
Type 5	Not analytic	Not analytic	Mixture family

to the untractable ideal information geometry. We have demonstrated the usefulness of our technique for a basic Bregman k-means clustering technique: Clustering statistical mixtures on a mixture family manifold. Although the MCIG structures are computationally convenient, we do not have in closed-form ∇F^* (nor F^*) because our Bregman generators are the sum of basic generators whose gradients are the sum of elementary gradients that cannot be inverted easily.[6] This step requires a numerical or symbolic technique [25].

We note that in the recent work of [27], Matsuzoe et al. defined a sequence of statistical manifolds relying on a sequential structure of escort expectations for non-exponential type statistical models.

Codes for reproducible results are available at:

https://franknielsen.github.io/MCIG/

7 Function $\mathrm{lse}_0^+(x)$ is a Bregman Generator

We give the proof of Theorem 3:

Proof Since $\mathrm{lse}_0^+(x_1, \ldots, x_d) = \log\left(1 + \sum_{i=1}^d \exp(x_i)\right)$ is twice continuously differentiable, it suffices to prove that $\nabla^2 \mathrm{lse}_0^+(x) \succ 0$. We have:

$$\partial_i \mathrm{lse}_0^+(x) = \frac{e^{x_i}}{1 + \sum_k e^{x_k}}, \tag{123}$$

$$\partial_j \partial_i \mathrm{lse}_0^+(x) \overset{j \neq i}{=} \frac{-e^{x_i} e^{x_j}}{(1 + \sum_k e^{x_k})^2}, \tag{124}$$

$$\partial_i \partial_i \mathrm{lse}_0^+(x) = \frac{e^{x_i}(1 + \sum_k e^{x_k}) - e^{x_i} e^{x_j}}{(1 + \sum_k e^{x_k})^2}. \tag{125}$$

[6]The Legendre conjugate of an infimal convolution of elementary functions is the sum of the elementary conjugate functions.

It follows that the Hessian $(\partial_j \partial_i \mathrm{lse}_0^+(x))_{ij}$ is a diagonally dominant matrix since:

$$e^{x_i}\left(1+\sum_k e^{x_k}\right) = e^{x_i} + e^{x_i}\sum_k e^{x_k} > \sum_{j\neq i}\left|-e^{x_i}e^{x_j}\right| = e^{x_i}\sum_{j\neq i} e^{x_j}. \tag{126}$$

To conclude that the Hessian matrix is SPD, we use Gershgorin circle theorem [55] to bound the spectrum of a square matrix: The eigenvalues of the Hessian matrix are thus real and fall inside a disk of center $(e^{x_i}(1+\sum_k e^{x_k}))_i$ and radius $e^{x_i}\sum_{j\neq i} e^{x_j}$. Therefore all eigenvalues are positive, and the Hessian matrix is positive definite.

For $x = (x_1, \dots, x_d) \in \mathbb{R}^d$, we have:

$$\nabla \mathrm{lse}(x) = \sigma(x), \tag{127}$$

where $\sigma(x)$ is the *softmax* function:

$$\sigma(x) := \left(\frac{e^{x_i}}{\sum_{k=1}^d e^{x_k}}\right)_{i\in\{1,\dots,d\}}. \tag{128}$$

By analogy, we may define for $x \in \mathbb{R}^d$:

$$\sigma_0^+(x) := \left(\frac{e^{x_i}}{1+\sum_k e^{x_k}}\right)_{i\in\{1,\dots,d\}}, \tag{129}$$

so that $\nabla \mathrm{lse}_0^+(x) = \sigma_0^+(x)$.

References

1. Amari, S.: Information Geometry and Its Applications. Applied Mathematical Sciences. Springer, Japan (2016)
2. Amari, Si, Cichocki, A.: Information geometry of divergence functions. Bull. Polish Acad. Sci.: Tech. Sci. **58**(1), 183–195 (2010)
3. Azoury, K.S., Warmuth, M.K.: Relative loss bounds for on-line density estimation with the exponential family of distributions. Mach. Learn. **43**(3), 211–246 (2001)
4. Banerjee, A., Merugu, S., Dhillon, I.S., Ghosh, J.: Clustering with Bregman divergences. J. Mach. Learn. Res. **6**(Oct), 1705–1749 (2005)
5. Bhattacharya, B.B., Mukherjee, S., et al.: Inference in Ising models. Bernoulli **24**(1), 493–525 (2018)
6. Boissonnat, J.D., Nielsen, F., Nock, R.: Bregman Voronoi diagrams. Discret. Comput. Geom. **44**(2), 281–307 (2010)
7. Boyd, S., Vandenberghe, L.: Convex Optimization. Cambridge University Press, Cambridge (2004)
8. Calin, O., Udriste, C.: Geometric Modeling in Probability and Statistics. Mathematics and Statistics. Springer International Publishing, Berlin (2014)

9. Cipra, B.A.: The Ising model is NP-complete. SIAM News **33**(6), 1–3 (2000)
10. Cobb, L., Koppstein, P., Chen, N.H.: Estimation and moment recursion relations for multimodal distributions of the exponential family. J. Am. Stat. Assoc. **78**(381), 124–130 (1983)
11. Cover, T.M., Thomas, J.A.: Elements of Information Theory. Wiley, New York (2012)
12. Critchley, F., Marriott, P.: Computational information geometry in statistics: theory and practice. Entropy **16**(5), 2454–2471 (2014)
13. Crouzeix, J.P.: A relationship between the second derivatives of a convex function and of its conjugate. Math. Programm. **13**(1), 364–365 (1977)
14. Davis, J.V., Dhillon, I.S.: Differential entropic clustering of multivariate gaussians. In: Advances in Neural Information Processing Systems, pp. 337–344 (2007)
15. Dawid, A.P.: The geometry of proper scoring rules. Ann. Inst. Stat. Math. **59**(1), 77–93 (2007)
16. Eguchi, S.: Second order efficiency of minimum contrast estimators in a curved exponential family. Ann. Stat. **11**, 793–803 (1983)
17. Eguchi, S.: Geometry of minimum contrast. Hiroshima Math. J. **22**(3), 631–647 (1992)
18. Fleisch, D.A.: A Student's Guide to Vectors and Tensors. Cambridge University Press, Cambridge (2011)
19. Frongillo, R., Reid, M.D.: Convex foundations for generalized MaxEnt models. In: AIP Conference Proceedings, vol. 1636, pp. 11–16. AIP (2014)
20. Gao, B., Pavel, L.: On the Properties of the Softmax Function with Application in Game Theory and Reinforcement Learning. ArXiv e-prints (2017)
21. Geman, S., Graffigne, C.: Markov random field image models and their applications to computer vision. In: Proceedings of the International Congress of Mathematicians, vol. 1, p. 2 (1986)
22. Grønlund, A., Larsen, K.G., Mathiasen, A., Nielsen, J.S.: Fast exact k-means, k-medians and Bregman divergence clustering in 1D (2017). arXiv:1701.07204
23. Kass, R.E., Vos, P.W.: Geometrical Foundations of Asymptotic Inference. Fisher-Rao metric of location-scale family is hyperbolic (and can be diagonalized), pp. 192–193. Wiley-Interscience (1997)
24. Lauritzen, S.L.: Statistical manifolds. Differential Geometry in Statistical Inference, p. 164 (1987)
25. Lauster, F., Luke, D.R., Tam, M.K.: Symbolic computation with monotone operators. Set-Valued and Variational Analysis, pp. 1–16 (2017)
26. Liu, Q., Ihler, A.T.: Distributed estimation, information loss and exponential families. In: Advances in Neural Information Processing Systems, pp. 1098–1106 (2014)
27. Matsuzoe, H., Scarfone, A.M., Wada, T.: A sequential structure of statistical manifolds on deformed exponential family. In: International Conference on Geometric Science of Information, pp. 223–230. Springer (2017)
28. Mitchell, A.F.S.: Statistical manifolds of univariate elliptic distributions. International Statistical Review/Revue Internationale de Statistique, pp. 1–16 (1988)
29. Nielsen, F.: A family of statistical symmetric divergences based on Jensen's inequality (2010). arXiv:1009.4004
30. Nielsen, F.: Legendre transformation and information geometry (2010)
31. Nielsen, F.: Hypothesis testing, information divergence and computational geometry. In: Geometric Science of Information, pp. 241–248. Springer (2013)
32. Nielsen, F.: An information-geometric characterization of Chernoff information. IEEE Signal Process. Lett. **20**(3), 269–272 (2013)
33. Nielsen, F.: Introduction to HPC with MPI for Data Science. Undergraduate Topics in Computer Science. Springer (2016). https://doi.org/10.1007/978-3-319-21903-5. https://doi.org/10.1007/978-3-319-21903-5
34. Nielsen, F., Boissonnat, J.D., Nock, R.: On Bregman Voronoi diagrams. In: Proceedings of the eighteenth annual ACM-SIAM symposium on Discrete algorithms. Society for Industrial and Applied Mathematics, pp. 746–755 (2007)
35. Nielsen, F., Boltz, S.: The Burbea-Rao and Bhattacharyya centroids. IEEE Trans. Inf. Theory **57**(8), 5455–5466 (2011)

36. Nielsen, F., Garcia, V.: Statistical exponential families: a digest with flash cards (2009) arXiv:0911.4863
37. Nielsen, F., Garcia, V., Nock, R.: Simplifying Gaussian mixture models via entropic quantization. In: 17th European Conference on Signal Processing (EUSIPCO), pp. 2012–2016. IEEE (2009)
38. Nielsen, F., Hadjeres, G.: Monte Carlo information geometry: the dually flat case (2018). CoRR arXiv:1803.07225
39. Nielsen, F., Nock, R.: On the smallest enclosing information disk. Inf. Process. Lett. **105**(3), 93–97 (2008)
40. Nielsen, F., Nock, R.: Sided and symmetrized Bregman centroids. IEEE Trans. Inf. Theory **55**(6), 2882–2904 (2009)
41. Nielsen, F., Nock, R.: Optimal interval clustering: application to Bregman clustering and statistical mixture learning. IEEE Signal Process. Lett. **21**(10), 1289–1292 (2014)
42. Nielsen, F., Nock, R.: Patch matching with polynomial exponential families and projective divergences. In: International Conference on Similarity Search and Applications, pp. 109–116. Springer (2016)
43. Nielsen, F., Nock, R.: On w-mixtures: finite convex combinations of prescribed component distributions (2017). CoRR arXiv:1708.00568
44. Nielsen, F., Nock, R.: On the geometric of mixtures of prescribed distributions. In: IEEE International Conference on Acoustics, Speech and Signal Processing (ICASSP) (2018)
45. Nielsen, F., Piro, P., Barlaud, M.: Bregman vantage point trees for efficient nearest neighbor queries. In: 2009 IEEE International Conference on Multimedia and Expo, ICME 2009, pp. 878–881. IEEE (2009)
46. Nielsen, F., Piro, P., Barlaud, M.: Tailored Bregman ball trees for effective nearest neighbors. In: Proceedings of the 25th European Workshop on Computational Geometry (EuroCG), pp. 29–32 (2009)
47. Nielsen, F., Sun, K.: Guaranteed bounds on information-theoretic measures of univariate mixtures using piecewise log-sum-exp inequalities. Entropy **18**(12), 442 (2016)
48. Nielsen, F., Sun, K., Marchand-Maillet, S.: On hölder projective divergences. Entropy **19**(3), 122 (2017)
49. Nock, R., Luosto, P., Kivinen, J.: Mixed Bregman clustering with approximation guarantees. In: Joint European Conference on Machine Learning and Knowledge Discovery in Databases, pp. 154–169. Springer (2008)
50. Notsu, A., Komori, O., Eguchi, S.: Spontaneous clustering via minimum γ-divergence. Neural Comput. **26**(2), 421–448 (2014)
51. Pelletier, B.: Informative barycentres in statistics. Ann. Inst. Stat. Math. **57**(4), 767–780 (2005)
52. Robert, C.P.: Monte Carlo methods. Wiley Online Library (2004)
53. Shima, H.: The Geometry of Hessian Structures. World Scientific, Singapore (2007)
54. Tang, Y., Salakhutdinov, R.R.: Learning stochastic feedforward neural networks. In: Advances in Neural Information Processing Systems, pp. 530–538 (2013)
55. Varga, R.S.: Geršgorin and His Circles, vol. 36. Springer Science & Business Media, Berlin (2010)
56. Watanabe, S., Yamazaki, K., Aoyagi, M.: Kullback information of normal mixture is not an analytic function. Technical report of IEICE (in Japanese) (2004-0), pp. 41–46 (2004)
57. Zhang, J.: Reference duality and representation duality in information geometry. In: AIP Conference Proceedings, vol. 1641, pp. 130–146. AIP (2015)

Information Geometry in Portfolio Theory

Ting-Kam Leonard Wong

Abstract We review some recent developments in stochastic portfolio theory motivated by information geometry, present illustrative examples and an extension of functional portfolio generation. Several problems are suggested for further study.

1 Introduction

In the first chapter of their influential monograph [1, p.1] Amari and Nagaoka explained the key idea of information geometry as follows:

> Information geometry ... allow[s] us to take problems from a variety of fields: statistics, information theory, and control theory; visualize them geometrically; and from this develop novel tools with which to extend and advance these fields.

In this chapter we show that this principle can be fruitfully applied to financial problems. We review some recent development in the field of stochastic portfolio theory (SPT) motivated by information geometry, present illustrative examples and an extension of functional portfolio generation (announced in [2] which is an early version of this paper), and suggest several problems for further study. It is hoped that the materials will be of interest to researchers in both information geometry and mathematical finance. The topics discussed are heavily influenced by the author's research interests. Other financial applications of information geometry are briefly reviewed in Sect. 1.2.

T.-K. L. Wong (✉)
University of Toronto, Toronto, Canada
e-mail: tkl.wong@utoronto.ca

F. Nielsen (ed.), *Geometric Structures of Information*, Signals and Communication Technology, https://doi.org/10.1007/978-3-030-02520-5_6

1.1 Main Ideas: Market Diversity and Volatility

To set the stage let us consider a universe of stocks represented by a capitalization-weighted index. A typical example is the S&P 500 Index which includes a significant portion of the US stock market. In a capitalization-weighted index, the influence of a stock is proportional to its market capitalization. Following the standard set up of stochastic portfolio theory (see for example [3, 4]), if we let $X_i(t) > 0$ be the market capitalization of stock i at time t, then

$$\mu_i(t) := \frac{X_i(t)}{X_1(t) + \cdots + X_n(t)} \tag{1}$$

is the *market weight* of the stock. Throughout this chapter we let time be discrete. Let $n \geq 2$ be the number of stocks in the market. The vector $\mu(t) = (\mu_1(t), \ldots, \mu_n(t))$ takes values in the open unit simplex Δ_n given by

$$\Delta_n := \{p = (p_1, \ldots, p_n) \in (0, 1)^n : p_1 + \cdots + p_n = 1\}.$$

Its closure in $\mathbb{R}^n$ is denoted by $\overline{\Delta}_n$. The vector $\mu(t)$ may also be regarded as the portfolio weights of the *market portfolio*.

A major objective of SPT is to construct investment strategies that beat the market under realistic conditions on market behaviors. These portfolios are called *relative arbitrages* with respect to the market portfolio. For more details about this important problem see for example [3–5] and their references. As a simple example, Fig. 1 plots the path of $\{\mu(t)\}$ for a hypothetical 3-stock market consisting of the US stocks Ford, IBM and Walmart. It should not be surprising that the relative performance of portfolios with respect to the market can be analyzed using appropriate geometries on the simplex.[1]

Regarding the market as a process in the simplex Δ_n, there are two natural quantities an investor may want to keep track of: *diversity* and *volatility*. Diversity refers to the degree of capital concentration in the equity market. For example, in Fig. 1, the market moves towards the vertex representing Walmart and thus becomes more concentrated. According to [6], many mutual funds tend to overweight small stocks and underweight large stocks (relative to the market index), so the change in market diversity is a significant predictor of their relative performance. To quantify diversity one introduces a positive *concave* function $\Phi : \Delta_n \to (0, \infty)$, and we say that the market is more diverse when $\Phi(\mu(t))$ is large. Typical examples include the Shannon entropy

$$\Phi(p) = -\sum_{j=1}^{n} p_j \log p_j$$

[1]In practice the number of stocks changes with time, and the market capitalization may fluctuate due to public offerings and other events. For simplicity these complications are neglected here.

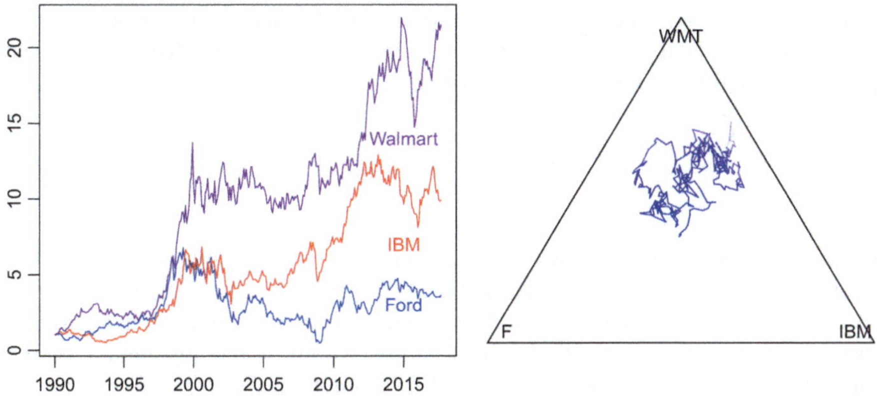

Fig. 1 Left: Monthly prices of the stocks (regarded as the capitalization and normalized to be 1 from January 1990). Right: Path of the market weight vector $\mu(t)$ in the simplex Δ_3

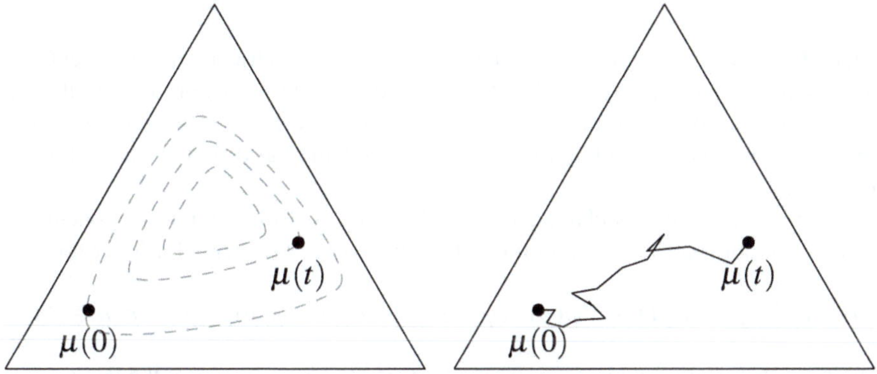

Fig. 2 Left: Change in market diversity measured by φ. The dashed curves represent level sets of φ. Right: Market volatility can be measured by the L-divergence $\mathbf{D}[\cdot \mid \cdot]$

as well as the λ-diversity function

$$\Phi(p) = \left(\sum_{j=1}^{n} (p_j)^{\lambda} \right)^{1/\lambda}, \tag{2}$$

where $0 < \lambda < 1$ is a parameter. More examples of measures of diversity can be found in [3, Chapter 3]. Note that we allow Φ to be asymmetric, so it can attain its maximum value at a point other than the barycenter $\overline{e} := \left(\frac{1}{n}, \ldots, \frac{1}{n}\right)$. As it turns out, it is mathematically more convenient to consider its logarithm $\varphi := \log \Phi$. Since $e^{\varphi} = \Phi$ is concave, we say that φ is *exponentially concave*. We remark that φ, being the logarithm of a concave function, is itself a concave function. Given φ, the time series of $\{\varphi(\mu(t))\}$ is an indicator of market diversity (see Fig. 2 (left)).

The volatility of the market weight $\mu(t)$ refers to the volatility of the stocks *relative to each other*. Anticipating the use of information geometry, note that the Euclidean quadratic variation (in discrete time) given by

$$\sum_{s=0}^{t-1} |\mu(s+1) - \mu(s)|^2 \tag{3}$$

may not be appropriate because the Euclidean norm on the simplex may not have a financial meaning (see however Example 3). In particular, the same displacement $v = \mu(s+1) - \mu(s)$ (approximated by a tangent vector) should have different norms on different portions of the simplex. Depending on the application, market volatility along a path should be quantified by a sum like

$$\sum_{s=0}^{t-1} \mathbf{D}\left[\mu(s+1) \mid \mu(s)\right],$$

where $\mathbf{D}[\cdot \mid \cdot] : \Delta_n \times \Delta_n \to [0, \infty)$ is possibly asymmetric in its arguments. In information geometry we say that $\mathbf{D}[\cdot \mid \cdot]$ a *divergence* (a rigorous definition is given in Definition 5). Intuitively, the asymmetry of $\mathbf{D}[\cdot \mid \cdot]$ reflects the effect of time: the time-reversed path $\tilde{\mu}(t) = \mu(T-t)$ probably has different impacts on the portfolio.

The main idea of this chapter is the following. A differentiable, exponentially concave function $\varphi : \Delta_n \to \mathbb{R}$ defines an *L-divergence* (L stands for logarithmic)

$$\mathbf{D}^{(1)}\left[q \mid p\right] := \log\left(1 + \nabla\varphi(p) \cdot (q-p)\right) - \left(\varphi(q) - \varphi(p)\right) \geq 0, \quad p, q \in \Delta_n, \tag{4}$$

which can be used to quantify market volatility. (The superscript will become clear in Definition 1.) An important example of L-divergence is the *excess growth rate* (also known as the *diversification return* [7]) defined for a fixed portfolio vector $\pi \in \overline{\Delta}_n$ by

$$\mathbf{T}_\pi\left[q \mid p\right] := \log\left(\sum_{i=1}^{n} \pi_i \frac{q_i}{p_i}\right) - \sum_{i=1}^{n} \pi_i \log \frac{q_i}{p_i}. \tag{5}$$

The corresponding exponentially concave function is $\varphi(p) = \sum_{i=1}^{n} \pi_i \log p_i$. Note that the L-divergence is different from the classical Bregman divergence of a concave function:

$$\mathbf{D}^{(0)}\left[q \mid p\right] := \nabla\varphi(p) \cdot (q-p) - \left(\varphi(q) - \varphi(p)\right). \tag{6}$$

The most important example of Bregman divergence is the relative entropy

$$\mathbf{H}\left(q \mid p\right) := \sum_{i=1}^{n} q_i \log \frac{q_i}{p_i}, \tag{7}$$

where the potential function φ is the Shannon entropy. Comparing (5) and (7), we see that the excess growth rate involves a nonlinear transformation of an integral, whereas the relative entropy is itself an integral. In Example 2 we show that the Rényi entropy generates the Rényi divergence in the sense of L-divergence and is related to the diversity function (2).

In Sect. 3, we will show that the L-divergence determines uniquely an investment strategy, called a *multiplicatively generated portfolio*, whose performance $V(t)$ relative to the market has the pathwise decomposition

$$\log V(t) - \log V(0) = \varphi(\mu(t)) - \varphi(\mu(0)) + \sum_{s=0}^{t-1} \mathbf{D}^{(1)}\left[\mu(s+1) \mid \mu(s)\right]. \quad (8)$$

From this decomposition, as long as $\varphi(\mu(t))$ remains bounded and the cumulative volatility grows at a steady rate, the portfolio will outperform the market in the long run (see Proposition 1). Furthermore, the dualistic geometry induced by the L-divergence (in the sense of [8, 9]; also see [10]) has interesting financial applications (see Sect. 5.2). In a continuous time framework and without using geometric concepts, these portfolios were first introduced by Fernholz [3, 11]. Here we adopt the discrete time, geometric approach established in [12, 13].

Following [2, 14], in Sect. 4 we will generalize the portfolio construction using the $L^{(\alpha)}$-divergence:

Definition 1 ($L^{(\alpha)}$*-divergence*) Let $\varphi : \Delta_n \to \infty$ be differentiable and α-exponentially concave, i.e, $e^{\alpha\varphi}$ is concave. The $L^{(\alpha)}$-divergence of φ is defined for $p, q \in \Delta_n$ by

$$\mathbf{D}^{(\alpha)}\left[q \mid p\right] := \frac{1}{\alpha} \log\left(1 + \alpha \nabla\varphi(p) \cdot (q - p)\right) - \left(\varphi(q) - \varphi(p)\right), \quad (9)$$

Our framework covers also the *additively generated portfolio* introduced recently in [15]. Note that the L-divergence is the $L^{(1)}$-divergence, and the Bregman divergence is equal to $\mathbf{D}^{(\alpha)}$ as $\alpha \downarrow 0$, so we will also call it the $L^{(0)}$-divergence. Using obvious notations, we have the identity

$$\mathbf{D}^{(\alpha)}_{\varphi}\left[\cdot \mid \cdot\right] \equiv \frac{1}{\alpha} \mathbf{D}^{(1)}_{\alpha\varphi}\left[\cdot \mid \cdot\right]. \quad (10)$$

According to the results obtained recently in [14], it appears that the $L^{(\alpha)}$-divergence is the canonical interpolation between the Bregman divergence and the L-divergence, and plays a fundamental role in information geometry.

1.2 Financial Applications of Information Geometry

Instead of conducting an extensive literature review, we content with giving a sample of other financial applications of information geometry. There are mainly two (overlapping) directions: (i) geometries on the state space of financial dynamics, and (ii) optimization using information-geometric quantities such as entropy and divergence.

Regarding the first direction, the paper [16] identifies a yield curve with a distribution function and studies its corresponding dynamics using the Fisher information metric. In option pricing, [17] applies Tsallis's deformed exponentials and generalized the Black-Scholes model to fat-tailed distributions. Though not directly related to finance, the paper [18] generalizes the concept of multiplicatively generated portfolio map (see Sect. 3) to a large class of optimal transport problems. The recent work [19] applies the Fisher metric in the study of systemic risk.

On the other hand, divergences are frequently useful as objective/cost functionals. In [20, 21], the authors generalize Markowitz's mean-variance model to a mean-divergence model, and show that the resulting portfolios have superior performance. Optimization of probabilistic functionals under a divergence constraint is studied [22] and applied to model risk.

1.3 Outline of the Paper

In Sect. 2 we present the market model and introduce two ways of representing a trading strategy and the associated value process. Section 3 reviews known results about multiplicatively generated portfolios with an emphasis on ideas and clarity. Motivated by these results and the recently introduced additively generated portfolio, in Sect. 4 we introduce a general framework for functional portfolio generation, where the $L^{(\alpha)}$-divergence arises naturally. Further properties of the L-divergence are discussed in Sect. 5 and several related problems are stated.

2 The Market Model

We work under the discrete time, pathwise setting used in our previous papers [12, 23, 24]. Let $n \geq 2$, the number of stocks in the market, be fixed. The data of our model is a sequence $\{\mu(t) = (\mu_1(t), \ldots, \mu_n(t))\}_{t=0}^{\infty}$ with values in the open unit simplex Δ_n. We regard $\mu(t)$ as the vector of market weights at time t. At this point we do not impose any condition on the sequence $\{\mu(t)\}_{t=0}^{\infty}$. In Proposition 1 we will give examples of path properties that lead to relative arbitrages.

We consider self-financing trading strategies in this market model. Let us express a strategy in terms of the number of shares held at each point in time. Furthermore, we use the market portfolio as the numéraire (i.e., unit of price). This means that the (relative) value of stock i is simply the market weight $\mu_i(t)$. We assume that trading is frictionless.

Definition 2 (*Self-financed trading strategy*) A self-financing trading strategy is a sequence $\eta = \{\eta(t)\}_{t=0}^{\infty}$, with values in $\mathbb{R}^n$, such that the self-financing identity

$$\sum_{i=1}^{n} \eta_i(t)\mu_i(t+1) \equiv \sum_{i=1}^{n} \eta_i(t+1)\mu_i(t+1) \tag{11}$$

holds for all time t. We always assume η is adapted in the sense that for each $t \geq 0$, $\eta(t)$ is a deterministic function of $\{\mu(s)\}_{0\leq s\leq t}$. The (relative) value process of η is defined by

$$V_\eta(t) = V_\eta(0) + \sum_{s=0}^{t-1} \eta(s) \cdot (\mu(s+1) - \mu(s)), \tag{12}$$

where $V_\eta(0) := \eta(0) \cdot \mu(0)$ and $a \cdot b$ is the Euclidean inner product.

In this paper all trading strategies are self-financed. Also, we only study the value of portfolios relative to the market portfolio, so for simplicity we may omit the words 'self-financed' and 'relative'. By the identity (11), the value (12) of the portfolio is equal to

$$V_\eta(t) = \eta(t) \cdot \mu(t).$$

Note that because we allow both long and short positions in the portfolio, the value $V_\eta(t)$ may take negative values. The self-financing identity (11) means that all changes in the portfolio value are due to price changes (but not addition or withdrawal of capital).

If the portfolio value $V_\eta(t)$ is strictly positive for all t, we may define the *portfolio weight vector* at time t by

$$\pi(t) = (\pi_1(t), \ldots, \pi_n(t)) = \left(\frac{\eta_1(t)\mu_1(t)}{V_\eta(t)}, \ldots, \frac{\eta_n(t)\mu_n(t)}{V_\eta(t)}\right). \tag{13}$$

The components of $\pi(t)$ represent the percentages of current capital invested in each of the stocks; clearly $\sum_{i=1}^{n} \pi_i(t) \equiv 1$. In this case, the value $V_\eta(t)$ can be expressed *multiplicatively* in the form

$$V_\eta(t) = V_\eta(0) \prod_{s=0}^{t-1} \left(\pi(s) \cdot \frac{\mu(s+1)}{\mu(s)}\right), \tag{14}$$

where $\frac{\mu(s+1)}{\mu(s)} = \left(\frac{\mu_i(s+1)}{\mu_i(s)}\right)_{1\leq i\leq n}$ is the vector of componentwise ratios. Compare this with the *additive* representation (12). If $\pi_i(t) \geq 0$ for all i and t, we say that the portfolio is *all-long*. It is clear that an adaptive sequence (in the sense of Definition 2) of portfolio weight vectors defines a self-financing all-long trading strategy for each initial value. The *market portfolio* corresponds to $\pi(t) \equiv \mu(t)$.

3 Multiplicatively Generated Portfolio

In this section we review the definition and main results of multiplicative functional generation using the approach of [12]. For simplicity of exposition we assume that the generating functions are smooth.

3.1 Pathwise Decomposition and Relative Arbitrage

Definition 3 (*Multiplicatively generated portfolio*) Let $\varphi : \Delta_n \to \mathbb{R}$ be smooth and exponentially concave, to be called a generating function. Given φ, we define a mapping $\boldsymbol{\pi} : \Delta_n \to \overline{\Delta}_n$, called the portfolio map, by

$$\boldsymbol{\pi}_i(p) = p_i \left(1 + D_{e_i - p}\varphi(p)\right), \quad i = 1, \ldots, n, \tag{15}$$

where $(e_1, \ldots, e_n)$ is the standard Euclidean basis and D_{e_i-p} is the directional derivative along the tangent vector $e_i - p$. It defines an all-long trading strategy η such that the portfolio weight at time t is

$$\pi(t) = \left(\frac{\eta_1(t)\mu_1(t)}{V_\eta(t)}, \ldots, \frac{\eta_n(t)\mu_n(t)}{V_\eta(t)}\right) = \boldsymbol{\pi}(\mu(t)). \tag{16}$$

We say that η (and $\boldsymbol{\pi}$) are generated multiplicatively by φ.

Here is a geometric interpretation of the formula (15). Consider the graph of the positive concave function $\Phi = e^{\varphi}$. Given $p \in \Delta_n$, let the tangent hyperplane to Φ at p be given by $q \mapsto \sum_{i=1}^n c_i q_i$ (see Fig. 3). We have

$$c_i = \Phi(p) + D_{e_i - p}\Phi(p) = \Phi(p)\left(1 + D_{e_i-p}\varphi(p)\right).$$

Since $\sum_{i=1}^n p_i(e_i - p) = 0$, the portfolio vector $\boldsymbol{\pi}(p)$ is given by

$$\boldsymbol{\pi}_i(p) = \frac{c_i p_i}{c_1 p_1 + \cdots + c_n p_n}, \quad i = 1, \ldots, n. \tag{17}$$

In particular, $\boldsymbol{\pi}(p)$ is an element of the closed simplex $\overline{\Delta}_n$ (so the portfolio is all-long), and the weight ratio $\boldsymbol{\pi}_i(p)/p_i$ is proportional to c_i. We say that the trading strategy is generated *multiplicatively* because the generating function φ specifies the weight ratios through its derivatives.

The following is the main result about multiplicatively generated portfolios. As this result is fundamental let us give a complete proof which also motivates the definition of the L-divergence. We also note that this proof is more transparent than the original proof (see [3, Theorem 3.1.5] which is formulated in continuous time).

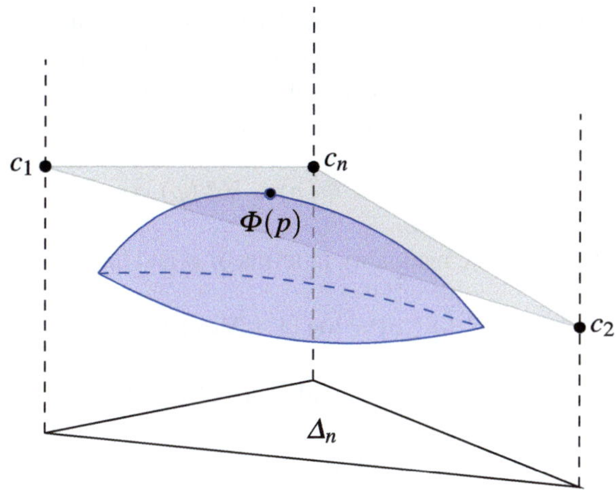

Fig. 3 Geometric interpretation of multiplicatively generated portfolio

Theorem 1 (Multiplicative decomposition [11, 12]) *Let η be the trading strategy generated multiplicatively by the exponentially concave function φ as in Definition 3. Then the value process of η satisfies the decomposition*

$$\log V_\eta(t) - \log V_\eta(0) = \varphi(\mu(t)) - \varphi(\mu(0)) + \sum_{s=0}^{t-1} \mathbf{D}^{(1)} \left[\mu(s+1) \mid \mu(s)\right], \quad (18)$$

where $\mathbf{D}^{(1)} [\cdot \mid \cdot]$ is the $L^{(1)}$-divergence of φ defined by (4).

Proof Using the multiplicative representation (14), we have

$$\frac{V_\eta(s+1)}{V_\eta(s)} = \sum_{i=1}^n \boldsymbol{\pi}_i(\mu(s)) \frac{\mu_i(s+1)}{\mu_i(s)}.$$

From (15), we have

$$\frac{\boldsymbol{\pi}_i(\mu(s))}{\mu_i(s)} = 1 + D_{e_i - \mu(s)} \varphi(\mu(s)),$$

so we get the useful identity

$$\begin{aligned} \frac{V_\eta(s+1)}{V_\eta(s)} &= 1 + \sum_{i=1}^n \mu_i(s+1) D_{e_i - \mu(s)} \varphi(\mu(s)) \\ &= 1 + D_{\mu(s+1)-\mu(s)} \varphi(\mu(s)) \\ &= 1 + \nabla \varphi(\mu(s)) \cdot (\mu(s+1) - \mu(s)). \end{aligned} \quad (19)$$

Here we think of the gradient $\nabla\varphi(\mu(s))$ as operating on tangent vectors of Δ_n. Financially, (19) says that the relative return $(V_\eta(s+1) - V_\eta(s))/V_\eta(s)$ of the portfolio is nothing but the directional derivative of φ.

By the concavity of $\Phi = e^\varphi$, for any $p, q \in \Delta_n$ we have

$$\Phi(p) + \nabla\Phi(p) \cdot (q - p) \geq \Phi(q).$$

Rewriting the inequality in terms of φ and taking logarithm on both sides, we have

$$\mathbf{D}^{(1)}[q \mid p] = \log(1 + \nabla\varphi(p) \cdot (q - p)) - (\varphi(q) - \varphi(p)) \geq 0.$$

Taking logarithm on both sides of (19) and rearranging, we get

$$\log V_\eta(s+1) - \log V_\eta(s) = \varphi(\mu(s+1)) - \varphi(\mu(s)) + \mathbf{D}^{(1)}[\mu(s+1)|\mu(s)].$$

Summing over time gives the decomposition (18).

From (18), the performance of the portfolio relative to the market can be attributed to two quantities. The first is the change in the market diversity $\varphi(\mu)$. It depends only on the beginning location $\mu(0)$ and the current location $\mu(t)$ of the market. Note that a change in $\varphi(\mu(t))$ is only caused by the component of market movement along the direction of $\nabla\varphi(\mu(t))$ which is perpendicular to the level set of φ. In particular, displacement along the same level set is not visible in this first term. The second term in (18) measures the volatility of the market, as it travels from $\mu(0)$ to $\mu(t)$, by the sum of $\mathbf{D}^{(1)}[\mu(s+1) \mid \mu(s)]$ over time. Intuitively, the functionally generated trading strategy η outperforms the market if and only if the volatility is greater than the change in market diversity. In SPT, this decomposition allows one to formulate conditions under which relative arbitrage (with respect to the market portfolio) exists. Here is the simplest version of this idea:

Proposition 1 (Relative arbitrage) *Fix a smooth, exponentially concave function $\varphi : \Delta_n \to \mathbb{R}$. Let $M > 0$ be a constant and let $T > 0$ be a finite time horizon. We say that a market weight sequence $\{\mu(t)\}_{t=0}^\infty$ satisfies property $\mathscr{P}$ if $\varphi(\mu(t)) - \varphi(\mu(0)) > -M$ for all t and $\sum_{s=0}^{T-1} \mathbf{D}^{(1)}[\mu(s+1)|\mu(s)] > M$, where $\mathbf{D}^{(1)}$ is the $L^{(1)}$-divergence of φ. Then there exists an all-long trading strategy η such that $V_\eta(T)/V_\eta(0) > 1$ (i.e., the strategy outperforms the market over the horizon $[0, T]$) for all market weight sequences satisfying property $\mathscr{P}$.*

Proof Let η be the trading strategy generated multiplicatively by φ. By Theorem 1, if the market weight sequence satisfies property $\mathscr{P}$, we have

$$\begin{aligned}\log V_\eta(T) - \log V_\eta(0) &= \varphi(\mu(t)) - \varphi(\mu(0)) + \sum_{s=0}^{t-1} \mathbf{D}^{(1)}[\mu(s+1) \mid \mu(s)] \\ &> -M + M = 0.\end{aligned}$$

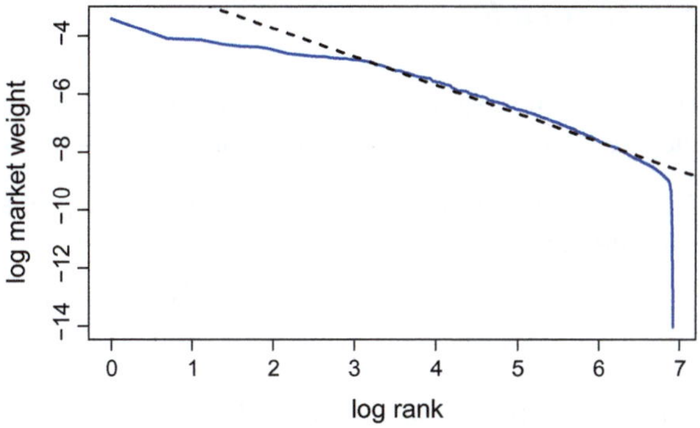

Fig. 4 Capital distribution of the Russel 1000 Index in June 2015 (taken from [25])

The proof of Proposition 1 is almost trivial because we already have the concept of multiplicatively generated portfolio. Without knowing this construction, it is not immediate why a relative arbitrage exists and how it is constructed. The usefulness of this result comes from the following observations (see [3]). Consider the *capital distribution* of the market defined by the reversed order statistics of the components of $\mu(t)$:

$$\mu_{(1)}(t) \geq \mu_{(2)}(t) \geq \cdots \geq \mu_{(n)}(t).$$

Empirically, it is found that if one plots $\log \mu_{(k)}(t)$ against $\log k$ (log of the rank), one gets an approximately linear curve (except the tail) which is relatively stable over time (see Fig. 4). This means that the capital distribution has an approximate Pareto distribution, and the market diversity $\varphi(\mu(t))$ is mean-reverting for φ suitably chosen. On the other hand, for a typical φ the market volatility $\sum_{s=0}^{T-1} \mathbf{D}[\mu(s+1)|\mu(s)]$ grows roughly linearly in time [26]. Thus, it appears that the market satisfies the conditions of (1).

The stability of the capital distribution has inspired many works on the construction and analysis of market models that exhibit such behaviors. Mathematically, these are systems of Brownian particles (representing the market capitalizations) where the drift and volatility coefficients depend on their relative rankings, and so they are called *rank-based models*. For more details we refer the reader to the papers [27–33] and their references.

Proposition 1 only addresses long term relative arbitrages. In practice, *short term* relative arbitrages are much more relevant and interesting. Naturally their constructions require more work and conditions (see for example [25, 34–36]). In particular, the paper [5] proves that market volatility alone does not imply the existence of short term relative arbitrage (this problem had been open in SPT for more than 10 years).

3.2 Examples

We give some examples of exponentially concave functions on Δ_n, the portfolios they generate as well as the corresponding $L^{(1)}$-divergences.

Example 1 (*Constant-weighted portfolio*) Fix a probability vector $\pi \in \overline{\Delta}_n$ and consider the function

$$\varphi(p) = \sum_{j=1}^{n} \pi_j \log p_j.$$

It is exponentially concave since $\Phi = e^{\varphi} = (p_1)^{\pi_1} \cdots (p_n)^{\pi_n}$, the geometric mean with weights $\pi_1, \ldots, \pi_n$, is concave on Δ_n. This function generates the constant-weighted portfolio $\boldsymbol{\pi}(p) \equiv \pi$, and the $L^{(1)}$-divergence is the excess growth rate $\mathbf{T}_{\pi}\left[q \mid p\right]$ given by (5). We also observe that

$$\begin{aligned}\varphi(\mu(t)) - \varphi(\mu(0)) &= \sum_{j=1}^{n} \pi_j \log \frac{\pi_j}{\mu_j(0)} - \sum_{j=1}^{n} \pi_j \log \frac{\pi_j}{\mu_j(t)} \\ &= H\left(\pi \mid \mu(0)\right) - H\left(\pi \mid \mu(t)\right)\end{aligned}$$

is the negative of the change in the relative entropy $H\left(\pi \mid \cdot\right)$. In [23], we call the decomposition (18) for this portfolio the *energy-entropy decomposition*.

Example 2 (*Diversity-weighted portfolio*) For $\lambda \in (0, 1)$ fixed, let φ be the function

$$\varphi(p) = \frac{1}{\lambda} \log \sum_{j=1}^{n} (p_j)^{\lambda},$$

which is the logarithm of the function Φ given by (2). Then φ is exponentially concave and generates the *diversity-weighted portfolio* where

$$\boldsymbol{\pi}_i(p) = \frac{(p_i)^{\lambda}}{\sum_{j=1}^{n} (p_j)^{\lambda}}, \quad i = 1, \ldots, n. \tag{20}$$

Note that this portfolio interpolates between the equal-weighted portfolio $\boldsymbol{\pi}(p) \equiv \overline{e}$ (when $\lambda \downarrow 0$) and the market portfolio $\boldsymbol{\pi}(p) \equiv p$ (when $\lambda \uparrow 1$). See [37] where the diversity-weighted portfolio is studied for negative values of λ. We remark that these cover, except for the log case, portfolios constructed using *Tukey's transformation ladder*; for more details see [38] where an extensive empirical study is given.

For $\lambda \in (0, 1)$ fixed, let $p^{(\lambda)} \in \Delta_n$ be given by $\boldsymbol{\pi}(p)$ as in (20). In information geometry, $p^{(\lambda)}$ is called the λ-*escort distribution* corresponding to the distribution p (see [10, Section 4.3]). Using this terminology we may interpret the portfolio in terms of the Rényi entropy and divergence.

Proposition 2 *Let* $0 < \lambda < 1$. *For* $p \in \Delta_n$, *we have*

$$\varphi(p) = \frac{1}{\lambda} \log \left(\sum_{j=1}^{n} (p_j)^{\lambda} \right) = (\alpha - 1)\mathbf{H}_{\alpha}(p^{(\lambda)}),$$

where $\alpha := \frac{1}{\lambda} \in (1, \infty)$, $p^{(\lambda)}$ *is the* λ*-escort distribution corresponding to* p, *and*

$$\mathbf{H}_{\alpha}(r) := \frac{1}{1-\alpha} \log \left(\sum_{j=1}^{n} (r_j)^{\alpha} \right)$$

is the Rényi entropy of order α.

Moreover, the $L^{(1)}$*-divergence of* φ *is given by*

$$\mathbf{D}^{(1)}[q \mid p] = (\alpha - 1)\mathbf{D}_{\alpha}(q^{(\lambda)} || p^{(\lambda)}), \tag{21}$$

where $\mathbf{D}_{\alpha}(\cdot||\cdot)$ *is the Rényi divergence of order* α *defined by*

$$\mathbf{D}_{\alpha}(p||q) := \frac{1}{\alpha - 1} \log \left(\sum_{j=1}^{n} (p_j)^{\alpha} (q_j)^{1-\alpha} \right). \tag{22}$$

Proof This is a direct computation and we only give the proof of the first statement. Using the fact that $p = \left(p^{(\lambda)}\right)^{(1/\lambda)}$, we have

$$\begin{aligned}
\frac{1}{\lambda} \log \left(\sum_{i=1}^{n} p_j^{\lambda} \right) &= \frac{1}{\lambda} \log \left(\sum_{i=1}^{n} \left(\frac{\left(p_i^{(\lambda)}\right)^{1/\lambda}}{\sum_{j=1}^{n} \left(p_j^{(\lambda)}\right)^{1/\lambda}} \right)^{\lambda} \right) \\
&= -\log \left(\sum_{j=1}^{n} \left(p_j^{(\lambda)}\right)^{\frac{1}{\lambda}} \right) \\
&= (\alpha - 1) \frac{1}{1-\alpha} \log \left(\sum_{j=1}^{n} \left(p_j^{(\lambda)}\right)^{\alpha} \right),
\end{aligned}$$

which is $\alpha - 1$ times the Rényi entropy.

Corollary 1 *Let* $\lambda \in (0, 1)$ *and* $\alpha := \frac{1}{\lambda} \in (1, \infty)$. *The relative value of the diversity-weighted portfolio with parameter* λ *satisfies*

$$\begin{aligned}
&\log V_{\eta}(t) V_{\eta}(0) \\
&= (\alpha - 1) \left[\mathbf{H}_{\alpha}\left(\mu^{(\lambda)}(t)\right) - \mathbf{H}_{\alpha}\left(\mu^{(\lambda)}(0)\right) + \sum_{s=0}^{t-1} \mathbf{D}_{\alpha}\left(\mu^{(\lambda)}(t+1) \middle|\middle| \mu^{(\lambda)}(t)\right) \right].
\end{aligned}$$

Consider a portfolio manager who tries to optimize over the parameter λ of a diversity-weighted portfolio. By Corollary 1, this means comparing the dynamics of the market weight $\mu(t)$ using different λ-escort geometries of the simplex. In particular, it is well-known that the Rényi divergence satisfies

$$\mathbf{D}_{\alpha}\left(r + tv||r\right) = \frac{\alpha}{2}t^2\|v\|_r^2 + O(|t|^3), \quad t \to 0, \tag{23}$$

where $\|v\|_r^2 := \sum_{i=1}^n v_i^2/r_i$ is the Fisher information metric at the point r. This geometric viewpoint may lead to new statistical methods and algorithms. In this regard, let us mention the recent work [39] which studies the dynamics of market diversity in the context of large rank-based models, as well as the paper [40] which proposes a model for predicting change in market diversity. More generally, optimization of functionally generated portfolio amounts to finding the geometry in which $\mu(t)$ has the least change in diversity and has the greatest cumulated volatility.

3.3 Multiplicative Cyclical Monotonicity

In this subsection we provide a financial argument, given in [12], that motivates the definition of multiplicatively generated portfolio.

Let us restrict to all-long trading strategies defined by portfolio maps, i.e., the portfolio weights satisfies $\pi(t) = \boldsymbol{\pi}(\mu(t))$ where $\boldsymbol{\pi} : \Delta_n \to \overline{\Delta}_n$ is a fixed deterministic function. When is $\boldsymbol{\pi}$ able to profit from market volatility? Intuitively it should satisfy the following property. Let O be a (small) neighborhood in the simplex Δ_n, and suppose $\mu(t) \in O$ for all t. From the discussion in Sect. 1.1 the capital distribution is stable. Then, we expect that the portfolio will outperform the market asymptotically as long as there is enough volatility. Specifically, it should outperform the market whenever it is periodic. This idea leads to the following definition.

Definition 4 (*multiplicative cyclical monotonicity (MCM)*) A portfolio map $\boldsymbol{\pi}$: $\Delta_n \to \overline{\Delta}_n$ is multiplicatively cyclical monotone if for any integer $m \geq 0$ and any cycle $\{\mu(t)\}_{t=0}^m$ with $\mu(0) = \mu(m)$ we have $V_\eta(m) \geq 1$, i.e.,

$$\prod_{t=0}^{m-1} \left(\boldsymbol{\pi}(\mu(t)) \cdot \frac{\mu(t+1)}{\mu(t)} \right) \geq 1. \tag{24}$$

In [12] we observed that this property characterizes the multiplicatively generated portfolio. The following result is the multiplicative analogue of Rockafellar's theorem which characterizes the subdifferentials of convex functions in terms of cyclical monotonicity [41, Section 24].

Theorem 2 *Suppose the portfolio map* $\boldsymbol{\pi} : \Delta_n \to \overline{\Delta}_n$ *is continuous. Then it is multiplicatively cyclical monotone if and only if there exists a differentiable, exponentially concave function* $\varphi : \Delta_n \to \mathbb{R}$ *which generates* $\boldsymbol{\pi}$ *in the sense of* (15).

Proof Let us provide a sketch of proof to illustrate the main idea. Continuity of $\boldsymbol{\pi}$ is included here only to simplify the statement (in the general case φ is not necessarily differentiable and we need to use supergradients). Suppose $\boldsymbol{\pi}$ is generated by φ. Consider a market weight sequence with $\mu(m) = \mu(0)$. By the decomposition (18), we have

$$\log V_\eta(m) - \log V_\eta(0) = \sum_{t=0}^{m-1} \mathbf{D}^{(1)}\left[\mu(t+1) \mid \mu(t)\right] \geq 0.$$

Thus $V_\eta(m) \geq 1$ and $\boldsymbol{\pi}$ is MCM.

Conversely, suppose that $\boldsymbol{\pi}$ is MCM. Consider the function φ defined by

$$\begin{aligned}\varphi(p) &= \varphi(p_0) + \inf\left\{\log V_\eta(t) - \log V_\eta(0)\right\} \\ &= \varphi(p_0) + \inf\left\{\sum_{s=0}^{t-1} \log\left(\sum_{i=1}^{n} \boldsymbol{\pi}_i(\mu(s)) \frac{\mu_i(s+1)}{\mu_i(s)}\right)\right\},\end{aligned}$$

where $p_0 \in \Delta_n$ is fixed, $\varphi(p_0) \in \mathbb{R}$ is arbitrary, and the infimum is taken over $t \geq 0$ and all market weight sequences $\{\mu(s)\}_{s=0}^{t}$ for which $\mu(0) = p_0$ and $\mu(t) = p$. Then it can be shown that φ is differentiable, exponentially concave, and generates the given portfolio map $\boldsymbol{\pi}$. It can also be shown that the function φ is unique up to an additive constant.

Using this characterization, in [12] we introduced a Monge-Kantorovich optimal transport problem and showed that the optimal coupling can be represented using exponentially concave functions and the portfolios they generate.

4 Generalized Functional Portfolio Generation

4.1 Motivations

As it turns out, Theorem 1 is not the only way to generate a portfolio such that a pathwise decomposition holds. In [15] Karatzas and Ruf introduced a novel *additive* generation and used it to construct relative arbitrages (see [42] for another extension involving an additional finite variation process). The following result uses the terminology of [43, Section 3.3] and adapts the construction to our discrete time setting. We omit the proof as it is contained (in the limit) in Theorem 5 below.

Theorem 3 (Additively generated portfolio [15]) *Let $\varphi : \Delta_n \to (0, \infty)$ be a smooth concave function and let $v_0 \in \mathbb{R}$ be an initial portfolio value. Then there is a self-financing trading strategy η satisfying $V_\eta(0) = v_0$ and*

$$\eta_i(t) = D_{e_i - \mu(t)}\varphi(\mu(t)) + V_\eta(t), \quad i = 1, \ldots, n. \tag{25}$$

Its relative value satisfies the decomposition

$$V_\eta(t) - V_\eta(0) = \varphi(\mu(t)) - \varphi(\mu(0)) + \sum_{s=0}^{t-1} \mathbf{D}^{(0)}\left[\mu(t+1) \mid \mu(t)\right], \tag{26}$$

where $\mathbf{D}^{(0)}[\cdot|\cdot]$ *is the Bregman (or* $L^{(0)}$*) divergence of* φ *as in* (6)*. We call* η *the strategy generated additively by* φ*.*

Note that the additively generated portfolio involves a concave rather than exponentially concave function.

Example 3 Consider the function

$$\varphi(p) = \frac{-1}{2}|p|^2 = \frac{-1}{2}\left(p_1^2 + \cdots + p_n^2\right).$$

It generates the trading strategy $\eta(t)$ given by $\eta_i(t) = |p|^2 - p_i + V_\eta(t)$. It is interesting to note that the Bregman divergence of φ is half of the squared Euclidean distance:

$$\mathbf{D}^{(0)}\left[q \mid p\right] = \frac{1}{2}|p - q|^2.$$

Thus the squared Euclidean distance indeed has a financial meaning for this specific trading strategy.

Observe that both the multiplicative and additive decompositions (18) and (26) can be written in the form

$$g(V_\eta(t)) - g(V_\eta(0)) = \varphi(\mu(t)) - \varphi(\mu(0)) + \mathbf{D}\left[\mu(t+1) \mid \mu(t)\right], \tag{27}$$

where g, φ and $\mathbf{D}[\cdot \mid \cdot]$ are suitable functions:

- (Multiplicative generation) $g(x) = \log x$ and $\mathbf{D}[\cdot \mid \cdot]$ is the $L^{(1)}$-divergence of the exponentially concave function φ.
- (Additive generation) $g(x) = x$ and $\mathbf{D}[\cdot \mid \cdot]$ is the $L^{(0)}$-divergence of the concave function φ.

It is natural to ask if there exist other portfolio constructions that admit pathwise decompositions of the form (27). To formulate this question rigorously we introduce the general concept of divergence.

Definition 5 (*Divergence on* Δ_n) A divergence on Δ_n is a non-negative functional $\mathbf{D}[\cdot \mid \cdot] : \Delta_n \times \Delta_n \to [0, \infty)$ satisfying the following conditions:

(i) $\mathbf{D}[q \mid p] = 0$ if and only if $p = q$.
(ii) It admits a quadratic approximation of the form

$$\mathbf{D}[p + \Delta p \mid p] = \frac{1}{2}\sum_{i,j=1}^{n} g_{ij}(p)\Delta p_i \Delta p_j + O(|\Delta p|^3) \tag{28}$$

as $|\Delta p| \to 0$, and the matrix $G(p) = \big(g_{ij}(p)\big)$ varies smoothly in p and is strictly positive definite in the sense that

$$\sum_{i,j=1}^{n} g_{ij}(p) v_i v_j > 0 \tag{29}$$

for all vectors $v \in \mathbb{R}^n$ that are tangent to Δ_n, i.e., $v_1 + \cdots + v_n = 0$.

If condition (i) is dropped and in (29) we do not strict inequality, we call $\mathbf{D}[\cdot \mid \cdot]$ a pseudo-divergence.

Example 4 Let Hess φ denote the Euclidean Hessian of φ. If $\alpha > 0$ and φ is α-exponentially concave, then its $L^{(\alpha)}$-divergence satisfies

$$\begin{aligned} &\mathbf{D}^{(\alpha)}[p + \Delta p \mid p] \\ &= \frac{-1}{2}(\Delta p)^{\top}\left(\operatorname{Hess}\varphi(p) + \alpha(\nabla\varphi(p))(\nabla\varphi(p))^{\top}\right)(\Delta p) + O(|\Delta p|^3). \end{aligned} \tag{30}$$

If φ is concave, then its $L^{(0)}$-divergence satisfies

$$\mathbf{D}^{(0)}[p + \Delta p \mid p] = \frac{-1}{2}(\Delta p)^{\top} \operatorname{Hess}\varphi(p)(\Delta p) + O(|\Delta p|^3).$$

It is easy to verify that the corresponding matrix $G(p)$ is semi-positive definite. They become true divergences if Hess $e^{\alpha\varphi}$ and Hess φ respectively are strictly positive definite.

Definition 6 (*General functional portfolio construction*) Let $\eta = \{\eta(t)\}_{t=0}^{\infty}$ be a self-financing trading strategy whose relative value process is $\{V_\eta(t)\}$, and let $\varphi, g : \Delta_n \to \mathbb{R}$ be functions on Δ_n where g is strictly increasing. We say that η is generated by φ with scale function g if there exists a pseudo-divergence $\mathbf{D}[\cdot : \cdot]$ on Δ_n such that (27) holds for all market sequences $\{\mu(t)\}_{t=0}^{\infty}$.

In this section we will introduce a new (α, C)-generation, and, after giving an empirical example, show that it characterizes all functional portfolio generation in the sense of Definition 6. For expositional convenience we always assume that the generating function is smooth. Extension of this construction to continuous time is left for future research.

4.2 A New Functional Portfolio Generation

Theorem 4 ((α, C)-generation) *Let $\alpha > 0$ and $C \geq 0$ be fixed parameters, and let $\varphi : \Delta_n \to \mathbb{R}$ be smooth and α-exponentially concave. Then, for any given initial* *value $v_0 \in \mathbb{R}$, there exists a unique self-financing trading strategy $\eta(t)$ such that $V_\eta(0) = v_0$ and*

$$\eta_i(t) = \alpha(C + V_\eta(t)) D_{e_i - \mu(t)} \varphi(\mu(t)) + V_\eta(t), \quad i = 1, \ldots, n, \tag{31}$$

for all t. We call η the strategy which is (α, C)-generated by φ.

Proof We will prove by induction on $T \geq 0$ that the statement holds on the time interval $[0, T]$. Consider $T = 0$. Let $V_\eta(0) = v_0$. If we define

$$\eta_i(0) = \alpha(C + v_0) D_{e_i - \mu(0)} \varphi(\mu(t)) + v_0,$$

then, since $\sum_{i=1}^n \mu_i (e_i - \mu) = 0$, we have

$$\eta(0) \cdot \mu(0) = \alpha(C + v_0) \sum_{i=1}^n \mu_i(0) D_{e_i - \mu(0)} \varphi(\mu(t)) + \sum_{i=1}^n \mu_i(0) v_0 = v_0.$$

Thus there is a unique strategy η which satisfies (31) at time 0.

Suppose by the induction hypothesis that the statement holds up to time T. Then $\eta(T)$ and $V_\eta(T)$ are uniquely defined, and the portfolio value at time $T + 1$ is given uniquely by

$$V_\eta(T + 1) := V_\eta(T) + \eta(T) \cdot (\mu(T + 1) - \mu(T)). \tag{32}$$

Thus there is a unique vector $\eta(T + 1)$ satisfying (31).

It remains to show that the strategy is self-financing at time $T + 1$, i.e., $\eta(T) \cdot \mu(T + 1) = \eta(T + 1) \cdot \mu(T + 1)$ (see (11)). Using (31), we have

$$\begin{aligned} \eta(T + 1) \cdot \mu(T + 1) &= \sum_{i=1}^n \mu_i(T + 1) \alpha(C + V_\eta(T + 1)) D_{e_i - \mu(T+1)} \varphi(\mu(T + 1)) \\ &\quad + \sum_{i=1}^n \mu_i(T + 1) V_\eta(T + 1) \\ &= V_\eta(T + 1) \quad (\text{since} \sum_{i=1}^n \mu_i(T + 1) D_{e_i - \mu(T+1)} = 0) \\ &= V_\eta(T) + \eta(T) \cdot (\mu(T + 1) - \mu(T)) \quad (\text{by (32)}) \\ &= \eta(T) \cdot \mu(T) + \eta(T) \cdot (\mu(T + 1) - \mu(T)) \\ &= \eta(T) \cdot \mu(T + 1). \end{aligned}$$

In the second last equality we used the self-financing property up to time T. This proves that the strategy is uniquely defined at all times.

In Theorem 5 we show that this trading strategy corresponds to the scale function given by

$$g(x) = \frac{1}{\alpha} \log(C + x). \tag{33}$$

Moreover, in Sect. 4.4 we show that up to an additive constant this function (together with $g(x) = x$) is the most general scale function. Comparing (31) with (15) and (25), we see that multiplicative generation corresponds to the case $C = 0$ and $\alpha = 1$, and additive generation corresponds to the limit when $\alpha = \frac{1}{C} \to 0$.

The trading strategy η given by (31) can be interpreted as follows.

Lemma 1 (Portfolio weight of η) *Let $\pi^{(\alpha)}$ be the portfolio process generated multiplicatively by the 1-exponentially concave function $\alpha\varphi$. If $V_\eta(t) > 0$, the portfolio weight vector $\pi(t)$ of the (α, C)-generated trading strategy η is given by*

$$\pi(t) = \left(\frac{\eta_1(t)\mu_1(t)}{V_\eta(t)}, \ldots, \frac{\eta_n(t)\mu_n(t)}{V_\eta(t)} \right) = \frac{C + V_\eta(t)}{V_\eta(t)} \pi^{(\alpha)}(t) - \frac{C}{V_\eta(t)} \mu(t). \tag{34}$$

In particular, $\eta(t)$ longs the multiplicatively generated portfolio $\pi^{(\alpha)}$ and shorts the market portfolio with weights depending on $V_\eta(t)$ and C.

Proof Direct computation using (31).

By increasing C, we may construct portfolios that are more aggressive than the multiplicatively generated portfolio. Note that we keep the parameter α so that we can generate different portfolios with the same generating function φ (as long as $e^{\alpha\varphi}$ is concave).

Next we show that the new portfolio generation admits a pathwise decomposition for the portfolio value.

Theorem 5 (Pathwise decomposition) *Consider an (α, C)-generated trading strategy η as in Theorem 4. If $V_\eta(\cdot) > -C$, then the value process satisfies the pathwise decomposition*

$$\frac{1}{\alpha} \log \frac{C + V_\eta(t)}{C + V_\eta(0)} = \varphi(\mu(t)) - \varphi(\mu(0)) + \sum_{s=0}^{t-1} \mathbf{D}^{(\alpha)} \left[\mu(s+1) \mid \mu(s) \right], \tag{35}$$

where $\mathbf{D}^{(\alpha)}$ is the $L^{(\alpha)}$-divergence of φ.

Proof The proof is similar to that of Theorem 1. By (31), for each time t we have

$$\begin{aligned}
&\frac{1}{\alpha} \log(C + V_\eta(t+1)) - \frac{1}{\alpha} \log(C + V_\eta(t)) \\
&= \frac{1}{\alpha} \log \frac{C + V_\eta(t) + \alpha(C + V_\eta(t))\nabla\varphi(\mu(t)) \cdot (\mu(t+1) - \mu(t))}{C + V_\eta(t)} \\
&= \frac{1}{\alpha} \log \left(1 + \alpha\nabla\varphi(\mu(t)) \cdot (\mu(t+1) - \mu(t))\right) \\
&= \varphi(\mu(t+1)) - \varphi(\mu(t)) + \mathbf{D}^{(\alpha)} \left[\mu(t+1) \mid \mu(t) \right].
\end{aligned} \tag{36}$$

This yields the desired decomposition. The condition $V_\eta(\cdot) > -C$ is imposed so that the logarithms make sense.

4.3 An Empirical Example

Consider a smooth and exponentially concave function φ. It is α-exponentially concave for all $0 < \alpha \leq 1$ and is concave (which corresponds to the case $\alpha \downarrow 0$). Thus both the additive and multiplicatively generated portfolios are well-defined. Unfortunately, while the $L^{(\alpha)}$-divergence is a natural interpolation, there does not seem to be a canonical choice for the constant C that connects the two basic cases.

In this example we consider instead the parameterized family $\{\eta^{(\alpha)}\}_{0 \leq \alpha \leq 1}$ where $\eta^{(\alpha)}$ is the trading strategy $(\alpha, \frac{1}{\alpha})$-generated by φ (so when $\alpha = 0$ it is the additively generated portfolio), and compare their empirical performance. We set $V_{\eta^{(\alpha)}}(0) = 1$ for all α. Note that $\eta^{(1)}$ is not the multiplicatively generated portfolio as it also shorts the market portfolio.

Consider as in Fig. 1 the (beginning) monthly stock prices of the US companies Ford, Walmart and IBM from January 1990 ($t = 0$) to September 2017 ($t = 332$). We normalize the prices so that at $t = 0$ the market weight is at the barycenter $\left(\frac{1}{3}, \frac{1}{3}, \frac{1}{3}\right)$. The path of the market weight $\mu(t)$ in the simplex Δ_3 is plotted in Fig. 1 (right).

We consider the 1-exponentially concave function

$$\varphi(p) = \sum_{i=1}^{3} \frac{1}{3} \log p_i \tag{37}$$

which generates multiplicatively the equal-weighted portfolio $\boldsymbol{\pi}(p) \equiv \overline{e} = \left(\frac{1}{3}, \frac{1}{3}, \frac{1}{3}\right)$. By (31), for each $\alpha \in [0, 1]$ the trading strategy is given by

$$\eta_i^{(\alpha)}(t) = \left(1 + \alpha V_\eta(t)\right)\left(\frac{1}{3\mu_i(t)} - 1\right) + V_\eta(t).$$

In terms of portfolio weights, we have

$$\pi^{(\alpha)}(t) = \frac{1 + \alpha V_\eta(t)}{V_\eta(t)} \overline{e} - \frac{1 + \alpha V_\eta(t) - V_\eta(t)}{V_\eta(t)} \mu(t).$$

Thus the portfolio longs more and more the equal-weighted portfolio as α increases. The corresponding $L^{(\alpha)}$-divergence is given by

$$\mathbf{D}^{(\alpha)}\left[q \mid p\right] = \frac{1}{\alpha} \log\left(1 + \alpha \sum_{i=1}^{n} \frac{1}{np_i}(q_i - p_i)\right) - \sum_{i=1}^{n} \frac{1}{n} \log \frac{q_i}{p_i}.$$

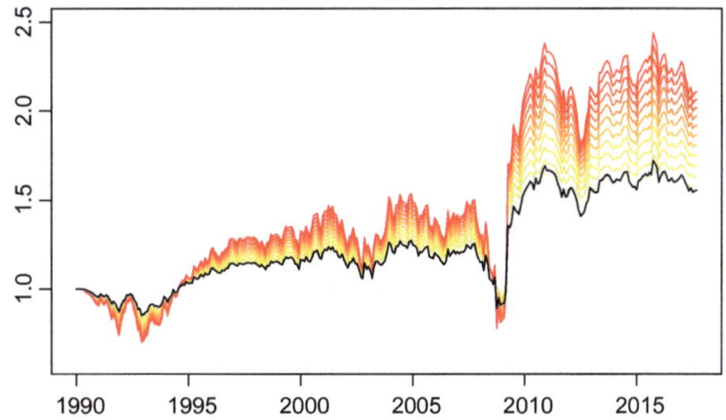

Fig. 5 Time series of the relative portfolio value $V_{\eta^{(\alpha)}}(t)$, from $\alpha = 0$ (yellow) to $\alpha = 1$ (red). The value of the equal-weighted portfolio is shown in black

The relative values of the simulated portfolios are plotted in Fig. 5. At the end of the period the portfolio value is increasing in α, and the additive portfolio ($\alpha = 0$) has the smallest value. It is interesting to note that the reverse is true at the beginning. Note that the values fluctuate widely in the period 2008–2009 corresponding to the financial crisis. For comparison, we also simulate the multiplicatively generated equal-weighted portfolio (i.e., $(\alpha, C) = (1, 0)$) and plot the result in Fig. 5. Surprisingly the additive and multiplicative portfolios have similar behaviors here. In this period, shorting the market by using a positive value for C gives significant advantage over both the additive and multiplicative portfolios. Dynamic optimization over our extended functionally generated portfolios is an interesting problem.

4.4 Characterizing Functional Portfolio Generation

Now we show that our (α, C)-generation is the most general one. Throughout this subsection we let η be a functionally generated trading strategy as in Definition 6. We assume that the scale function g is smooth and $g'(x) > 0$ for all x. We also require that the domain of g contains the positive real line $(0, \infty)$. Furthermore, we assume that φ is smooth, and η is non-trivial in the sense that for all $t \geq 0$ and all market weight paths $\{\mu(s)\}_{s=0}^{t}$ up to time t, the profit-or-loss

$$V_\eta(t+1) - V_\eta(t) = \eta(t) \cdot (\mu(t+1) - \mu(t))$$

is not identically zero as a function of $\mu(t+1) \in \Delta_n$. For technical reasons we also assume that for any $x > 0$, there exists $t \geq 0$ and a sequence $\{\mu(s)\}_{s=0}^{t}$ such that $V_\eta(t) = x$.

Theorem 6 *Under the above conditions, the scale function has one of the following forms. Either*

$$g(x) = c_1 x + c_2 \tag{38}$$

where $c_1 > 0$ *and* $c_2 \in \mathbb{R}$, *or*

$$g(x) = c_2 \log(c_1 + x) + c_3 \tag{39}$$

where $c_1 \geq 0$, $c_2 > 0$ and $c_3 \in \mathbb{R}$. In the first case φ is concave and η is additively generated by φ, whereas in the second case φ is α-exponentially concave with $c_2 = \frac{1}{\alpha}$ and η is (α, c_1)-generated by φ. The corresponding pseudo-divergence is the $L^{(\alpha)}$-divergence of φ.

Note that in (38) and (39) the additive constants are irrelevant and may be discarded. We will prove Theorem 6 with several lemmas. First we observe that the decomposition (27) already implies a formula of the trading strategy.

Lemma 2 *For any t and any tangent vector v of Δ_n (i.e., $v_1 + \cdots + v_n = 0$) we have*

$$\eta(t) \cdot v = \frac{1}{g'(V_\eta(t))} \nabla\varphi(\mu(t)) \cdot v. \tag{40}$$

In particular, for each $i = 1, \ldots, n$ we have

$$\eta_i(t) = \frac{1}{g'(V_\eta(t))} D_{e_i - \mu(t)}\varphi(\mu(t)) + V_\eta(t). \tag{41}$$

Proof From (27) we have the identity

$$g(V_\eta(t+1)) - g(V_\eta(t)) = \varphi(\mu(t+1)) - \varphi(\mu(t)) + \mathbf{D}\left[\mu(t+1) \mid \mu(t)\right] \tag{42}$$

which holds for all values of $\mu(t+1)$. Write

$$V_\eta(t+1) = V_\eta(t) + \eta(t) \cdot (\mu(t+1) - \mu(t))$$

and let $\mu(t+1) - \mu(t) = \delta v$, $\delta > 0$ sufficiently small, and compute the first order approximation of both sides of (42). Since $\mathbf{D}[\cdot \mid \cdot]$ is a pseudo-divergence, by (28) its first order approximation vanishes. Evaluating the derivatives and dividing by $\delta > 0$, we obtain (40).

Letting $v = e_i - \mu(t)$ in (41) for $i = 1, \ldots, n$, we get the formula (41).

Observe that (40) reduces to (25) when $g(x) = x$, and to (15) when $g(x) = \log x$. Also, we note that $\eta(t)$ depends only on $\mu(t)$ and the current portfolio value $V_\eta(t)$. Putting $v = \mu(t+1) - \mu(t)$ in (40), we have

$$V_\eta(t+1) - V_\eta(t) = \frac{1}{g'(V_\eta(t))} \nabla\varphi(\mu(t)) \cdot (\mu(t+1) - \mu(t)). \tag{43}$$

Consider the expression

$$\begin{aligned} & g(V_\eta(t+1)) - g(V_\eta(t)) \\ & = g\left(V_\eta(t) + \left[V_\eta(t+1) - V_\eta(t)\right]\right) - g(V_\eta(t)) \\ & = g\left(V_\eta(t) + \frac{1}{g'(V_\eta)} \nabla\varphi(\mu(t)) \cdot (\mu(t+1) - \mu(t))\right) - g(V_\eta(t)). \end{aligned} \tag{44}$$

By (27), this equals

$$\varphi(\mu(t+1)) - \varphi(\mu(t)) + \mathbf{D}[\mu(t+1)|\mu(t)],$$

which is a function of $\mu(t)$ and $\mu(t+1)$ only. Thus, the expression in (44) does not depend on the current portfolio value $V_\eta(t)$. From this observation we will derive a differential equation satisfied by g.

Lemma 3 *The scale function g satisfies the third order nonlinear ODE*

$$g' g''' = 2(g'')^2 \tag{45}$$

on the positive real line $(0, \infty)$.

Proof Given $x > 0$, write $x = V_\eta(t)$ for some $t \geq 0$ and market sequence $\{\mu(s)\}_{s=0}^t$. Let $\delta = \nabla\varphi(\mu(t)) \cdot (\mu(t+1) - \mu(t))$. From (44), for any δ, the expression

$$g(x + \frac{1}{g'(x)}\delta) - g(x) \tag{46}$$

does not depend on x.

Differentiating (46) with respect to x, we have

$$g'(x + \frac{1}{g'(x)}\delta)\left(1 - \delta\frac{g''(x)}{(g'(x))^2}\right) - g'(x) = 0.$$

Next we differentiate with respect to δ (since η is assumed to be non-trivial, this can be done by varying $\mu(t+1)$):

$$g''(x + \frac{1}{g'(x)}\delta)\frac{1}{g'(x)}\left(1 - \delta\frac{g''(x)}{(g'(x))^2}\right) + g'(x + \frac{1}{g'(x)}\delta)\frac{-g''(x)}{(g'(x))^2} = 0.$$

Differentiating one more time with respect to δ, we have

$$\begin{aligned} & g'''(x + \frac{1}{g'(x)}\delta)\frac{1}{(g'(x))^2}\left(1 - \delta\frac{g''(x)}{(g'(x))^2}\right) \\ & \quad + g''(x + \frac{1}{g'(x)}\delta)\frac{-g''(x)}{(g'(x))^3} - \frac{g''(x + \frac{1}{g'(x)}\delta)g''(x)}{(g'(x))^3} = 0. \end{aligned}$$

Setting $\delta = 0$, we get $\frac{g'''(x)}{(g'(x))^2} - 2\frac{(g''(x))^2}{(g'(x))^3} = 0$ which gives the ODE (45). (Note that we assumed that $g'(x) > 0$ for all x.)

With the differential equation in hand, it is not difficult to find the general solutions. They can be verified using direct substitution and the local uniqueness of the autonomous equation.

Lemma 4 *All solutions to the ODE* (45) *can be written in the form*

$$g(x) = c_0 + c_1 x \quad or \quad g(x) = c_2 \log(c_1 + x) + c_3, \tag{47}$$

where the c_i's are real constants. The constraints on the constants stated in Theorem 6 follow from our assumptions of g.

Now we are ready to complete the proof of Theorem 6. By Lemma 4, the scale function (up to an additive constant which is irrelevant) has the form (38) or (39). Consider the second case (the first case is similar). Then by (43), (44) and the third equality of (36) (which does not depend on α-exponential concavity of φ), for any $p = \mu(s)$ and $q = \mu(s+1)$, we have

$$\mathbf{D}[q \mid p] = \frac{1}{\alpha} \log(1 + \alpha \nabla\varphi(p) \cdot (q - p)) - (\varphi(q) - \varphi(p))$$

which is exactly the expression of the $L^{(\alpha)}$-divergence. By assumption $\mathbf{D}[\cdot \mid \cdot]$ is a pseudo-divergence so it is non-negative for all p, q. It is easy to check that this implies that φ is α-exponentially concave, and so η is the (α, c_1)-generated trading strategy.

5 Further Properties of L-divergence

In this section we gather some further properties of L-divergence and describe some related problems. For simplicity we focus on the $L^{(1)}$-divergence, and refer the reader to [14] for a systematic study of the $L^{(\alpha)}$-divergence. We always assume that the generating functions are smooth. It is clear that some of the problems make sense on domains other than the unit simplex.

5.1 Interpolation and Comparison

If $\varphi^{(0)}$ and $\varphi^{(1)}$ are exponentially concave functions on Δ_n, by the inequality of the arithmetic and geometric means, we have that

$$\varphi^{(\lambda)} := (1 - \lambda)\varphi^{(0)} + \lambda\varphi^{(1)}$$

is exponentially concave for any $0 < \lambda < 1$. If $\pi^{(0)}$ and $\pi^{(1)}$ are the portfolio maps generated multplicatively by $\varphi^{(0)}$ and $\varphi^{(1)}$ respectively, then $\varphi^{(\lambda)}$ generates the portfolio map

$$\pi^{(\lambda)}(\cdot) \equiv (1-\lambda)\pi^{(0)}(\cdot) + \lambda\pi^{(1)}(\cdot),$$

which is a constant-weighted portfolio of $\pi^{(0)}$ and $\pi^{(1)}$ [24, Lemma 4.4]. Thus, the spaces of exponentially concave functions and MCM portfolio maps (see Definition 3) are convex. Moreover, the L-divergence $\mathbf{D}^{(1)}[\cdot \mid \cdot]$ is concave in the function φ. In [13], this interpolation provides a new displacement interpolation for a logarithmic optimal transport problem.

Let $\mathbf{D}_\varphi$ and $\mathbf{D}_\psi$ be the L-divergences generated respectively by the exponentially concave functions φ and ψ, we say that $\mathbf{D}_\psi$ *dominates* $\mathbf{D}_\varphi$ if

$$\mathbf{D}_\psi^{(1)}[q \mid p] \geq \mathbf{D}_\varphi^{(1)}[q \mid p] \tag{48}$$

for all $p, q \in \Delta_n$. Financially, this means that the portfolio generated by ψ captures more volatility than the one generated by φ. An interesting problem is to find the maximal elements in this partial order, and the following result is obtained in [24] using the relative convexity lemma in [44].

Theorem 7 *Suppose φ is symmetric, i.e., $\varphi(p_1, \ldots, p_n) = \varphi(p_{\sigma(1)}, \ldots, p_{\sigma(n)})$ for any permutation σ of the coordinates. If*

$$\int_0^1 e^{-2\varphi((1-t)e_1 + t\bar{e})} dt = \infty,$$

then $\mathbf{D}_\varphi$ is maximal in the partial order (48)*: if $\mathbf{D}_\varphi$ is dominated by $\mathbf{D}_\psi$, then $\varphi - \psi$ is constant on Δ_n and $\mathbf{D}_\psi \equiv \mathbf{D}_\varphi$.*

As an example, the function $\varphi(p) = \frac{1}{n}\sum_{i=1}^n \log p_i$ (which generates the equal-weighted portfolio) is maximal. Another example is $\varphi(p) = \log\left(-\sum_{i=1}^n p_i \log p_i\right)$, the logarithm of the Shannon entropy.

Note that Theorem 7 is concerned with the global properties of the generating function. One can also study maximal exponentially concave functions over a local neighborhood; this idea is used in [18] to construct short term relative arbitrages.

5.2 Dualistic Geometry and the Generalized Pythagorean Theorem

Consider the $L^{(1)}$-divergence $\mathbf{D}[\cdot \mid \cdot]$ of an exponentially concave function φ on the simplex Δ_n. It defines a Riemannian metric g (as in (30)) and a dual pair of torsion-free affine connections (∇, ∇^*) (see [10, Chapter 6]). In [13] this geometry is derived and many interesting properties are shown. It generalizes the dually flat geometry of Bregman divergence (a unified framework is established recently in [14]).

Fig. 6 Generalized Pythagorean theorem for L-divergence

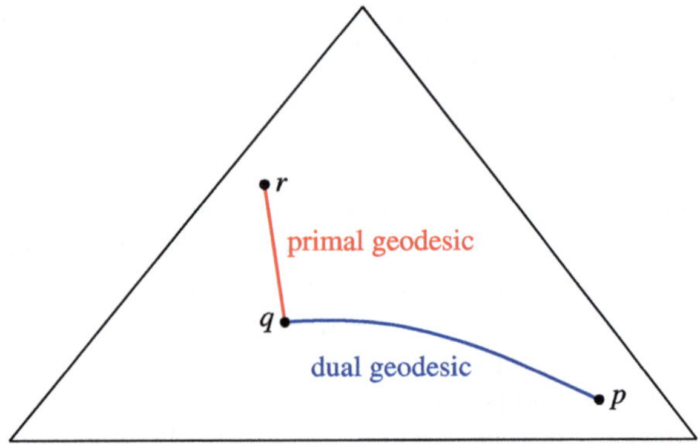

As before we let p denote a generic element of Δ_n. Now we regard this a global coordinate system of the manifold $M = \Delta_n$, and call it the *primal* coordinate system. Let $\boldsymbol{\pi}$ be the portfolio map generated by φ. It defines a *dual* coordinate system

$$p^* := \left(\frac{\pi_1(p)/p_1}{\sum_{j=1}^n \pi_j(p)/p_j}, \ldots, \frac{\pi_n(p)/p_n}{\sum_{j=1}^n \pi_j(p)/p_j} \right)$$

which also takes values in the unit simplex. The main properties of the geometry are summarized in the following theorem, and we refer the reader to [13, 14] for further properties related to the geodesic equations, gradient flows and connections with optimal transport.

Theorem 8 [13] *Consider the dualistic geometry induced by* $\mathbf{D}[\cdot \mid \cdot]$.

(i) The trace of a primal geodesic is a straight line under the primal coordinate system.
(ii) The trace of a dual geodesic is a straight line under the dual coordinate system.
(iii) The geometry has constant (primal and dual) sectional curvature -1 *(when* $n \geq 3$*) with respect to the induced Riemannian metric.*

In particular, the induced geometry is dually *projectively* flat but not flat. Furthermore, the $L^{(1)}$-divergence satisfies a generalized Pythagorean theorem.

Theorem 9 (Generalized Pythagorean theorem) *Given* $(p, q, r) \in (\Delta_n)^3$, *consider the dual geodesic joining* q *and* p *and the primal geodesic joining* q *and* r. *Consider the Riemannian angle between the geodesics at* q. *Then the difference*

$$\mathbf{D}[q \mid p] + \mathbf{D}[r \mid q] - \mathbf{D}[r \mid p] \tag{49}$$

is positive, zero or negative depending on whether the angle is less than, equal to, or greater than 90 *degrees (see Fig. 6).*

Further properties of the Pythagorean theorem can be studied. To give a flavor we present an interesting result for the excess growth rate $\mathbf{T}_\pi[\cdot \mid \cdot]$ defined by (5).

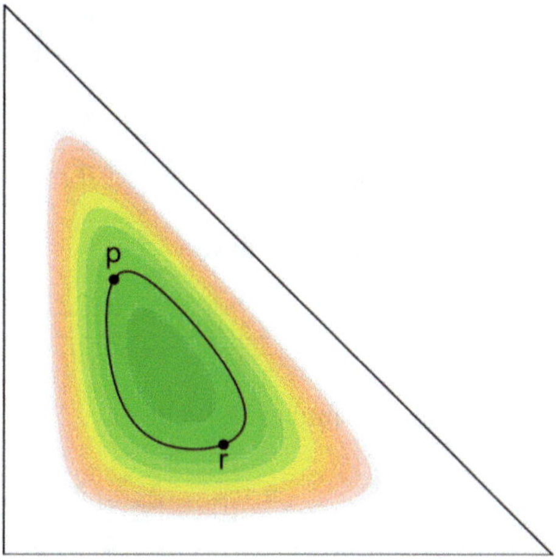

Fig. 7 The level sets of the map $q \mapsto \mathbf{T}_\pi[q \mid p] + \mathbf{T}_\pi[r \mid q] - \mathbf{T}_\pi[r \mid p]$ where p and r are fixed. Here $\pi = \overline{e}$ is the equal-weighted portfolio

Proposition 3 *Consider the excess growth rate $\mathbf{T}_\pi[\cdot \mid \cdot]$ for a fixed portfolio weight vector $\pi \in \overline{\Delta_n}$. For $p, r \in \Delta_n$ fixed, the mapping*

$$q \in \Delta_n \mapsto f(q) := \mathbf{T}_\pi[q \mid p] + \mathbf{T}_\pi[r \mid q] - \mathbf{T}_\pi[r \mid p] \tag{50}$$

is quasiconvex, i.e, the sublevel sets $\{q : f(q) \leq \lambda\}$ are convex (see Fig. 7).

Proof It suffices to show that the map

$$g(x) := \log\left(\pi \cdot \frac{x}{p}\right) + \log\left(\pi \cdot \frac{r}{x}\right)$$

is quasiconvex on Δ_n. We will use the following characterization of quasiconvex functions (see [45, Section 3.4.3]): g is quasiconvex if and only if

$$g(y) \leq g(x) \Rightarrow \nabla g(x) \cdot (y - x) \leq 0. \tag{51}$$

for any x and y.

Let $x, y \in \Delta_n$ be such that $g(y) \leq g(x)$. We have

$$\partial_i g(x) = \frac{\frac{\pi_i}{p_i}}{\pi \cdot \frac{x}{p}} + \frac{-\frac{\pi_i r_i}{x_i^2}}{\pi \cdot \frac{r}{x}}.$$

After some simplifications, we have

$$\begin{aligned}\nabla g(x) \cdot (y - x) &= \sum_{i=1}^{n} \left(\frac{\frac{\pi_i}{p_i}}{\pi \cdot \frac{x}{p}} + \frac{-\frac{\pi_i r_i}{x_i^2}}{\pi \cdot \frac{r}{x}} \right) (y_i - x_i) \\ &= \frac{\pi \cdot \frac{y}{p}}{\pi \cdot \frac{x}{p}} - \frac{\sum_{i=1}^{n} \pi_i \frac{r_i}{x_i} \frac{y_i}{x_i}}{\pi \cdot \frac{r}{x}}.\end{aligned} \tag{52}$$

Since $g(y) \leq g(x)$, we have

$$\frac{\pi \cdot \frac{x}{p}}{\pi \cdot \frac{x}{p}} \leq \frac{\pi \cdot \frac{r}{x}}{\pi \cdot \frac{r}{y}}.$$

Substituting this into (52), we get

$$\begin{aligned}\nabla g(x) \cdot (y - x) &\leq \frac{\pi \cdot \frac{r}{x}}{\pi \cdot \frac{r}{y}} - \frac{\sum_{i=1}^{n} \pi_i \frac{r_i}{x_i} \frac{y_i}{x_i}}{\pi \cdot \frac{r}{x}} \\ &= \frac{1}{\left(\pi \cdot \frac{r}{x}\right)\left(\pi \cdot \frac{r}{y}\right)} \left[\left(\pi \cdot \frac{r}{x}\right)^2 - \sum_{i,j=1}^{n} \pi_i \pi_j r_i r_j \frac{1}{x_i x_j} \frac{y_j}{x_j} \frac{x_i}{y_i} \right] \\ &= \frac{1}{\left(\pi \cdot \frac{r}{x}\right)\left(\pi \cdot \frac{r}{y}\right)} \sum_{i,j=1}^{n} \frac{\pi_i r_i}{x_i} \frac{\pi_j r_j}{x_j} \left(1 - \frac{x_i}{y_i} \frac{y_j}{x_j}\right).\end{aligned} \tag{53}$$

Let $A = \sum_{i=1}^{n} \frac{\pi_i r_i}{x_i}$ and let $\alpha_i = \frac{\pi_i r_i}{x_i} / A$. Note that α is a probability vector. Now we may write (53) in the form

$$C \left(1 - \sum_{i,j=1}^{n} \alpha_i \alpha_j \frac{x_i}{y_i} \frac{y_j}{x_j} \right),$$

where $C > 0$ is a constant. Let X and Y be independent and identically distributed random variables such that

$$\mathbb{P}\left(X = \frac{x_i}{y_i} \right) = \mathbb{P}\left(Y = \frac{x_i}{y_i} \right) = \alpha_i, \quad i = 1, \ldots, n.$$

By Jensen's inequality, we have

$$\sum_{i,j=1}^{n} \alpha_i \alpha_j \frac{x_i}{y_i} \frac{y_j}{x_j} = \mathbb{E}\left[X \cdot \frac{1}{Y} \right] = \mathbb{E}[X] \mathbb{E}\left[\frac{1}{Y} \right] \geq \mathbb{E}[X] \frac{1}{\mathbb{E}[X]} = 1.$$

Thus $\nabla g(\mathbf{x}) \cdot (y - x) \leq 0$ and we have proved that g is quasiconvex.

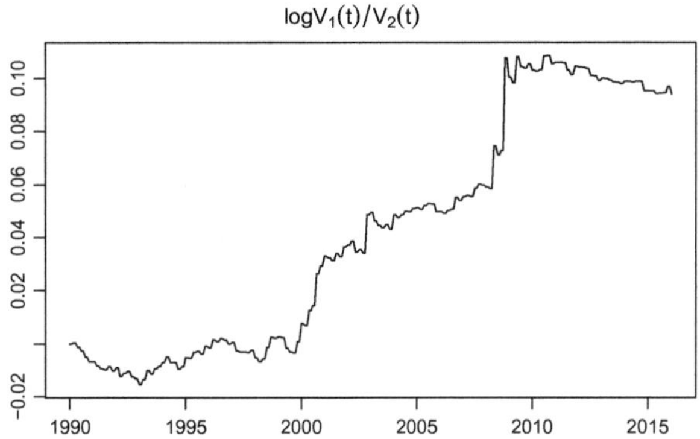

Fig. 8 Relative performance of the portfolio V_1 (rebalanced every month) versus V_2 (rebalanced every two months)

5.3 *Optimal Rebalancing Frequency*

Finally we mention a practical problem related to the L-divergence. The generalized Pythagorean theorem gives a geometric way to study the rebalancing frequency of a functionally generated portfolio. To give a simple example, consider the empirical example as in Fig. 1 and the equal-weighted portfolio $\boldsymbol{\pi}(p) \equiv \overline{e} = \left(\frac{1}{3}, \frac{1}{3}, \frac{1}{3}\right)$.

Consider two ways of implementing this portfolio: (i) rebalance every month; (ii) rebalance every two months. If V_1 and V_2 denote respectively the values of these implementations, Fig. 8 plots the time series of the relative performance $\log V_1(t)/V_2(t)$. For this data set, rebalancing every month boost the return by about 10% in log scale. Using the decomposition (18), we see that if T is the terminal time, then

$$\log \frac{V_1(T)}{V_2(T)}$$
$$= \sum_k \left(\mathbf{D}\left[\mu(2k+1) \mid \mu(2k)\right] + \mathbf{D}\left[\mu(2k+1) \mid \mu(2k+2)\right] - \mathbf{D}\left[\mu(2k+2) \mid \mu(k)\right]\right).$$

By Theorem 9, the sign of each term is determined by the Riemannian angle of the geodesic triangle. This angle summarizes in a single number the correlation among the stock returns that is relevant to the rebalancing frequency. Further work should study the joint relationship between the angle and the *size* of the geodesic triangle which determines the magnitude of (49).

In practice trading incurs transaction costs which have been neglected so far. Transaction costs create a drag of the portfolio value. A common setting is that the transaction cost is *proportional*, i.e., we pay a fixed percentage of the value exchanged. In our model, the transaction cost may be approximated by a functional $\mathbf{C}\left[q \mid p\right] \geq 0$ where p and q are the beginning and ending market weights of the holding period (Fig. 9). Since the transaction cost is proportional, the cost is of linear order when $q \approx p$. On the other hand, the L-divergence is approximately quadratic when $q \approx p$. Thus the net difference $\mathbf{D}\left[q \mid p\right] - \mathbf{C}\left[q \mid p\right]$ is negative when q is sufficiently close to p. Financially, this means that the investor should not rebalance

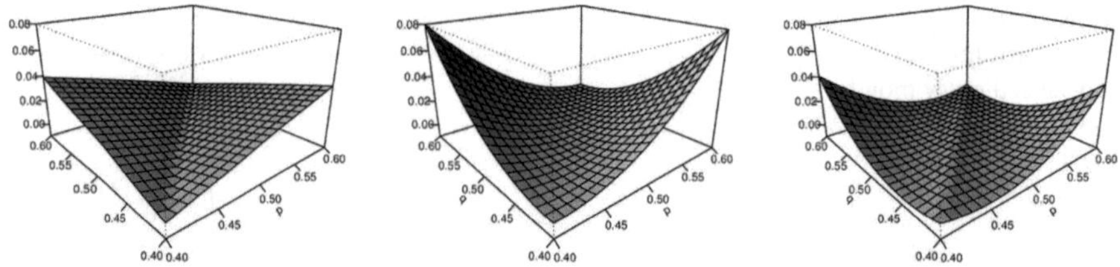

Fig. 9 Left: Transaction cost $\mathbf{C}[q \mid p]$. Middle: The L-divergence $\mathbf{D}[q \mid p]$. Right: The difference $\mathbf{D} - \mathbf{C}$. In these figures $n = 2$ and the x and y-axes are the first coordinates of p and q respectively

too often – at least when the increments of the market weights are 'small'. We end this paper with the following problem:

Problem 1 Design a robust strategy for rebalancing a given trading strategy.

References

1. Amari, S., Nagaoka, H., Harada, D.: Methods of Information Geometry (Translations of Mathematical Monographs). American Mathematical Society, Providence (2002)
2. Wong, T.-K.L.: On portfolios generated by optimal transport. arXiv preprint arXiv:1709.03169 (2017)
3. Fernholz, E.R.: Stochastic Portfolio Theory. Springer, Berlin (2002)
4. Fernholz, E.R., Karatzas, I.: Stochastic portfolio theory: an overview. In: Ciarlet, P.G. (ed.) Handbook of Numerical Analysis, vol. 15, pp. 89–167. Elsevier, Amsterdam (2009)
5. Fernholz, E.R., Karatzas, I., Ruf, J.: Volatility and arbitrage. Ann. Appl. Probab. **28**(1), 378–417 (2018)
6. Fernholz, R., Garvy, R., Hannon, J.: Diversity-weighted indexing. J. Portf. Manag. **24**(2), 74–82 (1998)
7. Booth, D.G., Fama, E.F.: Diversification returns and asset contributions. Financ. Anal. J. **48**(3), 26–32 (1992)
8. Eguchi, S.: Second order efficiency of minimum contrast estimators in a curved exponential family. Ann. Stat. **11**(3), 793–803 (1983)
9. Eguchi, S.: Geometry of minimum contrast. Hiroshima Math. J. **22**(3), 631–647 (1992)
10. Amari, S.-I.: Information Geometry and Its Applications. Springer, Berlin (2016)
11. Fenholz, R.: Portfolio generating functions. Quantitative Analysis in Financial Markets: Collected Papers of the New York University Mathematical Finance Seminar, vol. 1, pp. 344–367. World Scientific, Singapore (1999)
12. Pal, S., Wong, T.-K.L.: The geometry of relative arbitrage. Math. Financ. Econ. **10**(3), 263–293 (2016)
13. Pal, S., Wong, T.-K.L.: Exponentially concave functions and a new information geometry. Ann. Probab. **46**(2), 1070–1113 (2018)
14. Wong, T.-K.L.: Logarithmic divergences from optimal transport and Rényi geometry. Inf. Geom. **1**(1), 39–78 (2018)
15. Karatzas, I., Ruf, J.: Trading strategies generated by Lyapunov functions. Financ. Stoch. **21**(3), 753–787 (2017)
16. Brody, D.C., Hughston, L.P.: Interest rates and information geometry. In: Proceedings of the Royal Society of London A: Mathematical, Physical and Engineering Sciences, vol. 457, pp. 1343–1363. The Royal Society (2001)

17. Trivellato, B.: Deformed exponentials and applications to finance. Entropy **15**(9), 3471–3489 (2013)
18. Pal, S.: Embedding optimal transports in statistical manifolds. Indian J. Pure Appl. Math. **48**(4), 541–550 (2017)
19. Khashanah, K., Yang, H.: Evolutionary systemic risk: fisher information flow metric in financial network dynamics. Phys. A Stat. Mech. Appl. **445**, 318–327 (2016)
20. Nock, R., Magdalou, B., Briys, E., Nielsen, F.: On tracking portfolios with certainty equivalents on a generalization of Markowitz model: the fool, the wise and the adaptive. In: Proceedings of the 28th International Conference on Machine Learning (ICML-11), pp. 73–80 (2011)
21. Nock, R., Magdalou, B., Briys, E., Nielsen, F.: Mining matrix data with Bregman matrix divergences for portfolio selection. Matrix Information Geometry, pp. 373–402. Springer, Berlin (2013)
22. Breuer, T., Csiszár, I.: Information geometry in mathematical finance: model risk, worst and almost worst scenarios. In: 2013 IEEE International Symposium on Information Theory Proceedings (ISIT). IEEE (2013)
23. Pal, S., Wong, T.-K.L.: Energy, entropy, and arbitrage. arXiv preprint arXiv:1308.5376 (2013)
24. Wong, T.-K.L.: Optimization of relative arbitrage. Ann. Financ. **11**(3–4), 345–382 (2015)
25. Pal, S.: Exponentially concave functions and high dimensional stochastic portfolio theory. arXiv preprint arXiv:1603.01865 (2016)
26. Fernholz, R., Karatzas, I.: Relative arbitrage in volatility-stabilized markets. Ann. Financ. **1**(2), 149–177 (2005)
27. Banner, A.D., Fernholz, R., Karatzas, I.: Atlas models of equity markets. Ann. Appl. Probab. **15**(4), 2296–2330 (2005)
28. Dembo, A., Tsai, L.-C.: Equilibrium fluctuation of the Atlas model. Ann. Probab. **45**(6b), 4529–4560 (2017)
29. Fernholz, R., Ichiba, T., Karatzas, I.: A second-order stock market model. Ann. Financ. **9**(3), 439–454 (2013)
30. Ichiba, T., Pal, S., Shkolnikov, M.: Convergence rates for rank-based models with applications to portfolio theory. Probab. Theory Relat. Fields **156**(1–2), 415–448 (2013)
31. Ichiba, T., Papathanakos, V., Banner, A., Karatzas, I., Fernholz, R.: Hybrid Atlas models. Ann. Appl. Probab. **21**(2), 609–644 (2011)
32. Jourdain, B., Reygner, J.: Capital distribution and portfolio performance in the mean-field atlas model. Ann. Financ. **11**(2), 151–198 (2015)
33. Pal, S.: Analysis of market weights under volatility-stabilized market models. Ann. Appl. Probab. **21**(3), 1180–1213 (2011)
34. Banner, A.D., Fernholz, D.: Short-term relative arbitrage in volatility-stabilized markets. Ann. Financ. **4**(4), 445–454 (2008)
35. Fernholz, D., Karatzas, I.: On optimal arbitrage. Ann. Appl. Probab. **20**(4), 1179–1204 (2010)
36. Fernholz, R., Karatzas, I., Kardaras, C.: Diversity and relative arbitrage in equity markets. Financ. Stoch. **9**(1), 1–27 (2005)
37. Vervuurt, A., Karatzas, I.: Diversity-weighted portfolios with negative parameter. Ann. Financ. **11**(3–4), 411–432 (2015)
38. Ernst, P.A., Thompson, J.R., Miao, Y.: Tukeys transformational ladder for portfolio management. Financ. Mark. Portf. Manag. **31**(3), 317–355 (2017)
39. Monter, S.A.A., Shkolnikov, M., Zhang, J.: Dynamics of observables in rank-based models and performance of functionally generated portfolios. arXiv preprint arXiv:1802.03593 (2018)
40. Audrino, F., Fernholz, R., Ferretti, R.G.: A forecasting model for stock market diversity. Ann. Financ. **3**(2), 213–240 (2007)
41. Rockafellar, R.T.: Convex Analysis. Princeton Landmarks in Mathematics. Princeton University Press, New Jercy (1997)
42. Johannes, R. Kangjianan, X.: Generalised lyapunov functions and functionally generated trading strategies. arXiv preprint arXiv:1801.07817 (2018)
43. Vervuurt, A.: On portfolio construction through functional generation. Ph.D. thesis, Oxford Unviersity (2016)

44. Chuaqui, M., Duren, P., Osgood, B.: Schwarzian derivative criteria for valence of analytic and harmonic mappings. In: Mathematical Proceedings of the Cambridge Philosophical Society, vol. 143, pp. 473–486. Cambridge University Press, Cambridge (2007)
45. Boyd, S., Vandenberghe, L.: Convex Optimization. Cambridge University Press, Cambridge (2004)

Generalising Frailty Assumptions in Survival Analysis: A Geometric Approach

Vahed Maroufy and Paul Marriott

Abstract This paper uses Information Geometry in a practical and important applied statistical context: Cox regression in survival analysis. We explore the geometry of the corresponding model space including its potentially complex boundary. The exact manner that frailty terms in Cox's hazard model are specified has important implications for modelling. For example, it is very common to assume a gamma frailty for reasons of mathematical tractability and convenience. In this paper, we examine if there is a cost to having the gamma as a default option, without further scientific justification. We take a geometric approach to understanding the effect of precise model specification. We use a new, highly flexible but statistically well-behaved, way of specifying the frailty to calibrate modelling assumptions that are very commonly used in practice. We show that the gamma frailty assumption has the effect of considerably under-estimating standard errors when compared to our more general assumptions and, potentially, introducing bias. We comment on the implications of this. The survival times of adult acute myeloid leukaemia patients in northwest England are analyzed.

1 Introduction

One of the key insights of Information Geometry is the duality which links the exponential and mixture affine structures, [10]. Finite dimensional affine structures in the first are exponential families and affine structures in the second determine identification conditions in mixture models, [22]. This paper takes this fundamental insight and explores its implications in a practical and important applied statistical context.

V. Maroufy
Department of Biostatistics and Data Science, School of Public Health, University of Texas Health, Houston, USA
e-mail: vahed.maroufy@uth.tmc.edu

P. Marriott (✉)
Department of Statistics, University of Waterloo, Waterloo, ON, Canada
e-mail: pmarriot@uwaterloo.ca

F. Nielsen (ed.), *Geometric Structures of Information*, Signals and Communication Technology, https://doi.org/10.1007/978-3-030-02520-5_7

The motivating example of this paper is a study of the factors which affect the survival times of adult acute myeloid leukaemia patients. This is an example of a problem in lifetime data analysis, [20]. One of the most popular tools here is Cox's proportional hazard model, [9]. This models the time dependent hazard function, for subject i with covariates X_i, as a product

$$h_i(t) = h_0(t)\exp\{X_i\beta\},$$

where $h_0(t)$ is the time dependent baseline hazard common to all subjects. This will be treated as a non-parametric nuisance function. The second term is a time independent function of the explanatory variables, which is modelled parametrically. Under the proportional hazard assumption the interest parameters can be estimate using a partial likelihood. Often though there are unmeasured individual level covariates which need to be taken into account in the analysis. This is commonly done by adding subject level terms θ_i in the adapted hazard given by (1). Since this will give a parameter for each subject it is convenient to treat these as unobserved random effects, or *frailties*, and simply estimate parameters associated with the distribution of these terms.

This means that we work by mixing over different hazard functions and the key model choice issue is the specification of the frailty distribution. It is here that Information Geometry plays a role through consideration of the fundamental mixture geometry and its impact on inference of the parameters of interest. In particular we use tools associated with the *local mixture model*, see [2, 22]. Of special geometric interest is the potentially complex boundary that these models can have. For example [24] shows that local mixture models can have boundaries which are boundaries of polytopes or can be a non-smooth union of a finite number of smooth components.

Frailty models have been studied by many researchers; for example, [6, 12, 15, 19]. Various hazard models, including Cox's regression model, have been generalised by assuming a random frailty variable. The frailty model is commonly chosen to return a tractable marginal likelihood function; hence, gamma, inverse Gaussian and positive stable distributions with closed-form Laplace transformations are regular choices [7, p.77]. In [16] a discrete mixture model is form a gamma and an Inverse Gaussian, leading to a three-parameter model, is considered for the frailty. Among these the gamma is the most popular model choice, [19, 27, 28].

The goal of this paper is to investigate what might be the cost of specifying a particular parametric form, specifically the gamma, for the frailty distribution when this choice has been made purely for mathematical convenience. Model uncertainty is a critical problem in applied statistics, and – as we do here – its analysis can be treated in a geometric way. The use of geometry in the area is not new, for example the paper [8] provides an intriguing solution by proposing the 'double the variance' method for addressing the possibility of undetectably small departures from the model. Much more detail on the geometry of model specification can be found in [4].

We investigate the choice of a gamma frailty by calibrating it to a much more general space of mixtures, which can be used as a bench-marking tool. We use the excellent study of [13] to illustrate these effects in a real context. We show that making the gamma frailty assumption can have the effect of considerably under-estimating standard errors and its potential misspecification gives rise to important biases.

In general with mixture models, and specifically with the frailty models considered here, the analyst has to balance flexibility with inferential and computational tractability. In this paper we use the recently defined *discrete mixture of local mixture models*, introduced in [25], to achieve the calibration. This new model generates a finite dimensional and suitably parameterized space as a high quality approximation to a general mixture model with unknown mixing mechanism. The model is always identifiable and estimable, and its geometric and inferential properties allow for fast and efficient estimation algorithms. These properties make it suitable as a calibration tool, and we are currently exploring using it as a frailty model in its own right.

Notation, motivation and the main results of the paper, including our proposed method are presented in Sect. 2. Section 3 looks at the calibration, with Sect. 3.1 showing a simulation study, illustrating that the local mixture method returns similar biases, but larger – often more than double – standard deviations for the estimates compared to the Expected-Maximization method of [19] where a gamma frailty is assumed. This shows the considerable impact that fixing on a particular parametric form has on inference, and clearly illustrates that the reduction in standard error may be just an artefact of modelling choice rather than being real. In Sect. 3.2, the survival time of 1043 adults acute myeloid leukemia patients, recorded between 1982 and 1998 in northwest England, is analyzed, again with important differences in inferential conclusions being found. The paper closes with a short discussion in Sect. 4.

2 Methodology

Throughout this section, we follow the notation and definitions in [12, 20]. Let (T_i^0, C_i), for $i = 1, \ldots, n$, be the failure time and censoring time of the ith individual, and also let X be the $n \times p$ design matrix of the covariate vectors. Define $T_i = \min(T_i^0, C_i)$ and $\delta_i = I(T_i^0 < C_i)$, where $I(\cdot)$ is an indicator function. In addition, associated with the ith individual, an unobservable covariate θ_i, the frailty, is assumed, where θ_i's follow some distribution, Q. Adapting the proportional hazard model of [9] for the ith individual, conditional on the frailty θ_i, the hazard function is,

$$h_i(t) = \theta_i \, h_0(t) \exp\{X_i \beta\}, \tag{1}$$

where $h_0(t)$ is the base hazard function, X_i is the ith row of X and $\beta = (\beta_0, \ldots, \beta_{p-1})^T$ is a p-vector of regression coefficients. The cumulative hazard and survival functions are, respectively, defined as $H_i(t) = \int_0^t h_i(u)\, du$, and $S_i(t) = \exp\{-H_i(t)\}$, with the base cumulative hazard function $H_0(t)$. We also assume that the frailty θ is independent of X, and further that, given X and θ, censoring is independent and noninformative for θ and (h_0, β), see [12]. The full likelihood function for the parameter vector (β, H_0) is written as

$$L(\beta, H_0, Q) = \prod_{i=1}^{n} \int \left[\left(\theta\, h_0(T_i)\, e^{X_i\beta}\right)^{\delta_i} \exp\left\{-\theta\, H_0(T_i)\, e^{X_i\beta}\right\}\right] dQ(\theta). \tag{2}$$

In a similar way to a general mixture model problem, frailty survival models with an unknown frailty distribution, suffer from identification and estimability issues. Although, when all the covariates variables are continuous with a continuous distribution, [11], show that, given the distribution of the time duration variable, all the three multiplicative factors are, at least formally, identified. This theoretical result does not solve the identifiability issue in the general sense. For instance, when there is a discrete covariate then identifiability requires the corresponding regression coefficient to be limited to a known compact set [14, Ch. 2].

2.1 Local and Global Mixture Models

To allow full generality of the mixture structure it is tempting to only restrict Q to be a finite discrete distribution with an unknown number of components., i.e.

$$Q(\theta) = \sum_{i=1}^{N} \rho_i \delta_{\theta_i}(\theta),$$

where δ is the indicator function, and $\sum_{i=1}^{N} \rho_i = 1$ and $\rho_i > 0$. Indeed, as shown by [21], the non-parametric maximum likelihood estimate of Q lies in such a family. In that case the perturbation space would be 'parameterised' by N, the number of components, $(\theta_1, \ldots, \theta_N)$, the components, and $(\rho_1, \ldots, \rho_N)$, the mixing weights. However, this parameterization has many problems in implementation. Specifically, it is poorly identified and has complex boundaries. A key problem is that mixture components may be too close to one another to be resolved with a given set of data and so the order of the finite mixture is essentially not estimable, see [25].

The case where there is a single set of closely grouped components – or the much more general situation where Q is any small-variance distribution – is exactly the case which motivated the design of the local mixture model (LMM), see [1, 22]. The key idea is, essentially, to replace the unknown number of clustered mixing components with a fixed number of low order moments parameterised in an inferentially 'nice' way. The geometric intuition is that for a local perturbation all the mixing component distributions will lie close to a low dimensional linear (affine) space. This space is spanned by derivatives of the prior and parameterised with a small number of identified parameters.

Definition 1 For a density function, $f(y; \theta)$, belonging to the exponential family, the local mixture model of order k, centred at ϑ, is defined as

$$g_{\vartheta}(y; \lambda) = f(y; \vartheta) + \sum_{j=1}^{k} \lambda_j f^{(j)}(y; \vartheta), \tag{3}$$

where $\lambda := (\lambda_1, \ldots, \lambda_k) \in \Lambda_\vartheta, f^{(j)}(y; \vartheta) = \frac{\partial^j f(y;\theta)}{\partial \theta^j}|_{\theta=\vartheta}$. The parameter space, Λ_ϑ, is defined by

$$\Lambda_\vartheta := \{\lambda \mid g_\vartheta(y; \lambda) \geq 0, \ \forall y\}$$

is convex with a boundary determined by the non-negativity of (3). The structure of this boundary is the union of smooth manifolds and is studied in [24]. Also (3) is parameter invariant, and mean parameter is chosen only because of its clear interpretation, see [22].

In [25], local mixtures are generalised to discrete mixtures of local mixture models. In Definition 1 the point ϑ is fixed, selected by the analyst. This has the natural generalisation to have a set of possible centres, $\vartheta_1, \ldots \vartheta_L$, selected independently of the data. This gives rise to the following.

Definition 2 The discrete mixture of local mixture models is defined by

$$g(y; \underline{\vartheta}, \underline{\lambda}) = \sum_{l=1}^{L} \rho_l \, g_{\vartheta_l}(y; \lambda^l), \tag{4}$$

where $\underline{\lambda} = (\lambda^1, \ldots, \lambda^L)$, $\underline{\vartheta} = (\vartheta_1, \ldots, \vartheta_L)$, $\rho_l \geq 0$ and $\sum_{l=1}^{L} \rho_l = 1$.

The model selection task with such models is to select k, L and $\vartheta_1, \ldots \vartheta_L$ to balance estimability with the quality of approximation to a completely general mixture. We use the results of [25, Section 2.1] which show, for a given value of k, how to select the points $\vartheta_1 < \cdots < \vartheta_L$ to have a uniform L^1-bound on the difference between $g(y; \underline{\vartheta}, \underline{\lambda})$ and any mixture with mixing distribution Q whose support lies in a compact region containing $\{\vartheta_1, \ldots, \vartheta_L\}$.

We emphasis, before we show some of the details in an example, that in this paper, we are using the LMM structure as a calibration tool in order to evaluate the effect of selecting a particular parametric frailty model. We are generating a rich, but well behaved, class of mixtures which subsume the gamma assumption.

Example 1 To illustrate the flexibility of this framework, Fig. 1 shows some discrete mixtures of local mixture models for the exponential distribution on both the density and hazard scales. Using the methods of [25, Section 2.1] some straightforward numerical analysis shows that selecting points, in the rate parametrisation, at $\vartheta_1 = 0.5$, $\vartheta_2 = 1$, $\vartheta_3 = 2$ and $\vartheta_4 = 3.5$ and fixing $k = 4$ gives excellent L^1-approximations for all mixtures Q with support in $[0.3, 5]$. In the figure Model 1 is the unmixed exponential model, Model 2 is a single component local mixture centered at $\vartheta_2 = 1$, Models 3 and 4 are two components mixtures of local mixture centred at $(1, 3.5)$ and $(0.5, 1)$, respectively.

In [25] it was shown that an adaptation of the EM algorithm works well for fitting a finite mixture of local mixtures. We only consider here local mixture models of order $k = 4$. Increasing this degree – while mathematically possible – only adds a small marginal improvement to the local modelling performance, [23], at the cost of extra parameters.

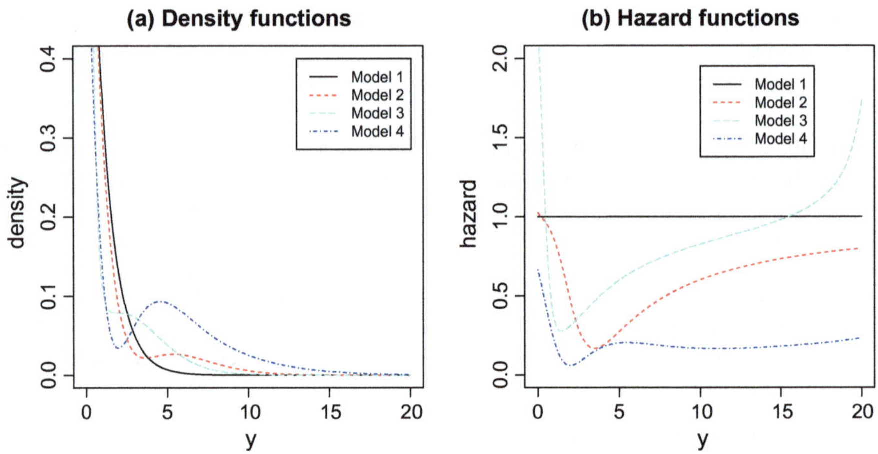

Fig. 1 The flexibility of the model class: **a** four models on a density scale, **b** the same models' hazard functions

The key, novel, issue in the estimation is to deal with the fact that the parameter space, Λ_ϑ, has a complex boundary. For this paper, for the Cox model, we need to consider the case where $f(y;\theta) = \theta \exp(-\theta y)$ and then the parameter space is characterized as the set of all λ's such that, for all $y > 0$,

$$\vartheta\lambda_4 y^4 - (4\lambda_4 + \vartheta\lambda_3)y^3 + (3\lambda_3 + \vartheta\lambda_2)y^2 - (2\lambda_2 + \vartheta\lambda_1)y + \lambda_1 + \vartheta \geq 0. \quad (5)$$

The boundary is characterised as being the cases where (5) has only repeated positive roots, and it can be shown, by direct calculation, that the parameter space is the union of smooth manifolds. This characterisation allows explicit parameterisations of the component manifolds to be computed and these are exploited in the numerical studies of this paper. See Appendix for details.

A cost of this boundary is that the maximum likelihood estimate may not exist in the regular sense as a turning point. Furthermore, asymptotic approximations in sample size, similar to that of [29], ultimately break down close to the boundary, irregardless of their order. In [3] a diagnostic approach is proposed which identifies when a first order asymptotic is appropriate, depending on the square distance of the MLE from the boundary, measured using the Fisher information.

3 Application to Frailty Modelling

In this section, we investigate the size of the effect on inference associated with fixing on a particular frailty distribution by comparing a standard gamma frailty model to our more general local mixture approach. In cases where the only reason for selecting a gamma frailty was convenience we argue differences, such as a reduction in standard error, are artefacts of model choice and not real.

In order to investigate the best possible case for the gamma frailty assumption we simulate, in Sect. 3.1, from that frailty distribution, and we only make the comparison to a one component local mixture with $k = 4$. With these restrictions we see that standard errors associated with the gamma assumption are considerably smaller than those of the LMM. Furthermore, in the real data example of Sect. 3.2, it turns out that the $k = 4$, one component local mixture is the appropriate choice for that data and we find important inferential differences which might be due to either misspecification bias, or artificially reduced standard errors, in the gamma based model.

Intuitively, for any fixed ϑ, the finite dimensional parameter vector λ represents the frailty distribution through its central moments. Local mixing also can be seen as a mechanism that extends a parametric model to a larger and more flexible space of densities which has nice geometric and inferential properties, allowing calibration of a particular parametric assumption. Substituting the local mixture expansion of Eq. (3) we obtain

$$\begin{aligned} l(\beta, H_0, \lambda) = & \sum\nolimits_{i=1}^{n} (\delta_i[\log h_0(T_i) + X_i\beta] + \log f(T_i, \beta, \vartheta)) \\ & + \sum\nolimits_{i=1}^{n} \log\left(1 + \sum\nolimits_{j=1}^{k} \lambda_j A_j(\delta_i, y_i)\right), \qquad \lambda \in \Lambda_\vartheta \end{aligned} \tag{6}$$

in which $A_j(\delta_i, y_i) = \frac{f^{(j)}(T_i,\beta,\vartheta)}{f(T_i,\beta,\vartheta)}$, and $y_i = H_0(T_i)e^{X_i\beta}$. We maximize Eq. (6) when estimating β, where H_0 and λ are considered as nuisance parameters. Thus, a profile likelihood optimization method is employed. That is, we first maximize for λ over Λ_ϑ to obtain $\hat{\lambda}$ and impute $\hat{H}_0$ for H_0, then maximize $l_p(\beta) = l(\beta, \hat{H}_0, \hat{\lambda})$ to estimate β.

To impute $h_0(t)$ and $H_0(t)$ we use the arguments in [12] to provide a recursive estimate of the cumulative hazard function using the fact that for two consecutive failure times, $T_{(i)}$ and $T_{(i+1)}$, we have $H_0(T_{(i+1)}) = H_0(T_{(i)}) + \Delta H_i$. Substituting this recursive equation into the log-likelihood function in (6), considering the conventions in [5] and taking partial derivative with respect to ΔH_i, we obtain

$$\frac{\partial l}{\partial \Delta H_i} = \frac{1}{\Delta H_i} - \sum_{\ell=i}^{n} e^{X_\ell\beta} + \frac{P'(e^{X_i\beta}[H_0(T_{(i)}) + \Delta H_i])}{P(e^{X_i\beta}[H_0(T_{(i)}) + \Delta H_i])}, \tag{7}$$

which is a function of just ΔH_i when $\hat{H}_0(T_{(i)})$ is given at time $T_{(i+1)}$, $P(\cdot)$ is a polynomial of degree four with its coefficients linear functions of $(\lambda_1, \lambda_2, \lambda_3, \lambda_4)$ and $P'(\cdot)$ is its derivative with respect to ΔH_i. When the denominator is not zero, Eq. (7) is a polynomial of degree five which can be solved numerically for ΔH_i. Note that when there is no frailty factor – that is $\lambda = (0, 0, 0, 0)$ – then the last term in Eq. (7) is zero, and the estimate of the cumulative hazard function reduces to the form in [18] which is the estimate in [19] with $\hat{\omega} = 1$.

Table 1 Bias and standard errors of coefficient estimates, when frailty is generated from $\Gamma(\frac{1}{\eta}, \eta)$. LMM is the local mixture method; Gamma the Expectation–Maximization method for the gamma frailty

			Gamma		LMM	
n	η	β	*bias*	*std*	*bias*	*std*
200	0.5	log 3	−0.057	0.18	−0.048	0.43
200	0.7	log 3	−0.056	0.21	−0.038	0.40
200	1	log 3	−0.117	0.22	−0.094	0.41

3.1 Simulation Study

In this section, a simulation study illustrates the effect of making the gamma frailty assumption when compared to the much richer local mixture method. We adapt the method in [19], which assumes a gamma model with mean 1 and variance η for the frailty, and apply the Expectation–Maximization algorithm.

We let $C = 0.01$, $\tau = 4.6$ and we follow a similar set-up as found in [17]. For each individual the event time is $T = [-\log(1 - U)\{\theta \exp\{\beta X\}\}^{-1}]^{-1/\tau} C^{-1}$, where $X \sim N(0, 1)$, $U \sim$ uniform[0, 1]. The censoring distribution is $N(100, 15)$, and frailty is assumed to follow a gamma distribution with mean 1 and variance η. Table 1 shows the bias and standard error for the estimates of the regression coefficient using both methods for three different values of η. It is clear that the local mixture method, which does not use any information about the frailty model, returns very similar biases as the Expectation–Maximization method for the gamma frailty. However, the standard deviation for the estimates in the local mixture method are almost twice as large as these for the Expectation–Maximization method.

Similar simulation studies are presented in Tables 2 and 3 where the frailty is generated from different models than the gamma distribution discussed above. In Table 2, the frailty is generated from uniform distributions with means set to one and different variances. The standard deviation for the LMM model is twice as large as that for Gamma model, similar to that in Table 1. In addition, since the gamma model is now misspecification, as the variation of the frailty gets larger, the LMM slightly beats the Gamma approach with respect to estimation bias.

Table 3 considers two different models for the frailty: the Inverse Gaussian and Inverse Gamma models. All have their mean set to one but with different variances. Although the LMM model still returns a bigger standard deviation, it is no longer twice as big as that for the Gamma model.

The results in Tables 1, 2 and 3 conveys three important messages. First, without assuming any specific model form for the unobserved frailty, our methodology does equally well in terms of the bias of estimation. Second, our method automatically addresses the possibility of departures from the gamma assumption by returning a bigger standard deviation, double in some cases. We argue that the apparent reduction of standard error, associated with the gamma, is an artifact of model choice which

Table 2 Bias and standard errors of coefficient estimates, when frailty is generated from uniform $U(1-\eta, 1+\eta)$ distributions. LMM is the local mixture method; Gamma the Expectation–Maximization method for the gamma frailty

			Gamma		LMM	
n	η	β	*bias*	*std*	*bias*	*std*
200	0.5	log 3	0.003	0.104	0.02	0.205
200	0.7	log 3	0.067	0.113	0.011	0.237
200	1	log 3	−0.23	0.144	−0.18	0.264

Table 3 Bias and standard errors of coefficient estimates, when frailty is generated from Inverse-gamma and Inverse-Gaussian. LMM is the local mixture method; Gamma the Expectation–Maximization method for the gamma frailty

			Gamma		LMM		
n	*Model*	*(mean, var)*	β	*bias*	*std*	*bias*	*std*
200	$IGamma(3, 2)$	$(1, 1)$	log 3	−0.106	0.112	−0.094	0.172
200	$IGamma(2, 1)$	$(1, 2)$	log 3	0.179	0.110	0.156	0.166
200	$IGauss(1, 0.5)$	$(1, 0.5)$	log 3	−0.130	0.115	−0.115	0.151
200	$IGauss(1, 1)$	$(1, 1)$	log 3	−0.198	0.104	−0.181	0.161

can only be justified if there were good, extra-data, reasons for trusting the gamma assumption. Third, in the case of model misspecification, the LMM dominates the Gamma model, as amount of frailty gets larger.

3.2 Data Example

We compare the two methods in the context of a study of the survival time of 1043 adult acute myeloid leukemia patients, recorded between 1982 and 1998 in northwest England [13]. In this example 16% of the survival times are censored and complete information is available for four covariates; age, sex, white blood cell count (WBC), and a measure of deprivation for the enumeration district of residence (Dep). [13], studied the data to investigate possible spatial variation in the survival time assuming a gamma marginal frailty with covariance structure for the frailty among the 24 districts. Although they find some indication of spatial variation between the districts, their analysis illustrates that, assuming covariance structure in the frailty variable does not affect the inference on the regression coefficients, and differences in the coefficient estimates are not significantly bigger than one standard deviation. The estimates and standard errors of the regression coefficients, $(\beta_1, \ldots, \beta_4)$, based on the independent gamma frailty are shown in the first two rows of Table 4. The frailty variance is estimated as $\hat{\eta} = 0.772$, indicating the existence of unobserved variation in the patients' survival time.

Table 4 Estimates and standard errors for covariate coefficient are presented. LMM, the local mixture method; Gamma; the estimates in *Henderson et al. (2002)* for the gamma frailty. Std are the associated standard errors for the gamma model

	Age	Sex	WBC	Dep
Gamma	0.0470	0.0563	0.0057	0.0547
Std	0.0045	0.0505	0.0008	0.0187
LMM	0.0422	0.0156	0.0114	0.0306

Applying the local mixture method we obtain $\hat{\lambda} = (-0.687, 0.017, 0.056, 0.023)$, also reflecting the existence of some unobserved variation. Furthermore, $\hat{\lambda}$ lie in the relative interior of Λ_{ϑ}, indicating that the one component local mixture is adequate. We note that if this estimate were to lay on the boundary the one component LMM approximation used here might still be a good L^1 approximation but perhaps poor in a KL-divergence sense. If this were to happen we would recommend trying a larger number of components.

The estimates for $(\beta_1, \ldots, \beta_4)$, using the LMM approach, are shown in the third row of Table 4. Comparing the differences between the estimates with the standard errors, we realize that all differences are inside the one standard deviation bound except for β_3, the coefficient for WBC, where the ratio is 7.125. This important difference in the estimate is interpreted as the result of a possible misspecification of the frailty model when a gamma distribution is imposed. It could be due to either misspecification bias or an under estimate of the standard error in the gamma frailty model. We note that our simulation results point to the fact that the bias in the LMM might be expected to be small when there is, in fact, a gamma frailty.

4 Discussion

In the context of frailty survival analysis, this paper applies an approximation of a general mixture model – with a completely unspecified mixing mechanism – by a novel family of models, the discrete mixture of local mixture models. These models are built in a geometrically and inferentially convenient way and have many advantages compared to finite mixture models. We do this to generate a calibration tool with which we can evaluate the effect of specifying a particular choice of frailty model. These geometric properties lead to inferential properties such as concavity of the log-likelihood function, identifiability and orthogonality in parametrization.

Our novel, highly flexible, but inferentially well-behaved, family gives a way of calibrating the effect of making a closed form parametric assumption on the frailty term. We see considerable differences in both the simulation and real data examples. Hence we conclude that making the gamma frailty assumption can have the effect of considerably under-estimating standard errors.

We note that, in this manuscript, we only use a one component LMM for the simulation and data example, and work under the assumption that the frailty is only a small component of the total variation. Nevertheless, the theory is quite general and allows for all mixing distributions in principle. For example we have recently applied it to an analysis of robustness in Bayesian conjugate and sparsity priors [26]. To apply Definition 2 here to its full potential, requires finding a good estimator for the cumulative hazard function and this is still an open problem.

Appendix

For the exponential distribution, the density of a single component local mixture has the form

$$\left\{\vartheta\lambda_4 y^4 - (4\lambda_4 + \vartheta\lambda_3)y^3 + (3\lambda_3 + \vartheta\lambda_2)y^2 - (2\lambda_2 + \vartheta\lambda_1)y + \lambda_1 + \vartheta\right\}\exp(-\vartheta y), \tag{8}$$

which is a density when the polynomial is non-negative for all $y > 0$. It will lie on the boundary of the parameter space when all the positive roots of the polynomial are of order 2 or 4 and $\lambda_4 > 0$. Polynomials of this form can be written as

$$\left\{a(y-r)^2(y^2+2by+c)\right\}, \tag{9}$$

then the roots of $(y^2 + 2by + c)$ must be repeated when they are positive, and a, b, c and r satisfy

$$\vartheta^5 - \vartheta^4 ar^2c - (2ar^2b - 2arc)\vartheta^3 - (2ac + 2ar^2 - 8arb)\vartheta^2 - (12ab - 12ar)\vartheta - 24a = 0. \tag{10}$$

This follows since a density integrates to one. Comparing (8) with (9) and imposing (10) results in an explicit parameterisation of the boundary in terms of b, c and r. The Jacobean of this has full rank as long as $r \neq 0, -b \pm \sqrt{b^2 - c}$. So, under these conditions the boundary is a smooth manifold. In the singular case where $r = -b \pm \sqrt{b^2 - c}$, since roots must have even order, we have $r = -b$ and $b < 0$. This singular set is given by a one dimensional manifold, again with an explicit parameterisation.

References

1. Anaya-Izquierdo, K., Marriott, P.: Local mixture models of exponential families. Bernoulli **13**, 623–640 (2007)
2. Anaya-Izquierdo, K., Marriott, P.: Local mixture of the exponential distribution. Ann. Inst. Math. Stat. **59**(1), 111–134 (2007)
3. Anaya-Izquierdo, K., Critchley, F., Marriott, P.: When are first order asymptotics adequate? a diagnostic. Stat **3**(1), 17–22 (2013)

4. Anaya-Izquierdo, K., Critchley, F., Marriott, P., Vos, P.: The geometry of model sensitivity: an illustration. In: Computational Information Geometry: For Image and Signal Processing. Springer, Berlin (2016)
5. Breslow, N.: Contribution to the discussion on the paper of Cox. Biom **34**, 216–217 (1972)
6. Clayton, D.G.: A model for association in bivariate life tables and its application in epidemiological studies of familial tendency in chronic disease incidence. Biometrika **65**(1), 141–151 (1978)
7. Cook, R.J., Lawless, J.F.: The statistical analysis of recurrent events (2007)
8. Copas, J., Eguchi, S.: Local model uncertainty and incomplete-data bias (with discussion). J. R. Stat. Soc. Ser. B (Stat. Methodol.) **67**(4), 459–513 (2005)
9. Cox, D.R.: Regression models and life-tables. J. R. Stat. Soc. B **34**, 187–200 (1972)
10. Critchley, F., Marriott, P.: Information geometry and its applications: an overview. In: Computational Information Geometry: For Image and Signal Processing, pp. 1–31. Springer, Berlin (2017)
11. Eleber, C., Ridder, G.: True and spurious duration dependence: the identifiability of the proportional hazard model. Rev. Econ. Stud. **49**, 403–409 (1982)
12. Gorfine, M., Zucher, D.M., Hsu, L.: Prospective survival analysis with general semiparametric shared frialty model: a pseudo full likelihood approach. Biometrika **93**(3), 735–741 (2006)
13. Henderson, R., Shimakura, S., Gorst, D.: Modeling partial variation in Leukemia survival data. J. Am. Stat. Assoc. **97**(460), 965–972 (2002)
14. Horowitz, J.L.: Semiparametric and Nonparametric Methods in Econometrics. Springer, Berlin (2010)
15. Hougaard, P.: Survival models for heterogeneous population derived from stable distributions. Biometrika **73**(2), 387–396 (1986)
16. Hougaard, P.: Frailty models for survival data. Lifetime Data Anal. **1**(3), 255–273 (1995)
17. Hsu, L., Chen, L., Gorfine, M., Malone, K.: Semiparametric estimation of marginal hazard function from case-control family studies. Biometrics **60**, 936–944 (2004)
18. Johansen, S.: An extension of Cox's regression model. Int. Stat. Rev. **51**, 258–262 (1983)
19. Klein, J.P.: Semiparametric estimation of random effects using cox model based on the EM algorithm. Biometrics **48**, 795–806 (1992)
20. Lawless, J.F.: Statistical Models and Methods for Lifetime Data. Wiley, New York (1981)
21. Lindsay, B.G.: Mixture Models: Theory, Geometry and Applications. Institute of Mathematical Statistics, Hayward (1995)
22. Marriott, P.: On the local geometry of mixture models. Biometrika **89**, 77–93 (2002)
23. Marriott, P.: Extending local mixture models. AISM **59**, 95–110 (2006)
24. Maroufy, V., Marriott, P.: Computing boundaries in local mixture models. Geometric Science of Information. Lecture Notes in Computer Science, vol. 9389, pp. 577–585 (2015)
25. Maroufy, V., Marriott, P.: Mixture models: building a parameter space. Stat. Comput. 1–7 (2016)
26. Maroufy, V., Marriott, P.: Local and global robustness in conjugate Bayesian analysis and sparsity models. Statistica Sinica. In press (2018)
27. Nielsen, G.G., Gill, R.D., Anderson, P.K., Sorensen, T.I.A.: A counting process approach to maximum likelihood estimation in frailty models. Scand. J. Stat. **19**(1), 25–43 (1992)
28. Vaupel, J.W., Manton, K.G., Stallard, E.: The impact of heterogeneity in individual frailty on the dynamics of mortality. Demography **16**, 439–454 (1979)
29. Zucker, D., Gorfine, M., Hsu, L.: Pseudo-full likelihood estimation for prospective survival analysis with a general semiparametric shared frailty model: asymptotic theory. J. Stat. Plan. Inference **138**(7), 1998–2016 (2008)

Some Universal Insights on Divergences for Statistics, Machine Learning and Artificial Intelligence

Michel Broniatowski and Wolfgang Stummer

Abstract Dissimilarity quantifiers such as divergences (e.g. Kullback–Leibler information, relative entropy) and distances between *probability distributions* are widely used in statistics, machine learning, information theory and adjacent artificial intelligence (AI). Within these fields, in contrast, some applications deal with divergences between *other-type* real-valued functions and vectors. For a broad readership, we present a correspondingly unifying framework which – by its nature as a "structure on structures" – also qualifies as a basis for similarity-based multistage AI and more humanlike (robustly generalizing) machine learning. Furthermore, we discuss some specificalities, subtleties as well as pitfalls when e.g. one "moves away" from the probability context. Several subcases and examples are given, including a new approach to obtain parameter estimators in continuous models which is based on noisy divergence minimization.

1 Outline

The goals formulated in the abstract are achieved in the following way and order: to address a wide audience, throughout the paper (with a few connection-indicative exceptions) we entirely formulate and investigate divergences and distances between

M. Broniatowski
Sorbonne Universite Pierre et Marie Curie, LPSM, 4 place Jussieu, 75252 Paris, France
e-mail: michel.broniatowski@sorbonne-universite.fr

W. Stummer (✉)
Department of Mathematics, University of Erlangen–Nürnberg, Cauerstrasse 11, 91058 Erlangen, Germany
e-mail: stummer@math.fau.de

W. Stummer
Affiliated Faculty Member of the School of Business and Economics, University of Erlangen–Nürnberg, Lange Gasse 20, 90403 Nürnberg, Germany

F. Nielsen (ed.), *Geometric Structures of Information*, Signals and Communication Technology, https://doi.org/10.1007/978-3-030-02520-5_8

functions, even for the probability context. In Sect. 2, we provide some non-technical background and overview of some of their principally possible usabilities for tasks in data analytics such as statistics, machine learning, and artificial intelligence (AI). Furthermore, we indicate some connections with geometry and information. Thereafter, in Sect. 3 we introduce a new *structured* framework (toolkit) of divergences between functions, and discuss their building-blocks, boundary behaviour, as well as their identifiability properties. Several subcases, running examples, technical subtleties of practical importance, etc. are illuminated, too. Finally, we study divergences between "entirely different functions" which e.g. appear in the frequent situation when for data-derived *discrete* functions one wants to find a closest possible continuous-function model (cf. Sect. 4); several corresponding noisy minimum-divergence procedures are compared – for the first time within a unifying framework – and new methods are derived too.

2 Some General Motivations and Uses of Divergences

2.1 *Quantification of Proximity*

As a starting motivation, it is basic knowledge that there are numerous ways of evaluating the proximity $d(p,q)$ of two real numbers p and q of primary interest. For instance, to quantify that p and q nearly coincide one could use the difference $d^{(1)}(p,q) := p-q \approx 0$ or the fraction $d^{(2)}(p,q) := \frac{p}{q} \approx 1$, scaled (e.g. magnifying, zooming-in) versions $d^{(3)}_m(p,q) := m\cdot(p-q) \approx 0$ or $d^{(4)}_m(p,q) := m \cdot \frac{p}{q} \approx 1$ with "scale" m of secondary (auxiliary) interest, as well as more flexible hybrids $d^{(5)}_{m_1,m_2,m_3}(p,q) := m_3 \cdot \left(\frac{p}{m_1} - \frac{q}{m_2}\right) \approx 0$ where m_i may also take one of the values p, q. All these "dissimilarities" $d^{(j)}(\cdot,\cdot)$ can principally take any sign and they are asymmetric which is consistent with the – in many applications required – desire that one of the two primary-interest numbers (say p) plays a distinct role; moreover, the involved divisions cause technical care if one principally allows for (convergence to) zero-valued numbers. A more sophisticated, nonlinear alternative to $d^{(1)}(\cdot,\cdot)$ is given by the dissimilarity $d^{(6)}_\phi(p,q) := \phi(p) - (\phi(q) + \phi'(q)\cdot(p-q))$ where $\phi(\cdot)$ is a strictly convex, differentiable function and thus $d^{(6)}_\phi(p,q)$ quantifies the difference between $\phi(p)$ and the value at p of the tangent line taken at $\phi(q)$. Notice that $d^{(6)}_\phi(\cdot,\cdot)$ is generally still asymmetric but always stays nonnegative independently of the possible signs of the "generator" ϕ and the signs of p,q. In contrast, as a nonlinear alternative to $d^{(4)}_m(\cdot,\cdot)$ one can construct from ϕ the dissimilarity $d^{(7)}_\phi(p,q) := q\cdot\phi(\frac{p}{q})$ (where $m=q$) which is also asymmetric but can become negative depending on the signs of p,q,ϕ. More generally, one often wants to work with dissimilarities $d(\cdot,\cdot)$ having the properties

(D1) $d(p,q) \geqslant 0$ for all p, q (nonnegativity),
(D2) $d(p,q) = 0$ if and only if $p = q$ (reflexivity; identity of indiscernibles[1]),

and such $d(\cdot,\cdot)$ is then called a *divergence* (or disparity, contrast function). Loosely speaking, the divergence $d(p,q)$ of p and q can be interpreted as a kind of "directed distance from p to q".[2] As already indicated above, the underlying directness turns out to be especially useful in contexts where the first component (point), say p, is always/principally of "more importance" or of "higher attention" than the second component, say q; this is nothing unusual, since after all, one of our most fundamental daily-life constituents – namely time – is directed (and therefore also time-dependent quantities)! Moreover, as a further analogue consider the "way/path-length" $d(p,q)$ a taxi would travel from point p to point q in parts of a city with at least one one-way street. Along the latter, there automatically exist points $p \neq q$ such that $d(p,q) \neq d(q,p)$; this non-equality may even hold for all $p \neq q$ if the street pattern is irregular enough; the same holds on similar systems of connected "one-way loops", directed graphs, etc. However, sometimes the application context demands for the usage of a dissimilarity $d(\cdot,\cdot)$ satisfying (D1), (D2) and

(D3) $d(p,q) = d(q,p)$ for all p, q (symmetry),

and such $d(\cdot,\cdot)$ is denoted as a *distance*; notice that we don't assume that the triangle inequality holds. Hence, we regard a distance as a symmetric divergence. Moreover, a distance $d(\cdot,\cdot)$ can be constructed from a divergence $\widetilde{d}(\cdot,\cdot)$ e.g. by means of either the three "symmetrizing operations" $d(p,q) := \widetilde{d}(p,q) + \widetilde{d}(q,p)$, $d(p,q) := \min\{\widetilde{d}(p,q), \widetilde{d}(q,p)\}$, $d(p,q) := \max\{\widetilde{d}(p,q), \widetilde{d}(q,p)\}$ for all p and q.

In many real-life applications, the numbers p, q of primary interest as well as the scaling numbers m_i of secondary interest are typically replaced by real-valued functions $x \to p(x)$, $x \to q(x)$, $x \to m_i(x)$, where $x \in \mathscr{X}$ is taken from some underlying set $\mathscr{X}$. To address the entire functions as objects we use the abbreviations $P := \{p(x)\}_{x\in\mathscr{X}}$, $Q := \{q(x)\}_{x\in\mathscr{X}}$, $M_i := \{m_i(x)\}_{x\in\mathscr{X}}$, and alternatively sometimes also $p(\cdot)$, $q(\cdot)$, $m_i(\cdot)$. This is conform with the high-level data processing paradigms in "functional programming" and "symbolic computation", where functions are basically treated as whole entities, too.

Depending on the nature of the data-analytical task, the function P of primary interest may stem either from a hypothetical model, or its analogue derived from observed/measured data, or its analogue derived from artificial computer-generated (simulated) data; the same holds for Q where "cross-over constellations" (w.r.t. to the origin of P) are possible.

The basic underlying set (space) $\mathscr{X}$ respectively the function argument x can play different roles, depending on the application context. For instance, if $\mathscr{X} \subset \mathbb{N}$ is a subset of the integers $\mathbb{N}$ then $x \in \mathscr{X}$ may be an index and $p(x)$ may describe the xth real-valued data-point. Accordingly, P is then a s-dimensional vector where s is the total number of elements in $\mathscr{X}$ with eventually allowing for $s = \infty$. In other

[1] See e.g. Weller-Fahy et al. [93].

[2] Alternatively, one can think of $d(p,q)$ as degree of proximity from p to q.

situations, x itself may be a data point of arbitrary nature (i.e. $\mathscr{X}$ can be any set) and $p(x)$ a real value attributed to x; this $p(x)$ may be of direct or of indirect use. The latter holds for instance in cases where $p(\cdot)$ is a density function (on $\mathscr{X}$) which roughly serves as a "basis" for the operationalized calculation of the "local aggregations over all[3] $A \subset \mathscr{X}$" in the sense of $A \rightarrow \sum_{x \in A} p(x)$ or $A \rightarrow \int_A p(x) \, \mathrm{d}\widetilde{\lambda}(x)$ subject to some "integrator" $\widetilde{\lambda}(\cdot)$ (including classical Riemann integrals $\mathrm{d}\widetilde{\lambda}(x) = \mathrm{d}x$); as examples for nonnegative densities $p(\cdot) \geqslant 0$ one can take "classical" (volumetric, weights-concerning) inertial-mass densities, population densities, probability densities, whereas densities $p(\cdot)$ with possible negative values can occur in electromagnetism (charge densities, polarization densities), in other fields of contemporary physics (negative inertial-mass respectively gravitational-mass densities) as well as in the field of acoustic metamaterials (effective density), to name but a few.

Especially when used as a set of possible states/data configurations (rather than indices), $\mathscr{X}$ can be of arbitrary complexity. For instance, each x itself may be a real-valued continuous function on a time interval $[0, T]$ (i.e. $x : [0, T] \rightarrow]-\infty, \infty[$) which describes the scenario of the overall time-evolution of a quantity of interest (e.g. of a time-varying quantity in an deterministic production process of one machine, of the return on a stock, of a neural spike train). Accordingly, one can take e.g. $\mathscr{X} = C\big([0, T],]-\infty, \infty[\big)$ to be the set of all such continuous functions, and e.g. $p(\cdot)$ a density thereupon (which is then a function on functions). Other kinds of functional data analytics can be covered in an analogous fashion.

To proceed with the proximity-quantification of the primary-interest functions $P := \big\{p(x)\big\}_{x \in \mathscr{X}}$, $Q := \big\{q(x)\big\}_{x \in \mathscr{X}}$, in accordance with the above-mentioned investigations one can deal with the pointwise dissimilarities/divergences $d_{\phi}^{(j)}(p(x), q(x))$, $d_{m_1(x),m_2(x),m_3(x)}^{(5)}(p(x), q(x))$ for fixed $x \in \mathscr{X}$, but in many contexts it is crucial to take "summarizing" dissimilarities/divergences

$$D_{\phi}^{(j)}(P, Q) := \sum_{x \in \mathscr{X}} d_{\phi}^{(j)}(p(x), q(x)) \cdot \lambda(x) \quad \text{or} \quad D_{\phi}^{(j)}(P, Q) := \int_{\mathscr{X}} d_{\phi}^{(j)}(p(x), q(x)) \, \mathrm{d}\lambda(x)$$

subject to some weight-type "summator"/"integrator" $\lambda(\cdot)$ (including classical Riemann integrals); analogously, one can deal with $D_{\phi,M_1,M_2,M_3}^{(5)}(P, Q) := \sum_{x \in \mathscr{X}} d_{m_1(x),m_2(x),m_3(x)}^{(5)}(p(x), q(x)) \cdot \lambda(x)$ or $D_{\phi,M_1,M_2,M_3}^{(5)}(P, Q) := \int_{\mathscr{X}} d_{m_1(x),m_2(x),m_3(x)}^{(5)}(p(x), q(x)) \, \mathrm{d}\lambda(x)$. Notice that the requirements (D1), (D2) respectively (D3) carry principally over in a straightforward manner also to these pointwise and aggregated dissimilarities between the functions (rather than real points), and accordingly one calls them (pointwise/aggregated) divergences respectively distances, too.

[3]Measurable.

2.2 Divergences and Geometry

There are several ways how pointwise dissimilarities $d(\cdot,\cdot)$ respectively aggregated dissimilarities $D(\cdot,\cdot)$ between two functions $P := \{p(x)\}_{x\in\mathscr{X}}$ and $Q := \{q(x)\}_{x\in\mathscr{X}}$ can be connected with geometric issues. To start with an "all-encompassing view", following the lines of e.g. Birkhoff [14] and Millmann and Parker [50], one can build from any set $\mathscr{S}$, whose elements can be interpreted as "points", together with a collection $\mathscr{L}$ of non-empty subsets of $\mathscr{S}$, interpreted as "lines" (as a manifestation of a principle sort of structural connectivity between points), and an arbitrary *distance* $\mathfrak{d}(\cdot,\cdot)$ on $\mathscr{S}\times\mathscr{S}$, an axiomatic constructive framework of geometry which can be of far-reaching nature; therein, $\mathfrak{d}(\cdot,\cdot)$ plays basically the role of a marked ruler. Accordingly, each triplet $(\mathscr{S},\mathscr{L},\mathfrak{d}(\cdot,\cdot))$ forms a distinct "quantitative geometric system"; the most prominent classical case is certainly $\mathscr{S}=\mathbb{R}^2$ with $\mathscr{L}$ as the collection of all vertical and non-vertical lines, equipped with the Euclidean distance $\mathfrak{d}(\cdot,\cdot)$, hence generating the usual Euclidean geometry in the two-dimensional space. In the case that $\mathfrak{d}(\cdot,\cdot)$ is only an *asymmetric divergence* but not a distance anymore, we propose that some of the outcoming geometric building blocks have to be interpreted in a direction-based way (e.g. the use of $\mathfrak{d}(\cdot,\cdot)$ as a marked directed ruler, the construction of points of equal divergence from a center viewed as distorted directed spheres, etc.). For $d(\cdot,\cdot)$ one takes $\mathscr{S}\subset\mathbb{R}$ whereas for $D(\cdot,\cdot)$ one has to work with $\mathscr{S}$ being a family of real-valued functions on $\mathscr{X}$.

Secondly, from any *distance* $\mathfrak{d}(\cdot,\cdot)$ on a "sufficiently rich" set $\mathscr{S}$ and a finite number of (fixed or adaptively flexible) distinct "reference points" s_i $(i=1,\dots,n)$ one can construct the corresponding Voronoi cells $V(s_i)$ by

$$V(s_i) := \{z\in\mathscr{S} : \mathfrak{d}(z,s_i) \leqslant \mathfrak{d}(z,s_j) \text{ for all } j=1,\dots,n\}.$$

This produces a tesselation (tiling) of $\mathscr{S}$ which is very useful for classification purposes. Of course, the geometric shape of these tesselations is of fundamental importance. In the case that $\mathfrak{d}(\cdot,\cdot)$ is only an *asymmetric divergence* but not a distance anymore, then $V(s_i)$ has to be interpreted as a directed Voronoi cell and then there is also the "reversely directed" alternative

$$\widetilde{V}(s_i) := \{z\in\mathscr{S} : \mathfrak{d}(s_i,z) \leqslant \mathfrak{d}(s_j,z) \text{ for all } j=1,\dots,n\}.$$

Recent applications where $\mathscr{S}\subset\mathbb{R}^d$ and $\mathfrak{d}(\cdot,\cdot)$ is a Bregman divergence or a more general conformal divergence, can be found e.g. in Boissonnat et al. [15], Nock et al. [64] (and the references therein), where they also deal with the corresponding adaption of k-nearest neighbour classification methods.

Thirdly, consider a "specific framework" where the functions $P := \widetilde{P}_{\theta_1} := \{\widetilde{p}_{\theta_1}(x)\}_{x\in\mathscr{X}}$ and $Q := \widetilde{P}_{\theta_2} := \{\widetilde{p}_{\theta_2}(x)\}_{x\in\mathscr{X}}$ depend on some parameters $\theta_1\in\Theta$, $\theta_2\in\Theta$, which reflect the strive for a complexity-reducing representation of "otherwise intrinsically complicated" functions P, Q. The way of dependence of the function (say) $\widetilde{p}_\theta(\cdot)$ on the underlying parameter θ from an appropriate space Θ

of e.g. manifold type, may show up directly e.g. via its operation/functioning as a relevant system-indicator, or it may be manifested implicitly e.g. such that $\widetilde{p}_\theta(\cdot)$ is the solution of an optimization problem with θ-involving constraints. In such a framework, one can induce divergences $D(\widetilde{P}_{\theta_1}, \widetilde{P}_{\theta_2}) =: f(\theta_1, \theta_2)$ and – under sufficiently smooth dependence – study their corresponding differential-geometric behaviour of $f(\cdot, \cdot)$ on Θ. An example is provided by the Kullback–Leibler divergence between two distributions of the same exponential family of distributions, which defines a Bregman divergence on the parameter space. This and related issues are subsumed in the research field of "information geometry"; for comprehensive overviews see e.g. Amari [3], Amari [1], Ay et al. [8]. Moreover, for recent connections between divergence-based information geometry and optimal transport the reader is e.g. referred to Pal and Wong [66, 67], Karakida and Amari [34], Amari et al. [2], Peyre and Cuturi [71], and the literature therein.

Further relations of divergences with other approaches to geometry can be overviewed e.g. from the wide-range-covering research-article collections in Nielsen and Bhatia [58], Nielsen and Barbaresco [55–57]. Finally, geometry also enters as a tool for visualizing quantitative effects on divergences.

2.3 Divergences and Uncertainty in Data

In general, data-uncertainty (including "deficiencies" like data incompleteness, fakery, unreliability, faultiness, vagueness, etc.) can enter the framework in various different ways. For instance, in situations where $x \in \mathscr{X}$ plays the role of an index (e.g. $\mathscr{X} = \{1, 2, \ldots, s\}$) and $p(x)$ describes the xth real-valued data-point, the uncertainty is typically[4] incorporated by adding a random argument $\omega \in \Omega$ to end up with the "vectors" $P(\omega) := \big\{p(x, \omega)\big\}_{x\in\mathscr{X}}$, $Q(\omega) := \big\{q(x, \omega)\big\}_{x\in\mathscr{X}}$ of random data points. Accordingly, one ends up with random-variable-type pointwise divergences $\omega \to d_\phi^{(j)}(p(x, \omega), q(x, \omega))$, $\omega \to d^{(5)}_{m_1(x),m_2(x),m_3(x)}(p(x, \omega), q(x, \omega))$ $(x \in \mathscr{X})$ as well as with the random-variable-type "summarizing" divergences $\omega \to D_\phi^{(j)}(P(\omega), Q(\omega)) := \sum_{x\in\mathscr{X}} d_\phi^{(j)}(p(x, \omega), q(x, \omega)) \cdot \lambda(x)$ respectively $\omega \to D_\phi^{(j)}(P(\omega), Q(\omega)) := \int_{\mathscr{X}} d_\phi^{(j)}(p(x, \omega), q(x, \omega))\, \mathrm{d}\lambda(x)$, as well as with $\omega \to D^{(5)}_{\phi,M_1,M_2,M_3}(P(\omega), Q(\omega)) := \sum_{x\in\mathscr{X}} d^{(5)}_{m_1(x),m_2(x),m_3(x)}(p(x, \omega), q(x, \omega)) \cdot \lambda(x)$, resp. $\omega \to D^{(5)}_{\phi,M_1,M_2,M_3}(P(\omega), Q(\omega)) := \int_{\mathscr{X}} d^{(5)}_{m_1(x),m_2(x),m_3(x)}(p(x, \omega), q(x, \omega))\mathrm{d}\lambda(x)$. More generally, one can allow for random scales $m_1(x, \omega), m_2(x, \omega), m_3(x, \omega)$.

In other situations with finitely-many-elements carrying $\mathscr{X}$, the state x may e.g. describe a possible outcome $Y(\omega)$ of an uncertainty-prone observation of a quantity Y of interest and $p(x), q(x)$ represent the corresponding probability mass functions ("discrete density functions") at x under two alternative probability mechanisms Pr, $\widetilde{Pr}$ (i.e. $p(x) = Pr[\{\omega \in \Omega : Y(\omega) = x\}]$, $q(x) = \widetilde{Pr}[\{\omega \in \Omega : Y(\omega) = x\}]$); as

[4] In a probabilistic approach rather than a chaos-theoretic approach.

already indicated above, $P := \{p(x)\}_{x\in\mathcal{X}}$ respectively $Q := \{q(x)\}_{x\in\mathcal{X}}$ serve then as a kind of "basis" for the computation of the probabilities $\sum_{x\in A} p(x)$ respectively $\sum_{x\in A} q(x)$ that an arbitrary event $\{\omega \in \Omega : Y(\omega) \in A\}$ $(A \subset \mathcal{X})$ occurs. Accordingly, the pointwise divergences $d_\phi^{(j)}(p(x), q(x)), d^{(5)}_{m_1(x),m_2(x),m_3(x)}(p(x), q(x))$ $(x \in \mathcal{X})$, and the aggregated divergences $D_\phi^{(j)}(P, Q) := \sum_{x\in\mathcal{X}} d_\phi^{(j)}(p(x), q(x))$, $D^{(5)}_{\phi,M_1,M_2,M_3}(P, Q) := \sum_{x\in\mathcal{X}} d^{(5)}_{m_1(x),m_2(x),m_3(x)}(p(x), q(x))$, $D^{(5)}_{\phi,M_1,M_2,M_3}(P, Q) := \int_{\mathcal{X}} d^{(5)}_{m_1(x),m_2(x),m_3(x)}(p(x), q(x))\,\mathrm{d}\lambda(x)$ can then be regarded as (nonnegative, reflexive) dissimilarities between the two alternative uncertainty-quantification-bases P and Q. Analogously, when e.g. $\mathcal{X} = \mathbb{R}^n$ is the n-dimensional Euclidean space and P, Q are classical probability density functions interpreted roughly via $p(x)\mathrm{d}x = Pr[\{\omega \in \Omega : Y(\omega) \in [x, x + \mathrm{d}x[\}$, $q(x)\mathrm{d}x = \widetilde{Pr}[\{\omega \in \Omega : Y(\omega) \in [x, x + \mathrm{d}x[\}$, then $d_\phi^{(j)}(p(x), q(x))$, $d^{(5)}_{m_1(x),m_2(x),m_3(x)}(p(x), q(x))$ $(x \in \mathcal{X})$, $D_\phi^{(j)}(P, Q) := \int_{\mathcal{X}} d_\phi^{(j)}(p(x), q(x))\,\mathrm{d}x$, $D^{(5)}_{\phi,M_1,M_2,M_3}(P, Q) := \int_{\mathcal{X}} d^{(5)}_{m_1(x),m_2(x),m_3(x)}(p(x), q(x))\,\mathrm{d}x$ serve as dissimilarities between the two alternative uncertainty-quantification-bases P, Q.

Let us finally mention that in concrete applications, the "degree" of intrinsic data-uncertainty may be zero (deterministic), low (e.g. small random data contamination and small random deviations from a "basically" deterministic system, slightly noisy data, measurement errors) or high (forecast of the price of a stock in one year from now). Furthermore, the data may contain "high unusualnesses" ("surprising observations") such as outliers and inliers. All this should be taken into account when choosing or even designing the right type of divergence which have different sensitivity to such issues (see e.g. Kißlinger and Stummer [37] and the references therein).

2.4 Divergences, Information and Model Uncertainty

In the main spirit of this book on geometric structures of information, let us also connect the latter with dissimilarities in a wide sense which is appropriate enough for our ambitions of universal modeling. In correspondingly adapting some conception e.g. of Buckland [20] to our above-mentioned investigations, in the following we regard a density function (say) $p(\cdot)$ as a fundamental basis of information understood as quantified real – respectively hypothetical – knowledge which can be communicated about some particular (family of) subjects or (family of) events; according to this information-as-knowledge point of view, pointwise dissimilarities/divergences/distances $d(p(x), q(x))$ $(x \in \mathcal{X})$ respectively aggregated dissimilarities/divergences/distances $D(P, Q)$ quantify the proximity between the two information-bases $P := \{p(x)\}_{x\in\mathcal{X}}$ and $Q := \{q(x)\}_{x\in\mathcal{X}}$ in a directed/nonnegative directed/nonnegative symmetric way. Hence, $d(\cdot, \cdot)$ respectively $D(\cdot, \cdot)$ themselves can be seen as a higher-level information on pairs of information bases.

Divergences can be used for the quantification of information-concerning issues for model uncertainty (model risk) and exploratory model search in various different ways. For instance, suppose that we search for (respectively learn to understand) a true unknown density function $Q^{true} := \{q^{true}(x)\}_{x\in\mathscr{X}}$ of an underlying data-generating mechanism of interest, which is often supposed to be a member of a prefixed class $\mathscr{P}$ of "hypothetical model-candidate density functions"; frequently, this task is (e.g. for the sake of fast tractability) simplified to a setup of finding the true unknown parameter $\theta = \theta_0$ – and hence $Q^{true} = Q_{\theta_0}$ – within a parametric family $\mathscr{P} := \{Q_\theta\}_{\theta\in\Theta}$. Let us first consider the case where the data-generating mechanism of interest Q^{true} is purely deterministic and hence also all the candidates $Q \in \mathscr{P}$ are (taken to be) *not* of probability-density-function type. Although one has no intrinsic data-uncertainty, one faces another type of knowledge-lack called model-uncertainty. Then, one standard goal is to "track down" (respectively learn to understand) this true unknown Q^{true} respectively Q_{θ_0} by collecting and purpose-appropriately postprocessing some corresponding data observations. Accordingly, one attempts to design a density-function-construction rule (mechanism, algorithm) $data \rightarrow P^{data} := \{p^{data}(x)\}_{x\in\mathscr{X}}$ to produce data-derived information-basis-type replica of a "comparable principal form" as the anticipated Q^{true}. This rule should theoretically guarantee that P^{data} converges – with reasonable "operational" speed – to Q^{true} as the number N^{data} of data grows, which particularly implies that (say) $D(P^{data}, Q^{true})$ for some prefixed aggregated divergence $D(\cdot, \cdot)$ becomes close to zero "fast enough". On these grounds, one reasonable strategy to down-narrow the true unknown data-generating mechanism Q^{true} is to take a prefixed class $\mathscr{P}^{hyp}$ of hypothetical density-function models and compute $infodeg := \inf_{Q\in\mathscr{P}^{hyp}} D(P^{data}, Q)$ which in the light of the previous discussions can be interpreted as an "unnormalized degree of informative evidence of Q^{true} being a member of $\mathscr{P}^{hyp}$", or from a reversed point of view, as an "unnormalized degree of goodness of approximation (respectively fit) of the data-derived density function P^{data} through/by means of $\mathscr{P}^{hyp}$". Within this current paradigm, if $infodeg$ is too large (to be specified in a context-dependent, appropriately quantified sense by taking into account the size of N^{data}), then one has to repeat the same procedure with a different class $\widetilde{\mathscr{P}^{hyp}}$; on the other hand, if (and roughly only if) $infodeg$ is small enough then $\widehat{Q^{data}} := \arg\inf_{Q\in\mathscr{P}^{hyp}} D(P^{data}, Q)$ (which may not be unique) is "the most reasonable" approximation. This procedure is repeated recursively as soon as new data points are observed.

In contrast to the last paragraph, let us now cope with the case where the true unknown data-generating mechanism of interest is prone to uncertainties (i.e. is random, noisy, risk-prone) and hence Q^{true} as well as all the candidates $Q \in \mathscr{P}$ *are* of probability-density-function type. Even more, the data-derived information-basis-type replica $\omega \rightarrow data(\omega) \rightarrow P^{data(\omega)} := \{p^{data(\omega)}(x)\}_{x\in\mathscr{X}}$ of Q^{true} is now a density-function-valued (!) random variable; notice that in an above-mentioned "full-scenario" time-evolutionary context, this becomes a density-function-on-functions-valued random variable. Correspondingly, the above-mentioned procedure for the deterministic case has to be adapted and the notions of convergence and smallness have to be stochastified, which leads to the need of considerably more advanced techniques.

Another field of applying divergences to a context of synchronous model and data uncertainty is Bayesian sequential updating. In such a "doubly uncertain" framework, one deals with a parametric context of probability density functions $Q^{true} = Q_{\theta_0}$, $\mathscr{P} := \{Q_\theta\}_{\theta\in\Theta}$ where the uncertain knowledge about the parameter θ (to be learnt) is operationalized by replacing it with a random variable ϑ on Θ. Based on both (i) an initial prior distribution $Prior_1[\cdot] := Pr[\vartheta \in \cdot\,]$ of ϑ (with probability density function pdf $\theta \rightarrow prior_1(\theta)$) and (ii) observed data $data_1(\omega), \ldots, data_{N^{data}}(\omega)$ of number N^{data}, a posterior distribution $Post_1[\cdot, \omega] := Pr[\vartheta \in \cdot \,|\, data_1(\omega), \ldots, data_{N^{data}}(\omega);\ prior[\cdot]\,]$ of ϑ (with pdf $\theta \rightarrow post_1(\theta, \omega)$) is determined with (amongst other things) the help of the well-known Bayes formula. This procedure is repeated recursively with new incoming data input (block) $data_{N^{data}+1}$, where the new prior distribution $Prior_2[\cdot, \omega] := Post_1[\cdot, \omega]$ is chosen as the old posterior and the new posterior distribution is $Post_2[\cdot, \omega] := Pr[\vartheta \in \cdot \,|\, data_1, \ldots, data_{N^{data}}, data_{N^{data}+1};\ Prior_2[\cdot, \omega]\,]$ (with pdf $\theta \rightarrow post_2(\theta, \omega)$), etc. The corresponding (say) aggregated divergence $D(P(\omega), Q(\omega))$ between the probability-density-valued random variables $\omega \rightarrow P(\omega) := \big\{prior_2(\theta, \omega)\big\}_{\theta\in\Theta}$, and $\omega \rightarrow Q(\omega) := \big\{post_2(\theta, \omega)\big\}_{\theta\in\Theta}$ serves as "degree of informativity of the new datapoint observation on the learning of the true unknown θ_0".

As another application in a "doubly uncertain" framework, divergences $D(P, Q)$ appear also in a dichotomous Bayesian testing problem between the two alternative probability densities functions P and Q, where $D(P, Q)$ represents an appropriate average (over prior probabilities) of the corresponding difference between the prior Bayes risk (prior minimal mean decision loss) and the posterior Bayes risk (posterior minimal mean decision loss). This, together with non-averaging versions and an interpretation of $D(P, Q)$ as a (weighted-average) statistical information measure in the sense of De Groot [29] can be found e.g. in Österreicher and Vajda [65]; see also Stummer [78–80], Liese and Vajda [42], Reid and Williamson [73]. In contrast of this employment of $D(P, Q)$ as quantifier of "decision risk reduction" respectively "model risk reduction" respectively "information gain", a different use of divergences $D(P, Q)$ in a "double uncertain" general Bayesian context of dichotomous loss-dependent decisions between arbitrary probability density functions P and Q can be found in Stummer and Vajda [81], where they achieve $D_{\phi_\alpha}(P, Q)$ (for some power functions ϕ_α cf. (5)) as upper and lower bound of the Bayes risk (minimal mean decision loss) itself and also give applications to decision making of time-continuous, non-stationary financial stochastic processes.

Divergences can be also employed to detect distributional changes in streams (respectively clouds) $(data_j)_{j\in\mathcal{T}}$ of uncertain (random, noisy, risk-prone) data indexed by j from an arbitrary countable set $\mathcal{T}$ (e.g. the integers, an undirected graph); a survey together with some general framework can be found in Kißlinger and Stummer [38]: the basic idea is to pick out two[5] non-identical, purpose-appropriately chosen subcollections respectively sample patterns (e.g. windows)

[5]Where one of them may e.g. stem from training data.

$data_{one}(\omega) := (data_{s_1}(\omega), \ldots, data_{s_{N_1}}(\omega))$, $data_{two}(\omega) := (data_{t_1}(\omega), \ldots, data_{t_{N_2}}(\omega))$, and to build from them data-derived probability-density functions $\omega \rightarrow data_{one}(\omega) \rightarrow P^{data_{one}(\omega)} := \big\{p^{data_{one}(\omega)}(x)\big\}_{x \in \mathscr{X}}$, $\omega \rightarrow data_{two}(\omega) \rightarrow P^{data_{two}(\omega)} := \big\{p^{data_{two}(\omega)}(x)\big\}_{x \in \mathscr{X}}$. If a correspondingly chosen (say) aggregated divergence $D\big(P^{data_{one}(\omega)}, P^{data_{two}(\omega)}\big)$ – which plays the role of a condensed change-score – is "significantly large" in the sense that it is large enough – compared to some sound threshold which within the model reflects the desired "degree of confidential plausibility" – then there is strong indication of a distributional change which we then "believe in". Notice that both components of the divergence $D(\cdot, \cdot)$ are now probability-density-function-valued random variables. The sound threshold can e.g. be derived from advanced random asymptotic theory.

From the above discussion it is clear that divergence-based model-uncertainty methods are useful tools in concrete applications for machine learning and artificial intelligence, see e.g. Collins et al. [25], Murata et al. [54], Banerjee et al. [9], Tsuda et al. [87], Cesa-Bianchi and Lugosi [21], Nock and Nielsen [63], Sugiyama et al. [85], Wu et al. [94], Nock et al. [62], Nielsen et al. [60], respectively Minka [51], Cooper et al. [26], Lizier [46], Zhang et al. [96], Chhogyal [22], Cliff et al. [23, 24].

3 General Framework

For the rest of this paper, we shall use the following

Main (i.e. non-locally used) Notation and Symbols

$\mathbb{R}$, $\mathbb{N}$, $\mathbb{R}^d$	Set of real respectively integer numbers respectively d-dimensional vectors
Θ, θ	Set of parameters, see p. 188
$\mathbb{1}$	Function with constant value 1
$\mathbf{1}_A(z) = \delta_z[A]$	Indicator function on the set A evaluated at data point z, which is equal to Dirac's one-point distribution on z evaluated at A
$\#A$	Number of elements in set A
$\mathscr{X}$; $\mathscr{X}_{\#}$	Space/set where data can take values in; space/set of countable size
$\mathscr{F}$	System of admissible events/data-collections (σ-algebra) on $\mathscr{X}$
λ	Reference measure/integrator/summator, see p. 160 & Sect. 3.1 on p. 165
λ-a.a.	λ-almost all, see p. 160
λ_L	Lebesgue measure ("Riemann-type" integrator), see p. 160, & Sect. 3.1
$\lambda_{\#}$	Counting measure ("classical summator"), see p. 160 & Sect. 3.1 on p. 165
$P := \big\{p(x)\big\}_{x \in \mathscr{X}}$	Function from which the divergence/dissimilarity is measured from, see p. 160
$Q := \big\{q(x)\big\}_{x \in \mathscr{X}}$	Function to which the divergence/dissimilarity is measured to, see p. 160
$M_i := \big\{m_i(x)\big\}_{x \in \mathscr{X}}$	Scaling function ($i = 1, 2$) respec. aggregation function ($i = 3$), see p. 161, (1) and paragraph (I1) thereafter, as well as Sect. 3.3 on p. 170
$p(\cdot), q(\cdot), m_i(\cdot)$,	Alternative representations of P, Q, M_i
$R := \big\{r(x)\big\}_{x \in \mathscr{X}}$	Function used for the aggregation function $m_3(\cdot)$, see Sect. 3.3.1 on p. 171
W_i	Connector function of the form $W_i := \big\{w_i(x, y, z)\big\}_{x,y,z \in \ldots}$, for adaptive scaling and aggregation functions $m_i(x) = w_i(x, p(x), q(x))$ $(i = 1, 2, 3)$, see e.g. Assumption 2 on p. 163 and Sect. 3.3.1.3 on p. 181

$\mathbb{P}, \mathbb{Q}, \mathbb{M}_i, \mathbb{W}_i$	Functions with $\mathbbm{p}(x) \geqslant 0$, $\mathbbm{q}(x) \geqslant 0$, $\mathbbm{m}_i(x) \geqslant 0$, $\mathbbm{w}_i(x) \geqslant 0$ for λ-a.a. $x \in \mathscr{X}$
$\mathbb{Q}^{\chi} := \{\mathbbm{q}^{\chi}(x)\}_{x \in \mathscr{X}}$	Function for the aggregation function $m_3(\cdot)$, see Sect. 4.2 on p. 184, (73)
$\boldsymbol{\mathbb{P}}, \boldsymbol{\mathbb{Q}}$	λ-probability density functions (incl. probability mass functions for $\lambda = \lambda_{\#}$), i.e. for which $\boldsymbol{\mathbbm{p}}(x) \geqslant 0$, $\boldsymbol{\mathbbm{q}}(x) \geqslant 0$ for λ-a.a. $x \in \mathscr{X}$ and $\int_{\mathscr{X}} \boldsymbol{\mathbbm{p}}(x)\,\mathrm{d}\lambda(x) = 1$, see Remark 2 on p. 172
$\boldsymbol{\mathbb{Q}}_{\theta} := \{\boldsymbol{\mathbbm{q}}_{\theta}(x)\}_{x \in \mathscr{X}}$	λ-probab. density function which depends on a parameter $\theta \in \Theta$, see p. 188
$\mathscr{R}(\frac{P}{M_1})$	Range (image) of the function $\{\frac{p(x)}{m_1(x)}\}_{x \in \mathscr{X}}$, see paragraph (I2) on p. 161
$\mathscr{R}(Y_1, \ldots, Y_N)$	Range (image) of the random variables $Y_1, \ldots, Y_N$, see p. 182
$\widetilde{\boldsymbol{\mathbb{Q}}}_{\theta} := \{\widetilde{\boldsymbol{\mathbbm{q}}}_{\theta}(x)\}_{x \in \mathscr{X}}$	λ-probab. density function (modification of $\boldsymbol{\mathbb{Q}}_{\theta}$) defined by $\widetilde{\boldsymbol{\mathbbm{q}}}_{\theta}(x) := \boldsymbol{\mathbbm{q}}_{\theta}(x) \cdot (1 - \mathbf{1}_{\mathscr{R}(Y_1(\omega), \ldots, Y_N(\omega))}(x))$, see p. 191
$\phi := \{\phi(t)\}_{t \in]a,b[}$	Divergence generator, a convex real-valued function on $]a, b[$, see p. 161, (1) and paragraph (I2), as well as Sect. 3.2 on p. 165
$\Phi(]a, b[)$;	Class of all such ϕ, see paragraph (I2) on p. 161
$\overline{\phi} := \{\phi(t)\}_{t \in [a,b]}$	Continuous extension of ϕ on $[a, b]$, with $\overline{\phi}(t) = \phi(t)$ for all $t \in]a, b[$, see (I2)
$\phi'_{+,c}(t)$	c-weighted mixture of left-hand and right-hand derivative of ϕ at t, see (I2)
$\Phi_{C_1}(]a, b[)$	Subclass of everywhere continuously differentiable ϕ, with derivative $\phi'(t)$ (being equal to $\phi'_{+,c}(t)$ for all $c \in [0, 1]$), see (I2) on p. 161
ϕ_{α}	α-power-function type divergence generator, see (5) on p. 166, (14), (18), (19)
ϕ_{TV}	Generator of total variation distance, see (31) on p. 169
ϕ_{ie}	Divergence generator with interesting effects, see (35) on p. 170
$\psi_{\phi,c}$	Function given by $\psi_{\phi,c}(s, t) := \phi(s) - \phi(t) - \phi'_{+,c}(t) \cdot (s - t) \geqslant 0$, see (I2)
$\overline{\psi_{\phi,c}}$	Bivariate extension of $\psi_{\phi,c}$, see (I2) on p. 161
$\overline{\int}_{\mathscr{X}} \ldots, \overline{\sum}_{\mathscr{X}} \ldots$	Integral/sum over extension of integrand/summand …, see (I2) & (2) on p. 165
$D^{c}_{\phi, M_1, M_2, \mathbb{M}_3, \lambda}(P, Q)$	Divergence between two functions P (scaled by M_1) and Q (scaled by M_2), generated by ϕ and weight c, and aggregated by $\mathbb{M}_3$ and λ, see (1) on p. 161
$D_{\phi, M_1, M_2, \mathbb{M}_3, \lambda}(P, Q)$	As above, but with $\phi \in \Phi_{C_1}(]a, b[)$ and obsolete c, see Sect. 3.2 on p. 165
$D_{\lambda}(P, Q)$	General λ-aggregated divergence, see p. 189, respectively pseudo-divergence, see Definition 2 on p. 195
$D_{\mathbb{M},\lambda}(\boldsymbol{\mathbb{P}}, \boldsymbol{\mathbb{Q}})$	Pointwise decomposable pseudo-divergence, scaled by $\mathbb{M}$ and aggregated by $\mathbb{M}$ and λ, see Sect. 4.6 on p. 200
NN0, NN1	Nonnegativity setup 0 respectively 1, see p. 166 resp. p. 171
$\mathfrak{P}^{\mathbb{R}\cdot\lambda}, \mathfrak{Q}^{\mathbb{R}\cdot\lambda}, \mathfrak{M}^{\mathbb{R}\cdot\lambda}$	Measures with λ-densities $\mathbbm{p}(\cdot) \cdot \mathbbm{r}(\cdot)$, $\mathbbm{q}(\cdot) \cdot \mathbbm{r}(\cdot)$, $\mathbbm{m}(\cdot) \cdot \mathbbm{r}(\cdot)$, see Remark 2 on p. 171
$\boldsymbol{\mathfrak{P}}^{\mathbb{1}\cdot\lambda}, \boldsymbol{\mathfrak{Q}}^{\mathbb{1}\cdot\lambda}$	Probability measures (distributions) with λ-densities $\boldsymbol{\mathbbm{p}}(\cdot)$, $\boldsymbol{\mathbbm{q}}(\cdot)$, see Remark 2
$\mathscr{Q}^{\lambda_2}_{\Theta}, \boldsymbol{\mathfrak{Q}}^{\mathbb{1}\cdot\lambda_2}_{\theta}$	Class of probability measures with λ_2-densities $\boldsymbol{\mathbbm{q}}_{\theta}(\cdot)$ with parameter $\theta \in \Theta$, see p. 188
$\boldsymbol{\mathfrak{P}}^{emp}_{N}, \boldsymbol{\mathbb{P}}^{emp}_{N}, \boldsymbol{\mathbbm{p}}^{emp}_{N}(\cdot)$	Data-derived empirical (probability) distribution, and probability mass function ($\lambda_{\#}$-density) thereof, see Remark 2 on p. 172
$\boldsymbol{\mathfrak{P}}^{\overline{emp}(\omega)}_{N}, \boldsymbol{\mathbb{P}}^{\overline{emp}(\omega)}_{N}$	Data-derived "extended" empirical (probability) distribution, and probability mass function thereof, see (85) on p. 190 and thereafter
DPD, CASD	Density-power divergences (see p. 174), Csiszar–Ali–Silvey divergences (see p. 177)
$\ell i_1, \phi^{*}(0), \ell i_2, \ell i_3$	Certain limits, see (50), (71), (72)
$\mathbb{P} \perp \mathbb{Q}$	The functions $\mathbb{P}$, $\mathbb{Q}$ are "essentially different", see (64) to (66) and thereafter
$\mathbb{P} \not\perp \mathbb{Q}$	Negation of $\mathbb{P} \perp \mathbb{Q}$, see p. 192
$\mathbb{P} \sim \mathbb{Q}$	The functions $\mathbb{P}$, $\mathbb{Q}$ are "equivalent" (concerning zeros), see (80)
$\mathbb{P} \nsim \mathbb{Q}$	Negation of $\mathbb{P} \sim \mathbb{Q}$, see p. 195
$\widehat{\theta}_{N, D_{\lambda_2}}$	Minimum-divergence estimator ("approximator") of the true unknown parameter θ_0, based on N data observations, see (82) on p. 189

$\widehat{\theta}_{N,D_{\lambda_{\#}}}, \widehat{\theta}_{N,D_{\lambda}}$	Certain minimum-divergence estimators, see (83), (86)
$\widehat{\theta}_{N,decD_{\lambda}}, \widehat{\theta}_{N,decD_{\overrightarrow{\mathbb{Q}}_{\tau},\lambda}}$	Certain minimum-divergence estimators, see (107), (123)
$\widehat{\theta}_{N,sup\mathscr{D}_{\phi,\lambda}}$	Certain minimum-divergence estimator, see (135)
$\mathscr{P}^{\lambda}$	Certain class of nonnegative, mutually equivalent functions, see p. 194
$\mathscr{P}^{\lambda\not\sim}, \widetilde{\mathscr{P}}^{\lambda}$	Certain classes of nonnegative functions, see p. 194
$\mathscr{P}^{\lambda}_{\Theta}, \mathscr{P}^{\lambda\perp}_{emp}, \mathscr{P}^{\lambda}_{\Theta,emp}$	Certain classes of nonnegative functions, see p. 195
$\mathfrak{D}^0, \mathfrak{D}^1, \rho_{\mathbb{Q}}$	Functionals and mapping for decomposable pseudo-divergences, see Definition 3 on p. 195
$\psi^{dec}, \psi^0, \psi^1, \rho$	Mappings for pointwise decomposable pseudo-divergences, see Definition 3 on p. 196
h_0, h_1, h_2	Mappings for pointwise decomposable pseudo-divergences, see Definition 3 on p. 196
ψ^{dec}_m	Perspective function of ψ^{dec}, see (120)

New Divergence Toolkit

In the above Sect. 2, we have motivated that for many different tasks within a broad spectrum of situations, it is useful to employ divergences as "directed distances", including distances as their symmetric special case. For the rest of the paper, we shall only deal with aggregated forms of divergences, and thus drop the attribute "aggregated" from now on. In the following, we present a fairly universal, flexible, multi-component system of divergences by adapting and widening the concept of scaled Bregman divergences of Stummer [81] and Stummer and Vajda [84] to the current context of arbitrary (measurable) functions. To begin with, let us assume that the modeled respectively observed (random) data take values in a state space $\mathscr{X}$ (with at least two distinct values), equipped with a system $\mathscr{F}$ of admissible events (σ-algebra) and a σ-finite measure λ (e.g. the Lebesgue measure, the counting measure, etc.). Furthermore, we suppose that $x \rightarrow p(x) \in [-\infty, \infty]$ and $x \rightarrow q(x) \in [-\infty, \infty]$ are (correspondingly measurable) functions on $\mathscr{X}$ which satisfy $p(x) \in]-\infty, \infty[$, $q(x) \in]-\infty, \infty[$ for λ-almost all (abbreviated as λ-a.a.) $x \in \mathscr{X}$.[6] To address the entire functions as objects we write $P := \{p(x)\}_{x \in \mathscr{X}}$, $Q := \{q(x)\}_{x \in \mathscr{X}}$ and alternatively sometimes also $p(\cdot), q(\cdot)$. To better highlight the very important special case of *λ-probability density functions* – where $p(x) \geqslant 0, q(x) \geqslant 0$ for λ-a.a. $x \in \mathscr{X}$ and $\int_{\mathscr{X}} p(x)\,\mathrm{d}\lambda(x) = 1$, $\int_{\mathscr{X}} q(x)\,\mathrm{d}\lambda(x) = 1$ – we use the notation $\overrightarrow{\mathbb{P}}, \overrightarrow{\mathbb{p}}, \overrightarrow{\mathbb{Q}}, \overrightarrow{\mathbb{q}}$ instead of P, p, Q, q (where $\overrightarrow{\cdots}$ symbolizes a lying 1). For instance, if $\lambda = \lambda_L$ is the Lebesgue measure on the s-dimensional Euclidean space $\mathscr{X} = \mathbb{R}^s$, then $\overrightarrow{\mathbb{P}}, \overrightarrow{\mathbb{Q}}$ are "classical" (e.g. Gaussian) probability density functions. In contrast, in the *discrete setup* where the state space (i.e. the set of all possible data points) $\mathscr{X} = \mathscr{X}_{\#}$ has countably many elements and $\lambda := \lambda_{\#}$ is the counting measure (i.e., $\lambda_{\#}[\{x\}] = 1$ for all $x \in \mathscr{X}_{\#}$), then $\overrightarrow{\mathbb{P}}, \overrightarrow{\mathbb{Q}}$ are probability mass functions and (say) $\overrightarrow{\mathbb{p}}(x)$ can be interpreted as probability that the data point x is taken by the underlying random (uncertainty-prone) mechanism. If $p(x) \geqslant 0$, $q(x) \geqslant 0$ for λ-a.a. $x \in \mathscr{X}$ (but not necessarily with the restrictions $\int_{\mathscr{X}} p(x)\,\mathrm{d}\lambda(x) = 1 = \int_{\mathscr{X}} q(x)\,\mathrm{d}\lambda(x)$) then we write $\mathbb{P}, \mathbb{Q}, \mathbb{p}, \mathbb{q}$ instead of P, p, Q, q.

---

[6]This means that there exists a $N \in \mathscr{F}$ with $\lambda[N] = 0$ (where the empty set $N = \emptyset$ is allowed) such that for all $x \in \mathscr{X} \backslash \{N\}$ (say) $p(x) \in]-\infty, \infty[$ holds.

Back to generality, we quantify the dissimilarity between the two functions P,Q in terms of divergences $D^c_\beta(P, Q)$ with $\beta = (\phi, M_1, M_2, \mathbb{M}_3, \lambda)$, defined by

$$0 \leqslant D^c_{\phi,M_1,M_2,\mathbb{M}_3,\lambda}(P, Q)$$
$$:= \int_{\mathscr{X}} \left[\phi\left(\tfrac{p(x)}{m_1(x)}\right) - \phi\left(\tfrac{q(x)}{m_2(x)}\right) - \phi'_{+,c}\left(\tfrac{q(x)}{m_2(x)}\right) \cdot \left(\tfrac{p(x)}{m_1(x)} - \tfrac{q(x)}{m_2(x)}\right)\right] \cdot \mathbb{m}_3(x)\,\mathrm{d}\lambda(x) \quad (1)$$

(see Stummer [81], Stummer and Vajda [84] for the case $c = 1, m_1(x) = m_2(x) = \mathbb{m}_3(x)$). Here, we use:

(I1) (measurable) *scaling functions* $m_1 : \mathscr{X} \to [-\infty, \infty]$ and $m_2 : \mathscr{X} \to [-\infty, \infty]$ as well as a nonnegative (measurable) *aggregating function* $\mathbb{m}_3 : \mathscr{X} \to [0, \infty]$ such that $m_1(x) \in]-\infty, \infty[$, $m_2(x) \in]-\infty, \infty[$, $\mathbb{m}_3(x) \in [0, \infty[$ for λ-a.a. $x \in \mathscr{X}$.[7] In accordance with the above notation, we use the symbols $M_i := \{m_i(x)\}_{x\in\mathscr{X}}$ respectively $m_i(\cdot)$ to refer to the entire functions, and $\mathbb{M}_\mathbb{i}$, $\mathbb{m}_\mathbb{i}(\cdot)$ when they are nonnegative as well as $\vec{\mathbb{M}}_i$, $\vec{\mathbb{m}}_i(\cdot)$ when they manifest λ-probability density functions. Furthermore, let us emphasize that we allow for / cover adaptive situations in the sense that all three functions $m_1(x)$, $m_2(x)$, $\mathbb{m}_3(x)$ (evaluated at x) may also depend on $p(x)$ and $q(x)$.

(I2) the so-called "divergence-generator" ϕ which is a continuous, convex (finite) function $\phi : E \to]-\infty, \infty[$ on some appropriately chosen open interval $E =]a, b[$ such that $[a, b]$ covers (at least) the union $\mathscr{R}(\frac{P}{M_1}) \cup \mathscr{R}(\frac{Q}{M_2})$ of both ranges $\mathscr{R}(\frac{P}{M_1})$ of $\{\frac{p(x)}{m_1(x)}\}_{x\in\mathscr{X}}$ and $\mathscr{R}(\frac{Q}{M_2})$ of $\{\frac{q(x)}{m_2(x)}\}_{x\in\mathscr{X}}$; for instance, $E =]0, 1[$, $E =]0, \infty[$ or $E =]-\infty, \infty[$; the class of all such functions will be denoted by $\Phi(]a, b[)$. Furthermore, we assume that ϕ is continuously extended to $\overline{\phi} : [a, b] \to [-\infty, \infty]$ by setting $\overline{\phi}(t) := \phi(t)$ for $t \in]a, b[$ as well as $\overline{\phi}(a) := \lim_{t\downarrow a} \phi(t)$, $\overline{\phi}(b) := \lim_{t\uparrow b} \phi(t)$ on the two boundary points $t = a$ and $t = b$. The latter two are the only points at which infinite values may appear. Moreover, for any fixed $c \in [0, 1]$ the (finite) function $\phi'_{+,c} :]a, b[\to]-\infty, \infty[$ is well-defined by $\phi'_{+,c}(t) := c \cdot \phi'_+(t) + (1 - c) \cdot \phi'_-(t)$, where $\phi'_+(t)$ denotes the (always finite) right-hand derivative of ϕ at the point $t \in]a, b[$ and $\phi'_-(t)$ the (always finite) left-hand derivative of ϕ at $t \in]a, b[$. If $\phi \in \Phi(]a, b[)$ is also continuously differentiable – which we denote by $\phi \in \Phi_{C_1}(]a, b[)$ – then for all $c \in [0, 1]$ one gets $\phi'_{+,c}(t) = \phi'(t)$ $(t \in]a, b[)$ and in such a situation we always suppress the obsolete indices c, $+$ in the corresponding expressions. We also employ the continuous continuation $\overline{\phi'_{+,c}} : [a, b] \to [-\infty, \infty]$ given by $\overline{\phi'_{+,c}}(t) := \phi'_{+,c}(t)$ $(t \in]a, b[)$, $\overline{\phi'_{+,c}}(a) := \lim_{t\downarrow a} \phi'_{+,c}(t)$, $\overline{\phi'_{+,c}}(b) := \lim_{t\uparrow b} \phi'_{+,c}(t)$. To explain the precise meaning of (1), we also make use of the (finite, nonnegative) function $\psi_{\phi,c} :]a, b[\times]a, b[\to [0, \infty[$ given by $\psi_{\phi,c}(s, t) := \phi(s) - \phi(t) - \phi'_{+,c}(t) \cdot (s - t) \geqslant 0$ $(s, t \in]a, b[)$. To extend this to a lower semi-continuous

[7] As an example, let $\mathscr{X} = \mathbb{R}$, $\lambda = \lambda_L$ be the Lebesgue measure (and hence, except for rare cases, the integral turns into a Riemann integral) and $\vec{\mathbb{m}}_1(x) := \frac{1}{2} \cdot x^{-1/2} \cdot \mathbf{1}_{[0,1]}(x) \geqslant 0$; since $\int_{\mathscr{X}} \vec{\mathbb{m}}_1(x)\,\mathrm{d}\lambda(x) = 1$ this qualifies as a probability density and thus is a possible candidate for $\vec{\mathbb{m}}_1(x) = \vec{\mathbb{q}}(x)$ in Sect. 3.3.1.2 below.

function $\overline{\psi_{\phi,c}} : [a,b] \times [a,b] \rightarrow [0,\infty]$ we proceed as follows: firstly, we set $\overline{\psi_{\phi,c}}(s,t) := \psi_{\phi,c}(s,t)$ for all $s,t \in]a,b[$. Moreover, since for fixed $t \in]a,b[$, the function $s \rightarrow \psi_{\phi,c}(s,t)$ is convex and continuous, the limit $\overline{\psi_{\phi,c}}(a,t) := \lim_{s\rightarrow a} \psi_{\phi,c}(s,t)$ always exists and (in order to avoid overlines in (1)) will be interpreted/abbreviated as $\phi(a) - \phi(t) - \phi'_{+,c}(t) \cdot (a-t)$. Analogously, for fixed $t \in]a,b[$ we set $\overline{\psi_{\phi,c}}(b,t) := \lim_{s\rightarrow b} \psi_{\phi,c}(s,t)$ with corresponding shorthand notation $\phi(b) - \phi(t) - \phi'_{+,c}(t) \cdot (b-t)$. Furthermore, for fixed $s \in]a,b[$ we interpret $\phi(s) - \phi(a) - \phi'_{+,c}(a) \cdot (s-a)$ as

$$\overline{\psi_{\phi,c}}(s,a) := \left\{\phi(s) - \overline{\phi'_{+,c}}(a) \cdot s + \lim_{t\rightarrow a} \left(t \cdot \overline{\phi'_{+,c}}(a) - \phi(t)\right)\right\} \cdot \mathbf{1}_{]-\infty,\infty[}\left(\overline{\phi'_{+,c}}(a)\right) + \infty \cdot \mathbf{1}_{\{-\infty\}}\left(\overline{\phi'_{+,c}}(a)\right),$$

where the involved limit always exists but may be infinite. Analogously, for fixed $s \in]a,b[$ we interpret $\phi(s) - \phi(b) - \phi'_{+,c}(b) \cdot (s-b)$ as

$$\overline{\psi_{\phi,c}}(s,b) := \left\{\phi(s) - \overline{\phi'_{+,c}}(b) \cdot s + \lim_{t\rightarrow b} \left(t \cdot \overline{\phi'_{+,c}}(b) - \phi(t)\right)\right\} \cdot \mathbf{1}_{]-\infty,\infty[}\left(\overline{\phi'_{+,c}}(b)\right) + \infty \cdot \mathbf{1}_{\{+\infty\}}\left(\overline{\phi'_{+,c}}(b)\right),$$

where again the involved limit always exists but may be infinite. Finally, we always set $\overline{\psi_{\phi,c}}(a,a) := 0$, $\overline{\psi_{\phi,c}}(b,b) := 0$, and $\overline{\psi_{\phi,c}}(a,b) := \lim_{s\rightarrow a} \overline{\psi_{\phi,c}}(s,b)$, $\overline{\psi_{\phi,c}}(b,a) := \lim_{s\rightarrow b} \overline{\psi_{\phi,c}}(s,a)$. Notice that $\overline{\psi_{\phi,c}}(\cdot,\cdot)$ is lower semi-continuous but not necessarily continuous. Since ratios are ultimately involved, we also consistently take $\overline{\psi_{\phi,c}}\left(\frac{0}{0},\frac{0}{0}\right) := 0$. Taking all this into account, we interpret $D^c_{\phi,M_1,M_2,\mathbb{M}_3,\lambda}(P,Q)$ as $\int_{\mathscr{X}} \overline{\psi_{\phi,c}}\left(\frac{p(x)}{m_1(x)},\frac{q(x)}{m_2(x)}\right) \mathrm{m}_3(x)\, \mathrm{d}\lambda(x)$ at first glance (see further investigations in Assumption 2 below), and use the (in lengthy examples) less clumsy notation $\overline{\int}_{\mathscr{X}} \psi_{\phi,c}\left(\frac{p(x)}{m_1(x)},\frac{q(x)}{m_2(x)}\right) \mathrm{m}_3(x)\, \mathrm{d}\lambda(x)$ as a shortcut for the implicitly involved boundary behaviour. □

Notice that despite of the "difference-structure" in the integrand of (1), the splitting of the integral into differences of several "autonomous" integrals may not always be feasible due to the possible appearance of differences between infinite integral values. Furthermore, there is non-uniqueness in the construction (1); for instance, one (formally) gets $D^c_{\phi,M_1,M_2,\mathbb{M}_3,\lambda}(P,Q) = D^c_{\tilde{\phi},M_1,M_2,\mathbb{M}_3,\lambda}(P,Q)$ for any $\tilde{\phi}(t) := \phi(t) + c_1 + c_2 \cdot t \quad (t \in E)$ with $c_1, c_2 \in \mathbb{R}$. Moreover, there exist "essentially different" pairs $(\phi,\mathbb{M})$ and $(\breve{\phi},\breve{\mathbb{M}})$ (where $\phi(t) - \breve{\phi}(t)$ is nonlinear in t) for which $D^c_{\phi,\mathbb{M},\mathbb{M},\mathbb{M},\lambda}(P,Q) = D^c_{\breve{\phi},\breve{\mathbb{M}},\breve{\mathbb{M}},\breve{\mathbb{M}},\lambda}(P,Q)$ (see e.g. [37]). Let us also mention that we could further generalize (1) by adapting the divergence concept of Stummer and Kißlinger [82] who also deal even with non-convex non-concave divergence generators ϕ; for the sake of brevity, this is omitted here.

Notice that by construction we obtain the following important assertion:

Theorem 1 *Let $\phi \in \Phi(]a,b[)$ and $c \in [0,1]$. Then there holds $D^c_{\phi,M_1,M_2,\mathbb{M}_3,\lambda}(P,Q) \geqslant 0$ with equality if $\frac{p(x)}{m_1(x)} = \frac{q(x)}{m_2(x)}$ for λ-almost all $x \in \mathscr{X}$. Depending on the concrete situation, $D^c_{\phi,M_1,M_2,\mathbb{M}_3,\lambda}(P,Q)$ may take infinite value.*

To get "sharp identifiability" (i.e. reflexivity) one needs further assumptions on $\phi \in \Phi(]a,b[)$, $c \in [0,1]$. As a motivation, consider the case where $\mathbb{m}_3(x) \equiv 1$ and $\phi \in \Phi(]a,b[)$ is affine linear on the whole interval $]a,b[$, and hence its extension $\overline{\phi}$ is affine-linear on $[a,b]$. Accordingly, one gets for the integrand-builder $\overline{\psi_{\phi,c}}(s,t) \equiv 0$ and hence $D^c_{\phi,M_1,M_2,\mathbb{M}_3,\lambda}(P,Q) = \int_{\mathscr{X}} \overline{\psi_{\phi,c}}\big(\frac{p(x)}{m_1(x)}, \frac{q(x)}{m_2(x)}\big)\,\mathrm{d}\lambda(x) = 0$ even in cases where $\frac{p(x)}{m_1(x)} \neq \frac{q(x)}{m_2(x)}$ for λ-a.a. $x \in \mathscr{X}$. In order to avoid such and similar phenomena, we use the following set of requirements:

Assumption 2 Let $c \in [0,1]$, $\phi \in \Phi(]a,b[)$ and $\mathscr{R}\big(\frac{P}{M_1}\big) \cup \mathscr{R}\big(\frac{Q}{M_2}\big) \subset [a,b]$. The aggregation function is supposed to be of the form $\mathbb{m}_3(x) = \mathbb{w}_3\big(x, \frac{p(x)}{m_1(x)}, \frac{q(x)}{m_2(x)}\big)$ for some (measur.) function $\mathbb{w}_3 : \mathscr{X} \times [a,b] \times [a,b] \to [0,\infty]$. Moreover, for all $s \in \mathscr{R}\big(\frac{P}{M_1}\big)$, all $t \in \mathscr{R}\big(\frac{Q}{M_2}\big)$ and λ-a.a. $x \in \mathscr{X}$, let the following conditions hold:

(a) ϕ is strictly convex at t;
(b) if ϕ is differentiable at t and $s \neq t$, then ϕ is not affine-linear on the interval $[\min(s,t), \max(s,t)]$ (i.e. between t and s);
(c) if ϕ is not differentiable at t, $s > t$ and ϕ is affine linear on $[t,s]$, then we exclude $c = 1$ for the ("globally/universally chosen") subderivative $\phi'_{+,c}(\cdot) = c \cdot \phi'_{+}(\cdot) + (1-c) \cdot \phi'_{-}(\cdot)$;
(d) if ϕ is not differentiable at t, $s < t$ and ϕ is affine linear on $[s,t]$, then we exclude $c = 0$ for $\phi'_{+,c}(\cdot)$;
(e) $\mathbb{w}_3(x,s,t) < \infty$;
(f) $\mathbb{w}_3(x,s,t) > 0$ if $s \neq t$;
(g) $\mathbb{w}_3(x,a,a) \cdot \psi_{\phi,c}(a,a) := 0$ by convention (even in cases where the function $\mathbb{w}_3(x,\cdot,\cdot) \cdot \psi_{\phi,c}(\cdot,\cdot)$ is not continuous on the boundary point (a,a));
(h) $\mathbb{w}_3(x,b,b) \cdot \psi_{\phi,c}(b,b) := 0$ by convention (even in cases where the function $\mathbb{w}_3(x,\cdot,\cdot) \cdot \psi_{\phi,c}(\cdot,\cdot)$ is not continuous on the boundary point (b,b));
(i) $\mathbb{w}_3(x,a,t) \cdot \psi_{\phi,c}(a,t) > 0$, where $\mathbb{w}_3(x,a,t) \cdot \psi_{\phi,c}(a,t) := \lim_{s\to a} \mathbb{w}_3(x,s,t) \cdot \psi_{\phi,c}(s,t)$ if this limit exists, and otherwise we set by convention $\mathbb{w}_3(x,a,t) \cdot \psi_{\phi,c}(a,t) := 1$ (or any other strictly positive constant);
(j) $\mathbb{w}_3(x,b,t) \cdot \psi_{\phi,c}(b,t) > 0$, where $\mathbb{w}_3(x,b,t) \cdot \psi_{\phi,c}(b,t)$ is analogous to (i);
(k) $\mathbb{w}_3(x,s,a) \cdot \psi_{\phi,c}(s,a) > 0$, where $\mathbb{w}_3(x,s,a) \cdot \psi_{\phi,c}(s,a) := \lim_{t\to a} \mathbb{w}_3(x,s,t) \cdot \psi_{\phi,c}(s,t)$ if this limit exists, and otherwise we set by convention $\mathbb{w}_3(x,s,a) \cdot \psi_{\phi,c}(s,a) := 1$ (or any other strictly positive constant);
(l) $\mathbb{w}_3(x,s,b) \cdot \psi_{\phi,c}(s,b) > 0$, where $\mathbb{w}_3(x,s,b) \cdot \psi_{\phi,c}(s,b)$ is analogous to (k);
(m) $\mathbb{w}_3(x,a,b) \cdot \psi_{\phi,c}(a,b) > 0$, where $\mathbb{w}_3(x,a,b) \cdot \psi_{\phi,c}(a,b) := \lim_{s\to a} \mathbb{w}_3(x,s,b) \cdot \psi_{\phi,c}(s,b)$ if this limit exists, and otherwise we set by convention $\mathbb{w}_3(x,a,b) \cdot \psi_{\phi,c}(a,b) := 1$ (or any other strictly positive constant);
(n) $\mathbb{w}_3(x,b,a) \cdot \psi_{\phi,c}(b,a) > 0$, where $\mathbb{w}_3(x,b,a) \cdot \psi_{\phi,c}(b,a) := \lim_{s\to b} \mathbb{w}_3(x,s,a) \cdot \psi_{\phi,c}(s,a)$ if this limit exists, and otherwise we set by convention $\mathbb{w}_3(x,b,a) \cdot \psi_{\phi,c}(b,a) := 1$ (or any other strictly positive constant). □

Under Assumption 2, we always interpret the corresponding divergence

$$D^c_{\phi,M_1,M_2,\mathbb{M}_3,\lambda}(P,Q) := D^c_{\phi,M_1,M_2,\mathbb{W}_3,\lambda}(P,Q) :=$$
$$:= \overline{\int}_{\mathscr{X}} \mathbb{w}_3\big(x, \frac{p(x)}{m_1(x)}, \frac{q(x)}{m_2(x)}\big) \cdot \big[\phi(\frac{p(x)}{m_1(x)}) - \phi(\frac{q(x)}{m_2(x)})$$
$$-\phi'_{+,c}(\frac{q(x)}{m_2(x)}) \cdot (\frac{p(x)}{m_1(x)} - \frac{q(x)}{m_2(x)})\big] \, \mathrm{d}\lambda(x)$$

as $\int_{\mathscr{X}} \overline{\mathbb{w}_3 \cdot \psi_{\phi,c}}\big(x, \frac{p(x)}{m_1(x)}, \frac{q(x)}{m_2(x)}\big) \, \mathrm{d}\lambda(x)$, where $\overline{\mathbb{w}_3 \cdot \psi_{\phi,c}}(x,s,t)$ denotes the extension of the function $\mathscr{X} \times]a,b[\times]a,b[\ni (x,s,t) \rightarrow \mathbb{w}_3(x,s,t) \cdot \psi_{\phi,c}(s,t)$ on $\mathscr{X} \times [a,b] \times [a,b]$ according to the conditions (g) to (n) above.

Remark 1 (a) We could even work with a weaker assumption obtained by replacing s with $\frac{p(x)}{m_1(x)}$ as well as t with $\frac{q(x)}{m_2(x)}$ and by requiring that then the correspondingly plugged-in conditions (a) to (n) hold for λ-a.a. $x \in \mathscr{X}$.
(b) Notice that our above context subsumes aggregation functions of the form $\mathbb{m}_3(x) = \tilde{\mathbb{w}}_3(x, p(x), q(x), m_1(x), m_2(x))$ with $\tilde{\mathbb{w}}_3(x, z_1, z_2, z_3, z_4)$ having appropriately imbeddable behaviour in its arguments x, z_1, z_2, z_3, z_4, the outcoming ratios $\frac{z_1}{z_3}, \frac{z_2}{z_4}$ and possible boundary values thereof. □

The following requirement is stronger than the "model-individual/dependent" Assumption 2 but is more "universally applicable" (amongst *all* models such that $\mathscr{R}(\frac{P}{M_1}) \cup \mathscr{R}(\frac{Q}{M_2}) \subset [a,b]$, take e.g. $E =]a,b[$ as $E =]0,\infty[$ or $E =]-\infty,\infty[$):

Assumption 3 Let $c \in [0,1]$, $\phi \in \Phi(]a,b[)$ on some fixed $]a,b[\in]-\infty,+\infty[$ such that $]a,b[\supset \mathscr{R}(\frac{P}{M_1}) \cup \mathscr{R}(\frac{Q}{M_2})$. The aggregation function is of the form $\mathbb{m}_3(x) = \mathbb{w}_3\big(x, \frac{p(x)}{m_1(x)}, \frac{q(x)}{m_2(x)}\big)$ for some (measurable) function $\mathbb{w}_3 : \mathscr{X} \times [a,b] \times [a,b] \rightarrow [0,\infty]$. Furthermore, for all $s \in]a,b[$, $t \in]a,b[$ and λ-a.a. $x \in \mathscr{X}$, the conditions (a) to (n) of Assumption 2 hold.

Important examples in connection with the Assumptions 2, 3 will be given in Sect. 3.2 (for ϕ) and Sect. 3.3 (for m_1, m_2, $\mathbb{w}_3$) below. With these assumptions at hand, we obtain the following non-negativity and reflexivity assertions:

Theorem 4 *Let the Assumption 2 be satisfied. Then there holds:*
(1) $D^c_{\phi,M_1,M_2,\mathbb{M}_3,\lambda}(P,Q) \geqslant 0$. *Depending on the concrete situation,* $D^c_{\phi,M_1,M_2,\mathbb{M}_3,\lambda}(P,Q)$ *may take infinite value.*
(2) $D^c_{\phi,M_1,M_2,\mathbb{M}_3,\lambda}(P,Q) = 0$ *if and only if* $\frac{p(x)}{m_1(x)} = \frac{q(x)}{m_2(x)}$ *for* λ*-a.a.* $x \in \mathscr{X}$.

Theorem 4 – whose proof will be given in the appendix – says that $D^c_{\phi,M_1,M_2,\mathbb{M}_3,\lambda}(P,Q)$ is indeed a "proper" divergence under the Assumption 2. Hence, the latter will be assumed for the rest of the paper, unless stated otherwise: for instance, we shall sometimes work with the stronger Assumption 3; thus, for more comfortable reference, we state explicitly

Corollary 1 *Under the more universally applicable Assumption 3, the Assertions (1) and (2) of Theorem 4 hold.*

Under some non-obvious additional constraints on the functions P, Q it may be possible to show the Assertions (1), (2) of Theorem 4 by even dropping the purely generator-concerning Assumptions 2(b) to (d); see e.g. Sect. 3.3.1.2 below. In the following, we discuss several important features and special cases of $\beta = (\phi, M_1, M_2, \mathbb{M}_3, \lambda)$ in a well-structured way. Let us start with the latter.

3.1 The Reference Measure λ

In (1), λ can be interpreted as a "governer" upon the *principle* aggregation structure, whereas the "aggregation function" $\mathbb{m}_3$ tunes the *fine* aggregation details. For instance, if one chooses $\lambda = \lambda_L$ as the Lebesgue measure on $\mathcal{X} \subset \mathbb{R}$, then the integral in (1) turns out to be of Lebesgue-type and (with some rare exceptions) consequently of Riemann-type. In contrast, in the *discrete setup* where $\mathcal{X} := \mathcal{X}_{\#}$ has countably many elements and is equipped with the counting measure $\lambda := \lambda_{\#} := \sum_{z \in \mathcal{X}_{\#}} \delta_z$ (where δ_z is Dirac's one-point distribution $\delta_z[A] := \mathbf{1}_A(z)$, and thus $\lambda_{\#}[\{z\}] = 1$ for all $z \in \mathcal{X}_{\#}$) then (1) simplifies to

$$0 \leqslant D^c_{\phi,M_1,M_2,\mathbb{M}_3,\lambda_{\#}}(P,Q)$$
$$:= \overline{\sum}_{z\in\mathcal{X}} \Big[\phi\big(\tfrac{p(z)}{m_1(z)}\big) - \phi\big(\tfrac{q(z)}{m_2(z)}\big) - \phi'_{+,c}\big(\tfrac{q(z)}{m_2(z)}\big) \cdot \big(\tfrac{p(z)}{m_1(z)} - \tfrac{q(z)}{m_2(z)}\big)\Big] \cdot \mathbb{m}_3(z)\,, \qquad (2)$$

which we interpret as $\sum_{z\in\mathcal{X}} \overline{\psi_{\phi,c}\big(\frac{p(z)}{m_1(z)}, \frac{q(z)}{m_2(z)}\big)} \cdot \mathbb{m}_3(z)$ with the same conventions and limits as in the paragraph right after (1); if $\mathcal{X}_{\#} = \{z_0\}$ for arbitrary $z_0 \in \widetilde{X}$, we obtain the corresponding one-point divergence over any space $\widetilde{X}$.

3.2 The Divergence Generator ϕ

We continue with the inspection of interesting special cases of
$\beta = (\phi, M_1, M_2, \mathbb{M}_3, \lambda)$ by dealing with the first component. For this, let $\Phi_{C_1}(]a,b[)$ be the class of all functions $\phi \in \Phi(]a,b[)$ which are also continuously differentiable on $E =]a,b[$. For divergence generator $\phi \in \Phi_{C_1}(]a,b[)$, the formula (1) becomes (recall that we suppress the obsolete c and subderivative index $+$)

$$0 \leqslant D_{\phi,M_1,M_2,\mathbb{M}_3,\lambda}(P,Q)$$
$$:= \overline{\int}_{\mathcal{X}} \Big[\phi\big(\tfrac{p(x)}{m_1(x)}\big) - \phi\big(\tfrac{q(x)}{m_2(x)}\big) - \phi'\big(\tfrac{q(x)}{m_2(x)}\big) \cdot \big(\tfrac{p(x)}{m_1(x)} - \tfrac{q(x)}{m_2(x)}\big)\Big] \cdot \mathbb{m}_3(x)\, \mathrm{d}\lambda(x)\,, \qquad (3)$$

whereas (2) turns into

$$0 \leqslant D_{\phi,M_1,M_2,\mathbb{M}_3,\lambda_\#}(P,Q) := \overline{\sum}_{x\in\mathscr{X}} \Big[\phi\big(\tfrac{p(x)}{m_1(x)}\big) - \phi\big(\tfrac{q(x)}{m_2(x)}\big) - \phi'\big(\tfrac{q(x)}{m_2(x)}\big)\cdot\big(\tfrac{p(x)}{m_1(x)} - \tfrac{q(x)}{m_2(x)}\big)\Big]\cdot \mathbb{m}_3(x).$$

Formally, by defining the integral functional $g_{\phi,\mathbb{M}_3,\lambda}(\xi) := \int_{\mathscr{X}} \phi(\xi(x))\cdot \mathbb{m}_3(x)\,\mathrm{d}\lambda(x)$ and plugging in e.g. $g_{\phi,\mathbb{M}_3,\lambda}\big(\frac{P}{M_1}\big) = \int_{\mathscr{X}} \phi\big(\frac{p(x)}{m_1(x)}\big)\cdot \mathbb{m}_3(x)\,\mathrm{d}\lambda(x)$, the divergence in (3) can be interpreted as

$$0 \leqslant D_{\phi,M_1,M_2,\mathbb{M}_3,\lambda}(P,Q) = g_{\phi,\mathbb{M}_3,\lambda}\big(\tfrac{P}{M_1}\big) - g_{\phi,\mathbb{M}_3,\lambda}\big(\tfrac{Q}{M_2}\big) - g'_{\phi,\mathbb{M}_3,\lambda}\big(\tfrac{Q}{M_2}, \tfrac{P}{M_1} - \tfrac{Q}{M_2}\big) \tag{4}$$

where $g'_{\phi,\mathbb{M}_3,\lambda}(\eta,\cdot)$ denotes the corresponding directional derivate at $\eta = \frac{Q}{M_2}$. If one has a "nonnegativity-setup" (NN0) in the sense that for all $x \in \mathscr{X}$ there holds $\frac{p(x)}{m_1(x)} \geqslant 0$ and $\frac{q(x)}{m_2(x)} \geqslant 0$ (but not necessarily $p(x) \geqslant 0$, $q(x) \geqslant 0$, $m_1(x) \geqslant 0$, $m_2(x) \geqslant 0$) then one can take $a = 0$, $b = \infty$, i.e. $E =]0,\infty[$, and employ the strictly convex power functions

$$\tilde{\phi}(t) := \tilde{\phi}_\alpha(t) := \tfrac{t^\alpha - 1}{\alpha(\alpha-1)} \in]-\infty,\infty[, \qquad t \in]0,\infty[,\ \alpha \in \mathbb{R}\backslash\{0,1\},$$
$$\phi(t) := \phi_\alpha(t) := \tilde{\phi}_\alpha(t) - \tilde{\phi}'_\alpha(1)\cdot(t-1) = \tfrac{t^\alpha - 1}{\alpha(\alpha-1)} - \tfrac{t-1}{\alpha-1} \in [0,\infty[, \quad t \in]0,\infty[,\ \alpha \in \mathbb{R}\backslash\{0,1\}, \tag{5}$$

which satisfy (with the notations introduced in the paragraph right after (1))

$$\phi_\alpha(1) = 0, \quad \phi'_\alpha(t) = \tfrac{t^{\alpha-1}-1}{\alpha-1}, \quad \phi'_\alpha(1) = 0, \quad \phi''_\alpha(t) = t^{\alpha-2} > 0, \quad t \in]0,\infty[, \tag{6}$$

$$\phi_\alpha(0) := \lim_{t\downarrow 0} \phi_\alpha(t) = \tfrac{1}{\alpha}\cdot \mathbf{1}_{]0,1[\cup]1,\infty[}(\alpha) + \infty\cdot\mathbf{1}_{]-\infty,0[}(\alpha), \qquad \phi_\alpha(\infty) := \lim_{t\uparrow\infty} \phi_\alpha(t) = \infty, \tag{7}$$

$$\phi'_\alpha(0) := \lim_{t\downarrow 0} \phi'_\alpha(t) = \tfrac{1}{1-\alpha}\cdot\mathbf{1}_{]1,\infty[}(\alpha) - \infty\cdot\mathbf{1}_{]-\infty,0[\cup]0,1[}(\alpha),$$
$$\phi'_\alpha(\infty) := \lim_{t\uparrow\infty} \phi'_\alpha(t) = \infty\cdot\mathbf{1}_{]1,\infty[}(\alpha) + \tfrac{1}{1-\alpha}\cdot\mathbf{1}_{]-\infty,0[\cup]0,1[}(\alpha) = \lim_{t\uparrow\infty}\tfrac{\phi_\alpha(t)}{t}, \tag{8}$$

$$\psi_{\phi_\alpha}(s,t) = \tfrac{1}{\alpha\cdot(\alpha-1)}\cdot\big[s^\alpha + (\alpha-1)\cdot t^\alpha - \alpha\cdot s\cdot t^{\alpha-1}\big], \quad s,t \in]0,\infty[, \tag{9}$$

$$\psi_{\phi_\alpha}(0,t) = \tfrac{t^\alpha}{\alpha}\cdot\mathbf{1}_{]0,1[\cup]1,\infty[}(\alpha) + \infty\cdot\mathbf{1}_{]-\infty,0[}(\alpha), \quad t \in]0,\infty[, \tag{10}$$

$$\psi_{\phi_\alpha}(\infty,t) = \infty, \quad t \in]0,\infty[,$$
$$\lim_{s\to\infty}\tfrac{1}{s}\cdot\psi_{\phi_\alpha}(s,1) = \tfrac{1}{1-\alpha}\cdot\mathbf{1}_{]-\infty,0[\cup]0,1[}(\alpha) + \infty\cdot\mathbf{1}_{]1,\infty[}(\alpha),$$
$$\psi_{\phi_\alpha}(s,0) = \tfrac{s^\alpha}{\alpha\cdot(\alpha-1)}\cdot\mathbf{1}_{]1,\infty[}(\alpha) + \infty\cdot\mathbf{1}_{]-\infty,0[\cup]0,1[}(\alpha), \quad s \in]0,\infty[, \tag{11}$$
$$\psi_{\phi_\alpha}(s,\infty) = \tfrac{s^\alpha}{\alpha\cdot(\alpha-1)}\cdot\mathbf{1}_{]-\infty,0[}(\alpha) + \infty\cdot\mathbf{1}_{]0,1[\cup]1,\infty[}(\alpha), \quad s \in]0,\infty[,$$

$\psi_{\phi_\alpha}(0,0) := 0$ (which is unequal to $\lim_{t\to 0}\lim_{s\to 0}\psi_{\phi_\alpha}(s,t)$ for $\alpha < 0$ and which is unequal to $\lim_{s\to 0}\lim_{t\to 0}\psi_{\phi_\alpha}(s,t)$ for $\alpha > 1$),

$\psi_{\phi_\alpha}(\infty,\infty) := 0$ (which is unequal to $\lim_{t\to\infty}\lim_{s\to\infty}\psi_{\phi_\alpha}(s,t)$ for $\alpha \in \mathbb{R}\backslash\{0,1\}$ and which is unequal to $\lim_{s\to\infty}\lim_{t\to\infty}\psi_{\phi_\alpha}(s,t)$ for $\alpha \in]0,1[\cup]1,\infty[$),

$$\psi_{\phi_\alpha}(0,\infty) := \lim_{s\to 0}\lim_{t\to\infty}\psi_{\phi_\alpha}(s,t) = \infty \tag{12}$$

$$\text{(which coincides with } \lim_{t\to\infty}\lim_{s\to 0} \psi_{\phi_\alpha}(s,t) \text{ for } \alpha \in \mathbb{R}\backslash\{0,1\}),$$

$$\psi_{\phi_\alpha}(\infty,0) := \lim_{s\to\infty}\lim_{t\to 0}\psi_{\phi_\alpha}(s,t) = \infty \tag{13}$$

$$\text{(which coincides with } \lim_{t\to 0}\lim_{s\to\infty} \psi_{\phi_\alpha}(s,t) \text{ for } \alpha \in \mathbb{R}\backslash\{0,1\}).$$

The perhaps most important special case is $\alpha = 2$, for which (5) turns into

$$\phi_2(t) := \tfrac{(t-1)^2}{2}, \quad t \in]0,\infty[= E, \tag{14}$$

having for $s, t \in]0, \infty[$ the properties (cf. (7)–(13))

$$\begin{aligned}
&\phi_2(1) = 0, \quad \phi_2'(1) = 0, \quad \phi_2(0) = \tfrac{1}{2}, \quad \phi_2(\infty) = \infty, \quad \phi_2'(0) = -\tfrac{1}{2}, \\
&\phi_2'(\infty) = \infty = \lim_{t\uparrow\infty} \tfrac{\phi_2(t)}{t}, \psi_{\phi_2}(s,t) = \tfrac{(s-t)^2}{2}, \qquad (15)\\
&\psi_{\phi_2}(0,t) = \tfrac{t^2}{2}, \quad \psi_{\phi_2}(\infty,t) = \infty, \quad \lim_{s\to\infty} \tfrac{1}{s}\cdot\psi_{\phi_2}(s,1) = \infty, \\
&\psi_{\phi_2}(s,0) = \tfrac{s^2}{2}, \quad \psi_{\phi_2}(s,\infty) = \infty, \quad \psi_{\phi_2}(0,0) := 0, \qquad (16)\\
&\psi_{\phi_2}(\infty,\infty) := 0, \quad \psi_{\phi_2}(0,\infty) = \infty, \quad \psi_{\phi_2}(\infty,0) = \infty.
\end{aligned}$$

Also notice that the divergence-generator ϕ_2 of (14) can be trivially extended to

$$\bar{\phi}_2(t) := \tfrac{(t-1)^2}{2}, \quad t \in]-\infty,\infty[= \bar{E}, \tag{17}$$

which is useful in a general setup (GS) where for all $x \in \mathscr{X}$ one has $\frac{p(x)}{m_1(x)} \in [-\infty, \infty]$ and $\frac{q(x)}{m_2(x)} \in [-\infty, \infty]$. Convex extensions to $]a, \infty[$ with $a \in]-\infty, 0[$ can be easily done by the shift $\bar{\phi}_\alpha(t) := \phi_\alpha(t-a)$.

Further examples of everywhere strictly convex differentiable divergence generators $\phi \in \Phi_{C_1}(]a,b[)$ for the "nonnegativity-setup" (NN0) (i.e. $a = 0$, $b = \infty$, $E =]0, \infty[$) can be obtained by taking the α-limits

$$\begin{aligned}
&\tilde{\phi}_1(t) := \lim_{\alpha\to 1}\phi_\alpha(t) = t\cdot\log t \in [-e^{-1},\infty[, \qquad t \in]0,\infty[,\\
&\phi_1(t) := \lim_{\alpha\to 1}\phi_\alpha(t) = \tilde{\phi}_1(t) - \tilde{\phi}_1'(1)\cdot(t-1) = t\cdot\log t + 1 - t \in [0,\infty[, \; t \in]0,\infty[, \; (18)\\
&\tilde{\phi}_0(t) := \lim_{\alpha\to 0}\phi_\alpha(t) = -\log t \in]-\infty,\infty[, \qquad t \in]0,\infty[,\\
&\phi_0(t) := \lim_{\alpha\to 0}\phi_\alpha(t) = \tilde{\phi}_0(t) - \tilde{\phi}_0'(1)\cdot(t-1) = -\log t + t - 1 \in [0,\infty[, \; t \in]0,\infty[, \; (19)
\end{aligned}$$

which satisfy

$$\phi_1(1) = 0, \quad \phi_1'(t) = \log t, \quad \phi_1'(1) = 0, \quad \phi_1''(t) = t^{-1} > 0, \quad t \in]0, \infty[,$$

$$\phi_1(0) := \lim_{t \downarrow 0} \phi_1(t) = 1, \quad \phi_1(\infty) := \lim_{t \uparrow \infty} \phi_1(t) = \infty, \tag{20}$$

$$\phi_1'(0) := \lim_{t \downarrow 0} \phi_1'(t) = -\infty, \quad \phi_1'(\infty) := \lim_{t \uparrow \infty} \phi_1'(t) = +\infty = \lim_{t \uparrow \infty} \tfrac{\phi_1(t)}{t}, \tag{21}$$

$$\psi_{\phi_1}(s,t) = s \cdot \log\left(\tfrac{s}{t}\right) + t - s, \quad s,t \in]0, \infty[, \tag{22}$$

$$\psi_{\phi_1}(0,t) = t, \quad \psi_{\phi_1}(\infty, t) = \infty, \quad \lim_{s \to \infty} \tfrac{1}{s} \cdot \psi_{\phi_1}(s,1) = \infty, \quad t \in]0, \infty[, \tag{23}$$

$$\psi_{\phi_1}(s,0) = \infty, \quad \psi_{\phi_1}(s,\infty) = \infty, \quad s \in]0, \infty[, \tag{24}$$

$$\psi_{\phi_1}(0,0) := 0 \text{ (which coincides with } \lim_{t \to 0} \lim_{s \to 0} \psi_{\phi_1}(s,t)$$
$$\text{but which does not coincide with } \lim_{s \to 0} \lim_{t \to 0} \psi_{\phi_1}(s,t) = \infty),$$

$$\psi_{\phi_1}(\infty,\infty) := 0 \text{ (which does not coincide with}$$
$$\lim_{t \to \infty} \lim_{s \to \infty} \psi_{\phi_1}(s,t) = \lim_{s \to \infty} \lim_{t \to \infty} \psi_{\phi_1}(s,t) = \infty,$$

$$\psi_{\phi_1}(0,\infty) := \lim_{s \to 0} \lim_{t \to \infty} \psi_{\phi_1}(s,t) = \infty$$
$$\text{(which coincides with } \lim_{t \to \infty} \lim_{s \to 0} \psi_{\phi_1}(s,t)),$$

$$\psi_{\phi_1}(\infty,0) := \lim_{s \to \infty} \lim_{t \to 0} \psi_{\phi_1}(s,t) = \infty$$
$$\text{(which coincides with } \lim_{t \to 0} \lim_{s \to \infty} \psi_{\phi_1}(s,t)),$$

as well as

$$\phi_0(1) = 0, \quad \phi_0'(t) = 1 - \tfrac{1}{t}, \quad \phi_0'(1) = 0, \quad \phi_0''(t) = t^{-2} > 0, \quad t \in]0, \infty[, \tag{25}$$

$$\phi_0(0) := \lim_{t \downarrow 0} \phi_0(t) = \infty, \quad \phi_0(\infty) := \lim_{t \uparrow \infty} \phi_0(t) = \infty, \tag{26}$$

$$\phi_0'(0) := \lim_{t \downarrow 0} \phi_0'(t) = -\infty, \quad \phi_0'(\infty) := \lim_{t \uparrow \infty} \phi_0'(t) = 1 = \lim_{t \uparrow \infty} \tfrac{\phi_0(t)}{t}, \tag{27}$$

$$\psi_{\phi_0}(s,t) = -\log\left(\tfrac{s}{t}\right) + \tfrac{s}{t} - 1, \quad s,t \in]0, \infty[, \tag{28}$$

$$\psi_{\phi_0}(0,t) = \infty, \quad \psi_{\phi_0}(\infty,t) = \infty, \quad \lim_{s \to \infty} \tfrac{1}{s} \cdot \psi_{\phi_0}(s,1) = 1, \quad t \in]0, \infty[, \tag{29}$$

$$\psi_{\phi_0}(s,0) = \infty, \quad \psi_{\phi_0}(s,\infty) = \infty, \quad s \in]0, \infty[, \tag{30}$$

$$\psi_{\phi_0}(0,0) := 0 \text{ (which does not coincide with}$$
$$\lim_{t \to 0} \lim_{s \to 0} \psi_{\phi_0}(s,t) = \lim_{s \to 0} \lim_{t \to 0} \psi_{\phi_0}(s,t) = \infty),$$

$$\psi_{\phi_0}(\infty,\infty) := 0 \text{ (which does not coincide with}$$
$$\lim_{t \to \infty} \lim_{s \to \infty} \psi_{\phi_0}(s,t) = \lim_{s \to \infty} \lim_{t \to \infty} \psi_{\phi_0}(s,t) = \infty),$$

$$\psi_{\phi_0}(0,\infty) := \lim_{s \to 0} \lim_{t \to \infty} \psi_{\phi_0}(s,t) = \infty$$
$$\text{(which coincides with } \lim_{t \to \infty} \lim_{s \to 0} \psi_{\phi_0}(s,t)),$$

$$\psi_{\phi_0}(\infty,0) := \lim_{s \to \infty} \lim_{t \to 0} \psi_{\phi_0}(s,t) = \infty$$
$$\text{(which coincides with } \lim_{t \to 0} \lim_{s \to \infty} \psi_{\phi_0}(s,t)).$$

An important, but (in our context) technically delicate, convex divergence generator is $\phi_{TV}(t) := |t-1|$ which is non-differentiable at $t = 1$; the latter is also the only point of strict convexity. Further properties are for arbitrarily fixed $s, t \in]0, \infty[$, $c \in [0, 1]$ (if not stated otherwise)

$$\phi_{TV}(1) = 0, \quad \phi_{TV}(0) = 1, \quad \phi_{TV}(\infty) = \infty, \tag{31}$$
$$\phi'_{TV,+,c}(t) = \mathbf{1}_{]1,\infty[}(t) + (2c-1)\cdot \mathbf{1}_{\{1\}}(t) - \mathbf{1}_{]0,1[}(t),$$
$$\phi'_{TV,+,1}(t) = \mathbf{1}_{[1,\infty[}(t) - \mathbf{1}_{]0,1[}(t),$$
$$\phi'_{TV,+,\frac{1}{2}}(t) = \mathbf{1}_{]1,\infty[}(t) - \mathbf{1}_{]0,1[}(t) = \text{sgn}(t-1)\cdot \mathbf{1}_{]0,\infty[}(t),$$
$$\phi'_{TV,+,c}(1) = 2c-1, \qquad \phi'_{TV,+,1}(1) = 1, \quad \phi'_{TV,+,\frac{1}{2}}(1) = 0, \tag{32}$$
$$\phi'_{TV,+,c}(0) = \lim_{t\to 0}\phi'_{TV,+,c}(t) = -1, \quad \phi'_{TV,+,c}(\infty) = \lim_{t\to\infty}\phi'_{TV,+,c}(t) = 1,$$
$$\psi_{\phi_{TV},c}(s,t) = \mathbf{1}_{]0,1[}(t)\cdot 2(s-1)\cdot \mathbf{1}_{]1,\infty[}(s) + \mathbf{1}_{]1,\infty[}(t)\cdot 2(1-s)\cdot \mathbf{1}_{]0,1]}(s)$$
$$+ \mathbf{1}_{\{1\}}(t)\cdot\Big[2(1-c)\cdot(s-1)\cdot \mathbf{1}_{]1,\infty[}(s) + 2c\cdot(1-s)\cdot \mathbf{1}_{]0,1]}(s)\Big],$$
$$\psi_{\phi_{TV},\frac{1}{2}}(s,1) = |s-1|, \tag{33}$$
$$\psi_{\phi_{TV},c}(0,t) = \lim_{s\to 0}\psi_{\phi_{TV},c}(s,t) = 2\cdot \mathbf{1}_{]1,\infty[}(t) + 2c\cdot \mathbf{1}_{\{1\}}(t),$$
$$\psi_{\phi_{TV},c}(\infty,t) = \lim_{s\to\infty}\psi_{\phi_{TV},c}(s,t) = \infty\cdot \mathbf{1}_{]0,1[}(t) + \ \infty\cdot \mathbf{1}_{\{1\}}(t)\cdot \mathbf{1}_{[0,1[}(c),$$
$$\lim_{s\to\infty}\tfrac{1}{s}\cdot\psi_{\phi_{TV},c}(s,1) = 2(1-c), \tag{34}$$
$$\psi_{\phi_{TV},c}(s,0) = \lim_{t\to 0}\psi_{\phi_{TV},c}(s,t) = 2(s-1)\cdot \mathbf{1}_{]1,\infty[}(s),$$
$$\psi_{\phi_{TV},c}(s,\infty) = \lim_{t\to\infty}\psi_{\phi_{TV},c}(s,t) = 2(1-s)\cdot \mathbf{1}_{]0,1]}(s),$$
$$\psi_{\phi_{TV},c}(0,0) := 0 \text{ (which coincides with both } \lim_{t\to 0}\lim_{s\to 0}\psi_{\phi_{TV},c}(s,t)$$
$$\text{and } \lim_{s\to 0}\lim_{t\to 0}\psi_{\phi_{TV},c}(s,t)),$$
$$\psi_{\phi_{TV},c}(\infty,\infty) := 0 \text{ (which coincides with both } \lim_{t\to\infty}\lim_{s\to\infty}\psi_{\phi_{TV},c}(s,t)$$
$$\text{and } \lim_{s\to\infty}\lim_{t\to\infty}\psi_{\phi_{TV},c}(s,t)),$$
$$\psi_{\phi_{TV},c}(0,\infty) := \lim_{s\to 0}\lim_{t\to\infty}\psi_{\phi_{TV},c}(s,t) = 2$$
$$\text{(which coincides with } \lim_{t\to\infty}\lim_{s\to 0}\psi_{\phi_{TV},c}(s,t)),$$
$$\psi_{\phi_{TV},c}(\infty,0) := \lim_{s\to\infty}\lim_{t\to 0}\psi_{\phi_{TV},c}(s,t) = \infty$$
$$\text{(which coincides with } \lim_{t\to 0}\lim_{s\to\infty}\psi_{\phi_{TV},c}(s,t)).$$

In particular, one sees from Assumption 2(a) that – in our context – ϕ_{TV} can only be potentially applied if $\frac{q(x)}{m_2(x)} = 1$ for λ-a.a. $x \in \mathscr{X}$ and from Assumption 2(c), (d) that we *generally* have to exclude $c = 1$ and $c = 0$ for $\phi'_{+,c}(\cdot)$ (i.e. we choose $c \in]0, 1[$); as already mentioned above, under some non-obvious additional constraints on the functions P, Q it may be possible to drop the Assumptions 2(c), (d), see for instance Sect. 3.3.1.2 below.

Another interesting and technically delicate example is the divergence generator $\phi_{ie}(t) := t - 1 + \frac{(1-t)^3}{3}\cdot \mathbf{1}_{[0,1]}(t)$ which is convex, twice continuously differentiable, strictly convex at any point $t \in]0, 1]$ and affine-linear on $[1, \infty[$. More detailed, one obtains for arbitrarily fixed $s, t \in]0, \infty[$ (if not stated otherwise):

$$\begin{aligned}
&\phi_{ie}(1) = 0, \quad \phi_{ie}(0) = -\tfrac{2}{3}, \quad \phi_{ie}(\infty) = \infty, \qquad (35)\\
&\phi_{ie}'(t) = 1 - (1-t)^2 \cdot \mathbf{1}_{]0,1[}(t),\\
&\phi_{ie}'(1) = 1, \quad \phi_{ie}'(0) = \lim_{t\to 0}\phi_{ie}'(t) = 0, \quad \phi_{ie}'(\infty) = \lim_{t\to\infty}\phi_{ie}'(t) = 1,\\
&\phi_{ie}''(t) = 2(1-t)\cdot \mathbf{1}_{]0,1[}(t), \quad \phi_{ie}''(1) = 0,\\
&\psi_{\phi_{ie}}(s,t) = \tfrac{(1-s)^3}{3}\cdot \mathbf{1}_{]0,1[}(s) + \ (1-t)^2 \cdot \Big[\tfrac{2}{3}\cdot(1-t) + (s-1)\Big]\cdot \mathbf{1}_{]0,1[}(t),\\
&\psi_{\phi_{ie}}(s,1) = \tfrac{(1-s)^3}{3}\cdot \mathbf{1}_{]0,1[}(s),\\
&\psi_{\phi_{ie}}(0,t) \ = \ \lim_{s\to 0}\psi_{\phi_{ie}}(s,t) \ = \ \tfrac{1}{3}\cdot \mathbf{1}_{[1,\infty[}(t) + \tfrac{1}{3}\cdot \Big[1 - (1-t)^2\cdot(1-2t)\Big]\cdot \mathbf{1}_{]0,1[}(t),\\
&\psi_{\phi_{ie}}(\infty,t) = \lim_{s\to\infty}\psi_{\phi_{ie}}(s,t) = \infty \cdot \mathbf{1}_{]0,1[}(t),\\
&\lim_{s\to\infty}\tfrac{1}{s}\cdot\psi_{\phi_{ie}}(s,1) = 0,\\
&\psi_{\phi_{ie}}(s,0) = \lim_{t\to 0}\psi_{\phi_{ie}}(s,t) = \big(s - \tfrac{1}{3}\big)\cdot \mathbf{1}_{[1,\infty[}(s) + s^2\cdot\big(1 - \tfrac{s}{3}\big)\cdot \mathbf{1}_{]0,1[}(s),\\
&\psi_{\phi_{ie}}(s,\infty) = \lim_{t\to\infty}\psi_{\phi_{ie}}(s,t) = \tfrac{(1-s)^3}{3}\cdot \mathbf{1}_{]0,1[}(s),\\
&\psi_{\phi_{ie}}(0,0) := 0 \text{ (which coincides with both } \lim_{t\to 0}\lim_{s\to 0}\psi_{\phi_{ie}}(s,t)\\
&\qquad\qquad \text{and } \lim_{s\to 0}\lim_{t\to 0}\psi_{\phi_{ie}}(s,t)),\\
&\psi_{\phi_{ie}}(\infty,\infty) := 0 \text{ (which coincides with both } \lim_{t\to\infty}\lim_{s\to\infty}\psi_{\phi_{ie}}(s,t)\\
&\qquad\qquad \text{and } \lim_{s\to\infty}\lim_{t\to\infty}\psi_{\phi_{ie}}(s,t)),\\
&\psi_{\phi_{ie}}(0,\infty) := \lim_{s\to 0}\lim_{t\to\infty}\psi_{\phi_{ie}}(s,t) = \tfrac{1}{3}\\
&\qquad\qquad \text{(which coincides with } \lim_{t\to\infty}\lim_{s\to 0}\psi_{\phi_{ie}}(s,t)),\\
&\psi_{\phi_{ie}}(\infty,0) := \lim_{s\to\infty}\lim_{t\to 0}\psi_{\phi_{ie}}(s,t) = \infty\\
&\qquad\qquad \text{(which coincides with } \lim_{t\to 0}\lim_{s\to\infty}\psi_{\phi_{ie}}(s,t)).
\end{aligned}$$

In particular, one sees from the Assumptions 2(a), (b) that – in our context – ϕ_{ie} can only be potentially applied in the following two disjoint situations:

(i) $\frac{q(x)}{m_2(x)} < 1$ for λ-a.a. $x \in \mathscr{X}$;
(ii) $\frac{q(x)}{m_2(x)} = 1$ and $\frac{p(x)}{m_1(x)} \leqslant 1$ for λ-a.a. $x \in \mathscr{X}$.

As already mentioned above, under some non-obvious additional constraints on the functions P, Q it may be possible to drop Assumption 2(b) and consequently (ii) can then be replaced by
$\widetilde{(ii)}$ $\frac{q(x)}{m_2(x)} = 1$ for λ-a.a. $x \in \mathscr{X}$;
see for instance Sect. 3.3.1.2 below.

3.3 *The Scaling and the Aggregation Functions m_1, m_2, $\mathbb{m}_3$*

In the above two Sects. 3.1 and 3.2, we have illuminated details of the choices of the first and the last component of $\beta = (\phi, M_1, M_2, \mathbb{M}_3, \lambda)$. Let us now discuss the *principal* roles as well as examples of m_1, m_2, $\mathbb{m}_3$, which widen considerably the divergence-modeling flexibility and thus bring in a broad spectrum of goal-oriented situation-based applicability. To start with, recall that in accordance with (1), the aggregation function $\mathbb{m}_3$ tunes the fine aggregation details (whereas λ can be interpreted as a "governer" upon the basic/principle aggregation structure); furthermore, the function $m_1(\cdot)$ scales the function $p(\cdot)$ and $m_2(\cdot)$ the function $q(\cdot)$. From a modeling perspective, these two scaling functions can e.g. be "purely direct" in the sense that $m_1(x), m_2(x)$ are chosen to directly reflect some dependence on the data-reflecting state $x \in \mathscr{X}$ (independent of the choice of P,Q), or "purely adaptive" in the sense that $m_1(x) = w_1(p(x), q(x))$, $m_2(x) = w_2(p(x), q(x))$ for some appropriate (measurable) "connector functions" w_1, w_2 on the product $\mathscr{R}(P) \times \mathscr{R}(Q)$ of the ranges of $\{p(x)\}_{x\in\mathscr{X}}$ and $\{q(x)\}_{x\in\mathscr{X}}$, or "hybrids" $m_1(x) = w_1(x, p(x), q(x))$ $m_2(x) = w_2(x, p(x), q(x))$. Also recall that in consistency with Assumption 2 we always assume $\mathbb{m}_3(x) = \mathbb{w}_3\big(x, \frac{p(x)}{m_1(x)}, \frac{q(x)}{m_2(x)}\big)$ for some (measurable) function $\mathbb{w}_3 : \mathscr{X} \times [a, b] \times [a, b] \to [0, \infty]$. Whenever applicable and insightfulness-enhancing, we use the notation $D^c_{\phi,W_1,W_2,\mathbb{W}_3,\lambda}(P, Q)$ instead of $D^c_{\phi,M_1,M_2,\mathbb{M}_3,\lambda}(P, Q)$.

Let us start with the following important sub-setup:

3.3.1 $\mathbf{m_1(x) = m_2(x) := m(x), \mathbb{m}_3(x) = r(x) \cdot m(x) \in [0, \infty]}$ for Some (meas.) Function $\mathbf{r : \mathscr{X} \to \mathbb{R}}$ Satisfying $\mathbf{r(x) \in]-\infty, 0[\cup]0, \infty[}$ for $\mathbf{\lambda}$-a.a. $\mathbf{x \in \mathscr{X}}$

As an interpretation, here the scaling functions are strongly coupled with the aggregation function; in order to avoid "case-overlapping", we assume that the function $r(\cdot)$ does not (explicitly) depend on the functions $m(\cdot)$, $p(\cdot)$ and $q(\cdot)$ (i.e. it is not of the form $r(\cdot) = h(\cdot, m(\cdot), p(\cdot), q(\cdot))$). From (1) one can deduce

$$0 \leqslant D^c_{\phi,M,M,R\cdot M,\lambda}(P, Q) \\ := \overline{\int}_{\mathscr{X}} \Big[\phi\big(\tfrac{p(x)}{m(x)}\big) - \phi\big(\tfrac{q(x)}{m(x)}\big) - \phi'_{+,c}\big(\tfrac{q(x)}{m(x)}\big) \cdot \big(\tfrac{p(x)}{m(x)} - \tfrac{q(x)}{m(x)}\big)\Big] \cdot m(x) \cdot r(x)\, \mathrm{d}\lambda(x)\,, \qquad (36)$$

which for the discrete setup $(\mathscr{X}, \lambda) = (\mathscr{X}_\#, \lambda_\#)$ (recall $\lambda_\#[\{x\}] = 1$ for all $x \in \mathscr{X}_\#$) simplifies to

$$0 \leqslant D^c_{\phi,M,M,R\cdot M,\lambda_\#}(P, Q) \\ = \overline{\sum}_{x\in\mathscr{X}} \Big[\phi\big(\tfrac{p(x)}{m(x)}\big) - \phi\big(\tfrac{q(x)}{m(x)}\big) - \phi'_{+,c}\big(\tfrac{q(x)}{m(x)}\big) \cdot \big(\tfrac{p(x)}{m(x)} - \tfrac{q(x)}{m(x)}\big)\Big] \cdot m(x) \cdot r(x)\,. \qquad (37)$$

Remark 2 (a) If one has a "nonnegativity-setup" (NN1) in the sense that for λ-almost all $x \in \mathscr{X}$ there holds $\mathbb{m}(x) \geqslant 0$, $\mathbb{r}(x) \geqslant 0$, $\mathbb{p}(x) \geqslant 0$, $\mathbb{q}(x) \geqslant 0$, then (36) (and hence also (37)) can be interpreted as scaled Bregman divergence $B_\phi(\mathfrak{P}, \mathfrak{Q} \,|\, \mathfrak{M})$ between the two nonnegative measures $\mathfrak{P}, \mathfrak{Q}$ (on $(\mathscr{X}, \mathscr{F})$) defined by $\mathfrak{P}[\bullet] := \mathfrak{P}^{\mathbb{R}\cdot\lambda}[\bullet] := \int_\bullet \mathbb{p}(x) \cdot \mathbb{r}(x)\, \mathrm{d}\lambda(x)$ and $\mathfrak{Q}[\bullet] := \mathfrak{Q}^{\mathbb{R}\cdot\lambda}[\bullet] := \int_\bullet \mathbb{q}(x) \cdot \mathbb{r}(x)\, \mathrm{d}\lambda(x)$, with scaling by the nonnegative measure $\mathfrak{M}[\bullet] := \mathfrak{M}^{\mathbb{R}\cdot\lambda}[\bullet] := \int_\bullet \mathbb{m}(x) \cdot \mathbb{r}(x)\, \mathrm{d}\lambda(x)$.

(b) In a context of $\mathbb{r}(x) \equiv 1$ and "λ-probability-densities" $\mathbb{p}$, $\mathbb{q}$ on general state space $\mathscr{X}$, then $\overset{\to}{\mathfrak{P}}^{\mathbb{1}\cdot\lambda}[\bullet] := \int_\bullet \mathbb{p}(x)\,\mathrm{d}\lambda(x)$ and $\overset{\to}{\mathfrak{Q}}^{\mathbb{1}\cdot\lambda}[\bullet] := \int_\bullet \mathbb{q}(x)\,\mathrm{d}\lambda(x)$ are probability measures (where $\mathbb{1}$ stands for the function with constant value 1). Accordingly, (36) (and hence also (37)) can be interpreted as scaled Bregman divergence $B_\phi\big(\overset{\to}{\mathfrak{P}}^{\mathbb{1}\cdot\lambda}, \overset{\to}{\mathfrak{Q}}^{\mathbb{1}\cdot\lambda} \,|\, \mathfrak{M}^{\mathbb{1}\cdot\lambda}\big)$ which has been first defined in Stummer [81], Stummer and Vajda [84], see also Kisslinger and Stummer [35–37] for the "purely adaptive" case $\mathbb{m}(x) = \mathbb{w}\big(\mathbb{p}(x), \mathbb{q}(x)\big)$ and indications on non-probability measures. For instance, if Y is a random variable taking values in the discrete space $\mathscr{X}_\#$, then (with a slight abuse of notation[8]) $\mathbb{q}(x) = \overset{\to}{\mathfrak{Q}}^{\mathbb{1}\cdot\lambda_\#}[Y = x]$ may be its probability mass function under a hypothetical/candidate law $\overset{\to}{\mathfrak{Q}}^{\mathbb{1}\cdot\lambda_\#}$, and $\mathbb{p}(x) = \frac{1}{N} \cdot \#\{i \in \{1, \ldots, N\} : Y_i = x\} =: \mathbb{p}_N^{emp}(x)$ is the probability mass function of the corresponding data-derived "empirical distribution" $\overset{\to}{\mathfrak{P}}^{\mathbb{1}\cdot\lambda_\#}[\bullet] := \overset{\to}{\mathfrak{P}}_N^{emp}[\bullet] := \frac{1}{N} \cdot \sum_{i=1}^N \delta_{Y_i}[\bullet]$ of an N-size independent and identically distributed (i.i.d.) sample $Y_1, \ldots, Y_N$ of Y which is nothing but the probability distribution reflecting the underlying (normalized) histogram. Typically, for small respectively medium sample size N one gets $\mathbb{p}_N^{emp}(x) = 0$ for some states $x \in \mathscr{X}$ which are feasible but "not yet" observed; amongst other things, this explains why density-zeros play an important role especially in statistics and information theory. This concludes the current Remark 2. □

In the following, we illuminate two important special cases of the scaling (and aggregation-part) function $m(\cdot)$, namely $\mathbb{m}(x) := 1$ and $m(x) := q(x)$:

3.3.1.1 $\mathbb{m}_1(\mathbf{x}) = \mathbb{m}_2(\mathbf{x}) := \mathbf{1}$, $\mathbb{m}_3(\mathbf{x}) = \mathbb{r}(\mathbf{x})$ for Some (Measurable) Function $\mathbb{r} : \mathscr{X} \to [\mathbf{0}, \infty]$ Satisfying $\mathbb{r}(\mathbf{x}) \in]\mathbf{0}, \infty[$ for $\boldsymbol{\lambda}$–a.a. $\mathbf{x} \in \mathscr{X}$

Accordingly, (36) turns into

$$\begin{aligned} 0 &\leqslant D^c_{\phi,\mathbb{1},\mathbb{1},\mathbb{R}\cdot\mathbb{1},\lambda}(P, Q) \\ &:= \overline{\int}_{\mathscr{X}} \Big[\phi\big(p(x)\big) - \phi\big(q(x)\big) - \phi'_{+,c}\big(q(x)\big) \cdot \big(p(x) - q(x)\big)\Big] \cdot \mathbb{r}(x)\,\mathrm{d}\lambda(x)\,, \end{aligned} \tag{38}$$

which for the discrete setup $(\mathscr{X}, \lambda) = (\mathscr{X}_\#, \lambda_\#)$ becomes[9]

$$\begin{aligned} 0 &\leqslant D^c_{\phi,\mathbb{1},\mathbb{1},\mathbb{R}\cdot\mathbb{1},\lambda_\#}(P, Q) \\ &:= \overline{\sum}_{x \in \mathscr{X}} \Big[\phi\big(p(x)\big) - \phi\big(q(x)\big) - \phi'_{+,c}\big(q(x)\big) \cdot \big(p(x) - q(x)\big)\Big] \cdot \mathbb{r}(x) \end{aligned} \tag{39}$$

---

[8] Respectively working with canonical space representation and $Y := id$.

[9] As a side remark, let us mention here that in the special case of continuously differentiable *strictly log-convex* divergence generator ϕ, one can construct divergences which are tighter than (38) respectively (39), see Stummer and Kißlinger [82]; in a finite discrete space and for differentiable *exponentially concave* divergence generator ϕ, a similar tightening (called L-divergence) can be found in Pal and Wong [66, 67].

Notice that for $\mathbb{r}(x) \equiv 1$, the divergences (38) and (39) are "consistent extensions" of the motivating pointwise dissimilarity $d_{\phi}^{(6)}(\cdot,\cdot)$ from Sect. 2. A special case of (38) is e.g. the rho-tau divergence (cf. Lemma 1 of Zhang and Naudts [95]).
Let us exemplarily illuminate the special case $\phi = \phi_\alpha$ together with $\mathbb{p}(x) \geqslant 0$, $\mathbb{q}(x) \geqslant 0$, for λ-almost all $x \in \mathscr{X}$ which by means of (9), (22), (28) turns (38) into the "explicit-boundary" version (of the corresponding "implicit-boundary-describing" $\overline{\int} \ldots$)[10]

$$0 \leqslant D_{\phi_\alpha,\mathbb{1},\mathbb{1},\mathbb{R}\cdot\mathbb{1},\lambda}(\mathbb{P},\mathbb{Q})$$
$$= \overline{\int}_{\mathscr{X}} \tfrac{\mathbb{r}(x)}{\alpha\cdot(\alpha-1)} \cdot \big[\mathbb{p}(x)^\alpha + (\alpha-1)\cdot\mathbb{q}(x)^\alpha - \alpha\cdot\mathbb{p}(x)\cdot\mathbb{q}(x)^{\alpha-1}\big]\,\mathrm{d}\lambda(x) \tag{40}$$
$$= \int_{\mathscr{X}} \tfrac{\mathbb{r}(x)}{\alpha\cdot(\alpha-1)} \cdot \big[\mathbb{p}(x)^\alpha + (\alpha-1)\cdot\mathbb{q}(x)^\alpha - \alpha\cdot\mathbb{p}(x)\cdot\mathbb{q}(x)^{\alpha-1}\big]\cdot \mathbf{1}_{]0,\infty[}\big(\mathbb{p}(x)\cdot\mathbb{q}(x)\big)\,\mathrm{d}\lambda(x)$$
$$+ \int_{\mathscr{X}} \mathbb{r}(x)\cdot\big[\tfrac{\mathbb{p}(x)^\alpha}{\alpha\cdot(\alpha-1)}\cdot\mathbf{1}_{]1,\infty[}(\alpha) + \infty\cdot\mathbf{1}_{]-\infty,0[\cup]0,1[}(\alpha)\big]\cdot\mathbf{1}_{]0,\infty[}\big(\mathbb{p}(x)\big)\cdot\mathbf{1}_{\{0\}}\big(\mathbb{q}(x)\big)\,\mathrm{d}\lambda(x)$$
$$+ \int_{\mathscr{X}} \mathbb{r}(x)\cdot\big[\tfrac{\mathbb{q}(x)^\alpha}{\alpha}\cdot\mathbf{1}_{]0,1[\cup]1,\infty[}(\alpha) + \infty\cdot\mathbf{1}_{]-\infty,0[}(\alpha)\big]\cdot\mathbf{1}_{]0,\infty[}\big(\mathbb{q}(x)\big)\cdot\mathbf{1}_{\{0\}}\big(\mathbb{p}(x)\big)\,\mathrm{d}\lambda(x),$$
$$\text{for } \alpha \in \mathbb{R}\backslash\{0,1\}, \tag{41}$$

$$0 \leqslant D_{\phi_1,\mathbb{1},\mathbb{1},\mathbb{R}\cdot\mathbb{1},\lambda}(\mathbb{P},\mathbb{Q})$$
$$= \int_{\mathscr{X}} \mathbb{r}(x)\cdot\big[\mathbb{p}(x)\cdot\log\big(\tfrac{\mathbb{p}(x)}{\mathbb{q}(x)}\big) + \mathbb{q}(x) - \mathbb{p}(x)\big]\cdot\mathbf{1}_{]0,\infty[}\big(\mathbb{p}(x)\cdot\mathbb{q}(x)\big)\,\mathrm{d}\lambda(x)$$
$$+ \int_{\mathscr{X}} \mathbb{r}(x)\cdot\infty\cdot\mathbf{1}_{]0,\infty[}\big(\mathbb{p}(x)\big)\cdot\mathbf{1}_{\{0\}}\big(\mathbb{q}(x)\big)\,\mathrm{d}\lambda(x)$$
$$+ \int_{\mathscr{X}} \mathbb{r}(x)\cdot\mathbb{q}(x)\cdot\mathbf{1}_{]0,\infty[}\big(\mathbb{q}(x)\big)\cdot\mathbf{1}_{\{0\}}\big(\mathbb{p}(x)\big)\,\mathrm{d}\lambda(x) \tag{42}$$

$$0 \leqslant D_{\phi_0,\mathbb{1},\mathbb{1},\mathbb{R}\cdot\mathbb{1},\lambda}(\mathbb{P},\mathbb{Q})$$
$$= \int_{\mathscr{X}} \mathbb{r}(x)\cdot\Big[-\log\big(\tfrac{\mathbb{p}(x)}{\mathbb{q}(x)}\big) + \tfrac{\mathbb{p}(x)}{\mathbb{q}(x)} - 1\Big]\cdot\mathbf{1}_{]0,\infty[}\big(\mathbb{p}(x)\cdot\mathbb{q}(x)\big)\,\mathrm{d}\lambda(x)$$
$$+ \int_{\mathscr{X}} \mathbb{r}(x)\cdot\infty\cdot\mathbf{1}_{]0,\infty[}\big(\mathbb{p}(x)\big)\cdot\mathbf{1}_{\{0\}}\big(\mathbb{q}(x)\big)\,\mathrm{d}\lambda(x)$$
$$+ \int_{\mathscr{X}} \mathbb{r}(x)\cdot\infty\cdot\mathbf{1}_{]0,\infty[}\big(\mathbb{q}(x)\big)\cdot\mathbf{1}_{\{0\}}\big(\mathbb{p}(x)\big)\,\mathrm{d}\lambda(x), \tag{43}$$

where we have employed (10), (11) (23), (24), (29), (30); notice that $D_{\phi_1,\mathbb{1},\mathbb{1},\mathbb{R}\cdot\mathbb{1},\lambda}(\mathbb{P},\mathbb{Q})$ is a generalized version of the Kullback–Leibler information divergence (resp. of the relative entropy). According to the above calculations, one should exclude $\alpha \leqslant 0$ whenever $\mathbb{p}(x) = 0$ for all x in some A with $\lambda[A] > 0$, respectively $\alpha \leqslant 1$ whenever $\mathbb{q}(x) = 0$ for all x in some $\tilde{A}$ with $\lambda[\tilde{A}] > 0$ (a refined alternative for $\alpha = 1$ is given in Sect. 3.3.1.2 below). As far as splitting of the first integral e.g. in (42) resp. (43) is concerned, notice that the integral $(\mathfrak{P}^{\mathbb{R}\cdot\lambda} - \mathfrak{Q}^{\mathbb{R}\cdot\lambda})[\mathscr{X}] := \int_{\mathscr{X}} \big[\mathbb{q}(x) - \mathbb{p}(x)\big]\cdot\mathbb{r}(x)\,\mathrm{d}\lambda(x)$ resp. $\int_{\mathscr{X}} \Big[\frac{\mathbb{p}(x)}{\mathbb{q}(x)} - 1\Big]\cdot\mathbb{r}(x)\,\mathrm{d}\lambda(x)$ may be finite even in cases where $\mathfrak{P}^{\mathbb{R}\cdot\lambda}[\mathscr{X}] = \int_{\mathscr{X}} \mathbb{p}(x)\cdot\mathbb{r}(x)\,\mathrm{d}\lambda(x) = \infty$ and $\mathfrak{Q}^{\mathbb{R}\cdot\lambda}[\mathscr{X}] = \int_{\mathscr{X}} \mathbb{q}(x)\cdot\mathbb{r}(x)\,\mathrm{d}\lambda(x) = \infty$ (especially in case of unbounded data space (e.g. $\mathscr{X} = \mathbb{R}$) when an additive constant is involved and $\mathbb{r}(\cdot)$ is bounded from above); furthermore, there are situations where $\mathfrak{P}^{\mathbb{R}\cdot\lambda}[\mathscr{X}] = \mathfrak{Q}^{\mathbb{R}\cdot\lambda}[\mathscr{X}] < \infty$ and thus $(\mathfrak{P}^{\mathbb{R}\cdot\lambda} - \mathfrak{Q}^{\mathbb{R}\cdot\lambda})[\mathscr{X}] = 0$ but $\int_{\mathscr{X}} \Big[\frac{\mathbb{p}(x)}{\mathbb{q}(x)} - 1\Big]\cdot\mathbb{r}(x)\,\mathrm{d}\lambda(x) = \infty$. For $\alpha = 2$, we obtain from (41) and (15) to (16)

[10]The first resp. second resp. third integral in (41) can be interpreted as divergence-contribution of the function-(support-)overlap resp. of one part of the function-nonoverlap (e.g. describing "extreme outliers") resp. of the other part of the function-nonoverlap (e.g. describing "extreme inliers").

$$0 \leqslant D_{\phi_2,\mathbb{1},\mathbb{1},\mathbb{R}\cdot\mathbb{1},\lambda}(\mathbb{P},\mathbb{Q}) = \int_{\mathscr{X}} \frac{\mathbb{r}(x)}{2} \cdot \big[\mathbb{p}(x) - \mathbb{q}(x)\big]^2 \, \mathrm{d}\lambda(x)\,, \tag{44}$$

where we can exceptionally drop the non-negativity constraints $\mathbb{p}(x) \geqslant 0, \mathbb{q}(x) \geqslant 0$. As for interpretation, (44) is nothing but half of the $\mathbb{r}(\cdot)$-weighted squared $L^2(\lambda)$-distance between $\mathbb{p}(\cdot)$ and $\mathbb{q}(\cdot)$.

In the special sub-setup of $\mathbb{r}(x) \equiv 1$ and "λ-probability-densities" $\breve{\mathbb{p}}$ $\breve{\mathbb{q}}$ on data space $\mathscr{X}$ (cf. Remark 2(b)), we can deduce from (41)–(43) the divergences

$$D_{\phi_\alpha,\mathbb{1},\mathbb{1},\mathbb{1}\cdot\mathbb{1},\lambda}(\breve{\mathbb{P}},\breve{\mathbb{Q}}) \tag{45}$$

which for the choice $\alpha > 0$ can be interpreted as "order$-\alpha$" density-power divergences DPD of Basu et al. [10] between the two corresponding probability measures $\breve{\mathfrak{P}}^{\mathbb{1}\cdot\lambda}$ and $\breve{\mathfrak{Q}}^{\mathbb{1}\cdot\lambda}$; for their statistical applications see e.g. Basu et al. [12], Ghosh and Basu [30, 31] and the references therein, and for general $\alpha \in \mathbb{R}$ see e.g. Stummer and Vajda [84]. In particular, the case $\alpha = 1$ corresponding divergence in (45) is called "Kullback–Leibler information divergence" between $\breve{\mathbb{P}}$ and $\breve{\mathbb{Q}}$, and is also known under the name "relative entropy". For $\alpha = 2$, we derive $D_{\phi_2,\mathbb{1},\mathbb{1},\mathbb{R}\cdot\mathbb{1},\lambda}(\breve{\mathbb{P}},\breve{\mathbb{Q}})$ from (44) with $\mathbb{r}(x) = 1$ which is nothing but half of the squared L^2-distance between the two "λ-probability-densities" $\breve{\mathbb{p}}$ and $\breve{\mathbb{q}}$.

For the special discrete setup $(\mathscr{X},\lambda) = (\mathscr{X}_\#, \lambda_\#)$ (recall $\lambda_\#[\{x\}] = 1$ for all $x \in \mathscr{X}_\#$), the divergences (41)–(44) simplify to

$$\begin{aligned}
& 0 \leqslant D_{\phi_\alpha,1,1,\mathbb{R}\cdot 1,\lambda}(\mathbb{P},\mathbb{Q}) \\
& = \textstyle\sum_{x\in\mathscr{X}} \frac{\mathbb{r}(x)}{\alpha\cdot(\alpha-1)} \cdot \big[(\mathbb{p}(x))^\alpha + (\alpha-1)\cdot(\mathbb{q}(x))^\alpha - \alpha\cdot\mathbb{p}(x)\cdot(\mathbb{q}(x))^{\alpha-1}\big] \\
& \qquad\qquad \cdot \mathbf{1}_{]0,\infty[}\big(\mathbb{p}(x)\cdot\mathbb{q}(x)\big) \\
& + \textstyle\sum_{x\in\mathscr{X}} \mathbb{r}(x) \cdot \big[\frac{\mathbb{p}(x)^\alpha}{\alpha\cdot(\alpha-1)} \cdot \mathbf{1}_{]1,\infty[}(\alpha) + \infty \cdot \mathbf{1}_{]-\infty,0[\cup]0,1[}(\alpha)\big] \cdot \mathbf{1}_{]0,\infty[}\big(\mathbb{p}(x)\big) \cdot \mathbf{1}_{\{0\}}\big(\mathbb{q}(x)\big) \\
& + \textstyle\sum_{x\in\mathscr{X}} \mathbb{r}(x) \cdot \big[\frac{\mathbb{q}(x)^\alpha}{\alpha} \cdot \mathbf{1}_{]0,1[\cup]1,\infty[}(\alpha) + \infty \cdot \mathbf{1}_{]-\infty,0[}(\alpha)\big] \cdot \mathbf{1}_{]0,\infty[}\big(\mathbb{q}(x)\big) \cdot \mathbf{1}_{\{0\}}\big(\mathbb{p}(x)\big)\,, \\
& \qquad\qquad \text{for } \alpha \in \mathbb{R}\backslash\{0,1\}, \qquad (46) \\
& 0 \leqslant D_{\phi_1,1,1,\mathbb{R}\cdot 1,\lambda}(\mathbb{P},\mathbb{Q}) \\
& = \textstyle\sum_{x\in\mathscr{X}} \mathbb{r}(x) \cdot \big[\mathbb{p}(x)\cdot\log\big(\frac{\mathbb{p}(x)}{\mathbb{q}(x)}\big) + \mathbb{q}(x) - \mathbb{p}(x)\big] \cdot \mathbf{1}_{]0,\infty[}\big(\mathbb{p}(x)\cdot\mathbb{q}(x)\big) \\
& + \textstyle\sum_{x\in\mathscr{X}} \mathbb{r}(x)\cdot\infty\cdot\mathbf{1}_{]0,\infty[}\big(\mathbb{p}(x)\big)\cdot\mathbf{1}_{\{0\}}\big(\mathbb{q}(x)\big) \\
& + \textstyle\sum_{x\in\mathscr{X}} \mathbb{r}(x)\cdot\mathbb{q}(x)\cdot\mathbf{1}_{]0,\infty[}\big(\mathbb{q}(x)\big)\cdot\mathbf{1}_{\{0\}}\big(\mathbb{p}(x)\big)\,, \\
& 0 \leqslant D_{\phi_0,1,1,\mathbb{R}\cdot 1,\lambda}(\mathbb{P},\mathbb{Q}) \\
& = \textstyle\sum_{x\in\mathscr{X}} \mathbb{r}(x)\cdot\big[-\log\big(\frac{\mathbb{p}(x)}{\mathbb{q}(x)}\big) + \frac{\mathbb{p}(x)}{\mathbb{q}(x)} - 1\big]\cdot\mathbf{1}_{]0,\infty[}\big(\mathbb{p}(x)\cdot\mathbb{q}(x)\big) \\
& + \textstyle\sum_{x\in\mathscr{X}} \mathbb{r}(x)\cdot\infty\cdot\mathbf{1}_{]0,\infty[}\big(\mathbb{p}(x)\big)\cdot\mathbf{1}_{\{0\}}\big(\mathbb{q}(x)\big) \\
& + \textstyle\sum_{x\in\mathscr{X}} \mathbb{r}(x)\cdot\infty\cdot\mathbf{1}_{]0,\infty[}\big(\mathbb{q}(x)\big)\cdot\mathbf{1}_{\{0\}}\big(\mathbb{p}(x)\big)\,, \\
& 0 \leqslant D_{\phi_2,1,1,\mathbb{R}\cdot 1,\lambda_\#}(\mathbb{P},\mathbb{Q}) = \textstyle\sum_{x\in\mathscr{X}} \frac{\mathbb{r}(x)}{2}\cdot\big[\mathbb{p}(x)-\mathbb{q}(x)\big]^2.
\end{aligned}$$

Hence, as above, one should exclude $\alpha \leqslant 0$ whenever $\mathbb{p}(x) = 0$ for all x in some A with $\lambda[A] > 0$, respectively $\alpha \leqslant 1$ whenever $\mathbb{q}(x) = 0$ for all x in some $\tilde{A}$ with $\lambda[\tilde{A}] > 0$ (a refined alternative for $\alpha = 1$ is given in Sect. 3.3.1.2 below).

In particular, take the probability context of Remark 2(b), with discrete random variable Y, hypothetical probability mass function $\mathbb{q}(x) := \vec{\mathbb{q}}(x) = \vec{\mathbb{Q}}^{\mathbb{1}\cdot\lambda_{\#}}[Y = x]$, and data-derived probability mass function (relative frequency) $\mathbb{p}(x) := \vec{\mathbb{p}}_N^{emp}(x) = \frac{1}{N} \cdot \#\{i \in \{1, \ldots, N\} : Y_i = x\}$ with sample size N. For $\mathbb{r}(x) \equiv 1$, the corresponding sample-size-weighted divergences $2N \cdot D_{\phi_\alpha, \mathbb{1}, \mathbb{1}, \mathbb{1}, \lambda_{\#}}(\vec{\mathbb{P}}_N^{emp}, \vec{\mathbb{Q}})$ (for $\alpha \in \mathbb{R}$) can be used as goodness-of-fit test statistics; see e.g. Kisslinger and Stummer [37] for their limit behaviour as the sample size N tends to infinity.

3.3.1.2 $\mathbf{m_1(x) = m_2(x) := q(x)}$, $\mathbf{\mathbb{m}_3(x) = r(x) \cdot q(x) \in [0, \infty]}$ for Some (meas.) Function $\mathbf{r} : \mathscr{X} \to \mathbb{R}$ Satisfying $\mathbf{r(x) \in]-\infty, 0[\cup]0, \infty[}$ for $\boldsymbol{\lambda}$−a.a. $\mathbf{x} \in \mathscr{X}$

In such a set-up, the divergence (36) becomes

$$
\begin{aligned}
0 &\leqslant D^c_{\phi, Q, Q, R\cdot Q, \lambda}(P, Q) \\
&= \overline{\int}_{\mathscr{X}} \left[\phi\left(\tfrac{p(x)}{q(x)}\right) - \phi(1) - \phi'_{+,c}(1) \cdot \left(\tfrac{p(x)}{q(x)} - 1\right)\right] \cdot q(x) \cdot r(x) \, \mathrm{d}\lambda(x) \qquad (47) \\
&= \overline{\int}_{\mathscr{X}} \left[q(x) \cdot \phi\left(\tfrac{p(x)}{q(x)}\right) - q(x) \cdot \phi(1) - \phi'_{+,c}(1) \cdot \left(p(x) - q(x)\right)\right] \cdot r(x) \, \mathrm{d}\lambda(x) \,, \qquad (48)
\end{aligned}
$$

where in accordance with the descriptions right after (1) we require that $\phi :]a, b[\to \mathbb{R}$ is convex and strictly convex at $1 \in]a, b[$ and incorporate the zeros of $p(\cdot), q(\cdot), r(\cdot)$ by the appropriate limits and conventions. In the following, we demonstrate this in a non-negativity set-up where for λ-almost all $x \in \mathscr{X}$ one has $\mathbb{r}(x) \in]0, \infty[$ as well as $\mathbb{p}(x) \in [0, \infty[$, $\mathbb{q}(x) \in [0, \infty[$, and hence $E =]a, b[=]0, \infty[$. In order to achieve a reflexivity result in the spirit of Theorem 4, we have to check for – respectively analogously adapt most of – the points in Assumption 2: to begin with, the weight $w(x, s, t)$ evaluated at $s := \mathbb{p}(x)$, $t := \mathbb{q}(x)$ has to be substituted/replaced by $\widetilde{w}(x, \widetilde{t}) := \mathbb{r}(x) \cdot \widetilde{t}$ evaluated at $\widetilde{t} = \mathbb{q}(x)$, and the dissimilarity $\psi_{\phi,c}(s, t)$ has to be substituted/replaced by $\widetilde{\widetilde{\psi}}_{\phi,c}(\widetilde{s}, \widetilde{t}) := \psi_{\phi,c}\left(\frac{\widetilde{s}}{\widetilde{t}}, 1\right)$ with the plug-in $\widetilde{s} = \mathbb{p}(x)$. Putting things together, instead of the integrand-generating term $w(x, s, t) \cdot \psi_{\phi,c}(s, t)$ we have to inspect the boundary behaviour of $\widetilde{w}(x, \widetilde{t}) \cdot \widetilde{\widetilde{\psi}}_{\phi,c}(\widetilde{s}, \widetilde{t})$ being explicitly given (with a slight abuse of notation) by the function $\widetilde{\psi}_{\phi,c} :]0, \infty[^3 \to [0, \infty[$ in

$$
\begin{aligned}
&\widetilde{\psi}_{\phi,c}\left(r, \widetilde{s}, \widetilde{t}\right) := r \cdot \widetilde{t} \cdot \psi_{\phi,c}\left(\tfrac{\widetilde{s}}{\widetilde{t}}, 1\right) = r \cdot \widetilde{t} \cdot \left[\phi\left(\tfrac{\widetilde{s}}{\widetilde{t}}\right) - \phi(1) - \phi'_{+,c}(1) \cdot \left(\tfrac{\widetilde{s}}{\widetilde{t}} - 1\right)\right] \\
&= r \cdot \widetilde{t} \cdot \left[\phi\left(\tfrac{\widetilde{s}\cdot r}{\widetilde{t}\cdot r}\right) - \phi(1) - \phi'_{+,c}(1) \cdot \left(\tfrac{\widetilde{s}\cdot r}{\widetilde{t}\cdot r} - 1\right)\right] = r \cdot \widetilde{t} \cdot \psi_{\phi,c}\left(\tfrac{\widetilde{s}\cdot r}{\widetilde{t}\cdot r}, 1\right). \qquad (49)
\end{aligned}
$$

Since the general right-hand-derivative concerning assumption $t \in \mathscr{R}\left(\frac{Q}{M_2}\right)$ has $\frac{\widetilde{s}}{t} = 1$ as its analogue, we require that the convex function $\phi :]0, \infty[\to]-\infty, \infty[$ is strictly convex (only) at 1 in conformity with Assumption 2(a) (which is also employed in

Assumption 3); for the sake of brevity we use the short-hand notation $2(a)$ etc. in the following discussion. We shall not need 2(b) to 2(d) in the prevailing context, so that the above-mentioned generator $\phi_{TV}(t) := |t-1|$ is allowed for achieving reflexivity (for reasons which will become clear in the proof of Theorem 5 in the appendix). The analogue of 2(e) is $\mathbb{r}(x) \cdot \widetilde{t} < \infty$ which is always (almost surely) automatically satisfied (a.a.sat.), whereas 2(f) converts to "$\mathbb{r}(x) \cdot \widetilde{t} > 0$ for all $\widetilde{s} \neq \widetilde{t}$" which is also a.a.sat. except for the case $\widetilde{t} = 0$ which will be below incorporated in combination with $\psi_{\phi,c}$-multiplication (cf. (50)). For the derivation of the analogue of 2(k) we observe that for fixed $r > 0, \widetilde{s} > 0$ the function $\widetilde{t} \rightarrow \widetilde{\psi}_{\phi,c}\big(r, \widetilde{s}, \widetilde{t}\big)$ is (the r-fold of) the perspective function (at $\widetilde{s}$) of the convex function $\psi_{\phi,c}\big(\cdot, 1\big)$ and thus convex with existing limit

$$\begin{aligned} &\ell i_1 := r \cdot 0 \cdot \psi_{\phi,c}\big(\tfrac{\widetilde{s}}{0}, 1\big) := \lim_{t \rightarrow 0} \widetilde{\psi}_{\phi,c}\big(r, \widetilde{s}, \widetilde{t}\big) = \\ &= -r \cdot \widetilde{s} \cdot \phi'_{+,c}(1) + r \cdot \widetilde{s} \cdot \lim_{\widetilde{t} \rightarrow 0} \big[\tfrac{\widetilde{t}}{\widetilde{s}} \cdot \phi\big(\tfrac{\widetilde{s}}{\widetilde{t}}\big)\big] = r \cdot \widetilde{s} \cdot (\phi^*(0) - \phi'_{+,c}(1)) \geqslant 0\,, \quad (50) \end{aligned}$$

where $\phi^*(0) := \lim_{u \rightarrow 0} u \cdot \phi\big(\frac{1}{u}\big) = \lim_{v \rightarrow \infty} \frac{\phi(v)}{v}$ exists but may be infinite (recall that $\phi'_{+,c}(1)$ is finite). Notice that in contrast to 2(k) we need not assume $\ell i_1 > 0$ (and thus do not exclude ϕ_{TV}). To convert 2(i), we employ the fact that for fixed $r > 0, \widetilde{t} > 0$ the function $\widetilde{s} \rightarrow \widetilde{\psi}_{\phi,c}\big(r, \widetilde{s}, \widetilde{t}\big)$ is convex with existing limit

$$r \cdot \widetilde{t} \cdot \psi_{\phi,c}\big(\frac{0}{\widetilde{t}}, 1\big) := \lim_{s \rightarrow 0} \widetilde{\psi}_{\phi,c}\big(r, \widetilde{s}, \widetilde{t}\big) = r \cdot \widetilde{t} \cdot (\phi(0) + \phi'_{+,c}(1) - \phi(1)) > 0\,,$$

where $\phi(0) := \lim_{u \rightarrow 0} \phi(u)$ exists but may be infinite. To achieve the analogue of 2(g), let us first remark that for fixed $r > 0$ the function $(\widetilde{s}, \widetilde{t}) \rightarrow \widetilde{\psi}_{\phi,c}\big(r, \widetilde{s}, \widetilde{t}\big)$ may not be continuous at $(\widetilde{s}, \widetilde{t}) = (0, 0)$, but due to the very nature of a divergence we make the 2(g)-conform convention of setting

$$r \cdot 0 \cdot \psi_{\phi,c}\big(\tfrac{0}{0}, 1\big) := \widetilde{\psi}_{\phi,c}\big(r, 0, 0\big) := 0$$

(notice that e.g. the power function ϕ_{-1} of (5) with index $\alpha = -1$ obeys $\lim_{\widetilde{t} \rightarrow 0} \widetilde{\psi}_{\phi_{-1}}\big(r, \widetilde{t}, \widetilde{t}\big) = 0 \neq \frac{r}{2} = \lim_{\widetilde{t} \rightarrow 0} \widetilde{\psi}_{\phi_{-1}}\big(r, \widetilde{t}^2, \widetilde{t}\big)$). The analogues of the remaining Assumptions 2(h),(j),(ℓ),(m),(n) are (almost surely) obsolete because of our basic (almost surely) finiteness requirements. Summing up, with the above-mentioned limits and conventions we write (47) explicitly as

$$
\begin{aligned}
0 \leqslant\ & D^{c}_{\phi,\mathbb{Q},\mathbb{Q},\mathbb{R}\cdot\mathbb{Q},\lambda}(\mathbb{P},\mathbb{Q}) \\
=\ & \int_{\mathscr{X}} \mathbb{r}(x)\cdot\big[\mathbb{q}(x)\cdot\phi\big(\tfrac{\mathbb{p}(x)}{\mathbb{q}(x)}\big) - \mathbb{q}(x)\cdot\phi\big(1\big) - \phi'_{+,c}\big(1\big)\cdot\big(\mathbb{p}(x)-\mathbb{q}(x)\big)\big] \\
& \qquad\qquad\qquad\qquad \cdot\mathbf{1}_{]0,\infty[}\big(\mathbb{p}(x)\cdot\mathbb{q}(x)\big)\,\mathrm{d}\lambda(x) \\
& +\big[\phi^{*}(0)-\phi'_{+,c}(1)\big]\cdot\int_{\mathscr{X}} \mathbb{r}(x)\cdot\mathbb{p}(x)\cdot\mathbf{1}_{]0,\infty[}\big(\mathbb{p}(x)\big)\cdot\mathbf{1}_{\{0\}}\big(\mathbb{q}(x)\big)\,\mathrm{d}\lambda(x) \\
& +\big[\phi(0)+\phi'_{+,c}(1)-\phi(1)\big]\cdot\int_{\mathscr{X}} \mathbb{r}(x)\cdot\mathbb{q}(x)\cdot\mathbf{1}_{]0,\infty[}\big(\mathbb{q}(x)\big)\cdot\mathbf{1}_{\{0\}}\big(\mathbb{p}(x)\big)\,\mathrm{d}\lambda(x) \\
=\ & \int_{\mathscr{X}} \mathbb{r}(x)\cdot\Big[\mathbb{q}(x)\cdot\phi\big(\tfrac{\mathbb{p}(x)}{\mathbb{q}(x)}\big) - \mathbb{q}(x)\cdot\phi\big(1\big) - \phi'_{+,c}\big(1\big)\cdot\big(\mathbb{p}(x)-\mathbb{q}(x)\big)\Big] \\
& \qquad\qquad\qquad\qquad \cdot\mathbf{1}_{]0,\infty[}\big(\mathbb{p}(x)\cdot\mathbb{q}(x)\big)\,\mathrm{d}\lambda(x) \\
& +\big[\phi^{*}(0)-\phi'_{+,c}(1)\big]\cdot\int_{\mathscr{X}} \mathbb{r}(x)\cdot\mathbb{p}(x)\cdot\mathbf{1}_{\{0\}}\big(\mathbb{q}(x)\big)\,\mathrm{d}\lambda(x) \\
& +\big[\phi(0)+\phi'_{+,c}(1)-\phi(1)\big]\cdot\int_{\mathscr{X}} \mathbb{r}(x)\cdot\mathbb{q}(x)\cdot\mathbf{1}_{\{0\}}\big(\mathbb{p}(x)\big)\,\mathrm{d}\lambda(x)\,. \qquad (51)
\end{aligned}
$$

In case of $\mathfrak{Q}^{\mathbb{R}\cdot\lambda}[\mathscr{X}] := \int_{\mathscr{X}} \mathbb{q}(x)\cdot\mathbb{r}(x)\,\mathrm{d}\lambda(x) < \infty$, the divergence (51) becomes

$$
\begin{aligned}
0 \leqslant\ & D^{c}_{\phi,\mathbb{Q},\mathbb{Q},\mathbb{R}\cdot\mathbb{Q},\lambda}(\mathbb{P},\mathbb{Q}) \\
=\ & \int_{\mathscr{X}} \mathbb{r}(x)\cdot\Big[\mathbb{q}(x)\cdot\phi\big(\tfrac{\mathbb{p}(x)}{\mathbb{q}(x)}\big) - \phi'_{+,c}\big(1\big)\cdot\big(\mathbb{p}(x)-\mathbb{q}(x)\big)\Big]\cdot\mathbf{1}_{]0,\infty[}\big(\mathbb{p}(x)\cdot\mathbb{q}(x)\big)\,\mathrm{d}\lambda(x) \\
& +\big[\phi^{*}(0)-\phi'_{+,c}(1)\big]\cdot\int_{\mathscr{X}} \mathbb{r}(x)\cdot\mathbb{p}(x)\cdot\mathbf{1}_{\{0\}}\big(\mathbb{q}(x)\big)\,\mathrm{d}\lambda(x) \\
& +\big[\phi(0)+\phi'_{+,c}(1)\big]\cdot\int_{\mathscr{X}} \mathbb{r}(x)\cdot\mathbb{q}(x)\cdot\mathbf{1}_{\{0\}}\big(\mathbb{p}(x)\big)\,\mathrm{d}\lambda(x) - \phi(1)\cdot\int_{\mathscr{X}} \mathbb{r}(x)\cdot\mathbb{q}(x)\,\mathrm{d}\lambda(x)\,. \qquad (52)
\end{aligned}
$$

Moreover, in case of $\phi\big(1\big)=0$ and $(\mathfrak{P}^{\mathbb{R}\cdot\lambda}-\mathfrak{Q}^{\mathbb{R}\cdot\lambda})[\mathscr{X}] = \int_{\mathscr{X}} \big(\mathbb{p}(x)-\mathbb{q}(x)\big)\cdot\mathbb{r}(x)\,\mathrm{d}\lambda(x) \in]-\infty,\infty[$ (but not necessarily $\mathfrak{P}^{\mathbb{R}\cdot\lambda}[\mathscr{X}] = \int_{\mathscr{X}} \mathbb{p}(x)\cdot\mathbb{r}(x)\,\mathrm{d}\lambda(x) < \infty$, $\mathfrak{Q}^{\mathbb{R}\cdot\lambda}[\mathscr{X}] = \int_{\mathscr{X}} \mathbb{q}(x)\cdot\mathbb{r}(x)\,\mathrm{d}\lambda(x) < \infty$), the divergence (51) turns into

$$
\begin{aligned}
0 \leqslant\ & D^{c}_{\phi,\mathbb{Q},\mathbb{Q},\mathbb{R}\cdot\mathbb{Q},\lambda}(\mathbb{P},\mathbb{Q}) = \int_{\mathscr{X}} \mathbb{r}(x)\cdot\mathbb{q}(x)\cdot\phi\big(\tfrac{\mathbb{p}(x)}{\mathbb{q}(x)}\big)\cdot\mathbf{1}_{]0,\infty[}\big(\mathbb{p}(x)\cdot\mathbb{q}(x)\big)\,\mathrm{d}\lambda(x) \\
& +\phi^{*}(0)\cdot\int_{\mathscr{X}} \mathbb{r}(x)\cdot\mathbb{p}(x)\cdot\mathbf{1}_{\{0\}}\big(\mathbb{q}(x)\big)\,\mathrm{d}\lambda(x) + \phi(0)\cdot\int_{\mathscr{X}} \mathbb{r}(x)\cdot\mathbb{q}(x)\cdot\mathbf{1}_{\{0\}}\big(\mathbb{p}(x)\big)\,\mathrm{d}\lambda(x) \\
& -\phi'_{+,c}\big(1\big)\cdot\int_{\mathscr{X}} \mathbb{r}(x)\cdot\big(\mathbb{p}(x)-\mathbb{q}(x)\big)\,\mathrm{d}\lambda(x)\,. \qquad (53)
\end{aligned}
$$

Let us remark that (53) can be interpreted as ϕ-divergence $D^{c}_{\phi}\big(\mu,\nu\big)$ between the two nonnegative measures μ,ν (on $(\mathscr{X},\mathscr{F})$) (cf. Stummer and Vajda [83]), where $\mu[\bullet] := \mathfrak{P}^{\mathbb{R}\cdot\lambda}[\bullet]$ and $\nu[\bullet] := \mathfrak{Q}^{\mathbb{R}\cdot\lambda}[\bullet]$. In the following, we briefly discuss two important sub-cases. First, in the "λ-probability-densities" context of Remark 2(b) one has for general $\mathscr{X}$ the manifestation $\mathbb{p}(x) := \vec{\mathbb{p}}(x) \geqslant 0$, $\mathbb{q}(x) := \vec{\mathbb{q}}(x) \geqslant 0$, and under the constraint $\phi(1)=0$ the corresponding divergence $D^{c}_{\phi,\vec{\mathbb{Q}},\vec{\mathbb{Q}},\mathbb{R}\cdot\vec{\mathbb{Q}},\lambda}(\vec{\mathbb{P}},\vec{\mathbb{Q}})$ turns out to be the ($\mathbb{r}$-)"local ϕ-divergence of Avlogiaris et al. [6, 7]; in case of $\mathbb{r}(x)\equiv 1$ this reduces – due to the fact $\int_{\mathscr{X}} \big(\vec{\mathbb{p}}(x)-\vec{\mathbb{q}}(x)\big)\,\mathrm{d}\lambda(x) = 0$ – to the classical Csiszar-Ali-Silvey ϕ-divergence CASD ([4, 27], see also e.g. Liese and Vajda [41], Vajda [89])

$$
\begin{aligned}
0 \leqslant D^{c}_{\phi,\vec{\mathbb{Q}},\vec{\mathbb{Q}},\mathbb{1}\cdot\vec{\mathbb{Q}},\lambda}(\vec{\mathbb{P}},\vec{\mathbb{Q}}) &= \int_{\mathscr{X}} \vec{\mathbb{q}}(x)\cdot\phi\big(\tfrac{\vec{\mathbb{p}}(x)}{\vec{\mathbb{q}}(x)}\big)\cdot\mathbf{1}_{]0,\infty[}\big(\vec{\mathbb{p}}(x)\cdot\vec{\mathbb{q}}(x)\big)\,\mathrm{d}\lambda(x)\\
&+\phi^{*}(0)\cdot\int_{\mathscr{X}}\vec{\mathbb{p}}(x)\cdot\mathbf{1}_{\{0\}}\big(\vec{\mathbb{q}}(x)\big)\,\mathrm{d}\lambda(x)+\phi(0)\cdot\int_{\mathscr{X}}\vec{\mathbb{q}}(x)\cdot\mathbf{1}_{\{0\}}\big(\vec{\mathbb{p}}(x)\big)\,\mathrm{d}\lambda(x)\\
&-\phi'_{+,c}(1)\cdot\int_{\mathscr{X}}\big(\vec{\mathbb{p}}(x)-\vec{\mathbb{q}}(x)\big)\,\mathrm{d}\lambda(x)\\
&=\int_{\mathscr{X}}\vec{\mathbb{q}}(x)\cdot\phi\big(\tfrac{\vec{\mathbb{p}}(x)}{\vec{\mathbb{q}}(x)}\big)\cdot\mathbf{1}_{]0,\infty[}\big(\vec{\mathbb{p}}(x)\cdot\vec{\mathbb{q}}(x)\big)\,\mathrm{d}\lambda(x)\\
&\quad+\phi^{*}(0)\cdot\mathfrak{P}^{\mathbb{1}\cdot\lambda}[\vec{\mathbb{q}}(x)=0]+\phi(0)\cdot\mathfrak{Q}^{\mathbb{1}\cdot\lambda}[\vec{\mathbb{p}}(x)=0]\,;
\end{aligned}
\tag{54}
$$

if $\phi(1)\neq 0$ then one has to additionally subtract $\phi(1)$ (cf. the corresponding special case of (52)). In particular, for the special sub-setup where for λ-almost all $x\in\mathscr{X}$ there holds $\mathbb{p}(x):=\vec{\mathbb{p}}(x)>0$, $\mathbb{q}(x):=\vec{\mathbb{q}}(x)>0$, $\mathbb{r}(x)\equiv 1$, $\phi(1)=0$, one ends up with the reduced Csiszar-Ali-Silvey divergence

$$
0\leqslant D^{c}_{\phi,\vec{\mathbb{Q}},\vec{\mathbb{Q}},\mathbb{1}\cdot\vec{\mathbb{Q}},\lambda}(\vec{\mathbb{P}},\vec{\mathbb{Q}})=\int_{\mathscr{X}}\vec{\mathbb{q}}(x)\cdot\phi\big(\tfrac{\vec{\mathbb{p}}(x)}{\vec{\mathbb{q}}(x)}\big)\,\mathrm{d}\lambda(x)
$$

which can be interpreted as a "consistent extension" of the motivating pointwise dissimilarity $d^{(7)}_{\phi}(\cdot,\cdot)$ from the introductory Sect. 2; notice the fundamental structural difference to the divergence (38) which reflects $d^{(6)}_{\phi}(\cdot,\cdot)$. For comprehensive treatments of statistical applications of CASD, the reader is referred to Liese and Vajda [41], Read and Cressie [72], Vajda [89], Pardo [68], Liese and Miescke [40], Basu et al. [13].

Returning to the general divergence setup (51), we derive the reflexivity result (to be proved in the appendix):

Theorem 5 *Let $c\in[0,1]$, $\mathbb{r}(x)\in]0,\infty[$ for λ-a.a. $x\in\mathscr{X}$, $\mathscr{R}\big(\frac{\mathbb{P}}{\mathbb{Q}}\big)\cup\{1\}\subset[a,b]$, and $\phi\in\Phi(]a,b[)$ be strictly convex at $t=1$. Moreover, suppose that*

$$
\int_{\mathscr{X}}\big(\mathbb{p}(x)-\mathbb{q}(x)\big)\cdot\mathbb{r}(x)\,\mathrm{d}\lambda(x)\;=\;0
\tag{55}
$$

(but not necessarily $\int_{\mathscr{X}}\mathbb{p}(x)\cdot\mathbb{r}(x)\,\mathrm{d}\lambda(x)<\infty$, $\int_{\mathscr{X}}\mathbb{q}(x)\cdot\mathbb{r}(x)\,\mathrm{d}\lambda(x)<\infty$). Then:
(1) $D^{c}_{\phi,\mathbb{Q},\mathbb{Q},\mathbb{R}\cdot\mathbb{Q},\lambda}(\mathbb{P},\mathbb{Q})\geqslant 0$. Depending on the concrete situation, $D^{c}_{\phi,\mathbb{Q},\mathbb{Q},\mathbb{R}\cdot\mathbb{Q},\lambda}(\mathbb{P},\mathbb{Q})$ may take infinite value.

(2) $D^{c}_{\phi,\mathbb{Q},\mathbb{Q},\mathbb{R}\cdot\mathbb{Q},\lambda}(\mathbb{P},\mathbb{Q})=0$ *if and only if* $\mathbb{p}(x)=\mathbb{q}(x)$ *for λ-a.a. $x\in\mathscr{X}$.* (56)

Remark 3 (a) In the context of non-negative measures, the special case $c=1$ – together with $\int_{\mathscr{X}}\mathbb{p}(x)\cdot\mathbb{r}(x)\,\mathrm{d}\lambda(x)<\infty$, $\int_{\mathscr{X}}\mathbb{q}(x)\cdot\mathbb{r}(x)\,\mathrm{d}\lambda(x)<\infty$ – of Theorem 5 was first achieved by Stummer and Vajda [83].
(b) Assumption (55) is always automatically satisfied if one has coincidence of finite total masses in the sense of $\mathfrak{P}^{\mathbb{R}\cdot\lambda}[\mathscr{X}]=\int_{\mathscr{X}}\mathbb{p}(x)\cdot\mathbb{r}(x)\,\mathrm{d}\lambda(x)=\int_{\mathscr{X}}\mathbb{q}(x)\cdot\mathbb{r}(x)\,\mathrm{d}\lambda(x)=\mathfrak{Q}^{\mathbb{R}\cdot\lambda}[\mathscr{X}]<\infty$. For $\mathbb{r}(x)\equiv 1$ this is always satisfied for λ-probability densities $\mathbb{p}(x):=\vec{\mathbb{p}}(x)$, $\mathbb{q}(x):=\vec{\mathbb{q}}(x)$, since $\vec{\mathfrak{P}}^{\mathbb{1}\cdot\lambda}[\mathscr{X}]=\vec{\mathfrak{Q}}^{\mathbb{1}\cdot\lambda}[\mathscr{X}]=1$.

(c) Notice that in contrast to Theorem 4, the generator-concerning Assumptions 2(b)–(d) are replaced by the "model-concerning" constraint (55). This opens the gate for the use of the generators ϕ_{ie} and ϕ_{TV} for cases where (55) is satisfied. For the latter, we obtain with $c = \frac{1}{2}$ explicitly from (49) and (33)

$$\widetilde{\psi}_{\phi_{TV},\frac{1}{2}}(r, \widetilde{s}, \widetilde{t}) := r \cdot \widetilde{t} \cdot \psi_{\phi_{TV},\frac{1}{2}}\big(\frac{\widetilde{s}}{\widetilde{t}}, 1\big) = r \cdot \widetilde{t} \cdot \big|\frac{\widetilde{s}}{\widetilde{t}} - 1\big| = r \cdot \big|\widetilde{s} - \widetilde{t}\big|,$$

and hence from (51) together with $\phi_{TV}(1) = 0$, $\phi_{TV}(0) = 1$ (cf. (31)), $\phi'_{TV,+,\frac{1}{2}}(1) = 0$ (cf. (32)), $\phi^*_{TV}(0) = \lim_{s\to\infty} \frac{1}{s} \cdot \psi_{\phi_{TV},\frac{1}{2}}(s, 1) = 1$ (cf. (34)) we get

$$0 \leqslant D^c_{\phi,\mathbb{Q},\mathbb{Q},\mathbb{R}\cdot\mathbb{Q},\lambda}(\mathbb{P}, \mathbb{Q}) = \int_{\mathscr{X}} \mathbb{r}(x) \cdot \big|\mathbb{p}(x) - \mathbb{q}(x)\big| \,\mathrm{d}\lambda(x) \tag{57}$$

which is nothing but the (possibly infinite) $\mathbb{r}(\cdot)$-weighted L_1-distance between the functions $x \to \mathbb{p}(x)$ and $x \to \mathbb{q}(x)$.

(d) In the light of (52), Theorem 4 (adapted to the current context) and Theorem 5, let us indicate that if one wants to use $\Xi := \int_{\mathscr{X}} \mathbb{q}(x) \cdot \phi\big(\frac{\mathbb{p}(x)}{\mathbb{q}(x)}\big) \cdot \mathbb{r}(x) \,\mathrm{d}\lambda(x)$ (with appropriate zero-conventions) as a divergence, then one should either employ generators ϕ satisfying $\phi(1) = \phi'_{+,c}(1) = 0$, or employ models fulfilling the assumption (56) together with generators ϕ satisfying $\phi(1) = 0$. On the other hand, if this integral Ξ appears in your application context "naturally", then one should be aware that Ξ may become negative depending on the involved set-up; for a counter-example, see Stummer and Vajda [83]. This concludes Remark 3.

As an important example, we illuminate the special case $\phi = \phi_\alpha$ with $\alpha \in \mathbb{R}\backslash\{0, 1\}$ (cf. (5)) under the constraint $(\mathfrak{P}^{\mathbb{R}\cdot\lambda} - \mathfrak{Q}^{\mathbb{R}\cdot\lambda})[\mathscr{X}] = \int_{\mathscr{X}} \big(\mathbb{p}(x) - \mathbb{q}(x)\big) \cdot \mathbb{r}(x) \,\mathrm{d}\lambda(x) \in]-\infty, \infty[$. Accordingly, the "implicit-boundary-describing" divergence (48) resp. the corresponding "explicit-boundary" version (53) turn into the generalized power divergences of order α (cf. Stummer and Vajda [83] for $\mathbb{r}(x) \equiv 1$)[11]

$$\begin{aligned}
&0 \leqslant D_{\phi_\alpha,\mathbb{Q},\mathbb{Q},\mathbb{R}\cdot\mathbb{Q},\lambda}(\mathbb{P}, \mathbb{Q}) \\
&= \overline{\int}_{\mathscr{X}} \tfrac{1}{\alpha\cdot(\alpha-1)} \cdot \Big[\big(\tfrac{\mathbb{p}(x)}{\mathbb{q}(x)}\big)^\alpha - \alpha \cdot \tfrac{\mathbb{p}(x)}{\mathbb{q}(x)} + \alpha - 1\Big] \cdot \mathbb{q}(x) \cdot \mathbb{r}(x) \,\mathrm{d}\lambda(x) \qquad (58)\\
&= \tfrac{1}{\alpha\cdot(\alpha-1)} \cdot \int_{\mathscr{X}} \mathbb{r}(x) \cdot \mathbb{q}(x) \cdot \Big[\big(\tfrac{\mathbb{p}(x)}{\mathbb{q}(x)}\big)^\alpha - \alpha \cdot \tfrac{\mathbb{p}(x)}{\mathbb{q}(x)} + \alpha - 1\Big] \cdot \mathbf{1}_{]0,\infty[}\big(\mathbb{p}(x) \cdot \mathbb{q}(x)\big) \,\mathrm{d}\lambda(x) \\
&+\phi^*_\alpha(0) \cdot \int_{\mathscr{X}} \mathbb{r}(x) \cdot \mathbb{p}(x) \cdot \mathbf{1}_{\{0\}}\big(\mathbb{q}(x)\big) \,\mathrm{d}\lambda(x) + \phi_\alpha(0) \cdot \int_{\mathscr{X}} \mathbb{r}(x) \cdot \mathbb{q}(x) \cdot \mathbf{1}_{\{0\}}\big(\mathbb{p}(x)\big) \,\mathrm{d}\lambda(x) \\
&= \tfrac{1}{\alpha\cdot(\alpha-1)} \int_{\mathscr{X}} \mathbb{r}(x) \cdot \Big[\mathbb{p}(x)^\alpha \cdot \mathbb{q}(x)^{1-\alpha} - \mathbb{q}(x)\Big] \cdot \mathbf{1}_{]0,\infty[}\big(\mathbb{p}(x) \cdot \mathbb{q}(x)\big) \,\mathrm{d}\lambda(x) \\
&+\tfrac{1}{1-\alpha} \cdot \int_{\mathscr{X}} \mathbb{r}(x) \cdot (\mathbb{p}(x) - \mathbb{q}(x)) \,\mathrm{d}\lambda(x) + \infty \cdot \mathbf{1}_{]1,\infty[}(\alpha) \cdot \int_{\mathscr{X}} \mathbb{r}(x) \cdot \mathbb{p}(x) \cdot \mathbf{1}_{\{0\}}\big(\mathbb{q}(x)\big) \,\mathrm{d}\lambda(x) \\
&+\big(\tfrac{1}{\alpha\cdot(1-\alpha)} \cdot \mathbf{1}_{]0,1]\cup]1,\infty[}(\alpha) + \infty \cdot \mathbf{1}_{]-\infty,0[}(\alpha)\big) \cdot \int_{\mathscr{X}} \mathbb{r}(x) \cdot \mathbb{q}(x) \cdot \mathbf{1}_{\{0\}}\big(\mathbb{p}(x)\big) \,\mathrm{d}\lambda(x),
\end{aligned}$$

[11] This can be interpreted analogously as in footnote 10.

where we have employed (8) and (7); especially, one gets for $\alpha = 2$

$$0 \leqslant D_{\phi_2,\mathbb{Q},\mathbb{Q},\mathbb{R}\cdot\mathbb{Q},\lambda}(\mathbb{P},\mathbb{Q}) = \overline{\int}_{\mathscr{X}} \tfrac{1}{2} \cdot \tfrac{(\mathbb{p}(x)-\mathbb{q}(x))^2}{\mathbb{q}(x)} \cdot \mathbb{r}(x)\,\mathrm{d}\lambda(x)$$
$$= \tfrac{1}{2} \int_{\mathscr{X}} \mathbb{r}(x) \cdot \tfrac{(\mathbb{p}(x)-\mathbb{q}(x))^2}{\mathbb{q}(x)} \cdot \mathbf{1}_{[0,\infty[}(\mathbb{p}(x)) \cdot \mathbf{1}_{]0,\infty[}(\mathbb{q}(x))\,\mathrm{d}\lambda(x)$$
$$+\infty \cdot \int_{\mathscr{X}} \mathbb{r}(x) \cdot \mathbb{p}(x) \cdot \mathbf{1}_{\{0\}}\big(\mathbb{q}(x)\big)\,\mathrm{d}\lambda(x)$$

which is called Pearsons's chisquare divergence. Under the same constraint $(\mathfrak{P}^{\mathbb{R}\cdot\lambda} - \mathfrak{Q}^{\mathbb{R}\cdot\lambda})[\mathscr{X}] \in]-\infty,\infty[$, the case $\alpha = 1$ leads by (18)–(22) to the generalized Kullback–Leibler divergence (generalized relative entropy)

$$0 \leqslant D_{\phi_1,\mathbb{Q},\mathbb{Q},\mathbb{R}\cdot\mathbb{Q},\lambda}(\mathbb{P},\mathbb{Q}) = \overline{\int}_{\mathscr{X}} \Big[\tfrac{\mathbb{p}(x)}{\mathbb{q}(x)} \cdot \log\big(\tfrac{\mathbb{p}(x)}{\mathbb{q}(x)}\big) + 1 - \tfrac{\mathbb{p}(x)}{\mathbb{q}(x)} \Big] \cdot \mathbb{q}(x) \cdot \mathbb{r}(x)\,\mathrm{d}\lambda(x)$$
$$= \int_{\mathscr{X}} \mathbb{r}(x) \cdot \mathbb{p}(x) \cdot \log\big(\tfrac{\mathbb{p}(x)}{\mathbb{q}(x)}\big) \cdot \mathbf{1}_{]0,\infty[}\big(\mathbb{p}(x)\cdot\mathbb{q}(x)\big)\,\mathrm{d}\lambda(x)$$
$$+ \int_{\mathscr{X}} \mathbb{r}(x) \cdot (\mathbb{q}(x) - \mathbb{p}(x))\,\mathrm{d}\lambda(x) + \infty \cdot \int_{\mathscr{X}} \mathbb{r}(x) \cdot \mathbb{p}(x) \cdot \mathbf{1}_{\{0\}}\big(\mathbb{q}(x)\big)\,\mathrm{d}\lambda(x)$$

(which equals (42)), and for $\alpha = 0$ one gets from (19), (25)–(27) the generalized reverse Kullback–Leibler divergence (generalized reverse relative entropy)

$$0 \leqslant D_{\phi_0,\mathbb{Q},\mathbb{Q},\mathbb{R}\cdot\mathbb{Q},\lambda}(\mathbb{P},\mathbb{Q}) = \overline{\int}_{\mathscr{X}} \big[-\log\big(\tfrac{\mathbb{p}(x)}{\mathbb{q}(x)}\big) + \tfrac{\mathbb{p}(x)}{\mathbb{q}(x)} - 1 \big] \cdot \mathbb{q}(x) \cdot \mathbb{r}(x)\,\mathrm{d}\lambda(x)$$
$$= \int_{\mathscr{X}} \mathbb{r}(x) \cdot \mathbb{q}(x) \cdot \log\big(\tfrac{\mathbb{q}(x)}{\mathbb{p}(x)}\big) \cdot \mathbf{1}_{]0,\infty[}\big(\mathbb{p}(x)\cdot\mathbb{q}(x)\big)\,\mathrm{d}\lambda(x)$$
$$+ \int_{\mathscr{X}} \mathbb{r}(x) \cdot (\mathbb{p}(x) - \mathbb{q}(x))\,\mathrm{d}\lambda(x) + \infty \cdot \int_{\mathscr{X}} \mathbb{r}(x) \cdot \mathbb{q}(x) \cdot \mathbf{1}_{\{0\}}\big(\mathbb{p}(x)\big)\,\mathrm{d}\lambda(x).$$

Notice that instead of the limit in (50) one could also use the convention $r \cdot 0 \cdot \psi_\phi\big(\tfrac{s}{0}, 1\big) := \widetilde{\psi}_\phi\big(r, s, 0\big) := 0$; in the context of λ-probability densities, one then ends up with divergence by Rüschendorf [75].

For the discrete setup $(\mathscr{X}, \lambda) = (\mathscr{X}_\#, \lambda_\#)$, the divergence in (51) simplifies to

$$\begin{aligned} 0 &\leqslant D^c_{\phi,\mathbb{Q},\mathbb{Q},\mathbb{R}\cdot\mathbb{Q},\lambda_\#}(\mathbb{P},\mathbb{Q}) \\ &= \textstyle\sum_{x\in\mathscr{X}} \mathbb{r}(x) \cdot \big[\mathbb{q}(x) \cdot \phi\big(\tfrac{\mathbb{p}(x)}{\mathbb{q}(x)}\big) - \mathbb{q}(x) \cdot \phi(1) - \phi'_{+,c}(1) \cdot \big(\mathbb{p}(x) - \mathbb{q}(x)\big)\big] \\ &\qquad \cdot \mathbf{1}_{]0,\infty[}\big(\mathbb{p}(x)\cdot\mathbb{q}(x)\big) \\ &+ \big[\phi^*(0) - \phi'_{+,c}(1)\big] \cdot \textstyle\sum_{x\in\mathscr{X}} \mathbb{r}(x) \cdot \mathbb{p}(x) \cdot \mathbf{1}_{\{0\}}\big(\mathbb{q}(x)\big) \\ &+ \big[\phi(0) + \phi'_{+,c}(1) - \phi(1)\big] \cdot \textstyle\sum_{x\in\mathscr{X}} \mathbb{r}(x) \cdot \mathbb{q}(x) \cdot \mathbf{1}_{\{0\}}\big(\mathbb{p}(x)\big) \end{aligned} \tag{59}$$

which in case of $\phi(1) = \phi'_{+,c}(1) = 0$ – respectively $\phi(1) = 0$ and (55) – turns into

$$0 \leqslant D^c_{\phi,\mathbb{Q},\mathbb{Q},\mathbb{R}\cdot\mathbb{Q},\lambda_\#}(\mathbb{P},\mathbb{Q}) = \textstyle\sum_{x\in\mathscr{X}} \mathbb{r}(x) \cdot \mathbb{q}(x) \cdot \phi\big(\tfrac{\mathbb{p}(x)}{\mathbb{q}(x)}\big) \cdot \mathbf{1}_{]0,\infty[}\big(\mathbb{p}(x)\cdot\mathbb{q}(x)\big)$$
$$+\phi^*(0) \cdot \textstyle\sum_{x\in\mathscr{X}} \mathbb{r}(x) \cdot \mathbb{p}(x) \cdot \mathbf{1}_{\{0\}}\big(\mathbb{q}(x)\big) + \phi(0) \cdot \textstyle\sum_{x\in\mathscr{X}} \mathbb{r}(x) \cdot \mathbb{q}(x) \cdot \mathbf{1}_{\{0\}}\big(\mathbb{p}(x)\big). \tag{60}$$

3.3.1.3 $\mathbf{m_1(x) = m_2(x) := w(p(x), q(x))}$, $\mathbb{m}_3(\mathbf{x}) = \mathbf{r(x) \cdot w(p(x), q(x))} \in [\mathbf{0}, \infty[$ for Some (Measurable) Functions $\mathbf{w} : \mathscr{R}(\mathbf{P}) \times \mathscr{R}(\mathbf{Q}) \to \mathbb{R}$ and $\mathbf{r} : \mathscr{X} \to \mathbb{R}$

Such a choice extends the context of the previous Sect. 3.3.1.2 where the "connector function" w took the simple form $w(u, v) = v$, as well as the setup of Sect. 3.3.1.1 dealing with constant $w(u, v) \equiv 1$. This introduces a wide flexibility with divergences of the form

$$0 \leqslant D^{c}_{\phi,W(P,Q),W(P,Q),R\cdot W(P,Q),\lambda}(P, Q)$$
$$:= \overline{\int}_{\mathscr{X}} \Big[\phi\big(\tfrac{p(x)}{w(p(x),q(x))}\big) - \phi\big(\tfrac{q(x)}{w(p(x),q(x))}\big)$$
$$-\phi'_{+,c}\big(\tfrac{q(x)}{w(p(x),q(x))}\big) \cdot \big(\tfrac{p(x)}{w(p(x),q(x))} - \tfrac{q(x)}{w(p(x),q(x))}\big)\Big] \cdot w(p(x), q(x)) \cdot r(x)\, \mathrm{d}\lambda(x), \quad (61)$$

which for the discrete setup $(\mathscr{X}, \lambda) = (\mathscr{X}_{\#}, \lambda_{\#})$ (recall $\lambda_{\#}[\{x\}] = 1$ for all $x \in \mathscr{X}_{\#}$) simplifies to

$$0 \leqslant D^{c}_{\phi,W(P,Q),W(P,Q),R\cdot W(P,Q),\lambda_{\#}}(P, Q) = \overline{\sum}_{x\in\mathscr{X}} \Big[\phi\big(\tfrac{p(x)}{w(p(x),q(x))}\big) - \phi\big(\tfrac{q(x)}{w(p(x),q(x))}\big)$$
$$-\phi'_{+,c}\big(\tfrac{q(x)}{w(p(x),q(x))}\big) \cdot \big(\tfrac{p(x)}{w(p(x),q(x))} - \tfrac{q(x)}{w(p(x),q(x))}\big)\Big] \cdot w(p(x), q(x)) \cdot r(x)\,. \quad (62)$$

A detailed discussion of this wide class of divergences (61),(62) is beyond the scope of this paper. For the λ-probability density context (and an indication for more general functions), see the comprehensive paper of Kisslinger and Stummer [37] and the references therein. Finally, by appropriate choices of $w(\cdot, \cdot)$ we can even derive divergences of the form (60) but with non-convex non-concave ϕ: see e.g. the "perturbed" power divergences of Roensch and Stummer [74].

3.3.2 Global Scaling and Aggregation, and Other Paradigms

Our universal framework also contains, as special cases, scaling and aggregation functions of the form $m_i(x) := m_{\ell,i}(x) \cdot H_i\big((m_{g,i}(z))_{z\in\mathscr{X}}\big)$ for some (meas., possibly nonnegative) functions $m_{l,i} : \mathscr{X} \mapsto \mathbb{R}$, $m_{g,i} : \mathscr{X} \mapsto \mathbb{R}$ and some nonzero scalar functionals H_i thereupon ($i = 1, 2, 3$, $x \in \mathscr{X}$). Accordingly, the components $H_i\big(\ldots\big)$ can be viewed as "global tunings", and may depend adaptively on the primary-interest functions P and Q, i.e. $m_{g,i}(z) = w_{g,i}(x, p(x), q(x))$. For instance, in a finite discrete setup $(\mathscr{X}_{\#}, \lambda_{\#})$ with strictly convex and differentiable ϕ, $m_1(x) \equiv m_2(x) \equiv 1$, $m_3(x) = H_i\big((w_{g,3}(q(x)))_{z\in\mathscr{X}}\big)$ this reduces to the conformal divergences of Nock et al. [64] (they also indicate the extension to equal nonunity scaling $m_1(x) \equiv m_2(x)$), for which the subcase $w_{g,3}(q(x)) := \big(\phi'(q(x))\big)^2$, $H_3\big((h(x))_{x\in\mathscr{X}}\big) := \big(1 + \sum_{x\in\mathscr{X}} h(x)\big)^{-1/2}$ leads to the total Bregman divergences of Liu et al. [44, 45], Vemuri et al. [91]. In contrast, Nock et al. [62] use $m_1(x) \equiv m_1 = H_1\big((p(x))_{z\in\mathscr{X}}\big)$, $m_2(x) \equiv m_1 = H_1\big((q(x))_{z\in\mathscr{X}}\big)$, $m_3(x) \equiv 1$. A more detailed discussion can be found in Stummer and Kißlinger [82] and Roensch and Stummer [74], where they also give versions for nonconvex nonconcave divergence generators. Let us finally mention that for the construction of divergence families, there are other recent paradigms which are essentially different to (1), e.g. by means of measuring the tightness of inequalities (cf. Nielsen et al. [60, 61]), respectively of comparative convexity (cf. Nielsen et al. [59]).

4 Divergences for Essentially Different Functions

4.1 Motivation

Especially in divergence-based statistics, one is often faced with the situation where the functions $p(\cdot)$ and $q(\cdot)$ are of "essentially different nature". For instance, consider the situation where the uncertainty-prone data-generating mechanism is a random variable Y taking values in $\mathscr{X} = \mathbb{R}$ having a "classical" (e.g. Gaussian) probability density $\mathbb{q}(\cdot)$ with respect to the one-dimensional Lebesque measure λ_L, i.e. $Pr[Y \in \bullet] := \vec{\mathfrak{Q}}^{\mathbb{1}\cdot\lambda_L}[\bullet] := \int_\bullet \mathbb{q}(x)\,\mathrm{d}\lambda_L(x)$ where the latter is almost always a Riemann integral (i.e. $\mathrm{d}\lambda_L(x) = \mathrm{d}x$); notice that we have set $\mathbb{r}(x) \equiv 1$ $(x \in \mathbb{R})$. As already indicated above, under independent and identically distributed (i.i.d.) data observations $Y_1, \ldots, Y_N$ of Y one often builds the corresponding "empirical distribution" $\vec{\mathfrak{P}}_N^{emp}[\bullet] := \frac{1}{N} \cdot \sum_{i=1}^N \delta_{Y_i}[\bullet]$ which is nothing but the probability distribution reflecting the underlying (normalized) histogram. By rewriting $\vec{\mathfrak{P}}^{\mathbb{1}\cdot\lambda_\#}[\bullet] := \vec{\mathfrak{P}}_N^{emp}[\bullet] = \int_\bullet \mathbb{p}(x)\,\mathrm{d}\lambda_\#(x)$ with empirical probability mass function $\mathbb{p}(x) := \frac{1}{N} \cdot \#\{i \in \{1, \ldots, N\} : Y_i = x\} =: \mathbb{p}_N^{emp}(x)$ one encounters some basic problems for a straightforward application of divergence concepts: the two aggregating measures λ_L and $\lambda_\#$ do not coincide and actually they are of "essentially different" nature; moreover, $\mathbb{p}(\cdot)$ is nonzero only on the range $\mathscr{R}(Y_1, \ldots, Y_N) = \{z_1, \ldots, z_s\}$ of distinguishable points $z_1, \ldots, z_s$ $(s \leqslant N)$ occupied by $Y_1, \ldots, Y_N$. In particular, one has $\lambda_L[\{z_1, \ldots, z_s\}] = 0$. Accordingly, building a "non-coarsely discriminating" dissimilarity/divergence $D(\vec{\mathbb{P}}, \vec{\mathbb{Q}})$ between such type of functions $\vec{\mathbb{P}} := \{\mathbb{p}(x)\}_{x\in\mathscr{X}}$ and $\vec{\mathbb{Q}} := \{\mathbb{q}(x)\}_{x\in\mathscr{X}}$, is a task like "comparing apples with pears". There are several solutions to tackle this. To begin with, in the following we take the "encompassing" approach of quantifying their dissimilarity by means of their common superordinate characteristics as "fruits". Put in mathematical terms, we choose e.g. $\mathscr{X} = \mathbb{R}$, $\lambda = \lambda_L + \lambda_\#$ and work with the particular representations $\mathbb{p}(x) := \widetilde{p}(x) \cdot \mathbf{1}_{\{z_1,\ldots,z_s\}}(x)$ with $\widetilde{p}(x) > 0$ for λ-almost all $x \in \{z_1, \ldots, z_s\}$ as well as $\mathbb{q}(x) := \widetilde{q}(x) \cdot \mathbf{1}_{\widetilde{A}\backslash\{z_1,\ldots,z_s\}}(x)$ with $\widetilde{q}(x) > 0$ for λ-almost all $x \in \widetilde{A}\backslash\{z_1, \ldots, z_s\}$ with some large enough (measurable) subset $\widetilde{A}$ of $\mathscr{X} = \mathbb{R}$ such that

$$1 = \int_{\mathscr{X}} \mathbb{p}(x)\,\mathrm{d}\lambda_\#(x) = \int_{\mathscr{X}} \mathbb{p}(x)\,\mathrm{d}\lambda(x) \text{ and } 1 = \int_{\mathscr{X}} \mathbb{q}(x)\,\mathrm{d}\lambda_L(x) = \int_{\mathscr{X}} \mathbb{q}(x)\,\mathrm{d}\lambda(x) \quad (63)$$

hold. In fact, with these choices one gets $Pr[Y \in \bullet] = \int_{\bullet} \vec{\mathbb{q}}(x)\, \mathrm{d}\lambda(x)$ and $\vec{\mathfrak{P}}_N^{emp}[\bullet] = \int_{\bullet} \vec{\mathbb{p}}(x)\, \mathrm{d}\lambda(x)$, as well as

$$\mathbb{p}(x) \cdot \mathbb{q}(x) = 0 \quad \text{for } \lambda\text{-almost all } x \in \mathscr{X}, \tag{64}$$

$$\mathbb{p}(x) \cdot \mathbf{1}_{\{0\}}\big(\mathbb{q}(x)\big) = \mathbb{p}(x) \quad \text{for } \lambda\text{-almost all } x \in \mathscr{X}, \tag{65}$$

$$\mathbb{q}(x) \cdot \mathbf{1}_{\{0\}}\big(\mathbb{p}(x)\big) = \mathbb{q}(x) \quad \text{for } \lambda\text{-almost all } x \in \mathscr{X} \tag{66}$$

for the special choices $\mathbb{p}(x) = \vec{\mathbb{p}}(x)$ and $\mathbb{q}(x) = \vec{\mathbb{q}}(x)$. By means of these and (63), the divergence (51) simplifies to

$$D^c_{\phi,\vec{\mathbb{Q}},\vec{\mathbb{Q}},\mathbb{1}\cdot\vec{\mathbb{Q}},\lambda}(\vec{\mathbb{P}}, \vec{\mathbb{Q}}) = \phi^*(0) + \phi(0) - \phi(1) \; > \; 0. \tag{67}$$

Since for arbitrary space $\mathscr{X}$ (and not only $\mathbb{R}$) and any aggregator λ thereupon, the formula (67) holds for all functions $\vec{\mathbb{P}} := \{\vec{\mathbb{p}}(x)\}_{x\in\mathscr{X}}$, $\vec{\mathbb{Q}} := \{\vec{\mathbb{q}}(x)\}_{x\in\mathscr{X}}$ which satisfy (63) as well as (64)–(66) for λ-almost all $x \in \mathscr{X}$, and since $\phi^*(0) + \phi(0) - \phi(1)$ is just a constant (which may be infinite), these divergences $D^c_{\phi,\vec{\mathbb{Q}},\vec{\mathbb{Q}},\mathbb{1}\cdot\vec{\mathbb{Q}},\lambda}(\vec{\mathbb{P}}, \vec{\mathbb{Q}})$ are not suitable for discriminating between such "essentially different" (basically orthogonal) λ-probability densities $\vec{\mathbb{P}}$ and $\vec{\mathbb{Q}}$. More generally, under the validity of (64)–(66) for λ-almost all $x \in \mathscr{X}$ – which we denote by $\mathbb{P} \perp \mathbb{Q}$ and which basically amounts to pair of functions of the type

$$\mathbb{p}(x) := \widetilde{p}(x) \cdot \mathbf{1}_A(x) \quad \text{with } \widetilde{p}(x) > 0 \text{ for } \lambda\text{-almost all } x \in A, \tag{68}$$

$$\mathbb{q}(x) := \widetilde{q}(x) \cdot \mathbf{1}_{B\setminus A}(x) \quad \text{with } \widetilde{q}(x) > 0 \text{ for } \lambda\text{-almost all } x \in B\setminus A, \tag{69}$$

with some (measurable) subsets $\widetilde{A} \subset B$ of $\mathscr{X}$ – the divergence (51) turns into

$$\begin{aligned} D^c_{\phi,\mathbb{Q},\mathbb{Q},\mathbb{R}\cdot\mathbb{Q},\lambda}(\mathbb{P}, \mathbb{Q}) &= \big[\phi^*(0) - \phi'_{+,c}(1)\big] \cdot \textstyle\int_{\mathscr{X}} \mathbb{r}(x) \cdot \mathbb{p}(x)\, \mathrm{d}\lambda(x) \\ &\quad + \big[\phi(0) + \phi'_{+,c}(1) - \phi(1)\big] \cdot \textstyle\int_{\mathscr{X}} \mathbb{r}(x) \cdot \mathbb{q}(x)\, \mathrm{d}\lambda(x) \; > \; 0 \end{aligned} \tag{70}$$

which now depends on $\mathbb{P}$ and $\mathbb{Q}$, in a rudimentary "weighted-total-mass" way. Inspired by this, we specify a statistically interesting divergence subclass:

Definition 1 We say that a divergence (respectively dissimilarity respectively distance)[12] $D(\cdot, \cdot)$ is encompassing for a class $\widetilde{\mathscr{P}}$ of functions if

- for arbitrarily fixed $Q := \{q(x)\}_{x\in\mathscr{X}} \in \widetilde{\mathscr{P}}$ the function $P := \{p(x)\}_{x\in\mathscr{X}} \to D(P, Q)$ is non-constant on the subfamily of all $P \in \widetilde{\mathscr{P}}$ with $P \perp Q$, and
- for arbitrarily fixed $P \in \widetilde{\mathscr{P}}$ the function $Q \to D(P, Q)$ is non-constant on the subfamily of all $Q \in \widetilde{\mathscr{P}}$ with $Q \perp P$.

Accordingly, due to (67) the prominently used divergences $D^c_{\phi,\vec{\mathbb{Q}},\vec{\mathbb{Q}},\mathbb{1}\cdot\vec{\mathbb{Q}},\lambda}(\vec{\mathbb{P}}, \vec{\mathbb{Q}})$ are not encompassing for the class of $\widetilde{\mathscr{P}}$ of all λ-probability densities; more gener-

[12] i.e. the properties (D1) and (D2) (respectively (D2) respectively (D1), (D2) and (D3)) are satisfied.

ally, because of (70) the divergences $D^c_{\phi,\mathbb{Q},\mathbb{Q},\mathbb{R}\cdot\mathbb{Q},\lambda}(\mathbb{P},\mathbb{Q})$ are in general encompassing for the class of $\widetilde{\mathscr{P}}$ of all λ-probability densities, but not for $\widetilde{\mathscr{P}} := \{\widetilde{P} := \{\widetilde{\mathbb{p}}(x)\}_{x\in\mathscr{X}} \mid \int_{\mathscr{X}} \mathbb{r}(x)\cdot\widetilde{\mathbb{p}}(x)\,\mathrm{d}\lambda(x) = \widetilde{c}\}$ for any fixed $\widetilde{c}$.

4.2 $\mathbb{m}_1(x) = \mathbb{m}_2(x) := \mathbb{q}(x)$, $\mathbb{m}_3(x) = \mathbb{r}(x)\cdot\mathbb{q}(x)^{\chi}\in[0,\infty]$ *for Some* $\chi > 1$ *and Some (Measurable) Function* $\mathbb{r}:\mathscr{X}\to[0,\infty[$

In the following, we propose a new way of repairing the above-mentioned encompassing-concerning deficiency for λ-probability density functions, by introducing a new divergence in terms of choosing a generator $\phi:]0,\infty[\to\mathbb{R}$ which is convex and strictly convex at 1, the scaling function $\mathbb{m}_1(x) = \mathbb{m}_2(x) := \mathbb{q}(x)$ as in the non-negativity set-up of Sect. 3.3.1.2, but the more general aggregation function $\mathbb{m}_3(x) = \mathbb{r}(x)\cdot\mathbb{q}(x)^{\chi}\in[0,\infty[$ for some power $\chi > 1$ and some (measurable) function $\mathbb{r}:\mathscr{X}\to[0,\infty[$ which satisfies $\mathbb{r}(x)\in]0,\infty[$ for λ-almost all $x\in\mathscr{X}$. To incorporate the zeros of $\mathbb{p}(\cdot)$, $\mathbb{q}(\cdot)$, $\mathbb{r}(\cdot)$ by appropriate limits and conventions, we proceed analogously to Sect. 3.3.1.2. Accordingly, we inspect the boundary behaviour of the function $\widetilde{\psi}_{\phi,c}:]0,\infty[^3\to[0,\infty[$ given by

$$\widetilde{\psi}_{\phi,c}\big(r,\widetilde{s},\widetilde{t}\big) := r\cdot\widetilde{t}^{\chi}\cdot\psi_{\phi,c}\big(\tfrac{\widetilde{s}}{\widetilde{t}},1\big) = r\cdot\widetilde{t}^{\chi}\cdot\big[\phi\big(\tfrac{\widetilde{s}}{\widetilde{t}}\big)-\phi(1)-\phi'_{+,c}(1)\cdot\big(\tfrac{\widetilde{s}}{\widetilde{t}}-1\big)\big]$$
$$= r\cdot\widetilde{t}^{\chi}\cdot\big[\phi\big(\tfrac{\widetilde{s}\cdot r}{\widetilde{t}\cdot r}\big)-\phi(1)-\phi'_{+,c}(1)\cdot\big(\tfrac{\widetilde{s}\cdot r}{\widetilde{t}\cdot r}-1\big)\big] = r\cdot\widetilde{t}^{\chi}\cdot\psi_{\phi,c}\big(\tfrac{\widetilde{s}\cdot r}{\widetilde{t}\cdot r},1\big).$$

As in Sect. 3.3.1.2, the Assumption 2(a) is conformly satisfied, for which we use the short-hand notation 2(a) etc. in the following discussion. Moreover, we require the validity of 2(b)–2(d) at the point $t = 1$. The analogue of 2(e) is $\mathbb{r}(x)\cdot\widetilde{t}^{\chi} < \infty$ which is always (almost surely) automatically satisfied (a.a.sat.), whereas 2(f) converts to "$\mathbb{r}(x)\cdot\widetilde{t}^{\chi} > 0$ for all $\widetilde{s}\neq\widetilde{t}$" which is also a.a.sat. except for the case $\widetilde{t} = 0$ which will be incorporated below. For the derivation of the analogue of 2(k) we observe that for fixed $r > 0$, $\widetilde{s} > 0$

$$\ell i_2 := r\cdot 0^{\chi}\cdot\psi_{\phi,c}\big(\tfrac{\widetilde{s}}{0},1\big) := \lim_{t\to 0}\widetilde{\psi}_{\phi,c}\big(r,\widetilde{s},\widetilde{t}\big) =$$
$$= r\cdot\widetilde{s}^{\chi}\cdot\lim_{\widetilde{t}\to 0}\big[\tfrac{\widetilde{t}^{\chi}}{\widetilde{s}^{\chi}}\cdot\phi\big(\tfrac{\widetilde{s}}{\widetilde{t}}\big)\big] = r\cdot\widetilde{s}^{\chi}\cdot\phi^*_{\chi}(0)\geqslant 0, \tag{71}$$

where $\phi^*_{\chi}(0) := \lim_{u\to 0}u^{\chi-1}\cdot u\cdot\phi\big(\frac{1}{u}\big) = \lim_{v\to\infty}\frac{\phi(v)}{v^{\chi}}$ exists but may be infinite. To convert 2(i), we employ the fact that for fixed $r > 0$, $\widetilde{t} > 0$ the function $\widetilde{s}\to\widetilde{\psi}_{\phi,c}\big(r,\widetilde{s},\widetilde{t}\big)$ is convex with existing limit

$$\ell i_3 := r\cdot\widetilde{t}^{\chi}\cdot\psi_{\phi,c}\big(\frac{0}{\widetilde{t}},1\big) := \lim_{s\to 0}\widetilde{\psi}_{\phi,c}\big(r,\widetilde{s},\widetilde{t}\big)$$
$$= r\cdot\widetilde{t}^{\chi}\cdot(\phi(0)+\phi'_{+,c}(1)-\phi(1)) > 0. \tag{72}$$

To achieve the analogue of 2(g), let us first remark that for fixed $r > 0$ the function $(\widetilde{s}, \widetilde{t}) \rightarrow \widetilde{\psi}_{\phi,c}\big(r, \widetilde{s}, \widetilde{t}\big)$ may not be continuous at $(\widetilde{s}, \widetilde{t}) = (0,0)$, but due to the very nature of a divergence we make the 2(g)-conform convention of setting

$$r \cdot 0^{\chi} \cdot \psi_{\phi,c}\big(\tfrac{0}{0}, 1\big) := \widetilde{\psi}_{\phi,c}\big(r, 0, 0\big) := 0\,.$$

The analogues of the Assumptions 2(h), (j), (ℓ), (m), (n) are obsolete because of our basic finiteness requirements. Putting together all the building-blocks, with the above-mentioned limits and conventions we obtain the divergence

$$\begin{aligned}
&0 \leqslant D^{c}_{\phi,\mathbb{Q},\mathbb{Q},\mathbb{R}\cdot\mathbb{Q}^{\chi},\lambda}(\mathbb{P},\mathbb{Q})\\
&:= \overline{\int}_{\mathscr{X}} \mathbb{r}(x) \cdot \Big[\mathbb{q}(x)^{\chi} \cdot \phi\big(\tfrac{\mathbb{p}(x)}{\mathbb{q}(x)}\big) - \mathbb{q}(x)^{\chi} \cdot \phi(1) - \phi'_{+,c}(1) \cdot \big(\mathbb{p}(x) \cdot \mathbb{q}(x)^{\chi-1} - \mathbb{q}(x)^{\chi}\big)\Big]\, \mathrm{d}\lambda(x)\\
&:= \int_{\mathscr{X}} \mathbb{r}(x) \cdot \Big[\mathbb{q}(x)^{\chi} \cdot \phi\big(\tfrac{\mathbb{p}(x)}{\mathbb{q}(x)}\big) - \mathbb{q}(x)^{\chi} \cdot \phi(1) - \phi'_{+,c}(1) \cdot \big(\mathbb{p}(x) \cdot \mathbb{q}(x)^{\chi-1} - \mathbb{q}(x)^{\chi}\big)\Big]\\
&\qquad\qquad \cdot \mathbf{1}_{]0,\infty[}\big(\mathbb{p}(x) \cdot \mathbb{q}(x)\big)\, \mathrm{d}\lambda(x)\\
&+\phi^{*}_{\chi}(0) \cdot \int_{\mathscr{X}} \mathbb{r}(x) \cdot \mathbb{p}(x)^{\chi} \cdot \mathbf{1}_{]0,\infty[}\big(\mathbb{p}(x)\big) \cdot \mathbf{1}_{\{0\}}\big(\mathbb{q}(x)\big)\, \mathrm{d}\lambda(x)\\
&+\big[\phi(0) + \phi'_{+,c}(1) - \phi(1)\big] \cdot \int_{\mathscr{X}} \mathbb{r}(x) \cdot \mathbb{q}(x)^{\chi} \cdot \mathbf{1}_{]0,\infty[}\big(\mathbb{q}(x)\big) \cdot \mathbf{1}_{\{0\}}\big(\mathbb{p}(x)\big)\, \mathrm{d}\lambda(x)\\
&= \int_{\mathscr{X}} \mathbb{r}(x) \cdot \Big[\mathbb{q}(x)^{\chi} \cdot \phi\big(\tfrac{\mathbb{p}(x)}{\mathbb{q}(x)}\big) - \mathbb{q}(x)^{\chi} \cdot \phi(1) - \phi'_{+,c}(1) \cdot \big(\mathbb{p}(x) \cdot \mathbb{q}(x)^{\chi-1} - \mathbb{q}(x)^{\chi}\big)\Big]\\
&\qquad\qquad \cdot \mathbf{1}_{]0,\infty[}\big(\mathbb{p}(x) \cdot \mathbb{q}(x)\big)\, \mathrm{d}\lambda(x)\\
&+\phi^{*}_{\chi}(0) \cdot \int_{\mathscr{X}} \mathbb{r}(x) \cdot \mathbb{p}(x)^{\chi} \cdot \mathbf{1}_{\{0\}}\big(\mathbb{q}(x)\big)\, \mathrm{d}\lambda(x)\\
&+\big[\phi(0) + \phi'_{+,c}(1) - \phi(1)\big] \cdot \int_{\mathscr{X}} \mathbb{r}(x) \cdot \mathbb{q}(x)^{\chi} \cdot \mathbf{1}_{\{0\}}\big(\mathbb{p}(x)\big)\, \mathrm{d}\lambda(x)\,. \qquad (73)
\end{aligned}$$

In case of $\mathfrak{Q}^{\mathbb{R}\cdot\lambda}_{\chi}[\mathscr{X}] := \int_{\mathscr{X}} \mathbb{q}(x)^{\chi} \cdot \mathbb{r}(x)\, \mathrm{d}\lambda(x) < \infty$, the divergence (73) becomes

$$\begin{aligned}
&0 \leqslant D^{c}_{\phi,\mathbb{Q},\mathbb{Q},\mathbb{R}\cdot\mathbb{Q}^{\chi},\lambda}(\mathbb{P},\mathbb{Q})\\
&= \int_{\mathscr{X}} \mathbb{r}(x) \cdot \Big[\mathbb{q}(x)^{\chi} \cdot \phi\big(\tfrac{\mathbb{p}(x)}{\mathbb{q}(x)}\big) - \phi'_{+,c}(1) \cdot \big(\mathbb{p}(x) \cdot \mathbb{q}(x)^{\chi-1} - \mathbb{q}(x)^{\chi}\big)\Big]\\
&\qquad\qquad \cdot \mathbf{1}_{]0,\infty[}\big(\mathbb{p}(x) \cdot \mathbb{q}(x)\big)\, \mathrm{d}\lambda(x)\\
&+\phi^{*}_{\chi}(0) \cdot \int_{\mathscr{X}} \mathbb{r}(x) \cdot \mathbb{p}(x)^{\chi} \cdot \mathbf{1}_{\{0\}}\big(\mathbb{q}(x)\big)\, \mathrm{d}\lambda(x)\\
&+\big[\phi(0) + \phi'_{+,c}(1)\big] \cdot \int_{\mathscr{X}} \mathbb{r}(x) \cdot \mathbb{q}(x)^{\chi} \cdot \mathbf{1}_{\{0\}}\big(\mathbb{p}(x)\big)\, \mathrm{d}\lambda(x)\\
&-\phi(1) \cdot \int_{\mathscr{X}} \mathbb{r}(x) \cdot \mathbb{q}(x)^{\chi}\, \mathrm{d}\lambda(x)\,. \qquad (74)
\end{aligned}$$

Moreover, in case of $\phi(1) = 0$ and $\int_{\mathscr{X}} \big(\mathbb{p}(x) \cdot \mathbb{q}(x)^{\chi-1} - \mathbb{q}(x)^{\chi}\big) \cdot \mathbb{r}(x)\, \mathrm{d}\lambda(x) \in [0,\infty[$ (but not necessarily $\int_{\mathscr{X}} \mathbb{p}(x) \cdot \mathbb{q}(x)^{\chi-1} \cdot \mathbb{r}(x)\, \mathrm{d}\lambda(x) < \infty$, $\int_{\mathscr{X}} \mathbb{q}(x)^{\chi} \cdot \mathbb{r}(x)\, \mathrm{d}\lambda(x) < \infty$), the divergence (73) turns into

$$\begin{aligned}
&0 \leqslant D^{c}_{\phi,\mathbb{Q},\mathbb{Q},\mathbb{R}\cdot\mathbb{Q}^{\chi},\lambda}(\mathbb{P},\mathbb{Q}) = \int_{\mathscr{X}} \mathbb{r}(x) \cdot \mathbb{q}(x)^{\chi} \cdot \phi\big(\tfrac{\mathbb{p}(x)}{\mathbb{q}(x)}\big) \cdot \mathbf{1}_{]0,\infty[}\big(\mathbb{p}(x) \cdot \mathbb{q}(x)\big)\, \mathrm{d}\lambda(x)\\
&+\phi^{*}_{\chi}(0) \cdot \int_{\mathscr{X}} \mathbb{r}(x) \cdot \mathbb{p}(x)^{\chi} \cdot \mathbf{1}_{\{0\}}\big(\mathbb{q}(x)\big)\, \mathrm{d}\lambda(x) + \phi(0) \cdot \int_{\mathscr{X}} \mathbb{r}(x) \cdot \mathbb{q}(x)^{\chi} \cdot \mathbf{1}_{\{0\}}\big(\mathbb{p}(x)\big)\, \mathrm{d}\lambda(x)\\
&-\phi'_{+,c}(1) \cdot \int_{\mathscr{X}} \big(\mathbb{p}(x) \cdot \mathbb{q}(x)^{\chi-1} - \mathbb{q}(x)^{\chi}\big) \cdot \mathbb{r}(x)\, \mathrm{d}\lambda(x)\,.
\end{aligned}$$

In contrast to the case $\chi = 1$ where for λ-probability-density functions $\mathbb{\tilde{p}}$, $\mathbb{\tilde{q}}$, the divergence (53) was further simplified due to $\int_{\mathscr{X}} \left(\mathbb{\tilde{p}}(x) - \mathbb{\tilde{q}}(x)\right) \mathrm{d}\lambda(x) = 0$, for the current setup $\chi > 1$ the latter has no impact for further simplification. However, in general, for the new divergence defined by (73) one gets for any $\mathbb{P} \perp \mathbb{Q}$ from (68), (69), (64)–(66) the expression

$$0 \leqslant D^{c}_{\phi,\mathbb{Q},\mathbb{Q},\mathbb{R}\cdot\mathbb{Q}^{\chi},\lambda}(\mathbb{P},\mathbb{Q})$$
$$= \phi^{*}_{\chi}(0) \cdot \int_{\mathscr{X}} \mathbb{r}(x) \cdot \mathbb{p}(x)^{\chi}\, \mathrm{d}\lambda(x) + \left[\phi(0) + \phi'_{+,c}(1) - \phi(1)\right] \cdot \int_{\mathscr{X}} \mathbb{r}(x) \cdot \mathbb{q}(x)^{\chi}\, \mathrm{d}\lambda(x) \quad (75)$$

which is encompassing for the class of λ-probability functions. By inspection of the above calculations, one can even relax the assumptions away from convexity:

Theorem 6 *Let* $\chi > 1$, $c \in [0,1]$, $\phi :]0,\infty[\rightarrow \mathbb{R}$ *such that both* $\phi'_{+,c}(1)$ *and* $\phi(0) := \lim_{s\rightarrow 0}\phi(s)$ *exist and* $\psi_{\phi,c}(s,1) = \phi(s) - \phi(1) - \phi'_{+,c}(1)\cdot(s-1) \geqslant 0$ *for all* $s > 0$*. Moreover, assume that* $\psi_{\phi,c}(s,1) = 0$ *if and only if* $s = 1$*. Furthermore, let the limits* $\ell i_2 \geqslant 0$ *defined by* (71) *and* $\ell i_3 \geqslant 0$ *defined by* (72) *exist and satisfy* $\ell i_2 + \ell i_3 > 0$*. Then one gets for the divergence defined by* (73)*:*
(1) $D^{c}_{\phi,\mathbb{Q},\mathbb{Q},\mathbb{R}\cdot\mathbb{Q}^{\chi},\lambda}(\mathbb{P},\mathbb{Q}) \geqslant 0$*. Depending on the concrete situation,* $D^{c}_{\phi,\mathbb{Q},\mathbb{Q},\mathbb{R}\cdot\mathbb{Q}^{\chi},\lambda}(\mathbb{P},\mathbb{Q})$ *may take infinite value.*

(2) $D^{c}_{\phi,\mathbb{Q},\mathbb{Q},\mathbb{R}\cdot\mathbb{Q}^{\chi},\lambda}(\mathbb{P},\mathbb{Q}) = 0$ *if and only if* $\mathbb{p}(x) = \mathbb{q}(x)$ *for* λ*-a.a.* $x \in \mathscr{X}$*.*
(3) For $\mathbb{P} \perp \mathbb{Q}$*, the representation* (75) *holds.*

Remark 4 (1) As seen above, if the generator ϕ is in $\Phi(]0,\infty[)$ and satisfies the Assumptions 2(a)–(d) for $t = 1$, then the requirements on ϕ in Theorem 6 are automatically satisfied. The case $\chi = 1$ has already been covered by Theorem 5.
(2) For practical purposes, it is sometimes useful to work with a sub-setup of choices $\chi > 1$, $c \in [0,1]$ and ϕ such that $\ell i_2 \in]0,\infty[$ and/or $\ell i_3 \in]0,\infty[$. □

Let us give some examples. To begin with, for $\alpha \in \mathbb{R}\backslash\{0,1\}$ take the power functions $\phi(t) := \phi_{\alpha}(t) := \frac{t^{\alpha}-1}{\alpha(\alpha-1)} - \frac{t-1}{\alpha-1} \in [0,\infty[$, $t \in]0,\infty[$, with the properties $\phi_{\alpha}(1) = 0$, $\phi'_{\alpha}(1) = 0$ (cf. (6)) and $\phi_{\alpha}(0) := \lim_{t\downarrow 0}\phi_{\alpha}(t) = \frac{1}{\alpha}\cdot \mathbf{1}_{]0,1]\cup]1,\infty[}(\alpha) + \infty \cdot \mathbf{1}_{]-\infty,0[}(\alpha)$. Then, for arbitrary $\chi \in \mathbb{R}$ one gets the representation

$$0 \leqslant D_{\phi_{\alpha},\mathbb{Q},\mathbb{Q},\mathbb{R}\cdot\mathbb{Q}^{\chi},\lambda}(\mathbb{P},\mathbb{Q})$$
$$:= \overline{\int}_{\mathscr{X}} \mathbb{r}(x) \cdot \left[\mathbb{q}(x)^{\chi} \cdot \phi_{\alpha}\big(\tfrac{\mathbb{p}(x)}{\mathbb{q}(x)}\big) - \mathbb{q}(x)^{\chi}\cdot\phi_{\alpha}(1) - \phi'_{\alpha}(1)\cdot\big(\mathbb{p}(x)\cdot\mathbb{q}(x)^{\chi-1} - \mathbb{q}(x)^{\chi}\big)\right] \mathrm{d}\lambda(x) \quad (76)$$

$$= \overline{\int}_{\mathscr{X}} \Big[\phi_{\alpha}\big(\tfrac{\mathbb{p}(x)}{w_{\widetilde{\chi}}(\mathbb{p}(x),\mathbb{q}(x))}\big) - \phi_{\alpha}\big(\tfrac{\mathbb{q}(x)}{w_{\widetilde{\chi}}(\mathbb{p}(x),\mathbb{q}(x))}\big)$$
$$-\phi'_{\alpha}\big(\tfrac{\mathbb{q}(x)}{w_{\widetilde{\chi}}(\mathbb{p}(x),\mathbb{q}(x))}\big)\cdot\big(\tfrac{\mathbb{p}(x)}{w_{\widetilde{\chi}}(\mathbb{p}(x),\mathbb{q}(x))} - \tfrac{\mathbb{q}(x)}{w_{\widetilde{\chi}}(\mathbb{p}(x),\mathbb{q}(x))}\big)\Big] \cdot w_{\widetilde{\chi}}(\mathbb{p}(x),\mathbb{q}(x)) \cdot \mathbb{r}(x)\, \mathrm{d}\lambda(x)$$
$$= D_{\phi_{\alpha},\mathbb{Q}^{\widetilde{\chi}},\mathbb{Q}^{\widetilde{\chi}},\mathbb{R}\cdot\mathbb{Q}^{\widetilde{\chi}},\lambda}(\mathbb{P},\mathbb{Q}) \quad (77)$$

with the adaptive scaling/aggregation function $w_{\widetilde{\chi}}(u,v) = v^{\widetilde{\chi}}$ and $\widetilde{\chi} := 1 + \frac{\chi-1}{1-\alpha}$; in other words, the divergence (76) can be seen as a particularly adaptively scaled

Bregman divergence of non-negative functions in the sense of Kißlinger and Stummer [37], from which their robustness and non-singularity-asymptotical-statistics properties can be derived as a special case (for the probability setup $\mathbb{P}, \mathbb{Q}, \mathbb{r}(x) \equiv 1$, and beyond). From (77), it is immediate to see that the case $\chi = 1$ corresponds to the generalized power divergences (58) of order $\alpha \in \mathbb{R}\backslash\{0,1\}$, whereas $\chi = \alpha$ corresponds to the unscaled divergences (40), i.e.

$$0 \leqslant D_{\phi_\alpha,\mathbb{Q},\mathbb{Q},\mathbb{R}\cdot\mathbb{Q}^\alpha,\lambda}(\mathbb{P},\mathbb{Q}) = D_{\phi_\alpha,\mathbb{1},\mathbb{1},\mathbb{R}\cdot\mathbb{1},\lambda}(\mathbb{P},\mathbb{Q}) \tag{78}$$
$$= \overline{\int}_{\mathscr{X}} \tfrac{\mathbb{r}(x)}{\alpha\cdot(\alpha-1)} \cdot \Big[\mathbb{p}(x)^\alpha + (\alpha-1)\cdot\mathbb{q}(x)^\alpha - \alpha\cdot\mathbb{p}(x)\cdot\mathbb{q}(x)^{\alpha-1}\Big]\,\mathrm{d}\lambda(x)\ (cf.\ (40))$$

which for $\alpha > 1, \mathbb{r}(x) \equiv 1, \mathbb{p} = \mathbb{p}, \mathbb{q} = \mathbb{q}$ is a multiple of the α-order density-power divergences DPD used by Basu et al. [10]; as a side remark, in the latter setup our divergence (77) manifests a smooth interconnection between PD and DPD which differs from that of Patra et al. [70], Ghosh et al. [32].

For (76), let us shortly inspect the corresponding ℓi_2 from (71) as well as ℓi_3 from (72). Only for $\alpha \in]0,1[\cup]1,\infty[$, one gets finite $\ell i_3 = \frac{r\widetilde{t}^\chi}{\alpha} \in]0,\infty[$ for all $\chi \in \mathbb{R}, r > 0, \widetilde{t} > 0$. Additionally, one obtains finite ℓi_2 only for $\chi = 1, \alpha \in]0,1[$ where $\ell i_2 = \frac{r\widetilde{s}}{1-\alpha}$ (PD case), respectively for $\chi > 1, \alpha \in]0,1[\cup]1,\chi[$ where $\ell i_2 = 0$, respectively for $\alpha = \chi > 1$ where $\ell i_2 = \frac{r\widetilde{s}^\alpha}{\alpha\cdot(\alpha-1)}$ (DPD case), for all $r > 0, \widetilde{s} > 0$.

Another interesting example for the divergence $D^c_{\phi,\mathbb{Q},\mathbb{Q},\mathbb{R}\cdot\mathbb{Q}^\chi,\lambda}(\mathbb{P},\mathbb{Q})$ in (73) is given for $\alpha \in \mathbb{R}\backslash\{0,1\}$ by the generators

$$\phi(t) := \widetilde{\widetilde{\phi}}_\alpha(t) := \tfrac{(\alpha-1)\cdot t^\alpha - \alpha\cdot t^{\alpha-1}+1}{\alpha\cdot(\alpha-1)}, \quad t > 0, \quad \widetilde{\widetilde{\phi}}_\alpha(1) = 0,\ \widetilde{\widetilde{\phi}}'_\alpha(1) = 0,$$

for which $t \to \widetilde{\widetilde{\phi}}_\alpha(t) = \widetilde{\widetilde{\phi}}_\alpha(t) - \widetilde{\widetilde{\phi}}_\alpha(0) - \widetilde{\widetilde{\phi}}'_\alpha(1)\cdot(t-1) = \psi_{\phi_\alpha}(t,1)$ is strictly decreasing on $]0,1[$ and strictly increasing on $]1,\infty[$. Hence, the corresponding assumptions of Theorem 6 are satisfied. Beyond this, notice that $\widetilde{\widetilde{\phi}}_\alpha(\cdot)$ is strictly convex on $]0,\infty[$ if $\alpha \in]1,2]$, respectively strictly convex on $]1-\frac{1}{\alpha-1},\infty[$ and strictly concave on $]0,1-\frac{1}{\alpha-1}[$ if $\alpha > 2$, respectively strictly convex on $]0,1+\frac{1}{1-\alpha}[$ and strictly concave on $]1+\frac{1}{1-\alpha},\infty[$ if $\alpha \in]-\infty,0[\cup]0,1[$. Furthermore, the corresponding ℓi_3 is finite only for $\alpha > 1$, namely $\ell i_3 = \frac{r\widetilde{t}^\chi}{\alpha\cdot(\alpha-1)} \in]0,\infty[$ for all $\chi \in \mathbb{R}$, $r > 0, \widetilde{t} > 0$. Additionally, if $\alpha > 1$ one gets finite ℓi_2 only for $\chi > \alpha > 1$ where $\ell i_2 = 0$, respectively for $\alpha = \chi > 1$ where $\ell i_2 = \frac{r\widetilde{s}^\alpha}{\alpha}$ for all $r > 0, \widetilde{s} > 0$. Notice that for $\chi = \alpha > 1$, the limits $\ell i_2, \ell i_3$ for the cases ϕ_α and $\widetilde{\widetilde{\phi}}_\alpha$ are asymmetric. Indeed, by straightforward calculations one can easily see that

$$0 \leqslant D_{\widetilde{\widetilde{\phi}}_\alpha,\mathbb{Q},\mathbb{Q},\mathbb{R}\cdot\mathbb{Q}^\alpha,\lambda}(\mathbb{P},\mathbb{Q}) = D_{\phi_\alpha,\mathbb{1},\mathbb{1},\mathbb{R}\cdot\mathbb{1},\lambda}(\mathbb{Q},\mathbb{P})$$
$$= \overline{\int}_{\mathscr{X}} \tfrac{\mathbb{r}(x)}{\alpha\cdot(\alpha-1)} \cdot \Big[\big(\mathbb{q}(x)\big)^\alpha + (\alpha-1)\cdot\big(\mathbb{p}(x)\big)^\alpha - \alpha\cdot\mathbb{q}(x)\cdot\big(\mathbb{p}(x)\big)^{\alpha-1}\Big]\,\mathrm{d}\lambda(x) \tag{79}$$

which is the "reversion" of the divergence (40).

4.3 Minimum Divergences - The Encompassing Method

So far, we have almost entirely dealt with aggregated divergences between functions $P := \{p(x)\}_{x \in \mathscr{X}}$, $Q := \{q(x)\}_{x \in \mathscr{X}}$ under the *same* aggregator (measure) λ. On the other hand, in Sect. 4.1 we have already encountered an important statistical situation where *two* aggregators λ_1 and λ_2 come into play. Let us now investigate such a context in more detail. To achieve this, for the rest of this paper we confine ourselves to the following probabilistic setup: the modeled respectively observed (random) data take values in a state space $\mathscr{X}$ (with at least two distinct values), equipped with a system $\mathscr{F}$ of admissible events (σ-algebra) and two σ-finite measures λ_1 and λ_2. Furthermore, let $\overset{\to}{\mathbb{P}} := \{\overset{\to}{\mathbb{p}}\}_{x \in \mathscr{X}}$, $\overset{\to}{\mathbb{Q}} := \{\overset{\to}{\mathbb{q}}\}_{x \in \mathscr{X}}$ such that $\overset{\to}{\mathbb{p}}(x) \geqslant 0$ for λ_1-a.a. $x \in \mathscr{X}$, $\overset{\to}{\mathbb{q}}(x) \geqslant 0$ for λ_2-a.a. $x \in \mathscr{X}$, $\int_{\mathscr{X}} \overset{\to}{\mathbb{p}}(x)\, \mathrm{d}\lambda_1(x) = 1$, and $\int_{\mathscr{X}} \overset{\to}{\mathbb{q}}(x)\, \mathrm{d}\lambda_2(x) = 1$; in other words, $\overset{\to}{\mathbb{P}}$ is a λ_1-probability density function and $\overset{\to}{\mathbb{Q}}$ is a λ_2-probability density function; the two corresponding probability measures are denoted by $\overset{\to}{\mathfrak{P}}^{\mathbb{1}\cdot\lambda_1}[\bullet] := \int_\bullet \overset{\to}{\mathbb{p}}(x)\, \mathrm{d}\lambda_1(x)$ and $\overset{\to}{\mathfrak{Q}}^{\mathbb{1}\cdot\lambda_2}[\bullet] := \int_\bullet \overset{\to}{\mathbb{q}}(x)\, \mathrm{d}\lambda_2(x)$. Notice that we henceforth assume $\mathbb{r}(x) = 1$ for all $x \in \mathscr{X}$.

More specific, we deal with a parametric framework of double uncertainty in the data and in the model (cf. Sect. 2.4). The former is described by a random variable Y taking values in the space $\mathscr{X}$ and by its probability law $\overset{\to}{\mathfrak{Q}}_{\theta_0}^{\mathbb{1}\cdot\lambda_2}[\bullet]$ which (as far as model risk is concerned) is supposed to be unknown but belong to a class $\mathscr{Q}_{\Theta}^{\lambda_2} = \{\overset{\to}{\mathfrak{Q}}_{\theta}^{\mathbb{1}\cdot\lambda_2}[\bullet] : \theta \in \Theta\}$ of probability measures on $(\mathscr{X}, \mathscr{F})$ indexed by a set of parameters $\Theta \subset \mathbb{R}^d$ (the non-parametric case works basically in analogous way, with more sophisticated technicalities). Accordingly, all $Pr[Y \in \bullet \,|\, \theta] = \overset{\to}{\mathfrak{Q}}_{\theta}^{\mathbb{1}\cdot\lambda_2}[\bullet] = \int_\bullet \overset{\to}{\mathbb{q}}_\theta(x)\, \mathrm{d}\lambda_2(x)$ $(\theta \in \Theta)$ are principal model-candidate laws, with θ_0 to be found out (approximately and with high confidence) by N concrete data observations described by the independent and identically distributed random variables $Y_1, \ldots Y_N$. Furthermore, we assume that the true unknown parameter θ_0 (to be learnt) is identifiable and that the family $\mathscr{Q}_{\Theta}^{\lambda_2}$ is (measure-theoretically) equivalent in the sense

$$\overset{\to}{\mathfrak{Q}}_{\theta}^{\mathbb{1}\cdot\lambda_2} \neq \overset{\to}{\mathfrak{Q}}_{\theta_0}^{\mathbb{1}\cdot\lambda_2} \quad \text{and} \quad \overset{\to}{\mathfrak{Q}}_{\theta}^{\mathbb{1}\cdot\lambda_2} \sim \overset{\to}{\mathfrak{Q}}_{\theta_0}^{\mathbb{1}\cdot\lambda_2} \quad \text{for all } \theta,\, \theta_0 \in \Theta \quad \text{with } \theta \neq \theta_0. \tag{80}$$

As usual, the equivalence $\overset{\to}{\mathfrak{Q}}^{\mathbb{1}\cdot\lambda_2} \sim \widehat{\overset{\to}{\mathfrak{Q}}}^{\mathbb{1}\cdot\lambda_2}$ means that for λ_2-a.a. $x \in \mathscr{X}$ there holds the density-function-relation: $\overset{\to}{\mathbb{q}}(x) = 0$ if and only if $\widehat{\overset{\to}{\mathbb{q}}}(x) = 0$; this implies in particular that $\overset{\to}{\mathbb{q}}(x) \cdot \mathbf{1}_{\{0\}}\big(\widehat{\overset{\to}{\mathbb{q}}}(x)\big) = 0$ and $\widehat{\overset{\to}{\mathbb{q}}}(x) \cdot \mathbf{1}_{\{0\}}\big(\overset{\to}{\mathbb{q}}(x)\big) = 0$ for λ_2-a.a. $x \in \mathscr{X}$, and by cutting off "datapoints/states of zero contributions" one can then even take $\mathscr{X}$ small enough such that $\overset{\to}{\mathbb{q}}(x) \cdot \widehat{\overset{\to}{\mathbb{q}}}(x) > 0$ (and hence, $\mathbf{1}_{]0,\infty[}\big(\overset{\to}{\mathbb{q}}(x) \cdot \widehat{\overset{\to}{\mathbb{q}}}(x)\big) = 1$) for λ_2-a.a. $x \in \mathscr{X}$. Clearly, since any λ_2-aggregated divergence $D_{\lambda_2}(\cdot,\cdot)$ satisfies (the aggregated version of) the axioms (D1) and (D2), and since θ_0 is identifiable, one gets immediately in terms of the corresponding λ_2-probability density functions $\overset{\to}{\mathbb{Q}}_\theta := \{\overset{\to}{\mathbb{q}}_\theta(x)\}_{x \in \mathscr{X}}$

$$\theta_0 = \operatorname{argmin}_{\theta \in \Theta} D_{\lambda_2}\big(\mathbb{Q}_{\theta_0}, \mathbb{Q}_\theta\big) \quad \text{for every } \theta_0 \in \Theta. \tag{81}$$

Inspired by this, one major idea of tracking down (respectively, learning) the true unknown θ_0 is to replace $\mathfrak{Q}_{\theta_0}^{\mathbb{1}\cdot\lambda_2}$ by a data-observation-derived – and thus noisy – probability law $\omega \to \mathfrak{P}_N^{obs(\omega);\mathbb{1}\cdot\lambda_1}[\bullet] := \int_\bullet \mathbb{p}^{Y_1(\omega),\ldots,Y_N(\omega)}(x)\,\mathrm{d}\lambda_1(x)$ where the λ_1-probability density function $\mathbb{P}_N^{obs(\omega)} := \{\mathbb{p}^{Y_1(\omega),\ldots,Y_N(\omega)}(x)\}_{x\in\mathscr{X}}$ depends, as indexed, on the outcome of the observations $Y_1(\omega), \ldots, Y_N(\omega)$. If $\mathfrak{P}_N^{obs(\omega);\mathbb{1}\cdot\lambda_1}$ converges in distribution to $\mathfrak{Q}_{\theta_0}^{\mathbb{1}\cdot\lambda_2}$ as N tends to infinity, then one *intuitively* expects to obtain the so-called *minimum-divergence estimator* ("approximator")

$$\widehat{\theta}_N(\omega) := \widehat{\theta}_{N,D_{\lambda_2}}(\omega) := \operatorname{arginf}_{\theta\in\Theta} D_{\lambda_2}\big(\mathbb{P}_N^{obs(\omega)}, \mathbb{Q}_\theta\big) \tag{82}$$

which estimates θ_0 consistently in the usual sense of the convergence $\theta_n \to \theta_0$ for $n \to \infty$. However, by the nature of our divergence construction, the method (82) makes principal sense only if the two aggregators λ_1 and λ_2 coincide (and if (82) is analytically respectively computationally solvable)! Remark that the minimum distance estimator (82) depends on the choice of the divergence $D_{\lambda_2}(\cdot,\cdot)$.

Subsetup 1. For instance, if by nature the set $\mathscr{X}$ of all possible data points has only countably many elements, say $\mathscr{X} = \mathscr{X}_\# = \{z_1, \ldots z_s\}$ (where s is an integer larger than one or infinity), then a natural model-concerning aggregator is the counting measure $\lambda_2 := \lambda_\#$ (recall $\lambda_\#[\{x\}] = 1$ for all $x \in \mathscr{X}$), and hence $\mathfrak{Q}_\theta^{\mathbb{1}\cdot\lambda_2}[\bullet] = \sum_{x\in\bullet} \mathbb{q}_\theta(x) = \sum_{x\in\mathscr{X}} \mathbf{1}_\bullet(x)\cdot \mathbb{q}_\theta(x)$ (where $\bullet$ stands for any arbitrary subset of $\mathscr{X}$). In such a context, a popular choice for the data-observation-derived probability law is the so-called "empirical distribution" $\omega \to \mathfrak{P}_N^{obs(\omega);\mathbb{1}\cdot\lambda_1}[\bullet] = \int_\bullet \mathbb{p}^{Y_1(\omega),\ldots,Y_N(\omega)}(x)\,\mathrm{d}\lambda_1(x) := \sum_{x\in\bullet} \mathbb{p}_N^{emp(\omega)}(x) =: \mathfrak{P}_N^{emp(\omega)}[\bullet]$, where $\lambda_1 := \lambda_\# = \lambda_2$ and $\mathbb{p}_N^{emp(\omega)}(x) := \frac{1}{N}\cdot \#\{i \in \{1,\ldots,N\} : Y_i(\omega) = x\}$ is the total number of x-observations divided by the total number N of observations. In other words, $\mathfrak{P}_N^{obs(\omega);\mathbb{1}\cdot\lambda_1}[\bullet] := \mathfrak{P}_N^{emp(\omega)}[\bullet] := \frac{1}{N}\cdot\sum_{i=1}^N \delta_{Y_i(\omega)}[\bullet]$, where $\delta_z[\bullet]$ is the corresponding Dirac (resp. one-point) distribution given by $\delta_z[A] := \mathbf{1}_A(z)$. Hence, in such a set-up it makes sense to solve the noisy minimization problem

$$\widehat{\theta}_N(\omega) := \widehat{\theta}_{N,D_{\lambda_\#}}(\omega) := \operatorname{arginf}_{\theta\in\Theta} D_{\lambda_\#}\big(\mathbb{P}_N^{emp(\omega)}, \mathbb{Q}_\theta\big) \tag{83}$$

where $\mathbb{P}_N^{emp(\omega)} := \{\mathbb{p}_N^{emp}(x)\}_{x\in\mathscr{X}}$ and $D_{\lambda_\#}(\cdot,\cdot)$ is the discrete version of any of the divergences above. Notice that – at least for small enough number N of observations – for some $x \in \mathscr{X}$ with $\lambda_\#[\{x\}] > 0$ one has $\mathbb{p}_N^{emp}(x) = 0$ but $\mathbb{q}_\theta(x) > 0$ (i.e. an "extreme inlier"), and hence, $\mathbb{q}_\theta(x)\cdot \mathbf{1}_{\{0\}}\big(\mathbb{p}_N^{emp}(x)\big) > 0$; this must be taken into account in the calculation of the explicit forms of the corresponding divergences.[13] By the assumed convergence, this effect disappears as N becomes large enough. □

Subsetup 2. Consider the "crossover case" where $\mathscr{X}$ is uncountable (e.g. $\mathscr{X} = \mathbb{R}$) and the family $\mathscr{Q}_{\Theta}^{\lambda_2}$ is assumed to be *continuous (nonatomic)* in the sense

$$0 = \vec{\mathfrak{Q}}_{\theta}^{\mathbb{1}\cdot\lambda_2}[\{z\}] = Pr[Y \in \{z\} \,|\, \theta] = \int_{\mathscr{X}} \mathbf{1}_{\{z\}}(x) \cdot \vec{\mathfrak{q}}_{\theta}(x)\, \mathrm{d}\lambda_2(x) \;\; \text{for all } z \in \mathscr{X}, \theta \in \Theta \quad (84)$$

(e.g. $\vec{\mathfrak{q}}_{\theta}(\cdot)$ are Gaussian densities with mean θ and variance 1), and the data-observation-derived probability law is the "extended" empirical distribution

$$\begin{aligned} \omega \to \vec{\mathfrak{P}}_N^{obs(\omega);\mathbb{1}\cdot\lambda_1}[\bullet] &= \int_{\bullet} \vec{\mathbb{p}}^{Y_1(\omega),\dots,Y_N(\omega)}(x)\, \mathrm{d}\lambda_1(x) \\ &:= \sum\nolimits_{x\in\bullet} \vec{\mathbb{p}}_N^{emp(\omega)}(x) \cdot \mathbf{1}_{\mathscr{R}(Y_1(\omega),\dots,Y_N(\omega))}(x) =: \vec{\mathfrak{P}}_N^{\overline{emp}(\omega)}[\bullet], \end{aligned} \quad (85)$$

where the extension on $\mathscr{X}$ is accomplished by attributing zeros to all x outside the finite range $\mathscr{R}(Y_1(\omega), \dots, Y_N(\omega)) = \{z_1(\omega), \dots, z_s(\omega)\}$ of distinguishable points $z_1(\omega), \dots, z_s(\omega)$ $(s \leqslant N)$ occupied by the observations $Y_1(\omega), \dots, Y_N(\omega)$; notice that the involved counting measure given by $\lambda_1[\bullet] := \sum_{z\in\mathscr{X}} \mathbf{1}_{\mathscr{R}(Y_1(\omega),\dots,Y_N(\omega))}(z) \cdot \delta_z[\bullet]$ puts 1 to each data-point z which has been observed. Because λ_1 and λ_2 are now essentially different, the minimum-divergence method (82) can not be applied directly (by taking either $\lambda := \lambda_1$ or $\lambda := \lambda_2$), despite of $\vec{\mathfrak{P}}_N^{\overline{emp}(\omega)}$ converging in distribution to $\vec{\mathfrak{Q}}_{\theta_0}^{\mathbb{1}\cdot\lambda_2}$ as N tends to infinity. □

There are several ways to circumvent the problem in Subsetup 2. In the following, we discuss in more detail our abovementioned new encompassing approach:

(Enc1) take the encompassing aggregator $\lambda := \lambda_1 + \lambda_2$ and the imbedding $\vec{\mathbb{P}}_N^{\overline{emp}(\omega)} := \{\vec{\mathbb{p}}_N^{\overline{emp}(\omega)}(x)\}_{x\in\mathscr{X}}$ with $\vec{\mathbb{p}}_N^{\overline{emp}(\omega)}(x) := \vec{\mathbb{p}}_N^{emp(\omega)}(x) \cdot \mathbf{1}_{\mathscr{R}(Y_1(\omega),\dots,Y_N(\omega))}(x)$;

(Enc2) choose a "sufficiently discriminating" (e.g. encompassing) divergence $D_{\lambda}(\cdot, \cdot)$ from above and evaluate them with the density-functions obtained in (Enc1);

(Enc3) solve the corresponding noisy minimization problem

$$\widehat{\theta}_N(\omega) := \widehat{\theta}_{N,D_\lambda}(\omega) := \operatorname{arginf}_{\theta\in\Theta} D_{\lambda}\big(\vec{\mathbb{P}}_N^{\overline{emp}(\omega)}, \check{\vec{\mathbb{Q}}}_{\theta}\big) \quad (86)$$

for $\check{\vec{\mathbb{Q}}}_{\theta} := \vec{\mathbb{Q}}_{\theta}$ respectively $\check{\vec{\mathbb{Q}}}_{\theta} := \widetilde{\vec{\mathbb{Q}}}_{\theta}$ (to be defined right below);

(Enc4) compute the noisy minimal distance $D_{\lambda}\big(\vec{\mathbb{P}}_N^{\overline{emp}(\omega)}, \check{\vec{\mathbb{Q}}}_{\theta}\big) > 0$ as an indicator of "goodness of fit" (goodness of noisy approximation");

[13] E.g. applying the divergence (46) for $\alpha \in \mathbb{R}\backslash\{0, 1\}$, the sum-entry $\mathbb{r}(x) \cdot \frac{\vec{\mathbb{q}}_{\theta}(x)^{\alpha}}{\alpha}$ appears, which can be viewed as penalty for the cell x being empty of data observations ("intrinsic empty-cell-penalty"); for divergence (60), the penalty is $\phi(0) \cdot \mathbb{r}(x) \cdot \vec{\mathbb{q}}_{\theta}(x)$.

(Enc5) investigate sound statistical properties of the outcoming estimator $\widehat{\theta}_N(\omega)$, e.g. show probabilistic convergence (as N tends to infinity) to the true unknown parameter θ_0, compute the corresponding convergence speed, analyze its robustness against data-contamination, etc.

Typically, for fixed N the step (Enc3) is not straightforward to solve, and consequently, the tasks described in the unavoidable step (Enc4) become even much more complicated; a detailed discussion of both is – for the sake of brevity – beyond the scope of this paper. As far as (Enc1) is concerned, things are non-trivial due to the generally well-known fact that "continuous" densities are only almost-surely unique. Indeed, consider e.g. the case where the θ-family of functions $\mathbb{Q}_\theta := \{\mathbb{q}_\theta(x)\}_{x\in\mathscr{X}}$ satisfies

$$\mathbb{q}_\theta(x) > 0 \text{ for all } x \in \mathscr{X} \text{ and } \mathfrak{Q}_\theta^{1\cdot\lambda_2}[\mathscr{X}] = \int_{\mathscr{X}} \mathbb{q}_\theta(x)\,\mathrm{d}\lambda_2(x) = 1 \text{ for all } \theta \in \Theta \tag{87}$$

and the alternative θ-family of functions $\widetilde{\mathbb{Q}}_\theta := \{\widetilde{\mathbb{q}}_\theta(x)\}_{x\in\mathscr{X}}$ defined by $\widetilde{\mathbb{q}}_\theta(x) := \mathbb{q}_\theta(x) \cdot (1 - \mathbf{1}_{\mathscr{R}(Y_1(\omega),\ldots,Y_N(\omega))}(x))$; for the latter, one obtains

$$\widetilde{\mathfrak{Q}}_\theta^{1\cdot\lambda_2}[\mathscr{X}] = \int_{\mathscr{X}} \widetilde{\mathbb{q}}_\theta(x)\,\mathrm{d}\lambda_2(x) = \int_{\mathscr{X}} \widetilde{\mathbb{q}}_\theta(x)\,\mathrm{d}(\lambda_1 + \lambda_2)(x) = 1 \text{ for all } \theta \in \Theta. \tag{88}$$

Furthermore, due to (85) one has

$$1 = \mathfrak{P}_N^{\overline{emp}(\omega)}[\mathscr{X}] = \int_{\mathscr{X}} \mathbb{p}_N^{\overline{emp}(\omega)}(x)\,\mathrm{d}\lambda_1(x) = \int_{\mathscr{X}} \mathbb{p}_N^{\overline{emp}(\omega)}(x)\,\mathrm{d}(\lambda_1 + \lambda_2)(x) \tag{89}$$

and the validity of (64)–(66) with $\mathbb{p}(x) := \mathbb{p}_N^{\overline{emp}(\omega)}(x)$, $\mathbb{q}(x) := \widetilde{\mathbb{q}}_\theta(x)$ and $\lambda = \lambda_1 + \lambda_2$; in other words, there holds the singularity (measure-theoretical orthogonality) $\mathbb{P}_N^{\overline{emp}(\omega)} \perp \widetilde{\mathbb{Q}}_\theta$ for all $\theta \in \Theta$. Accordingly, for the step (Enc2) one can e.g. take directly the (family of) encompassing divergences $D^c_{\phi,\mathbb{Q},\mathbb{Q},\mathbb{R}\cdot\mathbb{Q}^\chi,\lambda}(\mathbb{P}, \mathbb{Q})$ of (73) for $\mathbb{P} := \mathbb{P}_N^{\overline{emp}(\omega)}$, $\mathbb{Q} := \widetilde{\mathbb{Q}}_\theta$, $\lambda := \lambda_1 + \lambda_2$, $\mathbb{r}(x) \equiv 1$, and apply (75) to get

$$\begin{aligned} 0 \leqslant D^c_{\phi,\mathbb{Q},\mathbb{Q},\mathbb{1}\cdot\mathbb{Q}^\chi,\lambda}(\mathbb{P}, \mathbb{Q}) &= \phi^*_\chi(0) \cdot \textstyle\sum_{x\in\mathscr{R}(Y_1(\omega),\ldots,Y_N(\omega))} \left(\mathbb{p}_N^{emp(\omega)}(x)\right)^\chi \\ &+ \left[\phi(0) + \phi'_{+,c}(1) - \phi(1)\right] \cdot \int_{\mathscr{X}} \left(\mathbb{q}_\theta(x)\right)^\chi \mathrm{d}\lambda_2(x)\,; \end{aligned} \tag{90}$$

hence, the corresponding solution of (Enc3) does not depend on the data-observations $Y_1(\omega), \ldots, Y_N(\omega)$, and thus is "statistically non-relevant". As an important remark for the rest of this paper, let us mention that – only – in situations where no observations are taken into account, then $\widetilde{\mathbb{Q}}_\theta = \mathbb{Q}_\theta$, $\mathscr{R}(Y_1, \ldots, Y_N) = \emptyset$, and λ_1 collapses to the "zero aggregator" (i.e. $\lambda_1[\bullet] \equiv 0$).

In contrast, let us replace the alternative θ-family $\widetilde{\mathbb{Q}}_\theta$ by the original $\mathbb{Q}_\theta$, on which λ_1 acts differently. In fact, instead of (88) there holds

$$
\begin{aligned}
1 = \vec{\mathfrak{Q}}_\theta^{\mathbb{1}\cdot\lambda_2}[\mathscr{X}] = \int_{\mathscr{X}} \vec{q}_\theta(x)\,\mathrm{d}\lambda_2(x) < \int_{\mathscr{X}} \vec{q}_\theta(x)\,\mathrm{d}(\lambda_1+\lambda_2)(x) \\
= 1 + \sum_{x\in\mathscr{R}(Y_1(\omega),\ldots,Y_N(\omega))} \vec{q}_\theta(x) \qquad \text{for all } \theta\in\Theta\,;
\end{aligned}
\tag{91}
$$

moreover, one has for all $\theta \in \Theta$ the non-singularity $\vec{\mathbb{P}}_N^{\overline{emp}(\omega)} \not\perp \vec{\mathbb{Q}}_\theta$ but

$$
\mathbf{1}_{\{0\}}\big(\vec{q}_\theta(x)\big) = 0 \quad \text{for all } x\in\mathscr{X}, \tag{92}
$$

$$
\mathbf{1}_{\{0\}}\big(\vec{p}_N^{\overline{emp}(\omega)}(x)\big) = 1 - \mathbf{1}_{\mathscr{R}(Y_1(\omega),\ldots,Y_N(\omega))}(x) \quad \text{for all } x\in\mathscr{X}, \tag{93}
$$

$$
\mathbf{1}_{]0,\infty[}\big(\vec{p}_N^{\overline{emp}(\omega)}(x)\cdot\vec{q}_\theta(x)\big) = \mathbf{1}_{\mathscr{R}(Y_1(\omega),\ldots,Y_N(\omega))}(x) \quad \text{for all } x\in\mathscr{X}\,. \tag{94}
$$

Correspondingly, for the step (Enc2) one can e.g. take directly the (family of) encompassing divergences $D^c_{\phi,\mathbb{Q},\mathbb{Q},\mathbb{R}\cdot\mathbb{Q}^\chi,\lambda}(\mathbb{P},\mathbb{Q})$ of (73) for $\mathbb{P} := \vec{\mathbb{P}}_N^{\overline{emp}(\omega)}$, $\mathbb{Q} := \vec{\mathbb{Q}}_\theta$, $\lambda := \lambda_1+\lambda_2$, $\mathfrak{r}(x) \equiv 1$; the corresponding solution of the noisy minimization problem (Enc3) generally *does depend* on the data-observations $Y_1(\omega),\ldots,Y_N(\omega)$, as required. Let us demonstrate this exemplarily for the special subsetup where $\phi : [0,\infty[\to [0,\infty[$ is continuous (e.g. strictly convex on $]0,\infty[$), differentiable at 1, $\phi(1) = \phi'(1) = 0$, $\phi(t) \in]0,\infty[$ for all $t \in [0,1[\cup[1,\infty[$, $\chi > 1$, $\mathfrak{r}(x) \equiv 1$, and $\int_{\mathscr{X}} \vec{q}_\theta(x)^\chi\,\mathrm{d}\lambda_2(x) \in]0,\infty[$ for all $\theta\in\Theta$. Then, for each fixed $\theta\in\Theta$ we derive from (73) and (92)–(94) the divergence

$$
\begin{aligned}
&0 < D_{\phi,\vec{\mathbb{Q}}_\theta,\vec{\mathbb{Q}}_\theta,\mathbb{1}\cdot\vec{\mathbb{Q}}_\theta^{\chi},\lambda_1+\lambda_2}\big(\vec{\mathbb{P}}_N^{\overline{emp}(\omega)},\vec{\mathbb{Q}}_\theta\big) \\
&= \int_{\mathscr{X}} \vec{q}_\theta(x)^\chi\cdot\phi\Big(\frac{\vec{p}_N^{\overline{emp}(\omega)}(x)}{\vec{q}_\theta(x)}\Big)\cdot\mathbf{1}_{]0,\infty[}\big(\vec{p}_N^{\overline{emp}(\omega)}(x)\cdot\vec{q}_\theta(x)\big)\,\mathrm{d}(\lambda_1+\lambda_2)(x) \\
&+\phi(0)\cdot\int_{\mathscr{X}} \vec{q}_\theta(x)^\chi\cdot\mathbf{1}_{\{0\}}\big(\vec{p}_N^{\overline{emp}(\omega)}(x)\big)\,\mathrm{d}(\lambda_1+\lambda_2)(x) \\
&= \sum_{x\in\mathscr{R}(Y_1(\omega),\ldots,Y_N(\omega))} \vec{q}_\theta(x)^\chi\cdot\phi\Big(\frac{\vec{p}_N^{emp(\omega)}(x)}{\vec{q}_\theta(x)}\Big) + \phi(0)\cdot\int_{\mathscr{X}} \vec{q}_\theta(x)^\chi\,\mathrm{d}\lambda_2(x) < \infty\,.
\end{aligned}
\tag{95}
$$

When choosing this divergence (95) in step (Enc2), we call the solution $\widehat{\theta}_N(\omega)$ of the corresponding noisy minimization problem (86) of step (Enc3) a *minimum* (ϕ,χ)-*divergence estimator* of the true unknown parameter θ_0; in ML and AI contexts, the pair (ϕ,χ) may be regarded as "hyperparameter". Exemplarily, for the power functions $\phi := \phi_\alpha$ (cf. (5)) with $\alpha = \chi > 1$, we obtain from (95) (see also (78), (41)) the divergence

$$
\begin{aligned}
&]0,\infty[\ni D_{\phi_\alpha,\vec{\mathbb{Q}}_\theta,\vec{\mathbb{Q}}_\theta,\mathbb{1}\cdot\vec{\mathbb{Q}}_\theta^{\alpha},\lambda_1+\lambda_2}\big(\vec{\mathbb{P}}_N^{\overline{emp}(\omega)},\vec{\mathbb{Q}}_\theta\big) = \tfrac{1}{\alpha}\cdot\int_{\mathscr{X}} \vec{q}_\theta(x)^\alpha\,\mathrm{d}\lambda_2(x) \\
&+\sum_{x\in\mathscr{R}(Y_1(\omega),\ldots,Y_N(\omega))}\Big[\frac{(\vec{p}_N^{emp(\omega)}(x))^\alpha}{\alpha\cdot(\alpha-1)} - \vec{p}_N^{emp(\omega)}(x)\cdot\frac{\vec{q}_\theta(x)^{\alpha-1}}{\alpha-1} + \frac{\vec{q}_\theta(x)^\alpha}{\alpha}\Big] \\
&= \tfrac{1}{\alpha}\cdot\int_{\mathscr{X}} \vec{q}_\theta(x)^\alpha\,\mathrm{d}\lambda_2(x) \\
&+\tfrac{1}{N}\sum_{i=1}^N\Big[\frac{(\vec{p}_N^{emp(\omega)}(Y_i(\omega)))^{\alpha-1}}{\alpha\cdot(\alpha-1)} - \frac{\vec{q}_\theta(Y_i(\omega))^{\alpha-1}}{\alpha-1} + \frac{\vec{q}_\theta(Y_i(\omega))^\alpha}{\alpha\cdot\vec{p}_N^{emp(\omega)}(Y_i(\omega))}\Big],
\end{aligned}
\tag{96}
$$

where for the last equality we have used the representation

$$\sum_{x\in\mathscr{X}} \mathbb{p}_N^{emp(\omega)}(x) \cdot \mathbf{1}_{\mathscr{R}(Y_1(\omega),\ldots,Y_N(\omega))}(x) \cdot \delta_x[\bullet] \; = \; \tfrac{1}{N} \cdot \sum_{i=1}^{N} \delta_{Y_i(\omega)}[\bullet]; \quad (97)$$

notice that $\mathbb{p}_N^{emp(\omega)}(Y_i(\omega)) = \#\{j \in \{1,\ldots,N\} : Y_j(\omega) = Y_i(\omega)\}/N$. Clearly, the outcoming minimum (ϕ,χ)-divergence estimator of (95) (and in particular, the minimum (ϕ_α,α)-divergence estimator of (96)) depends on the data observations $Y_1(\omega),\ldots,Y_N(\omega)$, where for technical reasons as e.g. existence and uniqueness – as well as for the tasks (Enc4), (Enc5) – some further assumptions are generally needed; for the sake of brevity, corresponding details will appear in a forthcoming paper.

4.4 Minimum Divergences - Grouping and Smoothing

Next, we briefly indicate two other ways to circumvent the problem described in Subsetup 2 of Sect. 4.3, with continuous (nonatomic) $\mathscr{Q}_\Theta^{\lambda_2}$ and λ_2 from (84):

(GR) grouping (partitioning, quantization) of data: convert[14] everything into a purely discrete context, by subdividing the data-point-set $\mathscr{X} = \bigcup_{j=1}^{s} A_j$ into countably many – (say) $s \in \mathbb{N} \cup \{\infty\}\backslash\{1\}$ – (measurable) disjoint classes $A_1,\ldots,A_s$ with the property $\lambda_2[A_j] > 0$ ("essential partition"); proceed as in Subsetup 1 of Sect. 4.3, with $\mathscr{X}^{new} := \{A_1,\ldots,A_s\}$ instead of $\{z_1,\ldots,z_s\}$, and thus the ith data observation $Y_i(\omega)$ and the corresponding running variable x) manifest (only) the corresponding class-membership. For the subcase of Csiszar-Ali-Slivey divergences and adjacently related divergences, thorough statistical investigations (such as efficiency, robustness, types of grouping, grouping-error sensitivity, etc.) of the corresponding minimum-divergence-estimation can be found e.g. in Victoria-Feser and Ronchetti [92], Menendez et al. [47–49], Morales et al. [52, 53], Lin and He [43].

(SM) smoothing of the empirical density function: convert everything to a purely continuous context, by keeping the original data-point-set $\mathscr{X}$ and by "continuously modifying" (e.g. with the help of kernels) the empirical density $\mathbb{p}_N^{emp}(\cdot)$ to a function $\mathbb{p}_N^{emp,smo}(\cdot) \geqslant 0$ such that $\int_{\mathscr{X}} \mathbb{p}_N^{emp,smo}(x)\,\mathrm{d}\lambda_2(x) = 1$ and that for all $\theta \in \Theta$ there holds: $\mathbb{p}_N^{emp,smo}(x) = 0$ if and only if $\mathbb{q}_\theta(x) = 0$ (in addition to (80)). For the subcase of Csiszar-Ali-Slivey divergences, thorough statistical investigations (such as efficiency, robustness, information loss, etc.) of the corresponding minimum-divergence-estimation can be found e.g. in Basu and Lindsay [11], Park and Basu [69], Chapter 3 of Basu et al. [13], Kuchibhotla and Basu [39], Al Mohamad [5], and the references therein. Due to the "curse of dimensionality", such a solution cannot be applied successfully in a large-dimension setting, as required in the

[14] In several situations, such a conversion can appear in a natural way; e.g. an institution may generate/collect data of "continuous value" but mask them for external data analysts to group-frequencies, for reasons of confidentiality (information asymmetry).

so called "big data" paradigm. For instance (in preparation for divergence valuation), take $\mathscr{X} = \mathbb{R}^d$, λ_2 to be the $d-$dimensional Lebesgue measure and $\vec{\mathbb{P}}_N^{emp,smo}(x) := \frac{1}{N}\sum_{i=1}^{N} K(x, Y_i, h_n) = \int_{\mathscr{X}} K(x, y, h_n)\, \mathrm{d}\vec{\mathfrak{P}}_N^{emp}(y)$ where $K(\cdot, \cdot, \cdot)$ is an appropriate smooth kernel function with "bandwidth" h_n, e.g. $K(x, y, h_n) := \frac{1}{h_n}\widehat{K}\left(\frac{x-y}{h_n}\right)$ with appropriate nonnegative function $\widehat{K}(\cdot)$ satisfying $\int_{\mathbb{R}^d} \widehat{K}(y)\, \mathrm{d}\lambda_2(y) = 1$. Since such kernel smoothers KS use local averaging, and for large d most neighborhoods tend to be empty of data observations (because data often "live" on lower-dimensional manifolds, sparsity of data), a typical KS technique (choosing concrete kernels and bandwidths, etc.) needs then a huge amount N of data to provide a reasonable accuracy; for $d = 8$ one may need N to be 1 million. For background details, the reader is e.g. referred to DasGupta [28], Scott and Wand [77], Chapter 7 of Scott [76] and the references therein.

For the sake of brevity, a detailed discussion of (GR) and (SM) is beyond the scope of this paper.

4.5 Minimum Divergences - The Decomposability Method

Let us discuss yet another strategy to circumvent the problem described in Subsetup 2 of Sect. 4.3. As a motivation, for a divergence of the form

$$0 \leqslant D_\lambda(\mathbb{P}, \mathbb{Q}) = \int_{\mathscr{X}} f_1(x) \cdot \mathbf{1}_{]0,\infty[}\big(\mathbb{p}(x) \cdot \mathbb{q}(x)\big)\, \mathrm{d}\lambda(x)$$
$$+ \int_{\mathscr{X}} f_2(x) \cdot \mathbf{1}_{\{0\}}\big(\mathbb{p}(x)\big) \cdot \mathbf{1}_{]0,\infty[}(\mathbb{q}(x))\, \mathrm{d}\lambda(x)$$
$$+ \int_{\mathscr{X}} f_3(x) \cdot \mathbf{1}_{\{0\}}\big(\mathbb{q}(x)\big) \cdot \mathbf{1}_{]0,\infty[}(\mathbb{p}(x))\, \mathrm{d}\lambda(x) \tag{98}$$

with $f_1(x) \geqslant 0$, $f_2(x) \geqslant 0$, $f_3(x) \geqslant 0$, and an "adjacent" dissimilarity

$$\widetilde{D_\lambda}(\mathbb{P}, \mathbb{Q}) = \int_{\mathscr{X}} f_1(x) \cdot \mathbf{1}_{]0,\infty[}\big(\mathbb{p}(x) \cdot \mathbb{q}(x)\big)\, \mathrm{d}\lambda(x)$$
$$+ \int_{\mathscr{X}} g_2(x)) \cdot \mathbf{1}_{\{0\}}\big(\mathbb{p}(x)\big) \cdot \mathbf{1}_{]0,\infty[}(\mathbb{q}(x))\, \mathrm{d}\lambda(x)$$
$$+ \int_{\mathscr{X}} g_3(x)) \cdot \mathbf{1}_{\{0\}}\big(\mathbb{q}(x)\big) \cdot \mathbf{1}_{]0,\infty[}(\mathbb{p}(x))\, \mathrm{d}\lambda(x), \tag{99}$$

there holds $D_\lambda(\mathbb{P}, \mathbb{Q}) = \widetilde{D_\lambda}(\mathbb{P}, \mathbb{Q})$ for all equivalent $\mathbb{P} \sim \mathbb{Q}$ (where for both, the second and third integral become zero), but (in case that $g_2(\cdot), g_3(\cdot)$ differ sufficiently enough from $f_2(\cdot), f_3(\cdot)$) one gets $D_\lambda(\mathbb{P}, \mathbb{Q}) \neq \widetilde{D_\lambda}(\mathbb{P}, \mathbb{Q})$ for $\mathbb{P} \perp \mathbb{Q}$ and even for $\mathbb{P} \not\sim \mathbb{Q}$; in the latter two cases, depending on the signs of $g_2(\cdot)$, $g_3(\cdot)$, $\widetilde{D_\lambda}(\mathbb{P}, \mathbb{Q})$ may even become negative.

Such issues are of importance for our current problem where e.g. $\mathbb{P} := \vec{\mathbb{P}}_N^{\overline{emp}(\omega)} \perp \widetilde{\mathbb{Q}}_\theta =: \mathbb{Q}$. For further illuminations, and for the sake of a compact presentation, we use henceforth the notations $\mathscr{P}^\lambda$ for an arbitrarily fixed class of nonnegative, mutually equivalent functions (i.e. $\mathbb{P}_1 \sim \mathbb{P}_2$ for all $\mathbb{P}_1 \in \mathscr{P}^\lambda$, $\mathbb{P}_2 \in \mathscr{P}^\lambda$), and $\mathscr{P}^{\lambda\not\sim}$ for a corresponding class of nonnegative (not necessarily mutually equivalent) functions such that $\mathbb{P}_1 \not\sim \mathbb{P}_2$ for all $\mathbb{P}_1^\lambda \in \mathscr{P}^\lambda$, $\mathbb{P}_2 \in \mathscr{P}^{\lambda\not\sim}$. Furthermore, we employ $\widetilde{\mathscr{P}}^\lambda := \mathscr{P}^\lambda \cup \mathscr{P}^{\lambda\not\sim}$ and specify:

Definition 2 We say that a function $D_\lambda : \widetilde{\mathscr{P}}^\lambda \otimes \mathscr{P}^\lambda \to \mathbb{R}$ is a pseudo-divergence on $\widetilde{\mathscr{P}}^\lambda \times \mathscr{P}^\lambda$, if its restriction to $\mathscr{P}^\lambda \cup \mathscr{P}^\lambda$ is a divergence, i.e.

$$\begin{aligned} &D_\lambda(\mathbb{P},\mathbb{Q}) \geqslant 0 \ \text{ for all } \mathbb{P} \in \mathscr{P}^\lambda, \mathbb{Q} \in \mathscr{P}^\lambda, \quad \text{and} \\ &D_\lambda(\mathbb{P},\mathbb{Q}) = 0 \ \text{ if and only if } \mathbb{P} = \mathbb{Q} \in \mathscr{P}^\lambda . \end{aligned} \tag{100}$$

If also $D_\lambda(\mathbb{P},\mathbb{Q}) > 0$ for all $\mathbb{P} \in \mathscr{P}^{\lambda\not\sim}, \mathbb{Q} \in \mathscr{P}^\lambda$, then $D_\lambda(\cdot,\cdot)$ is a divergence.

As for interpretation, a pseudo-divergence $D_\lambda(\cdot,\cdot)$ acts like a divergence if both arguments are from $\mathscr{P}^\lambda$, but only like a dissimilarity if the first argument is from $\mathscr{P}^{\lambda\not\sim}$ and thus is "quite different" from the second argument. In the following, we often use pseudo-divergences for our noisy minimum-distance-estimation problem – cf. (81), (82) – by taking $\lambda = \lambda_1 + \lambda_2$, $\mathscr{P}^\lambda := \mathscr{P}^\lambda_\Theta := (\widetilde{\mathbb{Q}}_\theta)_{\theta\in\Theta} := (\{\widetilde{\mathbb{q}}_\theta(x)\}_{x\in\mathscr{X}})_{\theta\in\Theta}$ (cf. (87), (88)), and $\mathscr{P}^{\lambda\not\sim} := \mathscr{P}^{\lambda\perp}_{emp} := (\mathbb{P}_N^{\overline{emp}(\omega)})_{N\in\mathbb{N}} = (\{\mathbb{p}_N^{\overline{emp}(\omega)}(x)\}_{x\in\mathscr{X}})_{N\in\mathbb{N}}$ (cf. (85), (Enc1)) covering all numbers N of data observations (sample sizes), and the according $\widetilde{\mathscr{P}}^\lambda := \mathscr{P}^\lambda_{\Theta,emp} = \mathscr{P}^\lambda_\Theta \cup \mathscr{P}^{\lambda\perp}_{emp}$; notice that by construction we have even the function-class-relationship $\perp$ which is stronger than $\not\sim$. In such a setup, we have seen that for the choice $\mathbb{P} := \mathbb{P}_N^{\overline{emp}(\omega)}$, $\mathbb{Q} := \widetilde{\mathbb{Q}}_\theta$ the divergence $D^c_{\phi,\mathbb{Q},\mathbb{Q},\mathbb{1}\cdot\mathbb{Q}^\chi,\lambda}(\mathbb{P},\mathbb{Q}) > 0$ of (90) is unpleasant for (Enc3) since the solution does not depend on the data-observations $Y_1(\omega),\ldots,Y_N(\omega)$; also recall the special case of power functions $\phi := \phi_\alpha$ (cf. (5)) with $\alpha = \chi > 1$ which amounts to the unscaled divergences (78), (40) and thus to (41). In (95), for general ϕ we have repaired this deficiency by replacing $\mathbb{Q} := \widetilde{\mathbb{Q}}_\theta$ with $\mathbb{Q} := \mathbb{Q}_\theta$, at the cost of getting total mass larger than 1 but by keeping the strict positivity of the involved divergence; especially for $\phi := \phi_\alpha$, the divergence (41) has then amounted to (96).

In contrast, let us show another method to repair the (Enc3)-deficiency of (41), by sticking to $\mathbb{Q} := \widetilde{\mathbb{Q}}_\theta$ but changing the basically underlying divergence. In fact, we deal with the even more general

Definition 3 (a) We say that a pseudo-divergence $D_\lambda : \widetilde{\mathscr{P}}^\lambda \otimes \mathscr{P}^\lambda \to \mathbb{R}$ is decomposable if there exist functionals $\mathfrak{D}^0 : \widetilde{\mathscr{P}}^\lambda \mapsto \mathbb{R}$, $\mathfrak{D}^1 : \mathscr{Q} \mapsto \mathbb{R}$ and a (measurable) mapping $\rho_\mathbb{Q} : \mathscr{X} \mapsto \mathbb{R}$ (for each $\mathbb{Q} \in \mathscr{P}^\lambda$) such that[15]

---

[15]In an encompassing way, the part (a) reflects a measure-theoretic "plug-in" version of decomposable pseudo-divergences $D : (\mathscr{P}^{meas,\lambda_1} \cup \mathscr{P}^{meas,\lambda_2}) \otimes \mathscr{P}^{meas,\lambda_1} \mapsto \mathbb{R}$, where $\mathscr{P}^{meas,\lambda_1}$ is a family of mutually equivalent nonnegative measures of the form $\mathfrak{P}[\bullet] := \mathfrak{P}^{\mathbb{1}\cdot\lambda_1}[\bullet] := \int_\bullet \mathbb{p}(x)\,\mathrm{d}\lambda_1(x)$, $\mathscr{P}^{meas,\lambda_2}$ is a family of nonnegative measures of the form $\overline{\mathfrak{P}}[\bullet] := \overline{\mathfrak{P}}^{\mathbb{1}\cdot\lambda_2}[\bullet] := \int_\bullet \mathbb{q}(x)\,\mathrm{d}\lambda_2(x)$ such that any $\mathfrak{P} \in \mathscr{P}^{meas,\lambda_1}$ is not equivalent to any $\overline{\mathfrak{P}} \in \mathscr{P}^{meas,\lambda_2}$, and (101) is replaced with $D(\mathfrak{P},\mathfrak{Q}) = \mathfrak{D}^0(\mathfrak{P}) + \mathfrak{D}^1(\mathfrak{Q}) + \int_\mathscr{X} \rho_\mathfrak{Q}(x)\,\mathrm{d}\mathfrak{P}(x)$ for all $\mathbb{P} \in \mathfrak{P} \in \mathscr{P}^{meas,\lambda_1} \cup \mathscr{P}^{meas,\lambda_2}$, $\mathfrak{Q} \in \mathscr{P}^{meas,\lambda_2}$; cf. Vajda [90], Broniatowski and Vajda [18], Broniatowski et al. [19]; part (b) is new.

$$D_\lambda(\mathbb{P},\mathbb{Q}) = \mathfrak{D}^0(\mathbb{P}) + \mathfrak{D}^1(\mathbb{Q}) + \int_{\mathscr{X}} \rho_{\mathbb{Q}}(x)\cdot \mathbb{p}(x)\,\mathrm{d}\lambda(x) \quad \text{for all } \mathbb{P}\in\widetilde{\mathscr{P}}^\lambda,\ \mathbb{Q}\in\mathscr{P}^\lambda. \tag{101}$$

(b) We say that a pseudo-divergence $D_\lambda : \widetilde{\mathscr{P}}^\lambda \otimes \mathscr{P}^\lambda \to \mathbb{R}$ is pointwise decomposable if it is of the form $D_\lambda(\mathbb{P},\mathbb{Q}) = \int_{\mathscr{X}} \psi^{dec}(\mathbb{p}(x),\mathbb{q}(x))\,\mathrm{d}\lambda(x)$ for some (measurable) mapping $\psi^{dec} : [0,\infty[\times[0,\infty[\mapsto \mathbb{R}$ with representation

$$\begin{aligned}
&\psi^{dec}(s,t) := \psi^0\big(s + h_0(x,s)\cdot \mathbf{1}_{\{0\}}(t)\big)\cdot \mathbf{1}_{]\overline{c}_0,\infty[}(s)\cdot \mathbf{1}_{]c_0,\infty[}(t)\\
&+\psi^1\big(t + h_1(x)\cdot \mathbf{1}_{\{0\}}(t)\big)\cdot \mathbf{1}_{]c_1,\infty[}(t)\\
&+\rho\big(t + h_2(x)\cdot \mathbf{1}_{\{0\}}(t)\big)\cdot s \quad \text{for all } (s,t)\in[0,\infty[\times[0,\infty[\backslash\{(0,0)\}\,, \qquad (102)\\
&\psi^{dec}(0,0) := 0,
\end{aligned}$$

with constants $c_0, c_1, \overline{c}_0 \in \{0,1\}$, and (measurable) mappings $\psi^0, \psi^1, \rho : [0,\infty[\mapsto \mathbb{R}$, $h_1, h_2 : \mathscr{X} \mapsto [0,\infty[$, $h_0 : \mathscr{X}\times[0,\infty[\mapsto \mathbb{R}$, such that

$$\psi^{dec}(s,t) = \psi^0(s) + \psi^1(t) + \rho(t)\cdot s \geqslant 0 \quad \text{for all } (s,t) \in]0,\infty[\times]0,\infty[\,, \tag{103}$$

$$\psi^{dec}(s,t) = 0 \quad \text{if and only if} \quad s = t\,, \tag{104}$$

$$s + h_0(x,s) \geqslant 0 \quad \text{for all } s\in[0,\infty[\text{ and } \lambda\text{-almost all } x\in\mathscr{X}\,.$$

Remark 5 (a) Any pointwise decomposable pseudo-divergence is decomposable, under the additional assumption that the integral $\int_{\mathscr{X}} \ldots \,\mathrm{d}\lambda(x)$ can be split into three appropriate parts.
(b) For use in (Enc3), $\mathfrak{D}^1(\cdot)$ and $\rho_{\mathbb{Q}}(\cdot)$ should be non-constant.
(c) In the Definitions 2 and 3 we have put the "extension-role" to the first component $\mathbb{P}$; of course, everything can be worked out analogously for the second component $\mathbb{Q}$ by using (pseudo-)divergences $D_\lambda : \mathscr{P}^\lambda \times \widetilde{\mathscr{P}}^\lambda \to \mathbb{R}$.
(d) We could even extend (102) for bivariate functions $h_1(x,s), h_2(x,s)$. □

Notice that from (102) one obtains the boundary behaviour

$$\mathbb{R} \ni \psi^{dec}(s,0) = \psi^0(s + h_0(x,s))\cdot \breve{c_0} + \psi^1(h_1(x))\cdot \breve{c_1} + \rho(h_2(x))\cdot s \text{ for all } s>0, \tag{105}$$

$$\mathbb{R} \ni \psi^{dec}(0,t) = \psi^0(0)\cdot \breve{\overline{c}_0} + \psi^1(t) \quad \text{for all } t>0\,, \tag{106}$$

with $\breve{c_0} := \mathbf{1}_{]c_0,\infty[}(0)$, $\breve{c_1} := \mathbf{1}_{]c_1,\infty[}(0)$, $\breve{\overline{c}_0} := \mathbf{1}_{]\overline{c}_0,\infty[}(0)$. Notice that $\psi^{dec}(s,0)$ of (105) does generally not coincide with the eventually existent "(103)-limit" $\lim_{t\to 0}[\psi^0(s) + \psi^1(t) + \rho(t)\cdot s]$ $(s>0)$, which reflects a possibly "non-smooth boundary behaviour" (also recall (98), (99)). Moreover, when choosing a decomposable pseudo-divergence (101) in step (Enc2), we operationalize the solution $\widehat{\theta}_N(\omega)$ of the corresponding noisy minimization problem (86) of step (Enc3) as follows:

Definition 4 (a) We say that a functional $T_{D_\lambda} : \mathscr{P}^{\lambda}_{\Theta,emp} \mapsto \Theta$ generates a minimum decomposable pseudo-divergence estimator (briefly, min $-decD_\lambda$-estimator)

$$\widehat{\theta}_{N,decD_\lambda}(\omega) := T_{D_\lambda}\big(\vec{\mathbb{P}}_N^{\overline{emp}(\omega)}\big) \quad \text{for } \vec{\mathbb{P}}_N^{\overline{emp}(\omega)} \in \mathscr{P}^{\lambda\perp}_{emp} \tag{107}$$

of the true unknown parameter θ_0, if $D_\lambda(\cdot,\cdot) : \mathscr{P}^{\lambda}_{\Theta,emp} \otimes \mathscr{P}^{\lambda}_{\Theta} \mapsto \mathbb{R}$ is a decomposable pseudo-divergence and

$$T_{D_\lambda}(\vec{\mathbb{P}}) = \text{arginf}_{\theta\in\Theta}\big[\mathfrak{D}^1(\vec{\mathbb{Q}}_\theta) + \int_{\mathscr{X}} \rho_{\vec{\mathbb{Q}}_\theta}(x) \cdot \vec{\mathbb{p}}(x)\,\mathrm{d}\lambda(x)\big] \text{ for all } \vec{\mathbb{P}} \in \mathscr{P}^{\lambda}_{\Theta,emp}. \tag{108}$$

(b) If $D_\lambda(\cdot,\cdot)$ is a pointwise decomposable pseudo-divergence we replace (108) by

$$T_{D_\lambda}(\vec{\mathbb{P}}) = \text{arginf}_{\theta\in\Theta} \int_{\mathscr{X}} \psi^{dec}(\vec{\mathbb{p}}(x), \widetilde{\vec{\mathbb{q}}}_\theta(x))\,\mathrm{d}\lambda(x) \quad \text{for all } \vec{\mathbb{P}} \in \mathscr{P}^{\lambda}_{\Theta,emp},$$

but do not introduce a new notion (also recall that $\lambda = \lambda_2$ and $\widetilde{\vec{\mathbb{q}}}_\theta(\cdot) = \vec{\mathbb{q}}_\theta(\cdot)$ for the case of no observations, e.g. if $\vec{\mathbb{P}} \in \mathscr{P}^{\lambda_2}_{\Theta}$).

To proceed, let us point out that by (107) and (97) every min $-decD_\lambda$-estimator rewrites straightforwardly as

$$\widehat{\theta}_{N,decD_\lambda}(\omega) = \text{arginf}_{\theta\in\Theta}\big[\mathfrak{D}^1(\vec{\mathbb{Q}}_\theta) + \tfrac{1}{N}\sum_{i=1}^N \rho_{\vec{\mathbb{Q}}_\theta}(Y_i(\omega))\big] \tag{109}$$

and is Fisher consistent in the sense that

$$T_{\mathfrak{D}}(\vec{\mathbb{Q}}_{\theta_0}) = \text{arginf}_{\theta\in\Theta}\, \mathfrak{D}(\vec{\mathbb{Q}}_{\theta_0}, \vec{\mathbb{Q}}_\theta) = \theta_0 \quad \text{for all } \theta_0 \in \Theta\,. \tag{110}$$

Furthermore, the criterion to be minimized in (109) is of the form

$$\theta \;\mapsto\; \mathfrak{D}^1(\vec{\mathbb{Q}}_\theta) + \tfrac{1}{N}\sum_{i=1}^N \rho_{\vec{\mathbb{Q}}_\theta}(Y_i(\omega))$$

which e.g. for the task (Enc5) opens the possibility to apply the methods of the asymptotic theory of so-called M-estimators (cf. e.g. Hampel et al. [33], van der Vaart and Wellner [88], Liese and Mieske [40]). The concept of min $-decD_\lambda$-estimators (101) were introduced in Vajda [90], Broniatowski and Vajda [18] within the probability-law-restriction of the non-encompassing, "plug-in" context of footnote 15.

In the following, we demonstrate that our new concept of pointwise decomposability defined by (102) is very useful and flexible for creating new min $-decD_\lambda$-estimators and imbedding existing ones. In fact, since in our current statistics-ML-AI context we have chosen $\lambda[\bullet] := \lambda_1[\bullet] + \lambda_2[\bullet]$ with $\lambda_1[\bullet] := \sum_{z\in\mathscr{X}} \mathbf{1}_{\mathscr{R}(Y_1(\omega),\ldots,Y_N(\omega))}(z) \cdot \delta_z[\bullet]$ and $\lambda_2[\bullet]$ stemming from (87), we have seen that $\mathbb{P} := \vec{\mathbb{P}}_N^{\overline{emp}(\omega)} \perp \widetilde{\vec{\mathbb{Q}}}_\theta =: \vec{\mathbb{Q}}$ for all $\theta \in \Theta$. Hence, from (102), (105), (106) we obtain

$$
\begin{aligned}
&D_\lambda\Big(\vec{\mathbb{P}}_N^{\overline{emp}(\omega)}, \widetilde{\vec{\mathbb{Q}}}_\theta\Big) = \int_{\mathscr{X}} \psi^{dec}\Big(\vec{\mathbb{p}}_N^{\overline{emp}(\omega)}(x), \widetilde{\vec{\mathbb{q}}}_\theta(x)\Big)\,\mathrm{d}\lambda(x)\\
&= \int_{\mathscr{X}} \Big[\psi^0(0)\cdot \breve{\overline{c}}_0 + \psi^1\Big(\widetilde{\vec{\mathbb{q}}}_\theta(x)\Big)\Big]\cdot \mathbf{1}_{\{0\}}(\vec{\mathbb{p}}_N^{\overline{emp}(\omega)}(x))\,\mathrm{d}(\lambda_1+\lambda_2)(x)\\
&+ \int_{\mathscr{X}} \Big[\psi^0\Big(\vec{\mathbb{p}}_N^{\overline{emp}(\omega)}(x) + h_0\Big(x, \vec{\mathbb{p}}_N^{\overline{emp}(\omega)}(x)\Big)\Big)\cdot \breve{c}_0\\
&+\psi^1(h_1(x))\cdot \breve{c}_1 + \rho(h_2(x))\cdot \vec{\mathbb{p}}_N^{\overline{emp}(\omega)}(x)\Big]\cdot \mathbf{1}_{\{0\}}(\widetilde{\vec{\mathbb{q}}}_\theta(x))\,\mathrm{d}(\lambda_1+\lambda_2)(x)\\
&= \int_{\mathscr{X}} \Big[\psi^0(0)\cdot \breve{\overline{c}}_0 + \psi^1\Big(\vec{\mathbb{q}}_\theta(x)\Big)\Big]\,\mathrm{d}\lambda_2(x) + \sum_{x\in\mathscr{X}} \Big[\psi^0\Big(\vec{\mathbb{p}}_N^{\overline{emp}(\omega)}(x)\\
&+h_0\Big(x, \vec{\mathbb{p}}_N^{\overline{emp}(\omega)}(x)\Big)\Big)\cdot \breve{c}_0\\
&+\psi^1(h_1(x))\cdot \breve{c}_1 + \rho(h_2(x))\cdot \vec{\mathbb{p}}_N^{\overline{emp}(\omega)}(x)\Big]\cdot \mathbf{1}_{\mathscr{R}(Y_1(\omega),\dots,Y_N(\omega))}(x)\\
&= \int_{\mathscr{X}} \Big[\psi^0(0)\cdot \breve{\overline{c}}_0 + \psi^1\Big(\vec{\mathbb{q}}_\theta(x)\Big)\Big]\,\mathrm{d}\lambda_2(x) + \tfrac{1}{N}\sum_{i=1}^N \rho(h_2(Y_i(\omega)))\\
&+\tfrac{1}{N}\sum_{i=1}^N \frac{\psi^0\Big(\vec{\mathbb{p}}_N^{emp(\omega)}(Y_i(\omega)) + h_0\Big(Y_i(\omega), \vec{\mathbb{p}}_N^{emp(\omega)}(Y_i(\omega))\Big)\Big)\cdot\breve{c}_0 + \psi^1(h_1(Y_i(\omega)))\cdot\breve{c}_1}{\vec{\mathbb{p}}_N^{emp(\omega)}(Y_i(\omega))},
\end{aligned}
\tag{111}
$$

where we have employed (97); recall that $\vec{\mathbb{p}}_N^{emp(\omega)}(Y_i(\omega)) = \#\{j \in \{1,\dots,N\} : Y_j(\omega) = Y_i(\omega)\}/N$. Hence, we always choose $\mathfrak{D}^1(\vec{\mathbb{Q}}) = \mathfrak{D}^1(\widetilde{\vec{\mathbb{Q}}}_\theta) = \int_{\mathscr{X}} \Big[\psi^0(0) + \psi^1\Big(\widetilde{\vec{\mathbb{q}}}_\theta(x)\Big)\Big]\,\mathrm{d}\lambda_2(x) = \int_{\mathscr{X}} \Big[\psi^0(0) + \psi^1\Big(\vec{\mathbb{q}}_\theta(x)\Big)\Big]\,\mathrm{d}\lambda_2(x) = \mathfrak{D}^1(\vec{\mathbb{Q}}_\theta)$. Notice that the functions h_0, h_1, h_2 may depend on the parameter θ. Indeed, for $h_0(x,s) \equiv 0$, $h_1(x) \equiv 0$, $h_2(x) = \vec{\mathbb{q}}_\theta(x)$ $(\neq \widetilde{\vec{\mathbb{q}}}_\theta(x))$, the pseudo-divergence (111) turns into

$$
\begin{aligned}
&D_\lambda\Big(\vec{\mathbb{P}}_N^{\overline{emp}(\omega)}, \widetilde{\vec{\mathbb{Q}}}_\theta\Big) = \int_{\mathscr{X}} \Big[\psi^0(0)\cdot \breve{\overline{c}}_0 + \psi^1\Big(\vec{\mathbb{q}}_\theta(x)\Big)\Big]\,\mathrm{d}\lambda_2(x) + \tfrac{1}{N}\sum_{i=1}^N \rho\Big(\vec{\mathbb{q}}_\theta(Y_i(\omega))\Big)\\
&+\tfrac{1}{N}\sum_{i=1}^N \frac{\psi^0\Big(\vec{\mathbb{p}}_N^{emp(\omega)}(Y_i(\omega))\Big)\cdot\breve{c}_0 + \psi^1(0)\cdot\breve{c}_1}{\vec{\mathbb{p}}_N^{emp(\omega)}(Y_i(\omega))},
\end{aligned}
\tag{112}
$$

whereas for $h_0(x,s) \equiv 0$, $h_1(x) = \vec{\mathbb{q}}_\theta(x)$, $h_2(x) = \vec{\mathbb{q}}_\theta(x)$, (111) becomes

$$
\begin{aligned}
&D_\lambda\Big(\vec{\mathbb{P}}_N^{\overline{emp}(\omega)}, \widetilde{\vec{\mathbb{Q}}}_\theta\Big) = \int_{\mathscr{X}} \big[\psi^0(0)\cdot \breve{\overline{c}}_0 + \psi^1\big(\vec{\mathbb{q}}_\theta(x)\big)\big]\,\mathrm{d}\lambda_2(x)\\
&+\tfrac{1}{N}\sum_{i=1}^N \Big[\rho\Big(\vec{\mathbb{q}}_\theta(Y_i(\omega))\Big) + \frac{\breve{c}_1\cdot\psi^1(\vec{\mathbb{q}}_\theta(Y_i(\omega)))}{\vec{\mathbb{p}}_N^{emp(\omega)}(Y_i(\omega))}\Big] + \tfrac{1}{N}\sum_{i=1}^N \frac{\psi^0\Big(\vec{\mathbb{p}}_N^{emp(\omega)}(Y_i(\omega))\Big)\cdot\breve{c}_0}{\vec{\mathbb{p}}_N^{emp(\omega)}(Y_i(\omega))}.
\end{aligned}
\tag{113}
$$

The last sum in (112) respectively (113) is the desired $\mathfrak{D}^0(\vec{\mathbb{P}}_N^{\overline{emp}(\omega)})$. As an example, let us take $c_0 = c_1 = \overline{c}_0 = -1$ (and hence, $\breve{c}_0 = \breve{c}_1 = \breve{\overline{c}}_0 = 1$) and for $\alpha > 1$ the power functions $\phi(t) := \phi_\alpha(t) := \frac{t^\alpha - \alpha\cdot t + \alpha - 1}{\alpha\cdot(\alpha-1)}$ $(t \in]0,\infty[)$ of (6), for which by (9) and (103) one derives immediately the decomposition $\psi^0(t) := \psi^0_\alpha(t) := \frac{t^\alpha}{\alpha(\alpha-1)} > 0, \psi^1(t) := \psi^1_\alpha(t) := \frac{t^\alpha}{\alpha} > 0, \rho(t) := \rho_\alpha(t) := -\frac{t^{\alpha-1}}{\alpha-1} < 0\,(t \in]0,\infty[)$. Accordingly, (111) simplifies to

$$
\begin{aligned}
& D_\lambda\Big(\vec{\mathbb{P}}_N^{\overline{emp}(\omega)}, \widetilde{\vec{\mathbb{Q}}}_\theta\Big) := D_{\lambda,\alpha}\Big(\vec{\mathbb{P}}_N^{\overline{emp}(\omega)}, \widetilde{\vec{\mathbb{Q}}}_\theta\Big) \\
& = \tfrac{1}{\alpha} \int_{\mathscr{X}} \vec{q}_\theta(x)^\alpha \, \mathrm{d}\lambda_2(x) - \tfrac{1}{N\cdot(\alpha-1)} \sum_{i=1}^N \Big(h_2(Y_i(\omega))\Big)^{\alpha-1} \\
& + \tfrac{1}{N} \sum_{i=1}^N \frac{\Big(\vec{\mathbb{P}}_N^{emp(\omega)}(Y_i(\omega)) + h_0\Big(Y_i(\omega), \vec{\mathbb{P}}_N^{emp(\omega)}(Y_i(\omega))\Big)\Big)^\alpha + (\alpha-1)\cdot\Big(h_1(Y_i(\omega))\Big)^\alpha}{\alpha\cdot(\alpha-1)\cdot\vec{\mathbb{P}}_N^{emp(\omega)}(Y_i(\omega))},
\end{aligned}
\tag{114}
$$

and in particular the special case (112) turns into

$$
\begin{aligned}
& D_{\lambda,\alpha}\Big(\vec{\mathbb{P}}_N^{\overline{emp}(\omega)}, \widetilde{\vec{\mathbb{Q}}}_\theta\Big) = \tfrac{1}{\alpha} \int_{\mathscr{X}} \vec{q}_\theta(x)^\alpha \, \mathrm{d}\lambda_2(x) - \tfrac{1}{N\cdot(\alpha-1)} \sum_{i=1}^N \Big(\vec{q}_\theta(Y_i(\omega))\Big)^{\alpha-1} \\
& + \tfrac{1}{N\cdot\alpha\cdot(\alpha-1)} \sum_{i=1}^N \Big(\vec{\mathbb{P}}_N^{emp(\omega)}(Y_i(\omega))\Big)^{\alpha-1},
\end{aligned}
\tag{115}
$$

whereas the special case (113) simplifies to

$$
\begin{aligned}
& 0 < D_{\lambda,\alpha}\Big(\vec{\mathbb{P}}_N^{\overline{emp}(\omega)}, \widetilde{\vec{\mathbb{Q}}}_\theta\Big) = \tfrac{1}{\alpha} \int_{\mathscr{X}} \vec{q}_\theta(x)^\alpha \, \mathrm{d}\lambda_2(x) \\
& + \tfrac{1}{N} \sum_{i=1}^N \Big[\frac{\Big(\vec{q}_\theta(Y_i(\omega))\Big)^\alpha}{\alpha\cdot\vec{\mathbb{P}}_N^{emp(\omega)}(Y_i(\omega))} - \frac{\Big(\vec{q}_\theta(Y_i(\omega))\Big)^{\alpha-1}}{\alpha-1} \Big] + \tfrac{1}{N} \sum_{i=1}^N \frac{\Big(\vec{\mathbb{P}}_N^{emp(\omega)}(Y_i(\omega))\Big)^{\alpha-1}}{\alpha\cdot(\alpha-1)}.
\end{aligned}
\tag{116}
$$

Notice that (116) coincides with (96), but both were derived within quite different frameworks: to obtain (116) we have used the concept of decomposable pseudo-divergences (which may generally become negative at the boundary) together with $\vec{\mathbb{Q}} := \widetilde{\vec{\mathbb{Q}}}_\theta$ which leads to total mass of 1 (cf. (88)); on the other hand, for establishing (96) we have employed the concept of divergences (which are generally always strictly positive at the boundary) together with $\vec{\mathbb{Q}} := \vec{\mathbb{Q}}_\theta$ which amounts to total mass greater than 1 (cf. (91)). Moreover, choosing $h_0(x,s) \equiv 0$, $h_1(x) \equiv 0$, $h_2(x) \equiv 0$ in (114) gives exactly the divergence (90) for the current generator $\phi(t) := \phi_\alpha(t)$ with $\alpha > 1$; recall that the latter has been a starting motivation for the search of repairs. For $c_0 = c_1 = \overline{c}_0 = -1$ and the limit case $\alpha \to 1$ one gets $\phi(t) := \phi_1(t) := t \cdot \log t + 1 - t\,(t \in]0,\infty[)$ of (18), for which by (22) and (103) we obtain the decomposition $\psi^0(t) := \psi^0_1(t) := t \cdot \log t - t$, $\psi^1(t) := \psi^1_1(t) := t > 0$, $\rho(t) := \rho_1(t) := -\log t$. Accordingly, (111) simplifies to

$$D_\lambda\Big(\mathbb{P}_N^{\overline{emp}(\omega)},\widetilde{\widetilde{\mathbb{Q}}}_\theta\Big) := D_{\lambda,1}\Big(\mathbb{P}_N^{\overline{emp}(\omega)},\widetilde{\widetilde{\mathbb{Q}}}_\theta\Big)$$
$$= 1 - \tfrac{1}{N}\textstyle\sum_{i=1}^N \log\Big(h_2(Y_i(\omega))\Big)$$
$$+\tfrac{1}{N}\textstyle\sum_{i=1}^N \frac{\psi_1^0\Big(\vec{\mathbb{p}}_N^{emp(\omega)}(Y_i(\omega))+h_0\Big(Y_i(\omega),\vec{\mathbb{p}}_N^{emp(\omega)}(Y_i(\omega))\Big)\Big)+h_1(Y_i(\omega))}{\vec{\mathbb{p}}_N^{emp(\omega)}(Y_i(\omega))}, \tag{117}$$

and in particular the special case (112) turns into

$$D_{\lambda,1}\Big(\mathbb{P}_N^{\overline{emp}(\omega)},\widetilde{\widetilde{\mathbb{Q}}}_\theta\Big) = \frac{1}{N}\sum_{i=1}^N \log\Big(\vec{\mathbb{p}}_N^{emp(\omega)}(Y_i(\omega))\Big) - \frac{1}{N}\sum_{i=1}^N \log\Big(\vec{\mathbb{q}}_\theta(Y_i(\omega))\Big), \tag{118}$$

whereas the special case (113) becomes

$$0 < D_{\lambda,1}\Big(\mathbb{P}_N^{\overline{emp}(\omega)},\widetilde{\widetilde{\mathbb{Q}}}_\theta\Big) = \tfrac{1}{N}\textstyle\sum_{i=1}^N \log\Big(\vec{\mathbb{p}}_N^{emp(\omega)}(Y_i(\omega))\Big) - \tfrac{1}{N}\textstyle\sum_{i=1}^N \log\Big(\vec{\mathbb{q}}_\theta(Y_i(\omega))\Big)$$
$$+\tfrac{1}{N}\textstyle\sum_{i=1}^N \frac{\vec{\mathbb{q}}_\theta(Y_i(\omega))}{\vec{\mathbb{p}}_N^{emp(\omega)}(Y_i(\omega))}. \tag{119}$$

To end up this subsection, let us briefly indicate that choosing in step (Enc2) a decomposable pseudo-divergence of the form (respectively) (111)–(119), and in the course of (Enc3) minimize this over $\theta \in \Theta$, we end up at the corresponding min $-decD_\lambda$-estimator (109). For the special case (118) (i.e. $\alpha = 1$) this leads to the omnipresent, celebrated *maximum-likelihood-estimator* (MLE) which is known to be efficient but not robust. The particular choice (115) for $\alpha > 1$ gives the density-power divergence estimator DPDE of Basu et al. [10], where $\alpha = 2$ amounts to the (squared) L_2-estimator which is robust but not efficient (see e.g. Hampel et al. [33]); accordingly, taking $\alpha \in]1,2[$ builds a smooth bridge between the robustness and efficiency. The reversed version of the DPDE can be analogously imbedded in our context, by employing our new approach with $\phi(t) := \widetilde{\widetilde{\phi}}_\alpha(t)$ (cf. (79)).

4.6 Minimum Divergences - Generalized Subdivergence Method

One can flexibilize some of the methods of the previous Sect. 4.5, by employing an additional (a.s.) strictly positive density function $\mathbb{M}$ to define a pseudo-divergence $D_{\mathbb{M},\lambda} : \widetilde{\mathscr{P}}^\lambda \otimes \mathscr{P}^\lambda \to \mathbb{R}$ of the form $D_{\mathbb{M},\lambda}(\mathbb{P},\mathbb{Q}) =$
$\int_{\mathscr{X}} \psi^{dec}\big(\frac{\mathbb{p}(x)}{\mathbb{m}(x)}, \frac{\mathbb{q}(x)}{\mathbb{m}(x)}\big) \cdot \mathbb{m}(x)\, \mathrm{d}\lambda(x)$ for some (measurable) mapping
$\psi^{dec} : [0,\infty[\times[0,\infty[\mapsto \mathbb{R}$ with representation

$$\begin{aligned}
\psi^{dec}(s,t) &:= \psi^{0}\Big(s + h_0(x,s)\cdot \mathbf{1}_{\{0\}}(t)\Big)\cdot \mathbf{1}_{]\overline{c}_0,\infty[}(s)\cdot \mathbf{1}_{]c_0,\infty[}(t)\\
&\quad + \psi^{1}\Big(t + h_1(x)\cdot \mathbf{1}_{\{0\}}(t)\Big)\cdot \mathbf{1}_{]c_1,\infty[}(t)\\
&\quad + \rho\Big(t + h_2(x)\cdot \mathbf{1}_{\{0\}}(t)\Big)\cdot s \quad \text{for all } (s,t)\in[0,\infty[\times[0,\infty[\backslash\{(0,0)\}\,, \quad (cf.(102))\\
\psi^{dec}(0,0) &:= 0.
\end{aligned}$$

It is straightforward to see that $D_{\mathbb{M},\lambda}(\cdot,\cdot)$ is a pointwise decomposable pseudo-divergence in the sense of Definition 3(b), and one gets for fixed $m > 0$

$$\psi_m^{dec}(s,t) := m\cdot\psi^{dec}\Big(\tfrac{s}{m},\tfrac{t}{m}\Big) = m\cdot\psi^{0}\Big(\tfrac{s}{m}\Big) + m\cdot\psi^{1}\Big(\tfrac{t}{m}\Big) + \rho\Big(\tfrac{t}{m}\Big)\cdot s \geqslant 0 \quad \text{for all } (s,t)\in]0,\infty[\times]0,\infty[\,, \tag{120}$$

$$\begin{aligned}
&\psi_m^{dec}(s,t) = 0 \quad \text{if and only if} \quad s = t\,,\\
&\tfrac{s}{m} + h_0\Big(x,\tfrac{s}{m}\Big) \geqslant 0 \qquad \text{for all } s\in[0,\infty[\text{ and } \lambda\text{-almost all } x\in\mathscr{X}\,,\\
&\mathbb{R}\ni\psi_m^{dec}(s,0) = m\cdot\psi^{0}\Big(\tfrac{s}{m} + h_0\Big(x,\tfrac{s}{m}\Big)\Big)\cdot\breve{c}_0 + m\cdot\psi^{1}(h_1(x))\cdot\breve{c}_1 + \rho(h_2(x))\cdot s \quad \text{for all } s > 0\,,
\end{aligned}\tag{121}$$

$$\mathbb{R}\ni\psi_m^{dec}(0,t) = m\cdot\psi^{0}(0)\cdot\breve{\overline{c}}_0 + m\cdot\psi^{1}\Big(\tfrac{t}{m}\Big) \quad \text{for all } t > 0\,. \tag{122}$$

For each class-family member $\mathbb{M} := \vec{\mathbb{Q}}_{\tau}$ with arbitrarily fixed $\tau\in\Theta$, we can apply Definition 4 to $D_{\lambda}(\cdot,\cdot) := D_{\vec{\mathbb{Q}}_{\tau},\lambda}(\cdot,\cdot)$, and arrive at the corresponding $\min - decD_{\vec{\mathbb{Q}}_{\tau},\lambda}$-estimators

$$\widehat{\theta}_{N,decD_{\vec{\mathbb{Q}}_{\tau},\lambda}}(\omega) := T_{D_{\vec{\mathbb{Q}}_{\tau},\lambda}}\Big(\vec{\mathbb{P}}_N^{\overline{emp}(\omega)}\Big) \quad \text{for } \vec{\mathbb{P}}_N^{\overline{emp}(\omega)} \in \mathscr{P}_{emp}^{\lambda\perp} \tag{123}$$

of the true unknown parameter θ_0. Hence, analogously to the derivation of (111), we obtain from (102), (121), (122) for each $\tau\in\Theta$

$$\begin{aligned}
&D_{\vec{\mathbb{Q}}_{\tau},\lambda}\Big(\mathbb{P}_N^{\overline{emp}(\omega)},\widetilde{\mathbb{Q}}_{\theta}\Big) = \int_{\mathscr{X}} \psi^{dec}\Big(\tfrac{\vec{\mathbb{p}}_N^{\overline{emp}(\omega)}(x)}{\vec{\mathbb{q}}_{\tau}(x)},\tfrac{\widetilde{\vec{\mathbb{q}}}_{\theta}(x)}{\vec{\mathbb{q}}_{\tau}(x)}\Big)\cdot\vec{\mathbb{q}}_{\tau}(x)\,\mathrm{d}\lambda(x)\\
&= \int_{\mathscr{X}} \psi^{1}\Big(\tfrac{\vec{\mathbb{q}}_{\theta}(x)}{\vec{\mathbb{q}}_{\tau}(x)}\Big)\cdot\vec{\mathbb{q}}_{\tau}(x)\,\mathrm{d}\lambda_2(x) + \sum_{x\in\mathscr{X}}\Big[\vec{\mathbb{q}}_{\tau}(x)\cdot\psi^{0}\Big(\tfrac{\vec{\mathbb{p}}_N^{\overline{emp}(\omega)}(x)}{\vec{\mathbb{q}}_{\tau}(x)} + h_0\Big(x,\tfrac{\vec{\mathbb{p}}_N^{\overline{emp}(\omega)}(x)}{\vec{\mathbb{q}}_{\tau}(x)}\Big)\Big)\cdot\breve{c}_0\\
&+\vec{\mathbb{q}}_{\tau}(x)\cdot\psi^{1}(h_1(x))\cdot\breve{c}_1 + \rho(h_2(x))\cdot\vec{\mathbb{p}}_N^{\overline{emp}(\omega)}(x)\Big]\cdot\mathbf{1}_{\mathscr{R}(Y_1(\omega),\ldots,Y_N(\omega))}(x) + \psi^{0}(0)\cdot\breve{\overline{c}}_0\\
&= \int_{\mathscr{X}} \psi^{1}\Big(\tfrac{\vec{\mathbb{q}}_{\theta}(x)}{\vec{\mathbb{q}}_{\tau}(x)}\Big)\cdot\vec{\mathbb{q}}_{\tau}(x)\,\mathrm{d}\lambda_2(x) + \tfrac{1}{N}\sum_{i=1}^{N}\rho(h_2(Y_i(\omega))) + \psi^{0}(0)\cdot\breve{\overline{c}}_0\\
&+\tfrac{1}{N}\sum_{i=1}^{N}\frac{\psi^{0}\Big(\frac{\vec{\mathbb{p}}_N^{emp(\omega)}(Y_i(\omega))}{\vec{\mathbb{q}}_{\tau}(Y_i(\omega))} + h_0\Big(Y_i(\omega),\frac{\vec{\mathbb{p}}_N^{emp(\omega)}(Y_i(\omega))}{\vec{\mathbb{q}}_{\tau}(Y_i(\omega))}\Big)\Big)\cdot\breve{c}_0 + \psi^{1}(h_1(Y_i(\omega)))\cdot\breve{c}_1}{\vec{\mathbb{p}}_N^{emp(\omega)}(Y_i(\omega))}\cdot\vec{\mathbb{q}}_{\tau}(Y_i(\omega))\,.
\end{aligned}\tag{124}$$

Just as in the derivation of (112) respectively (113), reasonable choices for the "boundary-functions" in (124) are $h_0(x,s) \equiv 0$, $h_1(x) \equiv 0$, $h_2(x) = \frac{\vec{q}_\theta(x)}{\vec{q}_\tau(x)}$, respectively $h_0(x,s) \equiv 0$, $h_1(x) \equiv \frac{\vec{q}_\theta(x)}{\vec{q}_\tau(x)}$, $h_2(x) = \frac{\vec{q}_\theta(x)}{\vec{q}_\tau(x)}$. As for example, consider for all $\theta_0, \theta, \tau \in \Theta$ the scaled Bregman divergences in the sense of Stummer [81], Stummer and Vajda [84] (cf. Remark (2)(b)), for which we get from (36) with $r(x) \equiv 1$

$$0 \leqslant D^c_{\phi,\vec{\mathbb{Q}}_\tau,\vec{\mathbb{Q}}_\tau,\mathbb{1}\cdot\vec{\mathbb{Q}}_\tau,\lambda_2}(\vec{\mathbb{Q}}_{\theta_0},\vec{\mathbb{Q}}_\theta)$$
$$:= \int_{\mathscr{X}} \Big[\phi\big(\tfrac{\vec{q}_{\theta_0}(x)}{\vec{q}_\tau(x)}\big) - \phi\big(\tfrac{\vec{q}_\theta(x)}{\vec{q}_\tau(x)}\big) - \phi'_{+,c}\big(\tfrac{\vec{q}_\theta(x)}{\vec{q}_\tau(x)}\big)\cdot\big(\tfrac{\vec{q}_{\theta_0}(x)}{\vec{q}_\tau(x)} - \tfrac{\vec{q}_\theta(x)}{\vec{q}_\tau(x)}\big)\Big]\cdot \vec{q}_\tau(x)\,\mathrm{d}\lambda_2(x)\,,$$
$$=: D_{\phi,\vec{\mathbb{Q}}_\tau,\lambda_2}(\vec{\mathbb{Q}}_{\theta_0},\vec{\mathbb{Q}}_\theta)\,, \tag{125}$$

from which – together with (120) – one can identify immediately the pointwise decomposability with $\psi^0(s) := \psi^0_\phi(s) := \phi(s)$, $\psi^1(t) := \psi^1_\phi(t) := t\cdot\phi'_{+,c}(t) - \phi(t)$, $\rho(t) := \rho_\phi(t) := -\phi'_{+,c}(t)$; by plugging this into (124), one obtains the objective $D_{\phi,\vec{\mathbb{Q}}_\tau,\lambda_2}\Big(\mathbb{P}_N^{\overline{emp}(\omega)}, \widetilde{\vec{\mathbb{Q}}}_\theta\Big)$, which in the course of (Enc3) should be – for fixed $\tau \in \Theta$ – minimized over $\theta \in \Theta$ in order to obtain the corresponding τ-individual" $\min -decD_{\phi,\vec{\mathbb{Q}}_\tau,\lambda}$-estimator $\widehat{\theta}_{N,\tau}(\omega) := \text{arginf}_{\theta\in\Theta}\ D_{\phi,\vec{\mathbb{Q}}_\tau,\lambda}\Big(\mathbb{P}_N^{\overline{emp}(\omega)}, \widetilde{\vec{\mathbb{Q}}}_\theta\Big)$. Recall that this choice can be motivated by $0 = \min_{\theta\in\Theta} D_{\phi,\vec{\mathbb{Q}}_\tau,\lambda_2}(\vec{\mathbb{Q}}_{\theta_0},\vec{\mathbb{Q}}_\theta)$ and $\theta_0 = \text{argmin}_{\theta\in\Theta} D_{\phi,\vec{\mathbb{Q}}_\tau,\lambda_2}(\vec{\mathbb{Q}}_{\theta_0},\vec{\mathbb{Q}}_\theta)$. Furthermore, one gets even $0 = \min_{\theta\in\Theta}\min_{\tau\in\Theta} D_{\phi,\vec{\mathbb{Q}}_\tau,\lambda_2}(\vec{\mathbb{Q}}_{\theta_0},\vec{\mathbb{Q}}_\theta)$, $\theta_0 = \text{argmin}_{\theta\in\Theta}\min_{\tau\in\Theta} D_{\phi,\vec{\mathbb{Q}}_\tau,\lambda_2}(\vec{\mathbb{Q}}_{\theta_0},\vec{\mathbb{Q}}_\theta)$, and in case of $\max_{\tau\in\Theta} D_{\phi,\vec{\mathbb{Q}}_\tau,\lambda_2}(\vec{\mathbb{Q}}_{\theta_0},\vec{\mathbb{Q}}_\theta) < \infty$ also $0 = \min_{\theta\in\Theta}\max_{\tau\in\Theta} D_{\phi,\vec{\mathbb{Q}}_\tau,\lambda_2}(\vec{\mathbb{Q}}_{\theta_0},\vec{\mathbb{Q}}_\theta)$, $\theta_0 = \text{argmin}_{\theta\in\Theta}\max_{\tau\in\Theta} D_{\phi,\vec{\mathbb{Q}}_\tau,\lambda_2}(\vec{\mathbb{Q}}_{\theta_0},\vec{\mathbb{Q}}_\theta)$. This suggests the alternative, "τ-uniform" estimators $\widehat{\theta}_N(\omega) := \text{argmin}_{\theta\in\Theta}\ \min_{\tau\in\Theta} D_{\phi,\vec{\mathbb{Q}}_\tau,\lambda}\Big(\mathbb{P}_N^{\overline{emp}(\omega)}, \widetilde{\vec{\mathbb{Q}}}_\theta\Big)$, respectively $\widehat{\theta}_N(\omega) := \text{argmin}_{\theta\in\Theta}\ \max_{\tau\in\Theta} D_{\phi,\vec{\mathbb{Q}}_\tau,\lambda}\Big(\mathbb{P}_N^{\overline{emp}(\omega)}, \widetilde{\vec{\mathbb{Q}}}_\theta\Big)$. As a side remark, let us mention that in general, (say) $\min_{\tau\in\Theta} D_{\phi,\vec{\mathbb{Q}}_\tau,\lambda}\Big(\mathbb{P}_N^{\overline{emp}(\omega)}, \widetilde{\vec{\mathbb{Q}}}_\theta\Big)$ is not necessarily decomposable anymore, and therefore the standard theory of M-estimators is not applicable to this class of estimators.

With our approach, we can generate numerous further estimators of the true unknown parameter θ_0, by permuting the positions – but not the roles (!) – of the parameters (θ_0, θ, τ) in the (pseudo-)divergences of the above investigations. For the sake of brevity, we only sketch two further cases; the full variety will appear elsewhere. To start with, consider the adaptively scaled and aggregated divergence

$$0 \leqslant D^{rev}_{\phi,\overrightarrow{\mathbb{Q}}_\tau,\lambda_2}(\overrightarrow{\mathbb{Q}}_{\theta_0},\overrightarrow{\mathbb{Q}}_\theta) := D^c_{\phi,\overrightarrow{\mathbb{Q}}_{\theta_0}^2/\overrightarrow{\mathbb{Q}}_\tau,\,\overrightarrow{\mathbb{Q}}_\theta^2/\overrightarrow{\mathbb{Q}}_\tau,\,\mathbb{1}\cdot\overrightarrow{\mathbb{Q}}_{\theta_0},\lambda_2}(\overrightarrow{\mathbb{Q}}_{\theta_0},\overrightarrow{\mathbb{Q}}_\theta)$$

$$:= \int_{\mathscr{X}} \Bigg[\phi\Bigg(\frac{\overrightarrow{q}_{\theta_0}(x)}{\frac{\overrightarrow{q}_{\theta_0}(x)^2}{\overrightarrow{q}_\tau(x)}}\Bigg) - \phi\Bigg(\frac{\overrightarrow{q}_\theta(x)}{\frac{\overrightarrow{q}_\theta(x)^2}{\overrightarrow{q}_\tau(x)}}\Bigg) - \phi'_{+,c}\Bigg(\frac{\overrightarrow{q}_\theta(x)}{\frac{\overrightarrow{q}_\theta(x)^2}{\overrightarrow{q}_\tau(x)}}\Bigg)\cdot\Bigg(\Bigg(\frac{\overrightarrow{q}_{\theta_0}(x)}{\frac{\overrightarrow{q}_{\theta_0}(x)^2}{\overrightarrow{q}_\tau(x)}}\Bigg) - \Bigg(\frac{\overrightarrow{q}_\theta(x)}{\frac{\overrightarrow{q}_\theta(x)^2}{\overrightarrow{q}_\tau(x)}}\Bigg)\Bigg)\Bigg]$$

$$\cdot\overrightarrow{q}_\tau(x)\,\mathrm{d}\lambda_2(x)$$

$$= \int_{\mathscr{X}} \Big[\phi\big(\tfrac{\overrightarrow{q}_\tau(x)}{\overrightarrow{q}_{\theta_0}(x)}\big) - \phi\big(\tfrac{\overrightarrow{q}_\tau(x)}{\overrightarrow{q}_\theta(x)}\big) - \phi'_{+,c}\big(\tfrac{\overrightarrow{q}_\tau(x)}{\overrightarrow{q}_\theta(x)}\big)\cdot\big(\tfrac{\overrightarrow{q}_\tau(x)}{\overrightarrow{q}_{\theta_0}(x)} - \tfrac{\overrightarrow{q}_\tau(x)}{\overrightarrow{q}_\theta(x)}\big)\Big]\cdot\overrightarrow{q}_{\theta_0}(x)\,\mathrm{d}\lambda_2(x)$$

$$=: \int_{\mathscr{X}} \Big[\psi^{0,rev}_{\overrightarrow{q}_\tau(x)}(\overrightarrow{q}_{\theta_0}(x)) + \psi^{1,rev}_{\overrightarrow{q}_\tau(x)}(\overrightarrow{q}_\theta(x)) + \rho^{rev}_{\overrightarrow{q}_\tau(x)}(\overrightarrow{q}_\theta(x))\cdot\overrightarrow{q}_{\theta_0}(x)\Big]\,\mathrm{d}\lambda_2(x)$$

(indeed, by Theorem 4 and (80) this is zero if and only if $\theta = \theta_0$). By means of the involved mappings $\psi^0(s) := \psi^{0,rev}_m(s) := s\cdot\phi(\frac{m}{s})$, $\psi^1(t) := \psi^{1,rev}_m(t) := -m\cdot\phi'_{+,c}(\frac{m}{t})$, $\rho(t) := \rho^{rev}_m(t) := \frac{m}{t}\cdot\phi'_{+,c}(\frac{m}{t}) - \phi(\frac{m}{t}) =: \phi^{\circledcirc}(\frac{m}{t})$ $(s,t,m>0)$, the properties (103), (104) are applicable and thus $D^{rev}_{\phi,\overrightarrow{\mathbb{Q}}_\tau,\lambda_2}(\cdot,\cdot)$ can be extended to a pointwise decomposable pseudo-divergence on $\widetilde{\mathscr{P}}^\lambda\otimes\mathscr{P}^\lambda$ by using (102) with appropriate functions h_0,h_1,h_2 and constants c_0,c_1,$\overline{c}_0$. Furthermore, by minimizing over $\theta\in\Theta$ the objective (111) with these choices $\psi^{0,rev}_m(\cdot)$, $\psi^{1,rev}_m(\cdot)$, $\rho^{rev}_m(\cdot)$, in the course of (Enc3) we end up at the corresponding min $-decD^{rev}_{\phi,\overrightarrow{\mathbb{Q}}_\tau,\lambda}$-estimator. In particular, the corresponding special case $h_0(x,s)\equiv 0$, $h_1(x)\equiv 1$, $h_2(x) = \overrightarrow{q}_\theta(x)$ $(\neq \widetilde{\overrightarrow{q}}_\theta(x))$ leads to the objective (cf. (112) but with $\psi^1(1)$ instead of $\psi^1(0)$)

$$D^{rev}_{\phi,\overrightarrow{\mathbb{Q}}_\tau,\lambda_2}\Big(\mathbb{P}_N^{\overrightarrow{emp}(\omega)},\widetilde{\overrightarrow{\mathbb{Q}}}_\theta\Big) = \phi^*(0)\cdot\check{\overline{c}}_0 - \int_{\mathscr{X}}\overrightarrow{q}_\tau(x)\cdot\phi'_{+,c}\Big(\frac{\overrightarrow{q}_\tau(x)}{\overrightarrow{q}_\theta(x)}\Big)\,\mathrm{d}\lambda_2(x)$$
$$+\frac{1}{N}\sum_{i=1}^N \phi^{\circledcirc}\Big(\frac{\overrightarrow{q}_\tau(Y_i(\omega))}{\overrightarrow{q}_\theta(Y_i(\omega))}\Big)$$
$$+\frac{1}{N}\sum_{i=1}^N\Big[\phi\Big(\frac{\overrightarrow{q}_\tau(Y_i(\omega))}{\mathbb{p}_N^{\overrightarrow{emp}(\omega)}(Y_i(\omega))}\Big)\cdot\check{c}_0 - \frac{\overrightarrow{q}_\tau(Y_i(\omega))\cdot\phi'_{+,c}(\overrightarrow{q}_\tau(Y_i(\omega)))}{\mathbb{p}_N^{\overrightarrow{emp}(\omega)}(Y_i(\omega))}\cdot\check{c}_1\Big]$$

to be minimized over θ. As a second possibility to permutate the positions of the parameters (θ_0,θ,τ), let us consider

$$0\leqslant D^c_{\phi,\overrightarrow{\mathbb{Q}}_\theta,\overrightarrow{\mathbb{Q}}_\theta,\mathbb{1}\cdot\overrightarrow{\mathbb{Q}}_\theta,\lambda_2}(\overrightarrow{\mathbb{Q}}_{\theta_0},\overrightarrow{\mathbb{Q}}_\tau)$$
$$:= \int_{\mathscr{X}}\Big[\phi\big(\tfrac{\overrightarrow{q}_{\theta_0}(x)}{\overrightarrow{q}_\theta(x)}\big) - \phi\big(\tfrac{\overrightarrow{q}_\tau(x)}{\overrightarrow{q}_\theta(x)}\big) - \phi'_{+,c}\big(\tfrac{\overrightarrow{q}_\tau(x)}{\overrightarrow{q}_\theta(x)}\big)\cdot\big(\tfrac{\overrightarrow{q}_{\theta_0}(x)}{\overrightarrow{q}_\theta(x)} - \tfrac{\overrightarrow{q}_\tau(x)}{\overrightarrow{q}_\theta(x)}\big)\Big]\cdot\overrightarrow{q}_\theta(x)\,\mathrm{d}\lambda_2(x)\,; \quad (126)$$

this is a pointwise decomposable divergence between $\overrightarrow{\mathbb{Q}}_{\theta_0}$ and $\overrightarrow{\mathbb{Q}}_\tau$, but it is *not* a divergence – yet still a nonnegative and obviously *not* pointwise decomposable functional – between $\overrightarrow{\mathbb{Q}}_{\theta_0}$ and $\overrightarrow{\mathbb{Q}}_\theta$. Indeed, for $\theta = \theta_0 \neq \tau$ one obtains $D^c_{\phi,\overrightarrow{\mathbb{Q}}_{\theta_0},\overrightarrow{\mathbb{Q}}_{\theta_0},\mathbb{1}\cdot\overrightarrow{\mathbb{Q}}_{\theta_0},\lambda_2}(\overrightarrow{\mathbb{Q}}_{\theta_0},\overrightarrow{\mathbb{Q}}_\tau) > 0$. Notice that from (126) one gets

$$\int_{\mathscr{X}} \phi\Big(\tfrac{\vec{\mathbb{q}}_{\theta_0}(x)}{\vec{\mathbb{q}}_{\theta}(x)}\Big) \cdot \vec{\mathbb{q}}_{\theta}(x) \, \mathrm{d}\lambda_2(x) \geqslant \int_{\mathscr{X}} \Big\{ \Big[\phi\Big(\tfrac{\vec{\mathbb{q}}_{\tau}(x)}{\vec{\mathbb{q}}_{\theta}(x)}\Big) - \phi_{+,c}'\Big(\tfrac{\vec{\mathbb{q}}_{\tau}(x)}{\vec{\mathbb{q}}_{\theta}(x)}\Big) \cdot \tfrac{\vec{\mathbb{q}}_{\tau}(x)}{\vec{\mathbb{q}}_{\theta}(x)} \Big] \cdot \vec{\mathbb{q}}_{\theta}(x)$$
$$+\phi_{+,c}'\Big(\tfrac{\vec{\mathbb{q}}_{\tau}(x)}{\vec{\mathbb{q}}_{\theta}(x)}\Big) \cdot \vec{\mathbb{q}}_{\theta_0}(x) \Big\} \, \mathrm{d}\lambda_2(x) =: \mathscr{D}^c_{\phi,\vec{\mathbb{Q}}_{\tau},\lambda_2}(\vec{\mathbb{Q}}_{\theta_0}, \vec{\mathbb{Q}}_{\theta}) \,, \tag{127}$$

provided that the integral on the right-hand side exists and is finite. If moreover $\phi(1) = 0$, then by (54) the inequality (127) rewrites as

$$D^c_{\phi,\lambda_2}(\vec{\mathbb{Q}}_{\theta_0}, \vec{\mathbb{Q}}_{\theta}) := D^c_{\phi,\vec{\mathbb{Q}}_{\theta},\vec{\mathbb{Q}}_{\theta},\mathbb{1}\cdot\vec{\mathbb{Q}}_{\theta},\lambda}(\vec{\mathbb{Q}}_{\theta_0}, \vec{\mathbb{Q}}_{\theta}) \geqslant D^c_{\phi,\vec{\mathbb{Q}}_{\tau},\lambda_2}(\vec{\mathbb{Q}}_{\theta_0}, \vec{\mathbb{Q}}_{\theta}) \tag{128}$$

with (for fixed θ) equality if and only if $\theta_0 = \tau$; this implies that

$$D^c_{\phi,\lambda_2}(\vec{\mathbb{Q}}_{\theta_0}, \vec{\mathbb{Q}}_{\theta}) = \max_{\tau\in\Theta} \mathscr{D}^c_{\phi,\vec{\mathbb{Q}}_{\tau},\lambda_2}(\vec{\mathbb{Q}}_{\theta_0}, \vec{\mathbb{Q}}_{\theta}) \tag{129}$$
$$= \max_{\tau\in\Theta} \int_{\mathscr{X}} \Big[\psi^{1,sub}_{\vec{\mathbb{q}}_{\tau}(x)}(\vec{\mathbb{q}}_{\theta}(x)) + \rho^{sub}_{\vec{\mathbb{q}}_{\tau}(x)}(\vec{\mathbb{q}}_{\theta}(x)) \cdot \vec{\mathbb{q}}_{\theta_0}(x) \Big] \mathrm{d}\lambda_2(x) \tag{130}$$

with $\psi^0(s) := \psi^{0,sub}_m(s) \equiv 0$, $\psi^1(t) := \psi^{1,sub}_m(t) := t \cdot \phi(\frac{m}{t}) - m \cdot \phi_{+,c}'(\frac{m}{t})$, $\rho(t) := \rho^{sub}_m(t) := \phi_{+,c}'(\frac{m}{t})$ $(s, t, m > 0)$. In other words, this means that the Csiszar-Ali-Silvey divergence CASD $D^c_{\phi,\lambda_2}(\vec{\mathbb{Q}}_{\theta_0}, \vec{\mathbb{Q}}_{\theta})$ can be represented as the τ-maximum over – not necessarily nonnegative – pointwise decomposable (in the sense of (103), (104)) functionals $\mathscr{D}^c_{\phi,\vec{\mathbb{Q}}_{\tau},\lambda_2}(\vec{\mathbb{Q}}_{\theta_0}, \vec{\mathbb{Q}}_{\theta})$ between $\vec{\mathbb{Q}}_{\theta_0}$ and $\vec{\mathbb{Q}}_{\theta}$. Furthermore, from Theorem 5 and (130) we arrive at

$$0 = \min_{\theta\in\Theta} D^c_{\phi,\lambda_2}(\vec{\mathbb{Q}}_{\theta_0}, \vec{\mathbb{Q}}_{\theta}) = \min_{\theta\in\Theta} \max_{\tau\in\Theta} \mathscr{D}^c_{\phi,\vec{\mathbb{Q}}_{\tau},\lambda_2}(\vec{\mathbb{Q}}_{\theta_0}, \vec{\mathbb{Q}}_{\theta})$$
$$= \min_{\theta\in\Theta} \max_{\tau\in\Theta} \int_{\mathscr{X}} \Big[\psi^{1,sub}_{\vec{\mathbb{q}}_{\tau}(x)}(\vec{\mathbb{q}}_{\theta}(x)) + \rho^{sub}_{\vec{\mathbb{q}}_{\tau}(x)}(\vec{\mathbb{q}}_{\theta}(x)) \cdot \vec{\mathbb{q}}_{\theta_0}(x) \Big] \mathrm{d}\lambda_2(x) \,,$$
$$\theta_0 = \mathrm{argmin}_{\theta\in\Theta} \max_{\tau\in\Theta} \mathscr{D}^c_{\phi,\vec{\mathbb{Q}}_{\tau},\lambda_2}(\vec{\mathbb{Q}}_{\theta_0}, \vec{\mathbb{Q}}_{\theta}). \tag{131}$$

Accordingly, in analogy to the spirit of (81), (82), (86), respectively Definition 4 and (110), in order to achieve an estimator of the true unknown parameter θ_0 we first extend the "pure parametric case" $\mathscr{D}^c_{\phi,\vec{\mathbb{Q}}_{\tau},\lambda_2} : \mathscr{P}^{\lambda}_{\Theta} \otimes \mathscr{P}^{\lambda}_{\Theta} \mapsto \mathbb{R}$ to a singularity-covering functional $\mathscr{D}^c_{\phi,\vec{\mathbb{Q}}_{\tau},\lambda} : \mathscr{P}^{\lambda}_{\Theta,emp} \otimes \mathscr{P}^{\lambda}_{\Theta} \mapsto \mathbb{R}$, although it is not a pseudo-divergence anymore; indeed, by employing the reduced form of (102) we take

$$\mathscr{D}^c_{\phi,\vec{\mathbb{Q}}_{\tau},\lambda}(\vec{\mathbb{P}}, \vec{\mathbb{Q}}) := \int_{\mathscr{X}} \Big[\psi^{1,sub}_{\vec{\mathbb{q}}_{\tau}(x)}\Big(\vec{\mathbb{q}}(x) + h_1(x) \cdot \mathbf{1}_{\{0\}}(\vec{\mathbb{q}}(x))\Big) \cdot \mathbf{1}_{]c_1,\infty[}(\vec{\mathbb{q}}(x))$$
$$+\rho^{sub}_{\vec{\mathbb{q}}_{\tau}(x)}\Big(\vec{\mathbb{q}}(x) + h_2(x) \cdot \mathbf{1}_{\{0\}}(\vec{\mathbb{q}}(x))\Big) \cdot \vec{\mathbb{p}}(x) \Big] \mathrm{d}\lambda(x) \quad \text{for all } \vec{\mathbb{P}} \in \mathscr{P}^{\lambda}_{\Theta,emp}, \vec{\mathbb{Q}} \in \mathscr{P}^{\lambda}_{\Theta} \,. \tag{132}$$

Hence, analogously to the derivation of (111), we obtain from (132)

$$\sup_{\tau\in\Theta} \mathscr{D}^{c}_{\phi,\vec{\mathbb{Q}}_{\tau},\lambda}\Big(\mathbb{P}_N^{\overline{emp}(\omega)}, \widetilde{\vec{\mathbb{Q}}}_\theta\Big) = \sup_{\tau\in\Theta} \int_{\mathscr{X}} \psi^{1,sub}_{\vec{q}_\tau(x)}\Big(\vec{q}_\theta(x)\Big)\, \mathrm{d}\lambda_2(x)$$

$$+\sum_{x\in\mathscr{X}} \Big[\psi^{1,sub}_{\vec{q}_\tau(x)}\Big(h_1(x)\Big)\cdot \breve{c}_1 + \rho^{sub}_{\vec{q}_\tau(x)}(h_2(x))\cdot \vec{\mathbb{p}}_N^{\overline{emp}(\omega)}(x)\Big]\cdot \mathbf{1}_{\mathscr{R}(Y_1(\omega),\ldots,Y_N(\omega))}(x)$$

$$= \sup_{\tau\in\Theta} \int_{\mathscr{X}} \Big[\vec{q}_\theta(x)\cdot\phi\Big(\frac{\vec{q}_\tau(x)}{\vec{q}_\theta(x)}\Big) - \vec{q}_\tau(x)\cdot\phi'_{+,c}\Big(\frac{\vec{q}_\tau(x)}{\vec{q}_\theta(x)}\Big)\Big]\, \mathrm{d}\lambda_2(x) \qquad (133)$$

$$+\frac{1}{N}\sum_{i=1}^N \phi'_{+,c}\Big(\frac{\vec{q}_\tau(Y_i(\omega))}{h_2(Y_i(\omega))}\Big)$$

$$+\frac{1}{N}\sum_{i=1}^N \frac{h_1(Y_i(\omega))\cdot\phi\Big(\frac{\vec{q}_\tau(Y_i(\omega))}{h_1(Y_i(\omega))}\Big) - \vec{q}_\tau(Y_i(\omega))\cdot\phi'_{+,c}\Big(\frac{\vec{q}_\tau(Y_i(\omega))}{h_1(Y_i(\omega))}\Big)}{\vec{\mathbb{p}}_N^{emp(\omega)}(Y_i(\omega))}\cdot \breve{c}_1 \qquad (134)$$

to be minimized over $\theta\in\Theta$. In the view of (131), we can estimate (respectively learn) the true unknown parameter θ_0 by the estimator

$$\widehat{\theta}_{N,sup\mathscr{D}_{\phi,\lambda}}(\omega) := \operatorname{arginf}_{\theta\in\Theta} \sup_{\tau\in\Theta} \mathscr{D}^{c}_{\phi,\vec{\mathbb{Q}}_{\tau},\lambda}\Big(\mathbb{P}_N^{\overline{emp}(\omega)}, \widetilde{\vec{\mathbb{Q}}}_\theta\Big) \text{ for } \mathbb{P}_N^{\overline{emp}(\omega)} \in \mathscr{P}^{\lambda\perp}_{emp}, \qquad (135)$$

which under appropriate technical assumptions (integrability, etc.) exists, is finite, unique, and Fisher consistent; moreover, this method can be straightforwardly extended to non-parametric setups. Similarly to the derivation of (112) respectively (113), reasonable choices for the "boundary-functions" in (134) are $h_2(x) := \vec{q}_\theta(x)$ together with $h_1(x)\equiv 1$ respectively $h_1(x) := \vec{q}_\theta(x)$ (where the nominator in the last sum becomes $-\vec{q}_\tau(Y_i(\omega))\cdot\phi'_{+,c}(1))$). In the special case with $c_1 = 0 = \breve{c}_1$ – where the choice of $h_1(\cdot)$ is irrelevant – and $h_2(x) := \vec{q}_\theta(x)$, the estimator $\widehat{\theta}_{N,sup\mathscr{D}_{\phi,\lambda}}(\omega)$ was first proposed independently by Liese and Vajda [42] under the name *modified ϕ-divergence estimator* and Broniatowski and Keziou [16, 17] under the name *minimum dual ϕ-divergence estimator*; furthermore, within this special-case setup, Broniatowski and Keziou [17] also introduced for each fixed $\theta\in\Theta$ the related, so-called *dual ϕ-divergence estimator* $\widehat{\theta}_{N,\theta,\mathscr{D}_{\phi,\lambda}}(\omega) := \operatorname{argsup}_{\tau\in\Theta} \mathscr{D}^{c}_{\phi,\vec{\mathbb{Q}}_{\tau},\lambda}\Big(\mathbb{P}_N^{\overline{emp}(\omega)}, \widetilde{\vec{\mathbb{Q}}}_\theta\Big)$. The latter four references also work within a nonparametric framework. Let us also mention that by (128) and (129), $\widehat{\theta}_{N,\mathscr{D}_{\phi,\lambda}}(\omega)$ can be interpreted as *maximum sub-ϕ-divergence estimator*, whereas $\widehat{\theta}_{N,sup\mathscr{D}_{\phi,\lambda}}(\omega)$ can be viewed as *minimum super-ϕ-divergence estimator* (cf. Vajda [90], Broniatowski and Vajda [18] for the probability-measure-theoretic context of footnote 15).

Remark 6 Making use of the escort parameter τ proves to be useful in statistical inference under the model; its use under misspecification has been considered in Toma and Broniatowski [86], Al Mohamad [5], for Csiszar-Ali-Silvey divergences.

As a final example, consider $c_1 = 0$, $h_2(x) := \vec{q}_\theta(x)$, and $\phi(t) := t\log t + 1 - t$, for which we can deduce

$$\widehat{\theta}_{N,sup\mathscr{D}_{\phi,\lambda}}(\omega) = \widehat{\theta}_{N,\theta,\mathscr{D}_{\phi,\lambda}}(\omega) = \operatorname{argsup}_{\xi\in\Theta} \frac{1}{N}\sum_{i=1}^N \log\Big(\vec{q}_\xi(Y_i(\omega))\Big)$$

for all $\theta \in \Theta$, i.e. in this case all maximum sub-ϕ-divergence estimators and the minimum super-ϕ-divergence estimator exceptionally coincide, and give the celebrated maximum-likelihood estimator.

5 Conclusions

Motivated by fields of applications from statistics, machine learning, artificial intelligence and information geometry, we presented for a wide audience a new unifying framework of divergences between functions. Within this, we illuminated several important subcases – such as scaled Bregman divergences and Csiszar-Ali-Silvey ϕ-divergences – as well as involved subtleties and pitfalls. For the often desired task of finding the "continuous" model with best divergence-proximity to the observed "discrete" data, we summarized existing and also derived new approaches. As far as potential future studies is concerned, the kind of universal nature of our introduced toolkit suggests quite a lot of possibilities for further adjacent developments and concrete applications.

Acknowledgements We are grateful to three anonymous referees for their very useful suggestions and comments. W. Stummer wants to thank very much the Sorbonne Universite Pierre et Marie Curie Paris for its partial financial support.

Appendix: Proofs

Proof of Theorem 4. Assertion (1) and the "if-part" of (2) follow immediately from Theorem 1 which uses less restrictive assumptions. In order to show the "only-if" part of (2) (and the "if-part" of (2) in an alternative way), one can use the straightforwardly provable fact that the Assumption 2 implies

$$\overline{\mathbb{W}_3 \cdot \psi_{\phi,c}}(x,s,t) = 0 \qquad \text{if and only if} \qquad s = t \tag{136}$$

for all $s \in \mathscr{R}\big(\frac{P}{M_1}\big)$, all $t \in \mathscr{R}\big(\frac{Q}{M_2}\big)$ and λ-a.a. $x \in \mathscr{X}$. To proceed, assume that $D^c_{\phi,M_1,M_2,\mathbb{M}_3,\lambda}(P,Q) = 0$, which by the non-negativity of $\overline{\mathbb{W}_3 \cdot \psi_{\phi,c}}(\cdot,\cdot)$ implies that $\overline{\mathbb{W}_3 \cdot \psi_{\phi,c}}\big(\frac{p(x)}{m_1(x)}, \frac{q(x)}{m_2(x)}\big) = 0$ for λ-a.a. $x \in \mathscr{X}$. From this and the "only-if" part of (136), we obtain the identity $\frac{p(x)}{m_1(x)} = \frac{q(x)}{m_2(x)}$ for λ-a.a. $x \in \mathscr{X}$. □

Proof of Theorem 5. Consistently with Theorem 1 (and our adaptions) the "if-part" follows from (51). By our above investigations on the adaptions of the Assumptions 2 to the current context, it remains to investigate the "only-if" part (2) for the following four cases (recall that ϕ is strictly convex at $t = 1$):
(ia) ϕ is differentiable at $t = 1$ (hence, c is obsolete and $\phi'_{+,c}(1)$ collapses to $\phi'(1)$) and the function ϕ is affine linear on $[1, s]$ for some $s \in \mathscr{R}\big(\frac{P}{Q}\big)\backslash[a, 1]$;

(ib) ϕ is differentiable at $t = 1$, and the function ϕ is affine linear on $[s, 1]$ for some $s \in \mathscr{R}\big(\frac{P}{Q}\big)\backslash[1, b]$;
(ii) ϕ is not differentiable at $t = 1, c = 1$, and the function ϕ is affine linear on $[1, s]$ for some $s \in \mathscr{R}\big(\frac{P}{Q}\big)\backslash[a, 1]$;
(iii) ϕ is not differentiable at $t = 1$, $c = 0$, and the function ϕ is affine linear on $[s, 1]$ for some $s \in \mathscr{R}\big(\frac{P}{Q}\big)\backslash[1, b]$.
It is easy to see from the strict convexity at 1 that for (ii) one has $\phi(0) + \phi'_{+,1}(1) - \phi(1) > 0$, whereas for (iii) one gets $\phi^*(0) - \phi'_{+,0}(1) > 0$; furthermore, for (ia) there holds $\phi(0) + \phi'(1) - \phi(1) > 0$ and for (ib) $\phi^*(0) - \phi'(1) > 0$. Let us first examine the situations (ia) respectively (ii) under the assumptive constraint $D^c_{\phi,\mathbb{Q},\mathbb{Q},\mathbb{R}\cdot\mathbb{Q},\lambda}(\mathbb{P}, \mathbb{Q}) = 0$ with $c = 1$ respectively (in case of differentiability) obsolete c, for which we can deduce from (51)

$$\begin{aligned} 0 &= D^c_{\phi,\mathbb{Q},\mathbb{Q},\mathbb{R}\cdot\mathbb{Q},\lambda}(\mathbb{P}, \mathbb{Q}) \\ &\geqslant \int_{\mathscr{X}} \mathbb{r}(x)\cdot\big[\mathbb{q}(x)\cdot\phi\big(\tfrac{\mathbb{p}(x)}{\mathbb{q}(x)}\big) - \mathbb{q}(x)\cdot\phi(1) - \phi'_{+,c}(1)\cdot\big(\mathbb{p}(x)-\mathbb{q}(x)\big)\big] \\ &\qquad\qquad \cdot \mathbf{1}_{]0,\infty[}\big(\mathbb{p}(x)\big)\cdot\mathbf{1}_{]\mathbb{p}(x),\infty[}\big(\mathbb{q}(x)\big)\,\mathrm{d}\lambda(x) \\ &+\big[\phi(0)+\phi'_{+,c}(1)-\phi(1)\big]\cdot\int_{\mathscr{X}} \mathbb{r}(x)\cdot\mathbb{q}(x)\cdot\mathbf{1}_{\{0\}}\big(\mathbb{p}(x)\big)\cdot\mathbf{1}_{]\mathbb{p}(x),\infty[}\big(\mathbb{q}(x)\big)\,\mathrm{d}\lambda(x) \geqslant 0, \end{aligned}$$

and hence $\int_{\mathscr{X}} \mathbf{1}_{]\mathbb{p}(x),\infty[}\big(\mathbb{q}(x)\big)\cdot\mathbb{r}(x)\,\mathrm{d}\lambda(x) = 0$. From this and (55) we obtain

$$0 = \int_{\mathscr{X}} \big(\mathbb{p}(x)-\mathbb{q}(x)\big)\cdot\mathbb{r}(x)\,\mathrm{d}\lambda(x) = \int_{\mathscr{X}} \big(\mathbb{p}(x)-\mathbb{q}(x)\big)\cdot\mathbf{1}_{]\mathbb{q}(x),\infty[}\big(\mathbb{p}(x)\big)\cdot\mathbb{r}(x)\,\mathrm{d}\lambda(x)$$

and therefore $\int_{\mathscr{X}} \mathbf{1}_{]\mathbb{q}(x),\infty[}\big(\mathbb{p}(x)\big)\cdot\mathbb{r}(x)\,\mathrm{d}\lambda(x) = 0$. Since for λ-a.a. $x \in \mathscr{X}$ we have $\mathbb{r}(x) > 0$, we arrive at $\mathbb{p}(x) = \mathbb{q}(x)$ for λ-a.a. $x \in \mathscr{X}$. The remaining cases (ib) respectively (iii) can be treated analogously. □

References

1. Amari, S.-I.: Information Geometry and Its Applications. Springer, Japan (2016)
2. Amari, S.-I., Karakida, R., Oizumi, M.: Information geometry connecting Wasserstein distance and Kullback-Leibler divergence via the entropy-relaxed transportation problem. Info. Geo. (2018). https://doi.org/10.1007/s41884-018-0002-8
3. Amari, S.-I., Nagaoka, H.: Methods of Information Geometry. Oxford University Press, Oxford (2000)
4. Ali, M.S., Silvey, D.: A general class of coefficients of divergence of one distribution from another. J. R. Stat. Soc. **B–28**, 131–140 (1966)
5. Al Mohamad, D.: Towards a better understanding of the dual representation of phi divergences. Stat. Papers (2016). https://doi.org/10.1007/s00362-016-0812-5
6. Avlogiaris, G., Micheas, A., Zografos, K.: On local divergences between two probability measures. Metrika **79**, 303–333 (2016)
7. Avlogiaris, G., Micheas, A., Zografos, K.: On testing local hypotheses via local divergence. Stat. Methodol. **31**, 20–42 (2016)
8. Ay, N., Jost, J., Le, H.V., Schwachhöfer, L.: Information Geometry. Springer, Berlin (2017)
9. Banerjee, A., Merugu, S., Dhillon, I.S., Ghosh, J.: Clustering with Bregman divergences. J. Mach. Learn. Res. **6**, 1705–1749 (2005)

10. Basu, A., Harris, I.R., Hjort, N.L., Jones, M.C.: Robust and efficient estimation by minimizing a density power divergence. Biometrika **85**(3), 549–559 (1998)
11. Basu, A., Lindsay, B.G.: Minimum disparity estimation for continuous models: efficiency, distributions and robustness. Ann. Inst. Stat. Math. **46**(4), 683–705 (1994)
12. Basu, A., Mandal, A., Martin, N., Pardo, L.: Robust tests for the equality of two normal means based on the density power divergence. Metrika **78**, 611–634 (2015)
13. Basu, A., Shioya, H., Park, C.: Statistical Inference: The Minimum Distance Approach. CRC Press, Boca Raton (2011)
14. Birkhoff, G.D: A set of postulates for plane geometry, based on scale and protractor. Ann. Math. **33**(2) 329–345 (1932)
15. Boissonnat, J.-D., Nielsen, F., Nock, R.: Bregman Voronoi diagrams. Discret. Comput. Geom. **44**(2), 281–307 (2010)
16. Broniatowski, M., Keziou, A.: Minimization of ϕ-divergences on sets of signed measures. Stud. Sci. Math. Hungar. **43**, 403–442 (2006)
17. Broniatowski, M., Keziou, A.: Parametric estimation and tests through divergences and the duality technique. J. Multiv. Anal. **100**(1), 16–36 (2009)
18. Broniatowski, M., Vajda, I.: Several applications of divergence criteria in continuous families. Kybernetika **48**(4), 600–636 (2012)
19. Broniatowski, M., Toma, A., Vajda, I.: Decomposable pseudodistances in statistical estimation. J. Stat. Plan. Inf. **142**, 2574–2585 (2012)
20. Buckland, M.K.: Information as thing. J. Am. Soc. Inf. Sci. **42**(5), 351–360 (1991)
21. Cesa-Bianchi, N., Lugosi, G.: Prediction, Learning and Games. Cambridge University Press, Cambridge (2006)
22. Chhogyal, K., Nayak, A., Sattar, A.: On the KL divergence of probability mixtures for belief contraction. In: Hölldobler,S., et al. (eds.) KI 2015: Advances in Artificial Intelligence. Lecture Notes in Artificial Intelligence, vol. 9324, pp. 249–255. Springer International Publishing (2015)
23. Cliff, O.M., Prokopenko, M., Fitch, R.: An information criterion for inferring coupling in distributed dynamical systems. Front. Robot. AI **3**(71). https://doi.org/10.3389/frobt.2016.00071 (2016)
24. Cliff, O.M., Prokopenko, M., Fitch, R.: Minimising the Kullback-Leibler divergence for model selection in distributed nonlinear systems. Entropy **20**(51). https://doi.org/10.3390/e20020051 (2018)
25. Collins, M., Schapire, R.E., Singer, Y.: Logistic regression, AdaBoost and Bregman distances. Mach. Learn. **48**, 253–285 (2002)
26. Cooper, V.N., Haddad, H.M., Shahriar, H.: Android malware detection using Kullback-Leibler divergence. Adv. Distrib. Comp. Art. Int. J., Special Issue 3(2) (2014)
27. Csiszar, I.: Eine informationstheoretische Ungleichung und ihre Anwendung auf den Beweis der Ergodizität von Markoffschen Ketten. Publ. Math. Inst. Hungar. Acad. Sci. **A-8**, 85–108 (1963)
28. DasGupta, A.: Some results on the curse of dimensionality and sample size recommendations. Calcutta Stat. Assoc. Bull. **50**(3–4), 157–178 (2000)
29. De Groot, M.H.: Uncertainty, information and sequential experiments. Ann. Math. Stat. **33**, 404–419 (1962)
30. Ghosh, A., Basu, A.: Robust Bayes estimation using the density power divergence. Ann. Inst. Stat. Math. **68**, 413–437 (2016)
31. Ghosh, A., Basu, A.: Robust estimation in generalized linear models: the density power divergence approach. TEST **25**, 269–290 (2016)
32. Ghosh, A., Harris, I.R., Maji, A., Basu, A., Pardo, L.: A generalized divergence for statistical inference. Bernoulli **23**(4A), 2746–2783 (2017)
33. Hampel, F.R., Ronchetti, E.M., Rousseuw, P.J., Stahel, W.A.: Robust Statistics: The Approach Based on Influence Functions. Wiley, New York (1986)
34. Karakida, R., Amari, S.-I.: Information geometry of Wasserstein divergence. In: Nielsen, F., Barbaresco, F. (eds.) Geometric Science of Information GSI 2017. Lecture Notes in Computer Science, vol. 10589, pp. 119–126. Springer International (2017)

35. Kißlinger, A.-L., Stummer, W.: Some decision procedures based on scaled Bregman distance surfaces. In: Nielsen, F., Barbaresco, F. (eds.) Geometric Science of Information GSI 2013. Lecture Notes in Computer Science, vol. 8085, pp. 479–486. Springer, Berlin (2013)
36. Kißlinger, A.-L., Stummer, W.: New model search for nonlinear recursive models, regressions and autoregressions. In: Nielsen, F., Barbaresco, F. (eds.) Geometric Science of Information GSI 2015. Lecture Notes in Computer Science, vol. 9389, pp. 693–701. Springer International (2015)
37. Kißlinger, A.-L., Stummer, W.: Robust statistical engineering by means of scaled Bregman distances. In: Agostinelli, C., Basu, A., Filzmoser, P., Mukherjee, D. (eds.) Recent Advances in Robust Statistics - Theory and Applications, pp. 81–113. Springer, India (2016)
38. Kißlinger, A.-L., Stummer, W.: A new toolkit for robust distributional change detection. Appl. Stochastic Models Bus. Ind. **34**, 682–699 (2018)
39. Kuchibhotla, A.K., Basu, A.: A general setup for minimum disparity estimation. Stat. Prob. Lett. **96**, 68–74 (2015)
40. Liese, F., Miescke, K.J.: Statistical Decision Theory: Estimation, Testing, and Selection. Springer, New York (2008)
41. Liese, F., Vajda, I.: Convex Statistical Distances. Teubner, Leipzig (1987)
42. Liese, F., Vajda, I.: On divergences and informations in statistics and information theory. IEEE Trans. Inf. Theory **52**(10), 4394–4412 (2006)
43. Lin, N., He, X.: Robust and efficient estimation under data grouping. Biometrika **93**(1), 99–112 (2006)
44. Liu, M., Vemuri, B.C., Amari, S.-I., Nielsen, F.: Total Bregman divergence and its applications to shape retrieval. In: Proceedings of 23rd IEEE CVPR, pp. 3463–3468 (2010)
45. Liu, M., Vemuri, B.C., Amari, S.-I., Nielsen, F.: Shape retrieval using hierarchical total Bregman soft clustering. IEEE Trans. Pattern Anal. Mach. Intell. **34**(12), 2407–2419 (2012)
46. Lizier, J.T.: JIDT: an information-theoretic toolkit for studying the dynamcis of complex systems. Front. Robot. AI **1**(11). https://doi.org/10.3389/frobt.2014.00011 (2014)
47. Menendez, M., Morales, D., Pardo, L., Vajda, I.: Two approaches to grouping of data and related disparity statistics. Comm. Stat. - Theory Methods **27**(3), 609–633 (1998)
48. Menendez, M., Morales, D., Pardo, L., Vajda, I.: Minimum divergence estimators based on grouped data. Ann. Inst. Stat. Math. **53**(2), 277–288 (2001)
49. Menendez, M., Morales, D., Pardo, L., Vajda, I.: Minimum disparity estimators for discrete and continuous models. Appl. Math. **46**(6), 439–466 (2001)
50. Millmann, R.S., Parker, G.D.: Geometry - A Metric Approach With Models, 2nd edn. Springer, New York (1991)
51. Minka, T.: Divergence measures and message passing. Technical Report MSR-TR-2005-173, Microsoft Research Ltd., Cambridge, UK (2005)
52. Morales, D., Pardo, L., Vajda, I.: Digitalization of observations permits efficient estimation in continuous models. In: Lopez-Diaz, M., et al. (eds.) Soft Methodology and Random Information Systems, pp. 315–322. Springer, Berlin (2004)
53. Morales, D., Pardo, L., Vajda, I.: On efficient estimation in continuous models based on finitely quantized observations. Comm. Stat. - Theory Methods **35**(9), 1629–1653 (2006)
54. Murata, N., Takenouchi, T., Kanamori, T., Eguchi, S.: Information geometry of U-boost and Bregman divergence. Neural Comput. **16**(7), 1437–1481 (2004)
55. Nielsen, F., Barbaresco, F. (eds.): Geometric Science of Information GSI 2013. Lecture Notes in Computer Science, vol. 8085. Springer, Berlin (2013)
56. Nielsen, F., Barbaresco, F. (eds.): Geometric Science of Information GSI 2015. Lecture Notes in Computer Science, vol. 9389. Springer International (2015)
57. Nielsen, F., Barbaresco, F. (eds.): Geometric Science of Information GSI 2017. Lecture Notes in Computer Science, vol. 10589. Springer International (2017)
58. Nielsen, F., Bhatia, R. (eds.): Matrix Information Geometry. Springer, Berlin (2013)
59. Nielsen, F., Nock, R.: Bregman divergences from comparative convexity. In: Nielsen, F., Barbaresco, F. (eds.) Geometric Science of Information GSI 2017. Lecture Notes in Computer Science, vol. 10589, pp. 639–647. Springer International (2017)

60. Nielsen, F., Sun, K., Marchand-Maillet, S.: On Hölder projective divergences. Entropy **19**, 122 (2017)
61. Nielsen, F., Sun, K., Marchand-Maillet,S.: K-means clustering with Hölder divergences. In: Nielsen, F., Barbaresco, F. (eds.) Geometric Science of Information GSI 2017. Lecture Notes in Computer Science, vol. 10589, pp. 856–863. Springer International (2017)
62. Nock, R., Menon, A.K., Ong, C.S.: A scaled Bregman theorem with applications. Advances in Neural Information Processing Systems 29 (NIPS 2016), pp. 19–27 (2016)
63. Nock, R., Nielsen, F.: Bregman divergences and surrogates for learning. IEEE Trans. Pattern Anal. Mach. Intell. **31**(11), 2048–2059 (2009)
64. Nock, R., Nielsen, F., Amari, S.-I.: On conformal divergences and their population minimizers. IEEE Trans. Inf. Theory **62**(1), 527–538 (2016)
65. Österreicher, F., Vajda, I.: Statistical information and discrimination. IEEE Trans. Inf. Theory **39**, 1036–1039 (1993)
66. Pal, S., Wong, T.-K.L.: The geometry of relative arbitrage. Math. Financ. Econ. **10**, 263–293 (2016)
67. Pal, S., Wong, T.-K.L.: Exponentially concave functions and a new information geometry. Ann. Probab. **46**(2), 1070–1113 (2018)
68. Pardo, L.: Statistical Inference Based on Divergence Measures. Chapman & Hall/CRC, Boca Raton (2006)
69. Park, C., Basu, A.: Minimum disparity estimation: asymptotic normality and breakdown point results. Bull. Inf. Kybern. **36**, 19–33 (2004)
70. Patra, S., Maji, A., Basu, A., Pardo, L.: The power divergence and the density power divergence families: the mathematical connection. Sankhya 75-B Part 1, 16–28 (2013)
71. Peyre, G., Cuturi M.: Computational Optimal Transport (2018). arXiv:1803.00567v1
72. Read, T.R.C., Cressie, N.A.C.: Goodness-of-Fit Statistics for Discrete Multivariate Data. Springer, New York (1988)
73. Reid, M.D., Williamson, R.C.: Information, divergence and risk for binary experiments. J. Mach. Learn. Res. **12**, 731–817 (2011)
74. Roensch, B., Stummer, W.: 3D insights to some divergences for robust statistics and machine learning. In: Nielsen, F., Barbaresco, F. (eds.) Geometric Science of Information GSI 2017. Lecture Notes in Computer Science, vol. 10589, pp. 460–469. Springer International (2017)
75. Rüschendorf, L.: On the minimum discrimination information system. Stat. Decis. Suppl. Issue **1**, 263–283 (1984)
76. Scott, D.W.: Multivariate Density Estimation - Theory, Practice and Visualization, 2nd edn. Wiley, Hoboken (2015)
77. Scott, D.W., Wand, M.P.: Feasibility of multivariate density estimates. Biometrika **78**(1), 197–205 (1991)
78. Stummer, W.: On a statistical information measure of diffusion processes. Stat. Decis. **17**, 359–376 (1999)
79. Stummer, W.: On a statistical information measure for a generalized Samuelson-Black-Scholes model. Stat. Decis. **19**, 289–314 (2001)
80. Stummer, W.: Exponentials, Diffusions, Finance. Entropy and Information. Shaker, Aachen (2004)
81. Stummer, W.: Some Bregman distances between financial diffusion processes. Proc. Appl. Math. Mech. **7**(1), 1050503–1050504 (2007)
82. Stummer, W., Kißlinger, A-L.: Some new flexibilizations of Bregman divergences and their asymptotics. In: Nielsen, F., Barbaresco, F. (eds.) Geometric Science of Information GSI 2017. Lecture Notes in Computer Science, vol. 10589, pp. 514–522. Springer International (2017)
83. Stummer, W., Vajda, I.: On divergences of finite measures and their applicability in statistics and information theory. Statistics **44**, 169–187 (2010)
84. Stummer, W., Vajda, I.: On Bregman distances and divergences of probability measures. IEEE Trans. Inf. Theory **58**(3), 1277–1288 (2012)
85. Sugiyama, M., Suzuki, T., Kanamori, T.: Density-ratio matching under the Bregman divergence: a unified framework of density-ratio estimation. Ann. Inst. Stat. Math. **64**, 1009–1044 (2012)

86. Toma, A., Broniatowski, M.: Dual divergence estimators and tests: robustness results. J. Multiv. Anal. **102**, 20–36 (2011)
87. Tsuda, K., Rätsch, G., Warmuth, M.: Matrix exponentiated gradient updates for on-line learning and Bregman projection. J. Mach. Learn. Res. **6**, 995–1018 (2005)
88. van der Vaart, A.W., Wellner, J.A.: Weak Convergence and Empirical Processes. Springer, Berlin (1996)
89. Vajda, I.: Theory of Statistical Inference and Information. Kluwer, Dordrecht (1989)
90. Vajda, I.: Modifications of divergence criteria for applications in continuous families. Research Report No. 2230, Institute of Information Theory and Automation, Prague (2008)
91. Vemuri, B.C., Liu, M., Amari, S.-I., Nielsen, F.: Total Bregman divergence and its applications to DTI analysis. IEEE Trans. Med. Imag. **30**(2), 475–483 (2011)
92. Victoria-Feser, M.-P., Ronchetti, E.: Robust estimation for grouped data. J. Am. Stat. Assoc. **92**(437), 333–340 (1997)
93. Weller-Fahy, D.J., Borghetti, B.J., Sodemann, A.A.: A survey of distance and similarity measures used within network intrusion anomaly detection. IEEE Commun. Surv. Tutor. **17**(1), 70–91 (2015)
94. Wu, L., Hoi, S.C.H., Jin, R., Zhu, J., Yu, N.: Learning Bregman distance functions for semi-supervised clustering. IEEE Trans. Knowl. Data Engin. **24**(3), 478–491 (2012)
95. Zhang, J., Naudts, J.: Information geometry under monotone embedding, part I: divergence functions. In: Nielsen, F., Barbaresco, F. (eds.) Geometric Science of Information GSI 2017. Lecture Notes in Computer Science, vol. 10589, pp. 205–214. Springer International (2017)
96. Zhang, J., Wang, X., Yao, L., Li, J., Shen, X.: Using Kullback-Leibler divergence to model opponents in poker. Computer Poker and Imperfect Information: Papers from the AAAI-14 Workshop (2014)

Information-Theoretic Matrix Inequalities and Diffusion Processes on Unimodular Lie Groups

Gregory S. Chirikjian

Abstract Unimodular Lie groups admit natural generalizations of many of the core concepts on which classical information-theoretic inequalities are built. Specifically, they have the properties of shift-invariant integration, an associative convolution operator, well-defined diffusion processes, and concepts of Entropy, Fisher information, Gaussian distribution, and Fourier transform. Equipped with these definitions, it is shown that many inequalities from classical information theory generalize to this setting. Moreover, viewing the Fourier transform for noncommutative unimodular Lie groups as a matrix-valued function, relationships between trace inequalities, diffusion processes, and convolution are examined.

Keywords Haar measure · Convolution · Group theory · Harmonic analysis · Diffusion processes · Inequalities

1 Introduction

This paper explores aspects of information-theoretic inequalities that naturally extend from $\mathbb{R}^n$ to an arbitrary unimodular Lie group. The exposition is mostly a condensed summary of material that can be found in [1, 2], but also presents some new inequalities. The motivations for investigating such inequalities are twofold: (1) physical information-processing agents (such as mobile robots or microscopic organisms) often have configuration spaces with Lie-group structure, and their localization in the world is therefore inextricably connected to both information theory and geometry; (2) Due to the identical form of the entropy functional in continuous information theory and in statistical mechanics, results from one field carry over to the other, and so it becomes possible to make statements about the statistical mechanical entropy of passive objects such as DNA and loops in proteins using the results of Lie-theoretic information theory.

G. S. Chirikjian (✉)
Johns Hopkins University, Baltimore, MD 21218, USA
e-mail: gchirik1@jhu.edu
URL: http://www.rpk.lcsr.jhu.edu

F. Nielsen (ed.), *Geometric Structures of Information*, Signals and Communication Technology, https://doi.org/10.1007/978-3-030-02520-5_9

1.1 Mathematical Preliminaries

A unimodular Lie group, $(G, \circ)$, is one which possesses a bi-invariant integration measure. That is, it is possible to construct a measure μ and associated volume element $dg \doteq d\mu(g)$ around each $g \in G$ such that given any function $f : G \to \mathbb{C}$ whose measure

$$\mu(f) = \int_G f(g)\, dg$$

exists, the following invariance properties will hold

$$\int_G f(g)\, dg = \int_G f(g^{-1})\, dg = \int_G f(h \circ g)\, dg = \int_G f(g \circ h)\, dg \tag{1}$$

for arbitrary $h \in G$. Here, of course, g^{-1} is the inverse of $g \in G$, which is the unique element such that

$$g \circ g^{-1} = g^{-1} \circ g = e$$

with e being the identity element, which for every $g \in G$ satisfies

$$g \circ e = e \circ g = g\,.$$

This is the unique element of G with such properties.

The equalities in (1) are analogous to the properties of the Lebesgue integral

$$\int_{\mathbb{R}^n} f(\mathbf{x})\, d\mathbf{x} = \int_{\mathbb{R}^n} f(-\mathbf{x})\, d\mathbf{x} = \int_{\mathbb{R}^n} f(\mathbf{y} + \mathbf{x})\, d\mathbf{x} = \int_{\mathbb{R}^n} f(\mathbf{x} + \mathbf{y})\, d\mathbf{x}\,.$$

All compact Lie groups are unimodular, as are all nilpotent and semisimple Lie groups. When referring to Lie groups in this paper, the discussion is restricted to matrix Lie groups with elements that are square matrices, and group operation, $\circ$, being matrix multiplication and the identity element is the identity matrix in the case of a matrix Lie group. In this context, the set of $n \times n$ unitary matrices, $U(n)$, and all of its subgroups are compact Lie groups. An example of a noncompact semisimple Lie group is the much-studied $SL(2, \mathbb{R})$ consisting of all 2×2 matrices with real entries and unit determinant [3–6]. And an example of a nilpotent group is the Heisenberg group, $\mathcal{H}(3)$, consisting of matrices of the form

$$H(\alpha, \beta, \gamma) = \begin{pmatrix} 1 & \alpha & \beta \\ 0 & 1 & \gamma \\ 0 & 0 & 1 \end{pmatrix}$$

where $\alpha, \beta, \gamma \in \mathbb{R}$. Probability distributions, harmonic analysis, and diffusion processes on this group have been studied in detail [7, 8].

As a nontrivial example of a noncompact unimodular Lie group that arises frequently in engineering applications, and which is neither semisimple nor nilpotent, consider the Special Euclidean group, $SE(n)$, which consists of elements of the form $g = (R, \mathbf{t}) \in SO(n) \times \mathbb{R}^n$ with the semi-direct product group law

$$(R_1, \mathbf{t}_1) \circ (R_2, \mathbf{t}_2) = (R_1 R_2, R_1 \mathbf{t}_2 + \mathbf{t}_1).$$

Here $R_i \in SO(n)$, the special orthogonal group consisting of $n \times n$ rotation matrices, and the resulting semi-direct product group is denoted as

$$SE(n) = \mathbb{R}^n \rtimes SO(n).$$

Building on the classic works of Miller [9, 10] and Vilenkin [11–13], the author has published extensively on harmonic analysis and diffusion processes on this group in the context of applications in robotics and polymer science [14–16]. Detailed treatment of these topics can be found in [1, 2], and a recent concise summary can be found in [17]. In order to avoid repetition, examples in the present paper are instead illustrated with $\mathcal{H}(3)$ and $SO(3)$, though the general formulation is kept abstract and general, in the spirit of [18–25].

In the context of unimodular Lie groups, it then makes sense to consider probability density functions, i.e., $f : G \to \mathbb{R}$ with the properties

$$f(g) \geq 0 \text{ and } \int_G f(g)\, dg = 1.$$

Moreover, the concept of entropy of a pdf is simply

$$S(f) \doteq -\int_G f(g) \log f(g)\, dg, \tag{2}$$

and an entropy power is

$$N(f) \doteq \frac{1}{2\pi e} \exp\left(\frac{2}{\dim(G)} S(f)\right).$$

For unimodular Lie groups, a well-defined concept of convolution of functions in $(L^1 \cap L^2)(G)$ exists:

$$(f_1 * f_2)(g) \doteq \int_G f_1(h) f_2(h^{-1} \circ g)\, dh. \tag{3}$$

This inherits the associative property from G:

$$((f_1 * f_2) * f_3)(g) = (f_1 * (f_2 * f_3))(g).$$

Moreover, for broad classes of unimodular Lie groups, including all compact Lie groups and group extensions such as $SE(n)$, it is possible to define a concept of Fourier transform. This is based on the concept of an irreducible unitary representation (IUR) of G. An IUR is a unitary operator (which can be thought of as a square matrix of either finite or infinite dimension) with the properties

$$U(g_1 \circ g_2, \lambda) = U(g_1, \lambda)\, U(g_2, \lambda) \quad \text{and} \quad U(g^{-1}, \lambda) = U^*(g, \lambda)$$

where λ can be thought of as a frequency parameter and $*$ denotes the Hermitian conjugate. The space of all Λ values is called the unitary dual of G. In the case when G is Abelian, the unitary dual is also a group. In the case of compact Lie groups, Λ is a discrete space with countable elements. In the case of some noncompact unimodular Lie groups the space Λ has been fully characterized, and its description can be a quite complicated requiring a combination of continuous and discrete parameters.

Within this context the Fourier transform is defined as

$$\hat{f}(\lambda) = \int_G f(g)\, U(g^{-1}, \lambda)\, dg\,.$$

As with classical Fourier analysis, there is a convolution theorem

$$(\widehat{f_1 * f_2})(\lambda) = \hat{f}_2(\lambda)\, \hat{f}_1(\lambda)$$

and a reconstruction formula

$$f(g) = \int_\Lambda \mathrm{tr}\left[\hat{f}(\lambda)\, U(g, \lambda)\right] d\lambda\,.$$

Combining the above gives

$$(f_1 * f_2)(e) = \int_\Lambda \mathrm{tr}\left[\hat{f}_2(\lambda)\, \hat{f}_1(\lambda)\right] d\lambda = (f_2 * f_1)(e)\,.$$

Moreover, the Plancherel equality gives

$$\int_G f_1(g)\overline{f_2(g)}\, dg = \int_\Lambda \mathrm{tr}\left[\hat{f}_1(\lambda)\, \hat{f}_2^*(\lambda)\right] d\lambda\,.$$

When $f_1 = f_2 = f$, this becomes the familiar form of Parseval's equality

$$\int_G |f(g)|^2\, dg = \int_\Lambda \left\|\hat{f}(\lambda)\right\|_{HS}^2 d\lambda\,,$$

where $\|A\|_{HS}^2 = \mathrm{tr}(AA^*)$ is the familiar Hilbert-Schmidt norm.

In the context of Lie groups there are also natural generalizations of the concept of partial derivatives. Namely, if $X \in \mathcal{G}$ (the Lie algebra corresponding to the Lie group G), then a (left-invariant) directional derivative of $f(g)$ is computed as

$$(\tilde{X}f)(g) \doteq \frac{d}{dt} f\left(g \circ e^{tX}\right)\Big|_{t=0} .$$

If $\{E_i\}$ is a basis for $\mathcal{G}$, then $(\tilde{E}_i f)(g)$ can be thought of as partial derivatives, and the derivative in the direction $X = \sum_i x_i E_i$ can be written as

$$(\tilde{X}f)(g) = \sum_i x_i (\tilde{E}_i f)(g) .$$

Making a choice $\{E_i\}$ and defining an inner product by imposing orthonormality conditions $(E_i, E_j) = \delta_{ij}$ in effect fixes a metric for $\mathcal{G}$, which can be transported by left or right action to define a metric on G.

Operational properties of the Fourier transform include

$$\widehat{(\tilde{X}f)}(\lambda) = u(X, \lambda)\, \hat{f}(\lambda)$$

where

$$u(X, \lambda) \doteq \frac{d}{dt} U(e^{tX}, \lambda)\Big|_{t=0} .$$

This matrix function is linear in X, and so

$$u\left(\sum_i x_i E_i, \lambda\right) = \sum_i x_i u(E_i, \lambda) .$$

The concept of a Fisher information matrix with elements

$$F_{ij}(f) \doteq \int_G \frac{(\tilde{E}_i f)(g)(\tilde{E}_j f)(g)}{f(g)}\, dg$$

and a diffusion process on G can be described with an equation of the form[1]

$$\frac{\partial f}{\partial t} = -\sum_{i=1}^{dim(G)} h_i \tilde{E}_i f + \frac{1}{2} \sum_{i,j=1}^{dim(G)} D_{ij} \tilde{E}_i \tilde{E}_j f . \tag{4}$$

[1]The diffusion coefficients are D_{ij} and the drift coefficients are h_i. When $h_i = 0$ for all $i \in \{1, \ldots, dim(G)\}$ the diffusion process is called driftless.

With all of this in mind, it becomes possible to explore generalizations of inequalities from classical information theory. Two major hindrances to trivially extending these inequalities to the noncommutative case are: (1) for general functions on G,

$$(f_1 * f_2)(g) \neq (f_2 * f_1)(g);$$

and, (2) unlike in the case of $\mathbb{R}^n$ where the Gaussian distribution is simultaneously (a) the maximal entropy distribution subject to covariance constraints, (b) is closed under convolution and conditioning, and (c) solves the Euclidean-space version of (4), these attributes cannot in general be simultaneously satisfied for more general Lie groups. For example, the Entropy-Power Inequality (EPI) is not even true for the circle group $\mathbb{R}/\mathbb{Z} \cong SO(2)$.

This paper summarizes what is already known, introduces some new results, and poses some questions for future exploration. For additional background on convolution, entropy, and Fourier analysis (in the Euclidean on Lie group cases), see [26–38].

1.2 Structure of the Paper

Section 2 explains how to compute the integration measure for unimodular Lie groups. Section 3 reviews the concept of functions of positive type on unimodular Lie groups, and the resulting properties of Fourier matrices for such functions. Section 4 explains why trace inequalities are significant in the harmonic analysis of diffusion processes on noncommutative unimodular Lie groups, and reviews some of the most well-known trace inequalities and various conjectures put forth in the literature. Section 5 reviews how Fisher information arises in quantifying entropy increases under diffusion and reviews a generalization the de Bruijn identity, which is shown to hold for unimodular Lie groups in general. This too involves trace inequalities. Section 6 reviews definitions of mean of covariance of probability densities on unimodular Lie groups, and how the propagate under convolution. Section 7 illustrates the theory with specific examples ($SO(3)$ as an example of compact Lie groups and the Heisenberg group $\mathcal{H}(3)$ as an example of a noncommutative noncompact unimodular Lie group.)

2 Explicit Computation of the Bi-invariant Integration Measure

This section explains how to compute the bi-invariant integration measure for a unimodular Lie group, summarizing the discussion in [1, 2].

To begin, an inner product $(\cdot, \cdot)$ between arbitrary elements of the Lie algebra, $Y = \sum_i y_i E_i$ and $Z = \sum_j z_j E_j$, can be defined such that

$$(Y, Z) \doteq \sum_{i=1}^{n} y_i z_i \qquad \text{where} \qquad (E_i, E_j) = \delta_{ij}. \tag{5}$$

The basis $\{E_i\}$ is then orthonormal with respect to this inner product. The definition of the inner product together with the constraint of orthonormality of a particular choice of basis $\{E_i\}$ in (5) defines a metric tensor for the Lie group.

Let $\mathbf{q} = [q_1, \ldots, q_n]^T$ be a column vector of local coordinates. Then $g(t) = \tilde{g}(\mathbf{q}(t))$ is a curve in G where $\tilde{g} : \mathbb{R}^n \to G$ is the local parametrization of the Lie group G. Henceforth the tilde will be dropped since it will be clear from the argument whether the function $g(t)$ or $g(\mathbf{q})$ is being referred to. The right-Jacobian matrix[2] for an n-dimensional Lie group parameterized with local coordinates $q_1, \ldots, q_n$ is the matrix $J_r(\mathbf{q})$ that relates rates of change $\dot{\mathbf{q}}$ to $g^{-1}\dot{g}$, and likewise for $J_l(\mathbf{q})$ and $\dot{g}g^{-1}$, where a dot denotes d/dt. Specifically,

$$\dot{g}g^{-1} = \sum_j \omega_j^l E_j \qquad \text{and} \qquad \boldsymbol{\omega}^l = J_l(\mathbf{q})\dot{\mathbf{q}}$$

and

$$g^{-1}\dot{g} = \sum_j \omega_j^r E_j \qquad \text{and} \qquad \boldsymbol{\omega}^r = J_r(\mathbf{q})\dot{\mathbf{q}}.$$

In other words,

$$(\dot{g}g^{-1}, E_k) = \left(\sum_j \omega_j^l E_j, E_k \right) = \sum_j \omega_j^l \left(E_j, E_k\right) = \sum_j \omega_j^l \delta_{jk} = \omega_k^l.$$

The scalars ω_k^l can be stacked in an array to form the column vector $\boldsymbol{\omega}^l = [\omega_1^l, \omega_2^l, \ldots, \omega_n^l]^T$. Analogous calculations follow for the "r" case. This whole process is abbreviated with the "$\vee$" operation as

$$\left(\dot{g}g^{-1}\right)^\vee = \boldsymbol{\omega}^l \qquad \text{and} \qquad \left(g^{-1}\dot{g}\right)^\vee = \boldsymbol{\omega}^r. \tag{6}$$

Given an orthogonal basis $E_1, \ldots, E_n$ for the Lie algebra, projecting the left and right tangent operators onto this basis yields elements of the right- and left-Jacobian matrices[3]:

[2]Here 'right' and 'left' respectively refer to differentiation appearing on the right or left side in calculations. As such a 'right' quantity denoted with a subscript r is left invariant, and a 'left' quantity denoted with a subscript l is right invariant.

[3]The 'l' and 'r' convention used here for Jacobians and for vector fields is opposite that used in the mathematics literature. The reason for the choice made here is to emphasize the location of the "the most informative part" of the expression. In Jacobians, this is the location of the partial derivatives. In vector fields this is where the components defining the field appear.

$$(J_r)_{ij} = \left(g^{-1}\frac{\partial g}{\partial q_j}, E_i\right) \quad \text{and} \quad (J_l)_{ij} = \left(\frac{\partial g}{\partial q_j}g^{-1}, E_i\right). \tag{7}$$

In terms of the $\vee$ operation this is written as

$$\left(g^{-1}\frac{\partial g}{\partial q_j}\right)^{\vee} = J_r(\mathbf{q})\,\mathbf{e}_j \quad \text{and} \quad \left(\frac{\partial g}{\partial q_j}g^{-1}\right)^{\vee} = J_l(\mathbf{q})\,\mathbf{e}_j.$$

As another abuse of notation, the distinction between $J(\mathbf{q})$ and $J(g(\mathbf{q}))$ can be blurred in both the left and right cases. Again, it is clear which is being referred to from the argument of these matrix-valued functions.

Note that $J_r(h \circ g) = J_r(g)$ and $J_l(g \circ h) = J_l(g)$. For unimodular Lie groups,

$$|\det(J_r)(\mathbf{q})| = |\det(J_l)(\mathbf{q})| \quad \text{and} \quad dg = |\det(J_{r,l})(\mathbf{q})|\,d\mathbf{q}\,. \tag{8}$$

This dg has the bi-invariance property, and is called the *Haar measure*. Examples of how this looks in different coordinates are given for $\mathcal{H}(3)$ and $SO(3)$ in Sect. 7. In the compact case, it is always possible to find a constant c to normalize as $d'g \doteq c \cdot dg$ such that $\int_G d'g = 1$.

3 Functions of Positive Type

In harmonic analysis, a function $\varphi : G \to \mathbb{C}$ is called a *function of positive type* if for every $c_i \in \mathbb{C}$ and every $g_i, g_j \in G$ and any $n \in \mathbb{Z}_{>0}$ the inequality

$$\sum_{i,j=1}^{n} c_i \overline{c_j}\, \varphi(g_i \circ g_j^{-1}) \geq 0\,.$$

In some texts, such functions are also called *positive definite*, whereas in others that term is used only when the inequality above excludes equality except when all values of c_i are zero. Here a function of positive type will be taken to be one for which the matrix $M = [m_{ij}]$ with entries $m_{ij} \doteq \varphi(g_i \circ g_j^{-1})$ is Hermitian positive semi-definite (which can be shown to be equivalent to the above expression), and a positive definite function is one for which M is positive definite.

Some well-known properties of functions of positive type include [19, 21, 23]:

$$\begin{aligned} \varphi(e) &= \overline{\varphi(e)} \geq 0 \\ |\varphi(g)| &\leq \varphi(e) \\ \varphi(g^{-1}) &= \overline{\varphi(g)}\,. \end{aligned}$$

Moreover, if φ_1 and φ_2 are two such functions, then so are $\overline{\varphi_i}$, $\varphi_1 \cdot \varphi_2$, as are linear combinations of the form $a_1\varphi_1 + a_2\varphi_2$ where $a_i \in \mathbb{R}_{>0}$.

Clearly, if φ is a function constructed as

$$\varphi(g;\lambda) \doteq \mathrm{tr}\left[A^* U(g,\lambda) A\right]$$

when A is positive definite, then

$$\sum_{i,j=1}^{n} c_i \overline{c_j}\, \varphi(g_i \circ g_j^{-1}) = \sum_{i,j=1}^{n} c_i \overline{c_j}\, \mathrm{tr}\left[A^* U(g_i \circ g_j^{-1},\lambda) A\right]$$

$$= \left\| A^* \sum_{i=1}^{n} U(g_i,\lambda) \right\|_{HS}^{2} \geq 0$$

because

$$U(g_i \circ g_j^{-1},\lambda) = U(g_i,\lambda)\, U^*(g_j,\lambda)\,.$$

And hence $\varphi(g;\lambda)$ is a function of positive type. Moreover, by the same reasoning, if $f(g)$ is any functions for which $\hat{f}(\lambda)$ is a Hermitian positive definite matrix, then $f(g)$ will be a positive definite function. And, according to Hewitt and Ross [21, 23] (p. 683 Lemma D.12), if A and B are both positive definite matrices, then so is their product. This has implications regarding the positivity of the convolution of positive functions on a group.

In particular, if $\rho_t(g) = \rho(g;t)$ is the solution to a driftless diffusion equation with Dirac-delta initial conditions, then the Fourier-space solution is written as

$$\hat{\rho}(\lambda;t) = \exp\left[\frac{1}{2}\sum_{i,j=1}^{dim(G)} D_{ij} u(E_i,\lambda) u(E_j,\lambda)\right],$$

which is Hermitian positive definite, and hence $\rho_t(g)$ is a real-valued positive definite function for each value of $t \in \mathbb{R}_{\geq 0}$. Moreover,[4]

$$\rho_t(g) = \rho_t(g^{-1})\,.$$

It is not difficult to show that given two symmetric functions, $\rho_1(g) \doteq \rho_{t_1}(g;D_1)$ and $\rho_2(g) \doteq \rho_{t_2}(g;D_2)$, that

$$(\rho_1 * \rho_2)(g) = (\rho_2 * \rho_1)(g^{-1})\,.$$

Though this does not imply that $(\rho_1 * \rho_2)(g)$ is symmetric, it is easy to show that $(\rho_1 * \rho_2 * \rho_1)(g)$ is symmetric.

Moreover, if $f : G \to \mathbb{R}_{\geq 0}$ is a pdf which is not symmetric, it is not difficult to show that

---

[4]Here the dependence on $D = [D_{ij}]$ has been suppressed, but really $\rho_t(g) = \rho_t(g;D)$.

$$f'(g) \doteq \frac{f(g) + f(g^{-1})}{2}$$

and

$$f''(g) \doteq \frac{f(g) f(g^{-1})}{(f * f)(e)}$$

are symmetric pdfs.

For any positive definite symmetric pdf, the Fourier transform is a positive definite Hermitian matrix because

$$\begin{aligned}\hat{\rho}(\lambda) &= \int_G \rho(g)\, U(g^{-1}, \lambda)\, dg = \int_G \rho(g^{-1})\, U(g^{-1}, \lambda)\, dg \\ &= \int_G \rho(g)\, U(g, \lambda)\, dg = \int_G \rho(g)\, U^*(g^{-1}, \lambda)\, dg = \hat{\rho}^*(\lambda)\,.\end{aligned}$$

From positive definiteness, it is possible to write

$$\hat{\rho}(\lambda) = \exp H(\lambda)$$

where H is Hermitian, though not necessarily positive definite.

Moreover, every special unitary matrix can be expressed as the exponential of a skew-Hermitian matrix, and even more than that, if $g = \exp X$, then the IUR matrix $U(g, \lambda)$ can be computed as

$$U(g, \lambda) \;=\; \exp Z(\log(g), \lambda)$$

where

$$Z(X, \lambda) = \sum_{i=1}^{dim(G)} x_i u(E_i, \lambda) = -Z^*(X, \lambda)\,.$$

In analogy with the way that the exponential map for a matrix Lie group is simply the matrix exponential defined by the Taylor series, here and throughout this work the logarithm is the matrix logarithm defined by its Taylor series.

In this light, the Fourier inversion formula has in it the evaluation of

$$\mathrm{tr}\left[\exp H \exp Z\right]\,,$$

and the evaluation of $(\rho_1 * \rho_2)(e)$ has in it

$$\mathrm{tr}\left[\exp H_1 \exp H_2\right]\,.$$

Also, in the evaluation of probability densities for diffusion processes with drift, it is desirable to find approximations of the form

$$\operatorname{tr}\left[\exp(H+Z)\right] \approx \operatorname{tr}\left[\exp H' \exp Z'\right] .$$

For these reasons, there are connections between harmonic analysis on unimodular Lie groups and trace inequalities.

4 Trace Inequalities

In the Fourier reconstruction formula for a diffusion process on a Lie group, the trace of the product of exponentials of two matrices is computed. It is therefore relevant to consider: (1) when can the product of two exponentials be simplified; (2) even when the product cannot be simplified, it would be useful to determine when the trace operation has the effect of simplifying the result. For example, if A and B are bandlimited matrices and $\operatorname{tr}(e^A e^B) \approx \operatorname{tr}(\exp(A+B))$ then computing the eigenvalues of $A+B$, exponentiating each eigenvalue, and summing potentially could be much faster than directly exponentiating the matrices, and then taking the trace of the product. The statements that follow therefore may have some relevance to the rapid evaluation of the Fourier inversion formula for diffusion processes on Lie groups. Other matrix inequalities that may be applicable to extend the current analysis include [39–52].

4.1 Generalized Golden–Thompson Inequalities

For $n \times n$ Hermitian matrices A and B, the *Golden–Thompson inequality* [53–55] is

$$\varphi(e^A e^B) \geq \varphi(e^{A+B}) \tag{9}$$

where φ is one of a large number of so-called spectral functions. For the case when $\varphi(\cdot) = \operatorname{tr}(\cdot)$ (which is the case of primary interest in this chapter) this was proven in [54], and generalized in [56].

The Thompson Conjecture [57]: If H and K are Hermitian matrices, there exist unitary matrices U and V dependent on H and K such that

$$e^{iH} e^{iK} = e^{i(UHU^* + VKV^*)}. \tag{10}$$

The So-Thompson Conjecture [58]: If H and K are Hermitian matrices, there exist unitary matrices U and V dependent on H and K such that

$$e^{H/2} e^K e^{H/2} = e^{UHU^* + VKV^*}. \tag{11}$$

Interestingly, such conjectures have been proven [59] using techniques associated with random walks on symmetric space of Lie groups [60], thereby bringing the problem back to the domain of interest in this chapter.

In [61], Cohen et al. prove that *spectral matrix functions*,[5] $\varphi : \mathbb{C}^{n\times n} \to \mathbb{C}$ (including the trace) satisfy the inequality

$$\varphi(e^{(A+A^*)/2}e^{(B+B^*)/2}) \geq |\varphi(e^{A+B})| \tag{12}$$

when the following condition holds:

$$\varphi([XX^*]^s) \geq |\varphi(X^{2s})| \qquad \forall X \in \mathbb{C}^{n\times n} \qquad \text{for} \qquad s = 1, 2, \ldots$$

An interesting corollary to (12) is that if A is skew-Hermitian, and $B \in \mathbb{C}^{n\times n}$, then [61]:

$$\varphi(e^{(B+B^*)/2}) \geq |\varphi(e^{A+B})|. \tag{13}$$

Bernstein proved the following general statement for $A \in \mathbb{C}^{n\times n}$ [62]:

$$\mathrm{tr}(e^{A^*}e^{A}) \leq \mathrm{tr}(e^{A^*+A}) \tag{14}$$

Inequalities involving functions of products of exponentials have a long history (see e.g., [63, 64]) and remain an area of active investigation. A few recent papers include [65–67].

4.2 *Matrix Inequalities from Systems Theory*

Another sort of matrix inequality that may be useful would be extensions of results that comes from systems theory. For example, it is known that if $A, B \in \mathbb{R}^{n\times n}$ and $B = B^T > 0$ then [68]

$$\lambda_n(\hat{A})\mathrm{tr}(B) \leq \mathrm{tr}(AB) \leq \lambda_1(\hat{A})\mathrm{tr}(B) \tag{15}$$

where $\hat{A} = (A + A^T)/2$ and the eigenvalues are ordered as $\lambda_1 \geq \lambda_2 \geq \cdots \geq \lambda_n > 0$. This has been tightened by Fang et al. [69]:

$$\lambda_n(\hat{A})\mathrm{tr}(B) - \lambda_n(B)[n\lambda_n(\hat{A}) - \mathrm{tr}(A)] \leq \mathrm{tr}(AB) \leq \lambda_1(\hat{A})\mathrm{tr}(B) - \lambda_n(B)[n\lambda_1(\hat{A}) - \mathrm{tr}(A)] \tag{16}$$

Additional modifications have been made by Park [70].

Under the same conditions on A and B, Komaroff and Lasserre independently derived the inequality [71, 72]:

$$\sum_{i=1}^{n} \lambda_i(\hat{A})\lambda_{n-i+1}(B) \leq \mathrm{tr}(AB) \leq \sum_{i=1}^{n} \lambda_i(\hat{A})\lambda_i(B) \tag{17}$$

[5] These are functions that depend only the eigenvalues of a matrix, and are therefore invariant under similarity transformations.

and Lasserre tightened this with the result [73]:

$$f(-\epsilon) \leq \operatorname{tr}(AB) \leq f(\epsilon) \qquad \forall \epsilon > 0 \tag{18}$$

where

$$f(\epsilon) \doteq \frac{1}{\epsilon} \sum_{i=1}^{n} [\lambda_i(B + \epsilon \hat{A})\lambda_i(B) - \operatorname{tr}(B^2)].$$

Apparently, when $A \in \mathbb{C}^{n \times n}$ and $B = B^* \in \mathbb{C}^{n \times n}$ (17) holds with the substitutions $A^T \to A^*$ and $\operatorname{tr}(AB) \to \operatorname{Re}[\operatorname{tr}(AB)]$ [74]. Generalized formulas for products of arbitrary real matrices have been made recently [74–76].

4.3 Classical Matrix Inequalities

In a sense, inequalities of the form in the previous section can be traced back to work done as early as the 1930s. Mirsky [77] attributes the following result for arbitrary $A, B \in \mathbb{C}^{n \times n}$ to a 1937 paper by John von Neumann:

$$|\operatorname{tr}(AB)| \leq \sum_{i=1}^{n} \mu_i(A)\mu_i(B) \tag{19}$$

where $\mu_1(A) \geq \mu_2(A) \geq \cdots \geq \mu_n(A)$ are the singular values of A.

Hoffman and Wieland [78] states that for $n \times n$ normal matrices A and B (i.e., $A^*A = AA^*$ and $B^*B = BB^*$) permutations $\pi, \sigma \in \Pi_n$ can be found such that

$$\sum_{i=1}^{n} |\lambda_i(A) - \lambda_{\pi(i)}|^2 \leq \|A - B\|^2 \leq \sum_{i=1}^{n} |\lambda_i(A) - \lambda_{\sigma(i)}|^2 \tag{20}$$

where $\|A\|^2 = \operatorname{tr}(AA^*)$ is the Frobenius norm. For generalizations see [79]. In particular, Richter [80] and Mirsky [81] have shown that if A and B are both $n \times n$ Hermitian matrices,

$$\sum_{i=1}^{n} \lambda_i(A)\lambda_{n+1-i}(B) \leq \operatorname{tr}(AB) \leq \sum_{i=1}^{n} \lambda_i(A)\lambda_i(B) \tag{21}$$

and Marcus [82] showed that for normal matrices A and B, there exist permutations π and σ for which

$$\sum_{i=1}^{n} \lambda_i(A)\lambda_{\pi(i)}(B) \leq \operatorname{Re}[\operatorname{tr}(AB)] \leq \sum_{i=1}^{n} \lambda_i(A)\lambda_{\sigma(i)}(B) \tag{22}$$

4.4 The Arithmetic-Mean-Geometric-Mean (AM-GM) Inequality

The *arithmetic-geometric-mean inequality* states that the arithmetic mean of a set of positive real numbers is always less than the geometric mean of the same set of numbers:

$$\frac{1}{n}\sum_{i=1}^{n}\lambda_i \geq \prod_{i=1}^{n}\lambda_i^{\frac{1}{n}}. \tag{23}$$

This fact can be used to derive many useful inequalities. For example, Steele [83] uses the AM-GM inequality to derive a reverse Cauchy-Schwarz inequality of the form

$$\left(\sum_{k=1}^{n}a_k^2\right)^{\frac{1}{2}}\left(\sum_{k=1}^{n}b_k^2\right)^{\frac{1}{2}} \leq \frac{m+M}{2\sqrt{mM}}\sum_{k=1}^{n}a_k b_k \tag{24}$$

where $\{a_k\}, \{b_k\} \subset \mathbb{R}_{>0}$ and $0 < m \leq a_k/b_k \leq M < \infty$ for all $k \in \{1, \ldots, n\}$.

It is no coincidence that the numbers in (23) are denoted as λ_i because when they are interpreted as the eigenvalues of a positive definite Hermitian matrix, $A = A^* > 0$, and so

$$\frac{1}{n}\text{tr}(A) \geq |A|^{\frac{1}{n}}. \tag{25}$$

This is useful for bounding the trace of the matrix exponential of a not-necessarily-positive-definite Hermitian matrix, $H = H^*$, since $A = \exp H = A^* > 0$ can be substituted into (25). The determinant-trace equality for the matrix exponential $\det(\exp(H)) = e^{\text{tr}(H)}$ then gives $|\exp H|^{\frac{1}{n}} = |e^{\text{tr}(H)}|^{\frac{1}{n}}$ and so

$$\frac{1}{n}\text{tr}(\exp H) \geq e^{\text{tr}(H)/n}. \tag{26}$$

Though (25) is more fundamental than (26), the latter is directly useful in studying properties of diffusion processes on Lie groups.

It is also interesting to note that (25) generalizes in several ways. For example, if

$$\mu_p(\lambda_1, \ldots, \lambda_n) \doteq \left(\frac{1}{n}\sum_{i=1}^{n}\lambda_i^p\right)^{\frac{1}{p}}$$

then the AM-GM inequality can be stated as

$$\lim_{p\to 0}\mu_p(\lambda_1, \ldots, \lambda_n) \leq \mu_1(\lambda_1, \ldots, \lambda_n)$$

and more generally,

$$\mu_p(\lambda_1, \ldots, \lambda_n) \leq \mu_q(\lambda_1, \ldots, \lambda_n) \qquad p < q. \tag{27}$$

For each fixed choice of $\lambda_1, \ldots, \lambda_n$, the function $f : \mathbb{R} \to \mathbb{R}_{\geq 0}$ defined by $f(p) = \mu_p(\lambda_1, \ldots, \lambda_n)$ is an increasing function.

When $p = -1$, the *harmonic mean*

$$\mu_{-1}(\lambda_1, \ldots, \lambda_n) = n \cdot \left[\sum_{i=1}^{n} \frac{1}{\lambda_i}\right]^{-1}$$

results, and (27) for $p = -1$ and $q = 0$ implies that

$$n \cdot \left[\sum_{i=1}^{n} \frac{1}{\lambda_i}\right]^{-1} \leq \left[\prod_{i=1}^{n} \lambda_i\right]^{\frac{1}{n}}.$$

The implication of this for positive definite Hermitian matrices is that

$$n \cdot \mathrm{tr}[A^{-1}] \leq [\det A]^{\frac{1}{n}} \iff \mathrm{tr} A \leq \frac{1}{n}[\det A]^{-\frac{1}{n}}. \tag{28}$$

One generalization of (23) is the *weighted AM-GM inequality*

$$\frac{1}{\alpha}\sum_{i=1}^{n} \alpha_i \lambda_i \geq \left(\prod_{i=1}^{n} \lambda_i^{\alpha_i}\right)^{\frac{1}{\alpha}} \quad \text{where} \quad \alpha = \sum_{i=1}^{n} \alpha_i \;,\; \alpha_i \in \mathbb{R}_{>0}. \tag{29}$$

Another generalization is *Ky Fan's inequality* [84]:

$$\frac{\frac{1}{n}\sum_{i=1}^{n} \lambda_i}{\frac{1}{n}\sum_{i=1}^{n}(1-\lambda_i)} \geq \frac{\prod_{i=1}^{n} \lambda_i^{\frac{1}{n}}}{\prod_{i=1}^{n}(1-\lambda_i)^{\frac{1}{n}}} \tag{30}$$

which holds for $0 \leq \lambda_i \leq \frac{1}{2}$. If the numbers λ_i are viewed as the eigenvalues of a matrix, A, then Ky Fan's inequality can be written as

$$\frac{\mathrm{tr} A}{\mathrm{tr}(\mathbb{I} - A)} \geq \frac{|A|^{\frac{1}{n}}}{|\mathbb{I} - A|^{\frac{1}{n}}} \quad \text{or} \quad \mathrm{tr} A \geq \frac{n}{1 + |A|^{-\frac{1}{n}}|\mathbb{I} - A|^{\frac{1}{n}}}. \tag{31}$$

Of course this statement should then be restricted to those matrices that have real eigenvalues that obey $0 \leq \lambda_i(A) \leq \frac{1}{2}$.

4.5 Consequences for Harmonic Analysis and Diffusion Processes

If $\rho_{D^{(k)}}(g, t)$ denotes the solution to the driftless diffusion equation on G with diffusion coefficients $D_{ij}^{(k)}$, subject to initial conditions $\rho_{D^{(k)}}(g, 0) = \delta(g)$, then from the Golden–Thompson inequality

$$(\rho_{D^{(1)}} * \rho_{D^{(2)}})(e; t) \geq \rho_{D^{(1)}+D^{(2)}}(e; t) . \tag{32}$$

Alternatively, using the fact that $B^{\frac{1}{2}} A B^{\frac{1}{2}}$ is Hermitian positive definite whenever A and B are, the trace of the product $\text{tr}[AB] = \text{tr}[B^{\frac{1}{2}} AB]$ can be bounded using (25), resulting in

$$\frac{1}{n}\text{tr}(AB) \geq |A|^{\frac{1}{n}} |B|^{\frac{1}{n}} . \tag{33}$$

This can then be used to bound $(\rho_{D^{(1)}} * \rho_{D^{(2)}})(e; t)$ from below as well. But it can also be used in a different way. Given a diffusion with drift, the Fourier matrices will be of the form $\exp(H + Z)$ where $H = H^*$ and $Z = -Z^*$. Then the convolution of two diffusions with drifts being the negative of each other will be $\exp(H + Z) \exp(H - Z)$, which is Hermitian, and hence

$$\frac{1}{n}\text{tr}(\exp(H + Z) \exp(H - Z)) \geq |\exp(H + Z)|^{\frac{1}{n}} |\exp(H - Z)|^{\frac{1}{n}} . \tag{34}$$

Then from the determinant-trace equality, we can simplify

$$|\exp(H + Z)| = e^{\text{tr}(H+Z)} \quad \text{and} \quad |\exp(H - Z)| = e^{\text{tr}(H-Z)} ,$$

thereby giving that

$$\frac{1}{n}\text{tr}(\exp(H + Z) \exp(H - Z)) \geq e^{\frac{2}{n}\text{tr}(H)} . \tag{35}$$

These will be demonstrated in Sect. 7.

In the case when $G = \mathbb{R}^n$, covariances add under convolution and for a diffusion $\Sigma^{(k)} = D^{(k)} t$, and so

$$(\rho_{D^{(1)}} * \rho_{D^{(2)}})(\mathbf{0}; t) = \frac{1}{(2\pi t)^{n/2} \left|D^{(1)} + D^{(2)}\right|^{\frac{1}{2}}} = \rho_{D^{(1)}+D^{(2)}}(\mathbf{0}; t) .$$

This begs the question of how to define and propagate covariances on a unimodular Lie group, and what relationships may exist with Fisher information. Inequalities relating Fisher information and entropy are reviewed in Sect. 5, followed by definitions of covariance in Sect. 6 and the relationship be Fisher information and covariance.

5 Inequalities Involving Fisher Information and Diffusion Processes

This section connects trace inequalities with Fisher information and the rate of entropy increase under a diffusion process. The results presented here are an abridged version of those presented in [2].

5.1 *Rate of Increase of Entropy Under Diffusion*

The entropy of a pdf on a Lie group is defined in (2) If $f(g,t)$ is a pdf that satisfies a diffusion equation (regardless of the details of the initial conditions) then some interesting properties of $S_f(t)$ can be studied. In particular, if $\dot{S}_f = dS_f/dt$, then differentiating under the integral sign gives

$$\dot{S}_f = -\int_G \left\{ \frac{\partial f}{\partial t} \log f + \frac{\partial f}{\partial t} \right\} dg.$$

But from the properties of a diffusion equation,

$$\int_G \frac{\partial f}{\partial t}\, dg = \frac{d}{dt} \int_G f(g,t)\, dg = 0,$$

and so the second term in the above braces integrates to zero.

Substitution of

$$\frac{\partial f}{\partial t} = \frac{1}{2} \sum_{i,j=1}^{n} D_{ij} \tilde{E}_i^r \tilde{E}_j^r f - \sum_{k=1}^{n} h_k \tilde{E}_k^r f$$

into the integral for $\dot{S}_f$ gives

$$\begin{aligned}
\dot{S}_f &= -\int_G \left\{ \frac{1}{2} \sum_{i,j=1}^{n} D_{ij} \tilde{E}_i^r \tilde{E}_j^r f - \sum_{k=1}^{n} h_k \tilde{E}_k^r f \right\} \log f\, dg \\
&= -\frac{1}{2} \sum_{i,j=1}^{n} D_{ij} \int_G (\tilde{E}_i^r \tilde{E}_j^r f) \log f\, dg - \sum_{k=1}^{n} h_k \int_G (\tilde{E}_k^r f) \log f\, dg \\
&= \frac{1}{2} \sum_{i,j=1}^{n} D_{ij} \int_G (\tilde{E}_j^r f)(\tilde{E}_i^r \log f)\, dg + \sum_{k=1}^{n} h_k \int_G f\, (\tilde{E}_k^r \log f)\, dg \\
&= \frac{1}{2} \sum_{i,j=1}^{n} D_{ij} \int_G \frac{1}{f} (\tilde{E}_j^r f)(\tilde{E}_i^r f)\, dg + \sum_{k=1}^{n} h_k \int_G \tilde{E}_k^r f\, dg \\
&= \frac{1}{2} \sum_{i,j=1}^{n} D_{ij} \int_G \frac{1}{f} (\tilde{E}_j^r f)(\tilde{E}_i^r f)\, dg \\
&\geq 0
\end{aligned}$$

5.2 *The Generalized de Briujn Identity*

This section generalizes the de Bruijn identity, in which entropy rates are related to Fisher information.

Theorem 1 *Let* $f_{D,\mathbf{h},t}(g) = f(g, t; D, \mathbf{h})$ *denote the solution of the diffusion equation (4) with constant* $\mathbf{h} = [h_1, \ldots, h_n]^T$ *subject to the initial condition* $f(g, 0; D, \mathbf{h}) = \delta(g)$. *Then for any well-behaved pdf* $\alpha(g)$,

$$\frac{d}{dt} S(\alpha * f_{D,\mathbf{h},t}) = \frac{1}{2} \mathrm{tr}[D F^r(\alpha * f_{D,\mathbf{h},t})]. \tag{36}$$

Proof It is easy to see that the solution of the diffusion equation

$$\frac{\partial \rho}{\partial t} = \frac{1}{2} \sum_{i,j=1}^{n} D_{ij} \tilde{E}_i^r \tilde{E}_j^r \rho - \sum_{k=1}^{n} h_k \tilde{E}_k^r \rho \tag{37}$$

subject to the initial conditions $\rho(g, 0) = \alpha(g)$ is simply $\rho(g, t) = (\alpha * f_{D,\mathbf{h},t})(g)$. This follows because all derivatives "pass through" the convolution integral for $\rho(g, t)$ and act on $f_{D,\mathbf{h},t}(g)$.

Taking the time derivative of $S(\rho(g, t))$ gives

$$\frac{d}{dt} S(\rho) = -\frac{d}{dt} \int_G \rho(g, t) \log \rho(g, t)\, dg = -\int_G \left\{ \frac{\partial \rho}{\partial t} \log \rho + \frac{\partial \rho}{\partial t} \right\} dg. \tag{38}$$

Using (37), the partial with respect to time can be replaced with Lie derivatives. But

$$\int_G \tilde{E}_k^r \rho\, dg = \int_G \tilde{E}_i^r \tilde{E}_j^r \rho\, dg = 0,$$

so the second term on the right side of (38) completely disappears. Using the integration-by-parts formula[6]

$$\int_G f_1\, \tilde{E}_k^r f_2\, dg = -\int_G f_2\, \tilde{E}_k^r f_1\, dg,$$

with $f_1 = \log \rho$ and $f_2 = \rho$ then gives

[6]There are no surface terms because, like the circle and real line, each coordinate in the integral either wraps around or goes to infinity.

$$\frac{d}{dt}S(\alpha * f_{D,\mathbf{h},t}) = \frac{1}{2}\sum_{i,j=1}^{n} D_{ij}\int_G \frac{1}{\alpha * f_{D,\mathbf{h},t}}\tilde{E}_j^r(\alpha * f_{D,\mathbf{h},t})\tilde{E}_i^r(\alpha * f_{D,\mathbf{h},t})\,dg$$

$$= \frac{1}{2}\sum_{i,j=1}^{n} D_{ij}F_{ij}^r(\alpha * f_{D,\mathbf{h},t}) = \frac{1}{2}\mathrm{tr}\left[D\,F^r(\alpha * f_{D,\mathbf{h},t})\right].$$

The implication of this is that

$$S(\alpha * f_{D,\mathbf{h},t_2}) - S(\alpha * f_{D,\mathbf{h},t_1}) = \frac{1}{2}\int_{t_1}^{t_2}\mathrm{tr}\left[DF^r(\alpha * f_{D,\mathbf{h},t})\right]dt.$$

6 Mean, Covariance, and their Propagation Under Convolution

This section reviews concepts of mean and covariance for unimodular matrix Lie groups, and how they propagate under convolution. In these definitions, the concepts of Lie-theoretic exponential and logarithm play central roles. For a matrix Lie group, G, with corresponding Lie algebra, $\mathcal{G}$, the exponential map

$$\exp : \mathcal{G} \longrightarrow G$$

simply can be viewed as the matrix exponential defined by the Taylor series. In general, this map is neither surjective nor injective. However, it is possible to characterize the largest path-connected subset $\mathcal{G}^\circ \subset \mathcal{G}$ for which the image $G^\circ \doteq \exp(\mathcal{G}^\circ) \subset G$ has a well-defined inverse map

$$\log : G^\circ \longrightarrow \mathcal{G}.$$

This is also simply the matrix logarithm defined by its Taylor series.

For $SO(3)$, $SE(2)$, and $SE(3)$ which are three of the most common unimodular matrix Lie groups encountered in applications, the exponential map is surjective and G and G° differ only by a set of measure zero.

In what follows, it is assumed that all probability density functions $f : G \longrightarrow \mathbb{R}_{\geq 0}$ are either supported in G°, or that

$$\int_G f(g)\,dg = \epsilon + \int_{G^\circ} f(g)\,dg$$

where ϵ is an inconsequential probability. With this in mind, it becomes possible to blur the difference between G and G°.

6.1 Defining Mean

At least three different definitions for the mean of a pdf on a unimodular Lie group exist in the literature. The definitions reviewed here are all in the context of matrix-Lie-theoretic language which grew out of the author's applied work [85, 86]. For similar definitions expressed in differential-geometric terms see [87–89].

Directly generalizing the definition

$$\mathbf{m} = \int_{\mathbb{R}^n} \mathbf{x}\, f(\mathbf{x})\, d\mathbf{x}$$

to a Lie group is problematic because $\int_G g\, f(g)\, dg$ is not an element of the group. However, it is possible to define $m_0 \in G$ such that

$$\log m_0 = \int_G \log g\, f(g)\, dg\,.$$

Alternatively,

$$\int_{\mathbb{R}^n} (\mathbf{x} - \mathbf{m})\, f(\mathbf{x})\, d\mathbf{x} = \mathbf{0}$$

generalizes to searching for $m_1 \in G$ such that

$$\int_G \log\left(m_1^{-1} \circ g\right)\, f(g)\, dg \;=\; \mathbb{O}.$$

Thirdly,

$$\mathbf{m} = \underset{\mathbf{y} \in \mathbb{R}^n}{\text{argmin}} \int_{\mathbb{R}^n} (\mathbf{x} - \mathbf{y})^2\, f(\mathbf{x})\, d\mathbf{x}$$

generalizes as

$$m_2 = \underset{h \in G}{\text{argmin}} \int_G \left\|\log\left(h^{-1} \circ g\right)\right\|^2\, f(g)\, dg\,.$$

In general, no two of m_0, m_1, and m_2 are equal. However, in practice for distributions that are concentrated, they are quite close to each other. That said, if $f(g) = \rho(g)$ is a symmetric function, then all three reduce to the identity element, and hence are equal in this special case.

Though m_0 seems simple and straight forward, it has the undesirable property that shifting a symmetric pdf as $\rho(\mu^{-1} \circ g)$ does not automatically shift the mean from e to μ. m_2 has the problem that the norm $\|\cdot\|$ requires a choice of metric, and for noncompact unimodular Lie groups, a bi-invariant metric generally does not exist. Therefore, conjugating by an arbitrary $a \in G$ a symmetric pdf as $\rho(a^{-1} \circ g \circ a)$, which in the Euclidean setting would leave the mean fixed at e, results in a change to the value of the mean m_2 which depends on a.

In contrast, the mean m_1 shifts naturally with shifts of the pdf because

$$\int_G \log\left(m_1^{-1}\circ g\right)\,\rho(\mu^{-1}\circ g)\,dg = \int_G \log\left(m_1^{-1}\circ\mu\circ h\right)\,\rho(h)dh.$$

hence $m_1^{-1}\circ\mu = e$, or $m_1 = \mu$. Under conjugation of the pdf, the appearance of log linearly in the definition of m_1 means that

$$\int_G \log\left(m_1^{-1}\circ g\right)\rho(a^{-1}\circ g\circ a)\,dg = \int_G \log\left(m_1^{-1}\circ a\circ h\circ a^{-1}\right)\rho(h)\,dh = \mathbb{O}$$

can be written as

$$\int_G a^{-1}\,\log\left(m_1^{-1}\circ a\circ h\circ a^{-1}\right)\,a\,\rho(h)\,dh = a^{-1}\,\mathbb{O}\,a = \mathbb{O}.$$

But since

$$a^{-1}\,\log(g)\,a = \log(a^{-1}\circ g\circ a)\,,$$

then the mean m_1 of the conjugated pdf will be the conjugated mean. The implication of this general result in the special case when ρ is symmetric is

$$a^{-1}\circ m_1^{-1}\circ a = e \implies m_1 = e,$$

giving the desirable property of invariance of the mean of a symmetric function under conjugation.

For these reasons, m_1 is chosen here (and in the author's previous work) as the best definition of the mean, and this is what will be used henceforth, and denoted as μ. The value of μ can be obtained numerically with an iterative procedure using m_0 as the initial starting point.

6.2 Defining Covariance

Previously the concepts of $\log : G^\circ \to \mathcal{G}$ and $\vee : \mathcal{G} \to \mathbb{R}^n$ where defined. The composition of these maps is defined as

$$\log^\vee G^\circ \to \mathbb{R}^n\,.$$

That is, for any $g \in G^\circ$, $\log^\vee(g) \in \mathbb{R}^n$.

One way to define the covariance of pdf on a unimodular Lie group G is [1, 2, 85, 86]

$$\Sigma \doteq \int_G \log^\vee(\mu^{-1} \circ g)[\log^\vee(\mu^{-1} \circ g)]^T f(g)\, dg\,. \tag{39}$$

This definition is natural as a generalization of the concept of covariance in Euclidean space when the pdf of interest is relatively concentrated. Then, for example, a Gaussian distribution can be defined as

$$f(g;\mu,\Sigma) \doteq \frac{1}{(2\pi)^{d/2}|\Sigma|^{\frac{1}{2}}} \exp\left(-\frac{1}{2}[\log^\vee(\mu^{-1} \circ g)]^T \Sigma^{-1} \log^\vee(\mu^{-1} \circ g)\right).$$

This definition makes sense when the tails decay to negligible values inside a ball around μ for which the exponential and logarithm maps form a bijective pair. Otherwise, the topological properties of G become relevant.

Alternative definitions of scalar variance can be found in [2, 20, 90]. If the covariance as defined in (39) has been computed for pdfs f_1 and f_2, a convenient and accurate approximation for the covariance of $f_1 * f_2$ is known [1, 86]. This is known as a covariance propagation formula. In contrast, the scalar definitions in [20, 90] have exact propagation formulas, but these quantities do not have the form or properties that are usually associated with covariance of pdfs on Euclidean space.

An altogether different way to define covariance that does not involve any approximation is to recognize that for a Gaussian distribution with the mean serving as the statistic, the Cramér-Rao Bound becomes the equality

$$\Sigma_{gaussian} = F^{-1}_{gaussian}\,, \tag{40}$$

and since a Gaussian distribution with $\mu = e$ solves a driftless diffusion equation subject to Dirac delta initial conditions, it is possible to define a kind of covariance for such processes by computing the Fisher information and using (40). By generalization, an alternative definition of covariance can be taken as

$$\Sigma' \doteq F^{-1}\,.$$

The exact properties of this definition under convolution are unknown.

The covariance propagation formula for (39) involves the concept of the adjoint matrix, $Ad(g)$. This concept is reviewed in the following section.

6.3 The Adjoint Operators *ad* and *Ad*

Given $X, Y \in \mathcal{G}$, "little ad" operator is defined as

$$ad_Y(X) \doteq [Y, X] = YX - XY\,,$$

and "big Ad" is

$$Ad_g(X) \doteq gXg^{-1}.$$

Here "ad" is short for "adjoint". Both of these are linear in X. That is, for arbitrary $c_1, c_2 \in \mathbb{R}$ and $X_1, X_2 \in \mathbb{R}^{n\times n}$

$$ad_Y(c_1X_1 + c_2X_2) = c_1 ad_Y(X_1) + c_2 ad_Y(X_2)$$

and

$$Ad_g(c_1X_1 + c_2X_2) = c_1 Ad_g(X_1) + c_2 Ad_g(X_2).$$

Sometimes Ad_g is written as $Ad(g)$ and ad_Y is written as $ad(Y)$. It turns out that these are related as

$$Ad(\exp(Y)) = \exp(ad(Y)).$$

By introducing a basis for the Lie algebra of a Lie group, it is possible to express the Ad and ad operators as square matrices of the same dimension as the group. The distinction between operators and matrices can sometimes be confusing, which is why, for example, the matrices of $ad(X)$ and $Ad(A)$ are written as $[ad(X)]$ and $[Ad(A)]$ in [1, 2] where

$$[ad(X)]_{ij} = (E_i, ad(X)E_j) \quad \text{and} \quad [Ad(A)]_{ij} = (E_i, Ad(A)E_j)$$

computed using the inner product $(\cdot, \cdot)$.

6.4 Covariance Propagation

Given two pdfs with mean and covariance specified, i.e., $f_{(\mu_i,\Sigma_i)}(g)$ for $i = 1, 2$, one would like to be able to write expressions for μ_3, Σ_3 such that

$$f_{(\mu_1,\Sigma_1)} * f_{(\mu_2,\Sigma_2)} = f_{(\mu_3,\Sigma_3)}.$$

In the case when $G = \mathbb{R}^n$, the result is simply $\mathbf{m}_3 = \mathbf{m}_1 + \mathbf{m}_2$, and $\Sigma_3 = \Sigma_1 + \Sigma_2$. This result is nonparametric. That is, it does not require the pdfs to have a specific form, such as a Gaussian.

For general unimodular Lie groups, there is no simple exact formula. However, when the pdfs are concentrated, i.e., both $\|\Sigma_i\|$ are small, then it is possible to write [85]

$$\mu_3 \approx \mu_1 \circ \mu_2 \;\; \text{and} \;\; \Sigma_3 \approx \Sigma_1^{\mu_2} + \Sigma_2.$$

where

$$\Sigma_1^{\mu_2} \doteq [Ad_{\mu_2^{-1}}]\Sigma_1[Ad_{\mu_2^{-1}}]^T.$$

A higher-order approximation of the form [86]

$$\Sigma_3 \approx \Sigma_1^{\mu_2} + \Sigma_2 + \Phi(\Sigma_1^{\mu_2}, \Sigma_2) \tag{41}$$

is also possible, but in many applications $\Phi(\Sigma_1^{\mu_2}, \Sigma_2)$ is negligible because it depends quadratically on the elements of $\Sigma_1^{\mu_2}$ and Σ_2.

If $f_{(\mu,\Sigma)}(g) = f(g; \mu, \Sigma)$ denotes a Gaussian distribution in exponential coordinates on a d-dimensional Lie group, it will be of the form

$$f(g; \mu, \Sigma) = \frac{1}{(2\pi)^{d/2}|\Sigma_i|^{\frac{1}{2}}} \exp\left(-\frac{1}{2}[\log^\vee(\mu^{-1} \circ g)]^T \Sigma^{-1} \log^\vee(\mu^{-1} \circ g)\right).$$

Then

$$f_{(\mu_1,\Sigma_1)} * f_{(\mu_2,\Sigma_2)} \approx f_{(\mu_3,\Sigma_3)}. \tag{42}$$

The quality of these approximations, as well as those that are even more accurate, have been studied in [1]. This is also a nonparametric result.

In the case when the distributions are more spread out, it is possible to compute the covariance in a different way using the group Fourier transform using the convolution theorem. For example, if $\Sigma_i = D^{(i)}t$ and

$$f_{(\mu_i,\Sigma_i)}(g) \doteq \rho_{\Sigma_i}(\mu_i^{-1} \circ g; t),$$

and since

$$\hat{f}_i(\lambda) = U(\mu_i, \lambda) \exp\left(\sum_{j,k} \sigma_{jk}^{(i)} E_j E_k\right),$$

from the convolution theorem it is possible to write the Fourier version of (42) as

$$U(\mu_2, \lambda) \exp\left(\sum_{j,k} \sigma_{jk}^{(2)} E_j E_k\right) U(\mu_1, \lambda) \exp\left(\sum_{j,k} \sigma_{jk}^{(1)} E_j E_k\right) \approx U(\mu_3, \lambda) \exp\left(\sum_{j,k} \sigma_{jk}^{(3)} E_j E_k\right).$$

which can then be substituted in the reconstruction formula to reproduce (42), which produces an approximate expression involving traces, which is of a different type than the trace inequalities studied previously in the literature.

7 Examples

This section illustrates ideas presented earlier in this paper on the Heisenberg and rotation groups.

7.1 The Heisenberg Group, $\mathcal{H}(3)$

The Heisenberg group, $H(3)$, is defined by elements of the form

$$g(\alpha, \beta, \gamma) = \begin{pmatrix} 1 & \alpha & \beta \\ 0 & 1 & \gamma \\ 0 & 0 & 1 \end{pmatrix} \quad \text{where} \quad \alpha, \beta, \gamma \in \mathbb{R} \tag{43}$$

and the operation of matrix multiplication. Therefore, the group law can be viewed in terms of parameters as

$$g(\alpha_1, \beta_1, \gamma_1)g(\alpha_2, \beta_2, \gamma_2) = g(\alpha_1 + \alpha_2, \beta_1 + \beta_2 + \alpha_1\gamma_2, \gamma_1 + \gamma_2).$$

The identity element is the identity matrix $g(0, 0, 0)$, and the inverse of an arbitrary element $g(\alpha, \beta, \gamma)$ is

$$g^{-1}(\alpha, \beta, \gamma) = g(-\alpha, \alpha\gamma - \beta, -\gamma).$$

7.1.1 Lie Algebra and Exponential Map

Basis elements for the Lie algebra are

$$E_1 = \begin{pmatrix} 0 & 1 & 0 \\ 0 & 0 & 0 \\ 0 & 0 & 0 \end{pmatrix}; \quad E_2 = \begin{pmatrix} 0 & 0 & 1 \\ 0 & 0 & 0 \\ 0 & 0 & 0 \end{pmatrix}; \quad E_3 = \begin{pmatrix} 0 & 0 & 0 \\ 0 & 0 & 1 \\ 0 & 0 & 0 \end{pmatrix}. \tag{44}$$

A linear mapping between the Lie algebra spanned by this basis with $\mathbb{R}^3$ is defined by $E_i^\vee = \mathbf{e}_i$.

The Lie bracket is defined as $[E_i, E_j] = E_iE_j - E_jE_i = -[E_j, E_i]$, and so, as is always the case, $[E_i, E_i] = \mathbb{O}$. For these particular basis elements,

$$[E_1, E_2] = [E_2, E_3] = \mathbb{O} \quad \text{and} \quad [E_1, E_3] = E_2.$$

In addition, all double brackets involving the first two listed above are also zero,

$$[E_i, [E_1, E_2]] = [E_i, [E_2, E_3]] = \mathbb{O} \text{ for } i = 1, 2, 3$$

From these, and the bilinearity of the Lie bracket, it follows that for arbitrary

$$X = \sum_i x_i E_i \text{ and } Y = \sum_j x_j E_j$$

that

$$[X,Y]=\sum_{i,j} x_i y_j [E_i, E_j] = (x_1 y_3 - x_3 y_1)E_2\,. \tag{45}$$

If the inner product for the Lie algebra spanned by these basis elements is defined as $(X, Y) = \mathrm{tr}(XY^T)$, then this basis is orthonormal: $(E_i, E_j) = \delta_{ij}$.

The group $H(3)$ is nilpotent because $(x_1E_1 + x_2E_2 + x_3E_3)^n = 0$ for all $n \geq 3$. As a result, the matrix exponential is a polynomial in the coordinates $\{x_i\}$:

$$\exp\begin{pmatrix} 0 & x_1 & x_2 \\ 0 & 0 & x_3 \\ 0 & 0 & 0 \end{pmatrix} = g(x_1, x_2 + \frac{1}{2}x_1x_3, x_3). \tag{46}$$

The parametrization in (43) can be viewed as the following product of exponentials:

$$g(\alpha,\beta,\gamma) = g(0,\beta,0)g(0,0,\gamma)g(\alpha,0,0) = \exp(\beta E_2)\exp(\gamma E_3)\exp(\alpha E_1).$$

The logarithm is obtained by solving for each x_i as a function of α, β, γ. By inspection this is $x_1 = \alpha$, $x_3 = \gamma$ and $x_2 = \beta - \alpha\gamma/2$. Therefore,

$$\log g(\alpha,\beta,\gamma) = \begin{pmatrix} 0 & \alpha & \beta - \alpha\gamma/2 \\ 0 & 0 & \gamma \\ 0 & 0 & 0 \end{pmatrix}.$$

7.1.2 Adjoint Matrices for $\mathcal{H}(3)$

The adjoint matrix, defined by $[Ad(g)]\mathbf{x} = (gXg^{-1})^\vee$, is computed by evaluating

$$\begin{pmatrix} 1 & \alpha & \beta \\ 0 & 1 & \gamma \\ 0 & 0 & 1 \end{pmatrix}\begin{pmatrix} 0 & x_1 & x_2 \\ 0 & 0 & x_3 \\ 0 & 0 & 0 \end{pmatrix}\begin{pmatrix} 1 & -\alpha & \alpha\gamma - \beta \\ 0 & 1 & -\gamma \\ 0 & 0 & 1 \end{pmatrix} = \begin{pmatrix} 0 & x_1 & -\gamma x_1 + x_2 + \alpha x_3 \\ 0 & 0 & x_3 \\ 0 & 0 & 0 \end{pmatrix}.$$

Therefore,

$$(gXg^{-1})^\vee = \begin{pmatrix} x_1 \\ -\gamma x_1 + x_2 + \alpha x_3 \\ x_3 \end{pmatrix} \quad \text{and} \quad [Ad(g(\alpha,\beta,\gamma))] = \begin{pmatrix} 1 & 0 & 0 \\ -\gamma & 1 & \alpha \\ 0 & 0 & 1 \end{pmatrix}.$$

The fact that $\det[Ad(g)] = 1$ for all $g \in G$ indicates that this group is unimodular. This fact is independent of the parametrization. It can also be shown that for $X = \sum_{i=1}^{3} x_i E_i$ that

$$[Ad(\exp X)] = \begin{pmatrix} 1 & 0 & 0 \\ -x_3 & 1 & x_1 \\ 0 & 0 & 1 \end{pmatrix}. \tag{47}$$

7.2 Bi-invariant Integration Measure

The Jacobian matrices for this group can be computed in either parametrization. In terms of α, β, γ,

$$\frac{\partial g}{\partial \alpha} = E_1; \qquad \frac{\partial g}{\partial \beta} = E_2; \qquad \frac{\partial g}{\partial \gamma} = E_3.$$

A straightforward calculation then gives

$$g^{-1}\frac{\partial g}{\partial \alpha} = E_1; \qquad g^{-1}\frac{\partial g}{\partial \beta} = E_2; \qquad g^{-1}\frac{\partial g}{\partial \gamma} = E_3 - \alpha E_2.$$

Therefore

$$J_r(\alpha, \beta, \gamma) = \begin{pmatrix} 1 & 0 & 0 \\ 0 & 1 & -\alpha \\ 0 & 0 & 1 \end{pmatrix} \quad \text{and} \quad J_l(\alpha, \beta, \gamma) = \begin{pmatrix} 1 & 0 & 0 \\ -\gamma & 1 & 0 \\ 0 & 0 & 1 \end{pmatrix} \tag{48}$$

Then $|J_l(\alpha, \beta, \gamma)| = |J_r(\alpha, \beta, \gamma)| = 1$ and bi-invariant integration measure expressed in these coordinates is simply

$$dg = d\alpha d\beta d\gamma\,.$$

In exponential coordinates

$$J_r(\mathbf{x}) = \begin{pmatrix} 1 & 0 & 0 \\ x_3/2 & 1 & -x_1/2 \\ 0 & 0 & 1 \end{pmatrix} \quad \text{and} \quad J_l(\mathbf{x}) = \begin{pmatrix} 1 & 0 & 0 \\ -x_3/2 & 1 & x_1/2 \\ 0 & 0 & 1 \end{pmatrix} \tag{49}$$

and

$$dg = dx_1 dx_2 dx_3\,.$$

7.3 Covariance Propagation and the EPI for $\mathcal{H}(3)$

From (45)the only nonzero term in the second-order covariance propagation formula from [1, 86] is

$$\frac{1}{4}[X, Y]^{\vee}\left([X, Y]^{\vee}\right)^T = \frac{1}{4}(x_1 y_3 - x_3 y_1)^2 \mathbf{e}_2 \mathbf{e}_2^T\,.$$

Hence, for $\mathcal{H}(3)$, the second-order term in (41), which results from integrating the above as was done in [86], becomes

$$\Phi(\Sigma_1^{\mu_2}, \Sigma_2) = a\mathbf{e}_2\mathbf{e}_2^T$$

where

$$a = \frac{1}{4}\left(\sigma_{11}^{(1)}\sigma_{33}^{(2)} - \sigma_{13}^{(1)}\sigma_{31}^{(2)} - \sigma_{31}^{(1)}\sigma_{13}^{(2)} + +\sigma_{33}^{(1)}\sigma_{11}^{(2)}\right) \geq 0$$

with $\sigma_{ij}^{(1)} = \mathbf{e}_i^T \Sigma_1^{\mu_2} \mathbf{e}_j$ and $\sigma_{ij}^{(2)} = \mathbf{e}_i^T \Sigma_2 \mathbf{e}_j$.

Moreover, from the matrix identity for $A = A^T > 0$,

$$\det(A + \mathbf{u}\mathbf{v}^T) = (1 + \mathbf{v}^T A^{-1}\mathbf{u}) \det A\,,$$

it follows that

$$\det(\Sigma_{(1*2)}) = \det(\Sigma_1^{\mu_2} + \Sigma_2 + a\mathbf{e}_2\mathbf{e}_2^T) = (1 + a\mathbf{e}_2^T(\Sigma_1^{\mu_2} + \Sigma_2)^{-1}\mathbf{e}_2)\det(\Sigma_1^{\mu_2} + \Sigma_2)\,,$$

where $\Sigma_1^{\mu_2} = Ad_{\mu_2^{-1}} \Sigma_1 Ad_{\mu_2^{-1}}^T$. Consequently, from the Euclidean EPI and the unimodularity of G, which implies that

$$\det(\Sigma_1^{\mu_2}) = \det(\Sigma_1)\,,$$

it is not difficult to see that

$$\det(\Sigma_{(1*2)})^{\frac{1}{\dim(G)}} \geq \det(\Sigma_1^{\mu_2} + \Sigma_2)^{\frac{1}{\dim(G)}} \geq \det(\Sigma_1)^{\frac{1}{\dim(G)}} + \det(\Sigma_2)^{\frac{1}{\dim(G)}}.$$

That is, the entropy power inequality for pdfs from diffusion processes on $\mathcal{H}(3)$ follows in the small time limit from the classical EPI, and it is less restrictive than in the Euclidean case.

7.4 *The Case of $SO(3)$*

The group of rotations of three-dimensional space has elements that are 3×3 special orthogonal matrices, i.e., those satisfying

$$RR^T = \mathbb{I} \quad \text{and} \quad \det(R) = +1\,.$$

That is, they satisfy $RR^T = \mathbb{I}$ and $\det R = +1$. It is easy to see that closure of these properties under multiplication is satisfied because

$$(R_1R_2)^T(R_1R_2) = R_2^T R_1^T R_1 R_2 = R_2^T R_2 = \mathbb{I}$$

and

$$\det(R_1R_2) = \det(R_1)\det(R_2) = 1 \cdot 1 = 1.$$

7.4.1 The Lie Algebra

The Lie algebra $so(3)$ consists of skew-symmetric matrices of the form

$$X = \begin{pmatrix} 0 & -x_3 & x_2 \\ x_3 & 0 & -x_1 \\ -x_2 & x_1 & 0 \end{pmatrix} = \sum_{i=1}^{3} x_i E_i \,. \tag{50}$$

Every such matrix can be associated with a vector $\mathbf{x}$ by making the identification

$$E_i^\vee = \mathbf{e}_i \iff E_i = \hat{\mathbf{e}}_i \,.$$

For $SO(3)$ the adjoint matrices are

$$[Ad(R)] = R \quad \text{and} \quad [ad(X)] = X.$$

Furthermore,

$$[X, Y]^\vee = \mathbf{x} \times \mathbf{y}.$$

7.4.2 Exponential and Logarithm

It is well known that the exponential map $\exp : so(3) \to SO(3)$ is related to Euler's Theorem as

$$R = \exp(\theta N) = \mathbb{I} + \sin\theta\, N + (1 - \cos\theta)\, N^2 \,,$$

where $\theta \in [0, \pi]$ is the angle of rotation around the axis $\mathbf{n} \in S^2$, with N being the associated skew-symmetric matrix. Then $X = \theta N$ and $\mathbf{x} = \theta \mathbf{n}$. It is convenient to limit $\theta \in [0, \pi]$ and to allow $\mathbf{n}$ to take any value in the unit sphere, S^2. Moreover,

$$\text{tr}(R) = 1 + 2\cos\theta \quad \text{and} \quad N = \frac{R - R^T}{2\sin\theta} \,.$$

Then, since

$$\theta = \cos^{-1}\left[\frac{\text{tr}(R) - 1}{2}\right] \quad \text{and} \quad \sin(\cos^{-1} a) = \sqrt{1 - a^2} \,,$$

it follows that $\sin\theta$ can be written explicitly in terms of R as

$$\sin\theta = \sqrt{1 - \frac{(\text{tr}(R) - 1)^2}{4}} = \sqrt{\frac{3}{4} - \frac{(\text{tr}(R))^2}{4} + \frac{2\text{tr}(R)}{4}} \,.$$

Since $X = \theta N = \log R$, it follows that

$$\log(R) = \frac{\cos^{-1}\left[\frac{\mathrm{tr}(R)-1}{2}\right](R - R^T)}{\sqrt{3 - (\mathrm{tr}(R))^2 + 2\mathrm{tr}(R)}}. \tag{51}$$

This expression breaks down when $\theta = \pi$, which defines a set of measure zero, and hence is inconsequential when evaluating the logarithm under an integral.

7.4.3 Invariant Integration Measure

Two common ways to parameterize rotations are using the matrix exponential $R = \exp X$ and using Euler angles such as $R = R_3(\alpha)R_1(\beta)R_3(\gamma)$ where $0 \leq \alpha, \gamma \leq 2\pi$ and $0 \leq \beta \leq \pi$.

Relatively simple analytical expressions were derived by Park [91] for the Jacobian J_l when $R = \exp X$ as

$$J_l(\mathbf{x}) = \mathbb{I} + \frac{1 - \cos\|\mathbf{x}\|}{\|\mathbf{x}\|^2}X + \frac{\|\mathbf{x}\| - \sin\|\mathbf{x}\|}{\|\mathbf{x}\|^3}X^2 \tag{52}$$

The corresponding Jacobian J_r and its inverse are [1, 2]

$$J_r(\mathbf{x}) = \mathbb{I} - \frac{1 - \cos\|\mathbf{x}\|}{\|\mathbf{x}\|^2}X + \frac{\|\mathbf{x}\| - \sin\|\mathbf{x}\|}{\|\mathbf{x}\|^3}X^2$$

In terms of ZXZ Euler angles,

$$J_l(\alpha, \beta, \gamma) = [\mathbf{e}_3, R_3(\alpha)\mathbf{e}_1, R_3(\alpha)R_1(\beta)\mathbf{e}_3] = \begin{pmatrix} 0 & \cos\alpha & \sin\alpha\sin\beta \\ 0 & \sin\alpha & -\cos\alpha\sin\beta \\ 1 & 0 & \cos\beta \end{pmatrix}. \tag{53}$$

and

$$J_r = R^T J_l = [R_3(-\gamma)R_1(-\beta)\mathbf{e}_3, R_3(-\gamma)\mathbf{e}_1, \mathbf{e}_3] = \begin{pmatrix} \sin\beta\sin\gamma & \cos\gamma & 0 \\ \sin\beta\cos\gamma & -\sin\gamma & 0 \\ \cos\beta & 0 & 1 \end{pmatrix}. \tag{54}$$

From this we see that

$$dR = \frac{2(1 - \cos\|\mathbf{x}\|)}{\|\mathbf{x}\|^2} dx_1 dx_2 dx_3 = \sin\beta\, d\alpha d\beta d\gamma\,.$$

From these it can be shown that

$$\int_{SO(3)} dR = 8\pi^2\,.$$

7.4.4 Fourier Series

For $SO(3)$, irreducible unitary representations (IURs) [1] are enumerated by $l \in \mathbb{Z}_{\geq 0}$, and for any $R, A \in SO(3)$ these $(2l+1) \times (2l+1)$ IUR matrices have the fundamental properties

$$U^l(RA) = U^l(R)\, U^l(A) \;\text{ and }\; U^l(R^T) = U^l(R)^*$$

where $*$ is the Hermitian conjugate of a matrix. The explicit forms of these matrices when R is expressed in Euler angles are well known in Physics as the Wigner-D functions [92–96]

For functions $f \in L^2(SO(3))$, the Fourier coefficients are computed as

$$\hat{f}^l_{mn} = \int_{SO(3)} f(A) U^l_{mn}(A^{-1})\, dA\,. \tag{55}$$

The following orthogonality relation holds

$$\int_{SO(3)} U^l_{mn}(A)\overline{U^s_{pq}(A)}\, dA = \frac{1}{2l+1}\delta_{ls}\delta_{mp}\delta_{nq} \tag{56}$$

where dA is scaled so that $\int_{SO(3)} dA = 1$. The Fourier series on $SO(3)$ has the form

$$f(A) = \sum_{l=0}^{\infty}(2l+1)\sum_{m=-l}^{l}\sum_{n=-l}^{l}\hat{f}^l_{mn}U^l_{nm}(A)\,, \tag{57}$$

which results from the completeness relation

$$\sum_{l=0}^{\infty}(2l+1)\sum_{m=-l}^{l}\sum_{n=-l}^{l}U^l_{mn}(R^{-1})U^l_{nm}(A) = \delta(R^{-1}A)\,. \tag{58}$$

Another way to write (57) is

$$f(A) = \sum_{l=0}^{\infty}(2l+1)\text{trace}\left[\hat{f}^l\, U^l(A)\right]. \tag{59}$$

7.5 Diffusions on $SO(3)$

A diffusion process on $SO(3)$ commonly encountered in applications is of the form

$$\frac{\partial f}{\partial t} = \frac{1}{2} \sum_{i,j=1}^{3} D_{ij} \tilde{E}_i \tilde{E}_j f + \sum_{k=1}^{3} d_k \tilde{E}_k f \,. \tag{60}$$

By expanding the PDF in the PDE in (60) into a Fourier series on $SO(3)$, the solution can be obtained once we know how the differential operators X_i^R transform the matrix elements $U_{m,n}^l(A)$. Explicitly,

$$\tilde{E}_1 U_{mn}^l = \frac{1}{2} c_{-n}^l U_{m,n-1}^l - \frac{1}{2} c_n^l U_{m,n+1}^l; \tag{61}$$

$$\tilde{E}_2 U_{mn}^l = \frac{1}{2} i c_{-n}^l U_{m,n-1}^l + \frac{1}{2} i c_n^l U_{m,n+1}^l; \tag{62}$$

$$\tilde{E}_3 U_{mn}^l = -in U_{mn}^l; \tag{63}$$

where $c_n^l = \sqrt{(l-n)(l+n+1)}$ for $l \geq |n|$ and $c_n^l = 0$ otherwise. From this definition it is clear that $c_k^k = 0$, $c_{-(n+1)}^l = c_n^l$, $c_{n-1}^l = c_{-n}^l$, and $c_{n-2}^l = c_{-n+1}^l$.

By repeated application of these rules, it can be shown that [1]

$$\mathcal{F}\left(\frac{1}{2} \sum_{i,j=1}^{3} D_{ij} \tilde{E}_i \tilde{E}_j f + \sum_{i=1}^{3} d_i \tilde{E}_i f\right)_{mn}^{l} = \sum_{k=\max(-l,m-2)}^{\min(l,m+2)} \mathcal{A}_{m,k}^l \hat{f}_{k,n}^l,$$

where

$$\mathcal{A}_{m,m+2}^l = \left[\frac{(D_{11} - D_{22})}{8} + \frac{i}{4} D_{12}\right] c_{m+1}^l c_{-m-1}^l;$$

$$\mathcal{A}_{m,m+1}^l = \left[\frac{(2m+1)}{4}(D_{23} - i D_{13}) + \frac{1}{2}(d_1 + i d_2)\right] c_{-m-1}^l;$$

$$\mathcal{A}_{m,m}^l = \left[-\frac{(D_{11} + D_{22})}{8}(c_{-m}^l c_{m-1}^l + c_m^l c_{-m-1}^l) - \frac{D_{33} m^2}{2} - i d_3 m\right];$$

$$\mathcal{A}_{m,m-1}^l = \left[\frac{(2m-1)}{4}(D_{23} + i D_{13}) + \frac{1}{2}(-d_1 + i d_2)\right] c_{m-1}^l;$$

$$\mathcal{A}_{m,m-2}^l = \left[\frac{(D_{11} - D_{22})}{8} - \frac{i}{4} D_{12}\right] c_{-m+1}^l c_{m-1}^l;$$

Hence, application of the $SO(3)$-Fourier transform to (60) and corresponding initial conditions reduces (60) to a set of linear time-invariant ODEs of the form

$$\frac{d\hat{f}^l}{dL} = \mathcal{A}^l \hat{f}^l \quad \text{with} \quad \hat{f}^l(0) = \mathbb{I}_{2l+1}. \tag{64}$$

Here $\mathbb{I}_{2l+1}$ is the $(2l+1) \times (2l+1)$ identity matrix and the banded matrix $\mathcal{A}^l$ are of the following form for $l = 0, 1, 2, 3$:

$$\mathcal{A}^0 = \mathcal{A}^0_{0,0} = 0; \quad \mathcal{A}^1 = \begin{pmatrix} \mathcal{A}^1_{-1,-1} & \mathcal{A}^1_{-1,0} & \mathcal{A}^1_{-1,1} \\ \mathcal{A}^1_{0,-1} & \mathcal{A}^1_{0,0} & \mathcal{A}^1_{0,1} \\ \mathcal{A}^1_{1,-1} & \mathcal{A}^1_{1,0} & \mathcal{A}^1_{1,1} \end{pmatrix};$$

$$\mathcal{A}^2 = \begin{pmatrix} \mathcal{A}^2_{-2,-2} & \mathcal{A}^2_{-2,-1} & \mathcal{A}^2_{-2,0} & 0 & 0 \\ \mathcal{A}^2_{-1,-2} & \mathcal{A}^2_{-1,-1} & \mathcal{A}^2_{-1,0} & \mathcal{A}^2_{-1,1} & 0 \\ \mathcal{A}^2_{0,-2} & \mathcal{A}^2_{0,-1} & \mathcal{A}^2_{0,0} & \mathcal{A}^2_{0,1} & \mathcal{A}^2_{0,2} \\ 0 & \mathcal{A}^2_{1,-1} & \mathcal{A}^2_{1,0} & \mathcal{A}^2_{1,1} & \mathcal{A}^2_{1,2} \\ 0 & 0 & \mathcal{A}^2_{2,0} & \mathcal{A}^2_{2,1} & \mathcal{A}^2_{2,2} \end{pmatrix};$$

$$\mathcal{A}^3 = \begin{pmatrix} \mathcal{A}^3_{-3,-3} & \mathcal{A}^3_{-3,-2} & \mathcal{A}^3_{-3,-1} & 0 & 0 & 0 & 0 \\ \mathcal{A}^3_{-2,-3} & \mathcal{A}^3_{-2,-2} & \mathcal{A}^3_{-2,-1} & \mathcal{A}^3_{-2,0} & 0 & 0 & 0 \\ \mathcal{A}^3_{-1,-3} & \mathcal{A}^3_{-1,-2} & \mathcal{A}^3_{-1,-1} & \mathcal{A}^3_{-1,0} & \mathcal{A}^3_{-1,1} & 0 & 0 \\ 0 & \mathcal{A}^3_{0,-2} & \mathcal{A}^3_{0,-1} & \mathcal{A}^3_{0,0} & \mathcal{A}^3_{0,1} & \mathcal{A}^3_{0,2} & 0 \\ 0 & 0 & \mathcal{A}^3_{1,-1} & \mathcal{A}^3_{1,0} & \mathcal{A}^3_{1,1} & \mathcal{A}^3_{1,2} & \mathcal{A}^3_{1,3} \\ 0 & 0 & 0 & \mathcal{A}^3_{2,0} & \mathcal{A}^3_{2,1} & \mathcal{A}^3_{2,2} & \mathcal{A}^3_{2,3} \\ 0 & 0 & 0 & 0 & \mathcal{A}^3_{3,1} & \mathcal{A}^3_{3,2} & \mathcal{A}^3_{3,3} \end{pmatrix}.$$

The solution to (64) is then of the form of a matrix exponential:

$$\hat{f}^l(L) = e^{L\mathcal{A}^l}. \tag{65}$$

Since $\mathcal{A}^l$ is a band-diagonal matrix for $l > 1$, the matrix exponential can be calculated much more efficiently (either numerically or symbolically) for large values of l than for general matrices of dimension $(2l+1) \times (2l+1)$.

Given the explicit forms provided above, (32)–(35) can be verified.

7.5.1 Lack of an Entropy-Power Inequality

For all unimodular Lie groups, the EPI holds for concentrated Gaussian pdfs for which the first-order covariance propagation formula from [85] holds by application of the Euclidean EPI to Gaussians. However, for compact Lie groups (including the circle and n-torus) the EPI always breaks down. For example, the uniform distribution on the circle, $\rho(\theta) = 1/2\pi$, has entropy $S(\rho) = \log(1/2\pi)$. But since this distribution is stable under convolution, we have that $S(\rho * \rho) = S(\rho)$ and so the EPI cannot hold since $N(\rho * \rho) = N(\rho) < 2 \cdot N(\rho)$. Similarly, unlike for $\mathcal{H}(3)$, the EPI does not hold for $SO(3)$.

8 Conclusions

Many inequalities of information theory that are based on probability densities on Euclidean space extend to the case of probabilities on Lie groups. In addition to reviewing appropriate concepts of integration, convolution, partial derivative, Fourier transform, covariance, and diffusion processes on unimodular Lie groups, this paper also presents some new inequalities that extend to this setting those known in the classical Abelian case.

References

1. Chirikjian, G.S., Kyatkin, A.B.: Harmonic Analysis for Engineers and Applied Scientists. Dover Publications, Mineola, NY (2016)
2. Chirikjian, G.S.: Stochastic Models, Information Theory, and Lie Groups: Volume 2 - Analytic Methods and Modern Applications. Birkhäuser, Boston (2011)
3. Howe, R., Tan, E.C.: Non-abelian Harmonic Analysis. Springer, Berlin (1992)
4. Lang, S.: $SL_2(R)$. Addison-Wesley, Reading (1975)
5. Chandra, H.: Spherical functions on a semisimple Lie group II. Am. J. Math. **27**, 569–579 (1960)
6. Jorgenson, J., Lang, S.: Spherical Inversion on $SL_n(R)$. Springer, Berlin (2001)
7. Thangavelu, S.: Harmonic Analysis on the Heisenberg Group. Birkhäuser, Boston (1998)
8. Neuenschwander, D.: Probabilities on the Heisenberg Group: Limit Theorems and Brownian Motion. Lecture Notes in Mathematics, vol. 1630. Springer, Berlin (1996)
9. Miller Jr., W.: Lie Theory and Special Functions. Academic Press, New York (1968)
10. Miller Jr., W.: Some applications of the representation theory of the Euclidean group in three-space. Commun. Pure App. Math. **17**, 527–540 (1964)
11. Vilenkin, N.Ja., Klimyk, A.U.: Representation of Lie Groups and Special Functions, vol. 1–3. Kluwer Academic Publishers, Dordrecht, Holland (1991)
12. Vilenkin, N.J.: Special Functions and the Theory of Group Representations. American Mathematical Society, Providence (1968)
13. Vilenkin, N.J., Akim, E.L., Levin, A.A.: The matrix elements of irreducible unitary representations of the group of Euclidean three-dimensional space motions and their properties. Dokl. Akad. Nauk SSSR **112**, 987–989 (1957). (in Russian)
14. Wang, Y., Zhou, Y., Maslen, D.K., Chirikjian, G.S.: Solving the phase-noise Fokker-Planck equation using the motion-group Fourier transform. IEEE Trans. Commun. **54**(5), 868–877 (May, 2006)
15. Zhou, Y., Chirikjian, G.S.: Conformational statistics of semi-flexible macromolecular chains with internal joints. Macromolecules **39**(5), 1950–1960 (2006)
16. Chirikjian, G.S., Kyatkin, A.B.: An operational calculus for the Euclidean motion group with applications in robotics and polymer science. J. Fourier Anal. Appl. **6**(6), 583–606 (December, 2000)

17. Chirikjian, G.S.: Degenerate diffusions and harmonic analysis on SE(3): a tutorial. In: Albeverio, S., Cruzeiro, A., Holm, D. (eds.) Stochastic Geometric Mechanics, pp. 77–99. Springer, Berlin (2017)
18. Chirikjian, G.S.: Information-theoretic inequalities on unimodular Lie groups. J. Geom. Mech. **2**(2), 119–158 (June, 2010)
19. Folland, G.B.: A Course in Abstract Harmonic Analysis. CRC Press, Boca Raton, FL (1995)
20. Grenander, U.: Probabilities on Algebraic Structures. Dover Publications, Mineola (2008)
21. Gross, K.I.: Evolution of noncommutative harmonic analysis. Am. Math. Mon. **85**(7), 525–548 (1978)
22. Gurarie, D.: Symmetry and Laplacians. Introduction to Harmonic Analysis, Group Representations and Applications, Dover edn. Elsevier Science Publisher, The Netherlands, 1992 (2008)
23. Hewitt, E., Ross, K.A.: Abstract Harmonic Analysis I, and II. Springer, Berlin, 1963 and 1970. (Reprinted 1994)
24. Sugiura, M.: Unitary Representations and Harmonic Analysis, 2nd edn. North-Holland, Amsterdam (1990)
25. Taylor, M.E.: Noncommutative Harmonic Analysis. Mathematical Surveys and Monographs. American Mathematical Society, Providence, RI (1986)
26. Kunze, R.: L_p Fourier transforms on locally compact unimodular groups. Trans. Am. Math. Soc. **89**, 519–540 (1958)
27. Applebaum, D.: Probability on Compact Lie Groups. Springer, New York (2014)
28. Beckner, W.: Sharp inequalities and geometric manifolds. J. Fourier Anal. Appl. **3**, 825–836 (1997)
29. Beckner, W.: Geometric inequalities in Fourier analysis. Essays on Fourier Analysis in Honor of Elias M. Stein, pp. 36–68. Princeton University Press, Princeton (1995)
30. Blachman, N.M.: The convolution inequality for entropy powers. IEEE Trans. Inform. Theory **11**(2), 267–271 (1965)
31. Carlen, E.A.: Superadditivity of fishers information and logarithmic Sobolev inequalities. J. Funct. Anal. **101**, 194–211 (1991)
32. Cover, T.M., Thomas, J.A.: Elements of Information Theory, 2nd edn. Wiley-Interscience, Hoboken, NJ (2006)
33. Dembo, A., Cover, T.M., Thomas, J.A.: Information theoretic inequalities. IEEE Trans. Inf. Theory **37**(6), 1501–1518 (1991)
34. Varopoulos, N.T., Saloff-Coste, L., Coulhon, T.: Analysis and Geometry on Groups. Cambridge University Press, Cambridge (1992)
35. Maslen, D.K.: Fast Transforms and Sampling for Compact Groups, Ph.D. Dissertation, Department of Mathematics, Harvard University (May 1993)
36. Maslen, D.K., Rockmore, D.N.: Generalized FFTSA survey of some recent results. DIMACS Ser. Discret. Math. Theor. Comput. Sci. **28**, 183–237 (1997)
37. Hardy, G.H., Littlewood, J.E., Pólya, G.: Inequalities. Cambridge University Press, Cambridge (1932)
38. Pólya, G.: Isoperimetric Inequalities in Mathematical Physics. Princeton University Press, Princeton (1951)
39. Simon, B.: Trace Ideals and their Applications. Mathematical Surveys and Monographs, 2nd edn. American Mathematical Society, Providence (2010)
40. Bhatia, R.: Positive Definite Matrices. Princeton University Press, Princeton (2007)
41. Bernstein, D.S.: Matrix Mathematics: Theory, Facts, and Formulas with Application to Linear Systems Theory. Princeton University Press, Princeton (February 22, 2005)
42. Trotter, H.F.: On the product of semi-groups of operators. Proc. Amer. Math. Soc. 10545–10551 (1959)
43. So, W.: Equality cases in matrix exponential inequalities. SIAM J. Matrix Anal. Appl. **13**(4), 1154–1158 (October, 1992)
44. Mirsky, L.: A Note on normal matrices. Am. Math. Mon. **63**(7), 479 (August–September, 1956)

45. Reid, R.M.: Some eigenvalue properties of persymmetric matrices. SIAM Rev. **39**(2), 313–316 (June, 1997)
46. Schur, I.: Über die charakteristischen Wurzeln einer linearen Substitution mit einer Anwendung auf die Theorie der Integralgleichungen. Math. Annalen **66**, 488–510 (1909)
47. Thurston, H.S.: On the characteristic equations of products of square matrices. Am. Math. Mon. **38**, 322–324 (1931)
48. Scott, W.M.: On characteristic roots of matrix products. Am. Math. Mon. **48**(3), 201–203 (March, 1941)
49. Bhatia, R., Parthasarathy, K.R.: Positive definite functions and operator inequalities. Bull. Lond. Math. Soc. **32**, 214–228 (2000)
50. Andruchow, E., Corach, G., Stojanoff, D.: Geometric operator inequalities. Linear Algebra Appl. **258**, 295–310 (1997)
51. Bhatia, R., Kittaneh, F.: On singular values of a product of operators. SIAM J. Matrix Anal. Appl. **11**, 272–277 (1990)
52. Bhatia, R.: Matrix Analysis. Springer, Berlin (1996)
53. Thompson, C.J.: Inequalities and partial orders on matrix spaces. Indiana Univ. Math. J. **21**(5), 469–480 (1971)
54. Golden, S.: Lower bounds for the Helmholtz function. Phys. Rev. **137**, B1127–B1128 (1965)
55. Thompson, C.J.: Mathematical Statistical Mechanics. Macmillan, New York (1972); Reprint, Princeton University Press, Princeton (1979, 1992)
56. Lenard, A.: Generalization of the Golden-Thompson inequality $Tr(e^A e^B) \geq Tr(e^{A+B})$. Indiana Univ. Math. J. **21**, 457–467 (1971)
57. Thompson, R.C.: Special cases of a matrix exponential formula. Linear Algebra Appl. **107**, 283–292 (1988)
58. So, W., Thompson, R.C.: Products of exponentials of Hermitian and complex symmetric matrices. Linear and Multilinear Algebra **29**, 225–233 (1991)
59. So, W.: The high road to an exponential formula. Linear Algebra Appl. **379**, 69–75 (2004)
60. Klyachko, A.A.: Random walks on symmetric spaces and inequalities for matrix spectra. Linear Algebra Appl. **319**, 37–59 (2000)
61. Cohen, J.E., Friedland, S., Kato, T., Kelly, F.P.: Eigenvalue inequalities for products of matrix exponentials. Linear Algebra Appl. **45**, 55–95 (1982)
62. Bernstein, D.S.: Inequalities for the trace of matrix exponentials. SIAM J. Matrix Anal. Appl. **9**, 156–158 (1988)
63. Fan, K.: Maximum properties and inequalities for the eigenvalues of completely continuous operators. PNAS **37**, 760–766 (1951)
64. Fan, K.: On a theorem of Weyl concerning eigenvalues of linear transformations I. Proc. Nat. Acad. Sci. USA **35**, 652–655 (1949)
65. Bebiano, N., da Providência, J., Jr., Lemos, R.: Matrix inequalities in statistical mechanics. Linear Algebra Appl. **376**(1), 265–273 (January, 2004)
66. Friedland, S., So, W.: Product of matrix exponentials. Linear Algebra Appl. **196**, 193–205 (1994)
67. Friedland, S., Porta, B.: The limit of the product of the parameterized exponentials of two operators. J. Funct. Anal. **210**, 436–464 (2004)
68. Mori, T.: Comments on a matrix inequality associated with bounds on solutions of algebraic Riccati and Lyapunov equation. IEEE Trans. Autom. Control **AC-29**, 1088 (November, 1988)
69. Fang, Y., Loparo, K.A., Feng, X.: Inequalities for the trace of matrix product. IEEE Trans. Autom. Control **39**(12), 2489–2490 (December, 1994)
70. Park, P.-G.: On the trace bound of a matrix product. IEEE Trans. Autom. Control **41**(12), 1799–1802 (December, 1996)
71. Komaroff, N.: Bounds on eigenvalues of matrix products with an application to the algebraic Riccati equation. IEEE Trans. Autom. Control **35**(3), 348–350 (March, 1990)
72. Lasserre, J.B.: A trace inequality for the matrix product. IEEE Trans. Autom. Control **40**, 1500–1501 (1995)

73. Lasserre, J.B.: Tight bounds for the trace of a matrix product. IEEE Trans. Autom. Control **42**(4), 578–581 (April, 1997)
74. Zhang, F., Zhang, Q.: Eigenvalue inequalities for matrix product. IEEE Trans. Autom. Control **51**(9), 1506–1509 (September, 2006)
75. Xing, W., Zhang, Q., Wang, Q.: A trace bound for a general square matrix product. IEEE Trans. Autom. Control **45**(8), 1563–1565 (August, 2000)
76. Liu, J., He, L.: A new trace bound for a general square matrix product. IEEE Trans. Autom. Control **52**(2), 349–352 (February 2007)
77. Mirsky, L.: A trace inequality of John von Neumann. Monatshefte für Mathematik **79**, 303–306 (1975)
78. Hoffman, A.J., Wielandt, H.W.: The variation of the spectrum of a normal matrix. Duke Math. J. **20**, 37–40 (1953)
79. Cochran, J.A., Hinds, E.W.: Improved error bounds for the eigenvalues of certain normal operators. SIAM J. Numer. Anal. **9**(3), 446–453 (September, 1972)
80. Richter, H.: Zur Abschätzung von Matrizennormen. Mathematische Nachrichten **18**, 178–187 (1958)
81. Mirsky, L.: On the trace of matrix products. Mathematische Nachrichten **20**, 171–174 (1959)
82. Marcus, M.: An eigenvalue inequality for the product of normal matrices. Am. Math. Mon. **63**(3), 173–174 (March, 1956)
83. Steele, J.M.: The Cauchy-Schwarz Master Class : An Introduction to the Art of Mathematical Inequalities. Cambridge University Press, Cambridge, New York (2004)
84. Neuman, E., Sándor, J.: On the Ky Fan inequality and related inequalities I. Math. Inequalities Appl. **5**(1), 49–56 (2002)
85. Wang, Y., Chirikjian, G.S.: Error propagation on the Euclidean group with applications to manipulator kinematics. IEEE Trans. Robot. **22**(4), 591–602 (August, 2006)
86. Wang, Y., Chirikjian, G.S.: Nonparametric second-order theory of error propagation on the Euclidean group. Int. J. Robot. Res. **27**(11–12), 1258–1273 (2008)
87. Pennec, X.: *L'incertitude dans les problèmes de reconnaissance et de recalage–Applications en imagerie médicale et biologie moléculaire*, (Doctoral dissertation, Ecole Polytechnique X) (1996)
88. Pennec, X.: Intrinsic statistics on Riemannian manifolds: basic tools for geometric measurements. J. Math Imaging Vis. **25**, 127 (July, 2006)
89. Pennec, X., Arsigny, V.: Exponential barycenters of the canonical Cartan connection and invariant means on Lie groups. In: Barbaresco, F., Mishra, A., Nielsen, F. (eds.) Matrix Information Geometry, pp. 123–166. Springer, Berlin (May, 2012)
90. Heyer, H.: Probability Measures on Locally Compact Groups. Springer, New York (1977)
91. Park, F.C.: The Optimal Kinematic Design of Mechanisms. Ph.D. thesis, Division of Engineering and Applied Sciences, Harvard University, Cambridge, MA (1991)
92. Biedenharn, L.C., Louck, J.D.: Angular Momentum in Quantum Physics. Encyclopedia of Mathematics and Its Applications, vol. 8. Cambridge University Press, Cambridge (1985). (paperback version 2009)
93. Gelfand, I.M., Minlos, R.A., Shapiro, Z.Ya.: Representations of the Rotation and Lorentz Groups and their Applications. Macmillan, New York (1963)
94. Talman, J.: Special Functions. W. A. Benjamin Inc., Amsterdam (1968)
95. Varshalovich, D.A., Moskalev, A.N., Khersonskii, V.K.: Quantum Theory of Angular Momentum. World Scientific, Singapore (1988)
96. Wigner, E.P.: Group Theory and Its Applications to the Quantum Mechanics of Atomic Spectra. Academic Press, New York (1959)

Warped Riemannian Metrics for Location-Scale Models

Salem Said, Lionel Bombrun and Yannick Berthoumieu

Abstract The present contribution shows that warped Riemannian metrics, a class of Riemannian metrics which play a prominent role in Riemannian geometry, are also of fundamental importance in information geometry. Precisely, the starting point is a new theorem, which states that the Rao–Fisher information metric of any location-scale model, defined on a Riemannian manifold, is a warped Riemannian metric, whenever this model is invariant under the action of some Lie group. This theorem is a valuable tool in finding the expression of the Rao–Fisher information metric of location-scale models defined on high-dimensional Riemannian manifolds. Indeed, a warped Riemannian metric is fully determined by only two functions of a single variable, irrespective of the dimension of the underlying Riemannian manifold. Starting from this theorem, several original results are obtained. The expression of the Rao–Fisher information metric of the Riemannian Gaussian model is provided, for the first time in the literature. A generalised definition of the Mahalanobis distance is introduced, which is applicable to any location-scale model defined on a Riemannian manifold. The solution of the geodesic equation, as well as an explicit construction of Riemannian Brownian motion, are obtained, for any Rao–Fisher information metric defined in terms of warped Riemannian metrics. Finally, using a mixture of analytical and numerical computations, it is shown that the parameter space of the von Mises–Fisher model of n-dimensional directional data, when equipped with its Rao–Fisher information metric, becomes a Hadamard manifold, a simply-connected complete Riemannian manifold of negative sectional curvature, for $n = 2, \dots, 8$. Hopefully, in upcoming work, this will be proved for any value of n.

S. Said (✉)
Laboratoire IMS, CNRS, Université de Bordeaux, Bordeaux, France
e-mail: salem.said@u-bordeaux.fr

L. Bombrun
Laboratoire IMS, Bordeaux Sciences Agro, Bordeaux, France
e-mail: lionel.bombrun@u-bordeaux.fr

Y. Berthoumieu
Laboratoire IMS, Bordeaux INP, Bordeaux, France
e-mail: yannick.berthoumieu@u-bordeaux.fr

F. Nielsen (ed.), *Geometric Structures of Information*, Signals and Communication Technology, https://doi.org/10.1007/978-3-030-02520-5_10

1 Introduction

Warped Riemannian metrics are a class of Riemannian metrics which arise throughout Riemannian geometry [11, 40]. For example, the Riemannian metrics of surfaces of revolution, and of spaces of constant curvature (when restricted to polar coordinate charts), are warped Riemannian metrics. Closely similar to warped Riemannian metrics, warped semi-Riemannian metrics are very important in theoretical physics. Indeed, many gravitational models are given by warped semi-Riemannian metrics [38]. The present contribution shows that warped metrics, in addition to their well-known role in geometry and physics, play a fundamental role in information geometry, and have a strong potential for applications in statistical inference and statistical learning.

A unified definition of warped Riemannian metrics was first formulated in [11]. Here, only a special case of this definition is required. Precisely, let M be a complete Riemannian manifold, with length element ds^2_M , and consider the product manifold $\mathcal{M} = M \times (0\,,\infty)$, equipped with the length element $ds^2_{\mathcal{M}}$

$$ds^2_{\mathcal{M}}(z) = dr^2 + \beta^2(r)\, ds^2_M(x) \qquad \text{for } z = (x\,,r) \in \mathcal{M} \tag{1a}$$

where $\beta^2(r)$ is a strictly positive function. Then, the length element $ds^2_{\mathcal{M}}$ defines a warped Riemannian metric on $\mathcal{M}$. In Riemannian geometry, the coordinate r is a distance function, measuring the distance to some point or hypersurface [40]. In physics, r is replaced by the time t, and dr^2 is replaced by $-dt^2$ in formula (1a) (this is the meaning of "semi-Riemannian") [38]. In any case, the coordinate x can be thought of as a spatial coordinate which determines a position in M.

The intuition behind the present contribution is that warped Riemannian metrics are natural candidates for Riemannian metrics on location-scale models. Indeed, if $\mathcal{P}$ is a location-scale model on M, with location parameter $\bar{x} \in M$ and scale parameter $\sigma > 0$, then the parameter space of $\mathcal{P}$ is exactly $\mathcal{M} = M \times (0\,,\infty)$ with its points $z = (\bar{x}, \sigma)$. Thus, a warped Riemannian metric on $\mathcal{M}$ can be defined using (1a), after introducing a new scale parameter $r = r(\sigma)$ and setting $x = \bar{x}$.

As it turns out, this intuition is far from arbitrary. The main new result in the present contribution, Theorem 1 of Sect. 3, states that the Rao–Fisher information metric of any location-scale model is a warped Riemannian metric, whenever this model is invariant under the action of some Lie group. Roughly, Theorem 1 states that if M is a Riemannian symmetric space under the transitive action of a Lie group of isometries G, and if each probability density $p(x|\,\bar{x}, \sigma)$, belonging to the model $\mathcal{P}$, verifies the invariance condition

$$p(g \cdot x|\, g \cdot \bar{x}\,, \sigma) = p(x|\bar{x}, \sigma) \quad \text{for all } g \in G \tag{1b}$$

where $g \cdot x$ denotes the action of $g \in G$ on $x \in M$, then the Rao–Fisher information metric of the model $\mathcal{P}$ is a warped Riemannian metric.

A technical requirement for Theorem 1 is that the Riemannian symmetric space M should be irreducible. The meaning of this requirement, and the fact that it can be relaxed in certain cases, are discussed in Remarks 4 and 5 of Sect. 3. The proof of Theorem 1 is given in Appendix A.

A fundamental idea of information geometry is that the parameter space of a statistical model $\mathcal{P}$ should be considered as a Riemannian manifold [3, 15]. According to [7, 15], the unique way of doing so is by turning Fisher's information matrix into a Riemannian metric, the Rao–Fisher information metric. In this connection, Theorem 1 shows that, when the statistical model $\mathcal{P}$ is a location-scale model which is invariant under the action of a Lie group, information geometry inevitably leads to the study of warped Riemannian metrics.

In addition to stating and proving Theorem 1, the present contribution aims to explore its implications, with regard to the Riemannian geometry of location-scale models, and to lay the foundation for its applications in statistical inference and statistical learning.

To begin, Sect. 4 applies Theorem 1 to two location-scale models, the von Mises–Fisher model of directional data[16, 33], and the Riemannian Gaussian model of data in spaces of covariance matrices [14, 41, 42]. This leads to the analytic expression of the Rao–Fisher information metric of each one of these two models. Precisely, the Rao–Fisher information metric of the von Mises–Fisher model is given in Proposition 2, and that of the Riemannian Gaussian model is given in Proposition 3. The result of Proposition 2 is essentially already contained in [33], (see p. 199), but Proposition 3 is new in the literature.

Finding the analytic expression of the Rao–Fisher information metric, or equivalently of Fisher's information matrix, of a location-scale model $\mathcal{P}$ defined on a high-dimensional non-trivial manifold M, is a very difficult task when attempted by direct calculation. Propositions 2 and 3 show that this task is greatly simplified by Theorem 1. Precisely, if the dimension of M is d, then the dimension of the parameter space $\mathcal{M} = M \times (0\,,\infty)$ is $d+1$. Therefore, a priori, the expression of the Rao–Fisher information metric involves $(d+1)(d+2)/2$ functions of both parameters $\bar{x}$ and σ of the model $\mathcal{P}$. Instead of so many functions of both $\bar{x}$ and σ, Theorem 1 reduces the expression of the Rao–Fisher information metric to only two functions of σ alone. In the notation of (1a), these two functions are $\alpha(\sigma) = dr/d\sigma$ and $\beta(\sigma) = \beta(r(\sigma))$.

Section 5 builds on Theorem 1 to introduce a general definition of the Mahalanobis distance, applicable to any location-scale model $\mathcal{P}$ defined on a manifold M. Precisely, assume that the model $\mathcal{P}$ verifies the conditions of Theorem 1, so its Rao–Fisher information metric is a warped Riemannian metric. Then, the *generalised Mahalanobis distance* is defined as the Riemannian distance on M which is induced by the restriction of the Rao–Fisher information metric to M. The expression of the generalised Mahalanobis distance is given in Propositions 4 and 5. It was recently applied to visual content classification in [10].

The generalised Mahalanobis distance includes the classical Mahalanobis distance as a special case. Precisely, assume $\mathcal{P}$ is the isotropic normal model defined on $M = \mathbb{R}^d$, so each density $p(x\,|\,\bar{x},\sigma)$ is a d-variate normal density with mean $\bar{x}$ and covariance matrix σ^2 times the identity. Then, $\mathcal{P}$ verifies the invariance condition (1b) under the action of the group G of translations in $\mathbb{R}^d$. Therefore, by Theorem 1, its Rao–Fisher information metric is a warped Riemannian metric. This metric is already known in the literature, in terms of the length element [8, 9]

$$ds^2_{\mathcal{M}}(z) = \frac{2d}{\sigma^2}\,d\sigma^2 + \frac{1}{\sigma^2}\,\|d\bar{x}\|^2 \quad \text{for } z = (\bar{x}, \sigma) \in \mathcal{M} \tag{1c}$$

where $\|d\bar{x}\|^2$ denotes the Euclidean length element on $\mathbb{R}^d$. Now, the restriction of this Rao–Fisher information metric to $M = \mathbb{R}^d$ is given by the second term in (1c), which is clearly the Euclidean length element divided by σ^2, and corresponds to the classical Mahalanobis distance [36] — note that (1c) is brought into the form (1a) by letting $r(\sigma) = (2d)^{1/2}\log(\sigma)$.

Section 6 illustrates the results of Sects. 4 and 5, by applying them to the special case of the Riemannian Gaussian model defined on $M = \mathcal{P}_n$, the space of $n \times n$ real covariance matrices. In particular, it gives directly applicable expressions of the Rao–Fisher information metric, and of the generalised Mahalanobis distance, corresponding to this model. As it turns out, the generalised Mahalanobis distance defines a whole new family of affine-invariant distances on $\mathcal{P}_n$, in addition to the usual affine-invariant distance, which was introduced to the information science community in [39].

Section 7 provides the solution of the geodesic equation of any of the Rao–Fisher information metrics arising from Theorem 1. The main result is Proposition 6, which states that the solution of this equation, for given initial conditions, reduces to the solution of a one-dimensional second-order differential equation. This implies that geodesics with given initial conditions can be constructed at a reasonable numerical cost, which opens the possibility, with regard to future work, of a practical implementation of Riemannian line-search optimisation algorithms, which find the extrema of cost functions by searching for them along geodesics [1, 32].

Section 8 is a follow up to Sect. 7, focusing on the construction of Riemannian Brownian motion, instead of the solution of the geodesic equation. The main result is Proposition 7, which states that the construction of Riemannian Brownian motion reduces to the solution of a one-dimensional stochastic differential equation. This implies that Brownian paths can be constructed at a reasonable computational cost, which opens the possibility of a practical implementation of Riemannian stochastic search algorithms, which have the ability to avoid local minima and saddle points [5].

Section 9 is motivated by the special case of the isotropic normal model, with its Rao–Fisher information metric given by (1c). It is well-known that, after a trivial change of coordinates, the length element (1c) coincides with the length element of the Poincaré half-space model of hyperbolic geometry [8, 9]. This means that the parameter space of the isotropic normal model, when equipped with its Rao–Fisher information metric, becomes a space of constant negative curvature, and in particular a Hadamard manifold, a simply-connected complete Riemannian manifold of negative sectional curvature [13, 40]. One cannot but wonder whether other location-scale models also give rise to Hadamard manifolds in this way. This is investigated using Propositions 2 and 6, for the case of the von Mises–Fisher model. A mixture of analytical and numerical computations leads to a surprising new observation: the parameter space of the von Mises–Fisher model of n-dimensional directional data, when equipped with its Rao–Fisher information metric, becomes a Hadamard manifold, for $n = 2, \ldots, 8$. Ongoing research warrants the conjecture that this is true for any value of n, but this is yet to be proved.

Theorem 1, the main new result in the present contribution, has many potential applications, which will be developed in future work. Indeed, this theorem can provide the expression of the Rao–Fisher information metric, or equivalently of Fisher's information matrix, of a location-scale model, even if this model is defined on a high-dimensional non-trivial manifold. By doing so, it unlocks access to the many applications which require this expression, both in statistical inference and in statistical learning. In statistical inference, the expression of the Rao–Fisher information metric allows the computation of the Cramér-Rao lower bound, and the construction of asymptotic chi-squared statistics [30, 48]. In statistical learning, it allows the practical implementation of the natural gradient algorithm, which has the advantages of full reparameterisation invariance, of asymptotic efficiency of its stochastic version, and of the linear rate of convergence of its deterministic version [2, 12, 35].

A first step, towards developing the applications of Theorem 1, was recently taken in [50]. Using the expression of the Rao–Fisher information metric of the Riemannian Gaussian model, this derived and successfully implemented the natural gradient algorithm, for the problem of on-line learning of an unknown probability density, on the space $\mathcal{P}_n$ of $n \times n$ real covariance matrices. Future work will focus on extending this algorithm to more problems of statistical learning, including problems of on-line classification and regression in the space $\mathcal{P}_n$, and in other spaces of covariance matrices. In addition to its use in deriving the natural gradient algorithm for problems of statistical learning, the expression of the Rao–Fisher information metric of the Riemannian Gaussian model can be used in deriving so-called natural evolutionary strategies, for black-box optimisation in spaces of covariance matrices. These would generalize currently existing natural evolutionary strategies, mostly restricted to black-box optimisation in Euclidean space [37, 47].

The following Sect. 2 provides background on warped Riemannian metrics. While the present introduction expressed Riemannian metrics using length elements, Sect. 2 will use scalar products on the tangent space.

2 Background on Warped Riemannian Metrics

Assume M is a complete Riemannian manifold with Riemannian metric Q, and consider the manifold $\mathcal{M} = M \times (0, \infty)$. A warped Riemannian metric I on $\mathcal{M}$ is given in the following way [11, 38, 40]. Let α and β be positive functions, defined on $(0, \infty)$. Then, for $z = (x, \sigma) \in \mathcal{M}$, let the scalar product I_z on the tangent space $T_z\mathcal{M}$ be defined by

$$I_z(U,U) \;=\; (\alpha(\sigma)\,u_\sigma)^2 \;+\; \beta^2(\sigma)\,Q_x(u,u) \quad U \in T_z\mathcal{M} \tag{2a}$$

where $U \;=\; u_\sigma\,\partial_\sigma \;+\; u$ with $u_\sigma \in \mathbb{R}$ and $u \in T_xM$. The functions α and β are part of the definition of the warped metric I. Once these functions are fixed, it is possible to introduce a change of coordinates $r = r(\sigma)$ which eliminates α from (2a). Precisely, if $dr/d\sigma \;=\; \alpha(\sigma)$ then

$$I_z(U,U) \;=\; u_r^2 \;+\; \beta^2(r)\,Q_x(u,u) \tag{2b}$$

where $U \;=\; u_r\,\partial_r \;+\; u$ and $\beta(r) = \beta(\sigma(r))$.

The coordinate r will be called *vertical distance*. This is not a standard terminology, but is suggested as part of the following geometric picture. For $z = (x,\sigma) \in \mathcal{M}$, think of x as a horizontal coordinate, and of σ as a vertical coordinate. Accordingly, the points $z_0 = (x,\sigma_0)$ and $z_1 = (x,\sigma_1)$ lie on the same vertical line. It can be shown from (2b) that the Riemannian distance between z_0 and z_1 is

$$d(z_0,z_1) \;=\; r(\sigma_1) - r(\sigma_0) \quad \text{where} \quad \sigma_0 \,<\, \sigma_1 \tag{3a}$$

Precisely, $d(z_0,z_1)$ is the Riemannian distance induced by the warped Riemannian metric I.

The vertical distance r can be used to express a necessary and sufficient condition for completeness of the manifold $\mathcal{M}$, equipped with the warped Riemannian metric I. Namely, $\mathcal{M}$ is a complete Riemannian manifold, if and only if

$$\lim_{\sigma\to\infty} r(\sigma) - r(\sigma_0) = \infty \;\;\text{and}\;\; \lim_{\sigma\to 0} r(\sigma_1) - r(\sigma) = \infty \tag{3b}$$

where σ_0 and σ_1 are arbitrary. This condition is a special case of Lemma 7.2 in [11].

Let $K^{\mathcal{M}}$ and K^M denote the sectional curvatures of $\mathcal{M}$ and M, respectively. The relation between these two is given by the curvature equations of Riemannian geometry [18, 40]. These are,

$$\text{Gauss equation} \,:\, K_z^{\mathcal{M}}(u,v) \;=\beta^{-2}\,K_x^M(u,v) - \left(\beta^{-1}\partial_r\beta\right)^2 \tag{4a}$$

$$\text{Jacobi equation} \,:\, K_z^{\mathcal{M}}(u,\partial_r) \;=-\,\beta^{-1}\,\partial_r^2\beta \tag{4b}$$

for $u,v \in T_xM$. Here, the notations $K_z^{\mathcal{M}}$ and K_x^M mean that $K^{\mathcal{M}}$ is computed at z, and K^M is computed at x, where $z = (x,\sigma)$. Equation (4) are a special case of Lemma 7.4 in [11].

Note, as a corollary of these equations, that $\mathcal{M}$ has negative sectional curvature $K^{\mathcal{M}} < 0$, if M has negative sectional curvature $K^M < 0$ and β is a strictly convex function of r.

Remark 1 Equation (2) contain an abuse of notation. Namely, u denotes a tangent vector to M at x, and a tangent vector to $\mathcal{M}$ at z, at the same time. In the mathematical literature (for example, in [11, 38]), one writes $d\pi_z(U)$ instead of u, using the derivative $d\pi$ of the projection mapping $\pi(z) = x$, and this eliminates any ambiguity. In the present contribution, a deliberate choice is made to use a lighter, though not entirely correct, notation. ■

Remark 2 Consider the proof of Eq. (3). For (3a), let $\gamma(t)$ and $c(t)$ be curves connecting $z_0 = (x, \sigma_0)$ and $z_1 = (x, \sigma_1)$. Assume these are parameterised by $t \in [0, 1]$ as follows,

$$\gamma(t) \, : \, \begin{Bmatrix} x_\gamma(t) = x \quad \text{(constant)} \\ r_\gamma(t) = r(\sigma_0) + t\,(r(\sigma_1) - r(\sigma_0)) \end{Bmatrix} \quad ; \quad c(t) \, : \, \begin{Bmatrix} x_c(t) \\ r_c(t) \end{Bmatrix}$$

If $L(\gamma)$ and $L(c)$ denote the lengths of these curves, then from (2b),

$$L(c) = \int_0^1 \left((\dot{r}_c)^2 + \beta^2\, Q(\dot{x}_c, \dot{x}_c)\right)^{1/2} dt \geq \int_0^1 \dot{r}_c\, dt \; = \; r(\sigma_1) - r(\sigma_0) = L(\gamma)$$

where the dot denotes differentiation with respect to t, and the inequality is strict unless $\gamma = c$. This shows that $\gamma(t)$ is the unique length-minimising geodesic connecting z_0 and z_1. Thus, $d(z_0, z_1) = L(\gamma)$, and this gives (3a). For (3b), note that Lemma 7.2 in [11] states that $\mathcal{M}$ is complete, if and only if $(0\,,\infty)$ is complete when equipped with the distance $d_{(0,\infty)}(\sigma_0\,,\sigma_1) \; = |r(\sigma_1) - r(\sigma_0)|$. However, this is equivalent to (3b). ■

Remark 3 For each $\sigma \in (0\,,\infty)$, the manifold M can be embedded into the manifold $\mathcal{M}$, in the form of the hypersurface $M_\sigma = M \times \{\sigma\}$. Through this embedding, the warped Riemannian metric I of $\mathcal{M}$ induces a Riemannian metric Q^σ on M. By definition, this metric Q^σ is obtained by the restriction of I to the tangent vectors of M_σ [18, 40]. It follows from (2) that

$$Q^\sigma_x(u, u) \; = \; \beta^2(\sigma)\, Q_x(u, u) \tag{5}$$

The induced metric Q^σ will be called an extrinsic metric on M, since it comes from the ambient space $\mathcal{M}$. By (5), the extrinsic metric Q^σ is equal to a scaled version of the Riemannian metric Q of M, with scaling factor $\beta(\sigma)$. ■

3 Connection with Location-Scale Models

This section establishes the connection between warped Riemannian metrics and location-scale models. The main result is Theorem 1, which states that the Rao–Fisher information metric of any location-scale model is a warped Riemannian metric, whenever this model is invariant under the action of some Lie group.

To state this theorem, assume M is an irreducible Riemannian symmetric space, with invariant Riemannian metric Q, under the transitive action of a Lie group of isometries G [22]. Consider a location-scale model $\mathcal{P}$ defined on M,

$$\mathcal{P} = \{\, p(x|\bar{x},\sigma)\,;\ \bar{x} \in M\,,\ \sigma \in (0\,,\infty)\} \tag{6}$$

To each point $z = (\bar{x}, \sigma)$ in the parameter space $\mathcal{M} = M \times (0\,, \infty)$, this model associates a probability density $p(x|\bar{x}, \sigma)$ on M, which has a location parameter $\bar{x}$ and a scale parameter σ. Precisely, $p(x|\bar{x}, \sigma)$ is a probability density with respect to the invariant Riemannian volume element of M.

The condition that the model $\mathcal{P}$ is invariant under the action of the Lie group G means that,

$$p(\,g \cdot x|\,g \cdot \bar{x}\,,\sigma) \;=\; p(x|\bar{x},\sigma) \quad \text{for all } g \in G \tag{7}$$

where $g \cdot x$ denotes the action of $g \in G$ on $x \in M$.

The Rao–Fisher information metric of the location-scale model $\mathcal{P}$ is a Riemannian metric I on the parameter space $\mathcal{M}$ of this model [3]. It is defined as follows, for $z = (\bar{x}, \sigma) \in \mathcal{M}$ and $U \in T_z\mathcal{M}$,

$$I_z(U,U) \;=\; \mathbb{E}_z\left[(\,d\ell(z)\,U\,)^2\right] \tag{8}$$

where $\mathbb{E}_z$ denotes expectation with respect to the probability density $p(x|z) = p(x|\bar{x}, \sigma)$, and $\ell(z)$ is the log-likelihood function, given by $\ell(z)(x) = \log p(x|z)$.

In the following statement, $\nabla_{\bar{x}}\,\ell(z)$ denotes the Riemannian gradient of $\ell(z)$, taken with respect to $\bar{x} \in M$, while the value of σ is fixed.

Theorem 1 *If condition (7) is verified, then the Rao–Fisher information metric I of (8) is a warped Riemannian metric given by (2a), where*

$$\alpha^2(\sigma) = \mathbb{E}_z\,(\partial_\sigma \ell(z))^2 \quad \beta^2(\sigma) \;=\; \mathbb{E}_z\,Q\,(\nabla_{\bar{x}}\,\ell(z)\,,\,\nabla_{\bar{x}}\,\ell(z)\,)\,/\dim M \tag{9}$$

The expectations appearing in (9) do not depend on $\bar{x}$, so $\alpha(\sigma)$ and $\beta(\sigma)$ are well-defined functions of σ.

Remark 4 Recall the definition of an irreducible Riemannian symmetric space [22]. A Riemannian manifold M, whose group of isometries is denoted G, is called a Riemannian symmetric space, if for each $\bar{x} \in M$ there exists an isometry $s_{\bar{x}} \in G$, whose effect is to fix $\bar{x}$ and reverse the geodesic curves passing through $\bar{x}$. Further, M is called irreducible if it verifies the following condition. Let $K_{\bar{x}}$ be the subgroup of G which consists of those elements k such that $k \cdot \bar{x} = \bar{x}$. For each $k \in K_{\bar{x}}$, its derivative $dk_{\bar{x}}$ is a linear mapping of $T_{\bar{x}}M$. The mapping $k \mapsto dk_{\bar{x}}$ is a representation of $K_{\bar{x}}$ in $T_{\bar{x}}M$, called the isotropy representation, and M is called an irreducible Riemannian symmetric space if the isotropy representation is irreducible. That is, if the isotropy representation has no invariant subspaces in $T_{\bar{x}}M$, except $\{0\}$ and $T_{\bar{x}}M$. Irreducible Riemannian symmetric spaces are classified in [22] (Table I, p. 346 and Table II, p. 354). They include spaces of constant curvature, such as spheres and hyperbolic spaces, as well as spaces of positive definite matrices which have determinant equal to 1, and whose entries are real or complex numbers, or quaternions. ■

Remark 5 It is sometimes possible to apply Theorem 1, even when the Riemannian symmetric space M is not irreducible. For example, in Sect. 4.2, Theorem 1 will be used to find the expression of the Rao–Fisher information metric of the Riemannian Gaussian model [14, 41, 42]. For this model, when M is not irreducible, the Rao–Fisher information metric turns out to be a so-called multiply-warped Riemannian metric, rather than a warped Riemannian metric. The concrete case of $M = \mathcal{P}_n$, the space of $n \times n$ real covariance matrices, is detailed in Sect. 6. ■

Proof of Theorem 1 Recall the expression $U = u_\sigma \, \partial_\sigma + u$ with $u_\sigma \in \mathbb{R}$ and $u \in T_{\bar{x}}M$. Since the Rao–Fisher information metric I is bilinear and symmetric,

$$I_z(U, U) = I_z(\partial_\sigma, \partial_\sigma) \, u_\sigma^2 + 2I_z(\partial_\sigma, u) \, u_\sigma + I_z(u, u)$$

It is possible to show the following,

$$I_z(\partial_\sigma, \partial_\sigma) = \alpha^2(\sigma) \tag{10a}$$

$$I_z(\partial_\sigma, u) = 0 \tag{10b}$$

$$I_z(u, u) = \beta^2(\sigma) \, Q_{\bar{x}}(u, u) \tag{10c}$$

where $\alpha^2(\sigma)$ and $\beta^2(\sigma)$ are given by (9).
Proof of (10a): this is immediate from (8). Indeed,

$$I_z(\partial_\sigma, \partial_\sigma) = \mathbb{E}_z \left[(d\ell(z) \, \partial_\sigma)^2 \right] = \mathbb{E}_z \left[(\partial_\sigma \ell(z))^2 \right]$$

Proof of (10b): this is carried out in Appendix A, using the fact that M is a Riemannian symmetric space.
Proof of (10c): this is carried out in Appendix A, using the fact that M is irreducible, by an application of Schur's lemma from the theory of group representations [27].

The fact that the expectations appearing in (9) do not depend on $\bar{x}$ is also proved in Appendix A. Throughout the proof of the theorem, the following identity is used, which is equivalent to condition (7). For any real-valued function f on M,

$$\mathbb{E}_{g \cdot z} \, f = \mathbb{E}_z \, (f \circ g) \tag{11}$$

Here, $g \cdot z = (g \cdot \bar{x}, \sigma)$, and $f \circ g$ is the function $(f \circ g)(x) = f(g \cdot x)$, for $g \in G$ and $z = (\bar{x}, \sigma)$. ■

4 Examples: von Mises–Fisher and Riemannian Gaussian

This section applies Theorem 1 to finding the expression of the Rao–Fisher information metric of two location-scale models. These are the von Mises–Fisher model, which is widely used in the study of directional data [16, 33], and the Riemannian

Gaussian model, recently introduced in the study of data with values in spaces of covariance matrices [14, 41, 42].

The application of Theorem 1 to these two models is encapsulated in the following Proposition 1. Precisely, both of these models are of a common exponential form, which can be described as follows. Let M be an irreducible Riemannian symmetric space, as in Sect. 3. In the notation of (6), consider a location-scale model $\mathcal{P}$ defined on M, by

$$p(x|\bar{x}, \sigma) = \exp[\eta(\sigma) D(x, \bar{x}) - \psi(\eta(\sigma))] \tag{12a}$$

where $\eta(\sigma)$ is a certain parameter, to be called the natural parameter, and where $D : M \times M \to \mathbb{R}$ verifies the condition,

$$D(g \cdot x\,,\ g \cdot \bar{x}) = D(x\,,\ \bar{x}) \quad \text{for all } g \in G \tag{12b}$$

There is no need to assume that the function D is positive.

Proposition 1 *If the model $\mathcal{P}$ is given by Eq. (12), then the Rao–Fisher information metric I of this model is a warped Riemannian metric,*

$$I_z(U, U) = \psi''(\eta)\, u_\eta^2 + \beta^2(\eta)\, Q_{\bar{x}}(u, u) \tag{13a}$$

where $U = u_\eta\, \partial_\eta + u$ with $u_\eta \in \mathbb{R}$ and $u \in T_{\bar{x}}M$, and where

$$\beta^2(\eta) = \eta^2\, \mathbb{E}_z\, Q\,(\nabla_{\bar{x}} D, \nabla_{\bar{x}} D)\,/\dim M \tag{13b}$$

Proof For a model $\mathcal{P}$ defined by (12a), condition (12b) is equivalent to condition (7). Therefore, by application of Theorem 1, it follows that I is a warped Riemannian metric, of the form (2a),

$$I_z(U, U) = (\alpha(\sigma)\, u_\sigma)^2 + \beta^2(\sigma)\, Q_{\bar{x}}(u, u) \tag{14a}$$

where $\alpha^2(\sigma)$ and $\beta^2(\sigma)$ are given by (9). Consider the first term in (14a). By the change of coordinates formula [29], $u_\sigma = \sigma'(\eta)\, u_\eta$, where the prime denotes differentiation with respect to η. It follows that

$$(\alpha(\sigma)\, u_\sigma)^2 = \alpha^2(\sigma)\, \big(\sigma'(\eta)\big)^2\, u_\eta^2 \tag{14b}$$

However, by (9),

$$\alpha^2(\sigma)\, \big(\sigma'(\eta)\big)^2 = \mathbb{E}_z\, \big(\partial_\sigma \ell(z)\, \sigma'(\eta)\,\big)^2 = \mathbb{E}_z\, \big(\partial_\eta \ell(z)\,\big)^2 \tag{14c}$$

Here, the log-likelihood $\ell(z)$ is found from (12a),

$$\ell(z)(x) = \eta(\sigma)\, D(x, \bar{x}) - \psi(\eta(\sigma)) \tag{14d}$$

Therefore, the last expression in (14c) is

$$\mathbb{E}_z \left(\partial_\eta \ell(z) \right)^2 = -\mathbb{E}_z \, \partial^2_\eta \ell(z) = \psi''(\eta) \tag{14e}$$

where the first equality is the same as in [3], (see p. 28). Now, (14b) and (14c) imply

$$(\alpha(\sigma) \, u_\sigma)^2 = \psi''(\eta) \, u^2_\eta \tag{14f}$$

Replacing this in (14a), and writing $\beta(\eta) = \beta(\sigma(\eta))$, gives

$$I_z(U, U) = \psi''(\eta) \, u^2_\eta + \beta^2(\eta) \, Q_{\bar{x}}(u, u) \tag{14g}$$

which is the same as (13a). To prove the proposition, it remains to show that $\beta^2(\eta)$ is given by (13b). To do so, note that it follows from (14d),

$$\nabla_{\bar{x}} \, \ell(z) = \nabla_{\bar{x}} \left[\, \eta(\sigma) \, D(x, \bar{x}) \; - \; \psi(\eta(\sigma)) \right] = \eta(\sigma) \, \nabla_{\bar{x}} \, D(x, \bar{x})$$

Replacing this in (9) gives,

$$\beta^2(\eta) = \mathbb{E}_z \, Q \left(\nabla_{\bar{x}} \, \ell(z) \, , \nabla_{\bar{x}} \, \ell(z) \right) / \dim M \; = \eta^2 \; \mathbb{E}_z \, Q \left(\nabla_{\bar{x}} \, D, \nabla_{\bar{x}} \, D \right) / \dim M$$

and this is the same as (13b). ■

4.1 The von Mises–Fisher Model

The von Mises–Fisher model is a mainstay of directional statistics [16, 33]. In the notation of (12), this model corresponds to $M = S^{n-1}$, the unit sphere in $\mathbb{R}^n$, and to $G = O(n)$, the group of $n \times n$ real orthogonal matrices, which acts on $\mathbb{R}^n$ by rotations. Then, the expressions appearing in (12a) are

$$D(x, \bar{x}) = \langle x, \bar{x} \rangle \quad \psi(\eta) = \nu \log(2\pi) + \log \left(\eta^{1-\nu} I_{\nu-1}(\eta) \right) \tag{15}$$

for $\eta \in [\,0\,, \infty)$. Here, $\langle x, \bar{x} \rangle$ denotes the Euclidean scalar product in $\mathbb{R}^n$, so that condition (12b) is clearly verified, and $I_{\nu-1}$ denotes the modified Bessel function of order $\nu - 1$, where $\nu = n/2$. The natural parameter η and the scale parameter σ should be considered identical, in the sense that $\eta(\sigma) = \sigma$, as long as $\sigma \in (0\,, \infty)$. However, η takes on the additional value $\eta = 0$, which requires a special treatment.

Remark 6 The parameter space of the von Mises–Fisher model will be identified with the space $\mathbb{R}^n$. This is done by mapping each couple $(\bar{x}, \eta)$ to the point $z = \eta \, \bar{x}$ in $\mathbb{R}^n$. This mapping defines a diffeomorphism from the set of couples $(\bar{x}, \eta)$ where $\eta \in (0\,, \infty)$, to the open subset $\mathbb{R}^n - \{0\} \subset \mathbb{R}^n$. On the other hand, it maps all couples $(\bar{x}, \eta = 0)$, to the same point $z = 0 \in \mathbb{R}^n$. Note that each couple $(\bar{x}, \eta)$ where

$\eta \in (0\,,\infty)$ defines a distinct von Mises–Fisher distribution, which is a unimodal distribution with its mode at $\bar{x}$. On the other hand, all couples $(\bar{x}, \eta = 0)$ define the same von Mises–Fisher distribution, which is the uniform distribution on S^{n-1}. Therefore, it is correct to map all of these couples to the same point $z = 0$. ■

Proposition 1 will only provide the Rao–Fisher information metric of the von Mises–Fisher model on the subset $\mathbb{R}^n - \{0\}$ of the parameter space $\mathbb{R}^n$. Therefore, it is necessary to verify that this metric has a well-defined limit at the point $z = 0$. This is carried out in Proposition 2 below. In the statement of this proposition, a tangent vector $U \in T_z\mathbb{R}^n$, at a point $z \in \mathbb{R}^n - \{0\}$, is written in the form

$$U = u_\eta\,\bar{x} + \eta\,u \tag{16a}$$

where $z = \eta\,\bar{x}$, and where $u_\eta \in \mathbb{R}$ and $u \in T_{\bar{x}}S^{n-1}$. Here, u_η and u are unique, for a given U. Precisely,

$$u_\eta = \langle U, \bar{x}\rangle \quad u = \frac{1}{\eta}\,[\,U - \langle U, \bar{x}\rangle\,\bar{x}\,] \tag{16b}$$

as follows since $\bar{x}$ and u are orthogonal.

Proposition 2 *The Rao–Fisher information metric I of the von Mises–Fisher model is a well-defined Riemannian metric on the parameter space* $\mathbb{R}^n$. *On* $\mathbb{R}^n - \{0\}$, *it is a warped Riemannian metric of the form (13a), where*

$$\psi''(\eta) = \frac{1}{n} + \frac{n-1}{n}\frac{I_{\nu+1}(\eta)}{I_{\nu-1}(\eta)} - \frac{I^2_\nu(\eta)}{I^2_{\nu-1}(\eta)} \tag{17a}$$

$$\beta^2(\eta) = \frac{\eta^2}{n}\left(1 + \frac{I_{\nu+1}(\eta)}{I_{\nu-1}(\eta)}\right) \tag{17b}$$

and it extends smoothly to the value,

$$I_0(U, U) = \frac{1}{n}\,\|U\|^2 \tag{17c}$$

at the point $z = 0$. *Here,* $\|\cdot\|$ *denotes the Euclidean norm.*

Proof The Rao–Fisher information metric I on $\mathbb{R}^n - \{0\}$ is given by Proposition 1. This proposition applies because $M = S^{n-1}$ is an irreducible Riemannian symmetric space [22] (Table II, p. 354). Accordingly, for any point $z \in \mathbb{R}^n - \{0\}$, the metric I_z is given by (13a). Formulae (17) are proved as follows.
Proof of (17a): this is carried out in Appendix B, using the derivative and recurrence relations of modified Bessel functions [46].
Proof of (17b): this follows from (13b) and (15). By (15),

$$\nabla_{\bar{x}}\,D(x, \bar{x}) = x - \langle x, \bar{x}\rangle\,\bar{x}$$

which is just the orthogonal projection of x onto the tangent space $T_{\bar{x}}S^{n-1}$. Replacing in (13b) gives

$$\beta^2(\eta) = \frac{\eta^2}{n-1}\,\mathbb{E}_z\,\|\nabla_{\bar{x}}\,D\|^2 = \frac{\eta^2}{n-1}\,\mathbb{E}_z\left(1 - \langle x, \bar{x}\rangle^2\right) \tag{18a}$$

Here, in the first equality, $n-1$ appears because $\dim S^{n-1} = n-1$. The second equality follows by Pythagoras' theorem,

$$\|\, x - \langle x, \bar{x}\rangle\, \bar{x}\,\|^2 = \|\, x\,\|^2 - \|\,\langle x, \bar{x}\rangle\,\bar{x}\,\|^2 = 1 - \langle x, \bar{x}\rangle^2$$

since x and $\bar{x}$ belong to the unit sphere S^{n-1}. Formula (17b) is derived from (18a) in Appendix B, using the derivative and recurrence relations of modified Bessel functions [46].

Proof of (17c): for any point $z \in \mathbb{R}^n - \{0\}$, the metric I_z is given by (13a). This reads

$$I_z(U, U) = \psi''(\eta)\, u_\eta^2 + \beta^2(\eta)\,\|u\|^2 \tag{18b}$$

Consider the limit of this expression at the point $z = 0$. In (17a) and (17b), this corresponds to the limit at $\eta = 0$. This can be evaluated using the power series development of modified Bessel functions [46]. When replaced in (17a) and (17b), this gives the following developments,

$$\psi''(\eta) = \frac{1}{n} - \frac{12}{n^2(n+2)}\left(\frac{\eta}{2}\right)^2 + o\left(\eta^4\right)$$

$$\beta^2(\eta) = \frac{4}{n}\left(\frac{\eta}{2}\right)^2 + o\left(\eta^4\right)$$

which immediately imply that

$$\lim_{\eta\to 0}\ \psi''(\eta) = \frac{1}{n} \tag{18c}$$

$$\lim_{\eta\to 0}\ \beta^2(\eta) = 0 \tag{18d}$$

Replacing (18c) and (18d) in (18b) gives,

$$\lim_{z\to 0}\ I_z(U, U) = \frac{1}{n}\,u_\eta^2 \tag{18e}$$

Note that, from (16a),

$$\|U\|^2 = u_\eta^2 + \eta^2\,\|u\|^2$$

by Pythagoras' theorem, since $\bar{x}$ and u are orthogonal. At the point $z = 0$, one has $\eta = 0$, so that

$$\|U\|^2_{z=0} = u^2_\eta$$

This shows that (18e) is the same as

$$\lim_{z\to 0} I_z(U,U) = \frac{1}{n}\,\|U\|^2 \tag{18f}$$

This limit does not depend on the path along which z tends to $z = 0$. Therefore, I_z extends smoothly to I_0, which is given by (17c), at the point $z = 0$. This shows that I is a well-defined Riemannian metric throughout the parameter space $\mathbb{R}^n$. ■

4.2 The Riemannian Gaussian Model

The Riemannian Gaussian model was recently introduced as a means of describing unimodal populations of covariance matrices [14, 41, 42]. This model can be defined on any Riemannian symmetric space of non-positive sectional curvature. Let M be such a symmetric space and denote G its group of isometries. Then, the expressions appearing in (12a) are

$$D(x,\bar{x}) = d^2(x,\bar{x}) \quad \eta(\sigma) = -\frac{1}{2\sigma^2} \tag{19}$$

where $d(x,\bar{x})$ denotes the Riemannian distance in M, and condition (12b) is verified since each isometry $g \in G$ preserves this Riemannian distance. The function $\psi(\eta)$ is a strictly convex function of $\eta \in (-\infty\,,0)$, which can be expressed by means of a multiple integral [42], (see Proposition 1 in this reference). Precisely, $\psi(\eta)$ is the cumulant generating function of the squared Riemannian distance $d^2(x,\bar{x})$.

Proposition 1 cannot be applied directly to the Riemannian Gaussian model (19). This is because, in most cases of interest, the Riemannian symmetric space M is not irreducible. In such cases, before applying Proposition 1, it is necessary to introduce the De Rham decomposition theorem [22, 40].

Remark 7 Assume the Riemannian symmetric space M is moreover simply-connected. Then, the De Rham decomposition theorem implies that M is a Riemannian product of irreducible Riemannian symmetric spaces [22] (Proposition 5.5, p. 310). Precisely, $M = M_1 \times \cdots \times M_r$ where each M_q is an irreducible Riemannian symmetric space, and the Riemannian metric and distance of M can be expressed as follows,

$$Q(u,u) = \sum_{q=1}^{r} Q(u_q\,,u_q) \tag{20a}$$

$$d^2(x,y) = \sum_{q=1}^{r} d^2(x_q\,,y_q) \tag{20b}$$

where $x, y \in M$ are written $x = (x_1, \ldots, x_r)$ and $y = (y_1, \ldots, y_r)$ with $x_q, y_q \in M_q$, and where $u \in T_x M$ is written $u = u_1 + \cdots + u_r$ with $u_q \in T_{x_q} M_q$ naturally identified with an element of $T_x M$. Since M has non-positive sectional curvature, each M_q is either a Euclidean space, or a so-called space of non-compact type, having negative sectional curvature [22]. A concrete example of the De Rham decomposition is treated in Sect. 6, where $M = \mathcal{P}_n$ is the space of $n \times n$ real covariance matrices. ■

The following Proposition 3 gives the Rao–Fisher information metric of the Riemannian Gaussian model. Since this model is, in general, defined on a Riemannian symmetric space M which is not irreducible, the Rao–Fisher information metric turns out to be a multiply-warped Riemannian metric, rather than a warped Riemannian metric.

Remark 8 In the notation of (20), a multiply-warped Riemannian metric I is a Riemannian metric defined on $\mathcal{M} = M \times (0, \infty)$, in the following way [17, 45]

$$I_z(U, U) = (\alpha(\sigma)\, u_\sigma)^2 + \sum_{q=1}^{r} \beta_q^2(\sigma)\, Q_{\bar{x}}(u_q, u_q) \tag{21}$$

for $z = (\bar{x}, \sigma)$ and $U \in T_z\mathcal{M}$, where $U = u_\sigma\, \partial_\sigma + u$ with $u_\sigma \in \mathbb{R}$ and $u \in T_{\bar{x}} M$. Here, the functions α and β_q are positive functions defined on $(0, \infty)$. ■

Proposition 3 *The Rao–Fisher information metric of the Riemannian Gaussian model is a multiply-warped Riemannian metric. In terms of $\bar{x} \in M$ and $\eta = -1/2\sigma^2$, this metric has the following expression*

$$I_z(U, U) = \psi''(\eta)\, u_\eta^2 + \sum_{q=1}^{r} \left(4\eta^2 \psi_q'(\eta)/\dim M_q\right) Q_{\bar{x}}(u_q, u_q) \tag{22}$$

where $U = u_\eta\, \partial_\eta + u$, and where $\psi_q(\eta)$ is the cumulant generating function of the squared Riemannian distance $d^2(x_q, \bar{x}_q)$.

Proof Assume first that the Riemannian symmetric space M is irreducible, so that Proposition 1 applies directly, and the Rao–Fisher information metric is given by (13a),

$$I_z(U, U) = \psi''(\eta)\, u_\eta^2 + \beta^2(\eta)\, Q_{\bar{x}}(u, u) \tag{23a}$$

To obtain $\beta^2(\eta)$, replace into (13b) the fact that

$$\nabla_{\bar{x}}\, D(x, \bar{x}) = -2 \exp_{\bar{x}}^{-1}(x) \quad Q\left(\nabla_{\bar{x}}\, D, \nabla_{\bar{x}}\, D\right) = 4\, d^2(x, \bar{x})$$

where exp denotes the Riemannian exponential mapping, corresponding to the Riemannian metric Q of M [13], (see p. 407). It then follows from (13b) that,

$$\beta^2(\eta) = 4\eta^2\, \mathbb{E}_z\, d^2(x, \bar{x})/\dim M = 4\eta^2\, \psi'(\eta)/\dim M \tag{23b}$$

where the second equality holds since $\psi(\eta)$ is the cumulant generating function (log-moment generating function) of $d^2(x, \bar{x})$. From (23a) and (23b),

$$I_z(U, U) = \psi''(\eta)\, u_\eta^2 + \left(4\eta^2\psi'(\eta)/\dim M\right)\, Q_{\bar{x}}(u, u) \tag{23c}$$

which is the same as (22) with $r = 1$. This proves the proposition in the special case where M is irreducible. For the general case where M is not irreducible, write $U = u_\eta\, \partial_\eta + u$, with $u = u_1 + \cdots + u_r$ as in Remark 7. It is possible to prove that,

$$q \neq p \;: I_z(u_p\, , u_q) = 0 \tag{24a}$$

$$u = u_p \;: I_z(U, U) = \psi''(\eta)\, u_\eta^2 + \left(4\eta^2\psi_p'(\eta)/\dim M_p\right) Q_{\bar{x}}(u_p\, , u_p) \tag{24b}$$

Then, since the Rao–Fisher information metric I is bilinear and symmetric, (22) follows immediately, and the proposition is proved in the general case.

Proof of identities (24): this is carried out using the following properties (25). Note first that the probability density function of the Riemannian Gaussian model is given by (12a) and (19),

$$p(x|\bar{x}, \sigma) = \exp\left[\,\eta(\sigma)\, d^2(x, \bar{x}) - \psi(\eta(\sigma))\right] \tag{25a}$$

By substituting (20b) in this expression, it is seen that

$$p(x|\bar{x}, \sigma) = \prod_{q=1}^{r} \exp\left[\,\eta(\sigma)\, d^2(x_q\, , \bar{x}_q) - \psi_q(\eta(\sigma))\right] = \prod_{q=1}^{r} p(x_q|\bar{x}_q\, , \sigma) \tag{25b}$$

where $\psi_q(\eta)$ is the cumulant generating function of $d^2(x_q\, , \bar{x}_q)$, as stated after (22). The last equality shows that $(x_q\, ; q = 1, \ldots, r)$ are independent, and that each x_q has a Riemannian Gaussian density on the irreducible Riemannian symmetric space M_q, with parameters $z_q = (\bar{x}_q\, , \sigma)$. Now, identities (24) can be obtained from definition (8) of the Rao–Fisher information metric. To apply this definition, note from (25b), that the log-likelihood function $\ell(z)$ can be written,

$$\ell(z)(x) = \log p(x|z) = \sum_{q=1}^{r} \ell(z_q)(x_q) \quad \text{where } \ell(z_q)(x_q) = \log p(x_q|z_q) \tag{25c}$$

Proof of (24a): recall the polarisation identity, from elementary linear algebra [28], (see p. 29),

$$I_z(u_p\, , u_q) = \frac{1}{4}\, I_z(u_p + u_q\, , u_p + u_q) - \frac{1}{4}\, I_z(u_p - u_q\, , u_p - u_q)$$

By replacing (8) into this identity, it can be seen that,

$$I_z(u_p\,,u_q) \;=\; \mathbb{E}_z\left(\left(d\ell(z)\,u_p\right)\left(d\ell(z)\,u_q\right)\right) \tag{26}$$

Using (25c), it is then possible to write

$$I_z(u_p\,,u_q) \;=\; \mathbb{E}_z\left(\left(d\ell(z_p)\,u_p\right)\left(d\ell(z_q)\,u_q\right)\right) \;=\; \mathbb{E}_{z_p}\left(d\ell(z_p)\,u_p\right)\,\mathbb{E}_{z_q}\left(d\ell(z_q)\,u_q\right)$$

Here, the first equality follows from (26), since $u_p \in T_{\bar{x}_p}M_p$ and $u_q \in T_{\bar{x}_q}M_q$, and the second equality holds since x_p and x_q are independent. Now, each one of the two expectations appearing on the right-hand side is equal to zero, since the expectation of the derivative of the log-likelihood must be zero [3], (see p. 28). This shows that (24a) holds. ■

Proof of (24b): the condition $u = u_p$ implies $U = u_\eta\,\partial_\eta + u_p$. Replacing this in (8), it follows using (25c),

$$I_z(U,U) \;=\; \mathbb{E}_z\left(\sum_{q=1}^{r} d\ell(z_q)\,U\right)^2 = \mathbb{E}_z\left(\sum_{q=1}^{r} u_\eta\,\partial_\eta\ell(z_q) \;+\; d\ell(z_p)\,u_p\right)^2 \tag{27a}$$

where the second equality holds since $u_p \in T_{\bar{x}_p}M_p$. Since the x_q are independent, it is clear from (25c) that the $\ell(z_q)$ are independent. Accordingly, by expanding the right-hand side of (27a),

$$I_z(U,U) = \sum_{q\neq p} u_\eta^2\,\mathbb{E}_{z_q}\left(\partial_\eta\ell(z_q)\right)^2 + \mathbb{E}_{z_p}\left(d\ell(z_p)\,U\right)^2 \tag{27b}$$

Applying (14e) from the proof of Proposition 1 to each term in the sum over $q \neq p$, it follows that

$$I_z(U,U) = \sum_{q\neq p} \psi_q''(\eta)\,u_\eta^2 \;+\; \mathbb{E}_{z_p}\left(d\ell(z_p)\,U\right)^2 \tag{27c}$$

By (8), the expectation appearing in the second term is given by the Rao–Fisher information metric of the Riemannian Gaussian model on the irreducible Riemannian symmetric space M_p. This can be replaced from (23c), so that

$$\begin{aligned} I_z(U,U) &= \textstyle\sum_{q\neq p}\psi_q''(\eta)\,u_\eta^2 \;+\; \psi_p''(\eta)\,u_\eta^2 \;+\; \left(4\eta^2\psi_p'(\eta)/\dim M_p\right)\,Q_{\bar{x}}(u_p\,,u_p) \\ &= \textstyle\sum_q \psi_q''(\eta)\,u_\eta^2 \;+\; \left(4\eta^2\psi_p'(\eta)/\dim M_p\right)\,Q_{\bar{x}}(u_p\,,u_p) \end{aligned}$$

This immediately yields (24b), upon noting from (25a) and (25b) that $\psi(\eta) = \sum_q \psi_q(\eta)$. ■

Now, since identities (24) have been proved, (22) follows from the fact that the Rao–Fisher information metric I is bilinear and symmetric. ■

5 The Generalised Mahalanobis Distance

This section builds on Remark 3, made at the end of Sect. 2, in order to generalise the definition of the classical Mahalanobis distance, to the context of a location-scale model $\mathcal{P}$ defined on a Riemannian symmetric space M.

To begin, assume that, as in Theorem 1, the Riemannian symmetric space M is irreducible and the location-scale model $\mathcal{P}$ verifies condition (7). Then, according to Theorem 1, the Rao–Fisher information metric I of the model $\mathcal{P}$ is a warped Riemannian metric on the parameter space $\mathcal{M}$.

Recall from Remark 3 that this warped Riemannian metric I induces an extrinsic Riemannian metric Q^σ on M, for each $\sigma \in (0\,,\infty)$. The *generalised Mahalanobis distance* is defined to be the Riemannian distance on M which is induced by the extrinsic Riemannian metric Q^σ. The generalised Mahalanobis distance between $\bar{x}$ and $\bar{y}$ in M is denoted $d(\bar{x}, \bar{y}\,|\sigma)$. It is given by the following Proposition 4. The proof of Proposition 4 is omitted, since it is elementary.

Proposition 4 *The generalised Mahalanobis distance $d(\bar{x}, \bar{y}\,|\sigma)$ between $\bar{x}$ and $\bar{y}$ in M is given by*

$$d(\bar{x}, \bar{y}\,|\sigma) \;=\; \beta(\sigma)\, d(\bar{x}, \bar{y}) \tag{28}$$

where the function $\beta(\sigma)$ is given by (9), and where $d(\bar{x}, \bar{y})$ denotes the Riemannian distance in M.

Remark 9 The generalised Mahalanobis distance (28) reduces to the classical Mahalanobis distance, when $\mathcal{P}$ is the isotropic normal model on $M = \mathbb{R}^d$. In this case, the Rao–Fisher metric I is given by (1c) in the introduction, so that $\beta(\sigma) = 1/\sigma$. Replacing this in (28) yields

$$d(\bar{x}, \bar{y}\,|\sigma) \;=\; \frac{1}{\sigma}\, \|\bar{x} - \bar{y}\| \tag{29}$$

where $\|\bar{x} - \bar{y}\|$ is the Euclidean distance in $M = \mathbb{R}^d$. Now, (29) is the classical Mahalanobis distance [36]. ■

Expression (28) of the generalised Mahalanobis distance is valid only under the assumption that the Riemannian symmetric space M is irreducible. This assumption does not hold, when the model $\mathcal{P}$ is the Riemannian Gaussian model studied in Sect. 4.2. For this model, an alternative expression of the generalised Mahalanobis distance is given in Proposition 5 below.

As in Sect. 4.2, let $\mathcal{P}$ be the Riemannian Gaussian model on a Riemannian symmetric space M, where M is simply-connected and has non-positive sectional curvature. Proposition 3 states that the Rao–Fisher information metric I of the model $\mathcal{P}$ is a multiply-warped Riemannian metric on the parameter space $\mathcal{M}$. For each $\sigma \in (0\,,\infty)$, this multiply-warped Riemannian metric I induces an extrinsic Riemannian metric Q^σ on M. Precisely, Q^σ can be obtained from (22) of Proposition 3,

$$Q^{\sigma}_{\bar{x}}(u\,,u) \;=\; \sum_{q=1}^{r} \beta^2_q(\sigma)\, Q_{\bar{x}}(u_q\,,u_q) \quad \beta^2_q(\sigma) \;=\; 4\eta^2\psi^{\prime}_q(\eta)/\dim M_q \tag{30}$$

The generalised Mahalanobis distance $d(\bar{x}\,,\bar{y}\,|\sigma)$ is the Riemannian distance between $\bar{x}$ and $\bar{y}$ in M, induced by this extrinsic Riemannian metric Q^{σ}.

Proposition 5 *When $\mathcal{P}$ is the Riemannian Gaussian model, the generalised Mahalanobis distance $d(\bar{x}\,,\bar{y}\,|\sigma)$ between $\bar{x}$ and $\bar{y}$ in M is given by*

$$d^2(\bar{x}\,,\bar{y}\,|\sigma) \;=\; \sum_{q=1}^{r} \beta^2_q(\sigma)\, d^{\,2}(\bar{x}_q\,,\bar{y}_q) \tag{31}$$

where the notation is that of (20b).

Proof The proof hinges on the fact that the extrinsic Riemannian metric Q^{σ} of (30) is an invariant Riemannian metric on M. In other words, if G is the group of isometries of M, then

$$Q^{\sigma}_{g\cdot\bar{x}}(dg_{\bar{x}}\,u\,,dg_{\bar{x}}\,u) = Q^{\sigma}_{\bar{x}}(u\,,u) \quad \text{for all } g \in G \tag{32a}$$

where $dg_{\bar{x}}$ is the derivative of the isometry g at the point $\bar{x}$. The proof of (32a) is not detailed here. It follows since the Riemannian metric Q is also an invariant Riemannian metric on M, so that Q also verifies (32a), and since Q^{σ} is related to Q by (30). A general result in [22] (Corollary 4.3, p. 182), states that all invariant Riemannian metrics on M have the same geodesics. In particular, the metrics Q^{σ} and Q have the same geodesics, and therefore the same Riemannian exponential mapping exp. To find the generalised Mahalanobis distance between $\bar{x}$ and $\bar{y}$ in M, let $u = \exp^{-1}_{\bar{x}}(\bar{y})$, and note that

$$d^2(\bar{x}\,,\bar{y}\,|\sigma) \;=\; Q^{\sigma}_{\bar{x}}(u\,,u) \;=\; \sum_{q=1}^{r} \beta^2_q(\sigma)\, Q_{\bar{x}}(u_q\,,u_q) \tag{32b}$$

where the second equality follows from (30). Now, to prove (31) it is enough to prove that

$$Q_{\bar{x}}(u_q\,,u_q) \;=\; d^{\,2}(\bar{x}_q\,,\bar{y}_q) \tag{32c}$$

Indeed, (31) is then obtained by replacing (32c) into (32b). The proof of (32c) follows by writing, as in (32b),

$$d^2(\bar{x}\,,\bar{y}) \;=\; Q_{\bar{x}}(u\,,u) \;=\; \sum_{q=1}^{r} Q_{\bar{x}}(u_q\,,u_q) \;=\; \sum_{q=1}^{r} d^{\,2}(\bar{x}_q\,,\bar{y}_q) \tag{32d}$$

where the second equality follows from (20a), and the third equality follows from (20b). Since (32d) is an identity which holds for arbitrary $\bar{x} = (\bar{x}_1\,,\ldots,\,\bar{x}_r)$ and $\bar{y} = (\bar{y}_1\,,\ldots,\,\bar{y}_r)$, it follows that (32c) must hold true, as required. ■

Remark 10 The generalised Mahalanobis distance, whether given by (28) or by (31), has interesting properties, both geometric and statistical. From the geometric viewpoint, it is an invariant Riemannian distance on M,

$$d(g \cdot \bar{x}, g \cdot \bar{y}\,|\sigma) = d(\bar{x}, \bar{y}\,|\sigma) \quad \text{for all } g \in G \tag{33}$$

while, from the statistical viewpoint, just like the classical Mahalanobis distance, it can be used to build asymptotic chi-squared statistics, for hypothesis testing or classification [10]. This statistical aspect of the generalised Mahalanobis distance will be developed in future work. ■

6 A Concrete Example for the Riemannian Gaussian Model

The aim of this section is to illustrate the geometric concepts involved in Propositions 3 and 5, by applying these concepts to the concrete example of the Riemannian Gaussian model defined on $M = \mathcal{P}_n$, the space of $n \times n$ real covariance matrices.

The space $\mathcal{P}_n$ is a Riemannian symmetric space, which is simply-connected and has non-positive sectional curvature [22, 44]. It is usually equipped with its affine-invariant Riemannian metric [6, 44],

$$Q_{\bar{x}}(u, u) = \operatorname{tr}\left[\bar{x}^{-1}u\right]^2 \quad \bar{x} \in \mathcal{P}_n\,,\ u \in T_{\bar{x}}\mathcal{P}_n \tag{34a}$$

This metric is invariant under the action of the group of isometries $G = GL(n, \mathbb{R})$ on $\mathcal{P}_n$, which is given by affine transformations,

$$g \cdot \bar{x} = g\,\bar{x}\,g^t \tag{34b}$$

where ^{t} denotes the transpose. Moreover, this metric induces a Riemannian distance on $\mathcal{P}_n$, which is given by,

$$d^2(\bar{x}, \bar{y}) = \operatorname{tr}\left[\log\left(\bar{x}^{-1/2}\,\bar{y}\,\bar{x}^{-1/2}\right)\right]^2 \tag{34c}$$

This distance is also invariant under the action of the group $GL(n, \mathbb{R})$ on $\mathcal{P}_n$. In other words: $d(g \cdot \bar{x}, g \cdot \bar{y}) = d(\bar{x}, \bar{y})$.

The Riemannian Gaussian model on $\mathcal{P}_n$ is given by the probability density function [14, 41]

$$p(x|\bar{x}, \sigma) = Z^{-1}(\sigma)\,\exp\left[-\frac{d^2(x, \bar{x})}{2\sigma^2}\right] \tag{35a}$$

which is a probability density function with respect to the invariant volume element associated to the Riemannian metric (34a). The normalising factor $Z(\sigma)$ can be expressed as a multiple integral [41], (see Proposition 4 in this reference),

$$Z(\sigma) = \mathrm{C}_n \int_{\mathbb{R}^n} e^{-\|r\|^2/2\sigma^2} \prod_{i<j} \sinh\left(|r_i - r_j|/2\right) dr_1 \ldots dr_n \tag{35b}$$

where C_n is a numerical constant which only depends on n, and the integration variable is denoted $r = (r_1, \ldots, r_n) \in \mathbb{R}^n$. If $\eta(\sigma) = -1/2\sigma^2$, then $\psi(\eta) = \log Z(\sigma)$ is a strictly convex function of $\eta \in (-\infty, 0)$.

With (34) and (35) in mind, consider the application of Proposition 3 to the Riemannian Gaussian model on $\mathcal{P}_n$. This will lead to the expression of the Rao–Fisher information metric I of this model.

De Rham decomposition of $\mathcal{P}_n$: recall first that Proposition 3 uses the De Rham decomposition, introduced in Remark 7. For the Riemannian symmetric space $\mathcal{P}_n$, the De Rham decomposition states that $\mathcal{P}_n$ is a Riemannian product of irreducible Riemannian symmetric spaces $\mathcal{P}_n = \mathbb{R} \times S\mathcal{P}_n$, where $S\mathcal{P}_n$ is the set of $\bar{s} \in \mathcal{P}_n$ such that $\det(\bar{s}) = 1$. The identification of $\mathcal{P}_n$ with $\mathbb{R} \times S\mathcal{P}_n$ is obtained by identifying each $\bar{x} \in \mathcal{P}_n$ with a couple $(\bar{\tau}, \bar{s})$, where $\bar{\tau} \in \mathbb{R}$ and $\bar{s} \in S\mathcal{P}_n$ are given by

$$\bar{\tau} = \log\det(\bar{x}) \quad \bar{s} = e^{-\bar{\tau}/n}\, \bar{x} \tag{36a}$$

Note that the spaces $\mathbb{R}$ and $S\mathcal{P}_n$ are indeed irreducible Riemannian symmetric spaces. This is clear for $\mathbb{R}$, which is one-dimensional and cannot be decomposed into a product of lower-dimensional spaces. The fact that $S\mathcal{P}_n$ is irreducible can be found in [22], (Table II, p. 354). It will be convenient to write $\bar{x} = (\bar{x}_1, \bar{x}_2)$ where $\bar{x}_1 = \bar{\tau}$ and $\bar{x}_2 = \bar{s}$. If $u \in T_{\bar{x}}\mathcal{P}_n$, then $u = u_1 + u_2$,

$$u_1 = \frac{1}{n}\,\mathrm{tr}(\bar{x}^{-1}u)\,\bar{x} \quad u_2 = u - \frac{1}{n}\,\mathrm{tr}(\bar{x}^{-1}u)\,\bar{x} \tag{36b}$$

Here, $u_1 \in T_{\bar{x}_1}\mathbb{R}$, where $T_{\bar{x}_1}\mathbb{R} \subset T_{\bar{x}}\mathcal{P}_n$ is the one-dimensional subspace consisting of symmetric matrices v of the form $v = t\,\bar{x}$ with t any real number. On the other hand, $u_2 \in T_{\bar{x}_2}S\mathcal{P}_n$, where $T_{\bar{x}_2}S\mathcal{P}_n \subset T_{\bar{x}}\mathcal{P}_n$ is the subspace consisting of symmetric matrices v which satisfy $\mathrm{tr}(\bar{x}^{-1}v) = 0$. Using (36a) and (36b), (20a) and (20b) of Remark 7 can be written down,

$$Q_{\bar{x}}(u, u) = Q_{\bar{x}}(u_1, u_1) + Q_{\bar{x}}(u_2, u_2) \tag{36c}$$

$$d^2(\bar{x}, \bar{y}) = \frac{1}{n}\,|\bar{x}_1 - \bar{y}_1|^2 + d^2(\bar{x}_2, \bar{y}_2) \tag{36d}$$

where $Q_{\bar{x}}$ is the affine-invariant metric (34a) and $d(\bar{x}, \bar{y})$ or $d(\bar{x}_2, \bar{y}_2)$ is the Riemannian distance (34c). The proof of formulae (36c) and (36d) is a direct calculation, and is not detailed here.

The Rao–Fisher metric I : according to (22) of Proposition 3, the Rao–Fisher information metric I of the Riemannian Gaussian model on $\mathcal{P}_n$ is given by,

$$I_z(U,U) = \psi''(\eta)\,u_\eta^2 + \frac{4\eta^2\psi_1'(\eta)}{\dim \mathbb{R}}\,Q_{\bar{x}}(u_1\,,u_1) + \frac{4\eta^2\psi_2'(\eta)}{\dim S\mathcal{P}_n}\,Q_{\bar{x}}(u_2\,,u_2) \quad (37a)$$

for $z = (\bar{x}, \sigma)$ in the parameter space $\mathcal{M} = \mathcal{P}_n \times (0\,,\infty)$, and for $U = u_\eta\,\partial_\eta + u$ where $u = u_1 + u_2$ is given by (36b). Indeed, (37a) results from (22), by putting $r = 2$, as well as $M_1 = \mathbb{R}$ and $M_2 = S\mathcal{P}_n$. The functions appearing in (37a) are $\psi(\eta) = \log\, Z(\sigma)$ with $Z(\sigma)$ given by (35b), and, as shown in Remark 11 below,

$$\psi_1(\eta) = \frac{1}{2}\,\log(2\pi n) - \frac{1}{2}\,\log(-2\eta) \quad \psi_2(\eta) = \psi(\eta) - \psi_1(\eta) \quad (37b)$$

Moreover, $\dim \mathbb{R} = 1$ and $\dim S\mathcal{P}_n = \dim \mathcal{P}_n - 1 = n(n+1)/2 - 1$. Replacing into (37a) gives,

$$I_z(U,U) = \psi''(\eta)\,u_\eta^2 \; - \; 2\eta\,Q_{\bar{x}}(u_1\,,u_1) + \frac{8\eta^2\psi_2'(\eta)}{n^2+n-2}\,Q_{\bar{x}}(u_2\,,u_2) \quad (37c)$$

This expression of the Rao–Fisher information metric of the Riemannian Gaussian model on $\mathcal{P}_n$ can be computed directly from (34a), (36b) and (37b), once the function $\psi(\eta)$ is known. This function $\psi(\eta)$ has been tabulated for values of n up to $n = 50$, using a Monte Carlo method which was developed specifically for the evaluation of (35b) [49].

Remark 11 Assume x follows the Riemannian Gaussian probability density (35a) on $\mathcal{P}_n$. If $x = (x_1\,, x_2)$ where $x_1 \in \mathbb{R}$ and $x_2 \in S\mathcal{P}_n$, then the densities of x_1 and x_2 can be found by replacing (36d) into (35a). Precisely, this gives

$$p(x|\bar{x},\sigma) \propto \exp\left[-\frac{|x_1 - \bar{x}_1|^2}{2n\sigma^2}\right] \times \exp\left[-\frac{d^2(x_2\,,\bar{x}_2)}{2\sigma^2}\right]$$

It follows from this decomposition that x_1 and x_2 are independent, and that x_1 follows a univariate normal distribution of mean $\bar{x}_1$ and of variance $n\sigma^2$. In particular, the moment generating function $\psi_1(\eta)$ of the squared distance $|x_1 - \bar{x}_1|^2$ has the expression stated in (37b). ■

The generalised Mahalanobis distance on $\mathcal{P}_n$: applying Proposition 5 will yield the expression of the generalised Mahalanobis distance on $\mathcal{P}_n$. The Rao–Fisher information metric I as given by (37c) induces an extrinsic Riemannian metric Q^σ on $\mathcal{P}_n$, for each $\sigma \in (0\,,\infty)$,

$$Q^\sigma_{\bar{x}}(u,u) = -\,2\eta\,Q_{\bar{x}}(u_1\,,u_1) + \frac{8\eta^2\psi_2'(\eta)}{n^2+n-2}\,Q_{\bar{x}}(u_2\,,u_2) \quad \eta = -\frac{1}{2\sigma^2} \quad (38a)$$

The generalised Mahalanobis distance on $\mathcal{P}_n$ is the Riemannian distance induced on $\mathcal{P}_n$ by the extrinsic Riemannian metric Q^σ. If the generalised Mahalanobis distance between $\bar{x}$ and $\bar{y}$ in $\mathcal{P}_n$ is denoted $d(\bar{x}, \bar{y}\,|\sigma)$, then (31) of Proposition 5, along with (38a), imply

$$d^2(\bar{x}, \bar{y}\,|\sigma) = \frac{|\bar{x}_1 - \bar{y}_1|^2}{n\sigma^2} + \frac{4\psi_2'\left((-2\sigma^2)^{-1}\right)}{(n^2+n-2)\sigma^4}\, d^2(\bar{x}_2, \bar{y}_2) \tag{38b}$$

This distance can be computed directly from (34c), (36a) and (37b), once the function $\psi(\eta)$ has been tabulated using the Monte Carlo method of [49], or computed in any other way.

Affine invariance of the generalised Mahalanobis distance: the affine-invariant Riemannian metric Q of (34a) is well-known to the information science community, having been introduced in [39]. Besides the metric Q, a whole new family of affine-invariant Riemannian metrics Q^σ is provided by (38a). Indeed, to say that Q is affine-invariant means that it is invariant under affine transformations (34b). In other words

$$Q_{g\cdot\bar{x}}(dg_{\bar{x}}\, u, dg_{\bar{x}}\, u) \;=\; Q_{\bar{x}}(u, u) \quad \text{for all } g \in GL(n, \mathbb{R}) \tag{39a}$$

where $dg_{\bar{x}}$ denotes the derivative of the affine transformation (34b) at the point $\bar{x} \in \mathcal{P}_n$. On the other hand, a direct verification shows that each one of the metrics Q^σ also verifies (39a), so that

$$Q^\sigma_{g\cdot\bar{x}}(dg_{\bar{x}}\, u, dg_{\bar{x}}\, u) \;=\; Q^\sigma_{\bar{x}}(u, u) \quad \text{for all } g \in GL(n, \mathbb{R}) \tag{39b}$$

This means that each one of the metrics Q^σ is an affine-invariant Riemannian metric, as claimed. Furthermore, the fact that the metric Q^σ is invariant under affine transformations implies that the generalised Mahalanobis distance (38b) is also invariant under these transformations,

$$d(g\cdot\bar{x}, g\cdot\bar{y}\,|\sigma) \;=\; d(\bar{x}, \bar{y}\,|\sigma) \quad \text{for all } g \in GL(n, \mathbb{R}) \tag{39c}$$

This is because the generalised Mahalanobis distance (38b) is the Riemannian distance induced on $\mathcal{P}_n$ by Q^σ.

7 The Solution of the Geodesic Equation

The present section provides the solution of the geodesic equation of a multiply-warped Riemannian metric. The main result is the following Proposition 6. This proposition shows that the solution of the geodesic equation of a multiply-warped Riemannian metric, for given initial conditions, reduces to the solution of a one-dimensional second-order differential equation. As stated in Remark 8, warped Riemannian metrics are a special case of multiply-warped Riemannian metrics. Therefore, Proposition 6 also applies to the solution of the geodesic equation of a warped Riemannian metric. This special case of warped Riemannian metrics was treated separately in [38].

Let I be a multiply-warped Riemannian metric defined on $\mathcal{M} = M \times (0\,,\infty)$, in the notation of (21),

$$I_z(U,U) \,=\, (\alpha(\sigma)\,u_\sigma)^2 \,+\, \sum_{q=1}^{r} \beta_q^2(\sigma)\,Q_{\bar{x}}(u_q\,,u_q) \tag{40}$$

for $z = (\bar{x},\sigma)$ and $U \in T_z\mathcal{M}$, with $u = u_\sigma\,\partial_\sigma \,+\, u$ and $u = u_1 \,+\cdots+\, u_r$. As in (2b) of Sect. 2, introduce the vertical distance coordinate r, which is defined by $dr/d\sigma = \alpha(\sigma)$.

Proposition 6 *Let $\gamma(t)$ be a geodesic of the multiply-warped Riemannian metric I, with initial conditions $\gamma(0) = z$ and $\dot{\gamma}(0) = U$, and let $\gamma(t) = (\bar{x}(t),\sigma(t))$ and $r(t) = r(\sigma(t))$. Then, $r(t)$ verifies the second-order differential equation*

$$\ddot{r} \,=\, -\frac{1}{2}\frac{d}{dr}V(r) \quad V(r) \,=\, \sum_{q=1}^{r} \frac{\beta_q^2(r(0))}{\beta_q^2(r)}\,I_z(u_q\,,u_q) \tag{41a}$$

and $\bar{x}(t)$ is given by

$$\bar{x}(t) \,=\, \exp_{\bar{x}}\left[\sum_{q=1}^{r}\left(\int_0^t \frac{\beta_q^2(r(0))}{\beta_q^2(r(s))}ds\right)u_q\right] \tag{41b}$$

where exp *denotes the Riemannian exponential mapping of the metric Q on M.*

Proof The proof is given in Appendix C. It is a generalisation of the proof dealing with the special case of warped Riemannian metrics, which can be found in [38] (Proposition 38, p. 208). ■

Proposition 6 shows that the main difficulty, involved in computing a geodesic $\gamma(t)$ of the multiply-warped Riemannian metric I, lies in the solution of the second order differential equation (41a). Indeed, once this equation is solved, computing $\gamma(t)$ essentially reduces to an application of exp, which is the Riemannian exponential mapping of the metric Q on M. In the context of the present contribution, Q is an invariant metric on M, where M is a Riemannian symmetric space. Therefore, exp has a straightforward expression [22] (Theorem 3.3, p. 173). In particular, for the examples treated in Sect. 4, the expression of exp is well-known in the literature. For the von Mises–Fisher model, this expression is elementary, since geodesics on a sphere in Euclidean space are the great circles on this sphere. For the Riemannian Gaussian model, when this model is defined on the space $M = \mathcal{P}_n$ of $n \times n$ real covariance matrices, the expression of exp is widely used in the literature, as found in [39].

Remark 12 The differential equation (41a) is the equation of motion of a one-dimensional conservative mechanical system. As such, its solution can be carried out by quadrature [21], (see p. 11). Precisely,

$$t = \pm \int_{r(0)}^{r(t)} \frac{dr}{\sqrt{E - V(r)}} \tag{42a}$$

where the total energy E is a conserved quantity, in the sense that $\dot{E} = 0$, and it can be shown that $E = I_z(U, U)$. Recalling that $dr/d\sigma = \alpha(\sigma)$, this integral can be written

$$t = \pm \int_{\sigma(0)}^{\sigma(t)} \frac{\alpha(\sigma)}{\sqrt{E - V(\sigma)}} d\sigma \quad V(\sigma) = V(r(\sigma)) \tag{42b}$$

Here, if t is interpreted as time, then the integral on the right-hand side gives the time necessary to go from $\sigma(0)$ to $\sigma(t)$. In particular, replacing $\sigma(t)$ by ∞ and by 0 gives the two quantities

$$t_\infty = \int_{\sigma(0)}^{\infty} \frac{\alpha(\sigma)}{\sqrt{E - V(\sigma)}} d\sigma \quad t_0 = \int_{0}^{\sigma(0)} \frac{\alpha(\sigma)}{\sqrt{E - V(\sigma)}} d\sigma \tag{42c}$$

where t_∞ is the time necessary for $\sigma(t)$ to reach the value $\sigma = \infty$, and t_0 is the time necessary for $\sigma(t)$ to reach the value $\sigma = 0$. Since $\mathcal{M} = M \times (0, \infty)$, these two values $\sigma = \infty$ and $\sigma = 0$ are excluded from $\mathcal{M}$. Therefore, the geodesic $\gamma(t)$ cannot be extended beyond the time $t = \min(t_\infty, t_0)$, as it would then escape from $\mathcal{M}$. ■

Remark 13 A vertical geodesic is a geodesic $\gamma(t)$ for which $\dot{\gamma}(0) = U$ with $U = u_\sigma \, \partial_\sigma$. This means that all the u_q are zero. In (41a), this implies that $V(r) = 0$, so that $\ddot{r} = 0$ and $r(t)$ is an affine function of t. In (41b), this implies that $\bar{x}(t) = \bar{x}$ is constant. In Remark 2 of Sect. 2, it was shown that a vertical geodesic is a unique length-minimising geodesic. For a vertical geodesic, (42c) reads

$$t_\infty = \frac{1}{\sqrt{E}} \int_{\sigma(0)}^{\infty} \alpha(\sigma) \, d\sigma \quad t_0 = \frac{1}{\sqrt{E}} \int_{0}^{\sigma(0)} \alpha(\sigma) \, d\sigma \tag{43a}$$

These formulae provide another way of understanding conditions (3b) from Sect. 2. Precisely, since $dr/d\sigma = \alpha(\sigma)$, it is clear that t_∞ and t_0 are given by

$$t_\infty = \frac{1}{\sqrt{E}} \lim_{\sigma \to \infty} r(\sigma) - r(\sigma(0)) \quad t_0 = \frac{1}{\sqrt{E}} \lim_{\sigma \to 0} r(\sigma(0)) - r(\sigma) \tag{43b}$$

Thus, the first condition in (3b) is equivalent to $t_\infty = \infty$, which means that $\sigma(t)$ cannot reach the value $\sigma = \infty$ within a finite time, and the second condition in (3b) is equivalent to $t_0 = \infty$, which means that $\sigma(t)$ cannot reach the value $\sigma = 0$ within a finite time. ■

8 The Construction of Riemannian Brownian Motion

The present section provides an explicit construction of the Riemannian Brownian motion associated to any multiply-warped Riemannian metric. The main result is the following Proposition 7, which shows that this construction reduces to the solution of a one-dimensional stochastic differential equation. This is in analogy with Proposition 6 of the previous section.

Let I be a multiply-warped Riemannian metric on $\mathcal{M} = M \times (0, \infty)$, given by (40). Recall that a Riemannian Brownian motion associated to I is a diffusion process z in $\mathcal{M}$ which satisfies the identity, see [23, 24],

$$df(z(t)) = \frac{1}{2}\Delta_{\mathcal{M}} f(z(t)) + dm^f(t) \tag{44a}$$

for each smooth function f on $\mathcal{M}$, where $\Delta_{\mathcal{M}}$ is the Laplace–Beltrami operator of I, and where m^f is a local martingale with respect to the augmented natural filtration of z. In other words, z is a Riemannian Brownian motion associated to I, if z solves the martingale problem associated to $(1/2)\Delta_{\mathcal{M}}$. The Laplace–Beltrami operator $\Delta_{\mathcal{M}}$ is given by the following formula, which will be proved in Appendix D,

$$\Delta_{\mathcal{M}} f = \frac{1}{G}\,\partial_r\,(G\,\partial_r f) + \sum_{q=1}^{r} \frac{1}{\beta_q^2(r)}\,\Delta_{M_q} f \tag{44b}$$

where $G = G(r)$ is given by $G = \prod_{q=1}^{r} \beta_q^{\dim M_q}$. Equation (44a) defines a Riemannian Brownian motion process z, associated to the multiply-warped Riemannian metric I, through its relationship to the Laplace–Beltrami operator $\Delta_{\mathcal{M}}$. On the other hand, the following Proposition 7 shows how such a Riemannian Brownian motion process may be constructed explicitly.

Proposition 7 *Let z be a stochastic process with values in $\mathcal{M}$, and write $z(t) = (\bar{x}(t), \sigma(t))$ where $\bar{x}(t) = (\bar{x}_1(t), \ldots, \bar{x}_r(t))$ and $\bar{x}_q(t) \in M_q$. Let r and $(\theta_1, \ldots, \theta_r)$ be independent diffusion processes, where r is a one-dimensional diffusion process, which satisfies the stochastic differential equation*

$$dr(t) = \frac{1}{2}\frac{\partial_r G}{G}(r(t))dt + dw(t) \tag{45a}$$

with w a standard Brownian motion, and where θ_q is a Riemannian Brownian motion in M_q. If each $\bar{x}_q$ is a time-changed version of θ_q,

$$\bar{x}_q(t) = \left(\theta_q \circ \tau_q\right)(t) \quad \tau_q(t) = \int_0^t \frac{ds}{\beta_q^2(r(s))} \tag{45b}$$

and if $\sigma(t) = \sigma(r(t))$, then the process z is a Riemannian Brownian motion associated to the multiply-warped Riemannian metric I given by (40).

Proof The proof of this proposition is given in Appendix D. ■

Remark 14 The above Proposition 7 can be used for numerical simulation of a Riemannian Brownian motion z associated to a multiply-warped Riemannian metric I. According to Proposition 7, to simulate z, it is enough to simulate the solution r of the one-dimensional stochastic differential equation (45a), and then to independently simulate each θ_q as a Riemannian Brownian motion in the Riemannian symmetric space M_q. To simulate the solution r of (45a), it is enough to use any of the numerical schemes described in [26]. On the other hand, Riemannian Brownian motion in a symmetric space can be simulated using Lie group stochastic exponentials [4, 20, 31]. A detailed description of these methods falls outside the present scope, and is reserved for future work. ■

9 Surprising Observation: Hadamard Manifolds

In Sect. 2, the completeness and curvature of a warped Riemannian metric I were characterised by Formulae (3b) and (4), respectively. Here, based on Sect. 4.1, these formulae will be applied to the case where I is the Rao–Fisher information metric of the von Mises–Fisher model defined on S^{n-1}. Precisely, this application is carried out using a mixture of analytical and numerical computations, for the von Mises–Fisher model defined on S^{n-1} where $n = 2, \ldots, 8$. The result is a surprising observation: the parameter space of the von Mises–Fisher model, when equipped with the Rao–Fisher information metric I, becomes a Hadamard manifold, a simply-connected complete Riemannian manifold of negative sectional curvature [13, 40]. Since this observation is true for several values of n, it gives rise to a family of Hadamard manifolds. Part of this claim can be proved for any value of $n = 2, \ldots,$ as in the following proposition.

Proposition 8 *For any value of $n = 2, \ldots,$ the parameter space of the von Mises–Fisher model defined on S^{n-1} is a simply-connected manifold, which moreover becomes a complete Riemannian manifold when equipped with the Rao–Fisher information metric I.*

Proof Recall from Remark 6 that the parameter space of the von Mises–Fisher model defined on S^{n-1} is identified with $\mathbb{R}^n$. Of course, $\mathbb{R}^n$ is a simply-connected manifold [43]. Thus, to prove the proposition, it remains to prove that the parameter space $\mathbb{R}^n$ becomes a complete Riemannian manifold when equipped with the Rao–Fisher information metric I. This will be done by proving that all geodesics of the metric I which pass through the point $z = 0$ in $\mathbb{R}^n$ can be extended indefinitely. Then, a corollary of the Hopf–Rinow theorem [13] (Corollary I.7.2, p. 29) implies the required completeness of the parameter space $\mathbb{R}^n$.

First, note that the geodesics of the metric I which pass through the point $z = 0$ are exactly the radial straight lines in $\mathbb{R}^n$. Indeed, according to Remark 6, if $\gamma(t)$ is a geodesic of I where $\gamma(t) = (\bar{x}(t), \eta(t))$, then $\gamma(t)$ is identified with the curve

$z(t) = \eta(t)\,\bar{x}(t)$ in $\mathbb{R}^n$. Moreover, by Proposition 2, the restriction of I to $\mathbb{R}^n - \{0\}$ is a warped Riemannian metric of the general form (13a). Then, by Remark 13, the vertical geodesics $\gamma(t)$ of this warped Riemannian metric are parameterised by $r(t) =$ affine function of t and $\bar{x}(t) = \bar{x} =$ constant, where r is the vertical distance coordinate. Therefore, each vertical geodesic $\gamma(t)$ can be parameterised by $\eta(t) = \eta(r(t))$ and $\bar{x}(t) = \bar{x} =$ constant. This is identified with the curve $z(t) = \eta(t)\,\bar{x}$, which is a radial straight line in $\mathbb{R}^n$, in the direction $\bar{x}$. It remains to note that the geodesics of the metric I are just the geodesics of its restriction to $\mathbb{R}^n - \{0\}$, extended by continuity whenever they reach the point $z = 0$.

Let $z(t) = \eta(t)\,\bar{x}$ describe a geodesic of the metric I, as just explained. To say that this geodesic can be extended indefinitely is equivalent to saying that $\eta(t)$ cannot reach the value $\eta = \infty$ within a finite time. For the von Mises–Fisher model, $\eta(\sigma) = \sigma$ as long as $\sigma \in (0\,, \infty)$. Therefore, according to Remark 13, by evaluating the two conditions (3b), it is possible to know whether $\eta(t)$ can reach the two values $\eta = \infty$ and $\eta = 0$ within a finite time. These conditions now read

$$\lim_{\eta\to\infty} r(\eta) - r(\eta_0) \stackrel{?}{=} \infty \quad\text{and}\quad \lim_{\eta\to 0} r(\eta_1) - r(\eta) \stackrel{?}{=} \infty$$

where η_0 and η_1 are arbitrary. By (13a), $r(\eta)$ is defined by $dr/d\eta = \left(\psi''(\eta)\right)^{1/2}$. Therefore, the two conditions in (3b) are identical to

$$\int_{\eta_0}^{\infty} \left(\psi''(\eta)\right)^{1/2} d\eta \stackrel{?}{=} \infty \quad\text{and}\quad \int_{0}^{\eta_1} \left(\psi''(\eta)\right)^{1/2} d\eta \stackrel{?}{=} \infty \tag{46a}$$

where $\psi''(\eta)$ is given by (17a). For the first integral, recall the asymptotic expansion of modified Bessel functions at $\eta = \infty$ [46], (Sect. 7.23, p. 203. This formula appears with the wrong sign for the second term in parentheses, in [33]),

$$I_\nu(\eta) \;=\; \frac{e^\eta}{\sqrt{2\pi\eta}}\left(1 - \frac{4\nu^2 - 1}{8\eta} + \frac{(4\nu^2 - 1)(4\nu^2 - 3^2)}{2!(8\eta)^2}\right) + O\left(\eta^{-3}\right)$$

Using this asymptotic expansion, it follows by performing some direct calculations, and recalling that $\nu = n/2$,

$$\begin{aligned}\frac{I_{\nu+1}(\eta)}{I_{\nu-1}(\eta)} &= 1 - \frac{n}{\eta} + \frac{n(n-1)}{2\eta^2} + O\left(\eta^{-3}\right)\\ \frac{I^2_{\nu}(\eta)}{I^2_{\nu-1}(\eta)} &= 1 - \frac{n-1}{\eta} + \frac{(n-1)(n-2)}{2\eta^2} + O\left(\eta^{-3}\right)\end{aligned}$$

Replacing these expressions into (17a) immediately gives

$$\psi''(\eta) \;=\; \frac{n-1}{2\eta^2} + O\left(\eta^{-3}\right) \tag{46b}$$

Since $n > 1$, this implies that the first integral in (46a) is divergent, as required. The second integral in (46a) is actually convergent. Indeed, $\psi''(\eta)$ is a continuous function in the neighborhood of $\eta = 0$, as seen in the proof of Proposition 2, for the limit (18c). Thus, the first condition in (3b) is verified, while the second condition is not verified. This means that $\eta(t)$ cannot reach the value $\eta = \infty$ within a finite time, but that it can reach the value $\eta = 0$ within a finite time. The first of these two statements shows that the geodesic described by $z(t)$ can indeed be extended indefinitely. Now, any geodesic of the metric I which passes through the point $z = 0$ is described by some $z(t)$ of this form. ∎

The idea behind the proof of the completeness part of Proposition 8 can be summarised as follows. The restriction of the Rao–Fisher information metric I to $\mathbb{R}^n - \{0\}$ is a warped Riemannian metric. Thus, as stated in Sect. 2, $\mathbb{R}^n - \{0\}$ will be a complete Riemannian manifold, when equipped with this warped Riemannian metric, if and only if the two conditions in (3b) are verified. Once these conditions are evaluated, it turns out the first one is verified, but the second one is not. Thus, $\mathbb{R}^n - \{0\}$ is not a complete Riemannian manifold. However, this is only due to the fact that the point $z = 0$ is excluded. Once this point is included, the parameter space $\mathbb{R}^n$ is obtained, and this is a complete Riemannian manifold, when equipped with the Rao–Fisher information metric I. Precisely, a vertical geodesic in $\mathbb{R}^n - \{0\}$ can reach the point $z = 0$ within a finite time, but then all it does is pass through this point, and immediately return to $\mathbb{R}^n - \{0\}$. However, this vertical geodesic cannot escape to infinity within a finite time.

Proposition 8 established that the parameter space $\mathbb{R}^n$ of the von Mises–Fisher model is a simply-connected complete Riemannian manifold, for any value of n. To show that this parameter space is a Hadamard manifold, it remains to show that it has negative sectional curvature. This is done using numerical computation, for $n = 2, \ldots, 8$.

Precisely, let K_z denote the sectional curvature of $\mathbb{R}^n$ at a point $z \in \mathbb{R}^n$, with respect to the Rao–Fisher information metric I. Since, according to Proposition 2, I is a warped Riemannian metric, the sectional curvature K_z can be computed from formulae (4). To evaluate formula (4a), the Gauss equation, it is enough to note that for the von Mises–Fisher model, K^M is a constant equal to $+1$. Indeed, K^M is the sectional curvature of the unit sphere S^{n-1}. Then, formula (4a) reads

$$K_z(u, v) = \frac{1}{\beta^2} - \left(\frac{\partial_r \beta}{\beta}\right)^2 \quad u, v \in T_{\bar{x}} S^{n-1} \tag{47a}$$

where $z = \eta\,\bar{x}$. On the other hand, formula (4b), the Jacobi equation, can be copied directly,

$$K_z(u, \partial_r) = -\frac{\partial_r^2 \beta}{\beta} \tag{47b}$$

Here, $\beta(\eta)$ is given by (17b) of Proposition 2, and r is the vertical distance coordinate defined by $dr/d\eta = \left(\psi''(\eta)\right)^{1/2}$. In each one of formulae (47), the right-hand side

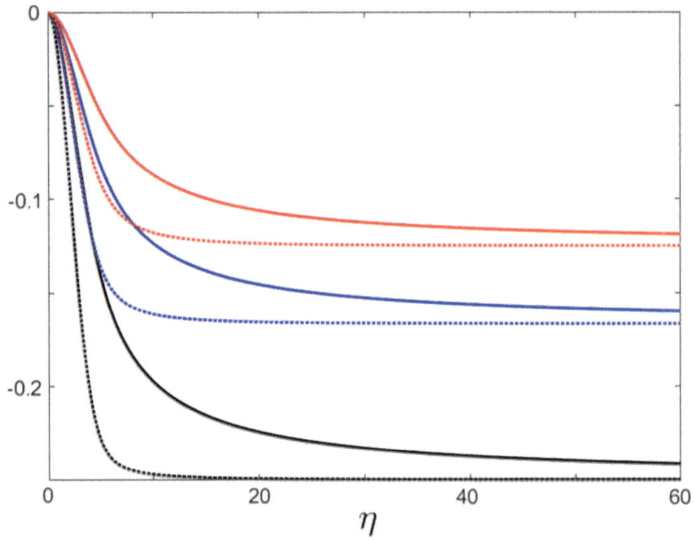

Fig. 1 Sectional curvature of the parameter space of the von Mises–Fisher model ($K^s(\eta)$ solid line, $K^r(\eta)$ dotted line, n = 3 black, n = 4 blue, n = 5 red)

is a real-valued function of η, independent of the vectors u, v. This will be denoted in the following way

$$K_z(u, v) = K^s(\eta) \quad K_z(u, \partial_r) = K^r(\eta) \tag{48}$$

Precisely, $K^s(\eta)$ is the sectional curvature of any section (u, v) tangent to the surface of a sphere centred at $z = 0$ and with Euclidean radius η. On the other hand, $K^r(\eta)$ is the sectional curvature of a radial section (u, ∂_r). In the following, $K^s(\eta)$ will be called the surface curvature, and $K^r(\eta)$ will be called the radial curvature.

Formulae (47) were computed numerically for $n = 2, \dots, 8$. It was systematically found that the sectional curvatures $K^s(\eta)$ and $K^r(\eta)$ are negative for all values of η. From these numerical results, it can be concluded with certainty that the sectional curvature of $\mathbb{R}^n$, with respect to the Rao–Fisher information metric I, is negative, when n ranges from $n = 2$ to $n = 8$. Figure 1 gives a graphic representation for $n = 3, 4, 5$. The sectional curvatures $K^s(\eta)$ and $K^r(\eta)$ behave in the same way, for all considered values of n. Precisely, they are equal to zero at $\eta = 0$, and decrease to a limiting negative value, as η becomes large. This limiting value, denoted $K^s(\infty)$ and $K^r(\infty)$, for $K^s(\eta)$ and $K^r(\eta)$, respectively, is given in the following Table 1. Remarkably, it appears from this table that $K^s(\infty)$ and $K^r(\infty)$ have the same first two digits after the decimal point.

Remark 15 Based on Proposition 8, and on the numerical results reported here, it has been found that the parameter space $\mathbb{R}^n$ of the von Mises–Fisher model becomes a Hadamard manifold, when equipped with the Rao–Fisher information metric I,

Table 1 Limiting value of the surface and radial curvatures

	$n = 2$	$n = 3$	$n = 4$	$n = 5$	$n = 6$	$n = 7$	$n = 8$
$K^s(\infty)$	−0.50	−0.25	−0.16	−0.12	− 0.10	−0.08	−0.07
$K^r(\infty)$	−0.50	−0.25	−0.16	−0.12	−0.10	−0.08	−0.07

for $n = 2, \ldots, 8$. Indeed, Proposition 8 shows that this parameter space is a simply-connected complete Riemannian manifold, for any value of $n = 2, \ldots,$ while the numerical results of Fig. 1 and Table 1 show that it has negative sectional curvature. Hopefully, future work will provide a mathematical proof of the proposition that the sectional curvature of the parameter space $\mathbb{R}^n$ is negative for $n = 2, \ldots,$ without restriction. ■

Remark 16 Preliminary results from ongoing research indicate that the Riemannian Gaussian model, which was studied in Sect. 4.2, when defined on $M = H^{n-1}$, the $(n-1)$-dimensional hyperbolic space, has similar properties to the von Mises–Fisher model, with regard to sectional curvature. Indeed, numerical computations show the sectional curvature of its parameter space $\mathcal{M} = H^{n-1} \times (0\,,\infty)$, equipped with the Rao–Fisher information metric I, is negative for $n = 3, 4, 5$. These numerical computations were carried out using formulae (4), for the sectional curvature of a warped Riemannian metric. This is justified because the hyperbolic space H^{n-1} is an irreducible Riemannian symmetric space, since it is a space of constant negative curvature, so the Rao–Fisher information metric (22) is a warped Riemannian metric. ■

Conclusion

The aim of the present contribution was to reveal a common geometric structure, shared by all location-scale models which are invariant under the action of some Lie group. Precisely, all of these location-scale models have a Rao–Fisher information metric which is a warped (eventually, multiply-warped) Riemannian metric. This provides a unified geometric framework for the study of a wide variety of location-scale models: von Mises–Fisher, and Riemannian Gaussian models, detailed in the above, or elliptically contoured distribution, generalised Wishart, and hyperboloid models, among many additional models. For such location-scale models, the rich yet tractable geometry of warped Riemannian metrics can be used to understand and solve important computational and theoretical problems. For example, future work will be able to address computational problems such as on-line estimation of mixture models on manifolds, regression between manifold-valued data sets, or black-box optimisation on manifolds. High-dimensional computations, with big data sets, as involved in these problems, are greatly simplified by the introduction of warped metrics, which afford exact computation of Riemannian gradients, Hessians, and geodesics, with a computational complexity largely independent of dimension. In particular, this means that exploiting the natural gradient algorithm, even with manifold-valued data and parameters, will be greatly streamlined. On the theoretical side, the connection between location-scale models and warped Riemannian metrics raises exciting new questions. Based on the theoretical and numerical results obtained above, it seems justified to make the conjecture that the parameter space of any invariant location-scale model, defined on a simply-connected symmetric space, turns into a Hadamard manifold, a simply-connected complete Riemannian manifold of negative curvature, when equipped with its Rao–Fisher information metric. Future work will focus on this conjecture, which seems to convey both geometric and statistical insight. Roughly, the fundamental example of a location-scale model is the isotropic normal model, whose parameter space turns into a space of constant negative curvature, indeed a Hadamard manifold, when equipped with its Rao–Fisher information metric. On the other hand, it is known to statisticians that any sufficiently regular location-scale model is locally normal, which means it is locally "similar" to an isotropic normal model. The question is then to know how this local statistical similarity translates into local geometric similarity. A glimpse of this statistical-geometric equivalence is seen from Table 1 in Sect. 9. For the von Mises–Fisher model, as η goes to infinity, the model converges to an isotropic normal model (see [33]). In this limit, it is seen in Table 1, that the two sectional curvatures $K^s(\eta)$ and $K^r(\eta)$ become equal, which means all sectional curvatures are equal, exactly as in a space of constant curvature. Intuitively, as the von Mises–Fisher model converges to an isotropic normal model, its geometry converges to that of a space of constant negative curvature.

Appendix A – Proof of Theorem 1

In order to complete the proof of Theorem 1, the following proposition is needed. The notation is that of Remark 4 and of (11).

Proposition 9 *Assume condition (7) holds. Then,*

$$\partial_\sigma \ell(z) \circ g \;=\; \partial_\sigma \ell(g^{-1} \cdot z) \quad \nabla_{\bar{x}}\, \ell(z) \circ g \;=\; dg_{\bar{x}}\, \nabla_{\bar{x}}\, \ell(g^{-1} \cdot z) \tag{49a}$$

In particular, if $g = s_{\bar{x}}$,

$$\partial_\sigma \ell(z) \circ s_{\bar{x}} \;=\; \partial_\sigma \ell(z) \quad \nabla_{\bar{x}}\, \ell(z) \circ s_{\bar{x}} \;=\; -\,\nabla_{\bar{x}}\, \ell(z) \tag{49b}$$

Proof Note that (49b) follows from (49a), by the definition of the geodesic-reversing isometry $s_{\bar{x}}$ [22]. Indeed, $s_{\bar{x}} \cdot \bar{x} = \bar{x}$ so $s_{\bar{x}}^{-1} \cdot z = z$. Moreover, $ds_{\bar{x}} = -\text{Id}$, as a linear mapping of $T_{\bar{x}}M$, where Id denotes the identity. To prove (49a), note that

$$(\partial_\sigma \ell(z) \circ g)\,(x) \;=\; \partial_\sigma \log p(g \cdot x|z) \;=\; \partial_\sigma \log p(x|g^{-1} \cdot z) \tag{50a}$$

where the second equality follows from condition (7). However,

$$\partial_\sigma \log p(x|g^{-1}\cdot z) \;=\; \partial_\sigma \ell(g^{-1}\cdot z)(x) \tag{50b}$$

Replacing (50b) in (50a) gives,

$$(\partial_\sigma \ell(z) \circ g)\,(x) \;=\; \partial_\sigma \ell(g^{-1}\cdot z)(x)$$

which is the first part of (49a). For the second part, a similar reasoning can be applied. Precisely, using condition (7), it follows,

$$(d\ell(z) \circ g)\,(x) \;=\; d\log p(g\cdot x|z) = d\log p(x|g^{-1}\cdot z) \;=\; d\ell^{(g)}(z)(x) \tag{51a}$$

where $d\ell(z)$ denotes the derivative of $\ell(z)$ with respect to $\bar{x}$, and $\ell^{(g)}(z) = \ell(g^{-1}\cdot z)$, so (51a) implies that,

$$d\ell(z) \circ g \;=\; d\ell^{(g)}(z) \tag{51b}$$

By the chain rule [29], for $u \in T_{\bar{x}}M$,

$$d\ell^{(g)}(z)\big|_{\bar{x}}\, u \;=\; d\ell(g^{-1}\cdot z)\, dg_{\bar{x}}^{-1}\, u$$

Replacing in (51b),

$$d\ell(z) \circ g|_{\bar{x}} \;=\; d\ell(g^{-1}\cdot z)\, dg_{\bar{x}}^{-1} \tag{51c}$$

The second part of (49a) can now be obtained as follows. By the definition of the Riemannian gradient [40],

$$Q\,(\nabla_{\bar{x}}\, \ell(z) \circ g\,, u) \;=\; d\ell(z) \circ g|_{\bar{x}}\, u \;=\; d\ell(g^{-1}\cdot z)\, dg_{\bar{x}}^{-1}\, u \tag{52a}$$

where the second equality follows from (51c). However,

$$d\ell(g^{-1}\cdot z)\, dg_{\bar{x}}^{-1}\, u \;=\; Q\left(\nabla_{\bar{x}}\, \ell(g^{-1}\cdot z), dg_{\bar{x}}^{-1}\, u\right)$$

Since g is an isometry of M, its derivative $dg_{\bar{x}}$ preserves the Riemannian metric Q. Therefore,

$$Q\left(\nabla_{\bar{x}}\, \ell(g^{-1}\cdot z), dg_{\bar{x}}^{-1}\, u\right) \;=\; Q\left(dg_{\bar{x}}\, \nabla_{\bar{x}}\, \ell(g^{-1}\cdot z),\; u\right) \tag{52b}$$

Replacing (52b) in (52a) gives,

$$Q\,(\nabla_{\bar{x}}\, \ell(z) \circ g\,, u) \;=\; Q\left(dg_{\bar{x}}\, \nabla_{\bar{x}}\, \ell(g^{-1}\cdot z),\; u\right)$$

To finish the proof, it is enough to note that the vector u is arbitrary. ■

Proof of (10b): recall the polarisation identity, from elementary linear algebra [28], (see p. 29),

$$I_z(\partial_\sigma, u) = \frac{1}{4}\, I_z(\partial_\sigma + u, \partial_\sigma + u) - \frac{1}{4}\, I_z(\partial_\sigma - u, \partial_\sigma - u)$$

by replacing (8) into this identity, it can be seen that,

$$I_z(\partial_\sigma, u) \;=\; \mathbb{E}_z\left((\partial_\sigma \ell(z))\,(d\ell(z)\,u)\right)$$

Then, by recalling the definition of the Riemannian gradient [40],

$$I_z(\partial_\sigma, u) \;=\; \mathbb{E}_z\left((\partial_\sigma \ell(z))\, Q\left(\nabla_{\bar{x}}\, \ell(z), u\right)\right) \tag{53a}$$

Denote the function under the expectation by f, and apply (11) with $g = s_{\bar{x}}$. Then,

$$\mathbb{E}_z\, f \;=\; \mathbb{E}_{s_{\bar{x}}\cdot z}\, f \;=\; \mathbb{E}_z\left(f \circ s_{\bar{x}}\right) \tag{53b}$$

since $s_{\bar{x}} \cdot z = z$. Note that (10b) amounts to saying that $\mathbb{E}_z\, f = 0$. To prove this, note that

$$f \circ s_{\bar{x}} = (\partial_\sigma \ell(z) \circ s_{\bar{x}})\, Q\left(\nabla_{\bar{x}}\ell(z) \circ s_{\bar{x}}\,, u\right) = -\partial_\sigma \ell(z) Q\left(\nabla_{\bar{x}}\ell(z), u\right) = -f \tag{53c}$$

where the second equality follows from (49b). Replacing in (53b) shows that $\mathbb{E}_z\, f = 0$. ■

Proof of (10c): the idea is to apply Schur's lemma to $I_z(u, u)$, considered as a symmetric bilinear form on $T_{\bar{x}}M$. First, it is shown that this symmetric bilinear form is invariant under the isotropy representation. That is,

$$I_z(u, u) \;=\; I_z\left(dk_{\bar{x}}\, u\,, dk_{\bar{x}}\, u\right) \quad \text{for all } k \in K_{\bar{x}} \tag{54a}$$

This is done using (11). Note from (8),

$$I_z(u, u) = \mathbb{E}_z\left[\left(\, Q\left(\nabla_{\bar{x}}\, \ell(z), u\right)\right)^2\right] \tag{54b}$$

Denote the function under the expectation by f. By (11),

$$\mathbb{E}_z\, f \;=\; \mathbb{E}_{k^{-1}\cdot z}\, f \;=\; \mathbb{E}_z\left(f \circ k^{-1}\right) \tag{54c}$$

since $k^{-1} \cdot z = z$ for $k \in K_{\bar{x}}$. To find $f \circ k^{-1}$, note that,

$$Q\left(\nabla_{\bar{x}}\, \ell(z) \circ k^{-1}, u\right) \;=\; Q\left(dk_{\bar{x}}^{-1} \nabla_{\bar{x}}\, \ell(z), u\right) \;=\; Q\left(\nabla_{\bar{x}}\, \ell(z), dk_{\bar{x}}\, u\right)$$

where the first equality follows from (49a) and the fact that $k \cdot \bar{x} = \bar{x}$, and the second equality from the fact that $dk_{\bar{x}}$ preserves the Riemannian metric Q. Now, by (54b) and (54c),

$$I_z(u, u) = \mathbb{E}_z\, f = \mathbb{E}_z\left(f \circ k^{-1}\right) = \mathbb{E}_z\left(Q\left(\nabla_{\bar{x}}\ell(z), dk_{\bar{x}}\, u\right)\right)^2 = I_z\left(dk_{\bar{x}}\, u\,, dk_{\bar{x}}\, u\right)$$

and this proves (54a).

Recall Schur's lemma, ([27], p. 240). Applied to (54a), this lemma implies that there exists some multiplicative factor β^2, such that

$$I_z(u,u) \;=\; \beta^2\, Q_{\bar{x}}(u,u) \tag{55a}$$

It remains to show that β^2 is given by (9). Taking the trace of (55a),

$$\text{tr}\, I_z = \beta^2\, \text{tr}\, Q_{\bar{x}} \;=\; \beta^2 \dim M \tag{55b}$$

If $e_1, \ldots, e_d$ is an orthonormal basis of $T_{\bar{x}}M$, then by (54b),

$$\text{tr}\, I_z = \mathbb{E}_z \sum_{i=1}^{d} \left(Q\left(\nabla_{\bar{x}}\, \ell(z), e_i\right)\right)^2 = \mathbb{E}_z\, Q\left(\nabla_{\bar{x}}\ell(z), \nabla_{\bar{x}}\ell(z)\right) \tag{55c}$$

Thus, (55b) and (55c) show that β^2 is given by (9). ■

To complete the proof of Theorem 1, it remains to show that the expectations appearing in (9) do not depend on $\bar{x}$.

For the first expectation, giving $\alpha^2(\sigma)$, note that,

$$\mathbb{E}_{g\cdot z}\left(\partial_\sigma \ell(g\cdot z)\right)^2 \;=\; \mathbb{E}_z\left(\partial_\sigma \ell(g\cdot z)\circ g\right)^2 \;=\; \mathbb{E}_z\left(\partial_\sigma \ell(z)\right)^2 \tag{56}$$

where the first equality follows from (11) and the second equality follows from (49a). Thus, this expectation has the same value, whether computed at $g \cdot z = (g \cdot \bar{x}, \sigma)$, or at $z = (\bar{x}, \sigma)$. Therefore, it does not depend on $\bar{x}$, since the action of G on M is transitive.

For the second expectation, giving $\beta^2(\sigma)$, note that by (11),

$$\mathbb{E}_{g\cdot z}\, Q\left(\nabla_{\bar{x}}\, \ell(g\cdot z)\,,\, \nabla_{\bar{x}}\, \ell(g\cdot z)\right) = \mathbb{E}_z\, Q\left(\nabla_{\bar{x}}\, \ell(g\cdot z)\circ g\,,\, \nabla_{\bar{x}}\, \ell(g\cdot z)\circ g\right) \tag{57a}$$

On the other hand, by (49a),

$$\nabla_{\bar{x}}\, \ell(g\cdot z)\circ g \;=\; dg_{\bar{x}}\, \nabla_{\bar{x}}\, \ell(z)$$

Moreover, since $dg_{\bar{x}}$ preserves the Riemannian metric Q,

$$\begin{aligned} Q(\nabla_{\bar{x}}\, \ell(g\cdot z)\circ g\,,\, \nabla_{\bar{x}}\, \ell(g\cdot z)\circ g) &= Q(dg_{\bar{x}}\, \nabla_{\bar{x}}\, \ell(z)\,,\, dg_{\bar{x}} \nabla_{\bar{x}}\, \ell(z)) \\ &= Q(\nabla_{\bar{x}}\, \ell(z)\,,\, \nabla_{\bar{x}}\, \ell(z)) \end{aligned}$$

Replacing in (57a) gives

$$\mathbb{E}_{g\cdot z}\, Q\left(\nabla_{\bar{x}}\, \ell(g\cdot z)\,,\, \nabla_{\bar{x}}\, \ell(g\cdot z)\right) \;=\; \mathbb{E}_z\, Q(\nabla_{\bar{x}}\, \ell(z)\,,\, \nabla_{\bar{x}}\, \ell(z)) \tag{57b}$$

so this expectation has the same value, at $g \cdot z$ and at z. By the same argument made after (56), it does not depend on $\bar{x}$. ■

Proof of (11): let dv denote the invariant Riemannian volume element of M, and note that,

$$\mathbb{E}_{g\cdot z}\, f \;=\; \int_M f(x)\, p(x|g\cdot z)\, dv(x) \;=\; \int_M f(x)\, p(g^{-1}\cdot x|z)\, dv(x) \tag{58a}$$

where the second equality follows from (7). Introduce the variable $y = g^{-1}\cdot x$. Since the volume element dv is invariant,

$$\int_M f(x)\, p(g^{-1}\cdot x|z)\, dv(x) \;=\; \int_M f(g\cdot y)\, p(y|z)\, dv(y) \tag{58b}$$

The last integral is the same as $\mathbb{E}_z\,(f\circ g)$. Therefore, (11) follows from (58a) and (58b). ∎

Appendix B – Proof of Proposition 2

It remains to prove (17a) and (17b). To do so, introduce the following notation, using (15),

$$t = \langle x, \bar{x}\rangle \quad Z(\eta) = e^{\psi(\eta)} = (2\pi)^{\nu}\, \eta^{1-\nu} I_{\nu-1}(\eta) \tag{59a}$$

Then, $Z(\eta)$ is the moment generating function of t, so

$$\mathbb{E}_z(t) = \frac{Z'(\eta)}{Z(\eta)} \quad \text{and} \quad \mathbb{E}_z\left(t^2\right) = \frac{Z''(\eta)}{Z(\eta)} \tag{59b}$$

where the prime denotes differentiation with respect to η. Recall the derivative and recurrence relations of modified Bessel functions [46],

$$\left(\eta^{-a} I_a(\eta)\right)' = \eta^{-a} I_{a+1}(\eta) \quad I_{a-1}(\eta) - I_{a+1}(\eta) \;=\; \frac{2a}{\eta}\, I_a(\eta) \tag{60}$$

where a is any complex number. By applying these relations to (59b), it is possible to show, through a direct calculation,

$$\mathbb{E}_z(t) = \frac{I_\nu(\eta)}{I_{\nu-1}(\eta)} \tag{61a}$$

$$\mathbb{E}_z\left(t^2\right) = \frac{1}{n} + \frac{n-1}{n}\,\frac{I_{\nu+1}(\eta)}{I_{\nu-1}(\eta)} \tag{61b}$$

Formulae (61) will provide the proof of (17a) and (17b).

Proof of (17a): since $\psi(\eta)$ is the cumulant generating function of t,

$$\psi''(\eta) \;=\; \mathrm{Var}_z(t) \;=\; \mathbb{E}_z\left(t^2\right) - \mathbb{E}_z\,(t)^2 \tag{62a}$$

where Var denotes the variance. Now, (17a) follows immediately by replacing from (61) into the right-hand side. ■

Proof of (17b): recall from (18a),

$$\beta^2(\eta) \;=\; \frac{\eta^2}{n-1}\,\mathbb{E}_z\left(1 - t^2\right) \tag{62b}$$

However, from (61b),

$$\mathbb{E}_z\left(1 - t^2\right) = \frac{n-1}{n}\left(1 + \frac{I_{\nu+1}(\eta)}{I_{\nu-1}(\eta)}\right) \tag{62c}$$

Now, (17b) follows by replacing (62c) into (62b). ■

Proof of (61a): using the derivative relation of modified Bessel functions, which is the first relation in (60), with $a = \nu - 1$, it follows that

$$Z'(\eta) \;=\; (2\pi)^{\nu}\,\eta^{1-\nu} I_\nu(\eta) \tag{63a}$$

Now, Formula (61a) follows by replacing this into (59b) and using (59a). ■

Proof of (61b): write (63a) in the form

$$Z'(\eta) \;=\; (2\pi)^{\nu}\,\eta\,\left(\eta^{-\nu} I_\nu(\eta)\right)$$

By the product rule

$$Z''(\eta) \;=\; (2\pi)^{\nu}\,\eta^{-\nu} I_\nu(\eta) + (2\pi)^{\nu}\,\eta\,\left(\eta^{-\nu} I_\nu(\eta)\right)'$$

The derivative in the second term can be evaluated from the derivative relation of modified Bessel functions, with $a = \nu$. Then,

$$Z''(\eta) \;=\; (2\pi)^{\nu}\,\eta^{-\nu} I_\nu(\eta) + (2\pi)^{\nu}\,\eta^{1-\nu} I_{\nu+1}(\eta)$$

Rearrange this formula as

$$Z''(\eta) \;=\; (2\pi)^{\nu}\,\eta^{1-\nu}\,\left(\eta^{-1} I_\nu(\eta) + I_{\nu+1}(\eta)\right)$$

By the recurrence relation of modified Bessel functions, which is the second relation in (60), with $a = \nu$, it then follows

$$Z''(\eta) = (2\pi)^{\nu}\,\eta^{1-\nu}\left(\frac{1}{2\nu}\,I_{\nu-1}(\eta) - \frac{1}{2\nu}\,I_{\nu+1}(\eta) + I_{\nu+1}(\eta)\right)$$

Recalling that $2\nu = n$, this can be written,

$$Z''(\eta) = (2\pi)^{\nu}\,\eta^{1-\nu}\left(\frac{1}{n}\,I_{\nu-1}(\eta) + \frac{n-1}{n}\,I_{\nu+1}(\eta)\right) \tag{63b}$$

Now, Formula (61b) follows by replacing this into (59b) and using (59a). ■

Appendix C – Proof of Proposition 6

The setting and notations are the same as in Sect. 7, except for the fact that $\bar{x}$ is written as x, without the bar, in order to avoid notations such as $\dot{\bar{x}}$ or $\ddot{\bar{x}}$. This being said, let $\tilde{\nabla}$ and ∇ denote the Levi-Civita connections of the Riemannian metrics I and Q, respectively. Thus, $\tilde{\nabla}$ is a connection on the tangent bundle of the manifold $\mathcal{M}$, and ∇ is a connection on the tangent bundle of the manifold M [13, 40]. Introduce the shape operator $S : T_xM \to T_xM$, which is given as in [40],

$$S(u) = \tilde{\nabla}_u\,\partial_r \quad u \in T_xM \tag{64}$$

for any $x \in M$. The following identities can be found in [40] (Sect. 2.4, p. 41),

$$\tilde{\nabla}_{\partial_r}\,\partial_r = 0 \tag{65a}$$

$$\tilde{\nabla}_{\partial_r}\,X = S(X) \tag{65b}$$

$$\tilde{\nabla}_X\,Y = \nabla_X\,Y - I(S(X), Y)\,\partial_r \tag{65c}$$

for any vector fields X and Y on M. Using these identities, it is possible to write the geodesic equation of the Riemannian metric I, in terms of the shape operator S. This is given in the following proposition.

Proposition 10 *Let $\gamma(t)$ be a curve in $\mathcal{M}$, with $\gamma(t) = (x(t), \sigma(t))$ and let $r(t) = r(\sigma(t))$. The curve $\gamma(t)$ is a geodesic of the Riemannian metric I if and only if it satisfies the geodesic equation*

$$\ddot{r} = I(S(\dot{x}), \dot{x}) \tag{66a}$$

$$\ddot{x} = -2\,\dot{r}\,S(\dot{x}) \tag{66b}$$

where $\ddot{x} = \nabla_{\dot{x}}\,\dot{x}$ is the acceleration of the curve $x(t)$ in M.

The shape operator S moreover admits a simple expression, which can be derived from expression (40) of the Riemannian metric I, using the fact that $\tilde{\nabla}$ is a metric connection [13] (Theorem I.5.1, p. 16).

Proposition 11 *In the notation of (40), the shape operator S is given by*

$$S(u) = \sum_{q=1}^{r} \frac{\partial_r \beta_q(r)}{\beta_q(r)}\, u_q \tag{67}$$

In other words, the decomposition $u = u_1 + \cdots + u_r$ provides a block-diagonalisation of S, where each block is a multiple of identity.

Combining Propositions 10 and 11, the geodesic equation (66) takes on a new form. Precisely, replacing (67) into (66) gives the following equations

$$\ddot{r} = \sum_{q=1}^{r} \beta_q(r)\partial_r\beta_q(r)\, Q(\dot{x}_q, \dot{x}_q) \tag{68a}$$

$$\ddot{x}_q = -2\,\dot{r}\, \frac{\partial_r \beta_q(r)}{\beta_q(r)}\, \dot{x}_q \tag{68b}$$

where $\dot{x} = \dot{x}_1 + \cdots + \dot{x}_r$ and $\ddot{x} = \ddot{x}_1 + \cdots + \ddot{x}_r$. The proof of Proposition 6 can be obtained directly from Eq. (68), using the following conservation laws.

Proposition 12 *Each one of the following quantities C_q is a conserved quantity,*

$$C_q \;=\; \beta_q^4(r)\, Q(\dot{x}_q\,, \dot{x}_q) \;\; for \;\; q = 1\,, \ldots, r \tag{69}$$

In other words, C_q remains constant when evaluated along any geodesic $\gamma(t)$ of the Riemannian metric I.

For now, assume that Propositions 10–12 are true. To prove Proposition 6, note the following.

Proof of (41a): it is enough to show that the right-hand side of (41a) is the same as the right-hand side of (68a). To do so, note from (41a) and (69) that

$$V(r) \;=\; \sum_{q=1}^{r} \frac{\beta_q^2(r(0))}{\beta_q^2(r)}\, I_z(u_q\,, u_q) \;=\; \sum_{q=1}^{r} \frac{C_q}{\beta_q^2(r)} \tag{70a}$$

Indeed, since $\dot{x}_q(0) = u_q$ and since C_q is a conserved quantity

$$\beta_q^2(r(0))\, I_z(u_q\,, u_q) \;=\; \beta_q^4(r)\, Q(\dot{x}_q\,, \dot{x}_q)\big|_{t=0} \;=\; C_q$$

Now, replacing the derivative of (70a) into the right-hand side of (41a) directly leads to the right-hand side of (68a). ■

Proof of (41b): recall from Remark 7 that M is the Riemannian product of the M_q. Therefore, the Riemannian exponential mapping of M is also the product of the Riemannian exponential mappings of the M_q. Precisely, (41b) is equivalent to

$$x_q(t) = \exp_{x_q(0)}\left[\left(\int_0^t \frac{\beta_q^2(r(0))}{\beta_q^2(r(s))} ds\right) u_q\right] \quad \text{for } q = 1, \ldots, r \tag{70b}$$

This means that the curve $x_q(t)$ in M_q is a reparameterised geodesic $\left(\delta_q \circ F\right)(t)$ where $\delta_q(t)$ is the geodesic given by $\delta_q(t) = \exp(t\, u_q)$ and $F(t)$ is the integral inside the parentheses in (70b). To prove (41b), it is sufficient to prove that (70b) solves Eq. (68b). Using the chain rule, (70b) implies that

$$\ddot{x}_q = \dot{F}^2\left(\ddot{\delta}_q \circ F\right) + F''\left(\dot{\delta}_q \circ F\right) = \dot{F}^2\left(\ddot{\delta}_q \circ F\right) + \frac{F''}{F'}\,\dot{x}_q = \frac{F''}{F'}\,\dot{x}_q \tag{70c}$$

where the third equality follows because δ_q is a geodesic, and therefore its acceleration $\ddot{\delta}_q$ is zero. By replacing the definition of the function $F(t)$, it is seen that (70c) is the same as (68b). It follows that (70b) solves (68b), as required. ■

Proof of Proposition 10: recall the geodesic equation is $\tilde{\nabla}_{\dot{\gamma}}\,\dot{\gamma} = 0$, which means that the velocity $\dot{\gamma}(t)$ is self-parallel [13, 40]. Here, the velocity $\dot{\gamma}(t)$ is given by $\dot{\gamma}(t) = \dot{r}\,\partial_r + \dot{x}$. Accordingly, the left-hand side of the geodesic equation is

$$\tilde{\nabla}_{\dot{\gamma}}\,\dot{\gamma} = \tilde{\nabla}_{\dot{\gamma}}\,\dot{r}\,\partial_r + \tilde{\nabla}_{\dot{\gamma}}\,\dot{x} = \ddot{r}\,\partial_r + \dot{r}\,\tilde{\nabla}_{\dot{\gamma}}\,\partial_r + \tilde{\nabla}_{\dot{\gamma}}\,\dot{x} \tag{71a}$$

where the second equality follows by the product rule for the covariant derivative [13, 40]. The second and third terms on the right-hand side of (71a) can be written in terms of the shape operator S. Precisely, for the second term,

$$\tilde{\nabla}_{\dot{\gamma}}\,\partial_r = \dot{r}\,\tilde{\nabla}_{\partial_r}\,\partial_r + \tilde{\nabla}_{\dot{x}}\,\partial_r = S(\dot{x}) \tag{71b}$$

where the second equality follows from (64) and (65a). Moreover, for the third term,

$$\tilde{\nabla}_{\dot{\gamma}}\,\dot{x} = \dot{r}\,\tilde{\nabla}_{\partial_r}\,\dot{x} + \tilde{\nabla}_{\dot{x}}\,\dot{x} = \dot{r}\,S(\dot{x}) + \ddot{x} - I(S(\dot{x}), \dot{x})\,\partial_r \tag{71c}$$

where the second equality follows from (65b) and (65c). Replacing (71b) and (71c) into (71a), the left-hand side of the geodesic equation becomes

$$\tilde{\nabla}_{\dot{\gamma}}\,\dot{\gamma} = (\ddot{r} - I(S(\dot{x}), \dot{x}))\,\partial_r + (\ddot{x} + 2\dot{r}\,S(\dot{x}))$$

Setting this equal to zero immediately gives Eq. (66). ■

Proof of Proposition 11 : recall the shape operator S is symmetric, since it is essentially the Riemannian Hessian of r [40] (Sect. 2.4, p. 41). Therefore, it is enough to evaluate $I(S(u), u)$ for $u \in T_x M$. Let X by a vector field on M, with $X(x) = u$. Then,

$$I(S(u), u) = I\,(S(X), X) = I\left(\tilde{\nabla}_{\partial_r}\,X\,,\,X\right) \tag{72a}$$

where the second equality follows from (65b). Using the fact that $\tilde{\nabla}$ is a metric connection [13] (Theorem I.5.1, p. 16), the right-hand side can be written as

$$I\left(\tilde{\nabla}_{\partial_r} X\,,X\right) = \frac{1}{2}\,\partial_r\, I(X,X) = \frac{1}{2}\,\partial_r \sum_{q=1}^{r} \beta_q^2(r)\, Q_x(u_q\,,u_q) \tag{72b}$$

where the second equality follows from (40). It remains to note that

$$\frac{1}{2}\,\partial_r\,\beta_q^2(r)\, Q_x(u_q\,,u_q) = \frac{\partial_r\,\beta_q(r)}{\beta_q(r)}\, I_z(u_q\,,u_q)$$

Accordingly, (72a) and (72b) imply

$$I(S(u),u) = I\left(\sum_{q=1}^{r}\frac{\partial_r\,\beta_q(r)}{\beta_q(r)}\,u_q\,,u_q\right) = I\left(\sum_{q=1}^{r}\frac{\partial_r\,\beta_q(r)}{\beta_q(r)}\,u_q\,,u\right) \tag{72c}$$

and (67) follows from the fact that S is symmetric. ■

Proof of Proposition 12: to say that C_q is a conserved quantity means that $\dot{C}_q = 0$. From (69),

$$\dot{C}_q = 4\dot{r}\,\beta_q^3(r)\,\partial_r\beta_q(r)\,Q(\dot{x}_q\,\dot{x}_q) + \beta_q^4(r)\,\frac{d}{dt}\,Q(\dot{x}_q\,\dot{x}_q) \tag{73a}$$

The last derivative can be expressed as

$$\frac{d}{dt}\,Q(\dot{x}_q\,\dot{x}_q) = 2\,Q(\ddot{x}_q\,,\dot{x}_q) = -4\dot{r}\,\frac{\partial_r\beta_q(r)}{\beta_q(r)}\,Q(\dot{x}_q\,,\dot{x}_q) \tag{73b}$$

where the second equality follows from (68b). By replacing (73b) into (73a), it follows immediately that $\dot{C}_q = 0$. ■

Appendix D – Proof of Proposition 7

The proof of Formula (44b), for the Laplace–Beltrami operator $\Delta_{\mathcal{M}}$, will introduce some useful notation.

Proof of Formula (44b): for any smooth function f on $\mathcal{M}$, it follows from (40) that the Riemannian gradient $\tilde{\nabla} f$ of f, with respect to the multiply-warped Riemannian metric I, is given by

$$\tilde{\nabla} f = (\partial_r f)\,\partial_r + \sum_{q=1}^{r} \beta_q^{-2}(r)\,\nabla_{\bar{x}_q} f \tag{74a}$$

where $\nabla_{\bar{x}_q} f$ is the Riemannian gradient of f with respect to $\bar{x}_q \in M_q$, computed with all other arguments of f being fixed. Expression (74a) can be verified by checking that

$$df\,U = I(\tilde{\nabla} f, U)$$

for any tangent vector U to $\mathcal{M}$. This follows directly from (40) and (74a). Now, by definition of the Laplace–Beltrami operator [34], (see p. 443),

$$\Delta_{\mathcal{M}} f = \operatorname{div} \tilde{\nabla} f \tag{74b}$$

where the divergence div V of a vector field V on $\mathcal{M}$ is found from

$$\mathcal{L}_V \operatorname{vol} = (\operatorname{div} V) \operatorname{vol} \tag{74c}$$

with the notation $\mathcal{L}$ for the Lie derivative, and vol for the Riemannian volume element of the metric I. This last formula can be applied along with the following expression of vol, which follows from (40),

$$\operatorname{vol} = G(r)\, dr \bigwedge_q \operatorname{vol}_q(\bar{x}_q) \tag{74d}$$

where the function $G(r)$ was defined after (44b), and where $\wedge$ denotes the exterior product, and vol_q is the Riemannian volume of M_q. From (74c) and (74d), applying the product formula of the Lie derivative [34] (Theorem 7.4.8, p. 414),

$$\operatorname{div} V = \frac{1}{G}\, \partial_r\, (G\, V_r) + \sum_{q=1}^{r} \operatorname{div}_{M_q} V_q \tag{74e}$$

where $V = V_r\, \partial_r + \sum_q V_q$ with each V_q tangent to M_q, and where $\operatorname{div}_{M_q} V_q$ denotes the divergence of V_q with respect to $\bar{x}_q \in M_q$. Formula (44b) follows directly from (74a), (74b) and (74e). Indeed, for the vector field $V = \tilde{\nabla} f$,

$$V_r = \partial_r f \quad V_q = \beta_q^{-2}(r)\, \nabla_{\bar{x}_q} f$$

as can be seen from (74a). ■

For the proof of Proposition 7, assume the process z is a Riemannian Brownian motion associated to I. Write $z(t) = (\bar{x}(t), \sigma(t))$ where $\bar{x}(t) = (\bar{x}_1(t), \ldots, \bar{x}_r(t))$ and each $\bar{x}_q(t)$ belongs to M_q. The proof consists in showing that the joint distribution of the processes $r(t) = r(\sigma(t))$ and $\bar{x}_q(t)$ is the same as described in Proposition 7. This is done through the following steps.

Step 1 – $r(t)$ **verifies** (45a): recall that, for any smooth function f on $\mathcal{M}$, the process z verifies (44a). If $f = r$, then by (44a) and (44b)

$$dr(t) = \frac{1}{2}\,\Delta_{\mathcal{M}} r(t)dt + dm^r(t) = \frac{1}{2}\frac{\partial_r G}{G}(r(t))dt + dm^r(t) \tag{75}$$

To prove that $r(t)$ verifies (45a), it is enough to prove that $dm^r(t) = dw(t)$ where w is a standard Brownian motion. Note that $dr^2(t)$ can be computed in two ways. The first way, by Itô's formula and (75)

$$dr^2 = r\Delta_{\mathcal{M}} r(t)dt + 2r(t)dm^r(t) + d[m^r](t)$$

where $[m^r](t)$ is the quadratic variation process of the local martingale $m^r(t)$ [25] (Theorem 17.16, p. 339). The second way, by (44a) with $f = r^2$,

$$\begin{aligned} dr^2 &= \left(I(\tilde{\nabla} r, \tilde{\nabla} r) + r\Delta_{\mathcal{M}} r(t)\right) dt + dm^{r^2}(t)dt \\ &= (1 + r\Delta_{\mathcal{M}} r(t))\, dt + dm^{r^2}(t)dt \end{aligned}$$

where the second line follows from (40) because $\tilde{\nabla} r = \partial_r$. By equating these two expressions of $dr^2(t)$, it follows that $dt - d[m^r](t)$ is the differential of a continuous local martingale of finite variation, and therefore identically zero [25] (Proposition 17.2, p. 330). In other words, $d[m^r](t) = dt$ and Lévy's characterisation implies that $dm^r(t) = dw(t)$, where w is a standard Brownian motion [25] (Theorem 18.3, p. 352). ■

Step 2 – $\bar{x}_q(t)$ **verifies** (45b): if f is a smooth function on $\mathcal{M}$, such that $f(z) = f(\bar{x}_q)$, then by (44a) and (44b)

$$df(\bar{x}_q(t)) = \frac{1}{2}\beta_q^{-2}(r(t))\,\Delta_{M_q} f(\bar{x}_q(t))\,dt + dm^f(t) \tag{76}$$

Define $l_q(t)$ to be the inverse of the time change process $\tau_q(t)$ defined in (45b), and let $\theta_q(t) = \left(\bar{x}_q \circ l_q\right)(t)$. By applying the time change $l_q(t)$ to (76)

$$df(\theta_q(t)) = \frac{1}{2}\,\Delta_{M_q} f(\theta_q(t))\,dt + d(m^f \circ l_q)(t)$$

where the first term on the right-hand side is obtained by replacing from the definition of $\tau_q(t)$ in (45b). Recall that a time change of a local martingale is a local martingale [25] (Theorem 17.24, p. 344). Therefore, $m^f \circ l_q$ is a local martingale, so θ_q solves the martingale problem associated to $(1/2)\Delta_{M_q}$. This means that θ_q is a Riemannian Brownian motion in M_q. ■

Step 3 – $r(t)$ **and** $\theta_q(t)$ **are independent**: it is required to prove that the processes $r(t)$ and $\theta_1(t), \ldots, \theta_r(t)$ are jointly independent. A detailed proof is given only of the fact that $r(t)$ and $\theta_q(t)$ are independent, for any fixed q. The complete proof is obtained by repeating similar arguments.

The following proof is modeled on [23] (Example 3.3.3, p. 84). Let $\left(\xi^i(t)\,;\ i = 1\,,\ldots,\,\dim M_q\right)$ be the stochastic anti-development of $\bar{x}_q(t)$. Precisely [19] (Definition 8.23, p. 119)

$$d\xi^i(t) \;=\; Q\left(e^i\,,d\bar{x}_q(t)\right) \;=\; \beta_q^{-2}(r(t))\,I\left(e^i\,,dz(t)\right) \tag{77a}$$

where the e^i form a parallel orthonormal moving frame above the stochastic process $\bar{x}_q(t)$ in M_q. On the other hand, it is possible to write, using Itô's formula [19] (Proposition 7.34, p. 109)

$$dr(t) \;=\; I\left(\partial_r\,,dz(t)\right) + \frac{1}{2}\,\Delta_{\mathcal{M}}r(t)\,dt \tag{77b}$$

Let $[\xi^i\,,r]$ denote the quadratic covariation of ξ^i and r. From [19] (Proposition 5.18, p. 63), since z is a Riemannian Brownian motion, it follows from (77a) and (77b) that

$$d[\xi^i\,,r](t) \;=\; \beta_q^{-2}(r(t))\,I(e^i\,,\partial_r)(z(t))\,dt \;=\; 0 \tag{77c}$$

as e^i and ∂_r are orthogonal with respect to I. This means that $(\xi^i(t))$ and $r(t)$ have zero quadratic covariation, and therefore $(\xi^i(t))$ and $w(t)$ have zero quadratic covariation. It follows as in [23] (Lemma 3.3.4, p. 85), that $r(t)$ and $\theta_q(t)$ are independent. ■

References

1. Absil, P.A., Mahony, R., Sepulchre, R.: Optimization Algorithms on Matrix Manifolds. Princeton University Press, Princeton (2008)
2. Amari, S.I.: Natural gradient works efficiently in learning. Neural Comput. **10**(2), 251–276 (1998)
3. Amari, S., Nagaoka, H.: Methods of Information Geometry. American Mathematical Society, Providence (2000)
4. Arnaudon, M.: Semi-martingales dans les espaces homogènes. Annales de l'I.H.P **29**(3), 269–288 (1993)
5. Arnaudon, M., Miclo, L.: A stochastic algorithm finding generalized means on compact manifolds. Stochastic. Process. Appl. **124**(10), 3463–3479 (2014)
6. Atkinson, C., Mitchell, A.: Rao's distance measure. Sankhya Ser. A **43**, 345–365 (1981)
7. Ay, N., Jost, J., Lê, H.V., Schwachhöfer, L.: Information geometry and sufficient statistics. Probab. Theory Relat. Fields **162**(1), 327–364 (2015)
8. Bensadon, J.: Black-box optimization using geodesics in statistical manifolds. Entropy **17**(1), 304–345 (2015)
9. Bensadon, J.: Applications of information theory to machine learning. Ph.D. thesis, Université Paris-Saclay (2016). https://tel.archives-ouvertes.fr/tel-01297163
10. Berthoumieu, Y., Bombrun, L., Said, S.: Classification approach based on the product of Riemannian manifolds from Gaussian parameterization space. In: International Conference on Image Processing (ICIP) (2017)
11. Bishop, R.L., O'NEILL, B.: Manifolds of negative curvature. Trans. Amer. Math. Soc. **145**, 1–49 (1969)

12. Bottou, L., Curtis, F.E., Nocedal, J.: Optimization methods for large-scale machine learning (2017). arXiv:1606.04838v2
13. Chavel, I.: Riemannian Geometry, A Modern Introduction. Cambridge University Press, Cambridge (2006)
14. Cheng, G., Vemuri, B.C.: A novel dynamic system in the space of SPD matrices with applications to appearance tracking. SIAM J. Imaging Sci. **6**(1), 592–615 (2013)
15. Chentsov, N.N.: Statistical Decision Rules and Optimal Inference. American Mathematical Society, Providence (1982)
16. Chikuse, Y.: Statistics on Special Manifolds. Lecture Notes in Statistics, 174. Springer Science+Business Media, LLC (2003)
17. Dobarro, F., Unal, B.: Curvature of multiply warped products. J. Geom. Phys. **55**, 75–106 (2005)
18. Do Carmo, M.P.: Riemannian Geometry, 2nd edn. Birkhauser, Boston (1992)
19. Emery, M.: Stochastic Calculus in Manifolds. Springer, Berlin (1980)
20. Estrade, A.: Exponentielle stochastique et intégrale multiplicative discontinues. Annales de l'I.H.P. **28**(1), 107–129 (1992)
21. Gallavotti, G.: The Elements of Mechanics. Springer Science+Business Media, LLC, New York (1983)
22. Helgason, S.: Differential Geometry and Symmetric Spaces. Academic Press, New York (1962)
23. Hsu, E.P.: Stochastic Analysis on Manifolds. American Mathematical Society, Providence (2002)
24. Ikeda, N., Watanabe, S.: Stochastic Differential Equations and Diffusion Processes. North-Holland Publishing Company, Amsterdam (1981)
25. Kallenberg, O.: Foundations of Modern Probability, 2nd edn. Springer, New York (2001)
26. Kloeden, P.E., Platen, E.: Numerical Solution of Stochastic Differential Equations. Springer, New York (2011)
27. Knapp, A.W.: Lie Groups Beyond an Introduction, 2nd edn. Birkhauser, Boston (2002)
28. Lang, S.: Introduction to Linear Algebra, 2nd edn. Springer, New York (1986)
29. Lee, J.M.: Introduction to Smooth Manifolds, 2nd edn. Springer Science+Business Media, New York (2012)
30. Lehmann, E.L., Romano, J.P.: Testing Statistical Hypotheses, 3rd edn. Springer Science+Business Media Inc, New York (2005)
31. Liao, E.: Lévy Processes on Lie Groups. Cambridge University Press, Cambridge (2004)
32. Luenbeger, D.G.: Linear and Nonlinear Programming, 2nd edn. Addison-Wesley publishing company, Reading (1973)
33. Mardia, K.V., Jupp, P.E.: Directional Statistics. Wiley, Chichester (2002)
34. Marsden, J.E., Ratiu, T.: Manifolds, Tensor Analysis, and Applications. Springer Publishing Company Inc., New York (2001)
35. Martens, J.: New insights and perspectives on the natural gradient method (2017). arXiv:1412.1193v8
36. McLachlan, G.J.: Discriminant Analysis and Statistical Pattern Recognition. Wiley, Hoboken (2004)
37. Ollivier, Y., Arnold, L., Auger, A., Hansen, N.: Information geometric optimization algorithms: a unifying picture via invariance principles. J. Mach. Learn. Res. **18**(18), 1–65 (2017)
38. O'Neill, B.: Semi-Riemannian Geometry with Applications to Relativity. Academic Press, San Diego (1983)
39. Pennec, X., Fillard, P., Ayache, N.: A Riemannian framework for tensor computing. Int. J. Comput. Vision **66**(1), 41–66 (2006)
40. Petersen, P.: Riemannian Geometry, 2nd edn. Springer Science+Business Media, LLC, New York (2006)
41. Said, S., Bombrun, L., Berthoumieu, Y., Manton, J.H.: Riemannian Gaussian distributions on the space of symmetric positive definite matrices. IEEE. Trans. Inf. Theory **63**(4), 2153–2170 (2017)

42. Said, S., Hajri, H., Bombrun, L., Vemuri, B.C.: Gaussian distributions on Riemannian symmetric spaces: statistical learning with structured covariance matrices. IEEE. Trans. Inf. Theory **64**(2), 752–772 (2018)
43. Spanier, E.H.: Algebraic Topology. Springer, New York (1966)
44. Terras, A.: Harmonic Analysis on Symmetric Spaces and Applications, vol. II. Springer, New York (1988)
45. Unal, B.: Doubly warped products. Differ. Geom. Appl. **15**, 253–263 (2001)
46. Watson, G.N.: A Treatise on the Theory of Bessel Functions. Cambridge University Press, Cambridge (1922)
47. Wierstra, D., Schaul, T., Glasmachers, T., Sun, Y., Peters, J., Schmidhuber, J.: Natural evolution strategies. J. Mach. Learn. Res. **15**(1), 949–980 (2014)
48. Young, G.A., Smith, R.L.: Essentials of Statistical Inference. Cambridge University Press, Cambridge (2005)
49. Zanini, P., Said, S., Congedo, M., Berthoumieu, Y., Jutten, C.: Parameter estimates of Riemannian Gaussian distributions in the manifold of covariance matrices. In: Sensor Array and Multichannel Signal Processsing Workshop (SAM) (2016)
50. Zanini, P., Said, S., Berthoumieu, Y., Congedo, M., Jutten, C.: Riemannian online algorithm for estimating mixture model parameters. In: Geometric Science of Information (GSI) (2017)

Clustering in Hilbert's Projective Geometry: The Case Studies of the Probability Simplex and the Elliptope of Correlation Matrices

Frank Nielsen and Ke Sun

Abstract Clustering categorical distributions in the probability simplex is a fundamental task met in many applications dealing with normalized histograms. Traditionally, differential-geometric structures of the probability simplex have been used either by (i) setting the Riemannian metric tensor to the Fisher information matrix of the categorical distributions, or (ii) defining the dualistic information-geometric structure induced by a smooth dissimilarity measure, the Kullback–Leibler divergence. In this work, we introduce for this clustering task a novel computationally-friendly framework for modeling the probability simplex termed *Hilbert simplex geometry*. In the Hilbert simplex geometry, the distance function is described by a polytope. We discuss the pros and cons of those different statistical modelings, and benchmark experimentally these geometries for center-based k-means and k-center clusterings. Furthermore, since a canonical Hilbert metric distance can be defined on any bounded convex subset of the Euclidean space, we also consider Hilbert's projective geometry of the elliptope of correlation matrices and study its clustering performances.

Keywords Fisher-Riemannian geometry · Information geometry
Multinoulli distribution · Hilbert simplex geometry · Polytope distance
Elliptope · Finsler geometry · k-means clustering · k-center clustering

F. Nielsen (✉)
Sony Computer Science Laboratories, Tokyo, Japan
e-mail: Frank.Nielsen@acm.org

K. Sun
CSIRO Data61, Sydney, Australia
e-mail: Ke.Sun@data61.csiro.au

F. Nielsen (ed.), *Geometric Structures of Information*, Signals and Communication Technology, https://doi.org/10.1007/978-3-030-02520-5_11

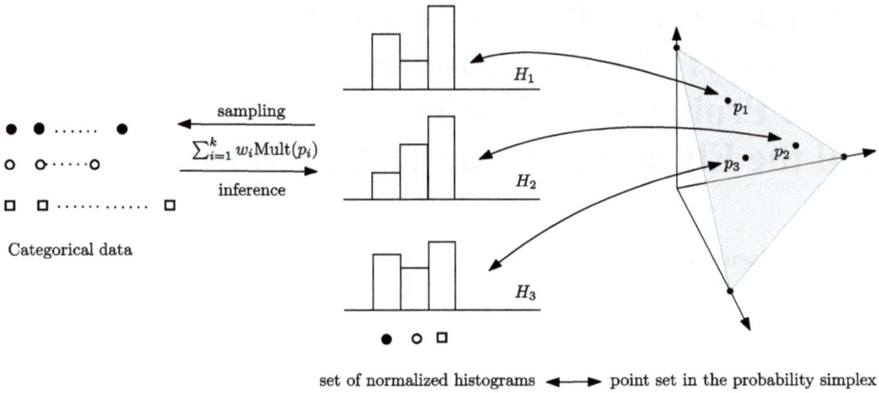

Fig. 1 Categorical datasets modeled by a generative statistical mixture model of multinoulli distributions can be visualized as a weighted set of normalized histograms or equivalently by a weighted point set encoding multinoulli distributions in the probability simplex Δ^d (here, $d = 2$ for trinoulli distributions — trinomial distributions with a single trial)

1 Introduction and Motivation

The categorical distributions and multinomial distributions are important probability distributions often met in data analysis [1], text mining [2], computer vision [3] and machine learning [4]. A multinomial distribution over a set $\mathscr{X} = \{e_0, \ldots, e_d\}$ of outcomes (e.g., the $d+1$ distinct colored faces of a die) is defined as follows: Let $\lambda_p^i > 0$ denote the probability that outcome e_i occurs for $i \in \{0, \ldots, d\}$ (with $\sum_{i=0}^d \lambda_p^i = 1$). Denote by m the total number of events, with m_i reporting the number of outcome e_i. Then the probability $\Pr(X_0 = m_0, \ldots, X_d = m_d)$ that a multinomial random variable $X = (X_0, \ldots, X_d) \sim \mathrm{Mult}(p = (\lambda_p^0, \ldots, \lambda_p^d), m)$ (where X_i count the number of events e_i, and $\sum_{i=0}^d m_i = m$) is given by the following probability mass function (pmf):

$$\Pr(X_0 = m_0, \ldots, X_d = m_d) = \frac{m!}{\prod_{i=0}^d m_i!} \prod_{i=0}^d \left(\lambda_p^i\right)^{m_i}.$$

The multinomial distribution is called a binomial distribution when $d = 1$ (e.g., coin tossing), a Bernoulli distribution when $m = 1$, and a “multinoulli distribution” (or categorical distribution) when $m = 1$ and $d > 1$. The multinomial distribution is also called a generalized Bernoulli distribution. A random variable X following a multinoulli distribution is denoted by $X = (X_0, \ldots, X_d) \sim \mathrm{Mult}(p = (\lambda_p^0, \ldots, \lambda_p^d))$. The multinomial/multinoulli distribution provides an important *feature representation* in machine learning that is often met in applications [5–7] as normalized histograms (with non-empty bins) as illustrated in Fig. 1.

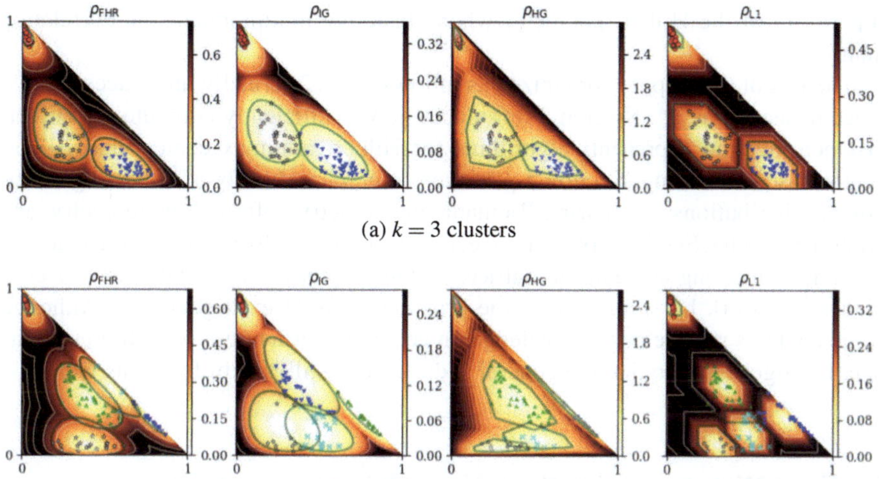

(a) $k = 3$ clusters

(b) $k = 5$ clusters

Fig. 2 Visualizing some k-center clustering results on a toy dataset in the space of trinomials Δ^2 for the considered four types of distances (and underlying geometries): Fisher-Hotelling-Rao metric distance (Riemannian geometry), Kullback–Leibler non-metric divergence (information geometry), Hilbert metric distance (Hilbert projective geometry), and total variation/L_1 metric distance (norm geometry). Observe that the L_1 balls have hexagonal shapes on the probability simplex (intersection of a rotated cube with the plane H_{Δ^d}). The color density maps indicate the distance from any point to its nearest cluster center

A multinomial distribution $p \in \Delta^d$ can be thought as a point lying in the probability simplex Δ^d (standard simplex) with coordinates $p = (\lambda_p^0, \ldots, \lambda_p^d)$ such that $\lambda_p^i = \Pr(X = e_i) > 0$ and $\sum_{i=0}^d \lambda_p^i = 1$. The open probability simplex Δ^d can be embedded in $\mathbb{R}^{d+1}$ on the hyperplane $H_{\Delta^d} : \sum_{i=0}^d x^i = 1$. Notice that observations with D categorical attributes can be clustered using k-mode [8] with respect to the Hamming distance. Here, we consider the different task of clustering a set $\Lambda = \{p_1, \ldots, p_n\}$ of n categorical/multinomial distributions in Δ^d [5] using center-based k-means++ or k-center clustering algorithms [9, 10], which rely on a dissimilarity measure (loosely called distance or divergence when smooth) between any two categorical distributions. In this work, we mainly consider four distances with their underlying geometries: (1) Fisher-Hotelling-Rao distance ρ_{FHR} (spherical geometry), (2) Kullback–Leibler divergence ρ_{IG} (dually flat geometry), (3) Hilbert distance ρ_{HG} (generalize Klein's hyperbolic geometry), and (4) the total variation/L1 distance (norm geometry). The geometric structures of spaces are necessary in algorithms, for example, to define midpoint distributions. Figure 2 displays the k-center clustering results obtained with these four geometries as well as the L^1 distance ρ_{L1} normed geometry on toy synthetic datasets in Δ^2. We shall now explain the Hilbert simplex geometry applied to the probability simplex, describe how to perform k-center clustering in Hilbert geometry, and report experimental results that demonstrate the superiority of the Hilbert geometry when clustering multinomials and correlation matrices.

The rest of this paper is organized as follows: Sect. 2 formally introduces the distance measures in Δ^d. Section 3 introduces how to efficiently compute the Hilbert distance. Section 4 presents algorithms for Hilbert minimax centers and Hilbert center-based clustering. Section 5 performs an empirical study of clustering multinomial distributions, comparing Riemannian geometry, information geometry, and Hilbert geometry. Section 6 presents a second use case of Hilbert geometry in machine learning: clustering correlation matrices in the elliptope [11]. Finally, Sect. 7 concludes this work by summarizing the pros and cons of each geometry. Although some contents require prior knowledge on geometric structures, we will present the detailed algorithms so that the general audience can still benefit from this work.

2 Four Distances with their Underlying Geometries

2.1 *Fisher-Hotelling-Rao Riemannian Geometry*

The Rao distance between two multinomial distributions is [6, 12]:

$$\rho_{\mathrm{FHR}}(p, q) = 2 \arccos \left(\sum_{i=0}^{d} \sqrt{\lambda_p^i \lambda_q^i} \right). \tag{1}$$

It is a Riemannian metric length distance (satisfying the symmetric and triangular inequality axioms) obtained by setting the metric tensor g to the *Fisher information matrix* (FIM) $\mathcal{I}(p) = (g_{ij}(p))_{d \times d}$ with respect to the coordinate system $(\lambda_p^1, \ldots, \lambda_p^d)$, where

$$g_{ij}(p) = \frac{\delta_{ij}}{\lambda_p^i} + \frac{1}{\lambda_p^0}.$$

We term this geometry the *Fisher-Hotelling-Rao (FHR) geometry* [13–16]. The metric tensor g allows one to define an inner product on each tangent plane T_p of the probability simplex manifold: $\langle u, v \rangle_p = u^\top g(p) v$. When g is everywhere the identity matrix, we recover the Euclidean (Riemannian) geometry with the inner product being the scalar product: $\langle u, v \rangle = u^\top v$. The geodesics $\gamma(p, q; \alpha)$ are defined by the Levi-Civita metric connection [17, 18] that is derived from the metric tensor. The FHR manifold can be embedded in the positive orthant of an Euclidean d-sphere in $\mathbb{R}^{d+1}$ by using the *square root representation* $\lambda \mapsto \sqrt{\lambda}$ [12]. Therefore the FHR manifold modeling of Δ^d has constant *positive* curvature: It is a spherical geometry restricted to the positive orthant with the metric distance measuring the arc length on a great circle.

2.2 Information Geometry

A divergence D is a smooth C^3 differentiable dissimilarity measure [19] that allows to define a dual structure in Information Geometry (IG), see [17, 18, 20]. A f-divergence is defined for a strictly convex function f with $f(1) = 0$ by:

$$I_f(p : q) = \sum_{i=0}^{d} \lambda_p^i f\left(\frac{\lambda_q^i}{\lambda_p^i}\right) \geq f(1) = 0.$$

It is a *separable* divergence since the d-variate divergence can be written as a sum of d univariate (scalar) divergences: $I_f(p : q) = \sum_{i=0}^{d} I_f(\lambda_p^i : \lambda_q^i)$. The class of f-divergences plays an essential role in information theory since they are provably the *only* separable divergences that satisfy the *information monotonicity* property [17, 21] (for $d \geq 2$). That is, by coarse-graining the histograms, we obtain lower-dimensional multinomials, say p' and q', such that $0 \leq I_f(p' : q') \leq I_f(p : q)$ [17]. The Kullback–Leibler (KL) divergence ρ_{IG} is a f-divergence obtained for the functional generator $f(u) = -\log u$:

$$\rho_{\mathrm{IG}}(p, q) = \sum_{i=0}^{d} \lambda_p^i \log \frac{\lambda_p^i}{\lambda_q^i}. \tag{2}$$

It is an asymmetric non-metric distance: $\rho_{\mathrm{IG}}(p, q) \neq \rho_{\mathrm{IG}}(q, p)$. In differential geometry, the structure of a manifold is defined by two independent components:

1. A *metric tensor* g that allows to define an inner product $\langle \cdot, \cdot \rangle_p$ at each tangent space (for measuring vector lengths and angles between vectors);
2. A *connection* ∇ that defines *parallel transport* $\prod_c^{\nabla}$, i.e., a way to move a tangent vector from one tangent plane T_p to any other one T_q along a smooth curve c, with $c(0) = p$ and $c(1) = q$.

In FHR geometry, the implicitly-used connection is called the Levi-Civita connection that is induced by the metric g: $\nabla^{\mathrm{LC}} = \nabla(g)$. It is a metric connection since it ensures that $\langle u, v \rangle_p = \langle \prod_{c(t)}^{\nabla^{\mathrm{LC}}} u, \prod_{c(t)}^{\nabla^{\mathrm{LC}}} v \rangle_{c(t)}$ for $t \in [0, 1]$. The underlying information-geometric structure of KL is characterized by a pair of *dual* connections [17] $\nabla = \nabla^{(-1)}$ (mixture connection) and $\nabla^* = \nabla^{(1)}$ (exponential connection) that induces a corresponding pair of dual geodesics (technically, ± 1-autoparallel curves, [18]). Those connections are said *flat* as they define two dual global affine coordinate systems θ and η on which the θ- and η-geodesics are (Euclidean) straight line segments, respectively. For multinomials, the *expectation parameters* are: $\eta = (\lambda^1, \ldots, \lambda^d)$ and they one-to-one correspond to the *natural parameters*: $\theta = \left(\log \frac{\lambda^1}{\lambda^0}, \ldots, \log \frac{\lambda^d}{\lambda^0}\right) \in \mathbb{R}^d$. Thus in IG, we have two kinds of midpoint multinomials of p and q, depending on whether we perform the (linear) interpolation on the θ- or the η-geodesics. Informally speaking, the dual connections $\nabla^{(\pm 1)}$ are said coupled to the FIM since we have $\frac{\nabla + \nabla^*}{2} = \nabla(g) = \nabla^{\mathrm{LC}}$. Those dual (torsion-free affine) connections are not metric connections but enjoy the following metric-compatibility property when used together as follows: $\langle u, v\rangle_p = \langle \prod_{c(t)} u, \prod^*{}_{c(t)} v\rangle_{c(t)}$ (for $t \in [0, 1]$), where $\prod := \prod^{\nabla}$ and $\prod^* := \prod^{\nabla^*}$ are the corresponding induced dual parallel transports. The geometry of f-divergences [19] is the α-geometry (for $\alpha = 3 + 2f'''(1)$) with the dual $\pm\alpha$-connections, where $\nabla^{(\alpha)} = \frac{1+\alpha}{2}\nabla^* + \frac{1-\alpha}{2}\nabla$. The Levi-Civita metric connection is $\nabla^{\mathrm{LC}} = \nabla^{(0)}$. More generally, it was shown how to build a dual information-geometric structure for *any* divergence [19]. For example, we can build a dual structure from the symmetric Cauchy–Schwarz divergence [22]:

$$\rho_{\mathrm{CS}}(p, q) = -\log \frac{\langle \lambda_p, \lambda_q\rangle}{\sqrt{\langle \lambda_p, \lambda_p\rangle\langle \lambda_q, \lambda_q\rangle}}. \tag{3}$$

2.3 Hilbert Simplex Geometry

In Hilbert geometry (HG), we are given a bounded convex domain $\mathscr{C}$ (here, $\mathscr{C} = \Delta^d$), and the distance between any two points M, M' of $\mathscr{C}$ is defined [23] as follows: Consider the two intersection points AA' of the line (MM') with $\mathscr{C}$, and order them on the line so that we have A, M, M', A'. Then the Hilbert metric distance [24] is defined by:

$$\rho_{\mathrm{HG}}(M, M') = \begin{cases} \left|\log \frac{|A'M||AM'|}{|A'M'||AM|}\right|, & M \neq M', \\ 0 & M = M'. \end{cases} \tag{4}$$

It is also called the Hilbert cross-ratio metric distance [25, 26]. Notice that we take the absolute value of the logarithm since the Hilbert distance is a *signed distance* [27]. When $\mathscr{C}$ is the unit ball, HG lets us recover the Klein hyperbolic geometry [26]. When $\mathscr{C}$ is a quadric bounded convex domain, we obtain the Cayley–Klein hyperbolic geometry [28] which can be studied with the Riemannian structure and the corresponding metric distance called the curved Mahalanobis distances [29, 30]. Cayley–Klein hyperbolic geometries have negative curvature. Elements on the boundary are called ideal elements [31].

In Hilbert geometry, the geodesics are *straight* Euclidean lines making them convenient for computation. Furthermore, the domain boundary $\partial\mathscr{C}$ needs not to be smooth: One may also consider bounded polytopes [32]. This is particularly interesting for modeling Δ^d, the d-dimensional open standard simplex. We call this geometry the *Hilbert simplex geometry* [33]. In Fig. 3, we show that the Hilbert distance between two multinomial distributions p (M) and q (M') can be computed by finding the two intersection points of the line $(1-t)p + tq$ with $\partial\Delta^d$, denoted as $t_0 \leq 0$ and $t_1 \geq 1$. Then

$$\rho_{\mathrm{HG}}(p, q) = \left|\log \frac{(1-t_0)t_1}{(-t_0)(t_1-1)}\right| = \log\left(1 - \frac{1}{t_0}\right) - \log\left(1 - \frac{1}{t_1}\right).$$

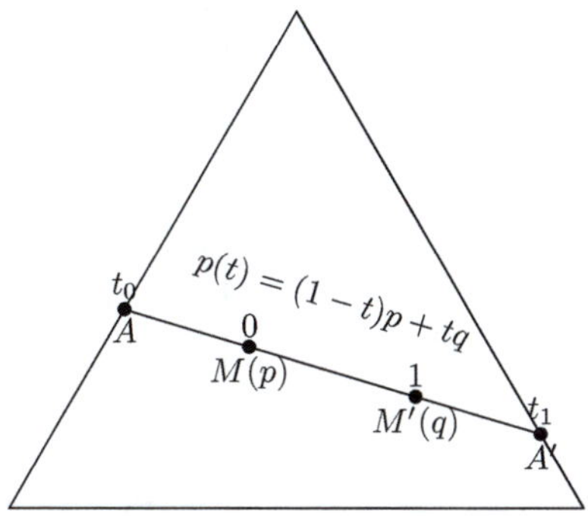

Fig. 3 Computing the Hilbert metric distance for trinomials on the 2D probability simplex as the logarithm of the cross ratio $(M, M'; A, A')$ of the four collinear points A, M, M' and A'

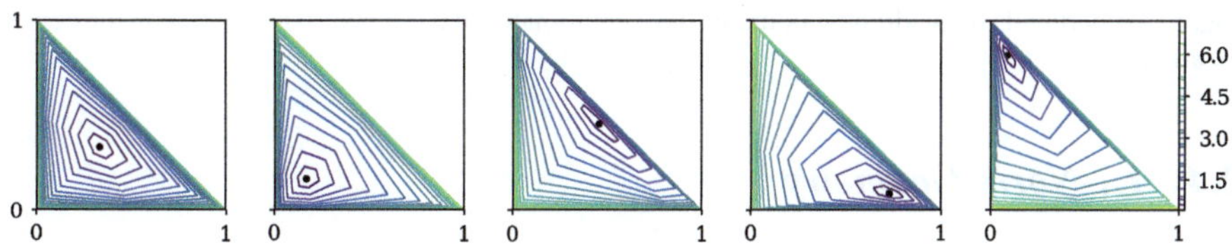

Fig. 4 Balls in the Hilbert simplex geometry Δ^2 have polygonal Euclidean shapes of constant combinatorial complexity. At infinitesimal scale, the balls have hexagonal shapes, showing that the Hilbert geometry is not Riemannian

The shape of balls in polytope-domain HG is Euclidean polytopes[1] [26], as depicted in Fig. 4. Furthermore, the Euclidean shape of the balls does not change with the radius. Hilbert balls have hexagons shapes in 2D [34], rhombic dodecahedra shapes in 3D, and are polytopes [26] with $d(d+1)$ facets in dimension d. When the polytope domain is not a simplex, the combinatorial complexity of balls depends on the center location [34], see Fig. 5. The HG of the probability simplex yields a non-Riemannian geometry, because, at an infinitesimal radius, the balls are polytopes and not ellipsoids (corresponding to squared Mahalanobis distance balls used to visualize metric tensors [35]). The isometries in Hilbert polyhedral geometries are studied in [36]. In Appendix 9, we recall that any Hilbert geometry induces a Finslerian structure that becomes Riemannian iff the boundary is an ellipsoid (yielding the hyperbolic Cayley–Klein geometries [27]). Notice that in Hilbert simplex/polytope geometry, the geodesics are not unique (see Figure 2 of [25]).

[1] To contrast with this result, let us mention that infinitesimal small balls in Riemannian geometry have Euclidean ellipsoidal shapes (visualized as Tissot's indicatrix in cartography).

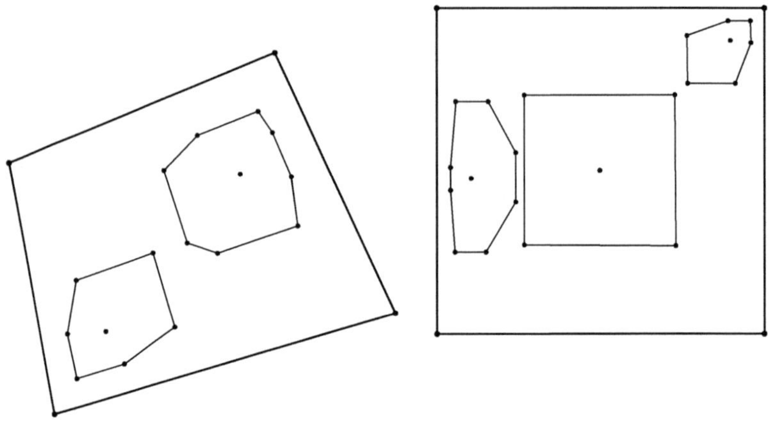

Fig. 5 Hilbert balls in quadrangle domains have combinatorial complexity depending on the center location

2.4 L_1-Norm Geometry

The Total Variation (TV) metric distance between two multinomials p and q is defined by:

$$\mathrm{TV}(p,q) = \frac{1}{2}\sum_{i=0}^{d} |\lambda_p^i - \lambda_q^i|.$$

It is a statistical f-divergence obtained for the generator $f(u) = \frac{1}{2}|u-1|$. The L_1-norm induced distance ρ_{L1} (L1) is defined by:

$$\rho_{L1}(p,q) = \|\lambda_p - \lambda_q\|_1 = \sum_{i=0}^{d} |\lambda_p^i - \lambda_q^i| = 2\mathrm{TV}(p,q).$$

Therefore the distance ρ_{L1} satisfies information monotonicity (for coarse-grained histograms p' and q' of $\Delta^{D'}$ with $D' < D$):

$$0 \leq \rho_{L1}(p',q') \leq \rho_{L1}(p,q).$$

For trinomials, the ρ_{L1} distance is given by:

$$\rho_{L1}(p,q) = |\lambda_p^0 - \lambda_q^0| + |\lambda_p^1 - \lambda_q^1| + |\lambda_q^0 - \lambda_p^0 + \lambda_q^1 - \lambda_p^1|.$$

The L_1 distance function is a polytopal distance function described by the dual polytope $\mathcal{Z}$ of the d-dimensional cube called the standard (or regular) d-cross-polytope [37], the orthoplex [38] or the d-cocube [39]: The cross-polytope $\mathcal{Z}$ can be obtained as the convex hull of the $2d$ unit standard base vectors $\pm e_i$ for

Table 1 Comparing the geometric modelings of the probability simplex Δ^d

	Riemannian Geometry	Information Rie. Geo.	Non-Rie. Hilbert Geo.
Structure	$(\Delta^d, g, \nabla^{LC} = \nabla(g))$	$(\Delta^d, g, \nabla^{(\alpha)}, \nabla^{(-\alpha)})$	(Δ^d, ρ)
	Levi-Civita $\nabla^{LC} = \nabla^{(0)}$	Dual connections $\nabla^{(\pm\alpha)}$ so that $\frac{\nabla^{(\alpha)}+\nabla^{(-\alpha)}}{2} = \nabla^{(0)}$	Connection of $\mathbb{R}^d$
Distance	Rao distance (metric)	α-divergence (non-metric) KL or reverse KL for $\alpha = \pm 1$	Hilbert distance (metric)
Property	Invariant to reparameterization	Information monotonicity	Isometric to a normed space
Calculation	Closed-form	Closed-form	Easy (Algorithm 1)
Geodesic	Shortest path	Straight either in θ/η	Straight
Smoothness	Manifold	Manifold	Non-manifold
Curvature	Positive	Dually flat	Negative

$i \in \{0, \ldots, d-1\}$. The cross-polytope is one of the three regular polytopes in dimension $d \geq 4$ (with the hypercubes and simplices): It has $2d$ vertices and 2^d facets. Therefore the L_1 balls on the hyperplane H_{Δ^d} supporting the probability simplex is the intersection of a $(d+1)$-cross-polytope with d-dimensional hyperplane H_{Δ^d}. Thus the "multinomial ball" $\mathrm{Ball}_{L_1}(p, r)$ of center p and radius r is defined by $\mathrm{Ball}_{L_1}(p, r) = (\lambda_p \oplus r\mathscr{Z}) \cap H_{\Delta^d}$. In 2D, the shape of L_1 trinomial balls is that of a regular octahedron (twelve edges and eight faces) cut by the 2D plane H_{Δ^2}: Trinomial balls have hexagonal shapes as illustrated in Fig. 2 (for ρ_{L1}). In 3D, trinomial balls are Archimedean solid cuboctahedra, and in arbitrary dimension, the shapes are polytopes with $d(d+1)$ vertices [40]. Let us note in passing, that in 3D, the L_1 multinomial cuboctahedron ball has the dual shape of the Hilbert rhombic dodecahedron ball.

Table 1 summarizes the characteristics of the three main geometries: FHR, IG, and HG. Let us conclude this introduction by mentioning the Cramér–Rao lower bound and its relationship with information geometry [41]: Consider an unbiased estimator $\hat{\theta} = T(X)$ of a parameter θ estimated from measurements distributed according to a smooth density $p(x; \theta)$ (i.e., $X \sim p(x; \theta)$). The Cramér–Rao Lower Bound (CRLB) states that the variance of $T(X)$ is greater or equal to the inverse of the FIM $\mathscr{I}(\theta)$: $V_\theta[T(X)] \succ \mathscr{I}^{-1}(\theta)$. For regular parametric families $\{p(x; \theta)\}_\theta$, the FIM is a positive-definite matrix and defines a metric tensor, called the Fisher metric in Riemannian geometry. The FIM is the cornerstone of information geometry [17] but requires the differentiability of the probability density function (pdf).

A better lower bound that does not require the pdf differentiability is the Hammersley–Chapman–Robbins Lower Bound [42, 43] (HCRLB):

$$V_\theta[T(X)] \geq \sup_\Delta \frac{\Delta^2}{E_\theta\left[\left(\frac{p(x;\theta+\Delta)-p(x;\theta)}{p(x;\theta)}\right)^2\right]}. \quad (5)$$

By introducing the χ^2-divergence, $\chi^2(P : Q) = \int \left(\frac{\mathrm{d}P - \mathrm{d}Q}{\mathrm{d}Q}\right)^2 \mathrm{d}Q$, we rewrite the HCRLB using the χ^2-divergence in the denominator as follows:

$$V_\theta[T(X)] \geq \sup_\Delta \frac{\Delta^2}{\chi^2(P(x;\theta+\Delta) : P(x;\theta))}. \tag{6}$$

Note that the FIM is not defined for non-differentiable pdfs, and therefore the Cramér–Rao lower bound does not exist in that case.

3 Computing Hilbert Distance in Δ^d

Let us start with the simplest case: The 1D probability simplex Δ^1, the space of Bernoulli distributions. Any Bernoulli distribution can be represented by the activation probability of the random bit x: $\lambda = p(x = 1) \in \Delta^1$, corresponding to a point in the interval $\Delta^1 = (0, 1)$. We write the Bernoulli manifold as an exponential family as

$$p(x) = \exp\left(x\theta - F(\theta)\right), \quad x \in \{0, 1\},$$

where $F(\theta) = \log(1 + \exp(\theta))$. Therefore $\lambda = \frac{\exp(\theta)}{1+\exp(\theta)}$ and $\theta = \log \frac{\lambda}{1-\lambda}$.

3.1 1D Probability Simplex of Bernoulli Distributions

By definition, the Hilbert distance has the closed form:

$$\rho_{\mathrm{HG}}(p, q) = \left|\log \frac{\lambda_q(1-\lambda_p)}{\lambda_p(1-\lambda_q)}\right| = \left|\log \frac{\lambda_p}{1-\lambda_p} - \log \frac{\lambda_q}{1-\lambda_q}\right|.$$

Note that $\theta_p = \log \frac{\lambda_p}{1-\lambda_p}$ is the canonical parameter of the Bernoulli distribution.

The FIM of the Bernoulli manifold in the λ-coordinates is given by: $g = \frac{1}{\lambda} + \frac{1}{1-\lambda} = \frac{1}{\lambda(1-\lambda)}$. The FHR distance is obtained by integration as:

$$\rho_{\mathrm{FHR}}(p, q) = 2 \arccos\left(\sqrt{\lambda_p \lambda_q} + \sqrt{(1-\lambda_p)(1-\lambda_q)}\right).$$

Notice that $\rho_{\mathrm{FHR}}(p, q)$ has finite values on $\partial\Delta^1$.

The KL divergence of the ± 1-geometry is:

$$\rho_{\mathrm{IG}}(p, q) = \lambda_p \log \frac{\lambda_p}{\lambda_q} + (1-\lambda_p) \log \frac{1-\lambda_p}{1-\lambda_q}.$$

The KL divergence belongs to the family of α-divergences [17].

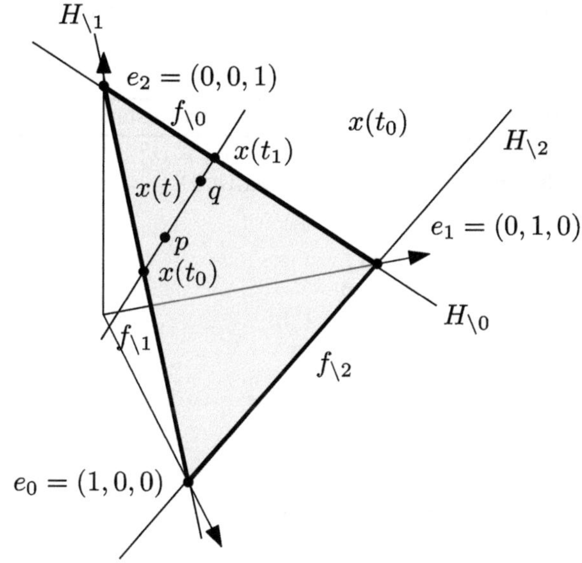

Fig. 6 Calculating the two intersection points $x(t_0)$ and $x(t_1)$ of the line (pq) with the boundary of the probability simplex Δ_d: For each facet $f_{\setminus i}$, we calculate the intersection point of line $x(t) = (1-t)p + tq$ with the d-dimensional hyperplane $H_{\setminus i}$ supporting the facet $f_{\setminus i}$

3.2 Arbitrary Dimension Case

Given $p, q \in \Delta^d$, we first need to compute the intersection of line (pq) with the border of the d-dimensional probability simplex to get the two intersection points p' and q' so that p', p, q, q' are ordered on (pq). Once this is done, we simply apply the formula in Eq. 4 to get the Hilbert distance.

A d-dimensional simplex consists of $d+1$ vertices with their corresponding $(d-1)$-dimensional facets. For the probability simplex Δ^d, let $e_i = (\underbrace{0, \ldots, 0}_{i}, 1, 0, \ldots, 0)$ denote the $d+1$ vertices of the standard simplex embedded in the hyperplane $H_\Delta : \sum_{i=0}^{d} \lambda^i = 1$ in $\mathbb{R}^{d+1}$. Let $f_{\setminus j}$ denote the simplex facets that is the convex hull of all vertices except e_j: $f_{\setminus j} = \text{hull}(e_0, \ldots, e_{j-1}, e_{j+1}, \ldots, e_d)$. Let $H_{\setminus j}$ denote the hyperplane supporting this facet, which is the affine hull $f_{\setminus j} = \text{affine}(e_0, \ldots, e_{j-1}, e_{j+1}, \ldots, e_d)$.

To compute the two intersection points of (pq) with Δ^d, a naive algorithm consists in computing the unique intersection point r_j of the line (pq) with each hyperplane $H_{\setminus j}$ $(j = 0, \ldots, d)$ and checking whether r_j belongs to $f_{\setminus j}$.

A much more efficient implementation given by Algorithm (1) calculates the intersection point of the line $x(t) = (1-t)p + tq$ with each $H_{\setminus j}$ $(j = 0, \ldots, d)$. These intersection points are represented using the coordinate t. For example, $x(0) = p$ and $x(1) = q$. Due to convexity, any intersection point with $H_{\setminus j}$ must satisfy either $t \leq 0$ or $t \geq 1$. Then, the two intersection points with $\partial\Delta^d$ are obtained by $t_0 = \max\{t : \exists j,\ x(t) \in H_{\setminus j} \text{ and } t \leq 0\}$ and $t_1 = \min\{t : \exists j,\ x(t) \in H_{\setminus j} \text{ and } t \geq 1\}$. Figure 6 illustrates this calculation method. This algorithm only requires $O(d)$ time and $O(1)$ memory.

Lemma 1 *The Hilbert distance in the probability simplex can be computed in optimal $\Theta(d)$ time.*

Algorithm 1: Computing the Hilbert distance

Data: Two points $p = (\lambda_p^0, \ldots, \lambda_p^d)$, $q = (\lambda_q^0, \ldots, \lambda_q^d)$ in the d-dimensional simplex Δ^d
Result: Their Hilbert distance $\rho_{\mathrm{HG}}(p, q)$

begin
  $t_0 \leftarrow -\infty$; $t_1 \leftarrow +\infty$;
  for $i = 0 \ldots d$ **do**
    if $\lambda_p^i \neq \lambda_q^i$ **then**
      $t \leftarrow \lambda_p^i / (\lambda_p^i - \lambda_q^i)$;
      if $t_0 < t \leq 0$ **then**
        $t_0 \leftarrow t$;
      else if $1 \leq t < t_1$ **then**
        $t_1 \leftarrow t$;
  if $t_0 = -\infty$ *or* $t_1 = +\infty$ **then**
    Output $\rho_{\mathrm{HG}}(p, q) = 0$;
  else if $t_0 = 0$ *or* $t_1 = 1$ **then**
    Output $\rho_{\mathrm{HG}}(p, q) = \infty$;
  else
    Output $\rho_{\mathrm{HG}}(p, q) = \left|\log(1 - \frac{1}{t_0}) - \log(1 - \frac{1}{t_1})\right|$;

Once an arbitrary distance ρ is chosen, we can define a ball centered at c and of radius r as $B_\rho(c, r) = \{x \,:\, \rho(c, x) \leq r\}$. Figure 4 displays the hexagonal shapes of the Hilbert balls for various center locations in Δ^2.

Theorem 1 (Balls in a simplicial Hilbert geometry [26]) *A ball in the Hilbert simplex geometry has a Euclidean polytope shape with $d(d+1)$ facets.*

Note that when the domain is not simplicial, the Hilbert balls can have varying combinatorial complexity depending on the center location. In 2D, the Hilbert ball can have $s \sim 2s$ edges inclusively, where s is the number of edges of the boundary of the Hilbert domain $\partial\mathcal{C}$.

Since a Riemannian geometry is locally defined by a metric tensor, at infinitesimal scales, Riemannian balls have Mahalanobis smooth ellipsoidal shapes: $B_\rho(c, r) = \{x \,:\, (x - c)^\top g(c)(x - c) \leq r^2\}$. This property allows one to visualize Riemannian metric tensors [35]. Thus we conclude that:

Lemma 2 ([26]) *Hilbert simplex geometry is a non-manifold metric length space.*

As a remark, let us notice that slicing a simplex with a hyperplane does not always produce a lower-dimensional simplex. For example, slicing a tetrahedron by a plane yields either a triangle or a quadrilateral. Thus the restriction of a d-dimensional ball B in a Hilbert simplex geometry Δ^d to a hyperplane H is a $(d-1)$-dimensional ball $B' = B \cap H$ of varying combinatorial complexity, corresponding to a ball in the induced Hilbert sub-geometry in the convex sub-domain $H \cap \Delta^d$.

3.3 Visualizing Distance Profiles

Figure 7 displays the distance profile from any point in the probability simplex to a fixed reference point (trinomial) based on the following common distance measures [18]: Euclidean (metric) distance, Cauchy–Schwarz (CS) divergence, Hellinger (metric) distance, Fisher-Rao (metric) distance, KL divergence and Hilbert simplicial (metric) distance. The Euclidean and Cauchy–Schwarz divergence are clipped to Δ^2. The Cauchy–Schwarz distance is projective so that $\rho_{CS}(\lambda p, \lambda' q) = \rho_{CS}(p, q)$ for any $\lambda, \lambda' > 0$ [44].

4 Center-Based Clustering

We concentrate on comparing the efficiency of Hilbert simplex geometry for clustering multinomials. We shall compare the experimental results of k-means++ and k-center multinomial clustering for the three distances: Rao and Hilbert metric distances, and KL divergence. We describe how to adapt those clustering algorithms to the Hilbert distance.

4.1 k-means++ Clustering

The celebrated k-means clustering [45] minimizes the sum of within-cluster variances, where each cluster has a center representative element. When dealing with $k = 1$ cluster, the center (also called centroid or cluster prototype) is the center of mass defined as the minimizer of

$$E_D(\Lambda, c) = \frac{1}{n} \sum_{i=1}^{n} D(p_i : c),$$

where $D(\cdot : \cdot)$ is a dissimilarity measure. For an arbitrary D, the centroid c may not be available in closed form. Nevertheless, using a generalization of the k-means++ initialization [9] (picking randomly seeds), one can bypass the centroid computation, and yet guarantee probabilistically a good clustering.

Let $C = \{c_1, \ldots, c_k\}$ denote the set of k cluster centers. Then the generalized k-means energy to be minimized is defined by:

$$E_D(\Lambda, C) = \frac{1}{n} \sum_{i=1}^{n} \min_{j \in \{1,\ldots,k\}} D(p_i : c_j).$$

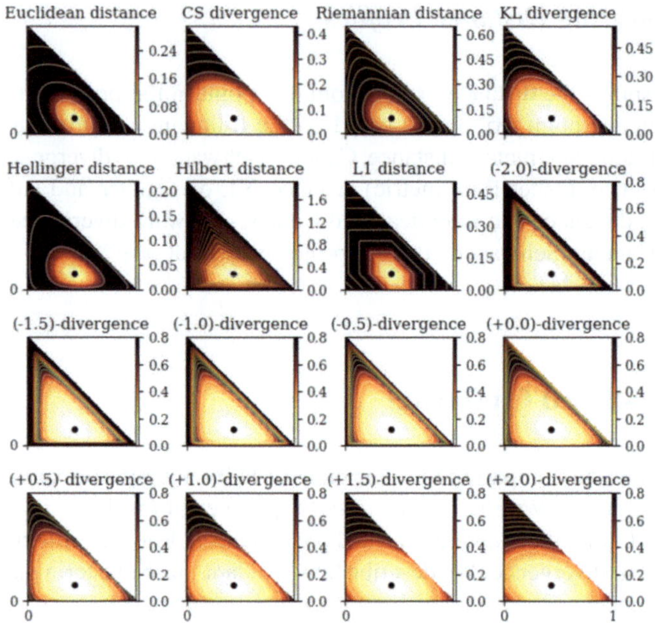

(a) Reference point (3/7,3/7,1/7)

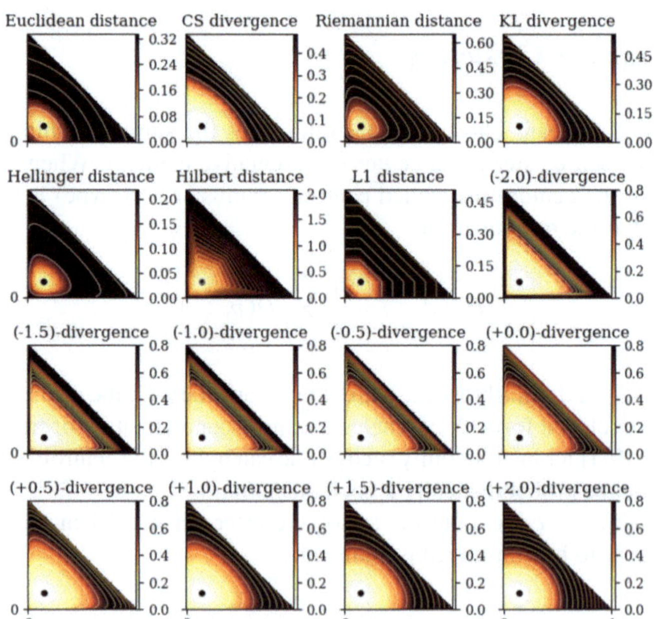

(b) Reference point (5/7,1/7,1/7)

Fig. 7 A comparison of different distance measures on Δ^2. The distance is measured from $\forall p \in \Delta^2$ to a fixed reference point (the black dot). Lighter color means shorter distance. Darker color means longer distance. The contours show equal distance curves with a precision step of 0.2

By defining the distance $D(p, C) = \min_{j\in\{1,\ldots,k\}} D(p : c_j)$ of a point to a set, we can rewrite the objective function as $E_D(\Lambda, C) = \frac{1}{n}\sum_{i=1}^n D(p_i, C)$. Let $E_D^*(\Lambda, k) = \min_{C\,:\,|C|=k} E_D(\Lambda, C)$ denote the global minimum of $E_D(\Lambda, C)$ wrt some given Λ and k.

The k-means++ seeding proceeds for an arbitrary divergence D as follows: Pick uniformly at random at first seed c_1, and then iteratively choose the $(k-1)$ remaining seeds according to the following probability distribution:

$$\Pr(c_j = p_i) = \frac{D(p_i, \{c_1, \ldots, c_{j-1}\})}{\sum_{i=1}^n D(p_i, \{c_1, \ldots, c_{j-1}\})} \quad (2 \leq j \leq k).$$

Since its inception (2007), this k-means++ seeding has been extensively studied [46]. We state the general theorem established by [47]:

Theorem 2 (Generalized k-means++ performance, [47]) *Let κ_1 and κ_2 be two constants such that κ_1 defines the quasi-triangular inequality property:*

$$D(x : z) \leq \kappa_1 \left(D(x : y) + D(y : z)\right), \quad \forall x, y, z \in \Delta^d,$$

and κ_2 handles the symmetry inequality:

$$D(x : y) \leq \kappa_2 D(y : x), \quad \forall x, y \in \Delta^d.$$

Then the generalized k-means++ seeding guarantees with high probability a configuration C of cluster centers such that:

$$E_D(\Lambda, C) \leq 2\kappa_1^2(1 + \kappa_2)(2 + \log k) E_D^*(\Lambda, k). \tag{7}$$

The ratio $\frac{E_D(\Lambda,C)}{E_D^*(\Lambda,k)}$ is called the *competitive factor*. The seminal result of ordinary k-means++ was shown [9] to be $8(2 + \log k)$-competitive. When evaluating κ_1, one has to note that squared metric distances are not metric because they do not satisfy the triangular inequality. For example, the squared Euclidean distance is not a metric but it satisfies the 2-quasi-triangular inequality with $\kappa_1 = 2$.

We state the following general performance theorem:

Theorem 3 (k-means++ performance in a metric space) *In any metric space $(\mathscr{X}, d)$, the k-means++ wrt the squared metric distance d^2 is $16(2 + \log k)$-competitive.*

Proof Since a metric distance is symmetric, it follows that $\kappa_2 = 1$. Consider the quasi-triangular inequality property for the squared non-metric dissimilarity d^2:

$$\begin{aligned} d(p, q) &\leq d(p, q) + d(q, r), \\ d^2(p, q) &\leq (d(p, q) + d(q, r))^2, \\ d^2(p, q) &\leq d^2(p, q) + d^2(q, r) + 2d(p, q)d(q, r). \end{aligned}$$

Let us apply the inequality of arithmetic and geometric means[2]:

$$\sqrt{d^2(p,q)d^2(q,r)} \leq \frac{d^2(p,q)+d^2(q,r)}{2}.$$

Thus we have

$$d^2(p,q) \leq d^2(p,q)+d^2(q,r)+2d(p,q)d(q,r) \leq 2(d^2(p,q)+d^2(q,r)).$$

That is, the squared metric distance satisfies the 2-approximate triangle inequality, and $\kappa_1 = 2$. The result is straightforward from Theorem 2.

Theorem 4 (k-means++ performance in a normed space) *In any normed space $(\mathscr{X}, \|\cdot\|)$, the k-means++ with $D(x:y) = \|x-y\|^2$ is $16(2+\log k)$-competitive.*

Proof In any normed space $(\mathscr{X}, \|\cdot\|)$, we have both $\|x-y\| = \|y-x\|$ and the triangle inequality:

$$\|x-z\| \leq \|x-y\| + \|y-z\|.$$

The proof is very similar to the proof of Theorem 3 and is omitted.

Since any inner product space $(\mathscr{X}, \langle\cdot,\cdot\rangle)$ has an induced norm $\|x\| = \sqrt{\langle x,x\rangle}$, we have the following corollary.

Corollary 1 *In any inner product space $(\mathscr{X}, \langle\cdot,\cdot\rangle)$, the k-means++ with $D(x:y) = \langle x-y, x-y\rangle$ is $16(2+\log k)$-competitive.*

We need to report a bound for the squared Hilbert symmetric distance ($\kappa_2 = 1$). In [26] (Theorem 3.3), it was shown that Hilbert geometry of a bounded convex domain $\mathscr{C}$ is isometric to a normed vector space iff $\mathscr{C}$ is an open simplex: $(\Delta^d, \rho_{\mathrm{HG}}) \simeq (V^d, \|\cdot\|_{\mathrm{NH}})$, where $\|\cdot\|_{\mathrm{NH}}$ is the corresponding norm. Therefore $\kappa_1 = 2$. We write "NH" for short for this equivalent normed Hilbert geometry. Appendix 8 recalls the construction due to [25], and shows the squared Hilbert distance fails the triangle inequality and it is not a distance induced by an inner product.

As an empirical study, we randomly generate $n = 10^6$ tuples (x, y, z) based on the uniform distribution in Δ^d. For each tuple (x, y, z), we evaluate the ratio

$$\kappa_1 = \frac{D(x:z)}{D(x:y)+D(y:z)}.$$

Figure 8 shows the statistics for four different choices of D: (1) $D(x:y) = \rho^2_{\mathrm{FHR}}(x,y)$; (2) $D(x:y) = \frac{1}{2}\mathrm{KL}(x:y) + \frac{1}{2}\mathrm{KL}(y:x)$; (3) $D(x:y) = \rho^2_{\mathrm{HG}}(x,y)$; (4) $D(x:y) = \rho^2_{\mathrm{L1}}(x,y)$. We find experimentally that κ_1 is upper bounded by 2 for ρ^2_{FHR}, ρ^2_{HG} and ρ^2_{L1}, while the average κ_1 value is smaller than 0.5. For all the compared distances, $\kappa_2 = 1$. Therefore ρ_{FHR} and ρ_{HG} have better k-means++ performance guarantee as compared to ρ_{IG}.

---

[2]For positive values a and b, the arithmetic-geometric mean inequality states that $\sqrt{ab} \leq \frac{a+b}{2}$.

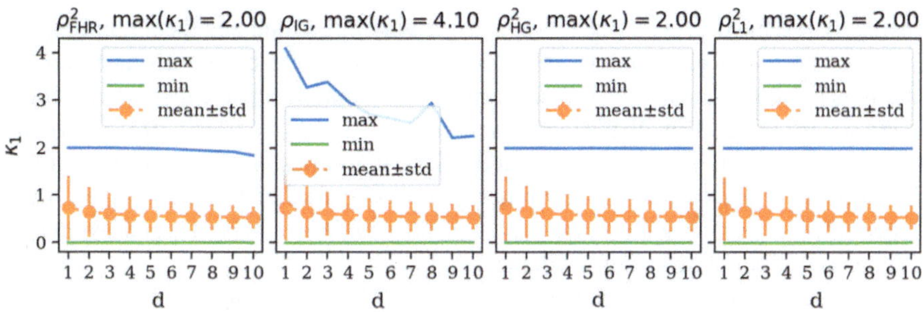

Fig. 8 The maximum, mean, standard deviation, and minimum of κ_1 on 10^6 randomly generated tuples (x, y, z) in Δ^d for $d = 1, \ldots, 10$

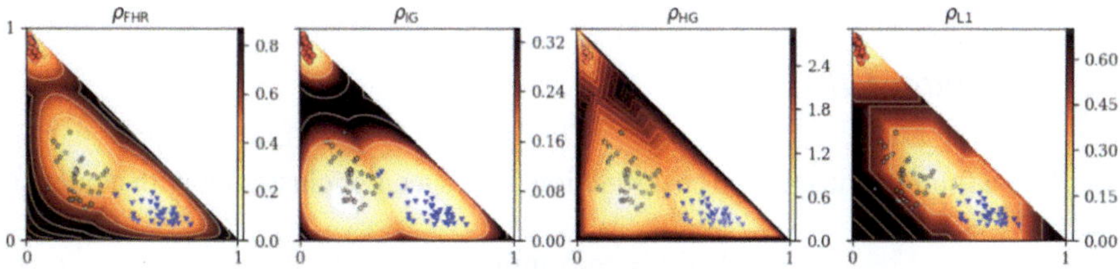

(a) $k = 3$ clusters

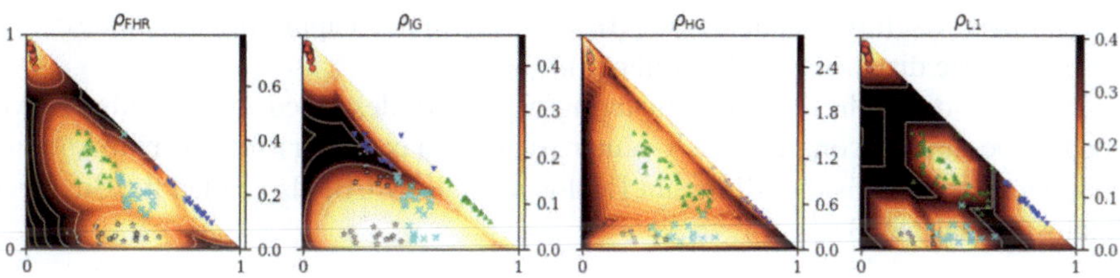

(b) $k = 5$ clusters

Fig. 9 k-Means++ clustering results on a toy dataset in the space of trinomials Δ^2. The color density maps indicate the distance from any point to its nearest cluster center

We get by applying Theorem 4:

Corollary 2 (k-means++ in Hilbert simplex geometry) *The k-means++ seeding in a Hilbert simplex geometry in fixed dimension is $16(2 + \log k)$-competitive.*

Figure 9 displays the clustering results of k-means++ in Hilbert simplex geometry as compared to the other geometries for $k \in \{3, 5\}$.

The KL divergence can be interpreted as a separable Bregman divergence [48]. The Bregman k-means++ performance has been studied in [48, 49], and a competitive factor of $O(\frac{1}{\mu})$ is reported using the notion of Bregman μ-similarity (that is suited for data-sets on a compact domain).

In [50], spherical k-means++ is studied wrt the distance $d_S(x, y) = 1 - \langle x, y\rangle$ for any pair of points x, y on the unit sphere. Since $\langle x, y\rangle = \|x\|_2\|y\|_2 \cos(\theta_{x,y}) =$

$\cos(\theta_{x,y})$, we have $d_S(x, y) = 1 - \cos(\theta_{x,y})$, where $\theta_{x,y}$ denotes the angle between a pair of unit vectors x and y. This distance is called the cosine distance since it amounts to one minus the cosine similarity. Notice that the cosine distance is related to the squared Euclidean distance via the identity: $d_S(x, y) = \frac{1}{2}\|x - y\|^2$. The cosine distance is different from the spherical distance that relies on the arccos function.

Since divergences may be asymmetric, one can further consider mixed divergence $M(p : q : r) = \lambda D(p : q) + (1 - \lambda)D(q : r)$ for $\lambda \in [0, 1]$, and extend the k-means++ seeding procedure and analysis [51].

For a given data set, we can compute κ_1 or κ_2 by inspecting triples and pairs of points, and get data-dependent competitive factor improving the bounds mentioned above.

4.2 k-Center Clustering

Let Λ be a finite point set. The cost function for a k-center clustering with centers C ($|C| = k$) is:

$$f_D(\Lambda, C) = \max_{p_i \in \Lambda} \min_{c_j \in C} D(p_i : c_j).$$

The farthest first traversal heuristic [10] has a guaranteed approximation factor of 2 for any metric distance (see Algorithm 3).

In order to use the k-center clustering algorithm described in Algorithm 2, we need to be able to compute the 1-center (or minimax center) for the Hilbert simplex geometry, that is the Minimum Enclosing Ball (MEB, also called the Smallest Enclosing Ball, SEB).

Algorithm 2: k-Center clustering

Data: A set of points $p_1, \ldots, p_n \in \Delta^d$. A distance measure ρ on Δ^d. The maximum number k of clusters. The maximum number T of iterations.

Result: A clustering scheme assigning each p_i a label $l_i \in \{1, \ldots, k\}$

begin
- Randomly pick k cluster centers $c_1, \ldots, c_k$ using the kmeans++ heuristic;
- **for** $t = 1, \ldots, T$ **do**
  - **for** $i = 1, \ldots, n$ **do**
    - $l_i \leftarrow \arg\min_{l=1}^{k} \rho(p_i, c_l)$;
  - **for** $l = 1, \ldots, k$ **do**
    - $c_l \leftarrow \arg\min_c \max_{i:l_i=l} \rho(p_i, c)$;
- Output $\{l_i\}_{i=1}^{n}$;

Algorithm 3: A 2-approximation of the k-center clustering for any metric distance ρ.

Data: A set Λ; a number k of clusters; a metric distance ρ.
Result: A 2-approximation of the k-center clustering

begin
  $c_1 \leftarrow \text{ARandomPointOf}(\Lambda)$;
  $C \leftarrow \{c_1\}$;
  for $i = 2, \ldots, k$ **do**
    $c_i \leftarrow \arg\max_{p \in \Lambda} \rho(p, C)$;
    $C \leftarrow C \cup \{c_i\}$;
Output C;

We may consider the SEB equivalently either in Δ^d or in the normed space V^d. In both spaces, the shapes of the balls are convex. Let $\Lambda = \{p_1, \ldots, p_n\}$ denote the point set in Δ^d, and $\mathcal{V} = \{v_1, \ldots, v_n\}$ the equivalent point set in the normed vector space (following the mapping explained in Appendix 8). Then the SEBs $B_{\text{HG}}(\Lambda)$ in Δ^d and $B_{\text{NH}}(\mathcal{V})$ in V^d have respectively radii r^*_{HG} and r^*_{NH} defined by:

$$r^*_{\text{HG}} = \min_{c \in \Delta^d} \max_{i \in \{1,\ldots,n\}} \rho_{\text{HG}}(p_i, c),$$

$$r^*_{\text{NH}} = \min_{v \in V^d} \max_{i \in \{1,\ldots,n\}} \|v_i - v\|_{\text{NH}}.$$

The SEB in the normed vector space $(V^d, \|\cdot\|_{\text{NH}})$ amounts to find the minimum covering norm polytope of a finite point set. This problem has been well-studied in computational geometry [52–54]. By considering the equivalent Hilbert norm polytope with $d(d+1)$ facets, we state the result of [54]:

Theorem 5 (SEB in Hilbert polytope normed space, [54]) *A $(1+\varepsilon)$-approximation of the SEB in V^d can be computed in $O(d^3 \frac{n}{\varepsilon})$ time.*

We shall now report two algorithms for computing the SEBs: One exact algorithm in V^d that does not scale well in high dimensions, and one approximation in Δ^d that works well for large dimensions.

4.2.1 Exact Smallest Enclosing Ball in a Hilbert Simplex Geometry

Given a finite point set $\{p_1, \ldots, p_n\} \in \Delta^d$, the SEB in Hilbert simplex geometry is centered at

$$c^* = \arg\min_{c \in \Delta^d} \max_{i \in \{1,\ldots,n\}} \rho_{\text{HG}}(c, x_i),$$

with radius

$$r^* = \min_{c \in \Delta^d} \max_{i \in \{1,\ldots,n\}} \rho_{\text{HG}}(c, x_i).$$

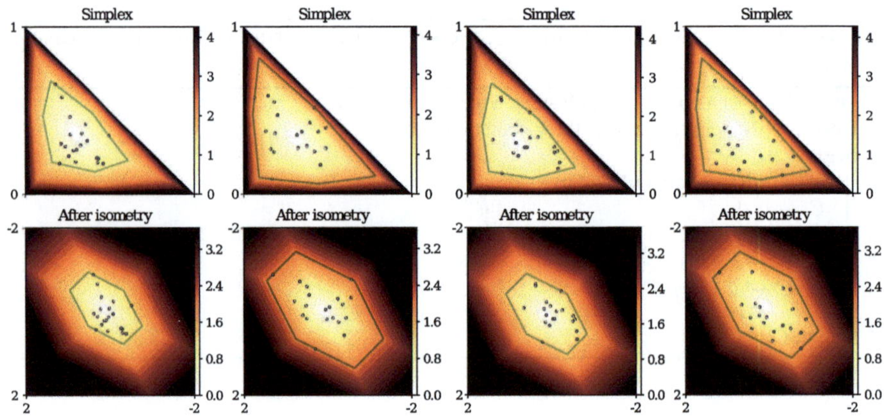

(a) 3 points on the border (b) 2 points on the border

Fig. 10 Computing the SEB in Hilbert simplex geometry amounts to compute the SEB in the corresponding normed vector space

An equivalent problem is to find the SEB in the isometric normed vector space V^d via the mapping reported in Appendix 8. Each simplex point p_i corresponds to a point v_i in the V^d.

Figure 10 displays some examples of the exact smallest enclosing balls in the Hilbert simplex geometry and in the corresponding normed vector space.

To compute the SEB, one may also consider the generic LP-type randomized algorithm [55]. We notice that an enclosing ball for a point set in general has a number k of points on the border of the ball, with $2 \leq k \leq \frac{d(d+1)}{2}$. Let $D = \frac{d(d+1)}{2}$ denote the varying size of the combinatorial basis, then we can apply the LP-type framework (we check the axioms of locality and monotonicity, [56]) to solve efficiently the SEBs.

Theorem 6 (Smallest Enclosing Hilbert Ball is LP-type, [56, 57]) *The smallest enclosing Hilbert ball amounts to find the smallest enclosing ball in a vector space with respect to a polytope norm that can be solved using a LP-type randomized algorithm.*

The Enclosing Ball Decision Problem (EBDP, [58]) asks for a given value r, whether $r \geq r^*$ or not. The decision problem amounts to find whether a set $\{r B_V + v_i\}$ of translates can be stabbed by a point [58]: That is, whether $\cap_{i=1}^n (r B_V + v_i)$ is empty or not. Since these translates are polytopes with $d(d+1)$ facets, this can be solved in linear time using *Linear Programming*.

Theorem 7 (Enclosing Hilbert Ball Decision Problem) *The decision problem to test whether $r \geq r^*$ or not can be solved by Linear Programming.*

This yields a simple scheme to approximate the optimal value r^*: Let $r_0 = \max_{i\in\{2,\dots,n\}} \|v_i - v_1\|_{\mathrm{NH}}$. Then $r^* \in [\frac{r_0}{2}, r_0] = [a_0, b_0]$. At stage i, perform a dichotomic search on $[a_i, b_i]$ by answering the decision problem for $r_{i+1} = \frac{a_i+b_i}{2}$, and update the radius range accordingly [58].

However, the LP-type randomized algorithm or the decision problem-based algorithm do not scale well in high dimensions. Next, we introduce a simple approximation algorithm that relies on the fact that the line segment $[pq]$ is a geodesic in Hilbert simplex geometry. (Geodesics are not unique. See Figure 2 of [25].)

4.2.2 Geodesic Bisection Approximation Heuristic

In Riemannian geometry, the 1-center can be arbitrarily finely approximated by a simple geodesic bisection algorithm [59, 60]. This algorithm can be extended to HG straightforwardly as detailed in Algorithm 4.

Algorithm 4: Geodesic walk for approximating the Hilbert minimax center, generalizing [60]

Data: A set of points $p_1, \dots, p_n \in \Delta^d$. The maximum number T of iterations.
Result: $c \approx \arg\min_c \max_i \rho_{\mathrm{HG}}(p_i, c)$

begin
- $c_0 \leftarrow \mathrm{ARandomPointOf}(\{p_1, \dots, p_n\})$;
- **for** $t = 1, \dots, T$ **do**
  - $p \leftarrow \arg\max_{p_i} \rho_{\mathrm{HG}}(p_i, c_{t-1})$;
  - $c_t \leftarrow c_{t-1} \#^{\rho}_{1/(t+1)} p$;
- Output c_T;

The algorithm first picks up a point c_0 at random from Λ as the initial center, then computes the farthest point p (with respect to the distance ρ), and then walk on the geodesic from c_0 to p by a certain amount to define c_1, etc. For an arbitrary distance ρ, we define the operator $\#^{\rho}_{\alpha}$ as follows:

$$p\#^{\rho}_{\alpha}q = v = \gamma(p, q, \alpha), \quad \rho(p : v) = \alpha\rho(p : q),$$

where $\gamma(p, q, \alpha)$ is the geodesic passing through p and q, and parameterized by α ($0 \leq \alpha \leq 1$). When the equations of the geodesics are explicitly known, we can either get a closed form solution for $\#^{\rho}_{\alpha}$ or perform a bisection search to find v' such that $\rho(p : v') \approx \alpha\rho(p : q)$. See [61] for an extension and analysis in hyperbolic geometry. See Fig. 11 to get an intuitive idea on the *experimental* convergence rate of Algorithm 4. See Fig. 12 for visualizations of centers wrt different geometries.

Furthermore, this iterative algorithm implies a core-set [62] (namely, the set of farthest points visited during the geodesic walks) that is useful for clustering large

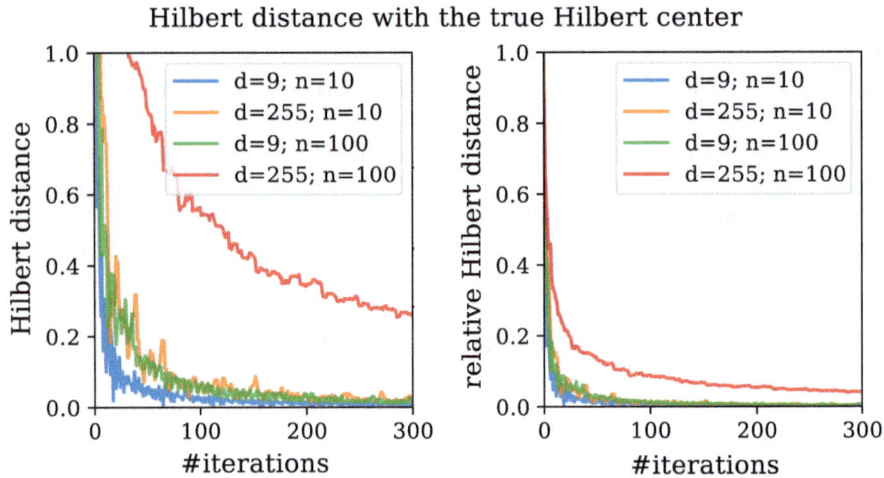

Fig. 11 Convergence rate of Algorithm 4 measured by the Hilbert distance between the current minimax center and the true center (left) or their Hilbert distance divided by the Hilbert radius of the dataset (right). The plot is based on 100 random points in Δ^9/Δ^{255}

data-sets [63]. See [52] for core-set results on containment problems wrt a convex homothetic object (the equivalent Hilbert polytope norm in our case).

A simple algorithm dubbed MinCon [53] can find an approximation of the Minimum Enclosing Polytope. The algorithm induces a core-set of size $O(\frac{1}{\varepsilon^2})$ although the theorem is challenged in [52].

Thus by combining the k-center seeding [10] with the Lloyd-like batched iterations, we get an efficient k-center clustering algorithm for the FHR and Hilbert metric geometries. When dealing with the Kullback–Leibler divergence, we use the fact that KL is a Bregman divergence, and use the 1-center algorithm ([64, 65] for approximation in any dimension, or [55] which is exact but limited to small dimensions).

Since Hilbert simplex geometry is isomorphic to a normed vector space [26] with a polytope norm with $d(d+1)$ facets, the Voronoi diagram in Hilbert geometry of Δ^d amounts to compute a Voronoi diagram wrt a polytope norm [66–68].

5 Experiments

We generate a dataset consisting of a set of clusters in a high dimensional statistical simplex Δ^d. Each cluster is generated independently as follows. We first pick a random center $c = (\lambda_c^0, \ldots, \lambda_c^d)$ based on the uniform distribution on Δ^d. Then any random sample $p = (\lambda^0, \ldots, \lambda^d)$ associated with c is independently generated by

Fig. 12 The Riemannian/IG/Hilbert/L_1 minimax centers of three point clouds in Δ^2 based on Algorithm 4. The color maps show the distance from $\forall p \in \Delta^2$ to the corresponding center

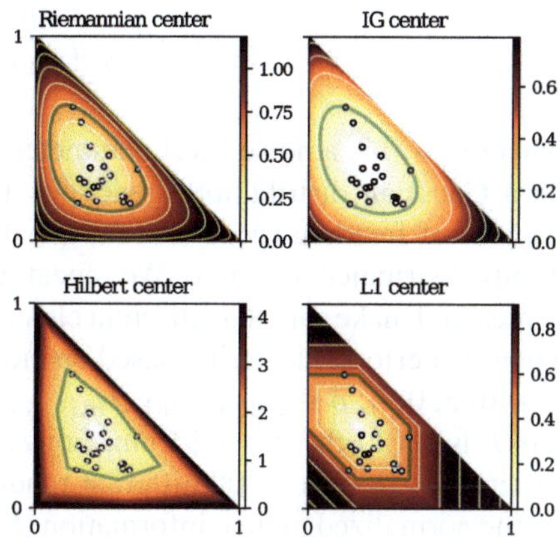

(a) Point Cloud 1

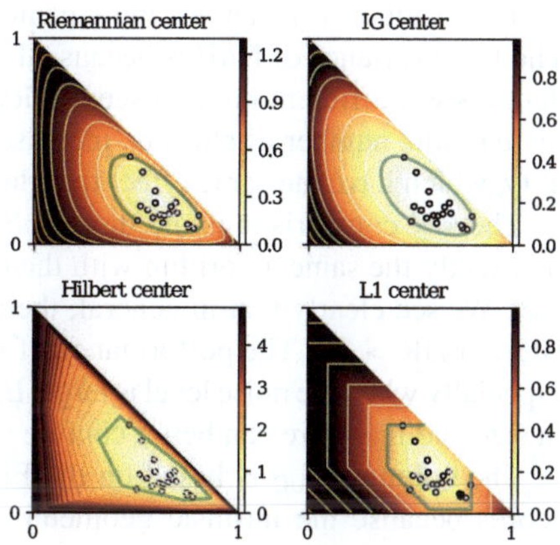

(b) Point Cloud 2

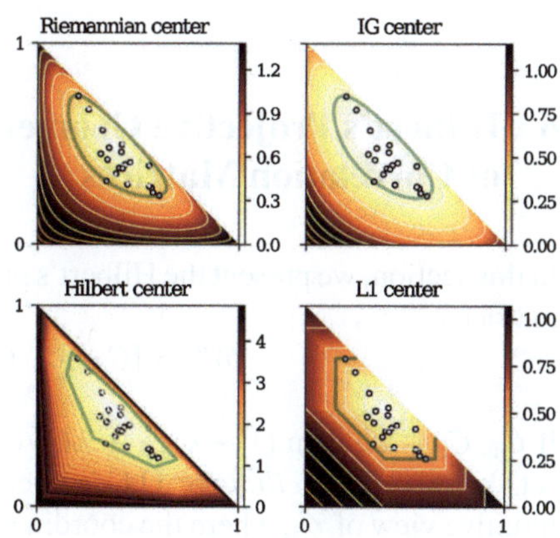

(c) Point Cloud 3

$$\lambda^i = \frac{\exp(\log \lambda_c^i + \sigma \varepsilon^i)}{\sum_{i=0}^{d} \exp(\log \lambda_c^i + \sigma \varepsilon^i)},$$

where $\sigma > 0$ is a noise level parameter, and each ε^i follows independently a standard Gaussian distribution (generator 1) or the Student's t-distribution with five degrees of freedom (generator 2). Let $\sigma = 0$, we get $\lambda^i = \lambda_c^i$. Therefore p is randomly distributed around c. We repeat generating random samples for each cluster center, and make sure that different clusters have almost the same number of samples. Then we perform clustering based on the configurations $n \in \{50, 100\}$, $d \in \{9, 255\}$, $\sigma \in \{0.5, 0.9\}$, $\rho \in \{\rho_{\mathrm{FHR}}, \rho_{\mathrm{IG}}, \rho_{\mathrm{HG}}, \rho_{\mathrm{EUC}}, \rho_{\mathrm{L1}}\}$. For simplicity, the number of clusters k is set to the ground truth. For each configuration, we repeat the clustering experiment based on 300 different random datasets. The performance is measured by the normalized mutual information (NMI), which is a scalar indicator in the range $[0, 1]$ (the larger the better).

The results of k-means++ and k-centers are shown in Tables 2 and 3, respectively. The large variance of NMI is because that each experiment is performed on random datasets wrt different random seeds. Generally, the performance deteriorates as we increase the number of clusters, increase the noise level or decrease the dimensionality, which have the same effect to reduce the inter-cluster gap.

The key comparison is the three columns ρ_{FHR}, ρ_{HG} and ρ_{IG}, as they are based on exactly the same algorithm with the only difference being the underlying geometry. We see clearly that in general, their clustering performance presents the order HG > FHR > IG. The performance of HG is superior to the other two geometries, especially when the noise level is large. Intuitively, the Hilbert balls are more compact in size and therefore can better capture the clustering structure (see Fig. 2).

The column ρ_{EUC} is based on the Euclidean enclosing ball. It shows the worst scores because the intrinsic geometry of the probability simplex is far from the Euclidean geometry.

6 Hilbert's Projective Geometry of the Space of Correlation Matrices

In this section, we present the Hilbert's projective geometry to the space of correlation matrices

$$\mathscr{C}^d = \{C_{d\times d} : C \succ 0; C_{ii} = 1, \forall i\}.$$

If $C_1, C_2 \in \mathscr{C}$, then $(1-\lambda)C_1 + \lambda C_2 \in \mathscr{C}$ for $0 < \lambda < 1$. Therefore $\mathscr{C}$ is a convex set, known as an *elliptope* [11] embedded in the p.s.d. cone. See Fig. 13 for an intuitive view of $\mathscr{C}_3$, where the coordinate system (x, y, z) is the off-diagonal entries of $C \in \mathscr{C}_3$.

In order to compute the Hilbert distance $\rho_{\mathrm{HG}}(C_1, C_2)$, we need to compute the intersection of the line (C_1, C_2) with $\partial\mathscr{C}$, denoted as C_1' and C_2', then we have

Table 2 k-means++ clustering accuracy in NMI on randomly generated datasets based on different geometries. The table shows the mean and standard deviation after 300 independent runs for each configuration. ρ is the distance measure. n is the sample size. d is the dimensionality of Δ^d. σ is noise level

k	n	d	σ	ρ_{FHR}	ρ_{IG}	ρ_{HG}	ρ_{EUC}	ρ_{L1}
3	50	9	0.5	0.76 ± 0.22	0.76 ± 0.24	$\mathbf{0.81 \pm 0.22}$	0.64 ± 0.23	0.70 ± 0.22
			0.9	0.44 ± 0.20	0.44 ± 0.20	$\mathbf{0.57 \pm 0.22}$	0.31 ± 0.17	0.38 ± 0.18
		255	0.5	0.80 ± 0.24	0.81 ± 0.24	$\mathbf{0.88 \pm 0.21}$	0.74 ± 0.25	0.79 ± 0.24
			0.9	0.65 ± 0.27	0.66 ± 0.28	$\mathbf{0.72 \pm 0.27}$	0.46 ± 0.24	0.63 ± 0.27
	100	9	0.5	0.76 ± 0.22	0.76 ± 0.21	$\mathbf{0.82 \pm 0.22}$	0.60 ± 0.21	0.69 ± 0.23
			0.9	0.42 ± 0.19	0.41 ± 0.18	$\mathbf{0.54 \pm 0.22}$	0.27 ± 0.14	0.34 ± 0.16
		255	0.5	0.82 ± 0.23	0.82 ± 0.24	$\mathbf{0.89 \pm 0.20}$	0.74 ± 0.24	0.80 ± 0.25
			0.9	0.66 ± 0.26	0.66 ± 0.28	$\mathbf{0.72 \pm 0.26}$	0.45 ± 0.25	0.64 ± 0.27
5	50	9	0.5	0.75 ± 0.14	0.74 ± 0.15	$\mathbf{0.81 \pm 0.13}$	0.61 ± 0.13	0.68 ± 0.13
			0.9	0.44 ± 0.13	0.42 ± 0.13	$\mathbf{0.55 \pm 0.15}$	0.31 ± 0.11	0.36 ± 0.12
		255	0.5	0.83 ± 0.15	0.83 ± 0.15	$\mathbf{0.88 \pm 0.14}$	0.77 ± 0.16	0.82 ± 0.15
			0.9	0.71 ± 0.17	0.70 ± 0.19	$\mathbf{0.75 \pm 0.17}$	0.50 ± 0.17	0.68 ± 0.18
	100	9	0.5	0.74 ± 0.13	0.74 ± 0.14	$\mathbf{0.80 \pm 0.14}$	0.60 ± 0.13	0.67 ± 0.13
			0.9	0.42 ± 0.11	0.40 ± 0.12	$\mathbf{0.55 \pm 0.15}$	0.29 ± 0.09	0.35 ± 0.11
		255	0.5	0.83 ± 0.14	0.83 ± 0.15	$\mathbf{0.88 \pm 0.13}$	0.77 ± 0.15	0.81 ± 0.15
			0.9	0.69 ± 0.18	0.69 ± 0.18	$\mathbf{0.73 \pm 0.17}$	0.48 ± 0.17	0.67 ± 0.18

(a) Generator 1

k	n	d	σ	ρ_{FHR}	ρ_{IG}	ρ_{HG}	ρ_{EUC}	ρ_{L1}
3	50	9	0.5	0.62 ± 0.22	0.60 ± 0.22	$\mathbf{0.71 \pm 0.23}$	0.45 ± 0.20	0.54 ± 0.22
			0.9	0.29 ± 0.17	0.27 ± 0.16	$\mathbf{0.39 \pm 0.19}$	0.17 ± 0.13	0.25 ± 0.15
		255	0.5	0.70 ± 0.25	0.69 ± 0.26	$\mathbf{0.74 \pm 0.25}$	0.37 ± 0.29	0.70 ± 0.26
			0.9	$\mathbf{0.42 \pm 0.25}$	0.35 ± 0.20	0.40 ± 0.19	0.03 ± 0.08	$\mathbf{0.44 \pm 0.26}$
	100	9	0.5	0.63 ± 0.22	0.61 ± 0.22	$\mathbf{0.71 \pm 0.22}$	0.46 ± 0.19	0.56 ± 0.20
			0.9	0.29 ± 0.15	0.26 ± 0.14	$\mathbf{0.38 \pm 0.20}$	0.18 ± 0.12	0.24 ± 0.14
		255	0.5	0.71 ± 0.26	0.69 ± 0.27	$\mathbf{0.75 \pm 0.25}$	0.31 ± 0.28	0.70 ± 0.27
			0.9	0.41 ± 0.26	0.33 ± 0.20	0.38 ± 0.18	0.02 ± 0.06	$\mathbf{0.43 \pm 0.26}$
5	50	9	0.5	0.64 ± 0.15	0.61 ± 0.14	$\mathbf{0.70 \pm 0.14}$	0.48 ± 0.14	0.57 ± 0.15
			0.9	0.31 ± 0.12	0.29 ± 0.12	$\mathbf{0.41 \pm 0.15}$	0.20 ± 0.09	0.26 ± 0.10
		255	0.5	0.74 ± 0.17	0.72 ± 0.17	$\mathbf{0.77 \pm 0.16}$	0.41 ± 0.20	0.74 ± 0.17
			0.9	0.44 ± 0.17	0.37 ± 0.16	0.44 ± 0.15	0.04 ± 0.06	$\mathbf{0.47 \pm 0.17}$
	100	9	0.5	0.62 ± 0.14	0.61 ± 0.14	$\mathbf{0.71 \pm 0.14}$	0.46 ± 0.13	0.54 ± 0.14
			0.9	0.30 ± 0.10	0.27 ± 0.11	$\mathbf{0.40 \pm 0.13}$	0.19 ± 0.08	0.25 ± 0.09
		255	0.5	0.73 ± 0.18	0.70 ± 0.18	$\mathbf{0.75 \pm 0.16}$	0.37 ± 0.20	0.73 ± 0.17
			0.9	0.43 ± 0.16	0.35 ± 0.14	0.41 ± 0.12	0.03 ± 0.06	$\mathbf{0.46 \pm 0.18}$

(b) Generator 2

Table 3 k-center clustering accuracy in NMI on randomly generated datasets based on different geometries. The table shows the mean and standard deviation after 300 independent runs for each configuration. ρ is the distance measure. n is the sample size. d is the dimensionality of the statistical simplex. σ is noise level

k	n	d	σ	ρ_{FHR}	ρ_{IG}	ρ_{HG}	ρ_{EUC}	ρ_{L1}
3	50	9	0.5	0.87 ± 0.19	0.85 ± 0.19	$\mathbf{0.92 \pm 0.16}$	0.72 ± 0.22	0.80 ± 0.20
			0.9	0.54 ± 0.21	0.51 ± 0.21	$\mathbf{0.70 \pm 0.23}$	0.36 ± 0.17	0.44 ± 0.19
		255	0.5	0.93 ± 0.16	0.92 ± 0.18	$\mathbf{0.95 \pm 0.14}$	0.89 ± 0.18	0.90 ± 0.19
			0.9	0.76 ± 0.24	0.72 ± 0.26	$\mathbf{0.82 \pm 0.24}$	0.50 ± 0.28	0.76 ± 0.25
	100	9	0.5	0.88 ± 0.17	0.86 ± 0.18	$\mathbf{0.93 \pm 0.14}$	0.70 ± 0.20	0.80 ± 0.20
			0.9	0.53 ± 0.20	0.49 ± 0.19	$\mathbf{0.70 \pm 0.22}$	0.33 ± 0.14	0.41 ± 0.18
		255	0.5	0.93 ± 0.16	0.92 ± 0.17	$\mathbf{0.95 \pm 0.13}$	0.88 ± 0.19	0.93 ± 0.16
			0.9	0.81 ± 0.22	0.75 ± 0.24	$\mathbf{0.83 \pm 0.22}$	0.47 ± 0.28	0.79 ± 0.22
5	50	9	0.5	0.82 ± 0.13	0.81 ± 0.13	$\mathbf{0.89 \pm 0.12}$	0.67 ± 0.13	0.75 ± 0.13
			0.9	0.50 ± 0.13	0.47 ± 0.13	$\mathbf{0.66 \pm 0.15}$	0.34 ± 0.11	0.40 ± 0.12
		255	0.5	$\mathbf{0.92 \pm 0.11}$	$\mathbf{0.91 \pm 0.12}$	$\mathbf{0.93 \pm 0.11}$	0.87 ± 0.13	$\mathbf{0.92 \pm 0.12}$
			0.9	0.77 ± 0.15	0.71 ± 0.17	$\mathbf{0.85 \pm 0.17}$	0.54 ± 0.19	0.74 ± 0.16
	100	9	0.5	0.83 ± 0.12	0.81 ± 0.13	$\mathbf{0.89 \pm 0.11}$	0.67 ± 0.11	0.76 ± 0.13
			0.9	0.48 ± 0.12	0.46 ± 0.12	$\mathbf{0.66 \pm 0.15}$	0.33 ± 0.09	0.39 ± 0.10
		255	0.5	$\mathbf{0.93 \pm 0.10}$	$\mathbf{0.92 \pm 0.11}$	$\mathbf{0.94 \pm 0.09}$	0.89 ± 0.11	0.92 ± 0.11
			0.9	0.81 ± 0.14	0.74 ± 0.15	$\mathbf{0.84 \pm 0.16}$	0.52 ± 0.19	0.79 ± 0.14

(a) Generator 1

k	n	d	σ	ρ_{FHR}	ρ_{IG}	ρ_{HG}	ρ_{EUC}	ρ_{L1}
3	50	9	0.5	0.68 ± 0.22	0.67 ± 0.22	$\mathbf{0.80 \pm 0.20}$	0.48 ± 0.22	0.60 ± 0.22
			0.9	0.32 ± 0.18	0.29 ± 0.17	$\mathbf{0.45 \pm 0.21}$	0.20 ± 0.14	0.26 ± 0.15
		255	0.5	0.79 ± 0.24	0.75 ± 0.24	$\mathbf{0.82 \pm 0.22}$	0.13 ± 0.23	$\mathbf{0.81 \pm 0.24}$
			0.9	0.35 ± 0.27	0.35 ± 0.21	$\mathbf{0.42 \pm 0.19}$	0.00 ± 0.02	0.32 ± 0.30
	100	9	0.5	0.66 ± 0.22	0.65 ± 0.22	$\mathbf{0.79 \pm 0.21}$	0.45 ± 0.19	0.59 ± 0.20
			0.9	0.30 ± 0.16	0.28 ± 0.14	$\mathbf{0.42 \pm 0.19}$	0.20 ± 0.12	0.26 ± 0.14
		255	0.5	0.78 ± 0.25	0.76 ± 0.24	$\mathbf{0.82 \pm 0.21}$	0.05 ± 0.14	0.77 ± 0.27
			0.9	0.29 ± 0.28	0.29 ± 0.20	$\mathbf{0.39 \pm 0.20}$	0.00 ± 0.02	0.22 ± 0.25
5	50	9	0.5	0.69 ± 0.14	0.66 ± 0.14	$\mathbf{0.77 \pm 0.13}$	0.50 ± 0.13	0.61 ± 0.14
			0.9	0.34 ± 0.12	0.30 ± 0.12	$\mathbf{0.46 \pm 0.15}$	0.22 ± 0.09	0.28 ± 0.10
		255	0.5	$\mathbf{0.80 \pm 0.15}$	0.76 ± 0.15	$\mathbf{0.82 \pm 0.14}$	0.24 ± 0.23	$\mathbf{0.81 \pm 0.14}$
			0.9	0.42 ± 0.21	0.38 ± 0.16	$\mathbf{0.46 \pm 0.15}$	0.00 ± 0.02	0.39 ± 0.22
	100	9	0.5	0.66 ± 0.13	0.64 ± 0.14	$\mathbf{0.77 \pm 0.14}$	0.47 ± 0.13	0.57 ± 0.13
			0.9	0.31 ± 0.11	0.28 ± 0.10	$\mathbf{0.44 \pm 0.13}$	0.21 ± 0.08	0.25 ± 0.09
		255	0.5	$\mathbf{0.80 \pm 0.16}$	0.76 ± 0.15	$\mathbf{0.82 \pm 0.13}$	0.12 ± 0.17	$\mathbf{0.81 \pm 0.16}$
			0.9	0.32 ± 0.19	0.30 ± 0.15	$\mathbf{0.41 \pm 0.13}$	0.00 ± 0.01	0.26 ± 0.18

(b) Generator 2

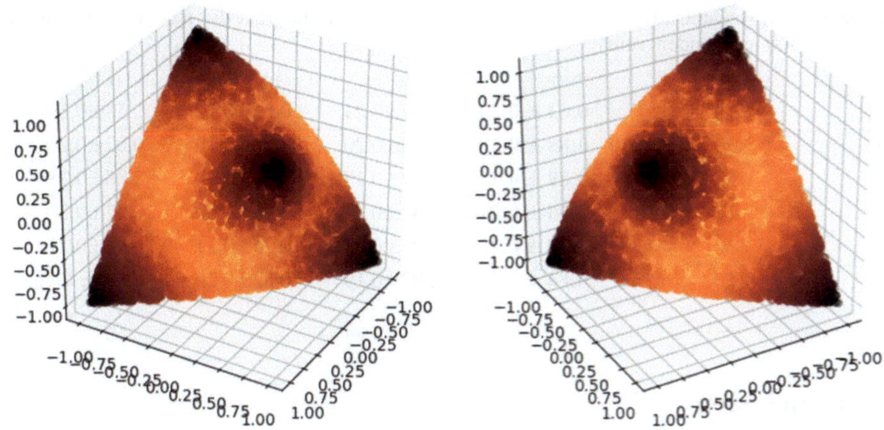

Fig. 13 The elliptope $\mathscr{C}_3$ (two different perspectives)

$$\rho_{\mathrm{HG}}(C_1, C_2) = \left|\log \frac{\|C_1 - C_2'\|\|C_1' - C_2\|}{\|C_1 - C_1'\|\|C_2 - C_2'\|}\right|.$$

Unfortunately there is no closed form solution of C_1' and C_2'. Instead, we apply a binary searching algorithm. Note a necessary condition for $C \in \mathscr{C}$ is that C has a positive spectrum. If C has at least one non-positive eigenvalue, then $C \notin \mathscr{C}$. To determine whether a given C is inside the elliptope requires a spectral decomposition of C. Therefore the computation of C_1' and C_2' is in general expensive.

We compare the Hilbert elliptope geometry with commonly used distance measures including the L_2 distance ρ_{EUC}, L_1 distance ρ_{L1}, and the square root of the log-det divergence

$$\rho_{\mathrm{LD}}(C_1, C_2) = \mathrm{tr}(C_1 C_2^{-1}) - \log|C_1 C_2^{-1}| - d.$$

Due to the high computational complexity, we only investigate k-means++ clustering. The investigated dataset consists of 100 matrices forming 3 clusters in $\mathscr{C}_3$ with almost identical size. Each cluster is independently generated according to

$$P \sim \mathscr{W}^{-1}(I_{3\times 3}, \nu_1),$$
$$C_i \sim \mathscr{W}^{-1}(P, \nu_2),$$

where $\mathscr{W}^{-1}(A, \nu)$ denotes the inverse Wishart distribution with scale matrix A and ν degrees of freedom, and C_i is a point in the cluster associated with P. Table 4 shows the k-means++ clustering performance in terms of NMI. Again Hilbert geometry is favorable as compared to alternatives, showing that the good performance of Hilbert clustering is generalizable.

Table 4 NMI (mean±std) of k-means++ clustering based on different distance measures in the elliptope (500 independent runs)

ν_1	ν_2	ρ_{HG}	ρ_{EUC}	ρ_{L1}	ρ_{LD}
4	10	**0.62 ± 0.22**	0.57±0.21	0.56±0.22	0.58±0.22
4	30	**0.85 ± 0.18**	0.80±0.20	0.81±0.19	0.82±0.20
4	50	**0.89 ± 0.17**	0.87±0.17	0.86±0.18	0.88±0.18
5	10	**0.50 ± 0.21**	0.49±0.21	0.48±0.20	0.47±0.21
5	30	**0.77 ± 0.20**	0.75±0.21	0.75±0.21	0.75±0.21
5	50	**0.84 ± 0.19**	0.82±0.19	0.82±0.20	**0.84 ± 0.18**

7 Conclusion

We introduced the Hilbert projective metric distance and its underlying non-Riemannian geometry for modeling the space of multinomials or the open probability simplex. We compared experimentally in simulated clustering tasks this geometry with the traditional differential geometric modelings (either the Fisher-Hotelling-Rao metric connection or the dually coupled non-metric affine connections of information geometry [17]).

The main feature of Hilbert geometry (HG) is that it is a metric non-manifold geometry, where geodesics are straight (Euclidean) line segments. This makes this geometry computationally attractive. In simplex domains, the Hilbert balls have fixed combinatorial (Euclidean) polytope structures, and HG is known to be isometric to a normed space [25, 69]. This latter isometry allows one to generalize easily the standard proofs of clustering (e.g., k-means or k-center). We demonstrated it for the k-means++ competitive performance analysis and for the convergence of the 1-center heuristic [60] (smallest enclosing Hilbert ball allows one to implement efficiently the k-center clustering). Our experimental k-means++ or k-center comparisons of HG algorithms with the manifold modeling approach yield superior performance. This may be intuitively explained by the sharpness of Hilbert balls as compared to the FHR/IG ball profiles.

Chentsov [70] defined statistical invariance on a probability manifold under Markov morphisms and proved that the Fisher Information Metric is the unique Riemannian metric (up to rescaling) for multinomials. However, this does not rule out that other distances (with underlying geometric structures) may be used to model statistical manifolds (e.g., Finsler statistical manifolds [71, 72], or the total variation distance — the only metric f-divergence [73]). Defining statistical invariance related to geometry is the cornerstone problem of information geometry that can be tackled from many directions (see [74] and references therein for a short review).

In this paper, we introduced Hilbert geometries in machine learning by considering clustering tasks in the probability simplex and in the correlation elliptope. A canonical Hilbert metric distance can be defined on any bounded convex subset of the Euclidean space with the key property that geodesics are straight Euclidean line segments thus making this geometry well-suited for fast and exact computations. Thus we may consider clustering in other bounded convex subsets like the simplotopes [75].

One future direction is to consider the Hilbert metric for regularization and sparsity in machine learning (due to its equivalence with a polytope normed distance).

Our Python codes are freely available online for reproducible research:
https://www.lix.polytechnique.fr/~nielsen/HSG/

8 Isometry of Hilbert Simplex Geometry to a Normed Vector Space

Consider the Hilbert simplex metric space $(\Delta^d, \rho_{\mathrm{HG}})$ where Δ^d denotes the d-dimensional open probability simplex and ρ_{HG} the Hilbert cross-ratio metric. Let us recall the isometry ([25], 1991) of the open standard simplex to a normed vector space $(V^d, \|\cdot\|_{\mathrm{NH}})$. Let $V^d = \{v \in \mathbb{R}^{d+1} \ : \ \sum_i v^i = 0\}$ denote the d-dimensional vector space sitting in $\mathbb{R}^{d+1}$. Map a point $p = (\lambda^0, \ldots, \lambda^d) \in \Delta^d$ to a point $v(x) = (v^0, \ldots, v^d) \in V^d$ as follows:

$$v^i = \frac{1}{d+1}\left(d \log \lambda^i - \sum_{j \neq i} \log \lambda^j\right) = \log \lambda^i - \frac{1}{d+1} \sum_j \log \lambda^j.$$

We define the corresponding norm $\|\cdot\|_{\mathrm{NH}}$ in V^d by considering the shape of its unit ball $B_V = \{v \in V^d \ : \ |v^i - v^j| \leq 1, \forall i \neq j\}$. The unit ball B_V is a symmetric convex set containing the origin in its interior, and thus yields a *polytope norm* $\|\cdot\|_{\mathrm{NH}}$ (Hilbert norm) with $2\binom{d+1}{2} = d(d+1)$ facets. Reciprocally, let us notice that a norm induces a unit ball centered at the origin that is convex and symmetric around the origin.

The distance in the normed vector space between $v \in V^d$ and $v' \in V^d$ is defined by:

$$\rho_V(v, v') = \|v - v'\|_{\mathrm{NH}} = \inf\left\{\tau \ : \ v' \in \tau(B_V \oplus \{v\})\right\},$$

where $A \oplus B = \{a + b \ : \ a \in A, b \in B\}$ is the Minkowski sum.

The reverse map from the normed space V^d to the probability simplex Δ^d is given by:

$$\lambda^i = \frac{\exp(v^i)}{\sum_j \exp(v^j)}.$$

Thus we have $(\Delta^d, \rho_{\mathrm{HG}}) \cong (V^d, \|\cdot\|_{\mathrm{NH}})$. In 1D, $(V^1, \|\cdot\|_{\mathrm{NH}})$ is isometric to the Euclidean line.

Note that computing the distance in the normed vector space requires naively $O(d^2)$ time.

Unfortunately, the norm $\|\cdot\|_{\mathrm{NH}}$ does not satisfy the parallelogram law.[3] Notice that a norm satisfying the parallelogram law can be associated with an inner product via the polarization identity. Thus the isometry of the Hilbert geometry to a normed vector space is not equipped with an inner product. However, all norms in a finite dimensional space are equivalent. This implies that in finite dimension, $(\Delta^d, \rho_{\mathrm{HG}})$ is *quasi-isometric* to the Euclidean space $\mathbb{R}^d$. An example of Hilbert geometry in infinite dimension is reported in [25]. Hilbert spaces are not CAT spaces except when $\mathcal{C}$ is an ellipsoid [76].

9 Hilbert Geometry with Finslerian/Riemannian Structures

In a Riemannian geometry, each tangent plane T_pM of a d-dimensional manifold M is equivalent to $\mathbb{R}^d$: $T_pM \simeq \mathbb{R}^d$. The inner product at each tangent plane T_pM can be visualized by an ellipsoid shape, a convex symmetric object centered at point p. In a *Finslerian geometry*, a norm $\|\cdot\|_p$ is defined in each tangent plane T_pM, and this norm is visualized as a symmetric convex object with non-empty interior. Finslerian geometry thus generalizes Riemannian geometry by taking into account generic symmetric convex objects instead of ellipsoids for inducing norms at each tangent plane. Any Hilbert geometry induced by a compact convex domain $\mathcal{C}$ can be expressed by an equivalent Finslerian geometry by defining the norm in T_p at p as follows [76]:

$$\|v\|_p = F_{\mathcal{C}}(p, v) = \frac{\|v\|}{2}\left(\frac{1}{pp^+} + \frac{1}{pp^-}\right),$$

where $F_{\mathcal{C}}$ is the *Finsler metric*, $\|\cdot\|$ is an *arbitrary norm* on $\mathbb{R}^d$, and p^+ and p^- are the intersection points of the line passing through p with direction v:

$$p^+ = p + t^+ v, \quad p^- = p + t^- v.$$

A geodesic γ in a Finslerian geometry satisfies:

$$d_{\mathcal{C}}(\gamma(t_1), \gamma(t_2)) = \int_{t_1}^{t_2} F_{\mathcal{C}}(\gamma(t), \dot{\gamma}(t))\mathrm{d}t.$$

In T_pM, a ball of center c and radius r is defined by:

$$B(c, r) = \{v \ : \ F_{\mathcal{C}}(c, v) \leq r\}.$$

[3] Consider $A = (1/3, 1/3, 1/3)$, $B = (1/6, 1/2, 1/3)$, $C = (1/6, 2/3, 1/6)$ and $D = (1/3, 1/2, 1/6)$. Then $2AB^2 + 2BC^2 = 4.34$ but $AC^2 + BD^2 = 3.84362411135$.

Thus any Hilbert geometry induces an equivalent Finslerian geometry, and since Finslerian geometries include Riemannian geometries, one may wonder which Hilbert geometries induce Riemannian structures? The only Riemannian geometries induced by Hilbert geometries are the *hyperbolic Cayley–Klein geometries* [27, 29, 30] with the domain $\mathcal{C}$ being an ellipsoid. The Finslerian modeling of information geometry has been studied in [71, 72].

There is not a canonical way of defining measures in a Hilbert geometry since Hilbert geometries are Finslerian but not necessary Riemannian geometries [76]. The Busemann measure is defined according to the Lebesgue measure λ of $\mathbb{R}^d$: Let B_p denote the unit ball wrt. to the Finsler norm at point $p \in \mathcal{C}$, and B_e the Euclidean unit ball. Then the Busemann measure for a Borel set $\mathcal{B}$ is defined by [76]:

$$\mu_{\mathcal{C}}(\mathcal{B}) = \int_{\mathcal{B}} \frac{\lambda(B_e)}{\lambda(B_p)} \mathrm{d}\lambda(p).$$

The existence and uniqueness of center points of a probability measure in Finsler geometry have been investigated in [77].

10 Bounding Hilbert Norm with Other Norms

Let us show that $\|v\|_{\mathrm{NH}} \leq \beta_{d,c}\|v\|_c$, where $\|\cdot\|_c$ is any norm. Let $v = \sum_{i=0}^d e_i x_i$, where $\{e_i\}$ is a basis of $\mathbb{R}^{d+1}$. We have:

$$\|v\|_c \leq \sum_{i=0}^{d} |x_i| \|e_i\|_c \leq \|x\|_2 \underbrace{\sqrt{\sum_{i=0}^{d} \|e_i\|_c^2}}_{\beta_d},$$

where the first inequality comes from the triangle inequality, and the second inequality is from the Cauchy–Schwarz inequality. Thus we have:

$$\|v\|_{\mathrm{NH}} \leq \beta_d \|x\|_2,$$

with $\beta_d = \sqrt{d+1}$ since $\|e_i\|_{\mathrm{NH}} \leq 1$.

Let $\alpha_{d,c} = \min_{\{v \,:\, \|v\|_c=1\}} \|v\|_{\mathrm{NH}}$. Consider $u = \frac{v}{\|v\|_c}$. Then $\|u\|_c = 1$ so that $\|v\|_{\mathrm{NH}} \geq \alpha_{d,c}\|v\|_c$. To find α_d, we consider the unit ℓ_2 ball in V^d, and find the smallest $\lambda > 0$ so that λB_V fully contains the Euclidean ball (Fig. 14).

Therefore, we have overall:

$$\alpha_d \|x\|_2 \leq \|v\|_{\mathrm{NH}} \leq \sqrt{d+1}\|x\|_2$$

In general, note that we may consider two arbitrary norms $\|\cdot\|_l$ and $\|\cdot\|_u$ so that:

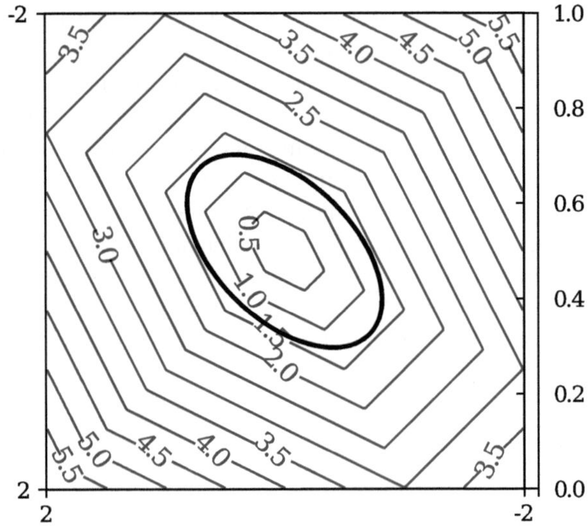

Fig. 14 Polytope balls B_V and the Euclidean unit ball B_E. From the figure the smallest polytope ball has radius ≈ 1.5

$$\alpha_{d,l}\|x\|_l \leq \|v\|_{\mathrm{NH}} \leq \beta_{d,u}\|x\|_u.$$

11 Funk Directed Metrics and Funk Balls

The Funk metric [78] wrt a convex domain $\mathscr{C}$ is defined by

$$F_{\mathscr{C}}(x, y) = \log\left(\frac{\|x - a\|}{\|y - a\|}\right),$$

where a is the intersection of the domain boundary and the affine ray $R(x, y)$ starting from x and passing through y. Correspondingly, the reverse Funk metric is

$$F_{\mathscr{C}}(y, x) = \log\left(\frac{\|y - b\|}{\|x - b\|}\right),$$

where b is the intersection of $R(y, x)$ with the boundary. The Funk metric is *not* a metric distance.

The Hilbert metric is simply the arithmetic symmetrization:

$$H_{\mathscr{C}}(x, y) = \frac{F_{\mathscr{C}}(x, y) + F_{\mathscr{C}}(y, x)}{2}.$$

It is interesting to explore clustering based on the Funk geometry, which we leave as a future work.

References

1. Agresti, A.: Categorical Data Analysis, vol. 482. Wiley, New Jercy (2003)
2. Aggarwal, C.C., Zhai, C.X.: Mining Text Data. Springer Publishing Company, Berlin (2012)
3. Messing, R., Pal, C., Kautz, H.: Activity recognition using the velocity histories of tracked keypoints. In: International Conference on Computer Vision, pp. 104–111. IEEE (2009)
4. Murphy, K.P.: Machine Learning: A Probabilistic Perspective. The MIT Press, Cambridge (2012)
5. Chaudhuri, K., McGregor, A.: Finding metric structure in information theoretic clustering. In: Conference on Learning Theory (COLT), pp. 391–402 (2008)
6. Lebanon, G.: Learning Riemannian metrics. In: Conference on Uncertainty in Artificial Intelligence (UAI), pp. 362–369 (2002)
7. Rigouste, L., Cappé, O., Yvon, F.: Inference and evaluation of the multinomial mixture model for text clustering. Inf. Process. Manag. **43**(5), 1260–1280 (2007)
8. Huang, Z.: Extensions to the k-means algorithm for clustering large data sets with categorical values. Data Min. Knowl. Discov. **2**(3), 283–304 (1998)
9. Arthur, D., Vassilvitskii, S.: k-means++: the advantages of careful seeding. In: ACM-SIAM Symposium on Discrete Algorithms (SODA), pp. 1027–1035 (2007)
10. Gonzalez, T.F.: Clustering to minimize the maximum intercluster distance. Theor. Comput. Sci. **38**, 293–306 (1985)
11. Tropp, J.A.: Simplicial faces of the set of correlation matrices. Discret. Comput. Geom. **60**(2), 512–529 (2018)
12. Kass, R.E., Vos, P.W.: Geometrical Foundations of Asymptotic Inference. Wiley Series in Probability and Statistics. Wiley-Interscience, New Jercy (1997)
13. Hotelling, H.: Spaces of statistical parameters. Bull. Amer. Math. Soc. **36**, 191 (1930)
14. Rao, C.R.: Information and accuracy attainable in the estimation of statistical parameters. Bull. Calcutta Math. Soc. **37**(3), 81–91 (1945)
15. Rao, C.R.: Information and the accuracy attainable in the estimation of statistical parameters. Breakthroughs in Statistics, pp. 235–247. Springer, New York (1992)
16. Stigler, S.M.: The epic story of maximum likelihood. Stat. Sci. **22**(4), 598–620 (2007)
17. Amari, Si: Information Geometry and Its Applications. Applied Mathematical Sciences, vol. 194. Springer, Japan (2016)
18. Calin, O., Udriste, C.: Geometric Modeling in Probability and Statistics. Mathematics and Statistics. Springer International Publishing, New York (2014)
19. Amari, Si, Cichocki, A.: Information geometry of divergence functions. Bull. Pol. Acad. Sci.: Tech. Sci. **58**(1), 183–195 (2010)
20. Shima, H.: The Geometry of Hessian Structures. World Scientific, Singapore (2007)
21. Liang, X.: A note on divergences. Neural Comput. **28**(10), 2045–2062 (2016)
22. Jenssen, R., Principe, J.C., Erdogmus, D., Eltoft, T.: The Cauchy–Schwarz divergence and Parzen windowing: connections to graph theory and mercer kernels. J. Frankl. Inst. **343**(6), 614–629 (2006)
23. Hilbert, D.: Über die gerade linie als kürzeste verbindung zweier punkte. Mathematische Annalen **46**(1), 91–96 (1895)
24. Busemann, H.: The Geometry of Geodesics. Pure and Applied Mathematics, vol. 6. Elsevier Science, Amsterdam (1955)
25. de la Harpe, P.: On Hilbert's metric for simplices. Geometric Group Theory, vol. 1, pp. 97–118. Cambridge University Press, Cambridge (1991)
26. Lemmens, B., Nussbaum, R.: Birkhoff's version of Hilbert's metric and its applications in analysis. Handbook of Hilbert Geometry, pp. 275–303 (2014)
27. Richter-Gebert, J.: Perspectives on Projective Geometry: A Guided Tour Through Real and Complex Geometry. Springer, Berlin (2011)
28. Bi, Y., Fan, B., Wu, F.: Beyond Mahalanobis metric: Cayley–Klein metric learning. In: IEEE Conference on Computer Vision and Pattern Recognition (CVPR), pp. 2339–2347 (2015)

29. Nielsen, F., Muzellec, B., Nock, R.: Classification with mixtures of curved Mahalanobis metrics. In: IEEE International Conference on Image Processing (ICIP), pp. 241–245 (2016)
30. Nielsen, F., Muzellec, B., Nock, R.: Large margin nearest neighbor classification using curved Mahalanobis distances (2016). arXiv:1609.07082 [cs.LG]
31. Stillwell, J.: Ideal elements in Hilbert's geometry. Perspect. Sci. **22**(1), 35–55 (2014)
32. Bernig, A.: Hilbert geometry of polytopes. Archiv der Mathematik **92**(4), 314–324 (2009)
33. Nielsen, F., Sun, K.: Clustering in Hilbert simplex geometry. CoRR arXiv: abs/1704.00454 (2017)
34. Nielsen, F., Shao, L.: On balls in a polygonal Hilbert geometry. In: 33st International Symposium on Computational Geometry (SoCG 2017). Schloss Dagstuhl–Leibniz-Zentrum fuer Informatik (2017)
35. Laidlaw, D.H., Weickert, J.: Visualization and Processing of Tensor Fields: Advances and Perspectives. Mathematics and Visualization. Springer, Berlin (2009)
36. Lemmens, B., Walsh, C.: Isometries of polyhedral Hilbert geometries. J. Topol. Anal. **3**(02), 213–241 (2011)
37. Condat, L.: Fast projection onto the simplex and the ℓ_1 ball. Math. Program. **158**(1–2), 575–585 (2016)
38. Park, P.S.: Regular polytopic distances. Forum Geom. **16**, 227–232 (2016)
39. Boissonnat, J.D., Sharir, M., Tagansky, B., Yvinec, M.: Voronoi diagrams in higher dimensions under certain polyhedral distance functions. Discret. Comput. Geom. **19**(4), 485–519 (1998)
40. Bengtsson, I., Zyczkowski, K.: Geometry of Quantum States: An Introduction to Quantum Entanglement. Cambridge University Press, Cambridge (2017)
41. Nielsen, F.: Cramér–Rao lower bound and information geometry. Connected at Infinity II, pp. 18–37. Springer, Berlin (2013)
42. Chapman, D.G.: Minimum variance estimation without regularity assumptions. Ann. Math. Stat. **22**(4), 581–586 (1951)
43. Hammersley, H.: On estimating restricted parameters. J. R. Stat. Society. Ser. B (Methodol.) **12**(2), 192–240 (1950)
44. Nielsen, F., Sun, K.: On Hölder projective divergences. Entropy **19**(3), 122 (2017)
45. Nielsen, F., Nock, R.: Further heuristics for k-means: the merge-and-split heuristic and the (k, l)-means. arXiv:1406.6314 (2014)
46. Bachem, O., Lucic, M., Hassani, S.H., Krause, A.: Approximate k-means++ in sublinear time. In: Proceedings of the Thirtieth AAAI Conference on Artificial Intelligence, pp. 1459–1467 (2016)
47. Nielsen, F., Nock, R.: Total Jensen divergences: definition, properties and k-means++ clustering (2013). arXiv:1309.7109 [cs.IT]
48. Ackermann, M.R., Blömer, J.: Bregman clustering for separable instances. Scandinavian Workshop on Algorithm Theory, pp. 212–223. Springer, Berlin (2010)
49. Manthey, B., Röglin, H.: Worst-case and smoothed analysis of k-means clustering with Bregman divergences. J. Comput. Geom. **4**(1), 94–132 (2013)
50. Endo, Y., Miyamoto, S.: Spherical k-means++ clustering. Modeling Decisions for Artificial Intelligence, pp. 103–114. Springer, Berlin (2015)
51. Nielsen, F., Nock, R., Amari, Si: On clustering histograms with k-means by using mixed α-divergences. Entropy **16**(6), 3273–3301 (2014)
52. Brandenberg, R., König, S.: No dimension-independent core-sets for containment under homothetics. Discret. Comput. Geom. **49**(1), 3–21 (2013)
53. Panigrahy, R.: Minimum enclosing polytope in high dimensions (2004). arXiv:cs/0407020 [cs.CG]
54. Saha, A., Vishwanathan, S., Zhang, X.: New approximation algorithms for minimum enclosing convex shapes. In: ACM-SIAM Symposium on Discrete Algorithms (SODA), pp. 1146–1160 (2011)
55. Nielsen, F., Nock, R.: On the smallest enclosing information disk. Inf. Process. Lett. **105**(3), 93–97 (2008)

56. Sharir, M., Welzl, E.: A combinatorial bound for linear programming and related problems. STACS **92**, 567–579 (1992)
57. Welzl, E.: Smallest enclosing disks (balls and ellipsoids). New Results and New trends in Computer Science, pp. 359–370. Springer, Berlin (1991)
58. Nielsen, F., Nock, R.: Approximating smallest enclosing balls with applications to machine learning. Int. J. Comput. Geom. Appl. **19**(05), 389–414 (2009)
59. Arnaudon, M., Nielsen, F.: On approximating the Riemannian 1-center. Comput. Geom. **46**(1), 93–104 (2013)
60. Bâdoiu, M., Clarkson, K.L.: Smaller core-sets for balls. In: ACM-SIAM Symposium on Discrete Algorithms (SODA), pp. 801–802 (2003)
61. Nielsen, F., Hadjeres, G.: Approximating covering and minimum enclosing balls in hyperbolic geometry. International Conference on Networked Geometric Science of Information, pp. 586–594. Springer, Cham (2015)
62. Bădoiu, M., Clarkson, K.L.: Optimal core-sets for balls. Comput. Geom. **40**(1), 14–22 (2008)
63. Bachem, O., Lucic, M., Krause, A.: Scalable and distributed clustering via lightweight coresets (2017). arXiv:1702.08248 [stat.ML]
64. Nielsen, F., Nock, R.: On approximating the smallest enclosing Bregman balls. In: Proceedings of the Twenty-Second Annual Symposium on Computational Geometry, pp. 485–486. ACM (2006)
65. Nock, R., Nielsen, F.: Fitting the smallest enclosing Bregman ball. ECML, pp. 649–656. Springer, Berlin (2005)
66. Deza, M., Sikirić, M.D.: Voronoi polytopes for polyhedral norms on lattices. Discret. Appl. Math. **197**, 42–52 (2015)
67. Körner, M.C.: Minisum hyperspheres, Springer Optimization and Its Applications, vol. 51. Springer, New York (2011)
68. Reem, D.: The geometric stability of Voronoi diagrams in normed spaces which are not uniformly convex (2012). arXiv:1212.1094 [cs.CG]
69. Foertsch, T., Karlsson, A.: Hilbert metrics and Minkowski norms. J. Geom. **83**(1–2), 22–31 (2005)
70. Cencov, N.N.: Statistical Decision Rules and Optimal Inference. Translations of Mathematical Monographs, vol. 53. American Mathematical Society, Providence (2000)
71. Cena, A.: Geometric structures on the non-parametric statistical manifold. Ph.D. thesis, University of Milano (2002)
72. Shen, Z.: Riemann-Finsler geometry with applications to information geometry. Chin. Ann. Math. Ser. B **27**(1), 73–94 (2006)
73. Khosravifard, M., Fooladivanda, D., Gulliver, T.A.: Confliction of the convexity and metric properties in f-divergences. IEICE Trans. Fundam. Electron. Commun. Comput. Sci. **90**(9), 1848–1853 (2007)
74. Dowty, J.G.: Chentsov's theorem for exponential families (2017). arXiv:1701.08895 [math.ST]
75. Doup, T.M.: Simplicial Algorithms on the Simplotope, vol. 318. Springer Science & Business Media, Berlin (2012)
76. Vernicos, C.: Introduction aux géométries de Hilbert. Séminaire de théorie spectrale et géométrie **23**, 145–168 (2004)
77. Arnaudon, M., Nielsen, F.: Medians and means in Finsler geometry. LMS J. Comput. Math. **15**, 23–37 (2012)
78. Papadopoulos, A., Troyanov, M.: From Funk to Hilbert geometry (2014). arXiv:1406.6983 [math.MG]

Jean-Louis Koszul and the Elementary Structures of Information Geometry

Frédéric Barbaresco

Abstract This paper is a scientific exegesis and admiration of Jean-Louis Koszul's works on homogeneous bounded domains that have appeared over time as elementary structures of Information Geometry. Koszul has introduced fundamental tools to characterize the geometry of sharp convex cones, as Koszul-Vinberg characteristic Function, Koszul Forms, and affine representation of Lie Algebra and Lie Group. The 2nd Koszul form is an extension of classical Fisher metric. Koszul theory of hessian structures and Koszul forms could be considered as main foundation and pillars of Information Geometry.

Keywords Koszul-Vinberg characteristic function · Koszul forms
Affine representation of lie algebra and lie group
Homogeneous bounded domains

1 Preamble

«La Physique mathématique, en incorporant à sa base la notion de groupe, marque la suprématie rationnelle…Chaque géométrie – et sans doute plus généralement chaque organisation mathématique de l'expérience – est caractérisée par un groupe spécial de transformations… Le groupe apporte la preuve d'une mathématique fermée sur elle-même. Sa découverte clôt l'ère des conventions, plus ou moins indépendantes, plus ou moins cohérentes» - **Gaston Bachelard, Le nouvel esprit scientifique, 1934**

In this article, I will pay tribute to a part of Professor Jean-Louis Koszul's work and fundamental and deep contributions of this great algebraist and geometer in the field of Information Geometry, which have many applications in the domain of applied mathematics, and in the emerging applications of Artificial Intelligence where the most efficient and robust algorithms are based on the natural gradient of the information geometry deduced from the Fisher matrix, as Yann Ollivier recently showed [1, 2].

F. Barbaresco (✉)
Thales Land & Air Systems, Voie Pierre-Gilles de Gennes, F91470 Limours, France
e-mail: frederic.barbaresco@thalesgroup.com

F. Nielsen (ed.), *Geometric Structures of Information*, Signals and Communication Technology, https://doi.org/10.1007/978-3-030-02520-5_12

After the seminal papers of Fréchet [3], Rao [4] and Chentsov [5], many mathematicians and physicists have studied Information Geometry. One can quote in mathematics, the works of Amari [6, 7] in the 80 s, which does not refer to the Koszul publications of the 50s and 60s where Koszul introduced the elementary structures of the Hessian geometries, and generalized the Fisher metric for homogeneous convex domains. In the physical field, many physicists have also addressed Information Geometry, without references to Koszul. Weinhold [8] in 1976 and Ruppeiner [9] in 1979 empirically introduced the inverse dual metric defined by the Hessian of Entropy, or Ingarden [10, 11] in 1981 in Statistical Physics. Mrugala [12, 13] in 1978, and Janyszek [14] in 1989, tried to geometrize Thermodynamics by jointly addressing Information Geometry and Contact Geometry. All these authors were not familiar with Representations Theory introduced by Kirillov, and more particularly the affine representation of Lie groups and Lie algebras, used and developed by Koszul in mathematics and by Souriau in statistical mechanics [79–84]. It thus appears that the first foundations of the information geometry goes back to Fréchet's paper of 1943 [3] (and his Lecture given during the winter of 1939 at the Institut Henri Poincaré), who first introduced the Clairaut(-Legendre) equation (fundamental equation in Information Geometry between dual potentials) and Fisher metric as the Hessian of a convex function. This Fréchet's seminal work was followed by Koszul's 50's papers [15, 16] which introduced new forms that generalize Fisher metric for sharp convex cones. It was not until 1969 that Souriau completed this extension in the framework of the Lie Group Thermodynamics with a cohomological definition of Fisher metric [17]. This last extension was developed by Koszul at the beginning of 80's in his Lecture "Introduction to Symplectic Geometry" [18]. I will conclude this survey by making reference to Balian [19], who has developed during 80's Information Geometry in Quantum Physics with a Quantum Fisher metric given by Von Neumann Entropy hessian [20].

Inspired by the French mathematical tradition, and the teachings of his master Elie Cartan (Koszul was PhD student of Henri Cartan but was greatly influenced by Elie Cartan), Jean-Louis Koszul was a real "avant-garde", if we take the definition given by Clausewitz «*An avant-garde is a group of units intended to move in front of the army to: explore the terrain to avoid surprises, quickly occupy the strong positions of the battlefield (high points), screen and contain the enemy the time the army can deploy*". Indeed, Jean-Louis Koszul was a pioneer, who explored and cleared many areas of mathematics, detailed in the book "Selected papers of JL Koszul" [21]. What I will expose, in this paper, is therefore only one part of his work which concerns homogeneous bounded domains geometry, from seminal Elie Cartan's earlier work on symmetric bounded domains. In a letter from André Weil to Henri Cartan, cited in the proceedings of the conference "*Elie Cartan and today's mathematics*" in 1984, it says "*As to the symmetrical spaces, and more particularly to the symmetric bounded domains at the birth of which you contributed, I have kept alive the memory of the satisfaction I felt in finding some incarnations in Siegel from his first works on quadratic forms, and later to convince Siegel of the value of your father's ideas on the subject*". At this 1984 conference, two disciples of Elie Cartan gave a conference, Jean-Louis Koszul [22] and Jean-Marie Souriau (Fig. 1).

Fig. 1 (on the left) Jean-Louis Koszul student at ENS ULM in 1940, (on the right) Jean-Louis Koszul at GSI'13 "Geometric Science of Information" conference at the École des Mines de Paris August 2013

In the book "*Selected papers of JL Koszul*" [21], Koszul summarizes the work, I will detail in the following: "*It is with the problem of the determination of the homogeneous bounded domains posed by E. Cartan around 1935 that are related [my papers]. The idea of approaching the question through invariant Hermitian forms already appears explicitly in Cartan. This leads to an algebraic approach which constitutes the essence of Cartan's work and which, with the Lie J-algebras, was pushed much further by the Russian School* [23–36]. *It is the work of Piatetski Shapiro on the Siegel domains, then those of E.B. Vinberg on the homogeneous cones that led me to the study of the affine transformation groups of the locally flat manifolds and in particular to the convexity criteria related to invariant forms*". In particular, J.L. Koszul source of inspiration is given in this last sentence of Elie Cartan's 1935 article [37]:

"*It is clear that if one could demonstrate that all homogeneous domains whose form* $\Phi = \sum_{i,j} \frac{\partial^2 \log K(z,z^*)}{\partial z_i \partial z_j^*} dz_i dz_j^*$ *is positive definite are symmetric, the whole theory of homogeneous bounded domains would be elucidated. This is a problem of Hermitian geometry certainly very interesting*". It was not until 1953 that the classification of non-Riemannian symmetric spaces has been achieved by Marcel Berger [38]. The work of Koszul has also been extended and deepened by one of his student Jacques Vey in [39, 40]. Jacques Vey has transposed the notion of hyperbolicity, developed

Fig. 2 (on the left) Professor Elie Cartan, (on the right) the Cartan family

by W. Kaup for Riemann surfaces, into the category of differentiable manifolds with flat linear connection (locally flat manifolds), which makes it possible to completely characterize the locally flat manifolds admitting as universal covering a convex open sharp cone of R^n, which had been studied by Koszul in [41]. The links between Koszul's work and those of Ernest B. Vinberg [23–30] were recently developed at the conference "*Transformation groups 2017*" in Moscow dedicated to the 80th anniversary of Professor EB Vinberg, in Dmitri Alekseevsky's talk on "*Vinberg's theory of homogeneous convex cones: developments and applications*" [42]. Koszul and Vinberg are actually associated with the concept of Koszul-Vinberg's characteristic function on convex cones, which I will develop later in the paper. Koszul introduced the so-called "*Koszul forms*" and a canonical metric given by the Hessian of the opposite of the logarithm of this Koszul-Vinberg characteristic function, from which I will show the links with Fisher's metric in Information Geometry, and its extension (Fig. 2).

Professor Koszul's main papers, which form the elementary structures of information geometry, are as follows:

- ***«Sur la forme hermitienne canonique des espaces homogènes complexes»*** [15] of 1955: Koszul considers the Hermitian structure of a homogeneous *G/B* manifold (*G* related Lie group and *B* a closed subgroup of *G*, associated, up to a constant factor, to the single invariant *G*, and to the invariant complex structure by the operations of *G*). Koszul says "*The interest of this form for the determination of* *homogeneous bounded domains has been emphasized by Elie Cartan: a necessary condition for G/B to be a bounded domain is indeed that this form is positive definite*". Koszul calculated this canonical form from infinitesimal data Lie algebra of *G*, the sub-algebra corresponding to *B* and an endomorphism algebra defining the invariant complex structure of *G/B*. The results obtained by Koszul proved that the homogeneous bounded domains whose group of automorphisms is semi-simple are bounded symmetric domains in the sense of Elie Cartan. Koszul also refers to André Lichnerowicz's work on Kählerian homogeneous spaces [43]. In this seminal paper, Koszul also introduced a left invariant form of degree 1 on *G*: $\Psi(X) = Tr_{g/b}[ad(JX) - J.ad(X)] \quad \forall X \in g$ with *J* an endomorphism of the Lie algebra space and the trace $Tr_{g/b}[.]$ corresponding to that of the endomorphism *g/b*. The Kähler form of the canonical Hermitian form is given by the differential of $-1/4\Psi(X)$ of this form of degree 1.

- ***«Exposés sur les espaces homogènes symétriques»*** [16] of 1959 is a Lecture written as part of a seminar held in September and October 1958 at the University of Sao Paulo, which details the determination of homogeneous bounded domains. He returned to [15] and showed that any symmetric bounded domain is a direct product of irreducible symmetric bounded domains, determined by Elie Cartan (4 classes corresponding to classical groups and 2 exceptional domains). For the study of irreducible symmetric bounded domains, Koszul refered to Elie Cartan, Carl-Ludwig Siegel and Loo-Keng Hua. Koszul illustrated the subject with two particular cases, the half-plane of Poincaré and the half-space of Siegel, and showed that with its trace formula of endomorphism *g/b*, he found that the canonical Kähler hermitian form and the associated metrics are the same as those introduced by Henri Poincaré and Carl-Ludwig Siegel [44] (who introduced them as invariant metric under action of the automorphisms of these spaces).
- ***«Domaines bornées homogènes et orbites de groupes de transformations affines»*** [45] of 1961 is written by Koszul at the Institute for Advanced Study at Princeton during a stay funded by the National Science Foundation. On a complex homogeneous space, an invariant volume defines with the complex structure the canonical invariant Hermitian form introduced in [15]. If the homogeneous space is holomorphically isomorphic to a bounded domain of a space C^n, this Hermitian form is positive definite because it coincides with the Bergmann metric of the domain. Koszul demonstrated in this article the reciprocal of this proposition for a class of complex homogeneous spaces. This class consists of some open orbits of complex affine transformation groups and contains all homogeneous bounded domains. Koszul addressed again the problem of knowing if a complex homogeneous space, whose canonical Hermitian form is positive definite is isomorphic to a bounded domain, but via the study of the invariant bilinear form defined on a real homogeneous space by an invariant volume and an invariant flat connection. Koszul demonstrated that if this bilinear form is positive definite then the homogeneous space with its flat connection is isomorphic to a convex open domain containing no straight line in a real vector space and extended it to the initial problem for the complex homogeneous spaces obtained in defining a complex structure in the variety of vectors of a real homogeneous space provided with an invariant

flat connection. It is in this article that Koszul used the affine representation of Lie groups and algebras. By studying the open orbits of the affine representations, he introduced an affine representation of G, written (f, q), and the following equation setting f the linear representation of the Lie algebra $\mathfrak{g}$ of G, defined by f and q the restriction to $\mathfrak{g}$ and the differential of q (f and q are differential respectively of f and q):

$$f(X)q(Y) - f(Y)q(X) = q([X, Y]) \quad \forall X, Y \in \mathfrak{g}$$
$$\text{with} \quad f : \mathfrak{g} \to gl(E) \quad \text{and} \quad q : \mathfrak{g} \mapsto E$$

- ***«Ouverts convexes homogènes des espaces affines»*** [46] of 1962. Koszul is interested in this paper by the structure of the convex open non-degenerate Ω (with no straight line) and homogeneous (the group of affine transformations of E leaving stable Ω operates transitively in Ω) in a real affine space of finite dimension. Koszul demonstrated that they can be all deduced from non-degenerate and homogeneous convex open cones built in [45]. He used for this the properties of the group of affine transformations leaving stable a non-degenerate convex open domain and an homogeneous domain.
- ***«Variétés localement plates et convexité»*** [41] of 1965. Koszul established the following theorem: let M *be* a locally related differentiable manifold. If the universal covering of M is isomorphic as a flat manifold with a convex open domain containing no straight line in a real affine space, then there exists on M a closed differential form α such that $D\alpha$ *(D* linear covariant derivative of zero torsion) is positive definite in all respects and which is invariant under every automorphism of M. If G is a group of automorphisms of M such that $G\backslash M$ is quasi-compact and if there exists on M a closed 1-differential form α invariant by G and such that $D\alpha$ is positive definite at any point, then the universal covering of M is isomorphic as a flat manifold with a convex open domain that does not contain a straight line in a real affine space.
- ***«Lectures on Groups of Transformations»*** [47] of 1965. This is lecture notes given by Koszul at Bombay "Tata Institute of Fundamental Research " on transformation groups. In particular in Chap. 6, Koszul studied discrete linear groups acting on convex open cones in vector spaces based on the work of C.L. Siegel (work on quadratic forms [48]). Koszul used what I will call in the following Koszul-Vinberg characteristic function on convex sharp cone.
- ***«Déformations des variétés localement plates»*** [49] of 1968. Koszul provided other proofs of theorems introduced in [41]. Koszul considered related differentiable manifolds of dimension n and TM the fibered space of M. The linear connections on M constitute a subspace of the space of the differentiable applications of the TMxTM fiber *product* in the space $T(TM)$ of the TM vectors. Any locally flat connection D (the curvature and the torsion are zero) defines a locally flat connection on the covering of M, and is hyperbolic when universal covering of M, with this connection, is isomorphic to a sharp convex open domain (without straight lines) in R^n. Koszul showed that, if M is a compact manifold, for a locally

flat connection on M to be hyperbolic, it is necessary and sufficient that there exists a closed differential form of degree 1 on M whose covariant differential is positive definite.
- ***«Trajectoires Convexes de Groupes Affines Unimodulaires»*** [50] in 1970. Koszul demonstrated that a convex sharp open domain in R^n that admits a unimodular transitive group of affine automorphisms is an auto-dual cone. This is a more geometric demonstration of the results shown by Ernest Vinberg [29] on the automorphisms of convex cones.

The elementary geometric structures discovered by Jean-Louis Koszul are the foundations of Information Geometry. These links were first established by Professor Hirohiko Shima [51–56]. These links were particularly crystallized in Shima book 2007 "*The Geometry of Hessian Structures*" [57], which is dedicated to Professor Koszul. The origin of this work followed the visit of Koszul in Japan in 1964, for a mission coordinated with the French government. Koszul taught lectures on the theory of flat manifolds at Osaka University. Hirohiko Shima was then a student and attended these lectures with the teachers Matsushima and Murakami. This lecture was at the origin of the notion of Hessian structures and the beginning of the works of Hirohiko Shima. Henri Cartan noted concerning Koszul's ties with Japan, "*Koszul has attracted eminent mathematicians from abroad to Strasbourg and Grenoble. I would like to mention in particular the links he has established with representatives of the Japanese School of Differential Geometry*". Shima's book [57] is a systematic introduction to the theory of Hessian structures (provided by a pair of a flat connection D and an Hessian metric g). Koszul studied flat manifolds with a closed 1-form α, such that $D\alpha$ be positive definite, where $D\alpha$ is a hessian metric. However, not all Hessian metrics are globally of the form $g = D\alpha$. Shima introduces the notion of Codazzi structure for a pair (D,g), with D a torsion-free connection, which verifies the Codazzi equation $(D_X g)(Y, Z) = (D_Y g)(X, Z)$. A Hessian structure is a Codazzi structure for which connection D is flat. This is an extension of Riemannian geometry. It is then possible to define a connection D' and a dual Codazzi structure (D',g) with $D' = \nabla - D$ where ∇ is the Levi-Civita connection. For a hessian structure (D, g) with $g = Dd\varphi$, the dual Codazzi structure $\left(D', g\right)$ is also a Hessian structure and $g = D'd\varphi'$, where φ' is the Legendre transform of $\varphi : \varphi' = \sum_i x^i \frac{\partial\varphi}{\partial x^i} - \varphi$. Shima observed that Information Geometry framework could be introduced by dual connections, and not only founded on Fréchet, Rao and Chentsov works [5]. A hessian structure (D, g) is of Koszul type, if there is a closed 1-form ω as $g = D\omega$. Using D and the volume element of g, Koszul introduced a 2nd form, which plays a similar role to the Ricci tensor for a Kählerian metric. Let υ be the volume element of g, we define a closed 1-form α such that $D_X \upsilon = \alpha(X)\upsilon$ and a symmetric bilinear form $\gamma = D\alpha$. In the following, α and γ forms are called 1st and 2nd form of Koszul for Hessian structure (D, g). We can consider the forms associated with the Hessian dual structure (D', g) by $\alpha' = -\alpha$ and $\gamma' = \gamma - 2\nabla\alpha$. In the case of a homogeneous regular convex cone Ω, with D the canonical flat connection of the ambient vector space, the Koszul forms α and γ for the canonical Hessian structure $(D, g = Dd\psi)$

Fig. 3 From left to right, Jean-Louis Koszul, Hirohiko Shima and Michel Nguiffo Boyom at GSI'13 (Geometric Science of Information) conference at the École des Mines of Paris in August 2013

are given by $\alpha = d \log \psi$ and $\gamma = g$. The volume element υ determined by g is invariant under the action of the group of automorphisms G of Ω.

Jean-Louis Koszul attended the 1st GSI "*Geometric Science of Information*" conference in August 2013 at the Ecole des Mines in Paris, where he attended the presentation of Hirohiko Shima, given for his honor on the topic "*Geometry of Hessian Structures* " [58]. In the photo below, we can see from left to right, Jean-Louis Koszul, Hirohiko Shima and Michel Nguiffo Boyom. Professor Michel Boyom has extensively studied and developed, at the University of Montpellier, Koszul models [59–66] in relation to symplectic flat affine manifolds and to the cohomology of Koszul-Vinberg algebras (KV Cohomology). Professor Boyom with his PhD student Byande [67, 68] have explored other links with Information Geometry. André Lichnerowicz worked in parallel on a closed topic about homogeneous Kähler manifolds [69] (Fig. 3).

2 Biographical Reminder of Jean-Louis Koszul Scientific Life

Jean Louis André Stanislas Koszul born in Strasbourg in 1921, is the child of a family of four (with three older sisters, Marie Andrée, Antoinette and Jeanne). He is the son of André Koszul (born in Roubaix on November 19th 1878, professor at the

Strasbourg university), and Marie Fontaine (born in Lyon on June 19th 1887), who was a friend of Henri Cartan's mother. Henri Cartan writes on this friendship "*My mother in her youth, had been a close friend of the one who was to become Jean-Louis Koszul's mother*" [70]. His paternal grandparents were Julien Stanislas Koszul and Hélène Ludivine Rosalie Marie Salomé. He attended high school in Fustel-de-Coulanges in Strasbourg and the Faculty of Science in Strasbourg and in Paris. He entered ENS Ulm in the class of 1940 and defended his thesis with Henri Cartan. Henri Cartan noted "*This promotion included other mathematicians like Belgodère or Godement, and also physicists and some chemists, like Marc Julia and Raimond Castaing*" [70] (for the anecdote, the maiden name of my wife Anne, is *Belgodère*, with a filial link with Paul Belgodère of the Koszul ENS promotion). Jean-Louis Koszul married on July 17th 1948 with Denise Reyss-Brion, student of ENS Sèvres, entered in 1941. They have three children, Michel (married to Christine Duchemin), Anne (wife of Stanislas Crouzier) and Bertrand. He then taught in Strasbourg and was appointed Associate Professor at the University of Strasbourg in 1949, and had for colleagues R. Thom, M. Berger and B. Malgrange. He was promoted to professor status in 1956. He became a member of Bourbaki with the 2nd generation, J. Dixmier, R. Godement, S. Eilenberg, P. Samuel, J. P. Serre and L. Schwartz. Henri Cartan remarked in [70] "*In the vehement discussions within Bourbaki, Koszul was not one of those who spoke loudly; but we learned to listen to him because we knew that if he opened his mouth he had something to say*". About this Koszul's period at Strasbourg University, Pierre Cartier [71] said "*When I arrived in Strasbourg, Koszul was returning from a year spent in Institute for Advanced Studies in Princeton, and he was after the departure of Ehresman and Lichnerowicz to Paris the paternal figure of the Department of Mathematics (despite his young age). I am not sure of his intimate convictions, but he represented for me a typical figure of this Alsatian Protestantism, which I frequented at the time. He shared the seriousness, the honesty, the common sense and the balance. In particular, he knew how to resist the academic attraction of Paris. He left us after 2 years to go to Grenoble, in a maneuver uncommon at the time of exchange of positions with Georges Reeb*". He became Senior Lecturer at the University of Grenoble in 1963, and then an honorary professor at the Joseph Fourier University [72] and integrated in Fourier Institute led by C. Chabauty. During this period, B. Malgrange [73] remembered Koszul seminar on "algebra and geometry" with his three students J. Vey, D. Luna and J. Helmstetter. In Grenoble, he practiced mountaineering and was a member of the French Alpine Club. Koszul was awarded by Jaffré Prize in 1975 and was elected correspondent at the Academy of Sciences on January 28th 1980. Koszul was one of the CIRM conference center founder at Luminy. The following year, he was elected to the Academy of São Paulo. Jean-Louis Koszul died on January 12th 2018, at the age of 97.

As early as 1947, Jean-Louis Koszul published three articles in CRAS of the Academy of Sciences, on the Betti number of a simple compact Lie group, on cohomology rings, generalizing ideas of Jean-Leray, and finally on the homology of homogeneous spaces. Koszul's thesis, defended in June 10th 1949 under the direction of Henri Cartan, dealt with the homology and cohomology of Lie algebras [74]. The jury was composed of M. Denjoy (President), J. Leray, P. Dubreil and H. Cartan.

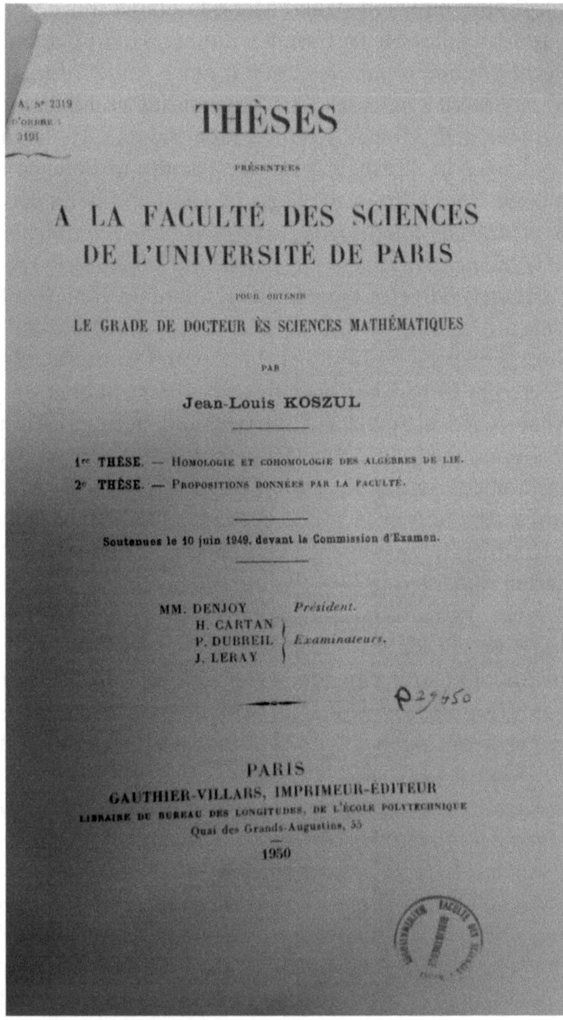
A, N° 2319
D'ORDRE :
3191

THÈSES

PRÉSENTÉES

A LA FACULTÉ DES SCIENCES
DE L'UNIVERSITÉ DE PARIS

POUR OBTENIR

LE GRADE DE DOCTEUR ÈS SCIENCES MATHÉMATIQUES

PAR

Jean-Louis KOSZUL

1re THÈSE. — Homologie et cohomologie des algèbres de Lie.
2e THÈSE. — Propositions données par la Faculté.

Soutenues le 10 juin 1949, devant la Commission d'Examen.

MM. DENJOY *Président.*
H. CARTAN
P. DUBREIL *Examinateurs.*
J. LERAY

PARIS
GAUTHIER-VILLARS, IMPRIMEUR-ÉDITEUR
LIBRAIRE DU BUREAU DES LONGITUDES, DE L'ÉCOLE POLYTECHNIQUE
Quai des Grands-Augustins, 55

1950

Fig. 4 Cover page of Koszul's PhD report defended June 10th 1949 with a Jury composed of Professors Arnaud Denjoy, Henri Cartan, Paul Dubreil and Jean Leray, published in [74]

Under the title "Works of Koszul I, II and III", Henri Cartan reported Koszul's PhD results to Bourbaki seminar [75–77]. See also, André Haefliger paper [78] (Fig. 4).

In 1987, an International Symposium on Geometry was held in Grenoble in honor of Jean-Louis Koszul, whose proceedings were published in "*les Annales de l'Institut Fourier*", Volume 37, No. 4. This conference began with a presentation by Henri Cartan, who remembered the mention given to Koszul for his aggregation [70]: "*Distinguished Spirit; he is successful in his problems. Should beware, orally, of overly systematic trends. A little less subtle complications, baroque ideas, a little more common sense and balance would be desirable*". About his supervision of Koszul's PhD, Henri Cartan writed "*Why did he turn to guide him (so-called)? Is it because he found inspiration in Elie Cartan's work on the topology of Lie groups?*

Perhaps he was surprised to note that mathematical knowledge is not necessarily transmitted by descent. In any case, he helped me to better know what my father had brought to the theory" [70]. On the work of Koszul algebrisation, Henri Cartan notes "*Koszul was the first to give a precise algebraic formalization of the situation studied by Leray in his 1946 publication, which became the theory of the spectral sequence. It took a good deal of insight to unravel what lay behind Leray's study. In this respect, Koszul's Note in the July 1947 CRAS is of historical significance.*" [70]. From June 26th to July 2nd 1947, CNRS, received an International conference in Paris, on "*Algebraic Topology*". This was the first postwar international diffusion of Leray's ideas. Koszul writes about this lecture "*I can still see Leray putting his chalk at the end of his talk by saying (modestly?) that he definitely did not understand anything about Algebraic Topology*". In writing his lectures at the Collège de France, Leray adopted the algebraic presentation of the spectral suite elaborated by Koszul. As early as 1950, J.P. Serre used the term "Leray-Koszul suite". Speaking of Leray, Koszul wrote "*around 1955 I remember asking him what had put him on the path of what he called the ring of homology of a representation in his Notes to the CRAS of 1946. His answer was Künneth's theorem; I could not find out more*". The sheaf theory, introduced by Jean-Leray, followed in 1947, at the same time as the spectral sequences.

In 1950, Koszul published an important book of 62 pages entitled "*Homology and Cohomology of Lie Algebras*" [74] based on his PhD work, in which he studied the links between homology and cohomology (with real coefficients) of a compact connected Lie group and purely algebraic problems of Lie algebra. Koszul then gave a lecture in São Paulo on the topic "*sheaves and cohomology*". The superb lecture notes were published in 1957 and dealt with the cohomology of Čech with coefficients in a sheaf. In the autumn of 1958, he again organized a series of seminars in São Paulo, this time on symmetric spaces [16]. R. Bott commented on these seminars "*very pleasant. The pace is fast, and the considerable material is covered elegantly. In addition to the more or less standard theorems on symmetric spaces, the author discusses the geometry of geodesics, Bergmann's metrics, and finally studies the bounded domains with many details*". In the mid-1960s, Koszul taught at the Tata Institute in Bombay on transformation groups [47] and on fiber bundles and differential geometry. The second lecture dealt with the theory of connections and the lecture notes were published in 1965. In 1986 he published "*Introduction to symplectic geometry*" [18] following a Chinese course in China (with the agreement of Jean-Louis Koszul given in 2017, this lecture given at the University of Nanjing will be translated into English by Springer and will be published in 2018). This book takes up and develops works of Jean-Marie Souriau [17, 79] on homogeneous symplectic manifolds and the affine representation of Lie algebras and Lie groups in geometric mechanics (another fundamental source of Information Geometry structures extended on homogeneous varieties [80–84]). Chuan Yu Ma writes in a review, on this latest book in Chinese, that "*This work coincided with developments in the field of analytical mechanics. Many new ideas have also been derived using a wide variety of notions of modern algebra, differential geometry, Lie groups, functional analysis, differentiable manifolds, and representation theory. [Koszul's book] emphasizes the differential-geometric and topological properties of symplectic manifolds. It gives a modern treatment of the subject that is useful for beginners as well as for experts*".

In 1994, in [21], a comment by Koszul explains the problems he was preoccupied with when he invented what is now called the "*Koszul complex*". This was introduced to define a theory of cohomology for Lie algebras and proved to be a general structure useful in homological algebra.

3 Koszul-Vinberg Characteristic Function, Koszul Forms and Maximum Entropy Density

Through the study of the geometry of bounded homogeneous domains initiated by Elie Cartan [37, 85], Jean-Louis Koszul discovered that the elementary structures are associated with Hessian manifolds on sharp convex cones [15, 16, 41, 45–47, 49, 50]. In 1935, Elie Cartan proved in [37] that the symmetric homogeneous irreducible bounded domains could be reduced to 6 classes, 4 canonical models and 2 exceptional cases. Ilya Piatetski-Shapiro [31–35], after Luogeng Hua [86], extended Siegel's description [44, 48] to other symmetric spaces, and showed by a counterexample that Elie Cartan's conjecture, that all transitive domains are symmetrical, was false. At the same time, Ernest B. Vinberg [23–30] worked on the theory of homogeneous convex cones and the construction of Siegel domains [44, 48]. More recently, the classical complex symmetric spaces were studied by F. Berezin [87, 88] in the context of quantification. In parallel, O.S. Rothaus [89] and Piatetski-Shapiro [31–35] with Karpelevitch, explored the underlying geometry of these complexes homogeneous fields, and more particularly the fibration areas on the components of the shilov boundary. In Italy, I note the work of E. Vessentini [90] and U. Sampieri [91, 92]. The Siegel domains, which fit into these classes of structures, nowadays play an important role in the processing of radar spatio-temporal signals and, more broadly, in learning from structured covariance matrices.

Jean-Louis Koszul and Ernest B. Vinberg have introduced a hessian metric invariant by the group of linear automorphisms on a sharp convex cone Ω through a function, called characteristic function ψ. In the following Ω is a sharp convex cone in a vector space E of finite size on R (a convex cone is sharp if there is no straight lines). In dual space E^* of E, Ω^* is the set of linear strictly positive forms on $\overline{\Omega}-\{0\}$. Ω^*, dual cone of Ω, is also a sharp convex cone. If $\xi\in\Omega^*$, then intersection $\Omega\cap\{x\in E/\langle x,\xi\rangle=1\}$ is bounded. $G=Aut(\Omega)$ is the group of linear transformation from *E that preserves* Ω (group of automorphisms). $G=Aut(\Omega)$ acts on Ω^* such that, $\forall g\in G=Aut(\Omega)$, $\forall\xi\in E^*$ then $\bar{g}.\xi=\xi\circ g^{-1}$. Koszul introduce an integral, of Laplace kind, on sharp dual convex cone, as:

Koszul-Vinberg Characteristic definition:

Let $d\xi$ Lebesque measure on E^*, following integral:

$$\psi_\Omega(x) = \int_{\Omega^*} e^{-\langle \xi, x \rangle} d\xi \quad \forall x \in \Omega \tag{1}$$

with Ω^* the dual cone, is analytical function on Ω, with $\psi_\Omega(x) \in]0, +\infty[$, called Koszul-Vinberg characteristic function of cone Ω.

Nota: the logarithm of the characteristic function is called « barrier function » for convex optimization algorithms. Yurii Nesterov and Arkadii Nemirovskii [93] have proved in modern theory of « *interior point* », using function $\Theta_\Omega(x) = \log(vol_n\{s \in \Omega^* / \langle s, x \rangle \le 1\})$, that all convex cones in R^n have a self-dual barrier, linked with Koszul characteristic function.

Koszul-Vinberg Characteristic function has the following properties:

- Bergman kernel of $\Omega + i R^{n+1}$ is written $K_\Omega(\mathrm{Re}(z))$ up to a constant. K_Ω is defined by integral:

$$K_\Omega(x) = \int_{\Omega^*} e^{-\langle \xi, x \rangle} \psi_{\Omega^*}(\xi)^{-1} d\xi \tag{2}$$

- ψ_Ω is an analytical function defined in the interior of Ω and $\psi_\Omega(x) \to +\infty$ when $x \to \partial\Omega$. If $g \in Aut(\Omega)$ then $\psi_\Omega(gx) = |\det g|^{-1} \psi_\Omega(x)$ and as $tI \in G = Aut(\Omega)$ for all $t > 0$, we have:

$$\psi_\Omega(tx) = \psi_\Omega(x)/t^n \tag{3}$$

- ψ_Ω is strictly log convex, such that $\phi_\Omega(x) = \log(\psi_\Omega(x))$ is strictly convex.

From this characteristic function, Koszul introduced two forms:

1st Koszul form α : Differential 1-form

$$\alpha = d\phi_\Omega = d \log \psi_\Omega = d\psi_\Omega / \psi_\Omega \tag{4}$$

is invariant with respect to all automorphisms $G = Aut(\Omega)$ of Ω. If $x \in \Omega$ and $u \in E$ then:

$$\langle \alpha_x, u \rangle = -\int_{\Omega^*} \langle \xi, u \rangle . e^{-\langle \xi, x \rangle} d\xi \text{ and } \alpha_x \in -\Omega^* \tag{5}$$

and

2nd Koszul form γ: Differential symmetric 2-form

$$\gamma = D\alpha = Dd \log \psi_\Omega \tag{6}$$

is a bilinear symmetric positive definite form invariant with respect to the action of $G = Aut(\Omega)$ and $D\alpha > 0$

Positivity is given by Schwarz inequality and:

$$Dd\ \log \psi_{\Omega}(u,v) = \int\limits_{\Omega^*} \langle \xi, u\rangle \langle \xi, v\rangle e^{-\langle \xi, u\rangle} d\xi \tag{7}$$

Koszul has proved that from this 2nd form, we can introduce an invariant Riemannian metric with respect to the action of cone automorphisms:
Koszul Metric: $D\alpha$ defines a Riemannian invariant structure by $Aut(\Omega)$, and the Riemannian metric is given by:

$$g = Dd\ \log \psi_{\Omega} \tag{8}$$

$$(Dd\ \log \psi(x))(u) = \frac{1}{\psi(u)^2}\left[\int\limits_{\Omega^*} F(\xi)^2 d\xi . \int\limits_{\Omega^*} G(\xi)^2 d\xi - \left(\int\limits_{\Omega^*} F(\xi).G(\xi) d\xi\right)^2\right] > 0$$

$$\text{with } F(\xi) = e^{-\frac{1}{2}\langle x, y\rangle} \quad \text{and} \quad G(\xi) = e^{-\frac{1}{2}\langle x, \xi\rangle}\langle u, \xi\rangle \tag{9}$$

The positivity could be proved by using Schwarz inequality, and the following properties for the derivative given by $d\ \log \psi = \frac{d\psi}{\psi}$ and $Dd\ \log \psi = \frac{Dd\psi}{\psi} - \left(\frac{d\psi}{\psi}\right)^2$ where $(d\psi(x))(u) = -\int\limits_{\Omega^*} e^{-\langle x, \xi\rangle}\langle u, \xi\rangle d\xi$ and $(Dd\psi(x))(u) = -\int\limits_{\Omega^*} e^{-\langle x, \xi\rangle}\langle u, \xi\rangle^2 d\xi$.

Koszul uses this diffeomorphism to define dual coordinates:

$$x^* = -\alpha_x = -d\ \log \psi_{\Omega}(x) \tag{10}$$

with $\langle df(x), u\rangle = D_u f(x) = \frac{d}{dt}\big|_{t=0} f(x+tu)$. When the cone Ω is symmetric, the map $x \mapsto x^* = -\alpha_x$ is a bijection and an isometry with only one fixed point (the manifold is a symmetric Riemannian space given by its isometry):

$$(x^*)^* = x, \left\langle x, x^*\right\rangle - n \text{ et } \psi_{\Omega}(x)\psi_{\Omega^*}(x^*) = cste \tag{11}$$

x^* is characterized by $x^* = \arg\min\{\psi(y)/y \in \Omega^*, \langle x, y\rangle = n\}$ and x^* is the gravity center of the transverse cut $\{y \in \Omega^*, \langle x, y\rangle = n\}$ of Ω^*:

$$x^* = \int\limits_{\Omega^*} \xi . e^{-\langle \xi, x\rangle} d\xi \Big/ \int\limits_{\Omega^*} e^{-\langle \xi, x\rangle} d\xi$$

$$\text{and} \quad \left\langle -x^*, h\right\rangle = d_h \log \psi_{\Omega}(x) = -\int\limits_{\Omega^*} \langle \xi, h\rangle e^{-\langle \xi, x\rangle} d\xi \Big/ \int\limits_{\Omega^*} e^{-\langle \xi, x\rangle} d\xi \tag{12}$$

In [94–97], Misha Gromov was interested by these structures. If we set $\Phi(x) = -\log \psi_{\Omega}(x)$, Gromov has observed that $x^* - d\Phi(x)$ is an injection where the image

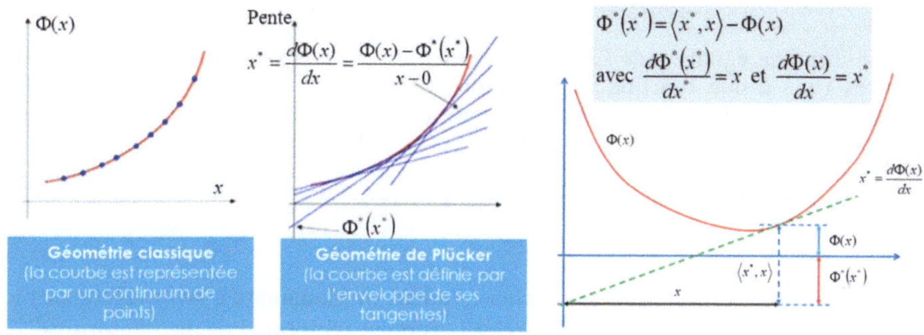

Fig. 5 Legendre transform and Plücker geometry

closure is equal to the convex envelop of the support and the volume of this envelop is the n-dimensionnel volume defined by the integral of hessian determinant of this function, $\Phi(x)$, where the map $\Phi \mapsto M(\Phi) = \int_{\Omega} \det(Hess(\Phi(x))).dx$ obeys a non-trivial inequality given by Brunn-Minkowsky:

$$[M(\Phi_1 + \Phi_2)]^{1/2} \geq [M(\Phi_1)]^{1/n} + [M(\Phi_2)]^{1/n} \tag{13}$$

These relations appear also in statistical physics. As the physicist Jean-Marie Souriau [17, 80–84, 98] did, it is indeed possible to define the concept of Shannon's Entropy via the Lengendre transform associated with the opposite of the logarithm of this Koszul-Vinberg characteristic function. Taking up the seminal ideas of François Massieu [99–102] in Thermodynamics (classmate of the Corps des Mines, it is François Massieu who influenced Henri Poincaré [103] who introduced the characteristic function in Probability, with a Laplace transform, and not a Fourier transform as did then Paul Levy), which were recently developed by Roger Balian in Quantum Physics [19, 20, 104–111], replacing Shannon Entropy by von Neumann Entropy. I will also note the work of Jean-Leray on the extensions of the Laplace transform in [112]. Starting from the characteristic function of Koszul-Vinberg, it is thus possible to introduce an entropy of Koszul defined as the Legendre transform of this function, which is the opposite of the logarithm of the characteristic function of Koszul-Vinberg (a logarithm lies the characteristic function of Massieu and the characteristic function of Koszul or Poincaré). Starting from the Koszul function, its Legendre transform gives a dual potential function in the dual coordinate system.x^* (Fig. 5):

$$\Phi^*(x^*) = \langle x, x^* \rangle - \Phi(x) \text{ with } x^* = D_x\Phi \text{ and } x = D_{x^*}\Phi^* \text{ where } \Phi(x) = -\log \psi_\Omega(x) \tag{14}$$

Concerning the Legendre transform [113], Darboux gives in his book an interpretation of Chasles: "*What comes back according to a remark of M. Chasles, to replace the surface with its polar reciprocal with respect to a paraboloid*". We have the same reference to polar reciprocal in "*Lessons on the calculus of variations*" by Jacques

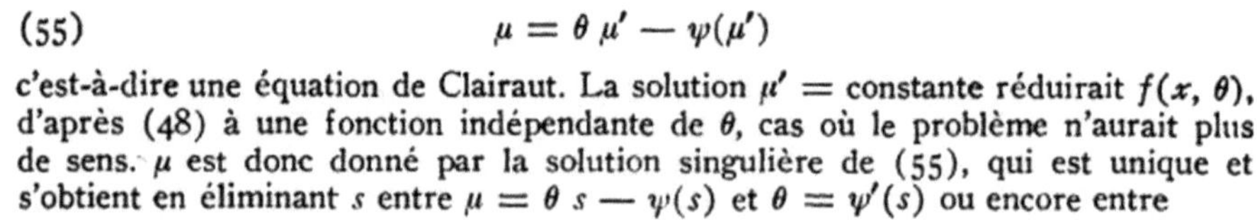
(55) $\mu = \theta\,\mu' - \psi(\mu')$

c'est-à-dire une équation de Clairaut. La solution μ' = constante réduirait $f(x, \theta)$, d'après (48) à une fonction indépendante de θ, cas où le problème n'aurait plus de sens. μ est donc donné par la solution singulière de (55), qui est unique et s'obtient en éliminant s entre $\mu = \theta\, s - \psi(s)$ et $\theta = \psi'(s)$ ou encore entre

Fig. 6 Legendre-Clairaut equation in 1943 Fréchet's paper

Hadamard, written by Maurice Fréchet (student of Hadamard), with references to M.E. Vessiot, which uses the "*figuratrice*", as polar reciprocal of the "*figurative*".

It is possible to express this Legendre transform only from the dual coordinate system x^*, using that $x = D_{x^*}\Phi^*$. We then obtain the Clairaut equation:

$$\Phi^*(x^*) - \left\langle (D_x\Phi)^{-1}(x^*), x^* \right\rangle - \Phi\left[(D_x\Phi)^{-1}(x^*)\right] \forall x^* \in \{D_x\Phi(x)/x \in \Omega\} \quad (15)$$

This equation was discovered by Maurice Fréchet in his 1943 paper [3] (see also in the appendix), in which he introduced for the first time the bound on the variance of any statistical estimator via the Fisher matrix, wrongly attributed to Cramer and Rao [4]. Fréchet was looking for "*distinguished densities*" [98], densities whose covariance matrix of the estimator of these parameters reaches this bound. Fréchet there showed that these densities were expressed while using this characteristic function $\Phi(x)$, and that these densities belong to the exponential densities family (Fig. 6).

Apparently, this discovery by Fréchet dates from winter of 1939, because Fréchet writes at the bottom of the page [3] "*The content of this dissertation formed part of our mathematical statistics Lecture at the Institut Henri Poincaré during the winter of 1939–1940. It is one of the chapters of the second edition (in preparation) of our ' Lessons in Mathematical Statistics', the first of which is 'Introduction: Preliminary Lecture on the Probability Calculation' (119 pages in quarto, typed in) has just been published at the University Documentation Center, Tournaments and Constans. Paris*". More details are given in appendix.

More recently Muriel Casalis [114, 115], the PhD student of Gérard Letac [116], has studied in her PhD, invariance of probability densities with respect to the affine group, and the links with densities of exponential families.

To make the link between the characteristic function of Koszul-Vinberg and Entropy of Shannon, we will detail the formulas of Koszul in the following developments. Using the fact that $-\langle \xi, x \rangle = \log e^{-\langle \xi, x \rangle}$, we can write:

$$-\left\langle x^*, x \right\rangle = \int_{\Omega^*} \log e^{-\langle \xi, x \rangle}.e^{-\langle \xi, x \rangle} d\xi \Big/ \int_{\Omega^*} e^{-\langle \xi, x \rangle} d\xi \quad (16)$$

and then developing the Legendre transform to make appear the density of maximum entropy in $\Phi^*(x^*)$, and also the Shannon entropy:

$$\Phi^*(x^*) = \langle x, x^*\rangle - \Phi(x) = -\int\limits_{\Omega^*} \log e^{-\langle \xi, x\rangle}.e^{-\langle \xi, x\rangle} d\xi \Big/ \int\limits_{\Omega^*} e^{-\langle \xi, x\rangle} d\xi + \log \int\limits_{\Omega^*} e^{-\langle \xi, x\rangle} d\xi$$

$$\Phi^*(x^*) = \left[\left(\int\limits_{\Omega^*} e^{-\langle \xi, x\rangle} d\xi \right). \log \int\limits_{\Omega^*} e^{-\langle \xi, x\rangle} d\xi - \int\limits_{\Omega^*} \log e^{-\langle \xi, x\rangle}.e^{-\langle \xi, x\rangle} d\xi \right] \Big/ \int\limits_{\Omega^*} e^{-\langle \xi, x\rangle} d\xi$$

$$\Phi^*(x^*) = \left[\log \int\limits_{\Omega^*} e^{-\langle \xi, x\rangle} d\xi - \int\limits_{\Omega^*} \log e^{-\langle \xi, x\rangle}. \frac{e^{-\langle \xi, x\rangle}}{\int\limits_{\Omega^*} e^{-\langle \xi, x\rangle} d\xi} d\xi \right]$$

$$\Phi^*(x^*) = \left[\log \int\limits_{\Omega^*} e^{-\langle \xi, x\rangle} d\xi. \left(\int\limits_{\Omega^*} \frac{e^{-\langle \xi, x\rangle}}{\int\limits_{\Omega^*} e^{-\langle \xi, x\rangle} d\xi} d\xi \right) - \int\limits_{\Omega^*} \log e^{-\langle \xi, x\rangle}. \frac{e^{-\langle \xi, x\rangle}}{\int\limits_{\Omega^*} e^{-\langle \xi, x\rangle} d\xi} d\xi \right]$$

with $\int\limits_{\Omega^*} \frac{e^{-\langle \xi, x\rangle}}{\int\limits_{\Omega^*} e^{-\langle \xi, x\rangle} d\xi} d\xi = 1$

$$\Phi^*(x^*) = \left[-\int\limits_{\Omega^*} \frac{e^{-\langle \xi, x\rangle}}{\int\limits_{\Omega^*} e^{-\langle \xi, x\rangle} d\xi}. \log \left(\frac{e^{-\langle \xi, x\rangle}}{\int\limits_{\Omega^*} e^{-\langle \xi, x\rangle} d\xi} \right) d\xi \right] \tag{17}$$

In this last equation, $p_x(\xi) = e^{-\langle \xi, x\rangle} / \int\limits_{\Omega^*} e^{-\langle \xi, x\rangle} d\xi$ plays the role of maximum entropy density as introduced by Jaynes [117–119] (also called, Gibbs density in Thermodynamics). I call the associated entropy, Koszul Entropy:

$$\Phi^* = -\int\limits_{\Omega^*} p_x(\xi) \log\, p_x(\xi) d\xi \tag{18}$$

with

$$p_x(\xi) = e^{-\langle \xi, x\rangle} / \int\limits_{\Omega^*} e^{-\langle \xi, x\rangle} d\xi = e^{-\langle x, \xi\rangle - \log \int\limits_{\Omega^*} e^{-\langle \xi, x\rangle} d\xi} = e^{-\langle x, \xi\rangle + \Phi(x)} \text{and}\; x^* = \int\limits_{\Omega^*} \xi. p_x(\xi) d\xi \tag{19}$$

This Koszul density $p_x(\xi) = \frac{e^{-\langle \xi, x\rangle}}{\int\limits_{\Omega^*} e^{-\langle \xi, x\rangle} d\xi}$ help us to develop the log likelihood:

$$\log\, p_x(\xi) = -\langle x, \xi\rangle - \log \int\limits_{\Omega^*} e^{-\langle \xi, x\rangle} d\xi = -\langle x, \xi\rangle + \Phi(x) \tag{20}$$

and deduce from the expectation:

$$E_\xi\left[-\log p_x(\xi)\right] = \langle x, x^*\rangle - \Phi(x) \tag{21}$$

We also obtain the equation about normalization:

$$\Phi(x) = -\log \int_{\Omega^*} e^{-\langle \xi, x\rangle} d\xi = -\log \int_{\Omega^*} e^{-[\Phi^*(\xi)+\Phi(x)]} d\xi = \Phi(x) - \log \int_{\Omega^*} e^{-\Phi^*(\xi)} d\xi$$
$$\Rightarrow \int_{\Omega^*} e^{-\Phi^*(\xi)} d\xi = 1 \qquad (22)$$

But we have to make appear the variable x^* in $\Phi^*(x^*)$. We have then to write:

$$\log\ p_x(\xi) = \log\ e^{-\langle x,\xi\rangle+\Phi(x)} = \log e^{-\Phi^*(\xi)} = -\Phi^*(\xi)$$
$$\Rightarrow \Phi^* = -\int_{\Omega^*} p_x(\xi)\log p_x(\xi)d\xi = \int_{\Omega^*} \Omega^*(\xi)p_x(\xi)d\xi = \Phi^*(x^*) \qquad (23)$$

Last equality is true, if we have:

$$\int_{\Omega^*} \Phi^*(\xi)p_x(\xi)d\xi - \Phi^*\left(\int_{\Omega^*} \xi . p_x(\xi)d\xi\right) \text{ with } x^* = \int_{\Omega^*} \xi . p_x(\xi)d\xi \qquad (24)$$

This last relation is associated to classical Jensen inequality. Equality is obtained for Maximum Entropy density for $x^* = D_x\Phi$ [120]:

$$\text{Legendre - Moreau Transform:}\quad \Phi^*(x^*) = \underset{x}{Sup}\left[\langle x, x^*\rangle - \Phi(x)\right]$$
$$\Rightarrow \begin{cases} \Phi^*(x^*) \geq \langle x, x^*\rangle - \Phi(x) \\ \Phi^*(x^*) \geq \int\limits_{\Omega^*} \Phi^*(\xi)p_x(\xi)d\xi \end{cases} \Rightarrow \begin{cases} \Phi^*(x^*) \geq E\left[\Phi^*(\xi)\right] \\ \text{equality if } x^* = \frac{d\Phi}{dx} \end{cases} \qquad (25)$$

We obtain for the maximum entropy density, the equality:

$$E\left[\Phi^*(\xi)\right] = \Phi^*(E[\xi]), \xi \in \Omega^* \qquad (26)$$

To make the link between this Koszul model and maximum entropy density [121–123] introduced by Jaynes [117–119], I use previous notation and I look for the density $p_x(\xi)$ that is the solution to this maximum entropy variational problem. Find the density that maximizes the Shannon entropy with constraint on normalization and on the knowledge of first moment:

$$\underset{p_x(.)}{Max}\left[-\int_{\Omega^*} p_x(\xi)\log p_x(\xi)d\xi\right] \text{ such that } \begin{cases} \int\limits_{\Omega^*} p_x(\xi)d\xi = 1 \\ \int\limits_{\Omega^*} \xi . p_x(\xi)d\xi = x^* \end{cases} \qquad (27)$$

If we consider the density $q_x(\xi) = e^{-\langle \xi, x\rangle} / \int\limits_{\Omega^*} e^{-\langle \xi, x\rangle} d\xi = e^{-\langle x, \xi\rangle - \log \int\limits_{\Omega^*} e^{-\langle \xi, x\rangle} d\xi}$ such that:

$$\begin{cases} \int\limits_{\Omega^*} q_x(\xi).d\xi = \int\limits_{\Omega^*} e^{-\langle \xi, x\rangle} d\xi / \int\limits_{\Omega^*} e^{-\langle \xi, x\rangle} d\xi = 1 \\ \log q_x(\xi) = \log e^{-\langle x, \xi\rangle - \log \int\limits_{\Omega^*} e^{-\langle \xi, x\rangle} d\xi} = -\langle x, \xi\rangle - \log \int\limits_{\Omega^*} e^{-\langle x, \xi\rangle} d\xi \end{cases} \quad (28)$$

By using the inequality $\log x \geq \left(1 - x^{-1}\right)$ with equality if $x = 1$, we can then write that:

$$-\int\limits_{\Omega^*} p_x(\xi) \log \frac{p_x(\xi)}{q_x(\xi)} d\xi \leq -\int\limits_{\Omega^*} p_x(\xi)\left(1 - \frac{q_x(\xi)}{p_x(\xi)}\right) d\xi \quad (29)$$

We develop the right term of the equation:

$$\int\limits_{\Omega^*} p_x(\xi)\left(1 - \frac{q_x(\xi)}{p_x(\xi)}\right) d\xi = \int\limits_{\Omega^*} p_x(\xi) d\xi - \int\limits_{\Omega^*} q_x(\xi) d\xi = 0 \quad (30)$$

knowing that $\int\limits_{\Omega^*} p_x(\xi) d\xi = \int\limits_{\Omega^*} q_x(\xi) d\xi = 1$, we can deduce that:

$$-\int\limits_{\Omega^*} p_x(\xi) \log \frac{p_x(\xi)}{q_x(\xi)} d\xi \leq 0 \Rightarrow -\int\limits_{\Omega^*} p_x(\xi) \log p_x(\xi) d\xi \leq -\int\limits_{\Omega^*} p_x(\xi) \log q_x(\xi) d\xi \quad (31)$$

We have then to develop the right term by using previous expression of $q_x(\xi)$:

$$-\int\limits_{\Omega^*} p_x(\xi) \log p_x(\xi) d\xi \leq -\int\limits_{\Omega^*} p_x(\xi)\left[-\langle x, \xi\rangle - \log \int\limits_{\Omega^*} e^{-\langle x, \xi\rangle} d\xi\right] d\xi \quad (32)$$

$$-\int\limits_{\Omega^*} p_x(\xi) \log p_x(\xi) d\xi \leq \left\langle x, \int\limits_{\Omega^*} \xi.p_x(\xi) d\xi \right\rangle + \log \int\limits_{\Omega^*} e^{-\langle x, \xi\rangle} d\xi \quad (33)$$

If we use that $x^* = \int\limits_{\Omega^*} \xi.p_x(\xi) d\xi$ and $\Phi(x) = -\log \int\limits_{\Omega^*} e^{-\langle x, \xi\rangle} d\xi$, then we obtain that the density $q_x(\xi) = e^{-\langle \xi, x\rangle} / \int\limits_{\Omega^*} e^{-\langle \xi, x\rangle} d\xi = e^{-\langle x, \xi\rangle - \log \int\limits_{\Omega^*} e^{-\langle \xi, x\rangle} d\xi}$ is the maximum entropy density constrained by $\int\limits_{\Omega^*} p_x(\xi) d\xi$ and $\int\limits_{\Omega^*} \xi.p_x(\xi) d\xi = x^*$:

$$-\int\limits_{\Omega^*} p_x(\xi) \log p_x(\xi) d\xi \leq \langle x, x^*\rangle - \Phi(x) \quad (34)$$

$$-\int_{\Omega^*} p_x(\xi)\log p_x(\xi)d\xi \leq \Phi^*(x^*) \tag{35}$$

In the following, we will write $x^* = \hat{\xi}$, to give to this variable the link with momentum $\hat{\xi} = \int_{\Omega^*} \xi . p_{\hat{\xi}}(\xi)d\xi$. To express the density with respect to the 1st moment as variable, we have to inverse $\hat{\xi} = \Theta(x) = \frac{d\Phi(x)}{dx}$, by writting $x = \Theta^{-1}(\hat{\xi})$ the inverse function (given by Legendre transform):

$$p_{\hat{\xi}}(\xi) = \frac{e^{-\left\langle \xi, \Theta^{-1}(\hat{\xi})\right\rangle}}{\int_{\Omega^*} e^{-\left\langle \xi, \Theta^{-1}(\hat{\xi})\right\rangle} d\xi} \text{ with } \hat{\xi} = \int_{\Omega^*} \xi . p_{\hat{\xi}}(\xi)d\xi \text{ and } \Phi(x) = -\log \int_{\Omega^*} e^{-\langle x, \xi\rangle} d\xi \tag{36}$$

We find finally the Maximum entropy density parametrized by 1st moment $\hat{\xi}$.

4 Links Between Koszul-Vinberg Characteristic Function, Koszul Forms and Information Geometry

Koszul Hessian Geometry Structure is the key tool to define elementary structures of Information Geometry, that appears as one particular case of more general framework studied by Koszul. In the Koszul-Vinberg Characteristic function $\psi_\Omega(x) = \int_{\Omega^*} e^{-\langle x, \xi\rangle} d\xi,\ \forall x \in \Omega$ where Ω is a sharp convex cone and Ω^* its dual cone, the duality bracket $<.,.>$ has to be defined. I will introduce it by using Cartan-Killing form $\langle x, y\rangle = -B(x, \theta(y))$ with $B(.,.)$ killing form and $\theta(.)$ Cartan involution. The inner product is then invariant with respect to automorphims of cone Ω. Koszul-Vinberg characteristic function could be developed as [124]:

$$\psi_\Omega(x + \lambda u) = \psi_\Omega(x) - \lambda\langle x^* + u\rangle + \frac{\lambda^2}{2}\langle K(x)u, u\rangle + \ldots \tag{37}$$

with $x^* = \frac{d\Phi(x)}{dx}$, $\Phi(x) = -\log \psi_\Omega(x)$ and $K(\mathrm{x}) = \frac{d^2\Phi(x)}{dx^2}$

In the following developments, I will write β, previous variable written x, because in thermodynamics, this variable corresponds to the Planck temperature, classically $\beta = \frac{1}{T}$. The variable β will be the dual variable of $\hat{\xi}$.

$$p_{\hat{\xi}}(\xi) = \frac{e^{-\left\langle \Theta^{-1}(\hat{\xi}),\xi \right\rangle}}{\int\limits_{\Omega^*} e^{-\left\langle \Theta^{-1}(\hat{\xi}),\xi \right\rangle}.d\xi} \quad \hat{\xi} = \Theta(\beta) = \frac{\partial \Phi(\beta)}{\partial \beta} \text{ with } \Phi(\beta) = -\log \psi_\Omega(\beta)$$

$$\psi_\Omega(\beta) = \int\limits_{\Omega^*} e^{-\langle \beta,\xi \rangle} d\xi, \quad S(\hat{\xi}) = -\int\limits_{\Omega^*} p_{\hat{\xi}}(\xi) \log p_{\hat{\xi}}(\xi).d\xi \text{ and } \beta = \Theta^{-1}(\hat{\xi})$$

$$S(\hat{\xi}) = \left\langle \hat{\xi}, \beta \right\rangle - \Phi(\beta) \tag{38}$$

Inversion of the function $\Theta(.)$ is given by $\beta = \Theta^{-1}(\hat{\xi})$ is achieved by Legendre transform using relation between Entropy $S(\hat{\xi})$ and the function $\Phi(\beta)$ (opposite of the logarithm of the Koszul-Vinberg characteristic function):

$$\begin{aligned} &S(\hat{\xi}) = \left\langle \beta, \hat{\xi} \right\rangle - \Phi(\beta) \\ &\text{with} \quad \Phi(\beta) = -\log \int\limits_{\Omega^*} e^{-\langle \xi,\beta \rangle} d\xi \quad \forall \beta \in \Omega \quad \text{and} \quad \forall \xi, \hat{\xi} \in \Omega^* \end{aligned} \tag{39}$$

We will prove that the 2nd Koszul form $-\frac{\partial^2 \Phi(\beta)}{\partial \beta^2}$ is linked with Fisher Metric of Information Geometry:

$$I(\beta) = -E\left[\frac{\partial^2 \log p_\beta(\xi)}{\partial \beta^2}\right] \tag{40}$$

To compute the Fisher metric $I(\beta)$, we use the following relations between variable

$$\begin{aligned} &\begin{cases} \log p_{\hat{\xi}}(\xi) = -\langle \xi, \beta \rangle + \Phi(\beta) \\ S(\hat{\xi}) = -\int\limits_{\Omega^*} p_{\hat{\xi}}(\xi). \log p_{\hat{\xi}}(\xi).d\xi = -E\left[\log p_{\hat{\xi}}(\xi)\right] \end{cases} \\ &\Rightarrow S(\hat{\xi}) = \langle E[\xi], \beta \rangle - \Phi(\beta) = \left\langle \hat{\xi}, \beta \right\rangle - \Phi(\beta) \end{aligned} \tag{41}$$

We can observe that the logarithm of the density is affine with respect to the variable β, and that the Fisher matrix is given by the hessian. We can then deduce that the Fisher Metric is given by the hessian.

$$I(\beta) = -E\left[\frac{\partial^2 \log p_\beta(\xi)}{\partial \beta^2}\right] = -E\left[\frac{\partial^2(-\langle \xi, \beta \rangle + \Phi(\beta))}{\partial \beta^2}\right] = -\frac{\partial^2 \Phi(\beta)}{\partial \beta^2} = \frac{\partial^2 \log \psi_\Omega(\beta)}{\partial \beta^2} \tag{42}$$

We can also identify the Fisher metric as a variance:

$$\log \Psi_\Omega(\beta) = \log \int\limits_{\Omega^*} e^{-\langle \xi,\beta \rangle} d\xi \Rightarrow \frac{\partial \log \Psi_\Omega(\beta)}{\partial \beta} = -\frac{1}{\int\limits_{\Omega^*} e^{-\langle \xi,\beta \rangle} d\xi} \int\limits_{\Omega^*} \xi.e^{-\langle \xi,\beta \rangle} d\xi \tag{43}$$

$$\frac{\partial^2 \log \Psi_\Omega(\beta)}{\partial \beta^2} = -\frac{1}{\left(\int\limits_{\Omega^*} e^{-\langle \xi,\beta \rangle} d\xi\right)^2}\left[-\int\limits_{\Omega^*} \xi^2.e^{-\langle \xi,\beta \rangle} d\xi.\int\limits_{\Omega^*} e^{-\langle \xi,\beta \rangle} d\xi + \left(\int\limits_{\Omega^*} \xi^2.e^{-\langle \xi,\beta \rangle} d\xi\right)^2\right] \quad (44)$$

$$\frac{\partial^2 \log \Psi_\Omega(\beta)}{\partial \beta^2} = \int\limits_{\Omega^*} \xi^2.\frac{e^{-\langle \xi,\beta \rangle}}{\int\limits_{\Omega^*} e^{-\langle \xi,\beta \rangle} d\xi} d\xi - \left(\int\limits_{\Omega^*} \xi.\frac{e^{-\langle \xi,\beta \rangle}}{\int\limits_{\Omega^*} e^{-\langle \xi,\beta \rangle} d\xi} d\xi\right)^2$$

$$= \int\limits_{\Omega^*} \xi^2.p_\beta(\xi) d\xi - \left(\int\limits_{\Omega^*} \xi.p_\beta(\xi) d\xi\right)^2 \quad (45)$$

$$I(\beta) = -E_\xi\left[\frac{\partial^2 \log p_\beta(\xi)}{\partial \beta^2}\right] = \frac{\partial^2 \log \psi_\Omega(\beta)}{\partial \beta^2} = E_\xi[\xi^2] - E_\xi[\xi^2] = Var(\xi) \quad (46)$$

In 1977, Crouzeix [125, 126] has identified the following relation between both hessian of entropy and characteristic function $\frac{\partial^2 \Phi}{\partial \beta^2} = \left[\frac{\partial^2 S}{\partial \hat{\xi}^2}\right]^{-1}$ giving a relation between the dual metrics with respect to their dual coordinate systems. The metric could be given by Fisher metric or given by the hessian of Entropy *S*:

$$ds_g^2 = d\beta^T I(\beta) d\beta = \sum_{ij} g_{ij} d\beta_i d\beta_j \quad \text{with} \quad g_{ij} = [I(\beta)]_{ij} \quad (47)$$

Thanks to Crouzeix relation [125] [126], we observe that 2 geodesic distances given by hessian of dual potential functions in dual coordinates systems, are equal:

$$ds_h^2 = d\hat{\xi}^T \left[\frac{\partial^2 S(\hat{\xi})}{\partial \hat{\xi}^2}\right] d\hat{\xi} = \sum_{ij} h_{ij} d\hat{\xi}_i d\hat{\xi}_j \quad \text{with} \quad h_{ij} = \left[\frac{\partial^2 S(\hat{\xi})}{\partial \hat{\xi}^2}\right]_{ij} \quad (48)$$

$$ds_h^2 = ds_g^2 \quad (49)$$

One can ask oneself the question of what is the most natural product of duality. This question has been treated by Elie Cartan in his thesis in 1894, by introducing a form called Cartan-Killing form, a symmetric bilinear form naturally associated with any Lie algebra. This form of Cartan-Killing is defined via the endomorphism ad_x of Lie algebra g via the Lie bracket:

$$ad_x(y) = [x, y] \quad (50)$$

The trace of the composition of these 2 endomorphisms defines this bilinear form by:

$$B(x, y) = Tr\left(ad_x ad_y\right) \quad (51)$$

The Cartan-Killing form is symmetric:

$$B(x, y) = B(y, x) \quad (52)$$

and verify associativity property:

$$B([x,y],z) = B(x,[y,z]) \tag{53}$$

given by:

$$\begin{aligned} B([x,y],z) &= Tr\left(ad_{[x,y]}ad_z\right) = Tr\left(\left[ad_x, ad_y\right]ad_z\right) \\ &= Tr\left(ad_x\left[ad_y, ad_z\right]\right) = B(x,[y,z]) \end{aligned} \tag{54}$$

Elie Cartan proved that if g is a semi-simple Lie algebra (the form of Killing is non-degenerate) then any symmetric bilinear form is a scalar multiple of the Cartan-Killing form. The Cartan-Killing form is invariant under the action of automorphisms $\sigma \in Aut(g)$ of the algebra g:

$$B(\sigma(x),\sigma(y)) = B(x,y) \tag{55}$$

This invariance is deduced from:

$$\begin{cases} \sigma[x,y] = [\sigma(x),\sigma(y)] \\ z = \sigma(y) \end{cases} \Rightarrow \sigma\left[x,\sigma^{-1}(z)\right] = [\sigma(x),z]$$

$$\text{by writting } \ ad_{\sigma(x)} = \sigma \circ ad_x \circ \sigma^{-1} \tag{56}$$

Then, we can write:

$$B(\sigma(x),\sigma(y)) = Tr\left(ad_{\sigma(x)}ad_{\sigma(y)}\right) = Tr\left(\sigma \circ ad_x ad_y \circ \sigma^{-1}\right) = Tr\left(ad_x ad_y\right) = B(a,y) \tag{57}$$

Cartan has introduced this natural inner product that is invariant by the automorphisms of the Lie algebra, from this Cartan-Killing form:

$$\langle x,y\rangle = -B(x,\theta(y)) \tag{58}$$

with $\theta \in g$ the Cartan involution (an involution on the Lie algebra g is an automorphism θ such that the square is equal to identity).

I summarize all these relations of information geometry from the characteristic function of Koszul-Vinberg, and the duality given via the Cartan-Killing form, as described in the figure below (Fig. 7):

Thanks to the expression of the characteristic function of Koszul-Vinberg and the Cartan-Killing form, one can express the maximum Entropy density in a very general way. For example, by applying these formulas to the cone Ω (self-dual: $\Omega^* = \Omega$) symmetric positive definite matrices $Sym^+(n)$, Cartan-Killing form gives us the product of duality:

$$\langle \eta,\xi\rangle = Tr(\eta^T\xi). \ \ \forall \eta,\xi \in Sym^+(n) = \left\{\xi/\xi^T = \xi, \xi > 0\right\} \tag{59}$$

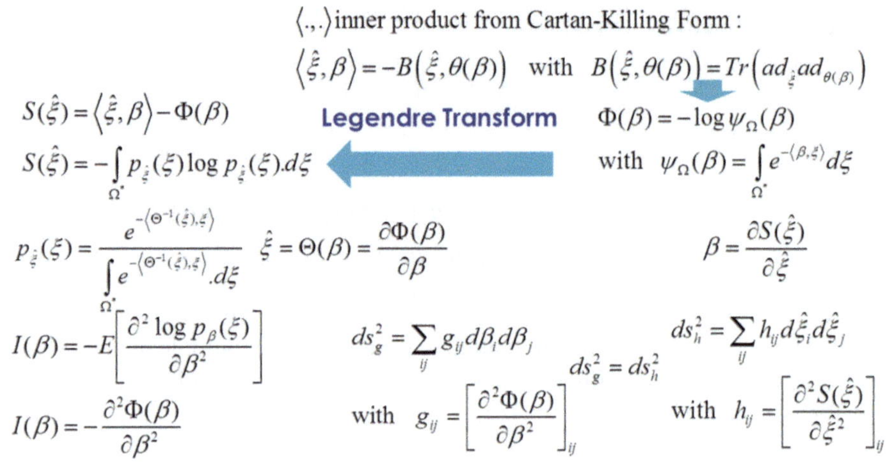

Fig. 7 Relations between cartan-killing form, koszul-vinberg characteristic function, potentials and dual coordinates, and metrics of information geometry

The maximum entropy density is given by:

$$\psi_\Omega(\beta) = \int_{\Omega^*} e^{-\langle \beta, \xi \rangle} d\xi = \det(\beta)^{-\frac{n+1}{2}} \psi_\Omega(I_d)$$

$$\text{and} \quad \hat{\xi} = \frac{\partial \Phi(\beta)}{\partial \beta} = \frac{\partial(-\log \psi_\Omega(\beta))}{\partial \beta} = \frac{n+1}{2} \beta^{-1} \tag{60}$$

From which, I can deduce the final expression:

$$p_{\hat{\xi}}(\xi) = e^{-\left\langle \Theta^{-1}(\hat{\xi}), \xi \right\rangle + \Phi\left(\Theta^{-1}(\hat{\xi})\right)} = \psi_\Omega(I_d).\left[\det\left(\alpha \hat{\xi}^{-1}\right)\right].e^{-Tr\left(\alpha \hat{\xi}^{-1} \xi\right)}$$
$$\text{with} \quad \alpha = \frac{n+1}{2} \tag{61}$$

We can apply this approach for multivariate Gaussian densities. In the case of multivariate Gaussian densities, as noted by Souriau [17, 79], the classical Gibbs expression can be rewritten by modifying the coordinate system and defining a new duality product [80–84, 98]. The multivariate Gaussian density is classically written with the following coordinate system (m, R), with m the mean vector, and R the covariance matrix of the vector z:

$$p_{\hat{\xi}}(\xi) = \frac{1}{(2\pi)^{n/2} \det(R)^{1/2}} e^{-\frac{1}{2}(z-m)^T R^{-1} (z-m)} \text{ with } \begin{cases} m = E(z) \\ R = E\left[(z-m)(z-m)^T\right] \end{cases} \tag{62}$$

By developing the term in the exponential:

$$\begin{aligned}\frac{1}{2}(z-m)^T R^{-1}(z-m) &= \frac{1}{2}\left[z^T R^{-1} z - m^T R^{-1} z - z^T R^{-1} m + m^T R^{-1} m\right] \\ &= \frac{1}{2} z^T R^{-1} z - m^T R^{-1} z + \frac{1}{2} m^T R^{-1} m \end{aligned} \tag{63}$$

I can write this density as a Gibbs density by introducing a new duality bracket between $\left(z, zz^T\right)$ and $\left(-R^{-1}m, \frac{1}{2}R^{-1}\right)$:

$$\begin{aligned} & p_{\hat{\xi}}(\xi) = \frac{1}{(2\pi)^{n/2}\det(R)^{1/2} e^{\frac{1}{2} m^T R^{-1} m}} e^{-\left[-m^T R^{-1} z + \frac{1}{2} z^T R^{-1} z\right]} = \frac{1}{Z} e^{-\langle \xi, \beta \rangle} \\ & \xi = \begin{bmatrix} z \\ zz^T \end{bmatrix} \text{ and } \beta = \begin{bmatrix} -R^{-1} m \\ \frac{1}{2} R^{-1} \end{bmatrix} = \begin{bmatrix} a \\ H \end{bmatrix} \\ & \text{with } \langle \xi, \beta \rangle = a^T z + z^T H z = Tr\left[z a^T + H^T z z^T\right] \end{aligned} \tag{64}$$

We can then write the density in Koszul form:

$$\begin{aligned} & p_{\hat{\xi}}(\xi) = \frac{1}{\int\limits_{\Omega^*} e^{-\langle \xi, \beta \rangle}.d\xi} e^{-\langle \xi, \beta \rangle} = \frac{1}{Z} e^{-\langle \xi, \beta \rangle} \\ & \text{with } \log(Z) = n\log(2\pi) + \frac{1}{2}\log\det(R) + \frac{1}{2} m^T R^{-1} m \\ & \xi = \begin{bmatrix} z \\ zz^T \end{bmatrix}, \hat{\xi} = E[\xi] = \begin{bmatrix} E[z] \\ E\left[zz^T\right] \end{bmatrix} = \begin{bmatrix} m \\ R + mm^T \end{bmatrix}, \beta = \begin{bmatrix} a \\ H \end{bmatrix} = \begin{bmatrix} -R^{-1} m \\ \frac{1}{2} R^{-1} \end{bmatrix} \\ & \text{with } \langle \xi, \beta \rangle = Tr\left[z a^T + H^T z z^T\right] \\ & R = E\left[(z-m)(z-m)^T\right] = E\left[zz^T - mz^T - zm^T + mm^T\right] = E\left[zz^T\right] - mm^T \end{aligned} \tag{65}$$

We are then able to compute the Koszul-Vinberg characteristic function whose opposite of the logarithm provides the potential function:

$$\begin{aligned} & \psi_\Omega(\beta) = \int\limits_{\Omega^*} e^{-\langle \xi, \beta \rangle}.d\xi \\ & \text{and } \Phi(\beta) = -\log\psi_\Omega(\beta) = \frac{1}{2}\left[-Tr\left[H^{-1} a a^T\right] + \log\left[(2)^n \det H\right] - n\log(2\pi)\right] \end{aligned} \tag{66}$$

that verifies the following relation given by Koszul and linked with 1st Koszul form:

$$\frac{\partial\Phi(\beta)}{\partial\beta}=\frac{\partial\left[-\log\psi_\Omega(\beta)\right]}{\partial\beta}=\int_{\Omega^*}\xi\frac{e^{-\langle\xi,\beta\rangle}}{\int_{\Omega^*}e^{-\langle\xi,\beta\rangle}.d\xi}=\int_{\Omega^*}\xi.p_{\hat{\xi}}(\xi).d\xi=\hat{\xi}$$

$$\frac{\partial\Phi(\beta)}{\partial\beta}=\begin{bmatrix}\frac{\partial\Phi(\beta)}{\partial\alpha}\\ \frac{\partial\Phi(\beta)}{\partial H}\end{bmatrix}=\begin{bmatrix}m\\ R+mm^T\end{bmatrix}=\hat{\xi} \tag{67}$$

The 2nd dual potential is given by the Legendre transform of $\Phi(\beta)$:

$$S(\hat{\xi})=\left\langle\hat{\xi},\beta\right\rangle-\Phi(\beta)\quad\text{with}\quad\frac{\partial\Phi(\beta)}{\partial\beta}=\hat{\xi}\quad\text{and}\quad\frac{\partial S(\hat{\xi})}{\partial\hat{\xi}}=\beta$$

$$S\left(\hat{\xi}\right)=-\int_{\Omega^*}\frac{e^{-\langle\xi,\beta\rangle}}{\int_{\Omega^*}e^{-\langle\xi,\beta\rangle}.d\xi}\log\frac{e^{-\langle\xi,\beta\rangle}}{\int_{\Omega^*}e^{-\langle\xi,\beta\rangle}.d\xi}.d\xi=-\int_{\Omega^*}p_{\hat{\xi}}(\xi)\log p_{\hat{\xi}}(\xi).d\xi \tag{68}$$

that is explicitly identified with the classical Shannon Entropy:

$$\begin{aligned}S(\hat{\xi})&=-\int_{\Omega^*}p_{\hat{\xi}}(\xi)\log p_{\hat{\xi}}(\xi).d\xi\\&=\frac{1}{2}\left[\log(2)^n\det\left[H^{-1}\right]+n\log(2\pi.e)\right]=\frac{1}{2}\left[\log\det[R]+n\log(2\pi.e)\right]\end{aligned} \tag{69}$$

The Fisher metric of Information Geometry is given by the hessian of the opposite of the logarithm of the Koszul-Vinberg characteristic function:

$$\begin{aligned}&ds_g^2=d\beta^T I(\beta)d\beta=\sum_{ij}g_{ij}d\beta_i d\beta_j\\&\text{with}\quad g_{ij}=[I(\beta)]_{ij}\text{ and }I(\beta)=-E_\xi\left[\frac{\partial^2\log p_\beta(\xi)}{\partial\beta^2}\right]=\frac{\partial^2\log\psi_\Omega(\beta)}{\partial\beta^2}\end{aligned} \tag{70}$$

Then, for the multivariate Gaussian density, we have the following Fisher metric:

$$ds^2=\sum_{ij}g_{ij}d\theta_i d\theta_j=dm^T R^{-1}dm+\frac{1}{2}Tr\left[\left(R^{-1}dR\right)^2\right] \tag{71}$$

Geodesic equations are given by Euler-Lagrange equations:

$$\begin{aligned}&\sum_{i=1}^{n}g_{ik}\ddot{\theta}_i+\sum_{i,j=1}^{n}\Gamma_{ijk}\dot{\theta}_i\dot{\theta}_j=0,\ k=1,\ldots,n\\&\text{with}\quad\Gamma_{ijk}=\frac{1}{2}\left[\frac{\partial g_{jk}}{\partial\theta_i}+\frac{\partial g_{jk}}{\partial\theta_j}+\frac{\partial g_{ij}}{\partial\theta_k}\right]\end{aligned} \tag{72}$$

that can be reduced to the equations:

$$\begin{cases} \ddot{R} + \dot{m}\dot{m}^T - \dot{R}R^{-1}\dot{R} = 0 \\ \ddot{m} - \dot{R}R^{-1}\dot{m} = 0 \end{cases} \tag{73}$$

I use a result of Souriau [17] that the component of « moment map » are constants (geometrization of Emmy Noether theorem), to identify the following constants [83]:

$$\frac{d\Pi_R}{dt} = \begin{bmatrix} \frac{d(R^{-1}\dot{R}+R^{-1}\dot{m}m^T)}{dt} & \frac{d(R^{-1}\dot{m})}{dt} \\ 0 & 0 \end{bmatrix} = 0$$
$$\Rightarrow \begin{cases} R^{-1}\dot{R} + R^{-1}\dot{m}m^T = B = cste \\ R^{-1}\dot{m} = b = cste \end{cases} \tag{74}$$

with Π_R the moment map introduced by Souriau [17]. This moment map could be computed if we consider the following Lie group acting in case of Gaussian densities:

$$\begin{bmatrix} Y \\ 1 \end{bmatrix} = \begin{bmatrix} R^{1/2} & m \\ 0 & 1 \end{bmatrix}\begin{bmatrix} X \\ 1 \end{bmatrix} = \begin{bmatrix} R^{1/2}X + m \\ 1 \end{bmatrix}, \begin{cases} (m, R) \in R^n \times Sym^+(n) \\ M = \begin{bmatrix} R^{1/2} & m \\ 0 & 1 \end{bmatrix} \in G_{aff} \end{cases}$$
$$X \approx \aleph(0, I) \to Y \approx \aleph(m, R) \tag{75}$$

$R^{1/2}$, square root of R, is given by Cholesky decomposition of R. $R^{1/2}$ is the Lie group of triangular matrix with positive elements on the diagonal. Euler-Poincaré equations, reduced equations from Euler-Lagrange equations, are then given by:

$$\begin{cases} \dot{m} = Rb \\ \dot{R} = R(B - bm^T) \end{cases} \tag{76}$$

Geodesic distance between multivariate Gaussian density is then obtained by "*geodesic shooting*" method that will provide iteratively the final solution from the tangent vector at the initial point:

$$\left(R^{-1}(0)\dot{m}(0), R^{-1}(0)\left(\dot{R}(0) + \dot{m}(0)m(0)^T\right)\right) = (b, B) \in R^n \times Sym^+(n) \tag{77}$$

From which, we then deduce the distance:

$$d = \sqrt{\dot{m}(0)^T R^{-1}(0)\dot{m}(0) + \frac{1}{2}Tr\left[\left(R^{-1}(0)\dot{R}(0)\right)^2\right]} \tag{78}$$

Geodesic shooting is obtained by using equations established by Eriksen [127, 128] for "*exponential map*" using the following change of variables:

$$\begin{cases} \Delta(t) = R^{-1}(t) \\ \delta(t) = R^{-1}(t)m(t) \end{cases} \Rightarrow \begin{cases} \dot{\Delta} = -B\Delta + bm^T \\ \dot{\delta} = -B\delta + \left(1 + \delta^T \Delta^{-1}\delta\right)b \\ \Delta(0) = I_p, \delta(0) = 0 \end{cases} \text{ with } \begin{cases} \dot{\Delta}(0) = -B \\ \dot{\delta}(0) = b \end{cases} \tag{79}$$

The method based on geodesic shooting consists in iteratively approaching the solution by geodesic shooting in direction $\left(\dot{\delta}(0), \dot{\Delta}(0)\right)$, using the following exponential map (Fig. 8):

$$\Lambda(t) = \exp(tA) = \sum_{n=0}^{\infty} \frac{(tA)^n}{n!} = \begin{pmatrix} \Delta & \delta & \Phi \\ \delta^T & \varepsilon & \gamma^T \\ \Phi^T & \gamma & \Gamma \end{pmatrix}$$
$$\text{with } A = \begin{pmatrix} -B & b & 0 \\ b^T & 0 & -b^T \\ 0 & -b & B \end{pmatrix} \tag{80}$$

The principle of geodesic shooting is the following. We consider one geodesic χ between θ_0 and θ_1 with an initial tangent vector V from the origin, and assume that V is modified by W, with respect to $V + W$. Variation of final point θ_1 could be obtained by Jacobi vector field $J(0) = 0$ and $\dot{J}(0) = W$:

$$J(t) = \frac{d}{d\alpha} \exp_{\theta_0}(t(V + \alpha W))|_{\alpha=0} \tag{81}$$

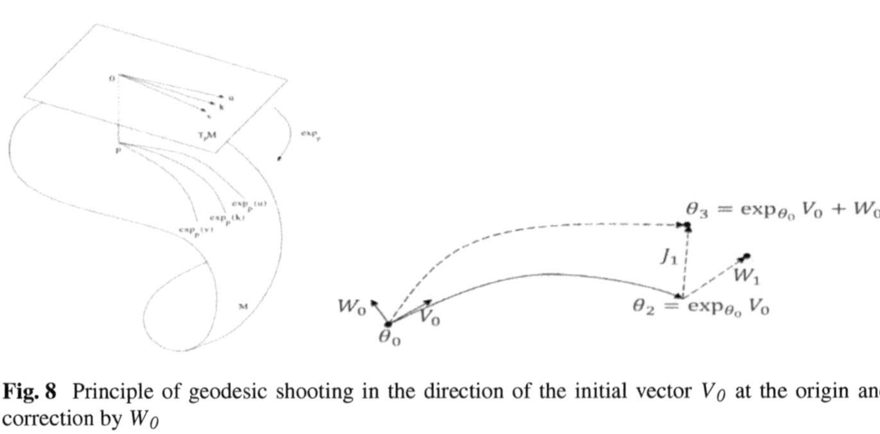

Fig. 8 Principle of geodesic shooting in the direction of the initial vector V_O at the origin and correction by W_O

5 Koszul's Study of Homogeneous Bounded Domains and Affine Representations of Lie Groups and Lie Algebras

Jean-Louis Koszul [15, 16, 41, 45–47, 49, 50] and his student Jacques Vey [39, 40] introduced new theorems with more general extension than previous results:

Koszul theorem [50]: Let Ω be a sharp convex open in an affine space of E of finite dimension on R. If a unimodular Lie group of affine transformations operates transitively on Ω, Ω is a cone.

Koszul-Vey Theorem [40]: Let M a hessian connected manifold associated with the hessian metric g. Assume that M has a closed 1-form α such that $D\alpha = g$ and that there is a group G of affine automorphisms of M preserving α, then:

- If M/G is almost compact, then the manifold, universal covering of M, is affinely isomorphic to a convex domain of an affine space containing no straight line.
- If M/G is compact, then Ω is a sharp convex cone.

Jean-Louis Koszul developed his theory, studying the homogeneous domains, in particular the homogeneous symmetric bounded domains of Siegel, which we note *DS* [44, 48]. He has proved that there is a subgroup G in the group of complex affine automorphisms of these domains (Iwasawa subgroup), so that G acts on *DS* in a merely transitive way. The Lie algebra g of G has a structure which is an algebraic translation of the Kähler structure *DS*.

Koszul considered on *G/B* an invariant complex structure tensor I. All the invariant volumes on *G/B*, equal up to a constant factor, define with the complex structure the same invariant Hermitian form on *G/B*, called Hermitian canonical form, denoted h. Let E be a differentiable fiber space of base M and let p be the projection of E on M, such that $p^*((pX).f) = X.(p^*f)$. The projection $p : E \to M$ defines an injective homomorphism p^* of the space of differential forms of M in the space of the differential forms of E such that for any form α of degree n on M and any sequence of n projectable vectors fields, we have $p^*(\alpha(pX_1, pX_2, \ldots, pX_n)) = (p^*\alpha)(X_1, X_2, \ldots, X_n)$. Let I be the tensor of an almost complex structure on the basis M, there exists on E a tensor J of type *(1,1)* and only one which possesses the following properties $p(JX) = I(pX)$ and $J^2X = -X \mod h, X \in g$ for any vector field X on E. Let G be a connected Lie group and B a closed subgroup of G, we note g the Lie algebra left invariant vector fields on G and b sub-algebra of g corresponding to B. The canonical mapping of G on *G/B* is denoted p (defining E as before). We assume that there exists on *G/B* an invariant volume by G, which consist in assuming that, for all $s \in B$, the automorphism $X \to Xs$ of g defines by passing to the quotient an automorphism of determinant *1* in *g/b*. Let r be the dimension of *G/B* and $(X_i)_{1 \le i \le m}$ a base of g such that $X_i \in b$, for $r \le i \le m$. Let $(\xi_i)_{1 \le i \le m}$ the base of the space of differential forms of degree *1* left invariant on G such that $\xi_i(X_j) = \delta_{ij}$. If ω is an invariant volume on *G/B*, then $\Omega = p^*\omega$ is equal, up to a constant factor, to $\xi_1 \wedge \xi_2 \wedge \ldots \wedge \xi_r$. We will assume the base (X_j) chosen so that this factor is equal to *1*, let $\Omega = \xi_1 \wedge \xi_2 \wedge \ldots \wedge \xi_r$. For any vector field that can be projected *X* on *G*, we have:

$$p^*(div(pX))\Omega = p^*((div(pX))\omega) = p^*((pX)\omega) = X\Omega = \sum_{j=1}^{r} \xi_j([X_j, X])\Omega \tag{82}$$

$$p^*(div(pX)) = \sum_{j=1}^{r} \xi_j([X_j, X]) \tag{83}$$

These elements being defined, Koszul calculates the Hermitian canonical form of *G/B*, denoted *h*, more particularly $\eta = p^*h$ on *G*. Let *X* and *Y* both right invariant vector fields on *G*. They are projectable and the fields pX and pY are conformal vector fields on *G/B* such that $div(pX) = div(pY) = 0$, because the volume and the complex structure of *G/B* are invariant under *G*. As a result, if κ is the Kähler form of *h* and if $\alpha = p^*\kappa$, then:

$$4\alpha(X, Y) = 4p^*(\kappa(pX, pY)) = p^*div(I[pX, pY]) \tag{84}$$

and as $p(J[X, Y]) = I[pX, pY]$, we obtain:

$$4\alpha(X, Y) = p^*div(J[X, Y]) = \sum_{i=1}^{2n} \xi_i([X_i, J[X, Y]]) \tag{85}$$

X and *Y* are two left invariant vectors fields on *G*. *X '* and *Y'* right invariant vectors fields coinciding with *X* and *Y* at the point *e*, neutral element of *G*. If $T = \left[X', Y'\right]$ is tight invariant vectors fields which coincide with $-[X, Y]$ on *e*, then:

$$[X, JT] = J[X, [X, Y]] - [X, J[X, Y]] \text{ at point e} \tag{86}$$

At point *e*, we have the equality:

$$4\alpha(X, Y) = \sum_{i=1}^{2n} \xi_i([J[X, Y], X_i] - J[[X, Y], X_i]) \tag{87}$$

As the form α is invariant on the left by *G*, this equality is verified for all points. For any endomorphism Θ of the space *g* such that $\Theta b \subset b$, we denote by $Tr_b\Theta$ the trace of the restriction of Θ to *b* and by $Tr_{g/b}\Theta$ the trace of the endomorphism of *g/b* deduced from Θ by passage to the quotient, with $Tr\Theta = Tr_b\Theta + Tr_{g/b}\Theta$. We have:

$$T_{r_{g/b}\Theta} = \sum_{i=1}^{2n} \xi_i(\Theta X_i) \tag{88}$$

Whatever $X \in g$ and $s \in B$, we have $J(Xs) - (JX)s \in b$. If $ad(Y)$ is the endomorphism of g defined by $ad(Y).Z = [Y, Z]$, we have $(Jad(Y) - ad(Y)J)g \subset b$ for all $Y \in b$. We can deduce, for all $X \in g$, the endomorphism $ad(JX) - Jad(X)$ leaves steady the subspace b. Koszul defines a linear form Ψ on the space g by defining:

$$\Psi(X) = Tr_{g/b}(ad(JX) - Jad(X)) \ , \forall X \in g \tag{89}$$

Koszul has finally obtained the following fundamental theorem:

Theorem of Koszul [15]:

The Kähler form of the Hermitian canonical form has for image by p^* the differential of the form $-\frac{1}{4}\Psi(X) = -\frac{1}{4}Tr_{g/b}(ad(JX) - Jad(X)), \forall X \in g$

Koszul note that the form Ψ is independent of the choice of the tensor J. It is determined by the invariant complex structure of *G/B*. The form Ψ is right invariant by B. For all $s \in B$, note the endomorphism $r(s) : X \to Xs$ of g. Since $J(Xs) = (JX)s \mod b$ and that $Tr_{g/b}ad(Y) = 0$, we have:

$$\Psi(Xs) = Tr_{g/b}(ad((JX)s) - Jad(Xs)), \quad \forall X \in g \,, \forall Y \in b \tag{90}$$

$$\Psi(Xs) = Tr_{g/b}(r(s)ad(JX)r(s)^{-1} - Jr(s)ad(X)r(s)^{-1}) \tag{91}$$

$$\Psi(Xs) = \Psi(X) + Tr_{g/b}((J - r(s)^{-1}Jr(s))ad(X)), \quad \forall X \in g, s \in B \tag{92}$$

As $\left(J - r(s)^{-1}Jr(s)\right)$ maps g in b, we get $\Psi(Xs) = \Psi(X)$. The form Ψ is not zero on b. This is not the image by p^* of a differential form of *G/B*. However, the right invariance of Ψ on B is translated, infinitesimally by the relation:

$$\Psi([b, g]) = (0) \tag{93}$$

Koszul proved that the canonical hermitian form h of a homogeneous Kähler manifold *G/B* has the following expression:

$$\begin{aligned} &\eta(X, Y) = \frac{1}{2}\Psi([JX, Y]) \\ &\text{with} \begin{cases} \Psi([X, Y]) = \Psi([JX, JY]) \\ \eta([JX, JY]) = \eta(X, Y) \end{cases} \forall X, Y \in g \end{aligned} \tag{94}$$

To do, the link with the first chapters, I can summarize the main result of Koszul that there is an integrable structure almost complex J on g, and for $l \in g^*$ defined by a positive J -invariant inner product on g:

$$\langle X, Y \rangle_l = \langle [JX, Y], l \rangle \tag{95}$$

Koszul has proposed as admissible form, $l \in g^*$, the form ξ:

$$\Psi(X) = \langle X, \xi \rangle = T_r[ad(JX) - J.ad(X)] \; \forall X \in g \tag{96}$$

Koszul proved that $\langle X, Y \rangle_\xi$ coincides, up to a positive multiplicative constant; with the real part of the Hermitian inner product obtained by the Bergman metric of symmetric homogeneous bounded domains DS by identifying g with the tangent space of DS. $\Psi(X)$ is the restriction to g of a differential form Ψ of degree 1, with left invariance on G. This form is fully defined by the invariant complex structure of G/B. This form is invariant to the choice of J. This form is invariant on the right by B. We have $\Psi([X, Y]) = 0$ with $X \in g$, $Y \in b$. The exterior differential $d\Psi$ of Ψ is the inverse image by the projection $G \to G/B$ of degree 2 form Ω. This form Ω is, up to a constant, the Kähler form h, defined by the canonical Hermitian form of G/B: $h(\pi.X, \pi.Y) = \frac{1}{2}(d\Psi)(X, J.Y), \forall X, Y \in G$ as it is proved in Bourbaki seminar by Koszul in [129].

The 1st Koszul form is then given by:

$$\alpha = -\frac{1}{4} d\Psi(X) \tag{97}$$

We can illustrate this structure for the simplest example of DS, the Poincaré upper half-plane $V = \{z = x + iy / y > 0\}$ which is isomorphic to the open $zz^* < 1$, which is a bounded domain. The group G of transformations $z \to az + b$ with a and b real values with $a > 0$ is simply transitive in V. We identify G and V by the application passing from $s \in G$ an element to the image $i = \sqrt{-1}$ by s.

Let's define vector fields $X = y\frac{d}{dx}$ and $Y = y\frac{d}{dy}$ which generate the vector space of left invariant vectors fields on G, and J an almost complex structure on V defined by $JX = Y$. As $[X, Y] = -Y$ and $ad(Y).Z = [Y, Z]$ then:

$$\begin{cases} Tr[ad(JX) - Jad(X)] = 2 \\ Tr[ad(JY) - Jad(Y)] = 0 \end{cases} \tag{98}$$

The Koszul forms and the Koszul metric are respectively given by:

$$\Psi(X) = 2\frac{dx}{y} \Rightarrow \alpha = -\frac{1}{4} d\Psi = -\frac{1}{2}\frac{dx \wedge dy}{y^2} \Rightarrow ds^2 = \frac{dx^2 + dy^2}{2y^2} \tag{99}$$

I note that $\alpha = -\frac{1}{4} d\Psi(X)$ is indeed the Kähler form of Poincaré's metric, which is invariant by the automorphisms of the upper half-plane.

The following example concerns $V = \{Z = X + iY / X, Y \in Sym(p), Y > 0\}$ the upper half-space of Siegel (which is the most natural extension of the Poincaré half-plane) with:

$$\begin{cases} SZ = (AZ + B)D^{-1} \\ A^T D = I,\; B^T D = D^T B \end{cases} \quad \text{with } S = \begin{pmatrix} A & B \\ 0 & D \end{pmatrix} \text{ and } J = \begin{pmatrix} 0 & I \\ -I & 0 \end{pmatrix} \tag{100}$$

We can then compute Koszul forms and the metric:

$$\Psi(dX + idY) = \frac{3p+1}{2} Tr(Y^{-1}dX)$$
$$\Rightarrow \begin{cases} \alpha = -\frac{1}{4}d\Psi = \frac{3p+1}{8} Tr\left(Y^{-1}dZ \wedge Y^{-1}d\bar{Z}\right) \\ ds^2 = \frac{(3p+1)}{8} Tr\left(Y^{-1}dZY^{-1}d\bar{Z}\right) \end{cases} \quad (101)$$

We recover Carl-Ludwig Siegel metric for the upper half space.

More recent development on Kähler manifolds are described in [130] et [131].

Koszul studied symmetric homogeneous spaces and defines the relation between invariant flat affine connections and the affine representations of Lie algebras and invariant Hessian metrics characterized by affine representations of Lie algebras. Koszul provides a correspondence between symmetric homogeneous spaces with invariant Hessian structures using affine representations of Lie algebras, and proves that a symmetric homogeneous space simply connected with an invariant Hessian structure is a direct product of a Euclidean space and of a homogeneous dual-cone. Let G be a connected Lie group and G/K a homogeneous space over which G acts effectively. Koszul gives a bijective correspondence between all planar G-invariantes connections on G/K and all of a certain class of affine representations of the Lie algebra of G. The main theorem of Koszul is:

Koszul's theorem: Let G/K be a homogeneous space of a connected Lie group G and be $\mathfrak{g}$ and $\mathfrak{k}$ the Lie algebras of G and K, assuming that G/K has G-invariant connection, then admits an affine representation (f, q) on the vector space E. Conversely, assume that G is simply connected and has an affine representation, then G/K admits a flat G-invariant connection.

In the foregoing, the basic tool studied by Koszul is the affine representation of Lie algebra and Lie group. To study these structures, Koszul introduced the following developments.

Let Ω a convex domain on R^n without any straight lines, and an associated convex cone $V(\Omega) = \{(\lambda x, x) \in R^n \times R / x \in \Omega, \lambda \in R^+\}$, then there exist an affine embedding:

$$\ell : x \in \Omega \mapsto \begin{bmatrix} x \\ 1 \end{bmatrix} \in V(\Omega) \quad (102)$$

If we consider η the group of homomorphism of $A(n, R)$ in $GL(n+1, R)$ given by:

$$s \in A(n, R) \mapsto \begin{bmatrix} \mathsf{f}(s) & \mathsf{q}(s) \\ 0 & 1 \end{bmatrix} \in GL(n+1, R) \quad (103)$$

and the affine representation of Lie algebra:

$$\begin{bmatrix} f & q \\ 0 & 0 \end{bmatrix} \quad (104)$$

with $A(n, R)$ the group of all affine representations of R^n. We have $\eta(G(\Omega)) \subset G(V(\Omega))$ and the pair (η, ℓ) of homomorphism $\eta : (G(\Omega) \rightarrow G(V(\Omega))$ and the application $\ell : \Omega \rightarrow V(\Omega)$ is equivariant.

If we observe Koszul affine representations of Lie algebra and Lie group, we have to consider G a convex Lie group and E a real or complex vector space of finite size, Koszul has introduced an affine representation of G in E such that:

$$\begin{aligned} & E \rightarrow E \\ & a \mapsto sa \, \forall s \in G \end{aligned} \tag{105}$$

is an affine representation. We set $A(E)$ the set of all affine transformation of a real vector space E, a Lie group called affine representation group of E. The set $GL(E)$ of all regular linear representation of E, a sub-group of $A(E)$.

We define a linear representation of G in $GL(E)$:

$$\begin{aligned} & \mathsf{f} : G \rightarrow GL(E) \\ & s \mapsto \mathsf{f}(s)a = sa - so \quad \forall a \in E \end{aligned} \tag{106}$$

and a map from G to E:

$$\begin{aligned} & \mathsf{q} : G \rightarrow E \\ & s \mapsto \mathsf{q}(s) so \quad \forall s \in G \end{aligned} \tag{107}$$

then, we have $\forall s, t \in G$:

$$\mathsf{f}(s)\mathsf{q}(t) + \mathsf{q}(s) = \mathsf{q}(st) \tag{108}$$

deduced from $\mathsf{f}(s)\mathsf{q}(t) + \mathsf{q}(s) = s\mathsf{q}(t) - s\mathsf{q} + so = s\mathsf{q}(t) = sto = \mathsf{q}(st)$.

Inversely, if a map q from G to E and a linear representation f from G to $GL(E)$ verifying previous equation, then we can define an affine representation from G in E, written by (f, q):

$$\mathrm{A}ff(s) : a \mapsto sa = \mathsf{f}(s)a + \mathsf{q}(s) \, \forall s \in G, \forall a \in E \tag{109}$$

The condition $\mathsf{f}(s)\mathsf{q}(t) + \mathsf{q}(s) = \mathsf{q}(st)$ is equal to the request that the following mapping is an homomorphism:

$$\mathrm{A}ff : s \in G \mapsto \mathrm{A}ff(s) \in A(E) \tag{110}$$

We write f the affine representation of Lie algebra g of G, defined by f and q the restriction to g to the differential of q (f and q differential of f and q respectively), Koszul proved the following equation:

$$\begin{aligned} & f(X)q(Y) - f(Y)q(X) = q([X, Y]) \forall X, Y \in \mathsf{g} \\ & \text{with} \quad f : \mathsf{g} \rightarrow gl(E) \quad \text{and}\, q : \mathsf{g} \mapsto E \end{aligned} \tag{111}$$

where $gl(E)$ the set of all linear endomorphisms of E, Lie algebra of $GL(E)$.

We use the assumption that:

$$q(Ad_s Y) = \left.\frac{d\mathsf{q}(s.e^{tY}.s^{-1})}{dt}\right|_{t=0} = \mathsf{f}(s)f(Y)\mathsf{q}(s^{-1}) + \mathsf{f}(s)q(Y) \tag{112}$$

We then obtain:

$$q([X,Y]) = \left.\frac{d\mathsf{q}(Ad_{e^{tX}}Y)}{dt}\right|_{t=0} = f(X)q(Y)\mathsf{q}(e) + \mathsf{f}(e)f(Y)(-q(X)) + f(X)q(Y) \tag{113}$$

where e is neutral element of G. Since $\mathsf{f}(e)$ is identity map and $\mathsf{q}(e) = 0$, we have the equality:

$$f(X)q(Y) - f(Y)q(X) = q([X,Y]) \tag{114}$$

A pair (f,q) of linear representation of f of a Lie algebra $\mathfrak{g}$ on E and a linear map q from $\mathfrak{g}$ in E is an affine representation of $\mathfrak{g}$ in E, if it satisfy:

$$f(X)q(Y) - f(Y)q(X) = q([X,Y]) \tag{115}$$

Inversely, if we assume that $\mathfrak{g}$ has an affine representation (f,q) on E, by using the coordinate systems $\{x^1,\ldots,x^n\}$ on E, we can express the affine map $v \mapsto f(X)v + q(Y)$ by a matrix representation of size $(n+1)\times(n+1)$:

$$aff(X) = \begin{bmatrix} f(X) & q(X) \\ 0 & 0 \end{bmatrix} \tag{116}$$

where $f(X)$ is a matrix of size $n \times n$ and $q(X)$ a vector of size n.

$X \mapsto aff(X)$ is an injective homomorphism of Lie algebra $\mathfrak{g}$ in Lie algebra of matrices $(n+1)\times(n+1)$, $gl(n+1,R)$:

$$\left| \begin{array}{l} \mathfrak{g} \to gl(n+1,R) \\ X \mapsto aff(X) \end{array} \right. \tag{117}$$

If we note $\mathfrak{g}_{aff} = aff(\mathfrak{g})$, we write G_{aff} linear Lie sub-group of $GL(n+1,R)$ generated by $\mathfrak{g}_{aff}$. One element of $s \in G_{aff}$ could be expressed by:

$$Aff(s) = \begin{bmatrix} \mathsf{f}(s) & \mathsf{q}(s) \\ 0 & 1 \end{bmatrix} \tag{118}$$

Let M_{aff} the orbit of G_{aff} from the origin o, then $M_{aff} = \mathsf{q}(G_{aff}) = G_{aff}/K_{aff}$ where $K_{aff} = \{s \in G_{aff}/\mathsf{q}(s) = 0\} = Ker(\mathsf{q})$.

We can give as example the following case. Let Ω a convex domain in R^n without any straight line, we define the cone $V(\Omega)$ in $R^{n+1} = R^n \times R$ by $V(\Omega) = \{(\lambda x, x) \in R^n \times R / x \in \Omega, \lambda \in R^+\}$. Then, there is an affine embedding:

$$\ell : x \in \Omega \mapsto \begin{bmatrix} x \\ 1 \end{bmatrix} \in V(\Omega) \tag{119}$$

If we consider η the group of homomorphisms of $A(n, R)$ in $GL(n+1, R)$ given by:

$$s \in A(n, R) \mapsto \begin{bmatrix} \mathrm{f}(s) & \mathrm{q}(s) \\ 0 & 1 \end{bmatrix} \in GL(n+1, R) \tag{120}$$

with $A(n, R)$ the group of all affine transformations in R^n. We have $\eta(G(\Omega)) \subset G(V(\Omega))$ and the pair (η, ℓ) of homomorphism $\eta : G(\Omega) \to G(V(\Omega))$ and the map $\ell : \Omega \to V(\Omega)$ are equivariant:

$$\ell \circ s = \eta(s) \circ \ell \ \textbf{and} \ d\ell \circ s = \eta(s) \circ d\ell \tag{121}$$

6 Koszul Lecture on Geometric and Analytics Mechanics, Related to Geometric Theory of Heat (Souriau's Lie Group Thermodynamics) and Theory of Information (Information Geometry)

Before that Professor Koszul passed away in January 2018, he gave his agreement to his book "Introduction to Symplectic Geometry" translation from Chinese to English by SPRINGER [18]. This Koszul's book translation genesis dates back to 2013. We had contacted Professor Jean-Louis Koszul, to deeper understand his work in the field of homogeneous bounded domains within the framework of Information Geometry. Professor Michel Boyom succeeded to convince Jean-Louis Koszul to answer positively to our invitation to attend the 1st GSI "*Geometric Science of Information*" conference in August 2013 at Ecole des Mines ParisTech in Paris, and more especialy to attend the talk of Hirohiko Shima, given for his honor on the topic "*Geometry of Hessian Structures*" (Fig. 9).

I was more particularly interested by Koszul's work developed in the paper «***Domaines bornées homogènes et orbites de groupes de transformations affines***» [45] of 1961, written by Koszul at the Institute for Advanced Studies at Princeton during a stay funded by the National Science Foundation. Koszul proved in this paper that on a complex homogeneous space, an invariant volume defines with the complex structure the canonical invariant Hermitian form introduced in [15]. It is in this article that Koszul uses the ***affine representation of Lie groups and Lie algebras***.

Fig. 9 Jean-Louis Koszul and Hirihiko Shima at GSI'13 "*Geometric Science of Information*" conference in Ecole des Mines ParisTech in Paris, October 2013

The use by Koszul of the affine representation of Lie groups and Lie algebras drew our attention, especially on the links of his approach with the similar one used by Jean-Marie Souriau in geometric mechanics in the framework of homogeneous symplectic manifolds. I have then looked for links between Koszul and Souriau works. I finally discovered, that in 1986, Koszul published this book "Introduction to symplectic geometry" following a Chinese course in China. I also observed that this book takes up and develops works of Jean-Marie Souriau on homogeneous symplectic manifolds and ***the affine representation of Lie algebras and Lie groups in geometric mechanics***.

I have then exchanged e-mails with Professor Koszul on Souriau works and on genesis of this Book. In May 2015, questioning Koszul on Souriau work on Geometric Mechanics and on Lie Group Thermodynamics, Koszul answered me "[*A l'époque où Souriau développait sa théorie, l'establishment avait tendance à ne pas y voir des avancées importantes. Je l'ai entendu exposer ses idées sur la thermodynamique mais je n'ai pas du tout réalisé à l'époque que la géométrie hessienne était en jeu.*] *At the time when Souriau was developing his theory, the establishment tended not to see significant progress. I heard him explaining his ideas on thermodynamics but I did not realize at the time that Hessian geometry was at stake*". In September 2016, I asked him the origins of Lie Group and Lie Algebra Affine representation. Koszul informed me that he attended Elie Cartan Lecture, where he presented seminal work on this topic: "*[Il y a là bien des choses que je voudrais comprendre (trop peut-être !), ne serait-ce que la relation entre ce que j'ai fait et les travaux de Souriau. Détecter l'origine d'une notion ou la première apparition d'un résultat est souvent difficile. Je ne suis certainement pas le premier à avoir utilisé des représentations affines de groupes ou d'algèbres de Lie. On peut effectivement imaginer que cela se trouve chez Elie Cartan, mais je ne puis rien dire de précis. A propos d'Elie Cartan: je n'ai pas été son élève. C'est Henri Cartan qui a été mon maître pendant mes années de thèse. En 1941 ou 42 j'ai entendu une brève série de conférences données par Elie à l'Ecole Normale et ce sont des travaux d'Elie qui ont été le point de départ de mon travail de thèse.] There are many things that I would like to understand (too much perhaps!), If only the relationship between what I did and the work of Souriau. Detecting the origin of a notion or the first appearance of a result is often difficult. I am certainly not the first to have used affine representations of Lie groups or Lie algebras. We can imagine that it is at Elie Cartan, but I cannot say anything specific. About Elie Cartan: I was not his student. It was Henri Cartan who was my master during my years of thesis. In 1941 or 42, I heard a brief series of lectures given by Elie at the Ecole Normale and it was Elie's work that was the starting point of my thesis work*".

After discovering the existence of this Koszul's book, written in Chinese based on a course given at Nankin, on "Introduction to Symplectic Geometry", where he made reference to Souriau's book and developed his main tools, I started to discuss its content. In January 2017, Koszul wrote me, with the usual humility "*[Ce petit fascicule d'introduction à la géométrie symplectique a été rédigé par un assistant de Nankin qui avait suivi mon cours. Il n'y a pas eu de version initiale en français.] This small introductory booklet on symplectic geometry was written by a Nanjing assistant who had taken my course. There was no initial version in French* ". I asked him if he had personal archive of this course, he answered "*[Je n'ai pas conservé de notes préparatoires à ce cours. Dites-moi à quelle adresse je puis vous envoyer un exemplaire du texte chinois.] I have not kept any preparatory notes for this course. Tell me where I can send you a copy of the Chinese text.* ". Professor Koszul then sent me his last copy of this book in Chinese, a small green book (Fig. 10).

I was not able to read the Chinese text, but I have observed in Chap. 4 "*Symplectic G-spaces*" and in Chap. 5 "*Poisson Manifolds*", that their equations content new original developments of Souriau work on moment map and affine representation of Lie Group and Lie Algebra. More especially, Koszul considered equivariance of moment map, where I recover Souriau theorem. Koszul shows that when $(M; \omega)$ is a connected Hamiltonian G-space and μ a moment map of the action of G, there exists an affine action of G on g^* (dual Lie algebra), whose linear part is the coadjoint action, for which the moment μ is equivariant. Koszul developed Souriau idea that this affine action is obtained by modifying the coadjoint action by means of a closed cochain (called cocycle by Souriau), and that $(M; \omega)$ is a G-Poisson space making reference to Souriau's book for more details.

About collaboration between Koszul and Souriau and another potential Lecture on Symplectic Geometry in Toulouse, Koszul informed me in February 2017 that: "*[J'ai plus d'une fois rencontré Souriau lors de colloques, mais nous n'avons jamais*

Fig. 10 Original small green Koszul's book "Introduction to symplectic geometry" in Chinese

collaboré. Pour ce qui est de cette allusion à un "cours" donné à Toulouse, il y erreur. J'y ai peut être fait un exposé en 81, mais rien d'autre.] I have met Souriau more than once at conferences, but we have never collaborated. As for this allusion to a "course" given in Toulouse, there is error. I could have made a presentation in 81, but nothing else. ". Koszul admitted that he had no direct collaboration with Souriau: "*[Je ne crois pas avoir jamais parlé de ses travaux avec Souriau. Du reste j'avoue ne pas en avoir bien mesuré l'importance à l'époque] I do not think I ever talked about his work with Souriau. For the rest, I admit that I did not have a good idea of the importance at the time*".

Considering the importance of this book for different communities, I tried to find an editor for its translation in English. By chance, I met Catriona Byrne from SPRINGER, when I gave a talk at IHES, invited by Pierre Cartier, on Koszul and Souriau works application in Radar. With help of Michel Boyom, we have convinced Professor Koszul to translate this book, proposing to contextualize this book with regard to the contemporary research trends in Geometric Mechanics, Lie Groups Thermodynamics and Geometric Science of Information. Professors Marle and Boyom accepted to check the translation and help me to write the forewords.

In the historical Foreword of this book, Koszul write "*The development of analytical mechanics provided the basic concepts of symplectic structures. The term symplectic structure is due largely to analytical mechanics. But in this book, the applications of symplectic structure theory to mechanics is not discussed in any detail*". Koszul considers in this book purely algebraic and geometric developments of Geometric/Analytic Mechanics developed during the 60th, more especially Jean-Marie Souriau works detailed in Chaps. 4 and 5. ***The originality of this book lies in the fact that Koszul develops new points of view, and demonstrations not considered initially by Souriau and Geometrical Mechanics community***.

Jean-Marie Souriau was the Creator of a new discipline called "*Mécanique Géométrique (Geometric Mechanics)*". Souriau observed that the collection of motions of a dynamical system is a manifold with an antisymmetric flat tensor that is a symplectic form where the structure contains all the pertinent information on the state of the system (positions, velocities, forces, etc.). Souriau said: "*[Ce que Lagrange a vu, que n'a pas vu Laplace, c'était la structure symplectique] What Lagrange saw, that Laplace didn't see, was the symplectic structure*". Using the symmetries of a symplectic manifold, Souriau introduced a mapping which he called the "*moment map*", which takes its values in a space attached to the group of symmetries (in the dual space of its Lie algebra). Souriau associated to this moment map, the notion of symplectic cohomology, linked to the fact that such a moment is defined up to an additive constant that brings into play an algebraic mechanism (called cohomology). Souriau proved that the moment map is a constant of the motion, and provided geometric generalization of Emmy Noether invariant theorem (invariants of E. Noether theorem are the components of the moment map). Souriau has defined in a geometrically way the Noetherian symmetries using the Lagrange-Souriau 2 form with the application map. Influenced by François Gallissot (Souriau and Galissot both attended ICM'54 in Moscow, and should have exchanged during this conference), Souriau has introduced in Mechanics the Lagrange 2-form, recovering seminal Lagrange ideas. Motivated by variational principles in a coordinate free formulation, inspired by Henri Poincaré and Elie Cartan who introduced a differential 1-form instead of the Lagrangian, Souriau introduced the Lagrange 2-form as the exterior differential of the Poincaré-Cartan 1-form, and obtained the phase space as a symplectic manifold. Souriau proposed to consider this Lagrange 2-form as the fundamental structure for Lagrangian system and not the classical Lagrangian function or the Poincaré-Cartan 1-form. This 2-form is called Lagrange-Souriau 2 form, and is the exterior derivative of the Lepage form (the Poincaré-Cartan form is a first order Lepage form). This structure is developed in Koszul book, where the authors shows that when $(M; \omega)$ is an exact symplectic manifold (when there exists a 1-form α on M such that $\omega = -d\alpha$), and that a symplectic action leaves not only ω, but α invariant, this action is strongly Hamiltonian ($(M; \omega)$ is a g-Poisson space). Koszul shows that a symplectic action of a Lie algebra g on an exact symplectic manifold $(M; \omega = -d\alpha)$ that leaves invariant not only ω, but also α, is strongly Hamiltionian.

In this Book in Chap. 4, Koszul calls symplectic G-space a symplectic manifold $(M; \omega)$ on which a Lie group G acts by a symplectic action (an action which leaves unchanged the symplectic form ω). Koszul then introduces and develop properties of the moment map μ (Souriau's invention) of a Hamiltonian action of the Lie algebra g. Koszul also defines the Souriau 2-cocycle, considering that the difference of two moments of the same Hamiltonian action is a locally constant application on M, showing that when μ is a moment map, for every pair *(a;b)* of elements of g, the function $c_{\mu}(a, b) = \{\langle\mu, a\rangle, \langle\mu, b\rangle\} - \langle\mu, \{a, b\}\rangle$ is locally constant on M, defining an antisymmetric bilinear application of *gxg* in $H^0(M;\ R)$ which verifies Jacobi's identity. ***This is the 2-cocycle introduced by Jean-Marie Souriau in Geometric Mechanics, that will play a fundamental role in Souriau Lie Groups Thermodynamics to define an extension of the Fisher Metric from Information Geometry (what I will call Fisher-Souriau metric in the following).***

To highlight the importance of this Koszul book, we will illustrate the links of the detailed tools, including demonstrations or original Koszul extensions, with Souriau's Lie Groups Thermodynamics, whose applications range from statistical physics to machine learning in Artificial Intelligence. In 1970, Souriau introduced the concept of co-adjoint action of a group on its momentum space, based on the orbit method works, that allows to define physical observables like energy, heat and momentum or moment as pure geometrical objects. In a first step to establish new foundations of thermodynamics, Souriau has defined a Gibbs canonical ensemble on a symplectic manifold M for a Lie group action on M. In classical statistical mechanics, a state is given by the solution of Liouville equation on the phase space, the partition function. As symplectic manifolds have a completely continuous measure, invariant by diffeomorphisms (the Liouville measure λ), Souriau has proved that when statistical states are Gibbs states (as generalized by Souriau), they are the product of the Liouville measure by the scalar function given by the generalized partition function $e^{\Phi(\beta)-\langle\beta,U(\xi)\rangle}$ defined by the energy U (defined in the dual of the Lie algebra of this dynamical group) and the geometric temperature β, where Φ is a normalizing constant such the mass of probability is equal to 1, $\Phi(\beta) = -\log \int\limits_M e^{-\langle\beta,U(\xi)\rangle} d\lambda$. Jean-Marie Souriau then generalizes the Gibbs equilibrium state to all symplectic manifolds that have a dynamical group. Souriau has observed that if we apply this theory for Galileo, the symmetry will be broken. For each temperature β, element of the Lie algebra $\mathfrak{g}$, Souriau has introduced a tensor $\tilde{\Theta}_{\beta}$, equal to the sum of the cocycle $\tilde{\Theta}$ and the heat coboundary (with [.,.] Lie bracket):

$$\tilde{\Theta}_{\beta}(Z_1, Z_2) = \tilde{\Theta}(Z_1, Z_2) + \langle Q, ad_{Z_1}(Z_2)\rangle \tag{122}$$

This tensor $\tilde{\Theta}_{\beta}$ has the following properties: $\tilde{\Theta}(X, Y) = \langle\Theta(X), Y\rangle$ where the map Θ is the symplectic one-cocycle of the Lie algebra $\mathfrak{g}$ with values in $\mathfrak{g}^*$, with $\Theta(X) = T_e\theta(X(e))$ where θ the one-cocycle of the Lie group G. $\tilde{\Theta}(X, Y)$ is constant on M and the map $\tilde{\Theta}(X, Y) : \mathfrak{g}\times\mathfrak{g} \rightarrow \Re$ is a skew-symmetric bilinear form, and is called the ***symplectic two-cocycle of Lie algebra*** $\mathfrak{g}$ associated to the *moment map* J, with the following properties:

$$\tilde{\Theta}(X, Y) = J_{[X,Y]} - \{J_X, J_Y\} \text{with } J \text{ the Moment Map} \tag{123}$$

$$\tilde{\Theta}([X,Y],Z)+\tilde{\Theta}([Y,Z],X)+\tilde{\Theta}([Z,X],Y)=0 \tag{124}$$

where J_X linear application from $\mathfrak{g}$ to differential function on $M : \mathfrak{g} \rightarrow C^\infty(M,R)$, $X \rightarrow J_X$ and the associated differentiable application J, called moment map:

$$J : M \rightarrow \mathfrak{g}^*, x \mapsto J(x) \quad \text{such that } J_X(x) = \langle J(x), X \rangle, X \in \mathfrak{g} \tag{125}$$

The geometric temperature, element of the algebra $\mathfrak{g}$, is in the kernel of the tensor $\tilde{\Theta}_\beta$:

$$\beta \in Ker\,\tilde{\Theta}_\beta \text{ such that } \tilde{\Theta}_\beta(\beta,\beta)=0, \quad \forall \beta \in \mathfrak{g} \tag{126}$$

The following symmetric tensor $g_\beta([\beta,Z_1],[\beta,Z_2])=\tilde{\Theta}_\beta(Z_1,[\beta,Z_2])$, defined on all values of $ad_\beta(.)=[\beta,.]$ is positive definite, and defines extension of classical Fisher metric in Information Geometry (as hessian of the logarithm of partition function):

$$g_\beta([\beta,Z_1],Z_2)=\tilde{\Theta}_\beta(Z_1,Z_2), \quad \forall Z_1 \in \mathfrak{g}, \forall Z_2 \in \mathrm{Im}\big(ad_\beta(.)\big) \tag{127}$$

$$\text{With } g_\beta(Z_1,Z_2) \geq 0, \quad \forall Z_1,Z_2 \in \mathrm{Im}\big(ad_\beta(.)\big) \tag{128}$$

***These equations are universal*, because they are not dependent on the symplectic manifold but only on the dynamical group *G*, the symplectic two-cocycle Θ, the temperature β and the heat *Q*. Souriau called it "*Lie groups thermodynamics*".**

This antisymmetric bilinear map (127) and (128), with definition (122) and (123) is exactly equal to the mathematical object introduced in Chap.* 4 *of Koszul's book by:

$$c_\mu(a,b)=\{\langle\mu,a\rangle,\langle\mu,b\rangle\}-\{\mu,\langle a,b\rangle\} \tag{129}$$

In this book, Koszul has studied this antisymmetric bilinear map considering the following developments. For any moment map μ, Koszul defines the skew symmetric bilinear form $\mathrm{c}_\mu(\mathrm{a,b})$ on Lie algebra by:

$$\mathrm{c}_\mu(\mathrm{a,b})=\big\langle d\theta_\mu(a),b\big\rangle, a,b \in \mathfrak{g} \tag{130}$$

Koszul observes that if we use:

$$\theta_\mu(st)=\mu(stx)-Ad^*_{st}\mu(x)=\theta_\mu(s)+Ad^*_s\mu(tx)-Ad^*_sAd^*_t\mu(x)=\theta_\mu(s)+Ad^*_s\theta_\mu(t)$$

by developing $d\mu(ax)={}^t ad_a\mu(x)+d\theta_\mu(a),\ x \in M, a \in \mathfrak{g}$, he obtains:

$$\langle d\mu(ax),b\rangle=\langle\mu(x),[a,b]\rangle+\big\langle \mathrm{d}\theta_\mu(a),b\big\rangle=\{\langle\mu,a\rangle,\langle\mu,b\rangle\}(x),\ x \in M, a,b \in \mathfrak{g} \tag{131}$$

We have then:

$$c_\mu(\mathrm{a},\mathrm{b}) = \{\langle\mu, a\rangle, \langle\mu, b\rangle\} - \langle\mu, [a,b]\rangle = \big\langle \mathrm{d}\theta_\mu(a), b\big\rangle, a, b \in \mathfrak{g} \tag{132}$$

and the property:

$$c_\mu([\mathrm{a},\mathrm{b}],\mathrm{c}) + c_\mu([b,c],a) + c_\mu([c,a],b) = 0,\ a, b, c \in \mathfrak{g} \tag{133}$$

Koszul concludes by observing that if the moment map is transform as $\mu' = \mu + \phi$ then we have:

$$c_{\mu'}(a,b) = c_\mu(a,b) - \langle\phi, [a,b]\rangle \tag{134}$$

Finally using $c_\mu(\mathrm{a},\mathrm{b}) = \{\langle\mu, a\rangle, \langle\mu, b\rangle\} - \langle\mu, [a,b]\rangle = \big\langle \mathrm{d}\theta_\mu(a), b\big\rangle, a, b \in \mathfrak{g}$, koszul highlights the property that:

$$\big\{\mu^*(a), \mu^*(b)\big\} = \{\langle\mu, a\rangle, \langle\mu, b\rangle\} = \mu^*\big([a,b] + c_\mu(a,b)\big) = \mu^*\{a,b\}_{c_\mu} \tag{135}$$

In Chap. 4, Koszul introduces the equivariance of the moment map μ. Based on the definitions of the adjoint and coadjoint representations of a Lie group or a Lie algebra, Koszul proves that when *(M; ω)* is a connected Hamiltonian *G*-space and $\mu : M \to \mathfrak{g}^*$ a moment of the action of *G*, there exists an affine action of *G* on g^*, whose linear part is the coadjoint action, for which the moment μ is equivariant. This affine action is obtained by modifying the coadjoint action by means of a cocycle. This notion is also developed in Chap. 5 for Poisson manifolds. Defining classical operation $Ad_s a = sas^{-1}, s \in G, a \in \mathfrak{g}$, $ad_a b = [a,b], a \in \mathfrak{g}, b \in \mathfrak{g}$ and coadjoint action given by $Ad_s^* =^t Ad_{s^{-1}}, s \in G$ with classical properties:

$$Ad_{\exp a} = \exp(-ad_a), a \in \mathfrak{g} \text{ or } Ad^*_{\exp a} = \exp^t(ad_a), a \in \mathfrak{g} \tag{136}$$

Koszul considers:

$$x \mapsto sx, x \in M, \mu : M \to \mathfrak{g}^* \tag{137}$$

From which, he obtains:

$$\langle d\mu(v), a\rangle = \omega(ax, v) \tag{138}$$

Koszul then study $\mu \circ s_M - Ad_s^* \circ \mu : M \to \mathfrak{g}^*$, and develops:

$$d\big\langle Ad_s^* \circ \mu, a\big\rangle = \big\langle Ad_s^* d\mu, a\big\rangle = \langle d\mu, Ad_{s^{-1}} a\rangle \tag{139}$$

$$\langle d\mu(v), Ad_{s^{-1}} a\rangle = \omega(s^{-1}asx, v) = \omega(asx, sv) = \langle d\mu(sv), a\rangle = (d\langle\mu \circ s_M, a\rangle)(v) \tag{140}$$

$$d\langle Ad_s^* \circ \mu, a\rangle = d\langle \mu \circ s_M, a\rangle \text{ and then proves that } d\langle \mu \circ s_M - Ad_s^* \circ \mu, a\rangle = 0 \quad (141)$$

Koszul considers the cocycle given by $\theta_\mu(s) = \mu(sx) - Ad_s^*\mu(x), s \in G$, and observes that:

$$\theta_\mu(st) = \theta_\mu(s) - Ad_s^*\theta_\mu(t), s, t \in G \quad (142)$$

From this action of the group on dual Lie algebra:

$$G \times \mathfrak{g}^* \to \mathfrak{g}^*, (s, \xi) \mapsto s\xi = Ad_s^*\xi + \theta_\mu(s) \quad (143)$$

Koszul introduces the following properties:

$$\mu(sx) = s\mu(x) = Ad_s^*\mu(x) + \theta_\mu(s), \forall s \in G, x \in M \quad (144)$$

$$G \times \mathfrak{g}^* \to \mathfrak{g}^*, (e, \xi) \mapsto e\xi = Ad_e^*\xi + \theta_\mu(e) = \xi + \mu(x) - \mu(x) = \xi \quad (145)$$

$$(s_1s_2)\xi = Ad_{s_1s_2}^*\xi + \theta_\mu(s_1s_2) = Ad_{s_1}^* Ad_{s_2}^*\xi + \theta_\mu(s_1) + Ad_{s_1}^*\theta_\mu(s_2)$$

$$(s_1s_2)\xi = Ad_{s_1}^*(Ad_{s_2}^*\xi + \theta_\mu(s_2)) + \theta_\mu(s_1) = s_1(s_2\xi), \forall s_1, s_2 \in G, \xi \in \mathfrak{g}^* \quad (146)$$

This Koszul study of the moment map μ equivariance, and the existence of an affine action of G on g^*, whose linear part is the coadjoint action, for which the moment μ is equivariant, is at the cornerstone of Souriau Theory of Geometric Mechanics and Lie Groups Thermodynamics. I illustrate this importance by giving Souriau theorem for Lie Groups Thermodynamics, and the link with, what I call, Souriau-Fisher metric (a covariant definition of Fisher metric):

Theorem (Souriau Theorem of Lie Group Thermodynamics). *Let Ω be the largest open proper subset of $\mathfrak{g}$, Lie algebra of G, such that $\int_M e^{-\langle \beta, U(\xi)\rangle} d\lambda$ and $\int_M \xi . e^{-\langle \beta, U(\xi)\rangle} d\lambda$ are convergent integrals, this set Ω is convex and is invariant under every transformation $Ad_g(.)$. Then, the fundamental equations of Lie group thermodynamics are given by the action of the group:*

$$Action\ of\ Lie\ group\ on\ Lie\ algebra : \beta \to Ad_g(\beta) \quad (147)$$

$$Characteristic\ function\ after\ Lie\ group\ action : \Phi \to \Phi - \langle \theta(g^{-1}), \beta\rangle \quad (148)$$

$$Invariance\ of\ entropy\ with\ respect\ to\ action\ of\ Lie\ group : s \to s \quad (149)$$

$$Action\ of\ Lie\ group\ on\ geometric\ heat : Q \to a(g, Q) = Ad_g^*(Q) + \theta(g) \quad (150)$$

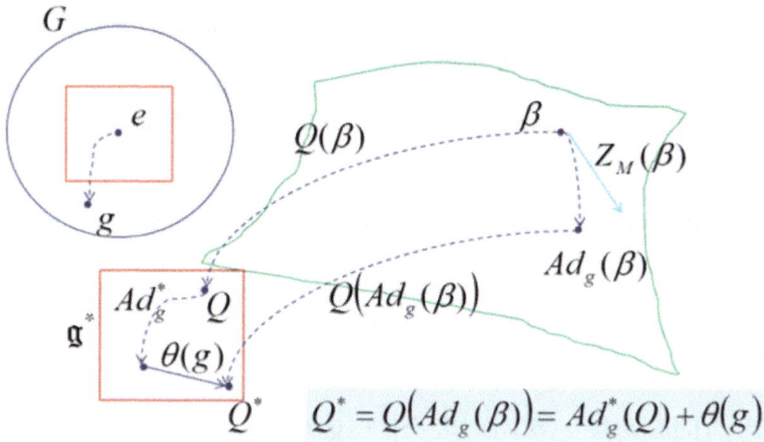

Fig. 11 Broken symmetry on geometric heat Q due to adjoint action of the group on temperature β as an element of the Lie algebra

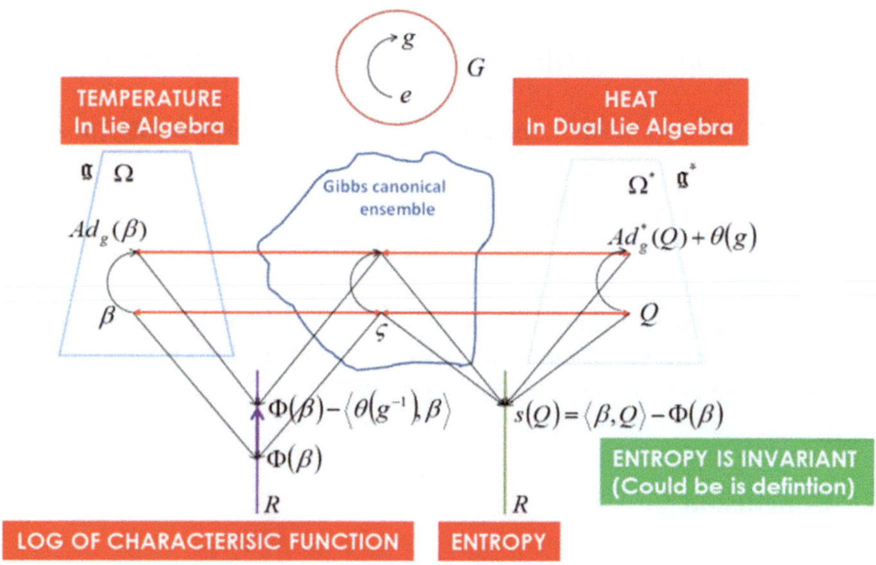

Fig. 12 Global Souriau scheme of Lie group thermodynamics, with entropy *s(Q)*, geometric heat Q element of dual Lie algebra and geometric temperature β element of Lie algebra

Souriau equations of Lie group thermodynamics, related to the moment map μ equivariance, and the existence of an affine action of G on g^*, whose linear part is the coadjoint action, for which the moment μ is equivariant, are summarized in the following figures (Figs. 11 and 12).

I finally observe that ***the Koszul antisymmetric bilinear map*** $c_\mu(a,b) = \{\langle\mu,a\rangle,\langle\mu,b\rangle\} - \langle\mu,\{a,b\}\rangle$ ***is equal to Souriau Riemannian metric***, introduced by mean of symplectic cocycle. I have observed that this metric is a generalization of the Fisher metric from Information Geometry, that I call the Souriau-Fisher metric, defined as a hessian of the partition function logarithm $g_\beta = -\frac{\partial^2\Phi}{\partial\beta^2} = \frac{\partial^2 \log\psi_\Omega}{\partial\beta^2}$ as in classical information geometry. This new definition of Fisher metric has the property to be covariant under the action of the group G. I have established the equality of two terms, between Souriau definition based on Lie group cocycle Θ and parameterized by "*geometric heat*" Q (element of dual Lie algebra) and "*geometric temperature*" β (element of Lie algebra) and hessian of characteristic function $\Phi(\beta) = -\log\Psi_\Omega(\beta)$ with respect to the variable β:

$$g_\beta([\beta,Z_1],[\beta,Z_2]) = \langle\Theta(Z_1),[\beta,Z_2]\rangle + \langle Q,[Z_1,[\beta,Z_2]]\rangle = \frac{\partial^2\log\psi_\Omega}{\partial\beta^2} \tag{151}$$

If we differentiate this relation of Souriau theorem $Q(Ad_g(\beta)) = Ad_g^*(Q) + \theta(g)$, this relation occurs:

$$\frac{\partial Q}{\partial\beta}(-[Z_1,\beta],.) = \tilde{\Theta}(Z_1[\beta,.]) + \left\langle Q, Ad_{z_1}([\beta,.])\right\rangle = \tilde{\Theta}_\beta(Z_1,[\beta,.]) \tag{152}$$

$$-\frac{\partial Q}{\partial\beta}([Z_1,\beta],Z_2.) = \tilde{\Theta}(Z_1,[\beta,Z_2]) + \left\langle Q, Ad_{.z_1}([\beta,Z_2])\right\rangle = \tilde{\Theta}_\beta(Z_1,[\beta,Z_2]) \tag{153}$$

$$\Rightarrow -\frac{\partial Q}{\partial\beta} = g_\beta([\beta,Z_1],[\beta,Z_2]) \tag{154}$$

The Souriau Fisher metric $I(\beta) = -\frac{\partial^2\Phi(\beta)}{\partial\beta^2} = -\frac{\partial Q}{\partial\beta}$ has been considered by Souriau as a *generalization of* "*heat capacity*". Souriau called it the "*geometric capacity*" and is also equal to "*geometric susceptibility*".

7 Conclusion

The community of "*Geometric Science of Information*" (GSI) has lost a mathematician of great value, who informed his views by the depth of his knowledge of the elementary structures of hessian geometry and bounded homogeneous domains. His modesty was inversely proportional to his talent. Professor Koszul built in over 60 years of mathematical career, in the silence of his passions, an immense work, which makes him one of the great mathematicians of the XX's century, whose importance will only affirm with the time. In this troubled time and rapid transformation of society and science, the example of Professor Koszul must be regarded as a model for future generations, to avoid them the trap of fleeting glories and recognitions too fast acquired. The work of Professor Koszul is also a proof of fidelity to his masters

Fig. 13 (on the left) Jean-Louis Koszul at Grenoble in December 1993, (on the right) last interview of Jean-Louis Koszul in 2016 for 50th birthday of Institut Joseph Fourier in Grenoble

and in the first place to Prof. Elie Cartan, who inspired him throughout his life. Henri Cartan writes on this subject "*I do not forget the homage he paid to Elie Cartan's work in Differential Geometry during the celebration, in Bucharest, in 1969, of the centenary of his birth. It is not a coincidence that this centenary was also celebrated in Grenoble the same year. As always, Koszul spoke with the discretion and tact that we know him, and that we love so much at home*". I will conclude by quoting Jorge Luis Borges "*Forgetfulness and memory are also inventive*" (Brodie's report). Our generation and previous one have forgotten or misunderstood the depth of the work of Jean-Louis Koszul and Elie Cartan on the study of bounded homogeneous domains. It is our responsibility to correct this omission, and to make it the new inspiration for the Geometric Science of Information. I will conclude by requesting you to listen to the last interview of Jean-Louis Koszul for 50th birthday of Joseph Fourier Institute [72], especially when Koszul he is passionate by "*conifers and cedars trees planted by Claude Chabauty*", or by the "*pretty catalpa tree*" which was at the Fourier Institute and destroyed by wind, "*the tree with parentheses*" he says, to which he seemed to be sentimentally attached. He also regrets that the Institute did not use the 1% artistic fund for the art mosaic project in the library. In this Koszul family of mathematicians, musicians, and Scientifics, there was a constant recollection of "beauty" and "truth". Our society no longer cares about timeless "beauty". We have then to extase ourself with Jean-Louis Koszul by observing beautiful "*Catalpa tree*" with "*Parenthese Mushroom*", before there is no longer people to contemplate them (Fig. 13).

> "*Seul la nuit avec un livre éclairé par une chandelle – livre et chandelle, double îlot de lumière, contre les doubles ténèbres de l'esprit et de la nuit. J'étudie ! Je ne suis que le sujet du verbe étudier. Penser je n'ose. Avant de penser, il faut étudier. Seuls les philosophes pensent avant d'étudier.* » - **Gaston Bachelard, La flamme d'une chandelle, 1961**

Appendix

Clairaut(-Legendre) Equation of Maurice Fréchet associated to "distinguished functions" as fundamental equation of Information geometry

Before Rao [4, 124], in 1943, Maurice Fréchet [3] wrote a seminal paper introducing what was then called the Cramer-Rao bound. This paper contains in fact much more that this important discovery. In particular, Maurice Fréchet introduces more general notions relative to "*distinguished functions*", densities with estimator reaching the bound, defined with a function, solution of Clairaut's equation. The solutions "envelope of the Clairaut's equation" are equivalents to standard Legendre transform without convexity constraints but only smoothness assumption. This Fréchet's analysis can be revisited on the basis of Jean-Louis Koszul's works as seminal foundation of "*Information Geometry*".

I will use Maurice Fréchet notations, to consider the estimator:

$$T = H(X_1, \ldots, X_n) \tag{155}$$

and the random variable

$$A(X) = \frac{\partial \log p_\theta(X)}{\partial \theta} \tag{156}$$

that are associated to:

$$U = \sum_i A(X_i) \tag{157}$$

The normalizing constraint $\int_{-\infty}^{+\infty} p_\theta(x)dx = 1$ implies that: $\int_{-\infty}^{+\infty} \ldots \int_{-\infty}^{+\infty} \prod_i p_\theta(x_i)dx_i = 1$

If we consider the derivative if this last expression with respect to θ, then $\int_{-\infty}^{+\infty} \ldots \int_{-\infty}^{+\infty} \left[\sum_i A(x_i)\right] \prod_i p_\theta(x_i)dx_i = 0$ gives:

$$E_\theta[U] = 0 \tag{158}$$

Similarly, if we assume that $E_\theta[T] = \theta$, then $\int_{-\infty}^{+\infty} \ldots \int_{-\infty}^{+\infty} H(x_1, \ldots, x_n) \prod_i p_\theta(x_i)dx_i = \theta$, and we obtain by derivation with respect to θ:

$$E[(T - \theta)U] = 1 \tag{159}$$

But as $E[T] = \theta$ and $E[U] = 0$, we immediatly deduce that:

$$E[(T - E[T])(U - E[U])] = 1 \tag{160}$$

From Schwarz inequality, we can develop the following relations:

$$\begin{aligned} &[E(ZT)]^2 \leq E\left[Z^2\right]E\left[T^2\right] \\ &1 \leq E\left[(T - E[T])^2\right]E\left[(U - E[U])^2\right] = (\sigma_T \sigma_U)^2 \end{aligned} \tag{161}$$

U being the summation of independent variables, Bienaymé equality could be applied:

$$(\sigma_U)^2 = \sum_i \left[\sigma_{A(X_i)}\right]^2 = n(\sigma_A)^2 \tag{162}$$

From which, Fréchet deduced the bound, rediscoved by Cramer and Rao 2 years later:

$$(\sigma_T)^2 \geq \frac{1}{n(\sigma_A)^2} \tag{163}$$

Fréchet observed that it is a remarkable inequality where the second member is independent of the choice of the function H defining the "*empirical value*" T, where the first member can be taken to any empirical value $T = H(X_1, \ldots, X_n)$ subject to the unique condition $E_\theta[T] = \theta$ regardless is θ.

The classic condition that the Schwarz inequality becomes an equality helps us to determine when σ_T reaches its lower bound $\frac{1}{\sqrt{n}\sigma_n}$.

The previous inequality becomes an equality if there are two numbers α and β (not random and not both zero) such that $\alpha\left(H' - \theta\right) + \beta U = 0$, with H' particular function among eligible H as we have the equality. This equality is rewritten $H' = \theta + \lambda' U$ with λ' a non-random number.

If we use the previous equation, then:

$$E[(T - E[T])(U - E[U])] = 1 \Rightarrow E\left[\left(H' - \theta\right)U\right] = \lambda' E_\theta\left[U^2\right] = 1 \tag{164}$$

We obtain:

$$U = \sum_i A(X_i) \Rightarrow \lambda' n E_\theta\left[A^2\right] = 1 \tag{165}$$

From which we obtain λ' and the form of the associated estimator H':

$$\lambda' = \frac{1}{nE\left[A^2\right]} \Rightarrow H' = \theta + \frac{1}{nE\left[A^2\right]} \sum_i \frac{\partial \log p_\theta(X_i)}{\partial \theta} \tag{166}$$

It is therefore deduced that the estimator that reaches the terminal is of the form:

$$H' = \theta + \frac{\sum_i \frac{\partial \log p_\theta(X_i)}{\partial \theta}}{n \int_{-\infty}^{+\infty} \left[\frac{\partial p_\theta(x)}{\partial \theta} \right]^2 \frac{dx}{p_\theta(x)}} \tag{167}$$

with $E\left[H'\right] = \theta + \lambda' E[U] = \theta$.

H' would be one of the eligible functions, if H' would be independent of θ. Indeed, if we consider:

$$E_{\theta_0}\left[H'\right] = \theta_0,\ E\left[\left(H' - \theta_0\right)^2\right] \le E_{\theta_0}\left[(H - \theta_0)^2\right] \forall H \text{ such that } E_{\theta_0}[H] = \theta_0 \tag{168}$$

$H = \theta_0$ satisfies the equation and inequality shows that it is almost certainly equal to θ_0. So to look for θ_0, we should know beforehand θ_0.

At this stage, Fréchet looked for "*distinguished functions*" ("*densités distinguées*" in French), as any probability density $p_\theta(x)$ such that the function:

$$h(x) = \theta + \frac{\frac{\partial \log p_\theta(x)}{\partial \theta}}{\int_{-\infty}^{+\infty} \left[\frac{\partial p_\theta(x)}{\partial \theta} \right]^2 \frac{dx}{p_\theta(x)}} \tag{169}$$

is independant of θ. The objective of Fréchet is then to determine the minimizing function $T = H'(X_1, \ldots, X_n)$ that reaches the bound. By deduction from previous relations, we have:

$$\lambda(\theta) \frac{\partial \log p_\theta(x)}{\partial \theta} = h(x) - \theta \tag{170}$$

But as $\lambda(\theta) > 0$, ***we can consider $\frac{1}{\lambda(\theta)}$ as the second derivative of a function*** $\Phi(\theta)$ such that:

$$\frac{\partial \log p_\theta(x)}{\partial \theta} = \frac{\partial^2 \Phi(\theta)}{\partial \theta^2} [h(x) - \theta] \tag{171}$$

from which we deduce that:

$$\ell(x) = \log p_\theta(x) - \frac{\partial \Phi(\theta)}{\partial \theta} [h(x) - \theta] - \Phi(\theta) \tag{172}$$

Is an independant quantity of θ. ***A distinguished function*** will be then given by:

$$p_\theta(x) = e^{\frac{\partial \Phi(\theta)}{\partial \theta}[h(x) - \theta] + \Phi(\theta) + \ell(x)} \tag{173}$$

with the normalizing constraint $\int_{-\infty}^{+\infty} p_\theta(x) dx = 1$.

These two conditions are sufficient. Indeed, reciprocally, let three functions $\Phi(\theta)$, $h(x)$ et $\ell(x)$ that we have, for any θ:

$$\int_{-\infty}^{+\infty} e^{\frac{\partial\Phi(\theta)}{\partial\theta}[h(x)-\theta]+\Phi(\theta)+\ell(x)}dx = 1 \tag{174}$$

Then the function is distinguished:

$$\theta + \frac{\frac{\partial \log p_\theta(x)}{\partial\theta}}{\int_{-\infty}^{+\infty}\left[\frac{\partial p_\theta(x)}{\partial\theta}\right]^2\frac{dx}{p_\theta(x)}} = \theta + \lambda(x)\frac{\partial^2\Phi(\theta)}{\partial\theta^2}[h(x)-\theta] \tag{175}$$

If $\lambda(x)\frac{\partial^2\Phi(\theta)}{\partial\theta^2} = 1$, when

$$\frac{1}{\lambda(x)} = \int_{-\infty}^{+\infty}\left[\frac{\partial \log p_\theta(x)}{\partial\theta}\right]^2 p_\theta(x)dx = (\sigma_A)^2 \tag{176}$$

The function is reduced to $h(x)$ and then is not dependent of θ.
We have then the following relation:

$$\frac{1}{\lambda(x)} = \int_{-\infty}^{+\infty}\left(\frac{\partial^2\Phi(\theta)}{\partial\theta^2}\right)^2[h(x)-\theta]^2 e^{\frac{\partial\Phi(\theta)}{\partial\theta}(h(x)-\theta)+\Phi(\theta)+\ell(x)}dx \tag{177}$$

The relation is valid for any θ, we can derive the previous expression from θ:

$$\int_{-\infty}^{+\infty} e^{\frac{\partial\Phi(\theta)}{\partial\theta}(h(x)-\theta)+\Phi(\theta)+\ell(x)}\left(\frac{\partial^2\Phi(\theta)}{\partial\theta^2}\right)[h(x)-\theta]dx = 0 \tag{178}$$

We can divide by $\frac{\partial^2\Phi(\theta)}{\partial\theta^2}$ because it doesn't depend on x.
If we derive again with respect to θ, we will have:

$$\int_{-\infty}^{+\infty} e^{\frac{\partial\Phi(\theta)}{\partial\theta}(h(x)-\theta)+\Phi(\theta)+\ell(x)}\left(\frac{\partial^2\Phi(\theta)}{\partial\theta^2}\right)[h(x)-\theta]^2dx = \int_{-\infty}^{+\infty} e^{\frac{\partial\Phi(\theta)}{\partial\theta}(h(x)-\theta)+\Phi(\theta)+\ell(x)}dx = 1 \tag{179}$$

Combining this relation with that of $\frac{1}{\lambda(x)}$, we can deduce that $\lambda(x)\frac{\partial^2\Phi(\theta)}{\partial\theta^2} = 1$ and as $\lambda(x) > 0$ then $\frac{\partial^2\Phi(\theta)}{\partial\theta^2} > 0$.

Fréchet emphasizes at this step, another way to approach the problem. We can select arbitrarily $h(x)$ and $l(x)$ and then $\Phi(\theta)$ is determined by:

$$\int_{-\infty}^{+\infty} e^{\frac{\partial\Phi(\theta)}{\partial\theta}[h(x)-\theta]+\Phi(\theta)+\ell(x)}dx = 1 \tag{180}$$

That could be rewritten:

$$e^{\theta.\frac{\partial\Phi(\theta)}{\partial\theta}-\Phi(\theta)} = \int_{-\infty}^{+\infty} e^{\frac{\partial\Phi(\theta)}{\partial\theta}h(x)+\ell(x)}dx \tag{181}$$

If we then fixed arbitrarily $h(x)$ and $l(x)$ and let s an arbitrary variable, the following function will be an explicit positive function given by $e^{\Psi(s)}$:

$$\int_{-\infty}^{+\infty} e^{s.h(x)+\ell(x)}dx = e^{\Psi(s)} \tag{182}$$

Fréchet obtained finally the function $\Phi(\theta)$ as solution of the equation:

$$\Phi(\theta) = \theta.\frac{\partial\Phi(\theta)}{\partial\theta} - \Psi\left(\frac{\partial\Phi(\theta)}{\partial\theta}\right) \tag{183}$$

Fréchet noted that this is the Alexis Clairaut Equation.

The case $\frac{\partial\Phi(\theta)}{\partial\theta} = cste$ would reduce the density to a function that would be independent of θ, and so $\Phi(\theta)$ is given by a singular solution of this Clairaut equation, that is unique and could be computed by eliminating the variable s between:

$$\Phi = \theta.s - \Psi(s) \text{ and } \theta = \frac{\partial\Psi(s)}{\partial s} \tag{184}$$

Or between:

$$e^{\theta.s-\Phi(\theta)} = \int_{-\infty}^{+\infty} e^{s.h(x)+\ell(x)}dx \text{ and } \int_{-\infty}^{+\infty} e^{s.h(x)+\ell(x)}[h(x)-\theta]dx = 0 \tag{185}$$

$\Phi(\theta) = -\log \int_{-\infty}^{+\infty} e^{s.h(x)+\ell(x)}dx + \theta.s$ where s is induced implicitly through the constraint given by the integral $\int_{-\infty}^{+\infty} e^{s.h(x)+\ell(x)}[h(x)-\theta]dx = 0$.

When we known the distinguished function, H' is among functions $H(X_1,\ldots,X_n)$ verifying $E_\theta[H] = \theta$ and such that σ_H reaches for each value of θ, an absolute minimum, equal to $\frac{1}{\sqrt{n}\sigma_A}$. For the previous equation:

$$h(x) = \theta + \frac{\frac{\partial \log p_\theta(x)}{\partial \theta}}{\int\limits_{-\infty}^{+\infty} \left[\frac{\partial p_\theta(x)}{\partial \theta} \right]^2 \frac{dx}{p_\theta(x)}} \tag{186}$$

We can rewrite the estimator as:

$$H'(X_1, \ldots, X_n) = \frac{1}{n}[h(X_1) + \ldots + h(X_n)] \tag{187}$$

and compute the associated empirical value:

$$t = H'(x_1, \ldots, x_n) = \frac{1}{n} \sum_i h(x_i) = \theta + \lambda(\theta) \sum_i \frac{\partial \log p_\theta(x_i)}{\partial \theta} \tag{188}$$

If we take $\theta = t$, we have as $\lambda(\theta) > 0$:

$$\sum_i \frac{\partial \log p_t(x_i)}{\partial t} = 0 \tag{189}$$

When $p_\theta(x)$ is a distinguished function, the empirical value t of θ corresponding to a sample $x_1, \ldots, x_n$ is a root of previous equation in t. This equation has a root and only one when X is a distinguished variable. Indeed, as we have:

$$p_\theta(x) = e^{\frac{\partial \Phi(\theta)}{\partial \theta}[h(x) - \theta] + \Phi(\theta) + \ell(x)} \tag{190}$$

$$\sum_i \frac{\partial \log p_t(x_i)}{\partial t} = \frac{\partial^2 \Phi(t)}{\partial t^2} \left[\frac{\sum_i h(x_i)}{n} - t \right] \text{ with } \frac{\partial^2 \Phi(t)}{\partial t^2} > 0 \tag{191}$$

We can then recover the unique root: $t = \frac{\sum_i h(x_i)}{n}$.

This function $T \equiv H'(X_1, \ldots, X_n) = \frac{1}{n} \sum_i h(X_i)$ can have an arbitrary form, that is a sum of functions of each only one of the quantities and it is even the arithmetic average of N values of a same auxiliary random variable $Y = h(X)$. The dispersion is given by:

$$\left(\sigma_{T_n}\right)^2 = \frac{1}{n(\sigma_A)^2} = \frac{1}{n \int\limits_{-\infty}^{+\infty} \left[\frac{\partial p_\theta(x)}{\partial \theta} \right]^2 \frac{dx}{p_\theta(x)}} = \frac{1}{n \frac{\partial^2 \Phi(\theta)}{\partial \theta^2}} \tag{192}$$

and T_n follows the probability density:

$$p_\theta(t) = \sqrt{n} \frac{1}{\sigma_A \sqrt{2\pi}} e^{-\frac{n(t-\theta)^2}{2.\sigma_A^2}} \quad \text{with} \quad (\sigma_A)^2 = \frac{\partial^2 \Phi(\theta)}{\partial \theta^2} \tag{193}$$

- **Clairaut Equation and Legendre Transform**

To summarize, Fréchet paper novelty, I have just observed that Fréchet introduced distinguished functions depending on a function $\Phi(\theta)$, solution of the Clairaut equation:

$$\Phi(\theta) = \theta.\frac{\partial\Phi(\theta)}{\partial\theta} - \Psi\left(\frac{\partial\Phi(\theta)}{\partial\theta}\right) \tag{194}$$

Or given by the Legendre Transform:

$$\Phi = \theta.s - \Psi(s) \text{ and } \theta = \frac{\partial\Psi(s)}{\partial s} \tag{195}$$

Fréchet also observed that this function $\Phi(\theta)$ could be rewritten:

$\Phi(\theta) = -\log \int\limits_{-\infty}^{+\infty} e^{s.h(x)+\ell(x)}dx + \theta.s$ where s is induced implicitly by the constraints given by integral $\int\limits_{-\infty}^{+\infty} e^{s.h(x)+\ell(x)}[h(x) - \theta]dx = 0$.

This equation is the fundamental equation of Information Geometry.

The "Legendre" transform was introduced by Adrien-Marie Legendre in 1787 to solve a minimal surface problem Gaspard Monge in 1784. Using a result of Jean Baptiste Meusnier, a student of Monge, it solves the problem by a change of variable corresponding to the transform which now entitled with his name. Legendre wrote: "*I have just arrived by a change of variables that can be useful in other occasions.*" About this transformation, Darboux in his book gives an interpretation of Chasles: "*This comes after a comment by Mr. Chasles, to substitute its polar reciprocal on the surface compared to a paraboloïd.*" The equation of Clairaut was introduced 40 years earlier in 1734 by Alexis Clairaut. Solutions "envelope of the Clairaut equation" are equivalent to the Legendre transform with unconditional convexity, but only under differentiability constraint. Indeed, for a non-convex function, Legendre transformation is not defined where the Hessian of the function is canceled, so that the equation of Clairaut only make the hypothesis of differentiability. The portion of the strictly convex function g in Clairaut equation $y = px - g(p)$ to the function f giving the envelope solutions by the formula $y = f(x)$ is precisely the Legendre transformation. The approach of Fréchet may be reconsidered in a more general context on the basis of the work of Jean-Louis Koszul.

References

1. Ollivier, Y., Marceau-Caron, G.: Natural Langevin dynamics for neural networks. In: Nielsen, F., Barbaresco, F. (eds.) Proceedings of the Conference on Geometric Science of Information (GSI 2017). Lecture Notes in Computer Science, vol. 10589, pp. 451–459. Springer, Berlin (2017) (Best paper award.)
2. Ollivier, Y.: True asymptotic natural gradient optimization manuscript disponible sur. http://arxiv.org/abs/1712.08449
3. Fréchet, M.: Sur l'extension de certaines évaluations statistiques au cas de petits échantillons. Rev. l'Institut Int. Stat. **11**(3/4), 182–205 (1943)
4. Rao, C.R.: Information and the accuracy attainable in the estimation of statistical parameters. Bull. Calcutta Math. Soc. **37**, 81–89 (1945)
5. Chentsov, N.N.: Statistical Decision Rules and Optimal Inferences, Transactions of Mathematics Monograph, vol. 53. American Mathematical Society, Providence, RI, USA, (1982)
6. Amari, S.I.: Differential Geometry of Statistical Models. SPRINGER Series Lecture Notes in Statistics, vol. 28 (1985)
7. Amari, S.I.: Information Geometry and Its Applications. SPRINGER Series Applied Mathematical Sciences, vol. 194 (2016)
8. Weinhold, F.: Thermodynamics and geometry. Phys. Today (1976)
9. Ruppeiner, G.: Thermodynamics: a Riemannian geometric model. Phys. Rev. A **20**, 1608 (1979)
10. Ingarden, R.S.: Information geometry in functional spaces of classical and quantum finite statistical systems. Int. J. Eng. Sci. **19**(12), 1609–1633 (1981)
11. Ingarden, R.S., Kossakowski, A., Ohya, M.: Information Dynamics and Open Systems: Classical and Quantum Approach. Springer, Berlin (1997). Accessed 31 March 1997
12. Mrugala, R.: Geometrical formulation of equilibrium phenomenological thermodynamics. Rep. Math. Phys. **14**, 419 (1978)
13. Mrugala, R., Nulton, J.D., Schön, J.C., Salamon, P.: Statistical approach to the geometric structure of thermodynamics. Phys. Rev. A **41**, 6, 3156 (1990)
14. Janyszek, H., Mrugala, R.: Riemannian geometry and the thermodynamics of model magnetic systems. Phys. Rev. A **39**(12), :6515–6523 (1989)
15. Koszul, J.L.: Sur la forme hermitienne canonique des espaces homogènes complexes. Can. J. Math. **7**, 562–576 (1955)
16. Koszul, J.L.: Exposés sur les Espaces Homogènes Symétriques. Publicação da Sociedade de Matematica de São Paulo, São Paulo, Brazil (1959)
17. Souriau, J.-M.: Structures des Systèmes Dynamiques. Dunod, Paris (1969)
18. Koszul, J.L.: Introduction to Symplectic Geometry. Science Press, Beijing (1986) (in chinese); translated by SPRINGER, with F.Barbaresco, C.M. Marle and M. Boyom forewords, SPRINGER, 2018
19. Balian, R., Alhassid, Y., Reinhardt, H.: Dissipation in many-body systems: a geometric approach based on information theory. Phys. Rep. **131**, 1–146 (1986)
20. Balian, R.: The entropy-based quantum metric. Entropy **16**, 3878–3888 (2014)
21. Selected Papers of J L Koszul, Series in Pure Mathematics, Volume 17, World Scientific Publishing, 1994
22. Koszul, J.L.: L'œuvre d'Élie Cartan en géométrie différentielle, in Élie Cartan, 1869–1951. Hommage de l'Académie de la République Socialiste de Roumanie à l'occasion du centenaire de sa naissance. Comprenant les communications faites aux séances du 4e Congrès du Groupement des Mathématiciens d'Expression Latine, tenu à Bucarest en 1969 (Editura Academiei Republicii Socialiste Romania, Bucharest, 1975) pp. 39–45
23. Vinberg, E.B.: Homogeneous cones. Dokl. Akad. Nauk SSSR. **133**, 9–12 (1960); Sov. Math. Dokl. **1**, 787–790 (1961)
24. Vinberg, E.B.: The Morozov-Borel theorem for real Lie groups. Dokl. Akad. Nauk SSSR. **141**, 270–273 (1961); Sov. Math. Dokl. **2**, 1416–1419 (1962)

25. Vinberg E.B., Convex homogeneous domains, Dokl. Akad. Nauk SSSR., 141 1961, 521–524; Soviet Math. Dokl., n°2, pp. 1470–1473, 1962
26. Vinberg, E.B.: Automorphisms of homogeneous convex cones. Dokl. Akad. Nauk SSSR. **143**, 265–268 (1962); Sov. Math. Dokl. **3**, 371–374 (1963)
27. Vinberg, E.B.: The Theory of Homogeneous Convex Cones, Trudy Moskovskogo Matematicheskogo Obshchestva, vol. 12, pp. 303–358 (1963); Trans. Moscow Math. Soc. **12**, 340–403 (1963)
28. Vinberg, E.B., Gindikin, S.G., Pyatetskii-Shapiro, I.I.: On the classification and canonical realization of bounded homogeneous domains. Trudy Moskov. Mat. Obshch. **12**, 359–388 (1963); Trans. Moscow Math. Soc. **12**, 404–437 (1963)
29. Vinberg, E.B.: The structure of the group of automorphisms of a convex cone. Trudy Moscov. Mat. Obshch. **13**, 56–83 (1964); Trans. Moscow Math. Soc. **13** (1964)
30. Vinberg, E.B.: Structure of the group of automorphisms of a homogeneous convex cone. Tr. Mosk. Mat. Obshchestva **13**, 56–83 (1965)
31. Pyatetskii-Shapiro, I.I.: Certain problems of harmonic analysis in homogeneous cones. Dokl. Akad. Nauk SSSR. 181–184 (1957)
32. Pyatetskii-Shapiro, I.I.: On a problem of E. Cartan Dokl. Akad. Nauk SSSR. **124**, 272–273 (1959)
33. Pyatetskii-Shapiro, I.I.: The geometry of homogeneous domains and the theory of automorphic functions, the solution of a problem of E. Cartan. Uspekhi Mat. Nauk. **14**(3), 190–192 (1959)
34. Pyatetskii-Shapiro, I.I.: On the classification of bounded homogeneous domains in n-dimensional complex space. Dokl. Akad. Nauk SSSR. **141**, 316–319 (1961); Sov. Math. Dokl. **141**, 1460–1463 (1962)
35. Pyatetskii-Shapiro, I.I.: On bounded homogeneous domains in n-dimensional complex space. Izv. Akad. Nauk SSSR. Ser. Mat. **26**, 107–124 (1962)
36. Gindikin, S.G.: Analysis in homogeneous domains. Uspekhi Mat. Nauk. **19**(4), 3–92 (1964); Russ. Math. Surv. **19**(4), 1–89 (1964)
37. Cartan, E.: Sur les domaines bornés de l'espace de n variables complexes. Abh. Math. Semin. Hambg. **1**, 116–162 (1935)
38. Berger, M.: Les espaces symétriques noncompacts. Ann. Sci. l'École Norm. Supérieure Sér. 3 Tome **74**(2), 85–177 (1957)
39. Vey, J.: Sur une notion d'hyperbolicité des variétés localement plates, Thèse de troisième cycle de mathématiques pures, Faculté des sciences de l'université de Grenoble (1969)
40. Vey, J.: Sur les automorphismes affines des ouverts convexes saillants, Annali della Scuola Normale Superiore di Pisa, Classe di Science, 3e série, tome **24**(4), 641–665 (1970)
41. Koszul, J.L.: Variétés localement plates et convexité. Osaka. J. Math. **2**, 285–290 (1965)
42. Alekseevsky, D.: Vinberg's Theory of Homogeneous Convex Cones: Developments and Applications, Transformation Groups 2017. Conference Dedicated to Prof. Ernest B. Vinberg on the occasion of his 80th birthday, Moscou, December (2017). https://www.mccme.ru/tg2017/slides/alexeevsky.pdf, http://www.mathnet.ru/present19121
43. Lichnerowicz, A.: Espaces homogènes Kähleriens. In: Colloque de Géométrie Différentielle. Publication du CNRSP, Paris, France, pp. 171–184 (1953)
44. Siegel, C.L.: Symplectic geometry. Am. J. Math. **65**, 1–86 (1943)
45. Koszul, J.L.: Domaines bornées homogènes et orbites de groupes de transformations affines. Bull. Soc. Math. France **89**, 515–533 (1961)
46. Koszul, J.L.: Ouverts convexes homogènes des espaces affines. Math. Z. **79**, 254–259 (1962)
47. Koszul, J.L.: Lectures on Groups of Transformations. Tata Institute of Fundamental Research, Bombay (1965)
48. Siegel, C.L.: Über der analytische Theorie der quadratischen Formen. Ann. Math. **36**, 527–606 (1935)
49. Koszul, J.L.: Déformations des variétés localement plates. Ann Inst Fourier **18**, 103–114 (1968)

50. Koszul, J.L.: Trajectoires convexes de groupes affines unimodulaires. In: Essays on Topology and Related Topics, pp. 105–110. Springer, Berlin (1970)
51. Shima, H.: Symmetric spaces with invariant locally Hessian structures. J. Math. Soc. Jpn. 581–589 (1977)
52. Shima, H.: Homogeneous Hessian manifolds. Ann. Inst. Fourier 91–128 (1980)
53. Shima, H.: Vanishing theorems for compact Hessian manifolds. Ann. Inst. Fourier 183–205 (1986)
54. Shima, H.: Harmonicity of gradient mappings of level surfaces in a real affine space. Geom. Dedicata 177–184 (1995)
55. Shima, H.: Hessian manifolds of constant Hessian sectional curvature. J. Math. Soc. Jpn. 735–753 (1995)
56. Shima, H.: Homogeneous spaces with invariant projectively flat affine connections. Trans. Am. Math. Soc. 4713–4726 (1999)
57. Shima, H.: The Geometry of Hessian Structures. World Scientific, Singapore (2007)
58. Shima, H.: In: Nielsen, F., Frederic, B. (eds.) Geometry of Hessian Structures. Springer Lecture Notes in Computer Science, vol. 8085, pp. 37–55 (2013) (planches: https://www.see.asso.fr/file/5104/download/25050), (vidéos GSI'13: https://www.youtube.com/watch?time_continue=139&v=6pyXxdIzDNQ, https://www.youtube.com/watch?time_continue=182&v=jG2tUjniOUs, https://www.youtube.com/watch?time_continue=6&v=I5kdMJvuNHA)
59. Boyom, M.N.: Sur les structures affines homotopes à zéro des groupes de Lie. J. Differ. Geom. **31**, 859–911 (1990)
60. Boyom, M.N.: Structures localement plates dans certaines variétés symplectiques. Math. Scand. **76**, 61–84 (1995)
61. Boyom, M.N.: Métriques kählériennes affinement plates de certaines variétés symplectiques. I Proc. Lond. Math. Soc. 3 **66**(2), 358–380 (1993)
62. Boyom, M.N.: The cohomology of Koszul-Vinberg algebras. Pac. J. Math. **225**, 119–153 (2006)
63. Boyom, M.N.: Some Lagrangian Invariants of Symplectic Manifolds, Geometry and Topology of Manifolds; Banach Center Institute of Mathematics, Polish Academy of Sciences, Warsaw, vol. 76, pp. 515–525 (2007)
64. Boyom, M.N., Byande, P.M.: KV Cohomology in Information Geometry Matrix Information Geometry, pp. 69–92. Springer, Heidelberg, Germany (2013)
65. Boyom, M.N.: Transversally Hessian foliations and information geometry I. Am. Inst. Phys. Proc. **1641**, 82–89 (2014)
66. Boyom, M.N., Wolak, R.: Transverse Hessian metrics information geometry MaxEnt 2014. In: AIP. Conference Proceedings American Institute of Physics (2015)
67. Byande, P.M., Ngakeu, F., Boyom, M.N., Wolak, R.: KV-cohomology and differential geometry of affinely flat manifolds. Information geometry. Afr. Diaspora J. Math. **14**, 197–226 (2012)
68. Byande, P.M.: Des Structures Affines à la Géométrie de L'information; European University Editions (2012)
69. Lichnerowicz, A.: Groupes de Lie à structures symplectiques ou Kähleriennes invariantes. In: Albert, C. (ed.) Géométrie Symplectique et Mécanique. Lecture Notes in Mathematics, vol 1416. Springer, Berlin (1990)
70. Cartan, H.: Allocution de Monsieur Henri Cartan, colloques Jean-Louis Koszul. Ann. l'Institut Fourier **37**(4), 1–4 (1987)
71. Malgrange, B.: Quelques souvenirs de Jean-Louis KOSZUL. Gazette des Mathématiciens **156**, 63–64 (2018)
72. Koszul, J.L.: Interview for "Institut Joseph Fourier" 50th birthday in 2016. https://www.youtube.com/watch?v=AzK5K7Q05sw
73. Barbaresco, F.: Jean-Louis Koszul et les structures élémentaires de la géométrie de l'information, revue MATAPLI 2018, SMAI (2018)
74. Koszul, J.M.: Homologie et cohomologie des algèbres de Lie. Bull. Soc. Math. Fr. Tome **78**, 65–127 (1950)

75. Cartan, H.: Les travaux de Koszul, I, Séminaire Bourbaki, Tome 1, Exposé **1**, 7–12 (1948–1951)
76. Cartan, H.: Les travaux de Koszul, II, Séminaire Bourbaki, Tome 1, Exposé **8**, 45–52 (1948–1951)
77. Cartan, H.: Les travaux de Koszul, III, Séminaire Bourbaki, Tome 1, Exposé **12**, 71–74 (1948–1951)
78. Haefliger, A.: Des espaces homogènes à la résolution de Koszul. Ann. l'inst. Fourier Tome **37**(4), 5–13 (1987)
79. Souriau, J.-M.: Mécanique statistique, groupes de Lie et cosmologie, Colloques int. du CNRS numéro 237, Géométrie symplectique et physique mathématique, pp. 59–113 (1974)
80. Barbaresco F., Koszul information geometry and Souriau Lie group thermodynamics. AIP Conference Proceedings 1641, 74, 2015, Proceedings of MaxEnt'14 Conference, Amboise, Septembre 2014
81. Barbaresco, F.: Koszul Information Geometry and Souriau Geometric Temperature/Capacity of Lie Group Thermodynamics, Entropy, vol. 16, pp. 4521–4565 (2014); Published in the book Information, Entropy and Their Geometric Structures, MDPI Publisher, September 2015
82. Barbaresco, F.: Symplectic structure of information geometry: fisher metric and Euler-Poincaré equation of Souriau Lie group thermodynamics. In: Geometric Science of Information, Second International Conference GSI 2015 Proceedings. Lecture Notes in Computer Science vol. 9389, pp. 529–540. Springer, Berlin (2015)
83. Barbaresco, F.: Geometric theory of heat from souriau lie groups thermodynamics and koszul hessian geometry: applications in information geometry for exponential families. Entropy **18**, 386 (2016)
84. Barbaresco, F.: Poly-symplectic Model of Higher Order Souriau Lie Groups Thermodynamics for Small Data Analytics, GSI 2017. Springer LNCS, vol. 10589, pp. 432–441 (2017)
85. Cartan, E.: Sur les invariants intégraux de certains espaces homogènes clos et les propriétés topologiques de ces espaces. Ann. Soc. Pol. De Math. **8**, 181–225 (1929)
86. Hua, L.K.: Harmonic Analysis of Functions of Several Complex Variables in the Classical Domains. American Mathematical Society, Providence, USA (1963)
87. Berezin, F.: Quantization in complex symmetric spaces. Izv. Akad. Nauk SSSR Ser. Math. **9**, 363–402 (1975)
88. Maliavin, P.: Invariant or quasi-invariant probability measures for infinite dimensional groups, Part II: unitarizing measures or Berezinian measures. Jpn. J. Math. **3**, 19–47 (2008)
89. Rothaus, O.S.: The Construction of Homogeneous Convex Cones. Annals of Mathematics, Series 2, vol. 83, pp. 358–376 (1966)
90. Vesentini, E.: Geometry of Homogeneous Bounded Domains. Springer, Berlin (2011); reprint of the 1st edn. C.IM.E., Ed. Cremonese, Roma 1968
91. Sampieri, U.: A generalized exponential map for an affinely homogeneous cone. Atti della Accademia Nazionale dei Lincei. Classe di Scienze Fisiche, Matematiche e Naturali. Rendiconti Lincei. Matematica e Applicazioni, Serie 8. **75**(6), 320–330 (1983)
92. Sampieri, U.: Lie group structures and reproducing kernels on homogeneous siegel domains. Ann. Mat. **152**(1), 1–19 (1988)
93. Nesterov, Y., Nemirovskii, A.: Interior-Point Polynomial Algorithms in Convex Programming. SIAM Series: Studies in Applied and Numerical Mathematics, pp. ix + 396 (1994)
94. Gromov, M.: Convex sets and kähler manifolds. In: Tricerri, F. (ed.) Advances in Differential Geometry and Topology, pp. 1–38. World Scientific, Singapore (1990)
95. Gromov, M.: In a Search for a Structure, Part 1: On Entropy. http://www.ihes.fr/~gromov/PDF/structre-serch-entropy-july5-2012.pdf (2012). Accessed 23 June 2012
96. Gromov, M.: Gromov Six Lectures on Probability, Symmetry, Linearity. October 2014, Jussieu, November 6th, 2014; Lecture Slides & video of Gromov lectures on youtube: http://www.ihes.fr/~gromov/PDF/probability-huge-Lecture-Nov-2014.pdf, https://www.youtube.com/watch?v=hb4D8yMdov4
97. Gromov, M.: Gromov Four Lectures on Mathematical Structures Arising from Genetics and Molecular Biology, IHES, October 2013; video of Lectures on youtube: https://www.youtube.com/watch?v=v7QuYuoyLQc&t=5935s

98. Barbaresco, F.: Les densités de probabilité «distinguées» et l'équation d'Alexis Clairaut: regards croisés de Maurice Fréchet et de Jean-Louis Koszul, GRETSI'17, Juan-Les-Pins (2017)
99. Massieu, F.: Sur les Fonctions caractéristiques des divers fluides. C. R. Acad. Sci. **69**, 858–862 (1869)
100. Massieu, F.: Addition au précédent Mémoire sur les Fonctions caractéristiques. C. R. Acad. Sci. **69**, 1057–1061 (1869)
101. Massieu, F.: Exposé des Principes Fondamentaux de la Théorie Mécanique de la Chaleur (note Destinée à Servir D'introduction au Mémoire de L'auteur sur les Fonctions Caractéristiques des Divers Fluides et la Théorie des Vapeurs); Académie des Sciences, Paris, 31 p. (1873)
102. Massieu, F.: Thermodynamique: Mémoire sur les Fonctions Caractéristiques des Divers Fluides et sur la Théorie des Vapeurs. Académie des Sciences, Paris, France, p. 92 (1876)
103. Poincaré, H.: Calcul des Probabilités. Gauthier-Villars, Paris, France (1896)
104. Balian, R., Balazs, N.: Equiprobability, inference and entropy in quantum theory. Ann. Phys. **179**, 97–144 (1987)
105. Balian, R.: On the principles of quantum mechanics. Am. J. Phys. **57**, 1019–1027 (1989)
106. Balian, R.: From Microphysics to Macrophysics: Methods and Applications of Statistical Physics, Vols. I and II. Springer, Heidelberg, Germany (1991, 1992)
107. Balian, R.: Incomplete descriptions and relevant entropies. Am. J. Phys. **67**, 1078–1090 (1999)
108. Balian, R., Valentin, P.: Hamiltonian structure of thermodynamics with gauge. Eur. Phys. J. B **21**, 269–282 (2001)
109. Balian, R.: Entropy, a protean concept. In: Dalibard, J., Duplantier, B., Rivasseau, V. (eds.) Poincaré Seminar 2003, pp. 119–144. Birkhauser, Basel, Switzerland (2004)
110. Balian, R.: Information in statistical physics. In: Studies in History and Philosophy of Modern Physics, Part B. Elsevier, Amsterdam (2005)
111. Balian, R.: François Massieu et les Potentiels Thermodynamiques, Évolution des Disciplines et Histoire des Découvertes. Académie des Sciences, Avril, France (2015)
112. Leray, J.: Un prolongement de la transformation de Laplace qui transforme la solution unitaire d'un opérateur hyperbolique en sa solution élémentaire. (Problème de Cauchy. IV.). Bulletin de la Société Mathématique de France **90**, 39–156 (1962)
113. Legendre, A.M.: Mémoire Sur L'intégration de Quelques Equations aux Différences Partielles; Mémoires de l'Académie des Sciences: Paris, France, pp. 309–351 (1787)
114. Casalis, M.: Familles exponentielles naturelles invariantes par un groupe de translations. C. R. Acad. Sci. Ser. I Math. **307**, 621–623 (1988)
115. Casalis, M.: Familles exponentielles naturelles invariantes par un groupe. Ph.D. Thesis, Thèse de l'Université Paul Sabatier, Toulouse, France (1990)
116. Letac, G.: Lectures on Natural Exponential Families and Their Variance Functions, Volume 50 of Monografias de Matematica (Mathematical Monographs); Instituto de Matematica Pura e Aplicada (IMPA). Rio de Janeiro, Brazil (1992)
117. Jaynes, E.T.: Information theory and statistical mechanics. Phys. Rev. Ser. II **106**, 620–630 (1957)
118. Jaynes, E.T.: Information theory and statistical mechanics II. Phys. Rev. Ser. **108**, 171–190 (1957)
119. Jaynes, E.T.: Prior probabilities. IEEE Trans. Syst. Sci. Cybern. **4**, 227–241 (1968)
120. Moreau, J.J.: Fonctions convexes duales et points proximaux dans un espace hilbertien. C. R. Acad. Sci. **255**, 2897–2899 (1962)
121. Dacunha-Castelle, D., Gamboa, F.: Maximum d'entropie et problèmes des moments. Ann. Inst. H. Poincaré Prob. Stat. **26**, 567–596 (1990)
122. Gamboa, F., Gassiat, E.: Maximum d'entropie et problème des moments: Cas multidimensionnel. Probab. Math. Stat. **12**, 67–83 (1991)
123. Dacunha-Castelle, D., Gamboa, F.: Maximum de l'entropie sous contraintes non linéaires. Ann. Inst H. Poincaré Prob. Stat. **4**, 567–596 (1990)
124. Burbea, J., Rao, C.R.: Entropy differential metric, distance and divergence measures in probability spaces: a unified approach. J. Multivar. Anal. **12**, 575–596 (1982)

125. Crouzeix, J.P.: A relationship between the second derivatives of a convex function and of its conjugate. Math. Progr. **3**, 364–365 (1977)
126. Hiriart-Urruty, J.B.: A new set-valued second-order derivative for convex functions. In: Mathematics for Optimization; Mathematical Studies Series, vol. 129. Elsevier: Amsterdam, The Netherlands (1986)
127. Eriksen, P.: Geodesics Connected with the Fisher Metric on the Multivariate Normal Manifold; Technical Report, pp. 86–13; Institute of Electronic Systems, Aalborg University, Aalborg, Denmark (1986)
128. Eriksen, P.S.: Geodesics connected with the Fisher metric on the multivariate normal manifold. In: Proceedings of the GST Workshop, Lancaster, UK (1987). Accessed 28–31 Oct 1987
129. Koszul, J.L.: Formes hermitiennes canoniques des espaces homogènes complexes. Sémin. Bourbaki Tome 3, Exposé **108**, 69–75 (1954–1956)
130. Bourguignon, J.P.: La géométrie kählérienne, domaines de la géométrie différentielle, séminaire Histoires de géométries, FMSH (2005). https://youtu.be/SDmMo4a1vbk
131. Gauduchon, P.: Calabi's extremal Kähler metrics: an elementary introduction. https://germanio.math.unifi.it/wp-content/uploads/2015/03/dercalabi.pdf
132. Cartier, P.: In memoriam Jean-Louis KOSZUL. Gazette des Mathématiciens **156**, 64–66 (2018)
133. Kosmanek, E.-E.: Hommage à mon directeur de thèse, Gazette des Mathématiciens **156**, 67, (2018)

MIX
Papier aus verantwortungsvollen Quellen
Paper from responsible sources
FSC® C105338
FSC
www.fsc.org

If you have any concerns about our products,
you can contact us on
ProductSafety@springernature.com

In case Publisher is established outside the EU,
the EU authorized representative is:
Springer Nature Customer Service Center GmbH
Europaplatz 3, 69115 Heidelberg, Germany

Printed by Libri Plureos GmbH
in Hamburg, Germany